SPACE SCIENCE SERIES

Tom Gehrels, General Editor

Planets, Stars and Nebulae, Studied with Photopolarimetry
Tom Gehrels, editor, 1974, 1133 pages

Jupiter
Tom Gehrels, editor, 1976, 1254 pages

Planetary Satellites
Joseph A. Burns, editor, 1977, 598 pages

Protostars and Planets
Tom Gehrels, editor, 1978, 756 pages

Asteroids
Tom Gehrels, editor, 1979, 1181 pages

Comets
Laurel L. Wilkening, editor, 1982, 766 pages

Satellites of Jupiter
David Morrison, editor, 1982, 972 pages

Venus
Donald M. Hunten et al., editors, 1983, 1143 pages

Saturn
Tom Gehrels and Mildred S. Matthews,
editors, 1984, 968 pages

Planetary Rings
Richard Greenberg and André Brahic,
editors, 1984, 784 pages

Protostars & Planets II
David C. Black and Mildred S. Matthews,
editors, 1985, 1293 pages

Satellites
Joseph A. Burns and Mildred S. Matthews,
editors, 1986, 1021 pages

The Galaxy and the Solar System
Roman Smoluchowski, John N. Bahcall,
and Mildred S. Matthews,
editors, 1986, 485 pages

Meteorites and the Early Solar System
John F. Kerridge and Mildred S. Matthews,
editors, 1988, 1269 pages

Mercury
Faith Vilas, Clark R. Chapman,
and Mildred S. Matthews,
editors, 1988, 794 pages

Origin and Evolution of Planetary and Satellite Atmospheres
S. K. Atreya, J. B. Pollack
and M. S. Matthews,
editors, 1989, 881 pages

Origin and Evolution of Planetary and Satellite Atmospheres

Origin and Evolution of Planetary and Satellite Atmospheres

S. K. Atreya
J. B. Pollack
M. S. Matthews

Editors

With 50 collaborating authors

THE UNIVERSITY OF ARIZONA PRESS
TUCSON

About the cover:

The front cover shows a view in the atmosphere of Titan which serves as a symbol of problems of planetary atmospheres. Its discovery by G. P. Kuiper in 1944 was the first recognition of an atmosphere of a satellite. Astronomical artists, beginning with Chesley Bonestell in the mid-1940s, delighted in showing Saturn in a blue sky from Titan's surface — a rare chance to show an extraterrestrial blue sky. Some decades later, culminating with Voyager imagery, a cloud or haze layer of significant optical depth was recognized. This painting shows Saturn in the sky from a position just above the cloud layer. This lowest layer from which Saturn is visible in a nonblack sky was whimsically named by artist Ron Miller the "Bonestellosphere." An exciting area of planetary research is the nature of the clouds and the atmospheric layer below them. Painting by William K. Hartmann.

The back cover shows a composite of six clear-filter images of Comet Halley taken by Giotto Halley Multicolor Camera on 13 March 1986. The peanut-shaped nucleus can be approximated by an ellipsoid of 16 × 8 × 8 km. The Sun was 29° above the horizontal on the left and 12° below the image plane on the average. The range to nucleus varied from 25,630 to 2730 km in this composite image. The strongest visible active area is located on the northern sunward tip of the nucleus. (Copyright Max Planck-Institut für Aeronomie, 1986. Courtesy H. U. Keller, Principal Investigator, Giotto/HMC.)

THE UNIVERSITY OF ARIZONA PRESS

This book was set in 10/12 Times Roman.
∞ This book is printed on acid-free, archival-quality paper.
Manufactured in the United States of America

93 92 91 90 89 5 4 3 2 1

Library of Congress Cataloging-in Publication Data

Origin and evolution of planetary and satellite atmospheres / S.K. Atreya, J.B. Pollack, M.S. Matthews, editors.
p. cm. — (Space science series)
Bibliography: p.
Includes index.
ISBN 0-8165-1105-5 (alk. paper)
1. Planets — Atmospheres. 2. Satellites — Atmospheres. I. Atreya, S. K. II. Pollack, James B. III. Matthews, Mildred Shapley. IV. Series.
QB603.A85075 1989
551.5'0999'2 — dc19 89-4651
CIP

British Library Cataloguing in Publication data are available.

CONTENTS

COLLABORATING AUTHORS

T. J. Ahrens, *328*
S. K. Atreya, *ix, 605*
P. Bodenheimer, *564*
A. P. Boss, *35*
P. Cerroni, *723*
A. F. Cheng, *682*
B. J. Conrath, *513*
A. Coradini, *723*
T. M. Donahue, *386*
G. Dreibus, *268*
C. Federico, *723*
B. Fegley, Jr., *78*
D. Gautier, *487*
R. Greenberg, *137*
R. A. Hanel, *513*
F. Herbert, *192*
W. B. Hubbard, *539*
D. M. Hunten, *386*
G. Igarshi, *306*
W. M. Irvine, *3*
E. K. Jessberger, *167*
R. E. Johnson, *682*
T. V. Johnson, *666*
T. D. Jones, *192*
J. F. Kasting, *386, 423*
J. Kissel, *167*
R. F. Knacke, *3*
M. A. Lange, *328*
L. A. Lebofsky, *192*
J. I. Lunine, *605*
G. Magni, *723*
D. L. Matson, *666*
G. E. Morfill, *35*
J. D. O'Keefe, *328*
T. Owen, *487*
M. Ozima, *306*
R. O. Pepin, *291*
J. B. Pollack, *ix, 564, 605*
R. G. Prinn, *78*
J. Rahe, *167*
R. E. Samuelson, *513*
G. Schubert, *450*
N. H. Sleep, *450*
S. C. Solomon, *450*
O. B. Toon, *423*
W. M. Tscharnuter, *35*
D. L. Turcotte, *450*
J. C. G. Walker, *386*
H. Wänke, *268*
P. R. Weissman, *230*

PREFACE

Nearly two decades of spacecraft exploration of the solar system, complemented by a variety of groundbased, orbital and suborbital observations, have resulted in an explosive growth in our knowledge of the formation of the planets and satellites and the origin and evolution of their atmospheres. The early exploratory missions, such as the Pioneers to Jupiter and Saturn, paved the way for more complex spacecraft missions like Voyagers 1 and 2 to Jupiter, Saturn, Uranus, Neptune and their moons and rings. The inner solar system has been extensively studied by orbiting spacecraft and *in situ* probes, deployed by the United States and the Soviet Union. The most noteworthy amongst these missions are the Soviet Venera series and the U.S. Pioneer Venus, and the Mariner and Viking missions to Mars. Recently, a gap in the knowledge of the history of solar-system formation and the origin and evolution of atmospheres was closed, to some extent, by a well-coordinated set of observations of a pristine body, Comet Halley, done from the European Space Agency's Giotto spacecraft, Soviet Union's VEGA 2, and Japan's Sakigake and Suisei spacecraft.

Results from the above spacecraft missions as well as Earth-based measurements have provided a solid set of constraints on hypotheses attempting to understand the origin and evolution of the atmospheres of solar-system objects. This data base includes measurements of the abundances of many of the major and minor constituents of these atmospheres as well as compositional information on asteroids and comets that may have served as volatile sources for the larger objects of the solar system. Comparisons of volatile abundances among objects of given type (e.g., terrestrial planets) and between one type and its possible sources (e.g., comets) is providing a powerful means to gain clues on the earliest history of atmospheres.

At the same time that this rich data set has been obtained and partially assimilated, new concepts and models have been developed that provide the modern paradigms for the origin and evolution of atmospheres. The classical concept that planets and satellites gained their atmospheres from a prolonged outgassing of their interiors has been supplemented by the possibility that these atmospheres may have begun forming during planetary and satellite accretion. In addition to focusing solely on ways in which these objects gained atmospheres, there has been a growing realization that loss processes may also have been very important. Examples of these processes include impact erosion, whereby a high-velocity meteoroid may cause part of the local atmosphere to be ejected to space by hydrodynamic escape, whereby lighter gases preferentially escape, and by charged-particle induced escape where removal of atmospheric constituents along the magnetic field lines occurs following their ionization.

Within the context of this intellectual environment, it seemed timely to organize a conference devoted to the history of atmospheres. In March 1987 a conference on the origin and evolution of the atmospheres of the planets and their satellites was convened in Tucson, Arizona to discuss the current status of the field and directions for future research—observational as well as theoretical. This book is the tangible product of that conference. It contains a set of comprehensive chapters based primarily on the review talks (although not limited to them). The chapters were critically reviewed by at least two referees, in addition to review by the editors. It is hoped that these chapters will serve as a valuable resource for advanced level graduate students and researchers in planetary science, especially those interested in the question of the origin and evolution of the solar system. In order to make the book affordable to as many scientists as possible, the cost has been kept at a reasonably low level through a grant from the U.S. National Aeronautics and Space Administration to defray some of the costs associated with the book's publication.

The five major sections in the book cover a wide range of topics, from the accretion of the planets, to the origin of volatiles for the inner and outer solar-system atmospheres. In addition, the present state of the relevant atmospheric properties, such as composition, heat balance and geochemical cycles, are reviewed from the perspective of the origin and evolution of the atmospheres. The objective of the first section of the book—early solar system—is to provide the context within which planets and their atmospheres formed by discussing the composition of the interstellar medium from which the solar system formed, the physical and chemical properties of the primordial solar nebula, and the means by which solid objects grew to large sizes in it. The next section—primitive bodies—discusses the composition of primitive bodies in the solar system both to gain insight into the composition of the bodies that helped to build planets and satellites and to investigate the possible importance of these objects as sources of volatiles for the bigger objects in the solar system that have atmospheres. In this context, this section also presents a discussion of the flux history of volatile-rich bodies to the inner solar system and what happens to them as they are accreted. The final sections deal with the three major classes of objects in the solar system that have atmospheres—the terrestrial planets, the giant planets, and satellites (including Pluto, although strictly speaking it probably formed in the solar nebula). Chapters in each of these sections range from ones that present the key observational constraints on the history of these atmospheres (especially on their composition), to ones describing models of the thermal and compositional evolution of the entire body, to ones discussing the merits of alternative models of the origin and evolution of these atmospheres. Finally, at the end of the book, a glossary is included as well as a comprehensive bibliography, to serve as a resource for further reading.

We wish to express our very deep appreciation to the many people who contributed to the making of this book, including the authors of the chapters and their reviewers. We are particularly grateful to Melanie Magisos for her tireless efforts in the preparation of this book and to Tom Gehrels for his enthusiastic support of the conference and the book. We also wish to thank all the participants of the conference for making it such a memorable event.

S. K. Atreya
J. B. Pollack
M. S. Matthews

PART I
Early Solar System

THE CHEMISTRY OF INTERSTELLAR GAS AND GRAINS

W. M. IRVINE
University of Massachusetts

and

R. F. KNACKE
State University of New York at Stony Brook

In the dense interstellar molecular clouds within which solar-type stars (and presumably planets) form, there is a complex chemistry which has a distinctive signature in terms of both composition and isotopic fractionation. Such clouds form a two-phase medium, with the dominant constituent by mass being gas-phase molecular material, but with a significant portion of the heavy elements being contained in particulate grains. Gas-phase constituents can be characterized quite precisely through their rotational spectra, provided that the molecules have a permanent electric dipole moment. At the present time molecular species have been identified with up to 13 atoms, and molecular weights up to 147. The grain component is less well characterized, and in fact there may not be a sharp dividing line between large molecules and small grains. The considerable uncertainty about the principal repository of volatiles in both the gas and grain phases is discussed in detail. It is noteworthy that strong isotopic fractionation is present, at least in the gas phase, with the D/H ratio at times enhanced by up to 1000 or more, relative to cosmic abundances.

I. INTRODUCTION

Although the direct observational evidence for the detection of protostars remains controversial, there seems to be no doubt that stars (and planetary systems, where they may exist) are born in the dense molecular clouds which form a major portion by mass of the interstellar medium in our Galaxy. Such clouds range from the $\sim$100 $M_\odot$ cold, dark clouds, some of which are as near as $\sim$100 pc, to the giant molecular clouds (GMCs) of 10^5 to 10^6 $M_\odot$, within which massive stars form. The chemistry of such dense interstellar clouds has distinctive characteristics, both in terms of composition and isotopic fractionation. Whether such characteristics are retained during the formation of a protostellar nebula is of fundamental importance to planetary science, since it may provide basic information on conditions in the nebula and links to the properties of the parental molecular cloud.

Molecular clouds contain both a gas and a solid phase, with the latter known as interstellar dust or grains. Many constituents of the gas phase can be unequivocally identified through their rotational (or in a few cases vibrational) spectra; in contrast, considerable controversy exists as to the interpretation of the much broader spectral features attributed to the solid grains. Although the grains constitute only about 1% by mass of these clouds, the percentage of elements heavier than helium incorporated in them may be comparable to or even greater than that present in the gas phase, at least for certain elements.

The last few years have seen considerable progress in both the determination of chemical compositions within molecular clouds and in an understanding of the processes leading to such compositions. Nonetheless, critical uncertainties remain, particularly concerning such questions as the major repositories for the volatile elements, the uniformity/diversity of composition under various conditions in molecular clouds, the heterogeneity of individual clouds, and the evolution with time of cloud chemistry. In what follows we present our view of the present knowledge concerning these matters and point, where possible, to future experiments which may resolve some of the uncertainties. Several recent reviews have appeared on these topics, which allow us to present a shorter story than would otherwise have been necessary (e.g., for the gas phase see Irvine et al. [1985,1987*a*], Guelin [1985] and Hjalmarson [1985]; discussions of interstellar grains have been presented by Martin [1978], Bussoletti [1985], Tielens and Allamandola [1987*a*] and Draine [1985]). We refer the reader to these earlier reviews for discussions of observational techniques and problems, of the physical characteristics of different types of interstellar clouds, and of the problems faced by theoretical models of the chemistry.

II. THE GAS PHASE

Qualitatively, the gas phase in dense molecular clouds consists almost entirely of molecular hydrogen, with trace constituents which include an interesting array of organic molecules (Table I). The richness of the rotational spectra observed at mm wavelengths is illustrated in Fig. 1, from a recent spectral survey of the Orion Molecular Cloud. Although quantitative abundance data is limited, it suggests that there is a basic uniformity in the com-

TABLE I
Identified Interstellar Molecules[a]

Simple Hydrides, Oxides, Sulfides, Halides, and Related Molecules				
H_2	CO	NH_3	CS	$NaCl$*
HCl	SiO	SiH_4*	SiS	$AlCl$*
H_2O	SO_2	CC	H_2S	KCl*
	OCS	CH_4*	PN	AlF*
	HNO ?			
Nitriles, Acetylene Derivatives, and Related Molecules				
HCN	$HC{\equiv}C{-}CN$	$H_3C{-}C{\equiv}C{-}CN$	$H_3C{-}CH_2{-}CN$	$H_2C{=}CH_2$*
H_3CCN	$H(C{\equiv}C)_2{-}CN$	$H_3C{-}C{\equiv}CH$	$H_2C{=}CH{-}CN$	$HC{\equiv}CH$*
$CCCO$	$H(C{\equiv}C)_3{-}CN$	$H_3C{-}(C{\equiv}C)_2{-}H$	HNC	
$CCCS$	$H(C{\equiv}C)_4{-}CN$	$H_3C{-}(C{\equiv}C)_2{-}CN$?	$HN{=}C{=}O$	
$HC{\equiv}CCHO$	$H(C{\equiv}C)_5{-}CN$		$HN{=}C{=}S$	
H_3CNC				
Aldehydes, Alcohols, Ethers, Ketones, Amides, and Related Molecules				
$H_2C{=}O$	H_3COH	$HO{-}CH{=}O$	H_2CNH	
$H_2C{=}S$	$H_3C{-}CH_2{-}OH$	$H_3C{-}O{-}CH{=}O$	H_3CNH_2	
$H_3C{-}CH{=}O$	H_3CSH	$H_3C{-}O{-}CH_3$	H_2NCN	
$NH_2{-}CH{=}O$	$(CH_3)_2CO$?	$H_2C{=}C{=}O$		
Cyclic Molecules				
C_3H_2	SiC_2	C_3H		
Ions				
CH^+	HCO^+	$HCNH^+$	H_3O^+?	
HN_2^+	$HOCO^+$	SO^+	HOC^+?	
	HCS^+		H_2D^+?	
Radicals				
OH	C_3H	CN	HCO	C_2S
CH	C_4H	C_3N	NO	NS
C_2H	C_5H	H_2CCN	SO	
	C_6H	$HSiCC$* or $HSCC$*		

[a]Table updated from Irvine et al. (1985,1987*a*). Recent detections have been made of C_6H (Suzuki et al. 1986; Cernicharo et al. 1987), C_2S (Saito et al. 1987), C_3S (Yamamoto et al. 1987*a*), $(CH_3)_2CO$ (Combes et al. 1987), PN (Ziurys 1987; Sutton et al. 1985), HC_2CHO (Irvine et al. 1987), a cyclic isomer of C_3H (Yamamoto et al. 1987*b*), CH_2CN (Saito et al. 1988), CH_3NC (Cernicharo et al. 1988), and several halides (Cernicharo and Guelin 1987).
(*) Indicates detection only in the envelope around the evolved stars.
(?) Claimed but not yet confirmed.

position of dense, quiescent clouds, with some striking departures from this uniformity that occur in regions of active star formation, and subtler differences both among and within clouds which may reflect differing formation conditions (such as elemental abundances in the gas phase) or possibly differing evolutionary states. The uniformity is reflected, for example, in Fig. 2, which compares selected abundances in two giant molecular clouds, a relatively nearby cold, dark cloud, and in the more diffuse, lower-density clouds which become visible at millimeter wavelengths when they absorb brighter background sources of continuum radiation. Some abundance differences presumably related to processes associated with star formation are illustrated in Fig. 3, which compares the composition of the quiescent Orion molecular cloud with that of material outflowing from the vicinity of the embedded

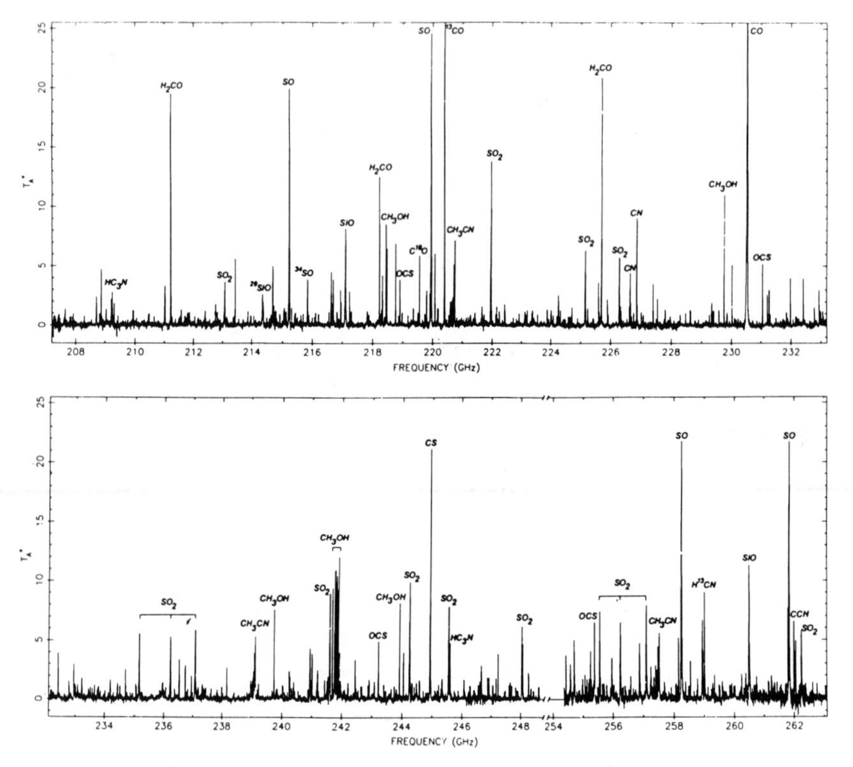

Fig. 1. Emission lines observed from gas phase molecules in the Orion Molecular Cloud during the spectral scan carried out at the Owens Valley Radio Observatory by Blake et al. (1987). Over 800 resolved spectral features are present and have been assigned to some 29 molecular species.

infrared source IRc2. The latter will be referred to below as the Orion outflow or "plateau" source. Note that the striking differences illustrated in Fig. 3 have thus far been found only in regions where massive stars are formed, and it is unclear whether the Sun formed in such a GMC, or instead in a much lower-mass dark cloud. T Tauri stars, which are believed to be of solar mass, are found in both types of region (see, e.g., Herbig 1982; Cohen and Kuhi 1979).

What are the distinctive chemical signatures of, first of all, the quiescent material? It is clear that this composition is very far from thermodynamic equilibrium. This is evident, for example, in the large abundance of the energetically unfavorable isomer HNC relative to the lower energy form HCN, and also in the chemically highly unsaturated nature of many of the organic constituents in an environment that is highly "reducing". The departure from equilibrium is illustrated by noting that there would be barely a single molecule of HNC present under thermochemical equilibrium in a cold molecular cloud. Such traits can be understood in terms of the kinetics of gas-phase ion-molecule chemistry in a low-temperature, low-density environment. Thus, the

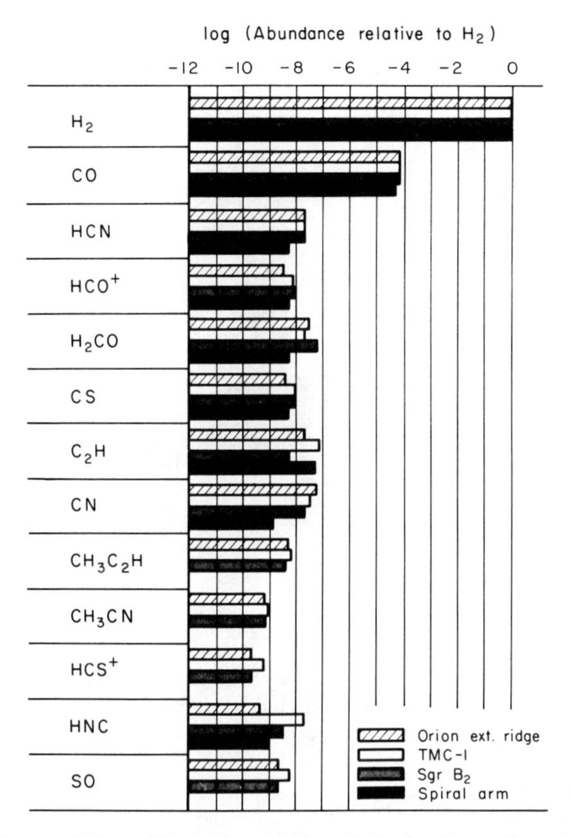

Fig. 2. Abundances relative to H_2 on a logarithmic scale for four interstellar molecular clouds: the Orion "extended ridge", Taurus Molecular Cloud One (TMC-1), the giant molecular cloud Sgr B2, and the "diffuse spiral arm clouds" towards W49N (figure from Irvine et al. 1987*a*).

barrier to isomerization from HNC to HCN is much higher than the typical energies available in molecular clouds, so that HNC which is formed will survive until it reacts chemically with another atom or molecule. Likewise, many hydrocarbon ions have barriers to reaction with molecular hydrogen, even when such a reaction is energetically favorable (Herbst et al. 1984), leading to a natural preference for unsaturated species in quiescent clouds.

Portions of molecular clouds that are heated by processes associated with massive star formation, such as shock waves and infrared radiation, may show a chemical composition that is closer to equilibrium (e.g., in the HNC/HCN ratio and/or degree of saturation of hydrocarbons). The Orion "hot core" is such a region, consisting of a dense clump or clumps apparently heated by a nearby star or protostar. The observed differences in degree of saturation among differing cloud regions were perhaps first stressed by Irvine and Hjalmarson (1984) and are illustrated in Table II, which compares the abundances of HC_3N, CH_2CHCN and CH_3CH_2CN in the cold cloud TMC-1, the GMC Sgr B2, the envelope of the evolved carbon star IRC+10216, and the

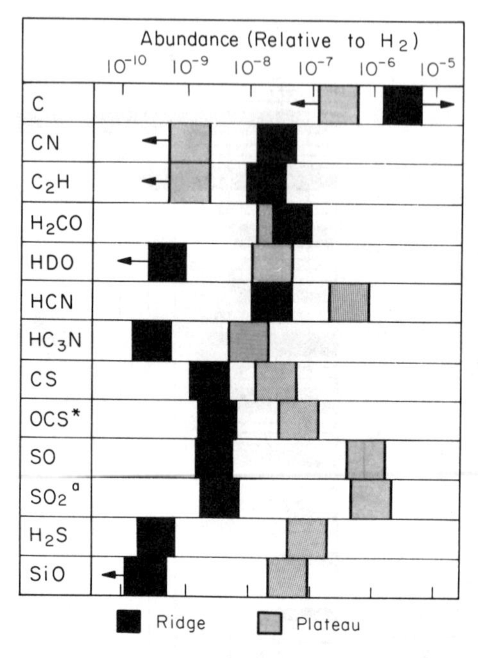

Fig. 3. Abundances relative to H_2 on a logarithmic scale for a quiescent cloud (the Orion Ridge) and for gas in the vicinity of active star formation (Orion Plateau). Values for OCS and SO_2 in the Ridge may not be typical of such material (figure from Blake et al. 1987 and Irvine and Hjalmarson 1987).

Orion "hot core". In the latter region both sufficient energy to overcome activation barriers for additional reaction pathways and the evaporation of material from grains may be important (see, e.g., Blake et al. 1987). The latter effect may be responsible for the enhanced abundances of water and ammonia that are also observed.

Among quiescent clouds, a major question concerns the reasons for variations in the abundances of the "carbon chains" (polyacetylene derivatives) and related species, such as the recently identified hydrocarbon ring C_3H_2.

TABLE II
Relative Abundances of Simple Nitriles with Varying Chemical Saturation[a]

Source	CH_3CH_2CN/HC_3N	CH_2CHCN/HC_3N
TMC-1	<0.2	0.03
Sgr B2	0.2	0.1
IRC + 10216	<0.02	<0.01
Orion hot core	6	1

[a]Table from Irvine et al. (1985,1987*a*).

The differences between the dark clouds in Taurus such as TMC-1 and those outside this region, such as L134N, are particularly noteworthy. The kinetic temperature and the density seem to be very similar in these two clouds, so that the origin of the chemical differences must be sought elsewhere. Although both age (evolutionary state) and the ultraviolet radiation environment have been proposed as causes (Stahler 1984; Millar and Freeman 1984), the local gas-phase C/O ratio may play a crucial role.

A. Isotopic Abundances

Probably the most distinctive signature of interstellar chemistry in cold clouds is the high degree of isotopic fractionation, particularly for hydrogen (see e.g., Watson and Walmsley 1982). The cosmic deuterium to hydrogen (D/H) ratio is estimated to be in the range $(0.8-2) \times 10^{-5}$ (Boesgaard and Steigman 1985). Values higher by three orders of magnitude are found for many interstellar molecules, as illustrated in Table III (some of the higher values reported in the literature for these species may reflect inaccurate estimates of the degree of saturation for the stronger isotopic lines [cf. Brown and Rice 1981]). Two of the species with reported D/H ratios $\gtrsim 0.1$ in cold clouds have multiple hydrogen atoms, and may have precursors subject to preferential loss of H relative to D by the mechanism suggested by Bell et al. (1987).

TABLE III
Deuterium Fractionation in Interstellar Clouds

Molecule	Observed Ratio[a]	Reference
DCN/HCN	0.002–0.02	Wootten 1987
DCO^+/HCO^+	0.004–0.02	Guelin et al. 1982
		Wootten et al. 1982
N_2D^+/N_2H^+	~0.01?	Guelin et al. 1982
DNC/HNC	0.01–0.04	Brown and Rice 1981
NH_2D/NH_3	0.003–0.14[b]	Olberg et al. 1985
		Walmsley et al. 1987
$HDCO/H_2CO$	~0.01	Guelin et al. 1982
C_2D/C_2H	0.01–0.05	Herbst et al. 1987
DC_3N/HC_3N	~0.02[c]	Langer et al. 1980*b*
		Wootten 1987
DC_5N/HC_5N	0.02[c]	Schloerb et al. 1981
		MacLeod et al. 1981
C_3HD/C_3H_2	0.03–0.15	Gerin et al. 1987
		Bell et al. 1987
HDO/H_2O	[d]	Olofsson 1984
CH_3OD/CH_3OH	[d]	Gottlieb et al. 1979

[a]Larger values for each ratio tend to occur in cold clouds and are probably rough upper limits to physically attainable values. Lower values than listed here may well occur in unobserved (particularly warmer) clouds.
[b]Different telescopes used for NH_3 and NH_2D.
[c]Only one source (TMC-1) observed.
[d]Both species detected, but ratio uncertain.

Note that even heavy (by interstellar standards) species such as HC_5N show significant fractionation. Partial preservation of such material (or of similarly fractionated, more refractory grain constituents) in the solar nebula would thus lead to significant D/H isotopic anomalies, such as have been observed in carbonaceous chondrites (see, e.g., Kerridge and Chang 1985; Chapter by Prinn and Fegley). Of course, enhancements in some species must be balanced by depletions elsewhere, but the latter occur primarily in the H_2 reservoir, which is so large relative to that represented by the constituents in Table III that the effect is negligible. Note that significant deuterium enhancement occurs for water as well as for hydrocarbons and nitrogen-containing species, but the degree is difficult to quantify because the H_2O abundance is quite uncertain (see below).

The gas-phase mechanism of deuterium fractionation is thought to be primarily the result of ion-molecule reactions such as

$$H_3^+ + HD \rightarrow H_2D^+ + H_2 + E_1 \tag{1}$$

and

$$CH_3^+ + HD \rightarrow CH_2D^+ + H_2 + E_2 \tag{2}$$

which proceed predominantly from left to right at interstellar cloud temperatures (10−100 K), since the degrees of exothermicity are (Smith et al. 1982*a*)

$$E_1 \sim 230 \text{ K}; E_2 \sim 370 \text{ K}. \tag{3}$$

The ions H_3^+ and CH_3^+ and their deuterated counterparts play key roles in interstellar chemistry, so that the fractionation which they acquire is transferred to heavier species by, e.g.,

$$\begin{aligned} H_2D^+ + CO &\rightarrow DCO^+ + H_2 \\ &\rightarrow HCO^+ + HD. \end{aligned} \tag{4}$$

There is some evidence for an inverse dependence of deuterium enhancement on cloud temperature for certain molecular species, as expected from reaction (1) (cf. Wootten 1987). In other cases, however, that effect is compensated by other aspects of the chemistry (e.g., C_2D/C_2H is higher in the warm Orion cloud than in the cold cloud TMC-1, and there is quite a high HDO abundance in Orion [cf. Herbst et al. 1987]). It has also been suggested that the D/H ratio may be enhanced for atomic hydrogen, which is a minor constituent but a potentially important one from the chemical point of view (Dalgarno and Lepp 1984). The possibility that the poorly understood reactions on grain surfaces might also lead to D/H fractionation must also be kept in mind (Tielens 1983).

For heavier elements the degree and importance (to the solar nebula) of isotopic fractionation is much less clear. First of all, it might be expected that stellar nucleosynthesis and mass loss would have caused the isotopic composition of interstellar material to have evolved during the 4.5 Gyr since the solar system formed, so that interstellar values measured today might be different to an uncertain degree from those in the precursor cloud of the solar nebula. Secondly, the isotopic composition of volatiles condensed on grains and entering the solar nebula would depend on the chemical composition of the condensate. This contrasts to the case of hydrogen, for which the large reservoir (H_2 and HD) is expected to remain in the gas phase. Finally, the effects for heavy elements are vastly less than for hydrogen, although still potentially enormous by terrestrial standards. As a result, observational uncertainties in the astronomical data become very significant.

Consider first the case of carbon. Anomalously heavy carbon has been reported in carbonaceous chondrites, with the suggestion that it may have an interstellar component (Kerridge and Chang 1985). This attribution is based on a series of observations which were interpreted as showing that the $^{12}C/^{13}C$ ratio for carbon monoxide in molecular clouds was significantly different from the terrestrial value of 89. Recent observations indicate, however, that the astronomical situation is far from clear. Thus, any underestimate of the degree of saturation of the strong ^{12}C lines will lead to a corresponding underestimate of the $^{12}C/^{13}C$ ratio. Careful studies of very rare isotopic species such as $^{13}C^{18}O$ have recently led to interstellar values that are closer to those for the solar system. Wilson et al. (1981) find from CO that $^{12}C/^{13}C$ = 75 $\pm$ 8 (1σ) for three nearby dark clouds, while Penzias (1983) obtains 100 $\pm$ 14 for the GMCs NGC 2264 and W3(OH). In contrast, Johansson et al. (1984) and Blake et al. (1987) both obtain $^{12}C/^{13}C \approx 40$ for several species in the Orion molecular cloud, whereas Scoville et al. (1983) find 96 $\pm$ 5 for this region. Henkel et al. (1982) find 80 $\pm$ 7 for the carbon isotope ratio in H_2CO for clouds in the solar neighborhood. A recent optical study of CH^+ in the solar neighborhood finds that $^{12}C/^{13}C = 43 \pm 4$ along several lines of sight, and attributes this low value to galactic evolution since the formation of the solar system (Hawkins and Jura 1987). There is general agreement that the carbon in the galactic center region is significantly heavier (factor of $\sim$4?) than that in the local gas, but this is not particularly relevant to questions concerning the early solar system. The situation is further complicated by the prediction, and apparent observational confirmation, of differential fractionation between the core and edge of clouds, as a result of differences in electron density and ultraviolet radiation field (see, e.g., Langer et al. 1980*a*). Models indicate that the *total* $^{12}C/^{13}C$ ratio should in most cases be between the values for CO and those for H_2CO (Langer et al. 1984). Bear in mind, however, that the carbon *grains* formed in the envelopes of evolved, carbon-rich stars are expected to have $^{12}C/^{13}C$ ratios that may easily be a factor of 2 less (i.e., heavier) than these gas-phase values. Thus, the meteorite carbon "anomalies" may indicate the inclusion of rather refractory carbonaceous grains of presolar origin, rather than icy, organic mantles condensed from the gas phase onto interstellar particles.

The situation is also complicated for oxygen. Concerning the present observational status, Wilson et al. (1981) find that gas phase $^{16}O/^{18}O$ is approximately terrestrial in nearby cold, dark clouds, while $^{18}O/^{17}O$ is roughly half the solar system value. The latter value agrees with that found in the Orion molecular cloud by Blake et al. (1987), i.e., $^{18}O/^{17}O = 3.5$ vs 5.5 for the solar system. In contrast, Penzias (1983) obtains a $^{16}O/^{18}O$ value of 900 $\pm$ 150 for two GMCs, compared to the terrestrial value of 500. All of the above results are for carbon monoxide.

Less data are available for nitrogen in the gas, but a recent careful study of ammonia in nearby GMCs by Güsten and Ungerechts (1985) gives $^{14}N/^{15}N \simeq 400$, compared with the terrestrial value of 270. It would be interesting to obtain corresponding values for dark clouds, but the lines would be very weak. In contrast, data for HCN are in reasonable agreement with the solar system value (Blake et al. 1987; Wannier et al. 1981). Interstellar nitrogen chemistry is less well understood than that of carbon and oxygen, partly because of uncertainties in reaction rates, so that it is unclear whether differential fractionation between NH_3 and HCN would be expected. For N_2, estimates of abundance must be made from the protonated ion N_2H^+. These show $^{14}N/^{15}N > 170$ (Linke et al. 1983).

Sulfur isotope ratios are also of interest for volatiles entering the solar system. Multitransition observations of SO_2 in Orion (Schloerb et al. 1983; Blake et al. 1987) lead to values for $^{32}S/^{34}S$ that are 30 to 40% below the terrestrial one. Observations of CS are consistent with a terrestrial ratio in many galactic sources, but the interpretation requires knowledge of the $^{12}C/^{13}C$ ratio (Frerking et al. 1980).

Even if all the above ratios were known exactly and could be extrapolated 4.6 Gyr into the past, the resulting isotopic composition of (say) a cometesimal forming in a molecular cloud would be dependent on presently unknown quantities such as the composition of the condensing volatiles, which depends in turn on the temperature of the interacting gas and grains, and the corresponding composition of refractory grain components. The composition of comets in such an environment has been considered in some detail by Yamamoto (1987; cf. also Vanýsek 1987).

B. Volatile Repositories

How is the cosmic complement of each of the volatile elements in the interstellar medium partitioned between the gas and the solid phases, and within each phase among the chemical constituents? In spite of our detailed knowledge of some aspects of astrochemistry, these major questions remain largely unanswered, or at least the answers remain highly controversial. In this section we review the situation for the gas phase.

Before examining each of the major volatile elements in turn, we must consider the fundamental problem of quantifying the cosmic abundances. To illustrate the problems, consider Table IV, which lists estimates of the abundance relative to hydrogen for C, N, O, S and P. In choosing values for comparison with interstellar observations, the average astrophysicist would probably rely on Allen (1976), while many planetary scientists would consult Cameron (1973*b*). Note that these sources and the more recent studies of Lambert (1978) and Grevesse (1984*b*) differ among themselves by 30 to 40% for carbon, nitrogen and oxygen. The difficulty is not primarily the observations themselves, but in their interpretation. Thus, the conversion of an equivalent width to an abundance requires accurate values of oscillator strengths for the weak lines which must be used in order to avoid saturation, and an accurate model of the solar atmosphere. This raises the more fundamental issue that these "cosmic abundances" are in fact *solar abundances* for the volatile elements. Even if the solar values were perfectly representative of the cosmic abundances in our portion of the Milky Way Galaxy one-third of a Hubble time ago, there is the possibility of evolutionary differences between such values and the present composition of the interstellar medium. The latter can be most directly approached via ultraviolet observations of diffuse interstellar clouds in front of bright background stars, such as those carried out by the Copernicus satellite. But these observations are dangerously incomplete, because they do not detect atoms incorporated in or on the particulate grains, which may include a significant portion of the volatile elements in the form of refractory polymeric mantles on grain cores of graphite or silicate (see below).

To continue our discussion, however, some values must be chosen to compare with observations of molecular clouds. For C, N and O we select the results of Lambert (1978), which (at least for oxygen) agree extremely well with those obtained by Grevesse et al. (1984) by a totally different approach.

TABLE IV
Cosmic Abundances of the Elements*

Element	a	b	c	d
H	$1.0(10)^{12}$	$1.0(10)^{12}$	$1.0(10)^{12}$	$1.0(10)^{12}$
O	$6.9(10)^{8}$	$6.6(10)^{8}$	$6.8(10)^{8}$	$8.3(10)^{8}$
C	$4.2(10)^{8}$	$3.3(10)^{8}$	$3.7(10)^{8}$	$4.7(10)^{8}$
N	$8.7(10)^{7}$	$9.1(10)^{7}$	$11.8(10)^{7}$	$9.8(10)^{7}$
S	$1.6(10)^{7}$	$1.6(10)^{7}$	$1.6(10)^{7}$	
P	$3.2(10)^{5}$	$3.2(10)^{5}$	$3.0(10)^{5}$	
Si	$4.5(10)^{7}$	$3.3(10)^{7}$	$3.1(10)^{7}$	
Fe	$3.2(10)^{7}$	$4.0(10)^{7}$		
Mg	$4.0(10)^{7}$	$2.6(10)^{7}$	$3.3(10)^{7}$	

*References: [a]Ross and Aller (1976); [b]Allen (1976); [c]Cameron (1973*b*); [d]Lambert (1978).

Note that this selection will reduce the deduced fraction of total carbon sequestered in gas-phase CO relative to an interpretation which relies on Allen's or Cameron's values for the cosmic C abundance. Lambert himself estimates the uncertainty in his *solar* values to be 20 to 25%.

1. Carbon. After H_2, the most abundant *detected* gas-phase constituent of molecular clouds is carbon monoxide. There remain uncertainties, however, concerning the fraction of the total carbon abundance represented by CO, as well as the abundance of other possible major carbon-containing species. The following discussion illustrates the types of questions that arise for all the volatile elements.

A basic problem is the determination of the total mass of material along a given line of sight, which is necessary in order to derive the fractional CO abundance. Since the mass in dense clouds is predominantly molecular hydrogen, the problem reduces to that of determining the absolute abundance of H_2 (in practice the column density in molecules per cm^2). Since this homonuclear molecule has no electric or magnetic dipole transitions between levels of different J (rotational quantum number) and v (vibrational quantum number) in the ground electronic state, detection of emission from cold clouds has not yet been achieved. Shock-excited H_2 in regions of massive star formation and fluorescent emission from material adjacent to hot stars has been seen (Shull and Beckwith 1982; Gatley and Kaifu 1987), as has ultraviolet absorption due to diffuse clouds in front of background stars. None of these observations applies directly to the ambient molecular gas in dense, quiescent clouds. Observation of the pure rotational electric quadrupole spectrum from such clouds, at 28 μm, should be feasible as detectors improve and can be placed above telluric absorption (Shull and Beckwith 1982).

A direct comparison between H_2 and CO has been made for the hot (post-shock?) gas in Orion (Watson 1984), yielding $CO/H_2 \approx 1.2 \times 10^{-4}$, and also for the cloud material in front of the strong infrared source in NGC 2024 (Black and Willner 1984), producing the *lower limit* $CO/H_2 \geq 8 \times 10^{-5}$. In general, however, this ratio is estimated indirectly from grain extinction and assumptions concerning the optical properties of grains and the gas-to-dust ratio (cf. Irvine et al. 1985). Such studies find values $CO/H_2 \sim 10^{-4}$, with little obvious dependence on cloud density (in contrast to some earlier work [see Frerking et al. 1982; Guelin 1987]). Scoville et al. (1983), however, report a value about five times higher for the Orion molecular cloud. For our assumed abundances (Table IV), a ratio of 10^{-4} corresponds to about 10% of the total carbon being in CO, which is probably the best current estimate in cold clouds. The fraction may be higher in warmer clouds such as Orion (see the recent detailed study by van Dishoeck and Black [1987]).

There is little observational evidence concerning the fraction of carbon in hydrocarbons in molecular clouds. A relevant result is the upper limit $CH_4/CO < 10^{-2}$ obtained by Knacke et al. (1985*a*) toward two infrared sources embedded in dense molecular clouds. Both methane (CH_4) and acetylene (HCCH) have been detected in the envelopes of evolved stars, but their lack of a permanent electric dipole moment precludes detection of pure rotational transitions between levels likely to be populated at typical cloud temperatures (10-100 K). A reported detection of CH_4 transitions resulting from a small rotationally induced dipole moment has been shown to result from misidentifications (Ellder et al. 1980). Models of gas-phase chemistry predict that $CH_4/CO << 1$, even in cases where the gas-phase C/O ratio is greater than unity (Leung et al. 1984; Watt 1985), consistent with the observations by Knacke et al. (1985*a*). In the early stages of chemical evolution, Leung et al. find that HCCH can be more abundant than CH_4, but not by a very large factor.

Other possible reservoirs for carbon include atomic carbon, CO_2, and the grains. Observations indicate that atomic carbon is vastly more abundant than some models predict for mature clouds, but that still $C/CO = 0.1$ to 0.2 (Keene et al. 1985). Carbon dioxide has not been observed directly because of its lack of an electric dipole moment, but estimates of its abundance can be made from that of its protonated form HCO_2^+. Observations of the latter species show that CO_2 is not a major carbon repository in typical interstellar clouds (Minh et al. 1987).

2. *Oxygen.* The chemical state of gas-phase oxygen depends critically on the gas-phase C/O ratio. Even though the total cosmic ratio $C/O \simeq 0.5$, scenarios exist in which the *gas-phase* ratio in dense clouds is greater than one; for example, due to the lower sublimation temperature of CH_4 relative to H_2O (see, e.g., Blake et al. 1987). In fact, the observed composition of quiescent clouds may be more consistent with models in which $C/O > 1$ than with the reverse, although this is by no means certain (Langer et al. 1984; Millar et al. 1987). If indeed $C/O > 1$, then the major repository of oxygen is predicted to be CO; in this case, most of the total oxygen would be in (or on) the grains.

For a gas-phase abundance $C/O < 1$, significant factions of the available oxygen may be in the form of O_2 or atomic oxygen, as well as in CO. Water is predicted to be somewhat less abundant, and NO considerably less so (Millar et al. 1987).

What do the observations tell us? For quiescent clouds, a fractional abundance of 10^{-4} for CO corresponds to 6% of the total oxygen (recall that at least one estimate of the CO abundance is $\sim$5 times this large [Scoville et al. 1983]). Molecular oxygen has no electric dipole rotational transitions, but it does have a strong magnetic dipole moment. The latter is in fact sufficient to render the atmosphere opaque, and thus preclude groundbased observations of these transitions. A corresponding transition of the rare isotopic form $^{16}O^{18}O$ is shifted out of the strong telluric lines and has been sought at millimeter wavelengths by two groups (Goldsmith et al. 1985; Liszt and Vanden Bout 1985). The molecular oxygen isotope was *not* detected, indicating that O_2/CO

< 1 in the sources observed, which included several GMCs, the Orion "plateau" source, and the Rho Oph dark cloud.

Atomic oxygen may be observed in hot, dense clouds via its $^3P_1 \rightarrow {}^3P_2$ ground state fine-structure transition at 63 μm, provided that the spectrometer can be flown above most of the atmospheric absorption. Using the Kuiper Airborne Observatory (KAO), Werner et al. (1984) detected atomic oxygen in the (presumably shocked) high-velocity outflow in the Orion molecular cloud and deduced an abundance relative to hydrogen $\sim 10^{-4}$. Unfortunately, this transition will not be easy to observe in cold interstellar clouds.

Although H_2O is not predicted to be the major oxygen reservoir in quiescent clouds, measurements of its abundance would certainly be desirable. Unfortunately, the low-energy rotational transitions of H_2O lie at submillimeter wavelengths, the last major region of the astronomical spectrum to remain largely unexplored. This will eventually be rectified, as detectors improve and instruments are flown above the strong telluric absorption. Both the $3_{13} \rightarrow 2_{20}$ and $4_{14} \rightarrow 3_{21}$ transitions, at 183 and 380 GHz, respectively, have been observed from the KAO in the Orion molecular cloud (Waters et al. 1980; Phillips et al. 1980), but both lines show evidence of maser action or at least large optical depths. As a result, abundance estimates are very uncertain (see Olofsson 1984). A reported transition of $H_2{}^{18}O$ (Phillips et al. 1978), which would surely be optically thin, has subsequently been shown actually to arise from SO_2 (N. Erickson, personal communication). Equipment has been designed to observe the lowest-energy transition of ortho-H_2O at 557 GHz from a spacecraft and would be a natural instrument for use with the Large Deployable Reflector (Schloerb 1987). (Note that the $6_{16} - 5_{23}$ water line at 22 GHz extensively observed by radio astronomers is too high in energy above the ground state to be excited, except in regions of active star formation. Even then, the uncertain excitation processes for this maser emission preclude reliable abundance estimates for water [cf. Rydbeck and Hjalmarson 1985].)

Three transitions of HDO have also been observed in molecular clouds, but it is very difficult to convert these to a total water abundance without knowing the degree of isotope fractionation. Olofsson (1984), however, suggests that in the Orion outflow (plateau source) the abundance of H_2O approaches that of CO, and Blake et al. (1987) attribute the clearly enhanced water abundance in this warmer Orion gas to evaporation of grain mantles.

3. Nitrogen. Although in early stages of the chemical evolution of an interstellar cloud theoretical models predict that atomic nitrogen may be quite abundant, as the composition approaches a steady state most gas-phase nitrogen appears as N_2 (see, e.g., Millar et al. 1987). Since N_2 has no permanent electric dipole moment, observational evidence in support of this conclusion must come from data on related species such as N_2H^+. The latter is widely observed in molecular clouds, and the $J = 1 \rightarrow 0$ transition often seems to be saturated and/or self-absorbed by low excitation cloud "halos." A search for the ^{15}N species was thus quite important, and such observations successfully detected both $^{15}NNH^+$ and $N^{15}NH^+$ in the cloud DR21(OH). Linke et al. (1983) interpret these results as showing that $N_2/CO \sim 1$ in this cloud, which implies that most nitrogen may exist as N_2. Only upper limits were obtained for the dark clouds TMC-1 and L134N, but these are consistent with abundances for N_2 comparable with or slightly smaller than that in DR21(OH). Thus, it is quite possible that most nitrogen in molecular clouds is in the gas phase as N_2.

Ammonia abundances seem to vary significantly both among cold clouds and, particularly, between such clouds and warmer regions (Irvine et al. 1987*a*). The latter effect may be due to evaporation of NH_3 from icy grain mantles (see, e.g., Walmsley et al. 1987). Even in such warm regions, however, the abundance ($\sim 10^{-5}$ relative to H_2) probably remains significantly below the point at which most nitrogen would be sequestered in ammonia.

The abundance of HCN also appears to show an interesting temperature dependence, which has been interpreted as the effect of additional production pathways becoming effective at warmer temperatures (Goldsmith et al. 1986). Nonetheless, $HCN/CO << 1$, so that HCN appears not to be the major nitrogen reservoir in molecular clouds.

4. *Sulfur.* In dense clouds with $O/C > 1$, Prasad and Huntress' (1982) model calculations predict that the major repositories for sulfur will be SO and SO_2. Sulfur monoxide has in fact been widely observed in both cold and warmer clouds, while SO_2 seems to be more restricted in its occurrence (see, e.g., Schloerb et al. 1983; Irvine et al. 1983). Nonetheless, the abundances of these species in cold clouds suggest gas-phase sulfur depletion by three orders of magnitude relative to cosmic abundances, in striking contrast to the near zero depletion observed in diffuse clouds (Jenkins 1987). Even under the influence of heating and/or grain sputtering associated with star formation, as in the Orion outflow (plateau) source, most sulfur ($\sim$90%) in molecular clouds may remain in the grains (Irvine et al. 1987*a*).

If the gas-phase oxygen abundance is low, CS or atomic S may be the major sulfur repositories (Prasad and Huntress 1982). Although CS is very widespread in molecular clouds, its abundance seems not to approach the value ($1\text{-}2 \times 10^{-5}$ relative to H_2) that would make it the major reservoir for sulfur (cf. Irvine et al. 1987*a*).

5. *Phosphorus.* The first interstellar phosphorus-containing molecule PN has only just been detected (Ziurys 1987; Turner and Bally 1987; cf. also Sutton et al. 1985). The reported abundance, however, corresponds to only a small fraction of the total cosmic phosphorus complement.

In the low-density, "diffuse" clouds observable in absorption against background stars, atomic phosphorus is either undepleted relative to its cosmic abundance or else (in colder regions) depleted by about a factor of 3 (Dufton et al. 1986). In dense clouds, calculations suggest that phosphorus may exist predominantly in atomic form, with the most abundant molecular form being PO (Thorne et al. 1984). Note that PH_3 appears not to be readily formed under interstellar (gas-phase) conditions. Searches for interstellar PO have thus far yielded only upper limits (Matthews et al. 1987), which, however, are not strikingly below the predicted values. Astronomical searches have likewise failed to detect HCP or PH_3 (see, e.g., Hollis et al. 1982).

III. THE GRAINS

A. The Physical State and Composition

The emerging picture of interstellar dust clearly shows that dust composition depends strongly on environment. Refractory materials survive in the diffuse clouds, while dense molecular clouds harbor both refractory and volatile solids. Circumstellar dust reflects physical and compositional properties of this special environment. Despite much recent progress, however, we still do not have a clear understanding of the gas and solids as interacting systems, nor do we know the major chemical state of even abundant elements like C, N and O.

Theoretical ion-molecule studies often de-emphasize the role of grains, except as a catalyst for the formation of H_2 molecules. The justification for this is reasonable agreement between observational abundances and gas-phase calculations, at least for most simple molecules (Irvine et al. 1985). Moreover, surface chemistry is a complicated subject and is uncertain because of the poorly understood properties of interstellar grain surfaces. On the other hand, H_2O and CO do certainly exist in both the gas and solid phases (see Sec. III.C below). Atoms should readily stick to grains; the problem is to understand how so much of the material returns to the gas (Martin 1978). On the grains energetic photons, cosmic rays and atomic ions can stimulate chemical reactions. Even the outlines of these grain processes are not yet clear, but there is laboratory and astronomical evidence that they influence interstellar chemistry (Draine 1985).

This section concentrates on the composition of interstellar dust and its relationship to the gas, with emphasis on observational results. The discussion is divided between grains in diffuse clouds and grains in molecular clouds. This roughly corresponds to the refractory and the volatile (plus refractory) dust components, although the separation is somewhat awkward for carbon. Evidence is strong that carbon compounds are major components of both volatile and refractory grain materials.

B. The Diffuse Cloud Component

1. Depletions. Ultraviolet observations of interstellar absorption lines in the spectra of hot stars behind diffuse clouds have been used to estimate depletions from cosmic elemental abundances in the gas and hence to reveal which species must be bound in grains. The observations sample gas where densities range from ~ 0.1 to $\sim 10^3$ cm^{-3}, the edges of molecular clouds (Spitzer 1985; Joseph et al. 1986). Figure 4 is a compilation of depletions in two moderately obscured sources (Snow and Joseph 1985), based on somewhat uncertain ($\sim$factor of 4?) estimates of the hydrogen column density. There is a general correlation between overall depletion and density (Joseph et al. 1986). Refractory elements including Fe, Mg and Si are strongly depleted. Most of the silicon must be in the grains, but even all of it (as silicates) could account for only about one-third of the observed continuum extinction (Martin 1978; Greenberg 1974).

C, N and O are much less depleted, making the measured depletions more sensitive to systematic effects such as uncertainties in oscillator strengths, questions concerning the corresponding hydrogen column densities, and possible differences between solar and interstellar elemental abundances (Jenkins 1987). Roughly half the oxygen may be incorporated in grains, although there are considerable variations from cloud to cloud. The most recent interpretations of the Copernicus data, including revised values of oscillator strengths, find virtually no depletion of nitrogen (Hibbert et al. 1985; cf. also Ferlet 1981). Towards many stars, the data are consistent with no depletion of gas-phase carbon, although some lines of sight show only one-third or so of the carbon is present in the gas (cf. York et al. 1983; Jenkins et

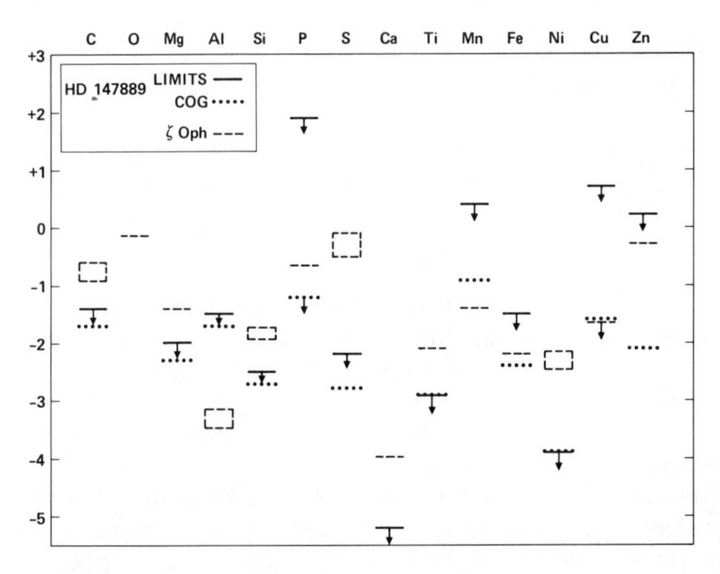

Fig. 4. Depletions toward HD 147889 and ζ Oph, defined as log (N/N_H) − log $(N/N_H)_\odot$ where N is the column density of the element in question, and N_H is the corresponding quantity for hydrogen. Upper limits and values derived from curve-of-growth column densities are shown for HD 147889, as well as values found earlier for ζ Oph (Snow and Joseph 1985).

al. 1983). The data are consistent with the main mass of the grains in diffuse clouds being an amorphous mixture of C, N and O (York et al. 1983).

Roughly 60% of the carbon in grains would be enough to give the observed mean extinction in the Galaxy. In this case, somewhat more gas might be missing than needed (Mathis et al. 1977). This might be evidence for large grains or for smaller than solar abundances in the interstellar medium (York et al. 1983).

2. *Silicates.* An Si-O band at 9.8 μm and an Si-O-Si bending mode near 18 μm identify silicate dust. Typical spectra of sources in molecular clouds are shown in Fig. 5; these spectra contain absorption bands of both the refractory silicates and of volatile ices (see Sec. III.C below). Silicates occur in diffuse and molecular clouds, and in circumstellar dust enshrouding late-type stars (cf. Merrill and Stein 1976). The only known cases of extinction without silicates are in the circumstellar environments of carbon stars, where silicon is bound in SiC.

Laboratory spectra of semi-amorphous silicates agree well with the position and shape of the 9.8 μm band (Krätschmer and Huffman 1979). Spectra of carbonaceous meteorites (Zaikowski and Knacke 1975; Forrest et al. 1979) and interplanetary dust particles (Sandford and Walker 1985) also agree reasonably well. Crystalline silicates like olivine or enstatite could only be present as minor constituents or in the amorphous forms. Negative searches for bound hydroxyl and water suggest that less than 25 to 50% of the silicates are hydrated (Knacke et al. 1985*b*).

If essentially all of the silicon is in grains, between 10 and 20% of the oxygen would be combined with it as silicates.

3. *Refractory Carbon Material.* The nature of the carbon compounds in diffuse clouds is still incompletely understood, but there has been encouraging progress recently. We discuss here several lines of evidence (another recent summary is given by Martin [1987]).

i. 2175 Å Carrier. A strong, interstellar absorption band at 2175 ± 30 Å is usually associated with carbon. For many years this feature was tentatively identified as graphite, although there were difficulties in explaining its position and shape (Gilra 1972; Mathis et al. 1977; Huffman 1977; Greenberg and Chlewicki 1983). The availability of laboratory optical constants may have contributed to the interest in graphite.

Opinion is swinging away from graphite and toward a more complex carbonaceous material of some kind, perhaps more reasonable on physical grounds (Mathis 1986). The absence of large amounts of graphite in meteorites implies that graphite is probably absent in interstellar dust also (Nuth 1985). Graphite should survive alteration processes affecting dust in the early solar system.

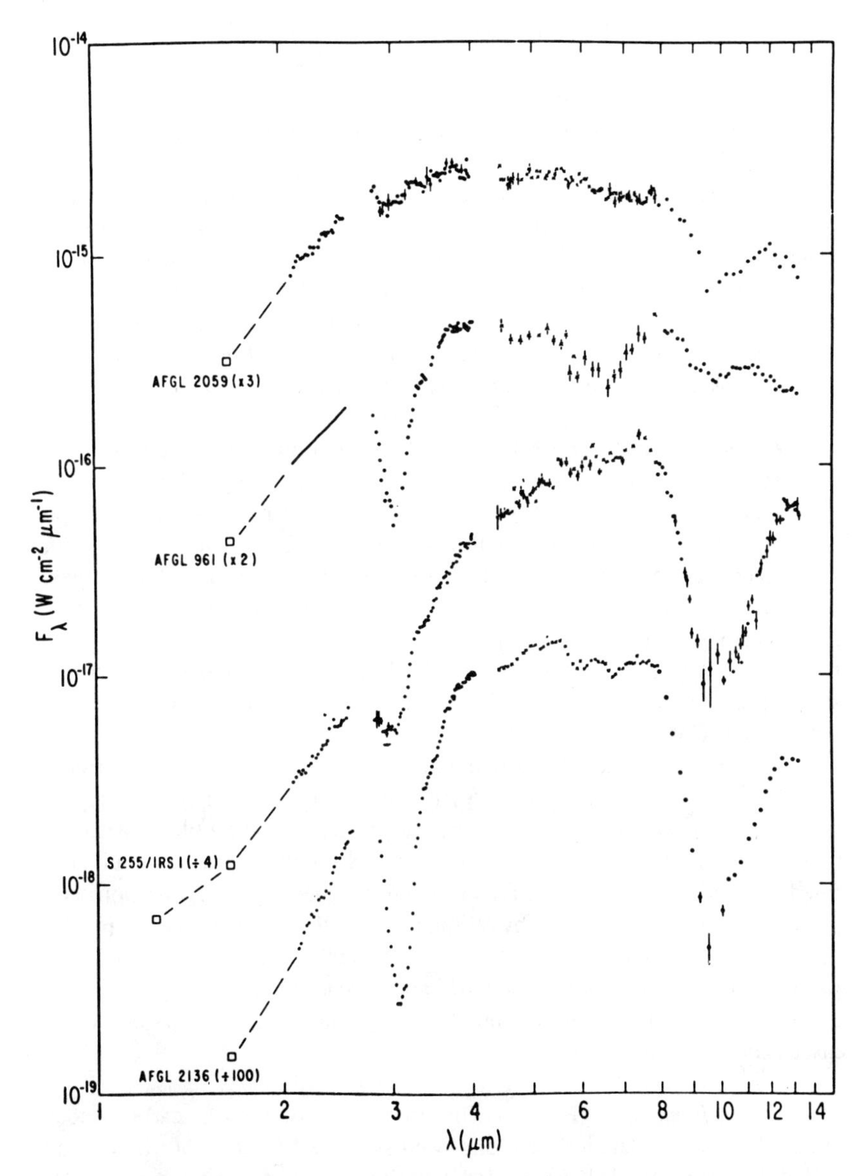

Fig. 5. Infrared spectra of protostars embedded in molecular clouds (Willner et al. 1982).

Sakata et al. (1983) found that a solid residue, "quenched carbonaceous composite (QCC)," which is formed in a discharge tube containing methane, absorbs at 2175 Å. The fit at 2175 Å is, however, only approximate; the feature in the laboratory spectrum is wider than the interstellar one. Leger and Puget (1984) proposed that "polycyclic aromatic hydrocarbons (PAHs)" could account for the 2175 Å band. These are organic molecules with benzene ring structures. Only a few percent of the interstellar carbon would be needed to account for the intensity of the ultraviolet band. Despite the spectral coincidences, the evidence for a carbon origin of the 2175 Å absorption is not conclusive. Steel and Duley (1987) recently found a 2175 Å feature in laboratory spectra of silicate particles. They assign the absorption to OH^- ions in the silicate lattice, and propose that this is the source of the interstellar band.

ii. Small Particle Emission. The IRAS satellite discovered extended dust emission, now named "infrared cirrus" (Low et al. 1984). This emission is strongest at 60 and 100 μm and often shows up at 12 and 20 μm also (Gautier 1986). It is correlated with high-latitude CO (Weiland et al. 1986). The rather high temperatures of 20 to 30 K inferred from the colors suggest small (a = 3-100 Å) grains. The abundances of the small particles must be greater than given by the power-law extinction derived by Mathis et al. (1977; Draine and Anderson 1985). In the Pleiades, small (~10 Å size) particles show up as scatterers in the ultraviolet (Witt et al. 1986) and as emitters in the far infrared (Cox and Leene 1987).

Sellgren (1984) identified a population of very hot (~1000 K) grains near HII regions from 2–5 μm spectra. Such temperatures also imply small grains (sizes of order 20 Å or less) which are transiently heated by ultraviolet photons. Emission bands at 3.28, 3.4, 6.2, 7.7, 8.6 and 11.3 μm, the unidentified infrared ("UIR") bands, are observed in or near many HII regions and planetary nebulae (cf. review by Willner 1984). Recent intermediate resolution spectra show that there are also weak UIR features at 3.46, 3.51 and 3.56 μm (de Muizon et al. 1986). The UIR bands occur in association with the hot continuum near HII regions and probably originate in the same particles (see discussion of PAHs below).

iii. Carbonaceous Grains. That there is some kind of carbonaceous material in interstellar dust has long been suspected (Platt 1956; Hoyle and Wickramasinghe 1962; Donn et al. 1966), but data were hard to accumulate. More recent infrared and ultraviolet observations provide more opportunities for spectral studies. Several of the UIR band positions are characteristic of carbon compounds (Knacke 1977; Sagan and Khare 1979). The bands at 3.28 and 3.4 μm coincide with characteristic C–H vibrations which are ubiquitous in organic compounds. Other bands have spectral similarities with organic substances found in carbonaceous chondrites and laboratory simulations of astrophysical chemistry. These materials consist of complicated and poorly defined mixtures of organic compounds sometimes characterized as tar or kerogen-like.

Duley and Williams (1981) and Duley (1987) proposed that hydrogenated amorphous carbon (HAC) is the dominant carbon condensate in clouds, and that PAH molecules evolve from it. Mathis (1986*a,b*) has summarized arguments for poorly ordered carbon as the source of visible extinction. Witt and Schild (1986) recently reported excess R- and I-band continuum emission near reflection nebulae. They interpret the spectrum as fluorescence by small, amorphous carbon particles. Borghesi et al. (1987) have reported infrared spectroscopy of amorphous carbon grains. They find bands at 3.4, 3.51, 5.78, 6.29, 6.85 and 11.3 μm in the spectra of their compounds. While the strong 3.28 μm feature is missing, the other features are suggestive of the UIR bands. Borghesi et al. argue that a mixture of amorphous carbon grains and PAHs may account for the interstellar emitter.

iv. QCCs. When examined in the infrared, the quenched carbonaceous composite (QCC) absorbed at 3.29, 3.24, 3.48, 6.25, 6.94, 11.40, 11.96 and 13.24 μm (Sakata et al. 1984). The wavelength positions are remarkably suggestive of the interstellar unidentified emission bands, although not exactly the same. In addition, an oxidized QCC shows features at 7.7 and 8.6 μm (Sakata et al. 1987). Since the QCCs also absorb at 2175 Å, this amorphous, carbonaceous material, or variations of it, is of great interest.

v. PAHs. Leger and Puget (1984) proposed that the hot grains near HII regions and the UIR emitters are polycyclic aromatic hydrocarbons (PAHs). In this interpretation, the UIR features are molecular emissions. Despite the newness of the PAH hypothesis (although Donn [1968] proposed them as a dust component before the UIRs were discovered), there are already reviews that strongly advocate the identification (Tielens and Allamandola 1987*a,b;* Leger and d'Hendecourt 1987) and a workshop on the subject (Leger et al. 1987).

Perhaps the best case for PAHs is in the 3.28 μm emission band, which is very diagnostic of C–H groups in ring structures (Bellamy 1975). The interstellar 6.2 and 7.7 μm features also agree with PAH features, but the 11.3 μm feature does not fit well (Fig. 6). Leger and d'Hendecourt (1987) propose that mixtures of weakly hydrogenated particles with a mean content of 50 atoms could remove the discrepancies. The features at 3.46, 3.51 and 3.56 μm are near C–H vibrations of molecular side chains like $-CH_3$ or $-C_2H_5$ attached to PAH rings (de Muizon et al. 1986), or could be evidence for anharmonicity in high vibrational transitions (Barker et al. 1987).

Puget et al. (1985) proposed that PAHs explain the IRAS cirrus emission at 100, 60 and 12 μm, but which is weak at 25 microns. In this model, the far-infrared flux is continuum small particle emission, and the flux in the 12 μm IRAS filter band is PAH band emission. If correct, this would show that the

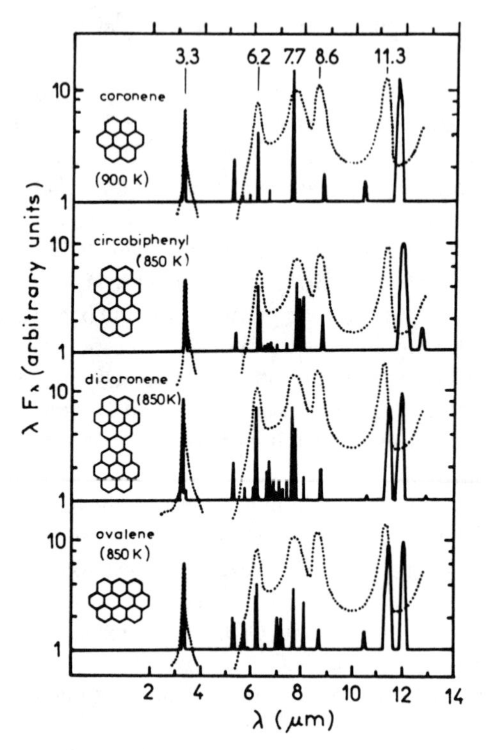

Fig. 6. Emission spectra of several PAHs calculated from their absorption spectra measured in the laboratory. The observed spectrum of the reflection nebula NGC 2023 is shown for comparison (dotted line) (figure from Leger and d'Hendecourt 1987).

infrared cirrus is carbonaceous material. A recent report of C_3H_2 (cyclopropenylidene, the first known interstellar hydrocarbon ring) emission from the cirrus may be relevant in this regard (Turner and Rickard 1987; see also Matthews and Irvine 1985). Band emission between 11.3 and 13.0 μm in spectra of HII regions could also be PAH resonances (Cohen et al. 1985).

There are difficulties with the PAH hypothesis also. Duley and Williams (1986) estimate that PAH chemical destruction lifetimes are less than 10^5 yr in diffuse clouds. If correct, this would make PAHs unlikely as a ubiquitous component of interstellar dust. However, the chemistry is complex and most processes are very uncertain. Omont (1986) appears to favor a more optimistic view of the destruction rates. Donn et al. (1987) argue that the spectroscopic mismatches of PAHs and the interstellar emitter are not easily reconciled, and that small PAH molecules could not emit the hot infrared continuum.

These results, as well as earlier measurements of the extinction shortward of 2200 Å (cf. review by Savage and Mathis 1979), mean that there are now several lines of evidence for a population of small (3-100 Å) grains. Approximately 5–15% of the carbon atoms would have to be incorporated in carbonaceous small particles emitting in the far infrared (Boulanger et al. 1985). While this is a small fraction of the carbon, interstellar abundances require that carbon, perhaps with some silicates, be a major grain constituent. Hot particles radiate strongly in emission bands. PAHs near hot nebulae require only 1 to 2% of the carbon (Puget et al. 1985), and QCCs or HACs should be similar. Despite the modest carbon requirements, the UIR data promise to give us key insights into the carbon chemistry.

vi. Interstellar 3.4 μm Absorber. There is a broad absorption at 3.4 μm (Fig. 7), characteristic of C-H bonds, observed toward the galactic center (Willner et al. 1979; Butchart et al. 1986). Most of the extinction toward the galactic center is produced in diffuse clouds in which ice is not observed and which should not contain volatile organic solids. Consequently, the 3.4 μm band is believed to be an indicator of a refractory component in the carbon chemistry (Greenberg 1976,1982). About 15–25% of the interstellar carbon in refractory grains would be needed to give this absorption (Draine 1985; Tielens and Allamandola 1987*a*).

The galactic center is thus far the only instance of this absorption in the diffuse medium (there are absorptions at 3.4 μm in molecular clouds; Sec. III.C). This may be a selection effect because the band would be difficult to see in most low-density clouds. Whether or not the galactic center 3.4 μm absorption is typical of diffuse clouds is not known.

4. Silicon Carbide, Carbonates. SiC forms in circumstellar shells of carbon-rich stars (Treffers and Cohen 1974; Merrill and Stein 1976). The absence of its characteristic 11.5 μm absorption in diffuse clouds shows that SiC is at most a minor constituent of refractory grains.

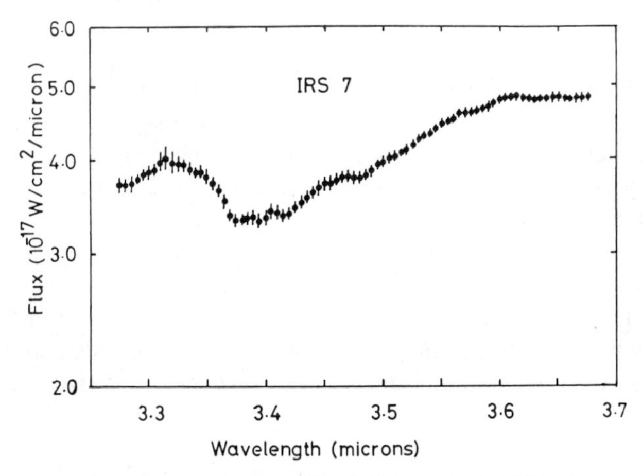

Fig. 7. Spectrum of the galactic center source IRS 7 (Butchart et al. 1986), showing the 3.4 μm feature that has been attributed to refractory organic material in grains.

Carbonates ($-CO_3$) would be observable through a strong band at 6.8–7.2 μm if they were abundant. Bands at these wavelengths occur in molecular clouds, but the absorption is weak or absent toward the galactic center where silicates absorb strongly (Willner et al. 1979). Hence, the refractory grains in this direction are not rich in carbonates, although a few percent abundance, typical of meteorite content, is not ruled out. The 6.8 μm molecular cloud band may arise from a volatile ice (Sec. III.C).

5. Summary. Debates about the nature of the grains in diffuse clouds are likely to continue for some time. For reasons of space, we have limited the discussion, but have included some ideas that are controversial. Amorphous or QCC material in small grains, rather than as molecules, may avoid the difficulty of the infrared continuum. There could be PAH molecules in addition to amorphous grains (Duley 1987; Borghese et al. 1987; Allamandola et al. 1987*b*). Indeed, all of the carbonaceous materials could contain PAH molecules. Interstellar carbonaceous material will surely reflect the complexity of carbon chemistry, and several, perhaps many, components may be present.

The recent results give intriguing clues to the carbon missing from the gas. A poorly ordered, complex carbonaceous material in the grains seems plausible based on spectroscopic and physical data, but the chemistry is still vague. The chemical state of oxygen and nitrogen in the refractory grains is even more uncertain.

C. The Volatile Component

1. Evidence for Volatiles. Rich infrared absorption spectra (Fig. 5) occur in sources that lie in or behind dense molecular clouds with strong extinctions ($A_V \geq 15$). These features are inferred to be absorptions of volatile ices. Some bands overlap, and the number of distinct volatile compounds is not clear.

2. Water Ice. The strongest volatile band is at the O-H stretch vibration of H_2O ice (3.07 μm). Water ice freezes out in the sheltered regions of dark clouds; it does not appear in diffuse clouds (cf. review by Martin 1978). The refractory grains must serve as nuclei for the volatiles, since ices would not nucleate in the clouds.

Only in rare cases do the 3.07 μm band shapes agree well with laboratory spectra of pure water ice. A source in which they do is OH 0739-14, an oxygen-rich star in a bipolar reflection nebula. In Fig. 8, the infrared spectrum is fitted with theoretical extinction by silicate core–ice mantle grains (Smith et al. 1986). Absorptions at 6.0 μm and 11.5 μm confirm the H_2O ice identification (Soifer et al. 1981).

3. Ice Mixtures. The spectra of IRS-2 and IRS-3 in Mon R2, shown in Fig. 8, are more typical of molecular cloud spectra and differ from those of

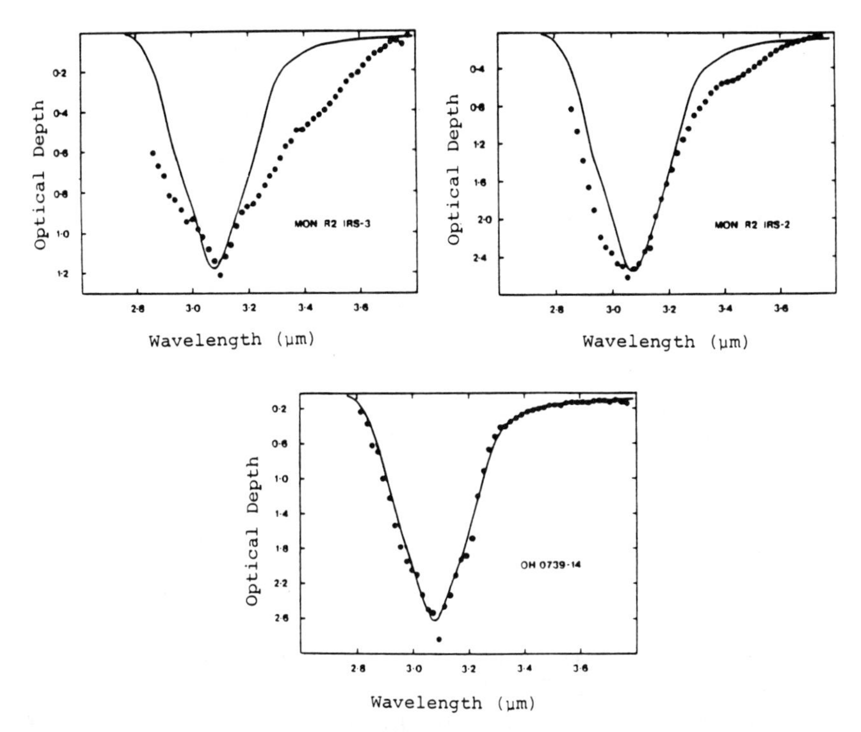

Fig. 8. Spectra of water ice sources. OH 0739-14 shows absorption of almost pure water ice. The Mon R2 sources have an ice band with additional absorption in the wings (Smith et al. 1986).

pure water ice; there is structure in the short and the long wavelength wings. This has led to some confusion, including occasional doubts that water ice is present at all. That the interstellar band is more complex than that of pure water ice is not surprising, since the functional groups C–H, N–H, and O–H all resonate in this spectral region. C–H groups absorb between 3.2 and 3.5 μm and N–H groups between 2.7 and 3.3 μm (Bellamy 1975). Consequently, mixtures of compounds of C, N and O have broader, more structured infrared bands than does H_2O.

Biological materials contain these groups also, and their spectra more or less resemble the interstellar band, as do many organic mixtures. However, the significance of this is limited. The fundamental vibrations are *near neighbor* interactions (O–H, C–H, etc.) and identify functional groups of a few atoms. It is questionable to infer the presence of *large-scale,* organized structures, such as bacteria (Hoyle et al. 1982), from band fits. Phosphorus abundances are also inconsistent with biological material (Duley 1984).

Laboratory ice mixtures of H_2O, CH_4, CH_3OH, H_2CO, NH_3, CO, CO_2 and other compounds show telling similarities with interstellar spectra (Fig. 9). Ultraviolet (Hagen et al. 1980) or proton irradiation (Strazzula 1985) of ice

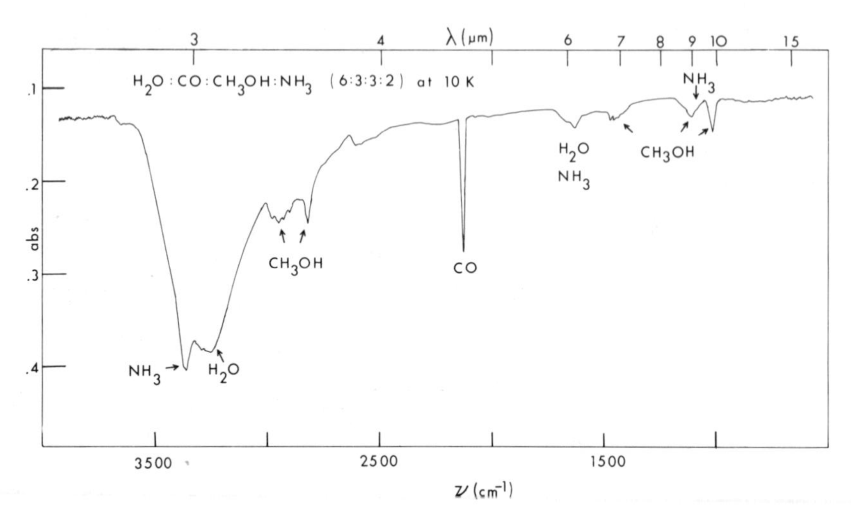

Fig. 9. Laboratory spectrum of an ice mixture containing H_2O, CO, CH_3OH and NH_3 (Hagen et al. 1980).

mixtures initiate chemical reactions that give products with spectra that often closely resemble the interstellar data.

As expected from the volatility of ices, the 3.07 μm band intensity does not correlate with the silicate band (Willner et al. 1982). Band intensities in clouds show that about 5–15% of the available oxygen is in H_2O ice (Gillett et al. 1975; Tielens 1983). Since about 6% is in gas phase CO and up to 20% in the silicates, this still leaves more than half of the oxygen unaccounted for.

4. Solid CO. Carbon monoxide freezes on grains and has been clearly detected in the denser clouds (Soifer et al. 1979; Lacy et al. 1984; Larson et al. 1985). No clear correlation between CO and H_2O ice abundance or the gas temperature has emerged (Geballe 1986).

Determining the relative amounts of CO in gas and dust is difficult. Column densities derived from radio measurements may not be of gas at the location of the infrared-absorbing material. The ratio of gas/solid CO is about 30 in NGC 2024 IRS2, the only source in which both phases have been observed by infrared absorption. However, because the temperature distribution and vapor pressure dependence are unknown, the *local* ratios of gas to solid CO remain uncertain (Geballe 1986).

5. Volatile Carbon Compounds: the 3.4 μm Wing. The long wavelength wing of the 3.07 μm band (Figs. 5 and 8) is at the 3.4 μm resonance of C–H groups (Capps et al. 1978; Hagen et al. 1980; Smith et al. 1986). Ammonia hydrates ($NH_3 \cdot H_2O$) could also contribute to the wing (Knacke et al. 1982).

The amount of organic material absorbing near 3.4 μm is uncertain because the substance is still poorly characterized (Draine 1985). Band strengths of C-H groups and optical constants for plausibly similar solids such as tholins (Khare et al. 1984) imply that of order 25% of the available carbon is required in the Orion sources (Knacke and McCorkle 1987). More measurements of optical constants of carbonaceous materials could firm up such estimates.

6.8 μm Absorption. An absorption at 6.8 μm could result from scissors vibrations of methyl ($-CH_3$) or methylene ($-CH_2$) groups (Soifer et al. 1979). Alcohols, primarily CH_3OH (d'Hendencourt and Allamandola 1986), or ammonium ions (NH_4^+) in clay minerals (Knacke et al. 1982), are also possibilities.

Absorptions Between 4 and 6 μm. A strong, broad absorption at 2165 cm^{-1} in a high-resolution spectrum of W33A may be the C≡N stretch band in nitriles (Lacey et al. 1984). Larson et al. (1985) pointed out that gas-phase CH_3NC, an isonitrile, gives better wavelength agreement with the interstellar absorber (in the gas phase CH_3CN is clearly present, and CH_3NC has probably been identified at about 5% the abundance of CH_3CN [Cernicharo et al. 1988; cf. also Irvine and Schloerb 1984]). Whether the 2165 cm^{-1} absorber is a gas or solid is not clear, nor have the proposed assignments been corroborated by other bands.

Absorption Band at 4.9 μm (2043 cm^{-1}). This band is also controversial. Sulfur compounds (Geballe et al. 1985), C_3, CN and CH_3OH (Larson et al. 1985), metal carbonyls and C=C groups (Tielens and Allamandola 1987*b*) are possibilities.

Summary for Carbon Compounds. Although only the CO identification is secure, there is now considerable circumstantial evidence for carbon compounds in both diffuse and dense clouds. Ultraviolet photolysis changes frozen volatiles into complex organic material that may be quite refractory and survive exposure to radiation in diffuse clouds (cf. Greenberg 1982). Cosmic rays (Strazulla 1985) and ion bombardment (Bibring and Rocard 1984) stimulate the reactions also. The products may be the refractory carbonaceous materials observed in diffuse clouds (Sec. III.B).

Carbonaceous substances are a difficult observational problem. Infrared spectroscopy tends not to be very specific when applied to complex organic mixtures, and the substances are hard to characterize, even in the laboratory. Coal, kerogen and petroleum chemistry and history are still controversial after more than a century of laboratory study (Durand 1980). The astronomical situation is not likely to be easier.

6. Nitrogen Compounds. Structure at 2.97 μm in the wing of the 3.07 μm band resembles solid ammonia absorption. Scattering effects complicate the interpretation, and the proposed identification needs to be corroborated with other bands (Knacke and McCorkle 1987).

7. *Sulfur Compounds*. Geballe et al. (1985) found absorptions at 3.9 μm and 4.9 μm (the 2043 cm^{-1} feature mentioned above) in W33A which they interpreted as bands of sulfur-containing molecules. The 3.9 μm feature lies near absorptions of H_2S that are shifted in the solid state. The 4.9 μm band resembles spectral features in ultraviolet-irradiated laboratory samples containing sulfur (see also Larson et al. 1985; Tielens and Allamandola 1987*a*).

D. Summary of Dust Composition

Tables V and VI summarize the known or suspected constituents of the grains. Progress in identifying the composition is accelerating, primarily because spectroscopy continues to improve. Nevertheless, our understanding is still very incomplete. Silicates in diffuse clouds and H_2O and CO in denser

TABLE V
Some Probable or Possible Refractories in Interstellar Dust

Component	Spectral Signature	Abundance[c]	Comments
Silicate[a]	9.8, 18 μm	100% Si, 10–20% O	observed in diffuse, molecular, and circumstellar clouds
Carbonacous[b]			
ultraviolet absorber	2175 Å	25% C	amount of C depends on carbon compound; silicate identification also proposed
hydrogenated amorphous carbon (HAC)	7000 Å	5–15% C	
quenched carbonaceous composite (QCC)	3.28, 3.4, 6.2, 7.7, 8.6, 11.3 μm		HAC, QCC and PAH may all be related
polycyclic aromatic hydrocarbon (PAH)		1–2% C	near ultraviolet sources[d]
organic refractory	3.4 μm	25% C	observed only toward galactic center
Silicon carbide[a]	11.4 μm		circumstellar only

[a]Secure identifications.
[b]Tentative identifications.
[c]Relative to total cosmic complement.
[d]If responsible for some IRAS cirrus emission, fraction of C would be higher.

TABLE VI
Some Probable or Possible Volatiles in Interstellar Dust

Component	Spectral Signature[a]	Abundance	Comments
H_2O[b]	3.07, 6.0, 12 μm	5–15% O	many molecular clouds
CO[b]	4.67 μm	few % of C	CO solid/gas ratio variable
Carbonaceous[c]			
Ice band "wing"	3.4 μm	10–25% C	common
Hydrocarbon or alcohol	6.8 μm		
"XCN, XNC"	4.62 μm		only in W33A
C_3, CN	4.9 μm		
Nitrogen[c]			
NH_3	2.97 μm	30% N	associated with H_2O ice
Sulfur[c]			
H_2S	3.9 μm	few % of S	only in W33A
"XS"	4.9 μm		

[a]Observed only in molecular clouds.
[b]Secure Identifications.
[c]Tentative Identifications.

regions are well established constituents of interstellar grains. There must be carbonaceous materials also, but they are complex and poorly understood. Other proposed molecular band identifications need corroboration.

IV. CONNECTIONS WITH THE SOLAR SYSTEM

Whether traces of relatively unaltered interstellar materials can be found in the solar system, particularly in primitive bodies, remains an intriguing question.

A. Isotopic Evidence

Some refractory dust certainly survives solar nebula processes and preserves the anomalous isotope ratios in meteorites (e.g., for oxygen: Clayton 1978). Geiss and Reeves (1981) have proposed that condensation onto interstellar particles will lock in the high deuterium fractionation observed in the gas phase (Table III) as the prenebular interstellar cloud collapses. Fractionation processes may also take place on the grains, although direct evidence for this is lacking (Tielens 1983). High D/H ratios occur in meteorites, particularly in the organic components (Kolodny et al. 1980; Robert and Epstein 1982) and in interplanetary dust particles (Zinner et al. 1983; Brownlee 1987). The only known fractionation mechanism is ion-molecule reactions at temperatures below 100 K. Because these occur in interstellar cloud conditions, this is one of the stronger lines of evidence for an interstellar remnant in the solar system. The possibility of similar reactions in the primitive solar nebula cannot be ruled out, but seems less likely (Kerridge and Chang 1985; Chapter by Prinn and Fegley).

Interstellar dust may also be present in comets. Arguments have been based partly on formation models that have comets accreting in the outer solar system where interstellar particles should survive, and on observed abundances (cf. Delsemme 1982; Bierman et al. 1982). The picture is supported by recent Comet Halley *in situ* measurements. Early results indicate that the D/H ratio is 3 to 24 times cosmic (Eberhardt et al. 1987*a*) in the comet water, perhaps paralleling the enhanced D/H abundances in the other "primitive" reservoirs, the carbonaceous chondrites and the interplanetary dust.

Likewise, the discovery of higher D/H abundance ratios in Titan, Uranus and Neptune than in Jupiter and Saturn may be direct evidence for distinct solid and gas reservoirs of solar nebula material (Owen et al. 1986), and for a connection with comets.

B. Identifiable Chemical Components

The high carbon abundances in the giant planets (C/H up to 4 times solar in Jupiter and Saturn and up to 20 times solar in Uranus) may mark the contributions of solid relative to gaseous material in the formation of these planets. These abundances may be the spoor of interstellar gas depletions (see the chapter by Pollack and Bodenheimer).

Silicate emission in many comets (Merrill 1974; Ney 1977) strikingly resembles interstellar silicate emission and absorption. There are also spectral similarities among comets, interstellar dust and the matrix of carbonaceous chondrites (Knacke 1978). However, recent observations of Comet Halley show features of crystalline silicates such as olivine or pyrexene that are not seen in the interstellar medium or around stars (Bregman et al. 1987).

The detection of S_2 in Comet IRAS-Araki-Alcock may be evidence for interstellar material. A'Hearn and Feldman (1985) suggest that the most probable formation mechanism for S_2 is irradiation of sulfur-containing molecules in an ice matrix. Because IRAS-Araki-Alcock is a recurrent comet, the S_2 must be distributed throughout the nucleus. Ultraviolet photons or cosmic rays do not penetrate deeply, so the reactions probably proceeded before the comet accreted, that is, in interstellar grains. A'Hearn and Feldman emphasize that the conclusion is still tentative, being based on extrapolations of complex chemical processes. There is evidence for sulfur compounds in interstellar dust (Sec. III.C).

Detection of emission near 3.36 μm in the spectrum of Comet Halley (Combes et al. 1986) may be evidence for a carbonaceous residue of interstellar material in comets. The emitter is almost certainly a complicated organic substance, and may be related to the carbon, hydrogen, oxygen and nitrogen "CHON" particles (Clark et al. 1987). The band shape resembles, but is not identical to, the galactic center absorption at 3.4 μm and to wing absorptions in molecular clouds. It is definitely different from PAH spectra, although PAHs could be among the absorbers (Knacke et al. 1987, and references therein).

Whether these materials evolved through the same processes and there is a direct interstellar dust-comet connection, is an open question. The present data are also consistent with parallel but different evolution in somewhat similar circumstances. What we know about dust, comets and primitive meteorites suggests intriguing similarities among them. It is hard to think of established differences between comets and interstellar grains other than size. Nevertheless, the spectroscopic data do not give the detailed mineralogy that may be necessary to investigate the relationships in detail. Comets may not remain pristine interstellar dust reservoirs because there will be continuing chemical change stimulated by energetic particles or photons, at least of the surface layers (cf. Strazzula 1986*b*). Interstellar grains will surely evolve continuously also. Therefore, a definitive chemical key linking interstellar grains to solar system objects may prove difficult to find. Evidence from elemental abundances, isotopes or dating techniques may provide alternate means.

V. CONCLUSION

It is clear from the foregoing discussion that there are major uncertainties regarding the chemical state of the principal volatile elements in dense interstellar clouds. Our view of the situation is indicated schematically in Fig. 10.

There is considerable indirect evidence that the bulk of the carbon is sequestered in grains, although the chemical state of this material remains controversial. Part of the evidence for this conclusion rests on the failure to detect the major portion of the carbon abundance in the gas phase, at least in quiescent clouds. More sensitive infrared observations of potentially abundant nonpolar molecules such as acetylene would be useful to confirm this point. Exploration of that portion of the infrared spectrum obscured by telluric absorption will also be critical to defining the chemical state of carbon in both the refractory and volatile grain components.

The situation for oxygen is less clear. A significant fraction must be contained in silicate grain cores and H_2O mantles, as well as in gas-phase CO. Nonetheless, more than half of the oxygen is presently unaccounted for. Clearly, spacecraft observations to determine the abundances of H_2O and O_2 in different environments are essential.

There is evidence that at least for one cloud the bulk of the nitrogen may be in the form of gas phase N_2. On the other hand, some fraction may be incorporated into poorly characterized, refractory organic material, and there is also the likelihood of ammonia and derivative species in the volatile grain component in dense clouds.

The low abundance of sulfur species in the gas phase argues strongly for the bulk of this element being contained in the grains, although again the chemical form of sulfur in the solid state is largely unknown. Now that phos-

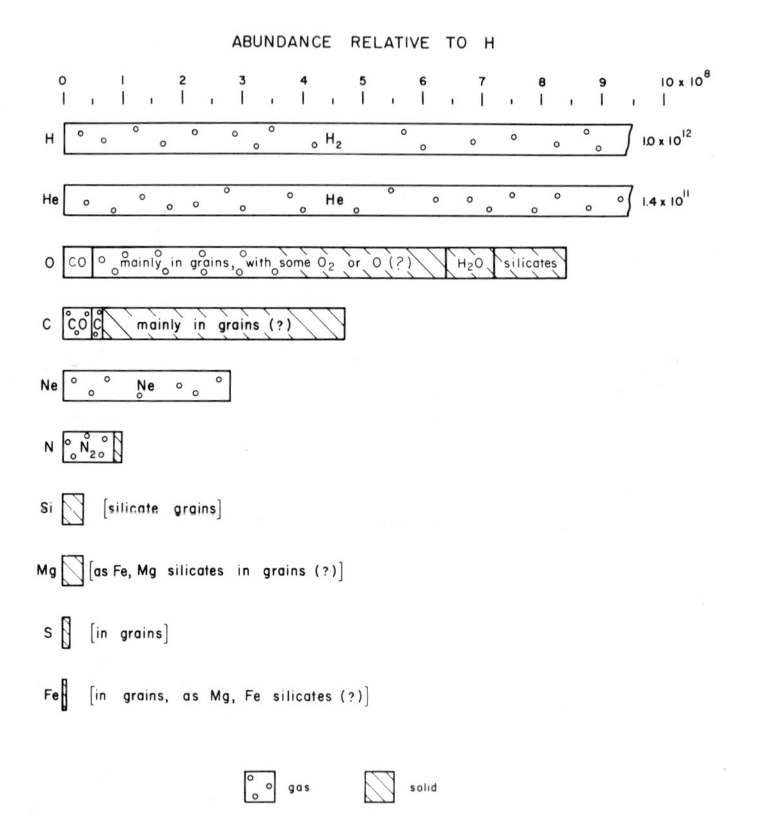

Fig. 10. Distribution of abundant elements in gas and dust in molecular clouds. Trace species (<1%) have not been included (although radio spectra of these are the source of much of our understanding of interstellar chemistry). The distributions are based on estimates discussed in the text for a "typical" dense molecular cloud. Abundances in specific clouds can vary by factors of 10 or more depending on physical conditions.

phorus has been detected in dense clouds, one may hope that the distribution of this element between possible phases will begin to be elucidated.

There is an increasing amount of highly suggestive, although not yet definitive, evidence for the survival of interstellar molecular material containing the volatile elements in primitive solar system objects, including comets, carbonaceous chondrites and interplanetary dust particles. Pursuit of this link promises to provide crucial information on the early state of the solar nebula.

Acknowledgments. The authors are grateful to several colleagues for helpful discussions, including Å. Hjalmarson, L. Ziurys, J. Mathis, A. Tokunaga and an anonymous referee. This research was supported in part by several grants from the National Aeronautics and Space Administration and the National Science Foundation.

MODELS OF THE FORMATION AND EVOLUTION OF THE SOLAR NEBULA

A. P. BOSS
Carnegie Institution of Washington

G. E. MORFILL
Max-Planck-Institut für Physik und Astrophysik

and

W. M. TSCHARNUTER
Universität Wien

We review observational and theoretical work relevant to the formation and evolution of the solar nebula, with emphasis on distinguishing which phases of planetary formation occurred in the presence of the solar nebula, a key concern for the origin and evolution of planetary atmospheres. Observations of regions of low-mass star formation provide important clues to the initial conditions for collapse of the presolar nebula. Energetic, episodic stellar winds (seen in FU Orionis stars and bipolar flows) are observed with mass-loss rates considerably higher than T Tauri stars. Removal of the gaseous portion of the solar nebula is thus probable within ≈1 to 10 Myr after formation of the proto-Sun, presenting a strong constraint on models of giant planet formation. Theoretical models of collapsing protostellar clouds largely agree with astronomical observations of young stellar objects. Models of rotating collapse possibly leading to the formation of the presolar nebula are described; formation of a single solar-type star apparently requires the collapse of a slowly rotating interstellar cloud. The primary uncertainty in solar nebula models is what physical process transported angular momentum outward and mass inward; leading candidates are turbulent viscosity and gravitational torques due to nonaxisymmetric structure. Convectively driven turbulent accretion disks have dominated the research of solar nebula modeling in the last decade, but severe problems with their basic physics have been uncovered. A new generation of solar nebula models is required before the detailed properties of the solar nebula (necessary for models of planetary formation) can be agreed upon. Because of large gaps in our knowledge, such as the angular momentum transport mechanism, not enough is known about the structure of the solar nebula to allow definitive answers even to fundamental questions such as the global stability of the gas in the nebula. Hence competing models of planetary formation must still be considered; in particular, the planetesimal accumulation and giant gaseous protoplanet models of planet formation are examined in the light of the time scale for nebula removal.

The basic paradigm of modern thinking on the origin of the solar system can be traced back to the nebular hypotheses advanced by I. Kant and P. S. Laplace in the eighteenth century: the planets formed out of the same rotating, flattened nebula of gas and dust that formed the Sun. Because of this cosmogonical link between the planets and the Sun, solar system formation is most naturally studied in the general context of the formation of solar-type stars in our Galaxy. With this point of view, it is also natural to employ a *forward modeling* approach, where one uses astronomical observations of star-forming regions to provide a starting point for evolutionary theoretical models which might lead to the formation of a solar system similar to our own. In this chapter we review current work on protostellar collapse and formation of preplanetary disks, and thus set the stage for the theme of this book, the formation of planetary atmospheres.

The primary relationship of this chapter to the origin and evolution of planetary and satellite atmospheres is through the question of the presence or absence of significant nebular gas during the various phases of planetary formation. Number densities at 1 AU of perhaps 10^{15} cm^{-3} during the solar-nebula phase are to be compared with densities of about 10 cm^{-3} in the present solar wind; at some time in the distant past, the solar system was swept nearly clean of gas. Whether the cosmic composition gas surrounding the growing Earth had a pressure of 10^{-5} bar or 10^{-19} bar is likely to influence strongly the elemental and isotopic composition of the Earth's present atmosphere, both directly through gas capture and blow-off, and indirectly through the effects of shielding the proto-Earth from early solar ultraviolet radiation. For some of the outer planets, this question is important not just for their atmospheres, but for the bulk of their mass: Jupiter and Saturn apparently must have formed largely in the presence of the solar nebula.

We will consider the formation of satellites to be a tertiary process, compared to planet formation, which is itself secondary to solar formation, and hence will concentrate on the latter two processes (see the Chapter by Coradini et al. for a description of current thinking about satellite formation).

I. OBSERVATIONS OF STAR FORMATION

The youngest stars are always associated with interstellar clouds of gas and dust. Their small spatial velocities and young ages require that they formed within the clouds. It is thus certain that stars form from the gravitational collapse of interstellar clouds.

The main assumption underlying this section is that present-day star formation is similar to star formation 4.5 Gyr ago when the presolar nebula was forming, and more specifically that single stars are currently being formed with planetary systems like the solar system. If this assumption is true, we can expect observations of present-day star formation to guide us in understanding the formation of the solar nebula. In order to maximize the probability of these assumptions being correct, we shall restrict our attention to the formation of low-mass stars $\approx M_{\odot}$ (= mass of Sun) with solar compositions in the disk of our Galaxy. Shu et al. (1987) have recently reviewed star formation in molecular clouds.

It has been thought that the presolar nebula was injected with the nucleosynthetic products of massive star evolution immediately prior to collapse (Cameron and Truran 1977). If this was the case, then the presolar nebula probably formed in a region of high-mass star formation, in which case the observational picture presented in this section, based largely on regions of low-mass star formation, may have to be somewhat altered. However, the interstellar medium at a given galactocentric radius is only isotopically homogeneous in species such as carbon isotopes to within the uncertainty of the measurements (about a factor of two in relative abundance), implying that the interstellar medium is well mixed at that level. Meteoritical evidence for isotopic anomalies in primitive solar system material involves anomalies of at most a few percent, however, which is well within the possible isotopic range of interstellar clouds. Furthermore, the recent detection of substantial amounts of radioactive ^{26}Al in interstellar clouds has removed much of the enthusiasm for a supernova trigger for the presolar nebula (Cameron 1984; Clayton 1984,1985), so an emphasis on low-mass star formation in this chapter seems appropriate.

Star Formation in Molecular Clouds

Star formation in the galactic disk occurs in dusty interstellar clouds composed of molecular hydrogen, He and molecules such as CO and NH_3. These molecular clouds contain most of the mass of the interstellar medium (Bally 1986). Molecular clouds have a hierarchy of structures extending from giant molecular clouds ($\approx$100 pc in size) to dense cores ($\approx$0.1 pc). Masses range from 10^6 $M_{\odot}$ to 0.1 $M_{\odot}$ for the smallest dense cores, while mean number densities range from 10 molecules cm^{-3} to 10^6 cm^{-3}. Temperatures of 10 K are common.

The dense cloud cores appear to be gravitationally bound and centrally condensed, making them prime candidates for low-mass star formation (Myers and Benson 1983). This hypothesis has been confirmed by observations showing the presence of infrared objects in nearly half of one large sample of cloud cores (Beichman et al. 1986). Of the infrared objects, less than one half are T Tauri stars, low-mass stars in the pre-main-sequence phase of evolution. The remaining infrared objects have no optically visible counterpart, and could be deeply embedded T Tauri stars or even younger protostellar objects still undergoing accretion.

Dense cores without infrared sources have been advanced as appropriate indicators of the initial conditions for low-mass star formation. In particular, the evidence for power-law density profiles in dark-cloud envelopes (Snell 1981; Arquilla and Goldsmith 1985) has been inferred to mean that protostellar collapse begins from the unstable equilibrium state of the singular isothermal sphere (Shu 1977), where $\rho \propto r^{-2}$. However, the fact that many of these clouds cores have already formed pre-main-sequence stars implies that gravitational collapse has long since started in these star-forming regions. Considering that further embedded infrared sources may someday be detected at lower infrared luminosities or other wavelengths in the remaining cloud cores, it may be that nearly all of these cloud cores have already undergone substantial collapse. If this is true, then their envelopes may not be indicative of the initial conditions for protostellar collapse, but rather of the envelope of a cloud which has already collapsed from some unknown initial configuration. We shall see in Sec. II that power-law density distributions arise in the collapse of clouds from a wide range of initial conditions.

Whether or not the dense cores have already or will form single or binary or even multiple protostars is unknown. This is quite important for solar nebula models, because the Sun is a single star. No binary embedded infrared objects have been seen to date, but this is not unexpected, largely because the spatial resolution of the Infrared Astronomical Satellite only allowed the visual separation of very wide binaries. In fact, many dense cores appear to be fragmented on slightly smaller (but poorly resolved) spatial scales than 0.1 pc (Heyer 1987), implying that multiple star formation may indeed result from dense cores. This is reassuring, considering that the majority of normal main-sequence stars are found in close binary systems which are themselves members of more widely separated systems (Abt 1983). While the binary frequency may need some revision (see, e.g., Morbey and Griffin 1987), the general trend seems to be for the frequency of binary and multiple systems to *increase* as observations improve (see, e.g., Wasserman and Weinberg 1987; Mayor and Mazeh 1987; McAlister et al. 1987). Unfortunately, if single stars are truly rare, then observations of the average dense core may not be relevant for the presolar nebula.

Angular Momentum of Dense Clouds

The primary determinant of whether a given dense cloud forms a single or multiple protostar may well be the angular momentum J of the cloud.

Binary systems of low-mass stars can have orbital angular momenta 10^3 times that of the solar system, and 10^5 times greater than the spin angular momentum of the Sun, which rotates at a rate typical of low-mass stars. Hence the rotation rates of dense clouds appear to be extremely important for determining the outcome of protostellar collapse. Rotation is very hard to measure, however, because even the most rapidly rotating dark clouds often have Doppler shifts which are barely measurable, and rotation can be intermingled with other large-scale motions such as relative motion between sub-clouds or small-scale motions such as turbulence. Goldsmith and Arquilla (1985) have reviewed the evidence for rotation in 16 dark interstellar clouds (Fig. 1). They found the specific angular momentum J/M to vary between $\approx 10^{21}$ cm^2 s^{-1} and $\approx 10^{24}$ cm^2 s^{-1}, with the higher values being found in the largest clouds. Their sample includes only the most rapidly rotating clouds, however. The rotation rates of 10^{-14} rad s^{-1} found in the dense cores studied by Heyer (1987) correspond to J/M about a factor of 10 lower for a comparable size cloud, implying that J/M may be smaller than $\approx 10^{20}$ cm^2 s^{-1} for cloud cores of solar mass.

Regardless of the process that produces dense cloud cores out of larger-scale structure, it is evident from Fig. 1 that the process does not proceed with conserved angular momentum, but more nearly with conserved angular velocity. The existence of dense cloud cores with masses comparable to low-

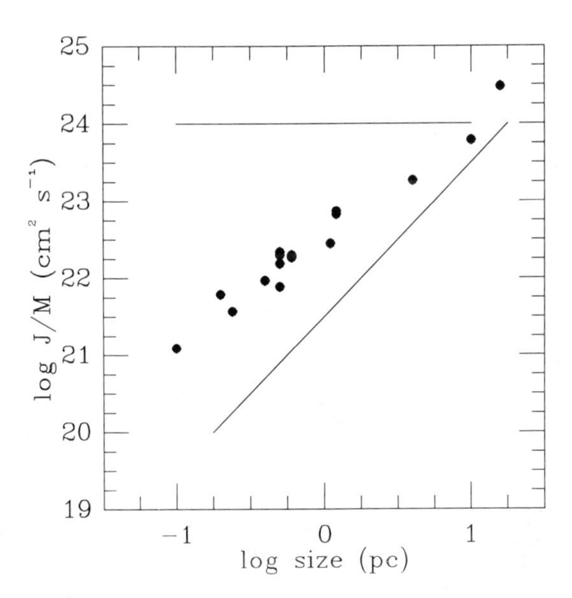

Fig. 1. Specific angular momentum J/M as a function of size for a sample of 16 rapidly rotating clouds (figure adapted from Goldsmith and Arquilla 1985). The horizontal line illustrates the expected distribution if cloud contraction and fragmentation occurs with constant angular momentum and mass, while the oblique line gives the trend for conserved angular velocity and mass.

mass stars and with relatively small specific angular momentum implies that a considerable amount of the fragmentation and angular momentum reduction necessary for low-mass star formation in molecular clouds is achieved during the transition from the quasi-static molecular cloud phase to the dense core phase, rather than during gravitational collapse to the protostellar phase.

Magnetic Fields

Magnetic fields appear to be quite important for the dynamics of diffuse and molecular interstellar clouds, and in particular for removing angular momentum from interstellar clouds as they form and contract from the general interstellar medium. The angular momentum removal is a consequence of the tendency of magnetic fields to preserve the angular velocity as the density of a cloud increases (Mouschovias 1977). Figure 1 is reasonably consistent with this scenario.

Magnetic fields are expected to be less important for the dynamics of dense clouds, because increasing cloud density shields the interior of the cloud from the ionizing radiation needed to maintain the fractional ionization (initially $\approx 10^{-8}$ to 10^{-7}) of the cloud. As the fractional ionization drops, the drag forces between the neutral gas and the ionized components decrease, allowing the neutral gas to contract relatively unimpeded by the magnetic field, a process known as ambipolar diffusion. One critical quantity is then the density n_d at which decoupling from the magnetic field occurs, in the sense of the magnetic field no longer being dynamically important; theoretical estimates of n_d have a wide range.

The critical density for decoupling has been observationally determined in one star-forming region by studies of the relative orientations of the ambient magnetic field (deduced by the polarization of light from embedded or background stars) and the rotation axes and minor axes of various clouds in the Taurus molecular cloud complex (Heyer et al. 1987). For relatively diffuse clouds with densities $<10^3$ to 10^4 atoms cm^{-3}, the clouds tend to be flattened with minor axes and/or rotational axes parallel to the magnetic fields. In the former case, this configuration implies that contraction preferentially down field lines has occurred, and hence that the magnetic field is strong enough to resist contraction across field lines. In the latter case, the magnetic field is evidently strong enough to transport to the surrounding envelope the component of angular momentum perpendicular to the field lines, leaving the component parallel to the field. However, for relatively dense cloud cores ($\gtrsim 10^3$ to 10^4 atoms cm^{-3}), the core and minor rotational axes are no longer aligned with the large-scale magnetic field (Fig. 2), implying that the field is no longer able to dominate the structure (Heyer 1987). Thus it appears that magnetic fields become dynamically insignificant at densities intermediate between diffuse clouds and dense cloud cores, i.e., $n_d \approx 3 \times 10^3$ cm^{-3}.

Considering that the collapse of the presolar nebula presumably started from a dense interstellar cloud core, magnetic fields apparently did not play an

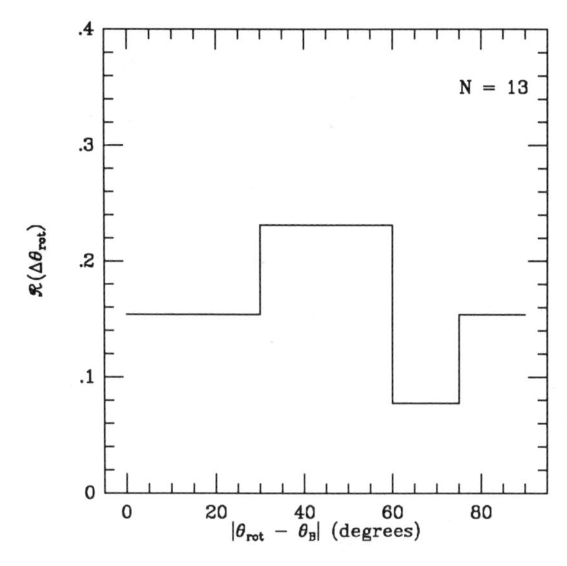

Fig. 2. Histogram of the distribution of the angle between the ambient magnetic-field direction and the rotation axis for a sample of 13 dense interstellar cloud cores in the Taurus-Auriga region (figure adapted from Heyer 1987). The lack of a clustering about zero angle implies that the magnetic field dynamically decouples from the gas at densities above about 3×10^3 cm^{-3}.

important role in the collapse, during the formation of the solar nebula, or in its earliest evolution. Once the proto-Sun began to generate a significant magnetic field through the dynamo mechanism, however, magnetic fields probably became important once again for angular momentum loss through the early solar wind.

Mass Loss from Young Stellar Objects

It has long been known that solar-type pre-main-sequence stars (T Tauris) lose mass in the form of stellar winds (Herbig 1958; Kuhi 1964), and that this mass loss could be significant for planet formation (Cameron 1973*a*). Estimates of the mass-loss rates for T Tauri stars are subject to substantial uncertainty, but typical rates appear to be on the order of about 10^{-8} $M_\odot$ yr^{-1} (Kuhi 1964; DeCampli 1981), about a factor of 10^6 higher than the current solar wind ($\approx 10^{-14}$ $M_\odot$ yr^{-1}). Some recent estimates of mass loss are as large as $\approx 10^{-6}$ $M_\odot$ yr^{-1} for one of the components of T Tauri itself (Schwartz et al. 1986). The ages of solar-type T Tauri stars are thought to be on the order of 1 Myr (Cohen and Kuhi 1979).

Strings of Herbig-Haro (HH) objects are found to be physically aligned with T Tauri stars, on the order of one pc distant, and to have spatial velocity vectors directed away from the central star. HH objects are significant clumps of gas shocked by high-velocity (100 to 400 km s^{-1}) stellar winds from their exciting stars (Fig. 3). At this speed, the stellar wind exciting the HH objects left the central star about 10^4 yr ago. A steady isotropic stellar wind from a normal T Tauri star is insufficient to accelerate these objects (Lada 1985). The association of HH objects and bipolar flows with T Tauri stars thus implies that T Tauri stars produce high-velocity outflows in their early evolution even more energetic than during their later pre-main-sequence evolution (Lada 1985).

Observations of the handful of FU Orionis stars also imply that low-mass stars experience an early phase of mass loss even more energetic than the T Tauri phase. FU Orionis stars are T Tauri stars that undergo episodic phases of greatly enhanced luminosity (factors of 10 to 100) for periods of about a decade (or more) at least every 10^4 yr (Herbig 1977). Accompanying this increased luminosity is increased mass loss, by a factor of 100 to 1000 times

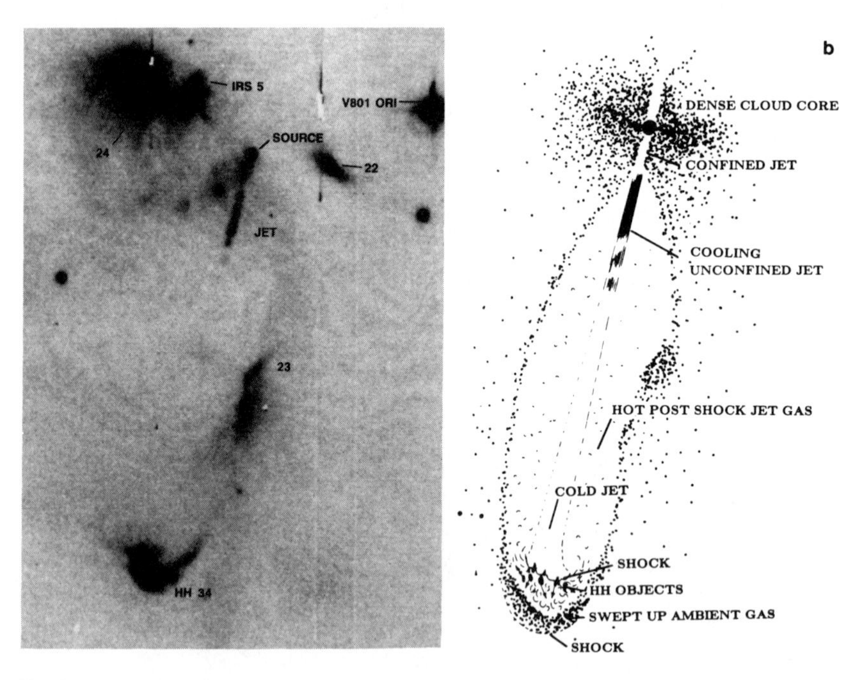

Fig. 3. An example of energetic mass loss from a low-mass pre-main-sequence star (figure adapted from Reipurth et al. 1986). (a) Optical CCD image showing Herbig-Haro object 34, located on the edge of a lobe of gas threaded by a narrow jet emanating from the exciting star (marked source). CO observations show that this system is embedded in a dense cloud core of size 0.1 pc and mass $\approx 4M_{\odot}$. (b) Theoretical interpretation of this region. A high-velocity wind from the source, collimated by the high-density cloud core, has evacuated an ovoidal cavity. HH 34 is produced by the bow shock between this wind and the ambient molecular cloud.

that of a T Tauri star (Croswell et al. 1987), to rates as high as 10^{-5} $M_{\odot}$ yr^{-1}. Considering that with more and longer observations we may find that FU Orionis outbursts last longer than a decade, and may recur more frequently than every 10^4 yr, the amount of mass lost during a FU Orionis phase may very well exceed that lost during the more nearly steady, but lower intensity, T Tauri wind phase. Graham and Frogel (1985) found an FU Orionis star associated with a HH object, tightening the suspicion that all of these mass-loss phenomena are linked together.

The underlying physical process driving T Tauri mass loss is usually thought to be a supercharged version of normal low-mass stellar winds, namely mechanical deposition of energy in the stellar atmosphere through Alfvén waves generated in the convection zone (see, e.g., DeCampli 1981). The process responsible for FU Orionis outbursts is more controversial (see Herbig 1977), and many suggestions have been made, such as rotational instability (Larson 1980), dissipation of rotational shear energy (Shu and Terebey 1984), and episodic mass accretion from a disk (Hartmann and Kenyon 1985).

Evidence for an even earlier phase of energetic mass loss comes from observations of bipolar molecular flows. These flows consist of very large amounts (on the order of 0.1 to 100 $M_{\odot}$) of high-velocity ($\approx$10 to 100 km s^{-1}), cold (10 K to 90 K) gas (Lada 1985). While it is uncertain what fraction of this gas is swept up from the surrounding medium, mass-loss rates on the order of 10^{-6} $M_{\odot}$ yr^{-1} are possible for solar-mass embedded objects. Perhaps most remarkable is that these outflows are often collimated in opposite directions centered on an exciting star. The central star tends to be an embedded infrared object rather than an optically visible pre-main-sequence star, implying that this phenomemon sometimes occurs prior to the T Tauri or FU Orionis phases. The dynamical ages of the outflows are on the order of 10^4 yr. Statistical arguments imply that bipolar outflows occur within the first 10^5 yr of pre-main-sequence evolution (Lada 1985). Their formation rate is comparable with that of solar-type stars, implying that, like the T Tauri and FU Orionis phases, many or perhaps all solar-type stars experience a similar phase.

The degree of collimation in bipolar flows varies substantially, but the fact that collimation is seen at all has led to a plethora of models for producing intrinsically collimated winds or for extrinsically collimating an initially isotropic wind. This point is clearly important for understanding the effect of such a wind on preplanetary matter in a circumstellar disk; if the wind is intrinsically collimated, such as by being formed at the surface of the disk (see, e.g., Torbett 1984), then the wind may not scour the preplanetary nebula at all. A more conventional picture is that the stellar wind originates as an isotropic flow (perhaps similar to an FU Orionis outburst) which is then collimated by larger-scale structure, such as a rotationally flattened disk (see, e.g., Boss 1987*b*, and references therein). Such a disk of gas and dust ($\approx$500 AU wide) has been detected about the T Tauri star HL Tauri (Grasdalen et al. 1984; Beckwith et al. 1986). In the extrinsic collimation model, collimation of the wind by the preplanetary disk almost certainly involves some loss of disk matter, and thus may remove much of the placental gas and dust. Continued wind erosion of the disk may lead to decreased collimation and eventually to nearly complete clearing of optically obscuring matter. These theoretical speculations are strongly supported by the observations to be described next.

Removal of Circumstellar Gas

The age at which young stars lost their circumstellar gas and dust is a critical question for models of planet formation. While theoretical models of the interaction of strong stellar winds with circumstellar matter have long implied that the solar nebula was removed within 1 to 10 Myr (see Sec. V below), observational confirmation is clearly crucial. Recent observations appear to confirm this theoretical time scale for nebula dispersal.

Direct evidence for circumstellar or nearby dust and gas comes from observations of infrared excesses in the first case and from Ca II and Na I absorption lines and concentrated CO emission in the second case. Let us first consider infrared excesses. While infrared excesses are common in T Tauri stars (Cohen and Kuhi 1979), no such excesses were found for a sample of 13 solar-type stars of age $\approx 3 \times 10^2$ Myr (Witteborn et al. 1982). Even more important for our purposes is the recent work by Walter (1986), who found no infrared excesses in two classes of pre-main-sequence solar-type stars. The first class, termed post-T Tauri stars, appears to be stars which have evolved beyond the T Tauri phase and are approaching the main sequence, with ages of about 10 Myr. The second class, termed naked T Tauri stars, appear to be normal T Tauri stars (with ages of 1 Myr) except for the absence of any discernible circumstellar envelope. This class places the most restrictive limit ($\approx$1 Myr) upon the lifetime of primordial circumstellar matter.

Consider now the evidence for gas. Hobbs (1986) searched a total of 16 stars with known infrared excesses for evidence of Ca II and Na I absorption, and found only one star with evidence of circumstellar gas, β Pictoris. However, even in the celebrated case of β Pic, where the infrared excess has an optical manifestation in the form of a flattened disk of dust extending 400 AU outward (Smith and Terrile 1984), the total mass of the gaseous disk is less than 2 $M_\oplus$ (=2 Earth masses) (Hobbs et al. 1985). Such a gas disk is so diffuse $<10^5$ cm^{-3}) compared to the solar nebula ($\approx 10^{15}$ cm^{-3}) that it is negligible for planet formation.

Hobbs et al. (1985) found no evidence at all for gas surrounding α Lyrae (Vega), in spite of the large infrared excess found by Aumann et al. (1984) to be caused by a shell of dust particles located 85 AU away from the star. While observations of both β Pic and α Lyrae have been hailed as direct evidence of planetary-system formation, because of the absence of appreciable gas, it seems much more likely that these observations are evidence of collisional processes leading to dust formation in a swarm of more aged comets and/or asteroids in the outer reaches of another planetary system. This is supported by the fact that Vega is a main-sequence A star of mass $\approx 3\ M_\odot$, with an age of $\sim 3 \times 10^2$ Myr, and the mass of β Pic is $\approx 2\ M_\odot$, yielding an age on the order of 1 Gyr, with both stars being so old as to no longer be experiencing planet formation in the usual sense. In fact, the spectral signature of the dust around these nearby stars implies that the mean grain size is quite small and that the closest analogue to the solar system may be the zodiacal light, rather than the solar nebula (Harper et al. 1984; Gradie et al. 1987).

Substantial gas disks have been directly detected through CO interferometric observations of the T Tauri star HL Tauri and the young star R Mon (NGC 2261) by Beckwith et al. (1986). HL Tauri also exhibits a large infrared excess implying at least 0.1 $M_\oplus$ of circumstellar dust. Recent observations by Sargent and Beckwith (1987*b*) imply a total gas mass of $\approx 0.1\ M_\odot$ within 1600 AU of HL Tauri, while R Mon is thought to be surrounded by a similarly sized gas disk with a mass of at least 0.01 $M_\odot$. Both of these stars appear to be very young. The luminosity and effective temperature of HL Tauri imply an age $< 10^5$ yr. while R Mon is the exciting star for a large bipolar flow, another indicator of extreme youth. The presence of gas disks in these two stars, as well as the other evidence presented in this section, is quite consistent with our scenario of energetic mass loss starting at $\approx 10^5$ yr and complete loss of circumstellar gas and dust by $\approx 10^6$ to 10^7 yr.

II. FORMATION OF THE SOLAR NEBULA

In the next two sections we describe attempts to model theoretically the formation of the presolar nebula from a collapsing interstellar cloud and to model the evolution of the solar nebula prior to its dispersal. Theoretical models are essential for learning about these events, in part because they occurred over 4.5 Gyr ago, but also because observations of present-day star formation provide little direct information about physical conditions in the cores of collapsing clouds or in circumstellar disks during the earliest phases of planetary formation.

The collapse of a fragment of an interstellar cloud to form the solar system is best studied within the context of attempts to understand star formation in general. Thus we will discuss models of protostellar collapse applicable to low-mass stars like the Sun, with special emphasis on calculations seeking to understand formation of the solar system. Theoretical work on protostellar collapse has been reviewed recently by Boss (1987*a*).

Nonrotating Collapse Models

The first detailed calculations of protostellar collapse (McNally 1964; Hayashi and Nakano 1965; Bodenheimer 1968; Larson 1969; Hunter 1969) involved the collapse of nonrotating clouds. The restriction to spherical symmetry simplifies the mathematics sufficiently to allow the evolution to be reliably calculated from the initial interstellar cloud through protostellar collapse to the formation of a pre-main-sequence star (Winkler and Newman 1980; Stahler et al. 1981).

The initial phases of protostellar contraction occur nearly isothermally (at $\approx$10 K) because of the transparency of the cloud in the infrared and the presence of dust grains, which radiate away the compressional energy produced by contraction. Collapse then occurs because thermal pressure is unable to counterbalance the increasingly larger gravitational forces that arise as the cloud becomes more compact. Isothermal collapse of a nearly uniform density cloud is characterized by the tendency for a cloud to collapse as a whole toward a $1/r^2$ density profile, with supersonic speeds first occurring toward the center of the cloud after roughly one free-fall time. The free-fall time $t_{ff} = (3\pi/(32G\rho_i))^{1/2}$ is the time for a pressureless, uniform-density sphere to collapse to a singularity, starting from density ρ_i. Isothermal collapse of a nearly equilibrium cloud (with density initially $\propto r^{-2}$) is characterized by collapse initiated from the center: an expansion wave moves out through the cloud envelope at the sound speed, and successive layers of the envelope begin to collapse once the wave has passed (Shu 1977). In the latter case, the density in the collapsing region tends toward an $r^{-3/2}$ density profile. In both cases, the most supersonic speeds are concentrated toward the center.

The first collapse phase is halted when the center of the cloud becomes dense enough to be opaque to infrared radiation, so that the compressional energy no longer freely exits the protostar. This defines the beginning of the nonisothermal regime of protostellar collapse. The rising central temperature halts the collapse in the central regions, while the outer regions continue to infall onto the first core. When the first core reaches temperatures of $\approx$2000 to 3000 K, the molecular hydrogen begins to dissociate into atomic hydrogen. The energy needed for dissociation is removed from the thermal support of the protostar, and a second collapse phase begins. The second collapse halts when all the molecular hydrogen at the center is dissociated, yielding the second core. The second core grows from an initial mass of $\approx$0.01 $M_\odot$ to its final mass by accretion of the infalling cloud envelope through an accretion shock of size a few solar radii. The process of collapse and accretion requires about 10^5 to 10^6 yr (roughly a few t_{ff}) for a nonrotating, solar-type protostar. Once the cloud envelope is accreted or blown away, the accretion shock turns into the optically visible photosphere of the pre-main sequence star. The pre-main-sequence phase consists of another 10^6 to 10^7 yr of contraction to higher densities and temperatures, until sustained thermonuclear fusion initiates the main-sequence phase.

Recently Tscharnuter (1987) developed a new numerical code for studying protostellar collapse in spherical and axial symmetry, and has encountered some remarkable results that may alter this picture of the main accretion phase of protostars. The new results suggest that the second protostellar core explosively re-expands following its initial formation, necessitating a delay before the central regions re-collapse and re-form the second core, followed by another re-expansion, and so on for perhaps 20 cycles.

The physics behind Tscharnuter's oscillations may be understood as follows. The second collapse is initiated by the dissociation of molecular hydrogen, which reduces the effective adiabatic exponent $\Gamma_1 = (\partial \ln P / \partial \ln \rho)_{ad}$ below the critical value of 4/3. $\Gamma_1 \approx 4/3$ deep in protostellar clouds was first noted by Bodenheimer (1968). The second collapse is halted by the completion of dissociation and the consequent rising of Γ_1 above 4/3. That a bounce ensues has long been known (Hayashi and Nakano 1965); the question is the amplitude of the bounce. Bodenheimer (1968) found no bounces, Larson (1969) found only a small rebound, while Winkler and Newman (1980) found a more vigorous rebound, but gave few details, and thereafter suppressed the bounces by using a large time step. Tscharnuter (1987) finds that with his new thermodynamical description, Γ_1 falls below 4/3 during the first rebound, producing an explosive expansion that results in the nearly complete evacuation of the center of the cloud; the central density decreases by a factor on the order of 10^{22}. Tscharnuter finds that this explosive expansion is independent of mass (between 1 and 100 $M_\odot$) and insensitive to the initial density distribution as long as the cloud is initially optically thin to infrared radiation.

Because of the cyclic behavior, formation of the final protostellar core is delayed in Tscharnuter's models until the envelope of the cloud has largely collapsed onto the oscillating central region ($\approx$3 to 4 initial t_{ff}). The protostellar cloud spends most of its life without a well-defined protostellar core; outflow is as likely to be seen as inflow. The final core is formed in one last collapse lasting on the order of 10^4 yr. The last collapse starts from a higher mean density, higher optical-depth configuration (cf. Hayashi and Nakano 1965) than is the case in the classical picture (see, e.g., Larson 1969), and consequently produces a protostellar core with a higher temperature. Tscharnuter's (1987) results imply that we may not yet fully understand the formation of the proto-Sun; until we do, models of the evolution of the rest of the solar nebula must be considered provisional.

Rotating Collapse Models

Spherically symmetrical models of protostellar evolution are consistent with observations of the luminosity and effective temperature of embedded infrared sources found in molecular cloud cores (Beichman et al. 1986) and of T Tauri stars (Stahler 1983*a*). The basic thermodynamics of star formation are thus reasonably well known. However, in order to account for the detailed spectra of embedded protostars (see, e.g., Adams et al. 1987) and the existence of binary and planetary systems, the effects of rotation must be included. The observational evidence presented in Sec. I suggests that nonmagnetic models of rotating clouds are appropriate for the protostellar collapse phase. The first nonmagnetic calculations including the effects of rotation were restricted to axisymmetry, that is, symmetry about the rotation axis.

Three different outcomes appear to be possible for the isothermal collapse of rotating, axisymmetric, uniform density clouds (Boss and Haber 1982). First, a rotating cloud which is initially close to thermal (Jeans) equilibrium may collapse only marginally, to form a Bonnor-Ebert spheroid (Stahler 1983*b*), a diffuse, equilibrium structure which is the rotating analogue of the spherically symmetrical, isothermal Bonnor-Ebert sphere. The outcome for a cloud which is initially farther from thermal equilibrium and undergoes significant collapse is dependent upon its angular momentum content and distribution. High J/M clouds undergo centrifugal rebounds during their isothermal collapse and so may form rings (Larson 1972; Black and Bodenheimer 1976), while in low J/M clouds, centrifugal forces play a lesser role and flattened disks form (Norman et al. 1980). Terebey et al. (1984) showed that in the collapse of a cloud with an initial $1/r^2$ density profile, rotation leads to the formation of a flattened disk. Low angular momentum clouds ($J/M < 10^{20}$ cm^2 s^{-1}) collapse into the nonisothermal regime to form pressure-supported, quasi-equilibrium disks (Safronov and Ruzmaikina 1978; Tscharnuter 1978; Boss 1984*a*); these disks are the rotating analogues of the cores found in spherically symmetric models of single star formation.

The formation of a ring in an axisymmetric calculation has often been inferred to imply that fragmentation into a multiple system would have occurred if the restriction to axisymmetry had been relaxed (see, e.g., Larson 1972). Indeed, the instability to fragmentation of isothermal rings has been demonstrated in three-dimensional calculations (Cook and Harlow 1978; Norman and Wilson 1978). However, in the case of nonisothermal rings, thermal pressure is able to stabilize some rings against fragmentation (Boss 1986). Furthermore, interstellar clouds may well fragment without experiencing an intermediate phase of ring formation (Boss and Bodenheimer 1979; Bodenheimer et al. 1980*b*). Hence axisymmetric calculations are not able to assess fully the possibilities for fragmentation; three-dimensional calculations of protostellar collapse are necessary.

The question of fragmentation is crucial, because one fundamental astronomical constraint is that our Sun is a single star. One could argue that the Sun is in a binary system with Jupiter as a sub-stellar companion, but considering that the Sun is 1047 times more massive than Jupiter, we shall hold to the convention that Jupiter is a planet formed by secondary processes attending the formation of the Sun. With this definition, binary systems require the mutual formation of both stars, presumably through fragmentation during collapse. Models of solar system formation that involve an intermediary multiple protostellar system, thus require either the ejection of a single protostar which could form the solar system, or the orbital decay of a multiple system into a single protostar. Formation of the presolar nebula by the former means, namely formation in a three-body protostellar system, which undergoes orbital decay and ejects a single protostar that could be the presolar nebula, does not appear likely. Because three-body formation requires the initial cloud to be strongly disposed to fragmentation, the ejected protostar will be similarly disposed, and is likely to fragment again during its own collapse (Boss 1982; Boss 1983). Thus the former scenario appears still to require the orbital decay of a binary system (the latter scenario) in order to obtain a single protostar.

Whether or not a given binary protostar system undergoes orbital decay is not well understood. The main cause of orbital decay appears to be loss of the binary's orbital angular momentum to surrounding matter through the action of gravitational torques. A criterion based on this process (Boss 1984*b*) can be used to predict which binary protostars should survive and which should merge together. When this criterion is used in concert with exploration of the parameter space available for the collapse of rotating interstellar clouds (Boss 1985,1986,1987*c*), it is possible to predict which initial clouds will collapse ultimately to form single stars like the Sun. These surveys are limited to fragmentation which occurs during the first collapse phase and during the formation of the first core; binary fragmentation or fission during subsequent contraction to a pre-main-sequence star cannot be ruled out as yet but appears to be unlikely because of the expected enhanced efficiency of orbital decay at higher temperatures and smaller binary separations. In agreement with the axisymmetric models, these surveys suggest that the solar system formed from a slowly rotating cloud with $J/M < 10^{20}$ cm^2 s^{-1}. Though their frequency is hard to assess in the absence of an unbiased survey of cloud rotation rates, interstellar clouds rotating this slowly apparently exist today (see Sec. I), giving important observational support to this theoretical picture of solar nebula formation.

Angular Momentum Transport During Collapse

A disk formed from a slowly rotating cloud is thus a reasonable candidate for the solar nebula. While this disk may not be susceptible to binary star formation, generally it will still be rotating too rapidly to allow sufficient matter to move inward and form a central proto-Sun. In the present solar system, the Sun contains about 99.9% of the total mass, but only about 2% of the total angular momentum. Clearly, considerable depletion of angular momentum in the presolar matter must have occurred, balanced by concentration of angular momentum in the preplanetary matter. Hence mechanisms have been sought which transport angular momentum outward and mass inward, both during the dynamic protostellar collapse phase and during the quasi-equilibrium, solar nebula phase.

In an axisymmetric calculation, angular momentum transport can occur by viscous stresses. While the molecular viscosity of protostellar clouds is too small to be of use, if the cloud is turbulent, the turbulent motions can result in an effective viscosity. There is abundant evidence for disordered motions in interstellar clouds, which lends support for the possible importance of turbulent viscosity during protostellar collapse. Early calculations of the effect of turbulent viscosity on protostellar collapse indicated that the turbulent stresses can indeed be important; for high J/M clouds, turbulent stresses transferred sufficient angular momentum outward to prevent rings from forming (see the calculations of Regev and Shaviv 1981); instead, disks resulted.

However, there are a number of difficulties with the hypothesis of the importance of turbulent stresses during protostellar collapse. First, the large-scale disordered motions in giant molecular clouds appear to consist of Alfvén waves (Morfill and Stenholm 1980; Myers and Goodman 1988) and strong stellar winds from newly formed stars (Norman and Silk 1980), so this observation may not be relevant for the smaller-scale process of collapse. Second, while interstellar clouds appear to be unstable to turbulence on the basis of the Reynolds number usually calculated for laboratory flows, this may be misleading, because of the absence of shear induced by the walls that accompany laboratory experiments. Shear in protostellar clouds may derive principally from differential rotation, in which case rotation probably acts to stabilize the flow. This follows because the Rayleigh criterion for instability in a rotating, incompressible fluid requires even more differential rotation than can be found in the extreme case of Keplerian rotation about a massive central object (Safronov 1969). Instability in a compressible, rotating fluid appears to require the Schwarzschild criterion for convective instability to be met, which does not occur during the early, nearly isothermal phases of protostellar collapse. Third, regardless of the torque-producing mechanism, it is unlikely that significant angular momentum transfer will occur during the collapse of a cloud whose rotational period P is longer than its collapse time $\approx t_{ff}$; dense cores have $P > 10^6$ yr and $t_{ff} < 10^6$ yr. Finally, even if turbulence does somehow occur, its ability to transport significant angular momentum during collapse is uncertain. For example, Tscharnuter (1978) found that with one assumption about the strength of turbulence (essentially an estimate of the critical Reynolds number Re_{cr}), turbulence had little effect on the collapse, while with another assumption (a smaller Re_{cr}), ring formation could be prevented. Vanajakshi and Jenkins (1985) found that with their formulation of turbulent stresses, likely strengths of the turbulent stresses could not prevent ring formation. Hence it may be that turbulent viscosity is negligible during the collapse phase. Turbulent stresses may still be important during the quasi-equilibrium solar nebula phase, however.

As discussed in Sec. I, magnetic fields appear to be an efficient means for redistributing angular momentum in diffuse and molecular interstellar clouds. Without magnetic fields, it is possible that slowly rotating, dense clouds could not be formed out of molecular clouds. We have assumed that protostellar collapse begins when magnetic fields cease to be dynamically significant, and by this reasoning, magnetic fields cannot be expected to transport significant angular momentum during protostellar collapse. However, this assumption is surely an oversimplification, and some degree of magnetic field interaction is likely at least during the initial phases of collapse (Mouschovias and Paleologou 1986).

A third mechanism for ameliorating the angular momentum transport problem during protostellar collapse is conversion of spin angular momentum of the initial cloud into the orbital angular momentum of a binary protostar system (see, e.g., Boss and Bodenheimer 1979). However, considering again that the Sun is not a member of a binary system, this mechanism cannot apply to the presolar nebula. It thus appears that the majority of the remaining angular momentum transport between presolar and preplanetary matter must occur within the solar nebula.

III. SOLAR NEBULA MODELS

A brief historical account of the development of solar nebula modeling is presented in the recent review by Wood and Morfill (1988). Here we concentrate on the viscous accretion disk models which have dominated the last decade of modeling, and on those aspects most relevant from the point of view of planetary atmospheres: time scales, nebular conditions and physical processes.

A simple criterion for the growth of a solar nebula around a proto-Sun can be derived as follows. Assume that the collapse of an extended cloud leads to an approximately constant mass influx $\dot{M}$ into the central cloud region for a time scale τ_i (of the order 1 or a few free-fall times). Let this collapsed gas be assembled in a small, dense, rotating disk (an accretion disk), that has the function of redistributing angular momentum, thus allowing the gas to evolve radially inward to form a star at the center. Clearly, the physical process responsible for the angular momentum transport determines the time scale τ_D with which the disk may get rid of its accumulated mass. Whatever the origin of this process, we may describe it with an effective viscosity ν and then the rate of change of the disk mass, M_d, is given by

$$\frac{\mathrm{d}M_d}{\mathrm{d}t} = \dot{M} - \frac{3\nu}{r_d^2} M_d \tag{1}$$

where r_d is the radius of the disk. (Note that we put $\dot{M}(t > \tau_i) \equiv 0$.) At the same time the embryo star in the center grows at a rate

$$\frac{\mathrm{d}M_*}{dt} = \frac{3\nu}{r_d^2} M_d \tag{2}$$

i.e., the mass loss from the disk goes into feeding the star. A "classical" accretion disk (see, e.g., von Weizsäcker 1948; Lüst 1952), which is dominated by the central mass M_* requires $M_d < M_*$, typically. The two equations can be solved subject to supplementary descriptions of ν and r_d from physical models (e.g., r_d is determined using conservation of angular momentum as in Cassen and Summers [1983]). The result is that initially the collapse invariably leads to the formation of a flattened, rotating polytrope in the center of the collapsing cloud, and that only later a protoplanetary accretion disk evolves. The exact cross-over depends on the ratio of infall time/viscous time:

$$\eta = \left(\frac{M_c}{\dot{M}} \right) / \left(\frac{r_d^2}{3\nu} \right) \tag{3}$$

where M_c is the cloud mass.

For small η, the disk configuration is dominated by the angular momentum of the infalling matter, while large η means that viscosity can redistribute angular momentum into some quasi-equilibrium configuration. The system may be highly susceptible to breakup for small η and during the early stages of the collapse, in which case a three-dimensional accretion disk model is necessary. However, even in this case the fragments of a broken-up solar nebula may merge again and migrate to the centrally growing star under the effect of tidal forces and gas drag, possibly validating the use of axisymmetric models of the later evolution of an accretion disk. These arguments, while plausible, have not yet been investigated by numerical experiment.

From these deliberations, it appears most likely then, that a *viscous accretion disk* (as opposed to a clumpy central cloud fragment) is an evolutionary stage of star and planet formation which may occur towards the end of the collapse and accretion phase. As such, it may not be responsible for the fast time scales (of $\sim 10^6$ yr) of star formation, but it may still carry a significance of its own: if it occurs, it quite likely determines the conditions in which planets and other solid bodies form (assuming that giant planet formation did not already occur through gravitional instability).

Axisymmetric Solar Nebula Models

A large number of axisymmetric viscous accretion-disk models exists in the literature (Lynden-Bell and Pringle 1974; Cameron 1978; Lin and Papaloizou 1980; Lin 1981; Cassen and Moosman 1981; Lin and Bodenheimer 1982; Cassen and Summers 1983; Morfill and Völk 1984; Ruden and Lin 1986; Cabot et al. 1987*a,b*). A number of recent review papers are also available (Cameron 1985; Lin and Papaloizou 1985; Morfill et al. 1985; Wood and Morfill 1988). Axisymmetric accretion-disk models have the advantage that their physics is clear (with one important exception, viscosity), and that they are mathematically simple and easy to derive and use.

The basic assumptions are reasonably well founded: hydrostatic pressure equilibrium is assumed in the vertical direction (perpendicular to the midplane of the nebula). Also, most models describe a phase when the Sun has already formed at the center, and the mass of the disk is a small fraction of a solar mass, as suggested by the 0.05 $M_\odot$ mass of the reconstituted solar nebula

(Weidenschilling 1977*a*). The main exceptions to the rule of minimum-mass disks are the solar mass-sized disks of Cameron and Pine (1973) and Cameron (1978). When the disk mass is only a minor contributor to the gravitational field, the gravitational field is simply that of a point mass. Because of this dominant central mass point, and the belief that pressure gradients in the radial direction are relatively small, most models are assumed to be in Keplerian rotation.

Conservation of energy is applied by assuming that the local heating due to viscous dissipation of shearing motions is balanced by radiative and convective energy loss in the vertical direction, followed by radiative loss from the surface of the nebula. This energy equation must be supplemented by expressions for the viscosity, pressure and the opacity: icy dust particles occur in the outer, cool regions of the disk, where the central temperature is less than 160 K, and a mixture of silicates and Fe particles occur inside the ice sublimation boundary.

Infalling matter from the remainder of the presolar cloud is usually ignored, thereby isolating the nebula from the difficult problem of protostellar collapse. Here, the exceptions are the work by Cameron (1978), who used a crude mapping between the cloud and the disk, and Cassen and Moosman (1981) and Cassen and Summers (1983), who calculated the infall of gas on hypersonic ballistic trajectories. These three studies thus included the buildup of the disk. The models of Morfill and Völk (1984) are based on a snapshot from Tscharnuter's (1978) axisymmetric protostellar collapse calculations, and thus did not evolve along with the collapse.

Vertically averaged hydrodynamical equations (i.e., the continuity equations for mass and momentum) appropriate for a thin disk then give the time evolution of the nebula. The dynamical evolution of a viscous accretion disk inevitably follows that first given by Lynden-Bell and Pringle (1974): viscous stresses move mass primarily inward, onto the central Sun, while transferring angular momentum always outward, to an expanding outer region of the disk. If no mass is being added to the nebula, or is only being added at a low rate, the density at fixed radius decreases with time, as the nebula is accumulated onto the Sun.

The most contentious issue in viscous accretion-disk models is the strength and formulation of the viscous stresses. The models are generally calculated using the Navier-Stokes equations viscosity formalism, with the coefficient of viscosity as a free parameter. Molecular viscosity is far too small (by some 10 orders of magnitude) to produce evolution on the time scale required for observations of star formation, so an effective viscosity is assumed to exist. Recent solar nebula work has concentrated on effective viscosities due to turbulence, particularly following the demonstration by Lin and Papaloizou (1980) of the instability of the solar nebula to turbulent convection in the vertical direction. Eddington-Sweet circulation currents were proposed by Cameron and Pine (1973) as an alternative mechanism for transporting angular momentum, but work by Cabot and Savedoff (1982) eliminated that possibility. Lin and Pringle (1987) suggested a formula for an effective viscosity due to gravitational instability; we will say more about gravitationally induced torques later in this section. Magnetic fields could also account for an effective viscosity, though the considerations of Sec. I tend to minimize their role.

The main problem is that the physics of turbulence, which involves nonlinear flows, onset criteria, damping and more, still is not well understood. In planetary atmospheres, there are thin sheet-like layers of turbulence separating more laminar flow regimes. In stars, we have convection regions which, when averaged, give rise to anomalous heat and radiation transport. Similarly one might expect convective flows in protostellar clouds and also some time and space dependent turbulence. While this fact is not in dispute (e.g., the highly resolved collapse calculations by Tscharnuter clearly show convective effects once a quasi-equilibrium disk is formed), the physical description at this stage is still not adequate and a great deal more work needs to be done. The continuing efforts of Canuto and Goldman (1985) and their collaborators have shown that progress can be made in refining our understanding of this critical process.

Steady Models

The simplest turbulent disk models to study (and hence the most widespread in the literature, with applications ranging over cataclysmic variables, binary neutron-star systems, protogalaxies, and planetary rings in addition to star and planet formation) are the *steady models*. In these models it is assumed that the ratio η, defined earlier, is much greater than unity, so that the disk may evolve with a steady supply of gas toward a steady state.

Following Shakura and Sunyaev (1973), the so-called α-model of turbulent viscosity is most often used, where the coefficient of viscosity is described by analogy to molecular viscosity as

$$\nu = \frac{1}{3} \alpha c_s L \tag{4}$$

where c_s is the sound speed, and L is the correlation length of the turbulent eddies. The dimensionless parameter α describes the strength of the turbulence and requires physical input, as does L. Most authors use the only relevant physical length scale for L, namely the thickness of the accretion disk. The disk thickness is obtained by balancing the gas pressure with the gravitational attraction of the central protostar, perpendicular to the equatorial plane. This assumes that the self-gravity of the disk is unimportant. In this case, the disk thickness is $L \sim c_s/\Omega_*$ where Ω_* is the Keplerian angular velocity around the central star, which varies with radial distance as $r^{-3/2}$. If the disk is massive, then Ω_* has to be replaced by a function $\Omega(r)$ of the disk mass distribution.

Figure 4, taken from the review by Wood and Morfill (1988), summarizes the physical structure of a "generic" steady accretion-disk model, calculated on the basis of the assumptions just outlined. While the power-law dependences of the various physical quantities are fixed, their absolute values can be changed by varying the assumptions of the model. For example, the midplane temperature is given by

$$T_m \approx 54600\, \alpha^{-1/3} M^{1/2} \dot{M}^{2/3}\, \kappa_0^{1/3} r^{-3/2} \tag{5}$$

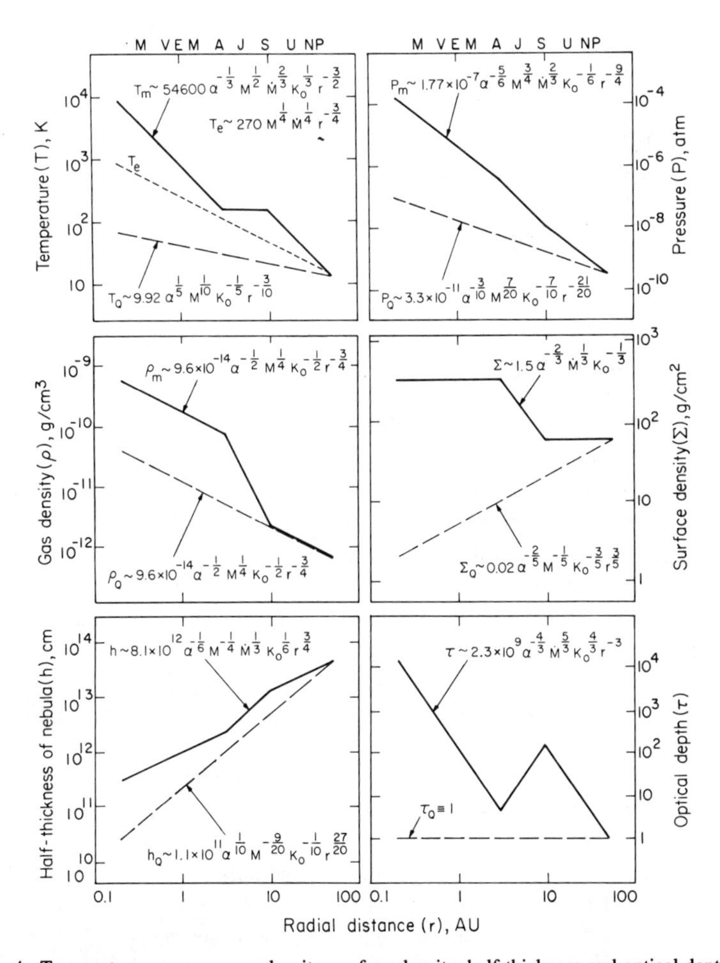

Fig. 4. Temperature, pressure, gas density, surface density, half-thickness and optical depth as a function of heliocentric distance in a simple steady-state viscous accretion-disk model of the solar nebula (figure from Wood and Morfill 1988). Subscripts *m, e* and *Q* stand for midplane, photospheric and quiescent (when the nebula is no longer being fed new matter). The middle segments of the curves represent conditions where both silicates and ices contribute importantly to opacity.

where M is the mass of the Sun in solar units, $\dot{M}$ is the mass accretion rate from the presolar nebula onto the disk and (under the steady-state assumption) from the disk onto the Sun (in units of $M_\odot/10^6$ yr), and κ_0 is a coefficient in the opacity expression $\kappa = \kappa_0 T^2$. Other parameters of interest include the total disk mass (in solar units)

$$M_d \approx 5.2 \times 10^{-5} \alpha^{-2/3} \dot{M}^{1/3} R^2 \tag{6}$$

and the disk lifetime (in years)

$$\tau \approx 10^6 \dot{M}^{-1} \tag{7}$$

where R is the overall disk radius in AU. Figure 4 was constructed using $\alpha = 0.3$, $M = 1$, $\dot{M} = 1$, $\kappa_0 = 2 \times 10^{-4}$ cm g^{-1} K^{-2} for $T < 160$ K (ices), $\kappa_0 = 10^{-6}$ cm g^{-1} K^{-2} for $160 < T < 1600$ K (Fe and silicates), and $R = 50$. By adopting different values for α, $\dot{M}$ etc. one can easily create a "new" solar nebula model.

Table I summarizes the structure of the solar nebula models derived in the axisymmetric accretion-disk models constructed to date. The table gives the exponents in the power-law dependences of the midplane (where possible) temperature and density on the radius, the disk mass, disk lifetime and α. Shown for reference are the assumptions of Safronov (1969) and Kusaka et al. (1970), the Cameron and Pine (1973) model, and Weidenschilling's (1977*a*) reconstituted nebula. For some models, the definition of α varies and so cannot be directly intercompared, but hopefully the tabulated values give a comparable estimate of the assumed (in most cases) or calculated (Cabot et al. 1987*a,b*) viscous strengths. It can be seen from Table I that within certain limits, nearly all models are in agreement about the temperature and density profiles: $T \propto r^{-1/2}$ to $r^{-3/2}$, and $\rho \propto r^{-1/2}$ to r^{-2}. For comparison, hydrodynamical calculations of the earliest phases of presolar nebula collapse (Boss 1986) yield inner regions with $T \propto r^{-1}$ and an exponential density profile that necessarily includes segments of all of the Table I model exponents. Table I also shows that as models have progressed, the estimates of disk lifetimes have generally increased, while the values of α have decreased substantially. Extrapolation of the latter trend implies that, like the notorious case of the mass of Pluto, with further research α will eventually reach zero!

It is worth noting at this stage that the surface density of dynamical models, such as those involving viscous disks, need not necessarily bear any relation to the reconstituted nebula derived from the augmented masses of the planets (see, e.g., Weidenschilling 1977*a*). The reason for this is obvious. The present planetary system only gives information about the matter left behind. It is possible, though perhaps unlikely, that practically the full solar mass originally passed through an accretion disk of varying mass and size.

TABLE I
Summary of Axisymmetric Solar Nebula Models

Authors	T exponent	ρ exponent	$M_d/M_\odot$	τ (yr)	α (turbulence)
		Reference Models			
Safronov 1969	−0.5 to −1	−2 to −3	0.05	—	—
Kusaka et al. 1970	−0.5	−2	0.05	—	—
Cameron and Pine 1973	−1	−1 to −3	1	10^2 to 10^5	—
Weidenschilling 1977	−1	−2	0.05	—	—
		Viscous Accretion-Disk Models			
Cameron 1978	−0.77	−1.9	1 to 2	10^5	0.7
Lin 1981	−1.5	−0.75	0.02 to 0.04	10^5	2
Lin and Bodenheimer 1982	−1.5	−0.75	0.02	10^5 to 10^7	1
Cassen and Summers 1983	−0.5 to −1	−1.5 to −3	0.015 to 1.5	10^5 to 10^6	(varied)
Morfill and Völk 1984	−1.5	−0.75	0.055 to 0.11	10^6	0.3
Ruden and Lin 1986	0 to −1	−0.5 to −1	0.05	10^6	0.01
Cabot et al. 1987*a,b*	−0.5 to −1	0 (steps)	>0.1	$>2 \times 10^6$	0.0001 to 0.01

What mass remained in orbit about the central star then depends on the physical processes and their strength and relative importance.

Turbulent Viscosity

As the previous discussion implied, the strength of the turbulence, i.e., the precise value of $\alpha(t,r)$, is still a matter for conjecture. Cabot et al. (1987*a,b*), making use of a new closure argument to improve on Heisenberg's turbulence theory (Canuto et al. 1984; Canuto and Goldman 1985), have sought actually to *calculate* α for the convectively driven instability found by Lin and Papaloizou (1980). Cabot et al. arrive at values of $\alpha \sim 10^{-4}$ to 10^{-2}, considerably lower than the assumptions of previous modelers. Reduced viscous strength means that the time scale for viscous evolution may become uncomfortably long, compared to the observational evidence for nebula dispersal in 1 to 10 Myr (Sec. I). If τ is not less than these values, then evidently viscous effects are not capable of transporting significant mass through the nebula before the nebula itself is removed by some other means.

The possibility that turbulence may be quite inefficient should not be overlooked. The results of Cabot et al. (1987*a,b*) are based on several improvements beyond the use of the Canuto turbulence theory. First, the latest studies of the opacity of dust grains (Pollack et al. 1985) imply a weaker dependence on temperature than was originally thought, restricting somewhat the regions where turbulence could be expected. Second, with a more accurate representation of the physics of turbulent convection in a rotating, anisotropic, radiating fluid (see also Cowling 1951), the effective strength of the viscous stresses drops by a large factor (e.g., vertical convective motions drive considerably weaker motions in the radial direction in a rotating, flattened nebula, and the radial motions are the ones responsible for angular momentum transport and hence viscous evolution). Perhaps most importantly, Cabot et al. found that such a viscous accretion disk is unstable to global formation of rings (and probably fragmentation as well), which destroys the smooth radial profile necessary for turbulent stresses to act efficiently. In this case, the quasistatic structure assumed at the outset of the model derivation is called into question, and hence the results as well.

If the evolutionary time scale of the disk is relatively long, the last stages of mass accretion onto the protostar may be quite protracted. This possibility itself is not without appeal, since it could maximize the possible time scale for planet formation within an environment which still contained an ample supply of gas—important for scenarios involving giant planets.

The free energy available to drive turbulence is huge, and it is not even clear that shock waves (effectively supersonic turbulence) can be clearly ruled out. Also, other possible driving mechanisms like angular momentum mismatch between infalling and disk material and magnetosonic waves, have not been investigated yet. If there is a way to dissipate free energy faster, most physical systems evolve in that direction.

Evolutionary Models

Clearly time-dependent, evolutionary models also need to be studied. Cameron (1978) was the first to compute an evolutionary model, though the time evolution was treated through a series of steady models, rather than in a mathematically rigorous fashion. Cassen and Moosman (1981) constructed mathematically rigorous models which followed the buildup of the nebula by mass infall from the envelope of the presolar cloud, the dumping of nebular matter onto the central Sun, and the expansion of the outer regions of the nebula as angular momentum was transported outward and as higher angular momentum cloud material arrived at the disk. Cassen and Moosman (1981) varied widely the strength of the viscous stress, and Cassen and Summers (1983) concluded that it is not possible to rule out either low mass, low angular momentum nebulae with strong viscosity, or massive, high angular momentum nebulae with weak viscosity.

Lin and Papaloizou (1985) constructed evolutionary models by starting off with a given surface density distribution, assuming a value for $\alpha \equiv 10^{-2}$, computing the convection speed from standard mixing-length theory, and then deriving the radial spreading and relaxation of the nebula. No mass addition was assumed, as may be appropriate for the late stages of solar nebula evolution or for a remnant protoplanetary disk. By choosing the initial mass in this disk to be sufficiently large (in this case about 8% of 1 $M_\odot$) and by picking a sufficiently large α, it was possible to reach relatively high temperatures (in excess of 2×10^3 K out to ~2 AU) during the early stages of the disk evolution. There is abundant evidence for high temperatures in this region of the nebula, gathered from meteorite data (see the review by Wood and Morfill [1988] and several other chapters in *Meteorites and the Early Solar System* [1988]). However, global temperatures of this magnitude, which at the prevailing pressures are in excess of the sublimation temperatures of most minerals, constitute a major problem in themselves.

Consider the consequences. Above 2000 K all the dust is vaporized quickly. The opacity in the absence of dust drops to a value of $\sim 3.5 \times 10^{-5}$ $cm^2 g^{-1}$ and the optical depth for a disk of surface density of ≈ 1000 g cm^{-2} decreases to ~0.035. The disk rapidly cools by radiation and temperature gradients disappear. If convective instability were the main agent for driving turbulence, it would cease also. We then have the case of cooler gas with no dust. In principle, supersaturation may then lead to homogeneous nucleation and fresh growth of grains, increase in opacity, establishment of a new vertical temperature gradient, and the re-establishment of convective instability. However, this has not been calculated in any detail so far, and in view of the uncertainties associated with the theory of homogeneous nucleation, this is not surprising. Coagulation of grains to form planetesimals also removes the opacity needed for convective instability (Weidenschilling 1984), and may cause turbulence to be intermittent.

Coagulation

Dust particles in viscous accretion disks are subject to stochastic motions caused by the turbulent convection, and to secular motions caused by the viscous evolution of the nebula itself and by gas drag forces (see, e.g., Weidenschilling 1977*b;* Morfill 1985). On a much longer time scale, adiabatic orbital changes due to growth of the central Sun are also important for large particles on Keplerian orbits.

Assuming a narrow particle distribution, we can combine analytically the effects of radial transport with the simultaneously occurring coagulation by collisions (Morfill 1985), yielding the normalized equation:

$$\frac{1+\eta}{\eta^{1/2}}\left(1+\frac{9\ 3^{1/2}\eta}{4\pi\alpha^2}\right)\left(1+\frac{6\eta^2}{\gamma_G\pi^2\alpha^2}\right)^{-1}\frac{d\eta}{d\xi} = \left(\frac{2\pi}{3}\right)^{1/2}\frac{\pi\gamma_F Q f_i \alpha R^2}{H^2}\ \xi^{-1/2} \tag{8}$$

where $\gamma_G = 1.4$ is the ratio of specific heats, γ_F is a parameter describing the width of the particle size distribution (we choose $\gamma_F = 0.7$ corresponding to a particle size distribution $\propto \eta e^{-\eta}$; Morfill 1985), Q is the sticking factor, $f_i = 10^{-2}$ for ice and 1.25×10^{-3} for Si and Fe is the mass fraction of dust, α is the ratio of the turbulent velocity to the sound speed (Morfill and Völk 1984), and R/H is the ratio of the radius/vertical height of the accretion disk's outer boundary. The dimensionless parameter $\xi = r/R$ describes position in the disk, and the parameter $\eta = \tau_f/\tau_{Ko}$ is the ratio of the frictional coupling time between a dust particle (of radius a, material density ρ_s) and the gas (density ρ, thermal gas velocity c_s) and the coherence time of the turbulence. These times are given by:

$$\tau_f = \frac{2\pi^{1/2}\rho_s a}{3\rho c_s} \tag{9}$$

$$\tau_{Ko} = \frac{3^{1/2}h}{\pi\alpha c_s}, \tag{10}$$

where $h = h(\xi)$ is the disk scale height at position ξ. We see that η is linearly proportional to particle radius and inversely proportional to disk column density. The solution of the coagulation equation is shown in Fig. 5, for $Q = 1$ and 0.1, $\alpha = 10^{-3}$, and $R/H = 21$, corresponding to a disk with a radius of 40 AU and a central mass of 1 $M_\odot$. Decoupling from the turbulence occurs when $\eta > 1$, and we see that even the extreme case of $Q = 1$ does not produce decoupling. This calculation does not include collisional disruption or nucleation of volatiles lost during motion inward in the nebula, and hence may be regarded as an upper limit.

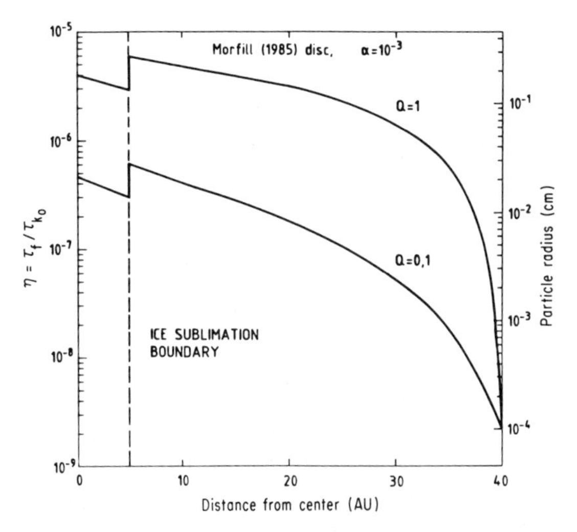

Fig. 5. Normalized dust-particle size η as a function of nebula position, with radial transport and turbulent coagulation (Morfill 1985). Decoupling from the turbulence occurs when $\eta > 1$. Two sticking coefficients Q were used.

For a strongly turbulent nebula, with $\alpha = 1/3$, a similar calculation shows that η may just reach unity in the extreme case of $Q = 1$. This result implies that gravitational settling of dust is all but impossible so long as the nebula is turbulent. Because gravitational settling is an important step in planet formation, we are faced with a dilemma. It is clearly necessary to have angular momentum transport in order to form the Sun out of the nebula, but turbulence acts to stifle planetesimal growth. Solutions to this dilemma include the possibility that angular momentum transport is caused by processes other than turbulence, such as spiral density waves (discussed later in this section). Another solution retains the assumption of convectively driven turbulence. Weidenschilling (1984) pointed out that coagulation of dust could lower the opacity such that the disk becomes optically thin, cools, and loses the vertical temperature gradient necessary for convective instability. Continuous mass influx from the surrounding cloud envelope will supply fresh dust to the disk, increasing the opacity and triggering convective instability again. Whether or not this works has not been convincingly demonstrated, and the possibility remains that the disk might "adiabatically" adjust to the opacity change by increasing its surface density (see, e.g., Morfill 1987).

Qualitatively, we can show how this idea works in the following way. We solve the coagulation equation (given above) in the limit $\eta << 1$

$$\frac{\mathrm{d}a}{\mathrm{d}t} = \frac{\alpha c_s Q \rho_d}{2\rho_s} \eta^{1/2} \tag{11}$$

together with the mass influx equation for the dust

$$\frac{d\rho_d}{dt} = \frac{f_i \dot{M}_p}{2\pi R^2 h} \tag{12}$$

where ρ_d is the space density of dust and $\dot{M}_p$ is the perpendicular mass inflow rate. If temperature variations are small, we treat c_s, h and α as constants, and calculate one cycle. The relevant time scale is the time taken for infalling matter to build up a dust layer of optical-thickness unity

$$t_r = \frac{8\pi\rho_s a R^2}{3 f_i \dot{M}_p Q_a} \tag{13}$$

where Q_a is the ratio of effective cross section to geometric cross section of the grains. For a disk of size 40 AU, particle size in μm (a_μ), and $\dot{M}_p = \epsilon M_\odot / 10^6$ yr, we get $t_r = 4.5 \times 10^8 a_\mu / Q_a \epsilon$ seconds.

Figure 6 plots the time development of the optical depth according to these equations, starting from unit optical depth. The calculations were made for different radial distances $\xi = r/40$ AU. As can be seen, the optical depth increases and then coagulation dominates and reduces it again. Typical time scales for turbulence (optical depth >1) in the Jupiter/Saturn region ($\xi = 1/8$ to $1/4$) are $\Delta t \approx 10 t_r$, i.e., ≈ 150 yr. Whether the disk can adjust thermally to these opacity changes depends on the thermal and cooling time scales. Certainly the cooling time scale is much smaller, and the thermal time scale $h/c_s \approx \Omega^{-1}$ is of order 5 yr, i.e., also much smaller. Hence this minimum set of equations is insufficient, and the full disk structure equations have to be included. The disk will adjust to the opacity changes in some way, and it is not clear whether intermittent turbulence can occur. However, if it does become intermittent, sedimentation of dust during the quiescent times (of the order of t_r) will be important.

The time scale for sedimentation of a dust particle is given by

$$\tau_{sed} = \frac{2R^3\xi^3}{GM_c\tau_f} \tag{14}$$

where M_c is the central mass, and is independent of the initial distance z above the nebula midplane. If $\tau_{sed} < t_r$, particles can sediment toward the midplane, where the dust mass density will exceed the gas density. In this case, the next turbulent phase (in the gas) will be unable to stir up the solids in the dust sub-disk, thereby allowing local gravitational instability of the solids. This condition translates into a minimum particle size for sedimentation $a > a_{min} \propto \xi^{3/2}$ cm. Because a_{min} increases with heliocentric distance, as dust moves inward in the nebula, the larger conglomerates will sediment out to the midplane, until in the vicinity of 1 AU only mm- to cm-sized particles remain. This result is independent of collisional disruption effects in the outer solar nebula, unless these processes restrict particle growth to even smaller values. The apparent "coincidence" of this particle size range with chondrule sizes and inclusions in meteorites may be thus explained.

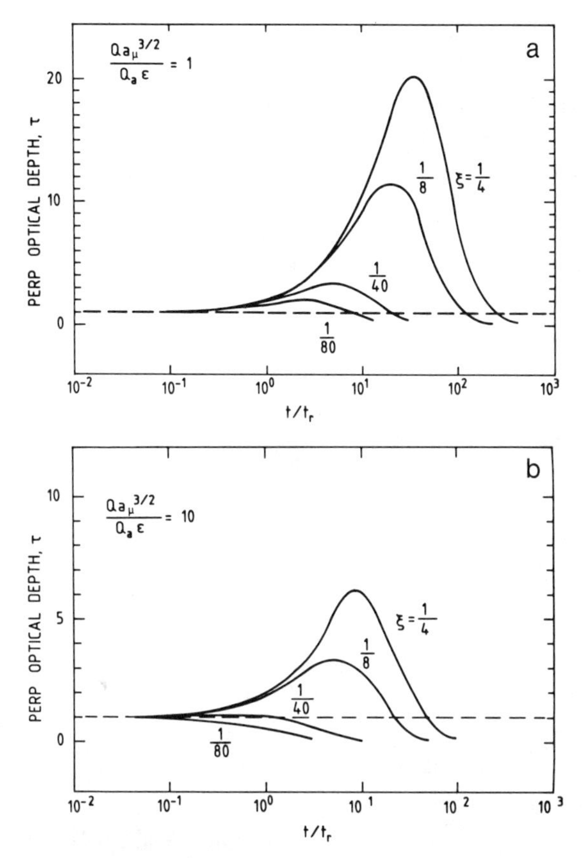

Fig. 6. Vertical optical depth τ in a turbulent disk caused by infalling matter from the cloud envelope, as a function of time. Time $t = 0$ corresponds to the onset of turbulence ($\tau = 1$). Accretion, vertical mixing and coagulation were included.

What happens to infalling interstellar dust particles during the turbulent time period Δt? Apart from possible heating during the traversal of the accretion shock, there is also the question of subsequent collisional growth. The final mean particle size of the infalling component (as opposed to one transported through the disk from the outer edge) after a given time period (either Δt or r^2/ν, whichever is smallest), is given by

$$a \approx \frac{3f_i \dot{M}_p Q_a}{4\pi\rho_s R^2} \min(\Delta t,\ R^2\xi^2/\nu). \tag{15}$$

Substituting standard values, we see that the final particle size in the region of the terrestrial planets is of the order of 1 to 10 μm. This second fine-grained component may perhaps be identified with the fine-grained matrix in meteorites.

Condensation Zones

In Sec. V, we will see that giant-planet formation presents a severe time scale problem for formation through accumulation of planetesimals. One means of solving this problem involves rapid formation of solid cores through "runaway accretion," which may require a high surface density in the giant-planet region and a relatively massive (>0.1 $M_\odot$) solar nebula. Morfill (1985) suggested that planets should preferentially form in regions where major species can condense, for the following reasons:

1. The total solid component is enhanced just outside condensation zones, according to the transport calculations of Morfill and Völk (1984);
2. Coagulation should be particularly efficient in regions of significant condensation, both because the surfaces of the particles should be covered with an irregular layer of condensing matter, and because particles which are loosely joined can be enveloped with condensates, which enhances the particle strength;
3. Turbulence will carry fluid elements into and out of condensation/sublimation zones many times, which may aid cold welding of coagulants;
4. In regions where the optical depth is decreased by dust coagulation, a turbulent nebula may adjust by increasing its surface density, thereby aiding coagulation.

While these arguments should be treated with caution, it is nevertheless interesting that viscous-disk models can have radical temperature profiles, e.g., $T \propto r^{-1}$ (Table I). With suitable choices for $\dot{M}$ and α, it is possible to have temperatures at the formation site of Jupiter of about 160 K (the sublimation temperature of H_2O ice) and at Saturn of about 60 K (the sublimation temperature of CO_2 and hydrated nitrogen $N_2 \cdot nH_2O$). These formation sites were at larger radii than the current location of the planets, if Jupiter and Saturn had to give up angular momentum in order to eject many small bodies from the solar system. This is a necessary requirement, particularly if the protoplanetary nebula contained as much as 0.1 $M_\odot$ of gas and dust. In this case the total dust component amounts to one Jupiter mass, whereas the sum of the giant-planet core masses is only about 1/5 this value. The rest has presumably been ejected outward (or lost to the Sun). For ejection, the kinetic energy of the ejected bodies must be doubled, which implies that Jupiter and Saturn moved inwards to a distance of about 0.6 of their formation distance. The formation sites can then be estimated as about 8 AU and 16 AU, respectively. Kinetic energy loss in the terrestrial planet region was perhaps less severe, but possibly not entirely negligible. In any case, accretion-disk models can be made consistent with Mercury forming at the Fe/Si condensation zone at about 1500 K, though recent work suggests that Mercury's composition can also be explained as a largely iron core whose mantle was lost through a head-on collision (see, e.g., Cameron and Benz 1987).

At first sight this accretion-disk scenario appears to be inconsistent with evidence (see, e.g., Wood and Morfill 1988) that the meteorites, formed in the asteroid belt, should have been born in a high-temperature ($\approx$1500 K) environment. However, it is unlikely that a cooling nebula, starting from such a high initial temperature, can explain the meteorites, because of the evidence for extremely rapid cooling ($\sim$10 to $\sim$100 K hr^{-1}) and for repeated heating and cooling cycles.

The composition of chondrites, which includes moderate volatiles, some hydrated silicates and magnetite, implies that the aggregation temperatures of the meteorite parent bodies were much lower, in the range of 400 to 500 K. The problem may then be reduced to mixing high temperature components, which bear the rapid cooling signature, with low-temperature material. Turbulent transport seems to be one way of effecting this, and maybe the rapid-temperature oscillations evident in chondrules and Ca-Al-rich inclusions could occur at the Fe/Si condensation zone (Morfill et al. 1987). This scenario implies that chondrules were formed in the hot inner nebula at $T \approx 1500$ K, and were there subject to rapid temperature oscillations. Subsequently, these particles (with their now characteristic signature) were dispersed and intermingled with other solids in the nebula, finally forming large aggregates and meteorite parent bodies. Because mixing by large-scale turbulence does not rapidly produce microscopic homogenization, and the turbulent length scale is much larger than the wavelength giving rise to gravitational instability in a dust disk, it may be expected that distinct "types" of parent bodies should form.

As an alternative, we may also consider a nonaxisymmetric nebula threaded by shock waves, triggered perhaps by Jupiter, resulting in rapid temperature variations as matter passes through the shocks. This would imply that Jupiter had grown to a substantial size before formation of the meteorite parent bodies. If rapid accretion occurred outside of the ice condensation zone, and coagulation inside this zone was much less efficient, this may have been possible, though the details remain to be studied.

Three-Dimensional Solar Nebula Models

While much work remains to be done on improving our understanding of the time evolution of viscous accretion disks, the problems encountered with this scenario so far suggest that other means of driving nebula evolution should be investigated as well. Three-dimensional models with nonaxisymmetric density distributions provide for a different means of transporting angular momentum in the solar nebula. Gravitational torques between nonaxisymmetric distributions of matter in a differentially rotating solar nebula can result in the desired outward transport of angular momentum (Larson 1984). The rate of transport can be comparable to that assumed in the most optimistic models of convectively driven turbulent accretion disks (Boss 1984*b*), so

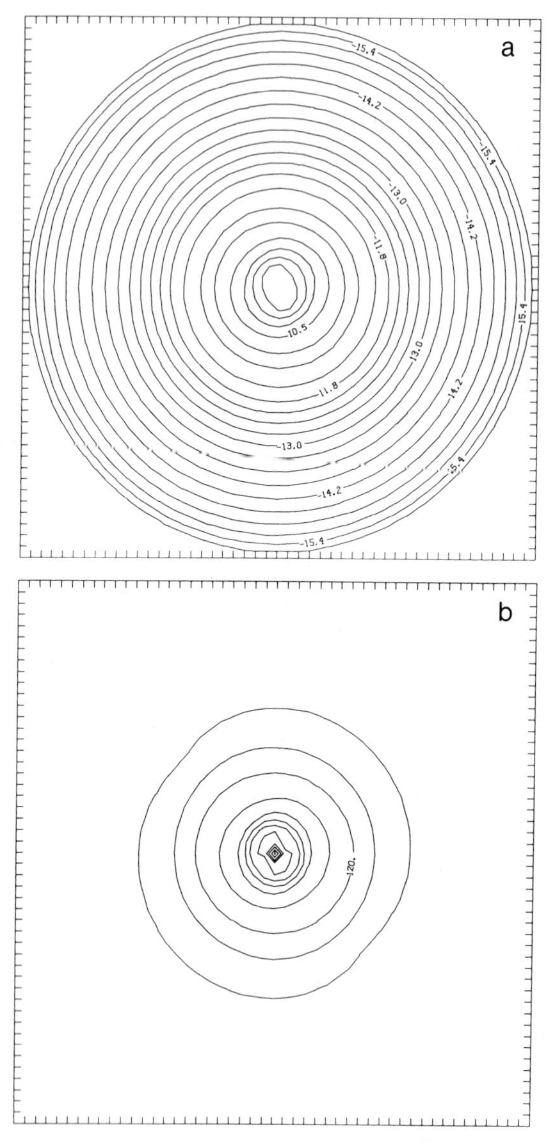

Fig. 7. Density (a) and temperature (b) contours in the equatorial plane of a three-dimensional model of the early solar nebula, prior to the formation of the proto-Sun (Boss in preparation). A region 80 AU across is shown. Density contours correspond to changes by factors of 2, temperature contours to changes of 50 K (minimum of 10 K at edges, maximum of 570 K at the center). In this model the disk has accreted a mass of 0.1 $M_\odot$ and has $J/M \approx 2 \times 10^{19}$ cm^2 s^{-1}.

gravitational torques are indeed a promising alternative to turbulent viscosity, provided that nonaxisymmetric structure occurs and is maintained.

Various types of nonaxisymmetric structure in the rotationally flattened proto-Sun and solar nebula have been proposed, including spiral density waves (Larson 1984), large-scale bars (Boss 1984*b*, 1985), and combinations of both (Yuan and Cassen 1985). Gravitational instability, possibly leading to giant-planet formation in the solar nebula (suggested by Cameron [1978] and calculated by Cassen et al. [1981]) is another possible source of nonaxisymmetric structure that could result in significant angular momentum transport.

In contrast to the uncertainties attending turbulent viscosity, angular momentum transport by gravitational torques simply requires significant nonaxisymmetric structure to be important in the solar nebula. Nonaxisymmetry is certain to be present at some level, and the determination of this level is just beginning to be explored. Recent three-dimensional calculations (Boss, in preparation) of the earliest phases of solar nebula formation and evolution show that angular momentum transport can occur on time scales on the order of 10^4 yr even for a mildly nonaxisymmetric nebula (Figs. 7 and 8). Shorter time scales are possible if the nebula is even more nonaxisymmetric. For example, binary fragments may decay inward and merge on the time scale of a few rotational periods (Boss 1984*b*), i.e., a few years.

While the early stages of presolar nebula collapse may not be as well understood as once thought (e.g., the oscillations found by Tscharnuter [1987]), and three-dimensional effects are likely to dominate the early stages,

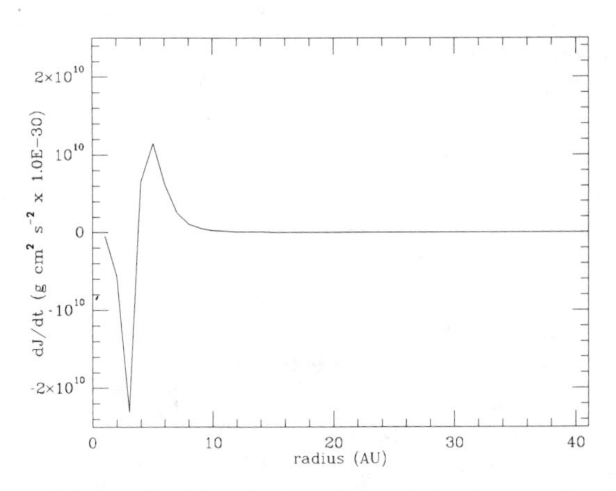

Fig. 8. Time rate of change of angular momentum by gravitational torques (on spherical shells), as a function of radius, for the model in Fig. 7. Because of the flattened nature of the nebula, spherical radius is roughly equivalent to cylindrical radius. Angular momentum is being lost from the inner 4 AU primarily to the region between 5 and 10 AU. In spite of the mild nonaxisymmetry of the model, the peak instantaneous angular-momentum transfer rate of $\approx 2 \times 10^{40}$ g cm^2 s^{-2} is capable of transporting the entire angular momentum of the model ($\approx 4 \times 10^{51}$ g cm^2 s^{-1}) in about 10^4 yr.

it cannot be ruled out that the final stages, once stabilized by a large central mass, could evolve somewhat more sedately in the form of a viscous protoplanetary disk.

Until we know the time evolution of the solar nebula, we cannot be sure of its structure at any instant; some steady-state models may be mathematically tractable but irrelevant to the solar nebula. Knowledge of the time evolution of this structure is tantamount to knowledge of the details of gas removal. Considering the present state of theoretical models, the observational value of 1 to 10 Myr (Sec. I) appears to be the most reliable estimate of the time scale for nebula removal.

IV. COSMOCHEMICAL IMPLICATIONS OF THE MODELS

Solar system samples from primitive bodies (mainly meteorites) primarily give information about the conditions within the nebula at the time the sample was preserved in the body. This means that remnants of the earliest phases of nebula evolution will be difficult to find—all the gas and dust present at the earliest times in the planetary region may have ended up in the Sun. Planetary surface studies (e.g., the spatial distribution of crater rates throughout the solar system) and analysis of giant-planet atmospheres (e.g., oxygen, carbon isotope ratios) and internal structures nevertheless are important constraints for models. Unfortunately, it is not easy to link solar nebula models with such information, in part because there are many stages and physical processes involved before solid bodies are formed, and in part because there may be a great deal of secondary alteration going on, postdating and masking the original nebula information.

Early work concentrated largely on the question of nebula temperature (see Cameron 1962) and the processes occurring in a slowly cooling nebula, in particular the condensation sequence of minerals and metallic compounds (see, e.g., Grossman 1972). The nebula was regarded as a "black box" without considering its dynamical evolution simultaneously with its thermal evolution.

The introduction of the concept of accretion disks changed this emphasis considerably; the importance of redistributing angular momentum, and the associated dissipation and heat production in the nebula as a necessary by-product of star formation became clearer. One of the early successes of accretion-disk theory was that the theoretical temperature gradient in the steady-state disk agreed quite well with the temperature profile derived by Lewis (1974*a*) on the basis of the composition of planets and their moons. However, the Lewis (1974*a*) profile can be fit by models with a wide range of viscous strengths, because temperatures vary only with small fractional powers of the assumed nebula parameters (Sec. III). Furthermore, this temperature profile need not be anymore representative of nebula properties at any given instant than the surface density, as was discussed in the last section.

Recently, cosmochemical consequences of turbulent disks have been analyzed in greater detail (Morfill 1983,1985; Morfill and Völk 1984; Morfill et al. 1985). (The chapter by Prinn and Fegley also deals with many aspects of the chemistry of the early solar nebula.) The important advance made by Morfill and colleagues was to couple a transport theory for dust and vapor products within a turbulent disk with simple physico-chemical processes (condensation and sublimation) and to calculate the spatial fractionation of minerals produced by the large-scale temperature gradient in the disk. In addition, coagulation was studied as well as statistical properties of the turbulent transport. So far, this work has only been applied to steady disks.

The results of these investigations can be summarized as follows.

(1) An ensemble average of a certain volume element of the nebula at a given radius r from the center contains condensates and residue. The condensates have been recycled through the hot inner region of the disk, and the residue is regarded as pure, unmodified interstellar material, which (depending on the position in the disk) may have lost volatiles, intermediates, etc. For the planetary region of interest (i.e., 1 AU to 4 AU), the ratio condensate/residue is about unity, decreasing with increasing distance from the center; that is, throughout the terrestrial planet region, roughly comparable amounts of pristine and thermally reprocessed grains exist. Thus a cosmochemical signature of a "hot environment" should be a ubiquitous feature of the nebula, even if the mean temperature is as low as 500 K—the value most accretion disk models assign to the Earth-Mars region. The reason is the diffusion associated with turbulent transport; the stochastic nature of turbulence produces a random walk of gas parcels and grains both upstream and downstream of the radial accretion flow onto the central Sun.

(2) While the above is true for an average description (in the sense of an ensemble average), it is also possible to study variations around this average. Large-scale turbulent eddies move large volumes of gas, which are dispersed gradually by smaller eddies. Complete mixing of gas, not as an ensemble average (which averages over many large-scale turbulent eddies), but homogeneously on much smaller (molecular) length scales, requires many coherence times. Thus a measure of the degree of inhomogeneity in a given sample can be obtained from comparing the turbulent-transport time scale $\tau_t \sim \Delta r^2/\nu$ with the coherence time $\tau_c \sim h/c_s$, where h is the disk half-thickness, and Δr is the distance to the nearest important (from the point of view of abundances) sublimation boundary. It is easy to show that this ratio is

$$\frac{\tau_t}{\tau_c} \simeq \left(\frac{\Delta r}{r}\right)^2 \frac{3\pi\Sigma}{\dot{M}} \sqrt{GM_* r} \tag{16}$$

where Σ is the disk surface density. For typical disk families (see the generic model shown in Sec. III), we get at 1 AU

$$\tau_t/\tau_c \simeq 10^3 \left(\frac{\Delta r}{r}\right)^2 . \tag{17}$$

Thus for $\Delta r/r \sim 1/2$, which is typical for the temperature gradients expected in disks, we may expect a variance in the "samples" (e.g., meteorites) of a few to $\sim$10% purely from statistical inhomogeneity. Thus some general "typecasting" into classes or categories is to be expected, associated with significant primordial (interstellar) inhomogeneities that survive turbulent mixing. It is impossible to quantify statistically these inhomogeneities beyond what has been indicated here. However, numerical Monte-Carlo experiments could be performed to simulate variations among meteorite types, based on simple fractionation as a function of position in a thermally structured protoplanetary accretion disk.

(3) Recently, Morfill and Clayton (1986) have investigated oxygen isotope fractionation in the protoplanetary accretion disk. This investigation highlights the predictive power of accretion-disk models. Originally, the viscous accretion-disk concept was introduced to overcome the angular momentum transport problem. Then, by deriving the transport theory, information could be obtained about bulk cosmochemical properties of meteorites and planets, as well as a measure of the variances. Now, by combining this with the Rayleigh distillation theory and gas-phase isotope exchange-rate information, it is possible to calculate isotope fractionation patterns and order the meteorite data within this new framework. Cross correlation with other information (e.g., mineralogical data) then allows tests of the applicability of the accretion-disk scenario. At the present time, indications are that some ordering of the meteorite data can be achieved this way, without taking recourse to "special effects." If these preliminary results are confirmed, this will have to be regarded as a major success. In the same way, application of isotope calculations to planetary atmospheres promises to be an important diagnostic tool in the future.

(4) A large-scale feature of solar nebula chemistry is the question of the missing CO. The problem is that radio observations show that a great deal of carbon in interstellar clouds resides in CO gas, whereas in planets CH_4 dominates. Somewhere, between the collapse and planet formation, processes must have occurred which changed the gas chemistry (Stevenson 1988). It appears that turbulent convective transport could be invoked to understand this problem, but the full implications have not yet been completely analyzed.

(5) Coagulation is of utmost importance in the solar nebula. We know that interstellar grains are typically in the size range $\sim$0.1 μm, and yet these are the building blocks of our planets. All researchers agree that initially coagulation is the fastest and single most important process leading to grain growth. This has a number of consequences, which have not yet been incorporated in accretion-disk models. These are:

a. The size spectrum of the dust particles becomes broader, as more large grains are formed;
b. The opacities change. First there may be a slight increase at *high* temperatures ($\gtrsim$200 K) but then there will be a significant drop as more dust surface area is lost through the coagulation. At large enough sizes the opacity is determined simply by the geometrical cross sections. Since viscous accretion-disk temperatures vary as the 1/3 power of the opacity, this effect is important;
c. Turbulence produces larger relative velocities between grains, and therefore speeds coagulation, assuming grain fragmentation is not also enhanced. However, if the optical depth (proportional to the opacity) decreases below $\sim$ unity, we have the same situation as that described in Sec. III in connection with grain evaporation: the origin of the turbulence, if the driving mechanism really is convective instability, could be inhibited (Weidenschilling 1984; Mizuno et al. 1987) and the disk becomes quiescent. Only if the disk is accreting matter, i.e., if the presolar cloud has not been exhausted, is it possible to increase the optical depth again, and start up turbulence anew. Otherwise the turbulence could only be maintained by a different driving agent altogether (e.g., frictional coupling with mass outflow). Turbulent coagulation, coupled with sedimentation during quiescent periods, does appear to be the fastest way to make a dust sub-disk, and therefore it is pleasing in some sense, to see that model calculations lead us in this direction.

The important task, however, is still to make a self-consistent disk model including coagulation and evaporation, as well as condensation, both for steady and for evolving disks. As a next step, nonaxisymmetric variations (e.g., density waves), should be studied (Fig. 7). Their effect will be to produce local "peaks" in temperature, which could be one of the "short-term events" appealed to by Wood and Morfill (1988).

This section has been devoted to the cosmochemical consequences of viscous accretion-disk models, the only models that are sufficiently developed dynamically for significant chemical consequences to be derived. One may speculate, however, that in mildly turbulent nebula models evolving primarily by gravitational torques, which may tend to preserve the radial ordering of fluid elements, considerably less mixing would occur than in a fully turbulent nebula.

V. REMOVAL OF NEBULAR GAS AND IMPLICATIONS FOR PLANET FORMATION

Observations of young stellar objects imply that the proto-Sun underwent several phases of energetic mass loss early in its evolution, with a peak mass loss rate of perhaps 10^{-5} $M_\odot$ yr^{-1} occurring within 10^5 to 10^6 yr after formation of the pre-main-sequence Sun (Sec. I). This strong stellar wind cleared away most of the optical obscuration surrounding the proto-Sun, i.e., it dissipated the gas and dust of the solar nebula. This role, usually assigned to the T Tauri phase (Cameron 1973*a*), may more properly belong to an early bipolar wind phase or an FU Orionis phase.

The details of how solar-nebula removal occurred are a subject of debate, in part because the details must depend on the structure and evolution of the nebula itself, which is poorly known. Turbulent viscosity and/or gravitational torques appear to have been responsible for the bulk of the mass transfer in the nebula, enough to form the central proto-Sun out of the nebula. In this section we consider the possibilities for removing the final vestiges of the nebula, after formation of the pre-main-sequence Sun, and the implications for planetary formation.

Horedt (1978) found that a T Tauri wind blowing over a concave solar nebula at 10^{-8} $M_{\odot}$ yr^{-1} could entrain gas and dust and carry it away from the solar system, removing about 0.1 $M_{\odot}$ of nebula matter in 10 Myr. Clearly a higher mass-loss rate, more in agreement with recent observations, would remove the nebula in a shorter interval. Elmegreen (1978) found that a stellar wind blowing over a disk would induce turbulence in the disk, and the effective viscosity thereby produced would result in the disk being accreted onto the central protostar. Attempts to reconcile these disparate points of view have not been successful (Elmegreen 1979; Horedt 1982*a*). Both theories neglect rotation and magnetic fields in the stellar wind.

Horedt (1982*b*) has also advanced ultraviolet radiation from the Sun as a means of heating the solar nebula sufficiently enough to cause its loss through thermal pressure. However, this mechanism appears to require ~1 Gyr to dissipate even a 0.01 $M_{\odot}$ nebula, which is much too long to explain the clearing of T Tauri stars within ≈1 Myr. Hayashi et al. (1985) suggest that a combination of stellar winds and extreme ultraviolet (EUV) radiation could have removed a solar nebula with a mass of about 0.01 $M_{\odot}$ within ≈2 Myr.

Unless the proto-Sun loses a large fraction of its mass through its early stellar wind, which is unlikely (Weidenschilling 1978), it appears that the most massive nebula that could be removed by a stellar wind is close to the minimum mass nebula, i.e., ≈0.01 $M_{\odot}$. This follows because the interaction of the outflowing wind with the disk gas is certain to be inefficient at removing the disk gas (e.g., the wind can flow outward preferentially along the rotational poles without interacting with the disk). Assuming a low efficiency of disk gas loss per unit stellar wind mass loss (perhaps on the order of 10%), a loss of 0.1 $M_{\odot}$ from the proto-Sun would only remove a 0.01 $M_{\odot}$ nebula. In order to remove an initially more massive solar nebula, it thus appears that mass transport processes in the nebula (Sec. III) must force the accretion of the majority of the matter onto the proto-Sun, leaving only a small residual for the solar wind and radiation to remove.

Regardless of the details of nebula removal, the observational evidence

(Sec. I) is that the bulk of the gas and dust is removed within 1 to 10 Myr, and theories of planet formation must meet this severe time constraint, which we will call the *removal constraint*. Dust grains sufficient to form the terrestrial planets and cores of the giant planets must be locked into bodies large enough to withstand the early solar wind; the gas necessary to form the envelopes of the giant planets must also be retained. The latter constraint would appear to require that the giant planets were essentially fully formed within 1 to 10 Myr after formation of the proto-Sun.

The compositional differences within the giant planet group (Uranus and Neptune have proportionately much less hydrogen and helium than Jupiter and Saturn) are important clues that the giant planets did not all form either in precisely the same manner or environment. For example, considering that planetary formation time scales by a given mechanism tend to increase with increasing radius and orbital period, it may be that significant nebula erosion and/or chemical evolution occurred between the formation of Jupiter/Saturn and Uranus/Neptune. While it is unclear whether strong stellar winds could preferentially remove the lighter gaseous species from the outer nebula, perhaps enhanced solar EUV radiation could perform this role.

We shall group theories of planet and satellite formation into two categories: formation by accumulation of a smaller population of bodies termed planetesimals, and gravitational instability of the gaseous portion of the nebula to form giant gaseous protoplanets.

Dust Instability and Planetesimal Accumulation

Coagulation induced by turbulent motions has already been discussed in the previous section. In addition, at the high densities of the solar nebula, Brownian motion can be another source of the relative motions needed to produce grain-grain collisions and coagulation. Once turbulent motions subside, grains can relatively rapidly sediment to the midplane of the nebula, requiring about 10^3 yr in the terrestrial planet region (Weidenschilling 1980; Nakagawa et al. 1981; Nakagawa et al. 1986). Growth to sizes on the order of 1 cm occurs during these phases. The dust settles into a very thin sub-disk at the midplane of the gaseous nebula.

The dust disk formed by the sedimenting dust is thought to have become gravitationally unstable at some point, rapidly breaking up into a large number of self-gravitating clumps, which could then contract into a swarm of planetesimals roughly of size 1 km (see, e.g., Safronov 1969; Goldreich and Ward 1973). The planetesimals accumulate into the planets, first through collisions with planetesimals on nearly identical orbits about the Sun, and then, more slowly, through collisions with planetesimals with significantly different orbits (for more details, see Safronov [1969], Wetherill [1980] and the Chapter by Greenberg).

While the terrestrial planets can be directly grown through this process, the giant planets must have begun accreting gaseous, hydrostatic atmospheres early in the formation of their rocky cores. By the time the cores reached about 10 $M_\oplus$, the atmospheres would have collapsed onto the cores, allowing the giant planets to accrete gas at a higher rate and rapidly to reach their present masses (Mizuno 1980). Mizuno's (1980) calculations were performed for giant planets embedded within the solar nebula defined by Kusaka et al. (1970). As can be seen in Table I, the Kusaka nebula is roughly consistent with viscous accretion-disk models, though physical quantities such as density and temperature may differ by factors of 10 and 2, respectively, compared to the generic accretion-disk model (Fig. 4). While even larger differences can be obtained by considering other possible disk models, the effects of different nebula models on the critical mass are unknown; Mizuno (1980) did not explicitly examine the dependence of the critical mass on the background nebula model. However, the fact that the critical mass came out to be very nearly constant throughout the giant-planet region implies that this effect may be small.

Because the earliest phases of the planetesimal accumulation process apparently occur on time scales much less than the nebula removal time scale, and because km-sized bodies should be resistant to even an energetic solar wind, formation of the terrestrial planets by this means is not affected by the removal constraint. Accumulation both in the presence of the solar nebula (Hayashi et al. 1985) and after nebula removal (Safronov 1969) have been advanced. In either situation, the time scale for final accumulation of the terrestrial planets is $\approx$100 Myr (Safronov and Ruzmaikina 1985). The final phases of terrestrial planet growth thus appear to have occurred well after the removal of the solar nebula, thereby obviating the need to consider final accumulation in the presence of gas.

Both theories encounter problems in giant-planet formation, unless an early phase of runaway accretion is invoked. Without an early runaway, and in the gas-rich approach, Hayashi et al. (1985) require 4 $\times$ 10 Myr for Jupiter and 6 $\times$ 100 Myr for Saturn to reach the critical core mass of 10 $M_\oplus$, and Neptune requires more than the age of solar system. Similar times are required in the accumulation models of Nakano (1987). In the gas-poor approach, Safronov (1969) and Wetherill (1980) require $\sim$1 Gyr to form a 10 $M_\oplus$ core for Jupiter by accumulation without a runaway. These time scales are inconsistent with the removal constraint; protogiant planets formed in this manner could not have accreted substantial gaseous envelopes.

It has long been suspected that forming the giant planets rapidly enough to satisfy the removal constraint requires special conditions in the outer solar system, such as enhanced gas density (Safronov 1969) or small relative velocities leading to runaway accumulation (Greenberg 1980). Lissauer (1987) has emphasized this approach to the time scale problem. Runaway accumulation of planetesimals in the outer solar nebula allows Jupiter's core to accumulate in $\sim$1 Myr, assuming also that the gas density in the nebula fell off less rapidly with heliocentric distance than has previously been thought likely

(Weidenschilling 1977*a*). Stevenson and Lunine (1987) have suggested that diffusively transported water vapor in a turbulent accretion disk would be preferentially frozen out at Jupiter's position, thereby enhancing the local nebula surface densities to values consistent with runaway accretion of Jupiter.

Lin and Papaloizou (1985) have asserted that both prototerrestrial planets and protogiant planets can form in the gas-rich disk of their convective accretion-disk models within $\approx 10^5$ to 10^6 yr. This would satisfy the removal constraint, but their growth times appear to be underestimates, because of the implicit assumption that only one body is growing; competition with other growing bodies means that the rapid growth by gas accretion will halt before planetary masses are formed. Weidenschilling (1984) found that the early grain growth necessary for terrestrial planet formation could not occur so long as turbulence was present in the Lin and Papaloizou accretion disk. Furthermore, because the underlying viscous accretion-disk model has been called into question (Cabot et al. 1987*a,b;* see Sec. III), these estimates must be viewed with caution.

There is also potentially a problem associated with the effects of the tidal force of a growing planet on the nearby nebula (Cameron 1979; also see Chapter by Pollack). Once the planet becomes large enough, the nebula will be tidally truncated inward and outward of the planet, so that gas can no longer flow onto the planet, thereby stifling further growth. On the other hand, this process does provide a means for cutting off growth at some point, and so may have determined the masses of the giant planets (Lin and Papaloizou 1980; Sekiya et al. 1987).

Giant Gaseous Protoplanets

Cameron (1978) suggested that the gaseous portion of a viscous accretion-disk model of the solar nebula was gravitationally unstable to the formation of rings, which would then break up and form giant gaseous protoplanets. Cassen et al. (1981) found that the instability could occur if the disks were cool and comparable in mass to the central proto-Sun. The time scale for this instability is very short, on the order of the rotational period, and so the removal constraint is well satisfied. Tidal truncation of the nebula may also occur in this scenario (Cameron 1979), but considering that the giant planets can start off with at least their final masses, tidal truncation does not present an impediment to growth. Instead, tidal truncation may simply affect the boundary conditions for giant planet evolution (Cameron et al. 1982; cf. Bodenheimer and Pollack [1986] for evolution in the planetesimal accumulation scenario).

While this theory is an attractive means for directly and rapidly forming the giant planets, substantial criticisms have been leveled at it (reviewed by Stevenson [1982*a*] and Pollack [1984]). For example, it appears to be difficult to account for the formation of rocky cores within the giant planets, and to explain the increase in the ratio of core mass to envelope mass with heliocentric radius. As a partial consequence, few people have seriously pursued this scenario. Nevertheless, Cameron (1985) has continued to support giant gaseous protoplanet formation in viscous accretion disks, at least until the work by Cabot et al. (1987) appeared. Until more detailed work is performed, however, on questions such as whether a gravitationally unstable disk actually forms distinct giant gaseous protoplanets, it appears premature to rule out the possibility that a relatively massive solar nebula underwent fragmentation into giant planets. Remembering recent observational evidence for substantial mass in some protostellar disks (Sargent and Beckwith 1987*b*; see Sec. I), and the trend toward more massive accretion-disk models (Cabot et al. 1987*b*), a minimum mass nebula incapable of local gravitational instability may not be the only possibility worth consideration.

VI. CONCLUSIONS

Models of the formation and evolution of the solar nebula, and the detailed link between the consequences and predictions of such models with solar system observations, are essential prerequisites for an understanding of star formation and planet formation. In this review, we have tried to summarize the current status and state of the art in solar nebula research.

Perhaps the most important factor for the formation of planetary atmospheres is the observational evidence in favor of nebula removal by 1 to 10 Myr after the collapse of the presolar nebula. Theoretical models of the protostellar collapse phase appear to have isolated the initial conditions necessary for formation of a single, solar-mass star, conditions which are in agreement with observations of star-forming regions. However, these models have not been developed enough to place firm constraints on the detailed physical properties of the solar nebula, and hence do not as yet constrain planetary atmosphere formation.

Many aspects of the dynamics of both steady and time-evolving viscous accretion disks have been well studied by now, and the thermodynamical and cosmochemical consequences predicted to date appear to be in accord with the solar system. These models obviously can also be used to constrain theories of planetary atmospheres. However, the basic hypothesis of a smoothly varying and evolving accretion disk has been threatened by sufficiently good physical arguments to shake the faith of even the firmest proponents of viscous disks.

Hence we do not at present have a generally accepted model of the solar nebula. Solar nebula models are in a state of flux from the steady viscous accretion-disk models of the last decade to a new generation which has not yet emerged. The new models should include more physical effects, such as the early evolution and accumulation of the proto-Sun, three dimensionality, and a self-consistent treatment of the coupled problems of grain coagulation, opacity and thermal evolution. Much remains to be learned.

Further laboratory effort is needed, too. The physics of coagulation and fragmentation in the velocity ranges of interest must be better understood. Chemical rate coefficients (especially at low temperatures), the role of "dust" as a catalyst for chemistry, and isotope fractionation are examples of areas where further, complementary research is required.

Acknowledgments. We thank P. Cassen and S. J. Weidenschilling for their detailed reviews of a difficult manuscript. APB thanks J. A. Graham for many discussions about pre-main-sequence stars. GEM thanks the University of Arizona for their hospitality during his visit.

SOLAR NEBULA CHEMISTRY: ORIGIN OF PLANETARY, SATELLITE, AND COMETARY VOLATILES

R. G. PRINN and B. FEGLEY, JR.
Massachusetts Institute of Technology

Complex interactions between chemical, physical and dynamical processes in the solar nebula are hypothesized to have played seminal roles in determining the reservoirs of volatile compounds and elements in the planets and their satellites and in comets. These processes include condensation, homogeneous and heterogeneous thermochemical and photochemical reactions, and disequilibration resulting from fluid transport, condensation, and cooling whenever they occur on time scales shorter than those for the chemical reactions. The ultimate potential starting materials were gases and grains in the pre-nebula interstellar cloud(s). As the nebula was forming the grains may have partially or completely evaporated or decomposed but not necessarily so. Extensive evidence from meteorites and inferences that asteroids are meteorite parent bodies implies that the nebula was extensively physically mixed and chemically reprocessed (but not necessarily chemically and isotopically homogenized) at least out to 4 AU. Possible energy sources for driving nebula chemical reactions include shock waves and photons from lightning discharges, radioactive decay processes, and solar and stellar photons, but we show that all of these are secondary to the thermal energy of the solar nebula itself. The enormous ultraviolet opacity of the gaseous nebula disk (due to H_2O, grains, etc.) caused solar ultraviolet photons to be absorbed only in the very hot inner nebula where rapid thermochemical reactions probably prevented significant photochemical disequilibration beyond some possible subtle effects on high temperature refractory material. Also, the high dust opacity in the immediate post-nebula solar system probably prevented the high ultraviolet flux from the early Sun being an effective energy input into early planetary atmospheres. Ultraviolet emission from the early Sun would still be remotely observable since the opacity for all photons emitted in directions other than toward the gaseous or dusty disk was probably very small. Surface catalysis was undoubtedly important in the nebula for (thermochemical) *organic synthesis. We discuss models for the chemical composition of the gases and grains as functions of space and time in the nebula and the implications of these models for the volatile contents of accreting planets, satellites and comets. The models include the pressure/temperature conditions appropriate to subnebulae around giant planets as well as those appropriate to the solar nebula itself. It is highly probable that as proposed by Lewis and Prinn the major gaseous C and N compounds in the solar nebula were CO and* N_2 *and that carbonaceous compounds formed from CO by Fischer-Tropsch-type reactions were present in amounts comparable to CO in order to explain recent outer solar system planetary and satellite observations. Vapor-phase hydration of silicates, like the chemical reduction of CO and* N_2*, was kinetically inhibited in the nebula. Formation of giant planetary satellites in well-mixed subnebulae created by* spin-off *as proposed by Pollack and Bodenheimer or by a massive impact as suggested by Cameron provides a natural explanation for the occurrence of* CH_4-rich *satellites like Titan and Triton using the chemistry described by Prinn and Fegley; satellite formation in planetary accretion disks which include nonlinear effects can also explain Titan but accretion-disk-products may be reprocessed in a later-forming* spin-off *or* impact-generated *subnebula. Certain volatile ratios (e.g.,* CO/CH_4, N_2/NH_3, H_20 *ice/silicate) in ice-rich bodies are diagnostic of their origin. For example, the observation of* CH_4 *and* NH_3 *in Halley implies that at least some of the material in this comet is neither of solar nebula nor interstellar origin but is material condensed in a subnebula of one of the giant planets.*

I. INTRODUCTION

A chemist investigating for the first time the distribution of the very volatile atmosphere-forming elements (H, O, C, N, Cl, S etc; He, Ne, Ar, etc.) and their compounds in the solar system would immediately ask why planets, satellites, comets and meteorites were endowed apparently with such different mixtures of these very volatile materials. Despite their often overemphasized similarities, the volatile element contents of the terrestrial planets Venus, Earth and Mars are significantly different and many of the differences are not attributable simply to (presently recognized) evolutionary processes; the other terrestrial planet Mercury is by comparison almost completely void of volatiles. The gaseous giant planets are all rich in H and He. But the heavier elements (O, C, N, Si, etc.) comprise only a small fraction of the masses of Jupiter and Saturn while they are a large fraction of Uranus and Neptune; and all four planets are enriched in these heavier elements relative to the Sun. The ice-rich satellites of the outer planets differ in uncompressed density (and thus implied ice/rock ratios) and, except for Titan and perhaps Triton, their atmospheres are thin and/or evanescent (e.g. Io) or essentially absent. Where data is available, comets (e.g., Halley) are more rich in CO than CH_4 whereas the converse appears to be true for ice-rich satellites (e.g., Titan, perhaps Triton?). Complex carbonaceous materials are major reservoirs of C, N, and H in carbonaceous chondrites and (by inference from reflection spectra) in many dark asteroids and some outer planetary satellites (e.g., small satellites of Jupiter and Uranus).

These intriguing patterns for volatile distribution are a result of a complex and as yet poorly understood sequence of events which began with the formation of solar system bodies in the primitive solar nebula and was followed by a wide range of both general and specific evolutionary processes on each body (see, e.g., Prinn 1982 for a review). The development of theories concerning the origin of the solar system, and more specifically the advancement of our ideas concerning the chemistry of the volatile elements in the gaseous and dusty solar nebula in which the planets formed, has been guided in large part by observations of the volatile contents of planets, planetary satellites, meteorites and comets. We will emphasize in this Chapter the origin of the volatile compounds formed from the abundant chemically reactive elements H, O, C, and N; the noble gases are discussed in the Chapter by Pepin. By way of an introduction to the chemical theories which we will address in this Chapter, we will first present here a review of some salient facts about chemically active volatiles in solar system objects observed to date.

A. Volatiles in Terrestrial Planets and Chondritic Meteorites

The observed inventories of the important volatiles expressed as CO_2, H_2O and N_2 on the three terrestrial planets which possess atmospheres and in the chondritic meteorites are summarized in Table I. No indigenous volatiles have been observed in Mercury's atmosphere at a pressure greater than 10^{-13} bar or on its surface (Lewis and Prinn 1984). The abundances in Table I illustrate that both the chondrites and Venus, Earth and Mars are depleted in C, N, and H_2O relative to solar abundances. Even the volatile-rich CI1 carbonaceous chondrites have retained only $\sim$ 5 to 10% of the solar abundances of C, N, and H_2O. The similar CO_2 and N_2 depletions for Earth and Venus indicate similar inventories of these volatiles on the two planets. Furthermore, on an absolute (g/g = gram/gram) basis, there is approximately 10 times more carbon than nitrogen on both Earth and Venus. However, Venus is apparently endowed with much less H_2O than the Earth, and Mars is apparently either volatile poor, or has outgassed less efficiently than Venus and Earth, or has lost a substantial portion of its volatile content through atmospheric erosion or escape at an earlier stage of its history.

However, comparisons of the planetary and chondritic volatile inventories are complicated by several factors. The apparently larger depletions for CO_2 and N_2 on the terrestrial planets relative to the chondritic meteorites may simply reflect partitioning of a significant fraction of their C and N inventories into iron cores and/or their silicate interiors. Similarly, the observed presence of CO_2(solid) in the Martian polar caps, and the suspected presence of subsurface water ice on Mars and of $CaCO_3$ on Venus means that the atmospheric inventories of these volatiles are likely to be underestimates of the planetary inventories (e.g., see Carr 1987; Prinn and Fegley 1987*a*).

The chondritic meteorites contain a variety of volatile-bearing phases whose occurrence and abundance vary from class to class. Carbon occurs as

TABLE I

Depletions of Important Volatiles in the Chondritic Meteorites and in Venus, Earth and Mars Relative to Solar Abundances [(g/g Si)/(g/g Si)][a]

Volatile	Venus[b]	Earth	Mars[c]	Chondritic Meteorites[d]		
				C	(H, L, LL)	E
CO_2	3×10^{-5}	3×10^{-5}	2×10^{-8}	$(0.8 - 6) \times 10^{-2}$	1×10^{-3}	$(4 - 5) \times 10^{-3}$
H_2O	2×10^{-9}	2×10^{-4}	4×10^{-12}	$(0.07 - 10) \times 10^{-2}$	$(0.2 - 2) \times 10^{-3}$	$(0.4 - 2) \times 10^{-3}$
N_2	1×10^{-5}	2×10^{-5}	5×10^{-9}	$(0.1 - 5) \times 10^{-2}$	$(4 - 6) \times 10^{-4}$	$(0.1 - 1) \times 10^{-2}$

[a]Solar abundances from Cameron (1982). The solar abundances of C, N and O (minus 15% of O assumed to be incorporated into rock as MgO + SiO_2) were expressed as CO_2/Si, N_2/Si, and H_2O/Si mass ratios for this normalization. Atmospheric inventories only were considered for Venus and Mars but atmospheric plus oceanic plus crustal inventories were considered for Earth. The abundance data for Venus, Earth and Mars are from Prinn and Fegley (1987*a* and references therein).

[b]Bulk composition models *V*2, *E*5, and *Ma*2 from the *Basaltic Volcanism Study Project* (1981*b*, p. 641) were used to determine Si contents for Venus, Earth and Mars, respectively. The Si contents for the chondritic meteorite classes are from Mason (1979).

[c]Present-day atmospheric inventories of CO_2, H_2O, and N_2 on Mars.

[d]The tabulated values are for CI1, CM2, C3 carbonaceous chondrites, EH4, EH5, EL6 enstatite chondrites, and H, L, LL ordinary chondrites. Carbon data: median values from Mason (1979). Nitrogen data: median values from Mason (1979) for the C and H, L, LL chondrites. Data of Grady et al. (1986) for the E chondrites ($\sim$100 to $\sim$1000 *ppm* N by weight). Mason (1979) gives a median value of 260 ppm N for E chondrites corresponding to a depletion factor of 3×10^{-3}. Water data: Mason (1971, 1979), Kolodny et al. (1980), Robert and Epstein (1982) and Yang and Epstein (1983) for the CI1, CM2, and C3 chondrites; Yang and Epstein (1983), McNaughton et al. (1981) and Robert et al. (1979) for H, L, LL chondrites; Boato (1954) and Yang and Epstein (1983) for the two EH4 chondrites Indarch and Abee, respectively. The range of water contents (wt. %) and the corresponding depletion factors for the carbonaceous chondrites are: CI1 chondrites, 3 – 10%, $(0.3 - 1) \times 10^{-1}$; CM2 chondrites, 1 – 16%, $(0.8 - 12) \times 10^{-2}$; and C3 chondrites, 0.1 – 2%, $(0.7 - 10) \times 10^{-3}$. If the higher values reported by Wiik (cited in Mason 1971) for the water contents of the CI1 chondrites are included, the upper limit is $\sim$20 % (wt.) corresponding to a depletion factor of $\sim 2 \times 10^{-1}$. However, some of this water is probably terrestrial contamination (see Boato 1954).

graphite, carbides, dissolved C in Fe,Ni alloy, in diamonds, in silicon carbide, disequilibrium organic matter, and carbonates (Mason 1971; Lewis et al. 1987; Bernatowicz et al. 1987). Nitrogen occurs as dissolved N in Fe,Ni alloy, disequilibrium organic matter, nitrides, and possibly as NH_4^+ salts (Mason 1971; Kung and Clayton 1978). However, the major N-carrier(s) in the ordinary chondrites is not well characterized. Finally, hydrogen occurs as water (adsorbed, hydrated in salts and in hydrous silicates) and as organic compounds. Water contents in carbonaceous chondrites may reach 10 wt. % (see Table I; Mason 1971) but are much lower (~ 100 to 1000 ppm by weight) in the ordinary and enstatite chondrites (Robert et al. 1979,1987*a,b;* McNaughton et al. 1981; Boato 1954; Yang and Epstein 1983). The hydrogen-bearing phases in these meteorites are apparently a combination of organic matter and trace hydrous phases (e.g., see Yang and Epstein 1983; Nagahara and El Goresy 1984). The stable isotope compositions of H, C, N and O in the chondritic meteorites have recently been reviewed by Pillinger (1984); some implications of the observed D/H ratios will be discussed later.

B. Volatiles in Ice-rich Satellites and Comets

With the exception of Io and Europa, all of the satellites of the outer planets for which we have reliable estimates of mass and radius appear to have low enough densities to conclude that ice must comprise a significant fraction of these satellites. Specifically, the densities of these *ice-rich* satellites range from ≈ 1.9 g cm^{-3} for Ganymede, Callisto and Titan down to ≈ 1.2 g cm^{-3} for Mimas, Iapetus and Enceladus (see Burns 1986 for a review). Models for Ganymede, Callisto and Titan suggest that they are 27 to 57, 34 to 58 and 30 to 58% ice, respectively; models for Mimas and Dione suggest they are 45 to 54% ice, and models for Enceladus, Tethys and Iapetus suggest in contrast a 65 to 78% ice composition with the remainder again being rock in each satellite (see Schubert et al. 1986 for a review). For the Uranian satellites and for the Saturnian satellite Rhea, Johnson et al. (1987) have constructed a range of models which imply that Titania is 35 to 53% ice, Oberon 40 to 58% ice and Rhea 53 to 73% ice, with the remainder again being rock in each satellite. We do not know the densities of the Neptunian satellites (Triton, Nereid) or of Pluto and its satellite Charon with sufficient accuracy to deduce useful ice/rock ratios.

In addition to the presence of ice implied from their densities, water ice has been identified spectroscopically on the surfaces of many satellites (Ganymede, Callisto, Rhea, Iapetus, Dione, Tethys, Enceladus, Mimas, Hyperion, Saturn's rings, Titania, Oberon, Umbriel, Ariel, Miranda, Charon), methane ice has been identified on the surfaces of Triton and Pluto, N_2 and CH_4 gas dominate the atmosphere of Titan, and liquid N_2 has been tentatively identified on the surface of Triton (see Burns 1986*b*; Cruikshank and Apt 1984; Cruikshank et al. 1984; Marcialis et al. 1987). In addition, there are many of the smaller satellites whose densities are not known but whose low albedos and reflection spectra imply the possible presence of dark carbonaceous material on their surfaces; e.g., Himalia, Elara, Pasiphae, Carme, Sinope, Lysithea, Ananke, Leda, Phoebe, and the rings of Uranus; also Iapetus has a spatially variable albedo suggesting a part water ice and part carbonaceous surface (see Cruikshank 1986).

Earth-based observations of cometary comae have led to identification of a large number of volatile species emanating from comets. Carbon, nitrogen, oxygen, sulfur and hydrogen are seen in molecules (CO, HCN, CH_3CN, H_2O), atoms (C, H, O, S), free radicals (CH, CN, CS, NH, NH_2, OH) and ions (C^+, CO^+, CO_2^+, CH^+, H_2O^+, N_2^+, OH^+, CN^+). The number abundances of H, C, N, O, S, Mg, and Fe (relative to Si = 1.0) in recent bright comets are 24.3, 3.73, 1.51, 22.3, 0.55, 1.06 and 0.9, respectively (Delsemme 1982). Except for H (which is depleted by a factor of $\approx 10^3$), these number abundances are not much different from the solar composition values particularly allowing for the factor-of-two uncertainties that accompany some of the cometary values.

Preliminary results from the Giotto mission to Halley suggest nuclear emanation rates for the gases CO, N_2, CH_4 and NH_3 relative to H_2O of 0.05 to 0.2, < 0.01 to 0.1, 0.02 and 0.01 to 0.02, respectively (Balsiger et al. 1986; Eberhardt et al. 1986; Allen et al. 1987). Thus $CH_4/CO \approx 0.1$ to 0.4 and $NH_3/N_2 > 0.1$. These abundance estimates for CH_4 and NH_3 are tentative (Allen et al. 1987) but even if they are 2 or 3 times less than estimated, their presence in comets has considerable cosmogonic implications which we will address later. The dust from comet Halley apparently consists of a predominantly *chondritic* core with an *organic* mantle composed mainly of highly unsaturated compounds (Kissel and Krueger 1987).

C. Volatiles in Giant Gaseous Planets

The four giant gaseous planets (Jupiter, Saturn, Uranus and Neptune) range in size from ~ 4 to 10 times the size of the Earth with masses that are ~ 15 to 300 times the Earth's mass ($M_\oplus$). These planets are composed of variable proportions of solar gas and heavy (atomic number > 2) elements, with the former fraction being H_2- and He- dominated and the latter fraction being made up of *icy* (e.g., H_2O, CH_4, NH_3) and *rocky* (e.g., MgO, SiO_2, Fe, etc.) material. The observed radii and mean densities of the four planets show that the gas/(ice + rock) ratio decreases monotonically from Jupiter ($R \sim 6.98 \times 10^4$ km,$\rho \sim 1.33$ g cm^{-3}), which is gas rich, to Neptune ($R \sim 2.45 \times 10^4$ km,$\rho \sim 1.67$ g cm^{-3}), which is (ice + rock) rich.

More detailed information about the bulk compositions of these four planets, and about the radial distribution of the gaseous and solid components, can be obtained by constructing interior structure models having several compositional zones and by fitting these models to the observed mass, radius and observed J_2 and J_4 gravitational moments. These modeling efforts which have been recently reviewed by Stevenson (1982), yield several important facts.

First, the estimated mass fractions of ice + rock for Jupiter, Saturn, Uranus and Neptune are $\sim 0.05 - 0.09$, $\sim 0.17 - 0.24$, $\sim 0.75 - 0.91$ and $\sim 0.81 - 0.96$, respectively (Pollack 1985). Second, although the *relative* proportion of ice + rock increases from Jupiter to Neptune, the *absolute* abundance of this material is approximately constant (within a factor of 2) while the *absolute* abundance of the solar gas component varies by about a factor of 100. Third, the ice/rock mass ratios for Uranus and Neptune are > 1 (Hubbard 1984*a*, and references therein). More details of interior modeling efforts are described in the chapter by Hubbard.

Although the solar gas component is H_2 and He dominated, remote sensing observations show that the relative proportions of H_2 to He and of several less abundant gases (e.g., CH_4, NH_3) to H_2 also vary from planet to planet (Gautier and Owen 1985). In particular, the H_2/He ratio is approximately solar (within the large observational uncertainties) on Uranus, is slightly enhanced over solar on Jupiter, and is enhanced by a factor of ~ 2 (or more) on Saturn (Conrath et al. 1987; Hanel et al. 1986; Gautier and Owen 1985). No accurate data are available for the H_2/He ratio on Neptune. The significantly increased H_2/He ratio on Saturn and the slightly increased H_2/He ratio on Jupiter if real are plausibly explained by an inhomogeneity in the radial distribution of He rather than by the segregation of He from H_2 in the region of the solar nebula where Saturn formed. Also, the CH_4/H_2 ratio for Jupiter and Saturn is slightly enhanced over the solar composition value by ~ 2 to 5 times while much larger enhancements of ~ 20 to 25 times are observed on Uranus and Neptune. Other hydrides (e.g., NH_3, H_2O, H_2S, etc.) are expected to exhibit similar behavior, but the observations of (or attempts to detect) these gases are complicated by condensation processes, photochemical destruction and vertical mixing in these planetary atmospheres. These topics are discussed at length in several other chapters (see those of Gautier and Owen, Pollack and Bodenheimer, and Hubbard).

II. NEBULA CHEMISTRY

A. Starting Materials: Interstellar Gases and Grains

The abundant volatile elements giving rise to atmospheres must have been present in the dust and gas of the interstellar cloud(s) which collapsed to form our primitive solar nebula. Telescopic observations of dense interstellar clouds indicate the presence of a rich assortment of carbon-, oxygen- and nitrogen-containing compounds in addition to molecular hydrogen which is the probable principal constituent. Knowledge of the dust in these clouds is less definitive but there is tentative evidence that silicates, magnetite, carbonaceous compounds including graphite and ices are all present. The dust thus bears some similarity to the *primitive* material in our present solar system (e.g., carbonaceous chondrites). A detailed review of interstellar clouds is given in the chapter by Irvine and Knacke.

The bulk elemental composition of the specific cloud(s) which collapsed to form our own solar system is inferred largely from data on the composition of the Sun for the volatile elements and of CI1 carbonaceous chondrites for the less volatile ones. A summary of the abundances of the most common elements in the solar system is given in Table II (Cameron 1982).

The collapse of the precursor interstellar cloud into a rotating disk-shaped primitive solar nebula was accompanied by increasing temperatures and densities within the nebula. A central question concerns the extent to which the interstellar dust and gas metamorphosed as it was accreted by the nebula and flowed inward toward the Sun. Extensive evaporation and re-equilibration of the interstellar material is possible to the extent that the material can be transported inwards to sufficiently warm temperatures, chemically reprocessed, and then either accreted locally or transported outwards again. The meteoritic evidence discussed by Niederer and Papanastassiou (1984) and Grossman and Larimer (1974) suggests that extensive chemical reprocessing occurred in at least the inner solar system and we will discuss later tentative evidence for some chemically reprocessed material even in Comet Halley.

B. Thermochemistry, Shock Chemistry, Photochemistry and Radiochemistry

Chemical reprocessing of interstellar material in the solar nebula requires in general a source of energy; either to overcome activation energy barriers or to drive otherwise endothermic reactions such as dissociations. There are several readily recognized sources of energy for chemical reactions in the nebula.

Thermochemistry Using Thermal Energy from Nebular Accretion. During the nebular collapse phase and prior to the formation of an active proto-Sun the major source of energy was undoubtedly conversion of gravitational potential energy to kinetic (thermal) energy. The net energy flux (radiative and convective) through the nebula is given roughly by $\phi_T = sT_e^4$ where T_e is the *effective* temperature of the nebula and s is Stefan's constant (for illustrative T_e of 100 to 500 K, $sT_e^4 = 5.7$ to 3544 J m^{-2}s^{-1}). Endothermic chemical reprocessing requiring an energy flux ϕ exceeding ϕ_T would lead to catastrophic

TABLE II
Ratios Relative to H_2 of the 10 Most Abundant Elements in a Solar-Composition Medium[a]

H_2	1.0	N	1.74×10^{-4}
He	0.135	Mg	7.97×10^{-5}
O	1.38×10^{-3}	Si	7.52×10^{-5}
C	8.35×10^{-4}	Fe	6.77×10^{-5}
Ne	1.95×10^{-4}	S	3.76×10^{-5}

[a]Table after Cameron (1982).

local cooling and thus could not be sustained. The availability of thermal energy for driving *thermochemical* reactions is, however, highly temperature dependent since the conversion efficiency varies roughly as the Boltzmann factor $b = \exp(-E/kNT)$ where N is Avogadro's number, E is the activation energy for the reaction of interest, T is the local temperature, and k is Boltzmann's constant (e.g., for $E = 5 \times 10^4$ J/mole,$b = 0.13$ at 3000 K and 2×10^{-9} at 300 K). When b values for all relevant reactions are sufficiently close to unity (i.e., when temperatures are sufficiently high), then the system essentially is in thermochemical equilibrium and there is no net conversion between molecular kinetic (thermal) and molecular internal energy. As the temperature is lowered, thermochemical equilibrium can be maintained provided the system cools at a sufficiently slow rate. Specifically, we need the chemical reaction time $t_{\mathrm{chem}} = -(\mathrm{dln}C/\mathrm{d}t)^{-1}$ to be $\leq$ the cooling time $t_{\mathrm{cool}} = -(\mathrm{dln}T/\mathrm{d}t)^{-1}$ where C is concentration. Once $t_{\mathrm{chem}} > t_{\mathrm{cool}}$, the reaction is quenched essentially and thermochemical equilibrium no longer applies. Since for a reaction with rate $R \propto \exp(-E/kNT)$ the chemical deceleration time in a cooling system $t_{\mathrm{decel}} = -(\mathrm{dln}R/\mathrm{d}t)^{-1} = (kNT/E)t_{\mathrm{cool}} << t_{\mathrm{cool}}$, then almost all cooling systems will ultimately reach disequilibrium states as $T \rightarrow 0$. For later discussion, we will assume that the usable thermal energy flux for endothermic thermochemical reactions $\phi \approx b\phi_T$; for thermo-neutral (equilibrium) and exothermic reactions there are of course no net external energy sources required but the activation energy barrier must still be overcome so the same formula can be justified.

Chemistry in Lightning Discharges and Thundershocks. The simultaneous presence of fluid motions and abundant particles in the solar nebula make charge separation (and hence lightning discharges and their accompanying thundershocks) a likely but at the present time speculative phenomenon. In effect, such discharges convert a (very small) fraction ϵ of the convective energy flux into intense electric currents, shock waves and ultraviolet photons all of which are then convertable with high efficiency into chemical potential energy through the breaking of chemical bonds. For Earth and Jupiter the lightning energy flux ϕ is approximately 4×10^{-7} and 4×10^{-5}, respectively, of the net flux in moist convective regions (Borucki et al. 1984); that is $\phi = \epsilon\, sT_e^4$ where $\epsilon \approx 4 \times 10^{-7}$ to 4×10^{-5}. It is arguable whether similar ϵ values apply in the nebula where the convection is largely dry but, if it does apply, then $\phi = (2.3 \text{ to } 1418) \times 10^{-5}$ J m^{-2}s^{-1} for nebular $T_e = 100$ to 500 K and $\epsilon = 4 \times 10^{-6}$. The temperatures attained in lightning are several thousand Kelvin (and hence the Boltzmann factors $b \rightarrow 1$) so that the number of chemical bonds broken per unit area per unit time $\phi_B \rightarrow \phi/D$ where D is a representative bond energy. However, a good deal of subsequent recombination occurs during cooling prior to quenching so the net ϕ_B is less than ϕ/D (e.g., on the Earth the net $\phi_B \approx 0.1 \times \phi/D$ using $D = 5 \times 10^5$ J/mole appropriate to O_2).

Photochemistry Driven by the Proto-Sun: Opacity Dilemmas. In the present-day solar system, ultraviolet light from the Sun provides a potent disequilibration mechanism in planetary atmospheres. The present-day flux F of the photochemically important Lyman-α radiation at 1 AU is not large (3.5×10^{11} photon cm^{-2} s^{-1}; 5.7×10^{-3} J m^{-2} s^{-1}), but the early Sun may have emitted a flux some 10^3 to 10^4 times greater than the present Sun (see, e.g., Zahnle and Walker 1982 for a comprehensive review). It has recently been conjectured that enhanced ultraviolet radiation from the proto-Sun served as an important disequilibration mechanism for nebula gases (Yung et al. 1987*a*), and (presumably after dissipation of the nebula) as both a disequilibration process and an energy source for hydrodynamic escape in evolving protoplanetary atmospheres (Zahnle and Walker 1982). Are these latter conjectures plausible?

First, it is arguable whether the proto-Sun was a powerful source of ultraviolet radiation while the dense *gaseous* nebula was present; current theories and observations do not enable definitive conclusions. To assess the viability of protosolar-driven photochemistry in the gaseous nebula, it is additionally important to determine the *ultraviolet* opacity of the nebula itself. In particular, if the opacity is very large the protosolar photons will be absorbed wholly in the very hot ($T \geq 2500$ K) relatively dense inner nebula where large thermochemical reaction rates involving abundant thermally produced radicals and atoms overwhelm or negate the dissociative effects of the ultraviolet photons. The solar ultraviolet photons then effectively become another source of heating to be added to the dominant source (namely conversion of the kinetic energy of infalling material to thermal energy). We note parenthetically that this situation in the gaseous inner nebula would be very different from that in the present-day terrestrial thermosphere where solar ultraviolet photons are the dominant source of energy for dissociation, where molecular diffusion, gravitational stratification and selective escape are important processes, and where pressures are much lower (and recombination reactions therefore much slower).

To calculate the minimum ultraviolet opacity we will deliberately use one of the lightest mass nebula models, specifically that of Hayashi (1981). The vertical optical depth (τ_z measured from midplane to outer surface of nebula at radius $r = r_1$) and radial optical depth (τ_r measured between radii r_1 and r_2) are:

$$\tau_z(r_1) = \frac{1}{2}(\Sigma_{gas}\sigma_{gas} + \Sigma_{dust}\sigma_{dust}) \quad (1)$$

$$\tau_r(r_1, r_2) = \int_{r_1}^{r_2} (\rho_{gas}\sigma_{gas} + \rho_{dust}\sigma_{dust})\mathrm{d}r. \quad (2)$$

We use the Hayashi (1981) expressions for vertical column mass densities for gas and dust (Σ_{gas}, Σ_{dust}) and midplane gas and dust mass density (ρ_{gas}, ρ_{dust}). For the Lyman-α wavelength ($\approx$ 122 nm), the minimum value of the gas absorption cross section σ_{gas} is given by considering the absorption cross section due to H_2O alone (1.59×10^{-17} cm^2 molecule^{-1}; Kley 1984). For a nebular H_2O molar mixing ratio of 10^{-3} and nebular mean molecular weight of 2.3, we thus have $\sigma_{gas} = 1.59 \times 10^{-17} \times 6 \times 10^{23} \times 10^{-3}/2.3 = 4.1 \times 10^3$ cm^2 g^{-1}. For the dust, we have $\sigma_{dust} \approx 3/(4aD)$ provided the particle radius $a >> 122$ nm. Specifically for $a = 10^{-4}$cm and particle density D = 4.4 g cm^{-3} (i.e., silicate plus metal), $\sigma_{dust} \approx 1.7 \times 10^3$ cm^{2}g^{-1}. Thus since $\sigma_{dust} \approx 0.4\, \sigma_{gas}$ and also $\Sigma_{dust} \approx 0.01\, \Sigma_{gas}$(Hayashi 1981), we can ignore the small additional ultraviolet opacity due to the dust (this is reasonable unless the dust has a much smaller scale height than the gas (e.g., 250 times smaller) in which case the dust and gas opacities in the central plane could be comparable). Hence we obtain:

$$\tau_z(r_1) \simeq \frac{1}{2} \times 1.7 \times 10^3 \left(\frac{r_1}{1\ \text{AU}}\right)^{-3/2} \times 4.1 \times 10^3$$

$$= 1.7 \times 10^7\ (r_1 = 0.35\ \text{AU}) \quad (3)$$

$$\tau_r(r_1, r_2) \simeq 1.4 \times 10^{-9} \times 4.1 \times 10^3 \times 1.5 \times 10^{13}$$

$$\times \left[\left(\frac{r_1}{1\ \text{AU}}\right)^{-7/4} - \left(\frac{r_2}{1\ \text{AU}}\right)^{-7/4}\right] /1.75$$

$$= 2.6 \times 10^8\ (r_1 = 0.35\ \text{AU},\ r_2 = 1\ \text{AU}). \quad (4)$$

These massive optical depths correspond to decreases in photochemically important photon fluxes ϕ from their midplane values at 0.35 AU due to absorption alone (i.e., not even including the $1/r^2$ geometric term) by 7 million and 110 million orders of magnitude, respectively. For photons beginning at the surface of the proto-Sun (i.e., $r_1 \rightarrow 0$), the τ_z and τ_r values are several orders of magnitude greater than their values computed above at 0.35 AU (we have not presented precise values because the Hayashi (1981) ρ and Σ expressions are strictly derived only for $r_1 \geq 0.35$ AU and because H_2O begins to be thermally dissociated to OH and H for $T \geq 2500$ K). It is almost academic to add that consideration of the more massive Cameron (1978) nebula would increase these numbers by more than another factor of 10. Thus, while the gaseous nebula was present, any protosolar-driven photochemistry must have been confined to the very hot gas near the proto-Sun where its ability to induce chemical disequilibration must have been small with surviving relicts, if any, restricted to high-temperature refractory material.

After the nebula gases and fine dust were dissipated (in a T-Tauri phase), the young Sun would then be initially irradiating a vast but thin disk of orbiting objects (boulder size to planet size?). High collision rates in this disk would have provided a potent source of dust (Greenberg et al. 1978*a*) which would not have abated until the planets as we know them today had largely accreted and collision rates had dropped dramatically: a process which probably took several 100 Myr (Safronov and Ruzmaikina 1985*a*). This is the very time during which the Sun was its most luminous in the ultraviolet (Zahnle and Walker 1982). Thus, to assess the viability of solar-ultraviolet driven photochemistry and hydrodynamic escape in early planetary atmospheres, we need to determine the ultraviolet opacity of the gas-free but very dusty, post-nebula disk. The radial optical depth τ_r can be related to the total mass $M(r)$ of dust inside radius r by

$$\tau_r = \frac{M\sigma_{\text{dust}}}{2\pi r h_{\text{dust}}} \simeq \frac{3M}{8\pi r h_{\text{dust}} a D} \tag{5}$$

where h_{dust} is the vertical scale height of the dust.

For an optical depth $\tau_r \geq 4 \ln 10 = 9.2$, a solar ultraviolet flux 10^4 times its present value will be reduced to $\leq$ present value. From Eq. (5) this is achieved provided

$$\begin{aligned} M(r) &\geq 77 \times r\, h_{\text{dust}}\, a\, D \\ &= 3.2 \times 10^{20}\ \text{g}\ (r = 1\ \text{AU}) \\ &= 5.3 \times 10^{-8}\ M_\oplus\ (r = 1\ \text{AU}) \end{aligned} \tag{6}$$

where we have used $h_{\text{dust}} \approx$ Earth radius $= 6380$ km, $a = 10^{-4}$ cm, $D = 4.4$ g cm^{-3}, and $r = 1.5 \times 10^{13}$ cm to obtain the numerical value. This is an extremely small value indeed. To place it in perspective, the numerical multi-body collision/accretion calculations of Greenberg et al. (1978*a*) resulted typically in about 1% of the total initial mass of colliding bodies being processed into unresolved (sub-31-m diameter) *dust* in 25,000 yr. This translates into a *dust* production rate inside 1 AU of 2.3×10^{21} g yr^{-1} or 3.8×10^{-7} $M_\oplus$ yr^{-1}. Given these massive production rates, it is difficult to argue that the $M(r)$ value required by Eq. (6) was not easily achieved while the terrestrial planets were accreting and collisions were common in the solar system. An important theoretical problem yet to be addressed is the precise duration of this dusty *high-opacity* epoch in the solar system and how the duration of this epoch compared to the duration of the enhanced ultraviolet emission from the young Sun. The latter problem is coincidentally common to Wetherill's (1981*b*) explanation for ^{36}Ar in terrestrial atmospheres.

As a postscript to this topic, it should be emphasized that the above opacity arguments refer to solar photons emitted in the direction of the (equatorial) nebula disk. Solar photons emitted in other (e.g., polar) directions may suffer little or no absorption and escape the protosolar system. Hence the fact that ultraviolet emission is observed from T-Tauri stars is not an argument in favor of low-opacity conditions and extensive solar-driven photochemical disequilibration in the nebula disk.

Photochemistry Driven by Stellar Photons. The maximum energy flux in interstellar photons with 90 nm $<$ wavelength $<$ 200 nm occurs in transparent parts of our Galaxy and is about 7×10^{-7} $Jm^{-2}s^{-1}$ (see, e.g., Duley and Williams 1984). Assuming a similar flux was applicable 4.6 Gyr ago, the energy flux for driving photochemistry in the outer skin of the solar nebula disk $\phi \leq 7 \times 10^{-7}$ $Jm^{-2}s^{-1}$ with the actual value depending on the opacity of interstellar space in the vicinity of the solar nebula. The interstellar flux is thus very small but it impinges the chemically perturbable (low-temperature) parts of the nebula. Hence it does not suffer from the same solar nebula opacity dilemmas which accompany solar-driven photochemistry and it may have been larger in the past than today. Specifically, the above ϕ value corresponds to a 90 to 200 nm photon flux of $\leq 5 \times 10^7$ photon $cm^{-2}s^{-1}$. In the inner nebula H_2O will be a major absorber of these photons (as already noted in the previous section). At 1 AU there are about 2.2×10^{23} H_2O molecule cm^{-2} on either side of midplane in the Hayashi (1981) model so that over the 10^5 to 10^6 yr lifetime of the nebula ≤ 0.07 to 0.7% of this H_2O will be dissociated by the above photon flux. Thus the interstellar flux would have had to have been at least 10^2 to 10^3 times greater than present in order to be important in nebula chemistry. In the cold outer nebula (e.g., $T < 150$ K), the stellar ultraviolet photons will mainly be absorbed by ice/rock grains and CO gas (H_2O is essentially totally condensed and we do not expect CH_4 in the solar nebula). Thus the H_2O photochemistry discussed by Yung et al. (1987*b*) could only have proceeded in the inner nebula and would require greatly enhanced interstellar ultraviolet fluxes over present-day values. Whatever photochemistry proceeds will also need to compete with shock chemistry and with the general inward transport and thermal reprocessing of material during the solar accretion phase.

Radiochemistry Driven by ^{26}Al. The short-lived radionuclide ^{26}Al was apparently present in dust particles in the early solar nebula with an average *mass* mixing ratio f relative to H_2 + He of $\approx 2 \times 10^{-9}$ (Lee et al. 1976). Consolmagno and Jokipii (1978) have noted that the 1.83 MeV γ photons and 1.16 MeV positrons (or the 1.02 MeV γ photons produced on their annihilation by electrons) which result from ^{26}Al decay were important sources of ionization in the solar nebula. Specifically, we can show that the power produced per unit area of the nebula disk by ^{26}Al decay (i.e., the energy flux available for radiochemistry) is

$$\phi \simeq jf\Sigma_{gas}\, d/26$$
$$= 1.2 \times 10^{-9} \left(\frac{r}{1\ \mathrm{AU}} \right)^{-3/2} \mathrm{J\ m^{-2}\ s^{-1}}$$
$$= 1.2 \times 10^{-9}\ \mathrm{J\ m^{-2}\ s^{-1}}\ (r = 1\ \mathrm{AU}) \quad (7)$$

where $j = 2.97 \times 10^{-14}\ s^{-1}$ is the ^{26}Al decay rate constant, $d = 3 \times 10^{11}$ J/mole is the average available ^{26}Al decay energy, and we have again used the Hayashi (1981) expression for Σ_{gas}. Evidently ϕ for radiochemistry is very small; indeed for ionization reactions even lightning may have been a more important (albeit spatially inhomogeneous) source of energy than ^{26}Al in the nebula.

C. Synopsis

To determine what types of chemistry will dominate in the nebula it is instructive to compare the *usable* energy fluxes ϕ available for chemical reactions to the net flux $\phi_T = sT_e^4$. Specifically for representative $T = 600$ K, $r = 1$ AU, $E = 5 \times 10^4$ J/mole, $\epsilon = 4 \times 10^{-6}$ and $T_e = 300$ K we have:

$$\phi(\text{thermochemistry})/\phi_T \approx \exp(-E/kNT) = 4 \times 10^{-5} \quad (8)$$

$$\phi(\text{thunder - shock chemistry})/\phi_T \approx \epsilon = 4 \times 10^{-6} \quad (9)$$

$$\phi(\text{solar - driven photochemistry})/\phi_T \approx 10^4 \times F \exp(-\tau_r)/\phi_T \approx 0 \quad (10)$$

$$\phi(\text{stellar - driven photochemistry})/\phi_T \approx 10^{-9} \quad (11)$$

$$\phi(\text{radiochemistry})/\phi_T \approx 3 \times 10^{-12} \quad (12)$$

Clearly thermochemistry is the expected dominant chemical process at $r = 1$ AU except for reactions with very large activation energies where shock chemistry (lightning) could be comparable. Solar photons and radioactivity are totally negligible at 1 AU. Stellar photons are not important in the inner nebula ($r \leq 1$ AU) but may be quite important in the cold, outer portions of the nebula ($r \geq 10$ AU). Specifically, for a low $T_e = 100$ K the value of ϕ/ϕ_T for stellar photochemistry increases to 10^{-7} (i.e., approaching thunder-shock chemistry and much greater than thermochemistry).

As we move closer to the Sun than 1 AU, thermochemistry becomes more and more dominant (e.g., at 1500 K, ϕ (thermochemistry)/$\phi_T = 0.018$ for the E value used in Eq. (8), while lightning becomes successively more important farther from the Sun (e.g., at 150 K, ϕ (thermochemistry)/$\phi_T = 4 \times 10^{-18}$ for the same E value while ϕ (thunder-shock chemistry)/ϕ_T remains at 4×10^{-6}). As a check, we note that if we apply Eqs. (8) through (12) to the present-day Earth atmosphere ($T_e = 255$ K, $T \simeq 300$ K, and most important relatively small τ_r value), we do predict correctly that solar-driven photochemistry is the dominant (abiotic) present-day atmospheric chemical process. Note also that when these equations are applied to specific chemical reactions the E value for that specific reaction must be used.

The fact that thermochemistry appears to have been the dominant type of chemistry overall in the solar nebula places important constraints on chemical reprocessing of cold interstellar material in the nebula. In particular, the cold material will need to be transported inwards to temperatures and pressures at which reactions can proceed within the lifetime of the nebula (10^5 to 10^6 yr). These temperatures and pressures are reached by approaching the proto-Sun in the solar nebula or approaching a giant protoplanet in a protoplanetary subnebula or atmosphere. This means that a discussion of nebula chemistry can only proceed in tandem with a discussion of nebula dynamics.

III. INTERPLAY BETWEEN NEBULA CHEMISTRY, DYNAMICS AND STRUCTURE

Advances in our understanding of nebula chemistry are coupled in several important ways to advances in our knowledge of nebula dynamics and evolution. The generally accepted evolutionary picture (see, e.g., Cassen et al. 1985; chapter by Boss et al.; Morfill et al. 1985) has the original interstellar cloud collapsing inwards with material of successively higher angular momentum accreting as time progresses thus forming a growing, rapidly rotating *accretion disk*. Mass inflow through the disk into the proto-Sun then continued until the proto-Sun ignited. Extensive chemical, petrographic and isotopic evidence from meteorites and inferences that asteroids are meteorite parent bodies suggest that the solar nebular accretion disk was extensively physically mixed (but not necessarily chemically and isotopically homogenized) out to at least 4 AU from the proto-Sun (Wilkening 1977; Niederer and Papanastassiou 1984; Grossman and Larimer 1974). The giant planets accreted in the nebula perhaps with their own accretion disks but we argue here that the probable site of formation of their regular satellites was in later-forming well-mixed subnebulae *spun off* during contraction of these initially very hot planets toward their present dimensions (chapter by Pollack and Bodenheimer) or *impact-generated* subnebulae formed by a massive collision (Cameron 1975). A simplified physical picture of the nebula is shown in Fig. 1.

It seems probable that an intense T-Tauri solar wind provided an important mode for the newly ignited proto-Sun to radiate away energy and that this intense wind first blew away the thinnest part of the nebular disk (over the protosolar poles) and also decelerated the inward gas flow in the protosolar equatorial plane; ultimately this wind then swept away all the remaining unaccreted gases and fine dust in the nebula. The time span from initial cloud collapse to proto-Sun ignition and nebular blow off was about 10^5 to 10^6 yr (Cameron 1985*a*).

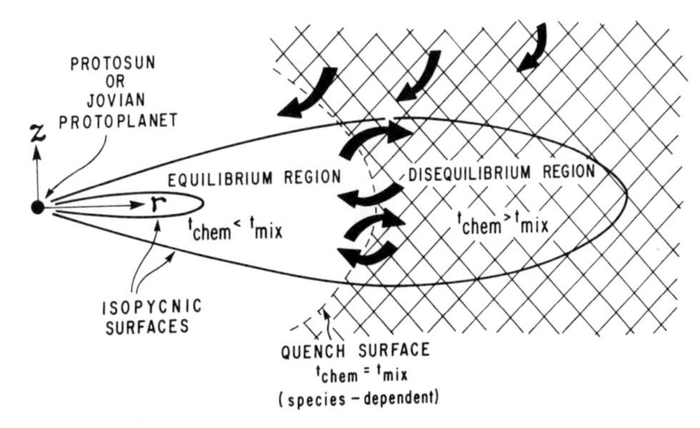

Fig. 1. Schematic illustrating the occurrence of a quench surface separating thermochemical equilibrium and transport-induced-disequilibrium regions in the solar nebula or a well-mixed protoplanetary subnebula. See text for details.

After the gaseous nebula was dissipated, the surviving bodies (ranging in size from giant gaseous protoplanets down to m-sized objects and perhaps smaller) began an evolutionary period of several 100 Myr duration characterized by high collision frequencies, accompanying accretion, fragmentation and degassing, evolution of terrestrial bodies, and further growth of the giant planets (see chapter by Greenberg). These processes proceeded in a largely gas-free but dust-rich (due to dust production in collisions) solar system environment.

Thermochemistry in both the solar nebula and protoplanetary subnebulae depended in two important ways on the pressure/temperature conditions in these nebulae. This is illustrated first in Figs. 2 and 3 where it is seen that the *P,T* profile for a representative solar nebula model (Lewis and Prinn 1980; Barshay 1981) implies that, in *thermochemical equilibrium,* CO and N_2 are much more abundant than CH_4 and NH_3 in the warm inner parts of the nebula. In contrast, CO is always less abundant than CH_4 and N_2 is at most comparable in abundance to NH_3 in the much higher-pressure (at a given *T*) protoplanetary subnebulae (Prinn and Fegley 1981). We should add that most solid/vapor and solid/solid transition temperatures in thermochemical equilibrium show much less pressure dependence than the above gas/gas transitions and thus they are much less sensitive to the choice of nebula model.

The nebula *P,T* profiles affect thermochemistry in a second important way. Because reaction rates generally increase rapidly with temperature and because the presence and abundance of efficient solid catalysts (e.g., Fe, Fe_xC particles) is temperature dependent, then the rates of the $CO \rightarrow CH_4$ and $N_2 \rightarrow NH_3$ reactions at the CO/CH_4 and N_2/NH_3 boundaries are much greater in the planetary subnebulae than in the solar nebula. As a result the $CO \rightarrow CH_4$ and $N_2 \rightarrow NH_3$ conversions are kinetically inhibited in the nebula (Lewis and Prinn 1980) but *not* in the subnebulae (Prinn and Fegley 1981*a*).

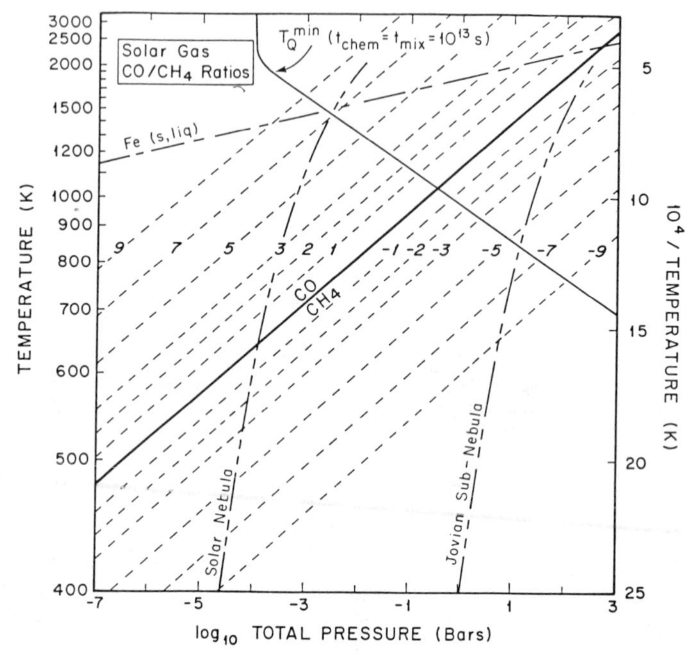

Fig. 2. Calculated ratios of CO/CH_4 at thermochemical equilibrium in a solar composition gas. The solid line labeled $CO - CH_4$ is the boundary where the abundances of the two gases are equal; CO is more abundant to the left and CH_4 is more abundant to the right. The dotted contours labeled 9,7,5, . . . −5,−7,−9 are constant $\log_{10} CO/CH_4$ contours. The line labeled Fe(s, liq) is the Fe condensation line; Fe(s, liq) is stable below this line. Two P,T profiles for an illustrative solar nebula model (Lewis and Prinn 1980; Barshay 1981) and an illustrative sub-nebula model (Prinn and Fegley 1981*a*) are also shown. The solid line labeled T_Q^{min} illustrates the minimum quench temperatures for homogeneous gas-phase conversion of CO to CH_4 assuming that the CO chemical lifetime (t_{chem}) = the maximum nebular mixing time (t_{mix}) of 10^{13} s (i.e., the nebular gases are mixed once during the nebula's lifetime). Shorter mixing times (i.e., more frequent mixing) result in higher quench temperatures. See Fig. 7 for details of Fe-catalyzed heterogeneous $CO \rightarrow CH_4$ conversion in the solar nebula.

Thermochemistry in the nebula potentially is profoundly influenced by nebula dynamics. Current models for the solar nebula accretion disk (see, e.g., Morfill et al. 1985) indicate that outward diffusive mixing is opposed by the advective inward flow. It is therefore difficult in these models for nebula material which might have been chemically processed in the hot inner nebula to diffuse outward (Stevenson 1987*b*). This conclusion is, however, at odds apparently with the observational evidence for extensive inner nebula processing and mixing contained in chondritic meteorites (see also Cabot et al. [1987*b*] for a discussion of other problems in current models). Perhaps the solution lies in the fact that the quadratically nonlinear momentum equation is replaced by a linear diffusive momentum equation in current accretion disk

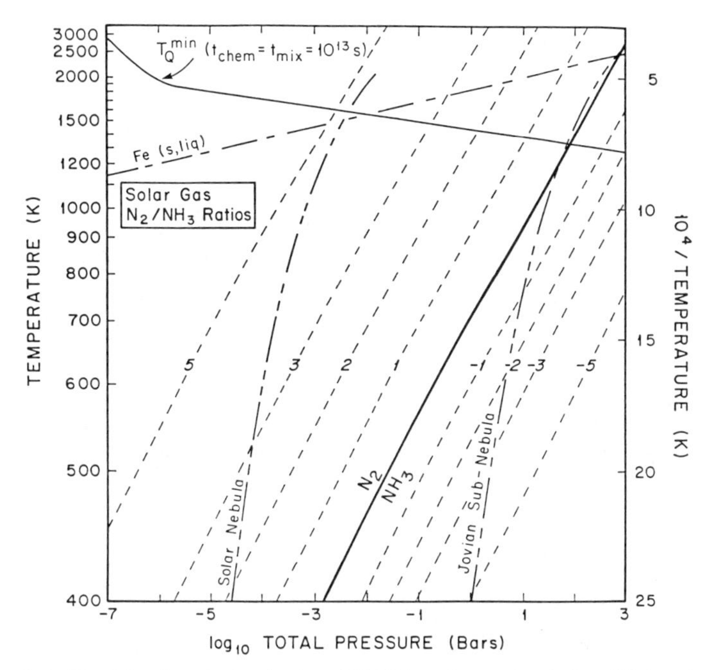

Fig. 3. As in Fig. 2 but for calculated ratios of N_2/NH_3 at thermochemical equilibrium in a solar composition gas. In this case, the line labeled T_Q^{min} illustrates the minimum quench temperatures for homogeneous gas-phase conversion of N_2 to NH_3.

models thus precluding in particular nonlinear counter-gradient momentum fluxes. These nonlinear fluxes are expected and when included in the models enable the shear to be maintained in an accretion disk in which diffusive mixing is strong enough to overcome the negative effect of the mean inward flow on outward trace constituent mixing (Prinn 1988).

Dynamics is also important in understanding the chemistry of subnebulae around the giant protoplanets. Prinn and Fegley (1981*a*) proposed a model for the chemistry in a convective circumplanetary subnebula involving inward mixing of accreted solar material (rich in CO and N_2), thermochemical conversion of the CO and N_2 to CH_4 and NH_3, and outward mixing of this processed material to form satellites rich in CH_4 and NH_3. If one assumes that the Prinn and Fegley (1981) subnebula is an accretion disk (its precise nature was in fact unspecified), then the efficiency for outward mixing in accretion disks which include nonlinear effects makes it probable that the above processed material could be mixed outward to form CH_4-rich satellites. Thus an accretion disk subnebula model may explain Titan. As mentioned earlier, it is, however, probable that the subnebulae in which outer planetary regular satellites formed were not accretion disks but later-occurring highly convective subnebulae spun off by the giant protoplanets as they contracted toward their present dimensions (chapter by Pollack and Bodenheimer). The reprocessing of CO and N_2 to yield CH_4 and NH_3 and the formation of CH_4-rich Titan-like satellites as described by Prinn and Fegley (1981*a*) is possible in this type of subnebula model with the difference being that the inward mixing occurs during the accretion phase and the outward mixing occurs during the spin-off process.

Besides accretion disks and spin-off disks, a third type of subnebula can be formed through collision of a massive object with a planet. This has been suggested as the origin of the Moon (Cameron 1985*b*). Since collision of a large object with Uranus is hypothesized to be causal in its large obliquity, the gases blown off during this collision may have formed a subnebula in which the Uranian regular satellites condensed (Cameron 1975). The volatile element chemistry in such an *impact-generated* Uranian disk again should be qualitatively similar to that discussed by Prinn and Fegley (1981*a*). In this case, CO to CH_4 reprocessing is first accomplished in the planet itself prior to the collision with some of the reprocessing being reversed (i.e., $CH_4 \rightarrow CO$) by atmospheric shock heating due to the impact itself.

A. Chemical-Dynamical Models

The general theory for disequilibration in a thermochemical system by rapid mixing was first proposed by Prinn and Barshay (1977) to explain the occurrence of CO in the CH_4-rich Jovian atmosphere and later applied to the solar nebula and to Jovian planetary subnebulae (Lewis and Prinn 1980; Prinn and Fegley 1981*a*). In the nebula context, we first determine for each chemical species the *quench radius* (designated r_q) inside which the temperatures are high enough (and thus thermochemical reactions fast enough) to ensure that the species concentration $[i]$ is essentially equal to that in thermochemical equilibrium $[i]_e$ (general or conditional depending on the circumstances). To first order this quench radius is defined as that radius at which the chemical destruction time $t_{chem} = -[i]/(d[i]/dt)$ for the species i equals the characteristic radial mixing time t_{mix}. Since T, P and hence $1/t_{chem}$ generally decrease with altitude z, the equilibrium region bounded by the *quench surface* is convex (see Fig. 1). Outside of the quench surface or radius we must solve the continuity equation for the chemical species. The convenient coordinate system is cylindrical (r, ϕ, z) centered on the Sun or protoplanet. If mixing is sufficiently rapid relative to chemistry for parcel motions in the vertical and around a (latitude) circle, the mixing ratio f_i of a noncondensing species i at radius r is approximately independent of altitude z and azimuth ϕ. If we also model the radial trace-species flux as an eddy-diffusive flux with eddy-diffusion coefficient K, we have for the steady-state continuity equation

$$\frac{\partial^2 f_i}{\partial r^2} + \left(\frac{1}{r} + \frac{1}{\rho} \frac{\partial \rho}{\partial r} \right) \frac{\partial f_i}{\partial r} = \frac{f_i}{K t_{chem}}. \tag{13}$$

In typical nebula models the radial density gradient $d\rho/dr = -\alpha\rho/r = -\rho/H$ where H is the radial density scale length, α is a dimensionless number of order unity, and $K = l^2_{mix}/t_{mix}$ where the mixing length l_{mix} is equated in diffusive accretion disks to the vertical density scale length. It is convenient also to define a chemical length l_{chem} which is the radial length for an e-fold change in the destruction rate of species i. That is

$$l_{chem} = \frac{-[i]/t_{chem}}{\frac{d}{dr}([i]/t_{chem})} \tag{14}$$

where $(d\ln t_{chem}/dr)^{-1} \leq l_{chem} \leq -(d\ln\rho/dr)^{-1} = H$. For a t_{chem} with a strong temperature dependence of exp (A/T) where $A >> T$, then l_{chem} varies approximately inversely as A (i.e., inversely as the activation energy). Making the coordinate change $R = r - r_q$ and using the fact that $Kt_{chem} = Kt_{mix} = l^2_{mix}$ at $R = 0$, Eq. (13) then becomes

$$\frac{\partial^2 f_i}{\partial R^2} + \left(\frac{1-\alpha}{R + r_q}\right)\frac{\partial f_i}{\partial R} \simeq \frac{f_i(o)}{l^2_{mix}} \exp\left[-R\left(\frac{1}{l_{chem}} - \frac{1}{H}\right)\right]. \tag{15}$$

In the vicinity of the quench surface $R + r_q \simeq r_q$ and l_{chem}, H and l_{mix} can be regarded as (positive) constants. Using two appropriate boundary conditions (f_i(o) equal to its thermochemical equilibrium value $f_i(r_q$, equil) and f_i finite and positive for large R), Eq. (15) then has the simple analytical solution

$$f_i = f_i\,(r_q, \text{equil})\left(1 - \frac{l^2_{chem}}{\beta l^2_{mix}}\left\{1 - \exp\left[-\left(\frac{1}{l_{chem}} - \frac{1}{H}\right)R\right]\right\}\right) \tag{16}$$

where $\beta = (1 - l_{chem}/H)/(1 - l_{chem}/r_q)$ and the second boundary condition requires $l^2_{chem} \leq \beta l^2_{mix}$ and $l_{chem} < H$. For $R >> l_{chem}$, the destruction of i is quenched and f_i has the constant value

$$f_i \simeq f_i(r_q, \text{equil})\left(1 - \frac{l^2_{chem}}{\beta l^2_{mix}}\right). \tag{17}$$

The interesting case occurs when $l^2_{chem} << \beta l^2_{mix} \simeq l^2_{mix}$ (that is, the air parcel is quenched chemically by moving a distance R much less than the mixing length) in which circumstance the mixing ratio of the species i in the disequilibrium region ($r > r_q$) is constant and is given simply by its equilibrium value at the quench radius. That is

$$f_i \simeq f_i(r_q, \text{equil}) \qquad \text{for } r \geq r_q. \tag{18}$$

Clearly, the larger the activation energy for destruction of the species i, then the more likely Eq. (18) is appropriate. Note also that when Eq. (18) applies, the convex shape of the quench surface (Fig. 1) means that air parcels are quenched for vertical as well as radial motions; that is, f_i is independent of z for $r \geq r_q$ as required in Eqs. (13) through (17).

The general approach is therefore first to calculate the thermochemical equilibrium mixing ratio for the species of interest throughout the nebula. Then appropriate kinetic expressions are obtained to describe the destruction rate of i and thus t_{chem} throughout the nebula. Finally, upper and lower limits to t_{mix} are defined (e.g., a *minimum* t_{mix} might be defined by $t_{\mathrm{mix}} = 3H/V_s$ where V_s is the sound speed and a *maximum* t_{mix} might be equated to the lifetime of the nebula). Thus a range of possible quench radii defined by t_{chem} is obtained and by comparing l^2_{chem} and l^2_{mix}, the validity of Eq. (18) is ascertained for each quench radius. The temperature at the quench radius is referred to as the *quench temperature* T_Q and it is often more convenient to define the quench position in the nebula by T_Q rather than r_q.

The reader should be aware that the above treatment does not take into account the mean inward flow of material toward the proto-Sun/protoplanet in accretion disks which we mentioned earlier. However, the aforementioned nonlinear effects allow diffusion to dominate this mean flow so the above treatment should hold approximately. This treatment is applicable without such qualifications to *spin-off* and *impact-generated* subnebulae (and for the present-day Jovian planets when cast in spherical coordinates).

B. Results of Chemical-Dynamical Models for Important C-, H-, N-, O-Bearing Volatiles

(i) Carbon and Nitrogen Compounds. The net reactions which relate CO, CH_4, CO_2, N_2 and NH_3 in a solar gas are

$$CO + 3H_2 = CH_4 + H_2O \tag{19}$$

$$N_2 + 3H_2 = 2NH_3 \tag{20}$$

$$CO + H_2O = CO_2 + H_2 \tag{21}$$

which all proceed to the right with decreasing temperature at constant pressure. However the extent to which these homogeneous gas-phase reactions approach quantitative conversions of CO to CH_4, CO to CO_2 and N_2 to NH_3 with decreasing temperature is severely limited by the rates at which the relevant rate-determining steps occur relative to the rates of nebular mixing and overall cooling.

The kinetic inhibition of reactions (19) to (21) in a solar gas can give dramatically different compositions than does the assumption of complete thermochemical equilibrium for the major carbon and nitrogen-bearing volatiles ultimately incorporated into the gases and grains accreted by planetary and cometary bodies. These differences are illustrated in Figs. 2 through 4

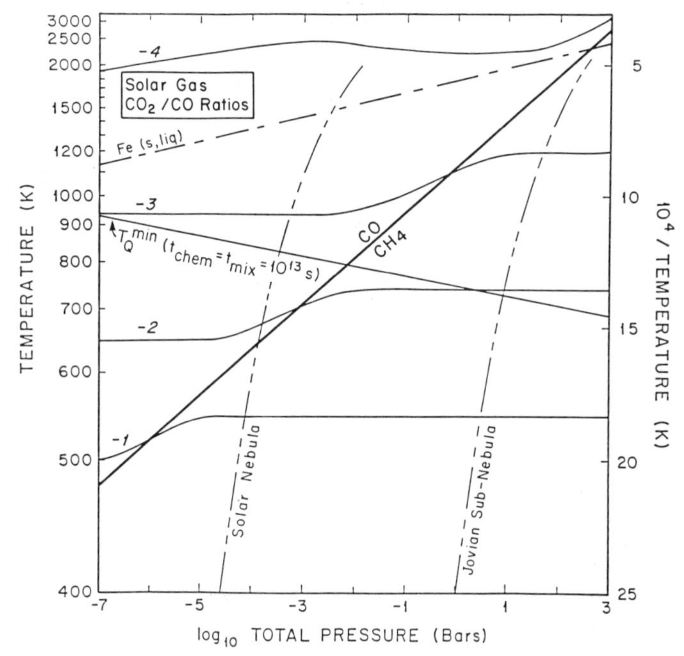

Fig. 4. As in Fig. 2 but for calculated ratios of CO_2/CO at thermochemical equilibrium in a solar composition gas. The inflections in the constant $\log_{10} CO_2/CO$ contours are caused by changes in the H_2O abundance from the CO- to CH_4- dominated regions. The T_Q^{min} line illustrates the minimum quench temperatures for homogeneous gas-phase conversion of CO to CO_2. If thermochemical equilibrium were maintained to lower temperatures and pressures, a CO_2- dominant field results (Lewis et al. 1979).

where the minimum quench temperatures for the homogeneous gas-phase conversions (Eqs. 19 to 21) are plotted along with the (CO/CH_4), (N_2/NH_3), and (CO_2/CO) ratios expected for maintenance of thermochemical equilibrium. The minimum quench temperatures, which are calculated by Prinn and Fegley (1981*a*) as described above, refer to the slowest possible nebular mixing rate of one overturn over the course of the solar nebula lifetime of $\sim 10^{13}$ s (Cameron 1985*a*; Lewis and Prinn 1980). Faster mixing rates (and thus shorter mixing times) will yield higher quench temperatures.

One very important point illustrated in Figs. 2 and 3 is that kinetic inhibition of homogeneous gas-phase $CO \rightarrow CH_4$ and $N_2 \rightarrow NH_3$ conversions almost certainly occurs at the low pressures commonly associated with solar nebula models. For the representative solar nebula model (Lewis and Prinn 1980; Barshay 1981) illustrated in Figs. 2, 3 and 4, the $CO \rightarrow CH_4$ and $N_2 \rightarrow NH_3$ conversions will quench at 1470 K and 1600 K, respectively, where $(CO/CH_4) \sim 10^7$ and $(N_2/NH_3) \sim 10^5$ (see Fig. 5). Furthermore, decreasing pressure leads to increases in these quench temperatures and to increases in the corresponding (CO/CH_4) and (N_2/NH_3) ratios at quench. Indeed at suffi-

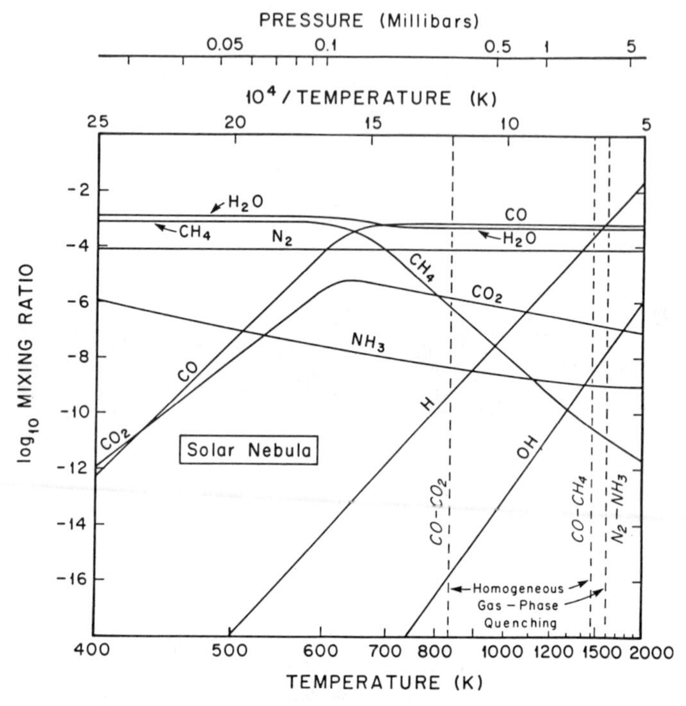

Fig. 5. Calculated equilibrium abundances of important H, C, N, O gases along the solar nebula P,T profile illustrated in Figs. 2,3 and 4. H_2 has an abundance close to 0 on a log scale and is not graphed for clarity. The vertical dashed lines illustrate the minimum quench temperatures for homogeneous gas-phase conversions of $N_2 \rightarrow NH_3$, $CO \rightarrow CH_4$, and $CO \rightarrow CO_2$ and correspond to the intersection of the T_Q^{min} lines with the solar nebula P,T profile in Figs. 2,3 and 4. The corresponding pressure scale for this model profile is also shown.

ciently low pressures of $\sim 10^{-6}$ bar, quenching will occur at such high temperatures that essentially no reprocessing of CO to CH_4 or of N_2 to NH_3 can occur even over the age of the solar system ($\sim 10^{17}$ s).

In contrast, Fig. 4 illustrates the facile nature of the homogeneous gas-phase $CO \rightarrow CO_2$ conversion, which proceeds down to $\sim$ 830 K and (CO_2/CO) $\sim 2 \times 10^{-3}$ for the representative solar nebula model shown. The relatively rapid interconversions of oxidized carbon species such as CO-CO_2-H_2CO (e.g., see Warnatz 1984) strongly suggest that equilibria among these species will be maintained even at relatively low pressures ($\sim 10^{-7}$ bar). The possible dominance of CO_2 and CO_2-bearing compounds over CH_4 in low temperature nebula condensates, which was first proposed by Lewis and Prinn (1980), is a consequence of this situation.

As indicated in Figs. 2, 3 and 4, increasing pressure leads to decreasing quench temperatures for reactions (19) to (21), although this effect is not as pronounced for the $CO \rightarrow CO_2$ conversion as it is for the $CO \rightarrow CH_4$ and N_2

$\rightarrow NH_3$ conversions. The lower quench temperatures lead to smaller (CO/CH_4) and (N_2/NH_3) ratios; indeed, for sufficiently high pressures of ~ 0.3 bar for $CO \rightarrow CH_4$ and ~ 80 bar for $N_2 \rightarrow NH_3$, quenching may produce equimolar mixtures of these gases. (Although uncertainties in the kinetic data may lead to small shifts in the specific pressure values, the relative positions of the CO - CH_4 and N_2 - NH_3 boundaries in *P, T* space require P_{total} for $N_2 = NH_3$ to be greater than P_{total} for $CO = CH_4$.) Still higher pressures will lead to even smaller (CO/CH_4) and (N_2/NH_3) ratios. For the representative Jovian subnebula model (Prinn and Fegley 1981*a*) illustrated in Figs. 2, 3 and 4, the relevant quench temperatures and gas ratios are (CO/CH_4) ~ 10^{-6} at ~ 840 K and (N_2/NH_3) ~ 1 at ~ 1370 K (see Fig. 6). However, in contrast to the $CO \rightarrow CH_4$ and $N_2 \rightarrow NH_3$ conversions, the slightly lower quench temperatures for the $CO \rightarrow CO_2$ conversion lead to slightly larger (CO_2/CO) ratios at higher pressures. Specifically, (CO_2/CO) ~ 10^{-2} at 740 K for the Jovian subnebula *(P, T)* profile shown. The much smaller (CO_2/CH_4) ratio in this case relative to that predicted for typical assumed solar nebula pressures of ~ 10^{-4} to ~ 10^{-7} bar leads to much smaller amounts of CO_2 and CO_2-bearing compounds incorporated into low-temperature condensates formed in

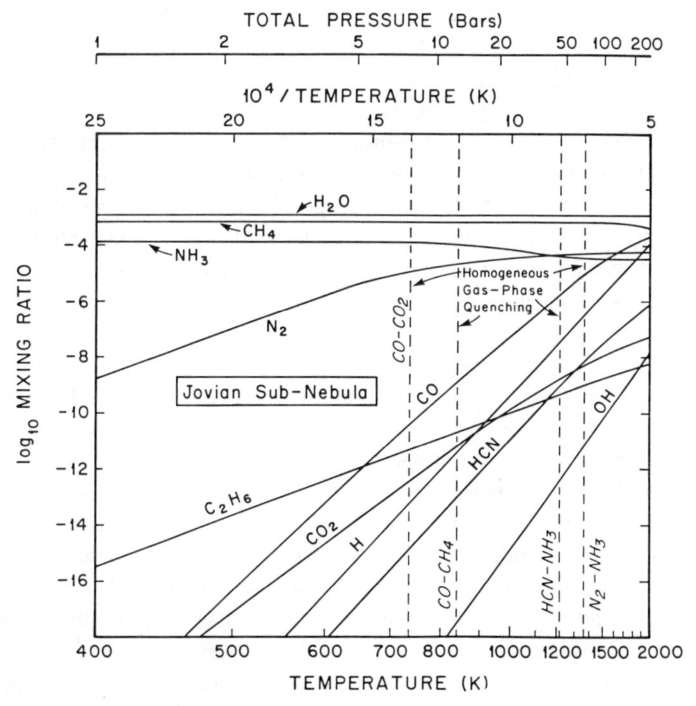

Fig. 6. As in Fig. 5 but for the Jovian subnebula *P,T* profile illustrated in Figs. 2,3 and 4. H_2 is not graphed for clarity.

spin off or impact-generated subnebulae around the giant planets than in solar nebula condensates (e.g., compare Prinn and Fegley 1981*a* and Lewis and Prinn 1980).

Finally, we note that small but significant amounts of HCN are predicted for homogeneous gas-phase quenching of the $HCN \rightarrow N_2$ and $HCN \rightarrow NH_3$ conversions over wide regions in *P, T* space covering plausible solar nebula and subnebula models. For the two representative nebular models shown in Figs. 5 and 6, quenching calculations done as described by Prinn and Fegley (1981*a*) (again assuming the slowest possible nebula mixing rate of one overturn in 10^{13} s) yield $(HCN/N_2) \sim 10^{-6}$ at $T_Q \sim 1460$ K along the solar nebula *(P, T)* profile and $(HCN/NH_3) \sim 10^{-5}$ at $T_Q \sim 1220$ K along the Jovian subnebula *(P, T)* profile. Somewhat larger amounts of HCN, which is an important precursor for the abiotic synthesis of complex organic molecules (Oro and Kimball 1961; Abelson 1966), may be produced in lightning discharges in nebular environments. This will be discussed in more detail below.

(ii) Heterogeneous Catalysis and Fischer-Tropsch-Type Reactions. The considerations discussed above for homogeneous gas-phase reactions may be modified somewhat by the presence of an abundant and active catalyst. In particular, if Fe metal grains are well mixed with the nebular (or subnebular) gases and are catalytically active, they may accelerate the rates of the $CO \rightarrow CH_4$, $N_2 \rightarrow NH_3$ and $CO \rightarrow CO_2$ conversions, and may also promote the synthesis of complex organic molecules from the CO and H_2 in the nebular gas via Fischer-Tropsch-type reactions. Indeed, the latter possibility was first anticipated by Urey (1953) who proposed that the $CO \rightarrow CH_4$ conversion "may well proceed through graphite or complex tarry compounds as intermediates." The feasibility of Urey's proposal was later experimentally demonstrated by Anders and coworkers (Hayatsu and Anders 1981; Studier et al. 1968) who synthesized complex organic compounds analogous to those in carbonaceous chondrites by iron meteorite-catalyzed Fischer-Tropsch-type reactions.

The heterogeneous Fe- catalyzed $N_2 \rightarrow NH_3$ conversion may occur over temperatures between the Fe metal condensation point (e.g., see Fig. 3) and a pressure-independent temperature of ~ 400 K where Fe metal is no longer stable and Fe_3O_4 forms. The incorporation of FeO into silicates may raise this latter temperature to a pressure-independent value of ~ 480 K. From Figure 3 it is clear that if the heterogeneous Fe catalysis of the $N_2 \rightarrow NH_3$ conversion is effective down to ~ 480 K where the last Fe metal is incorporated into FeO-bearing silicates, then for typical solar nebula pressures of $\sim 10^{-4}$ to $\sim 10^{-7}$ bar, the resulting (N_2/NH_3) ratio is much *greater* than unity. Furthermore, this conclusion is not altered if FeO-bearing silicate formation is neglected and Fe metal grains are assumed to be present (and catalytically active) in the nebular gas down to ~ 400 K where they will be deactivated by formation of Fe_3O_4 coatings. However, we note that if the Fe metal grains are not well mixed with the gas phase or are inactivated by thin coatings of FeS, which forms at a pressure-independent temperature of ~ 680 K, then heterogeneous catalysis may not be effective down to such low temperatures. The results of Lewis and Prinn (1980) for the solar nebula model illustrated in Figs. 3 and 5 indicate that the minimum quench temperature for the Fe catalyzed $N_2 \rightarrow NH_3$ conversion is ~ 530 K where $(N_2/NH_3) \sim 170$.

Similar arguments regarding the catalytic efficiency of Fe metal grains in higher pressure environments representative of those expected in subnebulae around the giant planets suggest that (N_2/NH_3) ratios much *less* than unity may be characteristic of these environments. Specifically for the Jovian subnebula *(P, T)* profile illustrated in Figs. 3 and 6, the results of Prinn and Fegley (1981*a*) suggest a minimum quench temperature of ~ 495 K and a (N_2/NH_3) ratio of $\sim 5 \times 10^{-4}$ for the Fe-catalyzed $N_2 \rightarrow NH_3$ conversion. However, it is clear from Fig. 3 that for subnebula pressures of $\sim 10^{-2}$ to $\sim 10^{2}$ bar, Fe catalysis down to ~ 400 K where Fe_3O_4 coatings will form will always yield (N_2/NH_3) ratios less than unity. The consideration of possible Fe catalysis of the $N_2 \rightarrow NH_3$ conversion thus does *not* alter our previous conclusions of N_2 dominance in the solar nebula and of NH_3 dominance in giant planet subnebulae that were reached on the basis of homogeneous gas-phase chemistry.

It is evident from Fig. 2 that if Fe catalysis of the $CO \rightarrow CH_4$ conversion is effective down to temperatures of ~ 400 to ~ 480 K, then over the entire range of pressures shown from $\sim 10^{-7}$ to $\sim 10^{3}$ bar, the resulting (CO/CH_4) ratios will be much *less* than unity. This contrasts with (CO/CH_4) ratios much *greater* than unity predicted for quenching of homogeneous gas-phase reactions at low pressures generally assumed for the solar nebula but reinforces conclusions regarding (CO/CH_4) ratios less than unity predicted for gas-phase quenching in subnebular environments at higher pressures.

The heterogeneous Fe-catalyzed $CO \rightarrow CH_4$ conversion in nebular and subnebular environments can be modeled using the results of laboratory studies of this reaction (e.g., see Vannice 1975,1982). The rate equation for the $CO \rightarrow CH_4$ conversion on metallic iron particles is

$$\frac{d}{dt}[CH_4] = -\frac{d}{dt}[CO] = [\text{sites}]k_{\text{site}}P_{H_2} \tag{22}$$

where $[i]$ is the number density per cm^3 of species *i*, P_{H_2} is the hydrogen pressure in bars and k_{site} is the experimental rate constant in units of CH_4 molecules produced per active site per s per bar on the Fe surface. The canonical value for the number of active sites per cm^2 surface is $\sim 10^{15}$. This is equivalent to a site number density of $\sim 10^5$ to 10^6 cm^{-3} in the solar nebula assuming iron present at solar abundance in the form of monodisperse, spherical ($r = 100$ μm) grains. A representative rate constant

$$k_{site} \sim 2.2 \times 10^7 \exp(-21{,}300/RT) \tag{23}$$

for the $CO \rightarrow CH_4$ conversion on *clean* Fe surfaces was taken from Vannice (1975). The chemical destruction time t_{chem} for the Fe-catalyzed $CO \rightarrow CH_4$ conversion is then given by

$$t_{chem} = -[CO] \Big/ \frac{d}{dt}[CO] \tag{24}$$

where the CO destruction rate is taken from Eq. (22).

Fig. 7 illustrates that the heterogeneous Fe catalysis of the $CO \rightarrow CH_4$ conversion in the solar nebula may lead to quenching of this reaction at ~ 900 K and $(CO/CH_4) \sim 10^{3.8}$ for the *minimum* radial mixing times implied by transport at 1/3 of sound speed (Cameron 1978; Lewis and Prinn 1980) or at

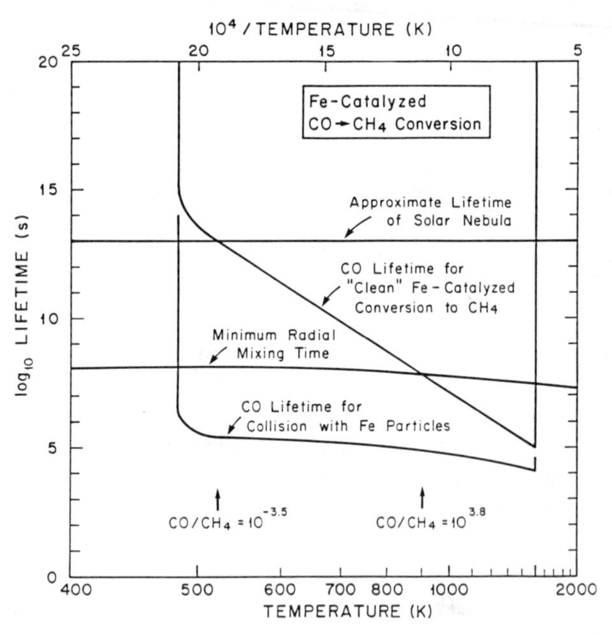

Fig. 7. Chemical and physical lifetimes associated with the Fe-catalyzed heterogeneous conversion of CO to CH_4 in the solar nebula. Fe catalysis first becomes possible at ~ 1500 K where Fe metal condenses (see Fig. 2) and becomes ineffective at ~ 480 K where the last metallic Fe is consumed by FeO incorporation into silicates. The shortest feasible nebular mixing time and the longest feasible nebular mixing time (see Sec. III.A) bracket the range of CO/CH_4 ratios produced by the heterogeneously catalyzed conversion on *clean* Fe surfaces. The expected inactivation of the Fe surface by rapidly forming carbonaceous coatings (Vannice 1982; Krebs et al. 1979), which may be similar to the "tar balls" observed in interplanetary dust particles (Bradley et al. 1984; Bradley and Brownlee 1986), will reduce the efficiency of CH_4 production.

temperatures as low as ~ 520 K and (CO/CH_4) ~ $10^{-3.5}$ for the *maximum* radial mixing time of one overturn in 10^{13} s. However, these results, which utilize Vannice's (1975) CO destruction rate constant for clean Fe surfaces, must substantially overestimate the efficiency of Fe catalysis because the inactivation of the Fe surface by rapidly forming carbonaceous coatings (Vannice 1982; Krebs et al. 1979), which may be similar to the "tar balls" observed in interplanetary dust particles (Bradley et al. 1984; Bradley and Brownlee 1986), is not considered. Therefore, the results in Fig. 7 should be viewed as an upper limit to the efficiency of the heterogeneous Fe catalyzed $CO \rightarrow CH_4$ conversion in the solar nebula.

The more likely course of events, which is indicated by the presence of carbonaceous material in some meteorites and of tar balls in interplanetary dust particles, is the Fe-catalyzed synthesis of complex organic molecules from CO and H_2 in the nebular gas by net reactions such as

$$(2n + 1)H_2 + nCO = C_nH_{2n+2} + nH_2O \tag{25}$$

$$(n + 1)H_2 + 2nCO = C_nH_{2n+2} + nCO_2 \tag{26}$$

$$2nH_2 + nCO = C_nH_{2n} + nH_2O \tag{27}$$

$$nH_2 + 2nCO = C_nH_{2n} + nCO_2 \tag{28}$$

$$2nH_2 + nCO = C_nH_{2n+1}OH + (n - 1)\, H_2O \tag{29}$$

$$(n + 1)H_2 + (2n - 1)CO = C_nH_{2n+1}OH + (n - 1)CO_2 \tag{30}$$

which exemplify the Fischer-Tropsch synthesis of alkanes, alkenes and alcohols, respectively. Such reactions, and similar reactions forming acetylenic and aromatic compounds, will proceed spontaneously in the presence of a suitable catalyst such as Fe or Ni (Bond 1962; Biloen and Sachtler 1981).

The possible significance of Fischer-Tropsch-type reactions for the formation of organic compounds found in carbonaceous chondrites has long been recognized (see, e.g., Urey 1953; Studier et al. 1968; Hayatsu and Anders 1981). However, quantitative discussions of the amount and type of organic compounds which may have been produced by such reactions operating in possible prenebular, nebular and subnebular environments requires a detailed knowledge of the effects of process parameters (e.g., pressure, temperature, time, reactant composition, catalyst chemical and physical state, presence of catalyst poisons, etc.) on the product composition and yield. This detailed knowledge is presently unavailable (e.g., as recently pointed out by Dictor and Bell [1986], the nature of the catalytically active iron phase is still controversial) but is essential for theoretical models of carbon chemistry and organic matter production by Fischer-Tropsch-type reactions in the solar nebula.

Insight into the probable effects of these important process parameters can nevertheless be obtained by qualitative comparisons of products of

Fischer-Tropsch-type reactions run under very different sets of conditions. Figure 8 illustrates that the hydrocarbons formed at *low* total conversions of CO to products using a clean Fe surface become progressively lighter and more C_1 rich as the reactant H_2/CO ratio increases from 4 to 100 (Krebs et al. 1979). Analogous trends were observed by Vannice (1975) for CO hydrogenation over clean Fe supported on Al_2O_3 as the H_2/CO ratio increased from 0.6 to ~ 15 (note that the nebular H_2/CO ratio is ~ 1200 if all carbon is present as CO).

In contrast, hydrocarbons produced at *high* total conversions of CO to products using industrial Fe catalysts are rich in large molecules. For example, an illustrative hydrocarbon composition from a commercial reactor (Dry 1981) is C_1 (2%), C_2 (2%), C_3 (5%), C_4 (5%), C_5 to C_{11} (18%), C_{12} to C_{18} (14%), C_{19} to C_{23} (7%), C_{24} to C_{35} (21%), and $> C_{35}$ (26%). Although the details of product selectivity (e.g. alkanes vs alkenes, carbon chain length distribution, degree of branching) vary somewhat, the commercial reactors operating at high total conversion of CO to products (with recycling of the outlet gas through the reactor) generally give larger molecules than the single-pass flow experiments operating at low total conversions and large H_2/CO ratios. The static experiments of Studier et al. (1968) also gave larger maximum carbon chain lengths at lower H_2/CO ratios.

These observations suggest that the conversion of nebular CO to organic material may yield a wide range of products from CH_4 to large molecules. Environmental conditions which allow high total conversions of CO (in dynamic or static conditions) and perhaps H_2/CO ratios smaller than the solar

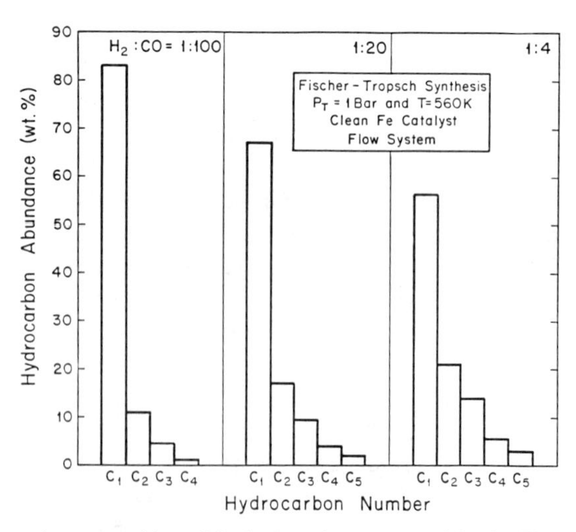

Fig. 8. Representative composition of the hydrocarbons produced by the Fischer-Tropsch synthesis on an Fe catalyst for a range of H_2/CO ratios from 1 : 100 to 1 : 4. The results are for low total conversion of CO to products in a flow system (from Krebs et al. 1979).

value of ~ 1200 may yield larger molecules than environments that allow low total conversions of CO and the solar H_2/CO ratio. In any case, the analogies between organic compounds in carbonaceous chondrites and the products of Fischer-Tropsch-type reactions (Hayatsu and Anders 1981), and the associations of carbonaceous material (tar balls) with Fe, Ni alloy, carbides and oxides (which have all been proposed as the catalytically active Fe phase [Dictor and Bell 1986]) in interplanetary dust grains (Bradley et al., 1984) suggest that Fischer-Tropsch-type reactions were operative in some environments in the solar nebula and early solar system.

(iii) Lightning and Shock Chemistry. We have mentioned earlier that lightning discharges and their accompanying thundershocks are a possible disequilibrating mechanism in nebular and subnebular environments favorable for charge separation. The high temperatures (several times 10^3 K) reached in lightning discharges lead to increasing degrees of molecular dissociation, atomization and ionization with increasing temperatures. These effects are more pronounced at lower total pressures than at higher total pressures. Recombination of these simple fragments during the rapid cooling of the shocked gas leads initially to the production of more complex fragments, then to thermally stable molecules. Sufficiently rapid cooling quenches these stable molecules at their high temperature abundances, which are generally enhanced over their equilibrium abundances at much lower temperatures (e.g., see the description by Borucki and Chameides [1984] of NO production in terrestrial lightning discharges via the shock heating and rapid quenching of $N_2 + O_2$). Repeated lightning discharges then lead to the buildup of these thermally stable species until a steady-state abundance is reached or they are lost by other processes (e.g., chemical reactions forming more complex molecules or vapor → solid condensation). Thus, lightning is a potentially significant source of disequilibrium products, especially in the cool, thermochemically inactive regions of the nebula near and beyond the water ice condensation point (T ~ 150 to 200 K over a wide pressure range).

Thermodynamic calculations of gas abundances presented in Figs. 9 and 10 and energetic considerations based on the First Law of Thermodynamics can be used to set upper limits on the production efficiencies and yields of important disequilibrium products such as HCN. Shock heating of a CO-, N_2-bearing gas parcel (e.g., in the solar nebula) produces HCN via the net reaction

$$2CO + N_2 + 3H_2 = 2HCN + 2H_2O \tag{31}$$

from which it can be easily deduced that the resultant HCN mixing ratio (X_{HCN}) is proportional to P_T the total pressure. This process is illustrated in Fig. 9 for the heating along an adiabatic (P,T) profile of an initially cool, H_2O-depleted (due to condensation) gas parcel in the outer solar nebula.

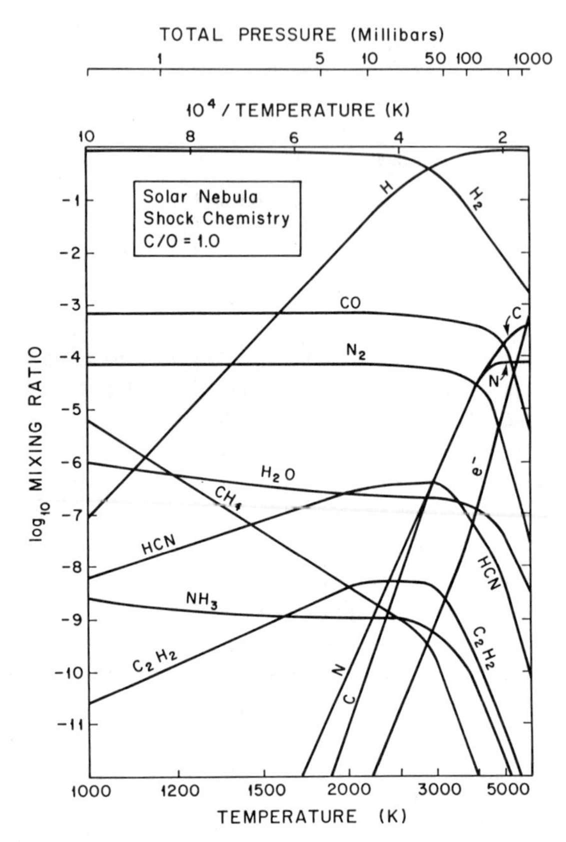

Fig. 9. Calculated equilibrium abundances of important H, C, N, O compounds in an initially cool, H_2O-depleted (by condensation) gas parcel which is shock heated by a lightning discharge in the outer region of the solar nebula. An adiabatic P,T profile is assumed; a Hugoniot or an isopycnic profile would give a lower pressure at a given temperature. The electron abundance shown is balanced by an identical abundance of positive ions (which are predominantly H^+) in order to preserve electroneutrality. However, the positive ions and many other radicals, atoms and negative ions are not graphed to simplify the figure.

On the other hand, shock heating of a CH_4-, N_2-bearing gas parcel or of a CH_4-, NH_3-bearing gas parcel (e.g., in the Jovian subnebula) produces HCN by the net reaction

$$2CH_4 + N_2 = 2HCN + 3H_2 \tag{32}$$

in the former case (where $X_{HCN} \propto P_T^{-1}$) or via

$$CH_4 + NH_3 = HCN + 3H_2 \tag{33}$$

in the latter case (where $X_{HCN} \propto P_T^{-2}$). Both of these two processes would contribute to HCN production during the heating along an adiabatic (P,T)

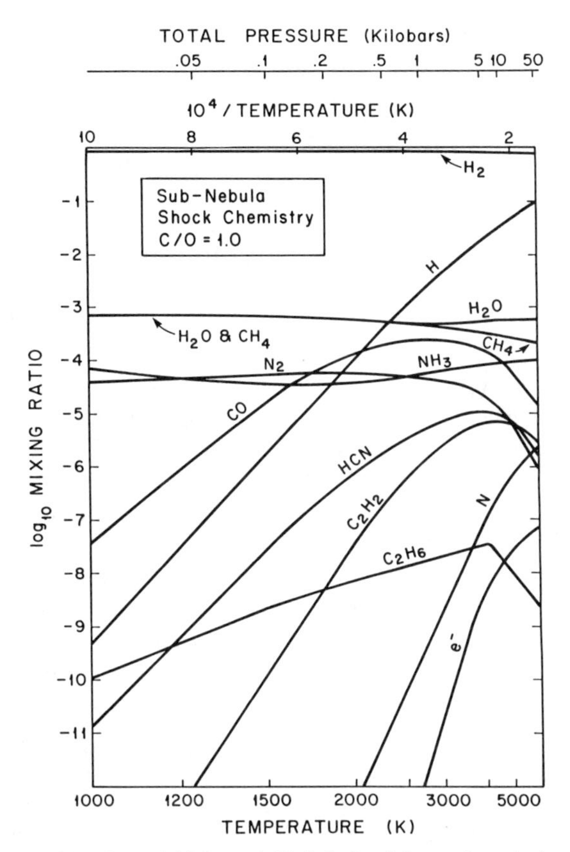

Fig. 10. As in Fig. 9 but for an initially cool, H_2O-depleted (by condensation) gas parcel which is shock heated by a lightning discharge in the outer region of the Jovian subnebula. The much higher total pressures along the assumed adiabatic *P,T* profile suppress molecular dissociation and ionization in comparison to the results illustrated in Fig. 9.

profile of an initially cool, H_2O-depleted (due to condensation) gas parcel in the outer regions of the hypothesized Jovian subnebula (see Fig. 10).

The maximum HCN concentrations produced by the adiabatic shock heating of nebular gases in these two different environments occur near temperatures of 3000 to 4000 K where $\sim$ (0.2 to 6) $\times$ 10^{18} HCN molecules are formed per mole of shocked gas. These concentrations correspond to *maximum* conversions of $\sim$ 0.3% (solar nebula) and $\sim$ 6.5% (Jovian subnebula) of the total nitrogen abundance into HCN. In the absence of a quantitative theory for lightning discharges (and their associated shock chemistry) in the solar nebula, it is instructive to use these maximum concentrations to deduce upper limits for HCN production efficiencies by lightning discharges.

If we (initially) neglect the dissociation of H_2 to H, the energy required for the reversible adiabatic heating and compression of H_2-rich nebular gas is given approximately by $\int C_v dT$ where C_v is the constant volume heat capacity of ideal $H_2(g)$. This energy is ~ 70 to 100 kJ mole^{-1} for heating of cold (T ~ 100 K) H_2-rich gas to the temperatures of 3000 to 4000 K that yield high HCN abundances (see Figs. 9 and 10). By contrast, reversible isobaric heating requires slightly more energy ($\int C_p dT \sim 90 - 120$ kJ mole^{-1}) for heating to the same temperatures. These two extremes bracket other possible (P,T) paths (e.g., isopycnic and Hugoniot) for shock heating by lightning.

Dissociation of H_2 to H requires another ~ 460 kJ per mole of H_2 dissociated. As illustrated by Figs. 9 and 10, H_2 dissociation is more important at the lower pressures commonly assumed for the solar nebula than at the higher pressures assumed for the Jovian subnebula. Furthermore, ionization of H → $H^+ + e^-$ does not appear to be significant in either environment in the 3000 to 4000 K range. Thus, energies of ~ 300 kJ mole^{-1} appear to be required for heating solar nebula gas to ~ 3000 K (where for this (P,T) path H_2 is ~ 50% dissociated to H) and energies of ~ 110 kJ mole^{-1} appear to be required for heating Jovian subnebula gas to ~ 4000 K (where along this (P,T) path H_2 is only a few percent dissociated to H).

The shock heating of 1 mole of nebular gas per m^2 per year thus requires a lightning energy flux ϕ on the order of 10^{-3} J m^{-2} s^{-1}; the estimates of ϕ deduced in Sec. II.B suggest that processing of 1 mole m^{-2} yr^{-1} is probably a reasonable upper limit for outer nebular regions. The corresponding *maximum* HCN yields of ~ $(0.2 - 6) \times 10^{18}$ molecules m^{-2} yr^{-1} yield *upper limits* to the HCN production efficiencies on the order of ~ 10^{13} to 10^{14} molecules J^{-1}. For comparison the NO production efficiencies in terrestrial lightning discharges are on the order of 10^{16} molecules J^{-1} (Borucki and Chameides 1984) and organic matter production efficiencies in laboratory sparking experiments on $CH_4 - NH_3 - H_2$ mixtures (see, e.g., Miller 1955) are ~ 10^{13} to 10^{15} molecules J^{-1} (10^{-7} g J^{-1}) for assumed product mean molecular weights of 10^4 to 10^2. Thus, lightning discharges in the solar nebula are probably less efficient sources of disequilibrium molecules (e.g., HCN) than terrestrial lightning or laboratory discharges. However, in the cool, thermochemically inactive outer regions of the solar nebula (where their disequilibrium products may be able to accumulate without rapidly being mixed inward to high-temperature regions where reequilibration can occur), lightning discharges may be a significant source of disequilibrium material. The production of HCN with its C ≡ N bond and potential for further reaction to form more complex organic molecules (Oro and Kimball 1961; Abelson 1966) is especially interesting in this regard.

(iv) Thermochemical Equilibrium/Disequilibrium for H_2O. Latimer (1950) and Urey (1952) both suggested that hydrated silicates were probable substances for retaining water in the solid planetary-forming materials that accreted to form the Earth and the other terrestrial planets. However, the lack of accurate thermodynamic data prevented them from calculating the stability

fields of hydrous silicates in the solar nebula. Despite this limitation, Urey (1952,1953) did calculate the thermodynamic conditions required for water retention as brucite $Mg(OH)_2$, water ice, and $NH_3 \cdot H_2O(s)$. This led him (Urey 1953) to conclude " that water and ammonia would have condensed in the solar nebula and the existence of the carbonaceous chondrites containing up to 10% of water confirms this conclusion . . ." However, the kinetic inhibition of the $N_2 \rightarrow NH_3$ conversion in the solar nebula, which was discussed earlier, will modify Urey's original prediction of $NH_3 \cdot H_2O(s)$ condensation.

More recent thermochemical calculations of the stabilities of H_2O-bearing phases in the solar nebula are presented in Table III and Fig. 11. These results illustrate that minor H_2O-bearing phases such as tremolite [$Ca_2Mg_5Si_8O_{22}(OH)_2$], Na-phlogopite [$NaMg_3AlSi_3O_{10}(OH)_2$], and hydroxyapatite [$Ca_5(PO_4)_3OH$] become stable in the 460 to 500 K temperature range. However, these minerals are seldom, if ever, observed in chondritic meteorites. On the other hand, major H_2O-bearing phases, which are observed in chondritic meteorites, exemplified by serpentine [$Mg_3Si_2O_5(OH)_4$], talc [$Mg_3Si_4O_{10}(OH)_2$], and brucite [$Mg(OH)_2$], do not become stable until temperatures of 400 K or below. However, significant formation of hydrous phases at these low temperatures in the solar nebula requires solid-solid (or gas-solid) chemical equilibrium to be approached during the lifetime of the nebula. Is this requirement satisfied or is attainment of equilibrium impossible at the (P,T) conditions where the major H_2O-bearing minerals are stable?

The formation of hydrated silicates by solid-solid reactions such as

$$\underset{\text{forsterite}}{Mg_2SiO_4} + \underset{\text{enstatite}}{MgSiO_3} + 2H_2O(g) = \underset{\text{serpentine}}{Mg_3Si_2O_5(OH)_4} \qquad (34)$$

requires the diffusion of Mg and Si between olivine and pyroxene at temperatures of 200 to 400 K. The characteristic diffusion time for this process can be estimated using the scaling relationship $t \sim r^2/D$ where r is the grain radius, D is the diffusion coefficient (cm^2s^{-1}), and t is the characteristic diffusion time. Freer (1981) has pointed out that under *dry* conditions, such as those expected in the solar nebula where $X_{H_2O} \leq 5 \times 10^{-4}$, divalent cations (e.g., Mg^{2+}, Fe^{2+}) in silicates generally diffuse faster than Si^{4+}. He further states that divalent cation diffusion in olivine is faster than in pyroxene.

In order to estimate the characteristic diffusion time for Mg and Si diffusion between olivine and pyroxene, which is the rate limiting step for hydrated silicate formation via reactions such as Eq. (34), we therefore make the following generous assumptions. First, we assume that both silicate reactants are in intimate contact (e.g., in the same grain) for long time periods. Second, we assume that the composite silicate grains are monodisperse, spherical grains with radii of 0.1 μm. This radius is comparable to the size of the very fine-grained matrix found in chondritic meteorites, but is significantly smaller than the size of the majority of silicate grains observed in chondrules. Third, we assume that the diffusion of Mg and Si between olivine and pyroxene is as rapid as divalent cation diffusion in olivine. Following Cohen et al. (1983), we take the data of Misener (1974) to model Fe − Mg diffusion in olivine. This diffusion coefficient is $D \sim 6.3 \times 10^{-3} \exp(-28{,}780/T)$ cm^{2}s^{-1}. We note that our assumptions all maximize the calculated diffusion rate and thus minimize the calculated diffusion time.

Despite these favorable assumptions, the characteristic diffusion time is estimated to be $\sim 10^{23}$s at 400 K, increasing to $\sim 10^{55}$ s at 200 K. These times are many orders of magnitude greater than the lifetime of the solar nebula ($\sim 10^{13}$ s) and the age of the solar system ($\sim 10^{17}$ s). It is easy to increase these characteristic diffusion times by making more realistic (and inherently less favorable) assumptions regarding silicate grain size and cation diffusion rates. Indeed, the assumed silicate grain radius of 0.1 μm, while comparable to the size of very fine-grained meteorite matrix and of olivines and pyroxenes found in interplanetary dust particles (e.g., see Peck 1983,1984; MacKinnon and Rietmeijer 1987), is significantly smaller than the size of the silicates commonly found in chondrules (grain size ≈ 10 to 100 μm). Furthermore, cation diffusion in pyroxene is significantly slower than cation diffusion in olivine (e.g., see Freer 1981). However, it is virtually impossible to decrease the estimated characteristic diffusion times.

Thus it is almost certain that reactions such as Eq. (34), which have been postulated for hydrated silicate formation in the solar nebula, simply do not have enough time to reach equilibrium because solid state diffusion is far too slow. As a consequence, the vapor phase hydration of monomineralic silicate grains, exemplified by the reactions

$$\underset{\text{forsterite}}{2Mg_2SiO_4} + 3H_2O(g) = \underset{\text{serpentine}}{Mg_3Si_2O_5\,(OH)_4} + \underset{\text{brucite}}{Mg(OH)_2} \tag{35}$$

$$\underset{\text{enstatite}}{4MgSiO_3} + 2H_2O(g) = \underset{\text{talc}}{Mg_3Si_4O_{10}\,(OH)_2} + \underset{\text{brucite}}{Mg(OH)_2} \tag{36}$$

may be the only kinetically feasible pathway for forming hydrated silicates in the solar nebula. However, as Table III and Fig. 11 show, formation of serpentine + brucite or of talc+ brucite via reaction (35) or (36) requires temperatures of ~ 200 to 300 K. The possible kinetic inhibition of gas-solid reactions at these low temperatures will now be considered.

The initial rate of reaction of water vapor with anhydrous silicate grains will depend on the collision rate of the H_2O molecules with the grain surfaces. This rate is given by

$$\sigma_w = 2.635 \times 10^{25}\,[P_w/(M_w T)^{1/2}] \tag{37}$$

where σ_w has units of cm^{-2}s^{-1}, P_w is the water vapor partial pressure, M_w is the molecular weight of water vapor, and T is temperature. The total number

TABLE III

Predicted H_2O-Bearing Phases Stable at Thermochemical Equilibrium in Solar Nebula Gas ($T \geq$ ice condensation point)[a]

Equilibration Temperature (K)	H_2O-Bearing Phase(s)	H_2O Content (w/o)	Reference
~500	Tremolite [$Ca_2Mg_5Si_8O_{22}(OH)_2$]	2.2	Lewis 1972*a*, *b*
<470[b]	Na − Phlogopite [$NaMg_3AlSi_3O_{10}(OH)_2$]	4.5	Hashimoto and Grossman 1987
~460[c]	Hydroxyapatite [$Ca_5(PO_4)_3OH$]	1.8	Fegley and Lewis 1980
~400[d]	Serpentine [$Mg_3Si_2O_5(OH)_4$]	13.0	Lewis 1974*a*
~230[d,e]	Serpentine [$Mg_3Si_2O_5(OH)_4$]	13.0	Barshay 1981
<274[f]	Serpentine [$Mg_3Si_2O_5(OH)_4$] + Brucite [$Mg(OH)_2$]	16.1	Hashimoto and Grossman 1987
~225	Serpentine [$Mg_3Si_2O_5(OH)_4$] + Brucite [$Mg(OH)_2$]	16.1	this work
~400[d]	Talc [$Mg_3Si_4O_{10}(OH)_2$]	4.8	Lewis 1972*a*, *b*
~340[d,e]	Talc [$Mg_3Si_4O_{10}(OH)_2$]	4.8	Barshay 1981
~160 − 280[g]	Talc [$Mg_3Si_4O_{10}(OH)_2$]	4.8	Larimer and Anders 1967
~250	Talc [$Mg_3Si_4O_{10}(OH)_2$] + Brucite [$Mg(OH)_2$]	8.3	this work
~160[h]	Water Ice [H_2O]	100	Lewis 1972*a*, *b*

[a]The temperatures listed are the highest temperatures at which the H_2O-bearing phase is stable along the illustrative solar nebula (P, T) profile in Fig. 11.

[b]Na-phlogopite forms at ~470 K at 10^{-3} bar total pressure (Hashimoto and Grossman 1987). It will form at slightly lower temperatures along the (P, T) profile shown in Fig. 11.

[c]The water content is calculated on the basis of 1 H_2O molecule per 2 hydroxyapatite molecules.

[d]These calculations assume solid-solid chemical equilibrium. However, the extremely slow rate of solid state diffusion at these low temperatures probably prevents the attainment of equilibrium over the lifetime of the solar nebula. See Sec III. B(iv) of the text for a discussion.

[e]Barshay's results are reproduced in Fig. 12 of Prinn and Fegley (1987*a*). He used a feldspar-nepheline silica buffer at $T < 600$ K in these calculations.

[f]Hashimoto and Grossman (1987) calculate that serpentine + brucite form at 274 K at 10^{-3} bar total pressure. This assemblage will form at lower temperatures along the (P, T) profile shown in Fig. 11.

[g]Larimer and Anders (1967) list 2 possible talc formation reactions: $3MgSiO_3 + SiO_2 + H_2O(g) \rightarrow Mg_3Si_4O_{10}(OH)_2$ and $3Mg_2SiO_4 + 5SiO_2 + 2H_2O(g) \rightarrow 2Mg_3Si_4O_{10}(OH)_2$ with condensation temperatures in the 160 to 280 K range for total pressures of ~10^{-6} to ~10^{-3} bar.

[h]The shaded region for H_2O (ice, liquid) in Fig. 11 illustrates the range of condensation temperatures assuming all carbon is CH_4 or CO for total pressures of 10^{-7} to 10^2 bar.

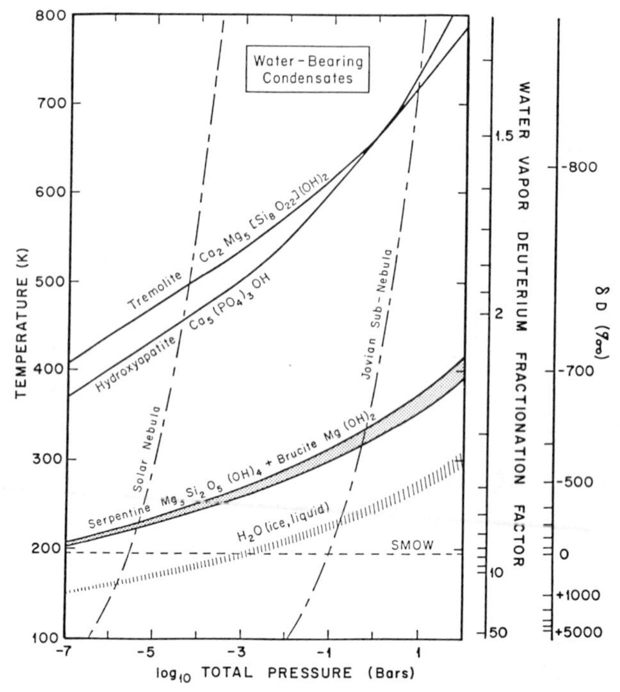

Fig. 11. The calculated stability fields for water-bearing condensates over a wide range of total pressures applicable to the solar nebula and giant planet subnebulae. The tremolite and hydroxyapatite curves are from Lewis (1974*a*) and Fegley and Lewis (1980). The serpentine + brucite curve represents hydration of forsterite by water vapor via reaction (35). The *P,T* profiles for the solar nebula and the Jovian subnebula are the same as shown in previous figures. The shaded regions for H_2O (ice, liquid) and serpentine + brucite illustrate the range of condensation temperatures if all carbon is present as CH_4 (top edge) or if all carbon is present as CO (lower edge). The amount of oxygen incorporated into *anhydrous* rock ($MgO + SiO_2$) is also taken into account. Equilibrium fractionation factors (Richet et al. 1977) for temperature-dependent deuterium partitioning between HD*(g)* and $H_2O(g)$ are also shown. These are independent of the absolute D/H ratio in either phase. Deuterium is enriched in H_2O (relative to HD) with decreasing temperature. However, δD = [(D/H)water - (D/H)SMOW]/[(D/H)SMOW] x 1000 where SMOW = standard mean ocean water with D/H = 1.557×10^{-4} (Pillinger 1984) depends on the absolute D/H value chosen for H_2 in the solar nebula. Following Cameron (1982), this is taken as 2.0×10^{-5}. Both the heterogeneous gas-solid equilibration of $H_2O(g)$ with anhydrous minerals (to form hydrous phases) and the homogeneous gas phase isotopic equilibration of $H_2O(g)$ and HD*(g)* are problematic at low temperatures and rapid mixing rates (i.e., short mixing times) in the solar nebula.

of collisions with all forsterite (or enstatite) grains in each cm^3 of the nebula is given by

$$\nu_w = \sigma_w A \tag{38}$$

where ν_w has units of cm^{-3}s^{-1} and A is the total surface area of all forsterite (or enstatite) grains in each cm^3 of the nebula. The grains are assumed to be monodisperse, spherical grains which are uniformly distributed at solar abundance in the gas. The present calculations also assume a grain radius of 0.1 μm consistent with the earlier diffusion calculations.

The collision time constant t_{coll} for water molecules to collide with forsterite (or enstatite) grains in each cm^3 of the nebula is then

$$t_{coll} \sim [H_2O]/\nu_w \tag{39}$$

where $[H_2O]$ is the molecular number density of water vapor. This abundance is taken as the stoichiometric amount of water vapor required for hydrated silicate formation (~ 20% of the total H_2O molecular number density). If every collision led to silicate hydration, Eq. (39) would also be the expression for the chemical time constant t_{chem}. However, only a small fraction of collisions which possess the necessary activation energy lead to hydration. This fraction is given by

$$f_w \sim \nu_w \exp(-E_a/RT) \tag{40}$$

where E_a is the activation energy and R is the ideal gas constant. The chemical time constant t_{chem} for silicate hydration is thus given by

$$t_{chem} \sim [H_2O]/f_w \sim t_{coll}/\exp(-E_a/RT). \tag{41}$$

Figure 12 illustrates the collision time constant t_{coll} for reaction (35), which exemplifies the vapor phase hydration of anhydrous silicates. The amount of H_2O that can be incorporated into serpentine + brucite (~ 20% of available water vapor in the CO-rich solar nebula) collides with 0.1 μm radius forsterite grains in less than an hour; this time increases linearly with the assumed grain radius. However, because only a small fraction of the collisions actually leads to chemical reaction, the chemical time constant t_{chem} is significantly longer than the collision time constant. In the absence of experimental data on the low pressure vapor phase hydration of forsterite, we utilize the results of Layden and Brindley (1963) and Bratton and Brindley (1965) on the vapor phase hydration of MgO to $Mg(OH)_2$ (which has an activation energy ~ 70 kJ $mole^{-1}$) in order to estimate t_{chem} for the vapor phase hydration of forsterite.

The estimated t_{chem} for forsterite hydration, which is displayed in Fig. 12, is $>10^4$ times longer than the solar nebula lifetime (i.e., $t_{chem} > 10^{17}$s) at the temperature (~ 225 K) where forsterite hydration to serpentine + brucite becomes thermodynamically favorable. Furthermore, this t_{chem} value should be regarded as a lower limit to the lifetimes (and thus an upper limit to the rates) for silicate hydration reactions in the solar nebula because reactions requiring the migration and diffusion of more than one metal cation will probably proceed slower than MgO hydration which involves only one metal cation. Thus, formation of serpentine, talc, brucite and of other related hydrous

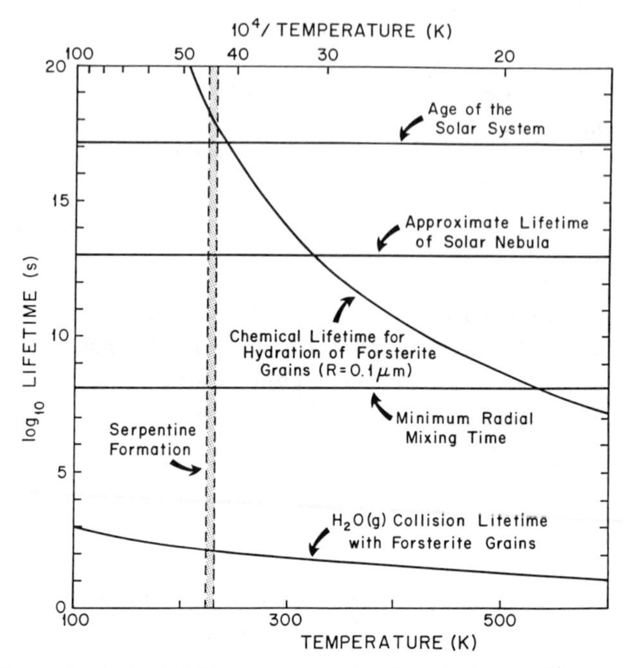

Fig. 12. Chemical and physical lifetimes associated with the hydration of forsterite Mg_2SiO_4 grains by $H_2O(g)$ to form serpentine $Mg_3Si_2O_5(OH)_4$ in the solar nebula. All Mg is assumed to be present in forsterite, which is a proxy for the more diverse suite of Mg- bearing minerals present in the solar nebula at low temperatures (e.g., see Barshay 1981). The assumed (spherical, monodisperse) forsterite grain diameter of 0.2 μm is consistent with the observations of 30 μm x 5 μm olivine plates and micron-sized olivine grains in C3V meteorite matrix (Peck 1983, 1984), which is believed to be a mixture of (relatively) unaltered nebular materials (McSween 1979*a*). The assumed grain size is also consistent with observed grain sizes for olivine and pyroxene in interplanetary dust particles (MacKinnon and Rietmeijer 1987). However most silicate grains found in chondrules in meteorites are significantly larger than 0.2 μm and thus will react significantly slower than the smaller assumed 0.2 μm grains. The shaded region for serpentine illustrates the range of condensation temperatures if all carbon is present as CH_4 (slightly higher *T*) or if all carbon is present as CO (slightly lower *T*). Note that the chemical lifetime for forsterite hydration (assuming it is as energetically favorable as MgO hydration) is longer than the lifetime of the solar nebula. Consequently, formation of serpentine (as well as talc and brucite which form at a similar temperature) is very probably kinetically inhibited.

minerals forming at these low temperatures $\leq$ 300 K is probably kinetically inhibited. The implications of this kinetic inhibition for water retention by the terrestrial planets, chondritic meteorites and asteroids will be examined in Sec. IV.

(v) Low temperature condensation. The nature of low temperature condensates (e.g., ice, hydrates, clathrates, etc.) in the nebula and subnebulae

depends sensitively on the degree of disequilibration for carbon, nitrogen and oxygen compounds. In particular, depending on the quench temperature T_Q for the $CO \rightarrow CH_4$ and $N_2 \rightarrow NH_3$ conversions (see Sec. III.B (i) above) we can proceed from the (low T_Q) *equilibrium* sequence (H_2O, $NH_3 \cdot H_2O$, $CH_4 \cdot 6H_2O$, etc.) to the (high T_Q) *disequilibrium* sequence (NH_4HCO_3, H_2O, NH_4COONH_2, CO_2, $CO \cdot 6H_2O$, $N_2 \cdot 6H_2O$, etc.). As already noted we expect the *disequilibrium* sequence in the solar nebula and the equilibrium sequence in protoplanetary subnebulae.

These two general sequences are further complicated in the solar nebula (but not in subnebulae) by probable catalytic conversion of CO to light and heavy hydrocarbons (Sec. III.B (ii)) and by probable kinetic inhibition of hydrated silicate formation (Sec. III.B (iv)). Catalytically formed hydrocarbons provide a mechanism for retention of cosmically important amounts of carbon at temperatures well above the $CO \cdot 6H_2O$ condensation point, while catalytically formed hydrocarbons and kinetic inhibition of silicate hydration lead to an increase in the H_2O/silicate (ice/rock) ratio (through conversion of $CO \rightarrow H_2O$ or prevention of $H_2O \rightarrow$ silicate OH).

A summary of the major low-temperature condensation sequences is provided in Table IV. The dramatic differences in the predicted condensation sequences in the solar nebula and the planetary subnebulae mean that certain volatile compound ratios (e.g., N_2/NH_3, CO/CH_4, H_2O ice/rock) may be diagnostic of origin for ice-rich bodies. We will address this point specifically in our discussion of ice-rich satellites and comets in the next section.

IV. IMPLICATIONS FOR SPECIFIC OBJECTS

A. Terrestrial Planets

The large depletions of the nonradiogenic rare gases relative to other, chemically reactive volatiles on the surface of the Earth (e.g., see Brown 1949) has led to the generally accepted hypothesis that the volatiles on Venus, Earth and Mars are almost entirely secondary and did *not* originate principally by gas capture from the solar nebula. Instead, the volatile endowments on the terrestrial planets are the result of chemical processes (e.g., for C, N and H_2O) and physical processes (e.g., for the rare gases) for retaining volatiles in solid grains. These volatile-bearing grains were either accreted in an essentially homogeneous fashion along with the bulk of planetary-forming materials or were delivered in an essentially inhomogeneous fashion to the terrestrial planets after they had finished (or were near the end of) their accretion. Some of the consequences of a simple two component inhomogeneous accretion model for planetary volatiles are discussed by Wänke (1981) and the chapter by Dreibus and Wänke.

The radial temperature gradient in the solar nebula was a major influence on the nature and abundance of the volatile-bearing grains. For the unlikely

TABLE IV
Low Temperature Condensation Sequences Predicted in the Protosolar Nebula and Protoplanetary Subnebulae*

Protosolar Nebula		Protoplanetary Subnebula
CO-rich	**CO plus Hydrocarbons**	
Largely anhydrous rock	Largely anhydrous rock Involatile carbonaceous compounds[a,b]	Largely hydrated rock
H_2O (150 K)	H_2O	H_2O (235 K)
NH_4HCO_3/NH_4COONH_2 (150, 130 K)[a]	NH_4HCO_3/NH_4COONH_2[a]	$NH_3 \cdot H_2O$ (160 K)
	Light hydrocarbons and hydrocarbon clathrates[c]	$CH_4 \cdot 6H_2O$ (94 K)
CO_2 (70 K)[a]	CO_2[a]	CO (20 K)[a]
$CO \cdot 6H_2O$ (60 K)[d]	$CO \cdot 6H_2O$[d]	N_2 (20 K)[a]
$N_2 \cdot 6H_2O$ (55 K)[d]	$N_2 \cdot 6H_2O$[d]	
CO (20 K)	CO (20 K)	
N_2 (20 K)	N_2 (20 K)	

[a]These compounds contain $\leq 10^{-2}$ of the total C or N.
[b]Exemplified by organic material in chondrites, and formed by nebular Fischer-Tropsch-type reactions and/or of interstellar origin.
[c]Exemplified by the light ($C_1 - C_5$) hydrobarbons formed as initial products in high $H_2 : CO$ environments (see Fig. 8). Precise condensation or enclathration temperature depends on the chain length and hydrocarbon abundance.
[d]There is insufficient H_2O to allow condensation of all C and N as CO and N_2 clathrates.
*Condensation temperatures in illustrative models (Lewis and Prinn 1980; Prinn and Fegley 1981*a*) are given for each condensate where available. For the nebula the "CO plus hydrocarbons" sequence is preferred with the condensation temperatures being slightly below those in the CO-rich case (the precise temperatures will depend on the degree of CO to hydrocarbon conversion).

case in which chemical equilibrium was completely maintained with decreasing temperature, the solid grains equilibrated at lower temperatures (i.e., farther from the proto-Sun) would be more volatile rich than the solid grains equilibrated at higher temperatures (i.e., closer to the proto-Sun). However, more plausibly, if chemical disequilibrium prevailed with decreasing temperature, processes such as Fischer-Tropsch-type reactions and lightning-induced shock heating would become more important and build up reservoirs of volatile-rich disequilibrium material. In turn, transport of such volatile-rich disequilibrium material inward to higher temperature regions would lead to its pyrolysis and to the eventual production of volatile-poor grains.

We also note that for both equilibrium and disequilibrium scenarios, decreasing temperatures may have allowed important exceptions to this general behavior. For example, if chemical equilibrium prevailed, the amounts of elemental carbon and nitrogen dissolved in Fe − Ni alloy would have gone through maxima in the inner regions of the solar nebula (see Figs. 13 and 14). On the other hand, if disequilibrium prevailed, decreasing temperatures would have kinetically inhibited water retention by solid grains (Fig. 12) or may have halted production of organic matter by deactivation of an essential catalyst (e.g., formation of FeS coatings on Fe particles at ~ 680 K).

A comparison of the observed C, N and H_2O inventories on Venus, Earth and Mars with the predicted inventories deduced using the idealized assumptions of complete chemical equilibrium and complete degassing of volatiles provide insight into the maximum contributions of equilibrium processes to the volatile inventories of these planets. This comparison is presented in Table

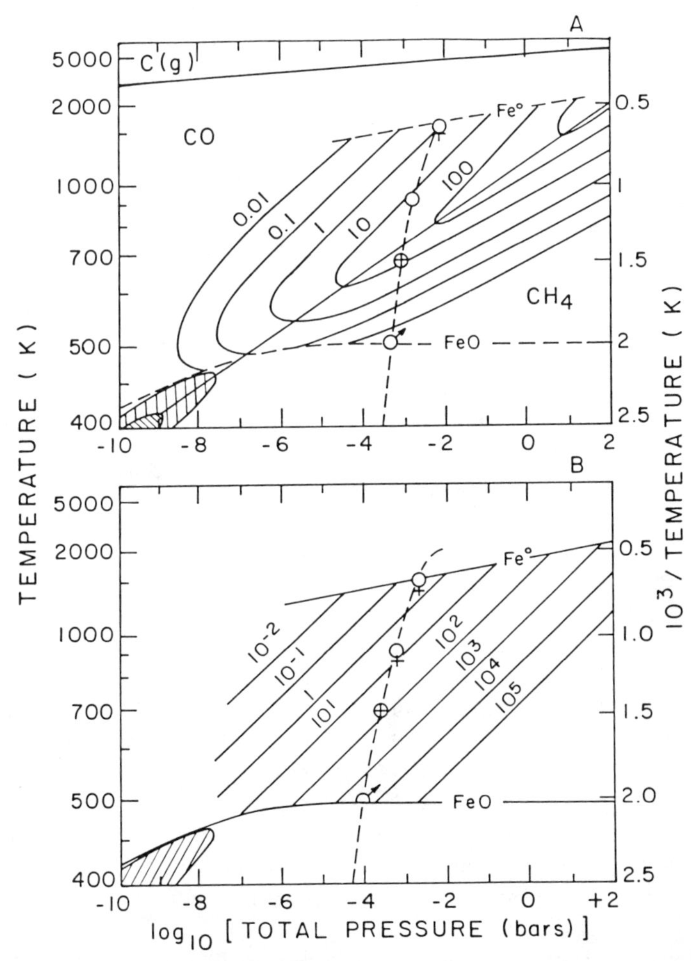

Fig. 13. Predicted concentrations (parts per million by mass) of elemental carbon dissolved in Fe - Ni alloy equilibrated with solar nebula gas. The assumed solar nebula *P,T* profile is also shown as a dashed line with the astrological symbols for Mercury, Venus, Earth and Mars from top to bottom. The predicted bulk carbon contents for Venus, Earth and Mars are calculated using metal contents from models *V*2, *E*5, and *Ma*2 from the *Basaltic Volcanism Study Project* (1981*b*, p. 641). (A) Complete thermochemical equilibrium between CO and CH_4 and all other species is maintained. (B) Kinetic inhibition of the CO → CH_4 conversion (figure after Lewis et al. 1979 and Lewis and Prinn 1980).

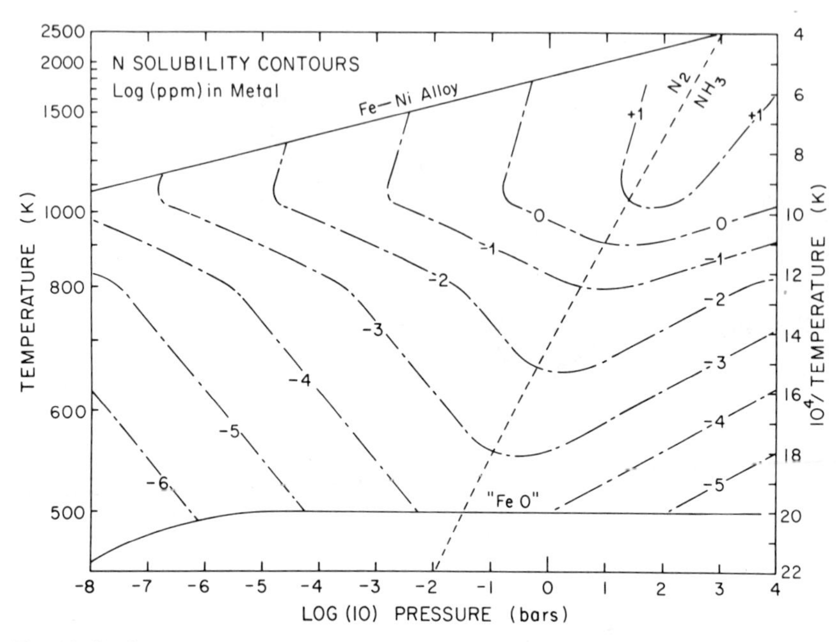

Fig. 14. Predicted amounts (parts per million by mass) of N dissolved in Fe - Ni alloy equilibrated with solar nebula gas (figure from Fegley 1983). Bulk nitrogen contents are calculated using the planetary composition models (for wt. % metal) given in the caption to Fig. 13.

V, where volatile inventories are tabulated on an absolute basis (g/g). Several trends are immediately apparent.

First, addressing carbon, we see that although the observed carbon inventories are approximately similar for Venus and the Earth, the bulk inventories predicted for maintenance of chemical equilibrium (between CO, CH_4, and Fe − Ni grains) decrease from Venus to Earth. The predicted inventories are also lower than the observed inventories for both Venus and the Earth. The observed carbon inventory on Mars (atmospheric CO_2), which is certainly a lower limit to the bulk inventory, is also larger than the predicted equilibrium value. Furthermore, since the degassing efficiency for carbon which is dissolved in Fe − Ni alloy is probably much less than 100%, the already significant discrepancies between the observed inventories and the (lower) predicted values will be increased.

However, as discussed previously in Sec. III.B, the $CO \rightarrow CH_4$ conversion was kinetically inhibited in the solar nebula (Fig. 2). As shown in Fig. 13, this leads to much larger amounts of elemental carbon dissolved in Fe − Ni alloy in the (*P*,*T*) region where CH_4(g) is thermodynamically stable (but kinetically inhibited from forming). In this case, the predicted carbon inventory for Venus remains the same while those for Earth and Mars are increased

TABLE V

Comparison (*g/g*) of Observed H_2O, C and N Volatile Inventories of Venus, Earth and Mars with Predictions of the Complete Equilibrium and Kinetic Inhibition Models[a]

Volatile	Venus		Earth		Mars	
	atmo.	bulk	atmo.	bulk	atmo.	bulk
CO_2 (obs.)	10^{-4}		4×10^{-10}	8×10^{-5}	4×10^{-8}	
(pred.)[b]		2×10^{-5}		$(0.07 - 3) \times 10^{-4}$		$(0.0001 - 8) \times 10^{-5}$
N_2 (obs.)	2×10^{-6}		6×10^{-7}	3×10^{-6}	7×10^{-10}	
(pred.)		4×10^{-9}		3×10^{-11}		10^{-13}
H_2O (obs.)	4×10^{-9}		5×10^{-9}	3×10^{-4}	5×10^{-12}	
(pred.)		0		10^{-5}		8×10^{-3}

[a]Atmospheric inventories based on Tables 2-4 of Prinn and Fegley (1987*a*). Terrestrial bulk inventory based on Table I. Note that this table is not normalized to Si and is simply on a (g/g) basis.

[b]Predictions based on Barshay (1981), and Figs. 13 and 14. The range of CO_2 values for Earth and Mars results from including (highest values) or excluding (lowest values) the effects of kinetic inhibition on dissolved carbon in grains.

over the equilibrium values. Indeed, depending on the extent to which CO(*g*) – Fe,Ni grain equilibrium is maintained with decreasing temperature and on the maintenance of the Fe – FeO equilibrium, Mars may have been endowed with as much (or more) carbon than the Earth.

Chemical analyses of the SNC (Shergottite-Nakhlite-Chassigny) meteorites, which are widely believed to come from Mars (reviewed by Laul 1986), may be instructive in constraining the bulk Martian carbon inventory. Although the carbon contents of the shergottite meteorites studied by Wright et al. (1986) were variable, the values found of $\sim 2 \times 10^{-5}$ to $\sim 6 \times 10^{-4}$g/g are much greater than the predicted equilibrium value for Mars but are compatible with those predicted from this CO/CH_4 disequilibrium scenario. The problem of degassing the carbon from the Fe – Ni alloy remains, however. Thus, while carbon dissolved in Fe – Ni alloy may have played some role in delivering carbon to the terrestrial planets, the probability of the Fischer-Tropsch-type reactions discussed in the previous section and the considerable evidence for ubiquitous organic material in the solar nebula discussed later in this section make organic material an equally probable source of carbon for Venus, Earth and Mars.

Second, addressing nitrogen, we see from Table V that the observed N_2 inventories for Venus, Earth and Mars are all significantly greater than the N_2 inventories predicted for N_2(g) – Fe,Ni grain equilibrium in the solar nebula (Fig. 14). Again, degassing efficiencies $< 100\%$ will increase the large discrepancies between the observed inventories and the predicted values. Kinetic inhibition of the $N_2 \rightarrow NH_3$ conversion in the solar nebula does not alter this situation because the N_2/NH_3 ratios are much greater than unity for typical nebular pressures down to ~ 400 K where Fe_3O_4 coatings will stop further nitrogen solution in the metal.

However, the nitrogen inventories of the terrestrial planets could have been supplied by other equilibrium sources. The highly reduced enstatite chondrites contain metal nitrides such as TiN (osbornite) and Si_2N_2O (sinoite); these minerals and other nitrides are also predicted to form in a solar gas having C/O ratios greater than the solar value of ~ 0.6 (Larimer and Bartholomay 1979). Accretion and degassing of only a trace amount of these materials (e.g., of all Zr as ZrN giving N $\sim 9 \times 10^{-7}$ g/g or of $\sim 1\%$ of all Ti as TiN giving N $\sim 2 \times 10^{-6}$ g/g in chondritic material) is sufficient to supply the observed nitrogen inventories on Venus and the Earth.

Alternatively, just as for carbon, accretion and degassing of a small amount of N-bearing disequilibrium organic material similar to that found in carbonaceous chondrites (e.g., see Hayatsu and Anders 1981) can also supply the nitrogen inventories on these planets. In this regard, we note that even the small amount of NH_3 predicted to be present in the solar nebula is more than sufficient to provide the amounts of nitrogen on the Earth and Venus. In fact, for $NH_3/N_2 \sim 2 \times 10^{-3}$ (which assumes Fe catalysis of the $N_2 \rightarrow NH_3$

conversion down to 480 K), accretion of only ~ 1% of this NH_3 as N-bearing organics will provide the required amounts of nitrogen in chondritic material. Hayatsu and Anders (1981) describe experiments showing NH_3 incorporation into organic material via Fischer-Tropsch-type reactions.

Third, examining H_2O in Table V, we see that increasing amounts of water are predicted to be retained in solid grains in the unlikely event that $H_2O(g)$-silicate grain equilibrium is maintained with decreasing temperature. Furthermore, the well known equilibrium fractionation of deuterium from HD to HDO (via the net reaction $HD + H_2O = H_2 + HDO$) with decreasing temperature (e.g., see Urey 1947) means that the D/H ratio of this bound water is also predicted to increase with decreasing temperature. This is illustrated in Fig. 11. The equilibrium model thus predicts that the Earth accreted significantly more hydrated phases (e.g., tremolite, hydroxyapatite, phlogopite, talc, serpentine) than did Venus because these phases only became thermodynamically stable in the cooler nebular region outside of 1 AU. Furthermore, Mars is predicted to have accreted even more of the hydrated minerals than the Earth did and is predicted to be even more water rich and more D rich. The (inferred) presence (by infrared reflection spectroscopy) of hydrated phases on the surface of the asteroid 1 Ceres (Lebofsky 1978), the (inferred) presence of abundant water on Mars (Carr 1987), and the larger than terrestrial D/H ratio on Mars (Owen et al. 1987), are in qualitative agreement with this trend but certainly are not unambiguous evidence in support of the equilibrium model. In particular, the higher than terrestrial D/H ratio on Mars may be due to D/H fractionation during hydrogen escape from Mars (Yung et al. 1987*b*).

The observation of D/H ~ 1.6×10^{-2} on Venus, measured by the Pioneer Venus mass spectrometer (Donahue et al. 1982), poses a potential problem for this equilibrium model of water retention. The observed D/H value, which is ~ 100 times higher than the terrestrial value of 1.557×10^{-4} SMOW (standard mean ocean water), is consistent with the depletion of an amount of water equivalent to ~ 0.3% of the terrestrial water inventory from Venus over geologic time (McElroy et al. 1982*a*; Donahue et al. 1982), with the water being primordial or added by ancient cometary impacts (Grinspoon 1987). This is a small amount of water which could have been supplied to Venus by the accretion of only trace amounts of hydrated phases (e.g., accretion of only 0.04% of all Ca if present as tremolite or accretion of only 0.002% of all Mg if present as serpentine). However, more water may have initially been present since Donahue et al. (1982) note that enhancement of the D/H ratio by hydrodynamic escape will only begin once the H_2O mixing ratio drops below ~ 2×10^{-2}; this is equivalent to the 0.3% of the terrestrial inventory mentioned above.

Supplying the very large water inventories on the Earth and Mars by equilibrium material is very problematical. As illustrated in Table III and Fig.

12, formation of the major water-bearing minerals serpentine, talc and brucite in the solar nebula was apparently kinetically inhibited inside the thermodynamic stability fields of these phases. Thus, unless hydration of Ca-bearing minerals to tremolite and subsequent accretion of the tremolite occurred with combined efficiencies $\sim$ 10%, the Earth for example must have received its water from some other source. This may have been in the fashion of an inhomogeneously accreted volatile-rich veneer (Anders 1968) or from ice-rich bodies gravitationally scattered into the inner solar system.

B. Chondritic Meteorites and Asteroids

The importance of disequilibrium processes for the production and preservation of organic matter in chondritic meteorites has recently been the focus of several studies. Hayatsu and Anders (1981) have summarized the evidence for and against the synthesis of different classes of organic compounds by Fischer-Tropsch-type reactions in the solar nebula. As discussed previously, the kinetic inhibition of the $CO \rightarrow CH_4$ conversion in the solar nebula, resulting in a supersaturated carbon-bearing gas, favors organic matter production by these reactions. Some investigators have even proposed that Fischer-Tropsch-type reactions may have synthesized organic compounds in the meteorite parent bodies. Although these latter proposals remain to be studied seriously, they are not inconsistent with our earlier observation that high total conversions of CO to organics and low H_2/CO ratios are more favorable for production of large molecules than are low total conversions of CO to organics and high H_2/CO ratios (see Sec. B (ii) and Fig. 8).

Isotopic studies of organic matter and coexisting phases (e.g., carbonates and hydrated silicates) in chondrites provide additional evidence for disequilibrium processing of the major volatiles C, N and H. The pattern of ^{13}C-light organic matter and ^{13}C-heavy carbonates, which is observed in the CI1 and CM2 chondrites (e.g., see Mason 1971), was reproduced in CO_2 and hydrocarbons generated from Fischer-Tropsch reactions of $CO + H_2$ done at 400 K (Lancet and Anders 1970). Unfortunately, this promising study has not been followed up in sufficient detail to determine the effects of processing variables (e.g., temperature, total pressure, H_2/CO ratio, type of catalyst, etc.) on the resultant $^{13}C/^{12}C$ ratios in the CO_2 and organic products.

On the other hand, the ^{15}N-rich soluble organic matter found by Becker and Epstein in CI1 and CM2 chondrites is much more fractionated than N-bearing organics produced via either Fischer-Tropsch-type or Miller-Urey reactions (summarized by Pillinger 1984). However, given the extensive evidence for $^{15}N/^{14}N$ heterogeneity between (and within) meteorites (Pillinger 1984), thus suggesting nebular heterogeneity, formation of the ^{15}N-rich soluble organics from a ^{15}N-rich reservoir with little (or no) fractionation during the synthesis cannot be ruled out.

However, the most dramatic isotopic variations in meteorites are displayed by hydrogen and deuterium. Figure 11 illustrates the well-known point that deuterium exchange between the major deuterium-bearing reservoir in the solar nebula (HD) and gaseous hydrides (e.g., H_2O in this case but the same is true for CH_4, NH_3, etc.) does not lead to large deuterium enrichments in the hydride until sufficiently low temperatures where the gas-phase exchange reaction will be kinetically inhibited. If we take CH_4 as a proxy for organic matter, we would not expect large D/H ratios in organics unless they had equilibrated with HD at similarly low temperatures (with larger D/H ratios requiring lower temperature equilibration).

Thus, the observation of D-rich organic matter having D/H ratios up to $\sim 1.05 \times 10^{-3}$ (as compared to $\sim 2 \times 10^{-5}$ for cosmic hydrogen and $\sim 16 \times 10^{-5}$ for terrestrial hydrogen) is very significant. As discussed at length by Pillinger (1984) and Yang and Epstein (1983), matter with such large D/H ratios is almost certainly of interstellar origin. This is a plausible conclusion because apparently only ion-molecule reactions, which occur in interstellar clouds, proceed with non-negligible rates at the low temperatures ($\leq$ 100 K) required to achieve such large D/H fractionations.

The survival of this D-rich material in meteorites poses important constraints on the severity and duration of thermal processing experienced by this material after its formation. The rate of isotopic re-equilibration back to lower D/H ratios (see Fig. 11) is clearly a temperature dependent process which is coupled in important ways to the rates of nebular transport and thermal evolution. Although no models of the survivability of D-rich material are currently available, the presence of high D phases in both ordinary and carbonaceous chondrites implies the kinetic inhibition of gas-solid reactions involved in the isotopic re-equilibration over a range of (P,T) conditions in the inner nebula. Indeed, such kinetic inhibition would not be surprising given the refractory nature of these D-rich phases (which are residues left after dissolution of the bulk of the meteorite in HCl-HF solutions) and the high C-H bond strength ($\sim$ 370 − 460 kJ mole^{-1} for a variety of organic compounds).

Our earlier discussion of the kinetic inhibition of H_2O (g)-silicate hydration reactions in the solar nebula implies that the hydrated silicates observed in CI1 and CM2 chondrites are not nebular products but are instead products of hydration reactions on the chondrite parent bodies. This interpretation is in accord with the extensive petrographic evidence summarized by Barber (1985) that implies an origin for the hydrated silicates by aqueous alteration on the chondrite parent bodies. More recent work by Tomeoka and Buseck (1985) also supports this model. Furthermore, aqueous activity on the CI1 and CM2 chondrite parent bodies is also required to explain the production of sulfate- and carbonate-bearing veins (see Barber 1985). In fact, as reviewed by Pillinger (1984), unidirectional oxidation in an aqueous environment of FeS (troilite) provides a plausible explanation for the observed sulfur isotopic compositions of FeS, elemental S, and the sulfate veins. Finally, Clayton and

Mayeda (1984) have shown that the oxygen isotopic compositions of CI1 and CM2 chondrites require $T < 293$ K and H_2O (liquid) volume fractions $> 44\%$ to produce the hydrated silicates in CM2 chondrites and aqueous alteration in a warmer, wetter environment to produce the hydrated silicates in CI1 chondrites.

Thus, a large number of chemical, isotopic and petrographic observations support the theoretical interpretation that the hydrated silicates in CI1 and CM2 chondrites are parent body and *not* nebular products. The kinetic inhibition of silicate hydration reactions in the solar nebula also implies that the hydrated silicates inferred to be present on the asteroid 1 Ceres are also aqueous alteration products. This interpretation is supported by the similarities in the infrared reflection spectra of the putative hydrated silicates on 1 Ceres and the hydrated phases in carbonaceous chondrites (Feierberg et al. 1981) and the inferences that asteroids are the source of the chondritic meteorites.

However, the interesting question of how water was originally retained on the asteroids and chondrite parent bodies must then be addressed. The simplest way to do this is by condensation of water ice. Figure 11 illustrates that water ice condenses at ~ 160 K along the solar nebula (P,T) profile used. This corresponds to $R \sim 3.4$ AU (near the outer edge of the present-day asteroid main belt) for the same model (Barshay 1981), which assumes $T \propto R^{-1}$ and takes $T \sim 550$ K at 1 AU. Higher nebular pressures lead to higher water ice condensation temperatures with $T \sim 200$ K at a total pressure of $\sim 10^{-3}$ bar (see Fig. 11). Assuming that HD and H_2O continued to equilibrate down to these temperatures, the implied δD (defined in the caption to Fig. 11) for the water ice (neglecting the smaller fractionation between water vapor and ice) is ~ -70 per mil for condensation at ~ 200 K up to $\sim +630$ per mil for condensation at ~ 160 K. The D/H ratio in this water ice should be reflected in the hydrated silicates in the CI1 and CM2 chondrites.

The inferred δD values for hydrated silicates in carbonaceous chondrites are ~ -100 per mil (Kolodny et al. 1980; Yang and Epstein 1983). This suggests that the precursor water ice either condensed at ~ 200 K or alternatively that HD $-$ H_2O equilibration was quenched at ~ 200 K. Thus, the available isotopic data are consistent with derivation of the hydrated silicates from water originally retained as water ice but do not rule out another (more complex) origin. We note that models of this type were first proposed by Dufresne and Anders (1962).

C. Giant Planets

Hypotheses for the origin of the giant planets generally involve either direct *collapse* of nebula gas and grains (e.g., see Cameron 1978) or *capture* of nebula gas and grains by a preformed solid core (see, e.g., Mizuno 1980). Because the directly collapsed planet can subsequently capture solid material, both scenarios predict planets which potentially are enriched by various amounts in those elements accreted in the solid material (ices, rock). The two models are therefore often considered equivalent for explaining bulk planetary composition but *not* for explaining the distribution of elements within a planet (clearly the capture model can potentially sequester more of the heavy elements in the core if interior mixing is weak).

Jupiter and Saturn were most likely formed where nebular temperatures exceeded 60 K but were less than the water ice condensation temperature. Thus examining the "CO plus hydrocarbons" case in Table IV, the *solid* material which comprised their precursor cores and/or accreted onto the gaseous planets (via an accretion disk or otherwise) contained the full solar complement of rock-forming elements except oxygen. Oxygen would be present as silicates, hydrated silicates and water ice but depleted relative to solar abundance by the factor d_O = [total O]/([total O] − [CO]) in the surrounding solar nebula (because CO remains in the gas phase). Clearly $d_O \simeq 1$ (i.e., no significant depletion) if catalytic conversion of CO to hydrocarbons plus H_2O was facile in the nebula while $d_O \simeq$ [total O]/([total O] − [total C]) = 2.5 if catalytic conversion was inhibited. Carbon would be depleted by the factor d_C = [total C]/([total C] − [CO]) which varies from $d_C \simeq 1$ for facile CO → hydrocarbon conversion up to $d_C \simeq$ [total C]/([NH_4HCO_3] + [NH_4COONH_2]) ≈ 100 for inhibited conversion. The observations force us to reject d_C values >> 1. In particular the Jovian and Saturnian ice-plus-rock mass fractions (Sec. I.C) combined with their 2 to 5 times enhancements in CH_4/H_2 lead us to conclude that C must have comprised 9 to 39% (Jupiter) and 3 to 12% (Saturn) of the mass of the ice-plus-rock fraction. Such high C contents require that at least partial CO → condensed hydrocarbon conversion occurred. Finally nitrogen in the solids would be depleted relative to solar abundance by the factor $d_N \simeq$ [total N]/([total N] − 2[N_2]) $\simeq$ [total N]/([NH_4HCO_3] + 2[NH_4COONH_2]) ≈ 100 (note that facile N_2 → involatile organic conversion is unlikely).

Uranus and Neptune most likely formed at temperatures below 60 K. In contrast to Jupiter and Saturn, the bulk of the mass of these planets was accreted as solid material rather than (solar-composition) gas. Fegley and Prinn (1986) have discussed a wide variety of generic models for these planets. We will focus again on the more plausible CO-plus-hydrocarbons case (see Table IV). The gas/solid ratio in material which formed Uranus as deduced from its bulk density and interior models varies from 11/89 to 25/75 by mass (Hubbard 1984*a*; Stevenson 1982*b*). To attain such ratios, enrichment factors *e* (relative to solar composition) for the ice- and rock-forming elements of 100 to 1000 are demanded. If the nebula temperature is below 20 K, then pure CO and N_2 solids will be present in the accreted condensed material and the ice/rock ratio in this material will be 2.3 (where "ice" here refers to all solids other than anhydrous rock; see Table IV). If the nebula temperature is instead about 50 K, then only a part of the nebular CO and N_2 is condensed (as clathrates) and the ice/rock ratio in accreted condensed material will be 0.6 to 2.3 with the precise value depending on the degree of CO $\rightarrow$ hydrocarbon conversion.

An important implication of the kinetic inhibition nebular models is that if Uranus and Neptune accreted large amounts of disequilibrium C and N (e.g., carbonaceous chondrite-like organic matter or CO, CO_2, N_2 in solids and clathrates), then planetary H_2 would be depleted by reactions such as

$$C_xH_yN_z + \frac{1}{2}(3z + 4x - y)H_2 \rightarrow xCH_4 + zNH_3 \quad (42)$$

$$CO + 3H_2 \rightarrow CH_4 + H_2O \quad (43)$$

$$CO_2 + 4H_2 \rightarrow CH_4 + 2H_2O \quad (44)$$

$$N_2 + 3H_2 \rightarrow 2NH_3 \quad (45)$$

and thus the H_2/He ratio would be decreased relative to its solar value of 7.4. The maximum H_2/He ratio decrease is by a factor $\approx 1 - 0.0028(e + 1) \geq 0.7$ which applies when nebular C and N are largely in CO and N_2 and nebular temperatures are $\leq$ 20 K. In contrast if a significant fraction of nebular C and N is in saturated organics and/or the nebular temperature $\approx$ 50 K then the H_2/He ratio is much less affected.

The Uranian J_2 and ϵ values favor ice/rock ratios > 1 (Hubbard 1984*a*, and references therein) and the Uranian H_2/He ratio is not different from the solar value within the measurement uncertainties (Conrath et al. 1987; Hanel et al. 1986). This argues again that conversion of some of the CO to hydrocarbons must have proceeded in the solar nebula as suggested (but not quantified) by Lewis and Prinn (1980); more specifically, the organic carbon was at least comparable to CO in abundance. It will be interesting to see if the Neptunian H_2/He ratio leads to a similar conclusion.

D. Satellites of Giant Planets and Pluto

The condensation sequence appropriate to satellite condensation in an equilibrium solar nebula was first elucidated by Lewis (1972*a*) who demonstrated the expected ubiquity of pure water ice (H_2O) as the major condensate in the outer solar system. In agreement with the suggestions by Urey (1952) and Miller (1961), he also pointed out the potential importance of retention of nitrogen and carbon in the form of *methane clathrate* ($CH_4 \cdot 6H_2O$) and *ammonia hydrate* ($NH_3 \cdot H_2O$). The later work of Lewis and Prinn (1980) showed, however, that the equilibrium assumption in the solar nebula was incorrect and that N_2 (*not* NH_3) and CO (*not* CH_4) were the dominant gaseous nitrogen and carbon compounds. This meant that the ice/silicate ratio was decreased from that in the equilibrium model (because CO not H_2O was the major oxygen reservoir) and also that $CH_4 \cdot 6H_2O$ and $NH_3 \cdot H_2O$ were *not* expected condensates in the solar nebula; how therefore could Titan's CH_4 be explained?

To help answer the latter query, Prinn and Fegley (1981*a*) proposed that the conditions in well-mixed subnebulae around the giant protoplanets (Pollack et al. 1976) provided precisely the chemical thermodynamic and chemical kinetic conditions necessary to make CH_4 and NH_3 (not CO and N_2) the dominant N and C gases in these subnebulae. They stated in the abstract of their paper that "satellites of the Jovian planets which accreted in sufficiently cool parts of their circumplanetary nebula are therefore predicted to retain large amounts of NH_3 and CH_4 in the form of clathrate/hydrates." More explicitly for Titan, they state "Titan is expected to have retained significant amounts of NH_3 (as $NH_3 \cdot H_2O$) and CH_4 (as $CH_4 \cdot 6H_2O$) and smaller amounts of HCN. If solar nebula temperatures at Saturn were equal to or even lower than 60 K, then retention of first N_2 (as $N_2 \cdot 6H_2O$) and then CO (as $CO \cdot 6H_2O$) is also possible." However an important cautionary statement was also made: namely mass-balance considerations indicated that complete condensation of CH_4 and NH_3 as $CH_4 \cdot 6H_2O$ and $NH_3 \cdot H_2O$ required more than *all* of the available water leaving *none* for forming the N_2 and CO clathrates. In addition, Lunine and Stevenson (1985*a*) give theoretical arguments and Davidson et al. (1987) give experimental results indicating CO clathrates are more stable than N_2 clathrates. Hence a model for the origin of Titan's atmosphere involving degassing of clathrates and hydrates would potentially yield copious amounts of NH_3 and CH_4 but clathration and subsequent degassing of CO and N_2 was problematical particularly for N_2.

Owen (1982*b*) later addressed the alternative possibility that Titan obtained its atmosphere by direct gravitational capture of N_2, CH_4 and other gases onto the primordial Titan. However, he ruled out this possibility due to the absence of the substantial amount of Ne (solar abundances of Ne and N are comparable) expected in a solar composition atmosphere after hydrogen loss. He reiterated the probable importance of methane clathrate suggested by Miller (1961), Lewis (1972*a*) and Prinn and Fegley (1981*a*) and also proposed specifically that the N_2 on Titan was derived from degassing of $N_2 \cdot 6H_2O$. From a cosmogonic viewpoint this is very difficult to justify. In particular, the high ice/rock ratio inferred for Titan from its density requires that much of the oxygen in the gaseous medium in which it formed be present as H_2O; this means that CO cannot be the major form of oxygen and hence $CO < CH_4$ as it is in the Prinn and Fegley (1981*a*) subnebula. However, the same chemical-dynamical conditions that yield $CO < CH_4$ also yield $N_2 < NH_3$ (see Figs. 2 and 3). As already noted, the formation of $N_2 \cdot 6H_2O$ in NH_3- and CH_4-rich subnebulae is improbable indeed and would need to be preceded by copious $CH_4 \cdot 6H_2O$, $NH_3 \cdot H_2O$ and $CO \cdot 6H_2O$ formation. These cosmogonic arguments point therefore to formation of Titan in a CH_4- and NH_3-rich subnebula by a process of accretion of rock and $NH_3 \cdot H_2O$ hydrate and $CH_4 \cdot 6H_2O$ clathrate as proposed by Prinn and Fegley (1981*a*) with the N_2 atmosphere being a photochemical and/or shock-chemical byproduct of outgassed NH_3 (see chapter by Lunine et al. for a discussion of relevant photochemical and shock-chemical $NH_3 \rightarrow N_2$ scenarios).

The identification of $CH_4(s)$ and tentatively $N_2(l)$—and hence $N_2(g)$—on Triton would by analogy with Titan, suggest condensation in a CH_4-rich subnebula with the surface $N_2(l)$ again being a photochemical byproduct of $NH_3 \cdot H_2O$ solid and/or NH_3 gas photolysis. Without a firm number for the density of this satellite it is, however, premature to form any definitive conclusions. The same is true for Pluto and Charon as for Triton.

The most extensive data we have on ice-rich bodies is on their densities and geometric albedos. The low geometric albedos of many of the smaller satellites, together with solar light reflection spectra of some of these low-albedo satellites, suggests that these *dark* satellites may be rich in organic material (Cruickshank 1986). From a cosmogonic viewpoint this organic material could either be: (1) unprocessed interstellar material, or (2) Fischer-Tropsch-type organics formed from CO in the solar nebula, or (3) photochemical decomposition products of $CH_4 \cdot 6H_2O$ resulting from solar ultraviolet irradiation of the satellite surfaces after dissipation of nebula gases and sweepup of solar system dust. Thus, the presence of organics alone is not very definitive in answering questions about origin.

Interior models of ice-rich satellites can be constructed using feasible chemical compositions and fits to the observed satellite bulk densities. In principal, the ratio of silicates to volatile condensates deduced from such models can be used to infer certain chemical properties of the medium in which these satellites formed. In particular, the ratio of H_2O to silicates in ice-rich satellites depends sensitively on the [CO]/[total C] ratio in the medium (nebula, subnebula) in which the satellite condensed. If [CO]/[total C] $\simeq$ 1, then CO is the major oxygen reservoir followed by H_2O and then oxygen in anhydrous rocks (modeled here as SiO_2 + MgO + FeO to a first approximation); specifically [CO] : [H_2O] : [2Si + Mg + Fe − S] is 3.2 : 1.1 : 1.0 in solar composition. Alternatively, if [CO]/[total C] $\simeq$ 0 (i.e., CH_4 and/or organics are the major forms of carbon), then H_2O is the major oxygen reservoir and [CO] : [H_2O] : [2Si + Mg + Fe − S] is then 0.0: 4.3: 1.0 in solar composition. Thus there is 4.3/1.1 = 3.9 times more water available for forming pure ice in a CH_4-rich solar-composition medium than in a CO-rich medium.

For application to ice-rich satellites, it is convenient to translate the above number ratios into *ice* (H_2O solid; NH_3 hydrate or solid; CH_4, CO and N_2 clathrates or solids) to anhydrous *rock* (SiO_2, MgO, FeO, FeS) ratios. For satellites condensing in a CH_4- and NH_3-rich protoplanetary subnebula, we have (Prinn and Fegley 1981*a*; Fegley and Prinn 1986):

$$\begin{aligned} \text{ice:rock} &= 0.61 : 0.39 \ (160 \text{ K} \leq T \leq 230 \text{ K}) \\ &= 0.64 : 0.36 \ (95 \text{ K} \leq T \leq 160 \text{ K}) \\ &= 0.66 : 0.34 \ (40 \text{ K} \leq T \leq 95 \text{ K}) \\ &= 0.74 : 0.26 \ (T \leq 40 \text{ K}) \end{aligned} \quad (46)$$

where the successive increases in ice mass fraction are due to successive formation of (i) H_2O (*s*), (ii) $NH_3 \cdot H_2O$ (*s*), (iii) $CH_4 \cdot 6H_2O$ (*s*) and (iv) CH_4 (*s*), respectively.

In constrast, for satellites condensing in a CO- and N_2-rich solar nebula, we have (Lewis and Prinn 1980; Fegley and Prinn 1986):

$$\begin{aligned} \text{ice:rock} &= 0.28 : 0.72 \ (60 \text{ K} \leq T \leq 140 \text{ K}) \\ &= 0.35 : 0.65 \ (20 \text{ K} \leq T \leq 60 \text{ K}) \\ &= 0.70 : 0.30 \ (T \leq 20 \text{ K}) \end{aligned} \quad (47)$$

where the successive increases in ice mass fraction are due to successive formation of: (i) $H_2O(s)$, (ii) $CO \cdot 6H_2O$ and $N_2 \cdot 6H_2O$ and finally (iii) CO and N_2 solids.

Schubert et al. (1986) and Johnson et al. (1987) have discussed a range of feasible interior models for the ice-rich satellites of the outer planets. Johnson et al. (1987) computed specifically the following *system-average* ice : rock ratios (where "ice" is H_2O only):

$$\begin{aligned} \text{ice:rock (Ganymede, Callisto)} &= 0.44 : 0.56 \text{ (differentiated)} \\ \text{ice:rock (Saturn satellites)} &= 0.44 : 0.56 \text{ (mass average, i.e., Titan essentially)} \\ &= 0.61 : 0.39 \text{ (object average)} \\ \text{ice:rock (Uranus satellites)} &= 0.41 : 0.59 \text{ (differentiated)} \\ &= 0.50 : 0.50 \text{ (homogeneous)} \end{aligned} \quad (48)$$

For the Saturn system, Titan only was considered differentiated while for the Uranus system, Titania and Oberon only were considered as possibly differentiated. It is evident that the icy Galilean satellites and Uranus satellites have ice-to-rock ratios which lie between the values expected in CO-rich and CH_4-rich nebulae/subnebulae. The same is true for Titan but *not* for the other Saturnian satellites whose ice-to-rock ratio is essentially identical to that expected in a CH_4-rich subnebula.

If we accept the hypotheses that: (a) the regular satellites of the outer planets formed in subnebulae, and (b) the presence of CH_4 on Titan and Triton means that CH_4 was present in appreciable amounts in these subnebulae, then ice-to-rock ratios which lie between the CO-rich and CH_4-rich nebula/subnebula ratios can be produced in at least three feasible ways:

1. Conversion of CO to CH_4 in the subnebula was incomplete (e.g., as may be possible in a poorly mixed subnebula);
2. Conversion of CO to CH_4 in the subnebula was complete but the subnebula (or the planetary disk system after the subnebula dissipated) subsequently accreted additional icy objects which formed in the CO-rich nebula;
3. Conversion of CO to CH_4 in the subnebula was complete but the larger ice-rich satellites lost some ice by blow off induced by post-accretional impacts as discussed in the chapter by Lunine et al.

Since all the outer planetary systems need to be augmented with elements heavier than H and He (by a process of accretion of solar nebula ice/rock solids without accretion of solar nebula gases), then methods (2) and (3) appear to us to be preferable at least in the absence of a quantitative demonstration of method (1). Note that based on our earlier discussion of Uranus, the ice-to-rock ratio in solar nebula objects will be increased (to lie between the CO-rich and CH_4-rich nebula values) to the extent that some of the solar nebula CO is converted to organics. Indeed, if it were not for the need to explain the regularity of the satellites and CH_4-rich satellites like Titan, the formation of the ice-rich satellites in a CO-plus organic-rich solar nebula (as opposed to a CH_4-rich subnebula) would appear quite feasible for explaining their observed ice-to-rock ratios (Johnson et al. 1987).

E. Comets

Prior to the Halley missions, the major parent volatile molecules inferred from both remote observations of cometary comae and analyses of stratospheric dust putatively of cometary origin were: (a) H_2O, (b) CO plus CO_2 with abundances $\approx$ 10 to 30% of H_2O, (c) HCN, NH_3, CH_3CN, etc. with abundances $\leq$ 1% of H_2O and (d) poorly characterized organic molecules $C_mH_nO_xN_y$ (which could serve as sources of coma C_2, C_3, CH, CN, etc.) with abundances perhaps a few % of H_2O (Delsemme 1982). Vega 1 analyses of the dust from Comet Halley indicated that the organic component of the dust consisted mainly of highly unsaturated compounds (Kissel and Krueger 1987).

Unfortunately this information on cometary volatiles by itself does not enable us to readily discriminate between two major possibilities: (a) interstellar material (dust, condensible gases) and (b) solar nebula condensates, as the parent material for comets. This is because the H_2O, CO clathrate, CO_2, NH_4COONH_2 (this contains a C $-$ N bond), NH_4HCO_3, HCN, and Fischer-Tropsch organics predicted in nebular condensates (Sec. III.B; also Lewis and Prinn 1980) bear too many similarities to the composition of (condensed or readily condensible) interstellar material inferred from astronomical observations, (CO, H_2O ice, HCN, carbonaceous compounds, etc; see the chapter by Irvine and Knacke). Nitrogen clathrate ($N_2 \cdot 6H_2O$) is predicted as a major low-temperature solar nebula condensate (Lewis and Prinn 1980) but the absence of N_2 in comets is plausibly explained by a greater stability of $CO \cdot 6H_2O$ over $N_2 \cdot 6H_2O$. Predicted kinetic difficulties in forming clathrates at low pressures (Lunine and Stevenson 1985*a*) are common to both the solar nebula and interstellar hypotheses for origin of cometary CO.

Early results from the Giotto Comet Halley mission provide what may prove to be definitive information on the (chemical) origin of Halley. The Giotto ion mass spectrometer data suggest production rates of CO and N_2 to be about 0.05 to 0.2 and < 0.01 to 0.1, respectively of H_2O production (Eberhardt et al. 1986; Balsiger et al. 1986); a result similar to that seen for previous (remotely-observed) comets. However, production rates of NH_3 and CH_4 are inferred from the same Giotto experiment tentatively to be as large as 0.01 to 0.02 and 0.02, respectively, of the H_2O production rates (Allen et al. 1987) implying $CH_4/CO \approx 0.1$ to 0.4 and $NH_3/N_2 > 0.1$ to 2. Both of these latter ratios are much greater than those seen in the interstellar medium and, if correct, argue strongly against accretion of Halley from pristine interstellar material (Allen et al. 1987).

Values of $NH_3/N_2 >> 1$ are however predicted in solar nebula solids (due to low-temperature condensation of NH_4HCO_3 and NH_4COONH_2) provided temperatures are not so low that N_2 clathrate or N_2 condense (see Table IV). A CO/H_2O ratio ≈ 0.15 is also predicted in solar nebula solids provided temperatures are low enough (and the pressure high enough) so that CO clathrate forms. Complete condensation of the solar abundance of C as CO clathrate would also use up all of the readily accessible cavities in the ice structure and thus inhibit N_2 clathrate formation. However, a CH_4/CO ratio as large as 0.1 in the solar nebula would require quenching of the $CO \rightarrow CH_4$ reaction in the solar nebula at temperatures as low as 700 K compared to a predicted quench temperature $\sim$ 1470 K and predicted CH_4/CO ratio $< 10^{-7}$ (see Fig. 2). Also the Halley NH_3/H_2O ratio of 0.01 to 0.02 significantly exceeds the *maximum* predicted NH_3/H_2O ratio in the solar nebula of 0.002 (the latter maximum ratio corresponds to the minimum quench temperature for $N_2 \rightarrow NH_3$ being 480 K due to total oxidation of Fe catalyst at lower temperatures). Therefore Halley does not appear to be a pristine solar nebula condensate either.

We have emphasized earlier that the thermodynamic and kinetic conditions necessary to produce CH_4 and NH_3 from CO and N_2 would occur in well-mixed subnebulae of the giant planets but not in the solar nebula (and not of course in the even cooler and less dense interstellar clouds). Table IV shows, however, that the CH_4/CO and NH_3/N_2 ratios in pristine subnebulae condensates are much greater than seen in Halley. An obvious working hypothesis to explain Halley is in fact a *heterogeneous mixture* of a small amount (e.g., 10%) of CH_4-rich subnebula condensate with a large amount (e.g. 90%) of CO-rich condensate of solar nebula or interstellar cloud origin.

We have two suggestions for producing such mixtures which are in fact analogous to our suggestions (1) and (2) for explaining the ice/rock ratios in ice-rich satellites. Let us first suppose that the giant planets possessed well-mixed subnebulae (accretion disk or spin off or impact generated). After the nebula and subnebulae gases are significantly dissipated, we are left in the outer solar system with CH_4-rich (subnebula-derived) objects orbiting around the giant planets and CO-rich (nebula-derived) objects orbiting around the Sun. Clearly some CO-rich solar system objects are expected to sweep through the giant planetary ring systems and accrete CH_4-rich material and *vice versa*. Alternatively, we can suppose that one or more of the giant planets possessed partially mixed subnebulae in which condensation of objects with small amounts of CH_4 (mixed outward from the inner disk) and large amounts of CO (mixed inward from the solar nebula) were a natural occurrence (since we must also explain CH_4-rich objects like Titan or Triton and since an impact-generated disk is likely for Uranus, it is unclear what the candidate giant planet is in this scenario). The dynamics of both suggestions need to be investigated to establish their viability and relative probability.

Nevertheless, the observation of such large amounts of CH_4 and NH_3 in Halley is persuasive evidence that at least some of the material in *this* comet (but not in CH_4-poor comets) is neither of solar nebula nor interstellar origin but is material condensed in a subnebula of one of the giant planets.

V. CONCLUSIONS

In this Chapter we have attempted to synthesize for the first time the wide variety of observational data and theories pertaining to the origin of the most volatile elements (C, N, H, O) in the solar system. It is clear that this subject is in an exploratory stage and seemingly *firm* conclusions may subsequently prove to be illusions. With this caveat in mind we offer the following as *firm conclusions:*

1. The starting materials were gases and grains in interstellar cloud(s).
2. As the nebula evolved, these gases and grains were at least partially reprocessed chemically; in particular the nebula was extensively physically mixed and chemically reprocessed (but not necessarily chemically and isotopically homogenized) at least out to 4 AU.
3. The major type of chemistry in the nebula was thermochemistry driven by the thermal energy of the nebula itself. Chemistry in lightning discharges and photochemistry driven by stellar photons were secondary but with their relative importance increasing with increasing radial distance. Solar photons played very little role if any because they were absorbed in the

very hot thermochemically controlled innermost nebula. Radiochemistry driven by ^{26}Al was minimal.

4. The gas-rich solar nebula and the dust-rich early post-solar nebula disk were highly opaque to ultraviolet radiation limiting severely the influence of the enhanced ultraviolet flux from the young Sun in the early solar system.
5. The reduction of N_2 and CO to NH_3 and CH_4 was kinetically inhibited in the solar nebula but not in the subnebulae of giant planets. Thus the nebula was (like interstellar material) rich in N_2 and CO (and organics formed from it) while subnebulae were rich in CH_4 and NH_3.
6. Vapor-phase hydration of silicates, like the chemical reduction of N_2 and CO, was very probably kinetically inhibited in the solar nebula.
7. The regular satellites of the outer planets originated in well-mixed subnebulae of these giant planets created by the accretion process, by spin off or by a massive impact.
8. Certain volatile ratios (e.g., CO/CH_4, N_2/NH_3, H_2O ice/silicate) in ice-rich bodies are diagnostic of their origin. For example, the CH_4 on Titan and Triton argues for their formation in a CH_4- and NH_3-rich giant-planetary subnebula while the observation of CH_4 and NH_3 (as well as the more-abundant CO and N_2) in Halley implies that at least some (see Conclusion 9 following) of the material in this comet is neither of solar nebula nor interstellar origin but was condensed also in a giant-planetary subnebula.
9. After the dissipation of the nebula, collisions between ice-rich solar system objects formed in the nebula and giant planetary satellite objects formed in subnebulae are inevitable and could produce hybrid objects characterized by a heterogeneous mixture of CO-, N_2-, CH_4-, and NH_3-containing ices.

How can our understanding of nebular chemistry and the origin of solar system volatiles be improved? Although we make no attempt at a comprehensive answer to this question, we think the following points worth emphasizing:

a. Current models of the solar nebula and protoplanetary subnebulae have recognizable important shortcomings. For example:
   (i) models of chemical processes are hampered severely by a lack of quantitative laboratory studies of the kinetics and mechanisms of a wide range of important gas/gas and gas/solid reactions (e.g., Fischer-Tropsch-type reactions) under conditions relevant to the nebula and subnebulae;
   (ii) current dynamical models ignore the quadratically nonlinear nature of momentum transport thus precluding expected counter-gradient momentum fluxes and leading to the erroneous conclusion that trace constituents in the nebula and subnebula must have been poorly mixed.

b. Current observations of many volatile-rich bodies in the solar system are insufficient to constrain meaningfully our knowledge of the chemistry of the solar nebula. We emphasize in particular the need to sample and analyze unperturbed material from comets, asteroids and ice-rich satellites and not to rely just on inferences from remote sensing and analyses of the (perturbed) material in cometary emanations.
c. Through improved observations, our knowledge of other protostellar systems needs to be expanded substantially so that it can provide definitive insights and constraints on the origin of the solar system.

If the need for these and other improvements is not already self-evident, it is worth closing by noting that, based on our current knowledge of solar nebula chemistry, we still do not know definitively in what form carbon, nitrogen and water were incorporated into the inner planets or how much interstellar material survives in the solar system today and where it is. We hope that one day such questions will have definitive answers.

Acknowledgments. We thank M. Allen and T. Johnson for discussions and manuscripts provided prior to publication, B. Vanderlaan for assisting with coding while working in the MIT UROP Program, G. Rodriguez for T_EXnique, and D. Souza for drafting. We also thank D. Stevenson and Y. Yung for their reviews of the original manuscript. Some of our stronger and clearer statements were made in response to their comments. This research was supported by grants from the National Science Foundation (Atmospheric Chemistry Program) and the National Aeronautics and Space Administration (Planetary Geochemistry and Materials Program, Planetary Astronomy and Atmospheres Program, and Innovative Research Program).

PLANETARY ACCRETION

R. GREENBERG
University of Arizona

Planets formed from three components of the solar nebula: rocky solids, ice and the nebular gas. Each component eventually contributed its own distinct complement of volatiles to planetary atmospheres. The terrestrial planets grew by accretion of rocky planetesimals; their atmospheres formed by subsequent outgassing. With additional icy solids, the Jupiter and Saturn embryos grew large enough to nucleate hydrodynamic collapse of nebular gas around them. In the less dense outer part of the nebula, Uranus and Neptune grew too late to capture nebular gas; the icy component dominates their atmospheric composition. Early analysis of this planetesimal scenario gave growth times for the cores of Jupiter and Saturn longer than the lifetime of the nebula, and for the outer planets longer than the age of the solar system. More recent studies show that much of the mass of the system of planetesimals remained in very small planetesimals, during early incipient runaway growth of substantial planetary embryos. Depending on the efficacy of radial transport mechanisms, subsequent evolution may have been: (a) by fairly quiescent accretion of small planetesimals; or (b) independent formation of many large embryos that went on to accrete one another in a violent phase controlled by random events. Case (a) would yield systematic compositional differences among planets; (b) would blur distinctions. In either case, growth times are uncertain.

I. INTRODUCTION

In recent years a consensus has emerged on a hypothesized sequence of events that is believed to have led to the formation of the planets. This scenario, the "planetesimal hypothesis," involves the gradual coalescence of many tiny bodies into planetary embryos, which eventually were able to sweep up much of the remaining material in the solar nebula and thus grow into planets. That a consensus has been achieved is remarkable because there remain large gaps in the story where the process of growth is simply not characterized or understood and because several physical processes with potentially strong effects have been identified but not yet incorporated into the model.

Despite these shortcomings, the consensus has emerged because other competing scenarios, most notably the "giant-gaseous-protoplanet" model, have now been shown to be inconsistent with the present, observable state of the solar system. In contrast, the favored hypothesis is still too amorphous to test. That very uncertainty may provide the leeway for adjusting the model until it fits observational constraints. In reviewing the state of our understanding of planetary accretion, it is as important to emphasize what we do not know as what we do know.

For constructing a first-order physical model of the formation of planets, three basic classes of material are available in the solar nebula: rocky solid condensates (including iron); icy solids (water, ammonia, methane and clathrates) which condensed in the cooler region beyond the present asteroid belt; and the nebular gas (predominantly hydrogen and helium). Later, when a more complete accretion model is available it may be possible to test and refine it with more detailed cosmochemical considerations, but this level of detail is appropriate at present.

The objective is to find a way to assemble these materials into three classes of planets. The terrestrial planets are made of rock. Their atmospheres formed by subsequent heating, outgassing and reprocessing of volatiles. It is quite possible that a late sprinkling of ice from the outer solar system (in the form of comets) also contributed to the terrestrial atmospheres. The Jovian planets are each made of ice and rock surrounded by an envelope of presumably nebular hydrogen and helium. And the outer planets consist of ice and rock with atmospheres dominated by volatiles from the icy component.

Until recently, a scheme for assembling the planets from giant gaseous protoplanets seemed to be a viable alternative to the planetesimal hypothesis. This scenario was largely developed by Cameron (1978), and was based conceptually on the earlier ideas of Kuiper (1951). In this model, as the solar nebula collapses gravitationally to form the Sun, subsidiary gravitational instabilities allow collapse to form giant protoplanets (Fig. 1). Instabilities on this scale require the nebula to contain an extra solar mass in addition to the mass that forms the Sun. Within each protoplanet, cooling permits condensation of grains of solid material, which rain inward toward the center of the body. Later a tremendous solar wind removes the gaseous envelopes from the inner planets, leaving only the rocky cores that we see as the terrestrial planets today. Farther from the Sun, removal of the gas leaves only partially coalesced material that we see as the asteroid belt. In the outer solar system, the gaseous envelopes are incompletely stripped, leaving the giant planets of today.

The giant-gaseous-protoplanet hypothesis has been generally abandoned

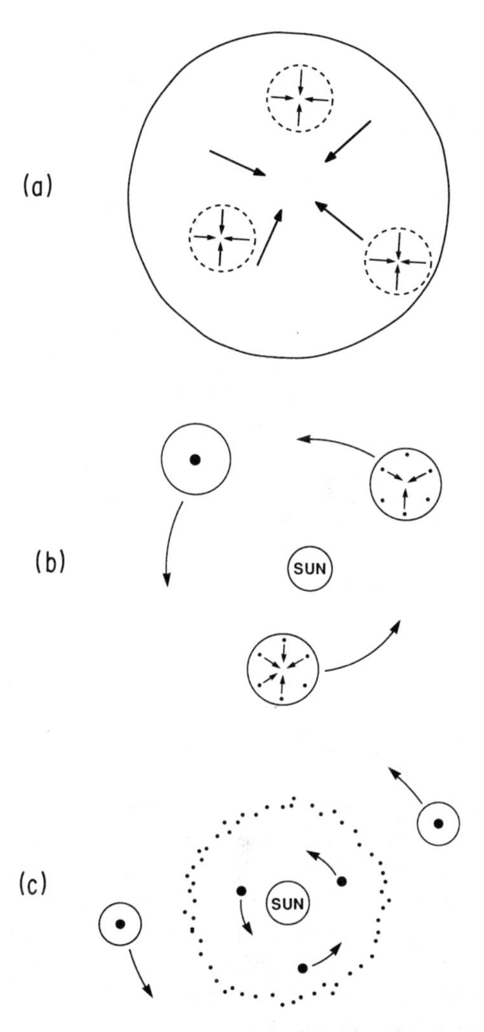

Fig. 1. The scenario of giant gaseous protoplanets. (a) Gravitational instability in a very massive nebula (twice the present mass of the Sun) yields giant protoplanets as well as the Sun. (b) Solids condense within the protoplanets and settle into cores. (c) Gaseous envelopes are blown away by a solar wind, leaving cores (the terrestrial planets) in the inner solar system, asteroids where core formation was incomplete, and giant planets farther out where gas removal was less effective. This scenario is now generally out of favor.

in recent years as close scrutiny showed some seemingly insurmountable shortcomings (Pollack 1985). In fact, most of the key steps in this scenario are suspect. For example, condensed material in such a protoplanet would probably not sink to form a core (Stevenson 1985), and it seems implausible that a solar wind could have blown away the great nebular mass as required.

The favored scenario, the planetesimal hypothesis as defined by Safronov (1969) and based on the earlier work of Schmidt (e.g. 1957), takes a very different perspective (Fig. 2). In this case, the nebula is assumed to be much less massive, only of few percent excess beyond the mass of the Sun $M_\odot$ itself. The density is insufficient for gravitational instability. As this nebula cools, solid rocky grains condense. Beyond the present position of the asteroid belt, temperatures decrease enough (<170 K) for ice grains to condense as well. While pressure support and turbulence keep the gas vertically extended, the solids settle to the equatorial midplane of the nebular disk.

Once the solids are concentrated at the midplane, gravitational instability clumps the material. In general, one can demonstrate from either a wave dynamics or energetics point of view that tidal effects stabilize long-wavelength disturbances (the differential force of the Sun prevents large-scale gravitational collapse), while random motion among the planetesimals stabilizes short wavelengths. Once the surface density σ at the midplane becomes great enough, there is a scale for gravitational instability between these extremes that clumps the material into bodies of mass

$$\text{mass} \simeq \sigma^3 \, (4\pi R^3/3 \, M_\odot)^2. \tag{1}$$

At distance from the Sun $R = 1$ AU, with surface density $\sigma \sim 10$ g cm^{-2} (a value obtained by smearing the Earth's mass over the terrestrial zone), gravitational collapse would yield bodies with radii of a few km. A process of collisional accretion is responsible for the subsequent agglomeration of these *planetesimals* into full-size planets.

The same calculation gives much larger bodies in the outer solar system, where the tidal effect of the Sun is weaker. In the Uranus-Neptune zone, with the relevant $\sigma \sim 1$ g cm^{-2}, these initial planetesimal radii would be 200 to 400 km, much further along the way to full-size planets. This result raises questions about the origin of comets, which are widely believed to come from this region, but are much smaller than the size given by this calculation. However, modifications to the instability theory (Sec. VI) can yield building blocks for the outer planets comparable in size to the original planetesimals in the terrestrial zone. Comets are probably samples of those original building blocks rather than fragments of larger bodies.

The outer planets are believed to have accreted from that population of dirty icy planetesimals (Greenberg et al. 1984). When Jupiter and Saturn reached $\sim$10 times the present Earth-mass, they became massive enough to nucleate gravitational collapse of nebular gas around themselves (see, e.g., Pollack 1985). The gas then constituted the bulk of those giant planets. Uranus and Neptune are assumed to have grown somewhat later, after the nebula was dispersed, so they were too late to acquire gaseous mantles. Their final sizes were right for scattering a significant fraction of the icy planetesimals into the Oort cloud (albeit with only $\sim$10% efficiency), while Jupiter and Saturn scattered almost all excess material out of the solar system.

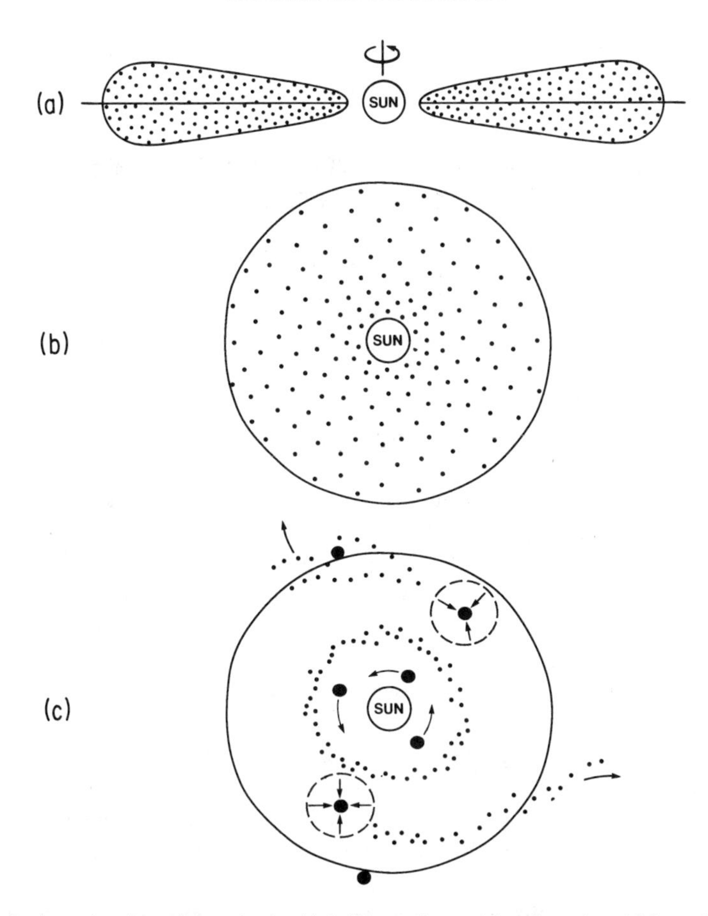

Fig. 2. Accretion from planetesimals. (a) Solids condense and settle to the midplane of a minimum mass nebula (edge-on view). (b) Gravitational instability in the thin disk clumps solids into planetesimals of radius ~3 km at 1 AU and hundreds of km in the outer solar system according to Eq. (1) (see text), but the latter value is more likely only a few km (see Sec. VI). (c) Planets grow by collisional accretion, outer planets nucleate collapse of nebular gas around themselves, Jupiter suppresses growth in the nearby asteroid zone, and icy planetesimals in the outer solar system are scattered into the Oort cloud and beyond.

Similarly, the terrestrial planets grew from the planetesimals in the inner solar system, but growth was limited because only rocky material had condensed in this region. Partial accretion took place in the asteroid belt, but was aborted due to the influence of nearby Jupiter, which may have scattered secondary bodies from its zone (some possibly Earth-size) into the asteroid zone (Safronov 1979; Davis et al. 1979). These large bodies would ultimately have been scattered by Jupiter out of the solar system, but not before they stirred up relative velocities in the asteroid belt to the present high values (~5 km s^{-1}).

The high velocities caused collisions to be not accretional, but destructive, grinding down and dismantling most of the small planets in that zone.

II. ACCRETION THEORY WITH A POWER-LAW SIZE DISTRIBUTION

Safronov's (1969) theory of accretion was based on a single fundamental assumption that was needed to make the analysis analytically tractable. The size distribution of planetesimals throughout the growth phase follows a simple power law, with the number of bodies dn in a mass range dm obeying the relation

$$\mathrm{d}N = m^{-q}\mathrm{d}m \tag{2}$$

where $1 < q < 2$. This distribution is not unreasonable for a collisionally evolved system. Dohnanyi (1969) showed that in a system where fragmentation dominates, an idealized equilibrium solution gave $q \sim 11/6$. Zvyagina and Safronov (1972) obtained $q \sim 3/2$ for a coagulation model (no comminution) and Zvyagina et al. (1974) found $q \sim 5/3$ for a simple model with both coagulation and fragmentation.

From our present perspective, we know these models were overly simplified. For example, if gravitational binding of bodies is taken into account, the law for fragmentation is significantly modified for the asteroid belt (Davis et al. 1979). Also these results assume that a steady state is reached, while we now have good reason to believe it is not. Nevertheless, as a starting point, the power-law assumption was reasonable and led to considerable insight into the relevant physical processes.

The most physically significant property of the power-law distribution assumed by Safronov was that with $q < 2$, most of the mass of the population resided in the largest bodies (Fig. 3a). Implicit in that assumption is the idea that there was a sharp upper cutoff to the population. Consider the alternatives. If the power law continued with no upper cutoff up to the size where $N = 1$ (Fig. 3b), most of the mass would already reside in a single largest body and there would be no further significant accretion. If there were a more gradual cutoff (Fig. 3c), most of the mass of the system would reside in bodies somewhat smaller than the largest bodies, and over that physically important size range q would be greater than 2. The simple power-law assumption would be violated unless the upper cutoff were sharp.

With the size distribution assumed to obey such a power law throughout the accretionary period, the evolution must proceed as shown schematically in Figs. 4 and 5. The largest bodies in the system always are accompanied by many others of the same size, and smaller bodies are negligible in terms of mass. The implicit picture is one of orderly hierarchical growth.

The other critical factor in the story of accretion, besides the size dis-

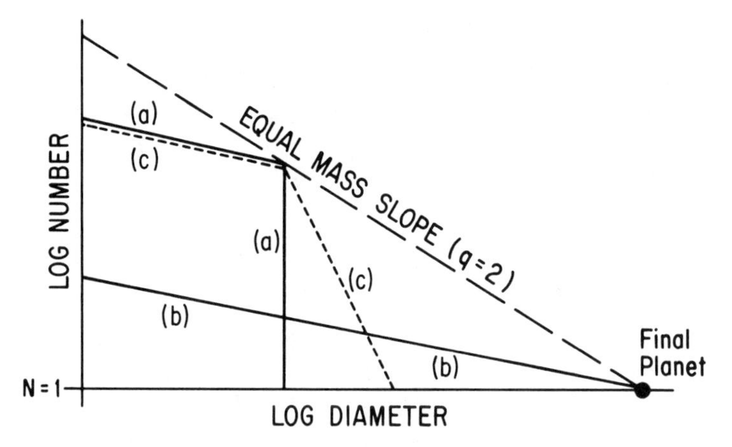

Fig. 3. (a) Size distribution (number N vs mass m) for planetesimals generally assumed by Safronov (1969). Most of the mass is in the largest bodies. (b) In that model, the same slope obtains until the final planet is formed. (c) With the upper mass cutoff less sharp, most of the mass resides in bodies smaller than the largest ones. This figure is schematic only.

tribution, is the relative velocity among planetesimals. The velocity controls the rate and probability of collisions because it determines both the rate at which each particle moves through the swarm of other particles and the size of its gravitational cross section. The velocity also determines the physical outcome of any collision. As in the asteroid belt, impacts too fast can yield disruption rather than accretion. Even if collisions are slow enough to allow accretion, some material (e.g., from craters) may be ejected.

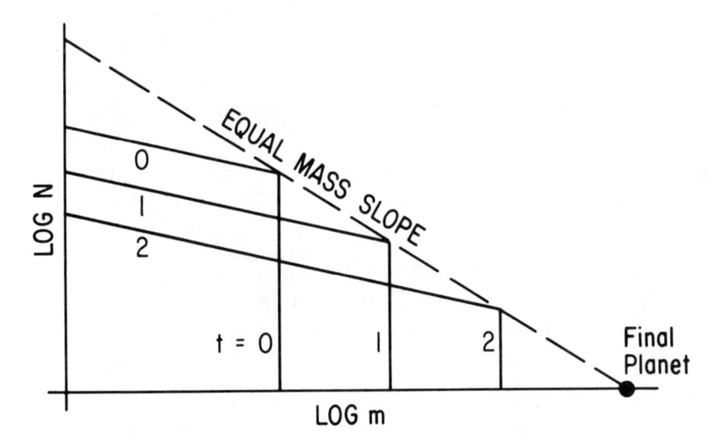

Fig. 4. If the size distribution obeyed Safronov's assumption throughout the accretionary period, the evolution would proceed as shown here schematically.

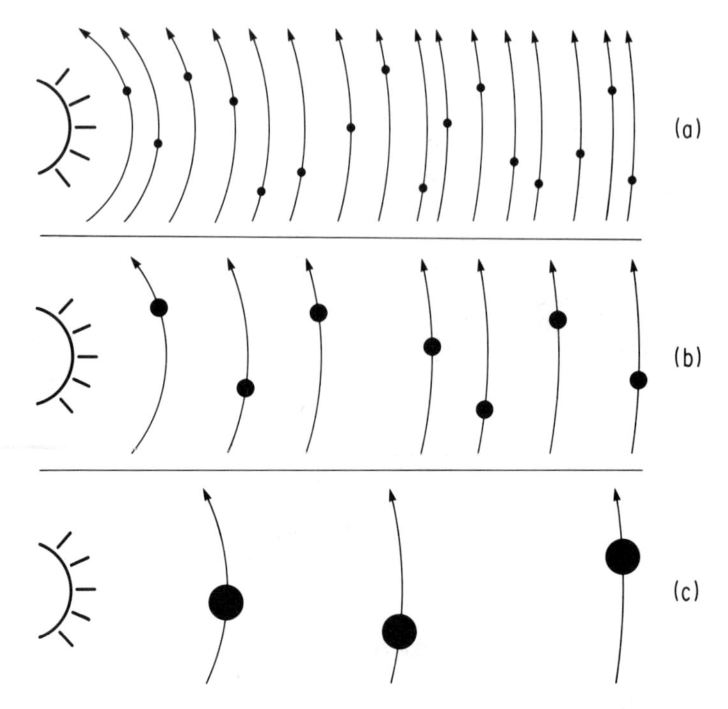

Fig. 5. In general, the process of accretion described by Safronov was one of orderly, hierarchical growth. The largest bodies were always all about the same size and dominated the system in terms of mass. As they grew (plot a to plot c) by mutual accretion, their feeding zones merged until the final isolated planets emerged.

Relative velocities were generally assumed to be due to the random (or *thermal* or *dispersion*) velocity components of the orbital motion, rather than the systematic motion of circular coplanar orbits. The out-of-plane random component represented by orbital inclination determines the thickness of the disk of planetesimals, which in turn affects the density of the swarm and frequency of collisions. The in-plane random component represented by orbital eccentricity helps determine the radial range over which material can be accreted, the all-important feeding zone.

Safronov showed that the random relative velocity in an equilibrium swarm of planetesimals is roughly the same as the escape velocity of the bodies that dominate the system. This result follows from the assumption of equilibrium between gravitational stirring and collisional damping; it can be demonstrated by the following simplified considerations.

Even if there is no random motion, planetesimals would pass by one another on circular orbits at speeds varying with distance from the Sun. Grav-

[1984]). On the other hand, more recent studies have shown that there are fundamental problems with the analytical theory. During some important portions of the accretion process, the size distribution differed significantly from the assumed power law, and the relevance of the power law is questionable throughout the evolution.

III. AN EVOLVING SIZE DISTRIBUTION

Some qualitative deviation from the idealized power law must have occurred according to the following argument (Greenberg and Rizk 1987). If the system did obey the sharply cut off power law at some time, any further growth of one of the larger bodies would have been dominated by accretion of other bodies of about the same size. The first such event would roughly double the mass of one body relative to all the others. Such discrete growth events, unlike the sort of continuous growth described by equations like Eq. (10), must lead to some spreading of the sizes of the largest bodies (as in curves 1 and 2 in Fig. 6). Without the sharp cutoff, most of the mass resides in bodies smaller than the largest ones. Random relative velocities can be much smaller than the escape velocity of the largest body. Runaway growth becomes a possibility. Some authors use the expression "runaway" to mean simply that the largest body grows faster than the second largest one, but that could be only a temporary condition, so I generally define runaway growth as separation of the largest planetesimal, or "planetary embryo" from the continuous portion of the size distribution as shown schematically in Fig. 6 (curve 3).

Numerical models of the accretion process allow simulation of the evolution of a planetesimal population without the artificial constraint of a power law. In this way, Greenberg et al. (1978*a*) showed that early evolution quickly led to incipient runaway of the largest planetary embryos. The size distribution was described by the number of bodies in each of a series of size intervals, and the random velocity was described by a characteristic eccentricity and inclination for each size.

The algorithm assumes that, within a given conceptual "box" moving in a circular Keplerian orbit around the Sun, the planetesimals move like molecules in a gas, with relative velocities controlled by the random components of their motion, which are much greater than the systematic Keplerian shear. Each planetesimal is assumed to move through a representative sample of the general population; none is spatially isolated.

The number of collisions that occur in each short time step between bodies of two sizes is proportional to the number of bodies of each size, their relative velocity and their collision cross section. The outcome of each collision in terms of sizes and velocities of debris or aggregations is determined from an algorithm based on experimental and theoretical results. Also in each time step, the average changes in random velocities due to gravitational in-

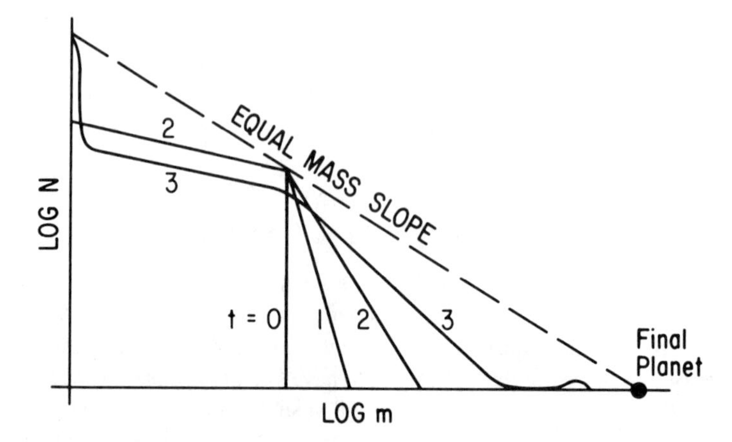

Fig. 6. A sharp upper cutoff to the planetesimal distribution could probably not be maintained. Evolution would proceed as shown schematically in time steps 0 to 2. If the largest body then accreted material much faster than the others, runaway growth could occur as shown in curve 3. Most mass remains in the originally dominant size range until the runaway planetary embryo finally reaches full size.

teractions are computed. After each time step, the size distribution and random velocity distribution is adjusted to reflect the effects of the collisions and gravitational interactions. In this simulation, the evolution of the size and velocity distributions are mutually dependent.

The evolution of the size distribution resulting from a typical simulation in the terrestrial planet zone is shown in Fig. 7. (Note that there were many computational time steps between the representative stages shown in Fig. 7.) Evolution qualitatively follows the scheme described in Fig. 6. In this case, all the planetesimals were assumed to have initial diameters of about 1 km. As time went on, a tail of small-particle debris was produced due to impact ejecta and fragmentation of some small bodies. A few bodies of ~1000 km were produced very quickly, because most of the mass of the system still resided in planetesimals not much larger than a few km, so relative velocities remained small with the gravitational cross sections of the larger bodies correspondingly large. Safronov's result that the dominant bodies would control the random velocities was correct, but his assumption that they would be the largest bodies does not always apply.

The time required for a few largest bodies to reach ~1000 km is somewhat uncertain, due to difficulties in numerical treatment of the statistics of the few largest bodies. Although it seems to agree quite well with the analytical results of Nakagawa (1978), the ~20,000 yr time scale in the simulations by Greenberg et al. may be artificially short to some degree. However, the principle seems secure that during this portion of the evolution, most of the mass of the system resided in particles much smaller than the largest bodies,

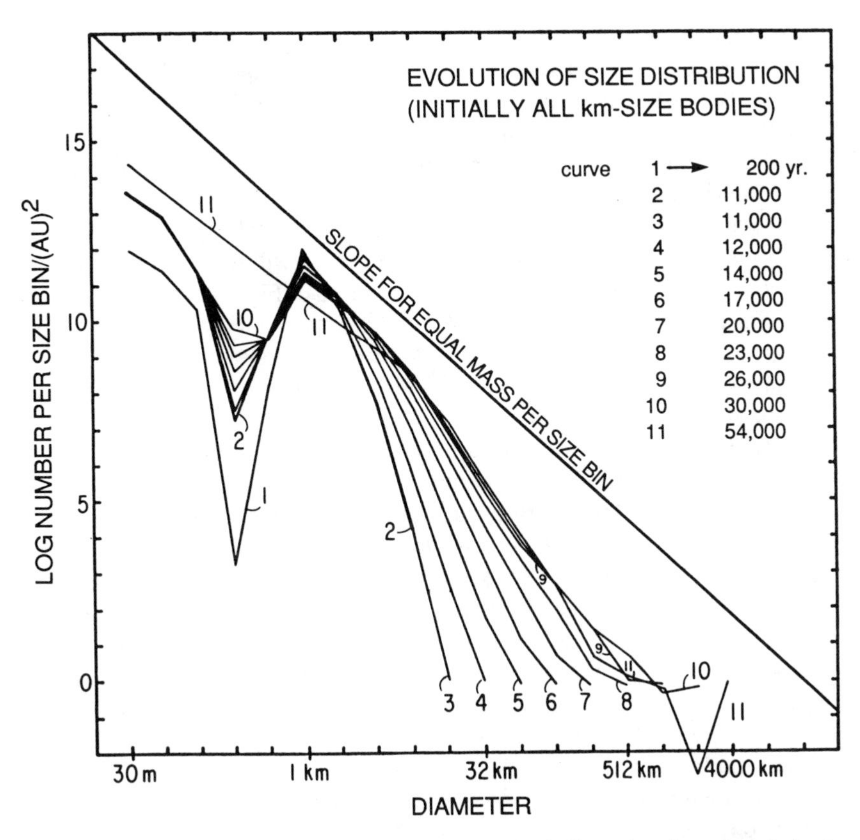

Fig. 7. Numerical simulations of planet growth gave results similar to the schematic evolution in Fig. 6. Initial planetesimals were taken as 1 km in diameter. A tail of small-particle debris forms with low slope so that it contains little mass. Most of the mass remains in the initial km-size bodies so random velocities stay small. Low velocities give large gravitational cross sections, which allow bodies hundreds of km in diameter to grow very quickly. Incipient runaway occurs by the time a planetary embryo reaches ~1000 km. Beyond that stage, the low random velocities mean that bodies are isolated on nearly circular orbits, so the particle-in-a-box approximations are invalid. The subsequent apparent runaway is therefore artificial.

with consequent accretion rates much faster than given by the Safronov theory.

Once the largest body reached ~1000 km, some of the underlying principles of the numerical simulation became invalid. First, because the larger bodies had very large gravitational cross sections and small orbital eccentricities, approach velocities began to be influenced by Keplerian shear rather than by random motion. It is not clear yet what is an acceptable statistical treatment of encounters under such conditions; the two-body encounter model has been called into question (see Sec. V.A). Second, because all bodies are on nearly circular orbits, these very important largest ones are effectively isolated from the soup of small particles that contains most of the mass of the system. The assumption that all bodies statistically sample the entire swarm is violated. Evolution in the special neighborhoods around the large bodies may be different from that elsewhere, but such spatial variations were not included in the numerical model.

For these reasons, as noted by Greenberg et al. (1978*a*), the simulation is only valid up to the point of incipient runaway growth (e.g., curve 9 in Fig. 7), not all the way to real runaway as in curve 11 of Fig. 7, or curve 3 in Fig. 6. At this point, there were a few 1000 km planetary embryos ($\sim$1 AU^{-2}), in a soup of small planetesimals which had most of the system's mass (Fig. 8). The nearly circular orbits mean that simple random motion is inadequate to feed the soup to the embryos.

Wetherill and Stewart (1987) have identified a number of factors that were important in giving this early runaway condition (or what I prefer to call incipient runaway). Most of these factors were included in the simulations of Greenberg et al. (1978*a*). The first is "lowering of eccentricity" by dynamical friction. In fact, dynamical friction had to be included in the numerical simulations in order to prevent artificial enhancement of random velocities. Second is the role of collisional damping and gas drag in lowering relative velocities. Collisional damping was an essential part of the numerical simulations, and unpublished numerical experiments with gas drag included showed that the latter was a negligible factor due to the low mass density of the gas. A related factor is the role of fragmentation, which produces small particles that help damp random motion; this was also an intrinsic element in the numerical simulations.

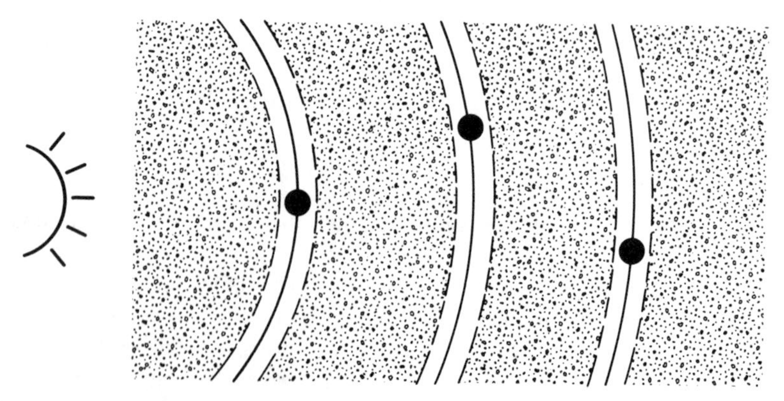

Fig. 8. The outcome of numerical simulations of early growth. A few 1000-km bodies (planetary embryos) are isolated on nearly circular orbits. Between those orbits, a soup of small planetesimals contains most of the system's mass.

Another factor, a so-called enhancement of the gravitational cross section at low encounter velocities due to the presence of the Sun is discussed in more detail in Sec. V.A. This factor was negligible while random motion governed relative velocities, i.e., it only became important after the part of the evolution where Greenberg et al.'s (1978*a*) numerical simulations were valid.

A final factor identified by Wetherill and Stewart is most critical in the sense that it represents the critical deviation from the Safronov model: the generation of planetesimal "seeds," bodies larger than the others that may proceed to runaway growth. A natural and seemingly inevitable way to generate such seeds is by stochastic merger of large planetesimals of similar mass (Wetherill and Stewart 1987). This process was identified by Greenberg and Rizk (1987) to have played a key role in the incipient runaway in the numerical simulations. Wetherill and Stewart suggest that such a seed must be discontinuously large compared with the rest of the size distribution, which sounds as if runaway must already have begun in order to continue. In fact, a seed need only be 2 to 3 times the mass of the next largest body (Wetherill and Stewart 1986; Greenberg and Rizk 1987). The work of Greenberg et al. (1978*a*) and Greenberg and Rizk (1987) has shown that such modest stochastic seeding would be practically unavoidable, leading to the incipient runaway that produces planetary embryos.

IV. WHAT HAPPENED NEXT?

The brief, early stage of accretion described above was quite different from what had been envisioned by Safronov, but it left a system still far from the final condition of full-size planets. What happened next remains an open question. Two possible end-member scenarios were described by Greenberg et al. (1978*b*), and these (or something between the two extremes) remain open hypotheses.

A. Isolated Embryos

The first scenario has the first-formed planetary embryos truly isolated on their circular orbits from the massive soup of planetesimals. Each embryo depletes its own local feeding zone of planetesimals as it grows. During growth its feeding zone expands, but only in proportion to its radius r (see Sec. V.A), while further growth requires amounts of mass proportional to r^2. Without access to enough material, the embryos stop growing at $r \sim 1000$ km. The soup between the depleted feeding zones is unaffected by the existence of those embryos, and more embryos emerge from the soup just as the first ones did. As each embryo reaches 1000 km, it depletes its local zone and stops growing. Additional embryos fill the orbital niches between the earlier ones. The size distribution returns to the Safronov condition (Fig. 9) because the spatial isolation imposes a sharp upper cutoff.

At this point, most of the mass of the system is in the largest bodies. For

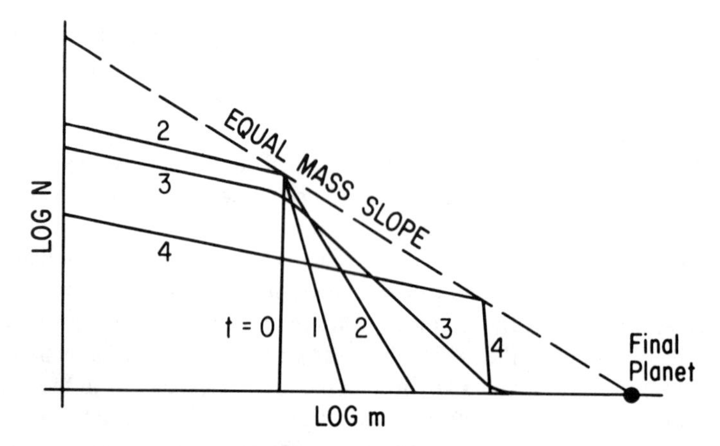

Fig. 9. If the first-formed embryos which follow from the numerical simulation (shown schematically as time steps 0 to 3 here) are truly isolated as shown in Fig. 8, then their growth would halt while other embryos of comparable size grew in the soup of planetesimals between the first embryos. In that case, the size distribution could return to that assumed by Safronov (shown here as step 4).

1000 km bodies in the Earth zone, this condition is reached when the orbital spacing is down to a few times 10^{-3} AU. The random velocities can now be pumped up toward $v_e \sim 500$ m s^{-1}. Corresponding orbital eccentricities give radial excursions comparable to the orbital spacing, so the isolation of these bodies comes to an end. With the system returned to the Safronov condition, one might imagine subsequent accretion (which is in fact most of the growth process, given that a 1000 km embryo is only 0.001 of the mass of the full-grown Earth) would follow the Safronov theory. Safronov and Ruzmaikina (1985*a*) describe a model in which adjacent feeding zones merge, allowing orderly accretion into ever-larger bodies, with the largest bodies in the separate feeding zones always about the same size.

However, Wetherill (1985,1986) showed that a system dominated by such large bodies would not evolve in that orderly, hierarchical way. He performed a Monte Carlo study of the behavior of a system of 500 bodies initially of radius ~1000 km. Results showed evolution characterized by high random velocities, violent collisions, random and abrupt accretion events and stochastic radial mass transport. The state of the system after 3 Myr in a characteristic experiment is shown in Fig. 10. Eccentricities are typically ~0.25, mass is distributed in a wide range of particle sizes and an embryo identified as Mercury according to its final position and size forms at 1.6 AU. The only characteristic of this evolution consistent with the Safronov theory is that accretion time is still ~100 Myr.

The results of this Monte-Carlo simulation of a discrete number of large planetesimals are only relevant in the context of this first scenario in which

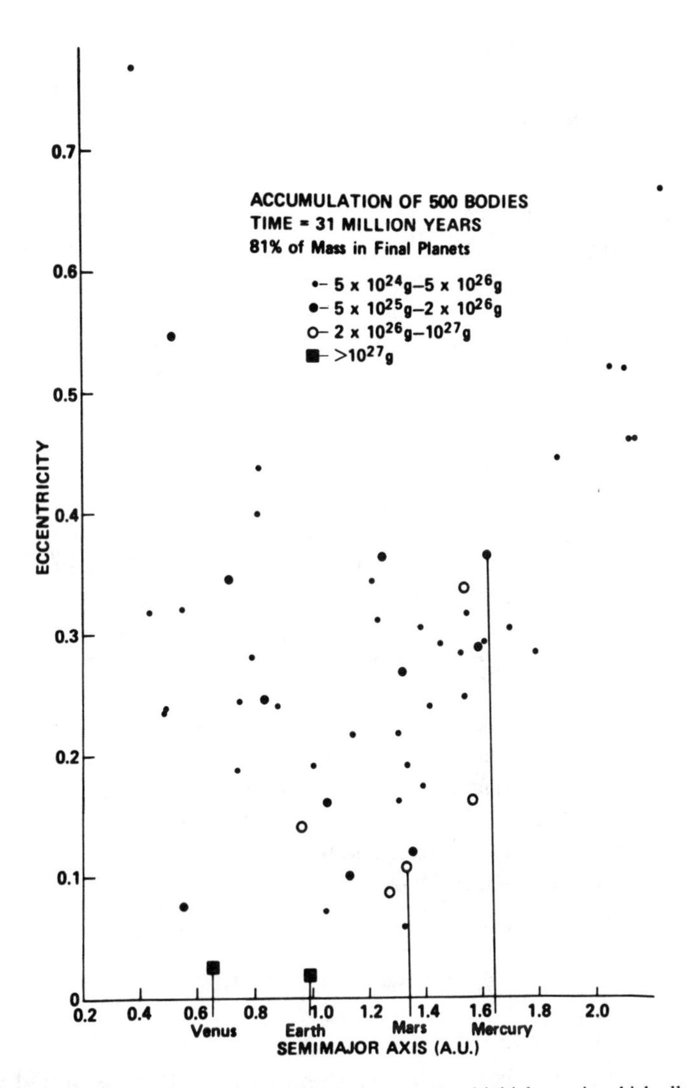

Fig. 10. Monte Carlo models show evolution from an assumed initial state in which all bodies were about 1000 km in radius (like curve 4 of Fig. 9). This plot (from Wetherill 1986) shows the distribution of bodies in size, location, and random velocity (given by orbital eccentricity e) at t = 30 Myr. The evolution is characterized by high velocity collisions between large bodies, which yield stochastic discrete growth events and random radial mass transport. At the stage shown, the body that is identified as Mercury, because of its ultimate position and size, is located beyond the other terrestrial planets at 1.6 AU. Although this simulation was predicated on a Safronovian size distribution, it leads to growth very different from the ordered, hierarchical evolution described by Safronov. Nevertheless, it does give growth of the terrestrial planets in ~100 Myr, just as Safronov's model did.

putative stalled growth of many bodies at 1000 km gives a size distribution consistent with Safronov's assumptions. Even in this case, we see that the orderly evolution style of the Safronov model never applied.

B. Runaway Growth

The second scenario assumes that the first-formed embryos were able to continue to feed on the soup of small planetesimals, rather than to become isolated in depleted feeding zones. Their incipient runaway growth was able to continue as shown schematically in Fig. 6. In this model, the rapidly formed, first large bodies were truly planetary embryos in the sense that their identities remained distinct as they grew by accreting smaller particles into the final full-size planets. Evolution would be rather quiescent, with random velocities far smaller than the escape velocities of the embryos, because most of the mass of the entire system remains in small bodies until the growth of planets is nearly complete. The positions of the growing planets in the solar system would not jump wildly as in the Wetherill (1985) simulation, although there could be considerable radial migration depending on the exchange of angular momentum with the soup during the feeding process.

This runaway scenario hinges on appropriate radial transport of material to supply mass for the growing embryos. Several radial transport processes have been considered to play a role in this process and are discussed in the next section. Most of these mechanisms have been elucidated to a certain degree by the study of planetary rings, which are similar continuous media of small particles interacting with discrete satellites (Ward 1984). How these processes work in presently observable systems is still incompletely understood, let alone their relative roles and interactions in evolution of the hypothetical planetesimal swarm.

V. RADIAL TRANSPORT

A. Range of Feeding Zones

The particle-in-a-box evaluation of impact rates becomes invalid when the feeding zone of an embryo gets so large that it extends to where Keplerian shear is faster than the random velocity. Equivalently, the feeding zone is wider than the range of random motion eR. Beyond this point in the evolution, delivery of material to where it may interact with the embryo is governed by the systematics of heliocentric orbits rather than by the random motion relative to circular orbits.

Growth rates beyond this stage are no longer governed by random encounters with particles on orbits that cross the embryo's. Instead, the rate of growth may depend on how effectively the embryo can extend its feeding zone by stirring the eccentricities of planetesimals from some range of semimajor axes so trajectories temporarily cross the embryo's orbit giving some chance of impact. The range of this extended feeding zone can be estimated analytically. For a planetesimal with semimajor axis differing from the embryo's a by an amount Δa, the embryo exerts an acceleration $\sim Gm/(\Delta a)^2$ for an encounter duration $\sim 4/n$. To give an embryo-crossing orbit, the change in velocity must be $\sim n\Delta a$, which implies $\Delta a \sim 4a(m/M_{\odot})^{1/3}$, i.e., a few times the radius of the embryo's Hill sphere of influence. The validity of this approximation has been demonstrated by numerical experiments as well (see, e.g., Dole 1962; Nishida 1983).

The planetesimal's trajectory may be only temporarily embryo crossing, but while it is, there is a chance of collision with the embryo. This probability can be evaluated using the two-body gravitational cross section and the particle-in-a-box methodology. Estimating the impact rate in this way, Greenberg et al. (1987*a*) find it to be remarkably close to (actually a few times faster than) the value one would get from naively continuing to use the two-body formula relevant to the case where random velocity governs encounters.

This same result has been obtained from Monte Carlo numerical experiments (Wetherill and Cox 1985). Wetherill and Cox interpret the difference between three-body Monte Carlo determinations of the impact rates and their invocation of a two-body formulation as representing a breakdown in the two-body approximation for embryo-planetesimal interactions as the embryo gets large and the random velocity stays small. However, a more physically meaningful interpretation is probably the explanation above, based on the transition from encounters dominated by random motion to ones controlled by systematic Keplerian relative velocities. Even in the latter case, impact rates can be estimated by properly invoking two-body approximations as described above (see also Greenberg et al. 1987*a*).

Because the impact rates continue to increase with embryo growth roughly as if random velocity still were in control, rapid runaway growth of the largest embryo can continue beyond the point where the particle-in-a-box numerical simulation broke down. But this growth cannot continue indefinitely. As described in Sec. IV.A, incremental growth $\mathrm{d}r$ in the embryo's radius r would require a mass increase proportional to $r^2\mathrm{d}r$, but would only extend the size of the feeding zone and the mass of available soup by an amount proportional to $\mathrm{d}r$. The feeding zone is compelled to grow fast enough to compensate for the depletion of soup. In the Earth zone this isolation of the embryo occurs by the time it reaches roughly the size of the Moon (see, e.g., Lissauer 1987), unless other processes can deliver planetesimals to the embryo.

B. Viscous Transport

One fundamental mechanism is viscous transport or collisional diffusion. This process can be understood in several ways. From a celestial mechanics point-of-view, each individual particle undergoes a series of collisions that change its velocity (as well as its mass). Its orbit undergoes a random walk in semimajor axis, such that a swarm of these particles must spread both inward toward the central body and outward. From an energetics point of view, the gradual dissipation of energy in collisional heating must come out of orbital motion, which tends on average to reduce semimajor axes, while the need to conserve angular momentum requires some particles to evolve outward. From a fluid dynamics point of view, the fast orbiting material on the inside of a swarm transports angular momentum to the slower outside by viscosity.

While this mechanism seems straightforward, there are some subtle complications. Viscous transport is accomplished on the microscopic scale by collisions of particles from the inside (fast-track) with those on the outside (slow-track). However, in Keplerian motion, when a particle with a fast mean motion moves out to the apocenter of its orbit, its angular velocity becomes slower than the mean motion of particles at this new location; this description would seem to suggest that the viscosity might be negative. In fact, evaluation of viscosity requires subtle accounting of statistics of collision sites and velocity changes. A recent review of the theory (Greenberg 1987*a*) shows that the rate of momentum transport can depend sensitively on the physical properties of the particles and the nature of their collisions. It is not clear that general formulae for viscosity can be lifted from the context of ring dynamics, or even that they are relevant for all rings applications.

C. Embryo-Soup Torques and Gap Formation

Another mechanism suggested by analogy with the dynamics of planetary rings is the torque due to gravitational interaction between an embryo and the soup. The small inner satellites of Saturn perturb the motion of ring particles at positions where the particles' orbital motion is in resonance with a satellite. These perturbations drive density waves in the ring, which propagate away from the site of the perturbation and are damped by interparticle collisions. The propagation and dissipation distorts the density distribution so it is not symmetrical relative to the driving satellite and a torque can be exerted on the satellite's orbit. Hourigan and Ward (1984) call this effect tidal drift because it is similar to the tidal torques that occur when a satellite distorts the figure of a planet, which in turn accelerates the satellite.

For a planetary embryo, the torques due to the swarms of planetesimals interior and exterior to its orbit tend to offset one another. However, if the density profile of the disk is not constant, there can be a net torque. Hourigan and Ward find that this effect could produce torques capable of sweeping an embryo across the entire disk of the solar system during the formation period. Such behavior would solve the embryo-isolation problem, but would overdo it.

On the other hand, the torques may be sufficient to repel the planetesimal swarm away from an embryo, just as shepherd satellites confine the edges of rings. This effect might open a wide gap in the swarm which would prevent further radial drift and stop further accretion. Hourigan and Ward have made a strong case that all these torque effects could play an important role in the planet-growth process, but they conclude that the results are still too quantitatively uncertain to be incorporated into a model of the early solar system.

D. Gas Drag

As long as the gaseous component of the nebula remained present, it exerted a drag force on the orbits of planetesimals (Adachi et al. 1976; Weidenschilling 1977*b*). The pressure which partially supported the gas against the pull of the Sun allowed the gas to rotate with an angular velocity slower than the Keplerian motion of the planetesimals. Thus the planetesimals experienced a headwind that removed orbital energy, causing them to spiral toward the Sun. (The gas also tended to damp random motion, but since the gaseous component was less concentrated at the midplane of the nebula, the mutual collisions of planetesimals probably dominated that damping.) Smaller particles, with their greater area-to-mass ratio, tend to be more susceptible to the effects of gas drag.

Nakagawa et al. (1983) estimate that accretion in an assumed gaseous environment (the Kyoto model), with the growth time largely controlled by orbital gas drag, would be about 10 Myr for the Earth, an order of magnitude faster than Safronov's widely quoted earlier value. Values farther out in the solar system remain unacceptably long however: 30 Myr for the Jovian core, 400 Myr for Saturn's core, and much greater than the age of the solar system for Uranus and Neptune. For Jupiter, this growth time is a significant improvement over that calculated in Sec. II, being marginally within a plausible lifetime for the nebular gas. But for Saturn, the core would still grow too late to acquire a gas mantle.

A mechanism for speeding the orbital decay of planetesimals, and thus accelerating planet growth, was proposed by Weidenschilling and Davis (1985). Planetesimals spiralling inward reach orbits that are in resonance with an embryo well before they reach the embryo itself. At such a resonance (where the embryo's gravitational perturbations are enhanced because the orbital periods are in a ratio of small whole numbers), continued orbital decay may be stopped (Greenberg 1978), while eccentricities would be enhanced. Weidenschilling and Davis suggested that planetesimals piled up at resonant positions, resulting in destructive collisions that reduced the mean size of the planetesimals. The greater drag on these smaller planetesimals pushed them through the resonance. They continued toward the embryo at a faster rate than if they had not been reduced in size.

The smaller size of these particles allowed gas drag then to deliver material to the embryo faster, thus speeding planetary growth. Also, accretion of material in the form of small particles would affect the thermal evolution of the primordial planetary atmospheres in a very different way than multi-km planetesimals. Accretion energy would be deposited higher in the atmosphere rather than the surface of the planet, and dust added to the atmosphere would increase its opacity. These effects would be important in determining the early compositional evolution of the atmospheres (see various chapters in this book). For example, it might help explain the low terrestrial abundance of neon (Mizuno et al. 1980).

Unfortunately there are serious problems with Weidenschilling and Davis' (1987) model. The time scale for pumping the eccentricities of planetesimals in resonance is so long, and the width of each resonance zone is so narrow, that mutual perturbations among the planetesimals may drive them through resonance before the desired destructive collisions take place. While this mechanism does not seem important in its latest incarnation, it does demonstrate the importance of subtle effects and the interrelationships among various mechanisms in the planet-growth process.

VI. GROWTH OF THE OUTER PLANETS

The literature on planetary accretion has directed considerable attention to the particular problem of the long time required to grow the outer planets. The approach has generally been to look for special fixes or adjustments to Safronov's basic theory to make it work beyond the asteroid zone. However, the above discussion suggests that concern with the single issue of time scale is premature, given the great uncertainty in even identifying the key mechanisms that govern growth.

The outer planets do provide important constraints. A successful theory should produce full-size solid cores for Jupiter and Saturn while the nebular gas is still present and available to collapse around them. Jupiter should form before the asteroid-zone can produce a planet, or else a different way must be found to suppress growth in that zone. Certainly, Uranus and Neptune must be formed in less than the age of the solar system. According to the principle of Occam's razor, we should try to satisfy these constraints as a natural continuation of the same theory that explains growth of the terrestrial planets, before invoking assumptions especially tailored to the outer solar system.

Nevertheless, Lissauer (1987) has pointed out that an *ad hoc* augmentation of the available solid mass would solve the Jovian accretion problem. If σ were >15 g cm^{-2}, the feeding zone of the embryo would have contained enough mass for accretion of the required core and would have allowed accretion well within the lifetime of the gaseous nebula. The more conventional value is $\sigma \sim 4$ g cm^{-2}, which comes from spreading the assumed ice and rock components of the planets over their respective zones (see, e.g., Weidenschilling 1977*a*). This procedure gives a general trend of $\sigma \propto R^{-3/2}$ for a minimum mass nebula, with a discontinuity of about a factor of 4 for the icy material that condensed beyond 3 AU. Lecar (1987) has given an elegant theoretical explanation for this $-3/2$ power law: if the original nebula was a uniform density, uniformly rotating sphere that flattened into a Keplerian disk with each element of mass conserving angular momentum, the disk would obey an $R^{-3/2}$ density law.

Lissauer's suggested augmentation of σ by a factor 4 in the Jovian formation zone does not fit such trends very well. It also means Jupiter would have subsequently had to have thrown considerable excess mass out of the solar system, which would have caused the planet to move roughly 1 AU closer to the Sun from its original formation site, but such modest movement is acceptable. Lissauer does not significantly augment the value of σ elsewhere in the nebula, because, for example, in the terrestrial planet region, it would be more difficult to remove excess mass. While there is no fundamental problem with tailoring the nebula in this way, it is not clear whether it will be necessary once the various radial transport issues are resolved.

Lissauer's fix for the growth of Jupiter is reminiscent of the fix discussed and rejected by Levin (1972*b*) for the Uranus-Neptune growth-time problem in Safronov's model. If σ were augmented by a factor of 30 over the conventional value (1 g cm^{-2}), then the growth time would be decreased from 100 Gyr to within the age of the solar system (see Eq. 11). In that case, those planets would have to eject so much mass from the solar system that their orbits would shrink by a factor of 10 from their original formation positions. That would mean that σ was 30 g cm^{-2} at >200 AU, implying a nebula mass ($\sim 2M_\odot$) too great to be consistent with any planetesimal hypothesis. Levin concluded that growth could be accelerated in the model by a modest increase of σ by a factor of a few, just as Lissauer has now proposed for Jupiter, but Uranus and Neptune would still grow an order of magnitude too slowly.

A more promising fix to the Safronov theory was to increase the gravitational cross sections of the planetesimals by decreasing the relative velocities (i.e., increase θ by an order of magnitude above the standard Safronov value of ~ 5). This approach also addressed another difficulty with the Safronov theory in the outer solar system: with the assumption that random velocities increased with the size of the largest planetesimals, most of the solid material in the outer solar system would have been ejected from the Uranus-Neptune zone before those planets could grow to 10% of their present masses. Safronov (1969,1972*a*) and Levin (1978) speculated with varying degrees of plausibility on rationales for damping the random motion in the outer solar system (reviewed by Greenberg et al. 1984). However, as shown in Sec. III we now have good reason to believe random velocities were much lower than assumed by Safronov, because most of the mass was in small bodies.

The size distribution of comets also indicates that much more mass resided in small bodies than in the growing embryos of Uranus and Neptune (Greenberg et al. 1984). Figure 11 shows the size distribution for observed comets (Whipple 1975) in the upper left, corrected for observational selection (open circles). If all cometary nuclei are as dark as Halley has proven to be, those points should be moved to the right in the figure, which would only serve to strengthen the following argument. There are about 10^{10} times as

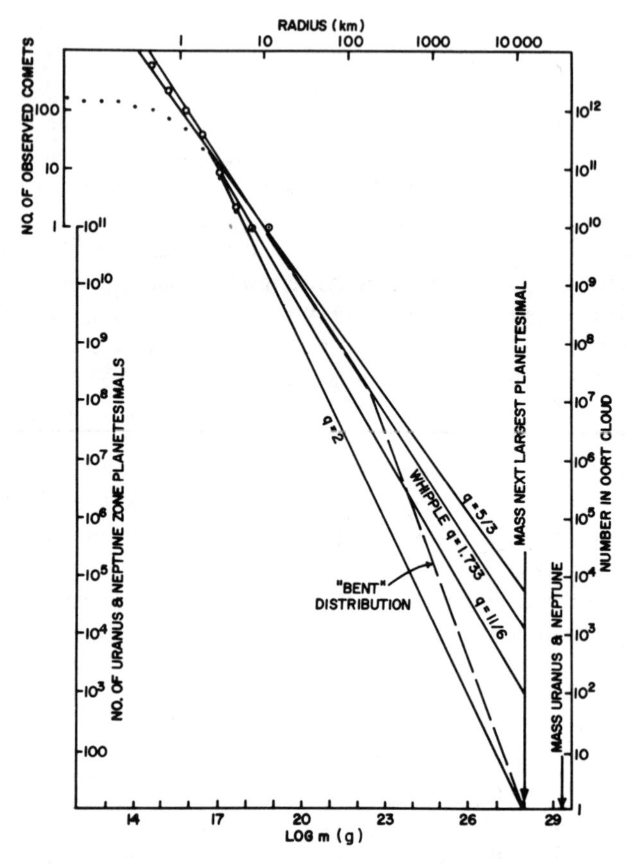

Fig. 11. Size distribution of observed comets (dots), corrected for observational selection (open circles), and related to the implied numbers in the Oort cloud and originally in the Uranus-Neptune zone. The latter population could not have extended as a single power law up to planetary sizes, because the total mass was not that great. A bent distribution is implied (figure from Greenberg et al. 1984).

many comets in the Oort cloud as in the observed sample (Weissman 1982), which gives the scale on the right. Also, there was a roughly 10% efficiency for delivering Uranus-Neptune zone planetesimals to the Oort cloud (Fernandez 1980), because most were ejected from the solar system entirely. Assuming that most Oort cloud comets came from this zone, there were originally 10 times as many planetesimals there, giving the scale on the left.

In that way, we can use the observed comet's size distribution as a representation of the distribution of Uranus-Neptune zone planetesimals. If we extrapolate this distribution to larger sizes using a best-fit power law or a Safronov power law, we find an implausibly large mass in that region. Only if the distribution bends more steeply downwards, as shown by the dashed line, would the total mass be within a few times the present planets' mass. Such a distribution could also include a planetesimal of about 0.1 of Uranus' mass to impact Uranus yielding the extreme obliquity.

These constraints define a size distribution similar to that found closer to the Sun in the simulation by Greenberg et al. (1978*a*). The bend at about 200 km corresponds to the initial planetesimal sizes expected from gravitational instability (Sec. I). The bulk of the mass of the distribution is in bodies of that size, so relative velocities should be damped relative to the Safronov assumption. By analogy with the earlier numerical model, the smaller bodies, including the observed sample of comets, would be fragmental debris.

However, application of the numerical simulation of collisional evolution showed that the comets could not be produced in this way (Greenberg et al. 1984). Figure 12 shows evolution in such a case. While an 8000 km embryo grew in about 100 Myr and random velocities stayed low, insufficient debris was produced in the km-size range to explain the present numbers of comets. The numbers were at least an order of magnitude small, and then began to decrease as mutual collisions ground these small bodies into even smaller pieces.

Those results led to an alternative hypothesis, that the original planetesimals were actually only a few km in radius, so that the present comets are

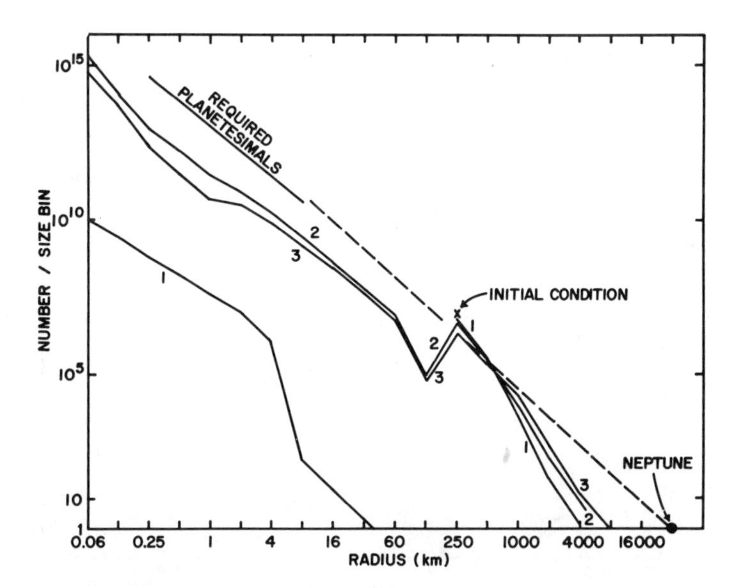

Fig. 12. If the initial population in the Uranus-Neptune zone was dominated by planetesimals hundreds of km in radius as given by Eq. (1) (see text), comet-sized bodies would form as collisional debris during the growth of the planets. However, the number of km-scale bodies formed this way would have been insufficient to populate the Oort cloud (figure from Greenberg et al. 1984).

survivors of that population, rather than collisional debris. In fact, there is good reason to question conventional use of the formula (Eq. 1) for the size of bodies produced by gravitational instability. Remember, Eq. (1) assumes that all the solid material has reached the midplane of the nebula. But the process of settling to the midplane is nonhomologous, because coagulation allows some larger particles to settle faster than others (Weidenschilling 1980).

At 30 AU, once about 0.5% of the solid material reaches the midplane, it goes gravitationally unstable, yielding planetesimals of radius ~3 km. These planetesimals would gravitationally stir their random velocities to a few m s^{-1}, thickening their layer to ~0.01 AU, and effectively removing themselves from the midplane. Subsequent settling of more small particles would repeat the process until all the material has clumped into planetesimals of ~3 km. If we assume that the initial population produced in this way had a distribution similar to the observed comets, subsequent collisional evolution (Fig. 13) would produce a rather large embryo in only 10^5 yr, thanks to the very low random velocity in a swarm dominated by such small (comet-size) bodies. The low random velocities also ensure that the comet-size bodies are not fragmented by collisions; they can survive until the embryo gets large enough to begin flinging them into the Oort cloud and beyond.

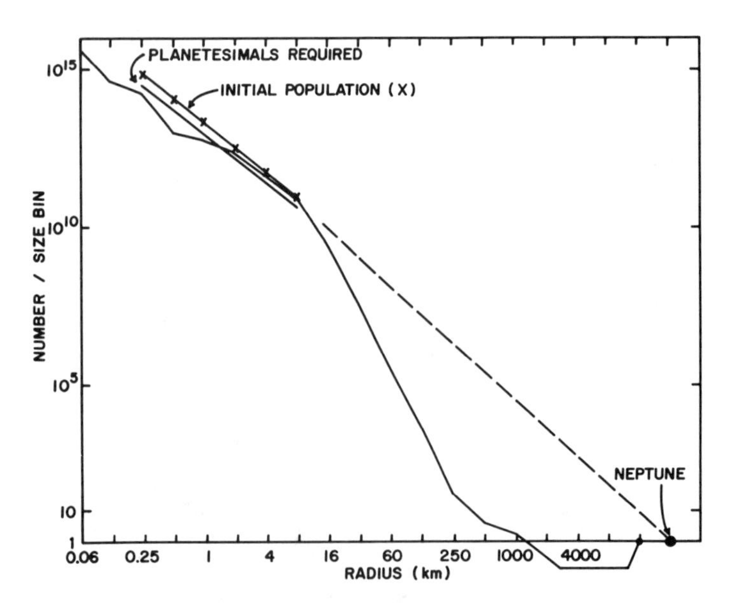

Fig. 13. If the initial population in the Uranus-Neptune zone consisted of comet-sized planetesimals, it could supply enough material to grow the planets and populate the Oort cloud. Therefore, comets are probably samples of the very early planetesimals, rather than collisional fragments (figure from Greenberg et al. 1984).

As in the case of the terrestrial planets, these particle-in-a-box calculations become invalid once random velocities cease to define the relative velocities between embryo and planetesimals. However, that computational problem, and the related physical problem of isolation of embryos on circular orbits, are less extreme in the outer solar system because relatively small random velocities correspond to relatively large orbital eccentricity. Whereas Keplerian shear began to influence relative velocity when embryos in the terrestrial zone were only ~10% of the final size, random motion continues to dominate until Uranus and Neptune are nearly full size (Greenberg et al. 1986). Only then would the mechanisms associated with isolation of embryos possibly slow the final accretion.

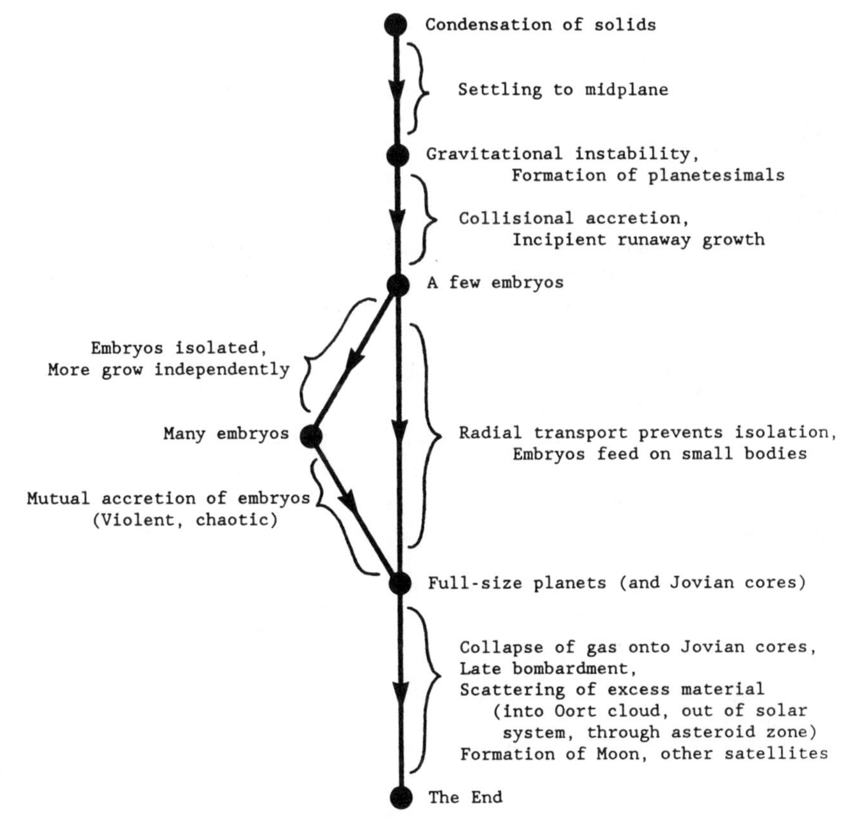

Fig. 14. Summary of the scenario of planetary accretion from planetesimals. The evolution after formation of the first few 1000 km embryos remains uncertain. Two possibilities are shown, but an intermediate course is also possible.

VII. CONCLUSIONS

In light of the above discussion, the often repeated concern about the specific problem of slow growth of the outer planets seems unwarranted. Rather, we need to understand the broader issue of growth from planetary embryos to full-size planets, or, in the case of Jupiter and Saturn, to full-size cores capable of nucleating gas collapse. Because embryos found themselves on nearly circular orbits, determination of their subsequent growth depends critically on how effectively material was transported radially.

If material was radially transported to the orbits of the embryos, accretion would have continued as a rather quiescent process of sweeping up small bodies. On the other hand, if many embryos were able to grow separately and independently, their final merging would have been very violent and chaotic. Thus the nature of the last stage of planet growth, including the final sweepup of planetesimals, the bombardment of primitive atmospheres, the formation of the Earth's Moon, the timing of gaseous collapse around the giant planets, and the stirring of the asteroid belt, all must remain speculative until either the accretion that preceded it is well understood, or unique constraints can be placed on the late stage by observed planetary properties (Fig. 14).

An example of the latter might be the existence of the Moon, which is now widely believed to result from a giant impact into the primitive Earth (see, e.g., Cameron 1986). If that belief is proven correct, it would argue for separate, independent embryo growth. Similarly, if studies of the evolution of primitive atmospheres yield hard requirements on the sizes of bombarding bodies or on the radial range of their provenance, they would help constrain accretion models (see, e.g., chapters by Ahrens and by Hunten et al.; Lange and Ahrens [1984]; Matsui and Abe [1986*a,b*]; Zahnle et al. [1988]).

There is no reason to believe that the present consensus scenario, in its most general terms that planets grew by accretion from a swarm of much smaller bodies, has any fatal difficulties. Important aspects of the planetary growth process are not yet understood, but the critical issues seem to be more clearly defined than ever, so we can be cautiously optimistic that a complete picture will emerge from the puzzle before long.

PART II
Primitive Bodies

THE COMPOSITION OF COMETS

E. K. JESSBERGER, J. KISSEL
Max Planck Institut für Kernphysik

and

J. RAHE
NASA Headquarters

This chapter summarizes recent results of studies dealing with the composition of comets, but focuses mainly on the results of the analysis of the in situ *measurements of Comet Halley's dust composition. Many discoveries, such as the presence of an unexpectedly large nucleus or the abundance of organic material of about 30% (in mass) of the dust, have lead to major changes in our ideas about the general nature of comets. The impact mass spectrometer onboard Vega 1 measured the elemental composition of Halley's dust and discovered an anorganic (mineral) fraction which appeared to be essentially CI chondritic, and an organic fraction consisting essentially of highly unsaturated hydrocarbons. The considerable carbon deficiency observed from the ground in other comets, can now be explained by a large amount of carbon present in the cometary dust and is spectroscopically invisible. The anorganic fraction generally forms a core which is embedded in essentially organic material. The density of pure organic grains (so-called CHON particles) is for the most part near 0.3 g* cm^{-3}. *All these measurements demonstrate a very fluffy nature of the cometary dust grains. The mass of most dust particles was found to be between* 10^{-12} *and* 10^{-14}*g. A major discovery is the observation that a considerable fraction of the dust grains serves as extended source of gas in the inner coma. It appears that, with the exception of a hydrogen depletion by about a factor of 1000, the light elements in Comet Halley show almost cosmic abundance; also the isotopic composition appears to be essentially solar. The ices consist of about 80% water and about 10% CO,* CO_2, H_2CO *and* H_2CO_2*; the remaining few percent appear to be made up of* CH_4, NH_3, *HCN and other parent molecules.*

I. INTRODUCTION

Cometary science has changed drastically following the spacecraft encounters of Comet P/Giacobini-Zinner in September 1985 by the International Cometary Explorer (ICE) and of Comet P/Halley by the international "Halley Armada" in March 1986, as well as by the comprehensive groundbased observational campaign focused on Halley's comet by the International Halley Watch (IHW). This unparalleled global scientific effort of exploration started immediately following Halley's recovery in 1982. Remote and *in situ* measurements were conducted from the ground, from Earth orbit, from Venus orbit, from interplanetary space, and from within the comet itself.

Comets are generally assumed to have formed from the same interstellar gas and dust cloud of which the primitive solar nebula was a part. However, many processes changed the interiors and exteriors of the planets and their satellites during their continuous evolution over the past 4.5 Gyr and have erased most information about the accumulation processes as well as the original state of the material. Comets, on the other hand, are expected to be remnants left over from the formation of the solar system which have preserved the chemical and physical characteristics of the condensing matter. They probably solidified at temperatures well below 150 K which are much lower than the corresponding temperatures for meteorites of more than 400 K. They are thought to be stored essentially unchanged at the cold outer edge of the solar system in the Oort cloud which extends some tens to thousands of AU from the Sun. Their small size and generally large distance from the Sun places them therefore among the most primitive objects remaining in our solar system. Comets have also been proposed as basic building blocks in the accumulation of the giant planets (Whipple 1964; Öpik 1973) and the discovery of large amounts of organic materials from Earth- and space-based measurements makes it appear likely that comets provided volatile and organic constituents for planetary atmospheres which may have been essential to the origin of life.

While the cometary nucleus is thus a unique witness of the remote past when it accumulated in the interstellar cloud of which the solar nebula was a part, the changing appearance of the coma and tail give us information about the present condition of interplanetary space with which the comet interacts.

II. THE NATURE OF THE VOLATILE COMPONENT OF COMETS

Long before the spacecraft flybys, it had been established that cometary material is an aggregate of ices and solid dust with weak cohesive strength. Independent observations had revealed the presence of volatiles, i.e., a mixture of molecules mainly made up of H, C, N and O atoms, and of refractory particles, i.e., dust grains of different sizes ranging from sub-μm, i.e., probably unaltered interplanetary dust grains, to particles with diameters more than centimeters.

When a comet comes close enough to the Sun, the outer layers of its nucleus are heated up, material vaporizes and is released into space while fresh material is exposed to solar radiation. The sublimated ices drag along dust and perhaps also ice particles and together they form the familiar comae and tails. The volatile component of the nucleus is made up of so-called parent molecules that are very difficult to detect. In fact, only a few have so far been identified (Table I). Most of the daughter species (atoms, ions, radicals) are visible through resonance fluorescence. The production of these species is phased with the water production rate. Not surprisingly, water appears to be the dominant volatile. Delsemme (1982,1985) noted already several years ago an interesting similarity in the spectra of the comets Bennett 1970 II, Kohoutek 1973 XII and West 1976 VI. He compared the elemental ratios of H, C, N, O and S and derived mean abundances in the volatile fraction for these comets (Table II), suggesting that these values might be typical for an average bright comet. Geiss' (1987) excellent review article on the composition measurements and the history of cometary matter lists the relative elemental abundances of the material released by Comet Halley at the time of the encounters and compares it with the abundances in the Sun (Grevesse 1984*b*),

TABLE I
Abundances of Probable Parent Molecules in the Coma of Comet Halley[a]

Species	Gas Production Rate Relative to H_2O[b]	Instrumental Technique[b]
CO	$0.05 \cdots 0.07$[1]	Giotto NMS, gas spectra[5]
	$0.17 \cdots 0.20$[2]	Rocket UV experiment[6]
	$0.13 \cdots 0.15$[2]	Giotta NMS, gas spectra[5]
CO_2	$\leqslant 0.035$	Giotta NMS, gas spectra[7]
	0.015	Vega 1 IKS, IR spectra[8]
CH_4	$\leqslant 0.07$	Giotto NMS, gas spectra[7]
	$\leqslant 0.04$	KAO, IR spectra[9]
	≈ 0.02[3]	Giotto IMS, ion spectra[10]
NH_3	$\leqslant 0.1$	Giotto NMS, gas spectra[7]
	$0.01 \cdots 0.02$[3]	Giotto IMS, ion spectra[10]
N_2	$\leqslant 0.02$[4]	Giotto NMS, gas spectra[5]
	< 0.02[3]	Giotto IMS, ion spectra[10]
saturated hydrocarbons	≈ 0.01	Vega 1 IKS, IR spectra[8]
unsaturated hydrocarbons	≈ 0.01	Vega 1 IKS, IR spectra[8]
H_2CO	≈ 0.01	Vega 1 IKS, IR spectra[8]
HCN	≈ 0.001	IRAM telescope, millimeter spectra[11]

[a]With few exceptions the data were obtained on March 6 and March 14, 1986 at heliocentric distances of 0.79 and 0.89 AU, respectively (table after Krankowsky and Eberhardt 1988).
[b]Notes: [1]released from nucleus; [2]released from nucleus and extended source; [3]inferred from modeling of coma; [4]data re-interpreted; [5]Eberhardt et al. (1986*b*, 1987*b*); [6]Woods et al. (1986); [7]Krankowsky et al. (1986); [8]Combes et al. (1986); [9]Drapatz et al. (1986); [10]Allen et al. (1987); [11]Bockelée-Morvan et al. (1986).

TABLE II
Mean Abundances in the Volatile Fraction of Bright Comets Normalized to Oxygen = 1.0[a]

Element	Atomic Mass	Mean Abundance
H	1.0	1.8 + 0.4
C	12.0	0.20 + 0.10
N	14.0	0.10 + 0.05
O	16.0	1.0
S	32.0	0.003 + 0.002

[a]Table after Delsemme 1985.

the solar nebula (Anders and Ebihara 1982) and carbonaceous meteorites (Kerridge 1985). Geiss assumes a gas/dust ratio of 2. Figure 1 (from Geiss 1987) clearly shows that the cometary oxygen, carbon and nitrogen are all at least one order of magnitude higher than C1 chondrites, and even more than in C2 or C3 chondrites. The cometary oxygen and carbon values are very similar to the solar values, but nitrogen is considerably less abundant in Halley than in the Sun; a fraction of nitrogen could be present in the form of N_2.

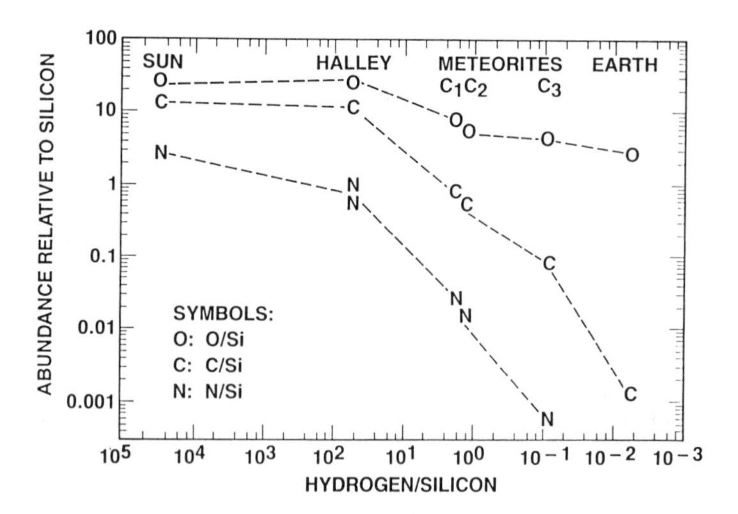

Fig. 1. Relative abundances of elements of the material released by Halley's comet. A gas/dust ratio of 2 at the source was assumed. For comparison, the abundances in the solar nebula (Anders and Ebihara 1982) and in carbonaceous meteorites (Kerridge 1985) are given together with estimates for the abundances on the Earth's crust plus mantle (Larimer 1971; Wänke, 1981; Wänke et al. 1984).

According to Spinrad (1987), the (O I) production rate at 1.5 AU for large comets such as Kohoutek, amounted to 130×10^{26} atoms s^{-1}; for medium-sized comets such as P/Halley or P/Giacobini-Zinner to 120 or 12×10^{26} atoms s^{-1}, respectively; and for small comets like P/Brooks to 1.2×10^{26} atoms s^{-1}. Halley produced near perihelion about 2×10^{29} mol s^{-1} or 4×10^{6} g s^{-1}. This amounts to about 2×10^{14} g/revolution which implies that Halley loses approximately a m-thick outer mantle every time it moves around the Sun. Assuming a constant gas production rate, one can estimate that the comet can make fewer than 10,000 revolutions.

Material which is more volatile than water is still present in Halley and other comets and causes pronounced activity at distances of 3 to 5 AU from the Sun where water ices are not affected. This might indicate that the nucleus has not been heated during and following its formation. Since some of the elements and compounds observed in Halley's comet, can have solidified only near absolute zero, the nucleus must have formed in an area reaching from near the outer planets to thousands of AU away from them. Although methane is a low-temperature condensate, it was found to be present in the outer surface layers, supporting the idea that the cometary material is made up of low-temperature pristine solar nebula matter.

Not all elements will appear in volatile form; the complex organic polymers, for example, remain solid and may only be detectable through *in situ* measurements. Solid organics were in fact measured by Kissel and Krueger (1987) and possibly by PICCA (positive ion cluster composition analyzer). Cosmic radiation could form solid organic material from water and carbon monoxide on grains that were aggregated into cometesimals, or also in the Oort cloud on the nucleus' outer layers (see Strazzulla et al. 1984, 1985*b*). A confirmation of this hypothesis would enhance the probability that more complex organic molecules exist which could have formed in the cometary environment.

III. THE COMPOSITION OF HALLEY'S DUST DERIVED FROM THE ANALYSIS OF THE *IN SITU* MEASUREMENTS

Even now, more than two years after the spacecraft encounters with Halley's comet, it is still too early to present a complete picture of the composition of the cometary dust. This chapter therefore only summarizes the present status of the continuing efforts to extract useful information on the dust from the mass spectra transmitted to Earth from the Giotto and the two Vega spacecraft. Earlier publications on this subject are by Kissel et al. (1986*a,b*), Clark et al. (1986), Jessberger et al. (1986), Solc et al. (1986), Sagdeev et al. (1986), Waesch (1986), Kissel and Krueger (1987), Brownlee et al. (1987), Jessberger et al. (1987), Krueger and Kissel (1987), Langevin et al. (1987), Mukhin et al. (1987) and Solc et al. (1987).

IV. THE EXPERIMENTS

For the *in situ* measurements of the chemical, isotopic and molecular composition of Halley's solid dust particles, the Particulate Impact Analyzers "PUMA 1 & 2" and "PIA" were installed onboard the three space probes Vega 1 and 2 and Giotto, respectively. The physical principle as well as the technical realization of the instruments have been described in detail elsewhere (Kissel 1986) so here we make only a few general comments.

A schematic view of the instrument is given in Fig. 2. Solid dust particles impinge on a clean target surface—Ag or Ag-doped Pt—with a relative velocity of 69 (for Giotto) and 79 km s^{-1} (for Vega), respectively. Upon impact the particle and some target material are vaporized and partly ionized. Positive ions are extracted by a grid arrangement, accelerated to 1 keV, and subsequently separated in a time-of-flight mass spectrometer according to the mass/charge ratio, $m/q = 2\, s^2\, U\, t^2$, where m = mass, q = charge, s = flight-path length, U = accleration voltage and t = flight duration. To increase the mass resolution to $m/\mathrm{d}m$ = 150, an ion reflector counterbalances possibly present initial ion energy differences. In the PUMA instruments two voltage settings alternate every 30 s to reflect ions with initial energies up to 50 eV ("long spectra") and up to 150 eV ("short spectra"), respectively (cf. Kissel and Krueger 1987). Ions with larger initial energies hit the wall and are lost for analysis. The remaining ions are detected by an open electron multiplier at the end of the drift tube.

Measured quantities are the flight duration (i.e., the time interval be-

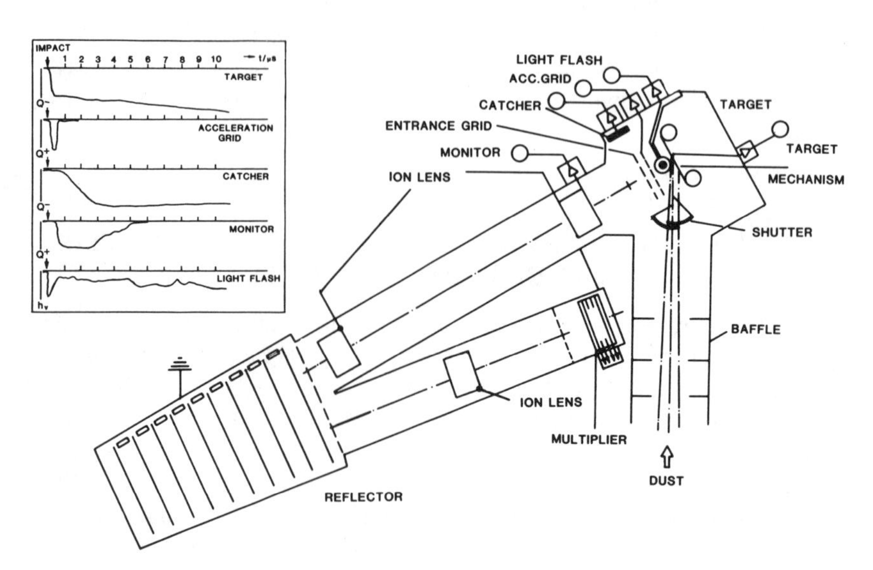

Fig. 2. Schematic view of the particulate impact analyzers onboard Giotto and Vega 1&2. The inset shows characteristic front-end signals.

tween the impact pulse at the front-end detectors [see Fig. 2] and the ion arrival time at the multiplier) and the peak height. The flight duration is proportional to $(m)^{1/2}$, and a mass scale can be calculated. The actual peak heights are onboard converted to their logarithms to cover a dynamic range of 5 decades. Because of the limited data transmission rate, the spectra are received on the Earth in differently compressed modes. In this chapter we discuss PUMA 1, mode 0 spectra, which are transmitted in a transient recorder-like mode, approximately 4 data points per amu. We use all 39 long and 40 short PUMA 1, mode 0 spectra which can be interpreted. The more compressed spectra of modes 1 to 3 are still being evaluated (Langevin et al. 1987; Solc et al. 1987).

V. THE MASS SPECTRA

All mass spectra discussed in this chapter (see Color Plates 1–5), have the same format: the original time-based full spectrum is plotted in yellow at the bottom of the graphs. The peaks at the right side are due to silver at 107 and 109 amu; they are well resolved in most cases. These peaks are used to define the mass scale and to estimate masses and densities of the dust particles (Kissel et al. 1986*a*). The same spectrum with the correct mass scale is shown in red upside down. In the upper part, a relevant portion of the mass spectrum is enlarged for clarity. Horizontal bars indicate intensity decades, vertical bars integer mass positions. The blue spectrum is a simulated spectrum and is calculated mimicking the actual peak shapes as closely as possible. It is important to note that we assume that only singly charged atomic ions are present and that the elements have a normal isotopic composition. The general validity of both assumptions will be demonstrated below. As a consequence, we have, strictly speaking, only upper limits for the ion abundances.

The (long) spectrum No. 54630 (Plate 1) is a prime example of a "well behaved" spectrum which exhibits nicely resolved peaks (note the logarithmic intensity scale). The peak ratios of 24/25/26 closely resemble the isotopic composition of magnesium (as indicated by the similarity of the yellow with the blue tracers). The ratio of the peaks at 35 and 37 is 3 : 1 as in chlorine, and there is only a slight excess at peak 30 when compared to the 28/29/30 ratio of silicon. Even the ratio of the peaks at 32 and 34 is not far from the isotopic composition of sulfur, and the 40/44 peak ratio is close to the calcium isotope ratio. At higher masses some small unidentified peaks at 64, 86 and 88 appear, but overall the spectrum is remarkably simple.

The (long) spectrum No. 51666 (Plate 2) shows unstructured background intensities below about mass 10, followed by an "intensity blend" which levels off at about 22 amu. The remainder of the spectrum is nearly as clean as the previous one. In the mass range 10 to 20, two peaks emerge from the blend. Although the minor isotopes of carbon and oxygen are partly covered by this mountain, we ascribe the peaks at 12 and 16 to these two elements.

The next (long) spectrum, No. 52069 (Plate 3), shows a similar pronounced blend, but besides well-resolved silver peaks, there is no other element discernible except perhaps a small contribution from iron. The three spectra discussed above (Plates 1, 2 and 3) demonstrate the wide range of features which were observed. Spectra showing no peaks are not reproduced here.

In the following, it will be demonstrated that the spectra are mostly composed of atomic ions and singly charged ions. This requires comparison of features observed in different spectra and one has therefore to assume that the ionization mechanism and related processes are the same for all particles. Normally, these assumptions would be tested by calibration experiments, but the velocities of particles, as they were encountered during the flybys, cannot yet be obtained in the laboratory. Simulations at lower impact speeds yield spectra with very different ion patterns and are of limited value for direct comparison with the *in situ* spectra (Krueger 1987).

We have already demonstrated that the pattern observed in spectrum No. 54630 (Plate 1) can be explained by atomic ions. The same holds especially for the features around mass 40; in spectrum No. 51666 (Plate 2), the intensity at mass 40 is low (65 counts, henceforth, cts) compared to that of ^{24}Mg (800 cts) and ^{16}O (12,100 cts). In No. 54981 (Plates 4*a,b*), the signal at mass 40 is considerably higher (550 cts) while the ^{24}Mg (1150 cts) and ^{16}O (2000 cts) signals are about the same as in the previous spectrum (Plate 3). This fact, as well as the presence of a pronounced signal at mass 44 in spectrum No. 54981 (Plates 4*a,b*), indicate that the intensity at mass 40 is not due to ^{24}Mg^{16}O^{+} but to ^{40}Ca^{+}. The absence of any signal at mass 20 in this spectrum indicates therefore the absence of doubly charged ^{40}CA^{++}. Preflight speculations that doubly and even triply charged ions will dominate the spectra, can therefore be discounted. Instead, predictions (Krueger 1984*a,b*) of the emission of singly charged atomic ions are confirmed.

VI. THE PARTICLES

The mass spectra demonstrate that various particle types were observed. The two (long) spectra No. 54630 (Plate 1) and No. 54981 (Plates 4*a,b*) show a considerable difference in the intensities of H, C, N and O relative to those of the heavier elements like Mg, which are about the same in both plates. Another more extreme example for HCNO-rich particles is shown in Plate 5, the (short) spectrum No. 51991, that is clearly dominated by peaks from the light elements. The presence of distinct particle types was noted very early in the data analysis (Kissel et al. 1986*a,b*). Particles like No. 51991, which show high abundances of light elements have been named *CHON* particles. Those like No. 54630, with no or only a minor content of these elements were called *silicates*. Particles such as No. 54981 were termed *mixed*. Figure 3 shows the ion abundance ratios C/Mg versus H/Mg separately for long and

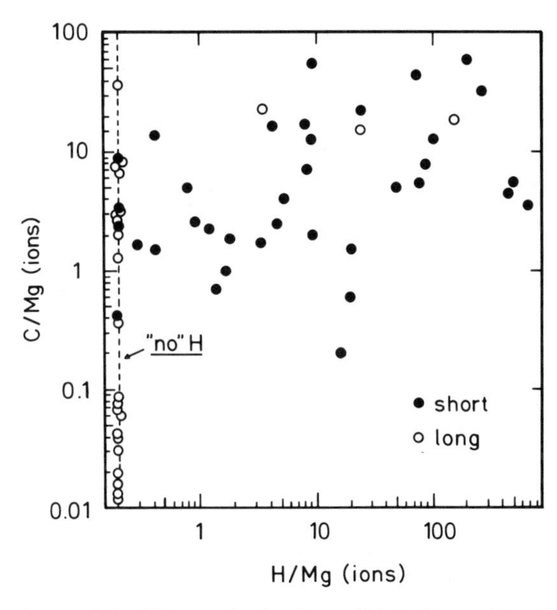

Fig. 3. Ion abundances of C and H normalized to those of Mg in short and long PUMA 1, mode 0 spectra.

short spectra. Both element ratios vary by 4 orders of magnitude, illustrating the variability of the particle compositions in terms of CHON/silicate ratios. In addition, it also demonstrates systematic differences in the ion composition of long and short spectra. Only 3 out of 39 long spectra contain all 3 elements while this holds for 32 out of 36 short spectra. All short spectra contain both, C and Mg, which is true for only 25 of the 39 long spectra. All C/Mg ion ratios from short spectra are >0.2. These differences must be due to different experiment conditions in the long and short modes, i.e., energy window settings in the ion reflector. The CHON ions do not pass the narrow energy window as easily as the silicate ions, which implies that their energy directly after impact is often larger than 50 eV. This may mean (Krueger and Kissel 1987) that the CHON ions stem from organic material surrounding a mineral core.

In Whipple's (1951,1963) classical model, the comet nucleus is composed of dust and ice, a dirty snowball. The presence of ices in the dust particles which were measured by Vega 1 at a distance of more than 8890 km from the nucleus, can be excluded on kinetic grounds; vaporization may leave a nonvolatile residue if the grain has been processed by irradiation (Hanner 1981). We conclude that the CHON material is refractory. Kissel and Krueger (1987) analyzed the remaining features which are left after the atomic ions—identified by the normal isotopic pattern of the elements—have been subtracted from the spectra. Applying statistical methods and organic chemistry systematics, they conclude that several types of organic molecules are present on the dust grains (Fig. 4). CN-containing compounds surprisingly are far more frequent than CO-containing compounds, a result which has important consequences for cosmochemistry and the evolution of life.

The density of pure anorganic grains is about 1 to 2 g cm^{-3} and that of pure organic CHON grains is mainly 0.3 to 1 g cm^{-3} (Kissel and Krueger 1987). These numbers indicate a very fluffy nature for cometary dust grains. The mass of most dust particles was found to be between 10^{-12} and 10^{-14} g (Kissel et al. 1986*a,b*; Solc et al. 1987). The dust-grain densities measured *in situ* are close to the average particle density of about 0.25 g cm^{-3}, deter-

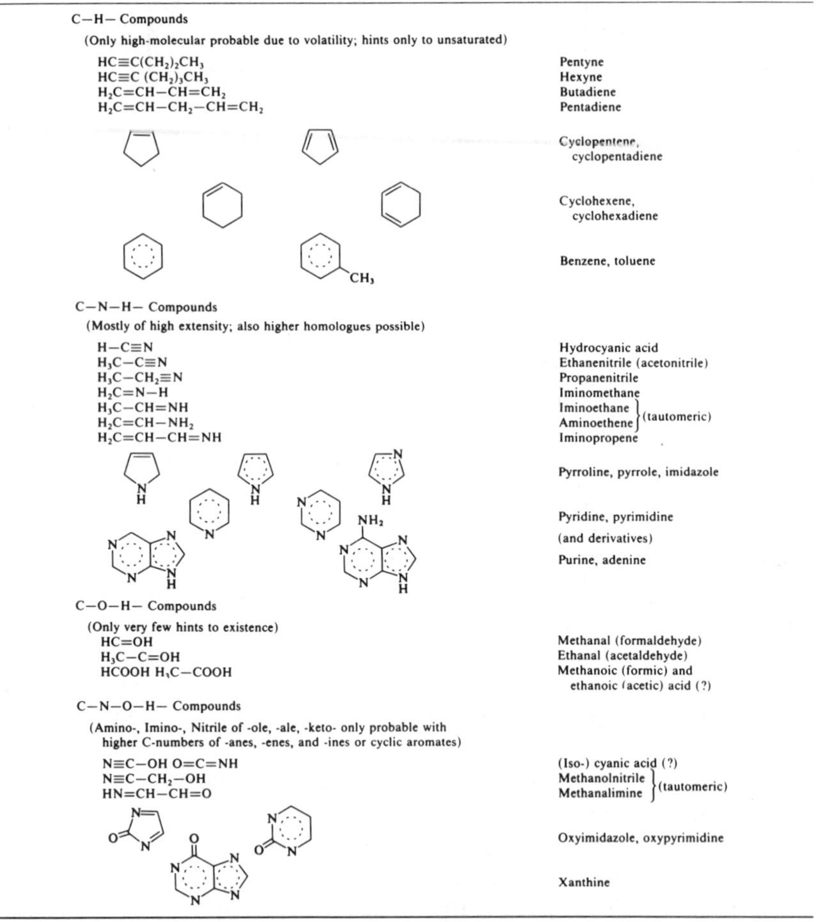

Structure isomers are additionally possible. Several types may form tautomers, mesomers and conformational isomers. Thus the molecules given here serve only as examples of the class of substances possibly present in the organic component of the dust. We are not sure yet if oxygen-containing species are present.

Fig. 4. Types of organic molecules as inferred from PUMA 1, mode 0 mass spectra (figure from Kissel and Krueger, 1987).

mined in the meteor streams of the Orionides and Aquarides which are both produced by Halley. Instead of speaking of the nucleus as a "dirty snowball," Whipple compared it with a "dirty snowdrift" which is made of freshly fallen snow and has somehow become compressed over time. The low packing density results in a very low heat conductivity between the outer dust layer and the ice-dust mixture underneath. A detailed model of a cometary nucleus that takes these and other results into consideration, has recently been published by Donn (1988) and Weissman (1988).

A major discovery is the observation that a considerable fraction of the dust grains serves as an extended source of gas in the inner coma. Eberhardt et al. (1986) found a large source of CO molecules in the inner coma that extended up to about 15,000 km from the nucleus and made up more than half the total CO flux; the total production of CO was estimated to be CO/H_2O = (10±3)%. The CN radical was found (A'Hearn et al. 1986; Eberhardt et al. 1987*c*; Krankowsky and Eberhardt 1988) in strange jets which remained narrow and well defined for up to about 50,000 km but which did not coincide with the jets made of visible dust. It is conceivable that the CN radicals did not come from the nucleus directly but are somehow related to streams of extremely tiny (and visually unobservable) particles consisting of CHON material. The CHON particles could form tar-like substances which may be a reason for the dark nucleus (see below). Greenberg et al. (1986) have shown in the laboratory that organic material can be formed out of cometary matter. He postulates that submicroscopic silicate cores are covered by organic coats and ice. They could form larger, loosely connected particles in agreement with the observed low density of the dust grains.

The observation that the CHON particles have almost cosmic abundance of carbon which is about eight times more abundant than in carbonaceous chondrites, may also solve a problem that has puzzled groundbased astronomers for years. Before the *in situ* measurements were made, groundbased observations had indicated that carbon (relative to oxygen) was about 4 times less abundant than in the cosmic abundance. Now it appears that the missing carbon is present in the CHON material.

Another surprise was the high abundance of very small particles. Groundbased optical and infrared observations had provided practically no information about particles smaller than 0.1 μm, and their number was expected to be very small. The dust detectors onboard the spacecraft found, however, that their abundance increased considerably towards smaller sizes, and Giotto measured tiny pieces of matter of only 10^{-17} g which were perhaps only about 100 atoms in diameter (Knacke 1987).

VII. THE ISOTOPES

Mass spectrometry of cometary dust particles allows also the determination of isotope ratios. Meteoritic studies have revealed important information on the origin, formation and evolution of the early solar system material. The study of isotopes in cometary dust should lead to equally important results. Cometary material is probably least affected by planetary processes that tend to dilute isotopic peculiarities. The mass spectra shown in Color Plates 1–5 (keeping their logarithmic intensity scale in mind) do not allow determination of small deviations from the normal isotopic composition. In meteorites, the largest isotope anomalies, with the exception of the noble gases, have been found in H, C and Mg. It is interesting to compare these data with those from Halley's dust particles.

The cosmic hydrogen isotope ratio is D/H= 0.00015. Derivations from this value cannot be determined from the positive ion spectra of Halley's comet since any intensity at mass 2 is due to H_2^+. Eberhardt et al. (1986) give a D/H ratio between 6 and 48×10^{-5}, indicating that the deuterium in Halley is enriched by at least a factor of 3 relative to the interstellar medium and the giant planets, but is comparable to the values for terrestrial oceans and the atmospheres of Titan and Uranus. The authors' $^{18}O/^{16}O$ ratio is comparable to the solar system value of 2×10^{-3}.

Carbon isotope data from PUMA 1 have been discussed by Solc et al. (1986) and Jessberger et al. (1987*a*). Wyckoff et al. (1988) inferred from spectroscopic data for CN a $^{12}C/^{13}C$ ratio of 65±8. In order to search for a relation of the $^{12}C/^{13}C$ ratio with the relative N content of the dust, the 12/13 ion intensity ratio is plotted in Fig. 5 versus the 16/14 ratio, which

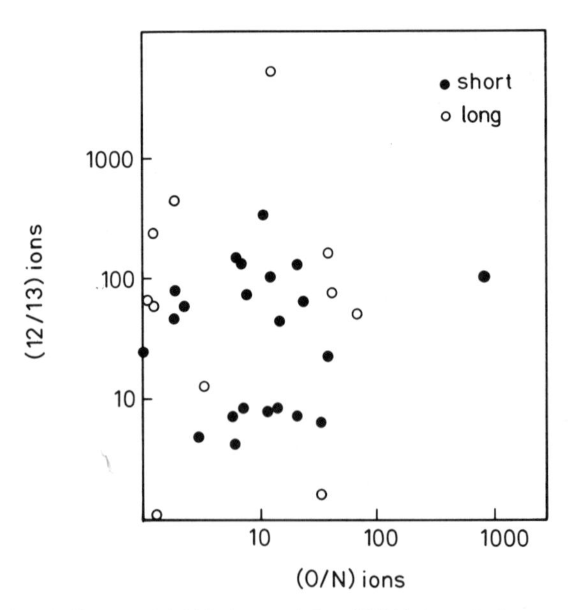

Fig. 5. Ion ratios 12/13 versus 16/14 in long and short PUMA 1, mode 0 spectra. Error bars are not available for individual data points.

correspond to O/N. It is remarkable that the measured 12/13 intensity ratios cover the range from 1 to 4000 while the galactic $^{12}C/^{13}C$ ratios range from 4 to 100 (Audouze 1977), and the terrestrial value is 89. In the graph no correlation is apparent that would correspond to the results of Wyckoff et al. (1988).

A special physical clue relating comets to the solar system comes from an analysis of the $^{12}C/^{13}C$ abundance ratio. The terrestrial ratio of 89 stems from the time when the solar system was formed out of the galactic gas and dust cloud, while this ratio in the interstellar medium reflects the present conditions. The observed interstellar ratio is still uncertain, but seems to be somewhat below the terrestrial value. If the cometary and interstellar $^{12}C/^{13}C$ ratios could be measured with high precision, the important question could perhaps be answered as to whether comets are intrinsic members of the solar system or intruders from interstellar space.

In Fig. 5 two 12/13 regimes are evident with the dividing line at 12/13 = 20 independent of the O/N ratio. Low 12/13 values could be due to an interference of $^{12}CH^{+}$ at mass 13. However, a correlation of the 12/13 ratio with the H intensity, which would then be expected, is not apparent (Jessberger et al. 1987). It is perhaps premature to assign the low 12/13 intensity ratios to carbon, although such values are very similar to those of extrasolar carbon found in certain components of meteorites (Yang and Epstein 1983) and in interplanetary dust particles (Zinner and Epstein 1988). The "normal" ratios of about 100 (Fig. 8) agree with those inferred for some comets from Earth-based observations (Vanýsek and Rahe 1978), but there are obvious uncertainties about these values which are mainly due to blends or systematic errors, as well as poorly understood fractionation processes. The much larger values of up to 4000 constitute a problem. Such ratios have never been observed and it seems too speculative at the present time to link them to theoretical models predicting such ^{12}C enrichments in portions of dense molecular clouds (Langer et al. 1984). Clearly, the $^{12}C/^{13}C$ value of Halley's dust deserves further attention.

The isotopic composition of magnesium attracted special attention in the last 14 yr following the discovery of the ^{26}Mg isotope anomaly in meteorites (Gray and Compston 1974; Lee et al. 1977). The ^{26}Mg excess of up to 15% has been attributed to extinct ^{26}Al ($T_{1/2}$=740,000 yr) which was probably live in the early solar system and could then have acted as a powerful heat source in primitive solar system bodies. In meteorites, large ^{26}Mg excesses are observed in minerals with high Al/Mg ratios of >100. Figure 6 shows the Al/Mg versus the Ca/Mg ion ratios of Halley's dust. The largest value obtained is about 1. Since Al and Mg have similar ion yields (Krueger 1987), a maximum Al/Mg atom ratio of 1 is implied. Large ^{26}Mg effects can therefore not be expected.

Among the species investigated, Mg is the only element with more than 2 isotopes. It is also the only element with more than 2 isotopes, for which the

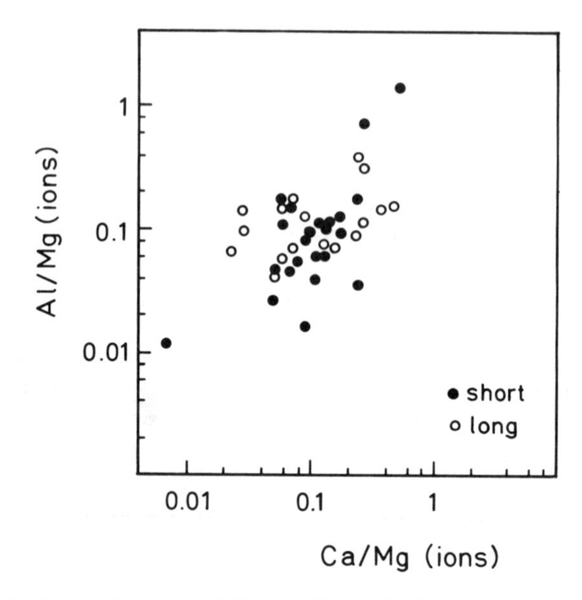

Fig. 6. Ion abundance ratios of Al and Ca normalized to Mg in PUMA 1, and mode 0 spectra.

abundance ratios of minor-to-major isotope are <10. The isotope diagram of Fig. 7 (Jessberger et al. 1987*a*), shows intensity ratios 24/25 plotted versus 26/25, where the symbol areas are proportional to the 25 signals. With the exception of a few low-intensity data, the 26/25 intensity ratios vary from 0.5 to 1.7 while the 24/25 ratios range from 3 to almost 100 irrespective of the peak height at mass 25.

Interestingly enough, the geometric averages of both ratios are about the same as the normal Mg isotope ratio, for long as well as short spectra. ^{26}Mg excesses of more than a factor of 2 would be detectable if they were present. 24/25 ratios below the normal value are accompanied by normal 26/25 ratios; they are either artifacts of the experiment of unknown origin or they are due to coincident molecular interferences on the peaks at masses 25 and 26. Such an interference, $^{12}C_2{}^+$, results in 24/25 ratios that are much larger than normal. A correlation of the 24/25 ratios with the 12 intensity or the 12/13 ratio could, however, not be established.

The most reliable isotopic information can perhaps be obtained for C1. In spectrum No. 54630 (Plate 1), the normal $^{35}C/^{37}C1$ ratio of about 3 is reproduced. This is also the case for many other spectra (Solc et al. 1986) and may demonstrate the instrument's capability to reproduce isotope ratios in mass ranges that are not overloaded with molecular ions. It should be noted, however, that it was recently shown by SIMS (secondary ion mass spectrometer) analysis of the PUMA (dust impact mass analyzer) Ag target that Cl is a terrestrial surface contamination (F. Buehler, personal communication).

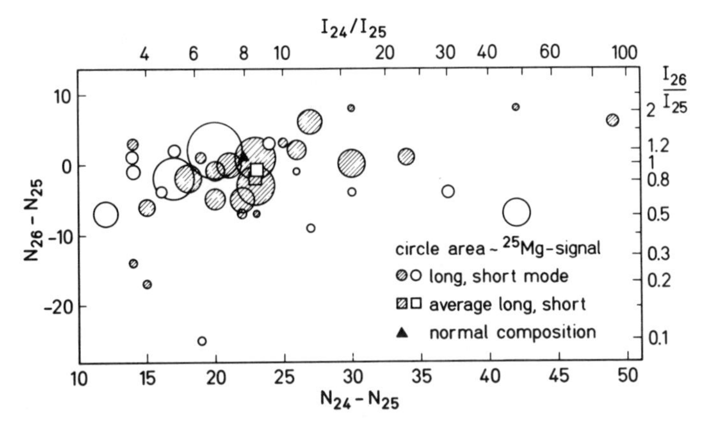

Fig. 7. Ion ratios 24/25 versus 26/25 in long and short PUMA 1, mode 0 spectra. The left-hand and bottom scales give the differences of the (logarithmic) signals at 24, 25 and 26 amu as received on the Earth. From these we calculate ion density ratios with the decade length $D = 25$ (cf. Jessberger and Kissel 1987), the scales of which are shown on the right-hand side and at the top.

VIII. THE BULK COMPOSITION

An excellent way to determine the bulk composition of a comet, would be to sample different sites and different depths of the nucleus and then fold the composition of many samples with the internal structure. The first part of this experiment is realized—at least to some extent—in the present data base. We have on hand the analyses of many grains coming from the nucleus from different sites and possibly from different depths below the immediate surface, especially when they are released in form of jets. We do not know, however, the location from which these samples come nor do we have reliable information on the internal structure of the nucleus. With these deficiencies, it appears justified to simply take the sum of all grains as representative for the whole comet, i.e., to sum the elemental ions identified in the spectra on the basis of their isotopic signature. In doing so, high-intensity spectra, i.e. "large" grains, dominate this sum in such a way that typically 60% of the total intensity of an element is contributed by 10% of the grains. Reasonable weighting factors are unfortunately difficult to derive.

The next step is to convert the ion numbers determined from the spectra to atomic numbers. Due to the lack of a good calibration, we used the relative ion yields derived for this type of instrument by Krueger (1987) which are believed to be correct within a factor of 2. Relative ion yields are sufficient since element abundances represent element ratios. For normalization, we choose Mg instead of the commonly used Si, because the identification of the ion abundance for Mg with the composition $24:25:26 = 7.3:0.9:1$ is better defined than the one for Si with $28:29:30 = 29.7:1.5:1$. The minor isotopes of Mg are less likely to be affected by background than those of Si.

TABLE III
Average Atomic Abundances of the Elements in Halley's Dust and Type 1 Carbonaceous (CI) Chondrites

Element	Halley's Dust[b]	CI[a]	Halley/CI
H[c]	2025	492	4.1
C	815	70.5	11.6
N	42	5.6	7.5
O	890	712	1.3
Na	10	5.3	1.9
Mg	= 100	= 100	= 1.0
Al	6.8	7.9	0.9
Si	185	93.0	2.0
S	72	47.9	1.5
K	0.2	0.35	0.5
Ca	6.3	5.68	1.2
Ti	0.4	0.22	1.9
Cr	0.9	1.25	0.7
Mn	0.5	0.88	0.6
Fe	52	83.7	0.6
Co	0.3	0.21	1.2
Ni	4.1	4.59	0.9

[a]Data from Anders and Ebihara 1982.
[b]The data on cometary dust derived in this work are based on a total of 79 high-quality mass spectra from PUMA-1.
[c]From short spectra only.

Table III gives the atomic element abundances as derived from the best 79, mode 0, spectra obtained from PUMA-1. The element abundances in CI chondrites are listed for comparison (Anders and Ebihara 1982). CI chondrites are the least metamorphosed rocks from the early solar system available for laboratory analysis. The abundances of the condensible elements in CI chondrites are the same as in the solar photosphere and by implication in the whole Sun and, by further implication, of the solar system as a whole. The CI chondrites thus provide the basis for the "cosmic" abundance of the elements. From Na upwards, the two abundances are the same within a factor of 2 which coincides with the uncertainty of the ion yields. The light volatile elements H, C and N (also O; cf. Jessberger et al. 1987*b*) are enriched relative to CI chondrites and are almost as abundant as in the solar photosphere (cf. Palme et al. 1981). The distinction of the *CHON* elements from the *silicates* has been noted before (Kissel et al. 1986*a,b*). The apparent enrichment of the CHON elements, that are present as refractory organics, identifies Halley's dust as the most unaltered early solar system material ever analyzed in the laboratory.

IX. VARIATIONS OF Mg, Si AND Fe

The discussion of element abundance variations from grain to grain will here be limited to Mg, Si and Fe since these are the three major rock-forming elements. When studying the form in which Fe is present (Anders 1986), e.g., in olivine or pyroxene, one notes that olivine and pyroxene differ by a factor of 2 in the (Mg+Fe)/Si ratio and again this coincides with the ion yield uncertainty. Brownlee et al. (1987) and later also Jessberger et al. (1987*b*) have argued that ion yields can be obtained by assuming a CI chondrite composition of the rock-forming elements in Halley's dust. The main argument is (Jessberger et al. 1987*b*) that the bulk Fe/Si ratio (Table III) is very low (only 0.3) which is not very common in early solar system bulk materials (Anders 1964; Mason 1971). Consequently, we correct the yields of these elements in such a way that the bulk compositions calculated from all spectra are those of CI chondrites.

The distribution of Fe/(Fe+Mg) obtained in the present study is shown in Fig. 8. As in Brownlee et al. (1987), who presented a similar plot from their data analysis, the ratios cover the whole range from 0 to 1. But, in contrast to their result, here some regimes are well separated: the two extremes with zero Fe (18 grains) and low Mg (4 grains), and two intermediate groups with Fe/(Fe+Mg) around 0.2 and around 0.5, respectively. The latter group extends to higher values which may indicate another group with Fe/(Fe+Mg) = 0.7. Carey et al. (1987) reported Fe/(Fe+Mg) microdistributions in three interplanetary dust particles and one Solar Max particle of different infrared (IR) types. Two particles of the layer-lattice-silicate IR type have ratios mostly in

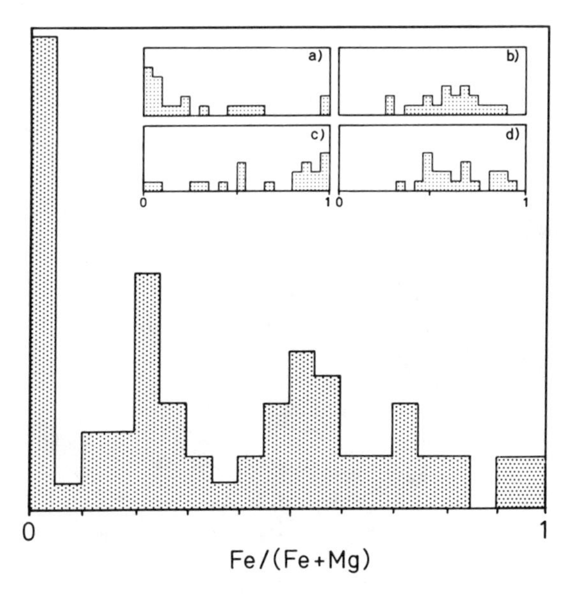

Fig. 8. Histogram of the Fe/(Fe+Mg) atom ratios in 74 grains of Halley dust. For comparison, the inserts show the variation of that ratio in interplanetary dust particles (Carey et al. 1987: (a) Viburnum, olivine IR type; (b) SM504-P4, serpentine layer-lattice-silicate IR class; (c) butterfly, pyroxene IR class; (d) W7029*A, layer lattice silicate.

the range 0.4 to 0.9. In an olivine IR-type particle, Fe/(Fe+Mg) tends towards low values and in a pyroxene IR-type particle towards high values. Comparing that with Halley's dust data one may identify the low values with forsteritic olivine IR types and those with Fe/(Fe + Mg) = 0.5 with layer-lattice-silicate and/or pyroxene IR-type grains.

Figure 9 shows Halley's dust data in the system Mg-Si-Fe (atom-%) as derived in the present study. The data are marked according to subjective quality criteria using the overall appearance of the spectrum and the clarity of the assignments of Mg, Si and Fe based on their isotopic signature. Type 1 lists excellent spectra and type 3 gives those which can just still be used. The composition in Fig. 9 shows considerable scatter, but the data points are not randomly distributed. The triangle ABC (A: Si = 100; B: Si = 50, Fe = 30; C: Fe = 100) is practically void of data, implying the lack of both Fe-rich pyroxenes as well as Fe-rich olivines. On the other hand, the triangle CDE (D: Mg = 60, Fe = 0; E: Mg = 35, Fe = 0) encompasses two thirds of all grains. The group of grains close to Fe = 0, Mg = Si = 50 indicates the survival of primary Mg-silicates which had not readily reacted with iron oxides to form olivine. This is again evidence for a low temperature, i.e., for the primitive nature of Halley's solid dust. It should be mentioned, however, that although the infrared spectra suggest the presence of noncrystalline grains, the identification of regions with pyroxine, olivine and enstatite requires crystalline grains. Nuth and Donn (1986) have shown that amorphous grains condense and produce structural peaks as they anneal.

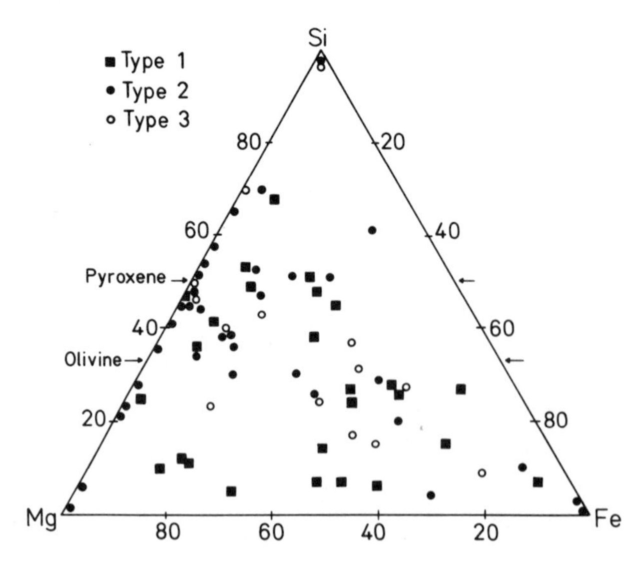

Fig. 9. Mg-Si-Fe diagram in Halley's dust grains. The ion yields are adjusted such that on the average Mg/Si/Fe is chondritic (see text). The types are a measure of quality of the spectra, type 1 being highest. Arrows indicate the loci of olivines and pyroxenes.

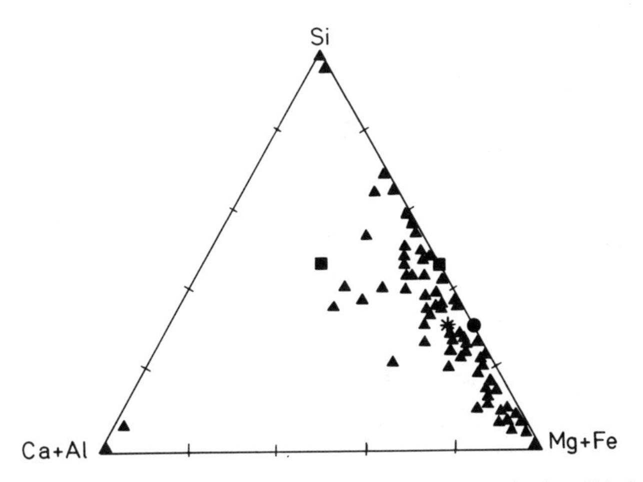

Fig. 10. (Ca+Al)-Si-(Mg+Fe) diagram in Halley's dust grains shown by the solid triangles. The star represents the chondritic composition, the solid circle the olivine point, and the 2 solid squares give the end member of pyroxenes.

Figure 10 shows the data in the diagram (Ca+Al)-Si-(Mg+Fe). The dust particles appear to have a composition away from the (Ca+Al)-rich side of the triangle. The compositions of the two grains with (Ca+Al) $>$ 90% are derived from spectra that are especially rich in unidentified mass lines and where the extracted atom numbers are uncertain. They are among the very few spectra with practically no Mg and Si. We should, however, point out that they are of lower quality (type 3) and that they should not be over interpreted. In summary, we can say that there is no grain in the sample which resembles the Ca, Al-rich inclusions found in carbonaceous chondrites and which is believed to have retained records from early solar system high temperature episodes.

At the present time, we cannot answer the question as to whether hydrated silicates are present in Halley's dust (Anders 1986) since the complex mixture of the silicate and CHON grain types makes the identification of OH-rich and C-free grains very difficult. Future studies of the PUMA and PIA measurements will also include a detailed analysis of the mode-1, -2, -3 data. The corresponding spectra will be compared with the mode-0, PUMA-1 spectra discussed in this chapter and cluster analytical techniques will be applied.

X. THE GAS/DUST RATIO

The gas/dust weight ratio can be derived from the measured composition of gas (Krankowsky and Eberhardt 1988) and dust (see Sec. III) by assuming that the primordial C/Mg ratio of 1126/100 (Anders and Ebihara 1982) is preserved for the whole (gas plus dust) comet. In the solid phase, the C/Mg

TABLE IV
Relative Atomic Element Abundances of the Material (Gas and Dust) Released by Halley's Comet

	This Chapter	Geiss (1987)[a]	Solar System[b]	CI[b]
H/Mg	31	39	25200	4.9
C/Mg	= 11.3	12	11.3	0.71
N/Mg	0.7	0.4 – 0.8	2.3	0.06
O/Mg	15	22.3	18.5	7.1
N/C	0.06	0.03 – 0.06	0.2	0.08
O/C	1.3	1.8	1.6	10.0

[a]These results normalized to Mg with the solar Mg/Si-ratio are given for comparison.
[b]The abundances in the primordial solar system and in CI chondrites are from Anders and Ebihara (1982).

ratio is 815/100 (Table III). If overall the primordial C/Mg ratio is maintained, we should then find for any 100 Mg atoms in the dust, 312 C atoms in the gas. Krankowsky and Eberhardt (1988) list the gas composition relative to 100 H_2O atoms as: CO = 17; CO_2 about 1.5; NH_3 about 1.5; CH_4 about 2; NH_3 about 1.5; $N_2 < 4$; all others < 3. We can therefore easily calculate that the 312 C atoms in the gas are accompanied by 3234 H atoms, 84 N atoms and 1826 O atoms. A fraction of the whole comet which contains 100 Mg atoms, has 42,442 amu in the solid and 37,370 amu in the volatile (gaseous) phase. The corresponding gas/dust weight ratio is then 0.9. This value is at the lower end of the range from 6.3 to 0.63 determined by McDonnell et al. (1986), using very different arguments.

We can now determine the total (gas and dust) composition of Halley's material which for the non-CHON elements has already been presented in Table III. Table IV gives the relative (to Mg) atomic abundances of the CHON elements for gas plus dust together with the result of a study by Geiss (1987) which is based on other lines of evidence. Considering the different approaches, the agreement of the results from the two studies is excellent. Again, one notes that the volatile elements H, C, N and O are strongly enriched in the cometary material relative to CI chondrites and are similarly abundant as in the primordial solar nebula.

XI. INTERNAL STRUCTURE AND EVOLUTION OF THE NUCLEUS

Cometary nuclei are so small and remain normally so far away from Earth that they were never optically resolved until the Vega 1 & 2 and Giotto flybys. All three spacecraft clearly revealed the long-suspected presence of a discrete nucleus. It has a highly irregular shape and its surface is covered by craters of different sizes (Fig. 11). Its battered body looks more like a com-

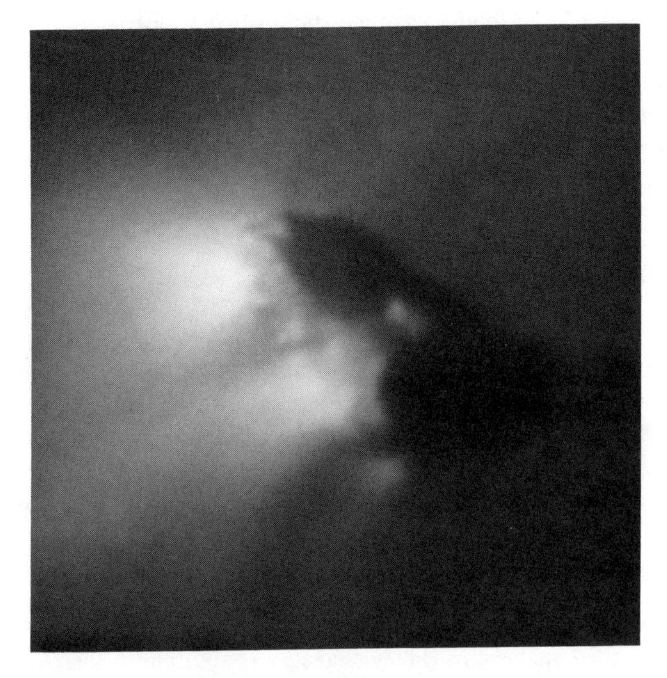

Fig. 11. This image of the nucleus of Comet Halley is composed of 7 images taken from Giotto on 13 March 1986. Several features such as a mountain, a chain of hills, a crater and jets can be seen in the originals. Photograph courtesy of H.U. Keller, Max Planck Institut für Aeronomie.

posite of many small cometesimals. The nucleus turned out to be at least twice as large (about 16 km long and 8 km across) as originally thought, and much darker, with an albedo lower than 4%, making it one of the darkest objects in the solar system. Only a few outer solar system objects, such as Jupiter's satellite Almethea, the dark side of Saturn's satellite Iapetus and the ring particles of Uranus are as black. The mean density is estimated to be 0.1 to 0.4 g cm^{-3}, the mass about 10^{18} g. Bright dust jets are emitted from relatively small active areas on the sunlit hemisphere.

While most astronomers were probably convinced that the comet nucleus must be mainly made up out of water ice, Halley was the first comet in which water was definitely identified (Mumma et al. 1986). Measurements from NASA's Kuiper Airborne Observatory (KAO) clearly revealed the presence of water and showed water vapor streaming from the nucleus with velocities between 0.8 and 1.4 km s^{-1}. At the time of the Vega 1 encounter, about 1.6 tons of water were emitted; when Vega 2 passed Halley, this amount had increased by a factor of 2.

An important new result is the detection of infrared radiation from the nucleus by the infrared spectrometer aboard Vega 1. The temperature of about

300 K is about 100 K higher than the sublimation temperature of water ice, indicating that the nucleus surface is covered by an insulating layer of dark, porous refractory material which is consistent with the observed low albedo (Combes et al. 1986). The thickness of the dust layer is unknown and could range from less than a cm to several m; it is also conceivable that it varies from one region to another.

Spinrad (1987) lists effective radii for selected comets (Table V). They range from a few to several 10 km. It is extremely difficult to derive rotational periods from the observation of variable light curves produced by a non-spherical rotating nucleus, since even at distances of several AU from the Sun, the nucleus is still hidden in an extended gas and dust coma. Rotational periods of short-period comets vary between a fraction of a day to about 1 day. Halley's nucleus appears to tumble along its orbit with a rotation period of 2.2 days and a superimposed precession of 7.4 days. The fact that asteroids of about the same size have rotation periods of approximately 0.3 days can perhaps be explained by a possible spin-up of asteroids as a consequence of collisions, and a deceleration of the cometary nucleus due to strong jet action of the rotating nucleus when the nucleus spins in a different sense as it orbits the Sun.

Ion mass spectrometer measurements revealed the presence of dissociation products of methane in Halley (Balsiger et al. 1986; Allen et al. 1988). The abundance of a low-temperature condensate such as methane in the near surface layers indicates that those layers have remained essentially unaltered until they are exposed to the Sun. This conclusion does not take into consideration that methane might be trapped in a water-ice lattice of a clathrate and thus have a higher vaporization temperature.

Although there is presently no direct evidence for the internal structure of the nucleus, several indirect arguments seem to favor an undifferentiated character (Donn and Rahe 1982; Delsemme 1982). The dust/gas ratio varies from comet to comet, but new and short-period (old) comets show no systematic

TABLE V
Effective Radii of Nuclei for Selected Comets[a]

Comet	*R* (km)	Reference
P/Encke	3.5	Mendis et al. 1985
P/Arend-Rigaux	5	A'Hearn 1986
P/Arend-Rigaux	5	Brooke and Knacke 1986
P/Neujmin 1	9.5	A'Hearn 1986
IRAS-Araki-Alcock	2.5–8	Harmon et al. 1988
Meier	13	Fernandez and Jockers 1983
Wirtanen	50	Mendis et al. 1985
Schwassmann-Wachmann 1	48	Mendis et al. 1985

[a]Table after Spinrad (1987) and Harmon et al. (1988).

difference in this ratio or the character of the solid particles. The continuum/ emission intensity ratio in their spectra appears also to have similar distributions (Donn 1977). Further, the emission spectra of new and old comets seem to be similar, in the visible as well as in the ultraviolet spectral region. For example, the CN/C_2 production-rate ratio turns out to be remarkably constant ($\pm$ 0.1 in the log) from comet to comet (A'Hearn and Schleicher 1980), and the relative production rates of different species such as CN, C_2, C_3 and OH appear to be unrelated to either the emission-to-continuum (gas-to-dust) ratio or the dynamical age of the comet. The fragmented comets studied by Sekanina (1980) and Kresák (1981) did not reveal any significantly different behavior of short- and long-period comets. Both groups fragmented at about the same rate, indicating a comparable structural strength against fragmentation, and the individual fragments showed about the same behavior. Sekanina (1977) also found that the relative nongravitational effects for the individual fragments due to the vaporizing gases, were proportional to their lifetime or mass, indicating the same vaporization pattern after fragmentation (Spinrad 1987).

While these arguments indicate an overall homogenous radial structure of the nucleus, it is also true that the behavior of the same comet changes noticeably during the same and also from one apparition to the next. This observation might be explained by changes in the nucleus' outer layers and also by a nucleus which was formed through an accretion of ice/dust aggregates.

It is generally assumed that the formation of comets occurred at the time when the solar system was formed (see, e.g., Whipple 1964; Öpik 1973; Delsemme 1977,1982). It has also been recognized that comets accumulated from a cloud of ice/dust aggregates at low temperature and low relative velocities (Donn 1963; Öpik 1973). What is not generally agreed upon is the region where comets formed. Dynamical as well as compositional arguments discussed above make it seem likely that the observed comets have always been connected with the Sun in some manner. Oort's (1950) proposal for a cloud of comets orbiting the Sun at 10,000 to 100,000 AU has been generally accepted, but the question of how the Oort cloud was formed is still wide open; several regions have been proposed for accretion of the nucleus prior to their residence in the Oort cloud, with the zone near Uranus and Jupiter at 20 to 30 AU (Whipple 1964) most commonly adopted (for a detailed discussion see, e.g., Delsemme [1977] Donn and Rahe [1982], Weissman [1982]).

On the other hand, arguments can be given to support the hypothesis that comets originated in the outer planetary region beyond the orbits of Uranus and Neptune. In the 4.5 Gyr since their formation, they are stored under nearly ideal conditions for preservation of their primitive nature. Several independent groundbased and space observations have demonstrated that CO is very abundant in comets and apparently distributed throughout the whole nucleus; it has about 10 to 20% of the water abundance but is at the same time several times more abundant than CO_2 (Krankowsky et al. 1986). CO and CO_2 vaporize at much lower temperatures than water ice. Their presence could explain why comets develop comae while still far away from the Sun.

It is interesting to note that a considerable fraction of the carbon monoxide does not appear to leave the nucleus in gaseous form; instead, it seems to come from fine dust particles in a region of about 15,000 km distance from the nucleus (Eberhardt et al. 1987*c*). Formaldehyde H_2CO and formic acid H_2CO_2 discovered in the dust grains, could be parent molecules for CO (Delsemme, personal communication). Part of the formaldehyde could also be in the form of polyformaldehyde (Huebner 1987). The expansion velocity at 2500 km distance from the nucleus was measured to be 800 m s^{-1}; at the distance of 20,000 km it had increased to about 1000 m s^{-1}. In March 1986, a coma brightness outburst in Comet Halley was followed by a large flare-up in CO^+, $CO_2{}^+$ and dust; the water production remained essentially constant during that time (Feldman et al. 1987). In case CO should be an abundant parent molecule, then Halley's nucleus cannot have been heated above 25 K (the sublimation temperature for CO) during its formation and this comet must have formed beyond 100 AU from the Sun.

On the other hand, CO^+ tail rays appear well within 10^4 km from the nucleus (Rahe and Donn 1968,1969), and it is very difficult to see how CO would be a parent species and being so volatile, not vaporize quickly unless trapped in grains. Its quoted high abundance would prevent efficient trapping, and also it would not be consistent with the hypothesis that pockets of CO, CO_2 and similar volatiles are present in the nucleus and are involved in outbursts as the one described above. Pirronello (1982) and Bar-Nun (1985) have published vaporization curves for CO_2/H_2O ice mixtures and obtained somewhat different results. Recent work by Donn (unpublished) has shown that the vaporization pattern of CO_2-water mixtures depends on the CO_2/H_2O ratio. It illustrates that the behavior of the minor volatile cometary constituents is quite complex and that more laboratory work is needed before the flyby results can be explained. If one explains a distributed source by the trapping of CO in grains, the other less volatile parent species would be expected to be more effectively trapped; however, nothing is reported about this.

Another argument comes from the detection of S_2 at 290 nm in the Sungrazing comet IRAS-Araki-Alcock 1983d (A'Hearn et al. 1983). S_2 has a lifetime against photodissociation of only 450 s at 1 AU (i.e., only 0.005 that of water) and a sublimation temperature of 20 K. The presence of free S_2 would point to an origin of the nucleus far beyond the orbit of Pluto (Greenberg et al. 1986).

XII. CONCLUSION

One of the major surprises resulting from the *in situ* measurements of Comet Halley, was the composition of the cometary dust. The analysis of the corresponding data sets is still not complete and new results are expected to be published in the near future. Fundamental new information about comets will be obtained by a mission such as NASA's planned Comet Rendezvous Asteroid Flyby Mission (CRAF). The chief goal of this mission is to determine the elemental, molecular, isotopic and mineralogical compositions and the physical, morphological and geological characters of the cometary nucleus, and to understand the changes that occur as functions of time and orbital positions. The present target for this mission is the short-period Comet P/Kopff. Unfortunately, it will take many years before data from this mission will be available for analysis.

Acknowledgments. It is a pleasure to thank H. Fechtig and H. Spinrad for many interesting suggestions. Special thanks are due to B. Donn for his careful reading of the manuscript and for his numerous constructive contributions.

ASTEROID VOLATILE INVENTORIES

L. A. LEBOFSKY
T. D. JONES
and
F. HERBERT
University of Arizona

Information from a variety of sources can now be used to aid our understanding of the formation and evolution of the asteroids. With improved observational techniques and detailed laboratory examinations of meteorites, we are beginning to unravel the mineralogy and physical properties of the asteroids. Taken together, these traits can be powerful discriminators of a body's chemical and physical history. In this chapter, we summarize briefly our general knowledge of asteroids in terms of size and spatial distribution, and describe the current observational and laboratory techniques employed in asteroid research. We then discuss specifically their compositions (with emphasis on major volatile content), the case for delivery of volatile-rich material to the terrestrial planets, the evolution and migration of volatile phases under heating, and current efforts to observe and inventory the surface materials of volatile-rich asteroids. The review concludes with a discussion of unresolved questions concerning volatile content of asteroids, and the prospects for their resolution by spacecraft measurements and new ground-based efforts.

The asteroids are a compositionally diverse group of objects. This diversity appears directly related to heliocentric distance, and thus an understanding of this distribution may give us some insight into the formation conditions of the entire solar system as well as the evolution of the asteroids. On theoretical as well as observational grounds, we suspect that the majority of asteroids are in part composed of volatile materials, such as water (in the form of water of hydration). Our purpose in this chapter is to examine the evidence for these volatiles and the role that the asteroids may have played in the formation of planetary atmospheres.

The largest asteroid, Ceres, was discovered on New Years Day, 1801, by Giuseppe Piazzi. It was at first thought that Ceres was the missing planet predicted by the Titius-Bode Law to orbit between Mars and Jupiter. When it became clear that Ceres was too small to play this role (Figs. 1 and 2), and was instead accompanied by a host of smaller objects, the natural conclusion was that the asteroids represented the shattered fragments of the missing body. The discovery and cataloging over the next 90 years of some 300 asteroids, which together possessed much less than a single lunar mass, quieted this speculation. With the advent of photographic plates in the early 1890s, the number of confirmed asteroids grew rapidly; there are currently almost 4000 numbered asteroids with well-determined orbits, with several thousand others observed and awaiting similar status. They range in size from a third of a lunar diameter to objects smaller than a kilometer (Figs. 1 and 2).

The asteroids became involved in our understanding of early terrestrial planet evolution when we began to realize the important role large impacts have played in shaping their surfaces. The obvious photographic evidence for the early, heavy bombardment of the terrestrial planets, coupled with studies of asteroid orbital dynamics and evolution, strongly suggests that the asteroids represent a much larger early population that may have been a significant volatile source for the terrestrial planets.

I. OBSERVATIONAL TECHNIQUES

Our understanding of asteroid physical properties, mineralogy and most importantly, composition comes from a variety of techniques. Ultraviolet, visual and near-infrared photometric and spectrophotometric observations record the reflective properties of asteroid surfaces. The spatial orientation (shape, period and pole position) of a given asteroid is revealed through study of the lightcurve obtained by broadband visual filter photometry (UBV) (see Taylor 1979). UBV photometry also served as a basis for the first efforts at building an asteroidal taxonomy (see Tholen 1984,p.9). With UBV data and slightly higher resolution eight-color photometry, detailed studies of the asteroid visual spectra were made, which, combined with albedo data, enabled the assignment of taxonomic classes reflecting different compositions. The obvious but crucial next step was to link carefully the resulting taxonomy with the spatial structure of the asteroid belt, revealing the distribution of types with heliocentric distance (Gradie and Tedesco 1982; see Fig. 9 below). Finally, with higher resolution spectrophotometry from about 0.3 to 3.5 μm, we can discern the individual mineralogy of some asteroids. This latter research benefits greatly from the laboratory comparison of asteroidal spectra with those of meteorites, most of which are considered asteroidal fragments (Dodd 1981). For more information on this widely used technique, we refer the

Fig. 1. Comparison of the sizes of four of the largest asteroids with the Moon. Relative geometric albedos are represented by the level of shading of the Moon and asteroids.

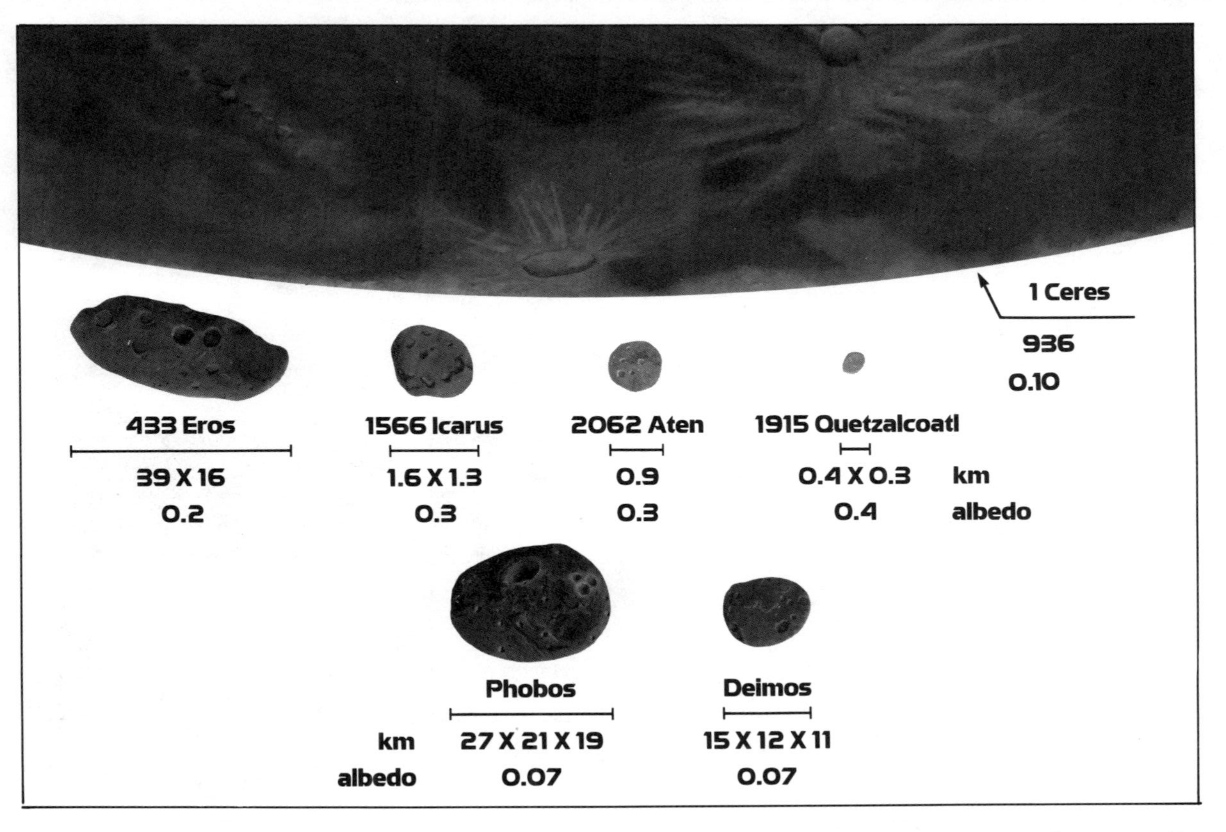

Fig. 2. Comparison of the sizes of a representative selection of the smallest known asteroids with the size of Ceres, the largest asteroid. Shown for comparison are the asteroid-like satellites of Mars, Deimos and Phobos. Relative geometric albedos are represented by the level of shading of the asteroids and satellites.

reader to various papers by Bell, Feierberg, Gaffey, Larson, Lebofsky, McCord and McFadden.

From about 3.5 μm out to millimeter and centimeter wavelengths, the radiation we observe from asteroids is dominated by thermal emission. This is of course solar radiation that has been absorbed primarily in the visual and near-infrared and reemitted at the longer wavelengths. Thermal infrared observations have been successfully used for determination of asteroid diameters and albedos and studies of the thermophysical properties of their surfaces. The most successful use of this technique was the survey carried out by the Infrared Astronomical Satellite (IRAS), which observed over 1800 asteroids in the 10- to 100-μm spectral region (IRAS Asteroid and Comet Survey 1986). For more detail, see the work of Morrison and Lebofsky (1979) and Lebofsky et al. (1985,1986).

Radar observation is a fairly new and powerful technique for the study of asteroids. By sending out a known signal and studying the returned echo, one

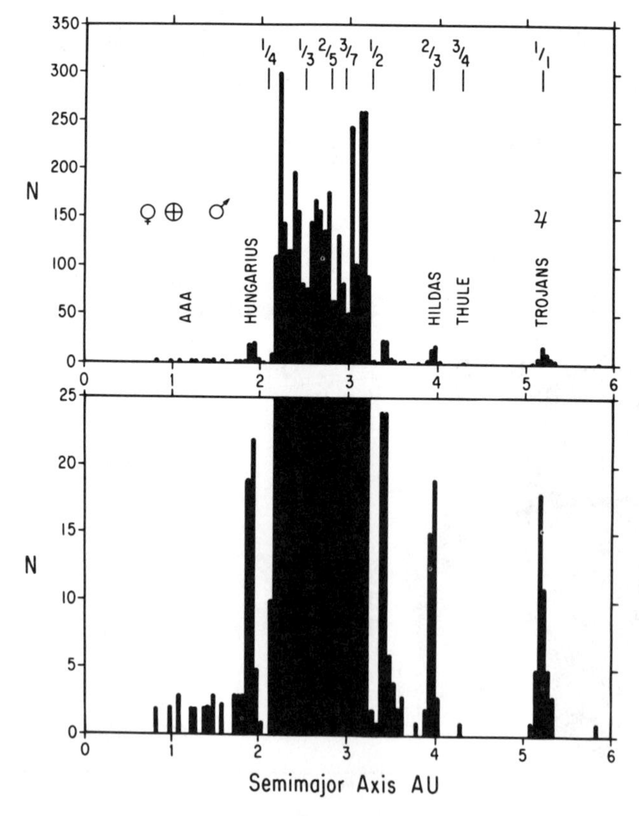

Fig. 3. Distribution of asteroids as a function of heliocentric distance in increments of 0.05 AU. The positions of the planets and the major asteroid groups are indicated. Fractions indicate ratios of orbital periods for the principal dynamic resonances with Jupiter (figure adapted from Gradie et al. 1979).

can characterize the size, shape and roughness of the surface. Also, because of the unique radar reflectance properties of metal, analysis of the reflected signal can determine the asteroid's metal content. Applications of radar observations are discussed by Ostro et al. (1985).

II. THE STRUCTURE OF THE ASTEROID BELT

In order to assess possible asteroid volatile contributions to the early terrestrial planet atmospheres, we must obtain in some way an estimate of the original asteroid compositions. Through groundbased observations combining the different approaches discussed above, researchers have constructed a broad compositional map of the asteroid belt that gives us our major insight into its history. This compositional structure is superimposed on the belt's very heterogeneous spatial distribution (see Fig. 3).

Asteroid Distribution

The asteroids can be divided into three major groups, related to their mean distances from the Sun: (1) the Apollo, Amor and Aten (AAA) asteroids whose orbits approach that of the Earth; (2) the main-belt asteroids with orbits between those of Mars and Jupiter (but which also include the Hungarias and

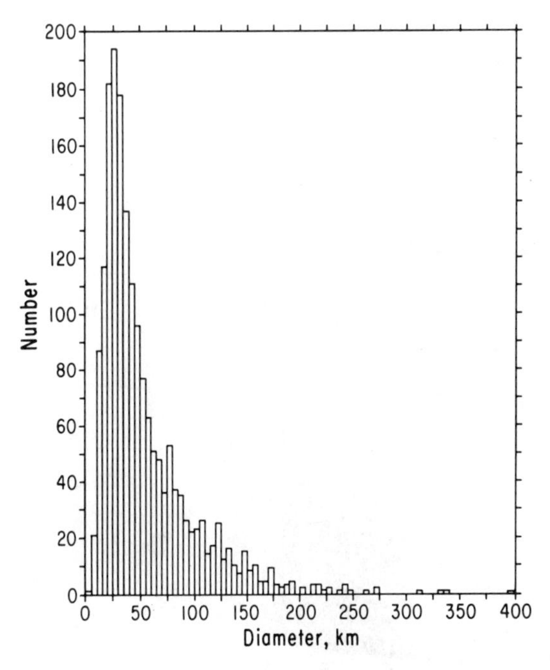

Fig. 4. The number of asteroids as a function of diameter. Diameters were obtained from IRAS measurements (figure from Veeder 1986). The three asteroids that have diameters larger than 400 km are not included.

Hildas, which actually lie just inside and outside the main belt, respectively); and (3) the Trojan asteroids, equidistant from the Sun and Jupiter at the L4 and L5 Jovian Lagrange points. In detail, the main-belt asteroids can be divided up into other "families," each with very similar orbital parameters. These distinct groups can be seen in Fig. 3, along with the resonance-associated gaps that demonstrate graphically Jupiter's great influence on main-belt asteroid distribution. The papers by Shoemaker et al. (1979), Gradie et al. (1979), and Zellner et al. (1985*b*) provide more detailed information on main-belt structure.

Asteroid Size and Albedo Dependence

Over the past ten years researchers have obtained thermal infrared observations of more than 200 asteroids. These have been coupled with the asteroid taxonomy in Gradie and Tedesco's preliminary analysis of the distribution of asteroid classes, discussed below. This work lately has been augmented significantly by the IRAS mission, which yielded albedos and diameters for over 1700 asteroids. Our goal here is to use the 1986 work of Veeder and the

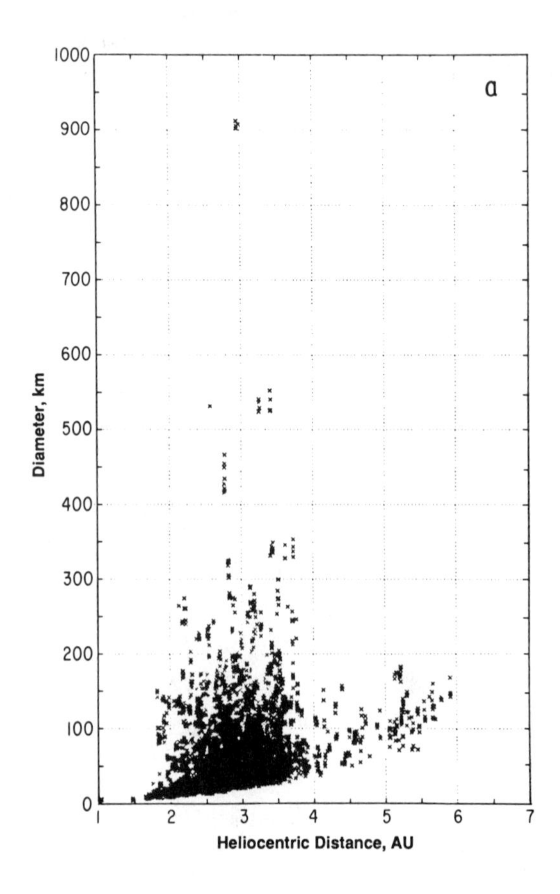

Asteroid Advisory Group to illustrate the asteroids' size and albedo distribution. When ultimately the IRAS observations are combined with other data sets, such as the Eight-Color Asteroid Survey (ECAS) (Zellner et al. 1985*a*), we will have a better picture of the overall compositional distribution in the asteroid belt (Gradie and Tedesco 1982; E. F. Tedesco and C. R. Chapman, personal communication).

Figure 4 shows the number of asteroids detected by IRAS as a function of diameter. The effect of observational limitations on the IRAS data is apparent in the drop in the number of asteroids below 30 km, where numbers should continue to rise. The IRAS limitations at smaller diameters are even more evident in Fig. 5, where we see the distribution of asteroid diameters as a function of solar distance. There is no clear trend other than this incomplete sampling of the more distant small asteroids; the threshold for detection increases to over 50 km at the distance of the Trojan asteroids. Of greater interest to our discussion of volatiles is the asteroid albedo distribution. Figure 6 plots the number of IRAS-observed asteroids vs geometric albedo. There is an obvious peak in the asteroid albedos at just under 0.05, with a tail that extends

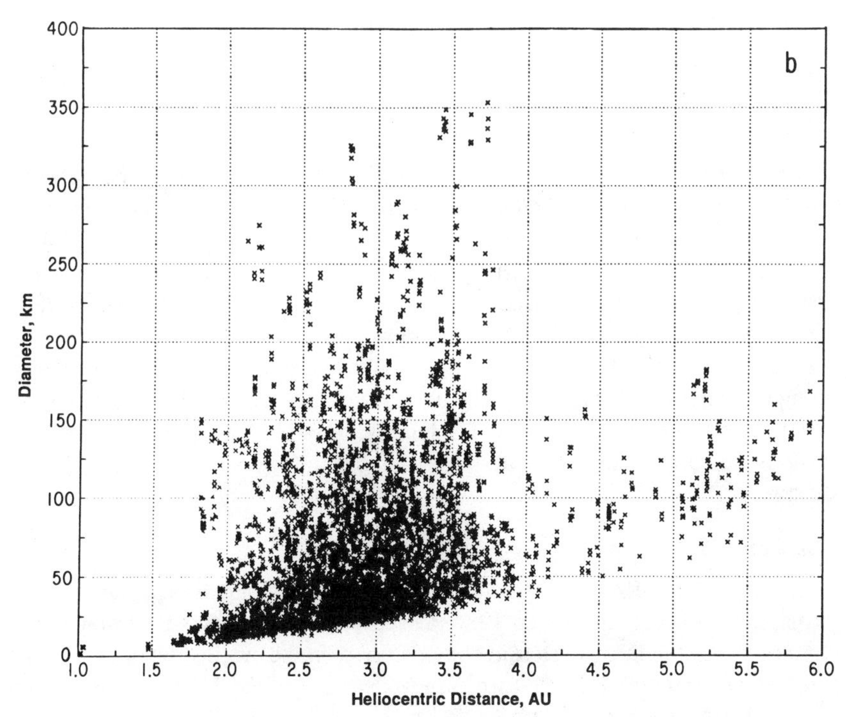

Fig. 5. (a) A plot of IRAS diameters of asteroids as a function of heliocentric distance. (b) An enlargement of (a). The sharp cutoff at lower diameters that rises with solar distance is due to the limiting flux observable by IRAS (figure from Veeder 1986).

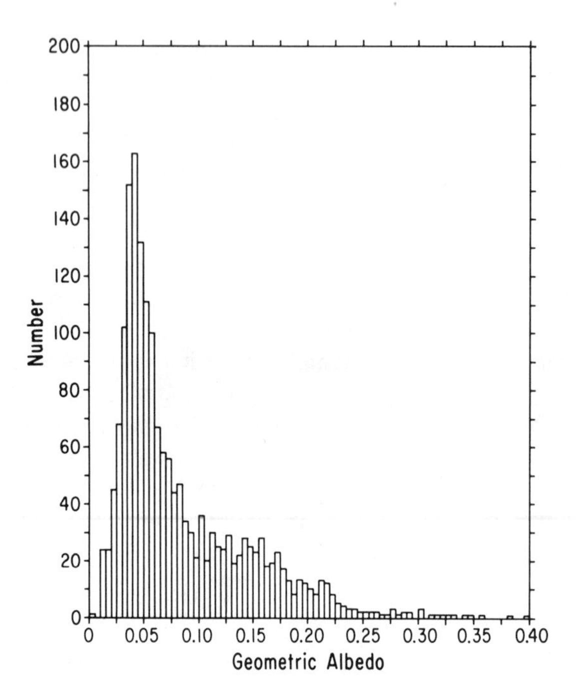

Fig. 6. The number of asteroids as a function of albedo. Albedos were obtained from IRAS measurements (figure from Veeder 1986).

out to 0.40. This distribution disagrees somewhat with older plots, which show a gap in the distribution around 0.10; this is perhaps due to uncertainty in IRAS albedo determinations at low flux levels. A plot of albedo vs solar distance is shown in Fig. 7. Here we see a clear trend; while higher-albedo asteroids are concentrated toward the inner belt, darker objects populate the entire belt and dominate the outer region from 3.5 AU all the way to the Trojans at 5.2 AU. This radial zoning of albedos, which arguably represents compositional gradation, is the key observation against which all theories of asteroid evolution are measured. Explaining how this gradation arose is crucial to our understanding of the asteroid belt's formation and subsequent history.

Asteroid Taxonomy

The ECAS of Zellner et al. (1985*a*) and the evolving asteroid taxonomy (Chapman et al. 1975; Tholen 1984; Chapman 1987) serve as the framework for our understanding of asteroid compositions. In the original work by Chapman et al., three major classes of asteroids—C, M and S—were identified from 24-color spectrophotometry (0.3 to 1.1 μm). Unfortunately, because of their spectral similarity to known meteorite types and a suggestive choice of letters, these original classes became inextricably linked in general usage to

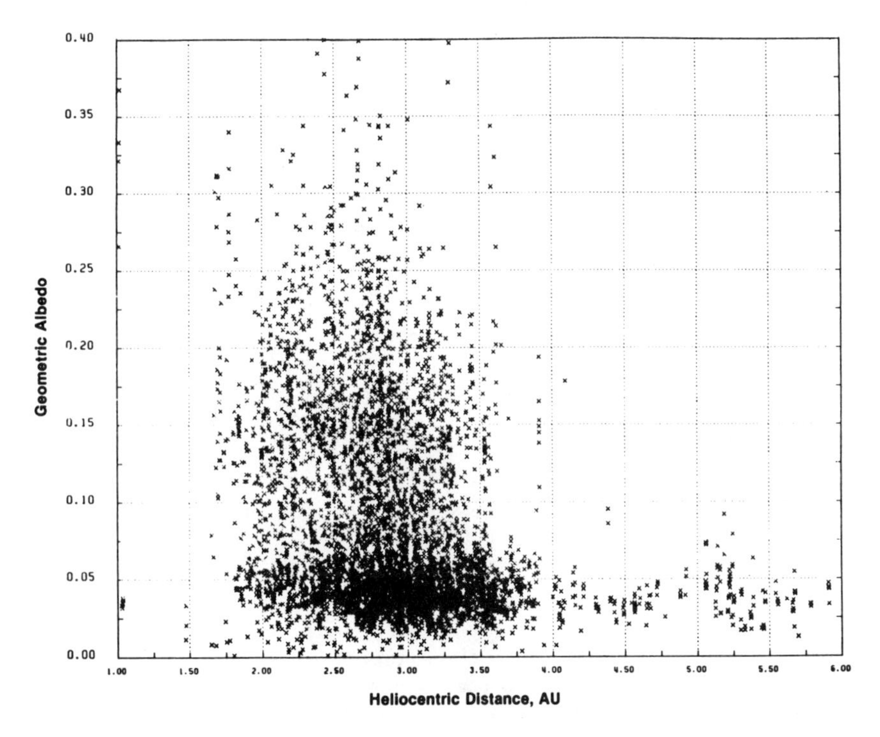

Fig. 7. A plot of IRAS geometric albedos of asteroids as a function of heliocentric distance (figure from Veeder 1986).

the corresponding meteorites: the carbonaceous chondrites, the metallic irons, and the stones. This clearly was not the intent of the first taxonomists, who recognized the speculative nature of comparing telescopic spectra to laboratory meteorite analogues. Few diagnostic features existed in the visual and near-infrared with which to make comparisons, except in rare cases (e.g., 4 Vesta). The later taxonomic classifications by Gradie and Tedesco (1982) and Tholen (1984) swung completely to the other end of the spectrum, avoiding any comparison of asteroid classes with possible meteorite analogues.

Tholen (1984) performed the most comprehensive classification to date of visual and near-infrared asteroid spectra. Using a cluster analysis based on 7 color indices measured in the ECAS, Tholen created a detailed taxonomy widely used today (see Fig. 8).

Radial Compositional Dependence in the Asteroid Belt

As noted above, the correlation between asteroid taxonomic class and probable composition based on meteorite types is a tentative one. It is quite probable that our meteorite collections do not represent the total range of asteroid compositions in the belt, nor is there any reason why all known mete-

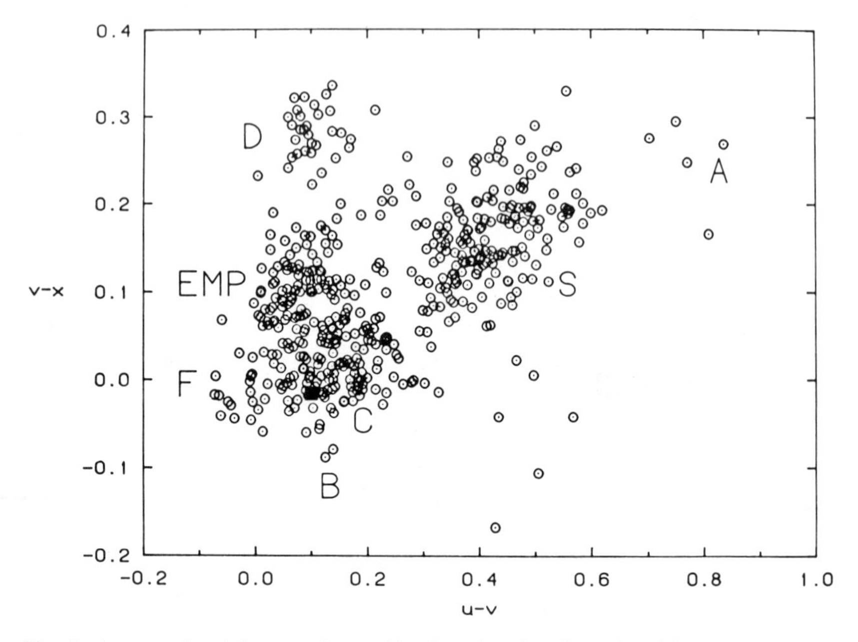

Fig. 8. An example of the several asteroid color-color plots from the Eight-Color Asteroid Survey, illustrating the separation of various classes in color space, used by Tholen (1984) to construct his taxonomy. Albedo information provides additional discrimination between classes.

orite assemblages should conveniently be exposed on asteroid surfaces. Also, because of the observational limitations inherent in defining the taxonomies, it is possible that asteroids assigned to the same class based on observable parameters may differ greatly in composition. For example, class Ms and class Ps look spectrally very similar, but can easily be distinguished by albedo and are thus likely to have very different compositions. With these caveats in mind, the distribution of asteroid classes (and thus composition) as a function of solar distance is still regarded as a first-order picture of the asteroid belt's original chemical state.

Gradie and Tedesco (1982), using a preliminary version of ECAS and still unpublished radiometric observations of asteroids, constructed a bias-corrected plot of the distribution of asteroid classes as a function of solar distance (Fig. 9). They demonstrated a marked radial distribution of the asteroids by class, consistent with the theory that the asteroids formed at or near their present belt locations with only slight redistribution. In the figure, it is quite apparent that the S asteroids dominate the inner belt, the Cs dominate the middle and outer belt, and the Ds dominate the Trojans. If one also equates the S, C and D classes, respectively, with volatile-poor materials, carbonaceous chondrite-like material, and "ultra-carbonaceous" material, then

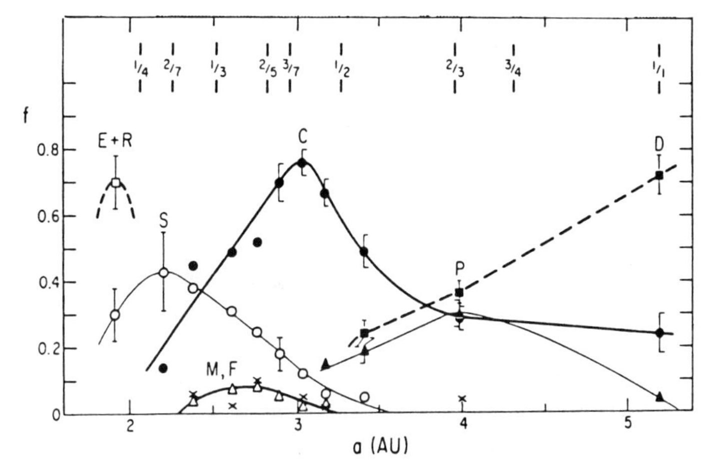

Fig. 9. Observed distribution of asteroid classes (relative to the number of asteroids in each distance increment) as a function of heliocentric distance. The Δ symbol is the fraction of class M, and the X is the fraction of class F. Smooth curves are drawn through the points of each class to delineate more clearly their distribution. Error bars represent ±1 standard-deviation uncertainties due to incomplete sampling (figure adapted from Gradie and Tedesco 1982).

this distribution may be indicative of formation temperatures or, alternatively, the effects of a radially dependent heating episode early in solar system history.

Recently, Bell (1986) has proposed three asteroid "superclasses" that relate the above taxonomy to the observed radial compositional variation. He views the variation as evidence for a strong electro-inductive heating episode in the belt that fell off sharply with heliocentric distance. The varying degrees of thermal processing created three superclasses: igneous, metamorphic and primitive asteroids. The igneous types dominate the belt sunward of 2.7 AU, the metamorphic objects lie in a narrow zone centered around 3.2 AU, and the primitive class dominates outside 3.4 AU. Bell's scenario also assigns the various meteorite types to asteroidal environments, controlled not only by heliocentric distance but also depth of origin in the various superclass parent bodies.

Meteorite Analogues and Belt Volatile Inventories

Mineralogical and spectral studies of meteorites, augmented by detailed asteroid spectroscopy and radiometric techniques, have encouraged comprehensive efforts to determine asteroidal compositions (see, e.g., Gaffey and McCord 1979). Not all spectral matches are as clear cut as the association between 4 Vesta and the basaltic achondrites (McCord et al. 1970; Feierberg and Drake 1980). For example, the C-class asteroids have no spectral twin among the carbonaceous chondrites, and the ordinary chondrites, the most common meteorite type, have few if any distinct asteroid analogues (McFadden and Vilas 1987). Spectral comparisons are complicated by meteorite weathering and spectral alteration during long terrestrial exposures (Gaffey and McCord 1979).

Nonetheless, we can use such an inferred asteroid-meteorite relationship to estimate the importance of volatiles (water and carbon, for example) among the present belt population. Water and carbon compounds are two meteoritic volatile species whose presence we can infer observationally in asteroidal material. (Rare gases and other volatiles may be present only in asteroidal regoliths.) The studies of Lebofsky and Feierberg (see Lebofsky et al. 1981; Feierberg et al. 1985*a*) in the 0.3- to 3.5-μm spectral region furnished convincing evidence of volatiles on main-belt asteroids. Spectral similarities to CI and CM chondrites (see Figs. 10 and 11) along with mean density determinations (Schubart and Matson 1979), indicated that Ceres' surface is similar in

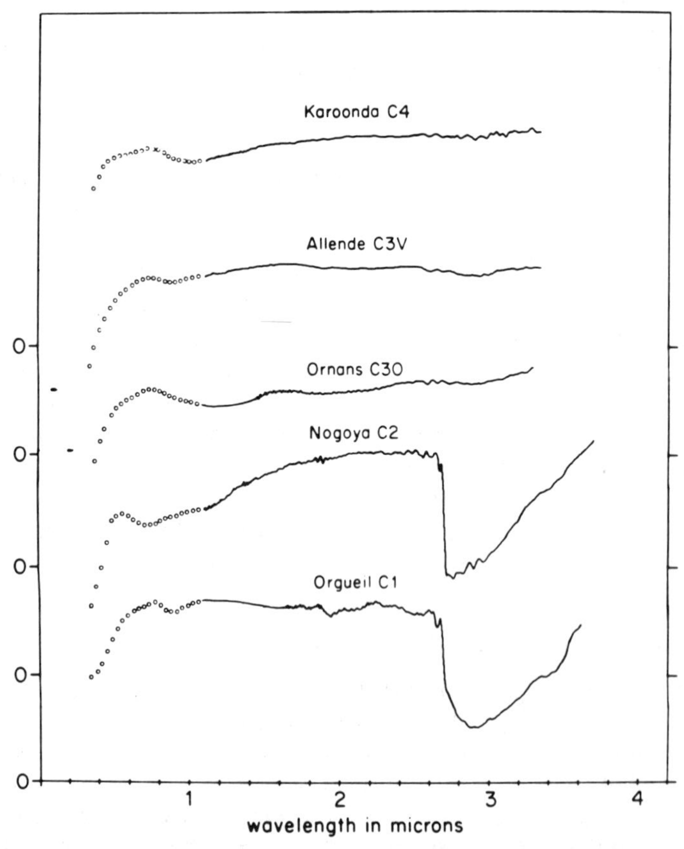

Fig. 10. Typical reflection spectra of carbonaceous chondritic meteorites. The deep 3-μm feature of Nagoya and Orgueil is caused by water of hydration in the hydrated silicate matrix.

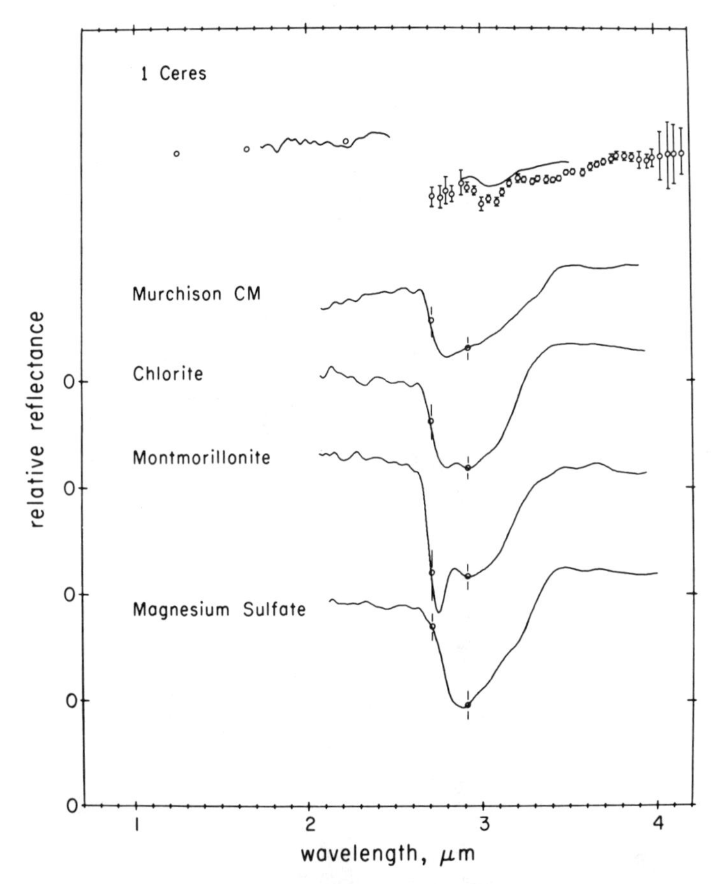

Fig. 11. Comparison of the spectral reflectance of Ceres with laboratory samples of hydrated minerals. These spectra show the absorption feature due to structural OH in clay minerals at 2.7 μm and the broad 3-μm band due to adsorbed water (figure from Lebofsky et al. 1981).

composition to these volatile-rich meteorites. In detail, and especially in light of slight albedo differences, Ceres may not be an exact analogue of any of our CI or CM samples, but at the least we can conclude that Ceres' surface material is composed primarily of hydrated silicates similar to those commonly found in carbonaceous chondrites. Table I gives approximate elemental abundances for these meteorites.

These abundances and the meteoritic associations discussed above enable us to estimate the approximate belt volatile inventory. Asteroids are numbered in order of discovery, the brightest obviously being discovered first. However, because the albedos of the C-class asteroids are generally much lower than the average S albedo, the Cs are larger for any given visual brightness (a function of diameter, geometric albedo and distance). Thus, even

TABLE I
Mineralogy of CI and CM Chondrites

Name	Formula	Abundance (wt.%)		Spectroscopic Signature (0.4–4.0 μm)
		CI	CM	
Phyllosilicates (layer-lattice silicates)	$Mg_?Fe_?Si_?OH_?$	60–80	50–80	>0.5 μm: Fe^{2+}-Fe^{3+} charge transfer band 1.4, 1.9, 3.0 μm: OH and H_2O absorptions
Carbon polymer	$C_?H_?O_?$	5	3	opaque material 3.4 μm: CH fundamental vibration bands
Metal	FeNi	rare	rare	no absorption bands, but can produce pronounced reddening
Sulfides				
Pyrrhotite	$Fe_{1-x}S$	rare	rare	opaque constituent
Troilite	FeS	~5	rare	
Pentlandite	$(Fe,Ni)_9S_8$	?	~3	
Pyrite	FeS_2	rare	—	
Sulfur	S	?	?	
Unidentified phase	$Fe_?Ni_?S_?O_?$	?	?	

Spinel group				
Magnetite	Fe_3O_4	10–15	~5	opaque constituent
Chromite	$FeCr_2O_4$	—	incl./Matrix	1 μm: Fe^{2+} electronic transition
Spinel	$MgAl_2O_4$	—	incl.	
Silicates				
Olivine	$(Mg,Fe)_2SiO_4$	rare }	5	1 μm: Fe^{2+} electronic transition in olivine, pyroxenes and felspars
Pyroxene	$(Mg,Fe,Ca)SiO_3$	rare }		
Feldspar	$NaAlSi_3O_8$-$CaAl_2Si_2O_8$	rare	rare	2 μm: Fe^{2+} band in orthopyroxenes
Glass	variable	rare	rare	
Salts				
Gypsum	$CaSO_4 \cdot 2H_2O$	rare	rare	3.5, 4.0 μm: CO_3 fundamental vibration bands
Magnesium sulfate	$MgSO_4 \cdot nH_2O$	5–15	—	3.0–4.0 μm: SO_4 overtone and combination bands
Calcite	$CaCO_3$	?	rare	1.4, 1.9, 3.0 μm: OH and H_2O absorption
Magnesite	$MgCO_3$	rare	rare	
Breunnerite	$(Fe,Mg)CO_3$	~3	?	
Dolomite	$CaMg(CO_3)_2$	rare	—	

though only 3 of the first 10 *numbered* asteroids are members of the C class or subclasses, Cs comprise 9 of the 13 *largest* asteroids. In fact, the total belt mass is dominated by the volatile-rich asteroid 1 Ceres (about 33%; Tedesco 1984). Since Ceres alone contains 1.8 times the mass of the next 10 largest asteroids, and CI or CM material contains approximately 10% water by mass (Lewis and Prinn 1984), we can conclude that volatile material is really quite abundant in the asteroid belt, on the order of 10^{23} g (the Earth's atmosphere, excluding the oceans, has a mass of about 5×10^{21} g). In absolute terms, however, the total mass of the asteroids is still small in comparison to the planets and their satellites: only about 5% of one lunar mass.

This overview of asteroid composition and distribution naturally leads to questions about their origin, their evolution over the age of the solar system and their possible volatile contributions to the planets. We proceed with a discussion of asteroid formation and the incorporation of their original complement of volatiles.

III. ASTEROID FORMATION

The present configuration of the asteroid belt, in terms of size, composition and radial distribution, may preserve information about the formation environment in this part of the solar nebula. In fact, the belt's existence serves to constrain models of solar system formation and accretion; they should not only predict a planetary system similar to our present configuration, but allow for the survival of the asteroids after planetary accretion is complete. How successful are current formation and accretion theories in explaining the radial compositional gradient and the other above-mentioned features of the asteroid belt?

Greenberg's chapter provides an excellent summary of the present state of planetary accretionary models; here we discuss only their relevance to the formation of the asteroid belt. Before our discussion of how the asteroids were assembled, though, we should list the materials that solar system formation models provide as plausible feedstocks for accretion in the asteroid belt.

Asteroid Building Supplies

The raw materials available for asteroid accretion were those condensates existing under the prevailing nebular conditions in the vicinity of the belt. According to a review by Lewis and Prinn (1984), temperatures in the belt ranged from roughly 400 to 200 K, with pressures on the order of 10^{-4} bar or less. The *equilibrium* condensates produced by cooling of the nebular gas to these temperatures were, in sequence, refractory oxides, enstatite, alkali aluminosilicates, FeO and FeS, the amphibole tremolite, magnetite and finally serpentine, talc and other hydrated silicates (Lewis 1972*a*; Grossman and Larimer 1974). At temperatures of approximately 170 K, water ice would have condensed, followed at 150 K by ammonia hydrates.

The hydrated silicates predicted by equilibrium reactions at 350 to 300 K include serpentine, talc and chlorite, formed when olivine grains react with nebular H_2O (OH radical). However, the equilibrium production of these silicates by nebular reactions of crystalline minerals with H_2O vapor at 10^{-8} bar, without liquid water, *and* at or only slightly above room temperature (Lewis and Prinn 1984; chapter by Prinn and Fegley), faces serious kinetic difficulties. If silicates escaped hydration in quantity, the condensation sequence would have been truncated, and the excess water vapor, if not condensed as ice, would have been lost as the nebula dissipated.

As Greenberg describes in his chapter, the dust grains formed by nebular condensates collapse into small bodies under mutual gravitation. We assume this process occurred on a time scale long in comparison to nebular condensation, the so-called homogeneous accretion scenario. Meteoritic evidence supports this contention; mineral assemblages predicted by inhomogeneous accretion schemes are not seen in meteorites (Fegley and Lewis 1980). Accretion of this material into small bodies may have yielded compositions quite different from the local condensates, however. At small scales, selective effects such as magnetic or electrical forces, and even the "stickiness" of the accreting grains, may have been more important than gravitational forces (Lewis 1974*b*).

Meteorite morphology and petrography imply that the asteroids were assembled from a wide variety of raw materials of disparate origin. The best examples of these unequilibrated assemblages are the carbonaceous chondrites, which contain refractory chondrules and calcium-rich inclusions intimately mixed with low-temperature hydrated silicates (phyllosilicates) and complex organic compounds (Wood and Chang 1985). Hydrated silicates are excellent volatile reservoirs: serpentinization of olivine and pyroxene can lead to the retention of up to 10% H_2O by mass (Lewis and Prinn 1984). The presence of deuterium-rich kerogen (a nonspecific term which denotes any organic mixture of ill-defined structure, composition and origin), stable only below 500 K (Kerridge and Chang 1985), suggests that surviving interstellar material could have influenced the isotopic content of protoasteroid and comet volatiles. If the numerous C-class asteroids are indeed analogous to carbonaceous chondrites, their kerogen-rich parent asteroids may have scattered this same material widely among the terrestrial planets (Kerridge 1983).

Initial Population and Collisional Evolution

Theory successfully explains the accretion of small planetesimals up to 1 km in size, but there is no satisfactory model detailing the construction of larger asteroids in the size range 10 to 100 km Greenberg (1978). To explain the growth of larger bodies from an initial population of small planetesimals, Greenberg favors a runaway growth scenario in which a few larger bodies emerge, though most of the mass still resides in objects on the order of 1 to 10 km in diameter. However, this is far from a plausible primordial size distribution for the asteroid belt.

While accretional models cannot produce a satisfactory picture of belt formation and original structure, the inverse approach, using the observed asteroid population to work backward, may be tried. Davis and coworkers (1985) produced a collisional model of the belt's evolution that, in providing a match to the present observed asteroid population, constrains the primordial population as well. According to their study, by the time mean velocities in the belt reached 5 km s^{-1}, the asteroids had evolved to the size distribution shown in Fig. 12. This so-called runaway growth population comes closest to reproducing the present one via collisional processing.

As Greenberg's chapter points out, once mean asteroid encounter velocities reached 5 km s^{-1}, mutual asteroid impacts yielded not accretional collisions, but destructive fragmentation which gradually eroded and dismantled most of the small planetesimals in the belt. The cause of the velocity

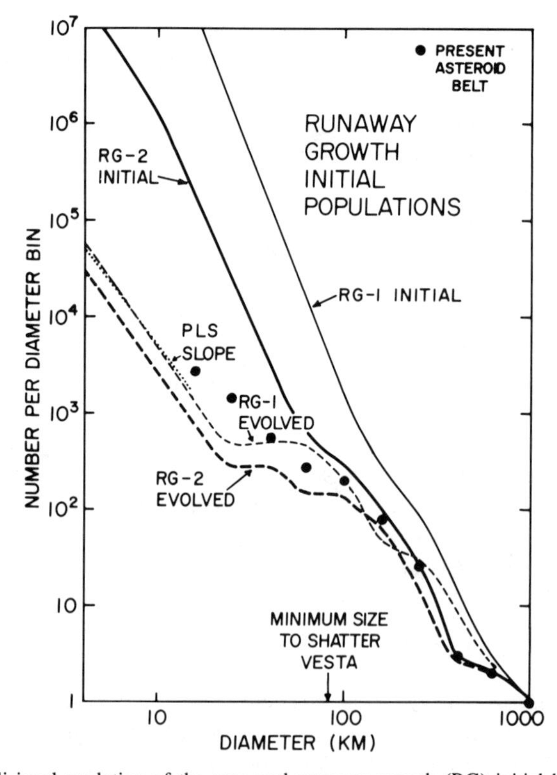

Fig. 12. Collisional evolution of the proposed runaway growth (RG) initial belt populations: RG-1 has a total mass 250 times that of the present belt, while RG-2 is 2 orders of magnitude smaller. Present belt population is denoted by filled circles. Most of the population mass is in small bodies (figure from Davis et al. 1985).

increase from smaller accretion values is still a subject for speculation. One possibility is that Jupiter's growth scattered planetesimals from its accretion zone inward through the main belt, so that frequent collisions and near-encounters increased relative velocities and effectively terminated asteroid accretion. Similarly, Wetherill (1975) has proposed that the stirring could have been accomplished by the passage through the belt of a single large Jupiter-scattered protoplanet, which later collided with a terrestrial planet or was ejected from the solar system. Alternatively, belt stirring may have resulted from the growth of Jupiter and the subsequent sweeping of its gravitational resonances through the belt.

What was the initial mass of the belt when the collisional environment halted accretion? The mean surface density of the belt was already depleted relative to other inner solar system locations when the erosional process began. However, asteroid accretion schemes and nebular reconstructions require substantial primordial masses; Weidenschilling (1975) placed an upper limit on the original belt mass of perhaps one Earth mass or more. By contrast, the total material present today is 1000 times smaller, comprising only about 5% of one lunar mass. The belt must have lost much of its original complement of material soon after formation. Davis et al. discuss possible mechanisms for explaining this belt attrition: (a) the same process that pumped up the mean asteroid velocities also eroded or ejected much of the belt mass; or (b) the collisions resulting from these enhanced velocities ground the debris into particles so small that they were swept away by solar radiation forces. Regardless of the pumping mechanism, the Davis et al. collisional model requires a total mass, at the time when velocities were pumped to 5 km s^{-1}, only slightly larger (by a factor of 2.5) than the present one.

If one accepts the Davis et al. model population, the attrition process mentioned above must have been quite vigorous in order to dispose of nearly an Earth mass of belt material. Since it takes only 10^5 to 10^6 yr to grow Ceres from km-sized planetesimals (Greenberg et al. 1978*b*), the time available for accretion before truncation by Jupiter-caused effects may have been very short. The ages of basaltic achondrites impose an upper limit on the length of vigorous mass loss by requiring that 4 Vesta (assuming it is indeed the parent body) have an intact crust about 10^7 yr after solar system formation. Even if most of this debris were ejected from the solar system by Jupiter, a substantial portion of it still must have been swept up by the terrestrial planets, potentially providing a large volatile source in the final stages of planetary accretion.

Once large-scale mass loss ended, mutual collisions acted primarily to modify the asteroids themselves rather than supply material outside the belt. Davis and his coworkers conclude that most of the present asteroids <250 km in diameter are shattered survivors of original bodies, reassembled after a catastrophic collision (see Fig. 13). Only a few of the largest objects are substantially intact. This prediction has a number of implications for observational interpretations of asteroid surfaces, which we address below.

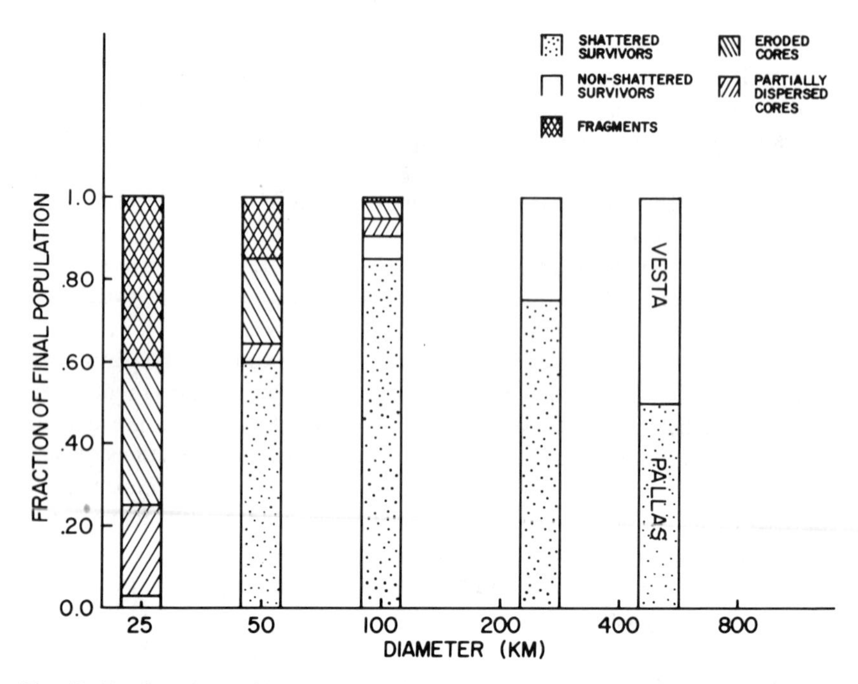

Fig. 13. Fraction of asteroids calculated to be either (1) nonshattered survivors, (2) shattered survivors, (3) eroded cores, (4) partially dispersed cores or (5) discrete fragments as a function of size on the RG-2 evolved population (figure from Davis et al. 1985).

Given an initial population comprising a few times the present asteroidal mass, the protoasteroids retained a substantial inventory of volatile elements. Assuming the present compositional stratification of the belt, the newly formed asteroids undoubtedly were dominated by carbonaceous, water-rich material. Water was abundant, comprising 10% or more of the C-class precursors. Ice may have been an important phase in both the early C class as well as the outer belt P and D objects. The latter may have been composed of roughly 20 to 40% water ice, with a maximum near the Galilean satellite value of 60% (Consolmagno and Lewis 1978*b*). Despite the subsequent collisional processing of these objects (shattering and reassembly, fragmentation, regolith overturn), the impact velocities were low enough for most volatiles to have survived, predominantly in the form of adsorbed and structural H_2O in hydrated silicates, and the C, H, O and N components of organic kerogens. The best spectral analogues of the C-class asteroids, CI and CM carbonaceous chondrites, serve to illustrate this point. CI chondrites, for example, are still rich in volatiles, despite long exposure to vacuum and mild impact brecciation on their parent bodies. The CIs contain about 10% H_2O, 3.5% C, 0.25% N, and 6.0% S by weight (Lewis and Prinn 1984). If the terrestrial planets intercepted even a fraction of this sort of material during late accretion and the heavy bombardment, the problem becomes not one of acquiring enough volatiles, but instead how to avoid getting too much. Only 0.02 Earth masses of CI-like material is sufficient to supply Earth with its entire volatile inventory (Anders 1968). The likelihood of supplying this asteroidal bonanza to the terrestrial planets via any of several delivery mechanisms is the subject of the next section.

Transport of Asteroidal Material to the Inner Solar System

We have mentioned that Jupiter's growth probably supplied a large amount of asteroidal material to the inner solar system, either by the action of Jupiter-zone debris scattered into the belt, thus increasing velocities and subsequent collisions, or alternatively, by sweeping its gap-clearing resonances through the belt as Jupiter grew. Wetherill (1978) showed that some main-belt debris perturbed by Jupiter into terrestrial planet-crossing orbits survives in a long-lived tail lasting up to 10^8 yr. Mars, undergoing frequent collisions with the debris leaving the belt, should have gained angular momentum and shifted its orbit outward, where its subsequent encounters with inner belt objects could send additional material sunward (Shoemaker 1977*b*). The initial stirring of the main belt would have also supplied a small, slowly declining stream of collision-generated fragments directly to planet-crossing orbits.

Since the end of the heavy bombardment period about 3.9 to 4.0 Gyr ago, delivery of asteroidal material to the terrestrial planets has continued at fairly steady rates (a recent estimate of the terrestrial accretion rate is 6.8×10^{10} g yr^{-1} [Kyte and Wasson 1986]). The existence of asteroid-free regions or *gaps* in the belt at distances associated with strong Jupiter commensurabilities suggests that the resonance itself might be clearing material from the belt. Indeed, analysis of orbits adjacent to the resonances revealed enhancements in eccentricity and inclination that oscillated about a forced value determined by the specific resonance involved (Greenberg and Scholl 1979). However, while unusual, these orbits apparently remain stable. Gravitational theories and simulations were unable to determine a mechanism which would actually *clear* the gaps until Wisdom's (1985*b*) work. He discovered the unusual chaotic nature of orbits near the Jupiter 3 : 1 commensurability (the Kirkwood Gap). Seemingly stable orbits near this zone can suddenly jump to much higher eccentricities and thus shift to planet-crossing orbits. Those objects leaving the gaps are swept up by collisions with Mars or Jupiter; some of these asteroids are thrown into the inner solar system, where they collide with the terrestrial planets or are ejected from the solar system.

This chaotic orbital behavior is the best candidate for providing the current flux of near-Earth or Earth-crossing asteroids. A spectral comparison of Earth-approaching asteroids with main-belt objects near the 3 : 1 resonance for similarities in composition would be a reasonable observational test of this mechanism. If resonances do transport belt material to the inner solar system, asteroids near prominent gaps should be represented in the near-Earth population. A corollary to this argument is that the most common meteorites arriving at Earth, the ordinary chondrites, should originate on spectrally similar parent bodies in the near-Earth population, in turn supplied from a similar population near the gaps. To date, however, only one small near-Earth asteroid with an ordinary chondrite spectral signature has been found, and none in the main belt. In a recent search of 11 asteroids near the 3 : 1 gap, no evidence for ordinary chondrite asteroids was found; the source of these ubiquitous meteorites remains unknown (McFadden and Vilas 1987).

Based on the main belt's composition alone, we would expect chaotic resonance effects to supply mainly C-class objects to the near-Earth or Earth-crossing population, though given the relatively small numbers so far observed, the distribution can be skewed dramatically by stochastic events in the belt (e.g., collisions). In fact, only about 20% of the two dozen or so Earth crossers characterized in McFadden's 1983 study were C class, while some 70% were from high albedo Ss. However, Tedesco and Gradie (1987) estimated that about 40% of the near-Earth population are Cs, with Ss comprising another 40%. Resonance effects, then, supply a constantly replenished stream of volatile-rich asteroids to the inner solar system; since the end of the heavy bombardment, their volatile contributions could have been significant. Moreover, because of the greater mass of the post-accretional belt, clearance of the gaps would have greatly increased the flux.

While the C-class asteroids may be thought of as relatively volatile-rich, the other asteroid classes and their (probable) meteoritic analogues range from greatly reduced specimens like the enstatite chondrites and the irons to the putative ultra-carbonaceous P and D asteroids, of which we have no demonstrable samples. The variation in volatile content can be attributed either to original compositional differences or to subsequent degassing under thermal processing. We next explore the possible heat sources that may have modified the protoasteroids and controlled their volatile content since formation.

IV. EARLY HEAT SOURCES

Long-Lived Radionuclides: Impact Heating

The potential of long-lived radionuclides and impact heating for thermally metamorphosing most bodies of asteroidal size is undoubtedly negligible. When the thermal diffusive time constant for a body ($\sim 10^4 R^2$ yr, where R is the radius in km) is short compared to the half-life of the radionuclide, its decay causes a very small temperature rise. Consequently, long-lived radionuclides (half-lives $\geq 10^9$ yr) have little effect on bodies smaller than a few 100 km in radius (see also the reference to Gaffey and Lazarewicz [1988] below).

The gravitational potential energy acquired by infalling material is equivalent to a temperature increase of about 5×10^{-4} R^2 for material with a density of 2 g cm^{-3}; this is only about 500 K for even a 1000 km object.

Random impacts due to orbital velocities typical of eccentricities and inclinations around 0.1 bring in kinetic energy equivalent to about 1000 K, but only a tiny fraction of this energy is expected to be retained, since it is released at the surface where it can radiate away immediately.

^{26}Al

Compared with longer-lived radioisotopes and impact energy, ^{26}Al is a potent heat source with a half-life of order 10^6 yr. Applying the relation given above, it can heat effectively bodies with radii roughly $\geq$10 km. Nevertheless, ^{26}Al suffers from two drawbacks as a heat source for asteroidal metamorphism. The first is that aluminum-correlated ^{26}Mg (the decay daughter of ^{26}Al) excesses are known only in very localized sites associated with rare calcium-aluminum-rich inclusions in meteorites; it seems particularly absent in CM-type material. The second is its lack of selectivity; it must have been distributed unevenly through the main belt in order to explain the radially dependent igneous activity inferred there by observation (Bowell et al. 1978; Lebofsky 1978). ^{26}Al must have been heterogeneously distributed even in individual parent bodies in order to have produced the complex partial melting scenarios observed in igneous meteorites (Mittlefehldt 1979; Hewins and Newsom 1988, ch. 3.2).

Electrical Induction Heating

Yet another mechanism for asteroid heating has hovered in the background for over 15 years, usually overshadowed by the ^{26}Al controversy, but steadily growing in favor. This electrical induction or ohmic heating scenario has received renewed attention because of improved measurements of rock and meteorite electrical conductivities and more exhaustive modeling of the many variables that influence its efficiency. This heat source has a dependence on heliocentric distance and asteroid size that seems consistent with the belt's compositional structure (Bell 1986). The mechanism's increasing popularity makes a short summary of its characteristics worthwhile.

Moving plasma with an embedded magnetic field impinging on an electrically conducting body will drive electrical currents through the body (Drell et al. 1965). Sonett et al. (1968) first proposed that an intense solar wind (suggested by observations of hydrogen outflows associated with T Tauri stars) during the late phases of stellar formation could induce currents in planetary bodies large enough to add significantly to their heat budgets. This scenario has been modeled with increasing sophistication for nearly two decades (Sonett et al. 1968,1970,1975; Herbert et al. 1977; Herbert and Sonett 1978,1979,1980). The results have confirmed the potential of the electrical induction mechanism for heating the Moon, asteroids and small asteroid-sized planetesimals. Since ^{26}Al radiogenic heating cannot explain all the early thermal activity on such a variety of different-sized objects, electrical induction heating is an attractive alternative.

Simply put, solar wind kinetic energy is deposited as heat in planetary interiors via electrical induction. Although the sizes of stellar wind energy fluxes (even those of proposed T Tauri-like outflows) are small compared to associated stellar luminosities, the low-heat conductivities of small planetary bodies lead to accumulation and retention of much of the intercepted energy flux. Over the course of the expected 10^4 to 10^7 yr duration of the solar T Tauri phase, objects of optimum size can receive enough heat to produce metamorphic temperatures.

Characteristics of Planetary Electrical Induction

With the possible exception of the Moon's case (see Sonnett et al. 1975), probably the most energetic induction mode is the so-called "unipolar" mode (the DC limit of the transverse magnetic mode). In this configuration the $-\mathbf{v} \times \mathbf{B}$ electric field in the moving plasma drives a current from the plasma through the body and back to the plasma (Fig. 14). As in any series circuit, the size of the current is sensitive to all impedances in the circuit. In particular, the plasma must possess enough charge carriers to support the current; the plasma must strike the body directly (and not be deflected completely by the pressure of the induced magnetic field, as the Earth's permanent field deflects the solar wind); and the body's surface layer must not be an insulator.

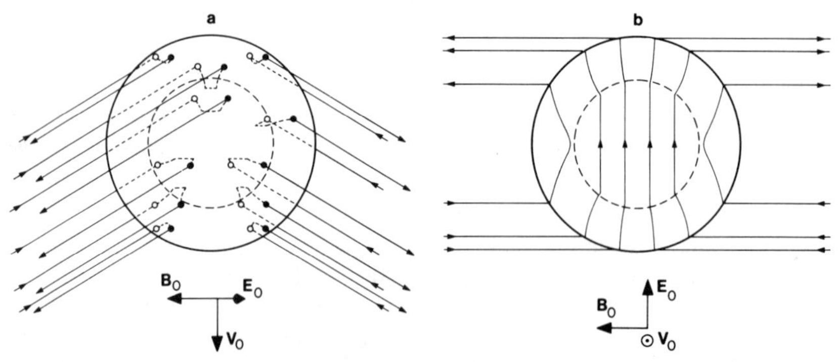

Fig. 14. Two views of typical current paths in an asteroid undergoing unipolar electromagnetic induction. Each is an isometric (i.e., without perspective) projection of three-dimensional current paths passing through the asteroid (assumed to have 2 layers of differing conductivity) and closing into a solar wind plasma. The plasma current paths do not terminate near the edge of the sketch but in reality continue on indefinitely. (A) "Top" view of the asteroid nearly antiparallel to **E** but rotated 10° clockwise around the $\mathbf{v}_0$ axis. $\mathbf{v}_0$ is the solar wind velocity vector and $\mathbf{B}_0$ and $\mathbf{E}_0$ the associated plasma magnetic and electric field vectors. The open circles show where the current paths enter the asteroid on the "bottom" of the object, and they emerge at the filled circles on the "top". The straight-line segments of the current paths within the asteroid are parallel to $\mathbf{E}_0$. (B) View looking up the solar wind stream (essentially sunward) with $\mathbf{B}_0$ directed left and $\mathbf{E}_0$ upward. Only current paths passing through a central asteroid cross section normal to the solar wind velocity vector $\mathbf{v}_0$ are shown. The current paths outside the asteroid are bent back towards the observer.

The principal factors determining the total heat input from induction are:

1. Plasma momentum flux;
2. Flow duration;
3. Magnetic back-pressure from the induced field;
4. Limitation of the current density by plasma sheath effects at the surface of the body;
5. Conduction of heat out of the body;
6. The possible electrical resistance (due to porosity) of any regolith that might exist;
7. Electrical conductivity of the material.

The interplay of factors (1) and (3) lead to an increase of heat input with decreasing body size; (2), (5) and (6) cause net heating to increase with increasing size, while (4) establishes a lower limit to the diameter at which substantial heating can occur. Also, for obvious reasons there is a decline in heating with increasing heliocentric distance. Thus any particular model scenario yields a particular object diameter at which heat deposition is a maximum.

The significance of the heating maximum at intermediate sizes has been discussed in a number of modeling studies (Herbert and Sonett 1978, 1979,1980). Those studies investigated the thermal evolution of asteroids in the main belt, and determined a range of conditions under which 4 Vesta would experience nearly complete melting while 1 Ceres and 2 Pallas would not. Vesta is believed to have undergone considerable igneous differentiation (Drake 1979; Bowell et al. 1978), while Ceres seems to have escaped exposure to high temperatures (Bowell et al. 1978; Lebofsky 1978). Pallas seems also to have remained unmelted (Bowell et al. 1978; Larson et al. 1983). Thus, what would otherwise be a puzzling divergence in the histories of these similarly sized asteroids seems naturally accounted for by the electrical induction hypothesis. Herbert and Sonett (1979,1980) also noted an instability of the electrical distribution that could cause spatially inhomogeneous "hot spots" in a parent body interior; this effect may help meet thermal history constraints imposed by igneous meteorite evidence (Mittlefehldt 1979).

Effect of Volatiles on Electrical Induction

The electrical conductivity of asteroidal material is a major factor affecting a minor planet's thermal evolution. For typical planetary materials, the variation of conductivity with respect to changes in temperature, bulk composition, physical state and trace contaminants is both extremely complex and poorly understood (see, e.g., Parkhomenko 1967). Laboratory studies have confirmed the strongly history-dependent conductivities of some likely materials (Schwerer et al. 1971; Duba and Boland 1984). One influential constituent is H_2O; with just over a ten-fold increase in relative moisture content, the bulk resistivity of common terrestrial igneous rock (normally highly insulating when dry) can be reduced by about six orders of magnitude (Parkhomenko 1967,p.123). Elemental carbon, which is deposited at grain boundaries when carbonaceous chondrite material is heated, strongly affects conductivity in a very history-dependent way (Duba and Boland 1984). Many sulfide minerals, such as FeS, are excellent conductors; when segregated along grain boundaries even as trace constituents, they can greatly increase bulk conductivity. Several of these volatile compounds are commonly found in chondritic meteorites (Dodd 1981; our Table I), and are also expected on theoretical grounds in protoplanetary materials (Lewis 1972*a*).

Volatile migration and loss is dependent on the thermal history of the parent body, but local volatile abundance also influences any electrically induced heat input. For example, water ice, carbon and reduced sulfur compounds driven from a heated interior and recondensed in an otherwise porous regolith may markedly reduce its series resistance to unipolar induction currents, prolonging the lifetime of a given heating episode. (The usually insulating surface layers tend to reduce heating rates markedly when the plasma flux begins to decrease.) Aside from this enhancement of electrical induction heating, volatile release and subsequent migration in asteroid-sized bodies probably had a marked effect on surface mineralogy and reflectance properties, which are the focus of asteroid remote-sensing efforts.

V. EFFECTS OF HEATING ON ASTEROID VOLATILE INVENTORIES

The chemical effects of the variety of heating scenarios outlined above are functions of composition, peak temperature, lithostatic pressure, duration of the episode, heating rate and the heat capacity of the material. With the early belt seemingly dominated by C-class, volatile-rich material, investigations of thermal effects on that type of assemblage would have the greatest relevance here. DuFresne and Anders (1962) pointed out that internal heating of a small body containing chemically bound H_2O would produce a narrow zone of liquid water; the layer would migrate upward under continued heating, resulting in a wet zone underlying a surficial icy layer analagous to permafrost. The most thorough study of the thermo-chemical changes associated with heating of volatile-rich parent bodies was conducted by Gaffey and Lazarewicz (1988). They modeled the evolution of several test objects up to 500 km in radius; the assumed starting composition was CM carbonaceous chondrite material. With conservative assumptions about the thermal input (long-lived radionuclides only, CI and CM abundances and no insulating surface layer), they hoped to predict the *minimum* thermo-chemical effects expected on originally homogeneous CM-type asteroids. The largest temperature rise, approximately 800°C, occurred in the core of the 500-km radius object, 1.5 Gyr after formation. Silicate melting did not occur. Dehydration of core minerals and mobilization of water and other volatiles created successive alteration zones outward from an Fe-rich olivine core: first a layer of olivine/pyroxene/magnetite, followed outward sequentially by layers of olivine/pyroxene/magnetite/phyllosilicate, phyllosilicates and FeO, and finally an unmodified surface layer. Liquid water was stable in two of the zones.

Of course, as dehydration of hydrous minerals and salts took place, water vapor and other volatiles were released, including CO_2 and CO from the oxidation of carbon compounds and the breakdown of carbonates. From C-chondrite heating experiments, the evolved gas is a mixture of H_2O, CO and CO_2 (90–95% by volume), with smaller amounts of CH_4 and SO_2. Peak gas release occurs between 600 and 900 K, delayed only 20 to 60 K at central pressures of 3 kbar. According to Gaffey and Lazarewicz' work (1987), H_2O and CO_2 vapor pressures can become so great in the heating zones ($\sim$2 kbar) that the released gases may exceed the confining pressures and fracture the surrounding rock. The H_2O released in the interior, propagating upward, could roughly double the water content in the 100 km layer nearest the surface. In this "damp" layer, a film of liquid would coat grain surfaces and provide an excellent environment for aqueous alteration and the formation of phyllosilicates.

Lewis and Prinn (1984) also discuss the chemical effects of heating carbonaceous chondrite material. The melting point of silicates with a large water content is fairly low; temperatures of perhaps 700 K could lead to separation of an FeS-rich core. As byproducts, such a high-temperature equilibration would produce H_2O, CO, H_2 and N_2, while lower temperatures would generate H_2O, H_2, CO_2 and N_2 or NH_3. On small bodies, the likely result of heating would be an H_2O–CO_2 fluid, bearing soluble materials such as NH_4^+, Na^+, K^+ and Ca^{2+} salts of $CO_3{}^{2-}$, $SO_4{}^{2-}$ and halide anions. Depending on the fluid composition, liquids could remain stable even at relatively low temperatures: the binary H_2O–NH_4Cl eutectic is 258 K, and that of H_2O–NH_3 is 173 K. Fluids reaching the surface layer would deposit their salts in pore spaces, creating CI-type assemblages. Interestingly, if a mobile CO-rich fluid gained access to the extremely low pressures of the surface environment, CO might very well decompose into graphite, depositing a dark phase that could dominate the surface optical properties (Gaffey and Lazarewicz 1988).

Wood (1979) suggested that the fragmented surface layer of most small bodies would possess a substantially reduced thermal conductivity; in such a case, or if radionuclides were augmented as a heat source by ^{26}Al or induction heating, the alteration zones outlined above would shift outward. Enhanced heating would be most important on smaller bodies, where heat loss is otherwise more efficient. Gaffey (personal communication) now believes that on relatively large bodies ($\geq$250 km radius), high central temperatures and alteration are virtually unavoidable, and any augmentation of the heat source can result in extensive internal metamorphism. An important implication for spectroscopic observations is that the outer layer of such an object could remain relatively undisturbed except for aqueous alteration effects (Hildebrand et al. 1987).

Aqueous Alteration

Plentiful meteoritic evidence for the aqueous alteration of asteroid surface layers, supplemented by telescopic observations, supports the existence of large volatile reservoirs. McSween (1987) notes that carbonaceous chondrites exhibit widespread aqueous alteration, probably because their fine-grained matrix made them permeable to fluids evolved from the interior. Kerridge and Bunch (1979) presented a thorough review of the morphological and mineralogical evidence for alteration in CI and CM chondrites. In the CIs, veins of Mg and Ca carbonates and sulfates permeate the hydrated Fe- and Mg-rich silicates which dominate the matrix (McSween 1979*a*). Both the veins and the characteristic magnetite morphology stem unequivocally from fluid deposition. Liquid water contact was maintained for a time $\geq$1000 yr, though leaching apparently did not transport original material over a scale $\gtrsim$1 cm.

The CM chondrites, though morphologically distinct from the CIs, also display mineral associations and textures characteristic of aqueous alteration. Analyses show that the matrix underwent addition of H_2O, CO_2 and water-soluble organic compounds during the process, which took place in a parent body regolith at temperatures <400 K (Bunch and Chang 1980). Carbon and H_2O were mobile phases during the process, but again bulk meteorite analyses were unaffected by liquid transport (McSween 1979*b*). Kerridge and Bunch (1979) conclude that both CI and CM chondrites were first subjected to low-velocity impacts and brecciation on a parent body surface, then altered by liquid water while regolith overturn and exposure continued. For chondrites of higher petrologic type, a few (of type CV, CO and LL3) exhibit aspects of sulfur oxide chemistry that also suggest an aqueous origin (Burgess et al. 1987).

Since we suspect C-class asteroid surfaces to be similar to carbonaceous chondrite material, spectroscopic searches might focus profitably on alteration products, but this approach quickly runs into a problem of nonuniqueness, noted by Kerridge and Bunch, that is still with us. The original water-bearing phase on the asteroids might as easily have been hydrated silicates instead of ice, and these precursor phyllosilicates would be spectroscopically indistinguishable from the similar products of aqueous alteration. Thus, we cannot tell if a given asteroid retains a primordial phyllosilicate surface or if it has undergone extensive internal heating and surface alteration. We can assume, however, based on kinetic inhibition of silicate hydration in the nebula (Prinn and Fegley 1987; see also their chapter), that asteroidal hydrated silicates are almost certainly of secondary origin. Spectral searches for hydrated silicates and other volatiles are important tools, then, for studies of asteroid composition, volatile content and thermal evolution.

VI. OBSERVATIONS OF ASTEROID VOLATILES

The thermal studies above predict that moderate heating of a carbonaceous asteroid would produce a relatively pristine outer layer, affected only by aqueous processes and perhaps minor outgassing (Fig. 15). The surface materials may contain structural, interlayer and adsorbed water in hydrated silicates, possibly underlain by an icy substrate. Deeper layers excavated by impact may be dehydrated or partially differentiated (Gaffey and Lazarewicz 1988; Hildebrand et al. 1987). Spectroscopic observations of carbonaceous chondrite matrix material (CCMM) are difficult; its opaque component, primarily organic kerogen but including some magnetite, gives rise not only to a low albedo but also to effective suppression of visual and near-infrared absorption features (Wilkening 1978). In mid-infrared reflectance, however, CCMM displays a broad 3-μm water absorption (Fig. 16), the subject of nearly a decade of spectroscopic investigation.

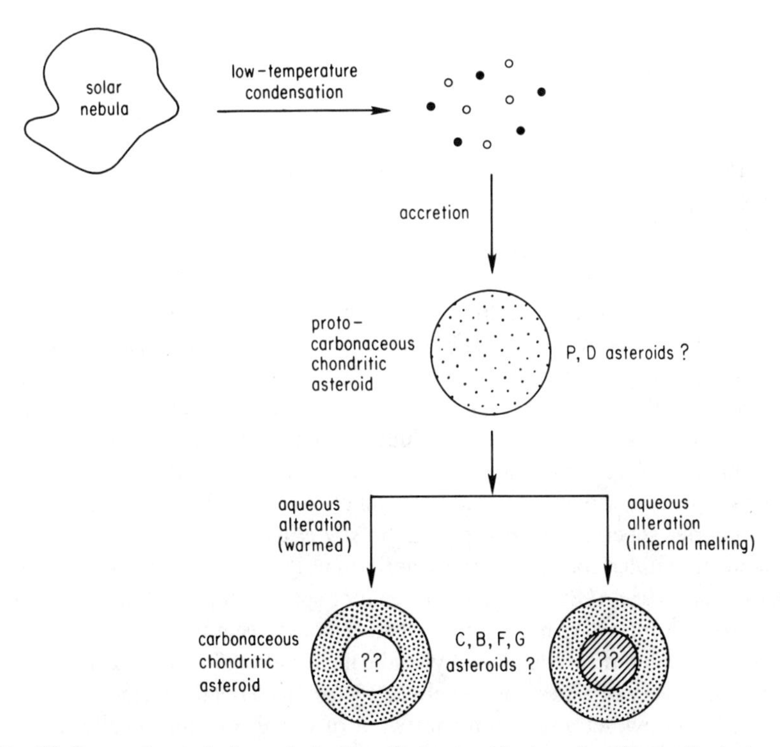

Fig. 15. Proposed scenario for producing low-albedo asteroids. Accretion of ice and/or hydrated silicates combined with high-temperature components formed very dark, ultracarbonaceous asteroids, of which the Ps and Ds may be relicts. Gentle or more intense internal heating produces aqueous alteration at the surface and either dehydration or partial differentiation near the core. It is presently impossible to tell whether the C,B,F and G classes have undergone more than mild heating in their interiors. None of our carbonaceous chondrite samples are good spectral matches for the C,B,F and G classes.

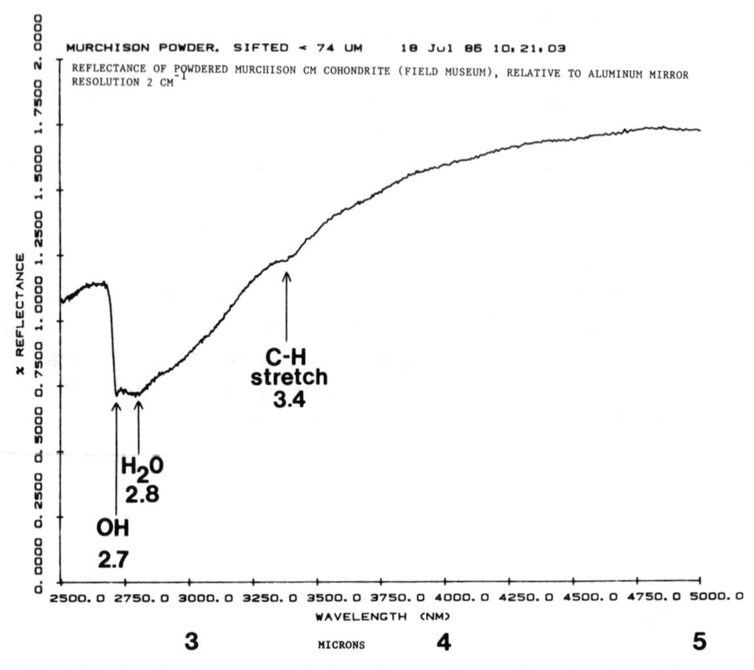

Fig. 16. Midinfrared reflectance of the Murchison CM chondrite, showing distinctive 3-μm water absorption and the subtle 3.4-μm hydrocarbon feature. This spectral region is diagnostic of several volatile species.

In the initial search for this asteroid "water of hydration" nearly a decade ago, Lebofsky used broadband photometry to identify the 3-μm band on 1 Ceres, the largest asteroid, then classified as a member of the C class (Lebofsky 1978; see Fig. 11). He later presented similar observations for seven additional objects, and concluded that both hydrated and anhydrous assemblages were present among the C class (Lebofsky 1980). Larson et al. (1979,1983) subsequently found both mineralogies on 2 Pallas. High-resolution Fourier spectra of Ceres from Lebofsky et al. (1981) showed the 3-μm feature resembled that of the montmorillonite group of clay minerals. Feierberg et al. (1981,1985*a*), using both Fourier spectroscopy and low-resolution spectrophotometry, searched more than a dozen asteroids for a variety of aqueous alteration products—phyllosilicates, hydrated salts and magnetite. Their discovery of a correlation between 3-μm band depth and the strength of an Fe^{3+} ultraviolet absorption shortward of 0.4 μm strengthened the compositional link between carbonaceous chondrites and the C-class asteroids.

Feierberg et al. (1985*a*) conclude that just over half of the ten largest C asteroids display the 3-μm signature of hydrated silicates. Additional water in the form of surface ice or frost was tentatively identified on Ceres by Lebofsky et al. (1981). Our current observation programs continue to make use of cir-

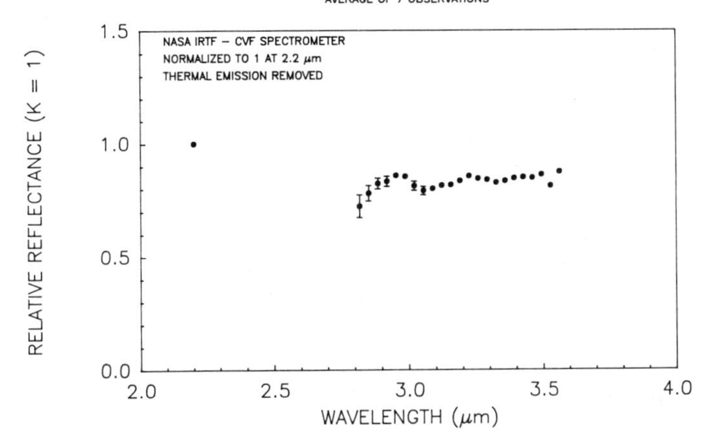

Fig. 17. CVF (circular variable filter) reflectance spectrum of 1 Ceres, showing the difficulty low-resolution spectrophotometry encounters in searching for subtle features. The 3.4-μm hydrocarbon absorption and the sharp band edge at 2.7 μm due to the OH group in clay minerals are both unresolved.

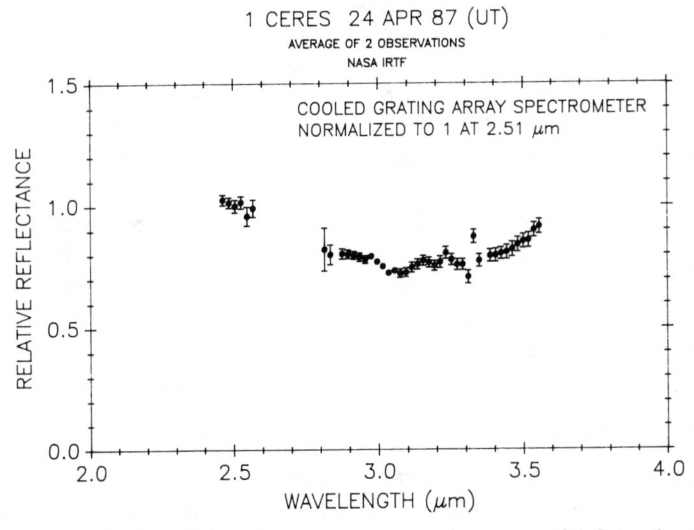

Fig. 18. Recent CGAS (cooled grating array spectrometer) spectrum of 1 Ceres showing increased resolution in the 3-μm region. CGAS spectra may yield the type and abundance of clay minerals, as well as confirmation of the presence of ice, frost or hydrocarbons. In this spectrum the 2.7-μm OH band edge emerges from the telluric H_2O absorption, and a 3.1-μm frost signature is visible. Note the absence of any obvious 3.4-μm C-H band. The spike at 3.35 μm is due to incomplete removal of atmospheric CH_4 extinction.

cular-variable-filter (CVF) spectrophotometry to search for hydrated asteroid surfaces (see Fig. 17), but we are turning to a new instrument at the NASA Infrared Telescope Facility (IRTF) for detailed studies of the 3-μm region. The Cooled Grating Array Spectrometer (CGAS) uses a linear 32-detector array to sample an asteroid's infrared spectrum simultaneously across a span of nearly 0.6 μm. The broad spectral range of the CGAS is especially useful in separating the hydration feature from the telluric H_2O absorption at 2.5–2.8 μm (Fig. 18). By measuring reflectance simultaneously on both sides of the telluric gap, we can obtain a much more accurate 3-μm band depth, crucial to characterizing the amount and type of hydrated silicates on a given surface. That, in turn, allows us to estimate the surface H_2O content. In addition, the CGAS resolution and high signal-to-noise ratio at and beyond 3 μm should be able to confirm the 3.1-μm frost feature, and perhaps even the C-H stretch band, diagnostic of hydrocarbons or kerogen at 3.4 μm.

VII. UNRESOLVED ISSUES IN ASTEROID VOLATILE OBSERVATIONS

In opening a new spectral region to scrutiny over the past few years, 3-μm spectroscopy has raised several new issues concerning the outer-belt volatile distribution. The observation of the 3-μm absorption on six of the ten largest C-class objects, and its seemingly random existence on smaller asteroids of that class, has been interpreted as the natural consequence of viewing collisional fragments of larger proto-asteroids (Feierberg et al. 1985*a*). Having undergone heating, the core fragments should now be anhydrous, while the outer mantle and surface fragments should be hydrated and aqueously altered. However, 2 Pallas, large enough to have survived 4.5 Gyr of collisions and so preserve its hydrated mantle intact, has a major anhydrous surface component (Larson et al. 1983). This may be ejecta from especially deep, penetrating impacts, or Pallas may actually have been shattered and reassembled, exposing random anhydrous and hydrated material on its surface. Both possibilities are consistent with a collision-induced origin for this body's unusually inclined orbit. Continued exploitation of the 3-μm spectral region will expand the sample size and reveal any correlations of the hydrated silicate absorption with either object size or heliocentric distance.

Phobos and Deimos

The morphology of the Martian moons provides further evidence that small carbonaceous "asteroids" possessed geologically significant amounts of volatiles. Viking images of Phobos, spectrally a C-class object, indicate that this small satellite may once have outgassed large amounts of water. Chains of small craters, aligned with linear surface fractures, may have formed when an impact vaporized bound water and forced it to the surface along the newly formed fracture planes (Veverka and Thomas 1979; Hartmann 1980). Admittedly Phobos and Deimos have had specialized histories in Martian orbit atypical of belt asteroids, but if the two moons actually formed in the inner belt, the rest of the C class should have incorporated at least as much water. Further, if even C-class objects had enough water to produce impact-induced devolatilization, the P and D classes may be remarkably wet objects.

The observed compositional zoning of the belt predicts a gradual shift to volatile-rich ultracarbonaceous surface material beyond 3 AU (Gradie and Veverka 1980; Vilas and Smith 1985). Gradie and Veverka proposed that the low albedos and reddish spectra of the Trojan (now classified as D) asteroids were due to the presence of kerogen-like organic compounds. Vilas and Smith observed P, D and C asteroids and explained a systematic steepening in their spectral slope in terms of a compositional or optical change in the organic material outward from the Sun. Such dark, kerogen-like materials are not unique to the outer asteroid belt or carbonaceous meteorites; many minor outer planet satellites are of spectral class C, and the Iapetus dark hemisphere is dominated by a silicate-organic mixture reminiscent of the D asteroids (Bell et al. 1985). More recently, the results from the Vega-Giotto Halley flybys confirmed a very low albedo (p_v = 0.04) and a volatile-rich, organic composition for many of the dust particles in the coma, the so-called carbon-hydrogen-oxygen-nitrogen or "CHON" particles. The spectral signature of extraterrestrial organic material is an important target of our current laboratory efforts (Jones et al. 1986), our CGAS study of carbonaceous asteroids, and planned spacecraft encounters with asteroids and comets (Galileo, CRAF).

Asteroid-Comet Relationships

The spectral similarities between many of these small bodies suggest a compositional link between at least some of the icy outer solar system bodies and outer-belt or more distant asteroids; this relationship has been discussed by Hartmann et al. (1985,1987). The major finding of their 1987 study is that asteroids with orbital elements typical of short-period comets exhibit spectral and albedo properties similar to outer solar system materials. From the spectrophotometric similarities between P, D and C asteroids, interplanetary dust particles, and comet nuclei or their associated dust, Hartmann et al. propose a genetic link between these hitherto disparate examples of solar system matter; they infer that the outer-belt asteroids, in particular the Ds, may be more similar in structure and composition to comet nuclei than other asteroids. They propose that small outer solar system objects all acquired their ice and carbonaceous dust as a natural consequence of formation, beginning in the main belt beyond roughly 3 AU out to at least the orbits of Uranus and Neptune (2060 Chiron and most comets fall into this category). A major implication of their thinking is that asteroids beyond 2.7 AU, and especially the Ps and Ds, may harbor significant quantities of ice. The Hildas and Trojans (at 4.0 and 5.2 AU, respectively) are particularly good prospects for subsurface ice, according to Hartmann and colleagues.

These conclusions should come with a few cautions. First, their dynamical selection of extinct comet candidates is subject to criticism; it is not inconceivable that asteroids perturbed from the outer belt (where low-albedo types do indeed dominate) may acquire short-period, comet-like orbits. In fact, it would be surprising if a perturbed asteroid did *not* acquire either an Earth-approaching or comet-like orbit, since observations have revealed no other reasonable storage areas outside the main belt, and Jupiter will truncate any asteroid or comet orbits with aphelia beyond 5 AU. Second, their claim of spectral similarity between cometary surfaces and the D asteroids may be premature if the nucleus is obscured by dust whose scattering properties are phase-dependent and thus different from those of a reflecting particulate surface. Hartmann and his coworkers recognize the difficulties in obtaining spectra of the nucleus free of coma effects, but cite studies of Halley which seem to minimize the effect of phase-related color changes.

Ice in the Outer Belt?

Because of their location and low albedos, the outer-belt P and D asteroids are generally regarded as water-rich ultracarbonaceous bodies, whose surfaces are dominated by clay minerals and organic compounds (Gradie and Veverka 1980). However, spectroscopic confirmation of this idea is still lacking: recent 3-μm observations of several P and D objects revealed no significant water of hydration absorption (Feierberg et al. 1985*b*; Jones et al. 1988). This is a troubling result, since shorter-wavelength spectra and albedo data, along with nebular condensation theories, predict a water-rich composition in the outer belt. The organic kerogens are apparently plentiful on P and D surfaces; early heating mechanisms seem to have been ineffective beyond 3 AU. Why, then, are hydrated silicates apparently absent? Surface dehydration due to external heating seems implausible, since the surface layer cools very efficiently by radiation, and any electro-inductive heating event would deposit its heat not at the surface but in deeper layers. Total dehydration from an interior heat source is hardly likely, given the great heat sink provided by water and other volatiles; we would also expect extensive thermal metamorphism of the entire body with associated spectral and albedo changes. Masking of the 3-μm water of hydration feature by surface organics is possible, but laboratory experiments suggest that suppressing such a strong feature is difficult: albedo decreases to some limiting value, say 0.04, but the H_2O band remains visible. If organics were produced in the nebula in grain-catalyzed reactions, perhaps individual phyllosilicate grains were coated with carbonaceous material, and this configuration might suppress the 3-μm absorption. (We are addressing this possibility in our current lab spectroscopy program.) We should note, though, that despite the intimate mixing of phyllosilicates and organic material in fine-grained CCMM, the water of hydration feature is quite strong.

The most reasonable explanation for the lack of the 3-μm feature on the ultracarbonaceous asteroids, it seems to us, is the storage of volatiles in the form of subsurface ice (as suggested by Bunch and Chang 1980). If the asteroids formed at temperatures lower than approximately 170 K, they may have incorporated water in the form of ice. Even if ice condensation was precluded in the asteroid formation region, an accretion zone that sampled the Jupiter region may have supplied ice-rich debris to outer-belt asteroids. The Ps and Ds may have formed originally as loose agglomerations of ice and anhydrous silicates; the outer layers would have quickly lost their ice content to sublimation or gentle external heating, while the interiors never warmed to high enough temperatures to drive H_2O vapor or fluid to the surface (Fig. 15). A lag deposit of carbonaceous material blanketing the surface, and the lack of subsequent aqueous alteration, might explain the lower albedos, the lack of a hydrated silicate spectral signature and the reddish spectra reminiscent of cometary surfaces. Interestingly, groundbased observations of P/Halley by Campins and Ryan (1987) and data from the Giotto mass spectrometer (Brownlee 1987) suggest that silicate grains released from the nucleus may indeed be anhydrous. Infrared mapping spectrometer images of a cometary nucleus from the projected CRAF mission should determine whether a carbonaceous surface can eliminate the 3-μm signature of any hydrated silicates or ice particles.

Interwoven with the problem of volatile storage on the outer-belt asteroids is the question of the origin of hydrated silicates (see chapter by Prinn and Fegley). CI and CM matrix material is composed primarily of Mg- and Fe-rich clay minerals similar to terrestrial serpentine; these silicates are both a predicted nebular condensate and a product of aqueous alteration. Their origin can tell us a great deal about the formation of the volatile content of the asteroids: if they originally acquired their volatiles from nebular hydrated silicates, they probably were never more than 10% H_2O by mass. However, if the phyllosilicates were formed by aqueous alteration in a fluid bath generated from melted water ice, the average protoasteroid could easily have been composed of up to 50% water.

Isotopic studies by Yang and Epstein (1983) supported a solar system origin for the carbonaceous chondrite phyllosilicates, but could not pinpoint the actual formation process. They proposed reactions with nebular water at 200 K as the likely source of the near-terrestrial D/H ratio for carbonaceous chondrite clay minerals, but kinetic obstacles pointed out by Grinspoon and Lewis (1987*a*) make such a fractionation unlikely. Some unknown process may have been responsible for a local enrichment of nebular water with deuterium; if hydrated silicate formation occurred in a region of the nebula which also supplied the Earth with its H_2O inventory, the "enhanced" (compared to solar) terrestrial and phyllosilicate D/H ratios could be explained. It is also quite possible that the similar D/H ratios are entirely coincidental and caused by two distinct processes. For example, the phyllosilicate D/H ratio could be determined when silicates and nebula-derived ice react during a parent body heating episode. If the ice had an enhanced D/H ratio (perhaps due to an outer solar system origin), the resultant hydrated silicates could have acquired the present near-terrestrial D/H ratio. An enhanced D/H value for outer solar system ices (exemplified by comets) is also consistent with the modification of terrestrial planet D/H ratios by long-term comet or asteroid bombardment, as discussed by Grinspoon and Lewis (1988). Thus deuterium-enriched ices may have something to do with producing the D/H ratios of both the asteroid hydrated silicates and the Earth's water inventory. Deuterium abundance measurements for cometary material and possible asteroid ices will be required to address this issue, probably obtained by spacecraft comet rendezvous or asteroid/comet sample return. It is worth recalling that comets may be quite heterogeneous and have no typical D/H ratio, so an early resolution of this problem is unlikely.

Spacecraft Observations

The present groundbased efforts at deciphering the composition and mineralogies of the various asteroid classes urgently require "ground truth" in the form of close-in, high-resolution spacecraft observations. Scientific experiments should include capabilities for imaging, mapping spectroscopy, remote sensing of chemical composition, polarimetry and radiometry. Ideally an asteroid as well as a comet rendezvous should be undertaken, but the proposed CRAF (Comet Rendezvous Asteroid Flyby) mission, which includes one or two asteroid flybys as well as the comet nucleus rendezvous, is a good beginning. The comet investigation, to include a surface penetrator, will yield information on carbonaceous, hydrated silicate and ice mixtures that can be applied to the more primitive asteroids. Mapping spectroscopy, if carried out at sufficient resolution during the asteroid encounters, can determine which mineral assemblages are present, and this in turn can provide us with a fairly good physico-chemical history of the object (Gaffey 1987). Radiometry and polarimetry of the surface can characterize the thermal and optical scattering properties of asteroidal and cometary material. Remote sensing of surface composition by gamma-ray or X-ray spectroscopy may not by possible during a flyby, but holds promise if done from orbit around the nucleus or from the penetrator. An aggressive development and launch schedule for CRAF will spur current groundbased asteroid research, and encourage new approaches to our studies of asteroid volatiles. Whether CRAF flies or not, the first observations of a volatile-rich, low-albedo asteroid surface will probably come from the Soviet Phobos orbiter/lander mission, now scheduled for a dual launch in 1988.

In anticipation of the Galileo and CRAF asteroid flybys, we are continuing laboratory spectral studies of meteoritic organics and other candidate low-albedo materials, especially under asteroidal surface temperatures and pressures. Organic absorption changes caused by the decline in temperature with heliocentric distance are of special interest. Mixtures of these opaques and various clay minerals will be matched spectrally with mid-infrared asteroid reflectance data in an attempt to characterize the supposedly more primitive outer-belt objects. The laboratory data will also support asteroid observations out to 5 μm by the Galileo and CRAF spectrometers. Looking farther ahead to the late 1990s, the Space Infrared Telescope Facility (SIRTF) may at last provide comprehensive mineralogical information on large numbers of asteroids; extensive laboratory and groundbased work remains to be done in support of any space-borne instrumentation.

VIII. CONCLUSION

Meteorite studies and telescopic observations continue to expand our knowledge of the asteroids. Once regarded as annoying debris cluttering the observer's field of view and posing a physical danger to spacecraft, these relatively unaltered bodies probably are the most accessible repositories of early solar system history. In the cooler regions of the outer asteroid belt, apparently unaffected by severe heating, the C, P and D populations undoubtedly harbor significant volatile inventories. The larger primordial belt population probably had an even greater percentage of volatile-rich, low-albedo asteroids, a potent reservoir for veneering early terrestrial planet atmospheres. The importance of the asteroid role in volatile delivery to the planets, relative to comets and other accretional debris, is obscured by our lack of knowledge about the dynamics of late-stage accretion and mass-loss mechanisms in the early belt. The volatile-rich asteroids contain carbon, structurally bound and adsorbed water, and undoubtedly some remnants of interstellar material predating the solar system. Continuing exploration of these bodies will provide direct clues to conditions in the solar nebula, the nature of planetesimal accretion, and the collisional and thermal environment in the early solar system. We should also not overlook the future potential of volatile-rich, near-Earth asteroids in providing raw materials for the further exploration of the solar system.

Acknowledgments. Much of the 3-μm laboratory spectroscopy that we have undertaken was made possible through the efforts of J. W. Salisbury, U.S. Geological Survey. The NASA Infrared Telescope Facility assisted us greatly in obtaining observing time and making available its expert staff. The authors thank M. Marley, D. Grinspoon, A. Hildebrand, and J. S. Lewis for their comments and suggestions. E. Tedesco and J. Bell contributed careful reviews which clarified several points. Thanks are due as well to K. Denomy and T. Schemenauer for their graphic skills in preparing the figures. The work was funded by the National Aeronautics and Space Administration.

THE IMPACT HISTORY OF THE SOLAR SYSTEM: IMPLICATIONS FOR THE ORIGIN OF ATMOSPHERES

P. R. WEISSMAN
Jet Propulsion Laboratory

Impacts play a dual role in the origin and evolution of planetary and satellite atmospheres. Small impacts of volatile-rich planetesimals can contribute to the body's volatile inventory, while large impacts blow away existing primordial atmospheres and cause outgassing of the main body. The conflict between these two processes is likely greatest during the end stage of accretion when giant impacts, the late heavy bombardment, and the final sweep up of planetesimals and clearing of the planetary formation zones occurred. The terrestrial planets are bombarded by both volatile-rich and volatile-poor sources, while the outer solar system satellites are dominated by volatile-rich impactors. Numerous uncertainties in the interpretation of the solar system cratering record and the chronology of events make it possible to obtain only order-of-magnitude estimates as to what actually occurred. The processes of zone clearing and dynamical exchange of material between zones are not fully understood, and have changed rapidly in recent years due to the discovery of new mechanisms and sources for delivering material into planet-crossing orbits. Temporal fluctuations in the rate of impacts may be quite large, particularly for cometary sources where a factor of two variation is common, and a factor of several hundred is possible. Extrasolar impactor sources such as Oort clouds around other stars probably play a very minor role in contributing to volatile inventories.

I. INTRODUCTION

Impacts have played a major role in the solar system's history, as evidenced by the cratered surfaces of the terrestrial planets, their moons and the satellite systems around the Jovian planets. The cratering record on those surfaces is the result of the complex interaction of impactor sources and dynamical delivery mechanisms, and their evolution over time. Moreover, the cratering record on many surfaces has been modified and/or obliterated by resurfacing events from both internal and external processes.

Deciphering this complex record is far from simple, and so far, highly incomplete. Absolute cratering rates have been determined on only two bodies, the Earth and the Moon, through radio-isotope dating of returned samples. In all other cases age estimates are only relative, and subject to numerous assumptions about the way the impactor flux varies radially and temporally through the planetary region. In addition, any record of the cratering flux prior to about 3.9×10^9 yr ago has been obscured by the "late heavy bombardment," a period of intense cratering in the terrestrial planets region that overlaid the earlier record. The magnitude of the late heavy bombardment itself is a subject of considerable controversy. Even the current estimates of the flux of potential impactors in Earth-crossing orbits is somewhat uncertain, by a factor of two or more. Lastly, our knowledge of impactor sources and dynamical mechanisms is continually evolving and (hopefully) improving, and thus any statements about the solar system's cratering history are a reflection of the state of our knowledge at any one time.

The problem is further complicated by the fact that the vast majority of impacts in the solar system, in particular those which we are concerned with in considering the origin of planet and satellite atmospheres, occurred during the first 10^9 yr of its history. We are forced to infer the state of the system during that period by extrapolating our present, incomplete knowledge of the planetary system backwards in time. But the age of the solar system exceeds the lifetimes of objects in planet-crossing orbits, i.e., potential impactors, by factors which range from 2×10^1 (Neptune) to greater than 4×10^3 (Jupiter). Thus, such extrapolations are easily susceptible to error.

Yet another problem is that of attempting to derive the original impactor populations from the observed cratering record. We have no direct experience with cratering events in the energy and size ranges typical for planetary bodies. What is known about required cratering energies is extrapolated from nuclear test explosions, the largest resulting crater of which is <200 meters in diameter (Nordyke 1977). Depending on which of several available crater scaling relations is used (Shoemaker et al. 1979; Holsapple and Schmidt 1982), vastly different results can be obtained.

With these cautions in mind, let us then consider what is known about the formation and early history of the solar system. That history can be divided into several distinct, though likely overlapping periods when different impact related processes were dominant. They are:

1. Nebula collapse, settling of solid material to the nebula plane, and formation of planetesimals;

2. Accretion of protoplanets from planetesimals;
3. Impact of the last, large bodies on the protoplanets;
4. Final planetesimal sweep up and zone clearing;
5. The late heavy bombardment;
6. Steady-state flux of comet and asteroid impactors.

Although they do in some sense involve impacts, the first two periods are outside the scope of this chapter, and the reader is referred to the chapters by Morfill and Boss, and by Greenberg for detailed discussions of those topics. Periods (4) and (5) likely overlap in time and may in fact be the same process. In addition, it should be recognized that the description of the periods above would be modified if one assumed a different scenario for the formation of the planetary system. For example, number (2) above might be changed to, "formation of gaseous protoplanets" which would imply a significantly different mechanism for the acquisition of the volatile reservoirs of the planets, but would not affect the following discussion of the solar system's impact history.

There are several major schools of thought (Lewis and Prinn 1984) with regard to the origin of planetary atmospheres: (1) that volatile-rich planetesimals constituted all or part of the originally accreted bodies and were outgassed as a result of the energy released from their own and subsequent impacts and from radio-isotope heating, providing the volatile reservoirs for the planets and satellites; (2) that the initial volatile reservoirs were partially or totally lost through a combination of steady-state and catastrophic processes, and that a veneer of volatiles was added later from a dynamically distant, volatile-rich reservoir; or (3) that the volatiles were directly captured from the gaseous component of the solar nebula when the Earth and other planets were nearly complete. If one accepts the first or third hypotheses, then the ensuing discussion of impacts in this chapter is superfluous. On the other hand, if one chooses the second hypothesis, then impacts are of paramount importance to the problem. The likely answer, however, is that some combination of the three hypotheses (and other variants on them) is the correct solution, and it remains to be decided what is the correct recipe for combining them all.

Additional factors which should be considered are the degree to which impacts erode planetary atmospheres due to blow off from the impact event, the degree to which they modify the atmospheric chemistry by their flight through them, and the effect of atmospheric heating from impacts, particularly during the early history of the solar system. There is also the possibility of extrasolar impactors and accretion sources which needs to be considered.

In the following sections each of the periods defined above, (3) through (6), will be discussed. However, first it is necessary to review the solar system's cratering record as observed on the surfaces of the terrestrial planets, the Moon, and the satellites of Mars and the outer planets. It is this record which has led to the definition of the different periods and impact-related processes cited above.

II. THE CRATERING RECORD IN THE SOLAR SYSTEM

As a result of planetary flyby and orbiter spacecraft, there presently exist data on cratering on all the terrestrial planets, the Moon, and on the satellite systems of Mars, Jupiter, Saturn and Uranus. Within the next decade that list can be expected to expand to include the satellites of Neptune and a few main-belt asteroids, and to extend the Galilean satellite imagery and the Venus radar coverage, each to close to 100% of their surface areas.

The cratering record is not complete at all the bodies. For Mercury, the resonance between its rotational and orbital periods resulted in the same face being illuminated during each flyby of the Mariner 10 spacecraft, and so only about 45% of the planet's surface was photographed. Radar mapping of Venus by the Soviet Venera 15 and 16 spacecraft covers only ¼ of the total surface, that north of 30°N latitude. On the Earth, geologic and weathering processes, in particular plate tectonics and sedimentation, have tended to destroy and/or obscure the cratering record for all but the largest impacts more than 600 Myr in age, and for all but the most recent craters <10 km in diameter. Identification of large impact structures on the Earth is still far from complete.

Orbiting spacecraft have yielded 100% coverage on the Moon and Mars. For Phobos and Deimos imaging by the Mariner 9 and Viking spacecraft covered 95% and 80% of each's surface, respectively (Thomas et al. 1986). For the outer planet satellites coverage is typically on the order of 30% to 50% at the resolution necessary for crater identification and counting, based on single close flybys by the Voyager spacecraft. Because of the varying distance of those flybys, the data sets are not uniform, often covering different ranges of crater sizes. In addition, as we have seen on the Moon and learned from the multiple spacecraft missions to Mars, the unimaged hemispheres of these satellites may display substantially different terrains and crater densities.

Analysis of the cratering record typically involves counting the craters covering a chosen area and then binning them by size. The counts are then plotted in a log-log format with log (diameter) as the abscissa and log (number/unit area) as the ordinate. The resulting size-frequency distributions usually show a large number of craters at small diameters, decreasing to steadily fewer craters as the diameter increases. The set of points is then fit with a power-law function: $\mathrm{d}N/\mathrm{d}D = cD^b$, where b, the slope of the curve fit on the log-log plot, is typically on the order of -3. For a slope of -3, craters in equal logarithmic intervals in diameter occupy the same fraction of the surface area. Size-frequency distributions are sometimes plotted as cumulative distribution where each bin counts all craters larger than that size range. Such plots are somewhat deceptive, as they imply better statistics than are actually present.

Because of the similar appearance of most such crater-count plots, the Crater Analysis Techniques Working Group (1979, chaired by A. Woronow) developed the "relative" size-frequency distribution plot, or "R-plot," where

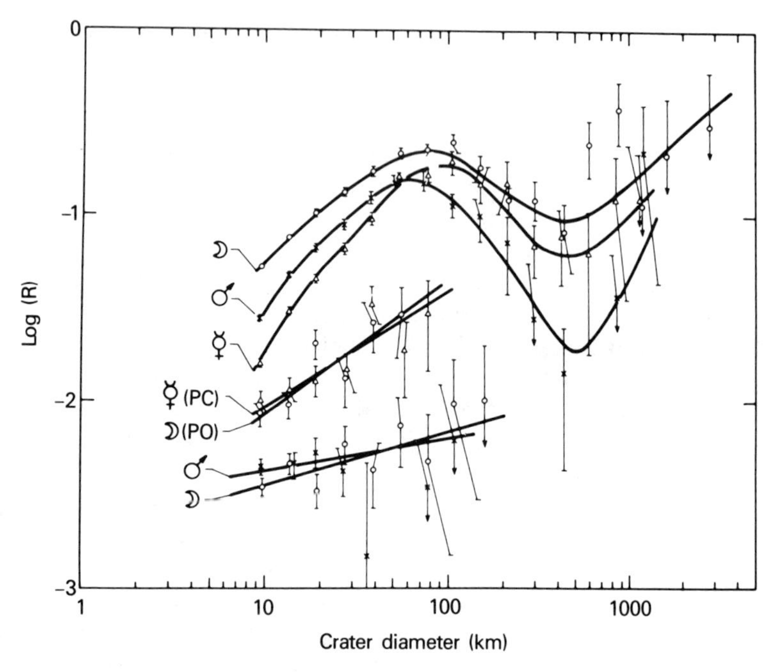

Fig. 1. Comparison of crater diameter/density distributions for the heavily cratered highlands on the Moon, Mars and Mercury (top three curves); for the post-Orientale and post-Caloris distributions (PO and PC) on the Moon and Mercury, respectively; and for the lunar post-mare and sparsely cratered areas of Mars (bottom two curves). (Figure from Carr et al. 1984.)

the crater frequencies are plotted versus their deviation from a $b = -3$ power law. An example of an R-plot is shown in Fig. 1 (Carr et al. 1984), which gives relative crater frequencies for different geologic units on the Moon, Mercury and Mars. R-plots are valuable in helping to infer something about the size distribution of the impacting projectile population.

Crater counting and the interpretation of size-frequency distribution plots has developed into as much an art as a science. There are numerous open questions with regard to how craters are identified and counted, and what exactly the plots mean. One of the major questions is that of saturation. Saturation is defined as the point in a surface's cratering history when new impactors create as many craters as the number of old craters they destroy, and thus the size-frequency distribution does not change. Hartmann (1984), for example, claims that many heavily cratered surfaces in the solar system are within a factor of a few of saturation. On the other hand, Woronow (1978) says that no known surfaces are saturated (the careful reader will note that the two views are not contradictory). The question is important because if saturation is achieved then much of the information on the size distribution of the impacting population is lost. Good reviews of these and other crater counting and analysis issues are given by Hartmann et al. (1981) and Chapman and McKinnon (1986).

The cratering record for the terrestrial planets zone, exhibited in Fig. 1, shows a remarkable similarity between the distributions for the heavily cratered terrains on Mercury, Mars and the Moon. This suggests that all three bodies, plus the Earth and Venus as well, were intensely bombarded during their early history by the same population of bodies. However, statistically significant differences in the crater populations do exist. These differences might be removed in part by considering the different impact velocities, gravity scaling, and gravitational focusing for each body. However, it is more likely that they reflect real differences in the times of surface solidification, differences in the absolute number of impactors perturbed into each class of planet-crossing orbits, and possibly resurfacing events on each body. Unfortunately, there is no way to unravel the effects of each of these mechanisms with the presently available data.

The crater densities on resurfaced areas are far lower than they are for the ancient cratered highlands on each body. Derived crater densities can be used to order geologic units in "relative" age. Fortunately, returned lunar samples from the Apollo and Luna missions also allow the surfaces to be dated "absolutely" through radio-isotope methods. By comparing these absolute ages with the counted crater density on each surface, a measure of the change of the cratering flux with time can be obtained. A plot of such data based on returned lunar samples is shown in Fig. 2 (Carr et al. 1984). There is a very sharp decrease in the cratering rate between 4.0 and 3.6×10^9 yr ago. This final phase of the intense period of impacts during the solar system's first Gyr is known as the "late heavy bombardment."

The late heavy bombardment is interesting because it comes at a relatively late time compared with the expected clearing time for the terrestrial planets zone. As will be discussed in Sec. III below, the formation time for the terrestrial planets is expected to be 200 to 300 Myr, yet the late heavy bombardment occurs roughly twice that value after the initiation of solar system formation. Possible sources for that bombardment will be considered in Sec. III.

The cratering record on resurfaced areas on the terrestrial planets, illustrated by the two lower pairs of lines in Fig. 1, show what appears to be a gradual evolution of the impactor flux. Whereas the crater size distribution (in the 10-to-100 km diameter size range) on the heavily cratered highlands implies a projectile population heavily weighted towards larger bodies, the younger post-Caloris and post-Orientale distributions are less so (though the differences are not yet quite statistically significant), and the post-mare lunar record and lightly cratered Martian units are distinctly closer to a $b = -3$ slope. This indicates that the projectile population before and during the late heavy bombardment is not the same as the population that provided the

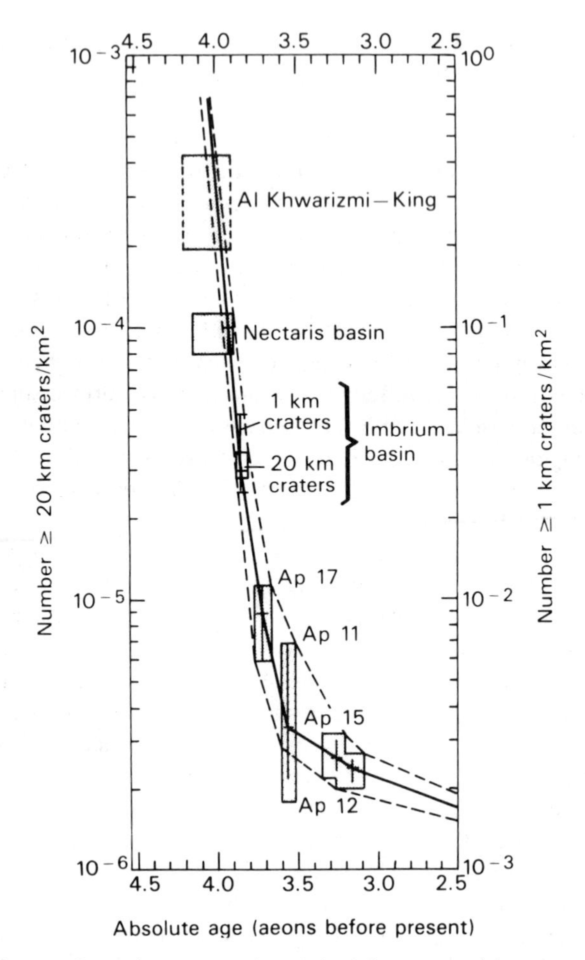

Fig. 2. Estimates of cratering rates vs time derived from returned lunar samples. The steep decline between 4.0 and 3.6 Gyr ago represents the end of the late heavy bombardment. (Figure from Carr et al. 1984.)

steady-state impactor flux during the last 3.5×10^9 yr of the solar system's history.

The cratering record in the terrestrial planets zone is relatively simple and straightforward when compared to that on the satellites of the outer planets. A comparison of the terrestrial record with the average cratering record for heavily cratered surfaces and resurfaced areas on the satellites of Jupiter, Saturn and Uranus is shown in Fig. 3 (Strom 1987). The actual variation of crater size distributions on individual satellites around each of the outer planets is even more diverse. For example, the heavily cratered terrain on Uranus's innermost satellite Miranda closely matches that of the Saturn aver-

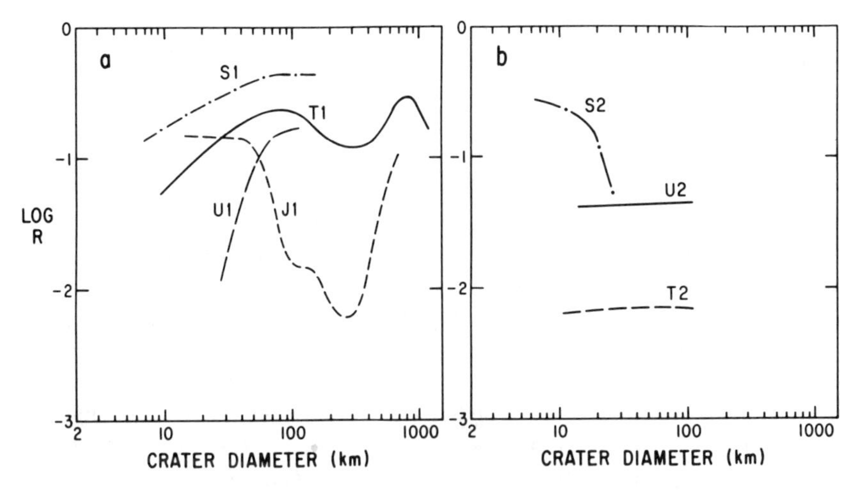

Fig. 3. Size-frequency distribution *R*-plots for (a) the old crater populations on the satellites of Jupiter (J1), Saturn (S1) and Uranus (U1), as compared with that in the terrestrial planets zone (T1), and (b) the young crater distributions on the satellites of Saturn (S2), Uranus (U2), and the terrestrial planets zone (T2). *R*-plots for individual satellites around each planet show considerable diversity from these averages. (Figure from Strom 1987.)

age shown, but the heavily cratered terrains on the other Uranian satellites are less dense and clearly deficient in large impactors.

Prior to the Voyager flybys, the low solar temperatures in the outer solar system led to expectations that the satellites of the outer planets would preserve the record of early intense cratering that characterized their formation, plus a continuous steady-state bombardment by comets. However, just before the first Jupiter flyby it was predicted (Peale et al. 1979) that tidal heating of satellites could supply sufficient energy to melt and resurface icy satellites. This was dramatically confirmed by satellite imagery. In the case of Io, the heating is so intense as to result in active volcanism on the satellite's surface. In addition, the cratering record on the Jupiter satellites that do preserve craters, Ganymede and Callisto, is radically different from that in the inner planets region. A suggested explanation is that the Jovian satellite craters reflect a population of planet orbiting impactors. But where then are the craters from the external impactors which must be passing through the Jovian system in abundance?

At Saturn the crater populations are very different from Jupiter. The ancient crater populations greatly resemble the crater populations in the inner solar system, while the young crater distributions on resurfaced areas bear some resemblance to the Jovian crater distributions. Thus, Saturn seems to be a mixture of ancient preserved crater record, plus younger craters caused by a planetocentric population of impactors. But exactly why the cratering projectiles at Saturn, presumably dominated by comets, should have the same size distribution as in the inner solar system, where asteroids and comets are mixed, is not understood.

For Uranus the cratering record again falls into older and younger populations, with considerable variation between individual satellites. The heavily cratered terrain on Miranda matches the terrestrial profile, though at even higher densities, and Oberon and Umbriel also show a strong similarity to the terrestrial cratering distribution. But Titania and Ariel have considerably different distributions, displaying a paucity of larger craters, interpreted in the other planet satellite systems as evidence of planetocentric impactors. Strangely, the young crater populations in the Uranus system resemble the young craters in the terrestrial planets zone.

It should be recognized that in the above discussion, the terms young and old should not be taken to refer to some sort of universal resurfacing event that occurred throughout the outer planets zone. The complex and individual history of tidal heating, impact and possibly disruption at each of the outer planets would in no way be expected to produce simultaneous resurfacing events between different satellite systems. In fact, there is no reason to suppose that resurfacing events within the same satellite system occurred simultaneously either.

Further analysis of the cratering record on the outer planet satellites, coupled with theoretical modeling of proposed projectile populations, may produce a more coherent picture of their respective impact histories. Or the problem may need an infusion of new and more complete data, as could be expected to come from the Galileo and Cassini missions at Jupiter and Saturn, respectively, and presumably other outer planet exploratory spacecraft missions.

The cratering record on the Earth consists of about 100 identified impact structures, ranging from a few tens of meters to 140 km in diameter (Grieve 1982). All but one of the craters are on land; a 45 km diameter impact crater buried under ocean sediment was recently identified in the North Atlantic off Nova Scotia (Jansa and Pe-Piper 1987). All but three of the impact structures are <600 Myr old; the exceptions are very large structures, 1.7 to 2.0 Gyr in age.

The counting of craters on accurately dated surfaces on the Earth and the Moon provides absolute measures of the average cratering flux in the inner solar system. Based on the known sample of North American astroblemes, Grieve and Dence (1979) estimate the terrestrial cratering rate over the past 600 Myr to be $1.4 \pm 0.4 \times 10^{-14}$ km^{-2}yr^{-1}, for craters larger than 10 km diameter. Given that the total surface area of the Earth is 5.1×10^{8} km^{2}, this translates to 7.1 craters >10 km diameter every Myr, of which 2 of the impacts would be expected to be on land (assuming a current ocean area of 72% of the Earth's surface).

A separate analysis of a subset of the North American craters by Shoemaker (1977) found a somewhat higher cratering rate of $2.2 \pm 1.1 \times 10^{-14}$

$km^{-2}yr^{-1}$, or about 11 craters >10 km per Myr. Shoemaker has argued that Grieve and Dence's larger sample (over a larger area) is observationally incomplete, and that it underestimates the total cratering rate. The error bars of the two estimates do overlap, and are thus better thought of as representative of the uncertainty of such estimates.

The lunar cratering rate estimated by Shoemaker (1977*a*) is $1.1 \pm 0.5 \times 10^{-14}$ $km^{-2}yr^{-1}$, based on 3.3 Gyr old surfaces. The estimate is in fairly good agreement with the terrestrial estimate by Grieve and Dence, allowing for differences in impact velocity, gravitational focusing, and gravity scaling, but falls somewhat short of Shoemaker's terrestrial estimate (though again, it is within the error bars). Further confounding the problem is a lunar cratering estimate by Neukum et al. (1975) who found a rate of only 0.3×10^{-14} $km^{-2}yr^{-1}$ for craters >10 km diameter on the lunar mare. The difference may in part be because Shoemaker's count is contaminated with craters from older surfaces, extending back to the higher cratering rates during the tail off of the late heavy bombardment.

Shoemaker et al. (1979) regard the difference in lunar and terrestrial cratering rates as significant. Since the two rates represent averages over different temporal periods, they conclude that the cratering rate has for some reason increased in the last 600 Myr, possibly due to an increase in the flux of long-period comets through the inner planets region. However, the decay time for major perturbations of the Oort cloud are typically only 2 to 3 Myr (Hut and Weissman 1985), and thus it is difficult to see how a factor of 2 (or more) increase in the cometary flux could be maintained over such an extended period of time. Shoemaker et al. also considered an asteroidal source due to the breakup of one or more large asteroids in the main belt. However, they rejected that idea because 600 Myr is considerably longer than the dynamical lifetime of Earth-crossing objects (30 Myr), and thus the same problem of sustaining the increased flux over an extended period again arises.

There has been some speculation in recent years on the possibility of periodicities in the terrestrial cratering record (Alvarez and Muller 1984) and a possible association with biological extinction events on the Earth. The biological extinctions have also been proposed to be periodic (Raup and Sepkoski 1984). A number of dynamical mechanisms, mostly involving cometary showers, have been suggested to explain the alleged periodicities. However, both the periodicities and the dynamical mechanisms have been challenged repeatedly on a variety of grounds (Shoemaker and Wolfe 1986; Weissman 1986; Sharpton et al. 1987), and are considered dubious, at best.

On the other hand, Schultz (1987) has suggested that recent craters on the Moon appear to cluster in a nonrandom fashion at certain ages: 30 ± 10 Myr and 10 ± 5 Myr. Since the age estimates are based on subjective measures of crater degradation, this is far from a conclusive result. However, it does point out the value of accurately dating a large sample of lunar craters, a most worthwhile project for future lunar explorers.

III. CRATERING MECHANISMS IN THE EARLY SOLAR SYSTEM: GIANT IMPACTS, PLANETARY ZONE CLEARING AND THE LATE HEAVY BOMBARDMENT

A. Giant Impacts on the Protoplanets

In the last few years it has been recognized that the final stages of planetary accretion likely involved the impact of fairly large bodies on the growing protoplanets. As each planetary embryo grew by runaway accretion of smaller planetesimals, it eventually could reach a point where it might encounter another growing embryo of only a slightly smaller size, and accreted it also. Given a massive enough secondary body, the energy of the impact could conceivably be great enough to disrupt the larger protoplanet.

The idea of giant impacts late in the planetary accretion phase has been around for some time, and has usually been associated with explanations for unusual planetary obliquities, namely that of Venus (180°) and Uranus (98°). Such explanations were always considered somewhat *ad hoc* as they involved invoking a singular event leading to an unusual planetary state. In fact, even the very modest obliquities of the other planets require some sort of late impactor bringing in a substantial amount of angular momentum; stochastic averaging of the angular momentum from many small impactors would lead to near-zero obliquities.

In the mid 1970s, Hartmann and Davis (1975) and Cameron and Ward (1976*b*) each proposed that the Moon might have formed from the impact of a Mars-sized planetesimal late in the Earth's accretional history. Such a hypothesis could explain the high mass ratio and high angular momentum of the Earth-Moon system, the lack of lunar volatiles and a substantial lunar core, and the similarity between lunar rocks and Earth mantle material. However, this idea did not really take hold until about a decade later when a strong consensus for it developed (Hartmann 1986; Kaula and Beachey 1986; Cameron 1986). The change in attitudes was in part due to the discovery of additional weaknesses in each of the other lunar origin hypotheses (capture, fission, co-accretion), but is also the result of a better understanding of planetary accretion. In addition, it is likely related to a generally increased acceptance of catastrophic theories, led by the asteroid impact hypothesis (Alvarez et al. 1980) for explaining biological extinction events on the Earth, in particular the Cretaceous-Tertiary boundary event.

Some understanding of the probability of large impactors during the final stages of planetary accretion can be obtained from the work of Wetherill (1986) who integrated the evolution of hypothetical samples of 500 planetesimals in the terrestrial planets zone, ranging in size from Ceres to the Moon, under a variety of initial assumptions and processes. Two steps in the early and later stages of one such integration, where the final configuration closely resembled the current terrestrial planets, are shown in Fig. 4. The evolving planetesimals are plotted as a function of their semimajor axes and eccentricity, and denoted by different symbols to indicate their masses. The growing protoplanets which eventually "become" the four terrestrial planets are identified.

In Fig. 4a, 10.9 Myr into the integration, more than half the mass of the system is already in the embryos of the final planets. Note the existence of a massive planetesimal close to "Venus" in the diagram. In the simulation, that planetesimal eventually impacted Venus with an energy equal to 65% of the gravitational potential energy of the combined body. Another interesting point is that at this time in the integration, at least four Mercury/Mars-sized planetesimals are still in the system, and "Mercury" is in an orbit outside that of Mars.

In Fig. 4b, 64 Myr into the integration, the system has cleared out considerably, and 90% of the mass is in the final planets. Mercury has assumed its rightful place in the solar system. Yet there is still one Mercury/Mars sized planetesimal just beyond the orbit of Mars. That planetesimal will either impact one of the nearly complete protoplanets, or it will be dynamically ejected from the system by a close planetary encounter.

Thus, giant impacts appear to be a likely and normal consequence of planetary accretion, and not a low probability, *ad hoc* solution to various problems. Recently, Wetherill and Cameron have each proposed that a grazing impact on the evolving Mercury embryo resulted in a partial stripping of that planet's silicate mantle, resulting in its unusually high mean density.

An example of a hypothetical impact of a Mars-sized planetesimal on the proto-Earth by Kipp and Melosh (1986) is shown in Fig. 5. The impact velocity is 15 km s^{-1}. In the first frame of the two-dimensional computer simulation of the oblique impact, 253 s after contact, jets of material are beginning to form on either side of the contact point. At 750 s, the impactor has totally disrupted and the jets are quite extended as material is accelerated into orbit and even to escape velocity. Meanwhile the core of the impactor moves towards an eventual merger with the planet's core.

What then are the effects of giant impacts on volatile retention in the growing protoplanets? All calculations to date have indicated a massive amount of heating, with jets of material blasted out of the impact site at a temperature of 10,000 K or more (Kipp and Melosh 1986). Cameron (1986) estimates that the entire proto-Earth would be heated in excess of 6000 K. Thus, any primordial atmosphere that had evolved up until the point of the impact would be expected to be blown away, and a substantial amount of new outgassing would result. Exactly how much remains to be seen when more detailed models and calculations are available.

Nor can the timing of the last giant impact on any of the planets be tied down very precisely. As shown by the Wetherill (1986) Monte Carlo simulations, such impacts can occur at almost any time in the first few hundred Myr of each planet's history. In the case of the Earth, the impact that is thought to have initiated the formation of the Moon must have come after core forma-

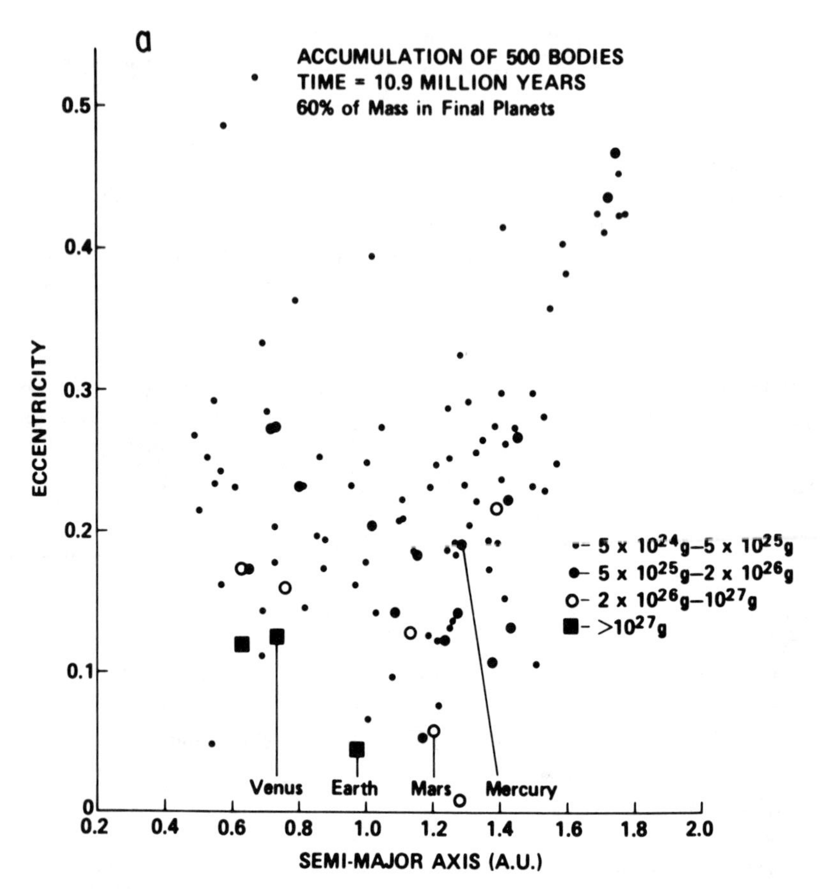

Fig. 4. Computer simulation of the accretionary evolution of a swarm of 500 planetesimals in the terrestrial planets zone by Wetherill (1986): (a) at 10.9 Myr more than half the mass of the system is already contained in the final growing embryos, but many large impactors remain in the system; (b) at 64 Myr more than 90% of the mass is in the final protoplanets and only a few large impactors remain to be swept up or dynamically ejected.

tion, but that can be anytime in the first 100 to 200 Myr of the Earth's history. The most likely history is that the evolving planetary embryos suffered several giant impacts each, and lost their primordial atmospheres repeatedly during their accretionary phase, with decreasing frequency after the first few tens of Myr, but possibly with greater individual effect as the mean size of the impacting planetesimals also grew.

B. Final Zone Clearing and Sweep Up of Planetesimals

During the early stages of the accretion of the planets the relative encounter velocities between planetesimals were initially low, so that inelastic collisions likely led to merger of bodies. As the planetary embryos grew, a few

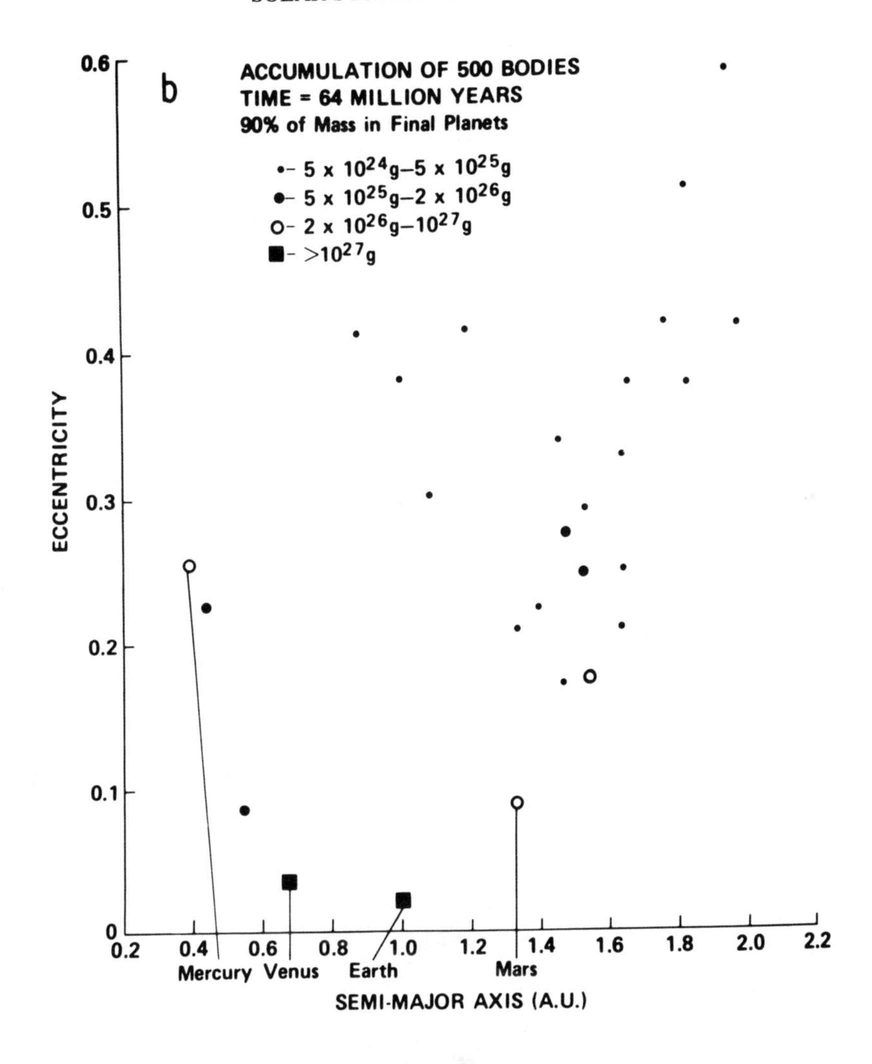

embryos began to "run away" and grow much faster than their companions. As a result, their gravitational cross sections increased and the probability of a scattering event became much greater than the probability of a collision. Thus, the relative velocity of collisions increased, typically to a value within a factor of 2 or 3 of the escape velocity of the largest planetesimal.

This pumping up of the encounter velocities for the small planetesimals resulted in increased orbital eccentricities which then gave them a probability of encountering more than one of the growing planetary embryos. Dynamical exchange of material between protoplanetary feeding zones tended to smear the original compositional gradient in the planetesimal populations resulting from the initial temperature gradient in the solar nebula. Planets close to the

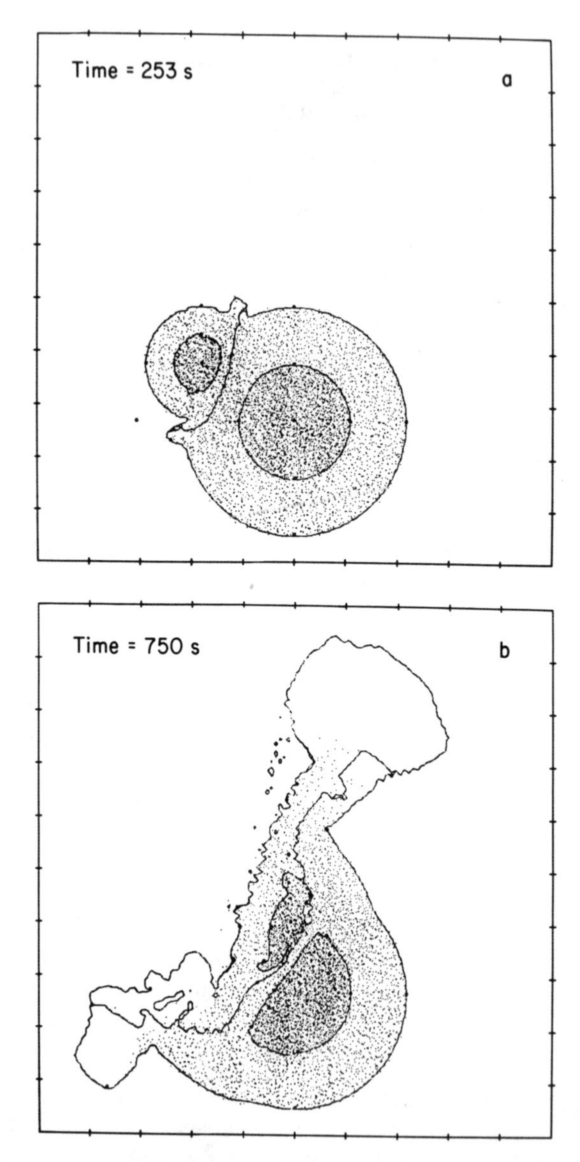

Fig. 5. Computer simulation of the impact of a differentiated, Mars-sized planetesimal on the differentiated proto-Earth by Kipp and Melosh (1986); (a) 253 s after impact as the vapor jets of mantle material begin to emerge at the sides of the contact; (b) 750 s after impact, large fractions of the mantles of both bodies are being ejected into orbit while the cores of the two bodies move towards an eventual merger.

Sun were bombarded by volatile-rich planetesimals from the outer asteroid belt and beyond, and vice versa. Planetesimals were passed between different planetary zones until they arrived at one of two possible end states: collision or ejection from the solar system. Two other end states: capture into satellite orbits or into stable reservoirs, were considerably less likely.

The time scale for all this to occur was a function of the masses of the scattering planets and their orbital periods. Thus, the larger protoplanets tended to scatter planetesimals out of their zones more rapidly than smaller ones, and planetesimals closer to the Sun had shorter lifetimes due to a higher probability of frequent close encounters with the protoplanets. The orbital evolution of a planetesimal typically required many planetary encounters, resulting in a random walk of its orbital elements. It is highly unlikely (except in the case of Jupiter and Saturn) that a single encounter alone could cause a sufficient orbital change to scatter it from one planetary zone to another.

Typical lifetimes for different planet-crossing orbits are shown in Table I (Wetherill 1975). It is seen that Jupiter- and Saturn-crossing orbits have the shortest lifetimes, because the large masses of those planets tend to eject objects from the planetary region on hyperbolic orbits. In most cases, the lifetimes for the other planets are determined by how long it takes each planet to change a particular orbit so that it becomes Jupiter and/or Saturn crossing. Jupiter and Saturn then quickly eliminate the object.

The final sweep up of material in the planetary system is not simply a function of the individual lifetimes for objects in each class of planet-crossing orbits. This happens in part because bodies might not spend all of their accretionary lifetime at the same heliocentric distance, as was seen for the case of the hypothetical "Mercury" in the Wetherill integration described in the previous section. In addition, new impactors may be introduced from a reservoir with a longer lifetime, after a particular planet's own zone has been swept clean. This is one proposed explanation for the late heavy bombardment, described in the next section.

TABLE I
Typical Lifetimes of Planet-Crossing Orbits[a]

Type of Orbit	Approximate Half-Life: 10^6 yr.
Earth crosser, aphelion = 3 AU	40
Earth crosser, aphelion = 4.5 AU	10
Mars crosser, perihelion = 1.2 AU	200
Mars crosser, perihelion = 1.5 AU	1200
Jupiter crosser	1
Saturn crosser	2
Uranus crosser	100
Neptune crosser	200

[a]Table from Wetherill (1975).

Most importantly, the precise sequence of planetary accretion is not well understood. Some hypotheses require that Jupiter and Saturn grew more rapidly than the terrestrial planets interior to them. One consequence of such a scenario is that Jupiter would scatter volatile-rich planetesimals out of its own zone and the outer asteroid belt, into the terrestrial planets zone while those planets were still accreting. Another consequence is that Jupiter would accelerate the planetesimals in the Mars zone, perhaps contributing to a decrease in the eventual mass of material that is allowed to accrete to form Mars.

Regardless of exactly when they accrete, once they do, Jupiter and Saturn act as a barrier to exchange of material between the zones interior and exterior to them. Rocky planetesimals from the terrestrial planets zone can be pumped up to Jupiter crossing, and then perhaps, in turn, Saturn-, Uranus- and Neptune-crossing orbits due to Jupiter perturbations. But given the smaller perturbations by Uranus and Neptune, it is unlikely that the planetesimals will detach their perihelia from Jupiter. Jupiter will eject them on a time scale of about 1 Myr, long before they would likely impact on the outer giant planets or their satellites.

The same rule would apply for icy material perturbed inward from the outer planets zone. Although the material might achieve crossing orbits with the terrestrial planets, their lifetimes in those orbits would still be predominantly controlled by Jupiter perturbations (unless they could somehow detach their aphelia from Jupiter), and would thus be on the order of 1 Myr.

The expected scenario therefore is that Venus and the Earth will clear the terrestrial planets zone in several typical crossing lifetimes, on the order of 150 to 300 Myr, with a long tail of material being supplied from Mars-crossing orbits (Wetherill 1981*a*). The time scale for completely clearing the Mars zone will be considerably longer, on the order of 1.5 to 3 Gyr for the longest-lived orbits.

Note that this "tail" of Mars zone material does not necessarily imply small bodies. It is entirely conceivable that fairly large planetesimals might survive many statistical lifetimes and still be present late in the evolution of the planets. Breakup of these large, late surviving planetesimals may provide periods of intense bombardment in the terrestrial planets zone. This point will be discussed further with regard to the late heavy bombardment, below.

In the outer planets region, Jupiter and Saturn will quickly eject or absorb all the planetesimals in their own zones, but will experience a continual stream of icy planetesimals from the Uranus and Neptune zones, lasting on the order of 1 to 2 Gyr. Monte Carlo simulations of the accretion of Uranus and Neptune from icy planetesimals by Fernandez and Ip (1981) showed that a substantial flux of material would pass through the Jupiter and Saturn zones, with between 0.2 and 3.4 Earth masses of icy planetesimals impacting on Jupiter and 0.8 to 2.7 Earth masses of icy planetesimals impacting on Saturn. Obviously, some small fraction of these planetesimals will also impact the

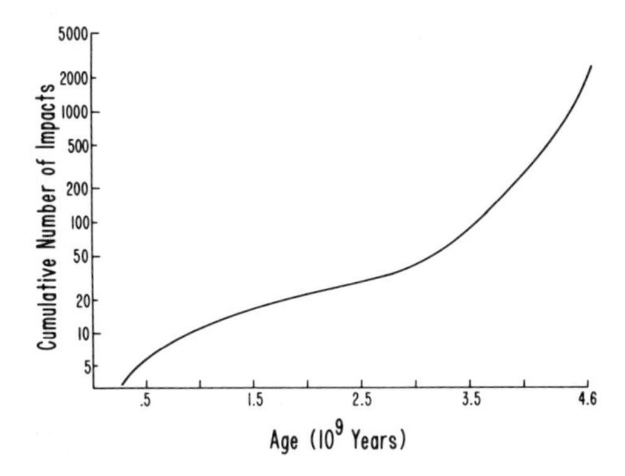

Fig. 6. Calculated time dependence of impacts in the inner planetary region by bodies initially in Uranus- and Neptune-crossing orbits. The cumulative number of impacts less than a given age are in arbitrary units. (Figure from Wetherill 1975.)

growing satellite systems around each planet, perhaps with catastrophic effect, given the expected high velocity of the encounters.

Fernandez and Ip also showed that a large fraction of the material from the Uranus-Neptune zone, on the order of 20 Earth masses or more, would wind up in short-period comet orbits that could encounter the terrestrial planets. Approximately 1% of that material would be expected to impact the Earth, 0.5% would impact Venus, and 0.2% would impact Mars (Ip and Fernandez 1988).

Thus, the clearing of the Uranus-Neptune zone provides a long-lived source of volatile material which could provide a volatile veneer on the terrestrial planets late in their evolution. An estimate of the flux versus time for this icy bombardment is shown in Fig. 6 (Wetherill 1975).

A somewhat different approach to the problem is by Chyba (1987) who estimated the total mass of late impactors on the Moon based on the sizes of the largest lunar impact basins, and then scaled the results to the Earth. He found that the Earth had accreted 1.1×10^{25} g of material. If 10% of those impactors were cometary objects, then the mass of accreted water would be equal to 40% of the mass of the Earth's current oceans. A comparable mass of volatiles would be expected to be delivered to Venus. Chyba's calculations should only be considered as order-of-magnitude estimates, but they do show that a significant fraction of volatiles could be delivered in this fashion.

C. The Late Heavy Bombardment

The lunar cratering chronology discussed in Sec. II shows a sharp decline in cratering rates between about 4.0 and 3.6 Gyr ago. This has been referred to in the literature as the late heavy bombardment. It occurs relatively late in the accretionary history of the terrestrial planets, at a time when the vast majority of that zone's planetesimals are already expected to have either impacted on the protoplanets, or been dynamically ejected from the inner planets region.

The similarity of cratering profiles on the terrestrial planets shown in the *R*-plot in Fig. 1 strongly suggests that the late heavy bombardment was not an isolated event in the Earth-Moon system. Rather, it appears that a flux of impactors flooded the terrestrial planets region at this point in the solar system's history, and is preserved in the cratering record of the heavily cratered terrain on each planet. Wetherill (1975) pointed out that the similarity in crater densities on the mare of the Moon and Mercury and the lightly cratered plains on Mars, suggests that the intense cratering may have been simultaneous on all three bodies, and only the more recent steady-state cratering by asteroids and comets is seen on these resurfaced areas. What then was the source of these impactors?

An essential requirement of any explanation for the late heavy bombardment is that the impactors be "stored" somewhere in the solar system until they are unleashed about 4.0 Gyr ago. Wetherill (1975) showed that assuming that the late heavy bombardment was simply the tail off of the accretion of the terrestrial planets implied too large an initial mass in the terrestrial planets zone. Wetherill examined a number of other possible impactor sources. They were: (1) disruption of a large asteroid in the asteroid belt; (2) breakup of small satellites around each planet; (3) disruption of a large planetesimal in the terrestrial planets zone; and (4) final clearing of planetesimals in the Uranus-Neptune zone.

The total mass of impactors required to explain the late heavy bombardment is estimated to be on the order of 10^{23} g, or roughly equal to a 200 km diameter asteroid. The expected lifetime of such a large asteroid against collisional disruption in the present-day asteroid belt is 10^{11} yr (Wetherill 1967). This is far longer than the age of the solar system. Thus, one must either invoke a very low probability event to explain such a source, or a much higher density of objects in the asteroid belt 4.0 Gyr ago.

Although breakup of planetocentric objects might provide an intense bombardment at one body, they would not provide a general impact signature throughout the terrestrial planets zone. Wetherill showed that such a source would result in far greater impact rates on the body around which it originated. Simultaneous breakup of satellites around several planets, yielding very similar impactor populations, is implausible.

At first glance, the breakup of a large planetesimal due to a close planetary approach seems to be a far more likely explanation. As seen in Sec. II, large planetesimals can persist late into the accretionary evolution of the bodies in the terrestrial planets zone, on the order of several hundred Myr or more. The most likely storage place for such a planetesimal is in a Mars-crossing orbit (Wetherill 1977) where the dynamical lifetime is comparable to (or even longer than) the storage time required, as shown in Table I.

A close approach to within the Roche limit of one of the nearly complete terrestrial planets would be expected to disrupt the body and unleash a flood of impactors throughout the terrestrial planets zone. A pass through the Roche zone of a major planet is more likely than an impact. Because of their larger cross sections, Venus or the Earth are the most likely source of the tidal disruption, rather than Mercury or Mars. The object would first have to be thrown into a Venus- or Earth-crossing orbit by one or more close Mars encounters. Having been stored in a Mars-crossing orbit, and upon being disrupted by a Venus or Earth encounter, the resulting debris would be in a sufficiently eccentric orbit to evolve rapidly to impact all the terrestrial planets.

The problem with this scenario arises from the recent work of Mizuno and Boss (1985) which showed that tidal disruption is highly unlikely to occur in dissipative planetesimals. This results because there is not sufficient time during the planetesimal's hyperbolic flyby of the Earth or Venus for it to physically disrupt, and because increased dissipation within the planetesimal lengthens the time scale for tidal disruption.

An alternative storage location with dynamical lifetimes similar to Mars-crossing orbits is the Uranus-Neptune zone. Large icy planetesimals could certainly grow in this zone, and obviously have, as evidenced by Pluto, Charon and Chiron. Although the storage time can be quite long, the objects must cross the "Jupiter barrier" to get into the inner solar system. As was seen in Sec. III.B above, the dynamical evolution accelerates as the object(s) is passed from the Uranus-Neptune zone to Saturn and then to Jupiter. But once in a Jupiter-crossing orbit, the object(s) would have only about 1 Myr to become Earth crossing and to be disrupted, before Jupiter would dynamically eject it (them) from the solar system. Jupiter itself might provide the source of the tidal disruption, but the resulting debris would still have the same dynamical problem. However, the short lifetime problem can be overcome if Uranus and Neptune supply new objects fairly continuously over an extended period, as would be expected, given the long dynamical lifetimes for objects in their heliocentric zones.

Regardless of the dynamical storage location, the physical difficulties with tidal disruption shown by Mizuno and Boss (1985), whether by Jupiter or by one of the terrestrial planets, remain. Thus, a plausible explanation for the late heavy bombardment remains something of a mystery. Given the fact that it appears to occur so late relative to expected planetary accretion scenarios, it seems likely that the late heavy bombardment is not the tail off of planetary accretion but rather is a late pulse superimposed on the tail off. Nor is there any reason to suppose that it was the only such pulse; it may have been preceded by several others which are not easily discernible from it in the cratering record (Wetherill 1981*a*).

IV. STEADY-STATE IMPACTS OF ASTEROIDS AND COMETS

A. Impactor Reservoirs

Once the initial population of planetesimals in planet-crossing orbits had been cleared out, about 3.6 Gyr ago, any additional impactors had to be supplied from existing reservoirs of small bodies. These reservoirs had to be in orbits that did not cross the orbits of any planets, so that the unperturbed dynamical lifetimes of the bodies were greater than the age of the solar system. The two major, known reservoirs are the asteroid belt and the Oort cloud. There is a third known reservoir which is available primarily to the Jovian satellites: the Trojan asteroids locked in Lagrange point librations ahead and behind Jupiter in its orbit.

The asteroid belt is a collection of some 20,000 rocky bodies brighter than absolute visual magnitude 18.0, corresponding to diameters of 0.9 km for typical S-type asteroids (albedo = 0.15), and 1.7 km for typical C-type asteroids (albedo = 0.04). They range in size up to about 10^3 km in diameter, and are primarily in moderately eccentric and inclined orbits between Mars and Jupiter (Chapman 1979). Precise orbits have been determined for approximately 3500 asteroids. Physical studies have shown that the population of the inner asteroid belt is dominated by metamorphosed surface compositions, primarily stony, stony-iron and iron asteroids, while the outer belt is typified by more primitive and volatile-rich carbonaceous chondrite surfaces. This compositional gradient presumably reflects the original compositional variation in the solar nebula, though that may have been blurred in part by dynamical mixing; it also likely reflects the effect of internal heat sources from short-lived radionuclides. The asteroids are a collisionally evolved population and some families of fragments from apparently recent collisions have been identified.

A small fraction of the asteroids are in unstable, planet-crossing orbits. Approximately 54 Earth-crossers, called Apollo asteroids, are currently known. They exhibit a diverse set of surface compositions, with examples of all major types found in the main belt (McFadden et al. 1985). A subset of the Apollos are also Venus and Mercury crossers. Asteroids which have periods <1 yr are known as Aten class objects, and those with perihelia >1.016 AU (the Earth's current aphelion distance) but <1.3 AU, close to the Earth's orbit, are known as Amor objects. A modest number of known asteroids, mostly discovered in the last 10 yr, are also in Jupiter-crossing orbits.

The Oort cloud is a roughly spherical cloud of 1 to 2×10^{12} comets surrounding the solar system and extending from approximately 2×10^4 to 2×10^5 AU (Weissman 1985*b*). The total mass of comets in the cloud is estimated to be between 15 and 25 Earth masses. Comets in the cloud are generally in eccentric, highly inclined orbits and those with perihelia close to the planetary system are randomly perturbed into the planetary region by passing stars, galactic tides, and giant molecular clouds (GMC's). The leading hypothesis is that the comets are icy planetesimals perturbed out of the Uranus-Neptune zone during the accretionary growth of those planets. The comets are preserved in deep freeze in their distant orbits, though there is some processing of their surface layers by cosmic-ray bombardment (Johnson 1985*a*), heating by nearby supernovae (Stern and Shull 1988), and erosion by hypervelocity interstellar dust impacts (Stern 1986).

There has been considerable speculation recently about the existence of a more massive inner Oort cloud (Fernandez 1985; Weissman 1985*b*), interior to the observed one, but not dynamically sampled because of a lack of perturbers passing through it. The number of comets in the inner cloud is likely between 5 and 20 times that in the outer cloud; the mass would be expected to scale accordingly. Star passages through the inner Oort cloud or close encounters with GMC's could cause massive showers of comets from the inner cloud to enter the planetary region, raising the impact rates on the planets and possibly resulting in biological extinction events on the Earth (Hut et al. 1987).

The Trojan asteroids are a collection of primitive composition asteroids locked in a 1 : 1 resonance with Jupiter, librating around the Lagrange L_4 and L_5 points in Jupiter's orbit. These are quasi-stable points located 60° ahead and behind Jupiter in its orbit, sometimes also referred to as the triangular libration points. Some 75 Trojans have been numbered with about ⅔ around the preceding libration point and the remainder following Jupiter. The Trojans could presumably be removed from the resonance by collisions with comets and/or each other, and would then have very high probabilities of making a close approach to Jupiter. Thus, their lifetimes once removed from the resonance can be expected to be very short, and they would be rapidly ejected from the solar system. They would have some small probability of impacting the Jovian satellites, but their impact probability for almost any other object (except Jupiter itself) would be essentially nil.

Other reservoirs may exist but there is little or no evidence of them. Gehrels (1957) and Kowal (1971) searched small areas around the Trojan points in the Saturn-Sun system, but detected nothing. However, the limiting magnitude of those searches would not have detected half of the numbered Jupiter Trojans if they were placed at Saturn's distance from the Sun, and none of them would likely have been in the small search fields covered. Weissman and Wetherill (1974) showed that stable orbits were possible for the Trojan points in the Earth-Sun system, but no objects have ever been found despite several searches. Lecar and Franklin (1973) suggested that there is a narrow stable region between the orbits of Jupiter and Saturn that could contain a residual belt of distant asteroids/planetesimals, but none have been detected. Conceivably, stable zones may also exist between the orbits of the other outer planets.

The only known planet-crossing asteroid in the outer solar system beyond Jupiter (not counting Pluto-Charon which are Neptune crossing, but are also locked in a 2 : 3 orbital libration which keeps them away from Neptune) is the asteroid 2060 Chiron. With a perihelion of 8.5 AU and an aphelion of 18.9 AU, it is in an orbit which crosses that of both Saturn and Uranus (Kowal 1979), and is thus dynamically unstable. The exact nature of Chiron is unknown. It may be the largest member of an undetected family of outer solar system objects (icy planetesimals) that has been perturbed out of its stable reservoir and is now slowly evolving inward. Or it may be an unusually large comet which is evolving into the planetary region from the inner Oort cloud. Or, it may even be a surviving planetesimal from the Uranus-Neptune zone, an explanation that in some sense encompasses the other two.

Regardless of how Chiron originated, it is a statistic of one from which it is not possible to infer very much about the existence or population of any outer planetary region reservoirs. Thus, any estimate of the steady-state flux of impactors on outer planet satellites must be an extrapolation of what is known about the impactor fluxes in the inner planetary region, and for that reason may only represent a lower limit on the true cratering flux.

B. Asteroidal Impactors

Studies of the cratering rate in the terrestrial planets zone has focused, for obvious reasons, on the Earth. Shoemaker et al. (1979) estimate that there is currently a steady-state population of 700 ± 300 Apollo asteroids, approximately 100 Aten asteroids, and approximately 500 Amor asteroids (about ¼ of the total estimated number of Amor asteroids) that will occasionally become Earth-crossing due to secular perturbations in their orbital elements. As in the previous section, the numbers given are for objects with absolute visual magnitude V(1,0) brighter than magnitude 18.0. Assuming a mix of compositional types in solving for the sizes and masses of these asteroids, Shoemaker et al. find a terrestrial impact rate for craters >10 km diameter of 2.3×10^{-14} km^{-2}yr^{-1}, with a probable error of about 50%. This is in good agreement with the derived cratering rate from known craters on the Earth by Shoemaker (1977*a*) and is within the error bars of the rate by Grieve and Dence (1979), both cited in Sec. II. However, it is not the total cratering rate, since the contribution by comets has not yet been included.

The estimate does, however, lend support to Shoemaker's contention that the recent cratering rate in the terrestrial planets zone is higher than the long-term average over the past 3 Gyr. The problem is that there does not appear to be any known mechanism to provide substantial, prolonged enhancements of the asteroidal flux among the terrestrial planets, other than that expected from short-term stochastic variations in the population.

In fact, a major problem with the asteroidal flux has been how it is supplied at all. As recently as 17 yr ago, Wetherill (1971) showed that the meteoroid flux and exposure age data at the Earth could only be explained if a substantial fraction of the Earth-crossing asteroids were in fact, extinct cometary nuclei. Until recently, no dynamical mechanisms for moving the required number of asteroids from the main belt to Earth-crossing orbits were known. That situation has only been remedied in the last few years.

Two mechanisms have now been identified which can rapidly move asteroids and asteroidal debris from the main belt into Earth-crossing orbits. They are secular resonances in the asteroid belt, first identified in the last century and studied in detail by Williams (1969), and chaotic motion near orbital commensurabilities, as studied by Wisdom (1983,1985*a*).

Commensurate orbits are those where the orbital period is a small number integer ratio of Jupiter's orbital period, i.e., 1/2, 1/3, 2/5, etc., and thus the orbits are said to be resonant with Jupiter. The distribution of asteroids versus semimajor axis shows gaps in the population at these commensurabilities, known as the Kirkwood gaps. Early analytical and numerical studies of commensurabilities showed that large variations in the orbits took place in or near the resonances, but that this was not sufficient to remove many objects from commensurate orbits.

Williams (1969) and Williams and Faulkner (1981) demonstrated that a different kind of commensurability, or resonance, also existed in the asteroid belt with similar gaps in the population, at least for the inner main belt. These resonances result from long-period terms in the motion of Jupiter, driven by perturbations from other planets. Three strong resonances in the inner asteroid belt arise from interactions between Jupiter and Saturn. Rather than being at a constant semimajor axis or orbital period, the secular resonances describe three-dimensional curves in a,e,i space, as shown in Fig. 7.

Debris from asteroids in orbits near the secular resonances can be thrown into the resonances by collisions with other asteroids. Wetherill and Williams (1979) and Wetherill (1987,1988) estimated the flux of meteorites and Earth-crossing asteroids that could be produced in this manner. Wetherill (1988) found that the secular resonances could provide a steady-state population of approximately 120 of the 800 $\pm$ 300 Apollo and Aten asteroids, and 195 of the approximately 2,000 Amor asteroids. The secular resonances will primarily sample asteroids at the inner edge of the asteroid belt, between 2.17 and 2.30 AU. The asteroid population at that distance range is dominated by stony, stony-iron and iron asteroids. Thus, the material perturbed into the terrestrial planets zone will be primarily volatile-poor asteroids.

The fraction of Earth-crossers supplied by the secular resonances is still relatively small. A larger source was analyzed by Wisdom (1983,1985*a*) who had been trying to understand the origin and width of the Kirkwood gaps. Wisdom found that the apparently periodic librations for orbits in the resonances with Jupiter's orbital period could suddenly turn chaotic, with huge increases in orbital eccentricity, up to 0.6 or more. An example of one such orbital evolution is shown in Fig. 8. The large eccentricities cause the asteroids to become planet crossing, and they can then be removed from the main belt by close encounters with the planets. The predicted areas of chaotic motion closely match the observed widths of the Kirkwood gaps.

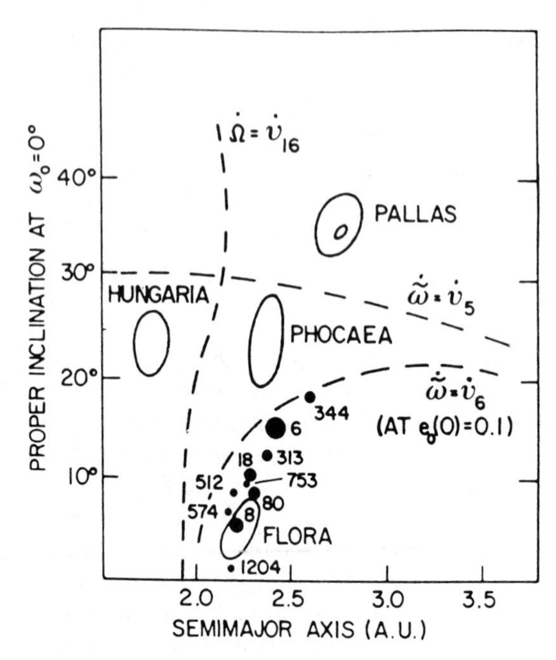

Fig. 7. Large asteroids near secular resonance, $\dot{\nu}_6$. A semimajor axis vs inclination (eccentricity = 0) plot of secular resonances in the inner asteroid belt, and the location of various asteroid families, as found by Williams (1969).

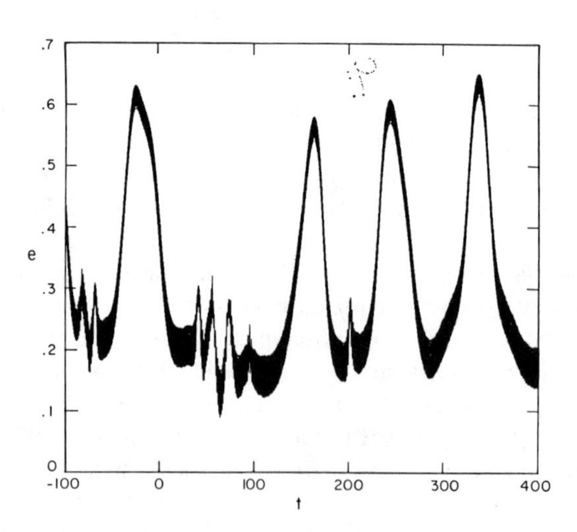

Fig. 8. Orbital eccentricity of a hypothetical asteroid at the 3/1 resonance showing variations in the chaotic motion, with the eccentricity ranging from relatively modest values, <0.15, to high values, >0.6, resulting in an Earth-crossing orbit. The orbital integration by Wisdom (1985*a*) is for a period of 5×10^5 yr.

At most of the major resonances, such as 2/1, and 5/2, the asteroids become Jupiter crossing but not Mars crossing. Because of the resonant motion, the asteroids avoid close approaches to Jupiter and thus can remain in the resonance indefinitely. However, at the 3/1 resonance at 2.50 AU the asteroids become Mars crossing as their eccentricities grow, then Earth crossing, but not Jupiter crossing. Close Mars encounters can remove the asteroids from the resonance. Thus, asteroids perturbed into orbits close to the 3/1 resonance can become chaotic in their motion and be thrown into orbits in the terrestrial planets zone. Wetherill (1988) estimated that the 3/1 resonance could produce about 330 of the observed steady-state population of Apollo and Aten asteroids, and about 1,400 of the Amor asteroids. At the distance of the 3/1 resonance, 2.5 AU, the asteroid belt is still dominated by stony and iron asteroids, though a substantial fraction of the population, on the order of 25%, appear to have primitive carbonaceous chondrite surfaces. Thus, the 3/1 resonance can supply some volatile-rich asteroids to the terrestrial planets zone, but the flux will still be predominately volatile-poor material.

With these two dynamical mechanisms, secular resonances and chaotic motion at orbital resonances, it now appears possible to supply over half of the expected population of Earth-crossing asteroids. It will be shown in the next section that the remaining fraction can likely be supplied by extinct cometary nuclei.

It is important to note that these two dynamical mechanisms have only been evaluated in detail relatively recently. It is entirely likely that there will be further refinements in the estimates above in the coming years. In addition, it is possible that as yet unrecognized dynamical mechanisms may exist for supplying main-belt asteroids to the terrestrial planets zone. The two dynamical mechanisms described above have resulted in a major shift in the thinking on this problem, and there is no reason to assume that the understanding of this problem will not continue to evolve in the near future.

Estimated cratering rates for asteroids and comets in both the terrestrial planets zone and for the Jovian and Saturnian satellite systems by Shoemaker (1982) are given in Table II. Note that the predicted total cratering rates for the Earth and Moon, based on estimates of the present Earth-crossing flux of small bodies, exceed the long-term average rates estimated from counted craters on dated surfaces on these bodies. This again provides some support for Shoemaker et al.'s contention that the present flux is anomalously high. Also, note that there is an approximately 20% discrepancy between the asteroid cratering rate for the Earth given in Table II with that cited above from Shoemaker et al. (1979); the reason for this discrepancy is not immediately obvious.

Asteroids are not expected to be a significant contributor to cratering rates among the outer planet satellites because their lifetimes in Jupiter-crossing orbits are so very short. Shoemaker and Wolfe (1982) estimated that asteroids constituted at most 10% of the Jupiter-crossing flux of small bodies,

TABLE II
Current Cratering Rates for Planets and Satellites, All in Units of 10^{-14} km^{-2}yr^{-1} [a]

	Mercury	Earth	Moon	Mars
Comets	1.3	1.0	0.9	0.3
Earth-crossing asteroids	1.0	1.9	1.0	0.4
Mars-crossing asteroids	0	0	0	2.0
Total	2.3	2.9	1.9	2.7

	Jovian Satellites			
	Io	Europa	Ganymede	Callisto
Long-period comets	0.6	0.7	0.5	0.38
Jupiter family comets	1.6	1.3	0.6	0.30
Extinct comets	3.0	2.5	1.2	0.55
Saturn family comets	0	0	0	0
Total	5.2	4.5	2.3	1.2

	Saturnian Satellites			
	Mimas	Tethys	Dione	Rhea
Long-period comets	0.5	0.3	0.2	0.12
Jupiter family comets and extinct comets	0.04	0.01	0.007	0.004
Saturn family comets	≈1.5	≈0.6	≈0.3	≈0.2
Total	≈2.0	≈1.0	≈0.5	≈0.3

[a]Table from Shoemaker (1982).

and even then expected that most of those so-called asteroids were, in fact, extinct cometary nuclei. However, dynamical mechanisms such as chaotic motion at the 2/1 and 5/2 resonances will produce Jupiter-crossing asteroids in the same fashion that the 3/1 resonance produces Earth-crossing asteroids. Some fraction of these asteroids may be perturbed out of the resonances and be allowed to make close approaches to Jupiter. Thus, it may be that this source requires more serious consideration in the future.

C. Cometary Impactors

Estimates of cometary impact rates on solar system bodies are highly uncertain because of a lack of detailed knowledge as to the size distribution and density of cometary nuclei, and the absolute flux of comets passing through the planetary region. Although the spacecraft flybys of Comet Halley have helped to clarify at least one part of the problem, that of cometary surface albedos, nucleus dimensions for other comets are still very poorly determined, and nucleus bulk densities are still a very open question. The uncertainties regarding observational completeness in the cometary flux estimates also remain.

Present attempts to determine the size distribution of cometary nuclei are based on observations of nuclei at large solar distances when they are presumably quiescent and their dust and gas comae have dispersed. High-quality observations of this type are available for only perhaps 30 to 40 comets. Even then, there is considerable evidence that comets can become and/or remain active at large heliocentric distances, and that they may always be accompanied by large clouds of dust and debris.

For example, Shoemaker and Wolfe (1982) attempted to estimate the cometary mass distribution in this manner, using the cratering rate on the Earth as an upper limit on their combined flux and mass estimates. But the resulting estimate was four times the measured cratering rate given in Sec. II. Given that the contribution from Earth-crossing asteroids could not be entirely ignored, Shoemaker and Wolfe concluded that 88% of the light in the distant cometary nucleus observations came from the unresolved coma. Though this is likely an overestimate of that effect, it does illustrate the problems associated with estimating nucleus sizes.

One area where Shoemaker and Wolfe were absolutely correct was that of the expected albedo of cometary surfaces. They reasoned that the comets, presumably formed in the Uranus-Neptune zone, were an extension of the asteroid compositional spectrum in the planetary region, and that the cometary surfaces would most resemble the dark, primitive surfaces of the Trojan asteroids. Although a variety of observations also pointed in this direction, it was not until the Halley spacecraft flybys that the very low albedo of cometary surfaces, on the order of 3 to 4%, was firmly established.

The other major area of uncertainty is that of the absolute flux of comets passing through the planetary region. Detailed studies of observational completeness by Everhart (1967) showed that a very large fraction of the long-period comets passing through the inner planets region were not detected. Taking a uniform sample of observations over a period of 127 yr, Everhart showed that only 337 comets with perihelia less than 4.0 AU and brighter than absolute magnitude 11.0 had been detected, out of an estimated flux of 8000 comets during that time. Although a number of factors discovered since 1967 would modify Everhart's estimate, it is obvious that the degree of incompleteness in cometary statistics must still be quite large.

The problem is yet further complicated by the discovery that the Halley nucleus was only active over about 30% of its sunlit surface area, and the expectation that the active fraction of surface area for most short-period comets might be only a few percent. Thus, the ability of comets to generate substantial comae may not be a direct function of the nucleus size.

Given all these uncertainties, estimates of the expected cometary cratering flux in the planetary region have been made, and are likely good as order-of-magnitude estimates. In the terrestrial planets zone the cratering is dominated by long-period comets, while among the outer planets both long- and short-period comets contribute significantly to the flux. Cratering rates calculated by Shoemaker (1982) for the terrestrial planets and outer planet satellites are shown in Table II. It is seen that the comets can contribute a significant fraction of the current cratering flux in the terrestrial planets zone, and virtually all of the cratering among the outer planets. Weissman (1987) finds a terrestrial cratering rate for long- and short-period comets combined of about half of Shoemaker's estimate. One factor contributing to the difference in the two estimates is a lower density of 0.6 g cm^{-3} assumed by Weissman, based on the more recent Halley mass estimates. However, the difference in the two estimates should primarily be considered as a reflection of the uncertainty inherent in making such estimates.

Unlike the asteroids, large temporal variations in the flux of comets passing through the planetary system are possible. Long-period comets are perturbed out of the Oort cloud by galactic perturbations and by the random passage of nearby stars, some of which will occasionally penetrate the cloud, greatly enhancing the flux of comets into the planetary region. A computer simulation of the number of dynamically new comets from the outer Oort cloud passing within 10 AU of the Sun over a period of 200 Myr by Heisler et al. (1987) is shown in Fig. 9. The very large spike at t = 75 Myr is caused by a star penetrating the Oort cloud to within 7200 AU of the Sun. Variations in the flux by a factor of 2 are common, and by a factor of 10 are possible.

Even greater variations in the flux are possible if one includes the denser and more populous inner Oort cloud. Hut and Weissman (1985) showed that increases in the average flux by a factor of several hundred or more are possible if a star passes through the inner comet cloud. Such intense comet showers would have a duration of 2 to 3 Myr and would result in many impacts on all the planets, as well as the larger satellites of the solar system. Weissman (1987) estimated that variations in the cometary flux due to major and minor comet showers would increase the average cometary cratering rate by an additional factor of 120%.

The problem is additionally complicated by the fact that there is no way to tell if the current flux of comets through the planetary region is close to the average long-term flux or not. The argument could be made that the apparent excess of current impactors over the long-term average cratering rates determined for the Earth and Moon indicates that the present cometary flux is enhanced. On the other hand, the uncertainties are so great that it is difficult to conclude anything definitive about perceived differences in observed and estimated cratering rates.

Another factor affecting the cometary impact rates is the perihelion distribution throughout the planetary region. Everhart (1967) showed that the perihelion distribution increased by a factor of about 2.5 between 0.1 and 1.0

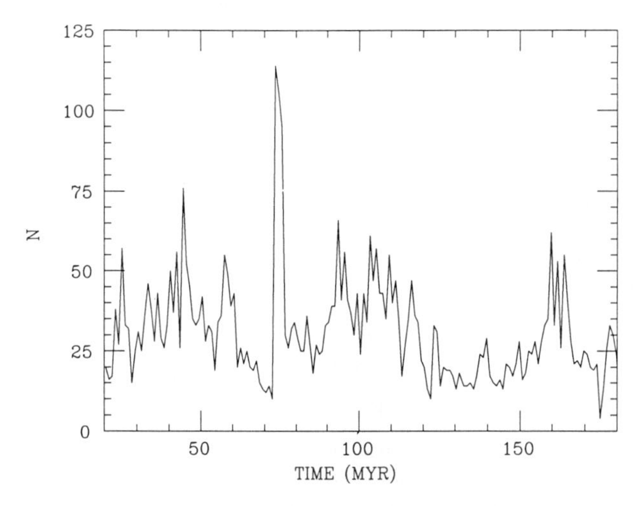

Fig. 9. Computer simulation of the temporal variation in the flux of dynamically new long-period comets entering the planetary region from the outer Oort cloud under the influence of random stellar perturbations and galactic perturbations over a period of 200 Myr (Heisler et al. 1987).

AU, and then either continued to increase or leveled off for perihelia greater than 1.0 AU. Dynamical modeling of the infusion of long-period comets into the planetary region by Weissman (1985*a*) showed that the perihelion distribution continues to increase throughout the planetary region and beyond, as shown in Fig. 10. This is caused by two factors: (1) the planetary region acts as a sink for Oort cloud comets and thus the supply of comets with perihelia closer to the planetary region is depleted; and (2) Jupiter and Saturn act as a dynamical barrier to the diffusion of comets inward, and many long-period comets are eliminated before they can reach the terrestrial planets zone.

The temporal and perihelion variations are recent developments in the understanding of long-period comet dynamics and thus are not included in Shoemaker's (1982) estimates of the cratering rates for the planets and satellites shown in Table II. A careful re-examination of this entire problem is clearly warranted.

Probably the greatest area of uncertainty with regard to cometary impact rates is the role of extinct cometary nuclei. There are two plausible physical end states for short-period cometary nuclei: (1) they eventually lose all their volatiles and disintegrate into meteoroid streams; or (2) they form thick lag deposits of nonvolatile, carbonaceous debris on their surfaces and become asteroidal in appearance. It is expected that most short-period comets are dynamically ejected from the solar system by Jupiter with a typical lifetime of 1 Myr (if they physically survive that long), as shown in Table I. But Marsden (1970) has pointed out that two very faint short-period comets, P/Arend-Rigaux and P/Neujmin I, are trapped in resonant librations that keep them

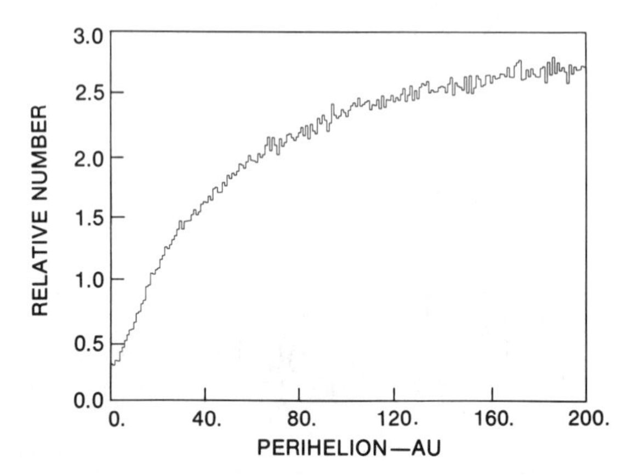

Fig. 10. Computer based estimate of the perihelion distribution of Oort cloud comets in the planetary region and beyond to 200 AU. The flux of planet-crossing comets monotonically increases as one moves outward from the Sun. (Figure from Weissman 1985*a*.)

away from Jupiter, thus greatly prolonging their dynamical lifetimes and allowing them to evolve physically toward asteroid-like objects.

In addition, it is possible that planetary perturbations and the effects of nongravitational forces caused by asymmetric jetting from the cometary nuclei can result in comets detaching their aphelia from Jupiter, again prolonging their dynamical lifetimes in the inner planetary region. This has somehow happened for periodic comet Encke which has an aphelion of 4.0 AU and a period of only 3.3 yr. Wetherill (1988) estimates that if only two short-period comets such as P/Encke evolve to asteroidal objects every 0.1 Myr, then this can provide about 450 of the steady-state population of Apollo asteroids, and about 330 of the Amor asteroids.

A number of observed Earth-crossing asteroids have been identified as possibly being extinct cometary nuclei. These include 3200 Phaeton which is in the same orbit as the Geminid meteor shower, 944 Hidalgo which is in an eccentric Jupiter-crossing orbit, 2101 Adonis which has an anomalous depolarized radar reflectivity similar to the icy Jovian satellite Callisto, and 2201 Oljato which is in a chaotic orbit and has been cited as unusual by a number of different observers. Study of these asteroids with reflectance spectroscopy is inconclusive: some appear to have surfaces typical of carbonaceous chondrites as would be expected for cometary bodies, but others do not.

Estimates of the mass of cometary volatiles delivered to the terrestrial planets by the steady-state flux of long- and short-period comets over the history of the solar system have been made. Weissman (1983) estimated delivered masses ranging from 10^{17} to 10^{20} g for each of the terrestrial planets, though those estimates were based on high albedos for cometary nuclei and would need considerable revision due to the several factors discussed above.

More recently, Ip and Fernandez (1988) have estimated a cometary contribution to the Earth of 3×10^{20} g over the last 4 Gyr, assuming a nucleus density of 0.5 g cm^{-3}. The cometary contributions to Venus, Mars and Titan over the same period of time are 4.2×10^{20} g, 1.0×10^{20} g and 9×10^{18} g, respectively. Note that these figures are far less than the cometary mass expected to impact each of these planets during the clearing of the Uranus-Neptune zone early in the solar system's history.

There has been considerable speculation in recent years about the role of impactors in biological extinction events on the Earth (Alvarez et al. 1980) and cometary sources have often been invoked to explain these. In particular, cometary showers have been invoked to explain alleged periodic extinctions. However, all of the proposed mechanisms run into a number of difficulties with regard to the dynamics of the shower-causing body (or bodies), and the expected cratering and siderophile record expected from them (Weissman 1986). The details of all the arguments involved are beyond the scope of this chapter. However, the most likely situation is that the alleged periodicity in the extinction record is spurious, and that any relationship between impacts and extinctions likely involves single, large, random impacts of comets or asteroids.

V. EXTRASOLAR IMPACTORS

The solar system does not exist apart from the rest of the Galaxy. It is moving slowly, and somewhat randomly, through a highly heterogeneous assemblage of stars and interstellar clouds. Although it is common to think about the solar system orbiting the galactic nucleus with a period of about 250 Myr, the true motion is far more complex. The Sun and planets currently oscillate up and down through the galactic plane with an amplitude of ± 85 parsecs and with a period of about 32 Myr (Bahcall and Bahcall 1985); this is referred to as the Sun's epicyclic motion. There is also a radial motion; the Sun is currently near the periapse of its galactic orbit (Innanen et al. 1978). In addition, gravitational encounters with giant molecular clouds (GMC's) change the Sun's random motion with a frequency of about once every 300 Myr.

There are two possible interstellar sources of material which may contribute to planetary and satellite atmospheres: interstellar comets and interstellar clouds. The comets likely result from dynamical ejection during planetary accretion and growth around other stars, and from the continuing dynamical evolution of their respective Oort clouds. This presupposes that the planetesimal formation processes which occurred in the solar nebula are a general feature of star formation. The detections of accretion disk-like structures around protostars such as HL Tau and R Monocerotis (Sargent and Beckwith 1987) and around young main-sequence stars (Gillett 1986) lend considerable credibility to that idea.

No comet has ever been observed passing through the solar system on a confirmed hyperbolic orbit. The most hyperbolic original orbit on record is that for comet 1976I with $1/a_o = -734 \times 10^{-6}$ AU^{-1}; this corresponds to a hyperbolic excess velocity of 0.81 km s^{-1}. This is quite small as compared with the Sun's 20 km s^{-1} velocity relative to the local group of stars, and is most likely a result of errors in the comet's orbit determination, or the effect of nongravitational forces on its motion. A truly hyperbolic comet approaching the solar system with a velocity of 20 km s^{-1} would have a $1/a_o$ value of 0.4504 AU^{-1}, more than 2 orders of magnitude greater than that for 1976I.

Sekanina (1976) used the fact that no comets with an interstellar origin had been observed to set an upper limit on the space density of interstellar comets in the solar neighborhood of 6×10^{-4} solar masses pc^{-3}. Sekanina used a mean mass for cometary nuclei of about 2.9×10^{17} g, yielding a maximum density of 4.1×10^{12} comets pc^{-3}, or about 4.7×10^{-4} AU^{-3}. This is about a factor of 5 less than the density of comets in the outer Oort cloud, so it represents a fairly modest limit.

The local density of interstellar comets increases if the solar system passes through the Oort cloud of another star. Stern (1987) estimated that the Sun has encountered 1.2×10^4 stars at ranges $<5 \times 10^4$ AU, the nominal radius of the outer Oort cloud, and about 500 stars at $<10^4$ AU, the radius of the inner Oort cloud. Planetary impacts are dominated by the inner Oort cloud passages, but the total number of extrasolar comet impacts on any planet is likely 2 to 4 orders of magnitude less than that for long-period comets from the Sun's own Oort cloud. The contribution from interstellar comets not gravitationally bound to other stars is even less. Thus, it is unlikely that there is any significant contribution to planetary atmospheres from interstellar impactors.

The effects of the passage of the solar system through an interstellar cloud has been considered by a number of authors (Shapley 1921; Begelman and Rees 1976; Talbot and Newman 1977; Frisch and York 1985), mostly with attention to possible climatic effects on the Earth. Accretion of interstellar material on the Sun might result in an increase in its luminosity. For interstellar cloud densities on the order of 10^2 to 10^3 cm^{-3}, the flux is enough to prevent the solar wind from reaching the Earth's orbit, leaving the Earth unprotected from low-energy galactic cosmic rays. A variety of magnetosphere-atmosphere interactions with possible climatic consequences have been speculated on.

Yabushita and Allen (1985) have estimated terrestrial accretion rates of 7×10^{13} g yr^{-1} for hydrogen and 6×10^{12} g yr^{-1} for dust and ice grains (though the latter figure does not quite reflect cosmic composition, and is likely somewhat high), based on a typical cloud density of 10^4 cm^{-3}. Assuming a typical GMC diameter of 20 parsecs and an encounter velocity of 20 km s^{-1}, the Earth would accrete material at this rate for a period of 1.2 Myr. Close passages to denser clumps within the GMC's may result in brief excursions in the accretion rate, by perhaps 4 orders of magnitude or more. The total intercepted mass for one such passage is thus on the order of 9.1×10^{19} g, mostly H_2. That is equivalent to one icy planetesimal with a diameter of 56 km (assuming a density of 1.0 g cm^{-3}).

Encounters with GMC's occur at an average interval of 300 Myr, so about 15 such encounters would be expected over the history of the solar system, equivalent to a total mass of 1.4×10^{21} g. This is still quite small compared to the total volatile inventory of the Earth, or the expected mass of impactors on the Earth during the clearing of the Uranus-Neptune zone. On the other hand, the climatological effects of intercepting such a mass of material, approximately 10^3 to 10^4 times the typical rate of meteoritic influx on the Earth, in a relatively brief but extended period of time, is obviously an interesting topic, worth pursuing in more detail in the future.

VI. IMPACT-RELATED PROCESSES ON ATMOSPHERIC EVOLUTION

There are a number of associated effects which need to be considered in studying the role of impacts in the evolution of planetary atmospheres. These are: impact erosion of atmospheres, effects of impactors on atmospheric chemistry, impact heating of the primordial atmosphere, and the effect of impact generated dust clouds in primitive planetary atmospheres.

The hypervelocity passage of a projectile through a planetary atmosphere will impart some of its kinetic energy to that atmosphere. This will manifest itself as a shock wave moving through the atmosphere ahead of the projectile. When the shock wave hits the ground it will be reflected and will impart considerable energy to the compressed volume of air between it and the projectile. In this manner, some of the atmosphere previously acquired by a planet or satellite will be blown off by later impacts.

This phenomenon has been modeled in some detail by Walker (1986*b*) who found that the volume of atmosphere which escapes the planet is roughly equal to the volume intercepted by the projectile on its downward flight. The impact with the solid planet surface does not contribute a significant amount of additional blow off because the velocity of the rebound rapidly decreases with distance from the point of impact (Melosh 1984).

Walker also found that atmospheric escape decreased with increasing planet size, and did not occur for planets with escape velocities $\gtrsim$10 km s^{-1}. No satellite in the solar system has an escape velocity >2.74 km s^{-1}. Given that fact, the atmosphere of Titan may be considered something of an oddity; this may indicate that Titan's atmosphere is primarily the result of outgassing due to an extended energy source such as tidal heating.

Among the terrestrial planets, only the Earth and Venus have escape velocities >10 km s^{-1}. In the case of Mars, Cameron (1983) pointed out that the longer dynamical lifetime for debris in the Mars zone probably led to many impacts long after that planet had cooled and its outgassing rate had decreased. Coupled with Mars' low escape velocity, the late impacts resulted in a substantial erosion of that early atmosphere at a time when replacement sources had already declined considerably.

The ultimate example of atmospheric erosion is a near-disruptive impact by a very large planetesimal as discussed in Sec. III.A. In that case one can expect that a very substantial fraction of the total atmosphere is lost (Cameron 1983).

The hypervelocity passage of a projectile through a planetary atmosphere can also have interesting effects on atmospheric chemistry. The severe shock heating of the atmosphere along the projectile path will result in chemical reactions between atmospheric constituents. Lewis et al. (1982) and Prinn and Fegley (1987*b*) showed that the passage of a 10 km asteroid or comet through the Earth's atmosphere would result in the conversion of NO to NO_2, and then to nitrous and nitric acid. The atmospheric mixing levels of NO_x would increase from approximately 10^{-8} to between 10^{-4} and 10^{-3}.

These atmospheric studies were made in an attempt to quantify one set of effects that could be associated with a major impact at the Cretaceous-Tertiary boundary, but it is obvious that the calculations are relevant to a much wider range of impact related problems. The resulting planet-wide effects include intense acid rain, chemical smog, and depletion of the ozone layer. The detailed environmental effects are difficult to predict but it is certainly likely that this could initiate a severe ecological crisis on the Earth, not even considering the effects of the impact itself.

In another study Fegley et al. (1986) considered the effects of an intense early bombardment on the atmosphere of the proto-Earth and concluded that it would result in a significant source of HCN and H_2CO, though much of it would be destroyed or would rain out of the atmosphere rather rapidly, on a time scale of hours.

The important point is that the evolving atmospheres around the growing protoplanets will have a large, near-continuous energy source from incoming projectiles, causing constant disruption of atmospheric chemistry. This will not stop when planetary accretion is completed, as Uranus-Neptune zone clearing and the constant flux of asteroids and comets will continue to provide a sporadic energy source for perturbing atmospheric chemistry. In addition to the examples worked for the Earth, one can also imagine interesting chemistry resulting in the atmospheres of Venus and Titan, if not also Mars.

Another interesting consequence of an early intense bombardment is the heating effect it might have on the proto-Earth. Matsui and Abe (1986*a*) showed that if the composition of projectiles was 0.1% to 1.0% water, then the resulting water atmosphere would serve as a thermal "blanket" and would prevent the planet's surface from radiating the impact energy to space. The atmosphere and surface temperature would continue to grow until the mass of water in the atmosphere was on the order of 10^{24} g and the temperature was 1500 K. At this point, the solubility of water in silicate melt would buffer the atmospheric water content. The result would be a magma ocean on the surface, covered by a thick, hot steam atmosphere. The atmosphere would slowly cool as the impact rate dropped off and the water would eventually condense to form the Earth's primitive oceans.

Interestingly, the maximum amount of water in this primitive atmosphere is of the same order of magnitude as the current mass of the Earth's oceans. Also, if the process was interrupted by a giant impact which removed the primordial water atmosphere, subsequent smaller impacts would rapidly recreate the same situation.

Zahnle et al. (1988) studied the same problem but with a wider range of parameters. They found that the water content of the projectiles did have some effect on the total mass of water in the atmosphere, either preventing the formation of a steam atmosphere if the impactors were too "dry," or raising the total atmospheric mass if the projectiles were "wet." Typical surface pressures for the steam atmosphere would be on the order of 30 bar. Although surface magma would outgas some of the dissolved water when the surface cooled, most of the dissolved water would remain buried in the mantle.

Matsui and Abe (1986*b*) showed that a similar chain of events could be expected on Venus, but as the impact rate decreased, the atmospheric cooling at Venus was not sufficient for water to condense out onto the surface. As a result Venus's primitive water remained in the hot atmosphere where it was subject to both photodissociation and atmospheric escape processes. Thus, a plausible explanation is provided for the tremendous difference in the water content of Venus and the Earth.

Another interesting point to consider is the effect of dust thrown up by impacts on the atmosphere. Planetary cooling from impact-induced dust is suggested to be the prime mechanism for initiating the Cretaceous-Tertiary extinction event (Toon et al. 1982). Grinspoon and Sagan (1987*b*) have suggested that the early Earth would be continuously shrouded in such dust clouds for the first few hundred Myr of its history, because the interval between impacts is less than the fall-out time for the dust. Even after impact rates decrease, sporadic impacts will result in new dust clouds and thus, continued fluctuations in the atmospheric and surface temperature of the planet.

It is clear that we are only first beginning to consider the many interacting mechanisms that might act during the impact accretion of a planetary atmosphere in the early solar system. The various papers discussed above consider each process separately, yet all must be occurring simultaneously, and will interact in some as yet not understood fashion. In addition, further consideration is certain to find additional mechanisms which need to be included in any study of the problem.

VII. DISCUSSION AND SUMMARY

The impact history of the solar system, as it relates to the origin of planet and satellite atmospheres, is that of a complex combination of interacting and conflicting processes. The most important part of that history is intimately involved with the formation of the planet and satellite systems, and the subsequent clearing of the planetary formation zones. Since then, a variety of volatile reservoirs and dynamical mechanisms have contributed to the delivery of both volatile-rich and volatile-poor impactors to the planets and satellites, while other physical processes related to impacts have modified and/or removed the evolving atmospheres.

Much has been learned about the impact record in the past 25 yr as a result of planetary exploration spacecraft and the Apollo missions to the Moon. Analyses of meteorites have also yielded important data on the early chronology of solar system formation and the composition of the solar nebula. At the same time, theoretical studies of solar nebula formation, planetary dynamics, accretionary processes, etc. have added greatly to the understanding of important physical processes in the early and present-day solar system. Further progress in all these areas can be expected to continue in coming years. However, the reduced pace of planetary exploration missions, in particular missions to the small bodies in the solar system, means that the new data which often results in quantum leaps forward in our knowledge of the solar system will be slower in coming.

The problem remains, however, that the bulk of the solar system's impact history, particularly as it applies to the accretion of planetary atmospheres, occurred in the first Gyr of that history. Impacts and impact-related processes since that time have played a relatively minor role in adding to, modifying, and/or eroding the existing atmospheres (though the brief perturbations caused by impacts are fascinating events to study in themselves). The record of early impacts has been obscured and/or obliterated by later cratering, be it the last major bombardment episode in the terrestrial planets zone, the late heavy bombardment, or subsequent episodes of resurfacing and planetocentric impactors on the satellites of the outer planets.

It is clear that the planets and satellites were bombarded early in their formation histories by the planetesimals that contributed to that formation. Because of dynamical scattering between planetary zones, the mix of impacting planetesimals included both volatile-rich and volatile-poor bodies. In fact, it is possible that the initial planetesimals formed in the gas-rich solar nebula were never volatile poor. The heating from impacts resulted in a continual outgassing, with catastrophic atmospheric loss and outgassing at the times of expected giant impacts. However, most of that discussion belongs rightfully to the chapters on solar system formation and planetary accretion, and will not be covered further here.

During the final stages of accretion and zone clearing, the planets and satellites were bombarded by a long-lived, volatile-rich source of planetesimals from the Uranus-Neptune zone. Although calculations of the expected mass of volatiles delivered to the Earth are comparable to the current mass of the oceans and atmosphere, it is not sufficient to say that those impacts alone can explain the origin of the Earth's atmosphere. A variety of loss mechanisms require that the total mass of volatiles brought to the Earth during its formation and subsequent impact history exceed the current volatile inventory by a factor of several, if not by an order of magnitude or more. More detailed estimates of the total mass of volatiles and their possible sources may be possible from further analyses of trace elements and isotopic ratios on the Earth. For the other terrestrial planets the important goal is to discover what happened to their volatile inventories, inventories that initially had to be comparable in size and composition to the primitive Earth's.

In the outer solar system, the important goal is to understand better the complex impact histories on the satellites, and the details of the impacting flux of comets, both from the inner and the outer Oort clouds. In addition, the search for other possible impactor reservoirs in the outer solar system should continue. Current and planned exploration spacecraft missions to the outer solar system can be expected to add greatly to our knowledge of the cratering distributions on outer planet satellites in the next 20 yr.

In the more distant future, it might be hoped that dating of returned samples from planetary surfaces will show whether the early chronology of impacts is the same on the other planets and satellites as it is on the Earth and Moon. In addition, detailed dating of many craters on each body could provide evidence of temporal variations in the flux of long-period comets through the planetary region, and possibly also asteroids. Analysis of impactor melts in craters might directly yield data on the compositional distribution of the impactors. However, for the moment such data is merely a distant dream. Our efforts must be concerned with finding ways to glean whatever additional understanding is possible out of the existing data that is already in our hands.

Acknowledgments. The author is very grateful to G. Wetherill, R. Strom, R. Greenberg and J. Williams for helpful discussions and comments, and for their reviews of an earlier draft of this paper. This work was supported by the NASA Planetary Geology and Geophysics Program and was performed at the Jet Propulsion Laboratory under contract with the National Aeronautics and Space Administration.

SUPPLY AND LOSS OF VOLATILE CONSTITUENTS DURING THE ACCRETION OF TERRESTRIAL PLANETS

G. DREIBUS and H. WÄNKE
Max-Planck-Institut für Chemie, Mainz

The different densities of the terrestrial planets are evidence for differences in their chemical composition. According to the two-component model, the composition of terrestrial planets can successfully be described as mixtures of a highly reduced volatile-free component A *and an oxidized, volatiles-containing component* B. *A mixing ratio of* A : B = *60 : 40 was estimated for Mars and 85 : 15 for the Earth. Mars obviously accreted homogeneously, while inhomogeneous accretion is favored for the Earth. Accretion is the most effective degassing stage of planetary matter; however, the early atmosphere formed in this way is also continuously removed by the accretion process itself. In addition, especially on Mars because of its homogeneous accretion, almost all H_2O added from component* B *reacted with metallic iron of component* A *and was reduced to H_2 which escaped. The huge quantities of H_2 not only greatly accelerated the extraction of gaseous species from the interior of the planet but also furthered the removal of these species due to hydrodynamic escape. This effect was somewhat smaller on Earth. Nevertheless, most of the water added to the Earth during accretion was lost in this way, except for a small portion that was added to the Earth after metallic Fe and Ni were no longer available. Large portions of Venus and the Earth, but not of Mars, became at least partially molten to great depth and, hence, had vigorously convecting mantles. Depending on their solubilities, the gases (HCl, H_2O, CO_2, rare gases, etc.) temporarily present in the early atmospheres entered these melts and were carried into the deep interiors. Due to the smaller mass of Mars, melting occurred on a smaller scale and, hence, the total amount of gases dissolved in its interior is expected to be lower. The present atmospheres of Venus, Earth and Mars are "secondary" atmospheres, while the "primary" atmospheres were almost completely lost. The huge differences observed in the concentrations of primordial rare gases per unit planetary mass on Venus, Earth and Mars are explained by* *differences in the amount of gases lost during accretion (inhomogeneous accretion for Venus and the Earth, but homogeneous accretion for Mars). The giant impact which probably formed the Moon also should have played an important role in the case of the Earth.*

I. INTRODUCTION

The Viking data revealed that the primordial rare-gas concentrations in the Martian atmosphere are 2 orders of magnitude below the terrestrial value. The amount of ^{36}Ar per g of the Martian mass is a factor of 5000 lower than in the carbonaceous chondrite Orgueil, and a factor of 25 lower even in the highly metamorphosed H chondrites. On the other hand, the ^{36}Ar content of Venus' atmosphere exceeds that of the Earth by a factor of 10^2 and that of Mars by a factor of 10^4. According to current models of the formation of the solar system (Larimer 1967; Lewis 1972*b*), the concentrations of volatiles in a given planet are expected to increase with distance from the Sun. A huge gas loss from the primitive atmosphere during the accretion of the planet seems to be the almost inevitable conclusion to explain the low primordial rare-gas concentrations in the Martian atmosphere.

Accretion is the most effective degassing stage in planetary evolution (Arrhenius et al. 1974; Lange and Ahrens 1983; Matsui and Abe 1986*a*). A model for the accretion of terrestrial planets from two chemically very different components *A* and *B* was postulated by Ringwood (1977*b*,1979*a*) and Wänke (1981). Based on estimates of the composition of the Earth and Mars, this chapter discusses the implications of this two-component model for the abundance of rare gases, H_2O and other volatile elements on Venus, Earth and Mars. The composition of Mars is based on the assumption that Mars is indeed the parent body of the SNC meteorites.

II. THE CHEMICAL COMPOSITION OF TERRESTRIAL PLANETS

The considerable differences in the density of the terrestrial planets indicate differences in their bulk chemical composition. Reliable estimates of the bulk composition of planetary bodies exist only for those of which we have samples for laboratory analyses. Besides the Earth and Moon, there are two other planetary objects from which we have samples in the form of meteorites. One of these objects, the eucrite parent body (EPB), has been studied for several years, and estimates of bulk composition have been published by various authors (Hertogen et al. 1977; Consolmagno and Drake 1977; Morgan et al. 1978; Dreibus and Wänke 1980). Eucrites are basaltic meteorites and two other groups of achondrites, the diogenites and howardites, are closely related to them. The asteroid Vesta has been proposed as their possible parent body because its reflectance spectra resemble those of eucrites (McCord et al. 1970) and howardites (Chapman 1976).

The other planetary object from which we have rocks in form of meteorites is in all probability the planet Mars. During the last 8 yr, a number of observations have been made which point towards Mars as the parent body of a special class of eight achondrites, the SNC meteorites (Wasson and Wetherill 1979; McSween and Stolper 1980; Dreibus et al. 1981; Bogard and Johnson 1983; Becker and Pepin 1984; Pepin 1985; Wiens et al. 1986). Geochemical studies (Dreibus et al. 1982; Burghele et al. 1983) as well as oxygen-isotope measurements (Clayton and Mayeda 1983) indicate that SNC meteorites come from one and the same parent body.

Chemical Composition of the Earth

The chemical composition of the Earth's mantle is known from investigations on mantle xenoliths which are frequently brought to the surface by volcanic eruption, as well as from komatiites and other mantle-derived rocks with limited fractionations (Sun 1982). Jagoutz et al. (1979) selected five nearly uncontaminated spinel-lherzolite nodules for detailed study. Trace element as well as isotope studies (Jagoutz et al. 1980) demonstrated the primitive (i.e., almost unfractionated) character of one of these nodules. From the study of these lherzolite nodules (Jagoutz et al. 1979; Wänke et al. 1984), the following abundance trends have been recognized (see Table I); the trends were obtained by normalizing the observed concentrations to Si and comparing them to Si-normalized abundances in the C1 carbonaceous chondrites (Palme et al. 1981).

Relative to Si and C1, all lithophile refractory elements including Mg have abundances of about 1.3. Corresponding to its slightly lower refractory behavior, Mg is somewhat less enriched. The enrichment of refractory ele-

TABLE I
Composition of the Earth's Mantle Compared with C1 Carbonaceous Chrondrites

Element Group	Si-normalized abundances relative to C1
1. Refractory oxyphile elements (Al, Ca, Ti, Sc and most refractory trace elements) and Mg *enriched*	× (1.3 ± 0.15)
2. V, Cr, Mn *depleted*	× (0.25 to 0.7)
3. Fe and moderately siderophile elements (Ga, Cu, W, Co, Ni) *depleted*	× (0.1 to 0.2)
4. Moderately volatile elements (Na, K, Rb, F, Zn) and the highly volatile element In *depleted*	× (0.1 to 0.2)
5. Highly siderophile elements (Ir, Os, Re, Au) *strongly depleted*	× (0.002)
6. Highly volatile elements (Cd, Ag, I, Br, Cl, Te, Se, C) *strongly depleted*	× (10^{-2} to 10^{-4})

ments might reflect a fractionation according to volatility. However, a uniform enrichment could also be interpreted in terms of a corresponding Si deficiency. The missing Si might have entered the core in metallic form (Ringwood 1958; Wänke 1981). Alternatively, the observed Si deficiency in the upper mantle could be compensated by a higher Si abundance in the lower mantle (Liu 1979).

The depletion of V, Cr and Mn, first noted by Ringwood (1966*b*), is striking. Depletion due to volatility could be expected in the case of Mn, the most volatile one of these three elements. However, under solar nebula conditions, Cr is only slightly more volatile than Si, while V is considerably less volatile than Si. Hence, Dreibus and Wänke (1979) argued that removal of these elements into the Earth's core was a more likely cause for their depletion than volatility. Vanadium, Cr and Mn might have partly entered the core in reduced form as metals or sulphides.

The high concentration of Ni in the Earth's mantle, which is incompatible with its high metal/silicate partition coefficient, has been the concern of many scientists for a long time. In a homogeneous accretion model with chemical equilibrium between mantle and core, the very similar abundances of the elements Ga, Cu, W, Ni and Co require not only lowering of the partition coefficient for Ni, but also conditions that make the partition coefficients of these five elements almost equal. In addition, the moderately volatile elements K, Na, Rb and F and the highly volatile In are similar in abundance to the moderately siderophile elements, as shown in Fig. 1.

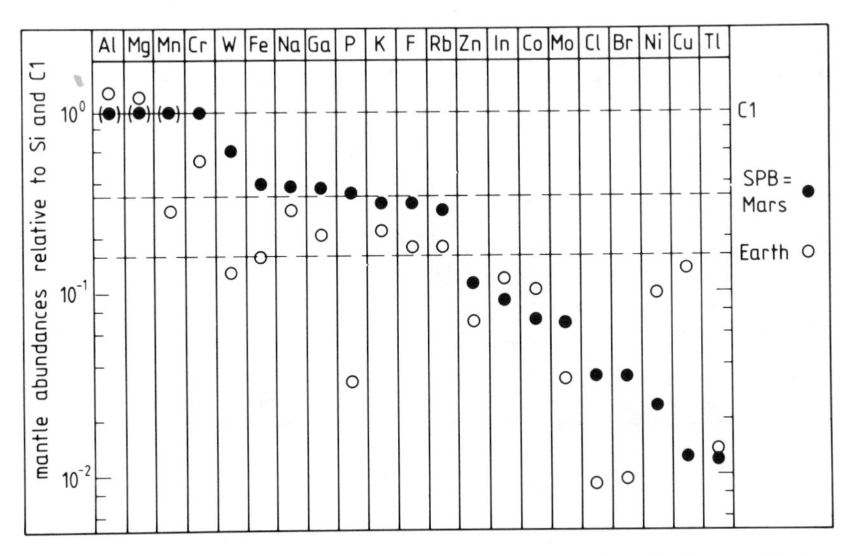

Fig. 1. Estimated elemental abundances in the Earth's mantle together with the respective data for the mantle of Mars (figure from Dreibus and Wänke 1984,1987).

TABLE II
Bulk Composition of the Earth

	Ringwood (1977*a*); Ringwood and Kesson (1977)	Morgan and Anders (1980)	Sun (1982)	Anderson (1983)	Wänke et al. (1984)
			Mantle + crust		
MgO%	38.1	34.1	38.0	34.0	36.85
Al_2O_3	3.9	4.1	4.3	3.8	4.20
SiO_2	45.9	47.9	44.5	47.9	45.95
CaO	3.2	3.2	3.5	3.1	3.54
TiO_2	0.3	0.20	0.22	0.20	0.23
FeO	8.1	8.9	8.36	7.9	7.58
Na_2O	0.1	0.25	0.39	0.27	0.39
P_2O_5	—	—	0.021	0.013	0.015
Cr_2O_3	0.3	0.9	0.44	0.3	0.44
MnO	0.14	0.11	0.11	0.13	0.13
K ppm	129	200	230	151	231
Rb	—	0.68	0.66	0.39	0.74
Cs	—	0.02	0.008–0.017	0.02	—
F	68	20	26	28	19.4
Cl	59	29	21–38	8	12
Br ppb	—	157	60–90	—	46
I	—	20		—	13
La ppm	—	0.56	—	0.57	0.52
Co ppm	98	—	110	101	105
Ni	1694	—	2000	1961	2108
Cu	34	46	30	29	28
Zn	50	109	56	37	49
Ga	6	5	4.8	4	3.8
W ppb	—	—	(21)	—	24.1
Th	75	76	—	77	—
U	21	21		20	21
			Core		
Fe%	86.2	84.5			80.27
Ni	4.8	5.6			5.46
Co					0.27
S	1.0	9.0			—
O	8.0	—			—
Si, Mn, Cr					14.00
Core mass %	31.2	32.4			33.5

The highly siderophile elements are again greatly overabundant compared with the concentrations expected from their partition coefficients, although less abundant than the moderately siderophile elements by more than an order of magnitude. The highly siderophile elements Pd, Ir, Re and Os appear again in almost chondritic abundances relative to each other.

Table II gives a compilation of the chemical composition of the Earth derived by various authors from geochemical, geophysical and cosmochemical constraints. The discussion in this chapter is based on the data of Wänke et al. (1984), but as seen from Table II, there are only minor deviations among the different estimates.

Chemical Composition of Mars

The discovery of trapped gases with element and isotope ratios characteristic of the Martian atmosphere in the shergottite EETA 79001 (Bogard and Johnson 1983; Becker and Pepin 1984; Wiens et al. 1986; Pepin 1985,1987*b*; Swindle et al. 1986; Hunten et al. 1987) and the low crystallization ages of these meteorites (Nyquist et al. 1979; Shih et al. 1982; Jagoutz and Wänke 1986) provide strong evidence that Mars is indeed the parent body of the SNC meteorites. Therefore these SNC meteorites allow very precise conclusions about the bulk composition of Mars (Wood and Ashwal 1981; Dreibus et al. 1982; Burghele et al. 1983; Dreibus and Wänke 1984,1985,1987; Jagoutz and Wänke 1986; Jagoutz 1987). The most recent estimates of this composition are given in Table III and Fig. 1.

Compared to the chemical composition of the Earth's silicate phase, the mantle of Mars has a higher content of moderately volatile (K, Rb, F, Na, etc.) and volatile elements (Cl, Br, I) and a higher concentration of moderately siderophile elements (Fe, W, P). However, elements with chalcophile tendencies are depleted in the Martian mantle according to their sulfide-silicate partition coefficients rather than their metal-silicate partition coefficients (Burghele et al. 1983; Dreibus and Wänke 1984).

III. THE TWO-COMPONENT MODEL FOR THE FORMATION OF TERRESTRIAL PLANETS

To explain the observed elemental abundance pattern of the Earth's and Mars' mantle, the following two-component model for the accretion of the terrestrial planets has been proposed (Ringwood 1977*b*,1979*a*; Wänke 1981). In this model, the composition of these planets can be successfully described as mixtures of:

Component A: Highly reduced and free of all elements with equal or higher volatility than Na, but containing all other elements in C1 abundance ratios. Fe and all siderophile elements are metallic, and even Si is present partly in metallic form; and

TABLE III
Bulk Composition of Mars

	Anderson (1972)	Morgan and Anders (1979)	Basalt. Volcan. Study Proj. (1981*b*)	Dreibus and Wänke[a] (1987)
		Mantle + crust		
MgO%	27.4	29.8	29.9	30.2
Al_2O_3	3.1	6.4	3.1	3.02
SiO_2	40.0	41.6	36.8	44.4
CaO	2.5	5.2	2.4	2.45
TiO_2	0.1	0.3	0.2	0.14
FeO	24.3	15.8	26.8	17.9
Na_2O	0.8	0.1	0.2	0.50
P_2O_5	—	—	—	0.16
Cr_2O_3	0.6	0.6	0.4	0.76
MnO	0.2	0.15	0.1	0.46
K ppm	573	77	218	315
Rb	—	0.258	—	1.12
Cs	—	0.026	—	0.07
F	—	24	—	32
Cl		0.88	—	38
Br ppb		4.7	—	145
I		0.59	—	32
Co ppm	—	—	—	68
Ni	—	—	—	480
Cu	—	—	—	2.6
Zn	—	42	—	74
Ga	—	2.4	—	6.6
W ppb	—	—	—	105
Th	77	113	60	56
U	17	33	17	16
		Core		
Fe%	72	88.1	63.7	77.8
Ni	9.3	8.0	8.2	7.6
Co	—	—	—	0.36
S	18.6	3.5	9.3	14.24
O	—	—	18.7	—
Core mass %	11.9	19.0	18.2	21.7

[a]For Shergottite parent body.

Component B: Oxidized and containing all elements, including the volatiles, in C1 abundances. Fe and all siderophile and lithophile elements are present mainly as oxides.

Based on their estimates of the bulk composition of Mars, Dreibus and Wänke (1987) concluded that the mixing ratio of component *A*/component *B* for Mars is 60 : 40, compared to a ratio of 85 : 15 for the Earth. Furthermore, they concluded that Mars accreted almost *homogeneously,* contrary to the *inhomogeneous* accretion of the Earth, as proposed by Jagoutz et al. (1979) and in more detail by Wänke (1981). According to this model, the accretion of the Earth started with the highly reduced component *A*. Segregation of the metallic FeNi phase, i.e. core formation, is supposed to be almost contemporaneous with accretion, leading to an FeO-free mantle during this stage. The present high mantle abundances of siderophile elements like Ni, Co, Cu (Fig. 1), which for Ni exceeds by far the amount to be expected from the metal-silicate partition coefficients exclude equilibration with a pure FeNi phase. That is the reason for the assumption, in this model, that the oxidized component *B* which supplied FeO, NiO, CoO, etc. to the mantle was only added in substantial amounts after the Earth had reached about two thirds of its present mass. In this inhomogeneous accretion model, the major portion of component *B* was never in equilibrium with larger amounts of metallic FeNi.

The argument for the homogeneous accretion of Mars is the obvious depletion of all chalcophile elements (Cu, Ni, Co., etc.) in the Martian mantle (Fig. 1). It was assumed that sulfur supplied by component *B* was responsible for formation of a sulfur-rich FeNi alloy, leading to a sulfide-silicate equilibrium. During core formation, extraction of elements from the mantle took place according to their sulfide-silicate partition coefficients as mentioned above (Burghele et al. 1983; Dreibus and Wänke 1984).

IV. PARENT BODIES OF PRIMITIVE METEORITES AS REMNANT BUILDING BLOCKS OF TERRESTRIAL PLANETS

Material of the reduced component *A* is, in several aspects, similar to that of enstatite chondrites. These meteorites contain Si as metal and most of their Mn and Cr as sulfides (Keil 1968), but they are depleted only slightly or not at all in volatile and moderately volatile elements. The composition of the oxidized component *B* is similar to that of carbonaceous chondrites. However, we do not wish to overemphasize such analogies because it might well be that the building blocks of planets are not found among the contemporary population of undifferentiated meteorites.

There are three important cosmochemical observations made during the last 10 yr which are not well understood:

1. The Moon, the Earth and probably all inner planets as well as the eucrite parent asteroid (i.e., all differentiated objects) are depleted in volatile and moderately volatile elements relative to C1 abundances. While small objects could have lost even moderately volatile elements during or after accretion, this seems to be impossible for large objects like the Earth or Moon.
2. In the oxygen isotope diagram of Clayton et al. (1976), all objects listed in (1) plot close to the same oxygen isotope fractionation line as do C1

chondrites and enstatite chondrites. Ordinary chondrites, however, plot differently.

3. Extinct ^{26}Al has been identified in certain meteorites in amounts sufficient to melt even objects in the km-size range (Lee et al. 1976). However, one may ask in this connection, why only a small fraction of asteroids were molten.

Long ago, Urey (1959*b*) introduced the idea of primary and secondary objects. The three observations described above could be connected in a scenario which we outline below in which differentiated bodies are primary objects and the undifferentiated bodies secondary objects.

If the time scale of accretion of km-sized objects was on the order of the half-life of ^{26}Al (0.72 Myr), the interiors of these objects were heated to silicate melting before they reached their final mass. Inside these rather loosely compacted bodies, mobilization of volatile and moderately volatile elements from hot interiors to cool surface layers would have started long before melting. However, fractionation (i.e., loss of volatilized elements) depends on the surface temperature of the planetesimals. Fractionation of even moderately volatile elements is possible only if the temperature exceeds a value of about 1000 K. Hence, loss of moderately volatile elements would not have been possible unless the planetesimals were disrupted by collision or internal pressure. Disruption of objects containing molten material will lead to formation of many droplets (chondrules) which would lose their volatiles very efficiently (Zook 1981; Wänke 1981).

Material not yet accreted as well as material dispersed from collisions of first-generation objects would continuously form new objects of asteroidal size. However, for objects that accreted at a time when ^{26}Al had decayed to a certain level, loss of volatiles was no longer possible. Thus, the last generation of objects (secondary objects) made in this scheme would have the highest concentrations of volatile elements. At the time of formation of secondary objects, the ambient gas may be more oxidized because of preferential removal of hydrogen in the direction perpendicular to the plane of the solar nebula, while escape of oxygen in form of water could have been prevented by formation of ice grains. Material of these secondary objects would have had time to react and equilibrate with H_2O. A different oxygen isotope composition of the H_2O would lead to the observed parallelism between the degree of oxidation and oxygen isotope composition of ordinary chondrites.

In the absence of water (in the inner part of the solar system) primary objects formed by unprocessed interstellar dust would have had extremely low oxygen fugacities due to reduction by its C and CH compounds upon its being heated. (High concentrations of carbon in solid particles were observed in Comet Halley; Jessberger et al. 1988.) Hence, not only FeNi etc., but also portions of Si, Cr, Mn, V, etc. could have converted to metals or sulfides. These primary objects would have the high C/O ratios required for the formation of enstatite chondrites (Larimer and Bartholomay 1979).

Leitch and Smith (1981) have found two types of olivine and pyroxene in E chondrites: one containing Mn and Cr, the other almost free of these elements. The latter might have been derived from extremely reduced objects in which Mn and Cr entered the sulfide phase. Thus E chondrites, as well as ordinary chondrites, could be mixtures of materials with different oxygen fugacities; i.e., they could be secondary objects. Achondrites might be derived from planetesimals of earlier generations which escaped destruction.

In intermediate regions of the solar nebula, small amounts of water added to the accreting planetesimals in the form of ice would, upon heating, lead to their partial oxidation. In the outer (cooler) regions of the solar system, the amount of water could have been large so that its complete evaporation exceeded the available energy supply. It is in these regions where C1 chondrites and comets formed.

In order to explain the different types of accretion (inhomogeneous and homogeneous) for the Earth and Mars, Dreibus and Wänke (1984) suggested that the volatile-free and highly reduced component *A* existed mainly at and inside the Earth's orbit, while the oxidized, volatile-containing component *B* dominated in the asteroidal belt. Mars, located near the transition of the region dominated by component *A* to that dominated by component *B*, was thought to have been fed from both components simultaneously and in approximately equal proportions. In the case of the Earth and Venus, component *B* material was added only during a late phase of accretion, since the transfer of material from the region outside of Mars' orbit required additional time. As this transfer may have been influenced by the perturbation effect of Jupiter on planetesimal orbits, it may also have been coupled to the formation time of Jupiter.

V. RARE GASES, WATER AND OTHER HIGHLY VOLATILE ELEMENTS IN PRIMITIVE METEORITES, COMPARED WITH THE CONCENTRATION OF GASES AND VOLATILES IN THE TERRESTRIAL PLANETS

Primordial Rare Gases

The amount of ^{36}Ar in the Martian atmosphere per g of the planet's mass is more than a factor of 100 lower than that in ordinary chondrites and about a factor of 5000 lower than that in the carbonaceous chondrite Orgueil (Table IV). On the other hand, the ^{36}Ar content of the Venus atmosphere exceeds that of the Earth by a factor of about 100 and even exceeds that of all primitive meteorites except the E4.5 chondrite South Oman, with its extraordinary high concentrations of primordial rare gases (Crabb and Anders 1981).

With respect to its state of reduction, component *A* in the models of Ringwood (1977*b*,1979*a*) and Wänke (1981) is similar to enstatite chondrites. As is obvious from their large metallic cores, all terrestrial planets are domi-

TABLE IV
Primordial Rare Gases, Water and Other Volatiles in Meteorites and Planets

	^{36}Ar 10^{-10}cm^3 g^{-1}	^{84}Kr 10^{-10}cm^3 g^{-1}	^{132}Xe 10^{-10}cm^3 g^{-1}	H_2O ppm	Cl ppm	Reference
C1 (Orgeuil)	7800	83	76	72900	678	a,l,h
C2 (Murray)	12500	93	100	65430	186	a,f,h
C3V (Vigarano)	5100	47	34	6170	149	a,f,h
C3O (Warrenton)	11500	59	46	—	226	a,h
E4 (Abee)	3700	15	8	—	560	b,h
E4-5 (S.Oman)	76000	170	29	—	—	b
H3 (Bremervörde)	760	13	13	—	76	c,i
H5 (Pultusk)	40	1.2	0.42	—	74	k,i
L3 (Krymka)	3630	38	34	—	—	c
L6 (Sultanpur)	130	1.5	1.6	—	—	k
Venus	21000*	—	—	—	—	d
Earth	210*	4.3*	0.16*	322	8	d,e
Mars	1.6*	0.055*	0.0075*	28	30	g
EETA 79001-C (glass)	1.34	0.047	0.0020			e,m

*Atmosphere only.
References (a) Mazor et al. 1970; (b) Crabb and Anders 1981; (c) Heymann and Mazor 1968; (d) Morgan and Anders 1980; (e) Becker and Pepin 1984; (f) Kerridge 1985; H_2O data related from the total H content; (g) Anders and Owen 1977; (h) Dreibus et al. 1979; (i) Quijano-Rico and Wänke 1969; (k) Zähringer 1968; (l) Boato 1954; (m) Hunten et al. 1987.

nated by the reduced component *A*. Oxygen isotope compositions of enstatite chondrites fall along the terrestrial fractionation line (Clayton et al. 1976). This is in agreement with the proposition that component *A* material similar to enstatite chondrites comprises 85% of the Earth. As expected from the higher abundance of component *B* on their parent body (Mars = 40% component *B*), the SNC meteorites in the oxygen isotope diagram of Clayton and Mayeda (1983) plot slightly above the terrestrial fractionation line, towards the more oxidized H and L chondrites.

It is also interesting to note that the highly reduced enstatite chondrites generally have primordial ^{36}Ar concentrations within the range of the oxidized carbonaceous chondrites, and that the ^{84}Kr/^{132}Xe ratio (Table IV) in the enstatite chondrite South Oman is closer to the ratios in the Earth's atmosphere (Wacker and Anders 1984) and the Martian atmosphere than to those in the C1 and other carbonaceous chondrites. Hence, one could even speculate that the rare gases on the terrestrial planets were not primarily supplied by carbonaceous type material. It might also be that the abundances of primordial rare gases in carbonaceous chondrites, but not in enstatite chondrites, were changed due to loss by diffusion, which should be largest in the case of Ne and Ar while Xe would be least affected. As indicated by their low K-Ar ages, C1 chondrites have suffered substantial loss of ^{40}Ar.

It is difficult to explain why Venus, Earth and Mars should have been formed from materials differing in their concentrations of primordial rare gases by 4 orders of magnitude. That is why Dreibus and Wänke (1985) proposed planet-specific differences in gas loss during accretion. As discussed by various authors, two quite different processes are thought to be of importance for the removal of primitive atmospheres:

1. Hydrodynamic escape due to the presence of large amounts of H_2 so that the mean atmospheric molecular weight remains low (Zahnle and Kasting 1986; Hunten et al. 1987);
2. Removal by large impacts (Watkins and Lewis 1984; Matsui and Abe 1986*a*; Walker 1986*b*; Ahrens and O'Keefe 1987; chapter by Hunten).

On Mars, because of its lower gravity, both processes will clearly act more efficiently than on the Earth and Venus; on these two planets, accretional energy leads to formation of huge magma oceans (Kaula 1979), especially if the blanketing effect of an early H_2O atmosphere (Matsui and Abe 1986*a*) is taken into account. Substantial portions of rare gases liberated during accretion will redissolve in the magma ocean and subsequently be carried into the solid regions of the highly convecting mantles. On Mars, melting occurred on a much smaller scale as the energy of accretion and core formation per unit mass is about 5 times lower (Basaltic Volcanism Study Project 1981*b*). In this respect, we note that the amounts of redissolved gases strongly depend on the thickness of the layers that equilibrate with the atmosphere.

The existence of H_2-rich primitive atmospheres on the terrestrial planets (Sekiya et al. 1980) is not generally accepted. However, in the two-component model for the formation of the terrestrial planets, reaction of metal from component *A* with H_2O from component *B* during accretion will produce large amounts of H_2. Assuming an H_2O content of component *B* equal to that of C1 chondrites (7.29% H_2O; see Table IV), 7×10^{24}g H_2 on the Earth and 2×10^{24}g H_2 on Mars would be generated in this way. Based on an accretion time scale of 100 Myr, escape fluxes of 2.6 and 2.7×10^{14} H_2 molecules cm^{-2} s^{-1} are obtained for Earth and Mars, respectively.

Impact-induced splash off seems the most likely model for the origin of the Moon (Hartmann and Davis 1975; Cameron and Ward 1976*b*; Ringwood 1979*a*,1986; Melosh and Sonett 1986; Wänke and Dreibus 1986; Wetherill 1986). The mass of the projectile probably considerably exceeded that of the Moon; its impact towards the end of terrestrial accretion would certainly have removed some fraction, perhaps a very large fraction, of any atmosphere present at that moment (see the chapter by Hunten et al). Clearly, gas loss has also to be expected in the case of Venus. However, unlike the Earth, Venus was obviously not hit late in its accretion history by an object of sufficient mass to form a moon. It has been suggested by Cameron (1983) that the primordial rare-gas concentrations of the atmospheres of Earth and Venus may have been similar prior to the removal of the terrestrial atmosphere by the giant moon-forming impact. Consequently all other atmospheric constituents (including H_2O) would also have been lost.

The low concentration of ^{36}Ar on Mars can be brought into agreement with the obviously high concentrations of moderately volatile and volatile elements such as K, Rb, Cl, Br and Pb, by assuming that the different rare-gas concentrations in Venus, Earth and Mars are due to different loss factors during accretion. The observed concentration of primordial rare gases on Venus can be taken as a lower limit of the concentrations initially added. In the model of Dreibus and Wänke (1985), relative to Venus, the Earth has lost 10^2, and Mars almost 10^4 times more of their primordial rare-gas contents.

Water and Halogens

Recently, Dreibus and Wänke (1985,1987) have estimated the halogen abundances on Mars derived from the correlation of the refractory element La with the volatile element Br in the SNC meteorites, assuming C1 abundance ratios of Cl/Br and I/Br as observed for the Earth (Wänke et al. 1984). In this way, they found for Mars an absolute Cl concentration of 30 ppm. Assuming the composition of component *B* to be identical to that of C1 meteorites (Palme et al. 1981), but with a correction for the mass of extraneous H_2O (Dreibus and Wänke 1987), 320 ppm Cl were actually added to the planet, of which all except the 30 ppm was lost during the accretion process together with H_2 and other gases. The 30 ppm Cl for the whole planet corresponds to 38 ppm Cl in the Martian mantle (Tables III and IV).

To estimate the water content on Mars, Dreibus and Wänke (1987) compared the solubility of H_2O and HCl in basaltic melts. They assumed that at the end of accretion the concentration of H_2O and other volatiles at the surface of the planet was zero (no atmosphere and hydrosphere). They further assumed that the solubility of H_2O and HCl is proportional to square root of pressure (Anderson 1975). In this way, they obtained an H_2O concentration in the Martian mantle of 36 ppm. Making the unrealistic assumption of 100% release, this would yield a water layer of about 130 m depth covering the whole planet.

A low water content of Mars is also reflected by the low water content of SNC meteorites. Fallick et al. (1983) have measured the N, C and H content and stable isotopic composition in Chassigny, Nakhla and Shergotty. The amount of water recovered from 440°C up to 1000°C were 0.9, 1.1 and 1.3 μmol g^{-1}, respectively. Yang and Epstein (1985) liberated 10.2 μmol H_2O g^{-1} by stepwise oxidation-pyrolysis of Shergotty from 350°C to 1100°C. The considerably lower value of 1.3 μmol g^{-1} obtained for Shergotty by Fallick et al. (1983) is probably due to the high precombustion temperature of 440°C compared to the 350°C usually used to exclude contamination from terrestrial organics. The 10.2 μmol H_2O g^{-1} found in Shergotty corresponds to 180 ppm, whereas in terrestrial MORB samples (Middle-Ocean-Ridge-Basalts) the mean water concentration is 2100 ppm (Delaney et al. 1978) (Table V).

TABLE V

Amounts of ^{132}Xenon, Water and Chlorine in Terrestrial and Martian Basalts

	^{132}Xe 10^{-12}cm^3 g^{-1}	H_2O ppm	Cl ppm	K ppm	La ppm	K/La	K/Cl	H_2O/Cl	H_2O/K	Reference
EARTH										
Basalts:										
MORB (Ave.)	0.46	2100	60					35		a,c
			25	650	2.8	230	26	—	—	d
Hawaii (Ave.)		6210	190	—	—	—	—	39	—	e
Arc (Ave.)		10660	1070	—	—	—	—	10	—	e
OIB (Kilauea)										
1701	1.68	3800	233	3000	10	300	13	16	1.3	b,d,f
1712	0.64	3800	200	3910	9.7	400	20	19	1.0	b,d,f
1699	1.16	5200	177	3750	—		21	29	1.4	b,d,f
Seawater		97%	1.94%	380				50	2550	
Atmosphere	16									g
Mantle		60*	0.5	127	0.35	360	250	160	0.6	h
Mantle + crust		480	11.8	231	0.52	440	20	42	2.1	h,i
MARS										
Basalts:										
Shergotty	5.94	180	108	1570	2.44	640	15	1.7	0.1	j,k,l
Atmosphere	0.75									
Mantle + crust		36	38	315	0.48	660	8	1.0	0.1	g,i
C1	7600	7.15%	678	517	0.245	2110	0.8	105	138	

*Estimated value assuming C1 Ir/H_2O ratio for the mantle and an Ir value of 3.2 ppb for Earth's mantle (Jagoutz et al. 1979) minus the crustal H_2O content of 6.95% (Larimer 1971).

References: (a) Allègre et al. 1986, 1987; (b) Staudacher and Allègre 1982; (c) Delaney et al. 1978; (d) Dreibus and Wänke 1987; (e) Garcia et al. 1979; (f) Kyser and O'Neil 1984; (g) Anders and Owen 1977; (h) Wänke et al. 1984; (i) Unpublished data of Dreibus and Wänke 1987; (j) Swindle et al. 1986; (k) Yang and Epstein 1985; (l) Laul et al. 1986.

Shergotty is enriched in La by a factor of 5 relative to the Martian mantle (Table V). Assuming a similar enrichment for H_2O (i.e., assuming that H_2O behaves as incompatibly as La), a mantle concentration of 180/5 = 36 ppm H_2O is obtained. This is exactly the value obtained by Dreibus and Wänke (1987) in a very indirect way. This exact match is, of course, purely fortuitous.

Compared to Martian basalts, the K/La ratio in terrestrial basalts is a factor of 2 lower (Fig. 1, Table V), reflecting a higher abundance of the moderately volatile element K on Mars (Dreibus and Wänke 1984,1985,1987). The similar K/Cl ratio in terrestrial and Martian basalts (Table V) indicates a similar enrichment of moderately volatile and volatile elements on Mars relative to Earth according to the higher portion of component *B* (40% vs 15%). However, the H_2O/Cl and H_2O/K ratios in terrestrial basalts exceed those on Mars by factors of 6 to 25. Even taking into account Martian Cl and K concentrations of about twice the terrestrial values, the poverty of water on Mars is evident.

According to the model of Wänke (1981), the accretion of the Earth started with the highly reduced and volatile-free material of component *A*. After the Earth had reached about 60% of its present mass, more and more of the oxidized component *B* was added. However, small amounts of metal still present were responsible for the extraction of highly siderophile elements (Ir, etc.). Towards the very end of accretion, metal became unstable; component *B* dominated and iron was added only in oxidized form. Hence, even the highly siderophile elements remained in the mantle in their C1 abundance ratios. As pointed out by Dreibus and Wänke (1987), the concentration of 3.2 ppb Ir in the Earth's mantle, equal to 2.1 ppb Ir for the whole Earth, corresponds to 0.44% C1 material. With a C1 H_2O content of 7.29% (Boato 1954), we find 320 ppm H_2O were added to the Earth after metallic Fe and Ni were no longer available. Dreibus and Wänke (1987) argued that only these 320 ppm remained in the form of water, while the much larger amount of water (~1.1% derived from 15% component *B*) which was added before metal became unstable was converted to H_2 and escaped (except for a few ppm dissolved in the magma ocean during accretion). From the estimate of 6.95% H_2O in the Earth's crust (Larimer 1971), we find that the crustal contribution of H_2O (mainly the oceans) to the whole Earth is 280 ppm. This would indicate that almost 90% of the total water inventory of the Earth resides in the crust.

Carbon, Nitrogen and Sulfur

Dreibus and Wänke (1987) argued that the large amounts of H_2 generated during planetary accretion by the reaction $Fe + H_2O = FeO + H_2$ should have removed all volatile species with high efficiency. The H_2 production ceased at that moment at which metallic iron was no longer available in the surface layers of the growing planets. Volatiles added after this stage were retained to a much larger extent. Hence, in the case of the Earth, the amounts of carbon,

Table VI
Carbon, Nitrogen and Sulfur on the Earth

	C ppm	N ppm	S ppm	Reference
Crust*	3600	350	880	a,b,c
Mantle	24	?	8	d
			350–1000	e
			317	f
Contribution to bulk Earth				
Crust*	14	1.4	3.5	
Mantle	16	?	235	
Mantle + crust	30	1.4	238	
C1 (Orgueil)	30900	1420	58000	g,h
C1 × 0.0044	136	6	255	

*Including atmosphere and hydrosphere
Reference: (a) Hunt 1972; (b) Turekian and Clark 1975; (c) Wedepohl 1981; (d) Wänke et al. 1984; (e) Sun 1982; (f) Ringwood and Kessen 1977; (g) Palme et al. 1981; (h) Kerridge 1985.

nitrogen and sulfur contained in the 0.44% C1 material, thought to have added to the Earth after metallic iron was no longer available, are the minimum concentrations to be expected in the Earth's mantle, crust and atmosphere. For the bulk Earth, this amounts to 136 ppm C, 6 ppm N and 255 ppm S.

Table VI shows the contribution of volatile elements C, N and S in the atmosphere, hydrosphere, crust and mantle to the bulk Earth. The mantle data derived from analyses of spinel-lherzolites (Wänke et al. 1984), probably do not reflect the true abundances. Partial loss of S, halogens and other volatiles during eruption of the ultramafic nodules as observed for subaerial basalts (Moore and Fabbi 1971; Unni and Schilling 1978; Sakai et al. 1982) cannot be excluded. The estimated higher sulfur concentrations of the Earth's mantle of 350 to 1000 ppm from Sun (1982) are derived from the observed sulfur concentrations in MORB and komatiites. The lower figure falls close to the values of 317 ppm given by Ringwood and Kesson (1977). Thus, crust and mantle contribute 238 ppm S to the bulk Earth, a value close to the 255 ppm S predicted from the late addition of 0.44% C1 material.

The amount of carbon to be expected from 0.44% C1 material exceeds the contribution of crust and mantle to the bulk Earth by a factor of 4 (Table VI). As in the case of nitrogen this difference is not really disturbing, considering the poor knowledge on the mantle and even the crustal concentrations of these elements. Like sulfur, carbon might have been lost from the ultramafic nodules, on which these mantle concentrations are based.

While the isotopic composition of carbon in C1 meteorites falls rather close to that of terrestrial carbon, C1 nitrogen is about 40 per mill heavier than the nitrogen of the Earth's atmosphere (see the chapter by Pepin). As the nitrogen of the undepleted Earth mantle seems to be even lighter than atmospheric nitrogen, the latter might be a mixture of heavy (C1) nitrogen and small amounts of light nitrogen brought to the Earth during earlier stages of accretion and held back in the mantle (Javay et al. 1986). In this context, it is interesting to note that the highly reduced enstatite chondrites which might be related to the reduced component *A*, could represent such a source. Alternatively, one might speculate that in C1 meteorites of today ^{15}N is enriched due to diffusion loss. Thus, nitrogen in the C1 material which was added to the Earth 4.5 Gyr ago might have been lighter and the absolute nitrogen concentrations correspondingly higher.

At present, we have no way to constrain the mantle and crustal abundances of C, N and S on Mars.

The Late Stage of the Inhomogeneous Accretion Sequence of the Earth

As pointed out by Dreibus and Wänke (1987), the 0.44% C1 material added at the very end of accretion after iron was no longer present in reduced form would bring about $34 \times 10^{-12}cm^3\ g^{-1}$ ^{132}Xe to the bulk Earth compared to $16 \times 10^{-12}cm^3\ g^{-1}$ ^{132}Xe present in the atmosphere (Tables IV and V). Since the portion of the Earth's inventory of ^{132}Xe which is in the atmosphere is probably not less than about 50% (Allègre et al. 1986,1987), this is in surprisingly good agreement. In the case of ^{36}Ar, the 0.44% C1 material could only deliver about 20% of the observed $210 \times 10^{-10}cm^3\ g^{-1}$ ^{36}Ar in the Earth's atmosphere (Table IV). As discussed above, it might be that C1 material had originally a higher concentration of light rare gases which later were lost by diffusion.

Comparing the concentrations of H, C, N and rare gases in the Earth's atmosphere and crust, Turekian and Clark (1975) have reached a similar conclusion. They postulated that the actual late veneer accumulating on the Earth had higher concentrations of rare gases as the sampled carbonaceous chondrites.

Furthermore, the agreement of the amount of heavier rare gases present in the Earth's atmosphere with that added by a late C1 veneer may be fortuitous. Pepin (see his chapter) underlines the fact that the isotopic composition of Kr and Xe rules out any direct acquisition of the terrestrial or Martian rare gases from the C1 reservoir. As seen from Figs. 2 and 3 of Pepin's chapter, terrestrial Kr is isotopically lighter relative to C1 Kr, and Martian Kr is even lighter while terrestrial Xe is isotopically heavier compared to C1 Xe, and Martian Xe is heaviest. Pepin (see his chapter) concludes that both terrestrial and Martian Xe could in principle be derived from C1 Xe by fractionation occurring during hydrodynamic escape of noble gases from the two planets, but he also stresses that neither the terrestrial nor the Martian Kr can be generated by fractionation of C1 Kr. The general problem is that it seems impossible to enrich the light isotopes in the residual reservoir. On the other hand, the problem might be solved if we except the idea of a loss of Kr and the lighter rare gases from the C1 matter by diffusion. Diffusion loss of Kr from C1 material would make it isotopically heavier. Hence one might speculate that C1 matter originally had an isotopic composition like the Martian Kr or even lighter. C1 material brought to the Earth during the accretion process could already have lost Kr and, hence, its Kr became heavier. It could then be expected that the gas loss and fractionation was even more severe for the C1 meteorite samples at the present time. Clearly, such a process would also affect the isotopic composition of the lighter gases, especially Ar and Ne. The expected trend is noticeable for Ne only, not for Ar. However, it might be marked by the dominance of Ar derived from highly retentive sites of C1 meteorites as well as by other processes (see the chapter by Pepin).

Assuming the diffusion constant D to vary as $m^{1/2}$ (m = atomic weight) (Zähringer 1962), Dreibus et al. (1988) estimate that in order to produce the heavier C1 Kr (Eugster et al. 1967), starting with Kr isotopically similar to the Martian Kr (chapter by Pepin), the original Kr content in C1 matter must have been 8 times higher than what is observed today.

A comparison of the $^{84}Kr/^{132}Xe$ ratio of C1 chondrites of 1.09 with the $^{84}Kr/^{132}Xe$ ratios of 27 for the Earth and 24 for Mars (EETA 79001 glass; Becker and Pepin 1984; Hunten et al. 1987) indicates that C1 meteorites must have lost Kr relative to Xe by a factor of about 23 to match the $^{84}Kr/^{132}Xe$ ratios of the Earth and Mars. This would be a three-fold higher gas loss than would be needed for the above mentioned fractionation of Martian Kr to C1 Kr.

Aside from the crude agreement of the amount of rare gases (especially of Xe) present in the Earth's atmosphere with that contributed by a late C1 veneer, there is the more general problem of where is the Xe of the C1 material which almost certainly was added in non-negligible amounts to the Earth and Mars. The portion of C1 Xe in the atmospheres of both planets could be very substantial in spite of the differences of the isotopic composition. If we assume, for example, that fractionation of Xe due to hydrodynamic escape on the Earth close to the end of accretion relative to C1 Xe was twice as high as for the Xe in the present atmosphere, the portion of C1 Xe added by the late C1 veneer could be 50%. The comparatively small difference of the isotopic composition of terrestrial Xe to Martian Xe must then be fortuitous.

VI. VOLATILE DEPLETION OF THE TERRESTRIAL PLANETS

Late addition of volatile-rich matter has been frequently discussed with respect to the origin of volatiles on the terrestrial planets. Such a model makes sense only if a complete loss of volatiles, especially of rare gases and water, occurred during the main stage of accretion. As seen from Table IV, all primitive meteorites except highly metamorphosed ordinary chondrites have higher concentrations of primordial rare gases (e.g., ^{36}Ar) than the Earth. Because of the possibility of terrestrial contamination, reliable data on the water content of ordinary chondrites do not exist. The Earth, however, with a bulk water content which in all probability does not exceed a value of 500 ppm (Dreibus and Wänke 1987), is a very dry object compared to any types of carbonaceous chondrites. Anders and Owen (1977) proposed an origin of the volatiles by late addition of a veneer of 6.3% CV3 material to the Earth and of 0.2% CV3 matter to Mars. As noted above, Dreibus and Wänke (1987) pointed out that all the water and perhaps also the primordial rare gases of the Earth can be explained by late addition of 0.44% C1 material. Clark (1986) has suggested contribution of cometary matter of which even smaller portions would suffice. We wish to emphasize that the major question is not the source of volatiles, but rather why are the terrestrial planets so poor in volatiles. If primitive meteorites play a non-negligible role as building blocks of terrestrial planets, huge gas losses during accretion must have occurred. The only undifferentiated meteorites with primordial rare-gas contents in the range of the Earth are highly metamorphosed ordinary chondrites of petrologic types 5 and 6. However, even if we find types of meteorites with primordial rare-gas contents similar to the Earth's, we still have to infer substantial gas loss in the case of Mars. Furthermore, we are faced with the problem of the rare-gas content of Venus, which exceeds that of the Earth by 2 orders of magnitude.

Loss of volatile constituents seems to be unavoidable in the case of halogens. According to the data obtained by Quijano-Rico and Wänke (1969) and Dreibus et al. (1979), there is a large variation of the chlorine content among ordinary chondrites, which is probably due to an inhomogeneous distribution of the chlorine-bearing mineral apatite. However, there is no clear-cut dependence of the chlorine concentration on petrologic type; mean values in H6 and L6 chondrites of 90 ppm and 101 ppm, respectively, fall close to the concentration of 107 ppm found in H3 chondrites. In contrast, bulk chlorine concentrations of 8 ppm for Earth and 30 ppm for Mars have been estimated by Dreibus and Wänke (1987).

The concentrations of primordial rare gases found in planetary atmospheres represent only lower limits on the total concentrations for the individual planets. There is no direct way to determine the degassing efficiency or the release factor R (rare-gas concentration in the atmosphere/total rare-gas content) of a particular rare-gas element for a given planet. For the Earth, a high release factor of about 0.5 for radiogenic ^{40}Ar is found from the estimated K content and the amount of ^{40}Ar in the Earth's atmosphere. The release factor for ^{36}Ar, and probably also those for other primordial rare-gas isotopes, can only be higher than that for ^{40}Ar, considering the 1.3 Gyr half-life of ^{40}K. Hence we can be reasonably certain that the primordial rare gases in the Earth's atmosphere represent at least 50% of their total planetary inventories.

The situation is very different in the case of Mars. From their estimates of the bulk concentration of K on Mars and the ^{40}Ar concentration in the Martian atmosphere (Anders and Owen 1977), Dreibus and Wänke (1987) obtained a release factor of 0.026 for ^{40}Ar. Again, the release factor for primordial rare gases is expected to be higher since the higher heat output in the early history of a planet should be coupled with a higher gas release.

Evidence from primordial rare-gas concentration in terrestrial rocks point to high release factors of primordial rare gases for the Earth. From Allègre et al. (1986,1987) we deduce a mean ^{132}Xe concentration of 0.46×10^{-12}cm^3 g^{-1} in MORB (see Table V), while concentrations in the glassy rims of submarine basalts from the East Rift zone of Kilauea are only slightly higher (i.e., 1.16×10^{-12}cm^3 g^{-1}; Staudacher and Allègre 1982). Compared with the 16×10^{-12}cm^3 g-planet^{-1} in the Earth's atmosphere, these are negligible amounts. On the other hand, Swindle et al. (1986) measured 5.94×10^{-12}cm^3 ^{132}Xe g^{-1} in Shergotty, an amount almost an order of magnitude above the amount contributed by the Martian atmosphere. Hiyagon and Ozima (1986) determined the noble-gas distribution coefficients for olivine-basalt melt systems; they found for Xe an olivine-basalt distribution coefficient $K_{Xe(olivine/melt)} = \leq 0.3$. If we assume that this observed Xe distribution between olivine and basaltic melt is also valid for the distribution of Xe between mantle and basalts of Mars, we calculate a ^{132}Xe concentration in the Martian mantle of 5.94×10^{-12}cm^3 g^{-1} ^{132}Xe $\times$ 0.3 = 1.8×10^{-12}cm^3 g^{-1} ^{132}Xe. Compared with the ^{132}Xe in the Martian atmosphere of 0.75×10^{-12}cm^3 g^{-1}, this rough estimate yields a release factor $R = 0.3$.

Qualitatively, the high concentrations of ^{132}Xe in Shergotty could indicate a low release factor for xenon and other primordial rare gases on Mars. The actual factor may be as low as $R = 0.1$. This would mean that a low release factor on Mars might account for about 1 order of magnitude of the 2 orders of magnitude lower concentrations of primordial rare gases in the atmosphere of Mars compared with terrestrial concentrations.

Dreibus and Wänke (1987) have argued that the release of ^{40}Ar may be governed by the degree of fractionation within the planet. They proposed a model in which ^{40}Ar is almost quantitatively released from the crust but is not released at all from the mantle. Such a model seems to be in line with the low release factor of ^{40}Ar on Mars, since, according to their estimates, the Martian crust is enriched in K relative to the mantle only by a factor of 2.5, while the Earth's crust is enriched in K by a factor of about 80. Dreibus and Wänke (1987) further argued that on Mars the release might be considerably higher for $^{129}Xe_{rad}$, if its parent ^{129}I was concentrated in the crust to a larger degree than potassium. Concentration of iodine in the surface layers of a planet during accretion seems plausible in light of the highly mobile character of this element. In contrast, fractionation of ^{244}Pu on Mars would be similar to, or even less than, that of potassium, and hence only a small fraction of the total ^{244}Pu would be expected to have entered the crust. In this way, it might be possible to explain the apparent contradiction of the high abundance of $^{129}Xe_{rad}$ in the Martian atmosphere but the absence of a noticeable contribution of ^{244}Pu fission xenon (Swindle et al. 1986). Both the high ^{40}Ar/^{36}Ar ratio as well as the high $^{129}Xe_{rad}$/^{132}Xe ratio are strong indications of the loss of considerably larger portions of primordial rare gases for Mars than for the Earth.

Acknowledgments. We wish to thank the reviewers T. Matsui and especially R. O. Pepin for their very constructive reviews, which have improved this chapter substantially. This work was carried out within the Forschergruppe Mainz, supported by the Deutsche Forschungsgemeinschaft.

PART III
Terrestrial Planets

ATMOSPHERIC COMPOSITIONS: KEY SIMILARITIES AND DIFFERENCES

R. O. PEPIN
University of Minnesota

Data bases for abundances and isotopic compositions of the noble gas, carbon and nitrogen inventories in the Sun and solar wind, in volatile-rich meteorites, and in the atmospheres and associated crustal reservoirs of Venus, Earth and Mars are briefly reviewed, evaluated, tabulated and discussed in the context of atmospheric origin and evolution. Characteristic mass distributions in these various volatile reservoirs provide crucial boundary conditions that must be satisfied in modeling the history of present-day atmospheres. Among these, in particular, are constraints posed by comparison of noble-gas isotope ratios in possible primordial volatile sources with the isotopic patterns that now exist on the terrestrial planets. Simple "veneer" scenarios in which volatiles are supplied from sources resembling contemporary meteorite classes fail to meet these constraints. It seems clear that the evolution of planetary atmospheres to their present states must have involved the action of processes capable of fractionating both elements and isotopes.

I. INTRODUCTION

About the most specific proposition one can make concerning planetary atmospheres, without challenge from some workers in the field, is that they derive from one or more source reservoirs of primordial solar system volatiles through the action of physical-chemical processes operating at some stage or stages in the pre- or early post-accretionary histories of the planets. It follows that the chemical and isotopic compositions of present-day atmospheres provide clues both to the characteristics of the source reservoir(s) and to the nature of the subsequent processing of its volatiles. Beyond these very general statements that atmospheric mass distributions are end-products of evolution from early primary volatile compositions, and that initial and final states are related by coherent processes, there is no consensus on the specifics of sources or mechanisms.

In current models for the origin and evolution of terrestrial planet atmospheres, primordial volatile sources may include the solar nebula, solar wind, comets, one or more known meteorite classes, or other bodies carrying different volatile distributions—with the *ad hoc* assumption that these represent meteorite classes that are either no longer extant in the solar system or exist but are unsampled. Processes invoked in various models range from (presumably) nonfractionating modes of gain or loss of atmospheric volatiles (which include impact degassing of meteorites or comets accreted by the planet; direct capture of ambient nebular gases; planetary outgassing; ejection of gases from planets by impact of accreting bodies into preexisting atmospheres) to mechanisms that fractionate either elements alone or both elements and isotopes (including adsorption of nebular gases, or implantation and diffusion of solar-wind gases, on dust grain surfaces prior to accretion; solution of ambient volatiles into the solid surfaces of planets during accretion; chemical partitioning; loss of atmospheric constituents from planetesimals or planetary bodies by Jeans or hydrodynamic escape or by nonthermal processes).

We have no model-independent ways to assess the relative importance of these, and perhaps other, primordial sources and evolutionary processes in establishing the various mass distributions that characterize contemporary planetary atmospheres. Assumptions in specific atmospheric models about the origins and compositions of primary reservoirs, their existence at particular times, the mechanisms responsible for transporting and fractionating their volatiles, and the astrophysical environments in which such processes operated are judged primarily by their perceived plausibility in the context of other models for the origin and evolution of the solar system. The one hard constraint, which all current scenarios fail to satisfy to varying degrees, is that they must account in detail for the elemental and isotopic compositions of present-day atmospheres.

The key diagnostic volatiles for tracing atmospheric origin and nonbiogenic evolution are the noble gases, nitrogen as N_2, and carbon as CO_2. This chapter briefly reviews and discusses the data base for abundances of these elements and the composition of their isotopes in terrestrial planet atmospheres and associated reservoirs, and in two candidate primordial reservoirs: the Sun-solar wind (≡ the solar nebula?) and volatile-rich meteorites. Helium is ephemeral in these atmospheres, and is excluded from consideration here. Noble-gas distributions in comets are still unknown.

II. VOLATILE INVENTORIES

Elemental abundances are set out in Table I, and isotopic data in Table II. Notes and comments on the tabulated data bases are given in the subsections below.

A. Solar System

Elemental abundances, except for ^{20}Ne, are averages of the separate compilations of solar system nuclidic abundances by Cameron (1982) and Anders and Ebihara (1982), with uncertainties that represent only the differences between the two estimates. The ^{20}Ne value is derived from the ^{36}Ar abundance and the solar-wind $^{20}Ne/^{36}Ar$ atom ratio of about 45 (Bochsler and Geiss 1977; Frick and Pepin 1981; Becker et al. 1986*a*), and is higher than Cameron's estimate from solar cosmic-ray data by $\sim$70%; the Anders-Ebihara value, which is actually a compromise between the solar-wind measurement and estimates from galactic stellar and molecular cloud observations, is intermediate. Ne and Ar isotope ratios are solar-wind composition measurements (Bochsler and Geiss 1977; Frick and Pepin 1981; Becker et al. 1986*a,b;* Frick et al. 1988). Extensive data exist for solar-wind carbon and nitrogen isotopes in lunar samples, with strong evidence for secular variation of N composition, but there is no consensus on what this may signify for the isotopic compositions of these elements in the young Sun or solar nebula (Kerridge 1980*a*,1982; Clayton and Thiemens 1980; Ray and Heymann 1980; Geiss and Bochsler 1982).

The extent to which this hybrid solar-solar wind data base actually represents the primordial solar nebula is an open issue. It is reasonable to suppose that the compositions of the Sun and the early nebula are closely related, but the unresolved question for the solar-wind data is that of fractionation of the wind with respect to solar photospheric abundances (Geiss 1982; Geiss and Bochsler 1985; Bochsler et al. 1986). There is some evidence that such fractionation may not be serious for the suite of elements considered here (Geiss and Bochsler 1985), but there are also observations that point to element-specific isotope effects in solar particle emission. One celebrated example is the large and unexplained discrepancy in the isotopic composition of Ne measured in the solar wind and in solar flares (Mewaldt et al. 1984). Taken at face value, this must mean that either the wind or flare isotopes are fractionated with respect to the true solar ratio.

B. Meteorites

Primitive, volatile-rich meteorites contain a rich and intriguing assortment of distinct noble-gas, carbon and nitrogen components, of both extra-solar and "local" (i.e., nebular) origin, associated with a variety of carrier phases (see Anders 1987,1988 for recent reviews). Here, however, we are interested in the possibility that meteoritic bodies accreted by planets were

TABLE I

Abundances of Carbon, Nitrogen, and Noble Gases in the Solar System (g/g-Solar Composition), Two Classes of Volatile-Rich Meteorites (g/g-Meteorite), and Terrestrial Planet Atmospheres (g/g-Planet)[a]

	^{12}C	^{14}N	^{20}Ne	^{36}Ar	^{84}Kr	^{130}Xe
[b]Solar system	3.90 ± 0.03(−03)	9.42 ± 0.03(−04)	2.24 ± 0.25(−03)	8.97 ± 0.36(−05)	5.84 ± 0.11(−08)	8.07 ± 1.36(−10)
Meteorites						
[c]CI Carbonaceous chondrites	3.70 ± 0.70(−02)	1.51 ± 0.31(−03)	2.89 ± 0.77(−10)	1.25 ± 0.10(−09)	3.57 ± 0.15(−11)	7.0 ± 1.9 (−12)
[d]Enstatite chondrites	1.50 − 7.00(−03)	1.33 − 9.46(−04)	~0 − 6.80(−11)	1.47 − 1220(−11)	5.35 − 639(−13)	1.05 − 27.8(−13)
Planetary Atmospheres						
[e]Venus	2.60 ± 0.04(−05)	2.20 ± 0.50(−06)	2.9 ± 1.3 (−10)	2.51 ± 0.97(−09)	$4.72\,^{+0.57}_{-3.40}$ (−12)	$8.9\,^{+2.5}_{-6.8}$ (−14)
[f]Earth	1.50 ± 0.40(−05)	1.45 ± 0.55(−06)	1.00 ± 0.01(−11)	3.45 ± 0.01(−11)	1.66 ± 0.02(−12)	1.40 ± 0.02(−14)
[g]Mars	1.11 ± 0.20(−08)	7.30 ± 1.90(−10)	4.38 ± 0.74(−14)	2.16 ± 0.55(−13)	1.76 ± 0.28(−14)	2.08 ± 0.41(−16)

[a]Power of ten multipliers in parentheses.
[b]Average of Anders and Ebihara (1982) and Cameron (1982), except for Ne; Ne from ^{36}Ar and $^{20}Ne/^{36}Ar \simeq 45$ in the solar wind (Bochsler and Geiss 1977).
[c]Mazor et al. (1970) for noble gases; Kerridge (1985) for C and N.
[d]Crabb and Anders (1981) for noble gases; Grady et al. (1986) and Kung and Clayton (1978) for C and N.
[e]von Zahn et al. (1983) for C, N, Ne and Ar; Donahue and Pollack (1983) for Kr; Donahue (1986) for Xe.
[f]Donahue and Pollack (1983, minimum outgassed inventories, Table IV) for C and N.
[g]Owen et al. (1977) for C and N; Hunten et al. (1987, and references therein) for SNC noble gas data.

TABLE II

Isotopic Compositions and Elemental Ratios of Carbon, Nitrogen, Neon and Argon in the Solar System, Two Classes of Volatile-Rich Meteorites, and Terrestrial Planet Atmospheres[a]

	$\delta^{13}C$	$\delta^{15}N$	[h] $^{14}N/^{12}C$ ($\times 10^{-2}$)	$^{20}Ne/^{22}Ne$	$^{36}Ar/^{38}Ar$	[h] $^{20}Ne/^{36}Ar$ ($\times 10^{-2}$)
[b]Solar system	—	—	24.15 ± 0.20	13.7 ± 0.4	5.6 ± 0.1	2500 ± 300
Meteorites						
[c]CI Carbonaceous	-10.3 ± 2.5	42 ± 11	4.1 ± 1.1	8.9 ± 1.3	5.30 ± 0.05	23.1 ± 6.4
chondrites	-8.4 ± 7.9	-32 ± 12	10.7 ± 7.0	$\sim 6.7 \pm 1.5$	5.46 ± 0.04	0.16 ± 0.02
[d]Enstatite chondrites						
Planetary atmospheres						
[e]Venus	-3 ± 18	$\sim 0 \pm 200$	8.5 ± 1.9	11.8 ± 0.7	5.56 ± 0.62	11.6 ± 6.8
[f]Earth	-6.4 ± 1.3	$\equiv 0$	9.7 ± 4.5	9.80 ± 0.08	5.320 ± 0.002	29.0 ± 0.3
[g]Mars	$<+50$	620 ± 160	6.6 ± 2.1	10.1 ± 0.7	4.1 ± 0.2	20.3 ± 6.2

[a]Units are permil deviations from standard compositions for $\delta^{13}C$ and $\delta^{15}N$, atom/atom for isotope ratios, and g/g for element ratios.
[b]Bochsler and Geiss (1977) for Ne composition; Frick et al. (1988) for Ar composition.
[c]Kerridge (1985) for $\delta^{13}C$ and $\delta^{15}N$; Mazor et al. (1970) for Ne and Ar composition (averages for Cl + CM classes).
[d]Grady et al. (1986) for $\delta^{13}C$ and $\delta^{15}N$; Deines and Wickman (1985) for $\delta^{13}C$; Kung and Clayton (1978) for $\delta^{15}N$; Crabb and Anders (1981) for Ne and Ar compositions (South Oman only).
[e]von Zahn et al. (1983) for $\delta^{13}C$ and $\delta^{15}N$; Donahue (1986) for Ne and Ar composition.
[f]Schwarcz et al. (1969) for $\delta^{13}C$; Eberhardt et al. (1965) for Ne composition; Nier (1950) for Ar composition.
[g]Nier and McElroy (1977) for $\delta^{13}C$ and $\delta^{15}N$; Wiens et al. (1986) for Ne and Ar compositions.
[h]Elemental abundance ratios from Table I data.

important primordial sources of atmospheric volatiles, and so the relevant measures are the average bulk inventories and isotopic compositions given in Tables I and II. This empirical approach avoids the question of how, when and from what sources the meteorites themselves acquired their gaseous elements, clearly a central issue for any truly comprehensive model for the origin and evolution of solar system volatiles.

In contrast to the relatively constant elemental abundances in CI chondrites (and even more constant elemental abundance ratios, a uniformity that extends to the CM, CV3 and CV4 carbonaceous chondrite classes as shown later in Fig. 1), the E chondrites span a broad abundance range. The largest E-chondrite noble-gas contents in Table I are for the South Oman meteorite (Crabb and Anders 1981), except for ^{20}Ne where the South Oman value is 1.98×10^{-11} g/g. The corresponding noble-gas isotope ratios in Table II are from South Oman alone, where the nonspallogenic ^{20}Ne/^{22}Ne ratio has been crudely estimated from a three-isotope correlation plot of the Crabb and Anders Ne data for South Oman and several other E chondrites.

C. Planetary Atmospheres

The compilations in Tables I and II refer to *atmospheric* inventories, except for terrestrial carbon and nitrogen. A relevant question is whether volatiles once in these atmospheres have been extracted by physical or chemical partitioning into surface materials, and perhaps later sequestered in planetary interiors by geologic cycling. Given the thermal state of Venus, the answer for that planet is likely to be no. On Earth, the hydrosphere, biosphere and sedimentary column are important subsidiary reservoirs for some atmophilic species; they contain fractions of total "atmospheric" inventories that range from probably minor for the nonradiogenic noble gases (including xenon; see below) to significant for nitrogen to dominant for carbon and hydrogen. To the extent that there has been a net transport of volatiles into the upper terrestrial mantle by subduction of sediments over geologic time, inventory estimates based on summation over all these "accessible" reservoirs may still be lower limits. For Mars, theoretical and experimental estimates of the carrying capacity of the cold megaregolith for H_2O and CO_2 are such that, for these species at least, the observable Martian volatile reservoirs (atmosphere-ice caps-layered deposits) are likely to contain only small fractions of the total inventories (Fanale 1976; Fanale and Jakosky 1982*a*; Fanale et al. 1982*a*).

Venus. Data from *in situ* compositional measurements of the Venus atmosphere by mass spectrometers and gas chromatographs on U.S. and Soviet spacecraft have been comprehensively reviewed and assessed by von Zahn et al. (1983). Abundances of C, N, ^{20}Ne and ^{36}Ar in Table I were calculated from the recommended mixing ratios given in their Table II. There is a profound difference of interpretation in the literature concerning the Kr and Xe concentrations measured by Venera 11-12 and by Pioneer Venus. The von

Zahn et al. review traces the derivation of the disparate ^{84}Kr mixing ratios from the two sets of instrumental data—their estimate of 0.4 ppm from Venera, based on analyses by Istomin et al. (1980) vs 0.025 ppm from Donahue et al.'s (1981) refinement of Pioneer Venus data—and sets out the authors' arguments in favor of the former. The contrary case is made by Donahue et al. (1981), and I have elected to use their lower value in Table I. Venusian Xe measurements are certainly in no better shape. The ^{132}Xe mixing ratio, reported as an upper limit of 10 ppb by Donahue and Pollack (1983), was apparently further refined from the Pioneer Venus data by Donahue (1986), where $^{84}Kr/^{132}Xe$ is given as 13.5 ± 3.5 implying a ^{132}Xe mixing ratio of about 1.9 ppb. This value has been used together with an assumed $^{130}Xe/^{132}Xe$ isotope ratio of 0.164 to calculate the ^{130}Xe abundance in Table I. Kr and Xe concentrations in this compilation thus consistently reflect the views of Donahue and his coworkers, but these views are controversial and do not compellingly rule out the possibility that the ^{84}Kr abundance is actually higher by a factor of 16 or more, and that Xe may not have been detected at all (von Zahn et al. 1983). The issue is substantive for atmospheric modeling, since the higher ^{84}Kr value converts relative abundances for Venus from a solar-like Ar : Kr : Xe distribution (see Fig. 1, below) to a Ne : Ar : Kr pattern closely resembling that for Earth, Mars and the carbonaceous meteorites.

The Ne and Ar isotopic data bear critically on this question, but do not resolve it. The $^{20}Ne/^{22}Ne$ ratio from Pioneer Venus was originally reported to be 14.3 ± 4.1 (Hoffman et al. 1980). The Table II value of 11.8 ± 0.7 is from a later compilation by Donahue (1986), and is significantly higher than the terrestrial and CI ratios. Thus there appears to be a solar signature in Venusian neon. It would seem, from the Pioneer Venus $^{36}Ar/^{38}Ar$ ratio of 5.56 ± 0.62 given in Table II (Hoffman et al. 1980; Donahue 1986), that this could also be the case for Ar (cf. the 5.6 ± 0.1 solar-wind ratio), though the uncertainty is large. However, this interpretation for Ar is at odds with the nominally precise value of 5.08 ± 0.05 derived from Venera data by Istomin et al. (1980), which is much lower than solar and even significantly below the terrestrial ratio. The Venera measurement is accepted by von Zahn et al. (1983) but is not considered in Donahue and Pollack's (1983) review or in Donahue's (1986) later data compilation. It is quite likely that two significantly different evolutionary tracks for the Venus atmosphere are implied by the elemental and isotopic noble gas distributions deduced from Venera data on the one hand, and from Pioneer Venus data on the other.

Earth. Abundances for nitrogen and carbon are estimates of "minimum outgassed inventories" in atmospheric and crustal reservoirs but excluding the upper mantle, taken from Table IV of Donahue and Pollack (1983). Terrestrial atmospheric nitrogen, with $^{14}N/^{15}N = 272$, is the standard reference gas for nitrogen isotopic measurements. The $^{12}C/^{13}C$ ratio in the *PDB* standard for $\delta^{13}C$ measurements is 88.99. The $\delta^{13}C_{PDB}$ value in Table II refers to integrated C from all reservoirs.

Mars. Mass spectrometric data on the composition of the Martian atmosphere from instruments on the Viking spacecraft were reported by Owen et al. (1977) and Nier and McElroy (1977). Uncertainties in mixing ratios for ^{20}Ne, ^{84}Kr and ^{132}Xe are about a factor of 3. Except for one very important ratio (^{129}Xe/^{132}Xe = 2.5 ± 70%), the isotopic compositions of Ne, Kr and Xe were not measured. The second data base used here for Mars atmospheric composition is that from the SNC meteorites, specifically from analyses of gases trapped in glassy nodules in the Antarctic shergottite EETA 79001. The geochemical case for origin of the SNC meteorites on Mars, and for trapping of an unfractionated sample of ambient Martian atmosphere in shock-generated melt phases in 79001, has been reviewed by Pepin (1985), Dreibus and Wänke (1985), Hunten et al. (1987) and Pepin (1987*b*). Detailed numerical comparisons of Viking and 79001 determinations for CO_2, N_2 and the noble gases are given in the last two of these reports. All abundances and isotopic compositions common to the two data sets agree within their respective uncertainties, and the generally more precise noble gas measurements from 79001 yield most of the isotopic ratios missing from Viking. Abundances of the two major atmospheric constituents, CO_2 and N_2, were determined more accurately by Viking; uncertainties in ^{36}Ar and ^{40}Ar from 79001 and Viking are comparable.

III. COMPARISONS AND IMPLICATIONS

Elemental abundance ratios calculated from the Table I data for meteoritic and atmospheric noble gases and normalized to the Table I solar composition are plotted in Fig. 1. To include all known classes of volatile-rich meteorites, relative abundances for CM, CV3, CV4 and CO3 carbonaceous chondrites (Mazor et al. 1970; Srinivasan et al. 1977; Alaerts et al. 1979; Matsuda et al. 1980) are also shown. The first three are indistinguishable from the CI distribution on this scale, and the CO3s fall within the E-chondrite range. The general pattern of progressively greater relative depletion with decreasing mass for Kr, Ar and Ne is suggestive of mass-dependent fractionation of solar-composition noble gases (but *not* of carbon and nitrogen: abundances of these elements, relative to ^{84}Kr as in Fig. 1, range from about 3 times solar for atmospheric N on Mars to 50,000 times solar for C in the E chondrites). Similar fractionation patterns have been generated in experimental investigations of noble-gas adsorption. Relative elemental compositions of gases adsorbed on various substrates in the laboratory tend to fall within the shaded area of Fig. 1, relative to ambient gas-phase compositions (Fanale and Cannon 1972; Frick et al. 1979; Niemeyer and Marti 1981; Yang et al. 1982; Yang and Anders 1982*a*,*b*; Wacker and Anders 1986). This process has been

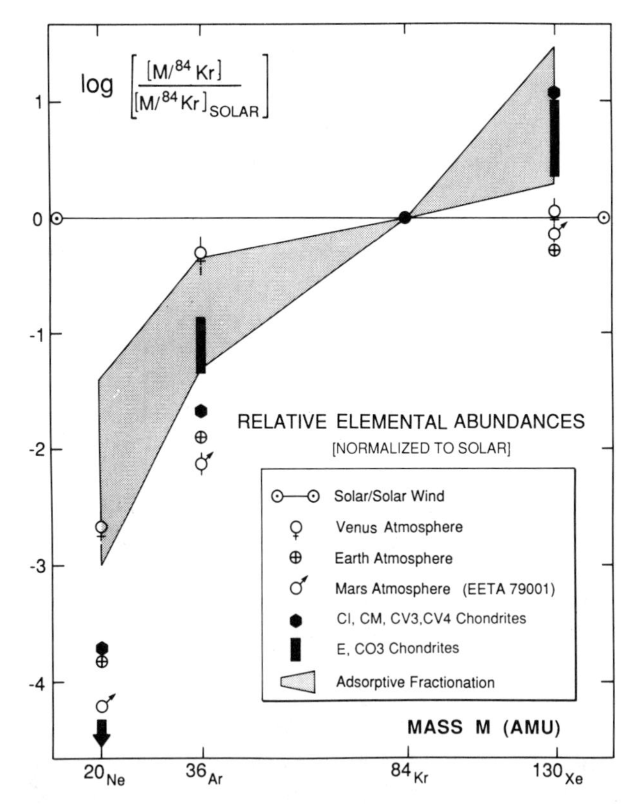

Fig. 1. Noble gas abundance ratios in terrestrial planet atmospheres and volatile-rich meteorites, plotted with respect to solar system relative abundances. Results from adsorptive fractionation laboratory experiments lie generally in the shaded area of the figure. Data are from Table I, and from references given in the text for CM, CV3, CV4 and CO3 chondrites and adsorption experiments.

postulated in some models for derivation of atmospheric volatiles directly from the solar nebula (see, e.g., Fanale and Cannon 1972; Pollack and Black 1982). However, the adsorptive fractionation process, while it generates elemental patterns similar to atmospheric distributions, cannot by itself satisfy certain isotopic constraints (see below and Hunten et al. 1988*a*). These constraints, and others that are important in evaluating the significance of apparent similarities and differences in solar system volatile compositions, are discussed in the following sections.

A. The "Planetary Component"

More than two decades ago, when only the meteoritic and terrestrial patterns in Fig. 1 were known, the term "planetary component" was introduced to designate CI-type noble gases, for what seemed then to be good reasons: the CI-Earth relative abundance distributions are virtually identical for Ne : Ar : Kr (Fig. 1); the terrestrial $^{20}Ne/^{22}Ne$ and $^{36}Ar/^{38}Ar$ ratios fall within the CI ranges (Table II); and the Kr isotopic compositions, shown later in Fig. 2, are rather similar. That the terminology was unfortunate is now clear. It created a wide-spread mind set toward an almost axiomatic presumption of a simple genetic relationship between "planetary" and terrestrial noble gases, and thereby established one glaring discrepancy (the obvious CI-Earth difference in Xe in Fig. 1) as the celebrated "Xe Problem" that somehow had to be solvable within the framework of this presumption. Moreover, "planetary" is a generic term, and its genetic implications could be assumed—and occasionally are, by workers not directly in the field—to apply to terrestrial planet atmospheres other than Earth's. Although it is not clear that a simple relationship between CI and planetary atmospheric noble gases in general can be dismissed on the basis of Fig. 1 alone (particularly if the Kr abundance on Venus is ~16 times higher than the Table I value, in which case the spread in the grouping of Ne : Ar : Kr ratios for all three planets and the CI meteorites is only about a factor of 4), the "Xe Problem" remains, now on at least two planets. More significantly, the isotopic data in Table II and in Figs. 2 and 3 below are inconsistent with this view. Use of "planetary" to designate meteoritic compositional patterns is dangerously misleading, and should cease.

B. Neon and Argon

The similarity of the $^{20}Ne/^{36}Ar$ ratios in the CI-type meteorite family and terrestrial planet atmospheres has prompted much discussion in the atmospheric modeling literature (see, e.g., review by Donahue and Pollack 1983). The central question is the implication of elemental ratios that show an extreme spread of a factor of only 2.5 (Table II) for primordial volatile compositions and possible subsequent modification of these compositions by evolutionary processes. Does this striking uniformity necessarily point to a single parent reservoir with a similar elemental ratio, or is it the coincidental result of derivation from compositionally different sources by fractionation, mixing, or both?

The former view would appear to favor scenarios in which atmospheric noble gases were acquired by accretion of planetesimals carrying a single type of volatile mass distribution, or of a more or less uniform mixture of accreted objects with different distributions (which would itself be a coincidence), on all three planets. Among the model classes of this type are the "veneer" theories (see, e.g., Anders and Owen 1977; see also Donahue and Pollack 1983), and to the extent that their late-accreting planetesimals are assumed to be represented among extant volatile-rich meteorite classes, they encounter the difficulties noted above in Sec. III.A. Isotopic compositions pose the most severe constraints. Venusian $^{20}Ne/^{22}Ne$ in Table II is more than 2σ higher than the terrestrial value and, more seriously, the Martian $^{36}Ar/^{38}Ar$ ratio is uniquely low among all known nonspallogenic solar system Ar compositions.

(It should be noted in this regard, since $^{36}Ar/^{38}Ar$ for Mars in Table II is an SNC value, that the report in Owen et al. [1977] that the atmospheric ratio is $5.3 \pm 10\%$ from Viking data is erroneous; spacecraft measurements indicate only a range of about 4 to 7 [Owen 1986].) Scenarios that aim to account for atmospheric noble-gas compositions by accretion of bodies or mixtures of bodies with volatile distributions resembling those in present-day meteorite populations cannot simultaneously accommodate these similar elemental but disparate isotopic compositions, and are thus unattractive.

We are then left with the second alternative, generation of coincidentally rather uniform Ne/Ar ratios as end-products of evolutionary tracks involving fractionating and mixing mechanisms, probably operating both prior to accretion and, to different degrees, in the separate post-accretional environments of the meteorites and the individual planets. The data clearly require processes that can fractionate isotopes, in precursor bodies, in the atmospheres of the assembled planets, or both. Jeans or hydrodynamic escape of early solar-composition atmospheres from planetesimals or planets are currently the most promising candidates (Donahue 1986; Zahnle and Kasting 1986; Sasaki and Nakazawa 1986; Hunten et al. 1987).

C. Krypton and Xenon

The relative underabundance of Xe in the Earth's atmosphere compared to the CI meteorites initiated a long search for the "missing" terrestrial Xe. Attention focused on crustal volatile reservoirs, and in particular on gases adsorbed on shales and slates in the sedimentary column (Canalas et al. 1968; Fanale and Cannon 1971). We now know that the Xe inventories in these reservoirs are much too small to account for the discrepancy (Bernatowicz et al. 1984,1985). Moreover, the "Xe Problem" was always much deeper than just the factor 25 difference in relative Xe/Kr abundances shown in Fig. 1. Isotopic distributions of both elements are also involved, and here the data convincingly rule out any simple acquisition of terrestrial gases directly from the CI reservoir.

Isotopic compositions of Kr and Xe in CI meteorites and the atmospheres of Earth and Mars (the latter from SNC meteorites data) are shown in Figs. 2 and 3. In the context of the preceeding discussion, the first-order observation is that the isotopic distributions are all *different* (subtly, but significantly, so for CI-Earth Kr and Earth-Mars Xe). However, there are coherent relationships among them. The curves and lines in the figures represent compositions derived by assuming that the various reservoirs are interrelated by mass-fractionating processes. Both terrestrial and CI Kr, for example, can be generated by fractionation of the Martian composition; perhaps more to the point, all three distributions could result from fractionation of a parent (solar?) composition that is isotopically lighter than Mars Kr. There is evidence from recent lunar sample studies that unfractionated solar-wind Kr is in fact isotopically light (Frick et al. 1988). For Xe in Fig. 3, the terrestrial and CI light isotope

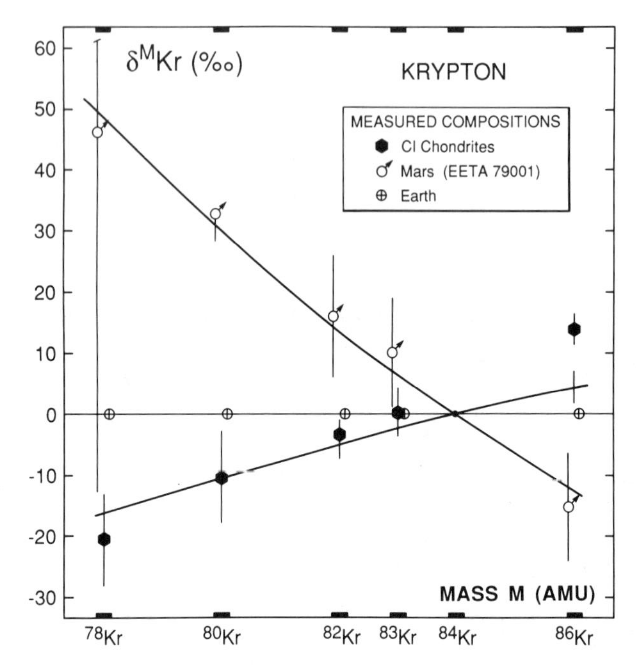

Fig. 2. Isotopic compositions of krypton in meteorites and planetary atmospheres, plotted relative to terrestrial atmospheric krypton. Units are permil deviation from the terrestrial composition. Data are from Basford et al. (1973) for Earth, Eugster et al. (1967) for the CI chondrites, and Becker and Pepin (1984) and Swindle et al. (1986) for the SNC meteorite EETA 79001.

patterns are fully consistent with fractionation of one to the other. The presence in chondritic and terrestrial Xe of nucleogenetic and fission products (Pepin and Phinney 1978) at the heavy isotope end of the spectrum is responsible for the departure of these isotopes from the fractionation curve; it is likely that the excess of chondritic ^{86}Kr in Fig. 2 also is due to a nucleogenetic Kr component. Because of the normalization to the terrestrial composition, the relationship of Martian Xe to the other compositions is not obvious from Fig. 3. Roughly speaking, it resembles Xe on Earth, but the significant deviations of the two sets of heavy-isotope ratios rule out the possibility of identical compositions. It turns out that the isotopic distribution of Martian Xe can be accurately reproduced, at all isotopes except ^{129}Xe where monoisotopic contributions from decay of ^{129}I are dominant, by mass fractionation of the CI chondrite composition (Swindle et al. 1986). So for Xe, one could advance the following scenario: fractionation of CI-Xe into Mars Xe, and fractionation of progenitor Xe resembling the CI composition, but without the nucleogenetic heavy-isotope component, into fission-free Earth Xe. In the latter case, the present-day meteoritic and terrestrial reservoirs would have acquired their additional nucleogenetic and fissiogenic Xe after the fractionating separation of the two reservoirs.

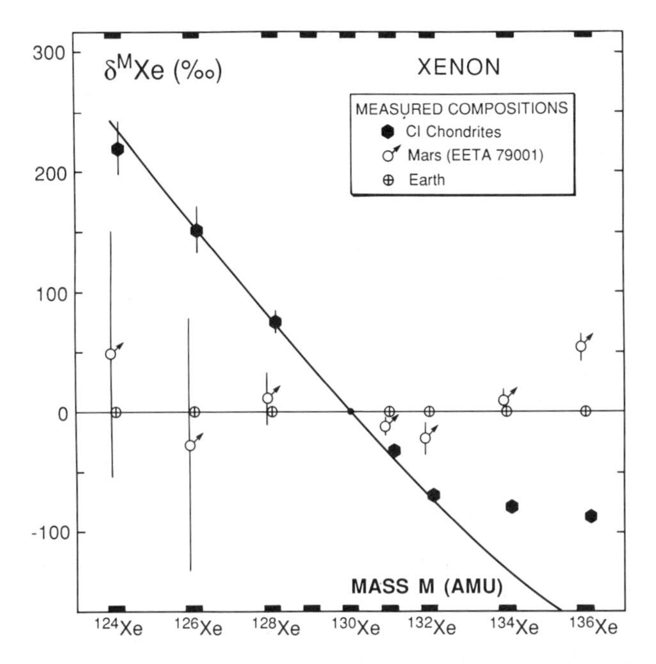

Fig. 3. Isotopic compositions of xenon in meteorites and planetary atmospheres, plotted relative to terrestrial atmospheric xenon. Units are permil deviation from the terrestrial composition. Data are from Basford et al. (1973) for Earth, Pepin and Phinney (1978) for the CI chondrites, and Swindle et al. (1986) for the SNC meteorite EETA 79001.

In terms of the "Xe Problem", it is clear that isotopic considerations destroy the presumption of a CI-Earth noble-gas relationship of the simple kind that led to the creation of the problem in the first place. They also point very firmly to the conclusion reached in Sec. III.B, that mechanisms capable of fractionating isotopes have been at work. For these heavy noble gases, Jeans escape probably played no role (Donahue 1986), but in the proper environment hydrodynamic escape could have been effective (Hunten et al. 1987*a,b*). The boundary conditions are still severe. Note that Figs. 2 and 3 imply a separation of Kr and Xe processing, in locale or time and perhaps also in type. While the terrestrial Xe composition can be generated by fractionation of a CI-like parent, the terrestrial Kr composition cannot. Derivation of Martian Xe directly from CI-Xe is appealing, since one might expect, on several grounds. a major carbonaceous chondrite component in materials accreted by Mars (Dreibus and Wänke 1985; see also their Chapter). But again, as for the CI-Earth pair, Martian Kr cannot be generated by fractionation of the CI composition. Realistic scenarios for generating meteoritic and planetary volatile distributions will have to accommodate this intriguing decoupling of the evolutionary histories of the two heaviest noble gases.

D. Carbon and Nitrogen

It was noted, in the introduction to Sec. III, that the relative depletions of these two elements in meteorites and planetary atmospheres are much less than for the noble gases. Even on Mars, where present-day atmospheric abundances of carbon and nitrogen probably constitute small to negligible fractions of their total contemporary or initial inventories, C/Kr and N/Kr ratios in the atmosphere are still higher than the corresponding solar ratios and exceed atmospheric Ar/Kr and Ne/Kr, which span roughly the same mass range as CO_2/Kr and N_2/Kr, by 2 to 5 orders of magnitude. Two general conclusions follow: contemporary C and N inventories could not have derived directly from a primordial source of solar composition, and they were probably not in the gas phase during most or all of the episodes of fractionating loss experienced by the noble gases. The first implies solid source materials, chemically fractionated from solar composition, and the second a sequestering of these elements in nonvolatile form in whatever planetary or preplanetary environments the major losses of noble gases occurred.

Elemental ratios and isotopic compositions are listed in Table II. The $^{14}N/^{12}C$ ratios could be identical on all three planets, with the CI value somewhat lower and the E chondrites spreading across this range. (For Mars, one would suspect that this identity is entirely fortuitous, since the atmospheric ratio, given the plausibility of losses to space and to the megaregolith, should not characterize the past or present planetary inventories. So coincidences do happen.) Carbon isotope ratios overlap where they have been measured; Martian $\delta^{13}C$ could be similar. Nitrogen composition, while nondiagnostic for Venus because of the 20% uncertainty, and almost certainly perturbed for Mars by heavy-isotope enrichment in nonthermal N loss (Nier and McElroy 1977; McElroy et al. 1977), is interesting for what it may say about derivation of terrestrial N—and by inference perhaps Venusian and Martian N as well—from meteoritic sources.

Consider a scenario in which carbon and nitrogen are supplied to Earth by accretion of a meteoritic veneer. In view of the discussion in Sec. III.B, one evident constraint is that the veneer material must have high enough contents of N and C relative to noble gases to supply the N-C inventories without contributing significantly to terrestrial noble-gas abundances. Enstatite chondrites can probably satisfy this requirement. The constraint is more stringent for the CI-CM classes, and they can moreover be ruled out on isotopic grounds. Though one may assume fractionating loss of N from the planetary atmosphere following veneer accretion, terrestrial $\delta^{15}N$ cannot be generated from the isotopically heavier CI (Table II) or CM (Kerridge 1985) compositions by such processes. The situation is clearly different for an E-chondrite source: as shown in Table II, $\delta^{15}N$ is isotopically lighter than terrestrial N.

The plausibility of an E-chondrite N and C source for Earth, and perhaps for Venus, depends on arguments for the existence of an E-type planetesimal population at the right time and place, for the operation of an appropriate atmospheric loss process, and for the timing of veneer accretion and outgassing to the atmosphere within the course of this process. The absence of significant carbon isotope fractionation between Earth and the E chondrites constrains the extent of CO_2 loss. Veneer accretion near the end of an initially intense episode of hydrodynamic atmospheric escape, when the heaviest mass that can be lifted out of the atmosphere by outflowing hydrogen (Hunten et al. 1987,1988*a*) has declined to the point where CO_2 loss is small but N_2 escape is not, seems to be one possibility. Noble gases are of course also lost, and so the escape process must generate their present-day atmospheric mass distributions as well.

For Mars, there is no way at present to apply isotopic constraints to candidate sources for N and C. This will require access to a Martian crustal reservoir whose N has not exchanged with ^{15}N-enriched atmospheric N, and a relatively precise measure of C composition. If geochemical modeling (Dreibus and Wänke 1985; see also their Chapter) and the signature of CI parent Xe (Sec. III.C) in Martian Xe are correctly signaling the presence of a major CI accretionary component in the planet, the initial C-N inventory would have been enormous. The problem, if there is one, might well lie in how to get rid of most of it.

TERRESTRIAL NOBLE GASES: CONSTRAINTS AND IMPLICATIONS ON ATMOSPHERIC EVOLUTION

M. OZIMA
and
G. IGARASHI
University of Tokyo

Isotopic data of terrestrial noble gas are reviewed and are compared with the two other principal noble gas components in the solar system, namely, the solar and planetary components. The characteristic features in terrestrial noble gases that are important in considering the evolution of the terrestrial atmosphere are that (1) primordial He and Ne in the Earth are solar-like; (2) while atmospheric Ne isotopic composition differs from that in the Earth's interior, Ar, Kr and Xe have almost identical compositions in the atmosphere and the Earth's interior; and (3) terrestrial (atmosphere and interior) and planetary noble gases are either identical or fractionated with enrichment of heavier isotopes relative to the solar noble gas component. We interpret these characteristics to indicate that (a) the primordial terrestrial noble gases are solar-type, and (b) while atmospheric Ne fractionation occurred after or during the formation of the terrestrial atmosphere, Xe fractionation predated the formation of the Earth. Examination of various models of terrestrial atmospheric evolution in the light of the above noble gas isotopic constraints suggests that an impact degassing model is least incompatible with these constraints.

Brown (1952) first noted that there is marked dissimilarity in relative abundance of volatiles, especially of the noble gases, between the Earth's atmosphere and that of the Sun. From this observation he concluded that the dissimilarity is due to a secondary origin of the terrestrial atmosphere, because a primary atmosphere consisting of a remnant of the solar nebula should have cosmic elemental abundance, which is represented by the solar atmosphere. Since then, noble gases have been a prime source of information on the origin and evolution of the terrestrial atmosphere.

There are two approaches in applying noble gas data to Earth science problems: the first one is to compare noble gas elemental abundances or their relative abundance pattern; the second is to examine the isotopic ratios. Since the successful attempt by Brown, the first approach has been yielding useful clues in understanding the origin of the planetary atmospheres (see, e.g., McElroy and Prather 1981; Wetherill 1981*b*; Pollack and Black 1982). However, elemental abundances, especially of diffusive elements like noble gases, are easily affected by fractionation and it is often difficult to ascertain whether the observed elemental abundances or their relative abundance pattern represent the initial state or merely reflect later alteration. In contrast, isotopic ratios are less susceptible to fractionation and it is generally safe to assume that they preserve a faithful record of the initial state. Hence, conclusions derived from isotopic ratios can be regarded as less ambiguous.

In the last few years, some interesting data of the isotopic compositions of terrestrial noble gases have been obtained. In Sec. I, we present the state of the art in terrestrial noble gases, especially the isotopic composition of the Earth's interior; the noble gas state of the paleoatmosphere is also briefly reviewed. In Sec. II, we discuss terrestrial noble gases as they relate to those in the solar system, and finally we examine various Earth-atmosphere evolution models in the light of noble gas constraints.

I. DATA BASE

A. Atmosphere

Air. It is very likely that the atmosphere is the major reservoir of terrestrial noble gases, accounting for most of the terrestrial noble gas inventory. This is inferred from comparison of noble gas contents in the atmosphere with those in various mantle-derived rocks (cf. Ozima and Podosek 1983, their Fig. 9.1). However, it should be noted that the mantle materials that have been analyzed so far are all from the upper mantle. Hence, one can argue that considerable amounts of the noble gases still remain in the lower mantle and even in the core. As will be discussed in Sec. I.B, very high $^{3}He/^{4}He$ ratios commonly observed in hot-spot volcanic rock seem to suggest the existence of primitive He in the deeper mantle; Xe may also be present in the deeper mantle (Ozima et al. 1985). Stevenson (1985*b*) has speculated that some amount of Xe could be incorporated in the core. However, since we have at present no way to estimate these amounts, we follow the conventional view in the following discussions, namely that the atmosphere approximates the total inventory of terrestrial noble gases.

Although the amount of noble gases in the atmosphere may approximate the total inventory in the Earth, isotopic compositions of some noble gases in the atmosphere and the Earth's interior such as Ne and He are distinctly differ-

TABLE I
Isotopic Compositional Data for Noble Gases in the Atmosphere

Isotope	Isotopic Ratios	Alternative Normalization	Percent Abundance (atomic)
Helium			
3	1.399 ± 13	1	0.000140
4	10^6	714800	~100
Neon			
20	100	9.800	90.50
21	0.296 ± 2	0.0290	0.268
22	10.20 ± 8	1	9.23
Argon			
36	0.3378 ± 6	1	0.3364
38	0.0635 ± 1	0.1880	0.0632
40	100	295.5	99.60
Krypton			
78	0.6087 ± 20	1.994	0.3469
80	3.9599 ± 20	12.973	2.2571
82	20.217 ± 4	66.23	11.523
83	20.136 ± 21	65.97	11.477
84	100	327.6	57.00
86	30.524 ± 25	100	17.398
Xenon			
124	0.3537 ± 11	2.337	0.0951
126	0.3300 ± 17	2.180	0.0887
128	7.136 ± 9	47.15	1.919
129	98.32 ± 12	649.6	26.44
130	15.136 ± 12	100	4.070
131	78.90 ± 11	521.3	21.22
132	100	660.7	26.89
134	38.79 ± 6	256.3	10.430
136	32.94 ± 4	217.6	8.857

ent. This difference is clearly a reflection of the particular evolutionary process of the Earth-atmosphere system. Hence, comparison of the noble gas isotopic compositions between the atmosphere and the Earth's interior would provide very useful information to explain the process (see, e.g., Mamyrin and Tolstikhin 1984; Ozima and Podosek 1983). Table I gives the isotopic compositions of noble gases in the atmosphere.

Paleoatmosphere. Evidence of time evolution for paleoatmospheric noble gas isotopic ratios such as $^{40}Ar/^{36}Ar$ or $^{129}Xe/^{130}Xe$ would impose a useful constraint on an Earth-atmosphere evolution model. So far, attempts have been made only for the case of $^{40}Ar/^{36}Ar$. Alexander (1975) first attempted to determine the $^{40}Ar/^{36}Ar$ ratio by analyzing several Precambrian cherts in the hope that ancient atmospheric Ar would have been incorporated during their deposition. The most difficult part in determining paleoatmospheric $^{40}Ar/^{36}Ar$ is in making correction for *in situ* decayed radiogenic ^{40}Ar. To solve this difficulty, Alexander used an ^{40}Ar-^{39}Ar isochron technique. Unfortunately, the attempt was not successful. Subsequently, Cadogan (1977) analyzed a 380 Myr old chert with the ^{40}Ar-^{39}Ar isochron technique and estimated $^{40}Ar/^{36}Ar = 291.0 \pm 1.0$ for tightly bound trapped Ar, that he concluded was the paleoatmospheric $^{40}Ar/^{36}Ar$ ratio. Jones (1982) analyzed several silicified plants as well as cherts, assuming that the former sample would yield a better record of paleoatmospheric Ar isotopic composition. However, the results were not very encouraging. Lately, Kelley et al. (1986) used an $^{36}Ar/^{40}Ar$ - $K/^{40}Ar$ - $Cl/^{40}Ar$ correlation diagram to infer the $^{40}Ar/^{36}Ar$ ratio in the trapped Ar in fluid inclusions in 268 Myr old authigenic mica, where the chlorine-correlated component would represent the indigenous trapped component from the contemporaneous sea water and the potassium-correlated component would correspond to *in situ* decayed radiogenic ^{40}Ar, thereby enabling them to discriminate each component. From this study they suggested a slightly smaller (a few percent) value of $^{40}Ar/^{36}Ar$ in the contemporaneous atmosphere. Conclusions so far obtained are still subject to considerable uncertainty owing to the difficulty in making accurate correction for radiogenic ^{40}Ar. It seems that reliable paleoatmospheric Ar isotopic ratios could be obtained only for samples that have negligible amount of potassium and hence require little correction for the *in situ* decayed radiogenic ^{40}Ar. Unfortunately, there is no such sample yet available.

B. The Earth's Interior

It is very likely that, except for radiogenic ^{40}Ar and ^{4}He, the noble gas amount in the crust is negligible in comparison with that in the mantle. Since we have no data for noble gas contents in the deeper mantle or in the core, we confine our discussion only to the upper mantle. However, this does not limit the general nature of our discussion of the evolution of the terrestrial atmosphere, since it is likely that the present atmosphere was derived from the upper mantle, or its constituting materials (see, e.g., Hamano and Ozima 1978).

He. Helium has two isotopes, ^{3}He and ^{4}He. While ^{4}He is produced from radioactive decay of U and Th, production of ^{3}He can be safely neglected in the total inventory of the Earth's interior (see, e.g, Mamyrin and Tolstikhin 1983). ^{3}He production becomes important only in very specific locales such as in Li-rich minerals (see, e.g., Mamyrin and Tolstikhin 1983) or in surface rocks at high altitude which have been exposed to cosmic ray radiation for a considerable time span ($>10^{5}$ yr) (Kurz 1986; Marti and Craig 1987). Consequently, we can safely assume that ^{3}He is essentially primordial and that the $^{3}He/^{4}He$ ratio in the Earth's interior has continuously decreased from the beginning of the Earth's formation to the present.

In Fig. 1, we summarize $^3He/^4He$ ratios in typical terrestrial materials. There is a large variation in the isotopic ratio, but we also notice several distinct aspects. First, the ratio in crustal materials is generally much smaller than it is in mantle-derived materials. This is easy to understand as U and Th are much more abundant in the crust than in the mantle. Second, the $^3He/^4He$ ratio in MORB (mid oceanic ridge basalt) falls in a rather narrow range from 1.1 to 1.2×10^{-5} (Kurz et al. 1982). Third, samples from hot-spot areas show the highest $^3He/^4He$ ratio among commonly occurring terrestrial materials (Kaneoka and Takaoka 1980; Kyser and Rison 1982; Allègre et al. 1987). Some diamonds have $^3He/^4He$ ratios, an order of magnitude higher than in any other terrestrial materials (Ozima and Zashu 1983*a;* Ozima et al. 1985; Honda et al. 1987).

The higher $^3He/^4He$ ratio in hot-spot materials is generally attributed to their deeper origin: hot-spot volcanic rocks are supposed to come from the deeper mantle, that is less differentiated and hence retains more ^3He (see, e.g., Hart et al. 1979; Kaneoka and Takaoka 1980; Staudacher and Allègre 1982; Allègre et al. 1987).

The extremely high $^3He/^4He$ ratios found in several diamonds lead to an interesting speculation as to the nature of the primordial He in the Earth (Ozima and Zashu 1983*a*). Since U and Th contents in diamonds are extremely low—so far there are several determinations giving U content <1 ppb (Kramers 1979)—the $^3He/^4He$ ratio is practically quenched in some diamond crystals. Therefore, diamonds with extremely low U and Th contents would have the $^3He/^4He$ ratio of the ambient mantle where the diamonds crystallized. Some diamonds are known to be extremely old, >3 Gyr (Kramers 1979; Richardson et al. 1984). Hence, because of their enormous chemical inertness and high temperature stability under condition of low O_2 fugacity and of low diffusivity of He (Honda et al. 1987), we may hope that

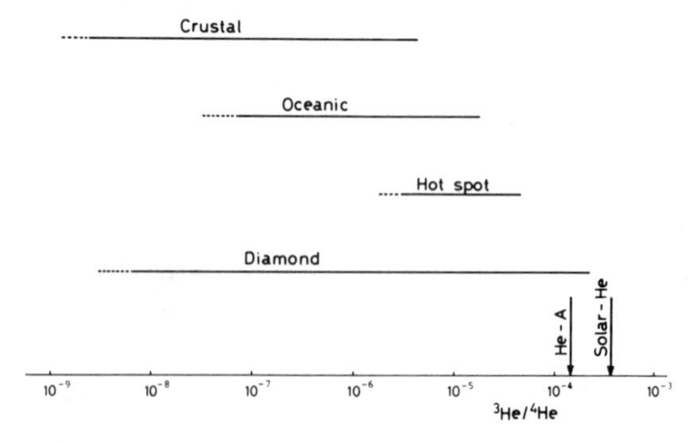

Fig. 1. $^3He/^4He$ ratios in various terrestrial materials. He-A (planetary He: $^3He/^4He = 1.4 \times 10^{-4}$) and solar He ($^3He/^4He = 4 \times 10^{-4}$) are also shown for comparison. Oceanic and hot spot denote oceanic volcanoes (except for hot-spot ones) and volcanoes from hot-spot regions. Diamond data are obtained from analyses of about 100 stones made by Ozima et al. (1985*a*, and unpublished data) and by Honda et al. (1987).

some diamonds would yield unique information of the He isotopic ratio in the ancient mantle, which cannot be inferred from other samples. Ozima et al. (1985*a*) found that several diamonds had a $^3He/^4He$ ratio of more than 10^{-4}, two of them even larger than the planetary He ratio (1.4×10^{-4}) up to 3.2×10^{-4}; a $^3He/^4He$ ratio higher than planetary He ratio is also found in an Australian diamond by Honda et al. (1987). Since the $^3He/^4He$ ratio decreases monotonically with time in the Earth's interior, the initial $^3He/^4He$ ratio at the time of the Earth's formation must have been even higher than the highest values observed in the diamond. Accordingly, Ozima and Zashu (1983*a*) concluded that primordial helium in the Earth cannot be planetary, but is likely to be solar ($^3He/^4He = 4 \times 10^{-4}$). The same conclusion was also suggested by Honda et al. (1987), who found solar-like neon in diamonds. We discuss this further below.

In summary, $^3He/^4He$ ratios in the Earth's interior vary by several orders of magnitude, which is attributed to addition of radiogenic ^4He from U and Th decays. From He isotopic compositions in some diamonds, we infer that primordial He in the Earth is very likely solar type, but not planetary.

Ne. Neon has three isotopes (^{20}Ne, ^{21}Ne and ^{22}Ne); it is customary to represent neon isotopic compositions in a three-isotope plot, i.e., a $^{20}Ne/^{22}Ne$-$^{21}Ne/^{22}Ne$ diagram (Fig. 2). In Fig. 2, we show three major neon

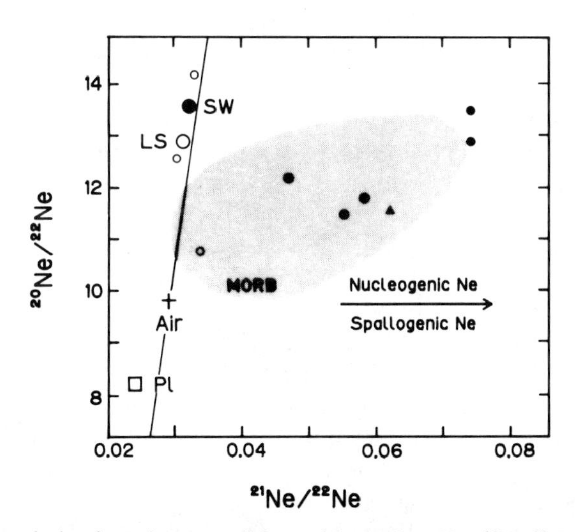

Fig. 2. Ne isotopic data for various terrestrial materials are plotted in a $^{20}Ne/^{22}Ne - ^{21}Ne/^{22}Ne$ diagram. The line passing through "Air" is a mass-fractionation line. Black circles: Zaire diamonds (Ozima and Zashu 1988); white circles: Australian and Arkansas diamonds (Honda et al. 1987); triangle: CO_2 well gas from New Mexico (Phinney et al. 1978); shaded area: MORB (Staudacher 1987; Poreda and di Brozolo 1984; Ozima and Zashu 1983*b*). Solar wind Ne (a large black circle denoted as SW), Ne in lunar breccia (a large white circle denoted as LS) and planetary Ne (a large white square) are shown for comparison.

isotopes commonly observed in the solar system, namely, solar wind, planetary and atmospheric Ne as well as the neon in various mantle-derived materials. We discuss genetic relations among these components in Sec. II.B. Unlike other noble gases such as Ar, Kr and Xe (except for nucleogenic isotopes), Ne in the Earth's interior seems to have a different isotopic composition from atmospheric Ne.

Poreda and di Brozolo (1984) observed $^{20}Ne/^{22}Ne$ ratios in several MORBs that were significantly higher (up to 9%) than that of atmospheric Ne. They ruled out that the difference is due to mass fractionation during magma generation and its transportation; they suggested instead a mixing of some unknown component of high $^{20}Ne/^{22}Ne$ ratio with atmospheric Ne. Staudacher (1987) suggested $^{20}Ne/^{22}Ne$ ratios ranging from 10 to 13 for MORB glasses. Phinney et al. (1978) observed a $^{20}Ne/^{22}Ne$ ratio of 11.6 ± 0.9 for CO_2 well gas from New Mexico. The latter sample also shows significantly higher $^{129}Xe/^{130}Xe$ and $^{131-136}Xe/^{130}Xe$ ratios than atmospheric Xe. Honda et al. (1987) first succeeded in analyzing Ne isotopic ratios in diamonds. They found four diamonds that show $^{20}Ne/^{22}Ne$ ratios significantly higher than atmospheric Ne, the highest being 14.2 ± 2.1 in a colorless diamond from Australia. Ozima and Zashu (1988) analyzed five diamonds from Zaire—very likely from a single geologic formation—for He and Ne isotopic ratios. All five stones show $^{20}Ne/^{22}Ne$ ratios higher than atmospheric Ne. The isotopic data for the five stones lie approximately on a line in a $^{20}Ne/^{36}Ar$–$^{22}Ne/^{36}Ar$ diagram, which they interpret as a mixing line between atmospheric Ne and "diamond Ne", the latter consisting of solar-like trapped Ne and *in situ* nucleogenic Ne.

Although nucleogenic ^{20}Ne and ^{22}Ne can be neglected in the Earth's interior, ^{21}Ne production may become significant (Kyser and Rison 1982). The nuclear reactions that are responsible for ^{21}Ne production are $^{18}O(\alpha,n)^{21}Ne$ and $^{24}Mg(n,\alpha)^{21}Ne$ (Wetherill 1954), where α and n are produced by decay of U and Th. Excess ^{21}Ne relative to atmospheric Ne is reported both in crust and mantle-derived materials, such as hydrothermal fluids (Hiyagon et al. 1986), volcanic gases (Craig and Lupton 1976; Poreda and di Brozolo 1984), mantle xenoliths (Kyser and Rison 1982) and some specific minerals (Tolstikhin 1978; Zadnik and Jeffery 1985; Ozima and Zashu 1987). They are unambiguously attributable to the above mentioned nucleogenic origin. These reactions would be more important in the crust, where U, and Th are highly concentrated.

In summary, there is ample evidence that Ne in the Earth's interior has different isotopic compositions ($^{20}Ne/^{22}Ne$, $^{21}Ne/^{22}Ne$) from atmospheric Ne. The excess in ^{21}Ne can be attributed to nucleogenic origin, i.e., $^{18}O(\alpha,n)^{21}Ne$ and $^{24}Mg(n,\alpha)^{21}Ne$, but the difference in $^{20}Ne/^{22}Ne$ ratio between air and the Earth's interior is more likely due to fractionation of the former during the evolution of the atmosphere. Ne in the Earth's interior appears to resemble solar Ne.

Ar and Kr. Except for radiogenic ^{40}Ar, that is a decay product of ^{40}K, Ar and Kr seem to have identical isotopic compositions in the atmosphere and the Earth's interior. In U-rich minerals, fissiogenic Kr may be recognizable, but they are barely important in the global inventory.

The Ar isotopic ratio $^{40}Ar/^{36}Ar$ in the mantle has been a prime source of information in constructing a mantle-degassing model. A number of investigations have been carried out to study the $^{40}Ar/^{36}Ar$ ratio in modern volcanic rocks, especially in submarine glasses which are least contaminated with atmospheric Ar, to yield a better representative of a pristine mantle Ar isotopic ratio (see, e.g., Fisher 1975; Hart et al. 1979; Ozima and Zashu 1983*b;* Sarda et al. 1985). These results show unequivocally a very high $^{40}Ar/^{36}Ar$ ratio of more than 10,000. While there is a general consensus among researchers that in the mantle source for MORB or in the upper depleted mantle, the present value of $^{40}Ar/^{36}Ar$ is higher than 10,000, some authors maintain that in the deeper mantle represented by hot-spot volcanic rock, $^{40}Ar/^{36}Ar$ is considerably lower (<1000) than that in the upper depleted mantle (Hart et al. 1979; Kaneoka and Takaoka 1980; Manuel and Sabu 1981; Allègre et al. 1983). However, Fisher (1985) argues that the low $^{40}Ar/^{36}Ar$ ratio is due to atmospheric Ar contamination. The issue is still open for future investigation.

Xe. Because of the very small amounts of xenon ($<10^{-11}$cm^{3}STP g^{-1}) in samples of mantle-derived materials, it is difficult to make precise isotopic measurements. However, apart from radiogenic (^{129}Xe) and fissiogenic ($^{131-136}Xe$) components, there seems no meaningful difference in the isotopic composition between atmospheric Xe and Xe in various mantle-derived materials. This is demonstrated in Fig. 3 where Xe isotopes are represented as a percentage deviation from atmospheric Xe.

Excess ^{129}Xe and ^{131}Xe-^{136}Xe relative to atmospheric Xe have been reported in various mantle-derived materials by several investigators, for example in MORBs (Staudacher and Allègre 1982; Allègre et al. 1987), in mantle xenoliths (Kaneoka and Takaoka 1978; Kyser and Rison 1982), in diamonds (Honda et al. 1987; Ozima and Zashu 1988), and in well gases (Butler et al. 1963). Excess ^{129}Xe relative to atmospheric Xe can be attributed to extinct nuclide ^{129}I, and $^{131-136}Xe$ excesses are due either to ^{238}U-spontaneous fission or to ^{244}Pu-spontaneous fission. Because of the short half-lives of 15.9 Myr of ^{129}I and of 82 Myr of ^{244}Pu, existence of those nucleogenic Xe isotopes must indicate very early separation of the atmosphere from their mantle sources, say, within a few 10 Myr after the formation of the Earth. Hence, their existence would impose crucial constraint on the atmospheric evolution, and establishment of the Xe isotopic anomalies bears prime importance in understanding the evolution of the atmosphere.

Although existence of excess ^{129}Xe in mantle-derived materials seem to be a rule rather than an exception, the extension and the provenance of the excess ^{129}Xe in the mantle are still not clear. Staudacher and Allègre (1982)

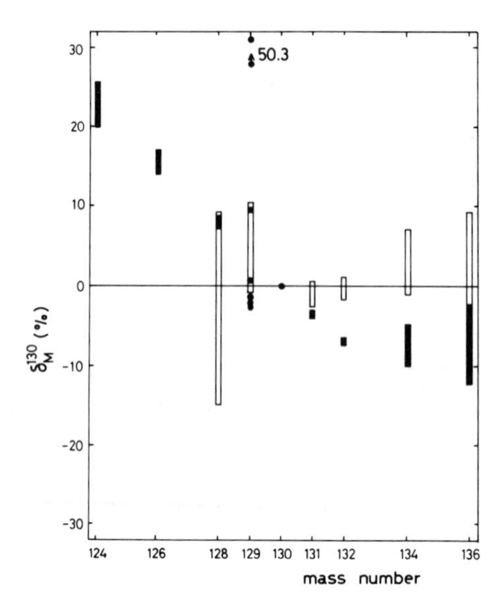

Fig. 3. Percentage deviation in Xe isotopic ratios (normalized to ^{130}Xe) from corresponding atmospheric Xe isotopic ratios are shown, where

$$\delta^{130}M \equiv [(^{i}M/^{130}M)_{\mathrm{sample}}/(^{i}M/^{130}M)_{\mathrm{air}} - 1] \times 100.$$

White columns represent various mantle-derived materials: volcanic rocks (Staudacher and Allègre 1982; Allègre et al. 1983; Ozima and Zashu 1983*b*); mantle xenoliths (Butler et al. 1963; Hennecke and Manuel 1975; Kaneoka et al. 1977; Kaneoka and Takaoka 1978; Saito et al. 1978); and diamonds (Honda et al. 1987; Ozima and Zashu 1988). Black columns and circles (^{129}Xe) represent chondrites (figure compiled by Pepin and Phinney 1978).

and Allègre et al. (1987) reported excess ^{129}Xe/^{130}Xe in MORB, some of which showed the highest isotopic ratios ever reported for terrestrial samples. However, such a large excess of ^{129}Xe in MORB has not been reported by other authors, though the absence of the isotopic anomaly may be due to atmospheric Xe contamination or to sample heterogeneity. Besides MORB, the largest excess ^{129}Xe was discovered in CO_2 well gas from New Mexico first by Butler et al. (1963) and later confirmed by several groups (Boulos and Manuel 1971; Phinney et al. 1978; Smith and Reynolds 1981). Excess ^{129}Xe reported in many mantle-xenoliths and diamonds may be real, though they are generally presented with large errors owing to the extremely small amount of Xe in these samples. Fissiogenic $^{131-136}$Xe are less well defined. Some samples show excesses in $^{131-136}$Xe isotopes relative to atmospheric Xe, but in most of these cases it is not possible to resolve unambiguously whether they are due to ^{238}U-fission or to ^{244}Pu-fission. Only in the case of the CO_2 well gas, excess $^{131-136}$Xe was identified to be of ^{238}U-fission origin without doubt (Phinney et al. 1978).

There has been no report which suggests the existence of ^{244}Pu-fissiogenic Xe components relative to atmospheric Xe. Excess $^{131-136}$Xe observed in some MORBs (Staudacher and Allègre 1982; Allègre et al. 1987) and in diamonds (Honda et al. 1987) are suggested to be of ^{238}U-fissiogenic origin. However, if we assume a value of ^{244}Pu/^{238}U = 0.015, estimated on meteorites (Podosek 1972), as the initial value and 32 ppb of U in the Earth, the fissiogenic ^{136}Xe from ^{244}Pu should be comparable with ^{136}Xe in the atmosphere. On the other hand, assuming 32 ppb of U in the whole Earth, the fissiogenic ^{136}Xe from U for 4.5 Gyr would be only about 1.7% of ^{136}Xe in the air (Bernatowicz and Podosek 1978). Recent estimates of ^{244}Pu/^{238}U in meteorites tend to show a significantly smaller value, ~0.005 (Marti et al. 1977; Hudson et al. 1988) than the previous estimated 0.015. Even taking the smaller estimate, ^{244}Pu-fissiogenic $^{131-136}$Xe would be larger or at least comparable with ^{238}U-fissiogenic components in the Earth as a whole.

From careful studies of meteorite and terrestrial data, Pepin and Phinney (1978) concluded that atmospheric Xe contains 6.7% of radiogenic ^{129}Xe decayed from ^{129}I and about 4.7% of fissiogenic ^{136}Xe from ^{244}Pu. With additional data on meteorites now available in the literature and with a refined analytical technique (multi-variate correlation analyses), Igarashi and Ozima (in preparation) confirmed the above conclusions, though the latter authors obtained a slightly smaller value of ^{244}Pu-derived fissiogenic ^{136}Xe (about 3.1% of atmospheric ^{136}Xe). These observational results, as well as the initial ^{244}Pu/^{238}U ratio inferred from meteorites, indicate that the Earth has acquired a significant amount of ^{244}Pu as well as ^{129}I. Hence, it is likely that the Earth contains non-negligible amount of both radiogenic ^{129}Xe and ^{244}Pu-derived fissiogenic Xe, some of which has been degassed into the atmosphere.

II. IMPLICATIONS AND CONSTRAINTS

A. Solar, Planetary and Terrestrial Noble Gases

In this section, we discuss characteristics of atmospheric noble gases with special reference to other noble gas components commonly occurring in the solar system. Here, the discussions are confined only to those relevant to terrestrial atmospheric evolution. For a more general account of noble gas cosmochemistry, especially those concerning extraterrestrial noble gas components, readers are referred to the Chapter by Pepin or to other literature (see, e.g., Ozima and Podosek 1983).

It is generally recognized that there are two widespread noble gas components in the solar system. The solar type is presumed to represent noble gases in the Sun and is observed in solar particle emissions such as the solar wind or solar flares (He and Ne), or defined as surface correlated components in lunar soils (Ar, Kr and Xe). The planetary type is widely observed in meteorites, typically in acid-resistant components called "Q" (Lewis et al. 1975) in chondrites, and has both elemental and isotopic compositions distinct from the former. Because of its widespread occurrence and fairly uniform elemental and isotopic compositions, this component is thought to come from some uniform and large-scale reservoir in the solar system, although its origin is still not clearly understood. While the existence of such major components is now hardly questioned, their characterization is still open to future investigation.

In the case of a solar component, it is now well known that, depending on the energy of the emitted particles, isotopic compositions of the noble gases vary considerably. For example, the $^{20}Ne/^{22}Ne$ ratio in the solar wind ($\sim$1 keV/n) is about 13.7 (Eberhardt et al. 1972), but ranges from 11 to 12 in solar flares ($E \sim$ 1 MeV/n: Black 1972; Etique et al. 1981; Nautiyal et al. 1981,1983; Rao et al. 1983). An even lower ratio is suggested for higher-energy particles (>10 MeV) from a satellite observation (Dietrich and Simpson 1979; Mewaldt 1984; Fan et al. 1984). Drastic change in an isotopic ratio is proposed in the case of He, where the $^{3}He/^{4}He$ ratio measured by a satellite ($E \sim$ 10 to 50 MeV/n) is about 10^{-2} in contrast to 4×10^{-4} in the solar wind ($\sim$1 keV/n: Dietrich and Simpson 1979). Hence, a question arises whether or not we could find a representative of the mean isotopic compositions of the noble gases in the Sun in any particular solar particle emissions. It has been argued that, because of its fairly constant elemental composition, especially of Fe/O, the solar wind represents the bulk chemical composition of the Sun (see, e.g., Meyer 1985*a*). However, on the basis of the observational fact that there is no fractionation in the isotopic ratios of C, N, O and Mg between the solar flare and terrestrial materials, Meyer (1985*a*) suggested that the solar flare is a better representative of the noble gas isotopic composition in the bulk Sun. Here, it should be noted that there are no observational data for these isotopic compositions in the solar wind: the solar wind may also have the same isotopic compositions as the Earth.

Because of such diversity in the isotopic ratios in solar particle emissions, it may appear to be hardly possible to assign any specific isotopic ratios or to choose a specific solar emission as the representative of the mean isotopic composition of noble gases in the Sun. In Fig. 4, we show $^{20}Ne/^{22}Ne$ ratios in various types of solar emissions with different energy ranges, together with the data for atmospheric Ne and planetary Ne. In spite of the considerable diversity in the isotopic ratios in different types of the solar emissions, it should be emphasized that $^{20}Ne/^{22}Ne$ ratios observed in solar emissions are distinctly larger than in atmospheric Ne or planetary Ne, although the higher energy particles (>10 MeV/n) have considerably lower $^{20}Ne/^{22}Ne$ ratios than atmospheric Ne, their contribution to the Ne isotopic ratio can be safely neglected because of the rapidly decreasing particle number with increasing energy. We therefore argue that solar Ne or Ne in the bulk Sun has a higher $^{20}Ne/^{22}Ne$ ratio than planetary and terrestrial Ne. As to ^{21}Ne, however, because of the addition of the nucleogenic component due to cosmic ray irradia-

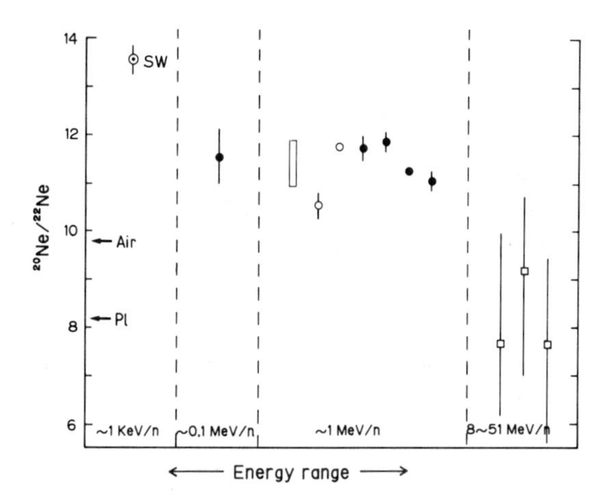

Fig. 4. $^{20}Ne/^{22}Ne$ ratios in solar particle emissions are plotted against their energy range. While the solar wind (SW) has a mean energy of a few keV/n, the solar flare is characterized by energy of more than 1 MeV/n. *Square:* satellite observation (Dietrich and Simpson 1979; Mewaldt 1984; Fan et al. 1984); *solid circle:* Ne implanted in lunar breccia (Etique et al. 1981; Nautiyal et al. 1981; Rao et al. 1983; Wieler et al. 1986); *hollow circle:* Ne in meteorites (Black 1972; Nautiyal et al. 1983); *rectangle:* Ne in cosmic dusts (Amari and Ozima 1988). "Air" and planetary Ne are indicated by arrows.

tion and its very small relative abundance, it is difficult to elucidate the original isotopic ratio; we therefore focus our discussion on the $^{20}Ne/^{22}Ne$ ratio.

For Ar and Kr, there seems to be no great difference in their isotopic ratios among terrestrial, planetary and solar components except for radiogenic ^{40}Ar. Black (1972) argued that $^{38}Ar/^{36}Ar$ was about 20% higher in solar flares than in the solar wind and in other components (planetary and terrestrial). However, Wieler et al. (1986) did not observe such a high ratio in solar flares, and argued for a nearly identical $^{38}Ar/^{36}Ar$ isotopic ratio for atmospheric Ar, planetary Ar and solar Ar. Planetary and solar Kr seem to be quite similar in the relative isotopic composition. Eberhardt et al. (1972) reported about −0.4%/amu fractionation (except for ^{86}Kr) in atmospheric Kr relative to solar or planetary Kr. However, Basford et al. (1973) observed two principal isotopic patterns in solar Kr, the first closely resembling the terrestrial Kr to within 10^{-3} at all isotopes and the second apparently fractionated with respect to the terrestrial isotopic composition by about −0.25%/amu, which is smaller than that observed by Eberhardt et al. (1972). It seems that the fractionation of atmospheric Kr relative to planetary or solar Kr is not well established and needs further confirmation. Since the degree of the fractionation, if real, is very small, we assume in the following discussion that Kr isotopic compositions are similar in atmospheric, planetary and solar components.

Both planetary and solar Xe, stripped from their fission-like heavy isotope components, are almost identical, thereby strongly suggesting that they represent a ubiquitous component in the solar system. The latter component was designated the primitive Xe by Takaoka (1972) and U-Xe by Pepin and Phinney (1978). Atmospheric Xe is precisely related to the U-Xe via mass-dependent fractionation. The fractionation is almost linear with mass number, and amounts to about 3.5%/amu (Fig. 5).

Below we summarize some major distinctions in isotopic compositions between the terrestrial atmospheric noble gases and the solar and planetary components:

(1) $^{20}Ne/^{22}Ne$ ratios are smallest (8.2 ± 0.4) for the planetary, intermediate (9.80 ± 0.08) for the atmosphere and highest (11 to 13.6) for solar wind and solar flares;

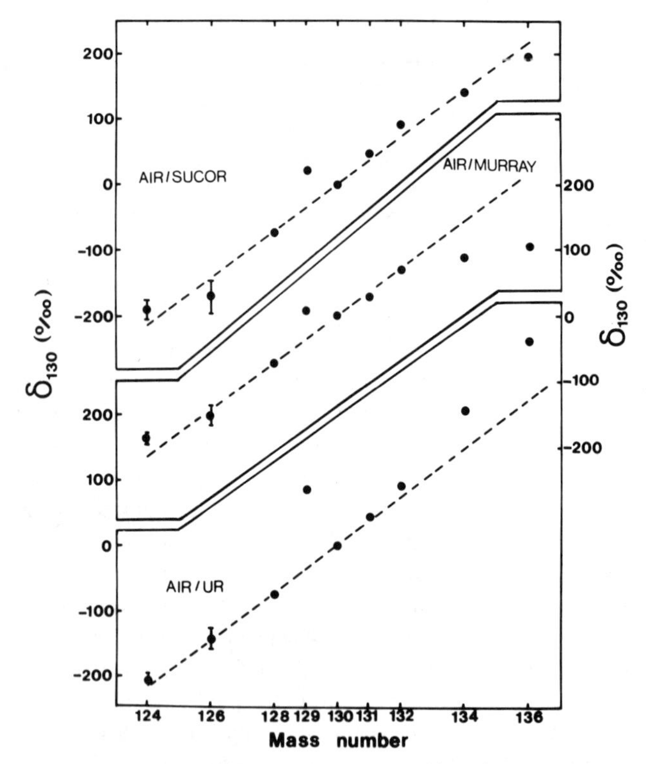

Fig. 5. A comparison of terrestrial Xe composition with three extraterrestrial compositions. The terrestrial data are plotted as per mil differences: $\delta_{130} = (Ra/R - 1) \times 1000$ where Ra and R correspond, respectively, to atmospheric Xe isotopic ratio and the extraterrestrial isotopic ratio (ratios are normalized to ^{130}Xe). SUCOR: surface correlated gas in lunar soil; Murray: carbonaceous chondrite; UR: U-Xe (Pepin and Phinney 1978). The dashed lines are straight lines drawn by eye through the data for isotopes ^{124}Xe, ^{126}Xe, ^{128}Xe and ^{130}Xe; they illustrate the approximate degree of fractionation of Xe relative to these extraterrestrial compositions (figure after Bernatowicz and Podosek 1978). The per mil scale at upper left is for the AIR-SUCOR composition, that at lower left is for AIR-UR, that on the right is for AIR-MURRAY.

(2) Ar and Kr are similar at all isotopes for atmosphere, solar and planetary components apart from nucleogenic isotopes;
(3) Planetary and solar Xe, apart from their fission-like isotopic components, have almost identical isotopic compositions. However, atmospheric Xe shows heavy and almost linearly mass-dependent fractionation relative to solar and planetary Xe.

Helium isotopic compositions are quite variable not only among terrestrial, planetary and solar ones, but also among the various terrestrial ones. Also, He escapes easily from the atmosphere, and the present atmospheric He is likely to be a transient rather than a permanent structure. Hence, we do not consider He isotopes in later discussions of atmospheric evolution.

B. Genetic Relation Among Noble Gas Components

In the preceding sections, we discussed the characteristic aspects of terrestrial noble gases with special emphasis on the distinction between atmospheric and mantle-derived components (I.B) and of atmospheric noble gases in reference to other major noble gas components in the solar system (II.A). Since the primitive solar nebula from which planets were formed is considered to be well homogenized, we may assume that terrestrial noble gases as well as the other two major noble gas components were derived originally from a uniform reservoir, that is, well-homogenized primordial noble gases in the primitive solar nebula. As the present Sun occupies more than 99.9% of the total mass in the solar system, it is reasonable to assume that, apart from later added nucleogenic components, the solar noble gas respresents the primordial component, and that both the terrestrial and planetary components were derived from this component. We discuss this point further below.

In Fig. 6, we show schematically isotopic compositions of all noble gases in the terrestrial (atmosphere and interior), planetary and solar components, where isotopic compositions are normalized to the solar component (represented by horizontal bars). A striking aspect is immediately apparent, that is, every fractionation relative to the solar component shows enrichment in the heavier isotopes.

This may be reasonably explained, if we assume that the solar component represents the primordial component and hence the others were derived from the former by some fractionation process(es). The fact that all the fractionations show enrichment in the heavier isotopes strongly suggests that the atmospheric and planetary noble gases are residual components in the fractionation process(es). This is easy to understand, since both the atmospheric and the planetary components are associated with solid (condensed) materials and condensation generally resulted in depletion of lighter elements and isotopes. However, the fact that the fractionation appears only in Xe in the case of terrestrial noble gases (both atmosphere and interior) requires an unusual fractionation process. We come back to this point later in this section.

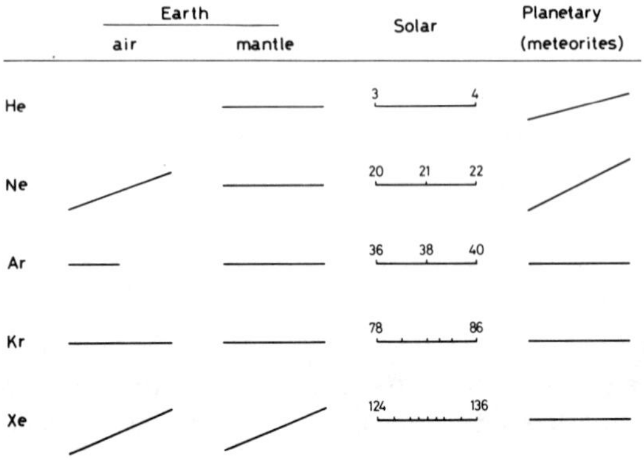

Fig. 6. Noble gas isotopic abundance pattern in the Earth (air and mantle) and in planetary compositions (carbonaceous chondrites) are shown relative to the solar abundance (horizontal line). Numerical figures shown for the solar abundance represent a mass number for an isotope. Note that noble gas abundance patterns in the Earth and meteorites (planetary) are either horizontal (no isotopic fractionation) or have a positive slope (isotopic fractionation with enrichment of heavier isotopes) relative to the solar abundance pattern. Ar in the Earth refers to the primordial component and ^{40}Ar is not shown because of its predominant radiogenic origin. The vertical axis is not to scale.

In contrast to Xe, fractionation appears to occur only in the atmospheric component in the case of Ne. We therefore argue that Ne fractionation occurred during or after the formation of the terrestrial atmosphere.

Considering all the observational facts, we propose the following scenario about the origin of terrestrial noble gases. The Earth originally acquired solar noble gases except for Xe. Prior to the Earth's accretion, Xe was fractionated during trapping by Earth-accreting planetesimals (Ozima and Nakazawa 1980). The atmosphere (including noble gases) was then derived from degassing of the Earth-accreting materials. During or after the formation of the terrestrial atmosphere, atmospheric Ne was fractionated. At present, we have no explanation for the mechanism of the Ne fractionation. This could be caused by hydrodynamic H_2 escape from the primitive atmosphere during a later stage in the Earth's accretion (see, e.g., Hunten et al. 1987; Sasaki and Nakazawa 1987; Chapter by Pepin).

As mentioned earlier, the fractionation process of Xe must be very specific: while the process resulted in large fractionation in Xe, the heaviest noble gas, the process left the lighter noble gases intact. In commonly encountered fractionation processes, the degree of fractionation is inversely related to a mass number, so that fractionation becomes larger for a small mass-number isotope. For example, for Rayleigh distillation and also for diffusion loss, a fractionation factor f [$f \equiv (N_2/N_1)/(N_2/N_1)_0$ where N_1 and N_2 are numbers of two isotopes and suffix 0 denotes the initial state] can be expressed as

$$f = (r)^{-\Delta M/2M} \tag{1}$$

where r denotes the residual fraction $N_1/(N_1)_0$, M a mass number and $\Delta M = M_2 - M_1$. Similarly for Jeans' escape, f can be expressed as

$$f = A \cdot e^{-\beta/M \cdot \ln r} \tag{2}$$

where A and β are constants (positive) independent of a mass M. In both cases we assumed $| \Delta M | / M << 1$. It can be seen from Eqs. (1) and (2) that, for the same value of r (<1) and ΔM, the fractionation factor f decreases with an increasing mass number M, i.e., less fractionation for a larger mass.

On the contrary, for a system which is gaining noble gases from an external system, the fractionation trend would be reversed. As such an example, Ozima and Nakazawa (1980) demonstrated that self-gravitational accomodation of noble gases into planetesimal structure results in more isotopic fractionation for larger mass atoms. The fractionation factor then is given:

$$f = a \exp (bM/RT) \tag{3}$$

where a and b are constants (positive) independent of a mass M, and R and T are a gas constant and temperature, respectively. As is obvious from Eq. (3), fractionation increases exponentially with increasing mass. Ozima and Nakazawa (1980) argued that gravitational capture of noble gases by a planetesimal can explain the most puzzling observational fact in terrestrial noble gases, namely, the larger fractionation occurs in the heaviest noble gases.

The above theory could explain the essential aspect of terrestrial Xe; however, there still remain some problems. First, although the gravitational capture theory can explain the large fractionation in Xe, it necessarily requires slight but non-negligible fractionation in the second heaviest noble gas Kr, and this is not observed. Second, Equation (3) results in a fractionation not only in isotopic ratio, but also in a much larger fractionation in elemental abundances. The elemental abundances and their relative pattern of noble gases implied from Eq. (3) give only a crude approximation of the atmospheric noble gases. Part of the difficulties may be resolved by assuming a specific size distribution of Earth-accreting planetesimals (Igarashi and Ozima 1988).

Another explanation of the large Xe isotopic fractionation was proposed lately by Hunten et al. (1987) and by Sasaki and Nakazawa (1988). Both papers appeal to hydrodynamic escape of H_2 from the primitive atmosphere, that would accompany fractionated escape of noble gases as well. The theory, however, fails to explain the observational facts that the Xe isotopic composition is identical both in the atmosphere and in the Earth's interior and that the fractionation is less for lighter noble gases.

C. Atmospheric Evolution Models

Since Brown's (1952) discovery that the terrestrial atmosphere is distinctly different from the solar in the relative abundance of volatile elements, it is now customary to assume that the present terrestrial atmosphere was degassed from solid terrestrial materials. The theory is in harmony with a currently accepted astrophysical scenario of planet formation, suggesting that planetesimals collided to form planets after the dissipation of the solar nebula, or that they formed in a vacuum (see, e.g., Safronov 1969). The present terrestrial atmosphere is therefore of a secondary origin. Consequently, the major issue in the origin of the atmosphere has been to resolve when and how the solid Earth was degassed. However, yet another astrophysical scenario has been proposed by the Kyoto group (see, e.g., Hayashi et al. 1985), who argue that planetesimals accreted to form the Earth in the presence of the solar nebula. As a necessary consequence of this scenario, we have to accept that the Earth was once surrounded by a solar nebula or that there was a primary atmosphere. Hayashi et al. (1977) concluded that the primary atmosphere which was attracted by the Earth's gravity amounted to about 50 atm. Eventually this thick primary atmosphere was dissipated by far-ultraviolet radiation during a T Tauri stage of the Sun (Sekiya et al. 1981). If the Earth did have such a dense primary atmosphere, even a small residual fraction becomes important in the atmospheric inventory.

In this section, we discuss various models of the origin of the atmosphere in the light of the noble gas constraints described in the preceeding sections. Here, we assume that noble gas evolution in the atmosphere paralleled the atmospheric evolution. This assumption is based on the fact that, except for radiogenic ^{40}Ar, noble gases were initially trapped together with other volatile components and large-scale liberation of the former from the solid Earth must also have accompanied the release of other major volatile components.

Mantle Degassing: The Case of $^{40}Ar/^{36}Ar$. Noting that the total output of volcanic gases and hot spring water over geologic time span is more than enough to account for the present mass of the atmosphere-hydrosphere, Rubey (1951) argued that the terrestrial atmosphere has been formed continuously throughout geologic history via fumaroles, hot springs and volcanoes. Fanale (1971) criticized the continuous degassing theory on the basis of many pieces of geologic evidences. It is also worth mentioning that Isua meta-sediments, the world's oldest rocks ($>$3.8 Gyr), suggest the existence of a considerable scale of oceans at the time of the formation of the original sediment, therefore favoring the early degassing ($>$3.8 Gyr) of the atmosphere.

Ozima and Kudo (1972) proposed that comparison of the Ar isotopic composition between the atmosphere and the Earth's interior would impose crucial constraints on mantle degassing. Since then, a number of mantle degassing models based on the ^{40}Ar/^{36}Ar ratio has been proposed by various authors (see, e.g., Schwartzman 1973; Fisher 1975; Ozima 1975; Bernatowicz and Podosek 1978; Hamano and Ozima 1978; Hart et al. 1979; Allègre et al. 1987). Their conclusions are more or less similar, suggesting early sudden degassing. As an example, we briefly review a mantle degassing model by Hamano and Ozima. They first assumed a two-stage atmospheric evolution model; that is, subsequent to early sudden degassing, the terrestrial atmosphere has evolved by continuous degassing from the mantle. Imposing constraints (1) $^{40}Ar/^{36}Ar > 5000$, and 80 ppm < K < 400 ppm in the present mantle, and (2) Ar degassing from the mantle to the atmosphere was at least as fast as K transport from the mantle to the crust, they showed that the initial sudden degassing had occurred before 3.8 Gyr and the degassing fraction was >75%. The remaining small fraction (<25%) has been degassed continuously to the present, most likely via volcanic action. This is illustrated in Fig. 7, where only the shaded area is permissible under the imposition of the above constraints (1) and (2). Consequently, terrestrial Ar isotopic data favor an early sudden degassing, not the continuous degassing proposed by Rubey. It is also interesting to note that the Ar isotope degassing model is compatible with extremely early degassing, that is, the degassing could be contemporaneous with the formation of the Earth. In the latter case, however, the fraction of the sudden degassing could become considerably less, say, about 75%.

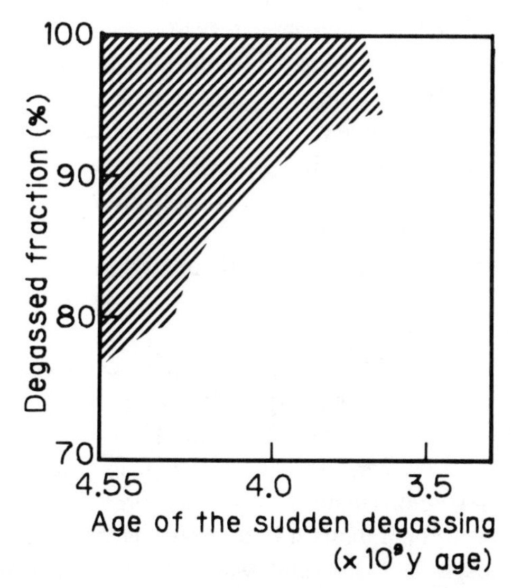

Fig. 7. Possible range for the time of sudden degassing (abscissa) and the degassed fraction (ordinate) in a mantle degassing model, where early sudden degassing and subsequent continuous degassing are assumed to have formed the Earth's atmosphere. With the constraints (1) $^{40}Ar/^{36}Ar > 5000$ and (2) 80 ppm < potassium < 400 ppm in the present mantle, only the shaded region is permissible in the degassing model (figure reconstructed from Hamano and Ozima (1978, Fig. 6).

Mantle Degassing: The Case of $^{129}Xe/^{130}Xe$. As discussed in the previous section, Ar isotopic data are consistent with an early sudden degassing case. The same conclusion has also been suggested from the observation of excess ^{129}Xe relative to atmospheric Xe in some mantle-derived materials by many authors (see, e.g., Boulos and Manuel 1971; Phinney et al. 1978; Thomsen 1981; Staudacher and Allègre 1982; Allègre et al. 1987), since the excess ^{129}Xe is most likely due to extinct ^{129}I and the existence of the former indicates that when the terrestrial atmosphere was degassed from the Earth's interior, there still remained ^{129}I; that is, the degassing occurred before the decay of ^{129}I. However, caution is in order in attributing the excess ^{129}Xe to a very early degassing as discussed below.

First, unlike Ar isotopic data where a large number of data obtained by many investigators show unequivocally the very high ^{40}Ar/^{36}Ar ratio in the upper mantle, Xe isotopic data are still very scanty and it would be premature to conclude whether or not the large excess ^{129}Xe reported in MORB by Staudacher and Allègre (1982) and Allègre et al. (1987) is indeed representative of the whole upper mantle. Hence, it is desirable to have more Xe isotopic data in MORB by other investigators. Second, even more difficulty in attributing the excess ^{129}Xe to the early mantle degassing is due to the fact that the excess ^{129}Xe in MORBs reported by Allègre et al. does not accompany excess ^{244}Pu-derived fissiogenic Xe components relative to atmospheric Xe. Because of the much longer half-life of ^{244}Pu than that of ^{129}I, the ratio of ^{244}Pu to ^{129}I remaining in the mantle after the mantle degassing must have been higher than at the time of the mantle degassing. Consequently, lack of ^{244}Pu-derived fissiogenic Xe in the MORBs studied by Allègre et al. indicates that the excess ^{129}Xe was not caused by mantle degassing, or its existence cannot be taken as an evidence of early mantle degassing. Instead Ozima et al. (1985*b*) argued that the lack of ^{244}Pu-derived fissiogenic components must indicate that the xenon in these materials was derived from a region(s) that has been isolated completely from the degassed mantle-atmosphere system since the formation of the Earth.

Impact Degassing. In the preceding section, we saw that the ^{40}Ar/^{36}Ar ratio in the Earth favors early catastrophic degassing, in which more than 75% of the volatiles were released into the atmosphere at an early stage in the Earth's history. Now, we consider what mechanism could have resulted in such effective degassing of volatiles from the Earth's interior. Solid diffusion is certainly too slow to be important in transporting volatiles from the Earth's interior to the atmosphere. If the Earth was once in a molten state and violent convective motion occurred, we may suppose such a very effective degassing as is imposed from the Ar isotopic data. However, it is not known whether a large part of the Earth had undergone violent convective motion. In this regard, impact degassing appears to offer a more realistic degassing mechanism.

An impact degassing origin of the terrestrial atmosphere was first proposed by Benlow and Meadows (1977) who suggested that shock heating of the Earth during the impacts of accreting dust particles resulted in degassing of volatiles. Lange and Ahrens (1982,1983) demonstrated experimentally that significant degassing of H_2O and CO_2 occurs when volatile-bearing minerals such as serpentine (H_2O) and calcite (CO_2) are subjected to impact shock (pressure) of up to a few 10 GPa. On the basis of these experiments, they suggested that the process of impact-induced devolatilization of volatile-bearing minerals during the Earth's accretion represents a major process in the evolution of a primitive atmosphere. Matsui and Abe (1986*a*) and Abe and Matsui (1987) examined the structure of the impact-degassed primary atmosphere consisting mainly of H_2O. They concluded that the impact-induced primary atmosphere can explain the essential characteristics of the present atmosphere-ocean system. Below we examine whether or not the impact degassing origin of the terrestrial atmosphere is compatible with the noble gas constraints.

As is obvious from the model, the impact degassing model imposes a severe constraint on the time of atmospheric evolution. Development of a primitive atmosphere must have occurred at the same time as the accretion of the Earth, or about 4.5 Gyr ago. Such extremely early atmospheric evolution is still compatible with the Ar isotopic constraints as described in Fig. 7. Neon isotopic constraint (Sec. II.A) requires that there was Ne isotopic fractionation between the impact-degassed primitive atmosphere and Earth's interior. However, the fractionation process should not have affected the heavier noble gases. As one such particular process, it may be possible that during the later stages of the Earth's accretion there could be noble gas escape from the atmosphere owing to the high temperature of the atmosphere. If there was substantial atmospheric escape from the Earth, we may also expect mass fractionation, which is more effective for lighter elements such as Ne. The escape process could be hydrodynamic escape (Hunten et al. 1987; Sasaki and Nakazawa 1988). It remains to be seen whether or not the required fractionation is indeed possible.

Other noble gas constraints (Sec. II.A) should be considered in the framework of the origin of the solar system, but not within the Earth-atmosphere system. This applies to all other atmospheric evolution models. Processes which are related to the noble gas constraints (1) and (2) (Sec. II.A) must have been completed before the accretion of the Earth.

Remnant of the Solar Nebula. Since Safronov's classical paper (1969) on the formation of planets, it has been customary to assume that the accretion of the terrestrial planets took place in a vacuum, or, more precisely, it occurred after the dissipation of the solar nebula. However, Hayashi et al. (1977) argued that in the absence of a gas drag force due to the solar nebula, accretion of planets would be too slow. For example, more than 10^{10} yr would be required for the outer planets—Uranus, Neptune and Pluto would not yet be born. Nakagawa et al. (1983) demonstrated that in the presence of the solar nebula, the gas drag force acted to increase the effective collision cross section of planetesimals due to gravitational attraction so that the accretion of planets proceeded much faster: for example, completion of the Earth's accretion took $<10^7$ yr. Since dissipation of the solar nebula is generally attributed to a T Tauri stage of the Sun, which is believed to have occurred about 10^7 yr after the formation of the Sun (Ezer and Cameron 1965), this scenario necessarily leads to a conclusion that the solar nebula still existed at the time of the Earth's formation. Consequently, the Earth attracted a thick primary atmosphere from the solar nebula by its own gravitational field.

Sasaki and Nakazawa (1988) took advantage of the thick primary atmosphere to explain the Xe isotopic fractionation. They argued that subsequent dissipation of the solar nebula during a T Tauri stage of the Sun would result in severe mass fractionation of elements including Xe. To produce the atmospheric Xe isotopic ratios, the primary atmosphere must have been dissipated to about 10^{-9} of the initial amount ($\sim$50 atm). This much removal of the primary atmosphere will result in almost complete dissipation of Ar and Kr, but relatively significant amount of Ne would remain in the primitive atmosphere together with Xe because of relatively large proportion of Ne in the solar nebula in comparison with that in the present atmosphere. In this process, Ne is also fractionated from the original isotopic ratio (assumed to be solar) close to the atmospheric one. In this scenario, all Ar and Kr in the atmosphere are assumed to be degassed from the Earth's interior.

Although the remnant nebula model would explain several important aspects of the terrestrial noble gases, the model contradicts a noble gas constraint, that is, Xe isotopic compositions of the Earth's atmosphere and interior are identical. Also, it is difficult to see how the scenario can explain the relative elemental abundance pattern observed in the present atmosphere. The above scenario assumes that the present atmosphere consists both of the remnant of the primary atmosphere (Ne and Xe) and of later degassed components (Ne, Ar, Kr and Xe). Hence, to produce the observed elemental abundance pattern, both the amount of the remnant and the degassed component must be specific, for which no reasonable mechanism was suggested.

III. SUMMARY

Terrestrial noble gas has some distinct features from the other major noble gas components in the solar system, i.e., solar and planetary components. While Ar and Kr are quite similar among the three components, terrestrial Ne (atmospheric Ne but not mantle Ne) and Xe are distinctly different from solar and planetary components. The distinction must be a reflection of a particular evolutionary process characteristic of the Earth. Below we summarize some of the characteristics which are directly related to the evolution of the terrestrial atmosphere.

(1) There is an almost ubiquitously observed extremely high $^{40}Ar/^{36}Ar$ ratio (>10,000) in MORB, in comparison with the atmospheric Ar isotopic ratio. This indicates that the depleted upper mantle, the source for MORB, must be characterized by an even higher $^{40}Ar/^{36}Ar$ ratio. Such a high $^{40}Ar/^{36}Ar$ ratio is consistent with a very early sudden degassing model.

(2) Atmospheric Ne has a very distinct isotopic ratio from Ne in the Earth's interior, the latter being quite similar to solar Ne, the putative primordial Ne. This suggests that the atmospheric Ne isotopic fractionation took place during or after the formation of the Earth's atmosphere. However, we do not know what the fractionation process was.

(3) Unlike other noble gases, both atmospheric Xe and Xe in the Earth's interior are quite similar (except for radiogenic ^{129}Xe and fissiogenic $^{131-136}Xe$), but they are distinctly different from solar and planetary Xe. This suggests that the distinction between terrestrial Xe, and solar and planetary Xe was established before the formation of the Earth. As a possible mechanism to produce the particular Xe isotopic composition in the Earth, we suggest gravitational trapping of noble gases from the environmental solar nebula by Earth-accreting planetesimals.

(4) Xe in some mantle-derived materials appears to be enriched in radiogenic ^{129}Xe and also in ^{238}U-derived fissiogenic $^{131-136}Xe$ compared to atmospheric Xe. However, these mantle-derived materials do not show ^{244}Pu-derived fissiogenic Xe, in contrast to atmospheric Xe, that is likely to have ~3 to 4% ^{244}Pu-derived Xe. Lack of the ^{244}Pu-derived Xe indicates that the source region from which the excess ^{129}Xe was derived has been isolated from the atmosphere-mantle (degassed) system for the whole history of the Earth. As a corollary, the excess ^{129}Xe observed in these mantle-derived materials cannot be taken as indicative of very early degassing as commonly assumed.

Acknowledgments. We thank D. Bogard and T. D. Swindle for helpful reviews of the manuscript. We are indebted to S. Amari for discussions on solar wind and solar flare noble gases.

FORMATION OF ATMOSPHERES DURING ACCRETION OF THE TERRESTRIAL PLANETS

T. J. AHRENS, J. D. O'KEEFE
California Institute of Technology

and

M. A. LANGE
Alfred Wegener Institute for Polar and Marine Research at Bremerhaven

Recent theories of accretion of planets assume they formed from a cloud of planetesimals which, in turn, formed from a protosolar accretionary disk according to Safronov, Wetherill and Kaula. Contemporaneous with the formation of the planets was the formation of proto-atmospheres during the accretionary period. Because the rare gas contents of the terrestrial planets are both similar to each other, and to meteorites, and are distinctly nonsolar, we assume the primitive meteorites are geochemical models of the planetesimals from which the planets accreted. We further assume that, upon impact devolatilization, the planetesimals produced proto-atmospheres largely of CO_2 *and* H_2O*. We describe the distribution of impact energy during accretion in the relevant impact velocity range from 5 to 45 km* s^{-1}*. We report the dependence of impact energy distribution in the presence of a regolith layer and the effect of silicate strength. Impact kinetic energy becomes divided into the following budgets: (1) kinetic energy of planetary material (10–30%); (2) kinetic energy of the projectile material (<5%); (3) internal energy of planetary material (60–90%); and (4) internal energy of a projectile material (5–20%). Laboratory experiments to measure the amount of shock devolatilization of serpentine (solid and porous), brucite, calcite and Murchison meteorite indicate that shock pressures between 8 and 60 GPa are required to drive* H_2O *and* CO_2 *from lattice sites in these materials. These results imply that impact devolatilization will occur on the*

Earth and Venus when these planets reach ~0.12 of the present value of their radius and that complete impact degassing of planetesimals occurs when these planets reach about 0.3 of their present radii. In contrast, impacts on Mars, because of its lower mass, do not begin to devolatilize infalling planetesimals until the planet is 0.3 of its final radius and, depending on the mineral, complete impact devolatilization may not occur even at the present size. Taking into account the CO_2, H_2O *and fine impact ejecta produced during accretion, Abe and Matsui have developed thermal models of accretion in which the proto-atmosphere acts as a thermal blanket and traps the internal energy budget due to impact at, or near the accreting planetary surface. Upon accreting the Earth or Venus to ~0.3 of their present radii, the surface temperature rapidly rises so as to induce dehydration regardless of shock pressure conditions, and eventually produces a magma ocean of a presently unspecified depth. Moreover, this magma ocean would persist over the remainder of the accretionary epoch and buffer the amount of* H_2O *in the atmosphere on account of the pressure effect of water on its solubility in liquid silicates. We quantitatively describe the impact of planetesimals onto a proto-atmosphere. This gives rise to what Cameron has named "atmospheric cratering." In principle, such cratering can eject and thereby erode co-accreting atmospheres. A wide range of models of planetary atmospheric accretion and erosion was studied in which the thermal structure, water content of planetesimals and mean planetesimal size were varied. For the Earth and Venus, depending upon the assumptions, a mass of water equivalent to several terrestrial oceans is ejected in the course of accreting these planets. In the case of Mars, calculations indicate that the process of atmospheric erosion is even more effective and this planet appears to have lost ~1 to* 10^2 *times the total* H_2O *inventory which accreted onto the planet.*

I. INTRODUCTION

During the last two decades, theories of the formation of the terrestrial planets and their atmospheres, which in the present discussion includes the Earth's ocean, have undergone a remarkable revolution. Prior to the understanding of impact processes and their role in accretion, it was assumed by many that the solid planets accreted by the process of infall of small solid objects as these condensed out of the solar nebula, as for example, in the inhomogeneous accretion models of Turekian and Clark (1969) and Grossman and Larimer (1974) or via a homogeneous process summarized in Ringwood (1979, Ch. 2). Extensive and in some cases, detailed calculations of thermal histories of the Earth and other planets were carried out in a series of papers in which it was assumed that most of the thermal energy of the planets which resulted in magmatism was a result of radiogenic heating and gravitational energy release during core formation (see, e.g., Lubimova 1958; McDonald 1959,1963; Hanks and Anderson 1969; Töksoz et al. 1978). Planetary outgassing, and hence, the formation of a planetary atmosphere was assumed to occur *after* the planet formed. This situation was unlike that assumed for the formation of the giant planets (see, e.g., Stevenson 1982*a*; Podolak and Cameron 1974) where the gravitational infall energy is usually assumed to provide a major component of the internal thermal energy within the planet. Until the ideas of Levin 1972*a*), Safronov (1969), Wetherill (1980) and Kaula (1979) were widely accepted, virtually all models of accretion utilized some form of equilibrium accretion equation in which the rate of growth with time t of a planetary radius r is given by

$$\frac{dr}{dt}\,\rho\,\frac{GM(r)}{r} = \epsilon\sigma\,(T_s^4 - T_p^4) \tag{1}$$

where ρ is the density of planetesimals, G is the gravitational constant, M is the mass of the planet, ϵ is the effective emissivity, σ is the Stefan-Boltzmann constant, T_s is the surface temperature of the planet and T_p is the temperature of planetesimals from which the planet is accreting. In Eq. (1) the left-hand side represents the rate of gravitational energy being delivered to the accreting planet, whereas the right-hand side equals the rate at which energy can be radiated to space as a gray body of emissivity ϵ. In these types of models, the accretion itself, does not heat the planetary object. Planetary heating solely depends on radioactivity and the release of gravitational energy upon core formation. The latter process represents the difference in gravitational energy, $\Omega_A - \Omega_B$, between a gravity differentiated planet such as the Earth which has an iron core in the center and a lighter outer silicate mantle (Fig. 1b), and a hypothetical undifferentiated object with iron blobs dispersed in a silicate matrix as indicated in Fig. 1a. Here

$$\Omega_B = 4\pi \int_0^R g r^3\, \rho \mathrm{d}r \tag{2}$$

where R is the planetary radius, g is gravity, and ρ local density at a radius r. A similar equation is written and evaluated for the configuration Ω_A of Fig. 1a. From our present knowledge of terrestrial structure, and taking into account knowledge of the compression properties of silicates and iron, the energy difference depending in detail on terrestrial models is in the range 1.7 to 2.3 J kg^{-1} of the Earth. This energy difference translates to a mean temperature rise of ~1500 K for the Earth. This would be similar for Venus, but consider-

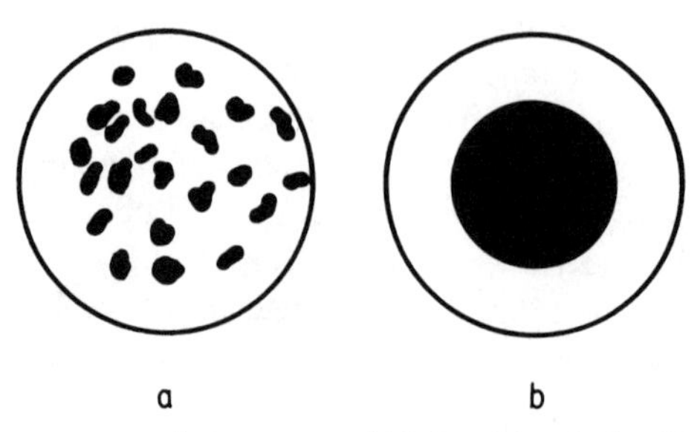

Fig. 1. Homogeneous accreted planetary model before (a) and after (b) core separation.

ably lower for Mars. The general picture that had emerged then in the 1970s (Lubimova 1958; McDonald 1959,1963; Hanks and Anderson 1969; Töksoz et al. 1978; Turekian and Clark 1969; Grossman 1972) was that the terrestrial planets accreted "gently" and heat generation subsequently occurred both from radioactivity and core formation which gave rise to early intense magmatism. This magmatic activity released the volatiles, including CH_4, CO_2 and H_2O which reacted, in part, in response to solar radiation. This, in turn induced loss of H via Jeans escape, and formed the present H_2O, CO_2 and N_2 atmospheres of Venus, Mars and Earth. This older concept of forming the planets first, and later, as a result of endogenic processes, forming the atmospheres and oceans was challenged in several early and thoughtful papers by Arrhenius et al. (1974), Benlow and Meadows (1977), and are included with several other concepts of accretion, in Ringwood's book (1979).

A key contribution to the acceptance of the impact accretion model was the work of Goldreich and Ward (1973) who quantitatively showed that protoplanetary dust condensed from a cloud around the proto-Sun and would coagulate into km-sized planetesimals which was an idea present in Safronov's (1969) work. Once this hypothesis was accepted, it followed that the scenario depicted in Fig. 2 could occur. Recently, very strong support for this type of scenario was forthcoming upon discovery via infrared and optical imagery of small-scale circumstellar gas and dust clouds around HL Tauri and R Monocerotis (Beckwith et al. 1986; Sargent and Beckwith 1987*a*), and

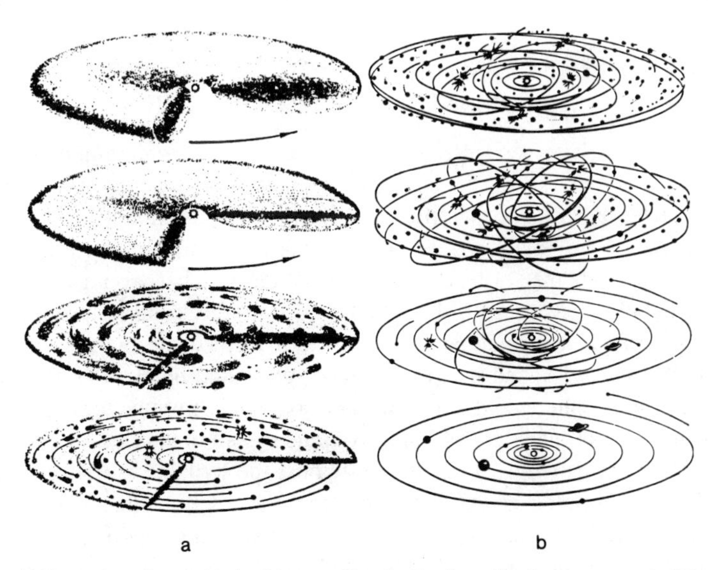

Fig. 2. (a) Formation of asteroid-sized intermediate bodies from the dust component of the solar nebula. (b) Gradual accretion of intermediate bodies into planets. The accretion of gas by giant planets is not shown. The initially flat system of intermediate bodies thickens due to their mutual gravitational perturbations (figure after Levin 1972*a*).

around the nearby star β-Pictoris by Smith and Terrile (1984). The presence of such clouds although very much in accord with the scenario of Fig. 2 have not yet been demonstrated to actually contain accreting planetesimals.

Lange and Ahrens (1982*a*) largely on the basis of shock-wave and thermochemical data, discussed the formation of a water-rich, impact-generated atmosphere on the Earth and other terrestrial planets. This work expanded on the early study of Jakosky and Ahrens (1979) who considered the burial of water within a planet upon reaction of impact-released water with surface silicates. In Sec. II, we consider briefly the basis of the chemistry of the various planetesimal models and discuss the chemical constraints that can be placed on the degree to which homogeneous accretion occurred on the terrestrial planets.

Our present understanding of the mechanisms which act to dissipate gas from a planetary atmosphere are briefly summarized in Sec. III. Our present knowledge of the partitioning of energy during planetary accretion processes is summarized in Sec. IV. Recent laboratory experiments, many of which were conducted by the present authors, define the regime of impact volatilization, and are summarized in Sec. V. We summarize in Sec. VI a description of the thermal effect resulting from impact accretion of a massive proto-atmosphere, largely developed by Abe and Matsui (1985,1986). Finally, in Sec. VII, we consider, for the first time, the combined effect of impact-induced atmospheric loss versus atmospheric gain upon impact of the planetesimals on the planets which already possess a protoplanetary atmosphere.

II. PROTOPLANETESIMAL MATERIALS; THE PROGENATORS OF PLANETARY ATMOSPHERES

It has long been recognized that there is a strong correlation between the abundance of elements in the solar photosphere with the elemental composition of meteorites as, for example, type I carbonaceous chondrites (Fig. 3). In Sec. VI we show how a model of terrestrial core, mantle and crust can be constructed from meteorite chemical abundances. The cooling of the protosolar nebula is sketched in Fig. 4. We assume this cooling resulted in the formation of planetesimals and these in turn formed the planets and meteorites. We are persuaded by the evidence, presented below, that the planets with their atmospheres did not form, approximately in their present orbit direct from the protonebular gas as for example shown in the classical Laplacian nebula hypothesis (Fig. 5). Evidence against the hypothesis that the planets and their atmospheres accreted directly from the solar nebula comes from the comparison of noble gas abundances of the Earth with solar composition (Fig. 6). Figure 7 indicates that the Earth is both very depleted in the light elements and represents a very different sample of the noble gases than the composition of the Sun. Pepin and Signer (1965) have pointed out that the inert gas in both the atmospheres of the planets and those evolved upon heating gas-bearing

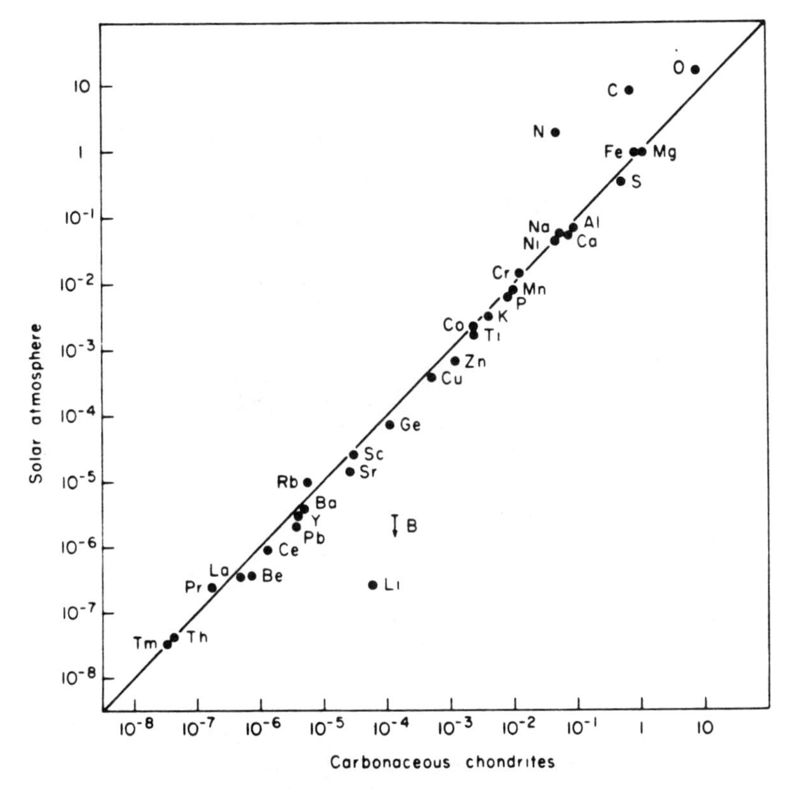

Fig. 3. Comparison of elemental abundances in type 1 carbonaceous chondrites with those in the solar photosphere. All abundances are normalized on the basis of Si = 10^6 (figure after Ringwood 1979).

meteorites are not similar to the Sun and are hence probably not the remnant gas of the solar nebula, but instead demonstrate a distinct "planetary" trend (Fig. 6).

Taking into account both the general agreement and elemental abundances among the Earth, meteorites and the Sun for the nongaseous elements (except those that require trapping in low-temperature ices or a fine glassy ground mass or on carbonaceous materials such as He, Ar, Kr, and Xe—demonstrated by Figs. 6 and 7), it is suggested that the planets and their atmospheres originated from planetesimals depleted in gaseous elements or accretion occurred inhomogeneously from at least two different types of planetesimals which condensed from the solar nebula. Possibly one type containing the primary nongaseous elements was solar-like in composition. The second, was depleted in light elements and gases. The terrestrial atmospheres are depleted in all elements which are likely to have been in proto-atmospheres if such proto-atmospheres are a direct remnant of the solar nebula that

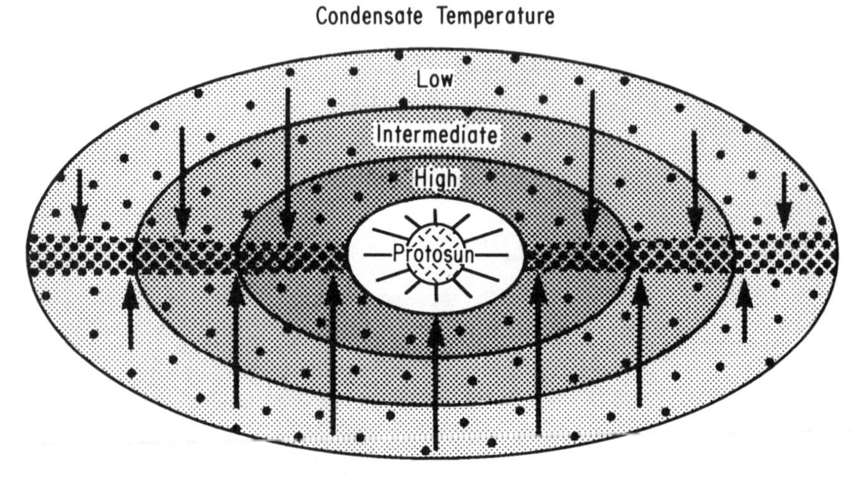

Fig. 4. Sketch showing notion of temperature distribution in a "cocoon nebula" surrounding the Sun, and prior to collapse into discoidal configuration. Small first-generation planetesimals form within the shell and spiral towards the ecliptic plane. Actual paths taken by sinking planetesimals are more complex than indicated and are controlled by a combination of gravity acting perpendicular to the ecliptic plane and gas-solid dissipation (figure after Ringwood 1979).

was present during accretion. What then is the evidence that the atmospheres are *not* residua of the protosolar nebula? They are two-fold:

First, as demonstrated in Figs. 6 and 7, the inert gas complement of the atmospheres of the terrestrial planets resembled the gas-rich components of the meteorites and not solar abundances. This correlation and the fact that the thermal and gravitational evolution of the planets and meteorites were distinctly different because of their different masses, indicate that the noble gases were taken up in similar objects when delivered to the planets. The solid phases containing noble gases presumably are still present in the asteroids and demonstratively occur in the small meteorites which fall on the Earth. These meteorites fall at sufficiently slow terminal velocity, that the noble gases are not released upon impact. They contain the complement of noble gases now present some 4.6 Gyr after they were emplaced prior to the planet-forming epoch (Fig. 2).

Second, the two key mass-selective, gas-loss mechanisms (Sec. III) for atmospheric escape from planets, Jeans loss and hydrodynamic escape, predict abundance patterns which would result from these processes starting from a solar noble gas pattern quite different from those observed in the atmospheres of Mars, Earth and Venus. The expected compositional patterns of gas loss which would occur if the solar nebula underwent such a gas-loss phase as described in T-Tauri type stars has not yet been studied (Cogley and Henderson-Sellers 1984).

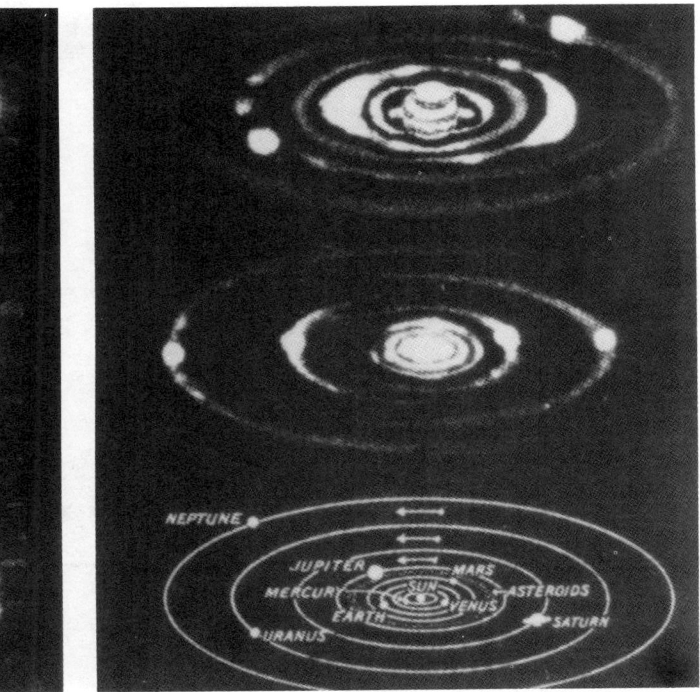

Fig. 5. A visual impression of the Laplacian nebula hypothesis. The young contracting and rotating proto-Sun sheds a concentric system of gaseous rings from which the planets later condensed (figure after Prentice 1978).

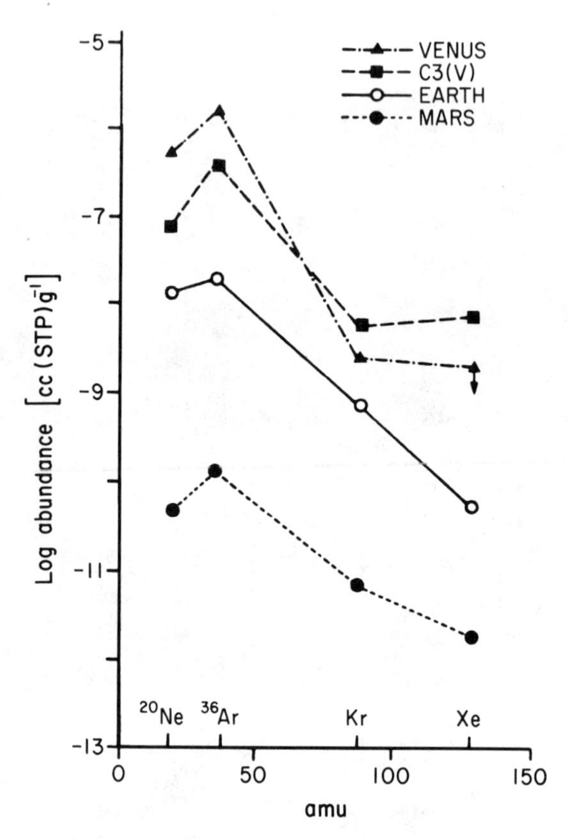

Fig. 6. Atmospheric abundances (per unit of whole-planet mass) of the noble gases for Venus, the Earth, Mars and a C3(V) chondrite meteorite (figure after Cogley and Henderson-Sellers 1984).

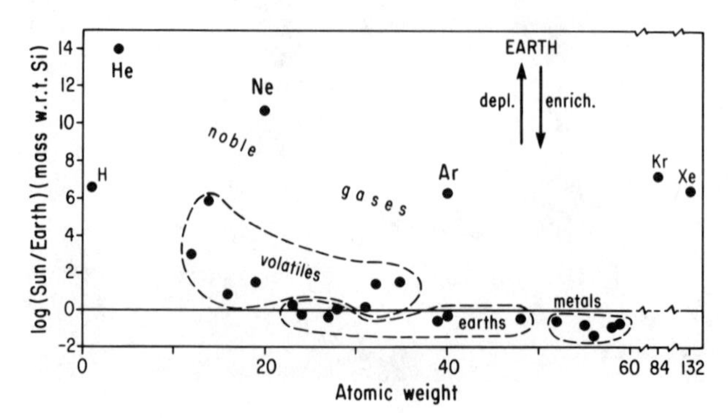

Fig. 7. Depletion factors for major elements in the Earth with respect to the Sun (based largely on data summarized by Walker [1977]).

Two broad classes of accretion models have been discussed in relation to accretion of planetary atmospheres; these are shown in Fig. 8 in cartoon form. Both models assume that a mixture of solar-system gas cools and solids condense from the gas via a sequence that has been worked out theoretically on the basis of thermodynamic properties of the gases and minerals as a function of temperatures for several hydrogen pressures (the dominant gas); this is summarized in Table I and in more detail in Fig. 9.

Heterogeneous accretion means that as solid grains condense from the protosolar cloud they immediately accrete onto the growing planets. Initially, a mineralogical profile like that shown in Fig. 10 (left) would result from a planet like the Earth with the refractory high temperature (HT) minerals in the center followed by Fe-Ni metal, Fe and silicates and topped off by the more volatile water-bearing and hydrocarbon-bearing minerals and the ices containing or reacting to form H_2O, CO_2 and NH_3. The latter materials are the raw materials required to produce (upon later photochemical reactions in the atmosphere, and reactions with the surface materials) the present planetary atmo-

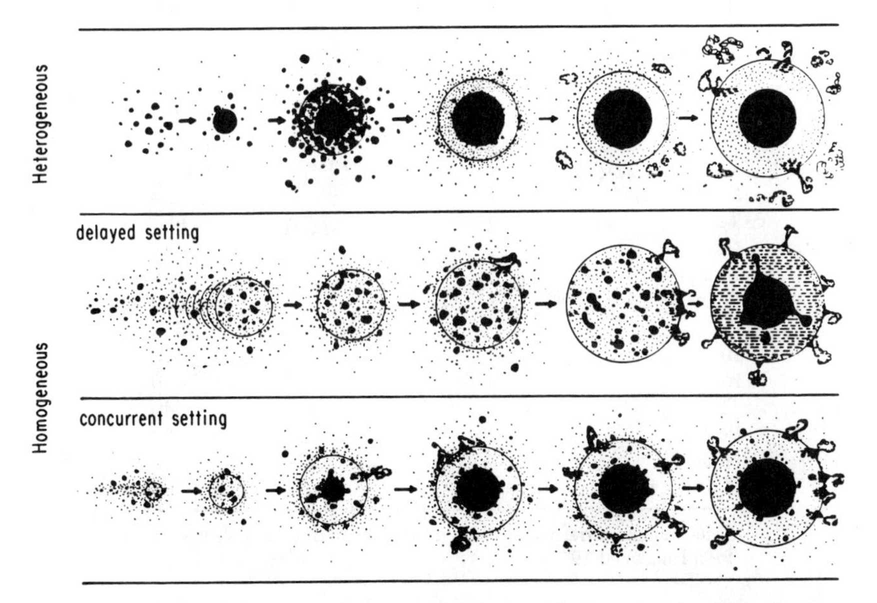

Fig. 8. Cartoons illustrating three possible models of terrestrial planet accretion. The adjectives "heterogeneous" and "homogeneous" refer to variations with time in the composition of accreting material. Most models of heterogeneous accretion argue that the upper mantle accreted as a volatile-rich late addition, giving rise to the crust and hydrosphere by differentiation; there may or may not have been a distinct episode in which the core separated from the lower mantle. Models of homogeneous accretion by delayed settling involve segregation and settling of the core long after the end of accretion; delayed settling is likely to be catastrophic and to melt much or all of the mass of the Earth. It is probable that the core settled out concurrently with accretion and that there were at least some deviations from homogeneity during accretion (figure after Cogley and Henderson-Sellers 1984).

TABLE I
Condensation Temperatures and Sequence of Phases and Elements Separating from Gas of Solar Composition at 10^{-4} atm Total Pressure[a]

Element or Compound	K
Ca, Al, Ti oxides and silicates; Platinum metals, W, Mo, Ta, Zr, REE	>1400
Mg_2SiO_4	~1360
Fe-Ni metal	~1360
Remaining SiO_2 (as $MgSiO_3$)	1200–1350
Cr_2O_3	—
P	1290
Au	1230
Li	1225
Mn_2SiO_4	1190
As	1135
Cu	1118
Ga	1075
$(K,Na)AlSi_3O_8$	~1000
Ag	952
Sb	910
F	855
Ge	812
Sn	720
Zn	660–760
Se	684
Te	680
Cd	680
S (as FeS)	648
FeO (10% ss. as $[Mg_{0.9}Fe_{0.1}]_2SiO_4$)	~600
Pb	520
Bi	470
In	460
Tl	440
Fe_3O_4	400
NiO	—
H_2O (as hydrated silicates)	~300

[a]Temperatures correspond to 50% condensation of a given element (table from Ringwood 1979*b*).

spheres. Late in the accretionary period, overturn of the interior to form the iron cores of the planets is predicted. Near-solar abundances of the elements, especially of the noble gases are predicted by this model for all the planets and as already indicated in Figs. 6 and 7, these patterns are not observed in detail. However, nearly chondritic abundances of siderophile elements are observed in the upper mantle of the Earth which agree with the predictions of this model.

Two variants of the older homogeneous accretion model are also shown in Fig. 8. One variant has the Earth accreting first and then forming a core.

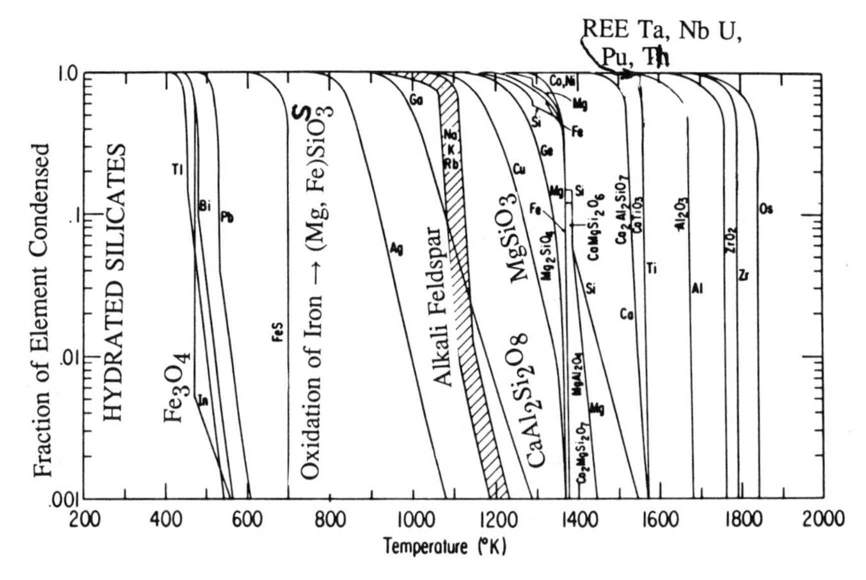

Fig. 9. Condensation of the elements from a gas of solar composition at 10^{-4} atm (figure from Ringwood 1979).

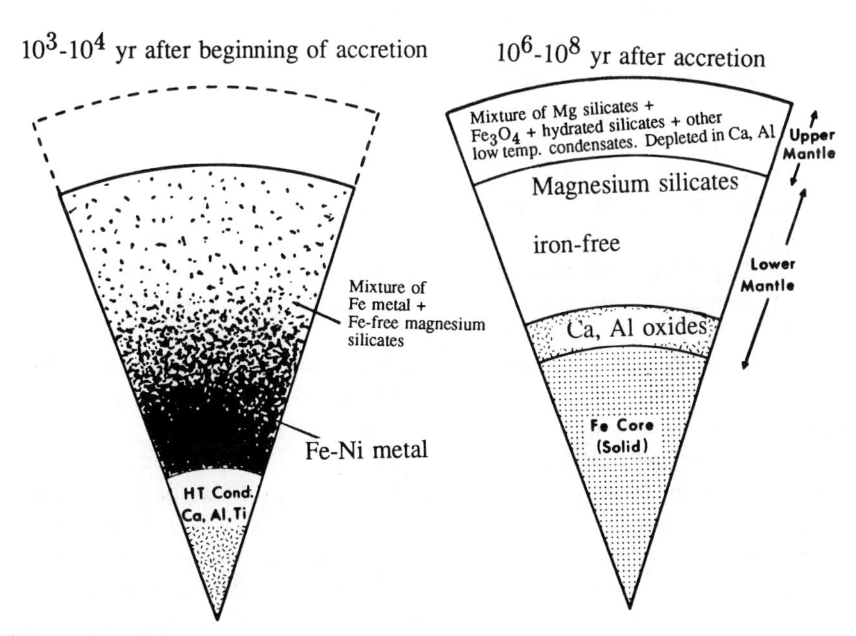

Fig. 10. Zonal structure of the Earth suggested by the heterogeneous accumulation hypotheses (figure after Ringwood 1979).

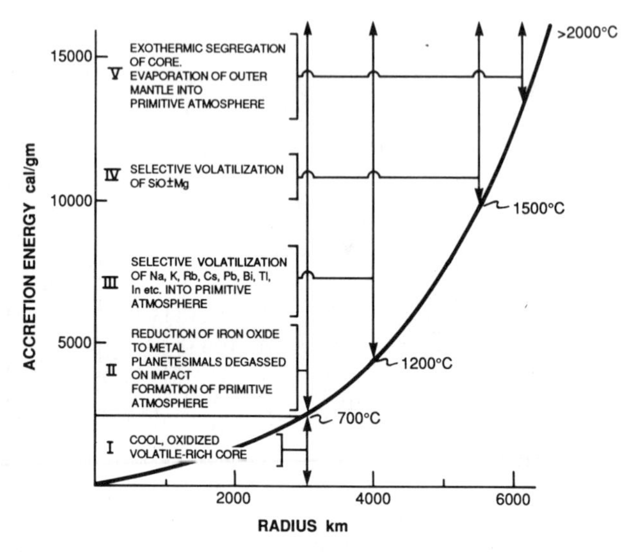

Fig. 11. Relationship between energy of accretion and radius of a growing Earth-sized terrestrial planet. The principal stages of accretion are also shown in relation to the energy of accretion and approximate surface temperatures (figure after Ringwood 1979).

This type of model is sketched in Fig. 11. In the lower model shown in Fig. 8, the core formation occurs simultaneously with accretion. In order to satisfy the two constraints (1) that the Earth's mantle appears to have a chondritic pattern of minor siderophile elements (e.g., Ni, Ir, Os, Au) which may not have been in equilibrium with a metallic core, and (2) that mantle xenoliths contain a small fraction, perhaps ~0.001 H_2O which appears to decrease with depth of origin, several workers have proposed a slightly inhomogeneous planetary accretion history. Moreover, the upper mantle contains a small chondritic fraction of the so-called incompatible elements such as La, Nd, K, Rb and U which generally partition strongly into the liquid upon partial melting. Also, the halogens in the Earth's mantle and in the oceans are found in their carbonaceous chondrite (C1) abundance ratios. Thus, Ringwood (1979), Wänke (1981), Anders (1968) and Smith (1982) propose models of planetary accretion in which the oxidation of materials increase as accretion of the planets occurs. For example, Wänke (1981) proposes that two-thirds of the Earth's mass accreted of phase A which is defined as a highly reduced material essentially free of Na, K, Rb, F, Zn, Cd, Ag, I, Br, Cl, Te, Se and C. He proposed that the remainder of the elements accreted in chondritic abundances with Fe, Cr, Mn and V as metals or sulfides, and Ga, Cu, W, Co, Ni and Si partially as metals. He then proposes that a second phase, phase B, provided a more oxidized veneer of volatile elements in chondritic abundances as well as Fe, Co, Ni and W in oxidized form. He suggests that the latter stages of accretion produced the atmosphere as does the precursor model of Anders (1968). As discussed below, this type of model appears to be required if one limits the concurrent infall upon accretion of minerals which contain both water and metallic iron. Very clearly, limits can be set in the case of the Earth, as to the amount of material which participated in the iron-water reaction

$$MgSiO_3 + Fe + H_2O \rightarrow \frac{1}{2} Mg_2SiO_4 + \frac{1}{2} Fe_2SiO_4 + H_2 \uparrow \quad (3)$$

and yet satisfy the observed budgets of water and Fe_2SiO_4 on the Earth and the other terrestrial planets (see, e.g., Ringwood 1979, p. 125). A nearly homogeneous model is specified in Table II. The higher-temperature components, phase A of Wänke (1981) which comprises some 85% of the Earth are given in column 3. The lower temperature components similar to Wänke's phase B are given in column 2. Stacey (1977) gives average values of the present Earth, which are shown in column 5.

Lange and Ahrens (1984) examined the question of the degree of the water-iron reaction and how this could have taken place on the Earth. Presumably, the water is released upon impact and reacts with both iron and silicate (represented by $MgSiO_3$) components of the accreting planetesimals. They considered in detail the water-iron reaction (Reaction 3) and infer that if this reaction were to continue during homogeneous accretion, since there is much more metallic iron than water, no free water would be present at the end of accretion. Moreover, assuming we know the FeO budget of the mantle from comparison of seismic and equation-of-state data, we can limit it to approximately 8% (mass) of the mantle or 3×10^{26} g. One could also assume that FeO is a light component in the Earth's core (45% by mass); this result is an upper bound of 1.2×10^{27} g of FeO in the Earth. Using these values to bound the degree of material processed via Reaction (3), Lange and Ahrens (1984) constrain the amount of homogeneous accretion seen by the Earth. As indicated in Table III, homogeneous accretion requires that planetesimals contain on the average 34.4 wt.% iron. By assuming that the Earth accreted slightly inhomogeneously with more metallic iron-rich planetesimals at the start of accretion, Lange and Ahrens found that initial planetesimals containing 36 wt.% Fe and 3% H_2O produce some 6×10^{26} g of FeO on the Earth during the first stage of planetary accretion. Such a model presumes that during the last 30% of accretion the iron-free material contains some 1.3×10^{25} g of water to form its present complement on the Earth. Greater inhomogeneous accretion (more Fe in the initial planetesimals) will decrease the FeO content of the Earth and this trades off with increased water budget as shown in Figs. 12 a,b and Table III.

III. THEORIES OF ATMOSPHERIC LOSS VIA NONIMPACT PROCESSES

The infall of the planetesimals, which themselves bring atmospheric constituents to planets, also causes the loss of some of the existing planetary atmosphere. This process has not previously been described quantitatively and

TABLE II

Compositions of Earth-forming Planetesimals[a] Compared with Bulk-Earth Composition[b] and Assumed Distribution of Planetesimal Types*

	Earth Models—Oxides and Elements			
	Ringwood (1979)			Stacey (1977)[b]
	Low-temperature comp. (15%) (wt.%)	High-temperature comp. (85%) (wt.%)	Mean[a]	
SiO_2	21.7	32.8	31.14	32
MgO	15.2	27.7	25.03	23
FeO	22.9	—	3.44	7.5
CaO	1.2	2.3	2.14	2
Al_2O_3	1.6	2.8	0.11	2
Na_2O	0.7	—	2.62	0.5
K_2O	0.07	—	0.01	
Cr_2O_3	0.35	0.2	0.22	
NiO	1.2	—	0.18	1 (K₂O through MnO, bracketed)
P_2O_5	0.3	—	0.05	
TiO_2	0.1	0.2	0.19	
MnO	0.2	0.1	0.12	
H_2O	19.2	—	2.88	—
Organic compounds	9.7	—	1.46	—

Metallic Fe	—	29.1	24.74	24
Metallic Ni	—	5.0	4.25	3
S	5.7	—	0.86	5

	Earth Models—Mineral Norms Normative Composition (wt.%)[c]		Present Model	
	Ringwood (mean)	Stacey	Planetesimal Type	Fraction (wt.%)
Plagioclase	7.6($Or_{0.01}Ab_{0.12}An_{0.87}$)	7.43($Or_{0.0}Ab_{0.57}An_{0.43}$)	Anorthite	7.7
Pyroxene	22.3($Wo_{0.07}Fs_{0.08}En_{0.85}$)	23.4($Wo_{0.12}Fs_{0.17}En_{0.71}$)	Enstatite[d]	22.5
Olivine	35.0($Fo_{0.91}Fa_{0.09}$)	36.2($Fo_{0.79}Fa_{0.21}$)	Forsterite[d]	35.4
Metallic nickel-iron	34.1	32.0	Iron[e]	34.4
Apatite	0.12			
Ilmenite	0.34	1.0		
Chromite	0.32			
Σ	99.8	100.0	Σ	100.0

[a]Composition of planetesimals according to model of Ringwood. Weighted mean composition (85% high-temperature condensates, 15% low-temperature condensates) used for norm calculation.

[b]Bulk-Earth composition (Stacey 1977).

[c]CIPW-norm on a water-free basis; organic compounds, metallic nickel-iron and sulfur also excluded from calculation.

[d]These phases assumed to contain hydrous phyllosilicates (by keeping fractions of planetesimal type constant), resulting in a total water content of 3 wt. %.

[e]Iron fraction to be varied from its homogeneous (present whole Earth) value (34.4 wt.%) to its extremal heterogeneous value (95%); up to when the total Earth's iron budget has been accreted. At that point the weight of iron in planetesimals is reduced to zero, and only silicate and hydrous phases are assumed to be accreted.

*Tables after Lange and Ahrens (1984).

TABLE III

Parameters for Model Calculations of Iron-Water Interactions during Accretion of the Earth[a]

Model No.	Water Content of Silicate Planetesimals (wt.%)	Initial Fraction of Iron Planetesimals (%)	Time for Termination of Iron Accretion (10^8 years)	Final FeO Budget (10^{26} g)	Final H_2O Budget (10^{25} g)	Remarks
1	3	34.4	1.6	7.16	0	homogeneous model
2	3	36	1.1	6.46	1.26	nominal model
3	3	50	0.7	5.80	4.92	heterogeneous model
4	3	95	0.4	2.43	11.57	heterogeneous model
5	6	34.4	1.6	14.30	0	homogeneous model
6	6	60	0.6	3.50	13.34	heterogeneous model
7	6	95	0.4	5.00	23.35	heterogeneous model

[a]See Fig. 12; table after Lange and Ahrens (1984).

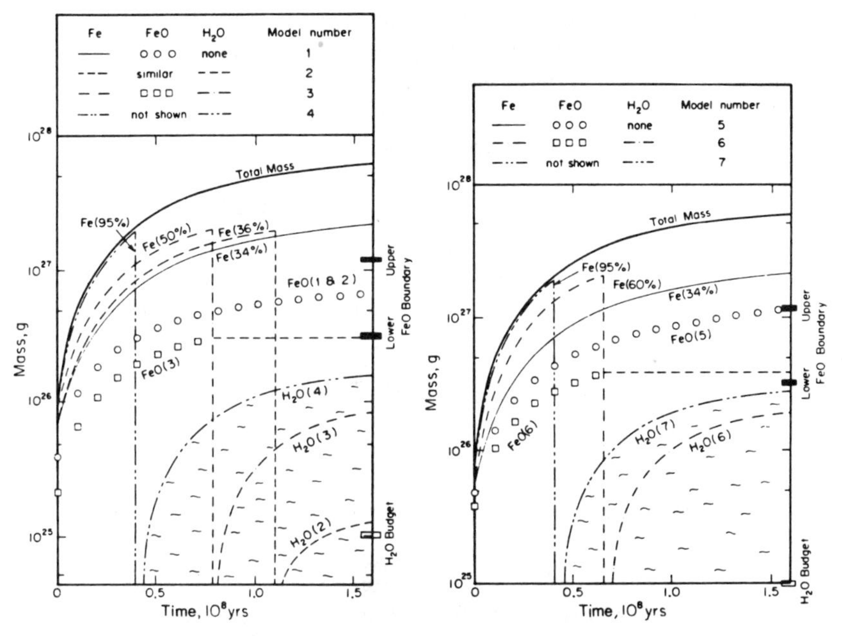

Fig. 12. (a) Accretion of the Earth (curve "Total Mass"), the growth of a core (curves labeled Fe), the formation of a terrestrial FeO budget (curves labeled FeO) and the growth of a terrestrial water reservoir (curves labeled H_2O) as a function of time (total accretion time, 1.6×10^8 yr). Numbers in parenthesis give the fraction of initially accreting iron. For heterogeneous accretion models (Fe,34%), iron accretion stops, once the mass of the core is reached (vertical lines) and from this point on, a water reservoir is built up. For these models, formation of FeO (due to interaction of iron with water) also ceases once the core is formed, and the FeO content remains constant from this point on (horizontal dashed line). The water content of infalling planetesimals for all of these models is 3 wt.%. Model numbers are described in Table III. (b) Results of model calculations for which the water content of planetesimals was increased to 6 wt.%.

is addressed in Sec. VII. The main purpose of this section is to summarize what has previously been recognized as the major mechanisms for atmospheric loss from planets. These two mechanisms are Jeans escape and hydrodynamic escape. Both are highly molecular specie, mass dependent. Moreover, in most cases, detailed description of photochemical reactions which occur at upper levels of the atmospheres are required to understand the roles of Jeans and hydrodynamic escape both during planetary accretion and later atmospheric evolution.

A. Jeans Escape

Jeans escape of species occurs at elevations of several scale heights in a planetary atmosphere as a result of the thermal energy ($kT \sim \frac{1}{2}mV^2$) overcoming the gravitational potential energy

$$\frac{1}{2}mV^2 > mMg/r \tag{4}$$

where m is the mass of the species, V is upwards velocity, M is the mass of the Earth, g is the gravitational acceleration and r is the distance from the center of the Earth. Net loss of the species occurs at altitudes such that the radius in Eq. (4) occurs at a height where the number of particles going upwards begins to exceed those going downwards. The mean expansion velocity of a species is

$$< v_c > = \frac{1}{2}\pi^{-\frac{1}{2}}\exp(-r_c/H_c)U(1 + r_c/H_c) \tag{5}$$

where r_c is the geometric distance of the critical radius for escape (Walker 1977; see also Table V below), n_c is the number density of particles and U is the most probable Maxwellian velocity:

$$U = \sqrt{2kT/m}. \tag{6}$$

Hcre

$$H_c = kT/mg(r) \tag{7}$$

is the characteristic scale height for escape at radius r_c. As is well known, the light molecules H and He escape from planets, whereas heavier molecules do not. This generalization may be quantified by calculating a characteristic escape time, τ_e in which $1/e$ of the gas of a single specie in the exosphere escapes. It is usually assumed that the gas which escapes the exosphere such as hydrogen becomes rapidly replenished from below. The characteristic escape time is given as

$$\tau_e = H_c/ < v_c >. \tag{8}$$

Present exobase conditions of the Earth are calculated and given in Table IV. Characteristic escape times for Jeans escape is very short on this time scale for

TABLE IV
Characteristic Time for the Reduction of Exospheric Density by Jeans Escape[a]

	H	H_2	He	O
Scale height, H (km)	1460	730	365	91
Most probable speed U (km s^{-1})	4.96	3.52	2.48	1.24
Mean expansion velocity (v_c) (km s^{-1})	7.26(−2)[b]	8.6(−4)	9.6(−32)	6.2(−32)
Exospheric escape time v_e (s)	2(4)	8.5(5)	3.8(9)	1.47(33)

[a]Table after Walker (1977).
[b]For example, 7.26(−2) ≡ 7.26 × 10^{-2}.

TABLE V
Characteristic Escape Times for the Planets[a]

	Moon[b]	Mercury[b]	Mars	Venus	Jupiter
T K	300	600	365	700	155
r_c km	1738	2439	3590	6255	69,500
$g(r_c)$ cm s^{-2}	162	376	332	827	2620
τ_e(H) s	3.55 (3)[c]	3.32 (3)	1.39 (4)	5.70 (5)	7.59 (617)
τ_c(He) s	2.03 (4)	1.40 (5)	2.66 (8)	2.87 (16)	9.59 (2455)
τ_e(O) s	2.25 (9)	7.32 (13)	1.02 (28)	7.61 (61)	1.20 (9818)
τ_e(A) s	3.35 (20)	2.64 (32)	2.09 (68)	3.24 (153)	3.04 (24532)
τ_e(Kr) s	3.97 (41)	1.10 (67)	4.34 (142)	1.07 (323)	9.99 (51509)

[a] Table after Walker (1977).
[b] For the Moon and Mercury the critical level is taken as the solid surface.
[c] For example, 3.55 (3) ≡ 3.55×10^3.

the species H, H_2 and He, and becomes very long for the heavier gases. These generalizations are also true for the other terrestrial planets as can be seen in Table V, but are not true for Jupiter where the relatively large gravitational acceleration offsets the somewhat higher temperatures of the surface gases. Jeans escape is given in Table V and in Fig. 13, in the units of the age of the Earth. Thus, we conclude that all the terrestrial planets retain their complete complement of all gases except H, He and H_2 against loss by the Jeans mechanism.

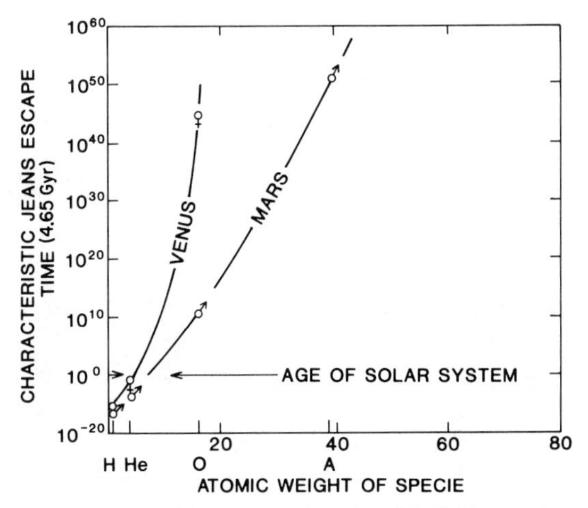

Fig. 13. Characteristic Jeans—loss escape time in units of 4.65 Gyr versus atomic weight for Venus and Mars.

B. Hydrodynamic Escape

Although theories of Jeans escape have been understood for some time (Jeans 1916; Chamberlain 1963; Walker 1977) recent progress has been made on theories of hydrodynamic escape (Hunten et al. 1987). At altitudes below the planetary exosphere where the mean free paths of molecules are only a fraction of the scale height, the outward flow of H, H_2 and He can sweep along the heavier gases essentially by the preferential upward motion of the light gases. The flux of light gas F_1 gives rise to the upward flux of the heavy gas F_2. Two simple models of hydrodynamic escape have been constructed. One is the Rayleigh case where at a given altitude a constant number N_1^0 of lighter molecules or atoms becomes constantly replenished via upward diffusion, and a second case in which the number of lighter molecules are initially fixed and decay with time. In the Rayleigh type model, the number density of the heavier species N_2 has an initial value of N_2^0 and decreases with time as

$$\ln\left(\frac{N_2}{N_2^0}\right) = -\left(\frac{m_c - m_2}{m_c - m_1}\right)\frac{F_1}{N_1^0}\,(t_2 - t_0) \tag{9}$$

where F_1 is the light element (usually hydrogen) escape flux, m_1 and m_2 are the light species (e.g., H) and heavy molecular mass, respectively. Here t_0 and t_2 is the initial time and time when the escape of the heavier components ceases. The species mass m_c is a critical mass and is the highest mass for which the hydrodynamic flow is capable of ejecting material (see, e.g., Hunten et al. 1987).

A formula similar to Eq. (9) may also be derived in the case that species (1) number density N_1 is not replenished from below in the atmosphere:

$$\ln\left(\frac{N_2}{N_2^0}\right) = -\left(\frac{m_c - m_2}{m_c - m_1}\right)\ln\left(\frac{N_1^0}{N_1}\right). \tag{10}$$

This expression is graphed in Fig. 14 and may be interpreted as follows. The lines labeled 2981, 55, etc. indicate the number of H atoms lost. Thus, the point A indicates that for 2981 H atoms lost, $1/10^{-2.5}$ or 316 species of having an amu of 30 are lost. Thus, 2981/316 or 9.43 H atoms are lost for each amu $= 30$ species.

IV. IMPACT PARTITIONING OF ENERGY

Two aspects of impact partitioning of energy are discussed in this section. First, we examine the question of how much of the incoming kinetic energy of a planetesimal can be deposited into the atmosphere of a planet during initial passage through the atmosphere and at what size a significant interaction with the solid planet occurs. Second, we discuss the impact of a

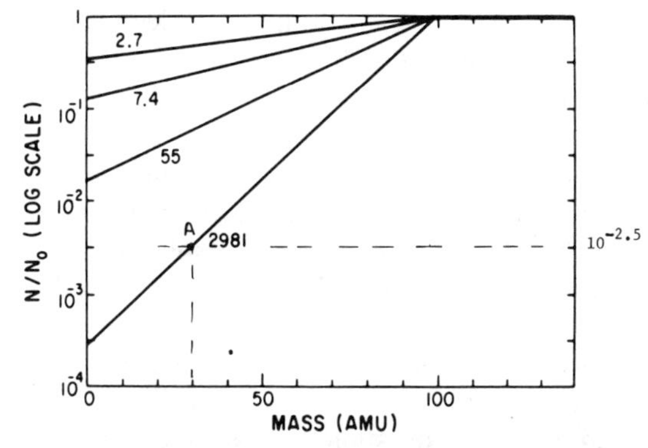

Fig. 14. The time evolution of the fractionation factor, which is the ratio of inventory remaining to initial inventory as a function of atomic mass. For purposes of illustration, a constant crossover mass m_c of 100 amu and constant hydrogen escape flux is assumed. The parameters on the curves are related to the time-integrated hydrogen loss, $\exp(F_1t/N_1)$ for the constant inventory, or N_1^0/N_1 for the Rayleigh case (figure after Hunten et al. 1987).

planetesimal onto an accreting planet for which the physical size of the impactor is large compared to the scale height of the atmosphere of the planet and/or the mean depth of the ocean. For the Earth these scale heights are respectively ~7 km and 4 km.

Both are important problems because in the Wetherill-Safronov type accretion scenario most of the planetary volatiles are thought to be delivered to the planet via impact of planetesimals. In any case, we are not dealing with a question of micrometeorites and as an approximation we will obtain an upper bound on the mass of an object that still delivers most of its energy to the atmosphere. This can be obtained by simply equating the mass of a column of air m_a to the mass of the projectile m_p or

$$m_a = (P_a)a/g = m_p. \tag{11}$$

Here P_a is the atmospheric pressure at the surface, a is the unit area, and g is gravity. This exercise yields $m_a \simeq 1$ kg. If the projectile has a density of 1 g cm^{-3}, a diameter of 0.12 m is obtained. Here we are ignoring meteorite breakup which will make this dimension even smaller. From both finite difference calculations and supersonic flow theory, O'Keefe and Ahrens (1982*a*) calculated that for a 72 km s^{-1} bolide impacting the atmosphere, energy-transfer considerations yielded a 170 m diameter object which would be just stopped by and hence deliver its energy to the atmosphere (before impacting the ground); again breakup was ignored.

In the case of an ocean impact of a 2.5 g cm^{-3} silicate object, the mass balance argument (Eq. 11) yields a projectile diameter of 0.7 m for a 4 km deep ocean. In contrast, from energy partitioning calculations for an impact at 25 km s^{-1}, O'Keefe and Ahrens (1982*a*) calculated that 15 bolide diameters is required just to stop a 2.93 g cm^{-3} bolide, which for a 4 km deep ocean translates to a 267 m diameter object.

Thus, we conclude that *upon initial passage* through the Earth's present-day atmosphere, objects with densities in the 1 to 3 g cm^{-3} range and smaller than 10^{-1} and 10^2 m in diameter, partition all or most of their energy into the atmosphere; objects in the same density range, with diameters <1 to 10^2 m deposit all or most of their energy into the ocean upon initial passage. O'Keefe and Ahrens (1982*a*) showed that these results are, surprisingly, quite insensitive to impact velocity.

We now need to ask the question, how does this size of planetesimals (small impactors), which can deposit energy in the atmosphere, compare with the planetesimals which accreted to form the terrestrial planets in the Safronov-Wetherill scenario? That is, how much of the terrestrial planet mass is delivered in what size impactors? Are the impactors contributing to most of the accreting mass much smaller or much larger than the above range?

According to the Safronov-Wetherill scenario, the number of impactors N of a given mass m can be estimated according to a statistical distribution of the form

$$\frac{\partial N(m)}{\partial m} = Cm^{-q} \tag{12}$$

where q (derived from the mass distribution of the asteroids) is believed to be in the range of 1.6 to 1.8; for examples, see Kaula (1979). The maximum size object m_1 to have accreted onto a planet (e.g., the Earth) is estimated to be in the range of 10^{24} to 10^{25} g or $\sim 10^{-3}$ to 10^{-4} the mass of the Earth. Evaluation of constants in various distributions and their implications as to the growth time of the Earth are summarized by Greenberg (1979). As C is calculated to be on the order of -10^{19} to 10^{20} $(g)^{1.8}$, it is found that the magnitude of masses of planetesimals which have come to accrete the Earth and Earth-sized planets falls off slowly. For example, using the above values which are known only very approximately, the 100th planetesimal to accrete onto the Earth still has a mass of $\sim$50% of m_1. These planetesimals have diameters of at least 10^2 greater than either the scale height of the atmosphere of any of the terrestrial planets, or the mean ocean depth, of the ocean on the Earth. *Thus the most significant transfer of energy of planetesimals forming a planet occurs upon impact on the surface*. Therefore, we will initially neglect the passage of a planetesimal through the atmosphere and ocean.

A typical scenario for the interaction of a large (100-km diameter) silicate planetesimal with a planetary surface is shown in Fig. 15. Details of the calculational method and summary of pertinent equations of state are given in O'Keefe and Ahrens (1977*a,b*) and Ahrens and O'Keefe (1977). Here we note

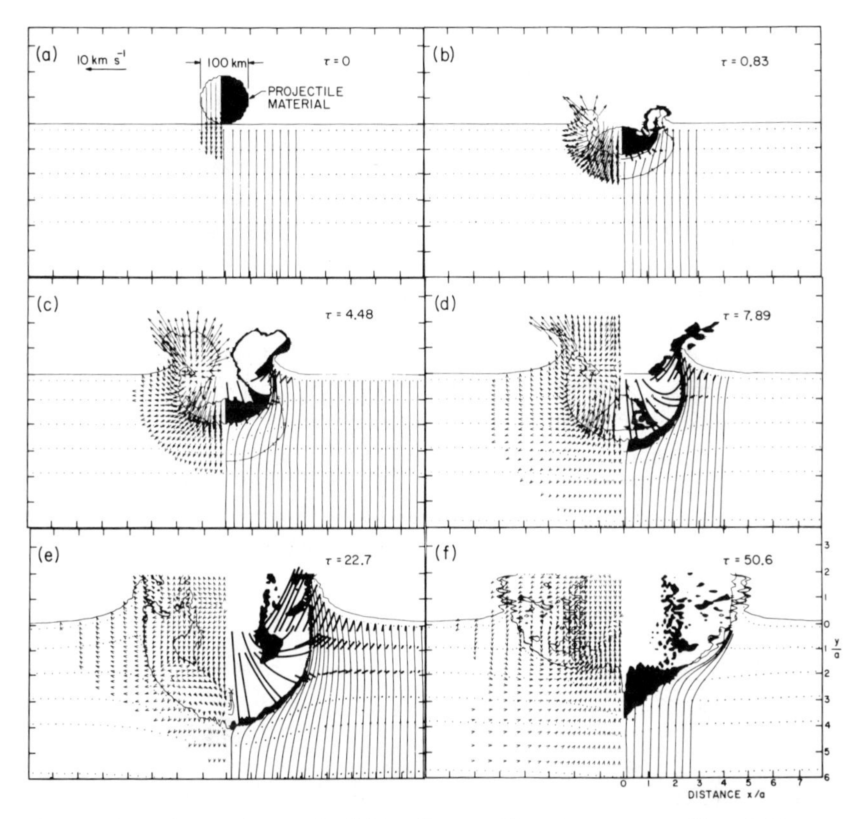

Fig. 15. Flow field upon impact of a 100-km diameter silicate impactor onto silicate half space with the Earth's gravity at 12 km s^{-1}. Panels (a), (b), (c), (d), (e) and (f) show flow at nondimensional times τ = 0, 0.83, 4.48, 7.89, 22.7 and 50.6. Particle velocity arrows in left panels indicate particle velocity motion scaled to 10 km s^{-1} vector of (a). Heavy curve in right-hand panels of (b), (c), (d) and (e) indicate trajectories of particle motion in transient cavity. Light vertical lines in right-hand panel, and horizontal aligned dots in both right- and left-hand panels are fixed to material in the target and demonstrate the total displacement. Note the gravity-driven rebound motion of the floor of a transient cavity between (e) and (f) resulting in a flat-floored crater and a reduction in displacement of vertical lines and horizontal dots.

that the characteristic, nondimensional time used in the calculation τ is

$$\tau = \frac{tv}{d} \tag{13}$$

where t is real time, v is impact velocity and d is projectile diameter. We note that the maximum projectile penetration for Earth gravity for craters whose growth are controlled by gravity, is 1.6 projectile diameters into the surface. Impactors having diameters $\gtrsim$10 km will give rise to gravity-controlled cratering on the Earth. This penetration occurs as approximately depicted in panel

(e) of Fig. 15. At later times, gravity drives the floor of the crater upward. This process eventually produces flat-floored craters.

The energy budget versus τ for 5 and 30 km s^{-1} impacts of silicate projectiles (2.9 g cm^{-3}) onto a 30% porous (2.0 g cm^{-3}) regolith half space is shown in Figs. 16a and b. The discontinuities in the energy versus non-dimensional time budgets are due to mismatches of energy budgets which occur when we rezone the numerical problem. The magnitude of these mismatches which occur at rezonings can range up to $\pm 5\%$. We believe that this is the numerical accuracy of these results. The peak in the internal energy of the projectile budget (at $\tau \sim 2$) is observed when a maximum amount of the projectile is shock compressed. This always occurs before the peak in the kinetic energy of the target budget which takes place at $\tau = \sim 3$. The latter always is seen when a maximal mass of target material has been set into motion by the resulting shock wave in the planet. After the shock in the projectile has decayed, the major transfer of energy occurs via work that the kinetic energy imparted to the target does on the target itself, producing an ever increasing internal energy budget. One convenient method of extrapolat-

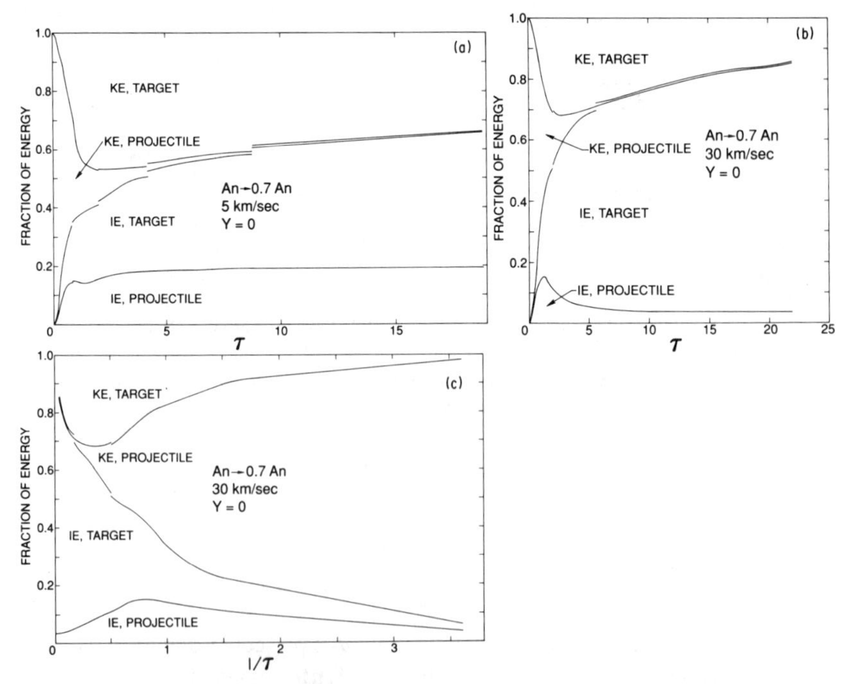

Fig. 16. (a) Fraction of impact energy versus τ for 2.9 g cm^{-3} (An) silicate projectile impacting porous, 2.0 g cm^{-3} (0.7An) regolith surface at 5 km s^{-1}. Zero strength (Y) is assumed. (b) Same as (a), except impact velocity is 30 km s^{-1}. (c) Same as (b), except $1/\tau$, rather than τ, is plotted.

ing to long times is via plot of $1/\tau$ (Fig. 16c) such as that shown for the same calculation as depicted in Fig. 16b. By extrapolating to $1/\tau = 0$, the late time values of energy partitioning such as given in the summary diagram (Fig. 17) may be obtained. As can be seen, the internal energy partitioned into the target varies from 0.7 to 0.85, of the total energy, as the projectile velocity increases from 5 to 30 km s^{-1}.

For larger projectiles (e.g., 100-m to 1000-km diameter), the impactor diameter becomes large compared to the planetary regolith thickness and the effective strength of the target Y is essentially negligible. Figs. 18a, b and c show the effect of impact energy partitioning for such interactions at 7.5, 15, and 45 km s^{-1}.

Whereas for regolith-free surfaces and small (<100-m diameter) projectiles, considerable target strength is appropriate to consider and the partitioning of energy versus τ is shown in Figs. 18d, e and f for impact at 7.5, 30 and 45 km s^{-1} against a target which has a von Mises strength of 3 kbars. A summary diagram for the above cases is shown in Fig. 19. As before, with a regolith, the final internal energy budget of the projectile decreases with impact velocity. The final kinetic energy partitioning between the target and internal energy of the target is a strong function of a target strength. This effect is discussed in more detail in O'Keefe and Ahrens (1977*a*). The final kinetic energy of the projectile remains less than a few percent and is not plotted on the summary diagram (Fig. 19). However, as the impact velocity increases to $\geq$ 45 km s^{-1}, much of the projectile is vaporized and the projectile does, in fact, retain a substantial (~5%) portion of the initial energy budget in the form of kinetic energy of expanding vaporized projectile.

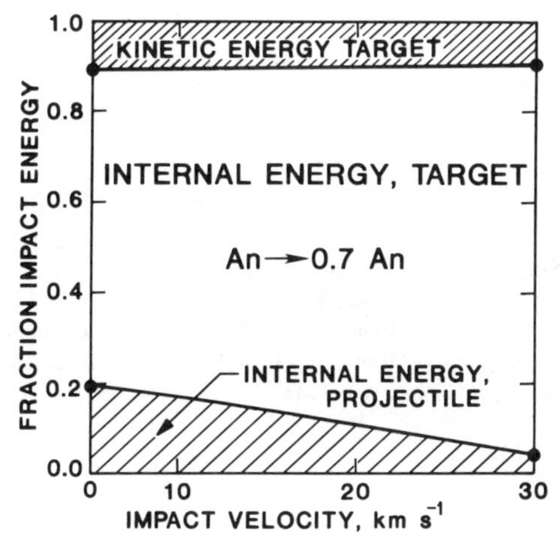

Fig. 17. Final energy budgets for 2.9 g cm^{-3} (An) silicate projectile imparting porous 2.0 g cm^{-3} (0.7An) regolith surface at speeds from 5 to 30 km s^{-1}.

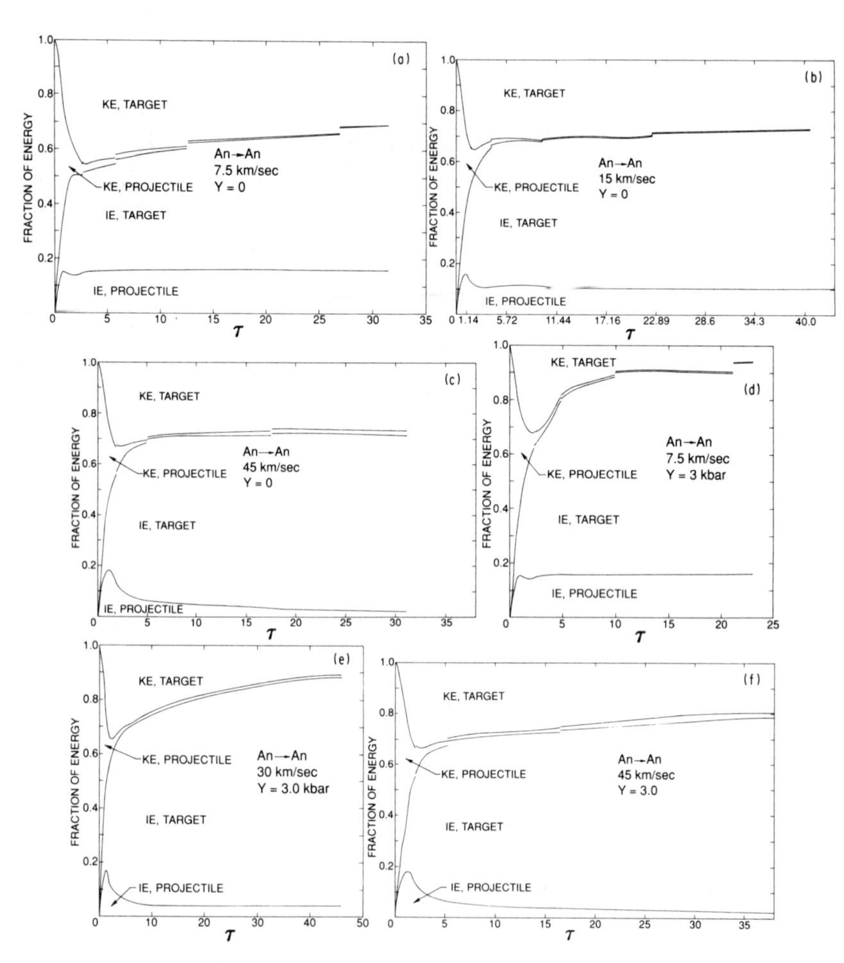

Fig. 18. (a) Fraction of impact energy versus τ for 2.9 g cm^{-3} (An) silicate projectile impacting solid 2.9 g cm^{-3} (An) silicate half space at 7.5 km s^{-1}. Zero strength Y is assumed. (b) Same as (a), except impact velocity is 15 km s^{-1}. (c) Same as (a), except impact velocity is 45 km s^{-1}. (d) Same as (a), except von Mises strength model ($Y = 3$ kbar) is assumed for silicate. (e) Same as (d), except impact velocity is 30 km s^{-1}. (f) Same as (e), except impact velocity is 45 km s^{-1}.

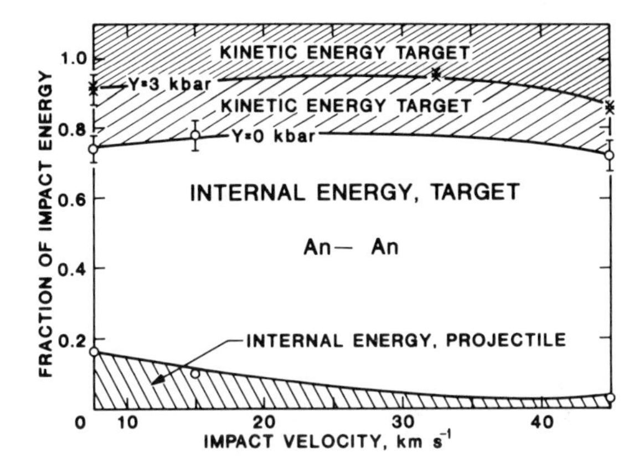

Fig. 19. Final fraction of impact energy versus impact velocity for nonporous silicate upon silicate impact. The difference in the fraction of energy, in the kinetic energy of target budget, depends strongly on assumed von Mises silicate strength *Y* indicated.

Although we have presented an extensive summary of our solid-planet impact-energy coupling calculations (summarized below in Table VI), calculation of actually how much of the ejecta energy either kinetic or internal may be transferred to an atmosphere is poorly known. As indicated in Figs. 17 and 19, a large fraction of the impact energy resides in the internal energy of the target; most of this is in ejecta. How much energy is transferred to the atmosphere is indicated in Table VII for a typical 30 km s^{-1} impact. Impact ejecta interaction with atmospheres, is at present only approximately modeled. Although there is considerable current research activity on the question of coupling impact energy to planetary atmospheres (see, e.g., Walker 1986*b*) some comments on the values given in Table VII may be of interest. The

TABLE VI
Partitioning of Energy upon Silicate Projectile Impacting Silicate Planet as a Function of Velocity

Target Porosity (%)	Velocity (km s^{-1})	Target Strength (kbar)	Fraction, Internal Energy Project (IE,P)	Fraction, Internal Energy Target (IE,T)	Fraction Kinetic Energy, Target (KE,T)
30	5	0	0.20	0.57	0.23
30	30	0	0.04	0.86	0.10
0	7.5	3	0.15	0.75	0.10
0	7.5	0	0.16	0.58	0.26
0	15	0	0.10	0.78	0.12
0	30	3	0.04	0.91	0.05
0	45	3	0.02	0.83	0.15
0	45	0	0.03	0.69	0.28

TABLE VII
Fraction of Total Energy that Ejecta Transfers to Atmosphere for a 30 km s^{-1} Asteroidal Impact[a]

	Fraction of Total Energy in the 0.5 km s^{-1} or Slower Ejecta	Efficiency of Transferring Energy to Atmosphere	Fraction Transferred to Atmosphere
Vapor-internal energy	0.12	1.0	0.12
Melt-internal energy	0.15	1.0	0.15
Solid-internal energy	0.03	0.1	0.00
Vapor-kinetic energy	0.01	1.0	0.01
Melt-kinetic energy	0.03	1.0	0.03
Solid-kinetic energy	0.05	1.0	0.05
Air shock			0.03
		Totals	0.39

[a]Table after O'Keefe and Ahrens (1982*a*).

seven energy budgets listed in this table were separately examined versus time in the course of this impact calculation and the peak values are the ones listed. In the Eulerian-type calculations, the amount of material that achieves sufficient entropy to remain molten or vapor upon pressure release increases with time as the impact-induced shock encompasses more material. However, as the shock expands, rezoning of the flow is carried out and the high entropy density material is mixed in with lower entropy density material. As a result, the mass of melt and vapor again decreases. The second column in Table VII is obtained from a calculation before rezoning. Thus the melt and vapor energy fractions are minimum value. The third, and hence fourth, columns are estimates. For the present, although we can assume ~0.4 of the impact energy is transferred to the planetary atmosphere, much of this energy is delivered late in the impact process and will, of course, heat the atmosphere, but little of this energy will result in atmospheric erosional loss.

V. IMPACT DEVOLATILIZATION OF MINERALS AND THEIR APPLICATION TO GENERATION OF PROTO-ATMOSPHERES DURING PLANETARY ACCRETION

Theoretical calculation of the entropy gain of minerals which are shocked and released to ambient pressure such as those reported by Ahrens and O'Keefe (1972), Kieffer and Simonds (1980), Lange and Ahrens (1980, 1982*a*) and Tyburczy and Ahrens (1986) for volatile-bearing minerals (containing CO_2, H_2O and SO_2), all suggest that shock pressures in the range of 30 to 50 GPa are required in nonporous minerals to induce partial devolatilization upon pressure release. These shock pressures can be induced upon impact of silicate projectiles at speeds of 2 to 3 km s^{-1} against these minerals.

It is difficult to detect partial vaporization by measuring release isentropes of volatile-bearing minerals. Substantial volatilization of the high pressure upon unloading would bring release isentropes to states which lie *above* the Hugoniot at a given density. Instead, the release isentropes measured both using the buffer techniques for porous $CaCO_3$ (chalk and limestone) and via velocity-interferometric methods for single crystal $CaCO_3$ (calcite) demonstrate that the isentropes lie at lower pressures than the respective Hugoniot in all cases (Figs. 20 and 21). This behavior was interpreted for the porous data (Tyburczy and Ahrens 1986) and single crystal data (Grady and Moody 1985) as arising from irreversible hysteretic behavior of polymorphic phase changes. Similar polymorphic phase changes also occur along the Hugoniot of other volatile-bearing minerals such as brucite, $Mg(OH)_2$ (Simakov et al. 1974); serpentine, $Mg_3Si_2O_5(OH)_4$ (Tyburczy, Ahrens and Lange, unpublished); and gypsum (Simakov et al. 1974).

Shock-recovery experiments conducted on single crystal serpentine, 20% porous serpentine, single crystal calcite and brucite all demonstrate that a fairly well-defined pressure range over which devolatilization (loss of H_2O or CO_2) occurs as indicated in Figs. 22a,b,c,d and Table VIII. In the experiments, samples were impacted in vacuum and as indicated in Fig. 23, were thoroughly shattered as expected for impact of planetesimals on planetary

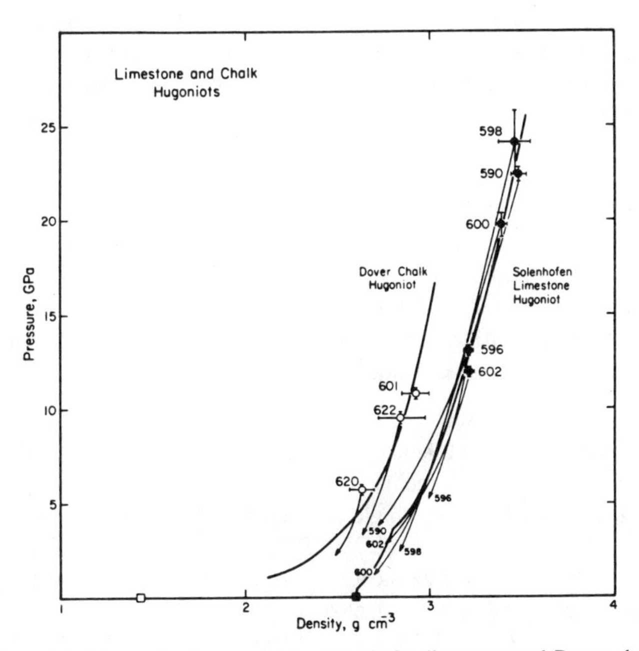

Fig. 20. Hugoniot states and release paths for Solenhofen limestone and Dover chalk. Solid symbols are for Solenhofen limestone; open symbols are for Dover chalk; squares represent initial densities (figure after Tyburczy and Ahrens 1986).

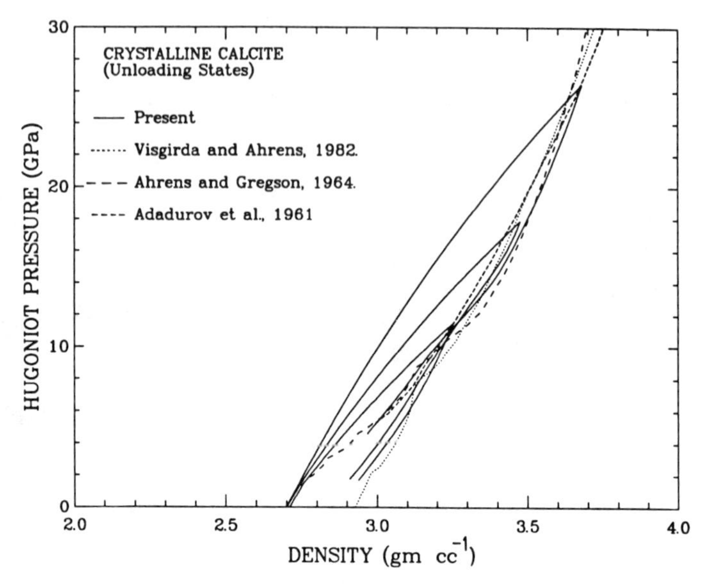

Fig. 21. Comparison of unloading curves with Hugoniot data for crystalline calcite (figure after Grady and Moody 1985).

surfaces. The amount of volatile loss was obtained by either measuring the total volatiles which could be extracted in recovered samples upon heating in a gravithermometric apparatus or in a water evolved analyzer. In the case of porous serpentine, both the solid products and gas evolved were analyzed in separate experiments as shown in Fig. 22b.

It is notable that in all cases shock-induced devolatization begins upon shock loading to much lower pressures, e.g., much lower impact velocities, than predicted for single crystals or even porous samples either from the thermodynamic calculation or release isentrope measurements. Two observations provide an explanation for the observed onset of impact devolatilization of CO_2- and H_2O-bearing minerals on planetary surfaces and within planetesimals at low impact velocities.

1. Accreting planetary surfaces are always covered by a porous regolith layer (Hartmann 1978) and hence a lower impact velocity criteria for devolatilization of porous media should be used.
2. Upon planar shock-loading single crystals of such volatile bearing minerals as calcite ($CaCO_3$), gypsum ($CaSO_4 \cdot 2H_2O$) and serpentine ($Mg_3Si_2O_5(OH)_4$), adiabatic shear bands induce local inhomogeneous regions of high temperature, which are observed both by photography and spectroscopically (Kondo and Ahrens 1983). These inhomogeneously high temperatures are indicated by small regions which are observed to comprise 10^{-2} to 10^{-4} of the area of the sample and appear to be 2000 to 3000 K above the continuum shock tem-

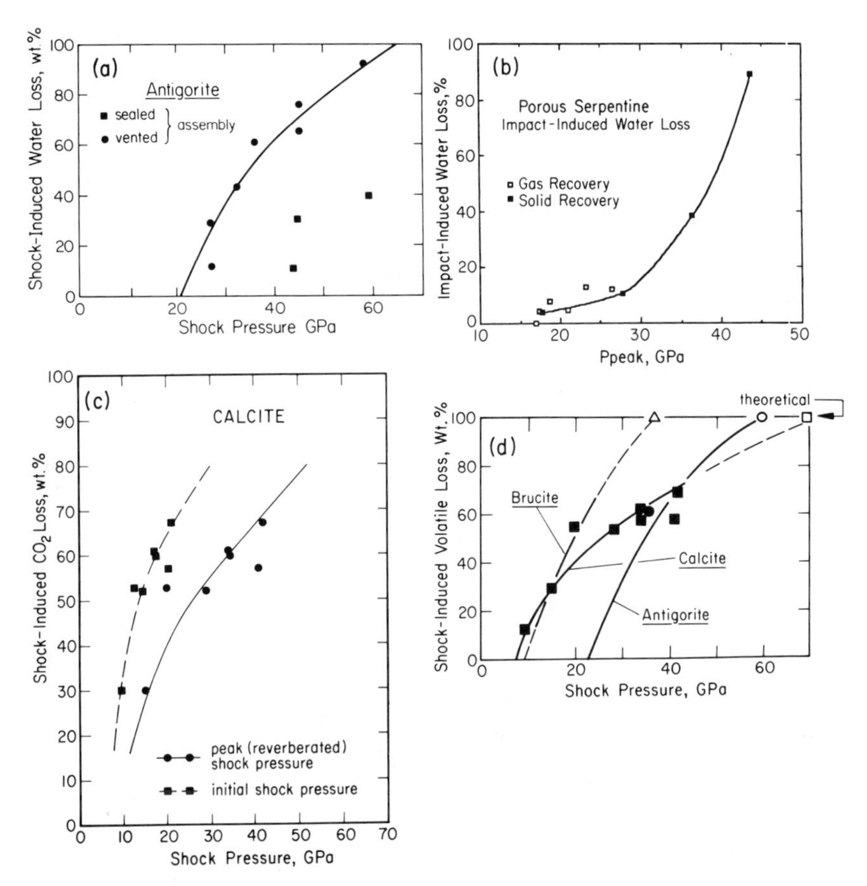

Fig. 22. (a) Shock-induced water loss as a function of shock pressure in antigorite serpentine. The solid line represents a fit to the vented-assembly data (figure after Lange et al. 1985). (b) Impact-induced water loss as a function of peak shock pressure for porous serpentine. Open symbols, gas recovery method; filled symbols, solid recovery method. The line is the empirical fit to the solid recovery data (figure after Tyburczy et al. 1987). (c) Maximum shock-induced CO_2 loss versus peak and initial shock pressure in a single crystal (figure after Lange and Ahrens 1986*b*). (d) Shock-induced water loss (in wt.% of total amount of initially present water) in brucite and antigorite-serpentine as a function of shock pressure in shock recovery experiments. Open circles indicate results for antigorite samples with 22% and 30% porosity (figure after Lange and Ahrens 1984).

TABLE VIII
Range of Shock Pressures for Devolatilization

Material	Specie Driven Out	Onset of Devolatilization (GPa)	Complete Devolatilization (GPa)
Serpentine			
$Mg_3Si_2O_5(OH)_4$ [a]	H_2O	20	65
20% Porous Serpentine [b]	H_2O	17	45
Brucite, $Mg(OH)_2$ [c]	H_2O	8	35
Calcite [d]	CO_2	10	40–50
$CaCO_3$			
Carbonaceous Chondrite [e]	H_2O, CO_2, SO_2		
Murchison meteorite	H_2S + . . .	12	30

[a]Lange et al. (1985).
[b]Tyburczy et al. (1987).
[c]Lange and Ahrens (1984).
[d]Lange and Ahrens (1986*a*).
[e]Tyburczy et al. (1986).

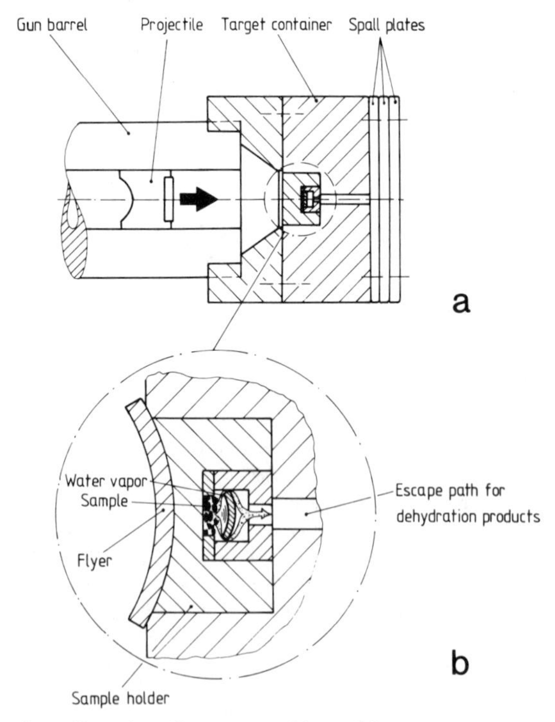

Fig. 23. Schematic configuration of target assembly used in present recovery experiments: (a) depicts the configuration immediately before projectile impact; (b) shows some of the details during shock loading. Escape of dehydration products (water vapor) is provided by the passage of vapor around the moving (away from the shock front) plate which is initially in contact with the shock-loaded sample (figure after Lange and Ahrens 1982*b*).

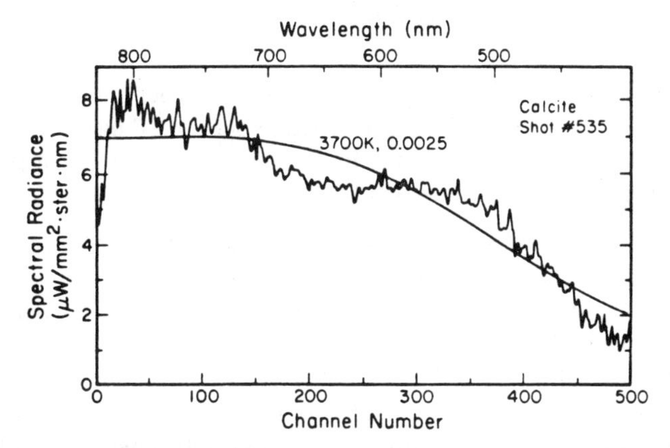

Fig. 24. Shock-induced spectral radiance versus channel number and wavelength for calcite as the shock wave of shock pressure of 40.4 GPa is propagating through the sample. The smooth curve for the spectrum is the best-fitted gray-body radiance. The color-temperature and emissivity are shown in the figure. The color-temperature is believed to indicate shear band temperature, whereas the emissivity of 0.0025 corresponds to the fraction of the surface radiating (figure after Kondo and Ahrens 1983).

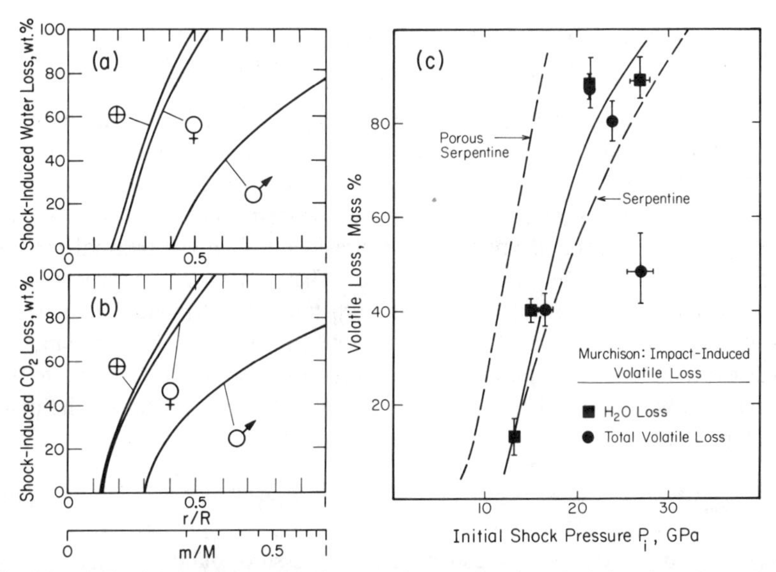

Fig. 25. (a) Shock-induced water loss in serpentine as a function of relative size r/R or relative mass m/M of the accreting terrestrial planets. Dehydration takes place when the impact pressures as given in Table VI are exceeded. This takes place at different stages in the accretional sequence for the Earth, Venus and Mars (figure after Lange and Ahrens 1984). (b) Maximum shock-induced CO_2 loss from impact on calcite (in wt.% of the total CO_2 content) versus relative size (r/R) and relative mass (m/M) for the Earth, Venus and Mars (figure after Lange and Ahrens 1986*b*). (c) Shock-induced mass loss as a function of initial shock pressure for Murchison. Squares represent H_2O loss; circles total volatile loss. H_2O and total volatile loss are similar with the exception of the erroneous datum for H_2O loss for shot 781. Dashed lines are fits to data for impact-induced dehydration of 20% porous serpentine and crystal-density serpentine (figure after Tyburczy et al. 1986).

perature (Fig. 24). Hence, because one has to consider porous regoliths and the fact that even for single crystal materials shear banding occurs, the onset of vaporization occurs at much lower impact velocities than previously predicted (see, e.g., Ahrens and O'Keefe 1972).

The experimental shock devolitilization data for water loss in porous serpentine and CO_2 loss in calcite may be applied in a very simple scenario. We assume that these materials impact the accreting planetary surface at just the infall velocity (equal to the growing planet escape velocity) (Figs. 25a and b) and the amount of the volatiles evolved corresponds to the shock-pressure, volatile-loss relations of Figs. 22a,b,c,d. These are then used to define the mass fraction of the planet growth interval which correspond to both the onset and complete vaporization of either the H_2O- or CO_2-bearing minerals. We use porous serpentine as representative of carbonaceous chondrite (Fig. 22b). These serpentine data compare favorably with experimental data from the Murchison meteorite shown in Fig. 25c as well as data for calcite which are calculated for vaporization of carbonate in Fig. 25b.

VI. THERMAL AND COMPOSITIONAL CONSEQUENCES OF CO-ACCRETION OF AN ATMOSPHERE

Although Lange and Ahrens (1982*a*) suggested that the impact-induced dehydration of water-bearing minerals would produce largely a water atmosphere on the growing planets, it was Abe and Matsui (1985) who first suggested the possibility that water in this atmosphere, and possibly the dust produced by planetesimal impact would drastically alter the thermal regime on the surface of growing planets (Fig. 26). They assumed that serpentine in planetesimals brought the Earth and the other terrestrial planets their water inventory during their accretion. Once impact pressures of infalling planetesimals exceeded $P \simeq 23$ GPa (for a porous regolith), the supply of water to the atmosphere was assumed to begin. Abe and Matsui (1985) estimate the peak shock pressure by

$$P = \rho[C_o + (K' + 1)V_i/8]V_i/2 \tag{14}$$

where ρ and C_o are density and bulk zero pressure-sound velocity respectively; K' is partial derivative of the bulk modulus with respect to pressure; and V_i the impact velocity. In all atmospheric accreting models this shock pressure is assumed to act on the entire impacting planetesimal and acts to release H_2O (or CO_2) in this mass; however, previous calculations have also included contributions to the atmosphere induced by shock-loading volatiles already present in the material of the planetary surface layer (see, e.g., Lange and Ahrens 1982*a*). The impact velocity is given as

$$V_i^2 = (2GM/R) + u^2 \tag{15}$$

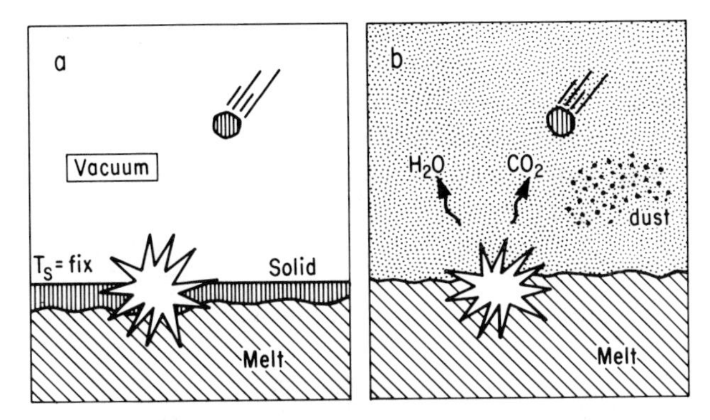

Fig. 26. Cartoon indicating the difference between the study of thermal regime of accretion via "thermal blanketing" and previous studies (figure from Abe and Matsui 1985).

where G, M, R and u are the gravitational constant, planetary mass, planetary radius and u is the velocity of planetesimals outside the region of influence of the growing planet. Here

$$u = (GM/R\theta)^{1/2} \tag{16}$$

where θ is the Safronov number. The values used by Abe and Matsui (1985) are similar to values used in Sec. VII below and by Lange and Ahrens (1987*a*), namely $\rho = 4.2$ g cm^{-3}, $C_o = 3$ km s^{-1}, $K' = 5$ and $\theta = 4$.

During accretion of the H_2O proto-atmosphere, as the Earth accretes from M to $M + \Delta M$, the atmospheric water increase is given by

$$\Delta M_a = \Delta M f_w f_i \tag{17}$$

where f_i is the amount of water in the planetesimal released by the impact and f_w is the mass fraction of water in the planetesimal. Abe and Matsui (1986) assume $f_w = 0.001$, 0.02 and 0.01, whereas Lange and Ahrens (1987*a*) assume a value of 0.01. However, as discussed later (Sec. VII.B), Lange and Ahrens (1987) assume major losses of water due to atmospheric cratering. If only the ocean water budget is assumed for the Earth, the present whole Earth value of f_w is 0.00027. In addition, Walker (1977)—and many others—have pointed out that water can be lost by a planet upon disassociation into H_2+O_2, and H_2 is then lost by Jeans escape.

Rehydration and reaction of shock-released water with iron or iron sulfides in planetesimals are also pathways with which water can be removed from the primordial atmosphere. These have been addressed, in part, by Jakosky and Ahrens (1979) and later by Lange and Ahrens (1984).

Matsui and Abe (1986*a*) conclude that once the radius of the Earth reaches between 0.2 to 0.4 of the present value ($R_\oplus$), thermal blanketing of the

Earth caused by a dense water atmosphere occurs. Although they did not specify the radius of the Earth at which this occurs, we expect that when the radius becomes >0.5 $R_\oplus$, the surface temperature becomes >900 K. At that point, surface rocks would become dehydrated regardless of impact additions. Matsui and Abe (1986*b*) point out that thermal blanketing is a more severe condition than the greenhouse effect. In a greenhouse effect, sunlight penetrates through the atmosphere (in the visible) but the thermal energy to be reradiated by the planetary surface is trapped because of the opacity in the infrared of the planetary atmosphere which contains abundant CO_2 and H_2O. Thermal blanketing is more severe. In this case, the effect of the Sun only determines the effective temperature at the top of the atmosphere. In thermal blanketing, solar radiation is incident on the top of the atmosphere but becomes completely scattered and absorbed by the greenhouse gases, H_2O and CO_2, and possibly the fine aerosol impact ejecta. As discussed in the previous section, for large impactors 60 to 90% of the energy of the impact is delivered as internal energy of the near surface material. The major effect of the proto-atmosphere then is to provide a partially insulating blanket to this flux of heat. These processes are described by the equation

$$(1 + 1/2\theta)\,\frac{GM}{R}\,\dot{M}\mathrm{d}t = 4\pi R^2\,(F_{\mathrm{atm}} - F_i)\mathrm{d}t + C_p\dot{M}\,(T_s - T_p)\,\mathrm{d}t + C_p M_s \dot{T}_s \mathrm{d}t \tag{18}$$

where the left-hand side is the rate of kinetic energy supplied to the surface provided by the impacting planetesimals. The first term on the right is the balance of heat flux of the atmosphere where F_{atm} is the energy flux escaping from the surface to interplanetary space and F_i is the energy flux from the interior to the surface layer. The second and third terms are the heat sinks to the planet as a result of heating a larger planetary material of mass M_s from planetesimal temperature T_p to the higher surface temperature T_s. For a gray-body radiative equilibrium atmosphere

$$F_{\mathrm{atm}} = 2(\sigma T_s^4 - S_o/4)/[(3kP_s/2g) + 2] \tag{19}$$

where σ is the Stefan-Boltzmann constant, S_o is the solar flux, k is the absorption coefficient in the atmosphere, P_s is the surface pressure, and g is surface gravity. Thermal blanketing as described by Eqs. (18) and (19), with the reasonable values of the constants chosen by Abe and Matsui (1985,1986*a*), quickly leads to temperatures above the solidus of crustal (basaltic) rocks (Figs. 27a and b). Moreover, as more planetesimals impact the planet, the additional water provided begins to dissolve in what is the start of a magma ocean. Matsui and Abe (1986) assume that the mass % of water dissolved in silicate X_w is given by the relation fit to data collected by Fricker and Reynolds (1968)

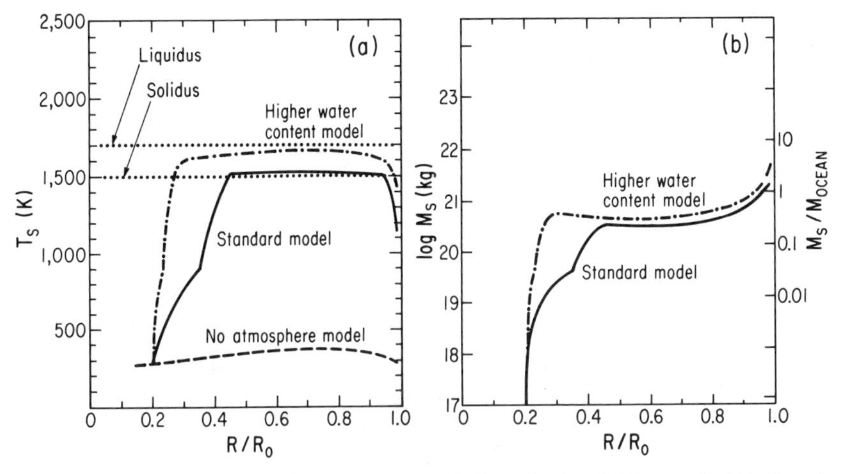

Fig. 27. (a) The evolution of surface temperature during accretion for the standard Earth and higher f_w models. The radius is normalized by the final value. The broken line traces the surface temperature of the model without an impact-generated atmosphere ($\tau_{acc} = 5 \times 10^7$yr). The surface temperature in the models including atmospheres begins to increase once the impact velocity exceeds a critical value. The rapid rise in the surface temperature for the standard model after the Earth grows to $\sim 0.3\ R_o$ is due to an increase in the total mass of the atmosphere because of initiation of the complete dehydration reaction of the surface layer. Once the surface temperature reaches the melting temperature, it becomes constant (figure after Matsui and Abe 1986*a*). (b) The total mass of the impact-generated H_2O atmosphere is plotted against the normalized radius for the standard ($f_w = 0.001$) Earth and higher water-content ($f_w = 0.01$) models. Note that the total mass remains nearly constant after the Earth grows to $0.4\ R_o$ and is very close to the present mass of the Earth's oceans ($\sim 1.4 \times 10^{21}$kg) (figure after Matsui and Abe 1986*b*).

$$X_w = 0.0002\ P_s^{0.54} \tag{20}$$

where P_s is the surface partial pressure (in pascals) of water. The growth of the atmosphere in this temperature range is then given simply by

$$\Delta M_a = \Delta M(f_w - \alpha X_w) \tag{21}$$

where α is a fraction of melt in the magma ocean. Equation 18 may be more simply written as

$$\Delta M V_i^2/2 = 4\pi R^2 F_{\mathrm{atm}} \Delta t + C_p \Delta M(T_s - T_p) \tag{22}$$

where Δt is the time interval associated in growing the atmosphere by ΔM. Because the solubility of water in silicate controls the water pressure, P_s, at the base of the atmosphere, the total water content of the atmosphere is buffered at a value close to that of the ocean water inventory in the case of the Earth (Fig. 28). As the impactor flux of kinetic energy is released at the sur-

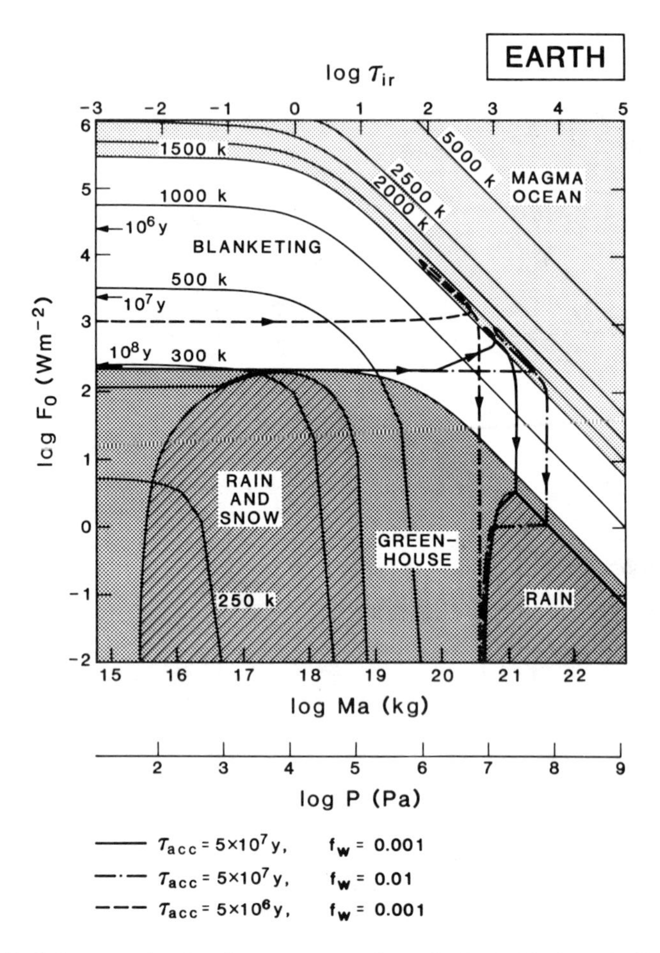

Fig. 28. Evolutionary tracks of surface temperature for an Earth-sized body at the Earth's distance from the Sun are plotted as functions of the energy flux F_o released at the surface and the mass M_a, surface pressure P, and optical depth τ_{ir} (for infrared radiation) of the H_2O atmosphere. The solid curves with numerals represent the curves of equal surface temperature (in K). The shaded region represents the greenhouse region in which the solar flux S_o is the dominant energy term controlling the thermal structure of the atmosphere. In the unshaded region, the blanketing region, the energy flux F_o released at the surface is the dominant term. The dotted region is the magma ocean region in which T_s exceeds the solidus temperature. The cross-hatched region depicts the range of conditions under which H_2O cannot exist in the gas phase and precipitation takes place. Arrows indicate the mean impact-energy fluxes for various accretion times of the planet. The evolutionary tracks for three Earth models are shown: the solid, dash-dot and broken lines represent the evolutionary tracks for the standard Earth model ($\tau_{acc} = 5 \times 10^7$ yr and $f_w = 0.001$), a higher water content model ($\tau_{acc} = 5 \times 10^7$ yr and $f_w = 0.01$) and a rapid accretion model (figure after Matsui and Abe, unpublished).

face, the impact energy flux decreases from 10^3 W m^{-2} to zero at the end of the accretionary period; the temperature at the surface decreases until the temperature falls below ~600 K at 100 bar (10^7 Pa); and the water in the atmosphere begins to condense (rains) and forms (in the case of the Earth) the ocean (Fig. 28). In the case of Venus, evolution ends with the greenhouse thermal regime in the atmosphere (Fig. 29), whereas in the case of Mars, the onset of H_2O precipitation occurs directly from the blanketing regime which may limit the amount of H_2O to be deposited from the proto-atmosphere onto

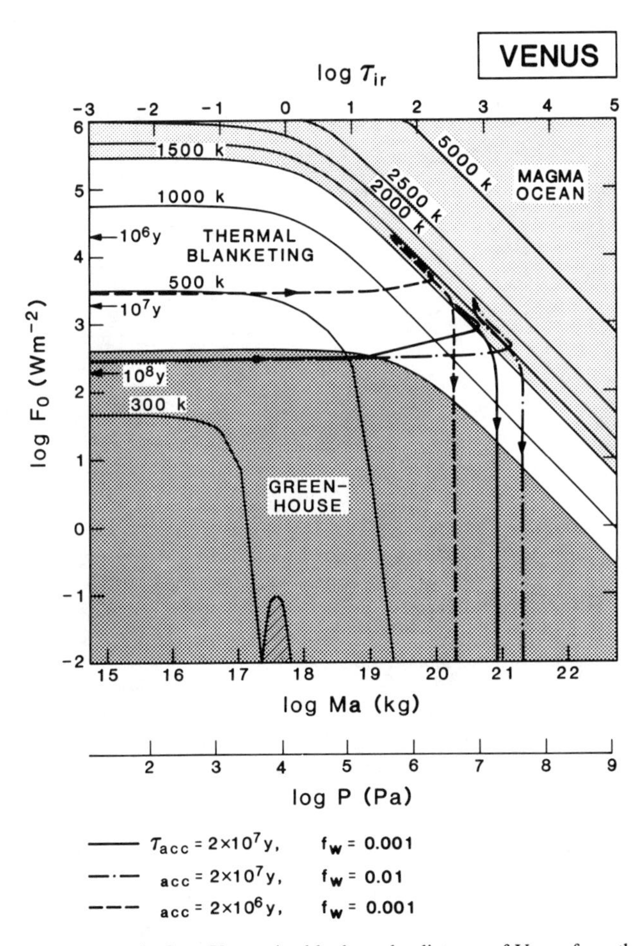

Fig. 29. Evolutionary tracks for a Venus-sized body at the distance of Venus from the Sun. See Fig. 28 for full explanation of symbols. Note that the area of the liquid H_2O region is much smaller than that of Fig. 28. The solid, dash-dot and broken lines represent the evolutionary tracks for the standard Venus model (τ_{acc} = 2 × 10^7 yr and f_w = 0.01), a high water-content model (f_w = 0.01), and a rapid accretion Venus model (τ_{acc} = 2 × 10^6 yr and f_w = 0.001), respectively (figure after Matsui and Abe, unpublished).

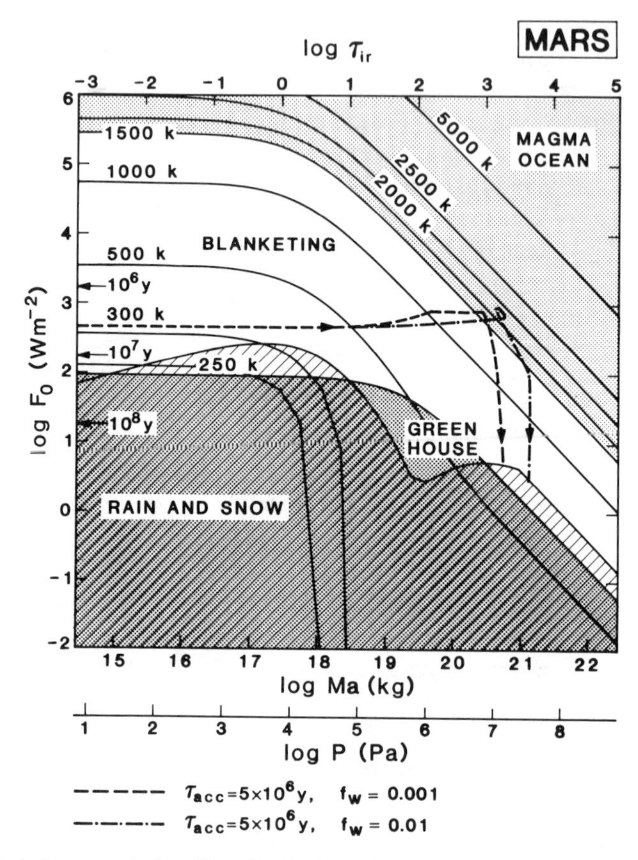

Fig. 30. Evolutionary tracks for a Mars-sized body at the distance of Mars from the Sun. See Fig. 28 for full explanation of symbols. Note that the area of the liquid H_2O region is much larger than that of Fig. 28. The broken and dash-dot lines represent the evolutionary tracks for rapid accretion models (τ_{acc} = 5 × 10^6 yr) with water contents, f_w of 0.001 and 0.01, respectively (figure after Matsui and Abe, unpublished).

the planet (Fig. 30). The effect of impact of planetesimals onto the proto-atmosphere is considered in the next section.

VII. IMPACT EROSION OF PLANETARY ATMOSPHERES

Although the effects of infall of large bolides onto the Earth was briefly considered by O'Keefe and Ahrens (1982*a,b*), Cameron (1983) first pointed out that significant loss of a planetary atmosphere could be a process which acted in the later stages of accretion. As already mentioned in Sec. IV, the initial passage of a planetesimal through a highly idealized planetary atmosphere delivers a small fraction (~10^{-1}) of its energy (Table VII) upon initial passage through the atmosphere. Primarily the interaction with the atmosphere is with the excavated ejecta (Table VII). In this section, we will first consider the impact into the present-day atmosphere of the Earth and then later consider how this process might have acted in protowater atmospheres of the Earth, Venus and Mars.

A. Impact into Air

Because of the strong mechanical impedance mismatch between the air and silicate, the calculation of the impact interaction of a projectile with an atmosphere, and the description of the ejecta interaction is difficult. Earlier, O'Keefe and Ahrens (1982*a,b*) studied the interaction of a 5 km s^{-1}, 10-km diameter silicate impactor interacting with a silicate half space covered with a 10-km thick layer of air. Partitioning of energy (Fig. 31) and the resulting flow field (Fig. 32) provide considerable insight into the physical processes which operate. As demonstrated in Fig. 31, the total fraction of energy initially partitioned promptly into the atmospheric layer because this speed impact is only ~0.08 of the bolide energy. However, a substantial fraction of the kinetic energy of the ejecta, is also later transferred to the air because of the air drag. This latter energy does not lead to atmospheric escape but just to the heating of the atmosphere. Thus, some 0.27 fraction of the internal energy of the resulting Earth and projectile ejecta is transferred to the atmosphere via radiation and conduction processes. O'Keefe and Ahrens estimated that some 0.39 energy fraction of the total energy of a very large bolide (~10^{28} ergs) would be transferred to the atmosphere (Table VII). Depending on the density, smaller (~10^2-m diameter) objects transfer virtually all of their energy into the atmosphere. Thus, impactors with energies >10^{28} erg impact the Earth

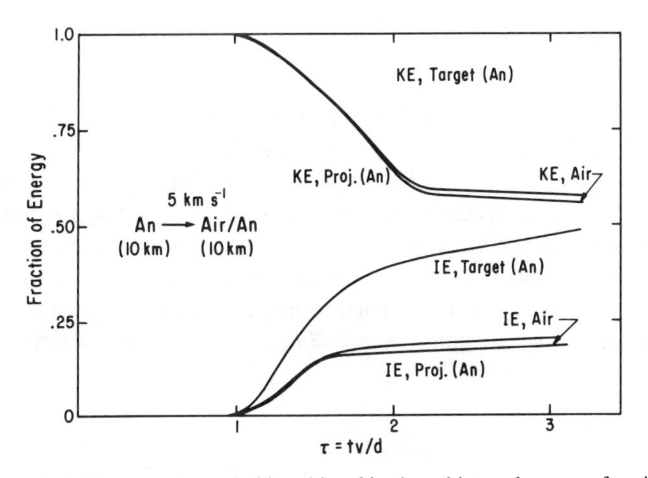

Fig. 31. Fraction of impact energy partitioned into kinetic and internal energy of projectile and target (air and silicate) versus dimensionless time for a 10-km diameter silicate projectile impacting a 10-km thick, 0.001 g cm^{-3} atmosphere overlying a silicate Earth (figure after O'Keefe and Ahrens 1982*a,b*).

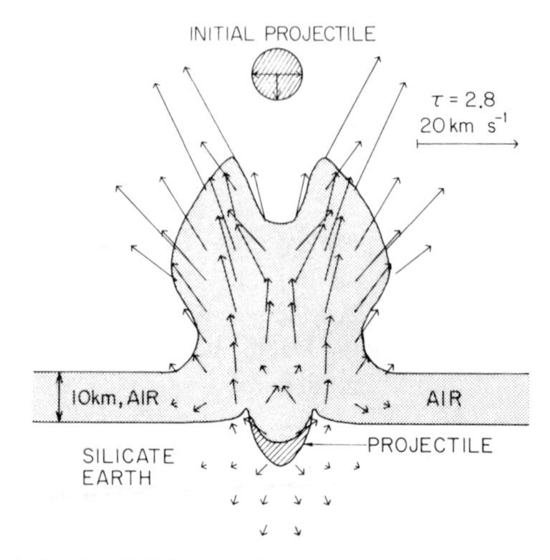

Fig. 32. Particle velocity flow field from a silicate projectile impacting a uniform 0.001 g cm^{-3} atmosphere overlying a silicate Earth at 5 s^{-1} at $\tau = 2.8$ when the projectile has delivered ~90% of its energy to the target(s) (figure after O'Keefe and Ahrens 1982*a,b*).

and virtually all interactions with the atmosphere occur very near the surface of the Earth.

Walker (1986*b*) has considered a simple model of impact of a projectile of radius R onto a planet with an atmosphere having a surface pressure P_a and surface gravity g and derived the following equation for the atmospheric mass ejected which is given by

$$M_e = \left(\frac{V_i}{V_e}\right)^2 \pi R^2 \frac{P_a}{g}\left(\frac{1}{1 + \Lambda V_i^2/2Q}\right) \tag{23}$$

where V_i and V_e are the impact and escape velocities, Λ is the ablation factor (0.1 to 0.2) and Q is a typical heat of vaporization (8×10^{10} erg g^{-1}). Aside from its simplicity, Eq. (23) has the virtue that some account is taken, by means of the ablation factor, of the additional gas produced by vaporization of the bolide upon atmospheric reentry which must be lofted to escape velocity along with the cylinder of gas intercepted by the bolide. The first term in Eq. (23) is the mass of the gas ejected and is proportional to the downward kinetic energy of the "plug" of atmosphere $\pi R^2 P_a g^{-1}$ (Fig. 33). The term in paren-

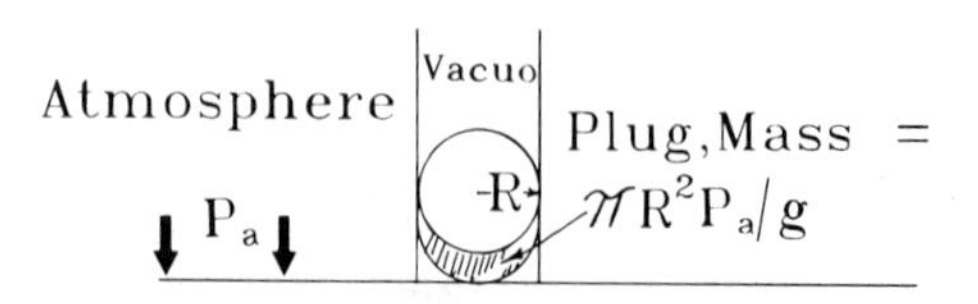

Fig. 33. Sketch of the Walker (1986*b*) atmospheric coupling model.

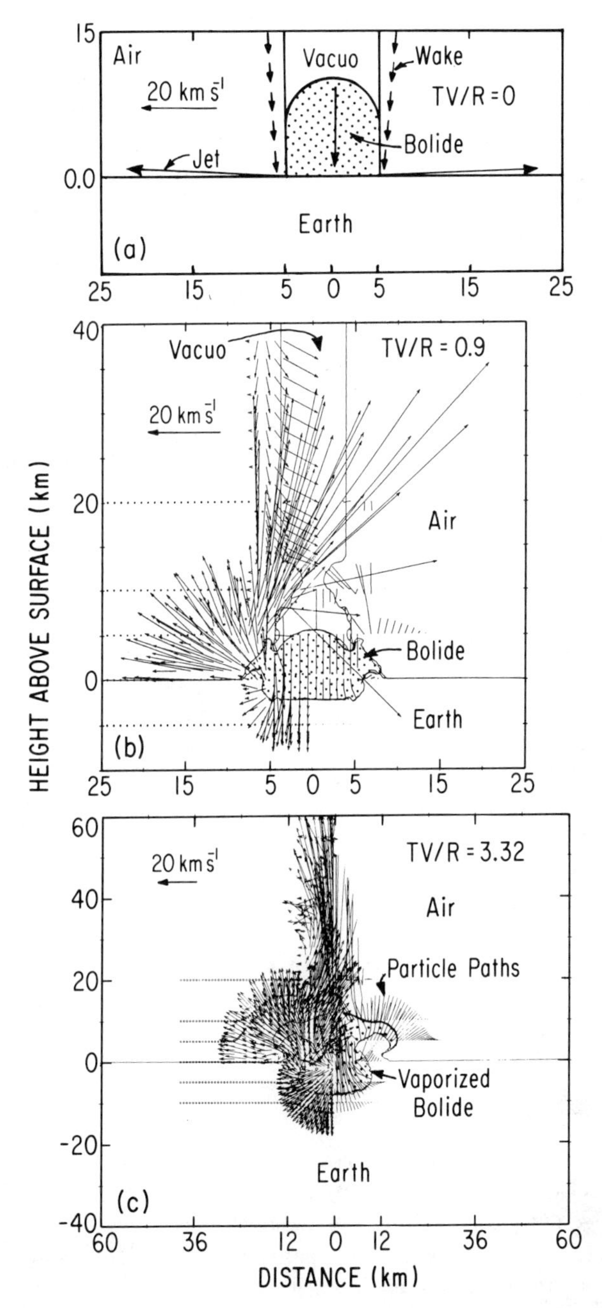

Fig. 34. (a) Initial conditions for calculation of numerical simulation of 10-km diameter impactor impinging on an exponential model of the Earth's atmosphere. The impact velocity is 20 km s^{-1} for silicate, 2.7 g cm^{-3}. The impactor radius is 5 km. The target is silicate. (b) Flow field at dimensionless time 0.9. Arrows on the left-hand side show particle velocity. Dots on the left indicate position of marker particles. Lines on the right-hand side indicate the path of marker particles. (c) Flow field at dimensionless time 3.32 (figure after O'Keefe and Ahrens 1988).

thesis takes into account the additional gas produced by ablation. The total energy in the plug might be expected to be somewhat lower than that contained in the high-velocity wake, as well as the very high-velocity jet and also high-enthalpy density jet which occurs upon impact of the gas cap with the planetary surface (Figs. 34a,b,c). Thus, we might expect Walker's (1986*b*) treatment to provide a minimum value for the energy coupling into an atmosphere, since it treats only the high-velocity parcels of gas which escape the planet and not the wake or jets. The latter may be seen to act on the impact-induced flow in numerical calculations such as those of O'Keefe and Ahrens (1988) for the impact of a bolide onto an exponential atmosphere. For example, for a 11 km s^{-1} (1×10^{30} erg) flow, some 0.2×10^{15} g of the atmosphere is ejected which requires an energy of 1×10^{25} erg. Thus, the ejected air of the Walker model represents 10^{-5} of the energy of the incoming bolide. Moreover, Walker (1986*b*) has pointed out that although the impact ejecta could put a sizeable fraction of the impact energy into the atmosphere as, for example, given in Table VII, the enthalpy density of the gas containing this contribution is so low as to virtually preclude that this coupling leads to significant atmospheric escape.

Actually, the fraction of energy that the ejecta can impart to the atmosphere is not as large as suggested in Table VII for large, gravity-controlled craters and may be estimated another way by using recent crater-scaling relations of Schmidt and Housen (1987). They give the relation of scaled crater radius π_r to the scaled energy π_2 as

$$\pi_r = 0.94\pi_2^{-0.22} \tag{24}$$

where

$$\pi_r = r(\rho/m)^{1/3} \tag{25}$$

and

$$\pi_2 = 3.22\ g\ a/V_i^2 \tag{26}$$

where r is crater radius, ρ is planetary density and m, V_i and a are bolide mass, impact velocity and radius. If the minimum kinetic energy of the ejecta is just sufficient to excavate the crater, and this ejecta, in turn, delivers all its kinetic energy to the atmosphere via particle drag, radiation and conduction, the minimum energy which could be imparted to the atmosphere is just the potential gravitational energy of crater excavation. This is given by

$$E_{xc} = 2\pi\rho g(\tan\theta)^2 r^4(3/4 - 2/3) \tag{27}$$

where θ is the angle of the downward slope (measured from the horizontal) of an assumed conical crater. Using Eqs. (24–27) and assuming tan $\theta = 0.4$, $V_i = 11$ km s^{-1} and the density of the planet and bolide is 3 g cm^{-3}, yields values of E_{xc}/E_{ke} which vary from 0.012 to 0.039 of the energy of the impactor, E_{ke}. This varies from 10^{20} to 10^{34} erg. For 11 km s^{-1} impactors these energies correspond to 0.082 to 355 km radii bolides. These values are considerably less than the maximum energy fraction inferred in Table VII to be coupled to the atmosphere by ejecta. Thus, we conclude that the prompt coupling of energy to the atmosphere is mostly due to projectile passage and this brings 3 to 8% of the bolide energy to the atmosphere. This is the energy available for impact erosion of planetary atmospheres. Later, ejecta coupling which induces local high temperatures but not sufficiently high-energy densities to produce atmospheric escape are loosely constrained to be $\lesssim 0.4$ of the initial energy of the bolide.

Let us examine the question of impact erosion in the present atmosphere. We know the approximate fraction of the energy of the impactor which is coupled to the atmosphere at the surface and we have recently (O'Keefe and Ahrens 1988) examined the shock-induced flow from such impacts on the Earth for an exponential atmosphere. The loss of the present atmosphere from a large explosion is an interesting problem and has been studied by Zeldovich and Raizer (1967); it is often called the Zeldovich flow. We assume that explosive-induced flow can approximate the impact-induced air flow because of the similarity of flow fields of Figs. 32, 34c and 35. Previous studies of the

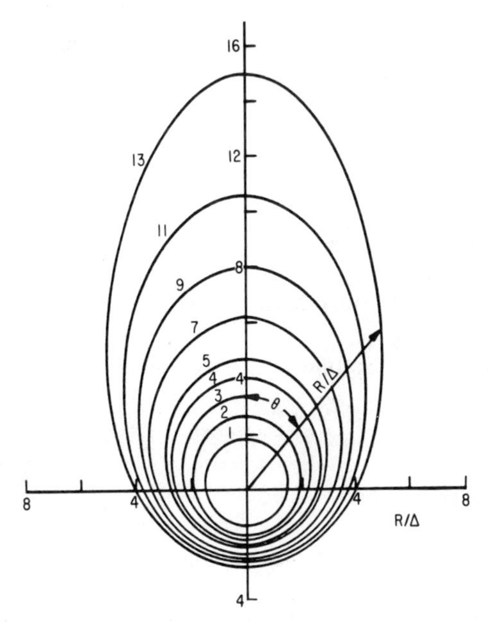

Fig. 35. Shock envelopes for various values of the nondimensional time scale $T(\gamma = 1.4)$ (figure after Bach et al. 1975).

Zeldovich flow were carried out by Kompaneets (1960), Troutman and Davis (1965), Grover and Hardy (1966), Hayes (1968), Lutzky and Lehto (1968), Laumbach and Probstein (1969) and Bach et al. (1975). These works demonstrate that in an exponential atmosphere the downward propagating shock (from a finite-height explosion) slows down, and the upward propagating shock accelerates both in shock velocity and particle velocity behind the shock front. The general idea of the Zeldovich flow is demonstrated in Fig. 35 in which the shock front is shown at various nondimensional times T. T is related to real time t by

$$T = t\left[\frac{E_o}{4\pi\Delta^5\rho_c J_o}\right]^{1/2} \equiv At \tag{28}$$

where E_o is the energy of the explosion, or in this case, the atmospheric coupled energy. Here Δ is the atmospheric scale height (7 km for the Earth), ρ_c is the density at the height of the explosion (taken here to be 10^{-3} g cm^{-3}) and $J_o = 0.4233$ is a constant which is a function of the polytropic exponent γ of the atmosphere (γ is taken to be 1.4). Thus, one can see that as a shock propagates upward, it accelerates. The strong shock conditions yield density and particle velocity at the shock front, in terms of the density ρ_a at a given height

$$\rho_a = \rho_c \exp(-R\cos\theta/\Delta) \tag{29}$$

where R is the radius to the shock front and θ is defined in Fig. 35. At the shock front, for strong shocks, the density is given as

$$\rho = \frac{2\rho_a}{\gamma + 1} \tag{30}$$

and also the shock velocity is given by

$$U = \Delta A/\{\xi^{3/2}[\exp(\lambda_1\omega/2) + \omega^2\lambda_2/4 + \omega^3\lambda_3/6]\} \tag{31}$$

where ξ and ω are related as follows

$$\xi \equiv \omega/\cos\theta \tag{32}$$

with ω the similarity variable

$$\omega \equiv \mathrm{dln}\ \rho_a/\mathrm{dln}\ \xi. \tag{33}$$

Here $\lambda_1 = -0.9085$, $\lambda_2 = 0.02902$ and $\lambda_3 = 0.00297$ are constants that apply for the $\gamma = 1.4$ case. The particle velocity u behind the shock front propagating at velocity U is

$$u = 2U/(\gamma + 1). \tag{34}$$

The particle velocity and density profiles as a function of ω given by Bach et al. (1975) were used to map the entire flow field via Eqs. (28–32). For altitudes which are a factor of ~6 greater than the atmospheric scale height, we assume that when the outward particle velocity of a zone is greater than the Earth's escape velocity of 11 km s^{-1}, the mass of gas in that zone (determined by the density and volume of the zone) escapes from the Earth.

Some typical calculated particle velocity profiles were obtained using the similarity solution of Bach et al. (1980) along the $\theta = 0°$ and $\theta = 80°$ directions for two times are shown in Figs. 36a and b. In the case of $\theta = 0°$, virtually all the gas above an elevation of 40 km, escapes from the Earth, whereas at $\theta = 80°$ at a much later time, no gas has sufficient velocity to escape.

For a value of $\Delta = 7$ km, $\gamma = 1.4$, $\rho_a = 10^{-3}$ g cm^{-3}, and an Earth with a surface area of 5.1×10^{18}cm^2, a total hemispherical mass of atmosphere of 1.8×10^{21} g is calculated upon integrating Eq. (29). This compares to the actual value of the Earth of 2.5×10^{21} g. Within the framework of the present theory of atmospheric escape for an atmosphere lying above a half space (no Earth curvature), the maximum mass of gas which can escape from a single explosion or impact will be the mass of gas *above* a plane tangent to the spherical Earth. For the above atmospheric parameters, this mass of gas is 2.6 $\times 10^{18}$ g and compares to a mass of 1.6×10^{16} g which is the largest mass that we calculate can escape from a single impact (Table IX, Fig. 37). These results were obtained by examining the particle velocity flow fields for 10^{26}- to 10^{28}-erg surface explosions in the Earth's atmosphere. The relative air mass accelerated upward to greater than a given velocity is shown in Fig. 37 and

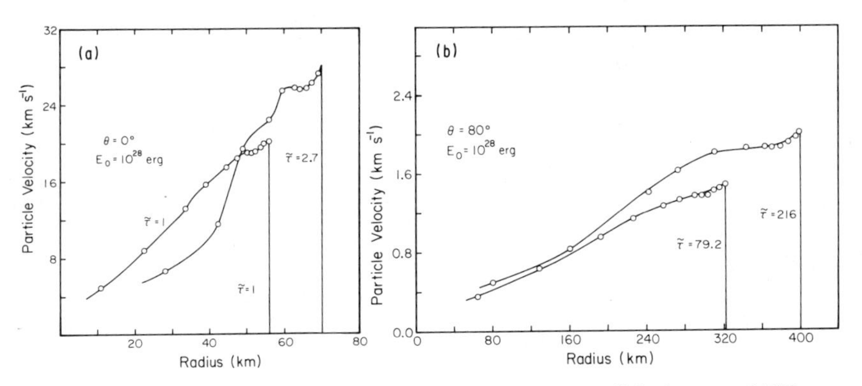

Fig. 36. (a) Particle velocity profile at two different times along $\theta = 0°$ for impacts of 10^{28} erg. (b) Particle velocity profile at two different times along $\theta = 80°$ for impacts of 10^{28} erg.

TABLE IX

Mass Fraction of Atmospheric Hemisphere Lofted to Escape vs Planetary Escape Velocity[a]

Coupled Energy (erg)	Velocity (km s^{-1}) 1	2	3	4	5	6	7	8	9	10	11	30	50
1.0×10^{26}	4.4-08[b]	7.0-09	0.0	0.0	0.0	0.0	0.0	0.0	0.0	0.0	0.0	0.0	0.0
0.5×10^{27}	4.4-07	1.1-07	2.2-08	8.8-09	3.5-09	5.6-11	0.0	0.0	0.0	0.0	0.0	0.0	0.0
1.0×10^{27}	5.6-07	1.1-07	4.4-08	4.4-08	1.8-08	8.8-09	3.5-09	2.2-10	0.0	0.0	0.0	0.0	0.0
0.5×10^{28}	8.8-06	5.6-07	4.4-07	1.1-07	1.1-07	1.1-07	4.4-08	4.4-08	4.4-08	1.8-08	1.8-08	0.0	0.0
1.0×10^{28}	8.8-08	5.6-07	5.6-07	4.4-07	4.4-07	1.1-07	1.1-07	1.1-07	1.1-07	4.4-08	4.4-06	0.0	0.0
0.5×10^{29}	8.8-06	8.8-06	8.8-06	5.6-07	5.6-07	5.6-07	5.6-07	5.6-07	4.4-07	4.4-07	4.4-07	2.2-08	3.5-09
1.0×10^{29}	8.8-06	8.8-06	8.8-06	8.8-06	8.8-06	1.1-06	5.6-07	5.6-07	5.6-07	5.6-07	5.6-07	4.4-08	4.8-08
0.5×10^{30}	8.8-06	8.8-06	8.8-06	8.8-06	8.8-06	8.8-06	8.8-06	8.8-06	8.8-06	8.8-06	1.8-06	4.4-07	1.1-07
1.0×10^{30}	8.8-06	8.8-06	8.8-06	8.8-06	8.8-06	8.8-06	8.8-06	8.8-06	8.8-06	8.8-06	8.8-06	5.6-07	4.4-07

[a]Table after Ahrens and O'Keefe (1987).
[b]As example, $4.4 - 08 = 4.4 \times 10^{-8}$.

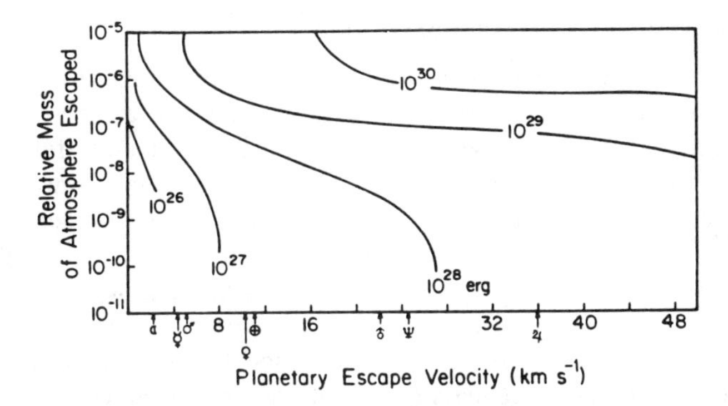

Fig. 37. Relative mass of a hemisphere of atmosphere, that will escape for an exponential atmosphere with various values of energy coupled into the atmosphere with γ = 1.4, $\rho_a = 10^{-3}$ g cm^{-3}, and Δ= 7km versus planetary escape velocity. For comparison, the escape velocity for the Moon, Mercury, Mars, Venus, Earth, Uranus, Neptune and Saturn is indicated.

Table IX. Thus, we conclude that within the framework of the Zeldovich flow problem, only some 5×10^{-6} of the total atmosphere of the Earth can be lost by a single impact. According to Table IX, the minimum coupled impact energy for this maximal atmospheric loss event is in the range of 0.5 to 1 $\times$ 10^{30} erg. We assume an atmospheric coupling factor of 0.08. This corresponds to a 2 g cm^{-3} density, 11 to 14 km radius, projectile infalling at 11 km s^{-1} into the present atmosphere. This maximal atmospheric loss projectile is only a factor of ~10 lower in energy than that inferred by Alvarez et al. (1980) and O'Keefe and Ahrens (1982*b*) for the Cretaceous-Tertiary extinction bolide.

We note that actually a very narrow energy regime range exists between the energy for maximal loss (10^{30} erg) versus that for impacts (or explosion) for which the result is total atmospheric retention which occurs for an escape velocity of 11 km s^{-1}. According to the equation of Fig. 38 for an escape velocity of 11 km s^{-1}, total atmospheric retention occurs at 10^{27} erg.

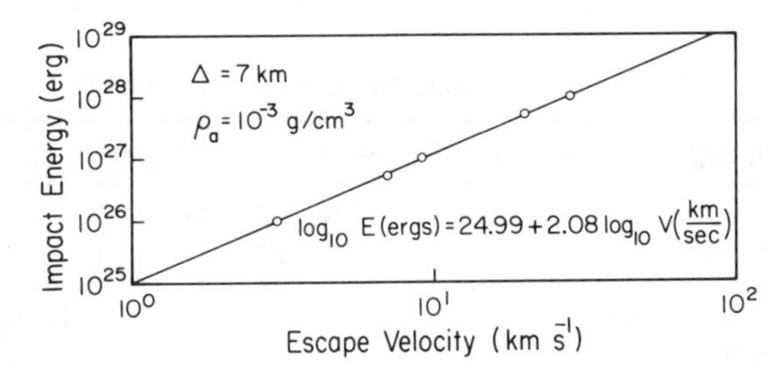

Fig. 38. Log$_{10}$ (impact energy, erg) versus log$_{10}$ (planetary escape velocity, *V*(km s^{-1} for total atmospheric retention). Linear fit to present calculations is indicated by a straight line.

B. Impact into a Proto-Atmosphere During Accretion

We model the formation of the planets and the growth of an impact-generated atmosphere in the framework of a Safronov-type accretional model. Once infall velocities of planetesimals exceed critical velocities for incipient to complete loss of structural water in water-bearing minerals, water vapor is added to an evolving proto-atmosphere. At that point, we consider the effect of bolide-atmosphere interactions in subsequent time steps of the model. We take the current mass of this atmosphere and its mean temperature into account to drive the relevant parameters (surface pressure, surface density) for the blow-off calculations. In our numerical modeling scheme at the end of each iteration, the evolving mass budget of the atmosphere (i.e., impact-induced minus blow-off) is obtained and is used as initial conditions for the following time step.

The accretion rate $\dot{M}$ assumed is that specified by Safronov (1969) and used by Abe and Matsui (1985,1986)

$$\dot{M} = \pi R^2(1 + 2\theta)\rho_p u \tag{35}$$

where R is the current radius of the solid protoplanet, θ the Safronov parameter, ρ_p is the spatial density of planetesimals in the region of the growing planet and u is the approach velocity or the velocity relative to the orbiting velocity of the planetesimals. Equation 35 can be transposed into

$$\dot{M} = \pi R^2(1 + 2\theta)\, \frac{4[M_f - M(t)]}{P_K \pi (a_o^2 - a_i^2)} \; . \tag{36}$$

where, M_f is the final mass of the planet, P_K is Keplerian period, and a_o and a_i are the radii of outer and inner edges of the feeding zone of planetesimals forming each planet (Safronov 1969; Lange and Ahrens 1976) (see Table X for values used).

Equation (36) is numerically solved and the time-dependent mass, radius and fractional mass $f(t) = M(t)/M_f$ is obtained. The (constant) mean radius of

TABLE X
Physical Parameters for Terrestrial Planets[a]

Planet	M_f (kg)	R_f (m)	$\bar{\rho}$ (kg m^{-3})	C_A (m^2)	f_w	T_e (K)
Venus	4.87×10^{24}	6.05×10^{6}	5.27×10^{3}	3.05×10^{22}	0.01	320
Earth	5.97×10^{24}	6.38×10^{6}	5.52×10^{3}	5.98×10^{22}	0.01	253
Mars	6.42×10^{23}	3.40×10^{6}	3.95×10^{3}	1.12×10^{23}	0.01	210

[a]M_f = Mass; R_f = radius; $\bar{\rho}$ = mean (compressed) density; $C_A = \pi(a_0^2 - a_I^2)$, a_0,a_i = radii of outer and inner feeding zones upon accretion; f_w = mass fraction of water in planetesimals; T_e = effective temperature of planet (table from Lange and Ahrens 1987*a*).

planetesimals r_{pl} determines the number of planetesimals, n_{pl} for each time step during accretion

$$n_{\text{pl}}(t) = \frac{\dot{M}(t)}{(4/3)\pi r_{\text{pl}}^3 \bar{\rho}_{pl}} \tag{37}$$

where $\bar{\rho}_{\text{pl}}$ is the mean density of infalling planetesimals (assumed for simplicity to be equal to the final mean density of the planet).

For each iteration, the current surface gravitational acceleration $g(t)$ and escape velocity $V_e(t)$ are computed. In the present study, we consider only dehydration of serpentine as the major source of atmospheric gases, for which P_{iv} = 12 GPa and P_{cv} = 33 GPa (Tyburczy et al. 1986) (see Sec. V).

Assuming that planetesimal composition is approximated in composition and shock impedance by the impacted surface material, the peak initial shock pressure P is calculated from Eq. (14) of Sec. VI. We further assume that the impact velocities can be approximated by the escape velocity of the planet. Here, values similar to those of Abe and Matsui are used with C_o (= 3750 m s^{-1}) and s (= 1.6) for equation-of-state parameters for pyroxene-bearing serpentinite as given by Marsh (1980). Thus, we effectively are using the equation of state of serpentinite centered at the mean density of the planet and planetesimals. These are all simplistic assumptions.

We now consider only release of water from infalling planetesimals and neglect the contribution from shocked partially hydrous surface materials. Our previous studies have shown (Lange and Ahrens 1982*a*,*b*,1984) that redistribution and reactions of free atmospheric water with surface minerals and metallic iron are probably not significant for the latter parts of the accretion process if the more iron-rich material accreted first. Thus, we consider impact devolatilization, and hence, the buildup of a primary water atmosphere only occurred after the point where the planet has achieved a fraction f of its final mass exceeding $f = f_{\text{bo}}$. For most models, atmospheric accretion occurred only when $f_{\text{bo}} \geq 0.75$ (or when 75% of the Earth's mass had already accreted) (Lange and Ahrens 1982*a*, 1984). The assumption $f_{\text{bo}} > 0.75$ also implies that during the earlier parts of the accretion process most of the delivered and shock-released water will not reside in a proto-atmosphere (cf. Jakosky and Ahrens 1979).

We assume the amount of shock-induced water released ΔM_a for a given impact is obtained by

$$\Delta M_a = \begin{cases} 0 & \text{for } P < P_{iv} \\ m_{\text{pl}} f_w \ [(P - P_{iv})/(P_{cv} - P_{iv})] & \text{for } P_{iv} \leq P \leq P_{cv} \\ m_{\text{pl}} f_w & \text{for } P > P_{cv} \end{cases} \tag{38}$$

where m_{pl} is the planetesimal mass, $f_w m_{\text{pl}}$ the mass fraction of structurally bound water within the planetesimals, and P_{iv} and P_{cv} the shock pressures required for incipient and complete vaporization of water from serpentine. We assume that f_w ranges between 0.001 and 0.01 (Lange and Ahrens 1987*a*).

At each time step we calculate the surface temperature either by using the radiative model, as in Eq. (19), or by assuming an adiabatic atmosphere with the surface temperature T_s fixed at either 800 K or 1200 K; this range can be thought of as either representing a possible range of magma ocean types or representing the typical effective temperatures observed on terrestrial lava lakes (Gradie et al. 1988). The former could also correspond to the solidi of a very H_2O- and Na_2O-rich low-temperature magma, the latter corresponding to a more common relatively alkali-rich composition. Notably, both temperatures are lower than the 1500 K used by Abe and Matsui which are close to the solidi of contemporary basaltic magmas.

The surface pressure is given as

$$P_s = M_a g / A_p \tag{39}$$

where A_p is the current surface area of the planet $4\pi R^2$ whereas the scale height is

$$\Delta = T_s R / (g\bar{m}) \tag{40}$$

where $\bar{m}$ is the mean molecular weight of the atmosphere (18 for water), T_s is the surface temperature, and the surface density ρ_s is from the ideal gas law given as

$$\rho_s = P_s \bar{m} / R T_s. \tag{41}$$

Thus, for impact of each planetesimal we first calculate atmospheric gas gain using Eq. (17) of Sec. VI and subject to the conditions of Eq. (38). We then calculate the flow field produced by the impact and determine if any atmosphere is lost as in the case of impact into an atmosphere as outlined in Sec. VII.A.

The time history of the Earth's atmospheric evolution for models with high water content, $f_w = 0.01$, low temperature and small planetesimal diameter is shown in Fig. 39 for constant surface low temperature; model (b) depicts the evolution with radiative feedback between surface temperature and the amount of atmospheric gas present. As can be seen, upon initiation of an atmospheric blowoff process at $f_{bo} = 0.75$, the atmosphere lost reaches a maximum at $\sim 1.58 \times 10^{15}$s. This is followed by a series of local maxima and minima in M_e, which are numerical artifacts and reflect the unsteady behavior resulting when the mass of atmospheric gases present M_a and the amount of escaped gas at the same time, are of approximately the same magnitude (M_a and M_e are current atmospheric mass and the amount of impact ejected atmo-

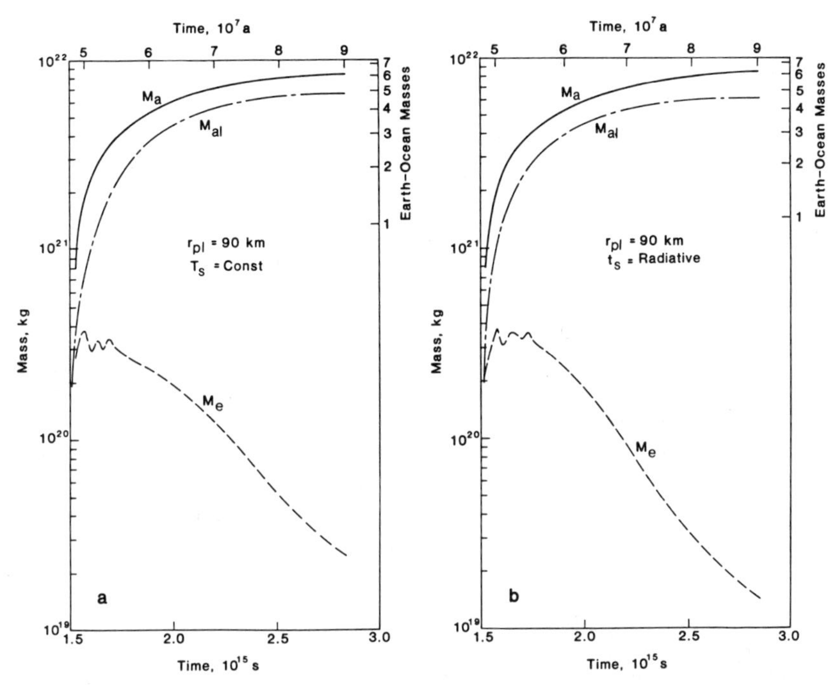

Fig. 39. Mass of atmospheric gases M_a, mass of blown-off atmosphere M_{al} and cumulative mass of escaped (dissipated) atmosphere M_e as a function of time during the final part of the Earth's accretion. The planetesimal radius $r_{\rm pl}$ used is 90 km. (a) constant surface temperature, 800 K; (b) radiative model.

sphere in each time step). Once M_a becomes significantly larger than M_e, the oscillations of M_e disappear and a steady decrease in M_e is observed. This is caused primarily by the fact that when $f(t)$ is approaching 1, the accretion rate $\dot{M}(t)$ and thus the number of planetesimals $n_{\rm pl}(t)$ and hence impactor rate has declined.

This decrease in M_e is somewhat more pronounced for model (b), resulting in less efficient impact cratering of atmospheric gases. This gives rise to a slightly lower value of the cumulative mass of escaped gas M_{al} at the end of accretion in Fig. 39b than Fig. 39a, caused by the higher surface temperatures T_s in this model. While for the Fig. 39a model the scale height Δ remains at an approximately constant value of 38 km, Δ values in model (b) are higher and vary significantly with time (Fig. 40a). As can be seen in Fig. 40a, due to the increase in atmospheric mass M_a, which is reflected in the increase in surface pressure P_s and surface density ρ_s, the optical depth and thus surface temperatures T_s increase correspondingly. This leads to an increase in atmospheric scale height Δ, which leads to *decreased* efficiencies of the atmospheric cratering process, mainly because of smaller atmospheric densities ρ_a for a given height (Eq. 29).

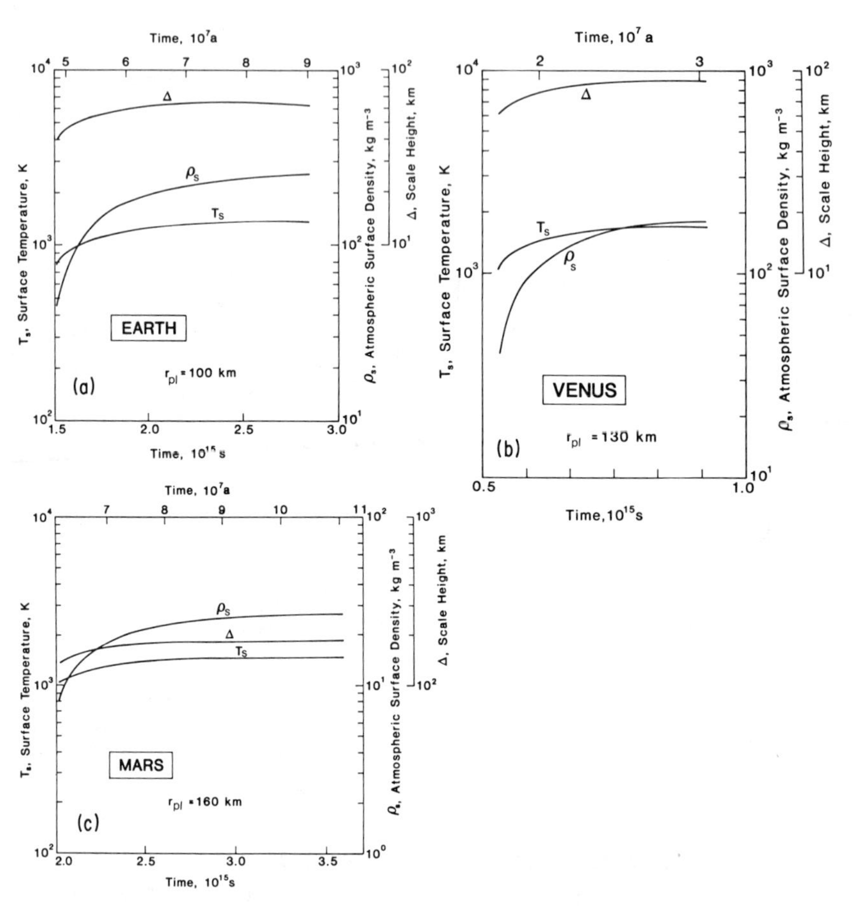

Fig. 40. (a) Earth surface temperature T_s, atmospheric surface density ρ_s and scale height Δ versus time. The planetesimal radius r_{pl} is 100 km. (b) Same as for (a), except for Venus, r_{pl} = 130 km. (c) Same as for (a) except for Mars, r_{pl} = 160 km.

Figure 41a and b shows the effect of varying planetesimal radius for a wet, $f_w = 0.01$ and cool Earth and Mars, respectively. This figure is a summary of a series of different calculations in which we varied planetesimal radius r_{pl} and found the radius of the planetesimal which induced the maximum atmospheric loss. The optimum planetesimal radius (which we define as r^*_{pl}) is thus that planetesimal size, which results in the maximum value of M_{al}, the cumulative mass of impact-dissipated (or eroded) gas at the end of accretion. The quantity M_{al} increases very slowly with increasing planetesimal radius for $r_{pl} \lesssim 15$ km. At $r_{pl} \sim 15$ km blowoff efficiencies increase dramatically with increasing r_{pl} until the optimum value r^*_{pl} is reached. From this point on, M_{al} decreases steadily with increasing r_{pl}, but at a slower rate than seen in the rising part of the curve.

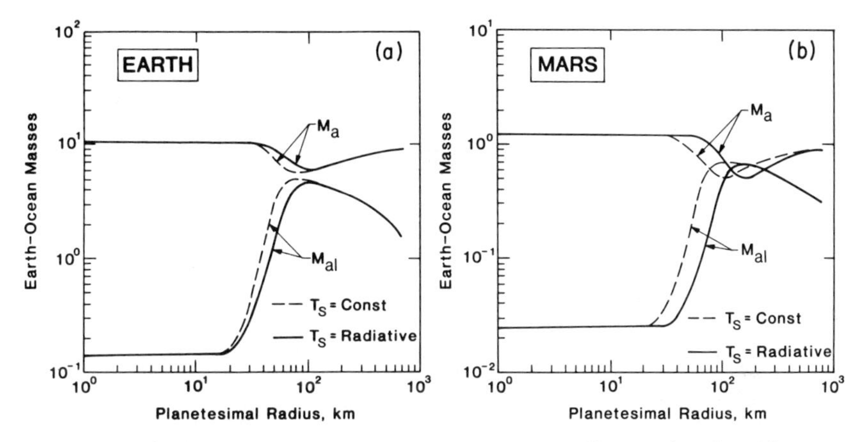

Fig. 41. (a) Mass of atmospheric gases M_a, and cumulative mass of impact-ejected gas M_{al} upon the accretion of the Earth as a function of planetesimal radii r_{pl}. Plotted are results for constant surface temperature T_s (dashed curves) and radiative models (solid curves); $f_{bo} = 0.75$ (b) Same as for (a), except for Mars.

This result parallels the result of Sec. VII.A for impact on air. There, we show that for small impactor energies (or impactor sizes), the efficiencies of impact cratering atmosphere loss (blowoff) are very small and become zero below a critical minimum energy. We also show that upon reaching maximum efficiencies, blowoff rates remain constant with increasing impactor energies.

This fact in conjunction with the decreasing planetesimal number as a function of size for a given accretion rate leads to the observed decrease in M_{al} for $r_{pl} > r^*_{pl}$. Thus, while the increase in impactor energy does not influence blowoff rates, the decrease in planetesimal number leads to a decreasing cumulative efficiency of the atmospheric cratering processes.

Shock-induced water release in infalling planetesimals starts, when the accreting planets are still small. For the Earth and Venus, this point lies at $f \simeq 0.16$ (Tyburczy et al. 1986). If we assume that time scales for processes, which consume atmospheric water (i.e., readsorbtion of water by anhydrous minerals [Jakosky and Ahrens 1979]) and iron water reactions (Lange and Ahrens 1984) require longer time scales than atmospheric cratering events (i.e., essentially the time span between subsequent planetesimal infalls), then atmospheric blowoff could be important in the earlier stages of accretion as well. In order to explore this effect, we conducted a series of calculations, where we varied f_{bo} between 0.001 and 0.9. For these models, we used the close-to-optimum planetesimal radius $r_{pl} = 10$ km (Fig. 42a) for the case of variable surface temperatures. Figure 42a gives the final values of M_a and M_e as a function of f_{bo}. For small values of f_{bo}, i.e., for models in which dehydration and blowoff is considered for earlier phases of the accretion process, the amount of atmospheric gases greatly exceeds the cumulative amount of blown-off gas (by up to a few orders of magnitude). This is due to the fact that

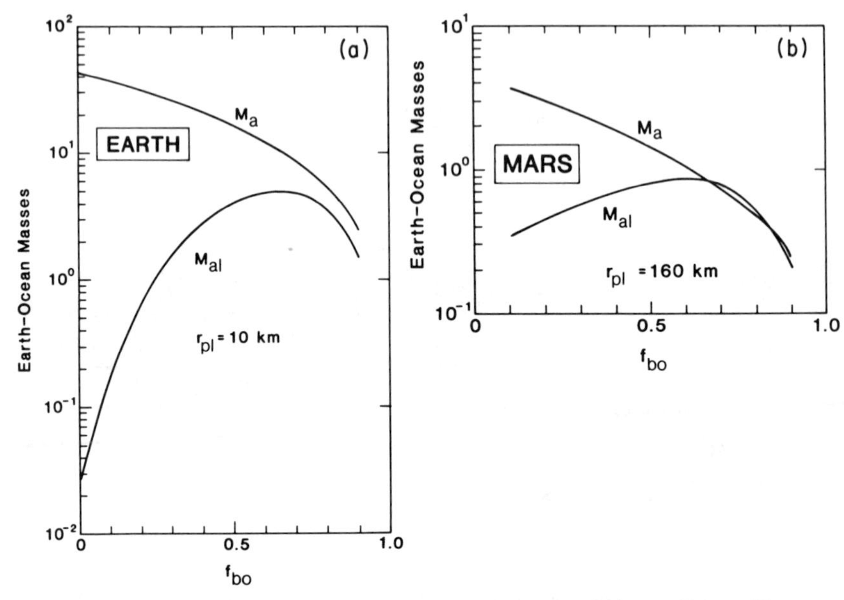

Fig. 42. (a) Mass of atmosphere M_a and final cumulative mass of blown-off gases M_{al} as a function of the mass fraction f_{bo} at which atmospheric cratering processes first act. Planetesimal radius r_{pl} = 10 km. (b) Same as for (a), except for Mars and r_{pl} = 160 km.

while infall energies are sufficient for shock-induced release of structural water, the fraction of this energy partitioned into the evolving atmosphere does not lead to appropriate blowoff efficiencies. A similar study is summarized in Fig. 42b for Mars.

The larger f_{bo} becomes, the smaller the difference between M_a and M_{al} gets; from about $f_{bo} = 0.65$, the curves are approximately parallel and for f_{bo} approaching 1.0, both curves approach 0. Thus, it can be seen (Figs. 41a and 42a) that atmospheric cratering is insufficient to remove completely an evolving proto-atmosphere, at least in the flat-planet approximation.

Several features of our models for Venus are similar to those of the Earth (see Fig. 40b for Venus). Because of the higher effective temperature T_e (Table X), surface temperatures are higher than on the Earth. This results in a greater atmospheric scale height, and, from the comparison of effects of temperature, predicts less efficient atmospheric cratering on Venus than on the Earth (e.g., Figs. 36a and b). Since for Venus, r^*_{pl} = 130 km for $f_w = 0.01$, the results shown represent the minimum feedback between surface temperature and mass of the atmosphere present. However, as in the case of the Earth, as the value assumed for f_w decreases, we expect r^*_{pl} to increase sharply and cratering efficiency to increase.

Mars demonstrates a smaller range of T_s, Δ and P_s (Fig. 40c) than the other terrestrial planets during the atmospheric accretion process. For Mars, we find $r^*_{\rm pl}$ = 160 km for f_w = 0.01 (Fig. 41b). We expect $r^*_{\rm pl}$ to increase sharply as f_w is decreased. The atmospheric mass M_a remains less for Mars than for Earth and Venus during accretion mainly because of the smaller value of g.

However, in contrast to the other planets, the cumulative mass of blown-off gas even for large values of $f_w \sim 0.01$ exceeds the amount of atmospheric gas present, but there is some atmosphere present at the end of accretion. Interestingly, the total atmospheric mass of dissipated gas exceeds this quantity. Thus, in contrast to the Earth and Venus where atmospheric cratering was less efficient and more atmospheric gas remained than was blown off, on Mars planetary cratering appears to be so effective that a large fraction of the volatiles delivered to the planet during accretion are probably lost (Fig. 41b). Other aspects of M_a and M_{al} versus $r_{\rm pl}$ are similar for the three planets, particularly the shift to a smaller value of $r^*_{\rm pl}$ for models with constant surface temperature and a decrease in M_a/M_{al}.

Figure 42b gives m_a and M_{al} as a function of $f_{\rm bo}$ for $r_{\rm pl}$ = 160 km (starting at $f_{\rm bo} = 0.1$, as discussed above for Mars). Similarly to the results of Fig. 41b, it can be seen that for $0.66 \lesssim f_{\rm bo} \lesssim 0.85$, the value of the functions M_{al} actually exceeds M_a. Thus, for models which treat atmospheric blowoff from planets with large values of $f_{\rm bo}$, the total rate of atmospheric loss by cratering can exceed the rate of the gas that is delivered to the planet.

Acknowledgments. This research was primarily supported by the National Aeronautics and Space Administration. M. A. Lange acknowledges the support of the Alfred Wegener Institute for Polar and Marine Research. We appreciate the technical discussions with Y. Abe about his work, and helpful comments from Y. Abe, R. Greenberg, K. Zahnle and an anonymous reviewer, and the assistance of S. Yamada in compiling this chapter.

Note Added in Proof: The question of energy coupling into the atmosphere upon impact of large bolides has been further studied. Recently, O'Keefe and Ahrens (1988) conducted a numerical finite-difference calculation with a geometry similar to that of Walker (1986*b*), except they also considered the high-velocity wake produced by passage of the projectile through an exponential atmosphere, as well as the jet induced by the strongly shocked air between the bolide and the surface.

ESCAPE OF ATMOSPHERES AND LOSS OF WATER

D. M. HUNTEN
University of Arizona

T. M. DONAHUE, J. C. G. WALKER
University of Michigan

J. F. KASTING
NASA Ames Research Center

The properties and limitations of several loss processes for atmospheric gases are presented and discussed. They include thermal loss (Jeans and hydrodynamic); nonthermal loss (all processes involve charged particles); and impact erosion, including thermal escape from a molten body heated by rapid accretion. Hydrodynamic escape, or "blowoff," is of particular interest because it offers the prospect of processing large quantities of gas and enriching the remainder in heavy elements and isotopes. In a second part, the water budgets and likely evolutionary histories of Venus, Earth and Mars are assessed. Although it is tempting to associate the great D/H enrichment on Venus with loss of a large initial endowment, a steady state with juvenile water (perhaps from comets) is equally probable.

I. INTRODUCTION

Historically, the study of atmospheric escape has focused on the loss of hydrogen and helium from the present atmosphere of the Earth, first suggested by Waterston in 1846 but given the name "Jeans escape" in honor of its quantitative formulation seventy years later (see Chamberlain and Hunten

1987, p. 400). It became almost a truism that light gases would be scarce in atmospheres like the Earth's, and that every gas would escape from a smaller body such as the Moon. However, not enough was known about other atmospheres, nor about the Earth's upper atmosphere, to test such ideas or to put actual numbers in them.

By 1949, the noble-gas abundances on the Earth and in the Sun had become reasonably well established. Suess (1949) and Brown (1949) independently concluded that these gases as a class are deficient by a factor of order 10^{-7}, in addition to a strong mass fractionation by another three orders of magnitude between neon and xenon. If the Earth ever had a primary atmosphere of solar composition, this is its only detectable remnant. Suess discussed the possibility that the effects might be due to Jeans escape, but quickly realized that this process cannot deliver a large fractionation except for tiny quantities of gas. Subsequently, the same point has been made by Shklovskii (1951), Öpik (1963) and Hunten (1973). However, Donahue (1986) has studied the possibility of carrying out a fractionation in planetesimal atmospheres, before the bodies were assembled into their final planets. This scenario gives a much larger area for the escape to work on, as well as smaller gravity wells to help make it work.

As discussed in Sec. II, a promising resolution of the difficulty has only recently been discovered in the process of "blowoff" or hydrodynamic escape. The papers of Shklovskii, Öpik and Hunten cited above all assumed that very large fluxes from a hydrogen-rich atmosphere implied indiscriminate, unfractionated loss of all heavier gases. Hunten (1979) and Sekiya et al. (1980*a,b*, 1981) showed that this is indeed true for sufficiently large hydrogen flux; Sekiya et al. also studied the hydrodynamics of the flow and suggested that enhanced solar EUV (extreme ultraviolet) radiation was required to drive it. More recently, as discussed in Sec. II, the fractionating effects of a more modest, though still large, flow have been quantitatively investigated. The mass fractionation stems from the competition between the upward drag forces and the downward force of gravity. In contrast to the Jeans process, the mass dependence is basically linear rather than exponential. Not enough time has passed since this discovery for the full implications to be studied, let alone for the generation of a consensus.

Nonthermal processes dominate the loss of hydrogen from the Earth and Venus at present, and perhaps ever since an initial transient state. On both planets, this dominant process is charge exchange between a hot proton and a thermal atom; the newly neutralized ion retains most of its original speed and can readily escape. On the Earth, the nonthermal loss rate is flexible enough to adjust to the thermal flux as it varies with solar activity; the sum remains equal to the rate controlled by diffusion in the lower thermosphere (cf. the review by Hunten and Donahue [1976]). A number of other nonthermal loss processes have been identified, one of which explains the anomalous isotopic ratio of nitrogen on Mars. Generally speaking, however, nonthermal escape rates are too small to affect the composition of a large reservoir such as the water in the Earth's oceans.

Hydrodynamic escape is not the only topic too young to have developed a consensus; another is atmospheric erosion during impacts of large bodies, up to asteroidal size, discussed in Sec. IV. It has been suggested that a "huge" impact by a Mars-sized body could have formed the Moon, but might also totally disrupt the Earth's atmosphere; however, there has been no quantitative study of the latter suggestion. Impact by a body of more modest size, on a planet as large as the Earth, cannot eject more than the fraction of atmosphere within line of sight, $\sim 5 \times 10^{-4}$. Typically, this fraction is multiplied by an efficiency factor no greater than 0.25 and much smaller for impacting bodies smaller than the scale height. The direct effect of the atmospheric passage is almost negligible for the cases studied so far; most of the loss is due to entrainment by ejecta from the surface impact. With such a low efficiency, it is quite possible for the planetesimal to bring in more volatiles than it ejects.

If accretion is rapid enough, it may be able to maintain a very hot surface and atmosphere. Among the likely consequences are an atmosphere containing all the available water as vapor, and rapid attainment of chemical equilibrium between interior and atmosphere. A large amount of water would react with reduced iron to produce free hydrogen, which could escape rapidly and drag along lighter gases.

Two principal scenarios for the initial environment of a planetary atmosphere are considered by workers in the field. The "Safronov" scenario assumes that the nebular gas has disappeared at an early stage of planetary accretion (Safronov 1969; Safronov and Ruzmaikina 1985). Thus, atmospheric material must be stored within the planetesimals in such forms as ices, water of hydration or chemically bound molecules. The "Kyoto" school (Hayashi et al. 1985) studies accretion in the presence of nebular gas and then considers the fate of the resulting massive proto-atmospheres. Their escape is necessarily hydrodynamic, and leads naturally to the associated mass fractionations. If the gas-poor scenario turns out to be correct, the most likely source for a hydrodynamic flow is ice from accreted low-temperature planetesimals.

The second major part of this review concerns loss of water from Venus, Earth and Mars. Here again, ideas are evolving rapidly. Discovery of the huge deuterium enrichment on Venus encouraged speculation that the planet started with an ocean equivalent of water, but suggestions of a steady state between selective escape and degassing or cometary accretion are becoming popular again. Both Venus and Mars are currently much drier than Earth, a fact that encourages ideas that they have lost most of their original endowment. It is generally believed that the Earth has retained most of its water, but this too is being questioned, partly because it too has a deuterium enrichment, at least relative to the value of Geiss and Reeves (1981) for the solar nebula. Moreover, if hydrodynamic escape is accepted as the cause of the observed noble-gas fractionation patterns, a substantial loss of hydrogen must have occurred.

II. THERMAL ESCAPE PROCESSES

In thermal escape, the loss of gas from a planet is powered by thermal energy resident in the gas or deposited there during the escape process. The case usually associated with the work of Jeans (1916,1925) makes sense only if the escape rate is small, so that the static structure of the atmosphere is not appreciably perturbed. Without this limitation, it is necessary to solve consistently for the structure, thermal balance and loss rate. To a smaller extent, this is already true for much more modest fluxes, such as that of hydrogen from the contemporary Earth. Here the structure and heat balance can be taken as given, but it is essential to consider the diffusive flow of the hydrogen through the background gas. The Jeans equation itself remains useful for rather large flows, but takes on a considerably different interpretation. General discussions may be found in Spitzer (1952), Chamberlain (1963), Tinsley (1974), Hunten (1973), Walker (1977) and Chamberlain and Hunten (1987).

Nearly all discussions of escape make use of the concept of an *exobase,* also called *critical level*. This is the level above which the integrated density of the atmosphere is equal to the mean free path, expressed in particles per unit area. If the collision cross section is Q and the region is isothermal with scale height H, the total density at the exobase n_c is given by

$$n_c QH = 1. \tag{1}$$

The region above the exobase is called the *exosphere,* and in it collisions are usually ignored. Thus, the escape flux can be considered to arise at the exobase. Again, these concepts make sense as long as the escape causes at most a small perturbation of the atmosphere. The most obvious limitation is the use of the plane-parallel approximation $n_c H$ in Eq. (1) for the integrated density. For an infinite, spherical atmosphere the actual integral diverges, and to obtain a reasonable result a more realistic model of the atmosphere is required. This topic has been considered by Herring and Kyle (1961), but the standard treatment is that of Chamberlain (1963). The latter is summarized, and the tables reproduced and extended, in Chamberlain and Hunten (1987).

Many of the properties of an exosphere and of escape are controlled by the ratio of the gravitational and thermal energies of an atom. Actually, the escape energy GMm/r is used, rather than the gravitational energy, which is negative; G is the Newtonian gravitational constant, M and m the masses of the planet and the atom, and r the radial distance. The translational thermal energy is $(3/2)kT$, where k is Boltzmann's constant and T the temperature. Omitting the 3/2, we obtain the escape parameter

$$\lambda = \frac{GMm}{kTr} = \frac{r}{H(r)} = \frac{v_e^2}{U^2}. \tag{2}$$

All these expressions are functions of r; if appropriate, they can be given the subscript c (critical level) or, less commonly, x (exobase). The escape velocity is v_e and the most probable speed of the Maxwellian is

$$U = \sqrt{2kT/m}. \tag{3}$$

Values of λ_c greater than 20 to 30 correspond to an atmosphere that is effectively plane-parallel for the mass m considered, although escape can be noticeably different from zero. For terrestrial H at modest exospheric temperature (800 K), λ_c is around 15, giving a substantial escape flux; however, the exospheric structure is controlled by O, for which λ_c is sixteenfold greater.

The situation $\lambda_c = 2$ corresponds to the solar wind (Parker 1963; Hunten 1973). The corresponding state in a planetary atmosphere is called hydrodynamic escape or, more colloquially, blowoff. The thermal structure and density profiles must be obtained as part of the solution. Normally the escape rate will be limited by the energy supply. For less extreme values of λ_c, vertical temperature gradients in the exosphere are small; the temperature is determined by a balance between ionospheric and auroral heating and downward conduction to the mesopause region. Typically, however, the temperature varies appreciably with time of day and latitude.

For an isothermal exosphere, the barometric equation may be written in the form

$$n(r) = n_c e^{\lambda - \lambda_c} \tag{4}$$

which reduces to the familiar $n_c e^{-z/H}$ when λ_c is sufficiently large. In a gas mixture, each constituent has its own copy of Eq. (4), since a static exosphere is in diffusive equilibrium.

As long as λ_c is not too small, the escape flux is given by the Jeans equation, which is obtained by integrating the upward flux in a Maxwellian distribution at the exobase, appropriately weighted in solid angle and with a lower limit at the escape speed. A factor B, typically ~0.5, is introduced to account for the fact that the high-velocity tail is somewhat depleted by the escape itself, as discussed below. The result is

$$F_J = Bn_c \frac{U}{\sqrt{\pi}} (1 + \lambda_c) e^{-\lambda_c}. \tag{5}$$

For atomic hydrogen on the Earth, typical values for low solar activity are $T =$ 900 K, $\lambda_c = 8$, $n_c = 10^5$ cm^{-3}, $B = 0.75$ and $F_J = 6 \times 10^7$ cm^{-2} s^{-1}. At this relatively low temperature, a considerably larger flux is produced by nonthermal processes, as discussed in the next section. If the total flux has been the same over the history of the Earth, the total quantity is equivalent to a few m of liquid water. However, a small change in atmospheric structure could increase the flux by a large factor, and even bigger differences during early history are readily conceivable (Sec. VI).

Although the use of the exobase level in the derivation of Eq. (5) seems arbitrary, the result is accurate as long as the velocity distribution is Maxwellian (Chamberlain 1963; Chamberlain and Hunten 1987), which also implies $B = 1$. As a partial justification, we note that the barometric Eq. (4) can be written $n_c e^{\lambda_c} = n e^{\lambda}$; substitution in Eq. (5) leaves the slowly varying factor λ_c as the only remnant of conditions at the exobase.

The reduction factor B is due to the fact that the high-velocity tail of the Maxwellian, depleted by the loss of atoms to space, can be restored only by collisions with other atoms. In the well-studied cases of H escaping from an ambient gas of O or CO_2, the restoration is inefficient because of the large mass ratio. The early attempts to model this phenomenon led to a confusing variety of results, summarized by Chamberlain and Campbell (1967), whose results themselves were correct only by compensation of two errors. Brinkmann (1971) and Chamberlain and Smith (1971) finally obtained convergence. Their Monte Carlo results have recently been confirmed by a treatment based on numerical solutions of the Boltzmann equation (Shizgal and Blackmore 1986; see Fig. 1).

For decades after its formulation, Eq. (5) was regarded as giving the flux F for a specified value of the exobase density n_c, although for most of the period too little was known about the upper atmosphere to say anything quantitative. This view is valid for helium and heavier atoms on the Earth, for which the flux is extremely small. However, for hydrogen the flux is in fact controlled by diffusion through the 100 to 130 km region, and n_c must adjust to give the flux. In the diffusion-controlled region, the H density falls off rapidly; at greater heights it takes on the normal very large scale height of diffusive equilibrium, around 800 km. Relatively small changes in the height at which these two regimes intersect are sufficient to give the necessary adjustment of n_c. The essence of this description is not changed by the fact that there are nonthermal losses in addition to the thermal ones described by the Jeans equation.

Thermal escape runs into two problems in trying to explain observed elemental fractionation patterns, both related to the negative exponential dependence of the flux on λ in Eq. (5). Over a substantial range of mass (say, from neon to xenon), the value of λ varies over the same range. If it has a reasonable value, say 10, for neon, it is so large for xenon that the flux of the latter is totally negligible. Second, the flux itself is always small if fractionation is to occur, and the composition of a substantial reservoir is difficult to affect. Both problems can be circumvented by postulating that the medium for the escape was a multitude of growing planetesimals (Donahue 1986). Taken

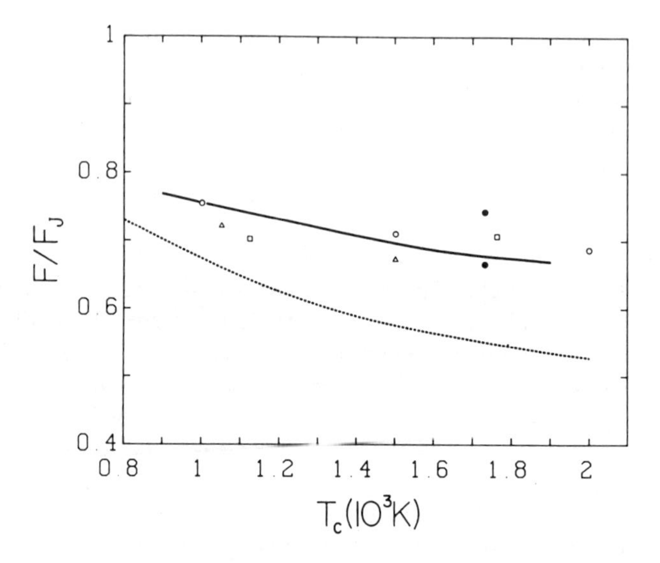

Fig. 1. Upper line: the Brinkmann factor *B* (see Eq. 5) as calculated by Shizgal and Blackmore (1986) for escape of H from the Earth. Earlier Monte Carlo results, shown by the symbols, are in good agreement, unlike the previous analytic result represented by the lower line.

together, they offer a very large area for the escape, and the growing masses replace the exponential tendency with a linear one.

Diffusion

The possibility that loss of gases might be inhibited by their slow diffusion through the stratosphere was raised by Harteck and Jensen (1948; see also Urey 1959*a*). These ideas were abandoned when it was learned that the atmosphere remains well mixed to ~100 km, but were then applied to the thermosphere (Bates and McDowell 1957,1959; Bates and Patterson 1961; Mange 1961; Kockarts and Nicolet 1962; Banks and Kockarts 1973). The following is based on the formulation of Hunten (1973), also summarized by Hunten and Donahue (1976) and Chamberlain and Hunten (1987).

The diffusion equations for a two-component gas of number density $n = n_1 + n_2$ are

$$w_1 - w_2 = \frac{F_1}{n_1} - \frac{F_2}{n_2} = -\frac{b}{n}\left(\frac{1}{n_1}\frac{dn_1}{dz} + \frac{m_1 g}{kT} + \frac{1+\alpha_1}{T}\frac{dT}{dz}\right) - \frac{K}{X_1}\frac{dX_1}{dz} \quad (6)$$

and a similar equation with interchanged subscripts. Vertical velocities are represented by *w*, and fluxes by *F*. The diffusion coefficient *D* is represented by *b*/*n*; *b* is nearly constant except for a temperature variation typically as $T^{0.7}$. The acceleration of gravity is *g*, and $\alpha_1 = -\alpha_2$ is the thermal diffusion factor. The equations are taken from Chapman and Cowling (1952), except for the last term, which represents vertical mixing as eddy diffusion, with coefficient K; $X_i = n_i/n$ is the mole fraction. In the parentheses, the second term is the reciprocal of the pressure scale height H_i, and the sum of the second and third terms is the reciprocal of the density scale height; however, to simplify the argument we shall ignore the distinction, or in other words ignore temperature gradients. We shall also omit the eddy term, an approximation that improves with altitude due to the $1/n$ factor in the molecular diffusion term. Moreover, in the cases of interest here the diffusion terms tend to enforce a height-independent mole fraction, so that the derivative in the eddy term will be small or zero. Finally, we omit the variation of gravity with height. More general treatments are given by Hunten (1973), Zahnle and Kasting (1986) and Hunten et al. (1987). We thus obtain the pair of equations

$$\begin{aligned} \frac{\mathrm{d}n_1}{\mathrm{d}z} &= -\frac{m_1 g}{kT}\, n_1 + \frac{1}{b}\,(n_1F_2 - n_2F_1) \\ \frac{\mathrm{d}n_2}{\mathrm{d}z} &= -\frac{m_2 g}{kT}\, n_1 + \frac{1}{b}\,(n_2F_1 - n_1F_2). \end{aligned} \tag{7}$$

Their sum is simply a hydrostatic equation for total mass density; the mutual diffusion effects are exhibited by differentiating the definition of the mole fraction X_2 and substituting, remembering that $X_1 + X_2 = 1$. The most interesting form of the result is for height-independent composition, that is X_1 and X_2 constant:

$$F_2 = \frac{X_2}{X_1} F_1 \left[1 - \frac{bgX_1}{kTF_1}(m_2 - m_1) \right] = \frac{X_2}{X_1} F_1 - \frac{bgX_2}{kT}(m_2 - m_1). \tag{8}$$

The two fluxes have been assumed to be independent of height (or zero), as required by continuity in the absence of chemical change. More of the algebraic details are given by Hunten et al. (1987).

Our first use of Eq. (8) is to derive the expression for the diffusion-limited, or "limiting," flux for a minor light gas, such as hydrogen in air. If subscripts are interchanged in the second form, and F_2 taken as zero, the result is

$$F_1 = \frac{bgX_1}{kT}(m_2 - m_1) = \frac{bX_1}{H_2}\left(1 - \frac{m_1}{m_2} \right). \tag{9}$$

Since m_1 is normally much smaller than m_2 the limiting flux is approximately bX_1/H_2. This flux has only a weak dependence on temperature, of order $T^{0.3}$, because b typically varies as $T^{0.7}$. Hydrogen loss from the Earth (and probably Mars) obeys Eq. (9) as accurately as the values are known (Hunten and

Donahue 1976; Walker 1977; Tinsley 1974; Breig et al. 1976). The number density at the exobase adjusts so that the major loss processes operate at the required rate. These processes are thermal escape and charge exchange with hot protons; the thermal part varies considerably with exospheric temperature, and the rest is taken up by the charge exchange.

In *hydrodynamic escape,* the flow is already supersonic before it reaches a density typical of an exobase; the latter level has lost most of its meaning under such conditions. Such flows pass the speed of sound at $\lambda = 2$ (Parker 1963). The planetary case has been studied by Sekiya et al. (1980*a*,*b*,1981), Watson et al. (1981), Kasting and Pollack (1983) and Zahnle and Kasting (1986). Before describing the nature of the flow, we shall consider its composition. The first form of Eq. (8) is rewritten in terms of a *crossover mass* m_c, defined as

$$m_c = m_1 + \frac{kTF_1}{bgX_1} \ . \tag{10}$$

This expression gives the mass in g; for amu, Boltzmann's constant can be replaced by the universal gas constant R. Equation (8) becomes

$$F_2 = \frac{X_2}{X_1} F_1 \left(\frac{m_c - m_2}{m_c - m_1} \right) . \tag{11}$$

Both Eqs. (8) and (11) can give a negative value of F_2 if F_1 is not large enough, or equivalently if $m_2 > m_c$. Such a downward flux will not exist in any realistic situation and must be replaced by zero; X_2 then falls off with increasing height. The fractionation expressed by Eq. (11) can be rather strong for masses less than m_c, as illustrated for some simple cases in Fig. 2, from Hunten et al. (1987), who also give other examples. This treatment is limited to a binary mixture; its extension to three or more components is unproven, though plausible as long as the light gas (subscript 1) dominates.

The fractionation patterns of Fig. 2 are obtained for a large, constant flux that gives a crossover mass of 200, with either of two assumptions: a constant or a declining atmospheric inventory. The fluxes must be proportional to the inventory for each constituent, a reasonable but restrictive assumption: the entire inventory must reside in the atmosphere, and allowance must be made for phenomena such as condensation and precipitation of water vapor in the Earth's atmosphere. Equation (11) does show flux to be proportional to mole fraction for the minor constituent. For the declining inventory, the evolution can be very simply treated as a Rayleigh process, which illustrates an important point even if it is not always realistic. If N_1 and N_2 are the inventories of the light and heavy gas, their loss rates are related by

$$\frac{dN_2}{dN_1} = \frac{N_2}{N_1(1 + y)} \tag{12}$$

where the *fractionation factor* $1 + y$, usually only slightly greater than unity, depends on the process. [Other terminology has been used: $1/R$ (McElroy and Yung 1976; McElroy et al. 1977); $1/x$ (Zahnle and Kasting 1986).] The integral of Eq. (12) is

$$\frac{N_2}{N_2^0} = \left(\frac{N_1}{N_1^0}\right)^{1/1+y}. \tag{13}$$

For Eq. (11), the exponent is simply $(m_c - m_2)/(m_c - m_1)$. Equation (13) gives the evolution of N_2 as a function of N_1 in terms of their initial values.

One well-known system that obeys Eq. (12) is the enrichment of ^{235}U by diffusion through a membrane, with diffusion coefficients inversely proportional to the square root of mass. Since the vapor used is the hexafluoride, the value of $1 + y$ is $\sqrt{352/349}$ and $y = 0.0043$. For n stages, the enrichment is $(1+y)^n$; if a mere factor of 2 is required, $n = 0.693/y = 161$. If Eq. (11) is obeyed, $y = (m_2 - m_1)/(m_c - m_1) \simeq 1/m_c$ for unit mass difference and large m_c. Curiously, the efficiency of separation of adjacent masses does not depend on the actual mass, unlike the diffusive case.

With the typical small values of y, a very large fraction of the original gas must be lost to get a large change of composition. The same principle is

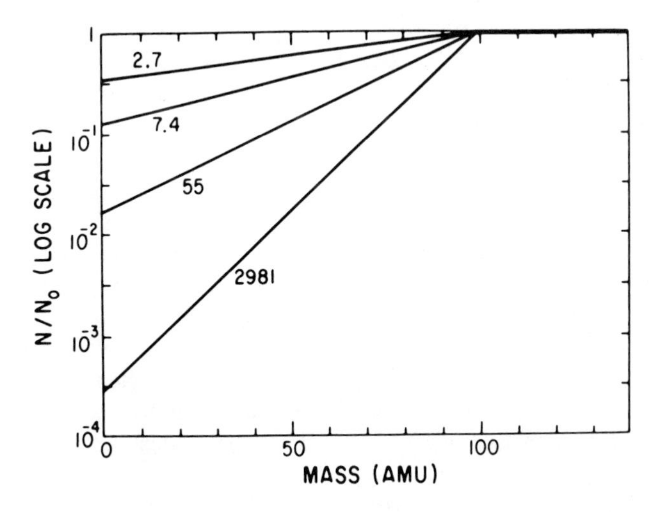

Fig. 2. Patterns of mass fractionation for hydrodynamic escape with a crossover mass of 100 amu. The ordinate is the ratio of remaining to initial inventory, and the numbers on the lines are proportional to time for Rayleigh fractionation (Eqs. 12 and 13). Similar patterns apply to the case of a constant atmospheric inventory maintained by degassing (figure from Hunten et al. 1987).

important even if fractionation is fully efficient, equivalent to an infinite value of y: to change the composition by a factor of 2, half the original gas must be lost.

Limits on the Flux

In principle there are three likely effects that can limit the escape flux of a constituent: the source strength (degassing, chemical change); diffusion; and energy input. Diffusion limits apply to a minor light gas escaping through a heavier one (for example, hydrogen on the Earth), and energy limits to a major light gas in or near the blowoff state. Actual cases do not always fall into a neat classification; citing hydrogen on the Earth again, the operation of the diffusion limit depends on the processes that limit the hydrogen mole fraction in the stratosphere and mesosphere, which are of the nature of source limits. Limitation by diffusion has been discussed above (see Eq. 9). We do not attempt a general discussion of source limits, although one example, by Kasting and Pollack (1983) is shown in Fig. 9 below. Factors determining the hydrogen mole fraction in the Earth's upper atmosphere are discussed in the chapter by Kasting and Toon.

Energy limits can appear in two different ways: directly affecting the possible flux, and indirectly through the magnitude of the exospheric temperature. A substantial thermal flux requires that λ be smaller than 10 to 15 and therefore that T/m be sufficiently large for the mass m under consideration. Direct energy limits for hydrodynamic flow were first discussed by Sekiya et al. (1980*b*) and Watson et al. (1981), and this work has been pursued by Kasting and Pollack (1983), Zahnle and Kasting (1986) and Hunten et al. (1987), from which the following is taken. Earlier attempts were made by Gross (1972) and Hunten (1979).

The detailed studies show that the limitation arises at very high altitudes, where the rapid expansion tends to cool the gas adiabatically; without a heat source, the temperature becomes so low that the escape is shut off. At such low densities the only significant heat source is solar ionizing radiation, which is absorbed with a cross section as large as 10^{-17} cm^2. The required energy per atomic mass unit is approximately the escape energy of a hydrogen atom GMm/r (definitions as for Eq. 2 above), evaluated at a level that may be considerably greater than the planetary radius a. For a global-mean energy input of ϕ, the energy-limited escape flux (amu s^{-1}) is $F_1 = \phi r/GMm$. Since $g = GM/r^2$, the crossover mass, Eq. (10), can be written

$$m_c - m_1 = \frac{kT}{bX_1}\frac{\phi r^3}{G^2M^2m} = \left(\frac{3}{4\pi G\rho}\right)^2 \frac{\phi kT}{bX_1 m}\frac{r^3}{a^6}\,. \tag{14}$$

The heating radius can be as great as $2a$, because the escaping atmosphere is very extended (Kasting and Pollack 1983), and the last term is then $8/a^3$.

TABLE I
Energy-Limited Fluxes (amu cm^{-2} s^{-1}) and Corresponding Crossover Masses for a Global-Mean Solar Input of 1 erg cm^{-2} s^{-1} [a]

	Earth	Mars	Moon	Planetesimal
Flux	1×10^{12}	5×10^{12}	2×10^{13}	5×10^{13}
m_c	2.5	20	197	1030
λ	300	61	14	4.5

[a]The planetesimal has the density of the Moon and a radius of 1000 km. The last line shows thermal escape parameters for a mass of 40 and a temperature of 1000 K.

Kasting and Pollack found a height around 2000 km, or a radius of 1.3 *a;* the examples in Table I use $r = a$, with $b = 2.2 \times 10^{19}$ cm^{-1} s^{-1} and $T = 400$ K. Values of λ for a mass of 40 and a temperature of 1000 K are included. The solar flux of 1 erg cm^{-1} s^{-1} is appropriate at 1 AU from the present Sun; quite different values exist at other distances and early in solar system history. Even for the present Earth the solar input could support a crossover mass of 2.5; with enhanced fluxes or smaller bodies this mass can easily attain 100 to 200.

The history of the solar flux has been discussed by Zahnle and Walker (1982*b*) and Canuto et al. (1982) on the basis of ultraviolet observations of young stars. For the region below 1000 Å the results of Zahnle and Walker can be represented by

$$\phi/\phi_0 = 10^8 t^{-5/6} \tag{15}$$

where ϕ_0 is the present flux and the time t is in yr. The enhancement factor is 150 at 10^7 yr; the corresponding crossover mass for the Earth is 225. Several examples of atmospheric fractionation, using Eq. (15) but with some variation of the exponent, are given by Hunten et al. (1987) and in the chapter by Pepin.

However, it is possible that the full energy (Eq. 15) may not have been available at the planets because of nebular absorption, which depends on the timing of the accretion process. Prinn and Fegley (see their chapter) point out that the EUV opacity would have been enormous in the presence of any nebular gas, and that dust from collisions between planetesimals might have maintained substantial opacity even after the gas had dissipated. Thus the availability of solar EUV energy could have been low until the accretion process was essentially over, some 10 to 100 Myr after it began (Safronov 1969). It must, however, be noted that Eq. (15) is based on observations of T Tauri stars, which cannot possibly have substantial opacity in the directions from which they are observed. It could still be argued that we are out of the plane of any preplanetary nebulae around these particular stars.

III. NONTHERMAL ESCAPE PROCESSES

Nonthermal processes dominate the present escape flux from Earth and probably Venus, and are responsible for the 75% enrichment of the $^{15}N/^{14}N$ ratio on Mars. Generally speaking, the actual fluxes are fairly small, though not as small as Jeans fluxes, and large compositional effects are therefore confined to small reservoirs. Large reservoirs include the Earth's oceans and the oxygen on Mars (as judged from the absence of measurable isotopic fractionation). With only one exception, the known nonthermal loss processes involve ions, and for all of them the production rates are limited by the smallness of the solar ionizing input and, in many cases, by competition for these photons by more abundant gases.

This section closely follows a survey by Hunten (1982), and much of it is directly quoted from the paraphrase in Chamberlain and Hunten (1987). A summary of the known nonthermal mechanisms is given in Table II. As mentioned above, the only one that does not directly involve ions or electrons is 3(b), photodissociation. Generally speaking, charged particles have difficulty escaping because they are trapped in magnetic fields; even a body that has no internal field is generally enveloped in a field of solar-wind origin. However, open field lines do exist, as in the Earth's polar regions (process 7). Charge exchange, process 1, can get rid of the charge while retaining most of the kinetic energy of the original ion. In processes 2–4, excess energy of reaction, or of the bombarding photon or electron, is converted into kinetic energy of the dissociated atoms. The incident fast particle in process 5 is normally an ion; fast atoms will work almost as well, but tend to be scarce. Acceleration of the ion is shown as process 8. The term "knock-on" tends to be applied when there is a single collision; "sputtering" refers to a multiple process, very common in electrical discharges, in which several atoms or molecules can be ejected in a backward direction. In process 6, an ion is produced far enough from the planet to be in the solar wind and be swept up in the flow. The same thing can happen to a satellite atmosphere in a rotating magnetospheric flow; Io is a prominent example.

A nonthermal atom can escape only if it is aimed upwards, has escape velocity, and is above the exobase, although a few downward velocities can be turned around in a collision. Slower atoms, including those that have lost speed in a collision, may remain bound and form a nonthermal *corona*. In turn, coronal atoms are particularly exposed to the sweeping process 6. Escape processes for bodies of Mars size or smaller tend to form coronae about larger planets. Table III indicates which processes are most active on the smaller planets and larger satellites; corona formation is indicated by (C). The abbreviations in the third column refer to Table II.

In most cases, nonthermal reactions generate rather high speeds, well above escape speed. There may be some mass dependence, since momentum conservation in a dissociation event favors the lighter atom. Nevertheless,

TABLE II
Nonthermal Processes Leading to Escape

Process	Examples[a]	Product[b]	Remarks
1. Charge exchange	(a) $H + H^{+*} \rightarrow H^+ + H^*$	N	
	(b) $O + H^{+*} \rightarrow O^+ + H^*$		
2. Dissociative recombination	(a) $O_2^+ + e \rightarrow O^* + O^*$	N	Energy divided equally
	(b) $OH^+ + e \rightarrow O + H^*$		H takes nearly all the energy
3. Impact dissociation	(a) $N_2 + e^* \rightarrow N^* + N^*$	N	e^* may be a photoelectron or an accelerated one
photodissociation	(b) $O_2 + h\nu \rightarrow O^* + O^*$		
4. Ion-neutral reaction	$O^+ + H_2 \rightarrow OH^+ + H^*$	N	
5. Sputtering or knock-on	(a) $Na + S^{+*} \rightarrow Na^* + S^{+*}$	N	Sputtering requires kilovolt or greater energies
	(b) $O^* + H \rightarrow O^* + H^*$		Knock-on requires much less.
6. Solar-wind pickup	$\{$ $O + h\nu \rightarrow O^+ + e$	I	Also electron impact
	O^+ picked up		Also magnetospheric wind for satellites
7. Ion escape	H^{+*} escapes	I	Requires open magnetic-field lines
8. Electric fields	$X^+ + eV \rightarrow X^{+*}$	I	Generates fast ions and electrons that participate in other processes

[a]An asterisk represents excess kinetic energy.
[b]N = neutral and I = ion.

TABLE III
Summary of Escape Mechanisms for the Terrestrial Bodies; Corona-Forming Processes (*c*) Are also Shown

Planet	Gas	Mechanisms[a]
Earth	H	1, CE; J; 7, IE
	D	same
	He	7, IE
	O(c)	2a, DR
Venus	H	2b, DR or 5b, KO
	H(c)	1, CE; 4, I–N
	O(c)	2a, DR
Mars	H	J
	H_2	J
	N	3a, ID; 2, DR
	O	2a, DR
Moon	H	J; 6, SWP
Mercury	He, Ar,---	6, SWP
Io	H, C, N,---	J?
	SO_2, S, O, O_2	?
	Na, K	5, Sp
E, G, C[b]	H, H_2	J
	O	2a, DR
Titan	H	J
	H_2	J; 2, DR
	N	3a, ID
Pluto	CH_4	J

[a]See Table II for abbreviations. J stands for Jeans or thermal escape.
[b]E, G, C are the Galilean satellites Europa, Ganymede, and Callisto.

there usually is a strong selectivity in the escape flux, because of diffusive separation. Solar ionizing photons tend to be absorbed at heights somewhat below the exobase but still well above the homopause (the level at which diffusive separation begins). Following Hunten (1982), we define the *exobase* as the bottom of the region from which escaping atoms originate, and Δz as the distance of the exobase above the homopause. Above the homopause, the two components have scale heights kT/m_1g, kT/m_2g, and the ratio of their densities changes with a "differential scale height" $H_d = kT/(m_2-m_1)g$. The fractionation factor is the same as the change in composition over a height range Δz, or

$$1 + y = \exp\left(\frac{\Delta z}{H_d}\right). \tag{16}$$

Since the exponent is usually small,

$$y \simeq \frac{\Delta z}{H_d} . \tag{17}$$

As noted above for cases described by Eq. (11), the magnitude of the fractionation depends on the difference of the masses, not the ratio.

A prominent example of enrichment described by Eq. (17) is nitrogen on Mars, discovered by Viking (Nier et al. 1976) and analyzed by McElroy et al. 1976,1977). The ratio $^{15}N/^{14}N$ is greater than the terrestrial value by a factor of 1.75. Appropriate values of Δz and H_d are 70 and 400 km, and thus $y \simeq 0.18$. Application of the Rayleigh formula (Eq. 13) suggests an initial endowment of about 2 mb, 10 to 15 times the present amount, and would require that all the nitrogen was degassed very early in planetary history. McElroy et al. consider several other scenarios of episodic or continuous degassing, all of which imply a larger initial endowment.

A very similar value of y applies to oxygen isotopes, but there is no detectable effect on the ratio. Presumably the total reservoir is much larger than that of nitrogen, and includes CO_2 and H_2O in the atmosphere, regolith and polar caps.

IV. IMPACT EROSION

Very large amounts of energy are released by the impact of an extraterrestrial body with a rocky planet. Erosion of an atmosphere can result if sufficient energy is imparted to atmospheric gas to blow it completely out of the gravitational field of the planet (O'Keefe and Ahrens 1982*a,b;* Lewis et al. 1982; Cameron 1983; Walker 1986*b;* Ahrens and O'Keefe 1987; chapter by Ahrens et al.). Quantitative study of the subject began only a few years ago, and in some areas has not even started. This section discusses several topics: the direct effect of the incoming object on the atmosphere (relatively minor in the terms defined below); acceleration of the local atmosphere by energy and ejecta from a large crater (which can happen for large events); accretionary heating of the whole planet and therefore the atmosphere; and the postulated effect of a huge impact such as the one currently favored for forming the Moon.

Event Sizes

Only a small fraction of the atmosphere is within line of sight from a point on the surface. We shall define a *large* event as one that ejects an amount comparable to this. A *small* event (though possibly very large for other purposes) ejects a much smaller quantity, and a *huge* event conceivably much more, perhaps even a significant fraction of the entire atmosphere. Quantitative information is available only for small events; the size of a large impact is not known.

An expression for the visible fraction of the atmosphere may be derived by noting that the height of the horizon plane at a distance x is approximately $z_h = x^2/2a$, where a is the planetary radius. For a constant scale height H, the density can be integrated upward from this level and horizontally with a weight $x dx$. The fraction of the atmospheric mass is found to be $H/2a$. For the Earth, H/a is almost exactly 10^{-3} and the visible fraction 5×10^{-4}. For other bodies of the same density, the scale height varies as $1/a$ and the visible fraction scales as $1/a^2$; it also scales inversely as the density. Thus, a body one-tenth the Earth's diameter and half as dense would have a visible fraction of 0.1, and even larger for a higher temperature or a lighter gas. This is among the factors that make impact erosion much more significant for a smaller body. For the Earth, 2000 large impacts would be required to reduce the atmospheric mass by $1/e$, or 10^4 impacts for a factor of 10^{-2}.

Atmospheric Interaction

A preliminary theory of the process of direct interaction has been developed by Walker (1986*b*); the conclusions are in accord with numerical studies and estimates by Ahrens and O'Keefe (1987), who then devote most of their attention to the indirect effects generated by the ejecta from the surface. Walker's treatment emphasizes that the response of an atmosphere to extraterrestrial impact is quite different from the response of the solid phase of the planet, mainly because of the large compressibility of the gas. Large compressibility implies that the gas is raised to very high temperatures and densities by compression, a process that plays much less of a role in the impact between solid bodies. On the other hand, the impact of a large extraterrestrial body imparts energy to the atmosphere over an extended area and over a period of some tens of seconds. For this reason, the response of an atmosphere to an impact event is quite different from the response of an atmosphere to a nuclear explosion, which occurs in a very short time and a very small volume, and is more relevant to the atmospheric effect of the actual surface impact. The most useful background information for the erosion problem appears to be that derived from the study of meteorite impacts on planetary atmospheres.

The first condition that must be met if impact erosion is to occur in large amounts is that the dimensions of the impacting body must be much larger than the atmospheric scale height. Shock-heated air can flow around a smaller impactor, dispersing impact energy into a relatively large volume of atmosphere. A sufficiently large impactor, on the other hand, imparts to the atmosphere that it encounters an energy per unit mass equal to the square of the impact velocity V_s. Since the impact velocity must exceed the planetary escape velocity V_e, the shock-heated gas could escape from the planet if the energy were not diluted. Dilution does occur, however, mainly because the very high temperatures (around 20,000 K) achieved by the shock-heated gas cause ablation, evaporation and dissociation (but little ionization) of the material of both the impactor and the solid planet. The energy that the impactor imparts to the gas it encounters must be shared between atmospheric gas and this vaporized material. Significant escape can occur only if there is enough energy in the gas to carry the mixture off into space.

This evaporative loading is described by the parameter

$$E = \Lambda \frac{V_s^2}{2Q} \tag{18}$$

where E is the relative contribution of evaporation to the mass of shock-heated gas, Λ describes the efficiency of the evaporation process and has typical values between 0.1 and 0.2 (Bronshten 1983) and Q is the latent heat of evaporation of the material of the impactor, with a typical value for meteors of 8×10^{10} erg g^{-1}.

The approximate theory of the interaction of impactor with atmosphere and the subsequent expansion of shock-heated gas outwards and upwards indicates that the mass of gas that can escape from a planet as a result of collision with an impactor of radius R is

$$M_e = \frac{\pi R^2 \sigma_a V_s^2}{V_e^2} \tag{19}$$

where σ_a is the mass of the atmosphere per unit area. The mass of gas that can escape is therefore the mass of the atmosphere intercepted by the impactor multiplied by the square of the ratio of impact velocity to escape velocity. The escaping gas, however, is made up of atmospheric gas and the vapor of the impactor and the planet in the proportions of 1 to E. The mass of atmospheric gas that escapes is therefore $M_e/(1+E)$. This is just the mass of atmospheric gas intercepted by the impactor multiplied by an *enhancement factor* equal to

$$\frac{V_s^2}{V_e^2(1 + E)} . \tag{20}$$

Significant escape occurs only when the enhancement factor is >1. A factor <1 implies that evaporative loading overwhelms impact heating so that almost none of the resultant gas mixture has enough energy to escape.

Illustrative values of the enhancement factor are plotted in Fig. 3 as a function of planetary mass and for various values of the ratio of impact velocity to escape velocity. For purposes of this calculation Λ was assumed equal to 0.16. Because this value is uncertain, the quantitative predictions of the theory may not be correct, but the qualitative behavior of the results should provide a useful guide.

Impact erosion does not occur from a planet of mass exceeding a limit

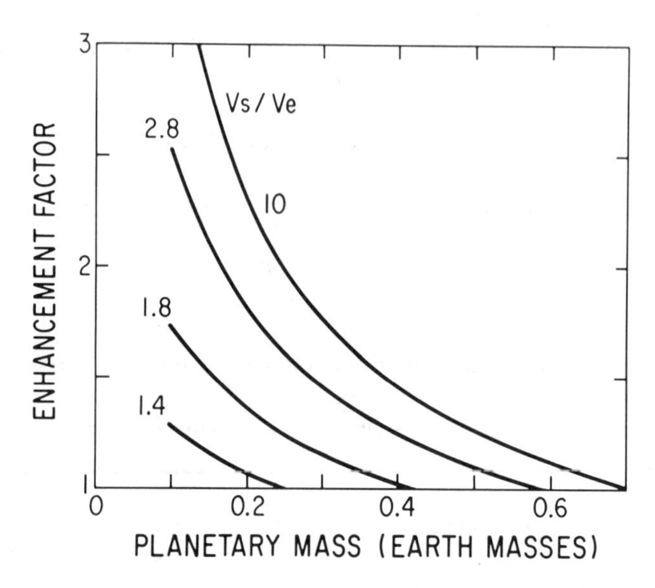

Fig. 3. Enhancement factors (Eq. 20) for impact erosion as a function of planetary mass. The curves are labeled with values of impactor velocity in units of the escape velocity.

possibly somewhat <1 Earth mass. Evaporative loading of the shock-heated gas increases with the square of the impact speed so that, for large planets, higher impact speeds simply yield more gas to be lifted out of the gravitational field of the planet. For the same reason, increasing impact velocity does not yield a very large increase of enhancement factor for fixed planetary mass. The rate of increase of enhancement factor with impact velocity is much less than the square of V_s/V_e. Even for quite small planets, significant impact erosion requires an impact velocity significantly larger than the escape velocity. Remember that there is much less escape when the enhancement factor (Eq. 20) is <1.

The theory of Walker (1986*b*) that is summarized here focuses on the initial interaction between impactor and planet, ignoring possible contributions to impact erosion by material rebounding from an impact crater. Most rebounding material will have velocities very much less than the original impact velocity and the energy imparted to a gas by a solid body traveling through it is a strong function of velocity, which is the justification for this neglect in a very simple theory. Such neglect would not, however, be justified for very large impact velocities, which could yield substantial amounts of rebounding material with velocity in excess of the escape velocity. The theory of Watkins (1983) and that of Ahrens and O'Keefe (1987) places most emphasis on interactions between rebounding material and the planetary atmosphere. The problem is sufficiently complex to call for much more study before these areas of disagreement can be resolved.

Figure 3 shows rather modest variations in enhancement factor (Eq. 20) for a wide range of planetary masses and impact velocities. In view of the approximate nature of the theory, it is probably sufficiently precise to conclude that if conditions for impact erosion are satisfied, the process leads to the escape of a mass of atmospheric gas approximately equal to the mass intercepted by the impactor. If we imagine, then, that the impactor contains atmophile elements, volatiles, we can ask whether or not a given impact event will add more volatiles to an accreting planet than it removes. Lange and Ahrens (1982*a*) have presented a more detailed analysis of this question. The answer depends on the radius of an assumed spherical impactor, its volatile content and the atmospheric mass per unit area. The comparison is presented in Fig. 4. Large impactors of large volatile content add mass to all but the densest atmospheres. Small impactors of low volatile content remove mass from most atmospheres. If we imagine that planetary accretion can be characterized by a single value of the effective radius of impactors and a single value for their volatile content and imagine that a large number of impact events occur, then Fig. 4 shows how the equilibrium mass of the atmosphere depends on the volatile content and effective radius of impactors. For example, 100-km impactors with a volatile content of 1% would yield a balance between impact erosion and accretion of atmosphere for an atmospheric mass per unit area approximately equal to that of the terrestrial ocean. These same 100 km impactors would yield an equilibrium atmosphere of only 1 terrestrial atmosphere if their volatile content were 0.01%. It might not be unreasonable to suppose that the Earth was built out of impactors with an average or effective radius of 100 km and a volatile content of 1%, so Earth's present volatile

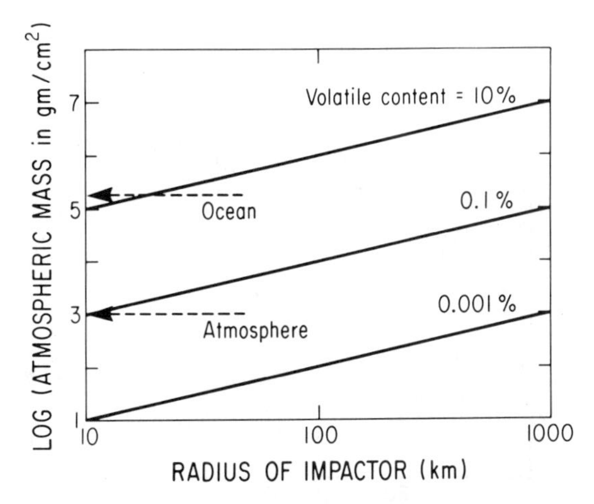

Fig. 4. The mass per unit area of a planetary atmosphere assumed to represent an equilibrium between the rate of addition of volatiles to the planet by accretion of material with the indicated volatile content and impact erosion of the atmosphere by impactors of the indicated radii. The arrows show the mass per unit area of the present terrestrial ocean and atmosphere.

content, approximately equal to the mass of the ocean, may represent a balance between impact erosion and accretion of volatiles. Parallel work by Matsui and Abe is described in the chapter by Ozima and Igarashi.

It seems that impact erosion is likely to be an important evolutionary process only if impactors with dimensions much larger than an atmospheric scale height are abundant, if impact velocities significantly exceed escape velocities, and if these impact events occur on planets that are not too large. Plainly, it is possible to imagine many accretion scenarios in which impact erosion would not play a major role. On the other hand, there are other aspects of the accretion process that are less easy to imagine away. The next section examines the possible influence of accretional energy on atmospheric evolution.

Interaction of Ejecta and Atmosphere

If the energy of a cratering event is large enough, a substantial fraction goes into atmospheric heating and into fast-moving ejecta that together can accelerate the ambient atmosphere to escape velocity or greater. At most, the effect is limited to the part of the atmosphere that is in line of sight from the impact. These effects have been modeled by Ahrens and O'Keefe (1987), building on earlier numerical work mainly concerned with the cratering itself and injection of material into the atmosphere (O'Keefe and Ahrens 1982*a,b*). Figure 5 illustrates the velocity profile directly above an impact of 10^{28} ergs

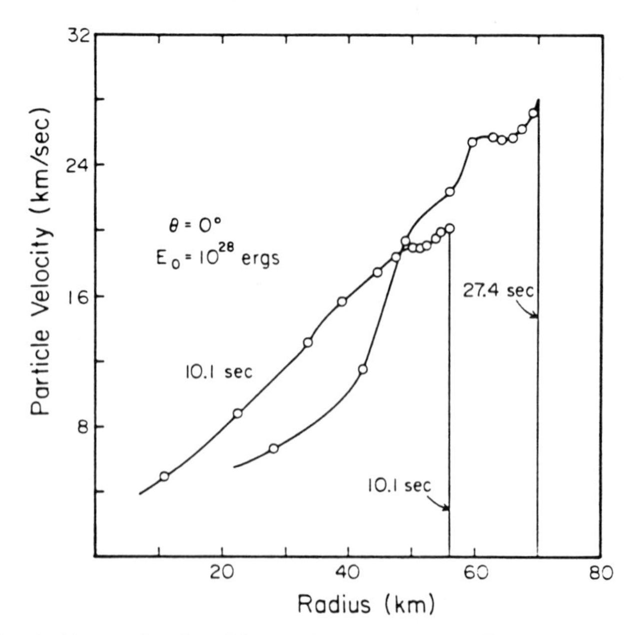

Fig. 5. Air velocities as a function of distance from an impact of 10^{28} ergs, calculated by Ahrens and O'Keefe (1987). The two curves refer to times of 10.1 and 27.4 s after the impact.

after times of 10 and 27 s; the Earth escape velocity is reached at around 35 km. For impact energy $>10^{30}$ ergs, the atmospheric mass ejected is fairly constant at 1.6×10^{16} g. This energy corresponds to a silicate body of 5 km diameter entering at 11 km s^{-1}.

The fraction of the entire atmosphere within the line of sight was estimated (using a somewhat crude model of the atmosphere) as 8×10^{-3}, and its mass as 2.6×10^{18} g (we find 5×10^{-4} above). Somewhat less than 1% of this, or 1.8×10^{-5} of the entire atmosphere, can escape as a consequence of the larger events. Just as with the direct influence of the incoming body on the atmosphere, this fraction is small enough to invite consideration of the balance between the loss and the gain due to volatiles brought in as part of the body. Such studies have been carried out by Lange and Ahrens (1984; see also the chapter by Ahrens et al.).

Once again, it appears that impact-induced escape is becoming inefficient for a body as large as the Earth, while it may be of much greater interest for smaller bodies.

Energy of Accretion

It is possible, as shown next, that accretional energy fluxes exceeded the solar energy flux at the surface of the Earth during at least the later stages of accretion, at a time when most of the atmospheric mass may have been released from the interior of the planet. Accretion of a planet of uniform density from an initially widely dispersed distribution releases gravitational energy equal to $(3/5)GM^2/R$ where M is the mass of the planet, G the universal constant of gravitation, and R the radius of the planet. For the Earth this gravitational energy is 2.3×10^{39} ergs. Imagine that this quantity of energy is released as heat in a period of D Myr and that a fraction B of this heat is buried beneath the surface of the Earth during accretion. Figure 6 presents the flux of gravitational energy per unit area of the Earth's surface as a function of the duration of the accretion event and for various values of the burial fraction. These fluxes are compared with the present-day solar constant, averaged over the globe, and the present-day value of the heat flow from the interior of the Earth. The flux of gravitational energy would exceed the present-day heat flow for any burial fraction in excess of 10^{-3}, even if accretion lasted for 1000 Myr. Burial fractions are likely to have been much larger than 10^{-3} and the duration of the accretion event much shorter than 1000 Myr, so it seems clear that the Earth must have started out with a hot interior in a state of rapid convection. These conditions could have promoted early and efficient degassing of the interior. For the comparison of the solar constant with the gravitational flux at the surface of the Earth, it is appropriate to consider just the line in Fig. 6 with a value of $B = 1$. The average flux of gravitational energy would have been comparable to the solar constant if D was shorter than 40 Myr. Even a longer duration of accretion very probably yielded fluxes of accretional energy larger than the solar constant near the end of accretion,

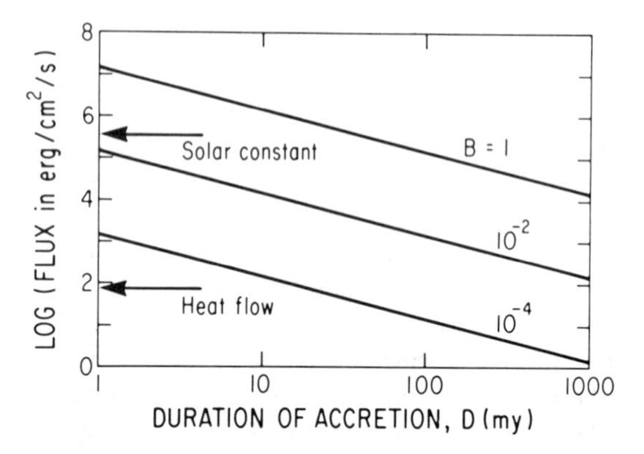

Fig. 6. The flux of gravitational energy at the surface of a planet as a function of the duration of the accretion event and the fraction of the accretional energy that is buried beneath the surface of the planet. Even very low burial fractions and long durations of accretion yield energy fluxes that exceed the present-day terrestrial heat flow from the interior. The line with $B = 1$ is appropriate for the heat flux leaving the surface of the planet; it exceeds the present-day solar constant for accretion durations $\lesssim$40 Myr.

because the rate of accretion and the gravitational energy released per unit mass added to the planet increase as the planetary mass increases.

Kasting (1988) has shown that this much additional energy at the surface of the Earth would lead to a runaway greenhouse, complete evaporation of the ocean, and surface temperatures under a steam atmosphere of 1500°C or more. Such high surface temperatures should have promoted the rapid achievement of thermodynamic equilibrium between the solid and gas phases. These conclusions agree with the relatively simple model of Matsui and Abe (1986*a,b*). A more detailed scenario is given by Zahnle et al. (1988), and the work is summarized in the chapter by Kasting and Toon. Of particular interest is the ratio of hydrogen to water vapor, which would have controlled the mean molecular weight of this atmosphere. If there was no metallic iron in the accreting material or exposed at the surface of the Earth, the hydrogen concentration would have been relatively low, and the mean molecular weight of the atmosphere would have been close to 18 (Holland 1984). Such an atmosphere could not have undergone bulk hydrodynamic escape into space, although the photolysis of water vapor in the upper atmosphere and the subsequent hydrodynamic escape of hydrogen might have been rapid (Walker 1982; Kasting and Pollack 1983; Watson et al. 1983).

It is, perhaps, more likely that free iron would have been present at the Earth's surface and in the accreting material. In this case, low oxygen fugacities would have yielded hydrogen partial pressures in excess of the water vapor partial pressure (Holland 1984). The mean molecular weight of the atmosphere may have been as low as 6 atomic mass units. Such an atmosphere would have been susceptible to hydrodynamic escape to space. The rate of escape is uncertain, but it may have been limited by the rate of supply

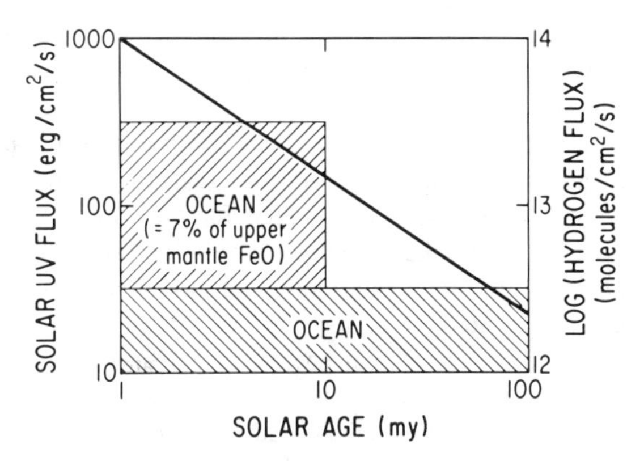

Fig. 7. An estimate of the extreme ultraviolet flux of energy from the Sun as a function of solar age and the corresponding rate of escape of hydrogen from the Earth if the escape rate were limited by the supply of EUV energy to the thermosphere. The cross-hatched areas show the escape rate that would remove all of the hydrogen in one terrestrial ocean over a period of 10 or 100 Gyr. The oxygen left behind could produce 7% of the iron oxide in the upper mantle.

of energy to the thermosphere by extreme ultraviolet radiation from the Sun. Figure 7 shows the solar ultraviolet flux as a function of solar age according to Zahnle and Walker (1982), and the corresponding escape flux of hydrogen, assuming zero nebular opacity and that all of the energy deposited by ultraviolet radiation in the thermosphere was carried off into space by hydrodynamic escape of the atmosphere. The cross-hatched areas in this figure represent hydrogen flux integrated over time to yield a total amount equal to the amount of hydrogen in the ocean. It seems possible that several oceanic masses could have been lost from the Earth as a result of hydrodynamic escape over a period of 100 Myr or so. Mass fractionation would have occurred during this escape event, with heavier gases preferentially retained by the Earth. The oxygen left behind by the escape of all of the hydrogen in 1 oceanic mass would be sufficient to produce 7% of the FeO in the upper mantle by oxidation of metallic iron.

Huge Impacts

A plausible mechanism for formation of the Moon is the impact of a Mars-sized body on a proto-Earth (Hartmann and Davis 1975; Cameron 1985*b*; Stevenson 1987*a*). Among several attractive features is that a plausible explanation is offered for the Moon's low density and extreme lack of volatiles, especially water. The very large impact could have yielded a massive and extended atmosphere of steam and vaporized silica. Rapid escape may have occurred from the top of this steam-silicate atmosphere. So far, however, this is only a matter of plausibility, because there has been no quantitative study of the rate at which the volatiles can be segregated from the more refractory components. If this segregation must occur on a diffusion time scale, the whole system might condense first, volatiles and all. More probably, the refractories rapidly formed small drops or grains, permitting rapid motion of the vapors to the surface of the system.

Equally little attention has been given to the loss or retention of volatiles by the proto-Earth. Reasoning back from the present state, we can conclude that such loss was much less efficient than for the Moon, but this hardly counts as a model. Despite the attractiveness of the basic idea, much more modeling is needed before it becomes anything more than an idea.

The ideas outlined here are preliminary and obviously in need of refinement. It is possible that accretional energy in the later stages of planetary growth could have resulted in a runaway greenhouse and very high surface temperatures under a steam atmosphere. Equilibrium reactions between metallic iron in accreting planetesimals and the steam atmosphere could have yielded a ratio of atmospheric hydrogen to water vapor exceeding 1, with a low mean molecular weight for the atmosphere. An atmosphere of low mean molecular weight could have suffered hydrodynamic escape to space with the possible loss of several oceanic masses of hydrogen during the later stages of accretion. Either because of this hydrodynamic escape or because of impact erosion during the course of planetary accretion, it is quite possible that the Earth and other inner planets retain only a small fraction of their initial volatile inventories.

V. LOSS OF WATER

Major issues that this section addresses are how much water each of the terrestrial planets retained (and outgassed) after accretion and what was the subsequent history of that water.

Initial Endowments

Cosmogonic models of condensation and accretion of the planets from the solar nebula lead to starkly different pictures of the early volatile inventories of the planets, particularly that of water and particularly with regard to Venus. One school of cosmogony (Holland 1963; Anders 1968; Lewis 1972*b*; Lewis and Prinn 1984) holds that Venus has always been very dry, whereas Earth and Mars both started with an ample supply of water. The other (Ingersoll 1969; Walker et al. 1970; Rasool and de Bergh 1970; Smith and Gross 1972; Walker 1975) arrives at the conclusion that, in effect, all of the terrestrial planets had essentially the same initial inventory of water. The concept of a dry primordial Venus has been elaborated in a long series of papers by Lewis and his colleagues (Lewis 1972*a,b;* Lewis 1973*a,b;* Barshay and Lewis 1975,1976,1978; Fegley and Lewis 1979,1980). It is based on a calculation of the equilibrium condensation state of material formed in regions of the solar nebula at various distances from the Sun. The condensates formed at a given heliocentric distance would have been those appropriate to the temperature and pressure at that place in the nebula. According to these calculations, water-bearing minerals, such as tremolite, serpentine and talc, could not have condensed where the present-day orbits of Mercury and Venus are located, whereas they could have formed at about 1 AU and beyond. If the initial planetary inventory of water is to reflect that in the early condensates, these models require that each planet was formed predominantly from material that had condensed in the neighborhood of its present orbit. Numerical models of accretion, such as that of Wetherill (1985), show all the terrestrial planets sharing essentially the same feeding ground of accreting material, one that extends from inside Mercury's orbit to beyond that of Mars, because of gravitational scattering of planetesimals into eccentric orbits after the planetesimals attain a certain size. In fact, it is argued that this kind of mixing is necessary if a few large planets are to be formed rather than a swarm of smaller ones.

The models of planetary formation developed by Turekian and Clark (1969) and by Walker and colleagues (Walker et al. 1970; Walker 1975,1977) postulate acquisition of volatiles by the planets from small amounts of volatile-rich material whose properties are like certain carbonaceous chondrites. The progenitors of planetary CO_2 and an appreciable amount of H_2O in these models are saturated alkanes $(CH_2)_n$ of the sort found as hydrocarbons in type I carbonaceous chondrites. Walker et al. (1970) point out that, if all the carbon dioxide found on Venus was derived from the reaction

$$CH_2 + 3Fe_3O_4 \rightarrow CO_2 + H_2O + 9FeO \tag{21}$$

about 10% of a terrestrial ocean would have been generated on Venus. Lewis and Prinn (1984) have vigorously criticized these proposals, partly on the grounds that there are virtually no known meteorites as rich in hydrocarbons as the mechanism would require, and partly because their thermodynamic-equilibrium models of nebular composition do not yield such compounds. In any event, even if the alkane source is capable of generating the CO_2 needed by the planets, it cannot produce as much water as there is in a terrestrial ocean. Additional water of hydration in minerals, such as talc, appears to be required, at least for the Earth. A fundamental issue is whether these minerals were equally available to all terrestrial planets. If vigorous gravitational mixing of planetesimals during accretion occurred, all four planets should have shared more or less equally the nebular material that condensed in the inner solar system. If water-bearing minerals in these planetesimals brought Earth its supply of water, they should also have brought comparable amounts to Venus and Mars. On the other hand, if volatile-rich interlopers from the outer solar system supplied the terrestrial planets with an important portion of their water, it is not easy to understand how any planet would have missed getting its share unless the planetesimals lost their supply of water before penetrating to 0.7 AU. (This might happen if the water existed as a relatively thin shell of ice on a rocky core.)

Present States and Their Evolution

Determination of the degree of deuterium enrichment has been regarded as the classical method of determining how much hydrogen has escaped from a planet and thus inferring its original endowment of water. For this to be feasible, the following information should be in hand: knowledge of the relative efficiency of all candidate mechanisms for loss of deuterium and hydrogen (and, perhaps, also of oxygen), knowledge of the value of the ratio of deuterium to hydrogen in juvenile water or its parent species, and an assessment of the amount of hydrogen the planet could have acquired since its formation from sources of water, such as comets, and the D/H ratio in those sources. Unfortunately, determining the "initial" value of D/H is a vexed matter. Geiss and Reeves (1981) have argued for a value of 2×10^{-5} in the primordial solar nebula. Models for the formation of Jupiter and Saturn predict a ratio close to the primordial one, and observations roughly confirm the prediction within a factor of 2 or so (Trauger et al. 1977; Beer and Taylor 1973; Macy and Smith 1978; Fink and Larson 1978; Owen et al. 1986). On the other hand, the ratio in Standard Mean Ocean Water (SMOW) on the Earth is 1.56×10^{-4}, and this is close to the value for hydrated minerals in meteorites (Geiss and Reeves 1981). Organic molecules in carbonaceous chondrites sometimes yield even higher values that can be as large as 2×10^{-3} (Kerridge, 1980*b*). These observations have led to suggestions that abundant ices in the solar system condensed from the solar nebula at such low temperatures that deuterium in molecular hydrogen was strongly partitioned into heavier molecules, such as water, methane and ammonia. The elevated value of D/H in SMOW and in CH_3D on Titan (Owen et al. 1986) could be accounted for by this model, if these ices were important sources of volatiles throughout the solar system. However, the enhancements in D/H predicted for Uranus and Neptune on the basis of their relatively high ice/gas ratio, have not been confirmed by observation (Macy and Smith 1978; de Bergh et al. 1986; Owen et al. 1986). Recently, Grinspoon and Lewis (1987*a*) have examined processes that might partition deuterium and have concluded that kinetic rates are too low to permit these processes to be effective in generating large values of D/H. They draw attention to proposals that comets may have become deuterium-rich by amassing large quantities of dust grains from interstellar clouds where large deuterium enrichment in molecules has been observed. The available measurements of D/H in comets—those for Comet Halley—are suggestive. Ip et al. (1987) have reported a value of 5.4×10^{-4} and Eberhardt et al. (1987*b*) give a range from 0.6×10^{-4} to 4.8×10^{-4}.

Indeed, some mechanism, such as the one investigated by Grinspoon and Lewis (1987*a*), may be required. One possibility that will be explored in greater detail in Sec. VI of this chapter is that enrichment occurred in the course of massive hydrogen escape from the young planets. If this did not occur it appears to be necessary to invoke a source of exotic, deuterium-rich material from the outer reaches of the solar system for the Earth's ocean. Venus and Mars should have had access to that same source. Thus, there is no way now known to assign definitive values of D/H to the primitive terrestrial planets. The best working hypothesis, adopted here, appears to be that the same enrichment mechanism, whatever it was, applied to all of them and probably to the meteorite parent bodies as well, and that the early value of D/H in each case was the value measured for SMOW: 1.6×10^{-4}.

Venus

The present-day value of D/H on Venus has been determined by two sets of measurements performed during the Pioneer Venus mission. An ion of mass 2 is observed in the ionosphere (Kumar et al. 1981). McElroy et al. (1982*a*) suggested that it is in fact D^+ rather than H_2^+. These authors also pointed out that, if this interpretation is correct, it implies that D/H is about 10^{-2} in the bulk atmosphere. This suggestion was pursued by Hartle and Taylor (1983) who performed a detailed analysis of the altitude dependence and local time variation of this ion. The ion behaved like D^+ rather than H_2^+ from which it can be distinguished because of the very different ion-molecular reactions that govern its distribution. Extrapolated to the lower atmosphere the value obtained for D/H was $(2.2 \pm 0.6) \times 10^{-2}$.

This value agrees with the ratio of $(1.6 \pm 0.2) \times 10^{-2}$ determined by Donahue et al. (1982) from water-related peaks in the spectra obtained by the Pioneer Venus Large Probe mass spectrometer. It is possible to differentiate between the contributions to these peaks of H_2O and HDO from the atmosphere of Venus and terrestrial contaminants transported from the Earth because of a serendipitous blockage between 50 and 26 km of the mass spectrometer inlet leak system by hydrated liquid cloud drops from the lower cloud layer. These droplets contained significant quantities of Venus water. They produced spectral peaks at 18 and 19 amu in which the contributions from this water were much larger in amplitude than those from either terrestrial water outgassing from the interior of the spectrometer or other minor constituents from the atmosphere. Comparison of data obtained during this period of enrichment with those obtained just before and just after blockage led to the determination of the quoted value of D/H.

Partitioning of the mass spectrometer water peaks between the Venus and Earth contributions is possible because the very different values of D/H permitted a determination of the mixing ratio of water vapor in the atmosphere of Venus below 25 km, where the leaks reopened. The volume mixing ratio f remained constant at about 100 (+100, −50) ppm between 25 and 10 km and then decreased by a factor of about 5 between 10 km and the surface. The gradient df/dz is therefore approximately -8×10^{-11} cm^{-1}. These results are similar to the spectrophotometer results obtained by Moroz on Veneras 11, 12, 13 and 14 in finding mixing ratios of about 15 ppm at the surface and about 60 ppm at 25 km (Moroz et al. 1980; Oyama et al. 1980; Schofield and

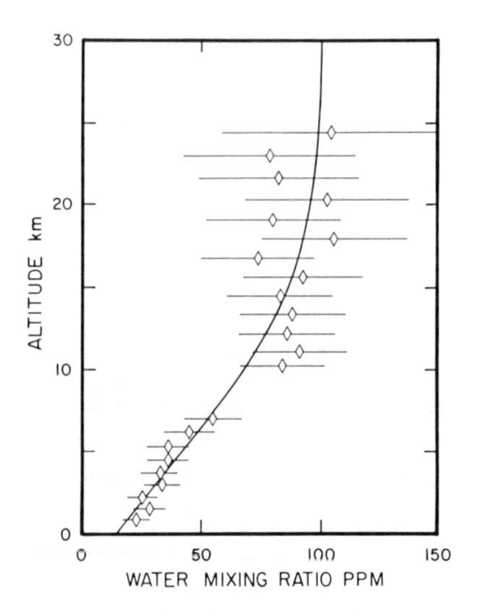

Fig. 8. Water-vapor mixing ratios (by volume) measured by the mass spectrometer on the Pioneer Venus Large Probe.

Taylor 1982; von Zahn et al. 1983). The details in the height variation are different, but both measurements agree that the ratio increases markedly with altitude (see Fig. 8).

Although the H_2O profile just described is adopted here as applying to Venus as a whole, it must be pointed out that the data are heavily weighted to low latitudes, and that there is strong evidence of variation with latitude. This evidence (Revercomb et al. 1985) comes from comparison of infrared net fluxes measured by all four Pioneer-Venus probes, and refers to considerably greater heights around 40 km. The mixing ratios (in ppm) are found to be 20 to 50 near 60°, 200 to 500 near 30°, and >500 near the equator. Furthermore the mixing ratios from the Pioneer-Venus mass spectrometer are based on a mixing ratio of 30 ppm for ^{36}Ar. If, as recommended by von Zahn et al. (1983) and von Zahn and Moroz (1984), the ^{36}Ar value is raised to 70 ppm, the water vapor must be adjusted accordingly. Thus, the globally averaged water-vapor mixing ratio is probably 200 ± 100 ppm.

If the mixing ratio of hydrogen in all forms remains essentially constant with altitude so that the deficiency in water vapor at low altitudes is made up by some other compound, these results would call for between 2.8 and 8.4 × 10^{23} hydrogen atoms per cm^2. At the present escape rate of 2×10^7 cm^{-2} s^{-1}, between 0.5 and 1.5 Gyr would be required to exhaust these atoms.

Dixon and Donahue (1987) have been attempting to determine whether the Pioneer Venus low-latitude profile can be explained in terms of a mechanism that calls for a flux of water vapor into the surface, followed by surface reactions that generate a return flux of a gas such as H_2 or H_2S and then by gas phase reactions that reconvert these species to H_2O below 10 km. Preliminary results of that exercise suggest a negative answer, so that there is a real possibility that the water vapor profile implies that the present surface of Venus is a sink for water vapor.

A one-dimensional expression for the flux Φ_0 into the surface is

$$\begin{aligned} K_0 n_0 \left(\frac{\mathrm{d}f}{\mathrm{d}z} \right)_0 &= \Phi_0 \\ 7 \times 10^{10} K_0 &= \Phi_0 \end{aligned} \tag{22}$$

in units of $\mathrm{cm}^{-2}\,\mathrm{s}^{-1}$. Here K_0 is the eddy diffusion coefficient ($\mathrm{cm}^2\,\mathrm{s}^{-1}$) and n_0 is the atmospheric number density, both at the surface. Obviously any reasonable value of the effective eddy diffusion coefficient will cause this flux to be very large compared to the hydrogen escape flux. The time required for the flux to exhaust the present water vapor inventory near the Pioneer Venus Large Probe entry site will be correspondingly short. There are 7×10^{22} water molecules cm^{-2} in the atmospheric reservoir. The $1/e$ time constant for emptying this reservoir is only

$$t_f = \frac{3.7 \times 10^4}{K_0}\ \mathrm{yr}.$$

This is so short that it seems there would have to be a strong source of water on Venus to balance the flow of vapor to the surface. Presumably the source would be a tectonic process, such as volcanism. If so, the water is probably juvenile, not recycled. There may be widespread volcanism on Venus, which could bring water from the mantle into the atmosphere. Recycling of crustal water, which occurs as a result of plate tectonics on Earth, is likely to be inhibited on Venus. Walker et al. (1970) have pointed out that, if the present escape flux of hydrogen on Venus is in dynamic balance with the input of juvenile water, the outgassing rate would be too weak by a factor of about 5×10^3 to produce the equivalent of a terrestrial ocean in 4.5 Gyr. If the present escape flux of water is in dynamical balance with an internal source (augmented, perhaps, by a cometary input), then early outgassing of water on Venus must have been very much more efficient than it is at present to have generated an ocean. What we know about present degassing rates for juvenile volatiles on Earth, compared to the inventories of species such as ^{36}Ar, suggests that the requirements on early compared to present terrestrial degassing rates are modest in comparison. Possibilities are that the present tectonic style of Venus severely inhibits degassing, that Venus never outgassed more than a few m of water (or both). Another possibility, suggested by the water vapor profile near the surface, is that there is a very large flux of water ($>10^{12}\ \mathrm{cm}^{-2}$

s^{-1}) into the surface of Venus that balances a similar flux of juvenile water from the interior.

The corollary—that the surface of Venus would become oxidized as the water reacts with it—creates problems in terms of sulphur chemistry (Prinn 1979; Lewis and Prinn 1984) but is in accord with conclusions reached by Pieters et al. (1986). These authors have analyzed multispectral images of the surface obtained by Venera 13 and interpret the results as suggesting the presence of ferric minerals in the basaltic surface of Venus. They conclude that the surface may, therefore, be relatively highly oxidized.

Mars

Anders and Owen (1977), after Viking, elaborated a model of Martian volatiles based on a suggestion that the volatiles were supplied by a late accretion of C3V-like carbonaceous chondrite material. From estimates of the global potassium abundance and measurements of the amount of ^{40}Ar in the atmosphere, they inferred that about 10% of the ^{40}Ar produced had been degassed. Using this value as a lower limit for the degassing factor of ^{36}Ar and the relative abundance of ^{36}Ar and other volatiles in C3V chondrites, they estimated the global inventory of C, N and H. The very low atmospheric inventory of the nonradiogenic noble gases on Mars led them to place correspondingly low values on other volatiles. In particular they concluded that only 6 m of water had been degassed.

Subsequent data revealed that nonradiogenic argon and neon are enhanced by two orders of magnitude on Venus, even though inventories of such volatiles as C and N are very similar to those of the Earth. Thus, it is not warranted to assume a constant ratio of noble gas volatiles and other volatiles on the terrestrial planets. In fact, McElroy et al. (1977) have exploited the observation of enrichment of Martian ^{15}N to argue that at least 120 m of water outgassed on Mars. This followed from a calculation of the amount of nitrogen that must have escaped to produce the observed enhancement of the $^{15}N/^{14}N$ ratio and the assumption that the Martian H/N ratio was the same as on Earth. Subsequent estimates of other sinks for nitrogen, including fixation by lightning, have led Yung and McElroy (1979) to conclude that as much as 100 mbar of nitrogen, 10 bar of CO_2 and 0.5 km of water may have outgassed on Mars.

Recently Carr (1986) has examined the geologic evidence for Martian water. He concludes, after a study of characteristics of terrain features of the cratered uplands at high and low latitudes, including outflow channels and valley networks, that Mars may have outgassed at least 100 m of water and perhaps as much as 500 m. He argues that the water (and other volatiles) are retained today in the cratered uplands, mostly at high latitudes and mainly in the megaregolith below 1 km.

Recently, also, Owen et al. (1988) have detected Martian HDO in the 3.67 μm band. Based on the estimate of HDO abundance from this measurement and a corresponding measurement of H_2O, they report a value for the ratio of HDO to H_2O in Martian water vapor in the range of 3 to 12 times terrestrial, with the average ratio being between 5 and 6 times terrestrial.

VI. LOSS OF WATER FROM VENUS, EARTH AND MARS

Hydrogen escape has probably played an important role in the evolution of all of the terrestrial planets, excepting Mercury which may have had little hydrogen from the start. Much, or all, of the hydrogen that has been lost from each planet was ultimately derived from water, either by photodissociation in the atmosphere or by direct reaction of water with reduced minerals present in the planet's crust. The oxygen released during these processes has in some cases been retained by the planet, while in other cases it too may have escaped along with the hydrogen. Here, we review the types of processes that are thought to have been operative and offer some estimates for the total amount of water that may have been lost from each planet.

Earth

The current rate at which hydrogen is escaping from the Earth's atmosphere is estimated to be 2.7×10^8 H atoms cm^{-2} s^{-1} (Hunten and Donahue 1976). During solar minimum, much of this escape occurs by way of nonthermal processes, most importantly charge exchange with hot H^+ ions, as discussed earlier. This mechanism is largely supplanted by Jeans escape as the exosphere heats up during solar maximum. The source of the escaping hydrogen is split between photodissociation of water vapor and oxidation of methane, each of which contributes approximately equal amounts of hydrogen to the lower stratosphere.

The amount of hydrogen that is presently escaping by these processes is small. If the current escape rate had remained constant for the last 4.5 Gyr, the quantity of water lost would be equivalent to a layer covering the Earth's surface to a depth of 5.7 m. By comparison, the 1.4×10^{24} g of water in the oceans is equivalent to an average depth of almost 3 km. Thus, hydrogen escape presently has a negligible effect on the Earth's water budget.

It is likely, however, that the importance of hydrogen escape was far greater in the distant past. There are at least two mechanisms that could have resulted in substantial loss of water. The first is escape of H_2 from an anoxic primitive atmosphere. Such an atmosphere was almost certainly present prior to the invention of photosynthesis (Walker 1977; Kasting and Walker 1981; Kasting and Ackerman 1986) and may have persisted until as recently as 2.4 Gyr ago (Walker et al. 1983). The hydrogen content of a low-O_2 atmosphere would have been determined by the balance between surface sources of reduced gases, escape of hydrogen to space and rainout of soluble trace species (Kasting et al. 1984*a*). To first order, one may neglect rainout and assume that all of the hydrogen that was vented into the lower atmosphere would have eventually escaped. Potentially important sources of hydrogen include volcanic outgassing of H_2O and CO (Walker 1977) and photostimulated oxidation of ferrous iron in the oceans (Braterman et al. 1983). The volcanic outgassing rate of reduced gases is equivalent to approximately 10^9 H atoms cm^{-2} s^{-1} at present (Kasting and Walker 1981) and was probably several times higher in the past. The rate at which hydrogen was produced from photostimulated iron oxidation was probably limited by availability of iron rather than photons and could have been as high as 10^{11} H atoms cm^{-2} s^{-1}, assuming that all of the iron that entered the early oceans was oxidized in this manner (Kasting et al. 1984*a*). While this latter assumption is unlikely to have been realized, it seems clear that a substantial amount of hydrogen could have been lost by this mechanism.

A second process by which large amounts of water may have been lost is by escape of hydrogen during the Earth's accretion. This possibility has already been discussed in Sec. IV. There it is concluded that several oceans' worth of water could have been lost during the first 100 Myr of the Earth's history by hydrodynamic escape of hydrogen from an impact-induced steam atmosphere. This estimate may be somewhat high because hydrogen may have escaped at less than the energy-limited rate, and because the solar EUV flux may have been attenuated by nebular gas or dust during part of the time. Nevertheless, consideration of this process leaves open the possibility that the deuterium enrichment in seawater relative to mantle water could have resulted from fractionation during hydrodynamic escape.

Mars

Mars presents a different picture with respect to water loss than does Earth, both because of its greater distance from the Sun and because it is much smaller. The higher D/H ratio on Mars relative to Earth (Sec. V) may imply that substantial amounts of water have been lost during the course of Mars' history. Exactly how much is unclear, since estimates of Mars' initial H_2O inventory range from 10 m (Anders and Owen 1977) to 1 km (Carr 1986). Furthermore, the interpretation of the present enhancement is complicated by the fact that Martian atmospheric water vapor can exchange with large reservoirs of ice in the polar caps and in the regolith.

Despite these difficulties one very interesting conclusion has been reached concerning water loss from Mars, namely, that water itself is currently escaping, rather than simply hydrogen (McElroy 1972; McElroy and Donahue 1972). The cause is related to Mars' low gravity. The escape energy for oxygen atoms from Mars' atmosphere is 1.99 eV. This is less than the energy available from dissociative recombination of O_2^+ in the Martian ionosphere (Table II, process 2), which can provide up to 3.5 and 5.3 eV per oxygen atom (depending on whether one atom is electronically excited). Consequently, recombination reactions that occur above the exobase can result in the loss of oxygen atoms to space. The rate at which O is presently being lost is about 6×10^7 cm^{-2} s^{-1} (McElroy et al. 1977).

Hydrogen escapes from the Martian atmosphere primarily by thermal evaporation (Jeans escape) of H atoms and H_2 molecules. The H atoms are produced by reactions involving H_2, notably

$$H_2 + O(^1D) \rightarrow H + OH$$
$$H_2 + CO_2^+ \rightarrow CO_2H^+ + H$$
$$CO_2H^+ + e \rightarrow CO_2 + H$$
$$H_2 + h\nu \rightarrow H + H \qquad (23)$$

(McElroy 1972). The abundance of H_2 is, in turn, controlled by the concentration of atmospheric O_2 through its effect on the odd-hydrogen photochemistry (McElroy and Donahue 1972). The feedback is such that an increase (decrease) in $[O_2]$ brought about by a decrease (increase) in the oxygen escape rate should cause a corresponding decrease (increase) in the concentration of H_2 and in the escape rate of hydrogen. The net result is that, over time scales of 10^5 yr or longer, H and O atoms are constrained to escape in a 2 to 1 ratio. The predicted hydrogen escape rate of 1.2×10^8 atoms cm^{-2} s^{-1} (McElroy et al. 1977) is in accord with fluxes inferred by Mariner 9 (Barth et al. 1972).

If one simply extrapolates the present water escape rate backward in time, the predicted amount of water lost is only enough to cover the Martian surface to a depth of 2.5 m. This estimate can be increased by a factor of 4 to 8 by including the variation in solar EUV flux (Eq. 15) and assuming that the oxygen escape rate scales proportionately. Even so, the predicted amount of water loss, 10 to 20 m, is not large by comparison to Carr's estimate for the initial H_2O inventory. If further observations were to confirm this estimate, and if the substantial deuterium enrichment found by Owen et al. (1988) proves correct, then additional mechanisms for hydrogen loss would need to be identified. The most likely possibilities are the same processes discussed above with respect to water loss from the Earth. Efficient reactions of oxygen with the Martian crust may, for example, have generated hydrogen much more rapidly than does the current atmospheric photochemistry. Such a scenario for hydrogen loss is given by Hunten et al. (1987).

One important distinction that should be noted is that Mars' small size would make its accretionary history quite different from that of Earth. Whether or not a steam atmosphere would have developed on Mars remains to be determined.

Venus

As discussed in Sec. V, Venus is the terrestrial planet on which loss of water is most apparent. The current hydrogen escape rate from Venus is lower than for Earth by about an order of magnitude. The important escape mechanisms are nonthermal. McElroy et al. (1982*a*) estimate an escape rate of 8×10^6 H atoms cm^{-2} s^{-1} from collisions of H with hot O atoms produced by dissociative recombination of O_2^+ (2a in Table II, followed by 5b). Charge exchange with hot H (1a in Table II) gives an escape rate of 1.2×10^7 H atoms cm^{-2} s^{-1} according to Kumar et al. (1983). The combined rate of hydrogen loss from these two mechanisms is sufficient to eliminate about 40 cm of water over the lifetime of the solar system. Although this is a small amount of water, it is by no means negligible. As we have seen, the time required for the present escape flux to exhaust the hydrogen in the contemporary atmosphere is uncertain (0.5 to 1.5 Gyr) because of uncertainty in the total inventory. At the low end of the range, the implication is that the water on Venus is being replenished at the same rate that it is lost. Possible sources for water include outgassing from the interior (Walker et al. 1970) and cometary impacts (Grinspoon and Lewis 1988). At the high end, it is consistent with the scenario of a gradual depletion over billions of years, either in the water left after massive early escape or in the slower loss over the lifetime of the planet of an amount less than a terrestrial ocean.

Venus may be similar to Mars in the sense that it, too, may be losing oxygen to space along with hydrogen (McElroy et al. 1982*b*). Because Venus lacks a strong magnetic field, oxygen ions produced above the plasmapause can be swept away by the solar wind. The estimated oxygen escape flux from this process is 6×10^6 atoms cm^{-2} s^{-1}, which could be taken to be equal to half the hydrogen escape rate, given the uncertainties in both calculations. The actual mechanism by which changes in oxygen might serve to regulate hydrogen escape is unclear. McElroy et al. suggest two possibilities: an increase in $[O_2]$ above the cloud deck might increase the rate of formation of hygroscopic aerosols, thereby decreasing the amount of water vapor in the upper atmosphere. Alternatively, changes in $[O_2]$ in the lower atmosphere might influence the abundances of H_2 and HCl. More detailed knowledge of the chemistry of Venus' atmosphere is needed to test these hypotheses.

The rate at which hydrogen is escaping today is negligible compared to the escape rate that must have prevailed in the past if Venus was indeed endowed with an Earth-like complement of water. The standard theory of how Venus could have lost this much water is the so-called "runaway greenhouse," first proposed by Hoyle (1955), but developed more fully by Ingersoll (1969). A description of this theory is given in the chapter by Kasting and Toon. Basically, the theory holds that Venus was always too hot for water to have condensed on its surface. Photodissociation of water vapor in the upper parts of this steam atmosphere produced copious amounts of hydrogen, which would have escaped by the hydrodynamic process described earlier in this chapter.

An alternative theory, proposed by Kasting et al. (1984*b*) and also described in the chapter by Kasting and Toon, is that Venus had a "moist greenhouse" atmosphere, in which oceans were present at the surface but the upper atmosphere was again wet, just as in the runaway greenhouse theory. The diffusion limit on escape was thereby overcome, and hydrogen could have escaped rapidly and hydrodynamically.

The most detailed model of hydrodynamic escape of hydrogen from Venus is that of Kasting and Pollack (1983). Their model included coupled, one-dimensional treatments of dynamics, photochemistry and radiative transfer. Figure 9 shows the hydrogen escape flux predicted by their model as a function of the H_2O mass mixing ratio at the cold trap, that is, the level at which the saturation mixing ratio of water vapor is at a minimum. The predicted hydrogen escape rate at the highest H_2O level studied was 2.7×10^{11} atoms cm^{-2} s^{-1}, which is sufficient to eliminate a terrestrial ocean of water in 2.4 Gyr. Kasting and Pollack's calculation, however, assumed the present-day, solar minimum EUV flux and a (minimal) 15% heating efficiency. For more liberal assumptions concerning solar EUV heating rates, the time required to lose an ocean of water would be much shorter. Figure 9 does demonstrate, however, that hydrodynamic escape can be limited by other factors besides energy. At low H_2O mixing ratios, the escape rate is limited by the finite supply of hydrogen at the base of the expansion. Even at high H_2O levels the predicted escape rate is lower than the energy-limited value by about a factor of 4.

The main objection to the proposed massive loss of water from Venus is the belief that it is difficult to dispose of the oxygen left behind. Lewis and

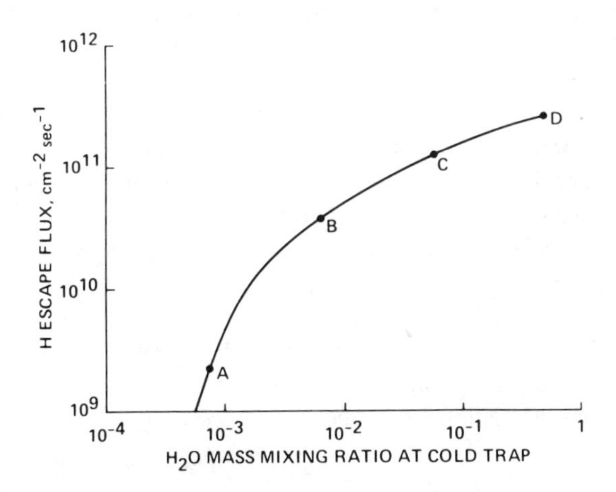

Fig. 9. Hydrodynamic escape flux of hydrogen from a hypothetical, water-rich atmosphere on early Venus. Four models, shown by the letters, were computed and the points joined by a smooth line. Each assumed a different stratospheric humidity; the solar flux was the present solar-minimum value, with a 15% heating efficiency. These models are all source-limited, although Model D is approaching the energy-limited situation (figure from Kasting and Pollack 1983).

Prinn (1984, p. 190) have pointed out that eliminating only one-third of the oxygen in Earth's oceans would entail the oxidation of some 80 km of Earth-like crust. This may not have been a problem if the surface was molten, as might have been the case under a runaway greenhouse atmosphere (chapter by Kasting and Toon). Alternatively, much of the oxygen in Venus' oceans may have escaped to space along with the hydrogen, if the crossover mass (Eq. 10) was above that of atomic oxygen. If escape rates were indeed this high, then loss of water by this process would have left little or no trace.

The loss of even several oceans' worth of water may have little to do with explaining the present D/H enrichment on Venus. During most of the hydrodynamic escape phase, the escape rate should have been high enough that little fractionation of deuterium would have occurred. The present deuterium enrichment was most likely produced by the much slower, nonthermal escape processes operating today. The D/H ratio expected depends on the initial D/H ratio, the fractionation factor $1 + y$, and on whether the water abundance in Venus' atmosphere is decreasing or is in steady state. If one assumes that water is in steady state, and if one further assumes a terrestrial D/H ratio for the H_2O source, then the predicted enrichment should be equal to $1 + y$. Krasnopolsky (1985) has calculated that $1 + y = 45$ for the charge-exchange mechanism. This value supersedes the 12.5 given by Kumar et al. (1983), who incorrectly defined it with respect to production of H and D atoms, instead of the actual D/H ratio in the lower atmospheric reservoir. By comparison, McElroy et al. (1982*a*) estimate that virtually no deuterium escapes via momentum transfer with hot O atoms, so that $1 + y \simeq \infty$ for this mechanism. If one weights these fractionation factors by their relative contributions to the total H escape rate, one obtains a mean value for $1 + y$ of 75, implying that D/H should be enriched by the same factor. Considering the uncertainties involved in the estimation of both fractionation factors and escape rates, we find this prediction to be in good accord with the data.

In summary, loss of water has apparently played an important role in the evolution of Venus, Earth and Mars. Further research will be required to better constrain the amounts of water involved and to use this knowledge to estimate loss rates of other volatiles.

Acknowledgments. Partial support in the preparation of this chapter was provided by several grants from the National Aeronautics and Space Administration and the National Science Foundation.

CLIMATE EVOLUTION ON THE TERRESTRIAL PLANETS

J. F. KASTING
and
O. B. TOON
NASA Ames Research Center

The long-term climate histories of Venus, Earth and Mars are compared and contrasted. It is argued that the Earth's climate has remained temperate over most of its history, despite a large secular increase in solar luminosity, because of a negative feedback cycle involving atmospheric CO_2 levels and climate. Prior to the emergence of the continents, the Earth may have possessed a dense (1 to 10 bar) CO_2 atmosphere and had a surface temperature as high as 100°C. However, a runaway greenhouse, that is, complete evaporation of the oceans, is excluded except during the initial accretion period. Mars may also have possessed a relatively warm, dense, CO_2 atmosphere early in its history. Mars lost most of its atmosphere and became cold because it was too small to continue recycling carbonate rocks. Venus may have started out with an Earth-like endowment of water, much of which may have initially existed as oceans. Venus, however, lost most of its water within a few 100 Myr by photodissociation followed by escape of hydrogen to space. Thereafter, carbonate formation was inhibited on the dry surface and CO_2 accumulated in Venus' atmosphere. Examination of the climate histories of the three planets suggests that an Earth-sized planet could maintain liquid water on its surface at orbital distances ranging from 0.95 AU to perhaps 1.5 AU—around the orbit of Mars. This, in turn, implies that there may be many other habitable planets within our Galaxy.

This chapter is concerned with changes in planetary climates that occur over time periods comparable to the age of the solar system. We restrict our attention to the very broadest of climatic variables, namely, global mean surface temperature and pressure. We do so in part because these are the most important variables on an evolutionary time scale and in part because little is known about long-term variations in other climatic parameters. Our goal is to present a unified view of climate evolution on the three outermost terrestrial planets—Venus, Earth and Mars.

We begin by summarizing what is known about the present climates of the terrestrial planets and about their climate histories. The reasons why those histories diverged so radically are then identified by considering each planet individually. A greater emphasis is placed on the Earth because much more is known about its past and because the Earth is more interesting than the other planets because we live there. We next touch briefly on the subject of atmospheric composition and climate during accretion: a topic that may have relevance to questions of hydrogen escape and noble gas fractionation discussed in Chapters by Hunten et al. and by Pepin. Finally, we discuss the implications of climate theory for the width of the continuously habitable zone around the Sun and consider the question of whether there are Earth-like planets orbiting other stars within our Galaxy.

I. OBSERVATIONAL CONSTRAINTS

This section summarizes the observational constraints that any theory of climate evolution must seek to explain. Since past surface temperatures and pressures are not themselves observable quantities, we are forced to rely on other indicators of past climates. An important paleoclimatic indicator for the Earth is the evidence for glaciation during various time periods. Hence, we include a brief discussion of the relationship between glaciation and global mean surface temperature.

A. Present Climates of the Terrestrial Planets

Because we have adopted a very narrow definition of climate, the present climates of the terrestrial planets are easy to describe. In terms of surface temperature, Venus is too hot (735 K), Mars is too cold (218 K), and Earth is just right (288 K) (Pollack 1979; Colin 1983). (This arrangement is sufficiently orderly that L. Margulis has termed it "the Goldilocks problem.") Venus has a 90 bar, largely CO_2 atmosphere; Earth has a 1 bar, N_2–O_2 atmosphere with about 3×10^{-4} bar of CO_2; and Mars has a 0.006 bar, CO_2 atmosphere (ibid). Interestingly, Venus, because of its high albedo (~0.75). actually absorbs significantly less solar energy than does the Earth, despite its position nearer the Sun. Venus' effective radiating temperature is about 230 K, compared to 253 K for Earth. The difference between the effective temperature and the surface temperature, 505 K for Venus, is a measure of the greenhouse effect of its dense CO_2 atmosphere. Other, less abundant gases (H_2O, SO_2 and CO) also contribute substantially to the greenhouse warming, as do the planet-covering, sulfuric acid clouds (Pollack et al. 1980). By comparison, the Earth's surface is warmed by about 35 K by the greenhouse effect of H_2O and CO_2, and Mars is warmed by 6 K by CO_2 alone.

B. Evidence for Climatic Change

Earth. The salient feature of the Earth's long-term climate history is the lack of any evidence for glaciation prior to about 2.7 billion years (Gyr) ago (Crowell 1983). This absence is all the more remarkable in light of the fact that solar luminosity was markedly lower at that time (see Sec. II). This first well-documented glacial episode, the Huronian glaciation, occurred between 2.0 and 2.5 Gyr ago (Frakes 1979; Crowley 1983). It was followed by another long period of apparent warmth that lasted until about 0.9 Gyr ago. This second period of warmth was, in turn, followed by a series of very severe glacial episodes between 0.9 and 0.6 Gyr ago that are collectively referred to as the Late Precambrian glaciations (ibid.). The last 600 Myr of the Earth's history, termed the Phanerozoic, has been marked by alternating periods of warmth and cold. The Earth is presently in the midst of an extended glacial epoch, the Pleistocene, that has lasted for several Myr. The extent of the ice cover has fluctuated with frequencies that are apparently related to variations in the Earth's orbital parameters, the so-called Milankovitch cycles (see, e.g., Imbrie and Imbrie 1981). We are currently experiencing a mild interglacial event that has lasted for about 10,000 yr.

Interpreting this climate record in terms of changes in global mean surface temperature T_s is not a straightforward procedure. Reasonably tight constraints can be inferred, however, during glacial periods. As discussed by Kasting (1987), T_s is unlikely to have ever been lower than about 5°C (278 K); otherwise, the oceans would have been in danger of freezing over, an event that is not thought to have occurred. By comparison, T_s is 15°C at present and was probably about 10°C during the peak of the last glacial episode, 18,000 yr ago (Schneider and Londer 1984). An upper limit on T_s during glacial periods can be inferred from the Earth's climate history over the last 100 Myr. Polar ice is known to have been absent during the Mesozoic Era, at which time T_s was ~10 deg warmer than today. Furthermore, despite being situated on the South Pole during most of this time, Antarctica was not glaciated until at least the Miocene (30 Myr ago), by which time surface temperatures had declined by about 5 deg (Frakes 1979). It therefore seems likely that T_s during glacial periods was no higher than about 20°C. This argument is not ironclad, since the glaciation of Antarctica may also have been affected by its separation from Australia and South America around 30 to 40 Myr ago and the subsequent establishment of the Antarctic circumpolar current (Crowley 1983). Our adopted upper limit on glacial T_s is probably accurate to within a few degrees, however, which is good enough for our present purposes.

Mean surface temperatures during nonglacial periods are harder to estimate. 20°C is a safe lower limit if one considers time periods long enough such that one or more of the continents have an appreciable chance of drifting over a pole. Upper limits on T_s are more speculative. Walker (1982*a*) has argued that the presence of evaporitic gypsum deposits precludes surface temperatures above 58°C during the last 3.5 Gyr. More stringent limits could probably be set during the latter part of this time by the continued presence of complex life. The fact that some simple organisms presently thrive in very warm environments, such as hot springs, makes it risky to use this criterion during the Precambrian, however. Surface temperatures prior to 3.5 Gyr ago are constrained only by the requirement that the Earth not lose all its water during this time by photodissociation and hydrogen escape (see Sec. III).

Attempts to determine paleoocean temperatures over the last 3.5 Gyr have been made using oxygen isotopes in cherts (Knauth and Epstein 1976) and chert-phosphate pairs (Karhu and Epstein 1986). Both studies yield extremely high temperatures (60° to 120°C) throughout most of the Precambrian. The latter study attempted to remove the effects of changes in ocean isotopic composition, which might otherwise have been responsible for the high temperatures reported in the earlier investigation. Unfortunately, these predictions can be reconciled with the evidence for Precambrian glaciations cited above only if the Earth's surface temperature is assumed to have fluctuated wildly, from very hot, to cold, and back to hot again, a very improbable climatic scenario. It seems more likely that the isotopic data reflect higher temperatures at some depth within the sediment, perhaps associated with a higher geothermal gradient at that time.

Mars. The climatic record on Mars is, understandably, more poorly resolved than on Earth. The discovery of fluvially-generated channels by the Mariner 9 spacecraft (McCauley et al. 1972; Masursky 1973; Milton 1973) provided strong evidence for flowing water on Mars' surface at some time in the past. An alternative hypothesis, that the channels were produced by liquid alkanes that condensed out of a methane-rich, primordial atmosphere (Yung and Pinto 1978), now seems unlikely because the Martian atmosphere was probably never that highly reducing. However, even if these channels were formed by water, such features are not necessarily indicative of warm surface temperatures. At least some of them, the so-called outflow channels, could have been generated by rapid release of near-surface water under present climatic conditions (Baker 1978; Wallace and Sagan 1979). This argument is harder to make for the valley networks, or runoff channels, found on the old, cratered terrain of the southern highlands. Most of these features appear to have been formed primarily by sapping of near-surface water, although precipitation and runoff remains a possibility in some cases (Squyres 1984). Even the sapping mechanism requires the presence of subsurface liquid water at shallow depths, which in turn implies much warmer surface temperatures in the distant past (ibid.). Whether or not an active hydrological cycle existed at some early epoch is presently a matter for speculation. This possibility may be testable by future *in situ* investigation of the Martian surface.

Venus. The climatic record on Venus, if it exists at all, has yet to be uncovered. The best prospect for learning about Venus' past climate is the upcoming Venus Radar Mapper Mission (Magellan), which is expected to map the surface of the planet with a resolution of roughly 100 m. If Venus had liquid water on its surface in the distant past, as we suggest in Sec. V, it is conceivable that the mapping mission will find evidence of it. It is also possible that any such evidence would have been completely obliterated by several billion years of volcanism and weathering.

One datum that has been suggested to bear indirectly on past climate is the factor of 100 enrichment, relative to terrestrial ocean water, in the D/H ratio in Venus' clouds (Donahue et al. 1982). If both planets formed with the same isotopic composition, and if the water in Venus' atmosphere has not been replenished over time, this measurement implies that Venus originally possessed at least 100 times more water than it has now. This observation is in accord with theories of Venus' evolution, discussed in Sec. V, that predict much larger water abundances in the past. A more recent interpretation, however, is that Venus' water is currently in steady state, with loss of hydrogen to space being balanced by continued outgassing from the interior or by cometary impacts (Grinspoon and Lewis 1987; see also Chapter by Hunten et al.). If this analysis is correct, then the present D/H enrichment tells us little about the planet's original water endowment, and models of Venus' past climate are at this point unconstrained by geochemical evidence.

Having summarized the evidence for climate change on the terrestrial planets, we now consider the mechanisms, both astronomical and atmospheric, that may have driven it.

II. ASTRONOMICAL INFLUENCES ON CLIMATE

A. The Faint Young Sun

One factor that must have influenced climate evolution on all of the terrestrial planets is the gradual increase in solar luminosity caused by the Sun's evolution. The reason for this increase is that, as the Sun converts hydrogen to helium, its mean molecular weight and density increase and its core becomes hotter. The rate of thermonuclear burning therefore increases and, hence, so must the outgoing energy flux. Most estimates place the initial solar luminosity at 25 to 30% below its present value (Newman and Rood 1977; Gough 1981). The increase in luminosity since that time may be described by the relation (Gough 1981)

$$L(t) = [1 + 0.4(1 - t/t_o)]^{-1}L_o. \tag{1}$$

Here, t_o is the present age of the Sun (4.6 Gyr) and L_o is the present luminosity. The variation in luminosity with time is shown in Fig. 1, along with the effective radiating temperature T_e for Earth, given by

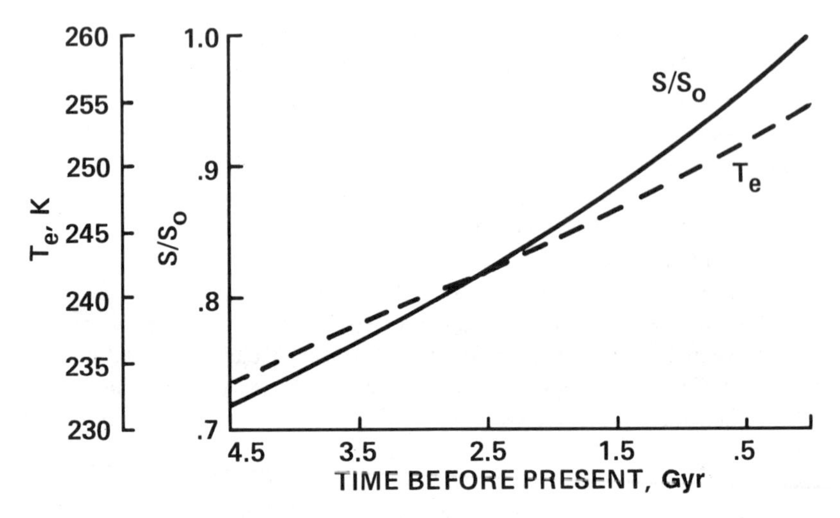

Fig. 1. Variation in solar luminosity S and effective temperature of the Earth T_e over geologic time. S_o is the present solar luminosity (figure from Kasting 1987).

$$\sigma T_e^4 = \frac{S}{4}(1 - A). \tag{2}$$

In this equation, σ is the Stefan-Boltzmann constant, S is the amount of sunlight intercepted by the Earth and A ($\simeq 0.3$) is the planetary albedo, assumed here to be constant. The change in effective temperature is nearly linear and amounts to about 20 deg over geologic time.

Although the precise magnitude of the increase in solar luminosity is model dependent, the basic concept depends only on the assumption that the Sun is converting hydrogen to helium. It is thus a prediction that is unlikely to go away, even if new discoveries are made about the mechanism by which the Sun produces its energy. Radical revisions in astrophysical theory could, of course, change the situation considerably. If, for example, the Sun has lost significant amounts of mass during its main-sequence lifetime, as recently suggested by Willson et al. (1987), then the early Sun could have been brighter, instead of dimmer, than today. In this case, the theory of climate evolution proposed here would need to be reconsidered.

B. Other Factors

Although the total energy flux from the young Sun was lower than today, its energy output in the extreme ultraviolet (EUV) was originally much higher (Canuto et al. 1982; Zahnle and Walker 1982). The reason is that the Sun was spinning faster early in its history. This increased the intensity of the magnetic fields in its atmosphere, which in turn increased the rate of excitation of various line emissions. The increased EUV emission would have been too small to affect climate directly, but it may have affected it indirectly by altering the chemical composition of the atmosphere (Canuto et al. 1982,1983; Kasting et al. 1984*a*). The EUV flux should have declined to more normal levels after the first few 100 Myr; thus, this factor should not have been important during most of solar system history.

Another astrophysical phenomenon that could conceivably have affected climate is encounters of the solar system with clouds of interstellar dust or gas (Thomas 1978). For example, an influx of interstellar hydrogen could have resulted in the formation of water-ice clouds at high altitudes in the Earth's atmosphere. Such clouds may, in turn, have triggered glaciations by increasing the planetary albedo (McKay and Thomas 1978; Paresce and Bowyer 1986). While we see no need to invoke such extraterrestrial influences to explain glaciations (see Sec. III), it is nevertheless difficult to rule them out as possible contributors to climatic change.

III. EARTH

A. The Faint Young Sun Problem

The secular increase in solar luminosity discussed in the previous section has important consequences for long-term climate evolution on the Earth. This was first brought to the attention of atmospheric scientists by Sagan and Mullen (1972) and has since come to be referred to as the "faint young Sun paradox." This problem has now been addressed by a number of investigators (see, e.g., Hart 1978; Owen et al. 1979; Rossow et al. 1982; Kuhn and Kasting 1983; Kasting et al. 1984*a;* Kasting 1987; Kiehl and Dickinson 1987). The discussion here parallels that of Kasting (1987).

The paradox associated with solar luminosity increases is that the Earth's mean surface temperature would have been below freezing prior to about 2.3 Gyr ago, provided that all other parameters (i.e., atmospheric composition, planetary albedo) remained unchanged (Sagan and Mullen 1972). Yet, the presence of sedimentary rocks indicates that liquid water was abundant back to the beginning of the geologic record, some 3.8 Gyr ago. Indeed, the absence of glacial deposits during the first half of the Earth's history indicates that the early Earth was most likely warmer than today (Sec. I.B).

The paradox disappears if one allows for the possibility that atmospheric composition has changed with time. Sagan and Mullen themselves suggested that enhanced concentrations of ammonia could have kept the early Earth warm. However, subsequent investigators (Kuhn and Atreya 1979; Kasting 1982) have shown that ammonia is photochemically converted to N_2 and hydrogen on a very short time scale. Methane could conceivably have augmented the greenhouse effect during the first half of the Earth's history, particularly if it reacted photochemically to produce additional, infrared-active hydrocarbons (Kasting et al. 1983; Kiehl and Dickinson 1987). But the most likely greenhouse gas is carbon dioxide. Several independent studies using radiative-convective climate models have shown that CO_2 concentrations of a few tenths of a bar would suffice to keep the Earth's surface temperature above freezing in the distant past (Owen et al. 1979; Kuhn and Kasting 1983; Kasting et al. 1984*a;* Kiehl and Dickinson 1987). These models have been intercompared for a particular CO_2-solar flux sequence suggested by Hart (1978). The Owen et al. calculations have since been shown to be in error (Kiehl and Dickinson 1987); the other models agree with each other quite closely.

A representative calculation of the greenhouse effect of large amounts of CO_2, from Kasting et al. (1984*a*), is shown in Fig. 2. A one-bar, N_2–O_2 background atmosphere with no O_3 was assumed in the calculation. The moist adiabatic lapse rate was used in the troposphere and the empirical relative humidity distribution of Manabe and Wetherald (1967) was assumed. The solid curve in this figure corresponds to the present solar luminosity and the temperature scale on the left; the dashed curve corresponds to a 30% reduced solar luminosity and the temperature scale on the right. The basic result in either case is that T_s increases approximately linearly with the logarithm of pCO_2: each factor of 10 increase in CO_2 produces roughly 10 deg of greenhouse warming. About one-tenth of a bar of CO_2 is required to keep the early Earth above freezing; several tenths of a bar would be required to raise T_s to its present value.

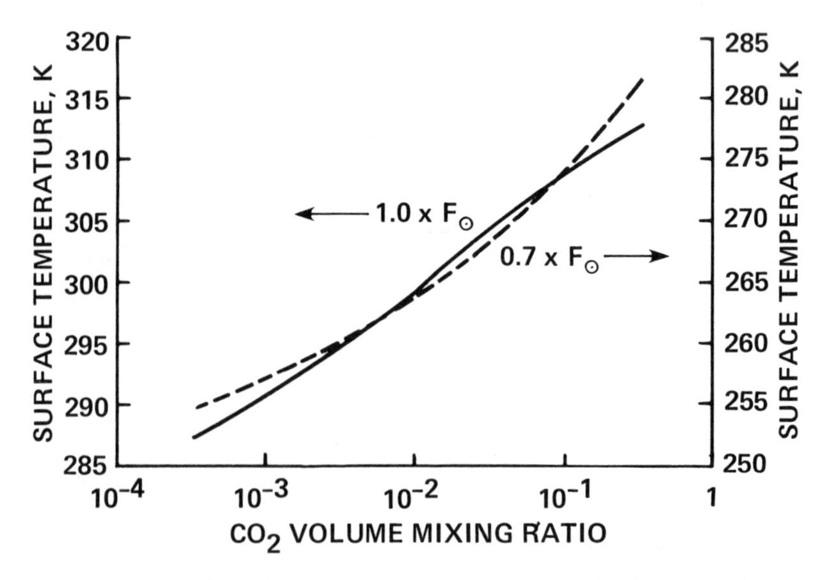

Fig. 2. Variation of the Earth's surface temperature with atmospheric CO_2 content for solar fluxes of 1.0 (solid curve) and 0.7 (dashed curve) times the present solar constant. A one-bar, N_2–O_2 background atmosphere is assumed (figure from Kasting et al. 1984*a*).

Similar one-dimensional climate calculations can be performed for various times in the Earth's history, as solar luminosity changes. Such calculations are most interesting for glacial periods when the Earth's surface temperature is bounded both above and below (Sec. I.B). Kasting (1987) has used this approach to infer atmospheric CO_2 concentrations during the Huronian glaciation 2.5 Gyr ago and during the Late Precambrian glaciation around 0.8 Gyr ago. Solar luminosity at these times was, respectively, 82% and 93.5% of the present value, according to Eq. (1). The corresponding bounds on CO_2 pressure are 0.03 to 0.3 bar during the Huronian and 10^{-3} to 10^{-2} bar during the Late Precambrian. These bounds may be used to estimate how pCO_2 has varied throughout geologic time (Fig. 3). The shaded area represents the range of CO_2 pressures that are consistent with the one-dimensional climate model. The general shape of the curve is dictated by the need to avoid cold temperatures during the Archean and mid-Proterozoic. The rationale for the upper limit of 10 bar of CO_2 at 4.5 Gyr ago is discussed in Sec. III.C. Perhaps the

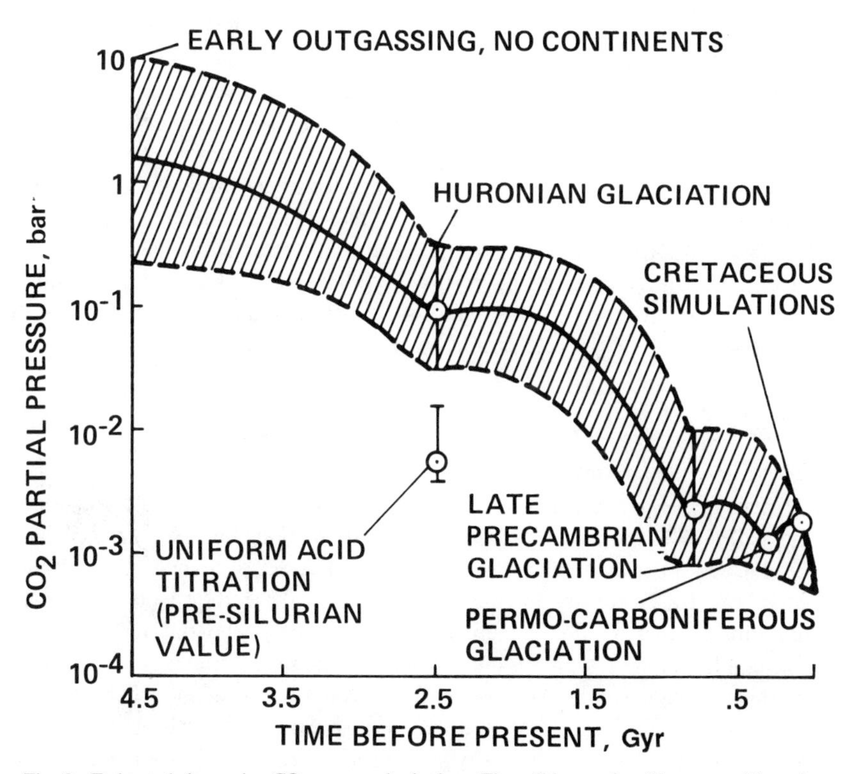

Fig. 3. Estimated change in pCO_2 over geologic time. The solid curve is a "best guess" based on one-dimensional climate model simulations. The shaded area represents the climatically reasonable range of pCO_2. The point labeled "uniform acid titration" is the pre-Silurian CO_2 pressure estimated by Holland and Zbinden (1986) from paleosol data (figure from Kasting 1987).

most striking feature of this curve is the steep downward trend in CO_2 predicted during the latter half of the Precambrian. This prediction could potentially be tested by careful analyses of weathering profiles in paleosols (ancient soils). The results of one such study (Holland and Zbinden 1986) are indicated by the datapoint labeled "uniform acid titration" in Fig. 3. This value for pCO_2 was derived from a suite of paleosols ranging in age from 1.5 to 2.5 Gyr. It is considerably below the mean CO_2 pressure predicted by the climate model for this interval, indicating that one or the other of these methods for estimating past CO_2 levels is significantly in error. It is furthermore noteworthy that the paleosol data show no evidence of the steep decline in pCO_2 predicted by the climate model (H. D. Holland, personal communication, 1986). Both discrepancies probably result from the relatively simple assumptions made in the original paleosol analysis (Pinto and Holland 1987).

Not all investigators agree that an increased greenhouse effect is required to solve the faint young Sun problem. Rossow et al. (1982) have suggested that cloud feedback alone could have kept surface temperatures above freezing even if atmospheric composition had remained unchanged. Cloudiness decreases with decreasing T_s in their model, so that the Earth absorbs more of the incident solar energy as it cools down. Their model, like other one-dimensional cloud models, is extremely speculative. It seems unlikely that cloud feedback by itself could have offset the drop in solar luminosity, since ice albedo feedback (which Rossow et al. did not consider) would have tended to further destabilize the climate system. Both their mechanism and the presence of other greenhouse gases could, however, have lowered the atmospheric CO_2 concentrations predicted in Fig. 3. On the other hand, neither of these additional mechanisms appears to be required. As discussed below, there are good reasons why past CO_2 levels should have been high enough to keep the early Earth warm.

B. The Carbonate-Silicate Cycle

To understand why atmospheric pCO_2 should have been higher in the past, it is necessary to consider the factors that control CO_2 levels over long time scales. The relevant processes are the interactions between atmospheric CO_2 and sedimentary rocks via the geochemical cycles of organic and carbonate carbon. Based on the apparent uniformity of the carbon isotopic record (Schidlowski et al. 1983), it seems that roughly 80% of the long-term cycling of carbon has always been through the latter of these two routes, the so-called carbonate-silicate geochemical cycle. By considering the details of this cycle, Walker et al. (1981) have proposed that atmospheric CO_2 levels are regulated over long time scales in such a way as to keep the Earth's surface temperature within the temperate regime. The basis for their argument is as follows. Carbon dioxide is removed from the atmosphere-ocean system by the process of silicate weathering. Dissolved CO_2 in rainwater reacts with silicate minerals in the soil, releasing calcium and magnesium ions and converting the CO_2

into bicarbonate. In the case of calcium, this process may be represented by the schematic reaction

$$CaSiO_3 + 2\,CO_2 + H_2O \rightarrow Ca^{++} + SiO_2 + 2\,HCO_3^-. \quad (3)$$

The dissolved species are transported by rivers to the ocean, where the calcium recombines with bicarbonate to form calcium carbonate

$$Ca^{++} + 2\,HCO_3^- \rightarrow CaCO_3 + CO_2 + H_2O. \quad (4)$$

In the present oceans, the task of precipitating calcium carbonate is performed by shelled organisms, most importantly the calcareous plankton. Prior to their appearance, this task may have been performed by mat-forming (stromatolitic) bacteria, or it may have occurred abiotically. The silica (SiO_2) released by weathering is precipitated by siliceous organisms in the form of opal.

Some mechanism must exist for replenishing CO_2; otherwise, carbonate deposition would deplete the present reservoir of carbon in the atmosphere and oceans in only 400,000 yr (Berner et al. 1983). Resupply of CO_2 is accomplished by the process of carbonate metamorphism or, equivalently, silicate reconstitution. According to the theory of plate tectonics, the seafloor spreads out from the midocean ridges across the ocean basins toward the margins of the continents, where it is subducted back downward into the mantle. When this happens carbonate sediments on the ocean floor are carried downward and subjected to high temperatures and pressures. Silicate minerals are reformed and gaseous carbon dioxide is released:

$$CaCO_3 + SiO_2 \rightarrow CaSiO_3 + CO_2. \quad (5)$$

Reaction (5) may be seen to be equivalent to the reverse of Reactions (3) plus (4). The CO_2 given off by this process reenters the atmosphere through volcanic outgassing along the plate margins. Some of the carbonate sediments may be carried downward into the upper mantle; this carbon is apparently recycled by outgassing at the midocean ridges (DesMarais 1985).

Self-regulation of the carbonate-silicate cycle arises from the fact that the weathering reactions (Reaction 3) are temperature dependent. More importantly, they depend on the amount of precipitation and runoff, and these factors themselves depend strongly on surface temperature (Walker et al. 1981; Berner et al. 1983). Increased T_s causes an increase in the weathering rate and hence an increase in the rate at which CO_2 is removed from the atmosphere-ocean system. But surface temperature, in turn, is related to the atmospheric CO_2 concentration by way of the greenhouse effect. Consider, then, what would happen if the system were to get out of balance and the CO_2 concentration were to become too high: T_s would increase, the weathering rate would increase; the rate of carbonate deposition would increase; and pCO_2 would decline. Conversely, if CO_2 concentrations were too low, T_s would decrease, the weathering rate would decrease, carbonate formation would decrease, and pCO_2 would build back up. Implicit in this discussion is the assumption that the rate of silicate reconstitution is unaffected by changes in surface temperature or in the rate of carbonate deposition. Since the bulk of the Earth's carbon is already in the carbonate rock reservoir, the rate of carbonate metamorphism should be independent of the other elements of the cycle.

The implications of this discussion for the faint young Sun problem are straightforward. Back when solar luminosity was lower, the Earth's surface temperature would have been lower as well, all other things being equal. Thus, the rate at which CO_2 was removed from the atmosphere by silicate weathering would have been slower than today. But the rate of tectonic cycling and, hence, of carbonate metamorphism should have been, if anything, higher than today as a consequence of a greater flow of heat from the Earth's interior. (The heat flow would have been higher because of enhanced heating from radioactive elements, in addition to a larger amount of residual heat left over from accretion.) The carbonate-silicate cycle would therefore have been out of balance and CO_2 would have begun to accumulate in the atsmosphere. The CO_2 cycle would have come back into balance only when sufficient carbon dioxide had accumulated to bring T_s up to something like its present value.

The feedback between weathering and climate in the carbonate-silicate cycle has been explored numerically by Walker et al. (1981) and, on shorter time scales, by Berner et al. (1983) and Lasaga et al. (1985). The predictions of such studies should be viewed qualitatively, rather than quantitatively, because the actual Earth system is much more complicated than any of the models. Nevertheless, it seems fair to state that the faint young Sun paradox is convincingly resolved by this feedback mechanism. This is most clearly illustrated by considering an extreme case. In the absence of additional greenhouse warming caused by CO_2 or other gases, the early oceans would have been completely frozen and silicate weathering should have virtually ceased. If carbon dioxide was released from volcanos at its present rate, a one-bar CO_2 atmosphere would have accumulated in only 20 Myr. The resulting greenhouse effect would have melted the ice and restored equable conditions within a geologically short time period.

C. Dense CO_2 Atmospheres and Hot Paleoclimates

While the early Earth should not have been overly cold, there is at least a possibility that it could have been quite hot. The total amount of carbon dioxide tied up in carbonate rocks today is some 3×10^{23} g, the equivalent of about 60 bar, were it all to be present in the atmosphere (Ronov and Yaroshevsky 1967; Holland 1978). This carbon, along with other volatile elements, was presumably brought to the Earth during accretion as a component of infalling planetesimals. A substantial fraction of these volatile compounds should have been released upon impact (Jakosky and Ahrens 1979; Holland 1984). Initially, reactions with iron in the infalling material may have kept the carbon in the form of CH_4 or CO. Once the rate of accretion had died down, however, these gases would have been oxidized to CO_2 by OH radicals produced from water vapor photolysis (Kasting et al. 1983).

How long this carbon dioxide remained in the atmosphere would have depended on its rate of removal by weathering. Silicate weathering today occurs primarily on the continents. But the continents were probably much smaller originally and may even have been entirely absent during the first few 100 Myr of the Earth's history (Walker 1986*a*, and references therein). Under such circumstances silicate weathering would have been restricted to the ocean floor. Since submarine weathering is slow, an appreciable fraction of the Earth's CO_2, perhaps as much as one-sixth, may have remained in the atmosphere-ocean system (Walker 1986*a*). If the *p*H of the ocean was less than 6, as seems likely in the presence of this much carbonic acid, the bulk of this CO_2 ($\sim$10 bar) would have resided in the atmosphere. Its removal would have awaited the emergence of the continents, a process that was slow to start and that did not approach completion until around 2 Gyr ago (Veizer 1983).

It is thus conceivable that a dense ($\sim$10 bar) CO_2 atmosphere could have enveloped the Earth for a period of several 100 Myr. Its existence, however, is far from certain. If the upper mantle was originally more reduced than at present, much of the Earth's CO_2 could have been sequestered in the mantle as graphite or diamond (J. C. G. Walker, personal communication, 1985). Some evidence that large amounts of carbon were tied up in the interior is provided by the apparent constancy of the ^{13}C content of marine carbonates back to at least 3.5 Gyr ago (Schidlowski et al. 1983, Fig. 7-3). If all the Earth's carbon was outgassed early, and if the organic carbon reservoir was initially small (as seems likely), then the initial ^{13}C content of carbonates should instead have been close to the average mantle value of -7 per mil. The fact that this is not observed implies one of two things: either, the organic carbon reservoir grew to its present size within the first Gyr of the Earth's history, in which case the biosphere must have achieved high productivity right from the start; or else, the surface reservoirs of carbonate and organic carbon grew simultaneously, indicating that the outgassing of carbon was gradual rather than abrupt (J. C. G. Walker, personal communication, 1985).

In view of the lack of convincing evidence in support of either position, we feel that it is not yet possible to constrain the time history of CO_2 outgassing. Thus, one cannot reliably estimate the CO_2 pressure in the Earth's primitive atmosphere. The possible climatic implications of very high CO_2 levels have nonetheless been explored by Kasting and Ackerman (1986), using a one-dimensional radiative-convective climate model. Their results are summarized in Fig. 4. The two curves show surface temperature as a function of CO_2 pressure for CO_2 pressures up to 100 bar. As in Fig. 2, the solid curve is

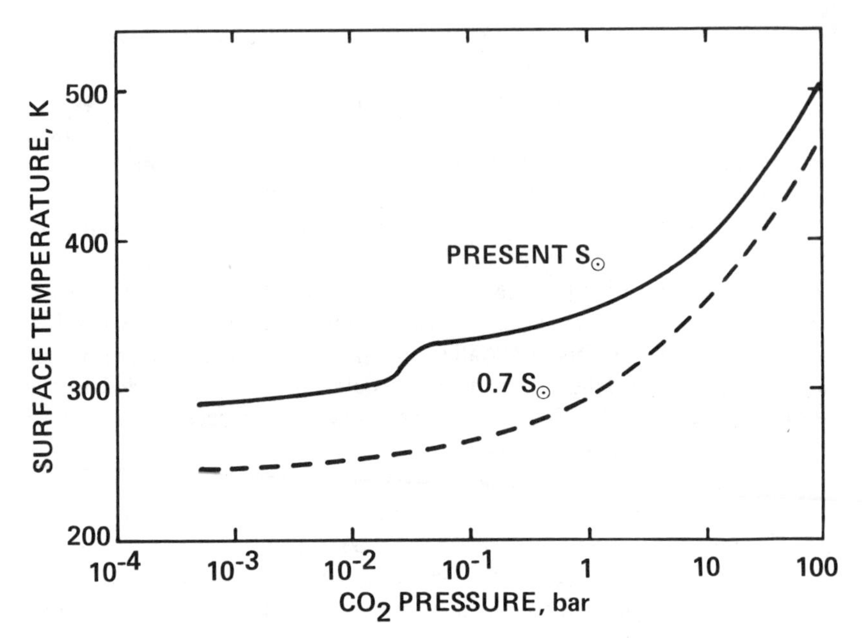

Fig. 4. Variation of the Earth's surface temperature with atmospheric CO_2 pressure for large CO_2 increases. The solid and dashed curves have the same meaning as in Fig. 2. One bar of N_2 plus O_2 is assumed, but the total surface pressure varies with CO_2 amount (figure from Kasting and Ackerman 1986).

for present solar luminosity and the dashed curve is for 30% reduced luminosity. Both calculations assume an enhanced value of the surface albedo (= 0.22) to compensate for the absence of clouds from the model. The moist adiabatic lapse rate was assumed in the troposphere, as before; however, the relative humidity was allowed to increase at high surface temperatures so as to produce an upper limit on T_s at a given CO_2 level.

A significant result of this calculation is that in neither case is the addition of CO_2 sufficient to trigger a runaway greenhouse, i.e., complete evaporation of the oceans. The predicted surface temperature for the present solar flux and 100 bar of CO_2 is just over 500 K. The corresponding staturation vapor pressure is about 29 bar; the remainder of the Earth's water (270 bar total) would reside in the ocean. The surface temperature required for runaway, by comparison, is 647.1 K, the critical point for water. Above the critical point, the phase change between gaseous and liquid water disappears and any water that was present could be considered to be part of the atmosphere. More will be said about this possibility in Sec. V. The predicted surface temperature for a plausible early Earth case, corresponding to reduced solar flux and 10 bar of CO_2, is 360 K. This is certainly warm compared to today, but perhaps not too warm to be consistent with the early appearance of life.

A related conclusion is that the early Earth was stable against loss of water via photodissociation followed by hydrogen escape. The hydrogen escape rate is limited under many circumstances by the mixing ratio (i.e., mole fraction) of hydrogen in the stratosphere in all of its chemical forms (Hunten 1973; Walker 1977). For a warm, humid atmosphere, the most important hydrogenous species is water vapor. Ingersoll (1969) has shown that, in a one-dimensional world, the mixing ratio of water vapor in the stratosphere is positively correlated with its mixing ratio at the surface. If the lower atmosphere contains more than about 20% water by volume, the stratosphere should be wet; otherwise, it should be relatively dry. The latter situation prevails on Earth today: the surface H_2O mixing ratio averages about 1% and the stratosphere contains only about 4 parts per million of water vapor. (In reality, the stratospheric H_2O concentration is also affected by dynamical processes, which cause the tropopause to be higher and colder at the equator than at high latitudes. Such dynamical effects could modify Ingersoll's (1969) prediction, but should not seriously undermine it.) A warm early atmosphere could conceivably have been much more moist. The model shown in Fig. 4, however, has a relatively low H_2O mixing ratio near the surface; the increased vapor pressure of water at high T_s is more than compensated by the increase in total pressure caused by the addition of CO_2. Thus, according to this model, the primitive stratosphere should have been even drier than today's.

D. Summary

The Earth was never cold enough to completely freeze its oceans, despite reduced solar luminosity in the past. The most likely mechanism for keeping the early Earth warm was an enhanced greenhouse effect caused by higher levels of atmospheric CO_2. Such elevated CO_2 concentrations are a natural consequence of the operation of the carbonate-silicate geochemical cycle on a tectonically active planet. The early Earth may, in fact, have possessed a warm, dense (1 to 10 bar) CO_2 atmosphere as a result of slower weathering rates caused by the absence of exposed land area. However, given its finite CO_2 inventory, the Earth was, and is, stable against a CO_2-induced runaway greenhouse and against the loss of its water.

IV. MARS

Having gone through the above discussion for Earth, one finds that the evolution of Mars' climate is relatively easy to understand. The problem of Mars' climate history closely parallels the faint young Sun problem on Earth. Mars is 1.52 times farther from the Sun than is Earth, so the incident solar flux is about 2.3 times smaller. Were Mars exactly like the Earth, this lower solar flux would presumably have been offset by much higher concentrations of atmospheric CO_2, and the resulting climate would have been roughly Earth-like. But Mars is much smaller than Earth; its mass is only about one-tenth as great. This difference in size, we shall argue, was primarily responsible for the two planets' divergent evolution.

A. Greenhouse Calculations

The first question that a climatologist might ask about Mars is: How much CO_2 would be required to raise its average surface temperature above freezing? Mars actually has about 30 times as much CO_2 in a vertical column in its atmosphere as does Earth, but this amount is clearly far from enough to keep its surface warm. Several different investigators have derived estimates for the greenhouse effect of large CO_2 amounts (Pollack 1979; Cess et al. 1980; Postawko and Kuhn 1986; Pollack et al. 1987*c*). (The Cess et al. calculations are, however, marred by the same error that invalidated the results of Owen et al. [1979]; see Sec. III.A.) The results of Pollack et al. (1987*c*) are exhibited in Fig. 5, which shows surface temperature T_s and planetary albedo A_p for a CO_2–H_2O atmosphere and present solar luminosity. As in the comparable Earth calculations (Fig. 2), the moist adiabatic lapse rate was used and the assumed relative humidity profile was that of Manabe and Wetherald (1967). Two features of Fig. 5 are of particular interest: first, the CO_2 pressure required to elevate T_s above freezing is approximately 2 bar, in good agreement with the results of Pollack (1979); and second, the planetary albedo increases at high CO_2 pressures because of an increase in diffuse reflection caused by Rayleigh scattering. This effect, which was first pointed out by Cess et al. (1980), is magnified by the fact that the Rayleigh scattering cross

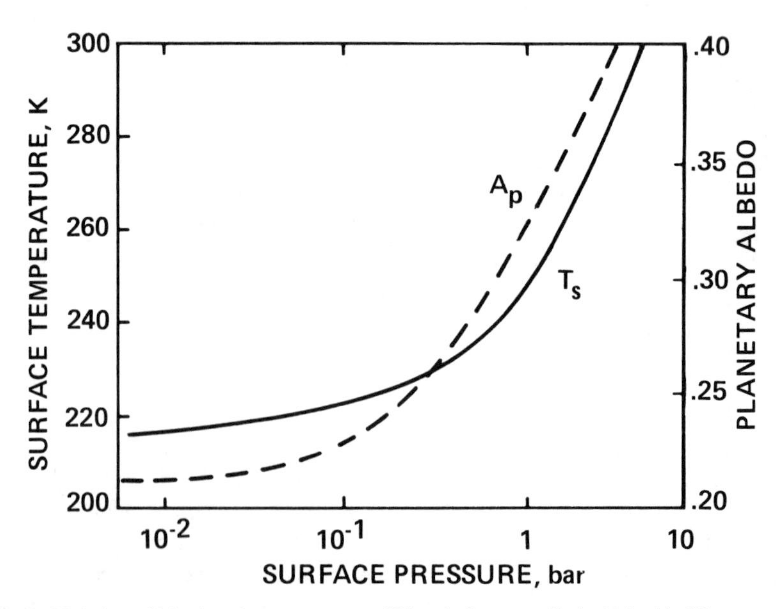

Fig. 5. Variation of Mars' surface temperature (T_s) and planetary albedo (A_p) with CO_2 pressure for present solar luminosity (figure from Pollack et al. 1987*c*).

section for CO_2 is about 2.5 times that of (terrestrial) air. The increase in planetary albedo at high surface pressures produces a negative feedback on surface temperature that increases the amount of CO_2 required to bring T_s above freezing. A similar effect is observed for the Earth calculations shown in Fig. 4, although in these much warmer atmospheres it competes with a decrease in A_p caused by absorption of solar near-infrared radiation by water vapor. While this albedo feedback is clearly significant, it must be remembered that albedo changes caused by clouds (which have been ignored in this model) could be equally important.

To apply such a model to early Mars, one must take several other factors into consideration. As noted previously (Sec. II.A), the solar flux was lower than today by about 30%. In addition, however, Mars' orbit is highly eccentric (e = 0.93); indeed, e varies quasi-periodically between 0 and 0.14 (Ward 1974). At times of maximum eccentricity, the solar insolation at perihelion can be as much as 35% higher than its orbitally averaged value. Furthermore, while the geologic record on Mars implies the existence of warm temperatures at some locales (Sec. I.B), it is not necessary that the planet's mean surface temperature have been above freezing. If one assumes that liquid water was confined to the equatorial regions, then the relevant value of the solar flux

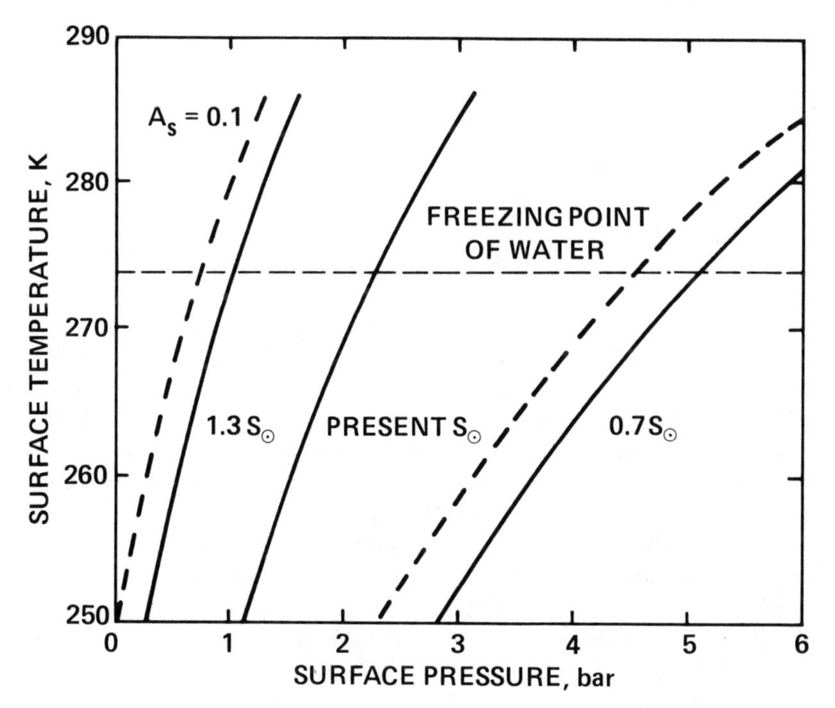

Fig. 6. Variation of Mars' surface temperature with CO_2 pressure for three different values of the solar luminosity. The dashed curves are for a surface albedo of 0.1 (figure from Pollack et al. 1987*c*).

could be 40% greater than the planetary average. Combining these three factors, one finds that the solar fluxes of interest range from 0.7 times the present value at Mars orbit (corresponding to globally and orbitally averaged conditions) to 1.3 times the present value (corresponding to equatorial conditions at perihelion during times of maximum eccentricity).

The consequences of allowing for such variations in insolation are exhibited in Fig. 6, again from Pollack et al. (1987*c*). The CO_2 pressure required to elevate T_s above freezing ranges from about 1 bar at maximum insolation to 5 bar at the minimum value. The dashed curves in Fig. 6 show the effect of lowering the surface albedo from its present value of 0.215 to 0.1 to account for a possible decrease in surface reflectivity on an unoxidized early Mars (Toon et al. 1980).

B. Partitioning of CO_2 Among the Atmosphere, Hydrosphere and Solid Planet

Two questions must be answered to determine whether a 1-to-5 bar CO_2 atmosphere could have existed in Mars' early history. The first is whether this much carbon dioxide was ever outgassed. Previous attempts to estimate Mars' CO_2 inventory have yielded values of 0.1 to several bars (Pollack and Yung 1980, and references therein). Many of these estimates, however, were based on scaling the terrestrial CO_2 reservoir by the ratio of various rare gas (e.g., ^{36}Ar) in the atmospheres of Mars and Earth. This procedure is probably invalid, because rare gases may have been incorporated into the planets by different mechanisms than were the other volatile elements, and because their abundance patterns may have been altered by fractionation occurring during hydrodynamic escape (Hunten et al. 1987; see also Chapters by Hunten et al. and by Pepin). McElroy et al. (1977) have estimated a total CO_2 inventory of about 1.5 bar, based on scaling by nitrogen abundances. Mars' initial N_2 inventory was itself estimated from the Viking measurement of the $^{15}N/^{14}N$ ratio in the Martian atmosphere. This procedure, too, is risky because it assumes that the vertical structure of Mars' atmosphere has remained unchanged with time and that the N_2/CO_2 ratio is the same for the two planets. A third method for estimating Mars' CO_2 inventory is simply to scale the Earth's inventory by the relative masses of the two planets, taking into account differences in surface area and gravity (Pollack et al. 1987*c*). This procedure yields a total inventory of about 10 (Mars) bar. This method is also unreliable because it does not account for possible differences in the availability of carbon and in the extent of outgassing. However, it does indicate that the hypothesized 1-to-5 bar early atmosphere should not be ruled out on this basis.

The second question that must be considered is whether the outgassed CO_2 would have resided in the atmosphere, in the hydrosphere or in carbonate rocks. On the Earth today, the bulk of the outgassed CO_2 ($\sim$60 bar) is in carbonates; a smaller amount ($\sim$0.02 bar) is dissolved in the oceans, predominately as bicarbonate; and only 3×10^{-4} bar is present in the atmosphere. Were these same ratios to apply to early Mars, the total amount of carbon dioxide required to maintain even a one-bar CO_2 atmosphere would be impossibly large. However, as in the case of early Earth (Sec. III.C), there is good reason to suppose that the relative partitioning should have been quite different. Mars' primitive ocean, if one existed at all, was presumably smaller than the Earth's. Carr (1986) estimates an average depth of 0.5 to 1 km, based on geomorphological evidence; most other estimates are considerably lower. The *p*H of this ocean would probably have been low because of the presence of carbonic acid; thus its total dissolved carbon content should have been smaller than the atmospheric reservoir (Pollack et al. 1987*c*). Carbonate rocks would have constituted a substantial sink for CO_2, but they would probably have been recycled by metamorphism on a time scale of a few 10 Myr. The recycling mechanism on early Mars may not have involved plate tectonics, as on Earth, but rather a process in which carbonate rocks were gradually buried by volcanic outflows (ibid.). In order for silicate weathering to compete with this process, T_s must have been at or above freezing; therefore, a substantial fraction, perhaps as much as one-half, of the total CO_2 inventory must have been present in the atmosphere.

The evidence on both of these counts indicates that a warm, dense CO_2 atmosphere could have been present on early Mars. Whether or not such an atmosphere actually existed remains to be determined. Close-up inspection of the valley networks would no doubt help to resolve this question. A relatively easy way perhaps to rule out this hypothesis (though not to confirm it) is to search spectroscopically for the presence of carbonate minerals in the crust. Such minerals ought to be abundant in at least some locales if this hypothesis is correct. The spectroscopic signature of carbonates has not yet been detected from Earth, but it ought to be observable by the upcoming Mars Observer Mission.

C. The Decline to Current Conditions

Even if a warm, dense CO_2 atmosphere did exist on Mars at some time in the past, it has evidently been absent for several billions of years. The reason for its disappearance is not hard to ascertain. The Earth has global-scale plate tectonics while Mars' does not. Thus, carbonate rocks, which are recycled on Earth, remain permanently locked up in the Martian crust. CO_2 has also been lost by other mechanisms, such as adsorption onto the regolith and formation of ice at the poles (Fanale 1976; Pollack 1979; Toon et al. 1980).

The reason why Mars does not exhibit plate tectonics is probably related to its smaller internal heat flow, about 30 mW m^{-2} compared to 100 mW m^{-2} for the Earth (Davies and Arvidson 1981). The interior heat flux on early Mars was probably 3 to 10 times larger than today because of an initially faster rate of heat release by radioactive elements and a larger amount of residual heat from accretion and core formation (Davies and Arvidson 1981). Such a high rate of heat flow, up to 3 times greater than the present terrestrial value, would likely have resulted in rapid resurfacing of the planet and efficient recycling of carbonate rocks. However, Mars lost its internal heat more rapidly than Earth because of its higher surface-to-volume ratio. Thus, the period of warm climate should have been relatively brief. The fact that the runoff channels are confined exclusively to the ancient, heavily cratered terrain indicates that the warm period did not extend beyond the end of the heavy bombardment, about 3.8 Gyr ago (Carr 1986). Even after the mean surface temperature had fallen below freezing, continued climatic deterioration could have been caused by carbonate formation in transient water pockets (Kahn 1985) and by uptake of CO_2 by the regolith. This overall cooling may have been interspersed with occasional warmer periods brought about by episodes of intense volcanism, acting in conjunction with orbital variations and a slowly warming Sun (Toon et al. 1980).

D. Summary

Mars could have started out with a dense (1 to 5 bar) CO_2 atmosphere, an Earth-like climate, and as much as a kilometer of water. On the other hand, if its initial volatile inventory was smaller than assumed here, Mars may always have been substantially cooler than the Earth, though not nearly so cold as it is today. In either case, such relatively warm climatic conditions prevailed for only a few 100 Myr. As the planet's interior cooled, the rate of crustal recycling decreased and CO_2 became trapped in carbonate rocks and in the regolith. Water was also trapped in the regolith as ice and in the form of hydrated minerals. The end result was a cold, alien, inhospitable planet.

V. VENUS

Venus is located at 0.72 times Earth's distance from the Sun and the incident solar flux is higher by a factor of 1.91. Even when the Sun was 30% less bright, the solar flux at Venus' orbit was 34% higher than the current flux for Earth. Consequently, Venus has always been faced with the opposite climatic problem from that of Earth and Mars: its surface temperature has probably always been near or above the maximum temperature at which liquid water can exist.

A. Original Water Endowment

One of the most remarkable aspects of Venus today is that it is incredibly dry. The amount of water in Venus' atmosphere, roughly 100 parts per million by volume (see Hunten et al.'s chapter), is about 10^5 times less than the amount in the Earth's oceans. Three clearly distinguishable theories have been offered to explain why this is so. The simplest is that Venus formed with very little water because the minerals that condensed from this relatively warm region of the solar nebula lacked water of hydration (Lewis and Prinn 1984).

However, if the CO_2 in Venus' present atmosphere was derived from oxidation of hydrocarbons, as seems likely, roughly one-sixth of a terrestrial ocean of water should have been produced at the same time (Walker et al. 1970). Furthermore, radial mixing of planetesimals between the zones of accretion of the Earth and Venus (Wetherill 1985) may well have provided similar amounts of volatiles to both planets. Much of this water could have been lost during the accretion process (Hunten et al.'s chapter); however, it is difficult to understand why Venus should have lost so much more of its water during this period than did Earth. Thus, it seems probable that an appreciable fraction of a terrestrial ocean of water was still present on Venus at the close of accretion.

B. The Runaway Greenhouse

The other two theories assume that Venus started out with a large amount of water, perhaps as much as Earth, but that it lost its water over the course of geologic time. The classical explanation for how this occurred is called the "runaway greenhouse" (Hoyle 1955; Sagan 1960; Gold 1964; Dayhoff et al. 1967; Ingersoll 1969; Rasool and DeBergh 1970; Pollack 1971; Goody and Walker 1972; Walker 1975; Watson et al. 1984). According to this theory, Venus was always too hot to allow water to condense upon its surface. Any water that the planet possessed was therefore present in the atmosphere as steam. Ingersoll (1969) showed that if the mixing ratio of water in such an atmosphere exceeded a critical value ($\sim$20% by volume), the stratospheric cold trap would be effectively "washed out" and water vapor would remain a major atmospheric constituent up to very high altitudes. Under these conditions, water vapor would be rapidly photodissociated by solar ultraviolet radiation, and the diffusion limit on hydrogen escape (Sec. II.C) would be overcome. Hydrogen would be lost by a rapid hydrodynamic escape process (Watson et al. 1981; Kasting and Pollack 1983), and oxygen would either be dragged along with it (Zahnle and Kasting 1986) or would react with reduced minerals in the crust (Walker 1975).

The basic concept of the runaway greenhouse is illustrated by Fig. 7, from Goody and Walker (1972). Their calculation was itself based on a similar computation by Rasool and DeBergh (1970). In this calculation, Venus, Earth and Mars were each presumed to have outgassed a pure water vapor atmosphere, starting from an initially airless state. Present solar luminosity was assumed and surface temperatures were calculated under the assumption that the atmosphere was grey and radiatively equilibrated at all levels. As the vapor pressure at the surface increased, Mars quickly encountered the saturation vapor pressure curve (solid line), and any water vapor released subsequently was frozen out as ice. The Earth encountered the saturation vapor pressure curve at a somewhat higher temperature, and water vapor condensed out to form oceans. On Venus, by contrast, the greenhouse effect became large before the saturation vapor pressure curve was reached, and all of the outgassed water accumulated in the atmosphere as steam.

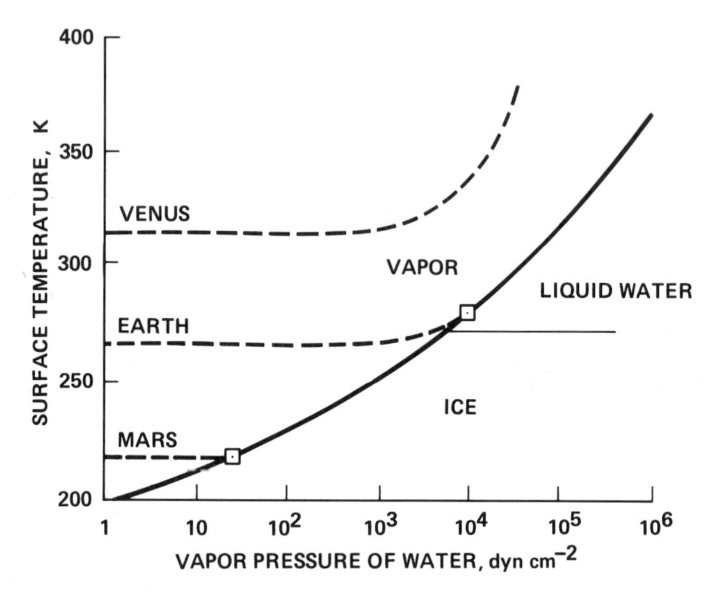

Fig. 7. Evolution of planetary surface temperatures with pressure for a pure water vapor atmosphere (figure from Goody and Walker 1972).

C. The Moist Greenhouse

The calculation presented in Fig. 7 has several obvious deficiencies. Foremost among these is the neglect of convection. When Goody and Walker's radiative equilibrium model is applied to the present Earth, it overpredicts the surface temperature by about 45 deg. This discrepancy is removed when heat transport by convection is taken into account. The stabilizing effect of convection on surface temperature is further increased when condensation is included because this lowers the rate at which temperature decreases with altitude in the troposphere (Lindzen et al. 1982; Kasting et al. 1984*b;* Lal and Ramanathan 1984).

Using a climate model which included the effects of moist convection, Kasting et al. (1984*b*) predicted that Venus may have originally had oceans and a surface temperature near 100°C. They termed their model atmosphere a "moist greenhouse" to distinguish it from the runaway case. The moist greenhouse model could account for the loss of Venus' water just as easily as could the runaway greenhouse model. The critical requirement, that the stratosphere be wet, was satisfied with a model atmosphere consisting of about one bar of N_2 plus O_2 and one bar of H_2O. The atmospheric CO_2 partial pressure was presumed to have been suppressed by formation of carbonate rocks. This model may even have an advantage over the runaway greenhouse model in explaining the subsequent evolution of Venus' atmosphere. Because the background pressure of noncondensable gases was smaller, the stratospheric cold trap would have developed only after nearly all of Venus' water had been lost. By contrast, a runaway greenhouse atmosphere containing some 90 bar of CO_2 should have been left with approximately 20 bar of residual water that would have been difficult to dispose of.

The model of Kasting et al. (1984*b*) was criticized by Vardavas and Carver (1985) because it did not account for possible variations in the planetary albedo. Vardavas and Carver presented a model in which planetary albedo decreased sharply at warmer surface temperatures in response to absorption of solar radiation by water vapor. For the same fully saturated conditions assumed by Kasting et al., they predicted that a runaway greenhouse could be triggered by a mere 2% increase in the solar flux at the Earth's orbit. This implies that Venus was always in the runaway greenhouse state.

Vardavas and Carver's model relied heavily on measurements of (purported) continuum absorption by H_2O in the visible and near infrared by Tomasi (1979*a,b*). Kasting (1988) criticized this approach and substituted one based on standard rotation-vibration band models. He also included an accurate treatment of moist and dry adiabatic lapse rates in nonideal atmospheres. The results of his calculation are summarized in Figs. 8 and 9. A fully saturated, N_2–O_2–low CO_2 atmosphere with a background surface pressure of one bar was assumed. Clouds were modeled by assuming an elevated surface albedo (A_s = 0.22). These assumptions should yield an upper limit on surface

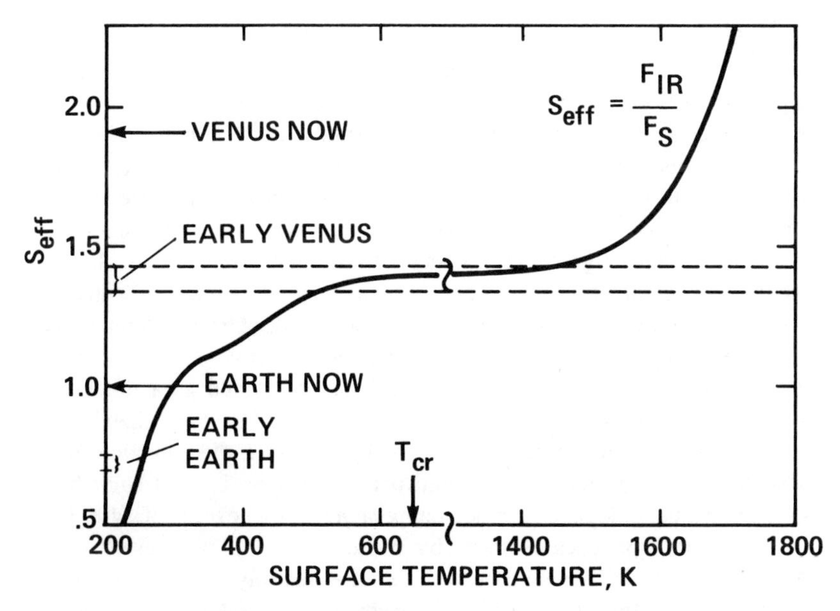

Fig. 8. Relationship between surface temperature and solar flux for an Earth-like planet. S_{eff} is the solar flux normalized to the present value at the Earth's orbit. The horizontal dashed lines bracket the solar flux expected for early Venus (figure from Kasting 1988).

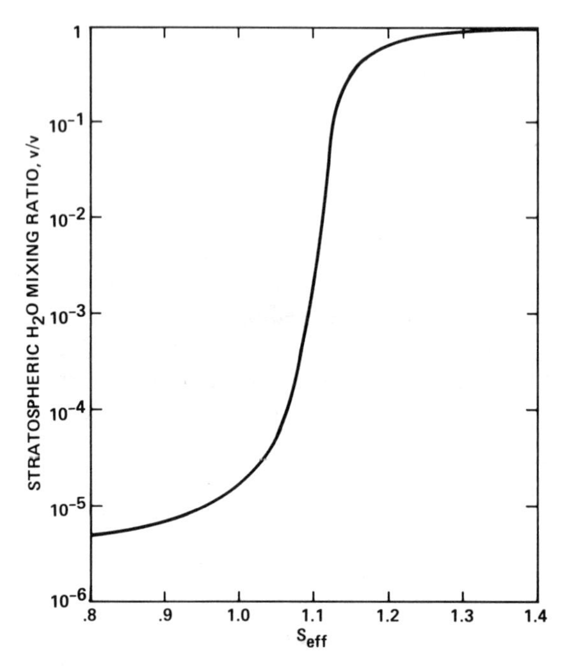

Fig. 9. H_2O volume mixing ratio in the stratosphere of an Earth-like planet as a function of normalized solar constant S_{eff} (figure from Kasting 1988).

temperature at a given solar flux (see below). The amount of water present at the planet's surface was assumed to be 1.4×10^{24} g (a full terrestrial ocean); this produces a surface pressure of 270 bar (under Earth gravity) when fully vaporized. This surface pressure may be compared with a saturation vapor pressure of 220.6 bar at the critical point for water (647.1 K). Since the pressure exerted by the ocean exceeds that at the critical point, the transition from a moist to a runaway greenhouse occurs at the critical temperature. By contrast, if the equivalent pressure of the oceans was lower than 220.6 bar, the transition to the runaway greenhouse would take place at a lower temperature.

For stability reasons, the usual radiative-convective approach was inverted in Kasting's model, and solar flux was calculated as a function of surface temperature (Fig. 8). Here, S_{eff} is the solar flux normalized to the present value at Earth's orbit. For this cloud-free model, the transition from a moist to a runaway greenhouse would occur at $S_{\mathrm{eff}} \simeq 1.4$. At solar fluxes above this value, calculated surface temperatures are in excess of 1400 K, in agreement with similar calculations by Watson et al. (1984). Present Venus, with S_{eff} equal to 1.91, would be well into the runaway greenhouse regime. Early Venus, on the other hand, was right on the borderline between the runaway and moist greenhouse states, according to this calculation. A reduction in solar flux by 25 to 30% (Sec. II.A) corresponds to values of S_{eff}

between 1.34 and 1.43, neatly bracketing the critical flux at which runaway is predicted to occur.

A real atmosphere would presumably have clouds distributed at various heights within the moist convective region. Sensitivity studies described in Kasting (1988) indicate that the effect of clouds in a warm, moist atmosphere would be to cool the surface, perhaps by a substantial amount. (The greenhouse effect of clouds is not important in an H_2O-dominated atmosphere because the gaseous infrared opacity is already very high.) Equivalently, clouds would have raised the solar flux required to maintain a given surface temperature. For example, an optically thick cloud layer located at a pressure of a few tenths of a bar could have raised the critical flux required for runaway to between 2 and 5 times the present terrestrial value, depending on the fractional amount of cloud cover. This implies that early Venus would almost certainly have been in the moist greenhouse state if it did indeed start out with an appreciable amount of water. Evidence of past oceans would be an intriguing thing to look for with the upcoming Venus Radar Mapping Mission.

Kasting (1988) also calculated stratospheric water vapor mixing ratio as a function of S_{eff} (Fig. 9). At low solar fluxes the predicted mixing ratio was a few parts per million, as on the present Earth. For S_{eff} greater than about 1.1, however, the stratosphere was wet, implying that water could be rapidly lost. The actual solar flux required for water loss should be somewhat higher than this value, because a real atmosphere should be only partly saturated and because clouds should help to keep it cool.

D. Summary

If the planetesimals that condensed from the solar nebula were radially mixed, Venus should have formed with a water endowment comparable to that of the Earth. This water may have stayed in the atmosphere as steam (the runaway greenhouse) or it may have condensed out to form oceans (the moist greenhouse). The latter hypothesis is favored from the standpoint of climate modeling. The moist greenhouse theory is also better able to explain how the last remnants of Venus' water were lost. In either case, the bulk of Venus' water was lost by photodissociation followed by hydrogen escape. Once the water was gone, there was no medium in which to form carbonate minerals, and CO_2 simply accumulated in the atmosphere. The end result was a planet which was hot, dry, and once again inhospitable toward life.

VI. CLIMATE DURING ACCRETION

Matsui and Abe (1986*a,b*) have suggested that the presence of an impact-induced steam atmosphere during Earth's accretion would have provided a sufficient blanketing effect to raise the surface temperature to approximately 1500 K, above the solidus for typical silicate rocks. The term "blanketing effect" is used in place of "greenhouse effect" to distinguish between an atmosphere that is heated primarily from below (by impacts), instead of from above. The predicted surface pressure in their model is approximately 100 bar throughout much of the accretion period. The amount of water in the atmosphere is buffered by the feedback between surface pressure, surface temperature and the degree of partial melting of the crust.

In performing this calculation, Matsui and Abe used a grey, radiatively equilibrated model similar to that of Goody and Walker (1972). As discussed in Sec. V.C, such a model may not yield a realistic estimate of surface temperature because it neglects the effect of convection. In this case, however, Matsui and Abe's results have been substantiated by the results of Kasting (1988), described in the previous section. The inclusion of an accretionary heat flux causes that model to switch to the runaway greenhouse regime, just as occurs in Fig. 8 for high values of the solar flux. The predicted surface temperature for a 100 bar steam atmosphere and a reasonable accretion rate ($\sim 10^{-8}$ Earth masses per year) is close to Matsui and Abe's estimate of 1500 K. Although the agreement is to some extent fortuitous, it indicates that Matsui and Abe's basic concept is sound. The Earth, and perhaps Mars and Venus as well, was probably enveloped in a dense steam atmosphere during the time that it was forming. This atmosphere would likely have been augmented by carbon-containing gases released from impacts and by H_2 produced from the reaction of steam with iron in the crust (Dreibus and Wänke 1984). Hydrogen would have escaped rapidly from the top of such an atmosphere by the hydrodynamic escape process described in the Chapter by Hunten et al. This period of planetary evolution may have played an important role in determining the abundance patterns of the rare gases (Chapter by Pepin).

VII. SUMMARY: THE CONTINUOUSLY HABITABLE ZONE

It is interesting to examine the implications of our climate history model for the width of the continuously habitable zone (CHZ) around the Sun. The concept of the CHZ was introduced by Hart (1978). He defined it as the region in space in which an Earth-like planet could maintain a climate conducive to life for a period long enough for complex life to have evolved, about 4 Gyr judging from our experience. Hart approached this question with a highly parameterized computer model, with which he attempted to calculate both atmospheric composition and surface temperature as a function of time. Most of his simulations encountered either the runaway greenhouse or runaway glaciation at some time during the Earth's history. From the few simulations that succeeded, Hart estimated that the CHZ extended from at most 0.95 to 1.01 AU. Elsewhere (Hart 1979), Hart went on to conclude that the chances of finding another habitable planet in our Galaxy might be rather remote.

We can make our own estimate for the width of the continuously habitable zone based on the discussion presented here. In Sec. V.C we argued that water could be rapidly lost from an Earth-like planet if the incident solar flux exceeds 1.1 times the present value at Earth's orbit. This calculation probably underestimates the critical solar flux to some extent because it assumes a fully saturated, cloud-free atmosphere. If we adopt this as a conservative estimate, however, then the inner edge of the CHZ is predicted to lie at 0.95 AU, in agreement with Hart's calculation. The agreement is accidental, since Hart based his limit on the distance at which a runaway greenhouse would have occurred. This distance, in our model, is about 0.85 AU.

Our model deviates significantly from Hart's in its placement of the outer edge of the continuously habitable zone. Hart's model encountered runaway glaciation at 1.01 AU because it did not include the stabilizing feedback provided by the carbonate-silicate geochemical cycle (Sec. III.B). If atmospheric CO_2 levels increase as surface temperatures approach freezing, as we believe they must, then the outer edge of the CHZ should be much farther away from the Sun than Hart calculated. According to our model, an Earth-sized planet, possessing the Earth's total CO_2 inventory, should be able to sustain liquid water out to well beyond the orbit of Mars. Indeed, the limiting constraint on the outer edge of the CHZ may be the difficulty in forming an Earth-sized object as one approaches the vicinity of the giant planets.

The implications of our theory for cosmic habitability are precisely the opposite of Hart's. If other planetary systems exist, there is good reason to believe that some of the planets will be habitable, at least in terms of surface temperature. Whether or not any of them are inhabited is, of course, an open question, but it would clearly be a mistake to dismiss this possibility on the grounds that the Earth is climatically unique.

COUPLED EVOLUTION OF THE ATMOSPHERES AND INTERIORS OF PLANETS AND SATELLITES

G. SCHUBERT
University of California

D. L. TURCOTTE
Cornell University

S. C. SOLOMON
Massachusetts Institute of Technology
and

N. H. SLEEP
Stanford University

The atmospheres and interiors of planets and satellites are highly coupled systems whose origins and evolutions are fundamentally interconnected. The evolution of a planet's atmosphere can be strongly influenced by its interior because the interior is potentially both a source and sink of atmospheric constituents. The evolution of a planet's interior can be strongly influenced by its atmosphere because mechanisms of heat loss from the interior depend on its volatile content and the atmosphere is potentially both a sink and source of these volatiles. Volatiles influence interior heat transport by decreasing the subsolidus viscosity and by promoting melting. Models of accretion and thermal evolution argue for formation of atmospheres on large planets (e.g., Earth and Venus) contemporaneously with accretion or shortly thereafter (a few 100 Myr) by impact degassing and/or highly vigorous overturning in a mantle heated by accretion and *core formation. Subsequent evolution of the atmosphere and interior may involve further degassing of volatiles retained within the planet at the end of accretion and regassing of the interior if there exist mechanisms such as plate tectonics for cycling volatiles back into the mantle. The dependence of mantle rheology on volatile content could provide a feedback mechanism that tends to keep regassing/degassing in balance and maintain relatively constant atmospheric mass. The temperature dependence of mantle viscosity compensates for the volatile dependence of viscosity in such a way as to keep mantle viscosity and heat flow at the values required to transfer the mantle's internal heat to the surface. Thus, as a consequence of degassing (regassing), the temperature of a mantle with volatile- and temperature-dependent viscosity is higher (lower) than the temperature of a mantle whose viscosity depends only on temperature. Consideration of the abundances of radiogenic and nonradiogenic noble gases in the Earth's atmosphere and of the fluxes of these gases from the mantle support a substantial degassing event early in the Earth's history, a decrease in degassing efficiency with time, and relatively inefficient outgassing over most of geologic time. Substantial regassing of the Earth's interior is suggested by recent estimates of the present CO_2 flux from the mantle. The ^{36}Ar and ^{40}Ar contents of Venus' atmosphere suggest less outgassing of Venus than of the Earth, consistent with the present lack of plate tectonics on Venus. These data also support the loss of ^{36}Ar on the Earth during a major Moon-forming collision event. Atmospheric erosion of Mars provides a mechanism for explaining its low content of the argon isotopes. With the exception of Io and Titan, the smaller outer planet satellites have no atmospheres of consequence. Accretional heating and impact degassing are less important for the icy satellites, while tidal heating is of unique importance for them. Io has a transient, volcanically controlled SO_2-S_2 atmosphere whose temperature, pressure, composition and surface distribution provide a direct connection to the internal sources of tidal dissipation. Buoyant NH_3 and CH_4 magmas may have substantially degassed Titan's interior and determined the composition and state of its present atmosphere.*

I. INTRODUCTION

The atmospheres and interiors of planets and satellites are highly coupled systems whose origins and evolutions can be fundamentally interconnected. Nevertheless, previous studies of the evolutions of planetary atmospheres and interiors have usually treated each of these systems as separate. While this review does not attempt any quantitative modeling of atmosphere-interior evolution (indeed, that is the subject of much needed research), it does attempt to discuss and emphasize the variety of mechanisms coupling the systems.

We will use the term atmosphere to refer to the gaseous envelope of volatile substances above the surface of a planet or satellite. Volatiles once in planetary interiors may end up in the atmosphere as a gas or on the surface as a liquid or a solid. These nongaseous reservoirs of volatiles (the Earth's oceans and ice caps, Mars' CO_2 ice caps, water ice on the outer-planet satellites, etc.) must, of course, be included in the inventory of a planet's volatile content.

The evolution of a planet's atmosphere can be strongly influenced by its interior because the interior is potentially both a source and sink of atmospheric volatiles. The evolution of a planet's interior can be strongly influenced by its atmosphere because mechanisms of heat loss from the interior depend on its volatile content and the atmosphere is potentially both a sink and source of these volatiles. Mechanisms of heat loss are responsible for surface volcanism and tectonics.

Not only are atmospheres and interiors coupled in their evolutions, but each often contains evidence relevant to deciphering the history of the other. For example, the atmospheric contents of noble gas isotopes provide clues to the timing and extent of mantle degassing. Also, volcanic surface expressions of interior activity can be used to infer the nature and volume of a past atmosphere.

Volcanic eruptions are a primary source of volatiles in the atmospheres of the terrestrial planets. The flux of volatiles to an atmosphere is dependent on the rate of volcanism, the concentrations of volatiles in the magmas, and the efficiencies of loss from magmas to the atmosphere. Volcanism is the direct result of heat loss from a planetary interior. Alternative tectonic mechanisms for heat loss provide greatly different rates of volcanism. The concentration of volatiles in a planetary interior is dependent on the rate of loss, the rate of recycling (if any) and primordial concentrations after accretion. The efficiency of transfer from magmas to atmosphere depends upon the volcanic style and secondary processes such as hydrothermal circulation and erosion.

A strong feedback component is also present. Styles of volcanic activity are certainly dependent on concentrations of volatiles in the magma. High volatile content leads to explosive volcanism which immediately transfers volatiles to the atmosphere. However, our understanding of this relationship is not well developed. Wilson and Mouginis-Mark (1987) have discussed the implications of pyroclastic deposits on Mars in terms of the volatile content of magmas.

It is expected that the presence or absence of plate tectonics on a planet can have a profound effect on interactions between planetary interiors and atmospheres. The 6 km of basalts that solidify at mid-ocean ridges to form the oceanic crust scavenge large quantities of volatiles from the Earth's interior. It is now well documented that the primary source of atmospheric helium on the Earth is the high temperature hydrothermal circulations adjacent to oceanic ridges (Lupton 1983*a*,*b*). Jenkins et al. (1978) have shown that there is a direct correlation between the helium flux and the heat flux. In fact, the global helium flux deduced from the geothermal heat flux in the oceans is in good agreement with the independently determined helium flux from the oceans to the atmosphere. O'Nions and Oxburgh (1983) and Oxburgh and O'Nions (1987) have discussed global correlations of heat flux and helium flux. Important questions are:

1. Can the fluxes of other volatiles be scaled with the heat flux?
2. Are hydrothermal circulations required to scavenge volatiles from extrusive and intrusive volcanic flows?

Plate tectonics can also influence global geochemical cycles of volatiles through subduction. Subduction recycles large quantities of volatiles into the interior of the Earth. Some, but not all, are returned to the surface in island-arc volcanics. The hydrologic cycle and the oceans play a very important role in the transfer processes. Hydrothermal circulations in the oceans hydrate the oceanic crust and deposit carbonates; the altered oceanic crust is subsequently subducted. Carbonates form a substantial fraction of the sediments that coat the oceanic crust. A substantial fraction of these sediments is subducted.

A number of studies of global geochemical cycles on the Earth have been carried out. Allègre et al. (1987) have given global balances for the rare gases. They show that inclusion of atmosphere-interior interactions is essential. Javoy et al. (1982) have given global balances for carbon; again, the storage and atmosphere are essential to the understanding of the system.

We will begin with a discussion of the physical processes that can couple atmosphere-interior evolution. This will be followed by a summary of what thermal history models of planetary interiors imply about atmospheric evolution. We will then review what atmospheric noble gases tell us about interiors and what volcanic styles and histories have to say about atmospheres. A final section will discuss the atmospheres and interiors of the outer-planet satellites.

II. MODES OF ATMOSPHERE-INTERIOR COUPLING

Atmosphere-interior coupling occurs in a variety of ways. These include the effects of volatiles on interior physical properties such as rock rheology and thermal conductivity, and the influence of volatiles on the degree of partial melting of the interior.

Effects of Volatiles on Rock Rheology and Physical Properties

Interior evolution depends on the ability of a planet to transfer heat from its deep interior to its surface via subsolidus convection or conduction. The selection of the heat transfer mechanism and the efficiency with which it operates depends on the physical properties controlling the modes of heat transport, primarily viscosity for convection and thermal conductivity for conduction. Subsolidus convection is believed to be the dominant process of heat transfer in the terrestrial planets and the Moon; either convection or conduction could control the transport of heat in the outer-planet satellites.

Since wet rock is more readily deformable than dry rock (Carter 1976; Tullis 1979), the presence of volatiles in a planet's interior reduces its viscosity μ at a specified temperature and pressure and thereby promotes convective heat transfer. Figure 1 illustrates the dependence of rock rheology on volatile content. Rock thermal conductivity k decreases with volatile content; this also tends to promote convective heat transfer relative to conduction.

The ways in which volatile content affect atmosphere-interior evolution are determined by the degassing-regassing history of the interior. In the clas-

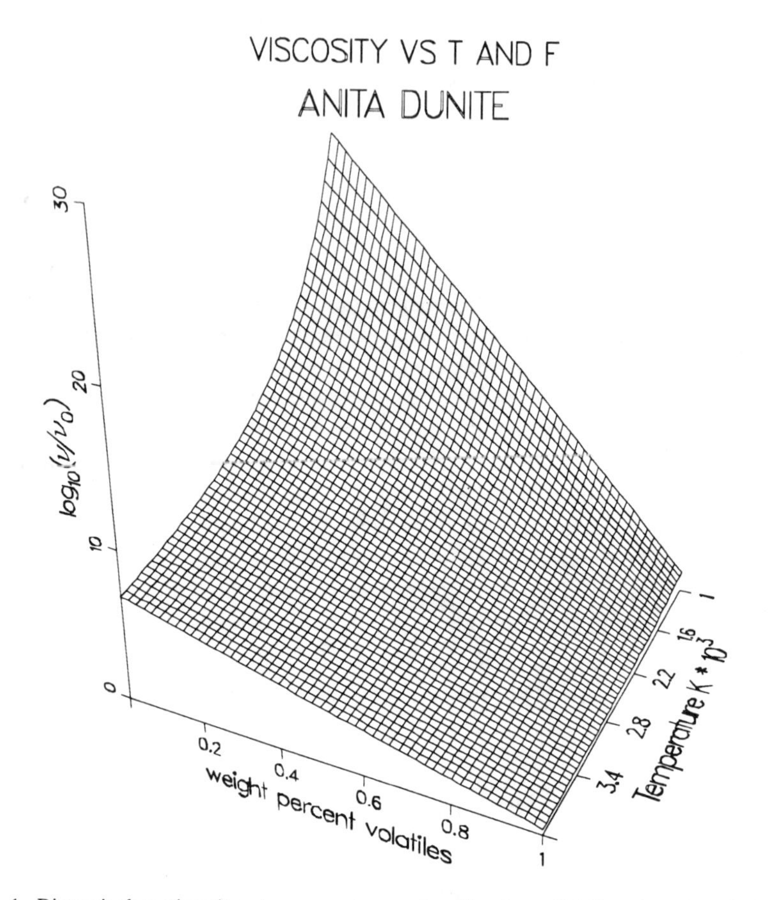

Fig. 1. Dimensionless viscosity ν/ν_o versus temperature T and wt. % F based on the relation $\nu/\nu_o = \exp(A_o - A'F/T)$, where $A_o = 64.4 \times 10^3$K and $A' = 61.1 \times 10^3$ K. The values of A_o and A' are from deformation data on Anita Bay dunite (Chopra and Paterson 1984).

sical view of Rubey (1951), planetary atmospheres are a consequence of the gradual degassing of planetary interiors throughout geologic time. However, especially in the cases of the larger planets, atmospheres may form contemporaneously or shortly after accretion by impact degassing and/or highly vigorous convection in a mantle heated by accretion and core formation. The interior may be highly depleted of volatiles at the end of the period of planetary accumulation. The subsequent evolution of the interior may involve regassing if mechanisms exist for cycling the atmosphere back into the interior.

On the Earth, plate tectonics provides such a mechanism through subduction. Extensive alteration of the oceanic crust by hydrothermal circulations adjacent to mid-ocean ridges introduces hydrous minerals and carbonates; these volatiles can be returned to the mantle at subduction zones. Subduction of sediments with the oceanic crust can also return volatiles to the mantle.

Competing processes, e.g., dehydration reactions in the descending slab, can return subducted volatiles to the atmosphere through island-arc volcanoes. That the potential for revolatilizing the Earth's mantle is large, follows from the observation that hydration of the oceanic crust can return the entire volume of the present oceans to the mantle over geologic time at present seafloor spreading rates. Even larger amounts of mantle recharging are feasible, given the likelihood of more vigorous mantle convection and plate motions in the past. Unfortunately, we are not certain about the direction or amount of any net volatile exchange between the present-day atmosphere and mantle. We discuss volatile cycling on the Earth in more detail in Sec. IV.

Interior regassing is more problematical for other planets and satellites that show no evidence of plate tectonics. However, volatiles may be returned to the mantle by delamination or foundering of the lithosphere, a process that could act on other planets as well as the Earth. Overthrusting is another tectonic process, familiar on the Earth, that could return volatiles to the uppermost mantles of other planets. Overthrusting can lead to conditions conducive to delamination. Shallow underthrusting and lithospheric foundering might occur in those regions of Venus showing evidence of compressional tectonics. Thus, possible mechanisms exist for recycling CO_2 back into Venus' upper mantle. Recycling of sulfur and SO_2, at least to shallow depths, is a strong possibility in the rapidly evolving volcanic environment of Io's surface and near-surface regions. Burial and subsidence would be the processes responsible for volatile recycling on Io.

Table I summarizes how the dependence of mantle viscosity μ on volatile content combines with the temperature dependence of μ to influence the thermal evolution of a convecting mantle in a degassing/regassing scenario. Outgassing dries out the interior and tends to increase its viscosity. However, the tendency for devolatilization to increase μ is compensated by the effect of temperature on viscosity. The mantle tends to maintain the required rate of heat loss by increasing temperature, reducing viscosity, and maintaining the

TABLE I
Rheological Influence of Mantle Devolatilization/Revolatilization on Interior Thermal History

Degassing (regassing)

↓

Decreases (increases) mantle volatile content

↓

Mantle viscosity tends to increase (decrease) from the volatile effect

↓

Mantle temperature increases (decreases) to compensate the volatile effect and maintain approximately constant mantle viscosity and heat flow. Vigor of mantle convection and quantity of heat transport are determined by internal heat production. Mantle temperature adjusts to accomplish the required heat transport.

level of convective vigor. The net result is a hotter mantle as a consequence of degassing, but mantle heat flow, viscosity and convective vigor are essentially the same as in a mantle with volatile-independent rheology. These effects are illustrated in Fig. 2, which compares the evolution of temperature with time in a mantle having a volatile- and temperature-dependent viscosity with one in which μ depends on temperature alone. The results are based on an evolution model of the mantle with parameterized convection, the details of which can be found in McGovern and Schubert (1988). The mantle is hotter and cools more slowly when μ depends on both volatile content and temperature (dotted curve) compared with the case in which μ is a function of only temperature (solid curve). Figure 2 also shows the degassing history. Degassing is very rapid in the first several 100 Myr, becoming more gradual with time. In this model, about 1.5 ocean masses of water are degassed over geologic time; one ocean mass is outgassed in about the first 500 Myr. Mantle viscosity and heat flow vs time are substantially the same for both the volatile-dependent and volatile-independent rheologies.

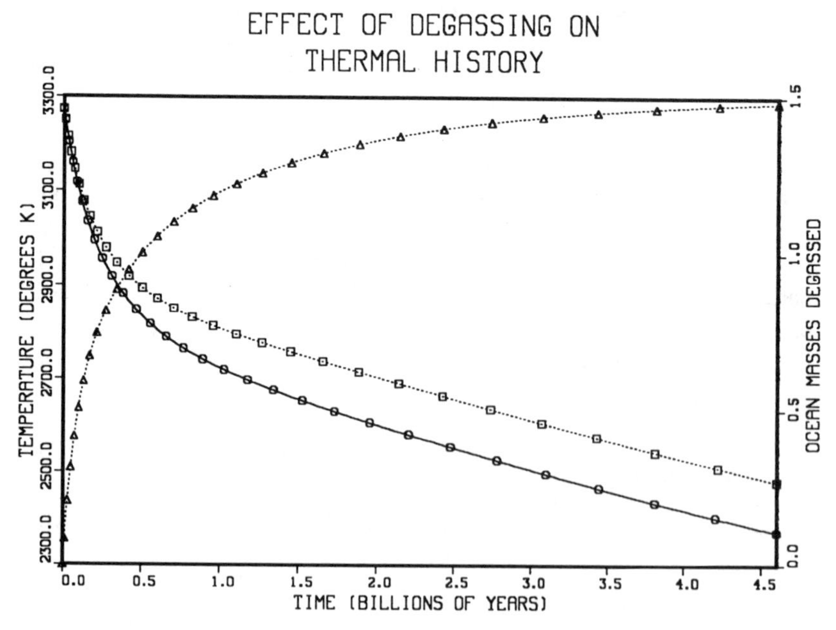

Fig. 2. Mantle temperature vs time for a volatile- and temperature-dependent viscosity (dotted curve) and a viscosity that depends on temperature only (solid curve). The amount of degassing (in terms of ocean masses) for the volatile-dependent mantle viscosity is shown by the monotonically increasing dotted curve. The mantle cools rapidly and most of the volatiles are outgassed during the first several 100 Myr of evolution. A volatile-dependent rheology results in a hotter mantle as a consequence of degassing. The figure is based on a parameterized convection model of the Earth's thermal evolution (McGovern and Schubert 1988).

Regassing increases the volatile content of the interior and tends to decrease its viscosity. However, as in the degassing case, the tendency for revolatilization to decrease μ is compensated by a reduction in mantle temperature so as to maintain viscosity, heat flow and convective vigor approximately constant. These effects are illustrated in Fig. 3 which shows mantle temperature vs time for a volatile- and temperature-dependent rheology (dotted curve) and a temperature-only dependence of rheology (solid curve). In addition, the figure shows the amount of water regassed into the mantle. The evolution, in terms of the amount of cooling and the quantity of water reabsorbed into the mantle, is rapid during the first several 100 Myr, becoming more gradual afterwards. In this particular model, about 3/4 of an ocean mass of volatiles (water) is re-injected into the mantle over geologic time, with the bulk of this occurring in the first Gyr. The main effect of the volatile-dependent mantle viscosity is a cooler mantle, compared with the case in which μ depends on temperature only. As in the degassing case, mantle viscosity and heat flow are essentially the same for both the volatile-dependent and volatile-independent viscosities. In both the regassing and degassing scenarios, the time rate of

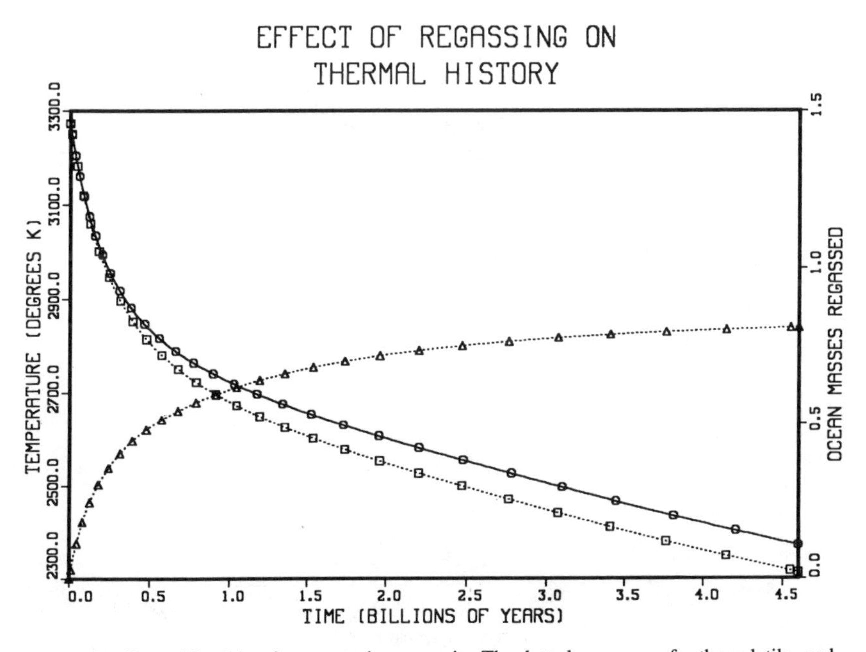

Fig. 3. Similar to Fig. 2 but for a regassing scenario. The dotted curves are for the volatile- and temperature-dependent rheology. The temperature curve decreases monotonically with time, while the amount of volatiles absorbed into the mantle increases monotonically with time. The solid curve is mantle temperature for a viscosity that depends on temperature only. The effect of a volatile-dependent viscosity is to make the mantle colder in this regassing situation (figure after McGovern and Schubert 1988).

change of temperature eventually tends to the same value for the volatile-dependent and volatile-independent rheologies; during the later stages of thermal evolution only a constant temperature offset distinguishes the volatile-dependent mantle cooling rate from the volatile-independent one.

Effects of Volatiles on Melting

Volatiles can have a profound effect on interior evolution by reducing the solidus temperature (Fig. 4) and promoting melting. Melt migration is more efficient than subsolidus convection in transporting heat and dissolved volatiles to the surface. Thus, melting promotes degassing. However, regassing promotes melting; an example may be found in island-arc volcanism. Dehydration of the descending oceanic crust and upward percolation of the volatiles may reduce the solidus temperature in the mantle wedge above the Benioff zone and lead to melting of mantle material and volcanism.

Volatiles can significantly alter heat transport from within a planet as attested to by hydrothermal circulation through oceanic ridge systems and geothermal regions on the Earth. Volcanic styles are a function of magma

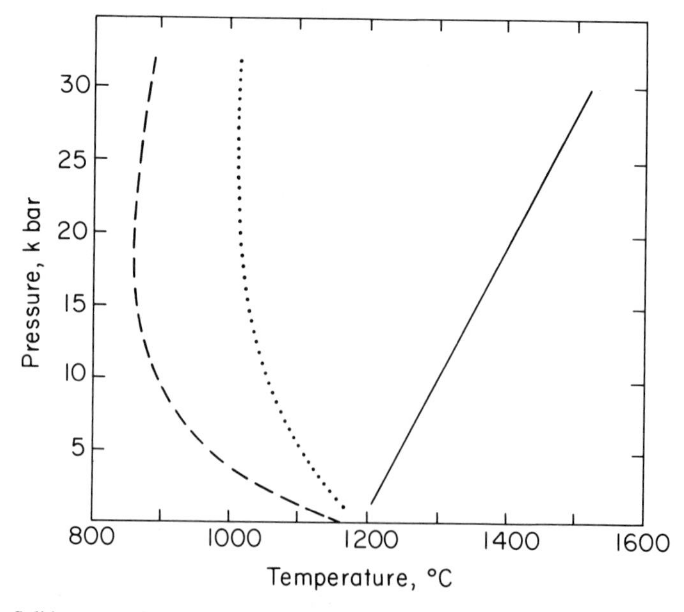

Fig. 4. Solidus curves for wet and dry peridotite. The dry solidus is after Kushiro et al. (1968). The wet solidi are from Mysen and Boettcher (1975). Dashed curve: experimentally determined solidus for peridotite + H_2O. Dotted curve: experimentally determined solidus for peridotite + H_2O + CO_2 with the mole fraction of H_2O in the vapor equal to 0.25.

volatile content so the record of volcanic activity on the planets can be informative about atmospheric evolution. The incorporation of minor constituent volatiles in the interiors of the icy satellites of the outer planets can have profound effects on the evolution of the satellites' interiors and surfaces. Volatiles can promote partial melting in the interiors at relatively low temperatures and they can provide mechanisms for weakening the rheology of the major constituent materials (Stevenson and Lunine 1986).

When volatiles promote melting they also influence interior evolution by changing the rheology of the partially molten rocks and by partitioning the heat-producing elements into the melts which can then redistribute the radioactives upward. By inducing partial melting, it might be expected that volatiles would further soften mantle material. However, the effect of partial melting on rheology might be just the opposite since small amounts of partial melt could dissolve volatiles and dry out the surrounding mantle rock (Karato 1986).

An example of the profound effects that volatile recharging can have on planetary interiors is provided by the late Archean-early Proterozoic granulite facies terrains on the Earth. It has been proposed that the deep crustal metamorphism exemplified by these terrains was the consequence of the upward migration through the lower continental crust of CO_2-rich fluids brought to the base of the continental lithosphere by subduction of marine carbonates (Newton 1987). Evidence for this process includes deformation-related alteration of gneisses, CO_2-rich fluid inclusions in quartz and other granulite minerals, and variable to extreme depletion of large ion lithophile elements (K, U, Rb—particularly Rb) relative to typical upper crustal rocks. The depletion of the heat-producing elements is especially important for the thermal regime of the lower continental crust. Wyllie (1980) has shown that a predominantly CO_2 fluid mixture of CO_2 and H_2O would initiate melting to form carbonatite or kimberlite liquids at the base of the continental lithosphere (at depths of 150 to 200 km). Upward migration of the melt could lead eventually to free streaming of CO_2 through the upper 100 km of the lithosphere or to capture of the CO_2 in mantle carbonate, depending on the thermal regime. The late Archean-Early Proterozoic thermal regime would favor the streaming of CO_2 (Newton 1987). Subducted carbonates may also be the source of many diamonds. Diamonds are often associated with eclogitic xenoliths; this eclogite may be subducted oceanic crust and the diamonds may be produced from associated subducted carbonates (Javoy et al. 1982).

The inability of a planet to recycle its volatiles can have equally profound implications for its evolution. Mars may be a prime example of the dire consequences to follow from the absence of volatile recycling. It has been proposed that Mars had a warm, dense, early CO_2 atmosphere which became trapped in carbonate rocks after $\sim$1 Gyr of evolution because of the absence of plate tectonic recycling on that planet (see the chapter by Kasting and Toon).

III. IMPLICATIONS FOR ATMOSPHERES OF INTERIOR THERMAL HISTORY MODELS

Terrestrial Planets

The generally accepted view of the thermal histories of the terrestrial planets is that immediately after accretion their interiors were hot, with temperatures essentially at the solidus, and that after a period of rapid cooling by a vigorous mantle convection system lasting several 100 Myr, the interiors adjusted to a more gradual cooling by convection lasting most of geologic time (see Figs. 2 and 3). The early heat source is gravitational potential energy made available by accretion and core formation contemporaneous with accretion or occurring shortly after planetary accumulation. Radiogenic heating contributes significantly on the geologic time scale, and possibly on a shorter time scale as well, if certain extinct radionuclides were incorporated into the planets. The dependence of mantle viscosity on temperature, which provides a thermostat-like control on internal temperature, has been the dominant factor in determining the evolutionary paths of thermal models (Tozer 1967; Schubert et al. 1979; Schubert 1979). The effect of volatiles on μ and the coupling with the atmosphere are only now being quantitatively incorporated into thermal history models of interiors. Jackson and Pollack (1987) took the dependence of μ on volatile content into account by *a priori* specifying the degassing history of the Earth. McGovern and Schubert (1988) have recently developed a self-consistent atmosphere-interior evolution model in which the degassing/regassing history is an outcome of the model.

The terrestrial planets are heated so severely by accretion and essentially contemporaneous core formation that their interiors must be substantially degassed upon completion of planet formation or within a few 100 Myr thereafter. Large amounts of early degassing characterize the results in Figs. 2 and 3. Thermal history models thus call for early and rapid degassing of the terrestrial planets in contrast to Rubey's (1951) view of gradual degassing of the Earth over geologic time, but in accordance with Fanale's (1971) view of early catastrophic degassing of the Earth. For the Earth, noble gas isotopic data support early atmospheric degassing (Hamano and Ozima 1978; chapter by Ozima and Igarashi) as will be discussed in more detail in Sec. VII.

Early degassing of the Earth and the terrestrial planets could have occurred contemporaneously with accretion or shortly thereafter by vigorous mantle overturning. The energy released in high-velocity impacts can liberate the water and carbon dioxide from volatile bearing minerals making an atmosphere an integral part of an accreting planet (Benlow and Meadows 1977; Lange and Ahrens 1982*a*,*b*,1984). Any volatiles retained within the accreting planet could be outgassed very rapidly by the energetic overturning of a mantle heated at least to its solidus.

Abe and Matsui (1985) and Matsui and Abe (1986*a*) have modeled the structure of an impact-induced atmosphere on the Earth (see also Zahnle et al.

1988). They assume that the energy of accretion is released continuously, clearly an approximation to the stochastic accretion process. As a consequence of this assumption, their model produces a steam-dominated H_2O-CO_2 accretionary atmosphere with a surface pressure on the order of 100 bar. A strong greenhouse effect in their model yields a surface temperature high enough ($\sim$1500 K) to maintain a partially molten crust. Formation of a magma ocean at the surface of the accreting Earth moderates the buildup of an atmosphere due to the solubility of water and carbon dioxide in the magma. The partitioning of H_2O and CO_2 between a magma ocean and the atmosphere introduces considerable uncertainty into the modeling of the accretionary atmosphere. The accretionary atmosphere would have been vulnerable to loss through hydrodynamic escape to space of hydrogen and other collisionally coupled constituents (Hunten et al. 1987; chapter by Hunten et al.). Many times the inventory of present atmospheric volatiles could have been lost during the accumulation of the Earth and shortly after the final stages of accretion, the H_2O vapor in the atmosphere would largely condense to form oceans.

Models of accretionary steam-dominated atmospheres have also been developed for Venus (Matsui and Abe 1986*b*). The early steam atmosphere on Venus is predicted to contain about as much water as there is in the present Earth's oceans. The major difference between the models of the steam atmospheres of the Earth and Venus is the eventual condensation of steam to form oceans in the former and the eventual dissociation of H_2O and escape to space of hydrogen in the latter (Matsui and Abe 1986*b*; chapter by Kasting and Toon). The contrasting evolutionary paths for the Earth's and Venus' steam atmospheres are due to the different amounts of solar radiation absorbed by the atmospheres after accretional energy deposition ceases to be the dominant energy source for the atmospheres.

The maintenance of very high surface temperatures during accretion of the Earth and Venus by the blanketing of a steam atmosphere would enhance temperatures throughout the planetary interiors promoting vigorous dynamic activity and core formation inside the growing planets.

Subsequent to a few 100 Myr after the end of accretion, the terrestrial planets undergo a slow and gradual cooling controlled by deep convective heat transport and near-surface conductive heat transport through a fixed or, in the case of the Earth, mobile lithosphere. The vigor of convection is determined by the internal heat production and the viscosity of the mantle. The strong temperature dependence of viscosity and the dependence of μ on volatile content exert a dominant rheological control on the thermal evolution (Figs. 2 and 3). Long-term sources of thermal energy include heat from the radioactive decay of ^{40}K, ^{238}U, ^{235}U and ^{232}Th, viscous dissipation in the mantle, ohmic dissipation in the core, the release of latent heat and gravitational potential energy upon inner core growth, and cooling of the interior at rates of order 100 K Gyr^{-1}.

During the slowly evolving phase of interior evolution there may be either a continuing gradual degassing of any volatiles left behind in the mantle after completion of accretion (Fig. 2) or a gradual regassing of the mantle by the post-accretion atmosphere (Fig. 3). Both regassing and degassing could occur simultaneously. The balance between the processes could be responsible for maintaining approximately constant ocean volume with geologic time. An imbalance between the processes might have changed the volume of the oceans in the past. The dependence of mantle rheology on volatile content can provide a feedback effect that tends to keep the processes in balance (McGovern and Schubert 1988). The volatile effect on rheology could moderate atmosphere-hydrosphere evolution in the same way that the temperature effect on rheology moderates interior thermal evolution. We discuss volatile exchange between the Earth's atmosphere and interior in Sec. IV; we note here that the present direction of net exchange is uncertain. Noble gases generated by radioactive decay in the Earth's mantle are gradually degassed over geologic time, but of course, they are only minor constituents of the atmosphere. Nevertheless, as we discuss in Sec. VII, the radiogenic noble gases are important tracers of atmospheric and interior evolution.

Degassing or regassing of certain portions of a planet's mantle can be limited by a chemical compositional stratification of the mantle that prevents material exchange between regions of different composition. It is conceivable that the Earth's upper and lower mantles are substantially isolated from each other because of differences in composition, although there is accumulating evidence to the contrary (Creager and Jordan 1984,1986). If the Earth's mantle is divided into separate upper and lower convective systems, then volatiles would have to diffuse across the 670 km interface and degassing or regassing of the lower mantle would occur slowly compared to the upper mantle. That the Earth is not completely degassed of original volatiles, is evident from the present-day escape of primordial nonradiogenic ^{3}He (Craig et al. 1975; Jenkins et al. 1978; Lupton 1983*a,b*). However, we cannot conclude from this that the ^{3}He has been stored in an *isolated* lower mantle. We can only conclude that the mantle has not been 100% degassed of its original volatiles. The retention of some volatiles in terrestrial planets would be facilitated by the isolation of a silicate core from impact degassing during planetary accretion. Impact degassing would not be effective until a planet has reached a critical size of about 10^3 km in radius.

Outer-Planet Satellites

Although many of the same considerations enter into the modeling of the thermal histories of outer-planet satellites, as occur in terrestrial-planet thermal evolution modeling, the sizes, compositions and environments of the satellites are so different from those of the inner planets that the satellite interior evolutionary paths are distinctive in many aspects (see Schubert et al. [1986] for a review of satellite thermal evolution). Of the major outer-planet satellites, only Io and Europa have rocky compositions similar to those of the terrestrial planets; the other satellites are predominantly water ice-rock mixtures with ice constituting about half of the satellite by mass. The icy composition of most of the satellites has two major effects on thermal history. There is a reduction in the radiogenic heat production proportional to the mass fraction of ice, and convective heat transport is controlled by the rheology of ice with some degree of stiffening by intermixed rock.

With the exception of Ganymede, Callisto and Titan, the satellites are lunar size and smaller. Accretional heating (Ellsworth and Schubert 1983; Squyres et al. 1988) and impact degassing are accordingly of much less significance for most of the satellites (accretional temperature profiles of Saturnian and Uranian satellites are shown in Fig. 5). Indeed, accretional heating is so marginal that a major question about the interiors of the satellites is the extent of ice-rock differentiation (Schubert et al. 1986). Models of the satellite interiors range from undifferentiated homogeneous mixtures of ice and rock to fully differentiated bodies with silicate cores and ice mantles. The smaller icy satellites of Saturn are not likely to have differentiated, whereas the larger satellites Ganymede, Callisto and Titan may be fully to partially differentiated.

While accretional heating is generally of reduced importance for the satellites, another heat source, namely tidal dissipation, is very significant. Tidal

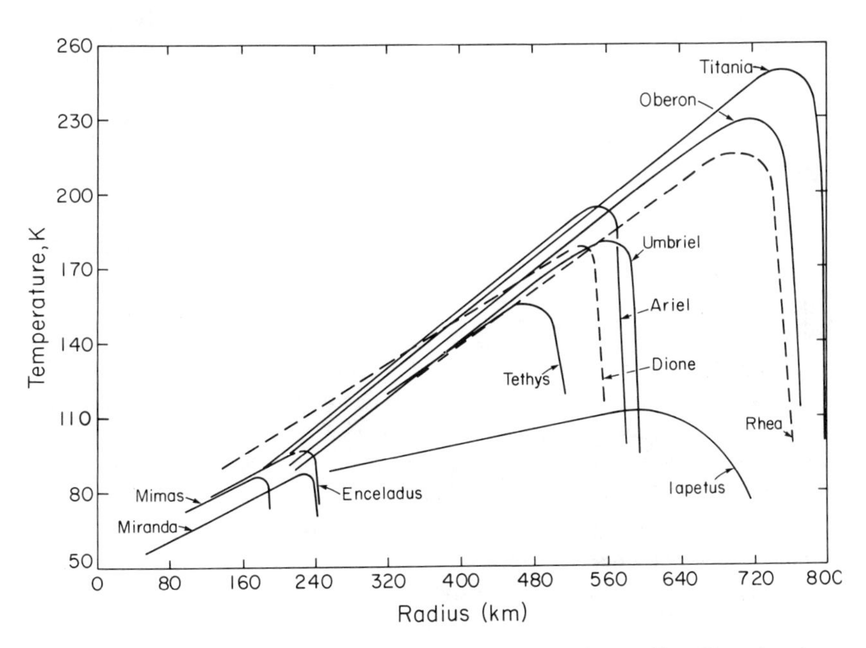

Fig. 5. Accretional temperature profiles of the Saturnian and Uranian satellites (figure based on the model of Squyres et al. 1988).

heating is a negligible heat source for terrestrial planet interiors, even for Mercury with its large eccentricity (tidal dissipation is directly proportional to the square of orbital eccentricity) (Kaula 1963; Peale and Cassen 1978; Schubert et al. 1988). However, tidal heating can be of great importance in the outer-planet satellites whenever orbital resonances among the satellites conspire to maintain large orbital eccentricities (Cassen et al. 1982; Schubert et al. 1986). Such is the case at present for Io (Peale et al. 1979; Schubert et al. 1981; Ross and Schubert 1985; Segatz et al. 1988) and perhaps for maintaining a subsurface liquid water ocean on Europa (Squyres et al. 1983*a*; Ross and Schubert 1987). Other satellites may have had significant tidal heating in the past. Both observations and models demonstrate that tidal heating is the source of energy for volcanism and degassing on Io.

The minor constituent ice $NH_3 \cdot H_2O$ can play a particularly crucial role in the evolutions of icy satellite interiors and atmospheres (Stevenson 1982*c*; Squyres et al. 1983*b*; Lunine and Stevenson 1985; Schubert et al. 1986). Subsolidus convection of H_2O ice generally keeps temperatures in icy satellite interiors below the melting point of H_2O ice. Ammonia hydrate ice melts at significantly lower temperature than does H_2O ice (ammonia hydrate ice melts at 173 K) and it can therefore form buoyant magma in icy satellite interiors and deliver volatile ammonia to the surface. If the melts of ammonia hydrate form interconnected pathways they can also drastically soften the interior rheology (Stevenson and Lunine 1986).

With the exceptions of Titan and Triton, the accretional and subsequent thermal histories of icy satellites have not produced any atmospheres of consequence. The accretion of the satellites led to varying amounts of internal ice-rock differentiation. Impact devolatilization of rock and impact vaporization of ices may have created transient atmospheres of H_2O, CH_4 and NH_3 during satellite accumulation, but these atmospheres, if they existed, have apparently been lost to space or have recondensed onto the satellite surfaces. Post-accumulation thermal histories are characterized by conductive cooling of the smallest undifferentiated satellites, and convective-conductive cooling of the larger differentiated/undifferentiated satellites (Schubert et al. 1986). If accretion or other sources of heat were ever sufficient to melt H_2O ice in any of the three larger satellites, then these internal oceans would have undergone rapid refreezing. Minor constituent ices may have melted during the course of a satellite's evolution with consequent modifications of surface morphology and internal structure, but only in the case of Titan did such melting lead to an atmosphere (Lunine and Stevenson 1987).

The silicate outer-planet satellites Io and Europa have degassed substantial amounts of volatiles. In the case of Europa, dehydration and outgassing of the interior have produced a cryosphere-hydrosphere (Cassen et al. 1982; Schubert et al. 1986). In the case of Io, tidal dissipation and volcanism have degassed the interior, producing transient SO_2 and perhaps S_2 atmospheres (Moreno et al. 1988*b*) and surface layers of condensed SO_2 and perhaps liquid and solid sulfur. Section VIII discusses the atmospheres and interiors of Europa, Io and Titan in more detail.

IV. CYCLING OF VOLATILES ON THE EARTH

We have discussed in general how degassing/regassing influences the evolution of atmospheres and interiors. It is only in the case of the Earth that we have some of the observations necessary to identify and quantify processes that transfer volatiles between the mantle and atmospheric or crustal reservoirs. Therefore, it is appropriate that we describe in more detail the volatile exchange mechanisms between atmosphere (crust) and mantle.

Liquid water, chlorine, carbon and sulfur are abundant in the crustal reservoirs of the Earth. Water occurs mainly in the oceans and in ice sheets and glaciers; a significant amount of water resides in the pores of sedimentary rocks. Chlorine is similarly concentrated in the oceans, but significant amounts occur in evaporite deposits. Carbon occurs as carbonate and organic carbon in sediments. The amounts of carbon in the atmosphere and the oceans, though volumetrically minor, are important for cycling between the major reservoirs. Sulfur occurs as sulfate in evaporite deposits in the oceans and in sedimentary sulfides.

Volatile Cycling Between Crust and Mantle

Volatiles are cycled between the crust and mantle by reactions in the oceanic crust at mid-oceanic ridges, by subduction of crustal material at trenches, and by emanations of island-arc volcanoes. Chlorine fluxes are constrained to some extent by measurements of the chlorine/water ratio in volcanic rocks and altered sea floor (Anderson 1974,1975; Ito et al. 1983). The processes at ridges are the most open to observation.

Volatile Fluxes Associated with Mid-Oceanic Ridges. New oceanic crust is formed at mid-oceanic ridges and a very large fraction is subsequently subducted back into the mantle. The current rate of crustal production or global spreading rate is 3.5 $km^2\ yr^{-1}$. An area of ocean crust equal to the Earth's surface area is produced every 150 Myr and an area equivalent to the ocean basins is produced every 90 Myr. Mid-oceanic ridge axes are important in the cycles of water and sulfur and are potentially important in the carbon and chlorine cycles.

The amount of water in mid-oceanic ridge basalts (MORB) is small enough, 0.2% by mass (Bryan and Moore 1977), and the amount of water in fully hydrated basalts is large enough, 3% by mass (see, e.g., Hart 1973), that water is recycled back into the mantle at ridges. The potential rate of water recycling is large. For example, hydration of the entire 5 km of oceanic crust would remove the oceans in 750 Myr. This amount is unacceptable, given the relatively constant freeboard of continents (Wise 1974; Schubert and Reymer 1985) unless this sink is balanced by sources.

Direct estimates of the amount of water added to the oceanic crust are obtained from deep drilling and oxygen isotopic studies of oceanic crust, including that preserved in ophiolites (Bowers and Taylor 1985). The actual hydration averaged over the crust is $< 1\%$ (equivalent to <1 km in average thickness of hydration). The cycling time for water at current plate velocities is thus comparable to the age of the Earth.

Sulfate is abundant in seawater but it is absent from 350°C hydrothermal fluids. Sulfide equivalent to 10 to 20% of the original sulfate is present. Significantly, the sulfide comes from sulfide originally in basalt (as FeS) rather than from seawater sulfate. This is indicated by sulfur isotopes (Kerridge et al. 1983) and enrichments of As, Se, Zn, Cu, Ag, Au and Pt which must be derived from the rock (Von Damm et al. 1985; Bischoff et al. 1983). Kerridge et al. (1983) prefer that 14% of the sulfide originate as seawater sulfate but their upper limit is 59%. Thus, at least 80% and maybe 98% of the sulfate which entered the hydrothermal system is not accounted for by sulfide at hydrothermal vents. The missing sulfate may react at depth to form sulfides (McDuff and Edmund 1982) or it may precipitate as anhydrite and later redissolve (Seyfried et al. 1984). These processes are suggested by isotopic studies of sulfide and by the presence of anhydrite crystals and pseudomorphs in deep core samples of young crust (Alt et al. 1986).

The chemical reactions in high-temperature axial systems differ from those in lower-temperature off-axial systems. In particular, reduction of sulfate is believed to occur only in the high-temperature hydrothermal environment. Thus, the global volume flux through these systems is needed to obtain the global sulfate loss. Methods of estimating the axial hydrothermal flux involve either thermal or geochemical considerations.

Geophysical arguments relating to the amount of axial hydrothermal circulation center on the mid-oceanic ridge heat balance. Both the heat content of the emplaced oceanic lithosphere and the heat per volume of the vented fluid are well constrained. An upper limit on the convective heat loss per area of oceanic crust of 71×10^6 MJ m^{-2} is obtained from the difference between the heat flow expected from the theory of a cooling plate and the measured conductive heat flow near mid-oceanic ridges (Sleep and Wolery 1978). This heat, if completely extracted at the ridge axis, would cool the crust and the upper mantle to seawater temperature to a depth of 15 km or cool the material along a linear temperature gradient to 30 km depth. These cooling depths are excessive because shallow axial magma chambers are often detected by seismic reflection at fast and intermediate rate ridges (Herron et al. 1978; Hale et al. 1982; Morton and Sleep 1985*a*; Detrick et al. 1987; Morton et al. 1987). Thermal modeling to match the observed magma chamber depths indicates that only 10 to 20% of the missing heat escapes by axial hydrothermal circulation (Morton and Sleep 1985*b*).

The fraction of hydrothermal circulation that is axial may be estimated using oxygen isotopes. The ocean and the oceanic crust are essentially uniform reservoirs before they react. The chemistry of exchange is well understood: ^{18}O is depleted from seawater at low temperatures and enriched at high temperatures by reaction with basalt. The $\delta^{18}O$ in seawater is "buffered" by a balance between the high- and low-temperature reactions (Muehlenbachs and Clayton 1976). The global amount of low-temperature exchange is constrained by rock samples. Calculations based on realistic paths of seawater through the ridge axis indicate that the effective water-rock (mass) ratio of high-temperature vents is about 0.5 and that the bulk of the reaction occurs at temperatures near the exit temperature (Bowers and Taylor 1985). Balance with hydration at shallow depths indicates that 9 to 20% of the hydrothermal circulation is axial, in agreement with the estimate based on thermal considerations.

The ratio of helium to heat appears to be relatively constant in high-temperature vent waters (Jenkins et al. 1978; Edmond et al. 1979*a,b;* Corliss et al. 1979*a,b;* Lupton 1983*a*). However, higher values are obtained at the East Pacific Rise at 13°N (Michard et al. 1984) and much lower values in the Guaymas basin (Lupton 1983*b*). The typical helium-to-heat ratio and the global ^{3}He flux (Craig et al. 1975; Welhan and Craig 1983) are often used to argue that nearly all hydrothermal circulation is axial.

The global amount of axial hydrothermal circulation (assuming that the heat carried by the circulation is 10^7MJ m^{-2} and that the vent temperature is 350°C) is 25 km^3 yr^{-1}, equivalent to cycling the volume of the ocean through axial vents in 54 Myr. The ocean is also cycled through off-axial vents on a time scale of $<$10 Myr. However, if all the circulation is axial, the ocean is cycled through axial vents every 8 Myr.

Cycle times for the global sulfur reservoir are obtained for various assumptions noting that 1/9 of the crustal sulfur is seawater sulfate (Ronov and Yaroshevskiy 1976). A maximum flux and a minimum cycle time of 70 Myr are obtained by assuming that all the circulation is axial and that all the sulfate reacts. An intermediate flux and cycle time of 600 Myr are obtained assuming our preferred axial flux and that all the sulfate reacts. Finally, a small flux and a long cycle time of 4000 Myr are obtained by assuming our preferred flux and that only an equivalent to the 15% of the original sulfate that is observed to vent as sulfide reacts.

Processes at Subduction Zones

Material is returned into the mantle with the down-going slab at subduction zones and added to the crust by island-arc volcanoes. Although the connection between these processes has been discussed since the early days of plate tectonics (Coats 1962), it is still difficult to quantify the extent to which volatiles in island-arc volcanoes come from the slab and the extent to which volatiles, particularly those in sediments, are subducted to great depths in the mantle. The difficulties are threefold:

1. The same rocks that are expected to be present in the slab—altered oceanic crust, serpentinite and marine sediments—are also present in the crust of island arcs. Assimilated sediment rafts are observed in gabbroic complexes formed deeper than 20 km in the roots of island arcs (see, e.g., Burns 1985). Much of the isotopic and chemical variability of island-arc volcanics can be explained by crustal processes without input from the slab (see, e.g., DePaolo 1980).
2. It is difficult to tell whether sediments are actually subducted into the mantle rather than just into the deep crust.
3. Volatile elements might move from the slab into the source region of volcanics independently of other components, in particular rare gases and radiogenic isotopes.

The composition of deep-sea sediments and the flux of volatiles from sediment subduction are strongly dependent on biological and oceanographic processes. Cretaceous and Tertiary pelagic sediments have abundant $CaCO_3$ formed by planktonic foraminifera which did not evolve until that time. The carbonate compensation depth above which carbonates enter pelagic sediments varies regionally and temporally depending on the amount of biological productivity and the chemistry of deep-water masses. The subduction of pelagic sediments is a significant sink for carbonate if it, in fact, occurs. About 4% of the crustal carbonate reservoir is in abyssal sediments (Ronov and Yaroshevskiy 1976; Wolery 1978). Since the oceanic crust is recycled every 90 Myr, the cycle time for carbonate from this process is 2250 Myr, much less time than it has been happening.

Constraints on Total Crust-Mantle Cycling from the Coupled Carbon and Sulfur Cycles

Constraints on the cycling of material between the crust and the mantle are obtained by considering the extent to which the crustal carbon and sulfur cycles behave as closed systems. Sediment subduction, hydrothermal processes at ridge axes, and ridge and arc volcanism are all involved in these cycles.

These cycles are strongly coupled by reactions of the form

$$2C + SO_4^{-2} = 2CO_2 + S^{-2}. \quad (1)$$

The partition of carbon between organic carbon and carbonate thus correlates with the partition of sulfur between sulfate and sulfide. The fractions of reduced and oxidized carbon and sulfur in the past are constrained by the carbon and sulfur isotope ratios in sediments. To see how isotope ratios are used, consider the carbon isotopic mass balance

$$f_{org}\delta^{13}C_{org} + f_{carb}\delta^{13}C_{carb} = \delta_{13}C_{bulk} \quad (2)$$

where f is the fraction in each reservoir, δ^{13} is the isotopic fractionation in units of per mil, org indicates the organic reservoir, carb indicates the carbonate reservoir, and bulk indicates the total quantity. Isotopic fractionation causes the reservoirs to differ in the isotopic ratio by a constant amount

$$\delta^{13}C_{org} - \delta^{13}C_{carb} = K \quad (3)$$

where K is biologically controlled. The mass balance is easily modified to distinguish the active material being deposited in sediments from the material that is buried and not active in the cycle at a given time (Garrels and Lerman 1984; Francois and Gérard 1986). Equations analogous to (2) and (3) apply to the partition of sulfur isotopes between sulfate and sulfide reservoirs.

The fraction of carbon in each reservoir is obtained from Eq. (2) by measuring isotopic ratios because the bulk Earth value is known to be about −6 per mil. In practice, the marine carbonate and marine sulfate reservoirs are better mixed and it is practical to measure carbon isotopes in limestones and sulfur isotopes in evaporites and compute the reservoir fractions from Eqs. (2) and (3). The correlation of sulfur and carbon isotopic ratios is predicted from Eq. (1) and the molar carbon and sulfur reservoir sizes. However, direct measurement of all the isotopic reservoirs is probably required for Archean sediments because the biological fraction was likely to have been different (Strauss 1986).

The correlation expected for closed crust reservoirs is in fact observed in the geologic record (Veizer et al. 1980; Garrels and Lerman 1984). In particular, the relative sizes of the reservoirs have remained constant. Systemic recycling of carbon and sulfur so that the recycled amounts balance the bulk Earth ratios, both in mass and isotopes, is conceivable but unlikely because carbon and sulfur behave quite differently both in igneous and sedimentary processes. A lower limit to the cycle time (and an upper limit to the cycling rate) of the smaller reservoir, sulfur, of 1 Gyr is indicated by the correlation of isotopic ratios. A similar upper limit is obtained by considering the balance of excess oxidants including sulfate (Wolery and Sleep 1976).

Des Marais (1985) and Marty and Jambon (1987) have combined measurements of the CO_2 to helium ratio in mid-oceanic ridge basalts (MORB) and high temperature vents with the ^{3}He flux of Craig et al. (1975) to estimate the present CO_2 flux from the mantle. The rate is sufficiently large to have built up the current carbon reservoir in 2 to 4 Gyr. This implies significant cycling of carbon back into the mantle especially if degassing rates were higher in the past. This rate is near the upper limit obtained above from isotopic considerations. However, carbon isotopes in MORB are generally near the bulk Earth ratio. Thus, either the mantle carbon reservoir is much larger than the recycled material, or little recycled material makes its way back to the ridge (Des Marais and Moore 1984; Marty and Jambon 1987).

V. VOLCANISM AND VOLATILES

We have noted how the interactions between planetary atmospheres and interiors are revealed in the elemental, chemical and isotopic compositions of the atmospheres and in the surface geological features. Section VII discusses what atmospheric radiogenic noble gases reveal about planetary interiors. This section deals with the opposite side of the coin, what volcanic landforms tell us about atmospheres and degassing of interiors.

Volcanic Styles vs Magma Volatile Content

Volcanism results if a planetary interior is sufficiently hot. The style of a volcanic eruption is controlled by a number of factors, including magma composition and temperature and surface atmospheric temperature and gravity. An important controlling factor is the volatile content of the magma, which determines whether the eruption is explosive or effusive. A second important factor is the surface atmospheric pressure, which can act to suppress explosive volcanism for a given magma volatile content. Different styles of volcanic eruptions lead to diagnostic differences in the volcanic landforms produced. Thus morphological studies of volcanic landforms on planetary surfaces can provide important indicators of the volatile budget of a planetary interior and the evolution of the planetary atmosphere.

A simple physical model of the eruption process (Wilson and Head 1983) illustrates the important role of magma volatile content. At sufficient depth so that all volatiles are dissolved in the melt, the vertical velocity u_d of magma rising in a volcanic vent is given by (Wilson and Head 1981)

$$u_d = w^2 g \, \Delta\rho / (12\eta) \tag{4}$$

where the magma is taken to be a Newtonian fluid of viscosity η, w is the width of the vent, g is the gravitational acceleration, $\Delta\rho$ is the effective density difference between the magma and the country rock (accounting for departures from hydrostatic pressure), and a laminar flow regime is assumed. The corresponding mass eruption rate M is given by

$$\dot{M} = w \, L \, \rho \, u_d \tag{5}$$

where L is the length of the volcanic vent. At depth, the primary effect of magma volatile content is to lower the viscosity, thus permitting larger rise and eruption rates. As the magma approaches the planetary surface, however, volatiles can exsolve from the magma and nucleate bubbles. The expansion of the gas releases energy, which causes the ascent velocity to increase (Wilson et al. 1980). If the volume fraction of gas bubbles exceeds about 0.75, the magma can disrupt, forming pyroclastic fragments and leading to an explosive eruption (Sparks 1978).

The conditions necessary for magma disruption and explosive volcanism depend on a number of physical properties of the magma and the planet (Wilson and Head 1983). Let the total weight fraction of the dominant volatile species in the magma before exsolution be n_t and the solubility of that species be n_s. Then the exsolved volatile weight fraction n_e, as a function of pressure p, is

$$n_e = n_t - n_s\,(p). \tag{6}$$

The condition in which gas bubbles occupy 3/4 of the volume (Sparks 1978) is satisfied by

$$3(1 - n_e)/\rho_\ell = n_e\,R\,T/(m\,p) \tag{7}$$

where ρ_ℓ is the density of the liquid magma, R is the gas constant, and T and m are the temperature and molecular weight of the gas, respectively. Substitution of Eq. (7) into Eq. (6) gives the pressure p_d at which disruption of the magma will occur, as long as p_d exceeds the atmospheric pressure p_a. The final eruption velocity u_f of the magma-gas mixture at the planetary surface is given approximately by (Wilson et al. 1980; Wilson and Head 1981,1983)

$$u_f^2/2 = u_d^2/2 + K'(n_d RT/m)\ln\,(p_d/p_a) \tag{8}$$

where n_d is $n_e(p_d)$ and K' is a constant near unity. Selected values of u_f and the corresponding surface mass flux $\dot{M}_c$ are given as functions of n_d for eruptions on the Earth, Mars and Venus in Table II (Wilson and Head 1983) under the assumption that H_2O or CO_2 is the dominant volatile phase.

As Table II demonstrates, the threshold volatile content for magma disruption and explosive volcanism varies considerably among the three terrestrial planets with atmospheres. An exsolved volatile content of about 0.01% is sufficient on Mars, while about 0.07% is necessary on the Earth and 3% is necessary on Venus (Wilson and Head 1983). These figures depend on the volatile constituent and on local elevation. A strong implication of Table II is that explosive volcanism should have been relatively common on Mars, as long as some volatiles were present in the magma source regions, while explosive volcanism on Venus, because of the high volatile contents required for magma disruptions, should have been comparatively rare.

Volcanic Histories of the Terrestrial Planets

A brief review of the volcanic history of each of the terrestrial planets indicates that the level of detailed information on volcanic styles and the consequent constraints on interior volatile contents and degassing differ considerably among the planets.

TABLE II

Minimum Mass Eruption Rate $\dot{M}_c$ for Eruption Column Collapse to Form Pyroclasts and Eruption Velocity u_f as Functions of Exsolved H_2O and CO_2 Content n_d[a]

H_2O	Earth		Mars		Venus lowland		Venus highland	
n_d (wt. %)	u_f (m s^{-1})	$\dot{M}_c$ (kg s^{-1})	u_f (m s^{-1})	$\dot{M}_c$ (kg s^{-1})	u_f (m s^{-1})	$\dot{M}_c$ (kg s^{-1})	u_f (m s^{-1})	$\dot{M}_c$ (kg s^{-1})
5	530	2.2×10^9	770	1.2×10^9	267	2.7×10^9	205	1.4×10^9
3	380	5.3×10^8	580	1.7×10^8	170	3.0×10^8	90	3.0×10^7
1	200	4.8×10^7	330	1.5×10^7	—	—	—	—
0.3	90	2.8×10^6	180	2.1×10^6	—	—	—	—
0.1	40	1.4×10^5	80	1.0×10^5	—	—	—	—
0.03	—	—	25	1.0×10^3	—	—	—	—
CO_2								
5	305	2.2×10^8	464	4.8×10^7	105	4.8×10^7	—	—
3	240	6.0×10^7	340	1.4×10^7	—	—	—	—
1	109	5.1×10^6	190	1.8×10^6	—	—	—	—
0.3	45	2.0×10^5	85	1.0×10^5	—	—	—	—
0.1	15	2.5×10^3	40	7.0×10^3	—	—	—	—
0.03	—	—	20	7.0×10^2	—	—	—	—

[a] Values are for eruptions at the 3 mbar level on Mars (appropriate to the upper parts of the Tharsis volcanoes) and at the 50 and 90 bar levels on Venus (appropriate to highland and lowland eruption sites, respectively). A blank entry means that magma disruption to form pyroclasts cannot occur. Table from Wilson and Head 1983.

Earth. Volcanic eruptions on the Earth display a wide range of styles, reflecting the diversity of magma properties (mafic to silicic, volatile-poor to volatile-rich) and eruption sites (subaerial, subglacial, submarine). Much of the volatile budget of magmas involved in the most violent explosive eruptions comes from the recycling process associated with sea-floor spreading and plate subduction (as discussed in the preceding section). Important variations in volcanic processes with time are known, such as the more common occurrence of ultramafic lavas in the Archean and the absence of well-developed ophiolite complexes preserved from that time, but no significant secular global changes in the volatile content of magma source regions appear to be required (Basaltic Volcanism Study Project 1981, Ch. 6).

Moon. The volcanic flux on the Moon is known to have decreased more or less monotonically between the time of heavy bombardment about 4 Gyr ago and the end of mare volcanism about 2.5 to 3 Gyr ago (Basaltic Volcanism Study Project 1981, Ch. 8). While the interior of the Moon has always been highly deficient in volatiles compared with the Earth, there is evidence that small amounts of carbon monoxide were present in lunar magmas (Sato 1976; Housley 1978). The large energy release per unit mass from this gas as it decompressed to the near-zero ambient lunar atmospheric pressure would have been sufficient to lead to magma disruption and pyroclast formation in most lunar eruptions (Wilson and Head 1981). Lunar dark mantle deposits may consist of such pyroclastic debris (Head 1974). Whether the styles of volcanic eruptions, and particularly the implied magma volatile content, varied with time on the Moon during the interval of mare volcanism is not known (Basaltic Volcanism Study Project 1981; Wilson and Head 1981).

Mercury. Like the Moon, Mercury is volatile poor. A volcanic origin has been suggested for the Mercurian smooth plains (Strom et al. 1975), and a number of possible volcanic landforms have been identified from the comparatively low-resolution Mariner 10 images of the surface (Basaltic Volcanism Study Project 1981, Ch. 5). The duration of plains volcanism was comparable to or somewhat shorter than that of the Moon (Basaltic Volcanism Study Project 1981, Ch. 8); no evolution of volcanic eruptive styles is discernible from available information.

Venus. Diverse and widespread volcanic landforms have been recognized in radar images of Venus' surface obtained from groundbased observatories (Campbell et al. 1984) and orbiting spacecraft (Barsukov et al. 1986). These landforms are generally less than several 100 Myr old, on the basis of the distribution of impact craters on Venus' surface (Campbell and Burns 1980; Barsukov et al. 1986). As noted above, because of the large atmospheric pressure on Venus, magma disruption and explosive volcanism will not occur unless the exsolved volatile content of the erupting magma exceeds several wt. %. A time-variable concentration of SO_2 in the upper atmosphere of Venus has been attributed (Esposito 1984) to episodic injection of volcanic eruption clouds. Consideration of volcanic landforms on Venus as viewed by radar and lander images, however, suggests that such explosive volcanic eruptions, and the implied high volatile contents of the erupting magmas, have been rare on Venus, at least over the interval of planetary history preserved in surface features (Head and Wilson 1986).

Mars. Of all the terrestrial planets, Mars preserves the temporally most extensive history of volcanic activity. As noted above, the Martian surface conditions favor explosive volcanism as long as the erupting magmas contain at least a modest weight fraction (0.01%) of volatiles. Pyroclastic deposits have been at least tentatively associated with a number of Martian volcanic features (Reimers and Komar 1979; Scott and Tanaka 1982; Mouginis-Mark et al. 1982). Francis and Wood (1982) have proposed that the highland paterae of Mars are mafic pyroclastic structures formed during a period of mantle outgassing, and that cessation of patera formation about 2 Gyr ago corresponded to a widespread reduction of mantle volatile levels below that necessary for explosive eruptive activity. Mouginis-Mark et al. (1987) have identified pyroclastic as well as effusive deposits on Alba Patera; because the youngest deposits are effusive lava flows, they suggest that this construct is intermediate in age and eruptive style between the highland paterae and the more recent effusive central-vent volcanoes such as the Tharsis Montes. A general decrease with time in the volatile content of the Martian mantle would be consistent with these results. Comparatively recent pyroclastic eruptions have been locally identified (Mouginis-Mark et al. 1982), however, so that either some heterogeneity in mantle volatile content or assimilation of near-surface water in shallow magma bodies is implied. Greeley (1987) has estimated the amount of water degassed through Martian volcanism to be equivalent to a global layer about 46 m thick, mostly during the first 2 Gyr of the planet's history. He did this by mapping volcanic materials, dating their surfaces by crater counts, inferring thicknesses of volcanic units from geologic relationships (e.g., crater burial), and estimating volatile contents in the lavas by analogy with terrestrial lavas. This exercise, though subject to considerable uncertainty, is an excellent illustration of how surface volcanism can be used to provide constraints on atmospheric evolution.

VI. EFFECTS OF VOLATILES ON TECTONIC STYLES

Role of Pore Fluids

Tectonic deformation of a planetary lithosphere is a response to stresses produced by internal dynamical processes. The form of the tectonic response depends on the mechanical properties of the lithosphere, particularly the mechanical strength as a function of depth. The strength of rock can be profoundly affected by the presence of a significant volatile component. Thus the distribution of volatiles at depth in a planet can be expected to have a signature in the geometry and style of tectonic behavior visible at the planetary surface.

Rock deformation in the upper lithosphere is predominantly by brittle fracture or frictional sliding on faults (Byerlee 1978; Brace and Kohlstedt 1980). Macroscopically, the empirical law governing both these modes of brittle behavior is given by the Coulomb criterion for shear traction τ on the fault at failure:

$$|\tau| = S + \mu(\sigma_n - p_f) \tag{9}$$

where S is the finite cohesive strength of the material, μ is the coefficient of fault friction, σ_n is the normal stress on the fault, and p_f is the pore fluid pressure. The presence of a significant fluid content at depth is felt in Eq. (9) through the lowering of the effective normal stress

$$\sigma^* = \sigma_n - p_f \tag{10}$$

and the consequent weakening of the surrounding material. On the Earth, p_f is typically equal to the hydrostatic pressure (the pressure at the base of a water column extending to the surface) at shallow depth. Below some point whose location is stratigraphically controlled, however, p_f often rises well above the hydrostatic value and follows a gradient somewhere between the hydrostatic and lithostatic gradients (lithostatic pressure equals the weight per unit area of the overlying rock) (Fertl 1976). Several possible relationships among p_f, σ^* and depth are illustrated in Fig. 6.

If p_f increases discontinuously at some depth h, the material below h will be substantially weaker than the material above. Such a situation can lead to a

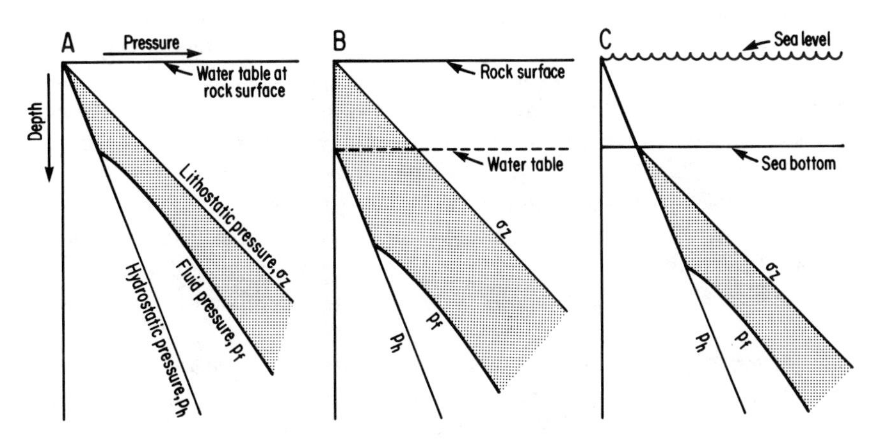

Fig. 6. Schematic of possible relationships among depth profiles of lithostatic pressure, fluid pressure and hydrostatic pressure (figure from Davis et al. 1983).

detachment surface, or decollement, on which the relatively weak and deformable overlying material can move laterally in response to gravitational or tectonic forces (Hubbert and Rubey 1959). The importance of detachment surfaces has been well recognized on the Earth in the formation of fold-and-thrust belts and accretionary wedges along compressive plate boundaries (see, e.g., Chapple 1978; Davis et al. 1983) and in the extension of continental crust and lithosphere (Wernicke and Burchfiel 1982). Detachment surfaces can develop for reasons other than a sharp increase in pore pressure, including a layer of reduced cohesive strength S or a layer (such as evaporite) which is ductile even at upper crustal conditions (Davis et al. 1983).

Detachment Tectonics

Volatiles have played little or no role in the tectonics of the Moon and Mercury. There is some evidence that detachment surfaces have been involved in tectonic deformation on Venus and Mars, and it is of interest to inquire whether tectonic features in such situations can constrain the nature or distribution of subsurface volatiles on those planets.

Venus. Whether volatiles are sufficiently abundant in the upper lithosphere of Venus to yield a value for p_f in excess of the atmospheric pressure has never been seriously examined, largely because of the low water content and high temperature of the lower atmosphere. However, the atmosphere D/H ratio has been interpreted to indicate that the water content of the atmosphere and, by inference, the upper crust was at one time considerably higher than at present (Donahue et al. 1982). If liquid water was ever present in any quantity on Venus' surface, there is also the possibility that layers of evaporite were formed; burial of such layers by younger basalts may yield a situation favorable to detachment tectonics (Wood and Amsbury 1986).

The tesserae, or parquet terrain units (Barsukov et al. 1986), on the basis of their detailed structure and the correlation of structure and topography, have been attributed to gravity spreading on a basal detachment (Kozak and Schaber 1986). The 10 to 20 km spacing between adjacent major fault structures as seen in radar images of compressional mountain belts and extensional rift valleys (Campbell et al. 1983,1984; Barsukov et al. 1986), however, suggests that the upper brittle portion of Venus' crust is only a few km thick (Solomon and Head 1984). Whether any ancient detachment surface in Venus' crust reflects the prior existence of volatiles or simply a general transition to ductile behavior in the mid to lower crust will thus be difficult to determine.

Mars. Numerous surface morphological features attest to the past presence of considerable volumes of subsurface water and ice on Mars (Squyres and Carr 1986; Carr 1987). Water-saturated layers at depth in the Martian crust may serve as zones of local weakness and detachment. Two types of tectonic features suggest that detachment zones may have been important on Mars.

First, mare-type ridges on the ridged plains have a characteristic spacing of 15 to 50 km; such a regular spacing has been explained in terms of a folding model under the assumption that the basalt units of the ridged plains are underlain by substantially weaker material (Saunders et al. 1981). Second, it has been suggested that the aureole deposits surrounding Olympus Mons are large-scale thrust sheets emplaced by sliding along a weak basal detachment (Harris 1977; Francis and Wadge 1983). Whether such weak zones imply a water-saturated layer with high p_f at the time of deformation has not been established.

VII. RADIOGENIC NOBLE GASES

We have discussed how surface geology provides clues about interior volatile content, past and present, and degassing. In this section we summarize how the atmosphere provides information about the structure and evolution of the interior. The radiogenic and nonradiogenic noble gas constituents of atmospheres are particularly useful in this regard and we focus largely on the radiogenic noble gases. A more thorough review of noble gases and how they constrain atmospheric evolution is given in the chapter by Ozima and Igarashi. Turcotte and Schubert (1988) have recently discussed the tectonic implications of the radiogenic noble gas contents of the atmospheres of the terrestrial planets.

The radiogenic noble gases ^{4}He and ^{40}Ar are particularly relevant to the study of atmosphere-interior coupling because they are produced by the decay of the long-lived heat producing elements ^{235}U, ^{238}U, ^{232}Th and ^{40}K that control the thermal evolution of the interior. The efficiency of escape of these gases provides insights about the transport processes and dynamics in planetary interiors. Diffusion of these gases is not likely to be important on a planetary scale. However, diffusion can transfer noble gases to grain boundaries or regions where other transport mechanisms are important. Partial melting is likely to be important in the transport of noble gases. Erosion and groundwater circulation are significant near-surface transport mechanisms on the Earth.

There are two possible approaches to the analysis and interpretation of noble-gas data. One involves the use of present-day fluxes to and from the atmosphere. However, fluxes are often difficult to determine. The other employs total gas content which is representative of time-integrated fluxes.

Earth

Argon. It is generally accepted that all ^{40}Ar in the Earth's atmosphere originated from the decay of ^{40}K in the Earth's interior; there is no primordial ^{40}Ar. The amount of ^{40}Ar in the present atmosphere is 6.6×10^{16} kg while the quantity of nonradiogenic ^{36}Ar is 2.1×10^{14} kg. The present atmospheric $^{40}Ar/^{36}Ar$ ratio is thus 295.5. Because of the heavy atomic mass of Ar there is no significant escape of Ar to space under normal circumstances. Thus, it is reasonable to assume that no atmospheric Ar has been lost from the atmosphere through geologic time, although it is possible that a significant amount of Ar could have been subducted into the mantle. The ^{40}Ar presently in the atmosphere is the accumulation throughout geologic time of radiogenic argon produced in the crust and mantle and transported to the surface by subsolidus convection and magma migration in the mantle and volcanism, hydrothermal circulation, and erosion in the crust.

Comparison of the ^{40}Ar content of the atmosphere with the time-integrated ^{40}Ar production rate in the crust and mantle can therefore be used to give the fraction of the mantle that has been degassed over geologic time if potassium concentrations in the crust and mantle can be prescribed. (Sleep [1979] has previously used the ^{40}Ar content of the atmosphere, together with a thermal history model of the Earth, to constrain the K concentration in the mantle.) Although the potassium content of the mantle is uncertain, the value is constrained by heat flow and geochemical data. For mantle potassium/ uranium and thorium/uranium mass ratios of 12,700 and 3.8, a continental crust potassium concentration (by mass) of 0.91%, and a mantle heat production equal to 70% of the heat flow from the mantle, the atmospheric ^{40}Ar content implies complete outgassing of the crust and either a 23% outgassing of the whole mantle or complete outgassing of the upper mantle plus 7% outgassing of the lower mantle (Turcotte and Schubert 1988). If the mantle concentration of potassium is somewhat lower than assumed above, or if internal heat production in the mantle is smaller than assumed, then the degassing efficiency of the mantle could be higher by perhaps a factor of about 2.

Consideration of the ^{40}Ar content of the atmosphere thus shows that the mantle has been only partially degassed of this radiogenic noble gas. It also demonstrates that a volume of the mantle significantly larger than that of the upper mantle must have been degassed. The lower mantle has therefore not been isolated from the upper mantle and the surface.

The ^{40}Ar and primordial nonradiogenic ^{36}Ar contents of the atmosphere can also be employed to infer how the mantle degassing rate has varied over geologic time. Hamano and Ozima (1978) have modeled atmospheric buildup of ^{36}Ar and ^{40}Ar assuming (1) transport of potassium from the mantle to the crust; (2) transport of argon from the mantle to the atmosphere; (3) transport of ^{40}Ar from the crust to the atmosphere; and (4) rates of transport proportional to the amounts of the constituents in the separate reservoirs. For reasonable assumptions regarding values of model parameters (for example, mantle ^{40}Ar/^{36}Ar, and potassium concentration in the mantle), their calculations imply a major degassing event prior to 3.8 Gyr ago in which 75% of the volatiles present in the mantle at that time were lost to the atmosphere. This result supports the scenario of rapid atmospheric formation contemporaneous with or shortly after accretion.

Substantial early degassing of the Earth is also supported by considerations of rare gas fluxes from the mantle (Holland 1984). The current degassing rates of ^{36}Ar and ^{20}Ne can be estimated from measurements of $^{3}He/^{36}Ar$, $^{20}Ne/^{36}Ar$ and the ^{3}He flux from the mantle to the oceans (discussed in the next section). These rates are found to be significantly less than the mean degassing rates of ^{36}Ar and ^{20}Ne based on the atmospheric content of these species divided by 4.5 Gyr. The implication is that the degassing rates of ^{36}Ar and ^{20}Ne were larger during the early evolution of the Earth. Models of degassing constrained by these data suggest an early substantial degassing event (Holland 1984). The models include one in which the degassing rate is assumed directly proportional to the mass of gas in the mantle and the secularly declining terrestrial heat flux.

The previous considerations argue for a major early loss of volatiles from the Earth's interior contemporaneous with accretion or within a few 100 Myr thereof by impact degassing or vigorous mantle overturning subsequent to accretion and core formation, followed by relatively inefficient partial degassing of the remaining mantle volatiles over most of geologic time.

Helium. Since helium readily escapes from the atmosphere, the atmospheric content of helium cannot be analyzed in the same way as was done for argon. However, Craig et al. (1975) have determined the flux of radiogenic ^{4}He from the oceans to the atmosphere and this can be compared with the production rate of ^{4}He in the mantle from the decay of ^{235}U, ^{238}U and ^{232}Th to obtain an estimate of the present mantle degassing efficiency. Implicit in this are the assumptions that the ^{4}He flux from the oceans to the atmosphere is the same as the flux from the mantle and the flux of primordial ^{4}He is negligible compared to the flux of radiogenic ^{4}He. If the $^{4}He/^{3}He$ ratio from carbonaceous chondrites enriched in gases is representative of the primordial mantle ratio then the latter assumption is justified.

The observed flux of ^{4}He from the mantle represents only about 9% of the ^{4}He produced in the mantle if ^{4}He is being extracted from the whole mantle with concentrations of uranium, thorium and potassium similar to those assumed in the case of ^{40}Ar production (Turcotte and Schubert 1988). If the ^{4}He flux is extracted only from the depleted upper mantle, then it is possible for the observed flux to be in balance with the production rate of ^{4}He for reasonable values of the concentrations of radioactive elements in the upper mantle (Turcotte and Schubert 1988). If the mantle is convecting as a whole, then the degassing efficiency of ^{4}He is rather low at present. One way to reconcile this low efficiency with the higher degassing efficiency of ^{40}Ar is to have larger degassing efficiencies in the past since the ^{40}Ar efficiency represents an average over geologic time while the ^{4}He efficiency is a present value. This is a reasonable possibility if the Earth's mantle has been cooling and decreasing its convective vigor since 4.5 Gyr ago. Indeed, the comparison of ^{4}He and ^{40}Ar degassing efficiencies adds to the evidence from noble gases for more rapid degassing early in the Earth's history. Another possible explanation for the difference between ^{4}He and ^{40}Ar deficiencies is that degassing of the mantle is controlled by solubility of gases in melt; the greater solubility of ^{40}Ar compared with ^{4}He would then account for the relative efficiencies.

Oxburgh and O'Nions (1987) have carried out a similar comparison of the observed ^{4}He flux with the mantle ^{4}He production rate and concluded that the measured flux is only 5% of that expected on the basis of the terrestrial heat flow. They conclude that either radiogenic helium is being retained within the mantle or that present models of the bulk chemistry of the mantle are in error and much of the terrestrial heat loss is nonradiogenic. We do not consider it necesary to question the estimates of mantle radiogenic heat production (recall that we took this to be 70% of the heat flow from the mantle) because it seems plausible that the mantle loses heat more readily than it loses ^{4}He. Most of the helium loss probably occurs upon partial melting at mid-ocean ridges. However, the upwelling is only partially melted and only a fraction of this melt will communicate with the surface. On the other hand, heat is lost not only directly by partial melting and upward magma migration at ridges but by conduction through the oceanic lithosphere everywhere beneath the ocean floor. Once mantle material moves laterally away from a ridge crest, helium diffusion will be too small for any significant loss to occur while a substantial heat loss occurs throughout the ocean basins. Greater retention of helium compared with heat by the mantle does seem consistent with our understanding of plate tectonics and heat transfer through the ocean basins.

Venus

The amounts of ^{36}Ar and ^{40}Ar in Venus' atmosphere are known and inferences about degassing of the planet's interior are thus possible. There is about 5 times less ^{40}Ar on Venus than on the Earth (1.4×10^{16} kg compared with 6.6×10^{16} kg; Pollack and Black 1982). Since measurements of radioactivity in Venus rocks (Vinogradov et al. 1973) give concentrations of radioactive elements similar to those in terrestial rocks, we attribute this difference to less efficient extraction of ^{40}Ar on Venus. If it is assumed that the concentration of potassium in Venus' mantle is the same as in the Earth's mantle, then the whole mantle degassing efficiency on Venus is only 8.5% (the upper mantle degassing efficiency is 28% assuming an upper mantle similar to the Earth's) (Turcotte and Schubert 1988). The absence of plate tectonics on Venus (Phillips et al. 1981) could lead to inefficient outgassing of its interior.

Venus' atmosphere contains about 50 times more ^{36}Ar than does the Earth's atmosphere (1.2×10^{16} kg compared to 2.1×10^{14} kg; Pollack and Black 1982). Wetherill (1981) and McElroy and Prather (1981) have attributed this to preferential solar wind implantation of ^{36}Ar into Venus' atmosphere. Cameron (1983) has suggested that the Earth has lost ^{36}Ar as a consequence of planetesimal collisions. The large collision hypothesized to be responsible for the origin of the Moon (Hartmann 1986; Cameron 1986) would have greatly eroded the Earth's primitive atmosphere. Thus the low concentration of

^{36}Ar on the Earth compared with Venus indirectly supports the collision hypothesis of lunar origin.

Mars

Mars has about 150 times less ^{40}Ar than Earth (4.5×10^{14} kg vs 6.6×10^{16} kg) and 1500 times less ^{36}Ar (1.4×10^{11} kg vs 2.1×10^{14} kg) (Pollack and Black 1982). Atmospheric erosion by planetesimal collisions can also explain the relative scarcity of ^{36}Ar on Mars (Cameron 1983). The small amount of ^{40}Ar in Mars' atmosphere can be obtained by degassing of only 640 m of a crust with terrestrial potassium concentration (Turcotte and Schubert 1988). Greater degassing of the Martian interior is implied by the numerous large volcanoes and other volcanic and fluvial landforms on Mars and by the atmospheric D/H ratio (Owen et al. 1988). Atmospheric erosion could also explain the small amount of ^{40}Ar retained in the present atmosphere. Dreibus and Wänke (1987) have attributed the low ^{40}Ar content of Mars' atmosphere to a relatively small crustal enrichment of even the most incompatible elements like K. They ascribe the low ^{36}Ar content of the atmosphere to more efficient loss by hydrodynamic escape during accretion and early atmosphere formation.

INTERIORS AND ATMOSPHERES OF SOME OUTER-PLANET SATELLITES

The volatiles at the surfaces of Europa, Io and Titan have intriguing connections with their interiors.

Europa

The source of water at Europa's surface was undoubtedly the hydrated silicates that originally accumulated to form this rocky satellite. Europa is sufficiently large that accretional heat could have completely dehydrated its interior. Radiogenic heating over geologic time could also have broken down the hydrated silicates. However, the extent to which interior dehydration has occurred is uncertain, and we do not know the thickness of the H_2O layer surrounding Europa's rocky core (Schubert et al. 1986). If Europa has a thick H_2O shell it is also possible for a liquid H_2O ocean, maintained by tidal heating, to exist beneath the ice. Thus there are three possibilities for the structure of Europa's outer layers: (1) a thin ice shell (thickness $\approx$ 10 km); (2) a thick ice shell (thickness $\approx$ 100 km); and (3) an outer ice shell surrounding a liquid water internal ocean.

Io

Io has a tenuous but nevertheless detectable atmosphere probably consisting primarily of sulfur dioxide and sulfur (sulfur has not been positively identified as an atmospheric gas). A model of the SO_2 atmosphere in equilibrium with a global surface deposit of SO_2 frost involves a supersonic flow from regions of surface evaporation due to solar insolation to areas of surface condensation (Ingersoll et al. 1985). In addition, transient atmospheres of SO_2 and S_2 are associated with the satellite's intense volcanic activity. Indeed, Io's atmosphere is probably volcanically controlled instead of being solar-insolation controlled through evaporation-condensation processes (Moreno et al. 1988*b*). Though volcanic eruptions can generate a global atmosphere on Io, the atmosphere is concentrated around the centers of volcanic activity. Since the locations and intensities of active plumes on Io vary with time, the atmospheric mass distribution and the total atmospheric mass will also be time variable. Supersonic atmospheric winds flow horizontally away from plume centers; condensation of SO_2 in the plume region (snow) and at the surface away from the plume deplete the plume-generated atmosphere (Moreno et al. 1988*b*). An oblique shock surrounding the plume center turns the falling gas parallel to the surface and causes enhanced condensation in the form of surface ring deposits of SO_2 around the plume (Moreno et al. 1988*b*). The volcanic atmosphere is probably the source of neutrals for Io's hot torus, cold torus and sodium cloud.

The volcanic control of Io's atmosphere provides a direct link between the instantaneous states of the atmosphere and the interior. Atmospheric composition, pressure, temperature and surface distribution depend on where the active volcanoes are, what they are erupting, the rates of eruption, and the physical states of the erupted gases. The location and intensity of volcanic activity on Io depend on how the interior is heated. The source of energy for Io's volcanism is tidal dissipation (Peale et al. 1979), but the magnitude of the heating and its distribution with depth, latitude and longitude depend on Io's internal structure. Segatz et al. (1988) have investigated two end-member possibilities, tidal dissipation in a partially molten thin global asthenosphere (magma ocean) and heating throughout a thick essentially solid mantle (Io probably has an iron core of radius about 1/2 the satellite's radius). Tidal dissipation in an internal magma ocean maximizes heat production near the equator (see Color Plate 6) while mantle dissipation concentrates heat production at the poles and near the core-mantle interface (Plate 7). The surface heat flow distributions corresponding to these two extreme internal heating modes are shown in Plates 8 and 9. Comparison of these distributions with plume and hotspot activity on Io does not allow for an unambiguous choice between the modes of internal heating, though a recent determination of topographic variations on Io (Gaskell et al. 1988) implies a strong role for asthenospheric heating (Segatz et al. 1988).

The extent to which volatile recycling occurs on Io is uncertain, but the high rate of resurfacing suggests burial and subsidence of volatiles and at least shallow recycling.

The volcanic control and variability of Io's atmosphere provide us with the opportunity to study first hand many of the atmosphere-interior connections that we can only infer indirectly for the terrestrial planets.

Titan

Although there is about 30 times more N_2 than CH_4 in Titan's gaseous atmosphere (Lindal et al. 1983), there could be as much as 7 times more CH_4 than N_2 in a combined atmosphere-ocean system if a methane-ethane-nitrogen ocean exists on Titan's surface (Lunine et al. 1983; Lunine and Stevenson 1987). We summarize here some of those features of the model proposed by Lunine and Stevenson (1987) for the origin of CH_4 in Titan's atmosphere hydrosphere that are particularly relevant to the atmosphere-interior interaction.

According to the model, at the close of accretion Titan had a NH_3, CH_4, H_2O atmosphere and a three-layer interior structure consisting of a NH_3-H_2O ocean about 600 km thick, an underlying rock layer about 500 to 1000 km thick, and an undifferentiated core of rock, water ice, methane ice and ammonia hydrate ice. The atmosphere and the outer two layers of the planet formed from the accretional melting and differentiation of core-type material. The core was undifferentiated during accretion since a critical radius of about 1000 km must be reached before accretional heating becomes effective in melting ices. The primordial and massive methane-dominated atmosphere was hypothesized to have been lost by extreme ultraviolet heating or atmospheric cratering sometime after accretion was completed.

The subsequent evolution of the interior involved core overturn and heating of core material, thereby releasing NH_3-H_2O and CH_4 magmas. The NH_3-H_2O ocean froze from the top and bottom forming an ice I outer layer, an intermediate residual liquid ocean and a lower NH_3-H_2O ice layer. Methane and ammonia hydrate magmas rose buoyantly to the surface to form the secondary atmosphere hydrosphere that survives to the present. Methane is more buoyant and less viscous than ammonia hydrate, allowing CH_4 to be the dominant contributor to the degassed atmosphere. The methane and ammonia volatiles must of course be capable of migrating through the freezing ocean and the CH_4, in particular, must survive possible enclathration in the ammonia-water ocean. Ammonia is later converted to N_2 through photolytic processes or shock-heating processes.

This model of Titan's atmosphere is speculative and we discuss it here not from the point of advocacy but as an example of the evolutionary processes that can occur in icy satellite interiors to give rise to an atmosphere by the buoyant rise of low melting point minor constituent volatiles.

Acknowledgments. This work was supported by several NASA grants. We are indebted to M. Segatz for permission to publish color plates 1–4.

PART IV
Outer Planets

THE COMPOSITION OF OUTER PLANET ATMOSPHERES

D. GAUTIER
Observatoire de Paris
and
T. OWEN
State University of New York at Stony Brook

This chapter reviews current information about elemental abundances and deuterium to hydrogen ratios in the atmospheres of the giant planets, together with contemporary hypotheses to explain the observations. Helium appears to be depleted (compared with the protosolar abundance) in the atmospheres of Jupiter and Saturn, but is in good agreement on Uranus. This behavior is consistent with theories for the internal structures of these planets. These theories predict that helium will be immiscible in metallic hydrogen at the temperatures and pressures occurring in the interiors of Jupiter (slightly), and Saturn (strongly), therefore precipitating toward the centers of these planets and depleting the atmospheric envelopes. But the pressure-temperature regimes inside Uranus and Neptune seem not to permit the formation of metallic hydrogen. The deuterium to hydrogen ratio is significantly larger on Uranus and Neptune than on Jupiter and Saturn. The latter two planets exhibit a value of $D/H \sim 2 \times 10^{-5}$*, consistent with the value deduced for the hydrogen gas in the primordial solar nebula. The higher values found on Uranus and Neptune indicate the influence of a second reservoir of deuterium,* viz. *the condensed, hydrogen-containing species that never equilibrated with the nebular gas. It appears that this enrichment is coming primarily from water in the planetary cores. Carbon and (to a less-well-determined degree) other heavy elements show an increasing enrichment in the sequence Jupiter, Saturn, Uranus, Neptune. This increase appears to be roughly correlated with the ratio of core mass to total planet mass. It may arise either from efficient upward convection of gases produced from the cores during the formation of these planets or from the dissolution of infalling planetesimals in the early atmospheres. The high values of D/H seem to impli-*

cate the cores as the primary source of the carbon enrichment for Uranus and Neptune. A number of observational tests are suggested to make further progress, including a better characterization of the solid material that was present in the outer solar nebula, and improved data on abundances and isotopic ratios that would result from atmospheric probes for all four planets.

The first indication that the giant planets had atmospheres very different from ours came from observations of absorption lines in their spectra by observers using their own eyes as detectors during the last century. The invention of the photographic plate allowed these bands to be recorded and more of them to be discovered, but it was not until 1932 that Wildt was able to show that methane and ammonia were responsible for the absorption bands in the red region of Jupiter's spectrum (Wildt 1932). Two years later, Adel and Slipher (1934) used their own laboratory spectra of methane at high pressure to show that this gas caused the visible absorption bands in spectra of Uranus and Neptune. Hydrogen was not detected in these atmospheres until Herzberg (1952) was able to show that a band found in the spectrum of Uranus near 8270 Å by Kuiper (1952*b*) was caused by the 3-0 pressure-induced dipole absorption of H_2. The 3-0 quadrupole spectrum of hydrogen was first found in the spectrum of Jupiter by Kiess et al. (1960) after which lines from the 4-0 system were identified in spectra of all four giant planets. The opening of the infrared brought many new discoveries, starting with C_2H_2 and C_2H_6 in the 10 μm window (Ridgway 1974). The next major step was the use of spacecraft, notably the infrared interferometer spectrometer (IRIS) instrument on the Voyager Spacecraft, that allowed the first definitive evaluations of helium in these atmospheres (Hanel et al. 1979*a*; Gautier et al. 1981).

Modern reviews of this subject have been published by Ridgway et al. (1976) and Prinn and Owen (1976) for Jupiter; Prinn et al. for Saturn (1984); Belton (1982*b*) for Uranus; Hunten (1984) for Uranus and Neptune; Trafton (1981) and Atreya (1987) for outer planets, while Larson (1980), Encrenaz (1984) and the present authors (1985) have given general surveys of the data from specific points of view.

In this chapter, we endeavor to bring the subject up to date by including results from the Voyager Uranus encounter and the latest groundbased observations. This new material has changed our perspective in several respects, while emphasizing some fundamental problems that still require additional data and modeling for their solution. We shall attempt to illustrate both of these aspects of the current state of our knowledge as we proceed.

I. SIMILARITIES AND DIFFERENCES AMONG THE ATMOSPHERES OF THE OUTER PLANETS

Two decades ago, the giant planets were commonly thought to have atmospheres with quite similar compositions and elemental abundances close to solar values. For instance, Cameron (1973*b*), wrote "there is no *a priori* reason to believe that any of the outer planets should contain a fractionated ratio of helium to hydrogen." This assumption was based on the concept that all giant planets were formed directly from the gaseous material of the primitive solar nebula. Subsequently, their high gravity and low exospheric temperatures would prevent the escape of even the lightest elements. Observations initially seemed to confirm these ideas. All giant planet atmospheres are mainly composed of hydrogen and helium; all exhibit minor components dominated by CNO compounds as expected from cosmic abundances.

However, the suspicion that the atmospheres of Uranus and Neptune might be enriched in heavy elements in keeping with the higher densities of the planets themselves was soon borne out by groundbased observations (Encrenaz et al. 1974; Lutz et al 1976). The exploration of the outer solar system by Pioneer and Voyager spacecraft and the improvement of measurements from groundbased and airborne telescopes have brought to light additional differences in composition, even in the ratios of He/H, that we summarize as follows:

1. Strong differences exist in the He/H ratio in the atmospheres of Jupiter, Saturn and Uranus. The Neptune value has not yet been determined; see Sec. IV.A;
2. Although the determination of the deuterium abundance is difficult and somewhat controversial, the D/H ratio seems to be equal to the protosolar value in Jupiter and Saturn and significantly higher in Uranus; see Sec. IV.B;
3. Atmospheres of *all* giant planets have been enriched in carbon; the enrichment increases when moving from Jupiter to Saturn to Uranus and Neptune. Jupiter and Saturn also exhibit above-solar N/H ratios. The case of O/H is still controversial in Jupiter and the ratio in the other giant planets is unknown; see Sec. IV.C.

In addition to differences in composition, differences in thermal structures must be mentioned since they can affect photochemical processes as well as horizontal and vertical distributions of elements. They are:

1. Jupiter, Saturn and Neptune exhibit substantial internal sources of energy, while Uranus exhibits only a very weak source of energy, if any. As a consequence, the convection must be rather small within Uranus so that disequilibrium species observed in Jupiter and Saturn (PH_3, GeH_4) are probably not present in the troposphere of Uranus (Lewis and Fegley 1984; Fegley and Prinn 1985,1986);
2. Due to differences in the orientations of their axes of rotation, the giant planets exhibit more or less strong seasonal effects resulting in latitudinal temperature variations. These affect the formation and distribution of species formed by photochemistry in the stratospheres, especially the hydrocarbons resulting from the photolysis of methane.

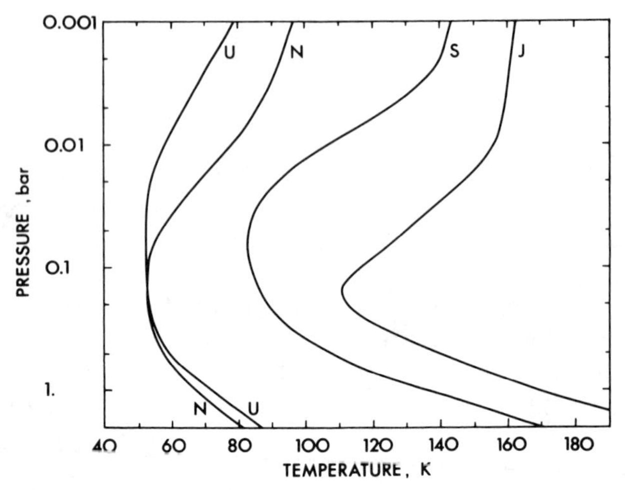

Fig. 1. Mean temperature profiles of the atmospheres of the giant planets. The Jovian profile is representative of the whole planetary disk (Bézard et al. 1983). Saturn's profile is the Voyager 2 ingress radio-occultation profile recorded at northern mid-latitudes (Tyler et al. 1982) and rescaled for the appropriate H_2/He ratio in Saturn (Conrath et al. 1984). Uranus and Neptune profiles were retrieved by Moseley et al. (1985) from far-infrared groundbased radiometric measurements and correspond to the whole planetary disk of each planet. Their stratospheres were modeled according to the work of Appleby (1986). The assumed temperature of Uranus at the 1 mbar level is somewhat cooler than that derived from Voyager radio-occultation measurements at the equator by Lindal et al. (1987) who indicate a temperature around 100 K at this level. The stratosphere of Neptune is still uncertain (figure adapted from Bézard et al. 1986*a*).

II. ATMOSPHERIC STRUCTURE

Mean atmospheric thermal structures of the four giant planets are compared in Fig. 1. More details and further discussion of Voyager results can be found in the chapter by Conrath et al.

The similarity of the thermal profiles in the atmospheres of the giant planets is not the result of chance. The thermal structure in the upper troposphere and in the stratosphere results from equilibrium between similar sources and sinks of radiative energy (Appleby and Hogan 1984; Appleby 1986). Sources of energy are incoming insolation and outgoing internal energy transferred by convection from the core of the planet up to the troposphere. The near infrared solar energy is absorbed at stratospheric levels by methane present in abundance in the four giant planets, and by aerosols, producing a pronounced thermal inversion on all of these planets except Uranus, where only a weak inversion occurs. Uranus is an exception because the condensation of methane in the Uranian upper troposphere limits its stratospheric abundance. A similar behavior was expected in the stratosphere of Neptune, but repeated groundbased infrared observations reveal the presence of stratospheric methane in amounts well above the value permitted by the saturation law at the tropopause temperature (Gillett and Rieke 1977; Macy and Sinton 1977; Courtin et al. 1979*a*; Tokunaga et al. 1983; Orton et al 1987*a*). The mechanism which permits methane to escape through the tropopause cold trap on Neptune, and not on Uranus is not known. It may be an indication of differences in the convecuive fluxes in the two planets, as could be suspected from a comparison of values of their respective internal sources of energy (see Table I). The envisaged process is simple transport of methane crystals through the tropopause into the stratosphere. The presence of large, easily visible cloud systems on Neptune, but not on Uranus, tends to support this idea (Smith 1984; Hammel and Buie 1987). Methane absorbs near-infrared solar energy, causing a more pronounced thermal inversion.

A substantial part of the incoming visible solar radiation is absorbed in the deep troposphere and re-emitted in the far-infrared spectral range where the opacity is caused by the pressure-induced absorption of H_2. This absorption is produced by H_2-H_2 and He-H_2 collisions, thereby providing the potential for determining the abundances of two main atmospheric components of all four giant planets. The upwelling radiation from the internal source of energy also contributes to the energy budget. The assumption of radiative equilibrium at all levels leads to the upper troposphere and stratospheric thermal structures similar to those shown in Fig. 1.

At pressure levels deeper than 0.6 to 1 bar, the transfer of energy is mainly by convection so that, under the assumption of a quasi-convective equilibrium, the thermal gradient is equal to the adiabatic lapse rate, $\Gamma = g/c_p$. Jupiter, because of its strong gravitational field, exhibits an adiabatic lapse rate of 2 K km^{-1}, almost three times the value for the three other giant planets. However, the very weak internal energy of Uranus may result in subadiabatic thermal gradients in the deep troposphere (Wallace 1983; Bézard and Gautier 1986*b*).

Thermal structures affect the vertical distributions of atmospheric components. In the deep troposphere, all noncondensable components are uni-

TABLE I
Energy Balance of Giant Planets[a]

	Jupiter	Saturn	Uranus	Neptune
T_{eff} (K)	124.4 ± 0.3	95.0 ± 0.4	59.1 ± 0.3	60.3 ± 2
Internal Energy over Solar Absorbed Energy	1.67 ± 0.09	1.78 ± 0.09	1.064 ± 0.062	2.52 ± 0.37

[a]References: for Jupiter, Hanel et al. (1981*a*); for Saturn, Hanel et al. (1983); for Uranus and Neptune, Conrath et al. (1988).

formly mixed by convection. Mixing ratios of condensable gases decrease rapidly with increasing altitude up to the tropopause. In principle, their abundances in the stratosphere are limited by the value reached at the tropopause level, with the remarkable exception mentioned above of CH_4 on Neptune. To first order, thermochemical equilibrium models calculated from plausible thermal models of the deep atmosphere provide a description of the composition of the upper troposphere in agreement with observations (Lewis 1969). However, the detection of PH_3 and GeH_4 on Jupiter and Saturn, also implies the existence of disequilibrating processes (Prinn and Owen 1976; Prinn et al. 1984).

A complex photochemistry mainly caused by solar irradiation occurs in the stratospheres and upper tropospheres of the giant planets and promotes dissociations of molecules coming from the bulk of the planet (H_2, CH_4, NH_3, PH_3) and formation of various new species such as hydrocarbons. Temperature structure affects the vertical distribution of disequilibrium components through kinetics of chemical reactions and through condensation processes as well.

Seasonal effects modulate temperature structures and stratospheric compositions. The low inclination of the Jovian axis of rotation explains the weak latitudinal temperature variations observed by Voyager between $-60°$ and $+60°$ of latitude (Hanel et al. 1979*b*). However, radiative models predict significant seasonal variations of stratospheric temperature at the Jovian poles (Bezanger et al. 1985). In contrast, a substantial seasonal effect is predicted

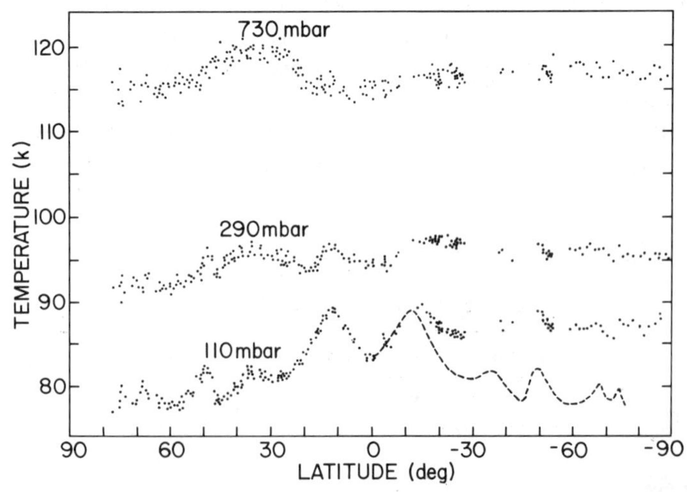

Fig. 2. Temperature-latitude profiles of Saturn retrieved from Voyager-IRIS measurements. The broken curve is a fit to the northern hemisphere 110-mbar temperatures which has been folded over the equator to permit a comparison with southern hemisphere results (figure from Conrath and Pirraglia 1983).

on Saturn, caused by the large obliquity of 26° (Cess and Caldwell 1979; Bézard et al. 1984; Bézard and Gautier 1985*b*), in reasonable agreement with Voyager observations which exhibit a clear, large-scale hemispheric thermal asymmetry (see Fig. 2). In spite of the 82° inclination (IAU convention) of the axis of Uranus, the large thermal inertia limits temporal variations of temperature (Bézard and Gautier 1986). A dynamical redistribution of heat precludes a substantial equator-to-pole temperature gradient in the troposphere (Hanel et al. 1986; Friedson and Ingersoll 1987) but not in the stratosphere where the ultraviolet occultation profile measured near 67° N is significantly warmer than the radio-occultation profile at the equator. A seasonal model for the lower atmosphere of Neptune has been elaborated by Wallace (1984). The latitudinal variation of the upper tropospheric temperature of this planet should be measured by Voyager in 1989.

III. ATMOSPHERIC COMPOSITION: OVERVIEW

We have gathered the available information on the composition of the atmospheres of the outer planets into three tables. Tables II and IV show the

TABLE II
Elemental Ratios, in Solar Units, Observed in the Outer Atmospheres of the Giant Planets*

	Sun	Jupiter / Sun	Saturn / Sun	Uranus / Sun	Neptune / Sun
He/H	0.09 [o]	0.7 ± 0.1 [p]	0.2 ± 0.1 [q]	1.0 ± 0.2 [r]	?
C/H	4.7×10^{-4} [a]	2.32 ± 0.18 [b]	2–6 [c,d]	~25 [e]	~35 [f]
				~25 [f]	$\sim 50 \pm 25$ [n]
				$\sim 20 \pm 10$ [n]	
N/H	9.8×10^{-5} [a]	~2 [g,h]	2–4 [g,h]	<<1 ? [h]	<<1 ? [h]
O/H	8.3×10^{-4} [a]	1/50 [i] (from H_2O)			
		>1/3 [j] (from CO)			
P/H	2.4×10^{-7} [k] (from carbonaceous chondrites)	1 ± 0.3 [l]	2.8 ± 1.6 [c]		
			~8 [m]		

*References: a : Lambert (1978); b : Gautier et al. (1982) revised by Gautier and Owen (1983a); c : Courtin et al. (1984); d : Buriez and de Bergh (1981); e : Lindal et al. (1987); f : Lutz et al. (1976); g : Marten et al. (1980); h : de Pater and Massie (1985); i : Bjoraker et al. (1986b); j : Noll et al. (1987) and Fegley and Prinn (1988*a,b*); k : Cameron (1982); l : Kunde et al. (1982); m : Bézard et al. (1987); n : Pollack et al. (1986*b*); o : Proto-Sun—Lebreton and Maeder (1986) and Cahen (1986); p : Gautier et al. (1981) revised by Conrath et al. (1984); q : Conrath et al. (1984); r : Conrath et al. (1987).

TABLE III
Minor Constituents of the Outer Planet Atmospheres

	Jupiter	Saturn	Uranus	Neptune
Troposphere	CH_4, NH_3, H_2O PH_3, GeH_4, CO[a], HCN	CH_4, NH_3 PH_3, CO,[a] GeH_4	CH_4, NH_3	CH_4, NH_3
Stratosphere	CH_4, CO[a] NH_3 (limited by tropopause cold trap)	CH_4 CO[a]	CH_4 (limited by trop. cold trap)	CH_4 (in excess compared to the trop. cold trap
	C_2H_2, C_2H_6, C_3H_4, C_6H_6(?), C_2H_4	C_2H_2, C_2H_6, C_3H_8(?), C_3H_4(?)	C_2H_2	C_2H_2, C_2H_6

[a]CO is predominantly present in the troposphere on Jupiter (Noll et al. 1988), but as it does not condense, some must occur in the stratosphere as well. On Saturn, the distribution is not yet known.

elemental abundance ratios derived from the abundances of major molecular constituents. A comparison of these ratios with those in the Sun demonstrates the basic points that we shall be discussing in detail: He/H is closest to the solar value in the atmosphere of Uranus; both Jupiter and Saturn show a depletion of helium resulting from the immiscibility of He in metallic hydrogen in their interiors. C/H is above the solar value everywhere. We shall argue that this results primarily from mixing gases liberated from the planetary cores during accretion with gases that collapsed onto the cores from the solar nebula. There may also be a contribution from infalling planetesimals (Stevenson 1983; Pollack et al. 1986). As the proportion of core mass to atmosphere mass increases, the value of C/H increases. The same should be true of other heavy elements except for neon. We discuss these issues in detail in Sec. IV. Table III contains the minor constituents. Values of the mixing ratios and the corresponding

TABLE IV
Helium Abundances in the Outer Atmospheres of the Giant Planets and in the Proto-Sun[a]

Y (per mass abundance)			
Jupiter	Saturn	Uranus	Proto-Sun
0.18 ± 0.04	0.06 ± 0.05	0.262 ± 0.048	0.27 ± 0.28

[a]References: for Jupiter, Gautier et al. (1981) revised by Conrath et al. (1984); for Saturn, Conrath et al. (1984); for Uranus, Conrath et al. (1987); for Proto-Sun, Lebreton and Maeder (1986) and Cahen (1986).

references can be found in Tables 1.2, 1.3, 1.4 and 1.5 of the recent book by Atreya (1986). Except for CH_4, NH_3 and H_2O, they consist of species that are out of equilibrium with their local surroundings, either because they are brought up from regions of high temperature rapidly enough to avoid chemical reactions (PH_3, CO, GeH_4), or they are formed in the upper atmosphere by photochemistry (C_2H_2, C_2H_6), or particle bombardment (C_6H_6) or lightning discharges (C_2H_2, HCN).

There is presently still a debate about the relative importance of solar ultraviolet, internal heat, and lightning discharges for the generation of some of these nonequilibrium species (Bar-Nun 1979; Lewis 1980*a,b;* Scarf et al. 1981; Lewis and Fegley 1984). However, a long-standing argument regarding the source of CO on Jupiter has just been settled. The issue was whether the CO was produced in the stratosphere by infalling H_2O or in the interior by the methane-water reaction. Measurement of the CO line halfwidths demonstrated that the latter explanation is correct (Noll et al. 1988).

Prinn and Olaguer (1981*b*) have called attention to the fact that N_2 produced in the deep atmosphere of Jupiter ($900 \leq T \leq 1700$ K) should be brought up to altitudes sounded by the Galileo probe in detectable amounts. They predict mixing ratios of N_2/H_2 in the range of 0.6 to 10 ppm for various assumptions regarding reaction rates and upward transport. This would make $N_2 \sim 70$ times more abundant than CO in the upper troposphere, within the sensitivity range of the mass spectrometer on the Galileo probe. Once detected, N_2 would become another important tracer of deep atmosphere chemistry and vertical convection in the troposphere.

For additional discussion of nonequilibrium processes in the deep atmospheres, see Fegley and Prinn (1985 for Saturn, 1986 for Uranus, 1988*a* for Jupiter); the chemistry of the upper atmospheres has been reviewed in a comprehensive book by Atreya (1986).

IV. BASIC PROBLEMS RELATED TO THE COMPOSITION OF THE OUTER PLANETS

A. Helium

The determination of the H_2/He ratio in the atmospheres of the giant planets is crucial for its cosmological implications as well as for understanding the internal structures and the evolution of these objects (see the chapters by Hubbard, and by Pollack and Bodenheimer). All planets as well as the Sun presumably formed from the primitive solar nebula, mainly composed of hydrogen and helium. Physical conditions prevailing even in the coldest part of the nebula exclude in principle any condensation or fractionation of these two gases which subsequently form the major part of the atmospheres of the giant planets. Contrary to the telluric planets, the escape of even the lightest elements was prevented by the large masses of these objects. Thus, one expects the ratio of H_2 to He in their atmospheres to be similar to the original value in the solar nebula. Within the framework of nucleosynthesis theories, the He abundance in the nebula 4.5 Gyr ago provides, through models of chemical evolution of galaxies, an estimate of the primordial He abundance which can then be compared to theoretical predictions of the standard model of the Big Bang (Gautier and Owen 1983*b;* Conrath et al. 1987). Unless helium differentiates from hydrogen within the interiors of the giant planets, as discussed below, the determination of the H_2/He ratio in the outer atmospheres of these objects may therefore provide a test of cosmological theories.

If helium separates from metallic hydrogen in the planetary interior and sinks towards the center, a depletion of He compared to the so-called solar abundance (i.e., the protosolar abundance) must be observed. The magnitude of the depletion provides a test of evolutionary models of giant planets.

Relative abundances of H_2 and He also affect the amplitude of the greenhouse effect in giant planet atmospheres through the far-infrared opacity, which is mainly caused by collisions between H_2-H_2 and H_2-He (Trafton 1967; Bachet et al. 1983).

The column density of hydrogen above the clouds in the troposphere can be estimated from groundbased observations of quadrupolar absorption lines in the visible range or of the pressure-induced vibration rotation spectrum in the near-infrared range (see reviews of Jupiter and Saturn H_2 observations by Prinn and Owen [1976], Trafton [1981] and Prinn et al. [1984]), but this information alone does not provide the value of the H_2/He ratio.

The He I resonance line at 584 Å is not observable from the Earth but has been detected from space on Jupiter by Pioneer 10 and Voyager (Carlson and Judge 1976; Broadfoot et al. 1981) and on Saturn and Uranus by Voyager (Sandel et al. 1982; Broadfoot et al. 1986). However, the line is formed in the upper atmosphere above the homopause in a region where H_2 and He are not mixed, and the precise inference of the H_2/He ratio in the mixing region far below the homopause level requires a knowledge of the vertical structure and of the eddy diffusion coefficient. Therefore, measurements of the 584 Å He line are not really helpful for determining the He abundance in the bulk of the planet.

The analysis of the thermal infrared spectrum of the giant planets provides a better approach. The far-infrared spectrum of Jupiter in the 200 to 700 cm^{-1} (14–50 μm) spectral range and of the other giant planets below 700 cm^{-1} ($\lambda > 14$ μm) consists almost exclusively of a continuum caused by the previously mentioned H_2-He collision-induced absorption (except at some frequencies where rotational lines of some minor constituents occur (Bézard et al. 1986*a*). This spectrum may be used to retrieve the thermal structure of the observed atmosphere provided the H_2/He ratio is known (Conrath and Gautier 1980).

The first method for inferring the H_2/He ratio, called the inversion method, takes advantage of the different spectral dependence of the H_2-H_2 and H_2-He absorption coefficients (Fig. 3). A set of temperature profiles is inferred

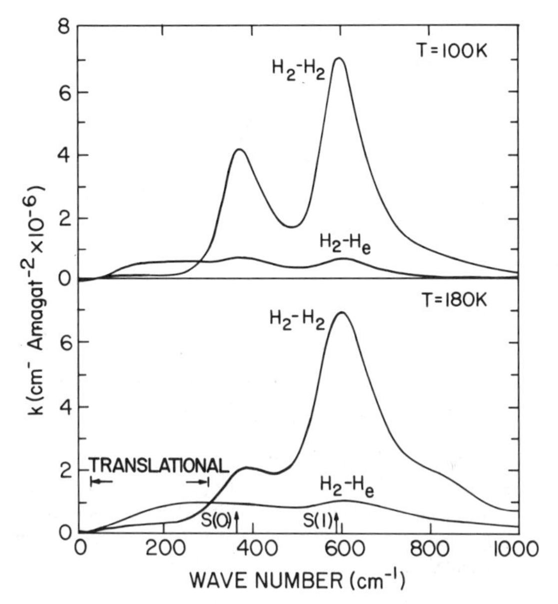

Fig. 3. Pressure-induced absorption coefficients k for H_2-H_2 collisions and H_2-He collisions at Saturn-like temperatures (top) and Jovian-like temperatures (bottom) in the troposphere (figure from Gautier et al. 1981).

from measured radiances for various values of the H_2/He ratio ρ. The solution for ρ is taken as that value for which the rms residual between measured and calculated radiances is a minimum (Gautier and Grossman 1972; Gautier et al. 1981). The disadvantage of the method is that it requires an extremely precise calibration of spectral measurements and a perfect knowledge of the spectral variation of the opacity, including the effects of clouds and aerosols and of any departure from the equilibrium value of the ortho-para ratio for H_2.

The second method for determining the H_2/He ratio combines infrared observations with radio-occultation data. This method avoids most of the difficulties inherent in the first approach. In this case, the major uncertainties in the determination of ρ come from uncertainties in the radio-occultation profile. The two methods were used simultaneously for the retrieval of H_2/He in Jupiter. For Saturn and Uranus, the analysis of the problem showed that the second method was more appropriate.

The observed helium abundances in the upper troposphere of Jupiter, Saturn and Uranus are indicated in Tables II and IV and are depicted graphically in Fig. 4. The value on Saturn is significantly less than the Jupiter value which is itself somewhat less than the Uranus value. For comparison with the solar value, it would be necessary in principle to calculate for each planet the effective He mass fraction $Y_{\rm eff}$ which would be obtained for a solar heavy-

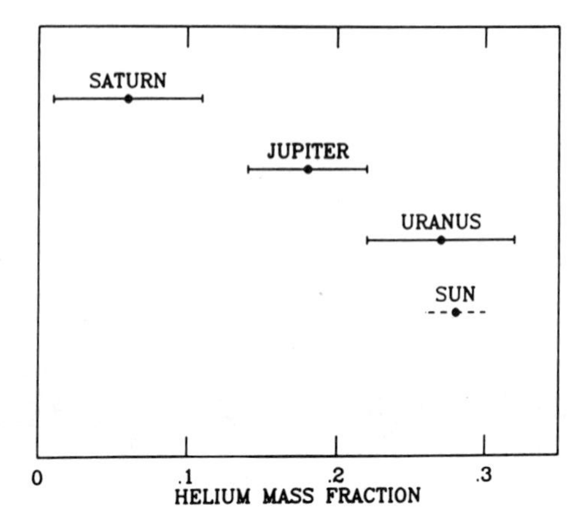

Fig. 4. Helium abundances (per mass) in the atmospheres of Jupiter, Saturn and Uranus compared to the protosolar value. He abundances in the giant planets were measured by Voyager and correspond to values given in Table II. The protosolar value is inferred from solar evolutionary models; uncertainties are not indicated by the modelers but all recent models propose values between 0.27 and 0.28 per mass (figure from Conrath et al. 1987).

element abundance Z_{eff} (since H_2 and He do not condense, the ratio of Y to the hydrogen mass fraction X is independent of altitude). Y_{eff} is given by:

$$Y_{eff} = \frac{Y}{X} \frac{(1 - Z_{eff})}{(1 + Y/X)} \tag{1}$$

where $Z_{eff} \simeq 0.02$ in the Sun (Grevesse 1984*a*; Meyer 1985*b*).

Considering the error bars, Y and Y_{eff} are presently indistinguishable. The most appropriate method for inferring the protosolar He abundance (i.e., the abundance of He in the early Sun) makes use of evolutionary models which are fit to the present mass, luminosity and age of the Sun. All the most recent models lead to a protosolar He mass fraction of about 0.27 to 0.28, in agreement with measurements of the helium abundance in the solar photosphere, that is expected never to have mixed with the interior (for a detailed discussion, see Conrath et al [1987]).

Considering the error bars, this solar value is in good agreement with the observed value in Uranus. On the contrary, Saturn and to a lesser extent Jupiter, exhibit a depletion of helium in their outer atmospheres compared to Uranus and the Sun.

This behavior is consistent with our present understanding of the internal structure and the evolution of the giant planets. At pressures higher than about 3 Mbar, which occur in the interiors of Jupiter and Saturn, hydrogen is expected to become metallic (see the chapter by Hubbard). At sufficiently low temperatures, helium may become immiscible in metallic hydrogen as shown

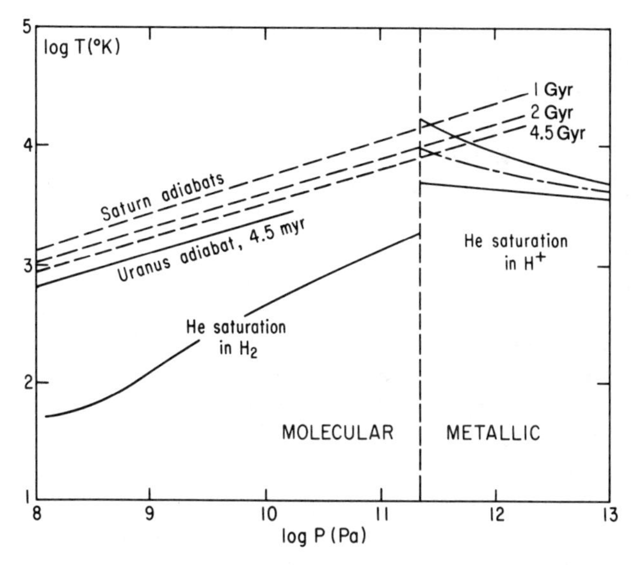

Fig. 5. Saturation temperature of helium for a mixture of H and He in "cosmic" abundance (75% H, 25% He), in a temperature-pressure diagram; pressure is in pascals (1 bar = 10^5Pa), and temperature in Kelvin. The vertical dashed line indicates the molecular metallic hydrogen transition. The upper and lower curves on the right represent the extreme possibilities due to theoretical uncertainties of helium solubilities in metallic hydrogen, while the dash-dot line represents the most plausible curve. The dashed line at the top represents three Saturn adiabats (line of constant entropy satisfying boundary conditions at the top of the atmosphere) at different ages, calculated from the origin of Saturn, as the planet cools down (4.5 Gyr corresponds to the present epoch). In the molecular H_2 region, adiabats are always above the saturation curve but in the metallic range the adiabat begins to intercept the dash-dot line about 2 Gyr after formation of the planet and helium raindrops form. Jovian adiabats, not shown for clarity, are located slightly above Saturn adiabats but should also intercept, somewhat later (i.e., more recently) the solubility curves. Uranus (and Neptune) adiabats are cooler than Saturn but they are believed to stop at the edges of the cores of both planets at about 2×10^{10} Pa and thus never reach the metallic H range (figure adapted from Stevenson 1982*b*).

in Fig. 5 (Smoluchowski 1973; Stevenson and Salpeter 1977*a*). When Jupiter and Saturn cooled from the initial high temperature acquired during their formation, the 3 Mbar pressure level may reach the saturation temperature for the H^+-He mixture. Helium droplets would then form and migrate towards the center of the planet, leading to a depletion of helium in the outer atmosphere in agreement with observations. Since Saturn is smaller than Jupiter in mass and size, its internal energy is expected to vanish earlier, so that the helium differentiation also starts earlier in Saturn, which explains the differences in He abundances in the outer atmospheres of the two biggest planets.

Interior models for Uranus and Neptune indicate that the hydrogen pressure at the edge of the core should not exceed around 200 kbar, (Hubbard 1984*b*), well below the pressure where the transition from molecular to metallic hydrogen occurs. Therefore, no helium depletion from differentiation is expected in these atmospheres, and the observed tropospheric H_2/He ratio should be equal to the value in the primitive solar nebula. This is indeed the case on Uranus (Conrath et al. 1987). Stevenson (1987*c*) has recently proposed a revised model for the interior of Uranus in which the gaseous envelope is much deeper than in previous models, so higher H_2 pressures could occur. Nevertheless, the similarity of H_2/He in the Uranian atmosphere with the solar value and the weakness of the planet's internal heat source (Table I) both argue against significant differentiation of helium in the deep interior.

Before leaving this topic, we should review the possibility that specific *chemical* processes could have modified the initial atmospheric He/H_2 ratio, at least on Uranus and Neptune. The planetesimals that presumably accreted to form the cores of the giant planets (see the chapter by Pollack and Bodenheimer) were initially composed of rocks, ices and/or clathrates containing CNO compounds. Heating during accretion would liberate most of these compounds into the nascent atmosphere of the forming planet. If the icy grains embedded in the nebula (from which the planetesimals themselves formed) contained large amounts of CO, CO_2, N_2 and complex hydrocarbons (see Sec. IV.C), these species would subsequently consume some amount of H_2 present in the initial planetary atmosphere to form CH_4 and NH_3 (see the chapter by Prinn and Fegley). As a result, the He/H_2 ratio would be increased compared to the value in the nebula. In fact, the effect would be negligible in Jupiter and Saturn where the H_2 reservoir is so large, but not in Uranus and Neptune. If, on the contrary, CNO compounds were mainly in the form of ices or clathrates of NH_3, CH_4, and H_2O, no significant reactions with H_2 would be expected.

Assuming the occurrence of kinetic inhibition in the nebula, and hence the prevalence of H-poor CNO compounds, Prinn and Fegley (see their chapter) predict a maximum increase of the He/H_2 ratio of 30% in Uranus and Neptune. Such a large effect is *not* observed in Uranus. Adopting a protosolar He abundance of 0.275 per mass, the maximum increase consistent with the upper limit of the Uranus value is on the order of 13%.

Pollack et al. (1986*a*) use a different approach: assuming that the carbon enrichment in the atmospheres of all giant planets is caused by the dissolution of infalling planetesimals, they elaborate a semi-empirical model fit to the observed carbon enhancement and derive the fraction of carbon that was in a condensed phase in the nebula in the regions of giant planet formation. Assuming in addition a carbonaceous chondrite-like composition for planetesimals, they predict a He/H_2 increase above the solar value on Uranus from 3 to 9%, an effect undetectable with the present uncertainties. The assumption of a chondritic composition for planetesimals is questionable in view of the prevalence of ices in the outer part of the nebula, but the essential requirement of the scenario of Pollack et al. (1986*a*) is simply that the infalling planetesimals contained a significant amount of unreduced carbon and nitrogen. As mentioned above, the composition of the solid phase material in the outer part of the nebula is still quite uncertain (see the chapter by Prinn and Fegley).

An effect in the opposite sense was predicted some time ago by MacFarlane and Hubbard (1982) who pointed out that CH_4 tends to dissociate at pressures above 200 kbar and temperatures above 2000 K, conditions that probably occur in the interior of Uranus. Such a process would enrich the outer atmosphere in H_2, resulting in a decrease of the He/H_2 ratio (Hubbard 1984*b*). Present Uranus observations limit the magnitude of the effect, if any, to 24%.

To summarize, it appears that the value of He/H in the present atmosphere of Uranus is a very close match to the value in the solar atmosphere. Models of solar evolution which are fitted to the present mass, luminosity and age of the Sun, have recently been revised by using improved radiative opacities and updated nuclear data. Results from various groups (see, for instance, Lebreton and Maeder 1986; Cahen 1986) lead to protosolar He abundances of 0.27 to 0.28. This new work is consistent with the He abundances in the vicinity of the solar system ($Y = 0.28 \pm 0.02$: Pagel 1982). Current models of chemical evolution of galaxies that try to interpret the destruction of deuterium between the Big Bang and the birth of the solar system predict a helium enrichment during this interval of about 3%; that leads to a primordial value $Y_p \approx 0.25$, close to the upper limit considered at the present time by most authors from measurements of the He abundance in very old galaxies (for a detailed discussion, see Conrath et al. [1987]). Primordial and solar nebula abundances of helium may therefore be consistent. However, should future observations of old galaxies lead to a value of $Y_p \lesssim 0.23$, both the standard model of the Big Bang and the models for the chemical evolution of galaxies would have to be revised.

Obviously the new data obtained by Voyager 2 at Neptune will provide additional tests of the various models that have been proposed to explain the variations in He/H observed in outer planet atmospheres thus far. If Neptune's helium abundance is found to be close to the Uranian value, it will confirm our preliminary conclusion that He/H in the atmospheres of these two planets is indeed the value from the primitive solar nebula. If on the contrary the Uranian and Neptunian values disagree, it will be necessary to invoke the scenario originally proposed by Stevenson (1983) and elaborated by Pollack et al. (1986*a*) or the effect proposed by MacFarlane and Hubbard (1982). Still greater improvements are probable. The next mission to the outer solar system will be Galileo, which will include a Helium Abundance Interferometer on its atmospheric probe, capable of an extremely precise *in situ* measurement of H_2/He in Jupiter's atmosphere. This measurement will permit an accurate evaluation of the helium differentiation within the planet. The Cassini Mission to the Saturn system will provide multiple radio occultations combined with highly sensitive infrared measurements which should certainly improve the existing accuracy for H_2/He on Saturn. A future Saturn probe could duplicate the precision of the He/H_2 determination on the Galileo mission. Similar probes have been strongly advocated for Uranus and Neptune by various mission planning groups (e.g., SSEC: Morrison et al. 1986).

B. Deuterium

The ratio of deuterium to hydrogen in the atmospheres of the outer planets has been the subject of considerable attention since the first detection of CH_3D on Jupiter by Beer and Taylor (1973). A comprehensive study of the thermochemical problems involved in deuterium-hydrogen exchange processes in the atmospheres of Jupiter and Saturn has just been completed by Fegley and Prinn (1988*b*). Here we shall simply give a brief review of the current status of D/H determinations for the outer solar system and our interpretation of them.

Both HD and CH_3D have been used as the deuterium-bearing molecules whose abundances are then ratioed to H_2 and CH_4, respectively, to obtain the desired value of D/H. Since hydrogen is the dominant gas in all four planetary atmospheres and since it is commonly assumed that vertical mixing will always provide contact with temperature levels that allow rapid exchange of D among various molecular species, the following equilibrium should apply:

$$\frac{D}{H} = \frac{1}{2}\frac{HD}{H_2} = \frac{1}{f}\frac{1}{4}\frac{CH_3D}{CH_4} \tag{2}$$

where f is the fractionation factor that represents the fact that D is a little more tightly bound than H ($\sim$ 0.5 K cal mol^{-1}) and therefore tends to become more concentrated in CH_4 than in H_2. Thus $f > 1$, and is a strong function of temperature. In fact, it varies between 1 and 2 for the range of temperatures appropriate to the outer planet atmospheres (Beer and Taylor 1973; Fegley and Prinn 1988*b*). Uranus may not be in equilibrium throughout at the present time, given that its upper troposphere is heated by the Sun more than by any internal source of energy. However, during the course of its evolution, this planet too must have passed through a stage in which convection kept its atmosphere well mixed, and the results of that epoch should be what we see today.

Observations of CH_3D and HD to date have not agreed with this assumption of equilibrium among hydrogen-bearing species (see Gautier [1985] for a review). In particular, the values of D/H obtained from observations of HD in spectra of Jupiter and Saturn have been systematically higher than the values deduced from CH_3D with $f = 1$, in violation of the chemical thermodynamics underlying Eq. (2): fractionation will *increase* the amount of deuterium in the methane.

This apparent paradox may be resolved in the near future by improved laboratory and observational data. According to W. H. Smith (personal communication, 1987), his most recent high-resolution observations indicate that a background of weak methane absorption lines distorts the local continuum around the HD lines in such a way that the HD abundances are overestimated. This perception substantiates the conclusion of Gautier (1985) and of Fegley and Prinn (1988*b*) that there is simply no plausible way to reconcile the higher values of D/H derived from HD for Jupiter and Saturn with the values obtained from CH_3D. We anticipate that the problem of evaluating the weak HD lines will only be worse on Uranus and Neptune, where the relative abundance of CH_4 is roughly 10 times higher, making the continuum even more uncertain. Future observations of the rotational spectrum of HD (Bézard et al. 1986*a*) as well as mass spectrometric measurements carried out in the atmospheres of both Jupiter and Saturn by instruments in entry probes should settle this issue. At the present time, we are inclined to favor D/H values obtained from CH_3D for the following additional reasons.

First, D/H ratios derived from HD abundances are not compatible with the protosolar deuterium abundance derived by Geiss and Boschler (1981) from ^{3}He solar-wind measurements ($D/H = 2\ ^{+1.5}_{-0.5} \times 10^{-5}$). We expect a similar value at least in Jupiter and Saturn because their hydrogen reservoirs, which originate from the primitive solar nebula, are too large to permit a significant enhancement of deuterium through the vaporization of ices of CNO compounds. Yet HD results from both planets yield values of D/H several times higher than the protosolar number.

Second, the comparison of D/H values derived from HD with present interstellar measurements $\sim 1 \pm 0.5 \times 10^{-5}$ (Vidal-Madjar 1987) imply a destruction of deuterium by a factor $\sim$ 10 between the birth of the solar system and the present era. Note that a large spread exists in the interstellar medium (ISM) measurements; some observations give values as low as 3×10^{-6} (Vidal-Madjar et al. 1983). The value quoted above proposed by Vidal-Madjar (1987) refers to a local average D/H ISM value, which is appropriate for comparison with the solar system abundance. It is expected that the abundance of deuterium in the ISM decreases with time, owing to nuclear burning in stars, and that the contemporary ISM value should be somewhat lower than the value in the solar nebula 4.5 Gyr ago, but only by a factor $\sim$ 2 at most (Vidal-Madjar 1987). Current models of the chemical evolution of galaxies (see, e.g., Delbourgo-Salvador et al. 1985) totally exclude a depletion factor of 10.

Therefore, we shall restrict our discussion to results from CH_3D. A summary of the current status of determinations of D/H in the outer solar system is given in Fig. 6. Here we are using the recent reestimates of fractionation factor values derived by Fegley and Prinn (1988*b*), $f = 1.205$ for Jupiter and 1.229 for Saturn. There is some spread in various determinations of the CH_3D mixing ratio in Jupiter. CH_3D/H_2 published values so far vary from $(2.3 \text{ to } 7) \times 10^{-7}$, each one with substantial uncertainties (see Gautier [1985] for details). We adopt the most recent value obtained by Bjoraker et al. (1986*b*) from 5 μm airborne observations made with a high spectral resolution (0.5 cm^{-1}), i.e.,

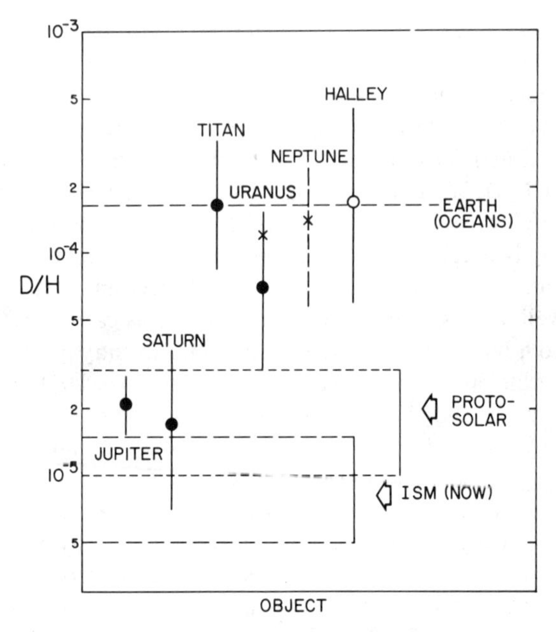

Fig. 6. Observed values of D/H as determined in methane in the outer planets and Titan, and in water in Halley's comet (see text for references). Predictions for Uranus and Neptune by Hubbard and MacFarlane (1980*b*) are marked with an X.

$CH_3D/H_2 = (2.2 \pm 0.45) \times 10^{-7}$. We believe, on the other hand, that the most precise determinations of the Jovian CH_4/H_2 ratios have been inferred from Voyager observations at 7.7 μm which provide a number of calibrated spectra with a high spatial resolution, permitting their selection in regions where the influence of clouds is negligible (Gautier et al. 1982). Adopting a C/H enrichment over the solar value of 2.32 ± 0.18 as indicated by these authors, we find $D/H = (2.1 \pm 0.6) \times 10^{-5}$. For Saturn, both CH_3D and CH_4 mixing ratios are uncertain. We conservatively adopt $CH_4/H_2 = (4 \pm 2) \times 10^{-3}$ from Buriez and de Bergh (1981). (This value is in good agreement with the Voyager determination of $4.5 ^{+2.4}_{-1.9} \times 10^{-3}$ [Courtin et al. 1984], but has somewhat less uncertainty.) With this mixing ratio for CH_4, far-infrared Voyager observations of CH_3D by Courtin et al. (1984) lead to $D/H = 2.0 ^{+2.4}_{-1.3} \times 10^{-5}$. Groundbased measurements of the CH_3D/CH_4 ratio by de Bergh et al. at 1.6 μm (see Owen et al. 1986) and of CH_3D at 5 μm by Bézard et al. (1987) lead to $D/H = 1.4 ^{+1.4}_{-0.9} \times 10^{-5}$ and $1.6 ^{+2.0}_{-0.7} \times 10^{-5}$, respectively. The average of these three independent determinations for Saturn is $D/H = 1.7 ^{+1.9}_{-1.0} \times 10^{-5}$.

Although the numbers seem to indicate that D/H on Saturn is systematically lower than on Jupiter, the uncertainties allow them to be identical, as we have noted before (Gautier and Owen 1983*a*; Gautier 1985; Owen et al. 1986). At this point, we conservatively prefer to think of the Jupiter and

Saturn results as two determinations of the same quantity, probably best expressed by D/H = $2.0 \pm 1.5 \times 10^{-5}$.

Turning to the smaller outer planets, we find a different situation. Whereas the heavy element cores of Uranus and Neptune must be approximately the same mass as the cores of Jupiter and Saturn (Stevenson 1982*b*; chapter by Hubbard), the smaller masses and higher densities of these planets imply much less massive hydrogen-helium envelopes. As Hubbard and MacFarlane (1980*b*) pointed out, this allows the possibility that the ratio of D/H set in the compounds that make up the cores could leave an imprint in the atmospheres.

To give a specific example, assume the minimum core size for Uranus is 10 $M_\odot$ (solar masses), leaving 4.5 $M_\odot$ for the atmosphere. (In fact, present models require the core mass to be a much larger fraction of the total [see the chapter by Hubbard].) With these proportions, as much as 35% of the atmospheric hydrogen could come from the core. But this ultra-extreme case, in which *all* the hydrogen in the core is released from the compounds that contain it is not only unrealistic, but it can be ruled out by the observed solar value of H/He (see above). Nevertheless, deuterium is clearly enriched in the Uranian atmosphere.

In fact, the observed value of D/H derived from CH_4D/CH_3 on Uranus is $9\ ^{+9.0}_{-4.5} \times 10^{-5}$, according to de Bergh et al. (1986). This number was determined without assuming equilibrium with H_2 via Eq. (2), i.e., setting $f = 1$. Given the lack of an agreed-upon model for the evolution of the planet, it is not possible to derive a convincing value for f, but it is probably between 1.2 and 1.4. We might therefore call this $7^{+8}_{-4} \times 10^{-5}$. This value is distinctly higher than the number derived for Jupiter and Saturn, although the error bars just overlap. Hence we do have evidence here for deuterium enrichment, presumably from the gases originally trapped or frozen in the core-forming ices. Both the methane in Titan's atmosphere (Owen et al. 1986; de Bergh et al. 1988), and the water in Halley's comet (Eberhardt et al. 1987*a*) suggest that D/H in the most abundant compounds in the core will be on the order of 2×10^{-4}.

This is exactly the scenario envisaged by Hubbard and MacFarlane (1980*b*) nearly ten years ago, and their predictions of D/H in the atmospheres of Uranus and Neptune are remarkably close to the values subsequently determined by de Bergh et al. (1986,1987,1988). The only substantive difference between more recent work and their original paper concerns their suggestion that low-temperature chemical equilibrium was responsible for the enrichment of D/H in the ice. The time scale for such equilibration is prohibitively long, inviting a search for alternative processes. An attractive solution is to assume that the enrichment took place before the formation of the solar nebula, as we discuss below.

It is easy to demonstrate that the core of Uranus must be invoked as a reservoir to provide the observed D/H enrichment. The mixing ratio of CH_4 in the atmosphere of Uranus is ~ 0.02 (Lutz et al. 1976; Lindal et al. 1987).

While this corresponds to ~ 20 times the solar value of C/H, it is too small to supply the expected enrichment of $D/H = 7 \times 10^{-5}$ in the entire atmospheric reservoir, starting with a value of $CH_3D/CH_4 = 8 \times 10^{-4}$ ($D/H = 2 \times 10^{-4}$ in the methane, $f = 1$; see Eq. 2). A much more abundant source of deuterium is available in the form of hydrogen compounds making up the core, where we expect water to dominate. To achieve the value of $D/H = 7 \times 10^{-5}$ that is observed in the atmosphere, we need a mixing ratio of H_2O to H_2 of approximately 0.5:

$$\frac{[D]}{[H_2O] + [H_2]} = 2\,\frac{D}{H} = 1.4 \times 10^{-4};\ \frac{[D]}{[H_2O]} = 4 \times 10^{-4} \tag{3}$$

which is the assumed starting value for H_2O in the core; so

$$\begin{gathered} \frac{4 \times 10^{-4}\,[H_2O]}{[H_2O] + [H_2]} = 1.4 \times 10^{-4} \\ 4 \times 10^{-4}\,[H_2O] = 1.4 \times 10^{-4}\,([H_2O] + [H_2]) \end{gathered} \tag{4}$$

then $2.6 \times 10^{-4}\,[H_2O] = 1.4 \times 10^{-4}\,[H_2]$, $H_2O/H_2 \sim 0.5$.
The water need not all be in the vapor phase at any given time to allow exchange of D between HDO and HD, as long as some of it is present as vapor and a phase-change cycle exists. Stevenson (1982*b*) has postulated the presence of an ionic "ocean" of water whose upper "surface" would be at a pressure of 200 bar and a temperature of 2500 K. This would do nicely as the required interface with the H_2-He atmosphere. (As previously mentioned, Stevenson [1987*c*] now favors a model excluding a discrete layering [rock core, ice and gas envelope] but including compositional gradients in the deep interior. However, an ionic H_2O - H_2 mixture still occurs at modest depths [$P \sim 105$ kbar and $T = 2000$ K].)

Taking our model of an atmosphere containing 4.5 $M_\oplus$ (Earth mass), approximately 3 $M_\oplus$ would be hydrogen. This means we need $0.5 \times (18/2) \times 3\ M_\oplus = 13.5\ M_\oplus$ of water, which contradicts the model because the core can only have a mass of $14.5\ M_\oplus - 4.5\ M_\oplus = 10\ M_\oplus$. However, if we reduce the mass of the atmosphere to any value below 3.5 $M_\oplus$, we are able to fit the atmosphere and the core within the total mass of the planet. As the mass of the atmosphere decreases, there is room for a progressively greater addition of "rock." For an atmosphere of mass 2.5 $M_\oplus$, for example, we have a total core mass of 12.0 $M_\oplus$. The atmosphere now contains 1.75 $M_\oplus$ of hydrogen, requiring only $0.5 \times 9 \times 1.75\ M_\oplus = 8\ M_\oplus$ of water. This makes the planet ~ 55% water by mass and leaves 4 $M_\oplus$ for other heavy element constituents.

This approach defines a family of models for $M_{atm} \leq 3.5\ M_\oplus$ in which the observed enrichment of D/H in the atmosphere can occur, starting with $D/H = 2 \times 10^{-4}$ in the water that forms the core. Obviously other hydrogen-containing compounds can be substituted for or added to the water. The range in core composition and mass fraction of core and atmosphere are consistent with current models for the interior of Uranus (Podolak and Reynolds 1987; Hubbard and Marley 1988; also see Hubbard's chapter).

This same line of reasoning (see Hubbard and MacFarlane 1980*b*) predicts that D/H on Neptune will be similarly enriched, in agreement with recent observations of CH_3D at 1.6 μm by de Bergh et al. (1987,1989). Their work suggests a higher value than the earlier result of Orton et al. (1987*a*) who found D/H $= 1.2 \pm 1 \times 10^{-5}$ from broadband measurements at 8.6 μm. Note that de Bergh et al. observed CH_3D absorption lines while Orton et al. derived their result from the fit of a continuum in emission that depends upon a number of elusive parameters.

It is useful to include a brief discussion of the abundance of deuterium on Titan in this review, since this observation provides some supporting evidence for the interpretation of apparent deuterium enrichments on Uranus and Neptune. De Bergh et al. (1986,1988) found D/H $= 1.6 \pm 0.8 \times 10^{-4}$ using the 1.6 μm band of CH_3D in Titan's spectrum. Investigating the reason for this large enrichment, Pinto et al. (1986) considered a variety of processes that could have increased the deuterium concentration on Titan during and after the satellite's formation. Aside from the extremely unlikely possibility (Grinspoon and Lewis 1987*a*) of catalysis of the reaction

$$HD + CH_4 \rightleftarrows CH_3D + H_2 \qquad (5)$$

on metallic grains at 500 K, they found the dominant process to be Jeans escape from Titan's upper atmosphere. Combined with other, less important processes, escape could increase D/H from an assumed starting value of $2 \pm 1.5 \times 10^{-5}$ to $4.4 \pm 3.3 \times 10^{-5}$, whereas the lower bound set by the observations was 8×10^{-5}. This enrichment factor of 2.2 is independent of the initial value of D/H (Pinto et al. 1986). Consequently, Owen et al. (1986) suggested that the observed enrichment was primordial, having been set prior to the formation of the satellite.

This possibility had been anticipated by Geiss and Reeves (1981) in their comprehensive discussion of deuterium in the solar system. Ion-molecule reactions in the interstellar medium are known to be capable of increasing D/H by as much as a hundred times, compared with the more modest ten fold observed on Titan. If the molecules carrying this enriched deuterium are then kept at a relatively low temperature ($T \lesssim 450$ K), the equilibrium expressed by Eq. (2) cannot be achieved in times shorter than the age of the galaxy. In this case, isotopic anomalies set in the ISM will be maintained through the formation of the solar nebula and incorporated in forming planetesimals. This idea seems substantiated by the high value of D/H in Halley's comet (Eberhardt et al. 1987*a*). Unless some additional process of deuterium enrichment on Titan can be identified, it appears that the prenebular isotope ratios in condensed matter were preserved in the Saturn subnebula as well (Owen et al. 1986). Chemical reactions during the process of accretion can then redistribute the deuterium among the volatile, hydrogen-containing molecules. This is what we believe to be responsible for the high D/H in the methane on Titan and in the atmospheres of Uranus and Neptune.

C. Carbon, Nitrogen, Oxygen and Sulfur

Carbon is the most useful calibrator of heavy element abundances, owing to the fact that methane does not condense at all in the atmospheres of Jupiter and Saturn. It does condense in the upper tropospheres of Uranus and Neptune, but we may still determine C/H in the deep atmospheres of these planets, below the clouds. The enrichment of carbon over the solar value as documented in Table II requires an explanation in terms of a model for the formation of the giant planets. A giant, gaseous protoplanet would be expected to maintain all the elemental abundances in the proportions found in the solar nebula, i.e., the cosmic ratios, obviously an unacceptable consequence.

These models are currently out of favor for other reasons as well (see the chapters by Hubbard and by Pollack and Bodenheimer). An attractive alternative is the idea of nucleation and collapse (Perri and Cameron 1974; Mizuno 1980) as described above. As the core is forming, it must be accumulating its own atmosphere as a result of degassing caused by accretional heating. The gases liberated from the ices during this phase come from three sources: (1) clathrate hydrates and/or adsorbed gases; (2) sublimation of ices and other low temperature volatiles; and (3) reaction products of these gases. Laboratory experiments show that H_2, He and Ne are not trapped in ice except at temperatures well below 25 K, whereas N_2, NH_3, CO, CH_4, Ar and the other heavier noble gases are (Bar-Nun et al. 1985,1987). Ratios as high as 1:1 for gas:ice can be obtained with sufficiently high gas pressure.

Thus, the atmosphere forming around a giant planet core will consist of two components: solar nebula gas that is depleted in condensible elements but with H, He and Ne in solar proportions, and gases emanating from the core itself. The net result of the mixing of these two reservoirs will be an enrichment in the resulting atmosphere of those gases carried in the core-forming materials.

This straightforward way to produce the observed enrichment of heavy elements (*viz.*, carbon) in the atmospheres of the outer planets has been challenged by Stevenson (1983). He pointed out that the upward mixing of core constituents by convection is inefficient and proposed instead that the enrichment is caused by a late influx of several $M_\oplus$ of planetesimals. As we have pointed out elsewhere (Gautier and Owen 1985), convective mixing may have been much more efficient in the past, particularly during the rapid collapse of the surrounding nebular gas onto the core and the subsequent high-luminosity phase of the cooling planet. Hence an additional source of heavy elements in the form of several $M_\oplus$ of late-accreting planetesimals may not be necessary.

Nevertheless, it is an approach to the problem of atmosphere formation that deserves consideration.

A detailed model of the planetesimal dissolution process has been developed by Pollack et al (1986*a*; also see the chapter by Pollack and Bodenheimer). The predictions of this model for the enhancement of carbon in the atmospheres of the outer planets are quite good for Jupiter and Saturn, but fail for Uranus. Neptune fits the model if the enrichment of C/H in its atmosphere is approximately 2.5 times the value found on Uranus as these authors assume (Pollack 1986*a,b*). However, other investigators have reported C/H to be more nearly equal in the two planets (Table II). The Voyager 2 encounter with Neptune in August of 1989 may allow a better determination of this important parameter, but an atmospheric probe will be required to obtain a definitive result.

Meanwhile, we point out that the deuterium enrichment predicted by Hubbard and MacFarlane (1980*b*) and observed by de Bergh et al. (1986, 1987,1988) in the atmospheres of Uranus and Neptune suggests that in these two planets, at least, chemical mixing between the cores and atmospheres is very good (see Sec. IV.B). To obtain this enrichment of deuterium by the dissolution of planetesimals would require a mass of planetesimals approximately equal to the mass of the core itself, which seems unlikely.

The above discussion has focused on carbon because of the low freezing point of methane. Obviously both models predict a similar enrichment for all of the other heavy elements except neon. Microwave observations that probe the regions of these atmospheres below the point at which ammonia condenses suggest that this is indeed true for NH_3 on Jupiter and Saturn (Table II).

The situation on Uranus and Neptune is more complex. Early observations of the microwave brightness temperatures of both of these planets suggested that ammonia was depleted (see Gulkis and de Pater [1984] for a review). Recent studies with the Very Large Array (VLA) appear to confirm this result (de Pater and Massie 1985). The reason for this deficiency may be that ammonia is dissolved in the massive water ocean-cloud system mentioned above, as Atreya and Romani (1985) have suggested. Further work on models of the interior of Uranus is required to settle the issue.

A similar problem exists with respect to H_2O on Jupiter, the only giant planet on which water has been detected so far. Measurements in the 5 μm window reveal the presence of H_2O lines in the Jovian spectrum, but the derived abundance is much lower than expected (Larson et al. 1975; Bjoraker et al. 1986*a*). The base of the region of the atmosphere sampled at 5 μm is at $P = 6$ bar and $T = 288$ K. At this level, the O/H ratio is $< 10^{-2}$ of the expected, above-solar value.

That this is not evidence for a global depletion of oxygen is demonstrated by the presence of CO in Jupiter's upper troposphere, and the stringent upper limit on silane (SiH_4), which require an oxygen abundance in the deep atmosphere that is within a factor 10 of the solar value (Noll et al. 1988; Fegley and

Prinn 1988*a*). Thus, we must once again be dealing with a cloud physics problem. A possible model that resolves this dilemma has been put forward by Lunine and Hunten (1987). They suggest that moist vertical convection on the planet is limited to narrow plumes, thereby keeping the upper atmosphere dry. Further study of the distribution of water vapor over the disk could provide a useful test of this model while providing valuable new data. However, the best determination of the vertical distribution of water will come from the Galileo probe when it reaches the 10 bar level, well below the condensation point for Jovian water clouds. (Given the solubility of NH_3 in water, these will actually be clouds of a dilute solution of ammonia [Lewis 1969].)

Sulfur is another element that seems underabundant. The dominant form of sulfur expected in these atmospheres is H_2S. The most careful search for this molecule has been made on Jupiter, where the results to date have been negative (Larson et al. 1984). Larson et al. (1984) established an upper limit of 5×10^{-4} of the expected enriched value from observations at 2.7 μm. Based on observations of methane lines, Larson et al. deduced that they were detecting reflected sunlight from a layer in the Jovian atmosphere with $P =$ 0.75 atm and $T =$ 150 K. This is just below the base of the uppermost cloud deck in their model for Jupiter's atmosphere. They interpreted their failure to detect water vapor in this same spectral region to mean that infrared photons at this wavelength go no lower than 1.2 bar, consistent with the lower boundary deduced from the methane lines. That means that ultraviolet photons obviously do not reach this region either. Larson et al. (1984) conclude that photolysis is not the explanation for the absence of detectable H_2S and that there is thus no role for photochemically generated sulfur compounds in the production of chromophores on Jupiter. Instead, they opt for the presence of NH_4SH clouds, as originally proposed by Lewis (1969), to serve as a sink for sulfur. The vapor pressure of H_2S above such clouds would be below the present upper limit. In their model, these clouds would be at a level with $P \sim$ 2 bar and $T \sim$ 200 K.

H_2S could be responsible for the additional opacity required to fit observations in the mm range (Bézard et al. 1983) but there are several other ways to interpret mm spectrum: cloud opacities, the appropriate shape factor for the absorption coefficient of NH_3 (de Pater and Massie 1985), etc. The solution of the problem requires high spectral resolution measurements, probably from space, or an entry probe that will get below the NH_4SH clouds, thought to be floating at $\sim$ 2 bar, where the H_2S should be well mixed. The Galileo probe will certainly accomplish this.

D. Noble Gases

No noble gas except helium has yet been detected in the atmospheres of the giant planets. They are obviously expected to be there: the question is in what amounts. The key point is that noble gases may be trapped in the form of clathrates in grains embedded in the primitive solar nebula. The composition of clathrates depends on the composition of the gas phase. As previously mentioned (Sec. IV.A), the region of formation of the giant planets may have been rich in CH_4 if chemical equilibrium prevailed or in CO if the $CH_4 \rightarrow CO$ reaction was inhibited (Lewis and Prinn 1980). The two possibilities lead to different compositions of the planetesimals which formed the cores of giant planets and subsequently enriched their atmospheres. Lunine and Stevenson (1985) have estimated in two cases the enrichment in noble gases in Jovian atmospheres as a function of the carbon enrichment with respect to the solar value. An example of such calculations for Jupiter is given in Fig. 7. Assuming that the primitive reservoir in the nebula contained structure II clathrates formed from a solar composition gas at equilibrium (Fig. 7, bottom panel), the present observed C/H ratio in the Jovian atmospheres ($\sim$ 2.3 times the solar value) would imply, for instance, a Kr/Ar atmospheric ratio 6 times higher than the solar ratio. Neutron diffraction results demonstrate that argon and krypton preferentially form structure II clathrates (for more details, see Lunine and Stevenson [1985]). On the other hand, should planetesimals have contained clathrates formed from disequilibrium material, the enhancement of Kr/Ar relative to the solar value would be as high as 8 (Fig. 7, top panel).

If clathrate formation indeed occurred, which is controversial, *in situ* measurements of noble gas abundances by means of mass spectrometers

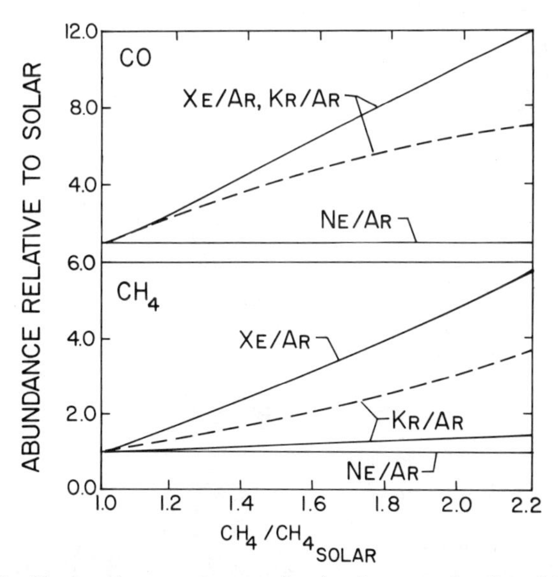

Fig. 7. Predicted noble gas enhancements over solar abundance as a function of CH_4 enhancement in the Jovian atmosphere. Top panel assumes CO-rich and bottom CH_4-rich clathrate. Essentially all CO brought into Jupiter is converted to CH_4. Formation temperature of CO clathrate is set at 60 K, and that of CH_4 clathrate at 100 K. Straight lines show enhancement for structure I clathrate and dashed lines show enhancement for structure II clathrate (figure from Lunine and Stevenson 1985).

aboard atmospheric probes could thus provide information on the nature and the composition of grains imbedded in the solar nebula in the region of formation of the giant planets.

V. CONCLUSIONS

We have stressed throughout this chapter that despite the advances of the last two decades, our knowledge of outer planet atmospheres is still sadly incomplete. Fortunately, help is on the way. Voyager 2 will reach Neptune in August of 1989. We have hopes for a probe into the atmosphere of Jupiter six years later, followed by a Titan probe in another six years. During this long interval, there will be many opportunities for improved observations from the ground and from Earth orbit, as new telescopes and detectors become available. The submillimeter range of the spectrum is still largely unexplored, and there are rich opportunities there for detecting minor species as well as evaluating the rotational lines of HD (Bézard et al. 1986).

Fifteen years ago, the Pioneer spacecraft were launched towards Jupiter on the first missions to the outer solar system. We are optimistic that a comparable growth of knowledge about the giant planets will occur during the next fifteen years. Thus, a review of this same subject written in 2003 should be able to describe solutions to many of the problems we are grappling with today, but we are equally certain that plenty of new enigmas will have been found to take their places.

THERMAL STRUCTURE AND HEAT BALANCE OF THE OUTER PLANETS

B. J. CONRATH, R. A. HANEL
and
R. E. SAMUELSON
Goddard Space Flight Center

Current knowledge of the thermal structure and energy balance of the outer planets is summarized. The Voyager spacecraft experiments have provided extensive new information on the atmospheric temperatures and energetics of Jupiter, Saturn and Uranus. All three planets show remarkably small global-scale horizontal thermal contrast, indicating efficient redistribution of heat within the atmospheres or interiors. Horizontal temperature gradients on the scale of the zonal jets indicate that the winds decay with height in the upper troposphere. This suggests that the winds are driven at deeper levels and are subjected to frictional damping of unknown origin at higher levels. Both Jupiter and Saturn have internal power sources equal to about 70% of the absorbed solar power. This result is consistent with the view that significant helium differentiation has occurred on Saturn. Uranus has an internal power no greater than 13% of the absorbed solar power, while Earth-based observations suggest Neptune has an internal power in excess of 100% of the absorbed solar power.

I. INTRODUCTION

One of the basic goals of modern planetary research is to describe the origin and evolution of the solar system. Evolutionary models of the major planets are central to this inquiry, partly because these bodies contain almost all the planetary mass and angular momentum of the solar system, and partly because they still retain signatures of their origin, such as their H_2 and He content and their excess internal thermal energy. Boundary conditions for the present epoch are provided by the observed thermal structure and heat balance of each planet.

The application of these boundary conditions is not necessarily straightforward. The thermal state of the atmosphere at any given time depends on the magnitude of the internal heat source, the vertical distributions of solar energy deposition and radiative loss to space, and the dynamical response of the atmosphere to these energy sources and sinks.

Some qualitative statements are possible. Deep in the atmosphere, solar deposition and radiative losses become negligible, and the heat flux, provided by an internal energy reservoir, is spherically symmetric except for possible dynamical perturbations due to rotation. Little latitudinal temperature variation along equal pressure surfaces should be evident. Closer to the outer layers, in regions of the atmosphere where absorption of sunlight and radiative losses to space become noticeable, the thermal inertia is still large and the radiative time constant long compared with the season, leading to a time-independent variation of temperature with latitude. Still higher in the atmosphere solar deposition and radiative losses will be greater and the radiative time constant short compared with a season, leading to a variation of latitudinal temperature structure with time of year. These temperature variations induced by radiative processes will be modified by dynamic activity, complicating the application of appropriate boundary conditions in evolutionary models.

One purpose of this Chapter is to review the thermal structure of the four major planets and discuss the diagnostic statements that can be made about their atmospheric dynamics. A second purpose is to summarize present knowledge of the heat balance of these planets and outline how the resulting boundary conditions lead to a better understanding of their internal structures and histories. The topics are interconnected and lead ultimately to important clues about the origin and evolution of the solar system.

II. THERMAL STRUCTURE

Knowledge of the thermal structure of the atmospheres of the giant planets is required if we are to achieve an understanding of the dynamics and energy transport processes in their outer layers. Such knowledge may also indirectly provide information on deeper layers which are not accessible to remote sensing. Groundbased infrared measurements have provided disk-averaged estimates of thermal structure in the upper tropospheres and stratospheres of all four outer planets (Gautier and Courtin 1979; Moseley et al. 1985; Orton et al. 1986). Temperature information at microbar pressure levels have been obtained from occultations of stars observed from Earth.

Recent measurements from spacecraft have provided much new information. Spatially resolved temperature structure was obtained for Jupiter using infrared radiometer data from Pioneers 10 and 11 (Orton 1975; Orton and Ingersoll 1977) and for Saturn from Pioneer 11 (Orton and Ingersoll 1980). Profiles were also obtained from the Pioneer spacecraft by means of the radio occultation method (Kliore and Woiceshyn 1976; Kliore et al. 1980). The Voyager spacecraft have flown by Jupiter, Saturn and Uranus with a Neptune encounter scheduled for August, 1989. The ultraviolet spectrometers carried on the Voyager spacecraft obtained estimates of mesospheric and exospheric temperatures of Jupiter, Saturn and Uranus (Atreya et al. 1979; Broadfoot et al. 1981; Smith et al. 1983; Broadfoot et al. 1986), while data from the infrared spectrometer (IRIS) and the radio occultation experiment have provided upper tropospheric and stratospheric thermal structure information.

In this review, we are primarily interested in the thermal structure of the upper troposphere and lower stratosphere since the dynamics and transport processes in these regions are most relevant to the energetics of the planetary interiors and atmospheres. We will concentrate primarily on the Voyager IRIS and radio occultation results.

Vertical Profiles

Analysis of the radio signal from the spacecraft passing behind the planet as viewed from Earth provides profiles of the temperature divided by the mean molecular weight. If the composition of the atmosphere is known then the mean molecular weight can be calculated, and a profile of temperature vs pressure can be determined. This approach has been used to obtain profiles for Jupiter and Saturn using measurements from Voyagers 1 and 2 and for Uranus using Voyager 2 data. Results were obtained from pressure levels of a few tenths of a millibar down to levels where the pressure exceeds 1 bar (Lindal et al. 1981*a*,1985,1986,1987). Although the number of profiles obtained in this way is limited to a maximum of two per spacecraft flyby, the results are quite detailed in vertical structure.

A second approach used to obtain vertical profiles uses measurements obtained with the infrared spectrometer (IRIS) carried on the spacecraft (Hanel et al. 1979*a,b*,1981*b*,1982,1986). If the opacity is known in the appropriate spectral regions, the measurements can be used in an inversion of the radiative transfer equation to obtain temperature as a function of atmospheric pressure level. On Jupiter and Saturn, measurements in the 1300 cm^{-1} CH_4 band provide temperature information between approximately 0.5 mbar and 10 mbar with a relatively low vertical resolution. Measurements between 200 and 650 cm^{-1}, where the opacity is dominated by the S(0) and S(1) pressure induced H_2 lines, provide information between approximately 100 and 600 mbar with a vertical resolution of about half of a pressure scale height. On Uranus, the cold stratospheric temperatures and the resulting lower signal-to-noise ratio prevent use of measurements in the CH_4 band. However, measurements in the pressure-induced H_2-dominated part of the spectrum between 200 and 400 cm^{-1} permit profiles to be obtained between about 60 mbar and 900 mbar. Although the IRIS derived profiles are relatively low in vertical resolution, they have been obtained from a variety of locations on the planets.

Profiles for Jupiter from both Voyagers 1 and 2 by the radio occultation method are shown in Fig. 1 (Lindal et al. 1981*a*). The dominant constituents, which contribute significantly to the mean molecular weight on Jupiter, are hydrogen and helium. The helium abundance used to obtain the profiles shown was derived from the Voyager IRIS spectra (Gautier et al. 1981). The profiles in the troposphere (the region below the temperature minimum) are essentially identical, and below about 500 or 600 mbar, they closely follow an adiabat. In the stratosphere (the region above the temperature minimum), the low-latitude profiles are highly structured and appear to be quite different from one another. Allison (1983) has suggested that the structure is a manifestation of vertically propagating equatorial waves. If this is truc, then wave propagation may play a significant role in transporting energy from the lower to the upper atmosphere, at least at low latitudes.

Profiles obtained by the radio occultation method for Saturn are shown in Fig. 2 (Lindal et al. 1985). Again the mean molecular weight is based on the determination of the helium abundance from Voyager measurements; in this case a combination of infrared spectroscopy and radio occultation data were used (Conrath et al. 1984). As on Jupiter, below about 500 mbar the profiles appear to follow an adiabat. In this case, we must carefully distinguish among

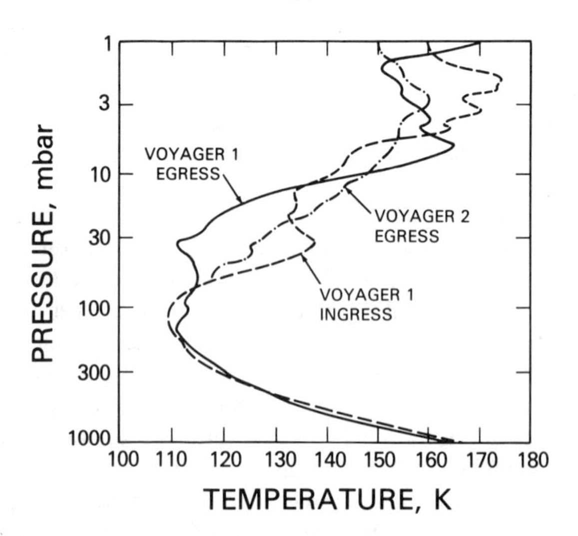

Fig. 1. Vertical temperature profiles of Jupiter obtained from Voyagers 1 and 2 using the radio occultation technique (Lindal et al. 1981*a*). The Voyager 1 ingress and egress profiles were acquired at 12°S lat and the equator, respectively, while the Voyager 2 egress profile was acquired at 60°S.

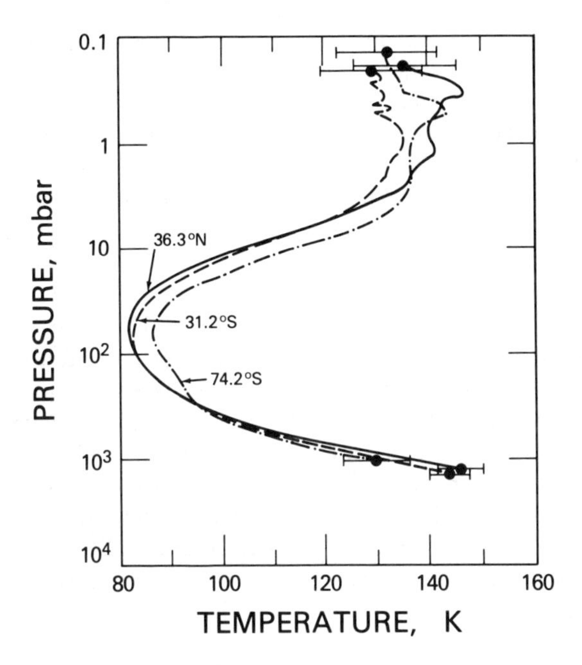

Fig. 2. Vertical temperature profiles of Saturn obtained from Voyagers 1 and 2 using the radio occultation technique (Lindal et al. 1985, 1986). The profiles labeled 74°.2 S, 36°.3 N, and 31°.2 S are from measurements taken at Voyager 1 ingress and Voyager 2 ingress and egress, respectively.

possible adiabats. In a predominantly H_2 atmosphere, the specific heat, and hence the abiabatic lapse rate, depends on the degree to which the ortho and para states of H_2 achieve thermodynamic equilibrium. Three possibilities have been suggested. If the ortho and para states are in equilibrium at the local temperature, an *equilibrium adiabat* may occur; this is the least steep of the three. If, on the other hand, the ortho-para ratio is the high temperature limit of three to one and transitions between ortho and para states do not occur, then a *normal adiabat* is expected. Finally, if the ortho-para ratio is the equilibrium value at the local temperature, but adiabatic parcel displacements occur on time scales short compared to the ortho-para equilibration time, then a *frozen equilibrium adiabat* will occur (Trafton 1966; Wallace 1980). In practice, the frozen equilibrium and normal adiabats are observationally indistinguishable. At the temperatures encountered on Jupiter, the frozen equilibrium adiabat is also essentially indistinguishable from the equilibrium adiabat. However, at the lower temperatures of Saturn, it may be marginally possible to distinguish between the two cases. Analysis of the occultation data by Bezard (1985) suggests that the Saturn profile may lie closer to the frozen equilibrium adiabat.

The Saturn profiles obtained by radio occultation at mid latitudes in both hemispheres appear to be quite similar. In contrast, profiles retrieved in these two regions from IRIS data are quite different. The northern hemisphere profile agrees quite well with the radio occultation profile, but the southern hemisphere profile does not. In addition, profiles retrieved in the southern hemisphere from Pioneer infrared radiometer measurements (Orton and Ingersoll 1980) do not agree with the Voyager occultation profile. The reason for this disagreement remains unknown.

Two temperature profiles for Uranus, both derived at low latitudes with the radio occultation method, are shown in Fig. 3 (Lindal et al. 1987). The stratosphere shows the presence of two warm regions, and it has been suggested by Lindal et al. (1986) that these are possibly due to absorption of sunlight by layers of particulates. Below about 400 mbar, the profiles appear to follow an adiabat. On Uranus, the temperatures are sufficiently low so that the equilibrium and frozen equilibrium adiabats are well separated. There seems to be little doubt that the Uranus lapse rate significantly exceeds that for an equilibrium adiabat even though IRIS spectra indicate an equilibrium ortho-para ratio in the upper troposphere (Hanel et al. 1986). The small dis-

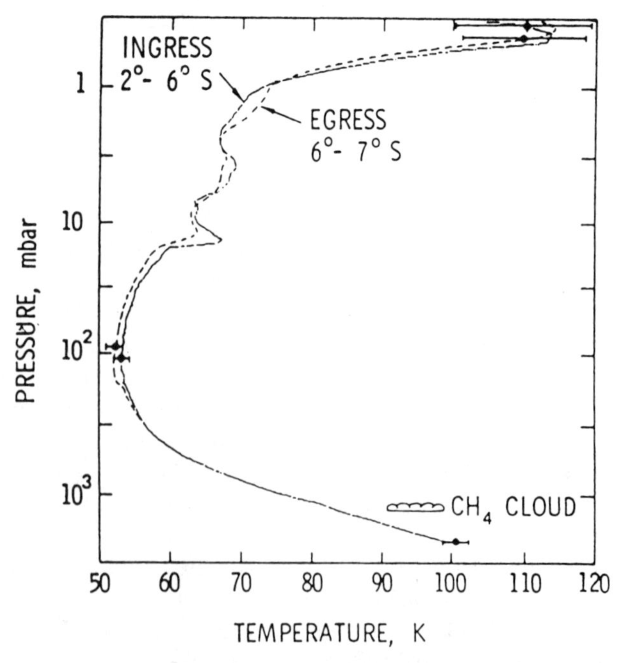

Fig. 3. Temperature profiles of Uranus obtained from Voyager 2 by the radio occultation technique (Lindal et al. 1987). The methane condensation level inferred from an abrupt lapse rate change is indicated.

continuity in the lapse rate near 1.2 bar is attributed to the condensation of CH_4 at that level.

The only information currently available on the vertical structure of the troposphere and stratosphere of Neptune has been derived from Earth-based disk averaged thermal emission measurements (Hildebrand et al. 1985; Moseley et al. 1985; Orton et al. 1986,1987*a*). These results indicate a profile remarkably similar to that of Uranus in the troposphere, but significantly warmer in the upper stratosphere. The similarity of tropospheric temperatures despite the greater distance of Neptune from the Sun is the result of the compensating effect of Neptune's much larger internal heat source. Enhanced absorption of solar energy is presumably responsible for the warmer upper stratosphere on Neptune and may result from enhanced stratospheric methane abundance, photochemically produced gases or smog particles.

Horizontal Structure

Groundbased spatially resolved measurements of Jupiter near 5 μm show substantial horizontal variations in brightness temperature (Westphal 1969; Terrile and Westphal 1977). However, these variations result primarily from cloud opacity gradients and the measurements provide little direct information on the horizontal temperature structure. Very-large-array measurements of Jupiter have recently been published by de Pater (1986). These spatially resolved measurements at millimeter and centimeter wavelengths show interesting latitudinal gradients at pressures greater than one bar which correlate with the belt-zone structure of the planet. NH_3 dominates the opacity at the wavelengths used so it is difficult to separate gradients of ammonia concentration from temperature gradients. Still another set of Earth-based measurements in the 1300 cm^{-1} CH_4 band by Cess et al. (1981) have provided information on upper stratospheric temperature structure. The Voyager IRIS measurements have provided the most detailed horizontal structure information on Jupiter, Saturn and Uranus, and we will confine our attention to this data set here.

Temperature vs latitude for Jupiter is shown in Fig. 4 for two layers approximately one half of a pressure scale height thick centered at 150 and 270 mbar (Gierasch et al. 1986). On the planetary scale, there are no significant horizontal differences between the equator and high latitudes. However, on the scale of the zonal jet system (or, equivalently, the scale of the belt-zone structure) there are latitudinal contrasts of 2 to 3 K. The observed longitudinal contrasts are usually less than this; exceptions occur in the regions of large discrete features such as the Great Red Spot, the white ovals, and the "barges" (Conrath et al. 1981).

Retrieved temperatures for the 150 and 290 mbar levels on Saturn are shown in Fig. 5 (Conrath and Pirraglia 1983). At the deeper level, there is little planetary scale contrast between the equator and high latitudes or between hemispheres. At the higher level there is a distinct asymmetry between hemispheres with little equator-to-pole contrast in the southern hemisphere

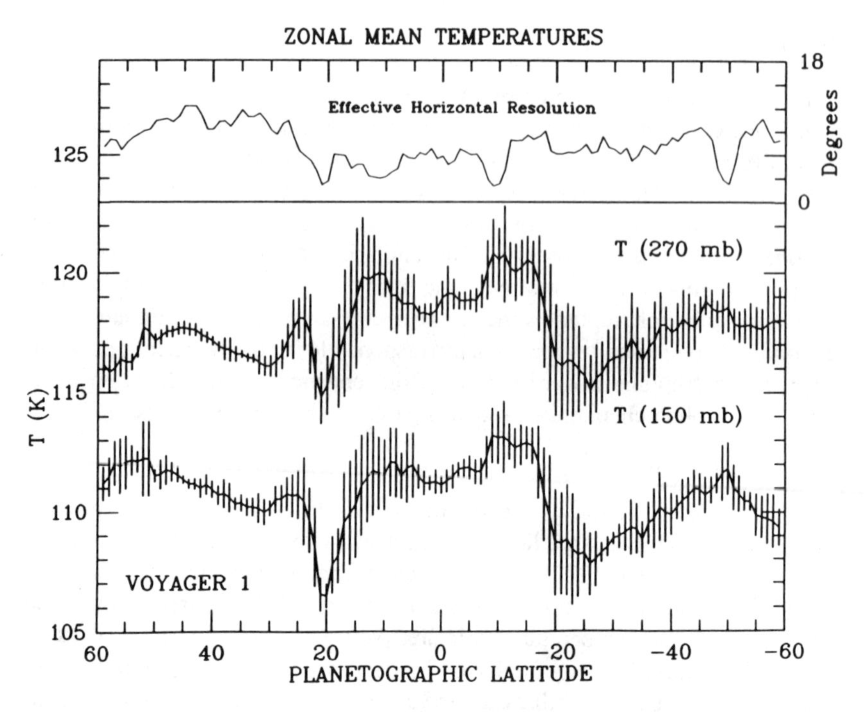

Fig. 4. Zonal mean atmospheric temperature vs latitude for Jupiter obtained from analysis of data from the Voyager 1 infrared spectroscopy experiment (Gierasch et al. 1986). The temperatures are means over layers approximately one-half of a pressure scale height thick centered at the indicated pressures. The vertical bars are the associated standard deviations. The effective spatial resolution of the measurements is indicated in the top panel.

but with a significant contrast in the north. Superposed on the planetary scale structure are thermal contrasts of 4 to 5 K on the scale of the jets.

Figure 6 shows temperatures for two atmospheric layers on Uranus retrieved from IRIS measurements (Flasar et al. 1987). Only small differences are found between both poles and the equator. Relatively weak temperature minima exist at mid latitudes in both hemispheres with stronger gradients at the upper level.

Dynamics

Both the vertical and horizontal temperature structures discussed above have implications for the dynamics of the atmospheres of these planets. The measured horizontal structure permits certain diagnostic statements to be made. On a rapidly rotating planet, the thermal wind relation is expected to hold (Holton 1972). The vertical derivative of the east-west component of the wind is proportional to the north-south temperature gradient on a constant

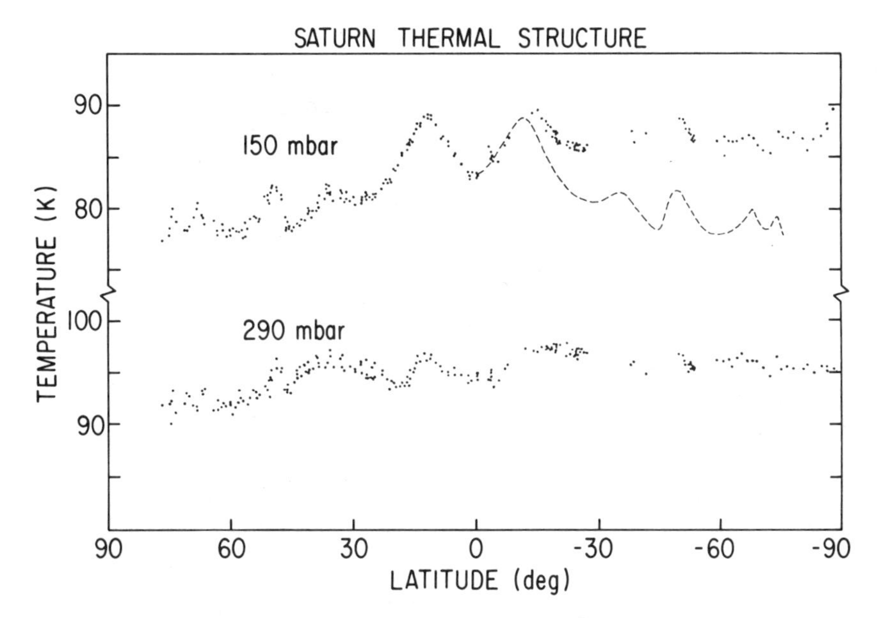

Fig. 5. Atmospheric temperature vs latitude for Saturn obtained from analysis of data from the Voyager 1 infrared spectroscopy experiment (Conrath and Pirraglia 1983). The temperatures are means over layers approximately one-half of a scale height thick. The broken curve presents the 150 mbar temperatures from the northern hemisphere replotted in the southern hemisphere for comparison.

pressure surface. This follows from the assumptions of hydrostatic balance in the vertical and geostrophic balance in the horizontal.

Thermal wind shear calculated from the measured latitudinal structure is shown in Figs. 7, 8 and 9 for Jupiter, Saturn and Uranus, respectively. In each case the results are compared with the zonal winds inferred from apparent cloud motions as observed with the Voyager imaging system. On Jupiter and Saturn the thermal wind shear is anticorrelated with the imaging-derived winds, indicating that the jets decay with height. This suggests that the jets are driven primarily at levels deeper than those observed, and we are seeing the decay of the jet system due to dissipation in the upper atmosphere. The source of this frictional dissipation is not known, but may be associated with the propagation of waves into this part of the atmosphere from deeper levels. On Uranus, the correlation is more difficult to establish because it has been possible to derive winds from cloud motion at only four latitudes (Smith et al. 1986). However, there is an indication that the wind field may also decay with height, at least at mid latitudes in the southern hemisphere. The reversal in sign of the thermal wind at low latitudes implies that this region is subrotating in contrast with Jupiter and Saturn whose equatorial regions superrotate.

The lack of large-scale thermal contrast on Jupiter suggests the existence

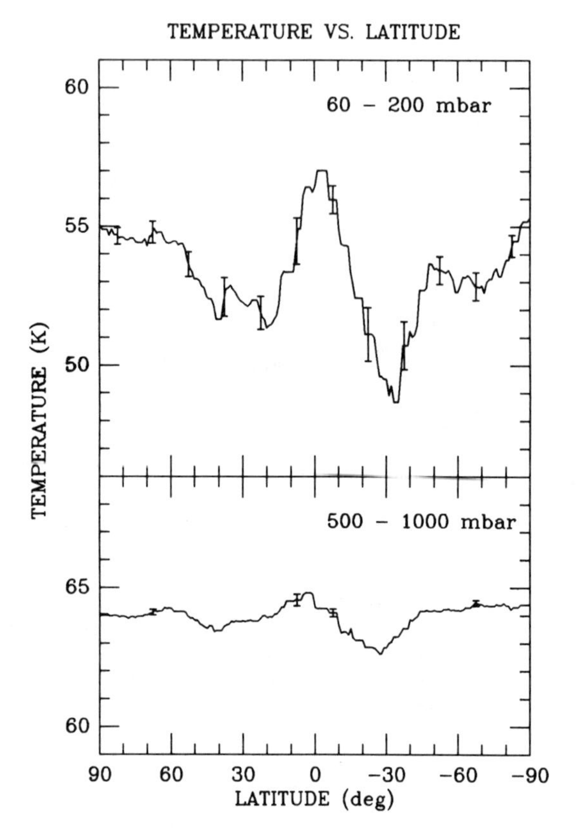

Fig. 6. Atmospheric temperature vs latitude for Uranus obtained from analysis of data from the Voyager 2 infrared spectroscopy experiment (Flasar et al. 1987). The temperatures are means over the layers indicated. The bars represent the formal errors due to instrument noise.

of an efficient mechanism for meridional heat transport. Since the solar energy is deposited preferentially near the equator, this energy must be redistributed or compensated for in some way. Ingersoll and Porco (1978) have suggested that the internal heat flux may adjust itself as a function of latitude so that it just compensates for the differential solar heating. In this view, the adiabatic interior with efficient eddy heat conduction effectively shorts out the atmospheric thermal gradient which would otherwise exist. According to Ingersoll and Porco, the internal heat flux must be at least 27% of the absorbed solar flux, or the mechanism does not work at all latitudes. This condition is met at Jupiter and Saturn.

The hemispheric temperature asymmetry observed in the upper troposphere on Saturn is apparently due to the seasonal asymmetry in solar heating in the statically stable upper troposphere. At the time of the Voyager

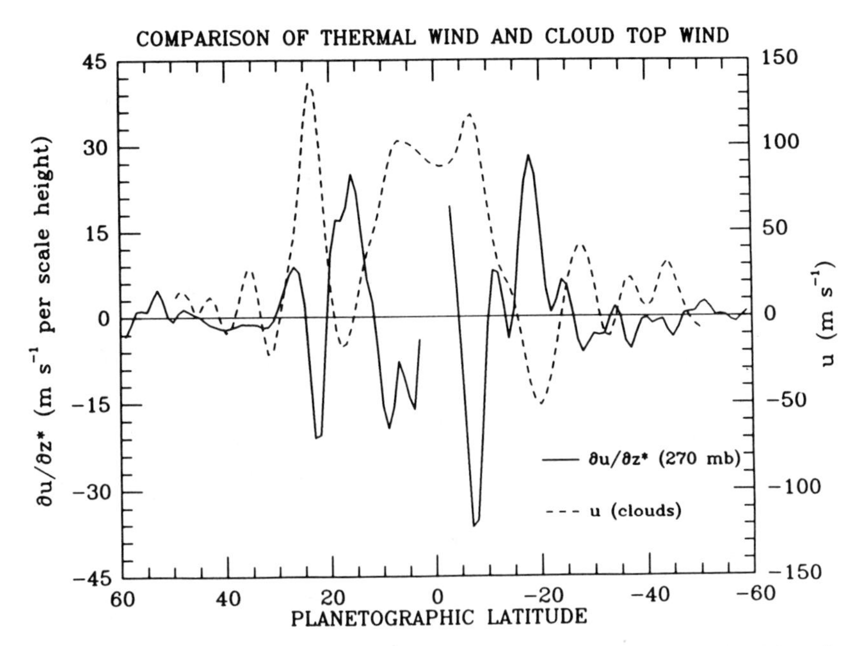

Fig. 7. Thermal wind shear (solid curve) for Jupiter at the 270 mbar level calculated from the latitudinal thermal structure shown in Fig. 5 (Gierasch et al. 1986). The broken curve is the cloud-top wind speed derived from cloud motion observed in Voyager images.

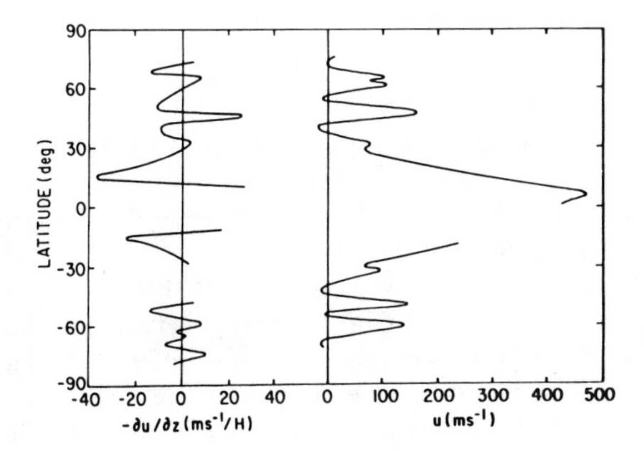

Fig. 8. Thermal wind shear (left) for Saturn at the 150 mbar level calculated from the temperatures shown in Fig. 6 (Conrath and Pirraglia 1983). For comparison, the cloud-top wind speed derived from Voyager images is shown on the right.

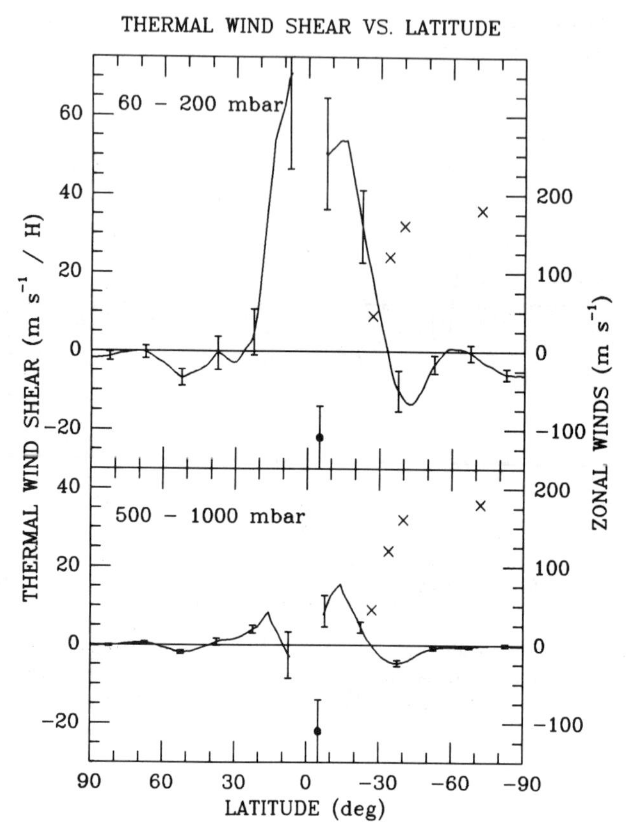

Fig. 9. Thermal wind shear for Uranus calculated from the temperatures shown in Fig. 7 (Flasar et al. 1987). The bars represent uncertainties due to instrument noise. The crosses are cloud-top zonal wind speeds derived from Voyager images. The solid circle with error bars is the zonal wind speed inferred from radio occultation data (Lindal et al. 1987).

encounters, the Sun had just crossed the equator into the northern hemisphere, and because of the long radiative time constant of the atmosphere, this hemisphere lags in its thermal response. The strength of the effect is apparently due to the relatively large obliquity (27°) of Saturn; Jupiter, on the other hand, has an obliquity of only about 3° so a strong seasonal signature is not expected. The implied differential absorption of solar energy in the upper troposphere of Saturn suggests that Jupiter should also experience differential solar heating in its upper troposphere; however, planetary-scale temperature gradients are not observed in this part of the atmosphere. Since the Ingersoll and Porco mechanism can operate only up to the top of the convective region, this implies the existence of some other dynamical means of redistribution of heat in this part of the atmosphere. Using Voyager IRIS observations of the para H_2 fraction as a tracer of motion in the Jovian troposphere, Conrath and Gierasch (1984) suggested the possibility of a weak equator to high latitude meridional cell which could redistribute heat poleward.

Uranus also shows little large-scale meridional thermal contrast. Because the internal heat flux lies well below 27% of the absorbed solar flux, the Ingersoll and Porco mechanism presumably cannot account for this behavior. Other dynamical processes within the atmosphere must be responsible for the redistribution of heat. Possible mechanisms include the existence of a baroclinic eddy regime (Friedson and Ingersoll 1987) or a zonally symmetric meridional circulation (Read 1986; Flasar et al. 1987).

A complete explanation has yet to be found for the observed lapse rates which exceed the equilibrium adiabatic lapse rates near the tops of the convective zones of the outer planets. Conrath and Gierasch (1984) applied mixing-length theory to the problem of buoyancy convection in the presence of para-to-ortho hydrogen conversion, and concluded that a regime of very slow overturning should exist in which much of the vertical heat flux is carried by the para hydrogen *latent* heat. However, if such a regime existed, then the lapse rate should be close to the equilibrium hydrogen lapse rate in disagreement with observations. One alternative considers that molecular weight stratification in the regions immediately above the condensation level of a relatively heavy constituent may enhance the static stability, permitting steeper lapse rates to exist. Flasar (1986) has suggested this as a possible mechanism operative above the CH_4 condensation level on Uranus. Another possibility is that rapid convective overturning occurs in very thin layers on a time scale short compared to the H_2 equilibration time. Within the layer this would result in a constant ortho-para ratio close to the equilibrium value for the mean temperature for the layer and the lapse rate would lie close to the frozen equilibrium value which is steeper than the equilibrium lapse rate. This possibility has been recently explored by Gierasch and Conrath (1987). This problem must be solved before a complete understanding of the energy transport through the upper part of the convective zone can be obtained.

Comparative studies of the dynamics of Uranus and Neptune should be of particular interest. The two planets are of approximately the same size, they have similar atmospheric composition and temperature, and their radiative relaxation times are comparable. However, there are major differences in the parameters that may control the dynamical forcing. Neptune has an obliquity of 29° and an internal heat source which is large relative to the solar energy absorbed by its atmosphere while Uranus has an extreme obliquity of 97° and a relatively small internal heat source. At present there is insufficient information to permit detailed diagnostic studies of the dynamics of Neptune's atmosphere, but the forthcoming Voyager encounter should substantially increase our knowledge in this area.

III. ENERGY BALANCE

All planets emit large amounts of thermal radiation; some radiate nearly twice as much as they receive from the Sun. Planetary size and effective temperature characterize the total emission, while cross section, bolometric Bond albedo and distance of the planet from the Sun determine the amount of absorbed solar energy. The bulk of the thermal energy is radiated in the middle and far infrared, while solar radiation is strongest in the visible and near infrared part of the spectrum. Planetary rings, and to a lesser degree nearby satellites, reflect and emit radiation as well and may affect the energy balance of a planet. Fluxes from stars, galaxies and the cosmic background are completely insignificant in this context.

The energy balance of a spherical planet is expressed by

$$4\pi R^2 \, \sigma T_{\mathrm{eff}}^4 = E_{\mathrm{int}} + \pi R^2 \,(1-A)\, \pi S/D^2. \qquad (1)$$

The term on the left represents emitted power; R is the planetary radius, σ the Stefan-Boltzmann constant, and T_{eff} the effective planetary temperature. A blackbody at that temperature and the same size as the planet emits the same spectrally integrated power as the planet. The first term on the right, E_{int}, represents internal heat sources. The last term in Eq. (1) characterizes absorbed solar power; πS is the solar constant (1374 watt m^{-2}), A the bolometric Bond albedo and D the distance of the planet from the Sun in astronomical units (1 AU $= 149.6 \times 10^9$ m).

On the Earth and on the other terrestrial planets radioactive decay of uranium, thorium, and potassium is the dominant internal heat source, but the magnitude of this source is only ~0.025% of the absorbed solar radiation and, in first order, may be neglected. For the terrestrial planets, emitted thermal and absorbed solar radiation are almost balanced. In contrast, the outer planets have retained varying degrees of their internal energy brought about by original gravitational collapse, and are far out of balance. Jupiter, Saturn and Neptune emit nearly twice as much energy as they absorb from the Sun, while Uranus has a relatively small internal heat source.

A measurement of the energy balance of a planet requires a precise determination of the radiative fluxes involved. This requires measurements of the spectrally integrated thermal emission and the absorbed solar radiation, characterized by the Bond albedo. The internal power is then found from Eq. (1). We discuss for each of the giant planets the thermal emission measurements, the observations that lead to the Bond albedo and finally the internal heat sources.

Thermal Emission Measurements

A measurement of the total thermal emission must take into account the spectrally integrated planetary flux in all directions. The spectral integration has to be carried out from a few wavenumbers (cm^{-1}) to an upper limit which depends on the effective temperature and on details of the emission spectrum of a particular planet. For Jupiter, the integration should be carried out to at least 2300 cm^{-1} to fully account for emission from the 5-μm hot spots. Above 2300 cm^{-1} thermal emission is rather insignificant and may be accounted for by an estimate derived from model calculations. On Saturn, it is sufficient to integrate up to 1400 cm^{-1}, just far enough to include stratospheric emission from the 1304 cm^{-1} CH_4 band. On Uranus and Neptune, integration up to 500 or 600 cm^{-1} is more than adequate due to the low temperatures of the emitting atmospheric layers and the steep slope of the Planck function at low temperatures and high wavenumbers.

The best approach to measure the emission spectrum uses spectrometers which have been calibrated on an absolute radiometric scale. Unfortunately, present instruments of this type, such as the IRIS on Voyager, only cover the spectral range above ~200 cm^{-1}, requiring extrapolation to lower wavenumbers. However, this extrapolation can be accomplished with reasonable confidence since the same spectra used for the spectral integration yield, simultaneously, atmospheric temperature profiles and composition information needed in the radiative transfer calculations of the extrapolation process. Integration of well-calibrated spectral data and of reliable extrapolation models has so far provided the best estimates of the thermal emission of the giant planets.

Groundbased measurements require a calibration source outside the Earth's atmosphere, and reference stars, as well as Mars, are sometimes used for that purpose. The stellar objects in particular have entirely different spectral characteristics than the giant planets; these characteristics must be well known if one wishes to use these sources for calibration purposes. The far infrared spectrum of Mars is not well known either and, moreover, is variable; the surface temperature and atmospheric opacity of Mars depend on the atmospheric dust content, which is subject to quasi-seasonal variations.

Integration over a wide spectral range is not the only requirement for obtaining the total infrared flux. One must also integrate planetary radiation over 4π sterradians. It is reasonable to expect the flux to depend on latitude but less strongly on longitude and solar phase angle. Because the radiative time constants of the atmospheric layers involved in the thermal emission (Jupiter ~6.6 yr; Saturn ~24 yr, Uranus and Neptune ~166 yr) are so much longer than the rotation periods of the giant planets, diurnal temperature variations must be insignificant. As far as latitude coverage is concerned, Earth-based observers are at a disadvantage, although a limited latitude range can be observed if one is willing to extend the observations over at least one half of the orbital period of a planet with a considerable obliquity. Uranus, for example, may be observed from Earth over all latitudes in a 42-year period. In all other cases spacecraft measurements are required to observe emission from high latitudes. Fortunately, the limited information available from spatially resolved latitudinal measurements from Pioneer and Voyager indicates that Jupiter, Saturn and Uranus have only small latitudinal dependences of atmospheric temperatures on constant pressure levels. Thus, the emitted flux is also essentially independent of latitude. As discussed previously, it has been suggested that internal mechanisms are at work, which equalize the emission over the surfaces of Jupiter and Saturn (Ingersoll and Porco 1978). Therefore, the directional dependence of the thermal flux seems to be rather weak and it has been neglected at the present stage of precision of thermal flux measurements. For Uranus, the measured latitudinal temperatures are more uniform than predicted from radiative transfer models without incorporation of latitudinal energy transport (Bézard and Gautier 1985*a*; Hanel et al. 1986). The observed gradients were included in the determination of the total thermal emission (Pearl et al. 1988). In the rapidly rotating outer planets the deviation from spherical shape has to be considered as well as special circumstances, such as the rings of Saturn or the orientation of Uranus' spin axis (Hanel et al. 1983; Pearl et al. 1988).

The first measurement of the radiative temperature of Jupiter in the 8 to 12 μm atmospheric window was carried out by Menzel et al. (1926). They obtained 130 ± 10 K, a remarkable result, considering the relatively primitive instrumentation available at that time. However, they did not recognize the presence of an internal energy source. Low (1965) pointed out that Jupiter radiates as much energy between 5 and 25 μm as it receives from the Sun, which implies a substantial internal energy source. Results of effective-temperature measurements of Jupiter are summarized in Table I. The clustering effect near 135 K, noticeable in all measurements between 1969 and 1974, may be due to systematic errors introduced in the calibration process. Aumann et al. (1969) and Armstrong et al. (1972) used stars (α Ori, α Boo and γ Dra) in that process. The spacecraft measurements (Pioneer, 1976 and

TABLE I
Thermal Emission Measurements of Jupiter

Effective Temperature, K	References
130 ± 10 at 8–12 μm	Menzel et al. 1926
128 ± 2.3	Murray and Wildey 1963
132 ± 6 at 8–12 μm	Low 1965
127 ± 6 at ~ 20 μm	
134 ± 4	Aumann et al. 1969
134 ± 4	Armstrong et al. 1972
135 ± 4	Trafton and Wildey 1974
136 ± 5	Murphy and Fesen 1974
125 ± 3	Ingersoll et al. 1975
123 ± 2	Erickson et al. 1978
124.4 ± 0.3	Hanel et al. 1981*a*

TABLE II
Thermal Emission Measurements of Saturn

Effective Temperature, K	References
97 ± 4	Aumann et al. 1969
99 ± 4	Nolt et al. 1974
± 3	
85 ± 2	Wright 1976
89 ± 3	Ward 1977
97 ± 3	Erickson et al. 1978
~95	Courtin et al. 1979*b*
99.3 ± 4.6	Gautier and Courtin 1979
96.5 ± 2.5	Orton and Ingersoll 1980
96.8 ± 2.5	Haas et al. 1982
96.1 ± 1.6	Melnick et al. 1983
95.0 ± 0.4	Hanel et al. 1983

Voyager, 1981) are in agreement with each other and with the aircraft measurements of Erickson et al. (1978). The spacecraft measurements depend for calibration on radiation from deep space and an on-board blackbody in the case of Pioneer (Ingersoll et al. 1975), and on deep space and on an isothermal, well-thermostated instrument in the case of Voyager (Hanel et al. 1980). Erickson et al. (1978) used Mars as a calibration source. The historical evolution of the Saturnian effective-temperature measurements is shown in Table II. Here again the most recent Voyager measurements are slightly lower than some earlier airborne and groundbased observations, but in many cases within the quoted error bars. Table III shows the same type of information for Uranus

TABLE III
Thermal Emission Measurements of Uranus

Effective Temperature, K	References
58.3 ± 3.0	Fazio et al. 1976
58.0 ± 2.0	Loewenstein et al. 1977
58.5 ± 2.0	Stier et al. 1978
57.0 ± 2.5	Courtin et al. 1979*b*
58.6 ± 2.0	Hildebrand et al. 1985
57.7 ± 1.8	Moseley et al. 1985
57.7 ± 2.0	Pollack et al. 1986*b*
59.1 ± 0.3	Pearl et al. 1988

TABLE IV
Thermal Emission Measurements of Neptune

Effective Temperature, K	References
55.0 ± 2.3	Loewenstein et al. 1977
59.7 ± 4.0	Stier et al. 1978
60.3 ± 2.0	Hildebrand et al. 1985
58.2 ± 1.9	Moseley et al. 1985

and Table IV for Neptune. The most recent estimates of the effective temperature and the total emitted power (that is, the left-hand term of Eq. 1) are summarized in Table V.

Bolometric Bond Albedo

A determination of the planetary albedo also requires many individual flux measurements. As for thermal emission, integrations have to be accomplished in the spectral domain and over 4π sterradian, but in this case there is a strong dependence on the solar angle. To obtain the absorbed solar power (that is, the second term on the right-side of Eq. 1), the bolometric Bond albedo A needs to be found. The other quantities in this expression, R, S and D, are relatively well known. However, a measurement of A, that is, of the fraction of the solar power intercepted by the planet and scattered into 4π sterradian is a nontrivial task. The reflected, or scattered radiation is

$$P_{\text{refl}} = \pi R^2 \int_{o}^{2\pi} \int_{o}^{\pi} \bar{I}(\theta,\phi) \sin\theta \, d\theta \, d\phi \tag{2}$$

TABLE V
Thermal Emission Parameters

Parameter	Jupiter[a]	Saturn[a]	Uranus[a]	Neptune[b]
Effective Temperature (K)	124.4 ± 0.3	95.0 ± 0.4	59.1 ± 0.3	58.3 ± 2.0
Total Emitted Power (10^{16} watt)	83.65 ± 0.84	19.77 ± 0.32	0.563 ± 0.012	0.484 ± 0.066

[a]Data from Voyager.
[b]Groundbased data (Table IV).

where $\bar{I}(\theta,\phi)$ is the mean disk intensity measured from the direction given by the Sun angle θ and an azimuth angle ϕ. In reality, it is very difficult to obtain a sufficiently dense set of full-disk measurements $\bar{I}$ to evaluate Eq. (2). Radiative transfer calculations may be used to find $\bar{I}$, provided one can specify sufficiently well the reflection function of all planetary surface elements and the dependences of that function on the Sun, emission and azimuth angles. Such calculated values of $\bar{I}$ may serve as an interpolation function between sparce data points provided by measurements; however, such calculations, involving multiple scattering, are very time consuming, although the physical principles are understood. What is not well known is how to describe a real atmosphere containing haze and clouds by a limited set of parameters and to derive a realistic reflectivity function. Only comparisons of radiative transfer models with planetary measurements promise progress in this area.

It has been common to assume $\bar{I}$ to be independent of the azimuth angle ϕ. This assumption seems to be justified in first order for the outer planets. It is less well justified for dark planets with bright polar caps, such as the Earth or Mars, for example. With the assumption that the disk intensity does not depend on the azimuth angle ϕ, the reflected power is

$$P_{\text{refl}} = 2\pi^2 R^2 \int_{o}^{\pi} \bar{I}(\theta) \sin\theta \, d\theta. \tag{3}$$

The incident solar power intercepted by the planet is simply the planetary cross section times the solar flux at the particular heliocentric distance,

$$P_{\text{incid}} = \pi R^2 \, \pi S/D^2. \tag{4}$$

The Bond albedo may be defined as the ratio

$$A = \frac{P_{\text{refl}}}{P_{\text{incid}}} = \frac{2}{S/D^2} \int_{o}^{\pi} \bar{I}(\theta) \sin\theta \, d\theta. \tag{5}$$

It has become customary to multiply and divide by $\bar{I}(o)$, and express the albedo as a product,

$$A = p \; q = \frac{\bar{I}(0)}{S/D^2} \; 2 \int_{o}^{\pi} \frac{\bar{I}(\theta)}{\bar{I}(0)} \sin\theta \, d\theta. \tag{6}$$

The first factor p, the geometric albedo, is accessible to Earth-based measurements, even for the outer planets. The second factor q, the phase integral, requires measurements of the phase function $\bar{I}(\theta)/\bar{I}(0)$ which can be accomplished only from space-based platforms. If the measurements of p and q are performed over narrow spectral intervals one obtains the spectral albedo. To find the Bond albedo one has to integrate the spectral albedo over the solar spectrum. This integration may be performed analytically if the spectral albedo values are available or, alternatively, at the detector of the measuring instrument, provided the response function of that flux detector is spectrally flat over the solar spectrum from approximately 0.2 to 4 μm. Radiometers on Pioneer and on Voyager fall somewhat short of the uniform response requirement. In that case, the actual instrument response function and a knowledge of the relative planetary spectrum have to be used to apply corrections.

Equally important for finding the proper level of the geometric albedo is the absolute calibration. To eliminate the effect of atmospheric absorption from groundbased measurements, it is necessary to compare the planetary radiation to that from a standard source. Ultimately this source must be the Sun; however, the difference in brightness between planet and Sun is so immense that the same instrument cannot be employed directly for both. The calibration has to be transferred via auxiliary objects. For groundbased observations, it is customary to use stars, hopefully of the same spectral type as the Sun, and certain standard areas on the Moon (Neff et al. 1984; McCord et al. 1972). Pioneer used groundbased measurements of Jupiter and Saturn for calibration purposes. The Voyager spacecraft carried a large, flat calibration target. The surface of this aluminum plate was chemically etched to produce a good Lambert diffusor over the wavelength range of interest. The spectral characteristic of this plate was measured before launch, but this calibration is not without problems. Although the IRIS radiometers have not changed significantly between the first calibrations before the Jovian encounters and the most recent calibration after the encounter with Uranus, changes in the characteristics of the diffusor plate between the laboratory measurements and the first space-borne measurements cannot be ruled out.

The geometric albedo values for the outer planets are shown in Table VI. The data for Jupiter, Saturn and Uranus have been obtained by Voyager (Hanel et al. 1981*a*, 1983; Pearl et al. 1988). The value of p for Neptune is obtained from groundbased measurements (Neff et al. 1985). Groundbased measurements of Uranus (Lockwood et al. 1983; Neff et al. 1985) give a somewhat higher value (0.28 $\pm$ 0.02) of p than the Voyager-IRIS measurements (0.208 $\pm$ 0.048), indicating a possible systematic difference in the calibrations between these observations.

A measurement of the phase integral is more difficult. Radiometric data must be obtained over a wide range of phase angles. For Voyager, the IRIS radiometer could not obtain raster scans of the planets at many phase angles. With a flyby trajectory and many demands on the pointing of the spacecraft platform, not enough time was available to measure the phase function properly. In this respect the photopolarimeter on the spinning Pioneer spacecraft was better suited for this particular measurement. Unfortunately, the spectral range of the Pioneer instrument was limited to two relatively narrow intervals

TABLE VI
Parameters Necessary to Derive the Absorbed Solar Radiation

Parameter	Jupiter[a]	Saturn[a]	Uranus[a]	Neptune[b]
Geometric Albedo	0.274 ± 0.013	0.242 ± 0.012	0.208 ± 0.048	0.25 ± 0.02[c]
Phase Integral	1.25 ± 0.10	1.42 ± 0.10	1.40 ± 0.14	1.25 ± 0.10[d]
Bond Albedo	0.343 ± 0.032	0.342 ± 0.030	0.290 ± 0.051	0.31 ± 0.04
Absorbed Power (10^{16} watt)	50.14 ± 2.48	11.14 ± 0.50	0.533 ± 0.038	0.192 ± 0.010

[a]Voyager, Pioneer.
[b]Groundbased data.
[c]Neff et al. 1985.
[d]Pollack et al. 1986*b*.

in the blue and red part of the spectrum (Tomasko et al. 1978). On objects that are more uniform than Jupiter, such as Uranus, determination of the geometric albedo and the phase function from several data sets, including IRIS radiometer measurements, is possible (Pearl et al. 1988). The Voyager imaging system can also be used to obtain phase functions. Even from a great distance, when Uranus illuminated only a single pixel of the Voyager 1 camera, a determination of the phase function was obtained (Pollack et al. 1986*b*). Again the spectral interval of the camera and the photometric calibration of the Vidicon tube set limits to the precision of the measurements.

The phase integrals shown in Table VI are from Pioneer (Tomasko et al. 1978), Voyager (Pearl et al. 1988) and for Neptune from cloud model calculations (Pollack et al. 1986*b*). The Bond albedo is simply the product $p\ q$ except for Uranus where it was determined directly. The error bar quoted by the individual authors for the Bond albedo treat the errors of the geometric albedo and that of the phase integral as statistically independent. This procedure may be questioned when the same instrument is used to derive both quantities. However, the main sources of errors in the determination of the geometric albedo are concerned with the absolute calibration, which does not enter the determination of the phase function. Clearly, the discrepancy between the groundbased and Voyager estimates of the albedo of Uranus points towards unrecognized systematic differences in the respective calibrations. With the Bond albedo established, along with the solar constant, the planetary dimensions and the mean heliocentric distances, the total solar power absorbed by the planet is obtained; it is given in the last line of Table VI. More elaborate evaluation of presently available data, more sophisticated radiative transfer models, and carefully planned future measurements are required to resolve present inconsistencies and decrease existing uncertainties.

TABLE VII
Energy Balance

Parameter	Jupiter	Saturn	Uranus	Neptune
Internal Power (10^{16} watt)	33.5 ± 2.6	8.63 ± 0.60	0.027 ± 0.040	0.29 ± 0.07
Energy Balance	1.67 ± 0.09	1.78 ± 0.09	1.05 ± 0.08 (<1.13)	2.52 ± 0.37
Internal Power/Mass (10^{-12}watt kg^{-1})	176 ± 14	152 ± 11	3.09 ± 4.60	29.6 ± 7.0
Luminosity log ($L/L_\odot$)	-9.062 ± 0.034	-9.651 ± 0.030	$-12.159^{+0.396}_{-\infty}$	-11.125 ± 0.094

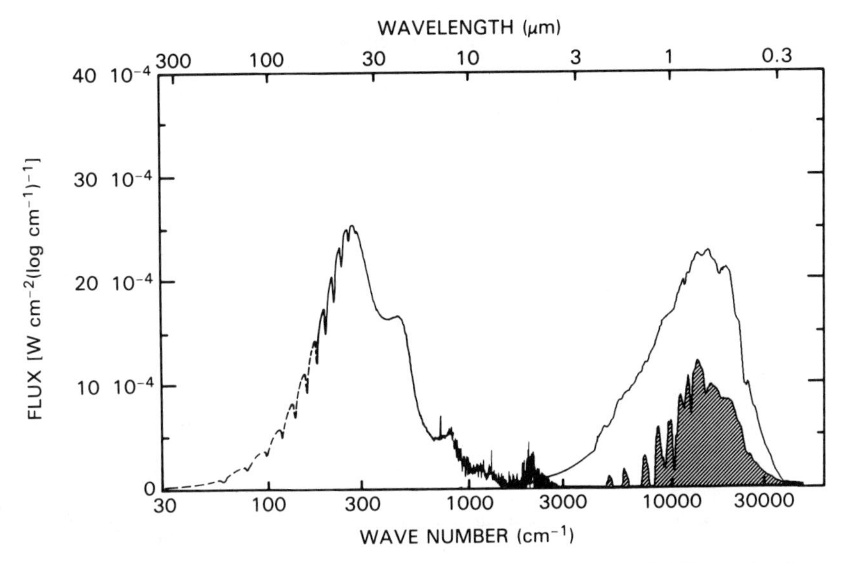

Fig. 10. Energy balance of Jupiter. The thermal emission between 230 and 2300 cm^{-1} is from Voyager-IRIS; below 230 cm^{-1} the emission is derived from model calculations constrained to match the observations between 200 and 300 cm^{-1}. The reflected solar spectrum (lower curve above 3000 cm^{-1}) was calculated from the solar spectrum at the distance of Jupiter (upper curve) and a Bond albedo of 0.343 (Hanel et al. 1981*a*).

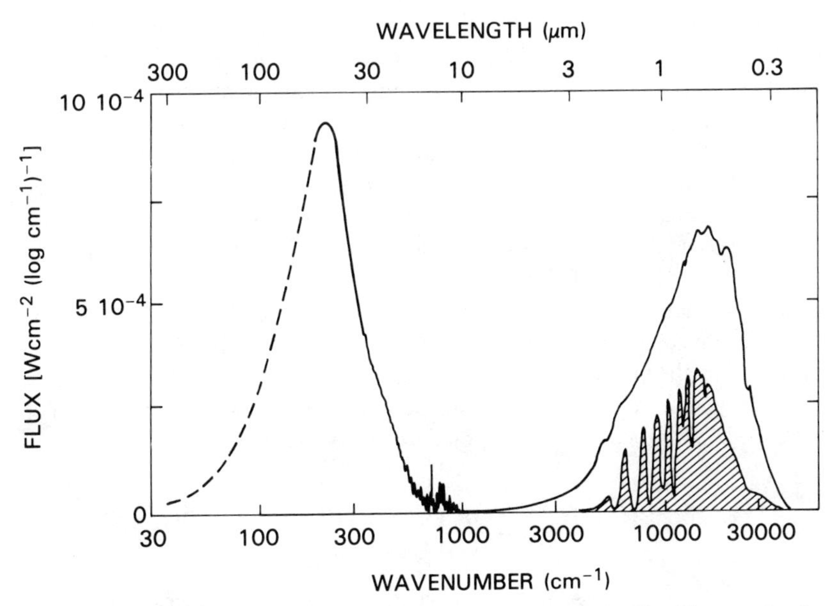

Fig. 11. The energy balance of Saturn. The format is the same as for Fig. 10 except for the vertical scale (Hanel et al. 1983).

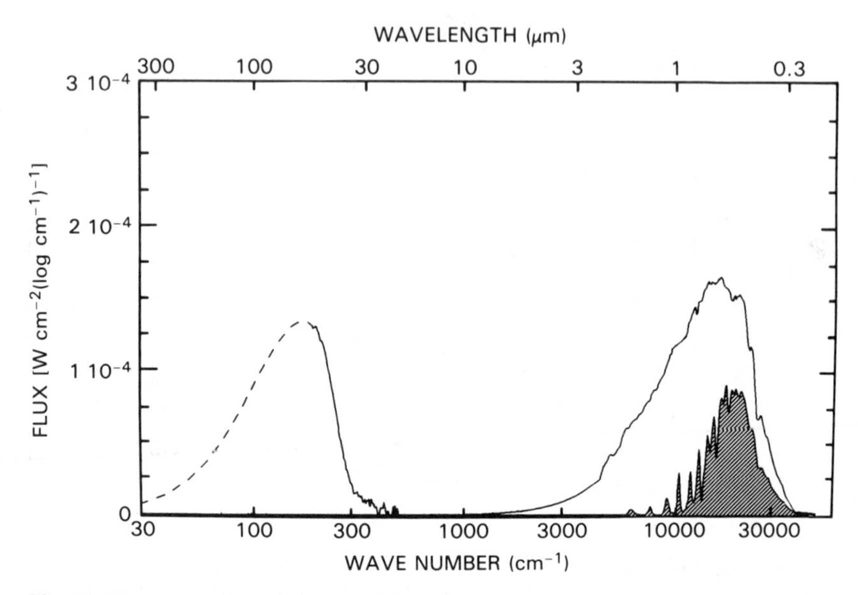

Fig. 12. The energy balance of Uranus. The format is the same as for Fig. 10 except for the vertical scale (Pearl et al. 1988).

Internal Energy

As shown in Eq. (1), the total emitted (Table V) and absorbed (Table VI) power determine the energy balance terms. Subtracting the absorbed from the emitted power yields the internal power; this is shown in the first line of Table VII. The second line of the same Table gives the ratio of the emitted radiation to the absorbed solar radiation, generally referred to as the energy balance of a planet. The third line of the Table provides the internal power per planetary mass and the last line, the planetary luminosity compared to that of the Sun.

The energy balance terms for Jupiter, Saturn and Uranus are displayed in Figs. 10 to 12. The difference between the area representing the solar flux available at the heliocentric distance of the planet (solid curve on the right) and the reflected solar radiation (hatched area on the right) is the solar flux absorbed by the planet. For Jupiter and Saturn this area is clearly smaller than the area representing the emitted flux (solid and dashed curve on the left). For Uranus the absorbed and emitted flux are nearly the same.

IV. IMPLICATIONS FOR INTERIOR STRUCTURE AND EVOLUTION

Several parameters of importance for testing evolutionary models are available from studies of the heat balance and thermal structure of the major planets. Both the helium abundance and the temperature at some reference level (1 bar, say) follow from an analysis of the infrared spectra and the radio occultation data. The excess thermal flux is inferred from a measure of the total emitted and total absorbed flux. These quantities serve as boundary conditions for the present epoch in the construction of evolutionary models.

Jupiter, Saturn and Neptune emit about twice as much thermal radiation as they would if they were in equilibrium with the Sun. This excess of thermal radiation originates in the interior of the planet, and is almost entirely derived from the release of gravitational potential energy. Initially, the protoplanet undergoes hydrodynamic collapse. As contraction continues, gravitational potential energy is converted into internal energy and stored as heat. The interior is highly compressible during this stage. Later, as the interior becomes less compressible, contraction slows down, and the stored internal energy begins to be liberated at a faster rate than it is being created by conversion from gravitational potential energy (Flaser 1973; Cameron and Pollack 1976; Pollack et al. 1977). At this point the interior begins to cool.

If the hydrogen envelope extends deeply enough, high pressures will cause the lower layers to dissociate and ionize, and hydrogen takes on the character of a fluid metal. If cooling continues and the temperature becomes low enough, the helium present becomes partially immiscible and will begin to separate from the metallic hydrogen; helium droplets will begin to form. Once they have reached about a centimeter in size, they will precipitate to lower, hotter levels where they will be reabsorbed into the surrounding medium. The net effect is to enrich the lower levels with helium at the expense of the upper levels (Salpeter 1973; Smoluchowski 1973; Hubbard and Smoluchowski 1973; Stevenson and Salpeter 1976). Because helium is more dense than hydrogen, there will be an increase of gravitational potential energy available for conversion to heat. This, in turn, will lengthen the time it takes to cool the interior.

On Jupiter the hydrogen envelope, mixed with helium, extends almost to the center of the planet. Hydrogen is in the metallic phase in the interior, but probably at too high a temperature to separate from helium. Evolutionary models that ignore chemical differentiation can be found that are consistent with the observed Jovian heat flux, implying that the interior is still too warm for helium separation to have begun (Graboske et al. 1975*b*). However, as discussed below, the most recent measurements of the helium abundance of Uranus suggest that separation may already have begun on Jupiter.

On Saturn hydrogen also extends to high enough pressures to enter into the metallic phase. However, the interior temperatures are sufficiently low to allow the onset of helium separation. Evolutionary models that ignore chemical differentiation appear to predict a flux excess about a factor three lower than observed for Saturn (Grossman et al. 1980). Apparently helium separation is required to extend the cooling time in order to raise the emitted flux predicted for the present epoch to the observed value. Considerable uncertainty exists, however, due in part to inadequately known equations of state at the temperatures and pressures of interest.

Hydrogen apparently does not extend deeply enough on Uranus and Neptune to ionize, and helium separation cannot occur. Even so, Neptune appears to have a substantial excess thermal flux while Uranus does not, implying there is some fundamental difference between these two otherwise similar planets. Hubbard (1978,1980) has used a convective-cooling, evolutionary model and concludes the larger amount of solar radiation absorbed in Uranus' atmosphere has provided an effective mechanism to enhance the escape of interior heat relative to that of Neptune. He shows the model to be consistent with the observed thermal balance of the two planets.

In addition to the internal heat flux, evolutionary models predict a planet's radius at any epoch. By increasing the size of the "silicate" core (comprised of SiO_2, MgO, Fe and Ni, for example) the radius can be made smaller for a given internal heat flux. Grossman et al. (1980), using pre-Voyager data to infer the temperature at 1 bar, derive core masses for Jupiter and Saturn ~19 times the mass of the Earth. Hubbard and McFarlane (1980*a*) suggest all outer planets except Pluto contain dense cores ~15 times the Earth's mass. Whether or not the formation of such similar dense cores throughout the outer solar system has an important bearing on solar system evolution is an intriguing, if open, question. Clearly, questions of this type can be better answered as more precise observational constraints become available.

Observed values of the helium mass fractions (in percent) for the outer layers of Jupiter, Saturn and Uranus are $Y_j = 18 \pm 4$, $Y_s = 6 \pm 5$ and $Y_u = 26.7 \pm 4.8$; the value for Uranus is very close to that of the Sun (Conrath et al. 1987). Helium depletion is clearly advanced for Saturn and may have started for Jupiter, although the listed uncertainties weaken the case for the latter planet. If helium separation is taken into account in Jovian models, the outgoing flux calculated for the present epoch may exceed the observed flux unless the theoretical treatment is modified in other ways. Two areas where modeling advances are needed are in improved equations of state for hydrogen-helium mixtures at high temperatures and densities, and in quantification of the way in which the vertical distribution of helium develops once helium separation begins.

STRUCTURE AND COMPOSITION OF GIANT PLANET INTERIORS

W. B. HUBBARD
University of Arizona

In the simplest model of a Jovian planet atmosphere, the atmospheric abundances are identical to the bulk interior abundances, except as modified by atmospheric condensation processes. This model is now known to be generally inadequate, on the basis of comparisons between detailed atmospheric composition measurements and (less-detailed) determinations of interior composition. The latter are primarily deduced by integrating high-pressure equations of state of plausible constituents to obtain interior models which satisfy observational constraints such as mass, radius, gravitational moments, luminosity and age. This chapter reviews the status of discrepancies between such determinations of interior and atmospheric composition, and reviews possible explanations via interior processes such as hydrogen-helium immiscibility and phase transitions in major constituents. We discuss the proposed structure of the core, mantle and deep atmosphere for each of the four giant planets.

I. INTRODUCTION

The atmospheres of the four giant planets were, until relatively recently, believed to be very similar in elemental composition to the solar atmosphere (Weidenschilling and Lewis 1973). This similarity was taken to be a consequence of the high surface gravities and low atmospheric temperatures of these planets, affording them the possibility of retaining hydrogen-rich atmospheres without significant loss since their time of formation. In this initial and oversimplified picture, any observed differences in the elemental abundances of giant planet atmospheres were to be understood in terms of precipitation of minor species in regions of the atmospheres where these species become condensible, with condensation regions changing systematically and predictably as a result of changing temperature structure.

Although condensation of minor species does affect the observed atmospheric abundances, after this is allowed for, one finds that the four giant planets have atmospheric compositions that differ intrinsically from solar, and from each other. Since atmospheric distillation over the age of the solar system cannot be responsible for these differences, we must attribute them to (a) elemental sorting processes that occurred at the time of formation of the planets, leading to bulk differences in initial composition between the giant planets, and (b) elemental sorting processes occurring within the planets subsequent to formation, preserving the bulk composition but causing the atmosphere and perhaps deeper layers to deviate from it.

There is now substantial evidence that both processes (a) and (b) are significant in the giant planets. It has been clear for some time that the bulk compositions of the giant planets are different from each other and from initial solar composition. Jupiter is closest to solar composition, followed by Saturn; but in Uranus and Neptune, hydrogen is a minor component by mass. These conclusions are secure and are based upon relatively gross properties of the hydrogen equation of state. On the other hand, processes of internal differentiation are more sensitive to details of the thermodynamics of the constituents, and are therefore more difficult to calculate. In this review, we discuss the current status of calculations of the thermodynamics of the major constituents of the giant planets, and their use in the calculations of interior models. The inferred bulk compositions of the giant planets are then discussed and compared with atmospheric compositions, leading to conclusions about the significance of sorting processes.

II. EQUATIONS OF STATE AND PHASE DIAGRAM OF PURE HYDROGEN

In primordial solar-composition matter, hydrogen comprises 74% by mass, helium 24%, and the remaining elements only about 2% (Anders and Ebihara 1982). Thus the behavior of hydrogen at high pressures plays a pivotal role in understanding the structure of the giant planets. Experimental data on hydrogen are available from diamond-cell investigations at low temperatures (Bell et al. 1986), and from single- and double-shock compression of hydrogen at high temperatures (Nellis et al. 1983). The shock compression data yield pressure and density points that are directly applicable to the calculation of pressure-density relations in giant planet interiors. Diamond-cell experiments provide more indirect data at present and hence are less useful for planetary modeling purposes, although they have achieved higher pressures than the shock experiments (1.6 Mbar vs 0.8 Mbar). Neither approach has reached the degree of compression necessary to produce pressure-ionized hydrogen (metallic hydrogen). Although Grigor'ev et al. (1972) reported achieving metallic hydrogen in an adiabatic compression experiment, their experiment has not been duplicated, and the precision was not useful for constraining theoretical equations of state. In fact, the pressures claimed to have been achieved in the experiments of Grigor'ev et al. (maximum value was 8 Mbar) were not actually measured but were calculated from a hydrodynamic code. For all practical purposes, constraints on hydrogen thermodynamics at pressures $\gtrsim 1$ Mbar are derived largely from theory.

A. Equations of State of Molecular Hydrogen: Experimental and Theoretical Results

The high-precision shock compression experiments on hydrogen and deuterium of Nellis et al. are at present the most relevant data set for testing planetary interior models. These experiments achieved a peak pressure $P =$ 0.76 Mbar and mass density $\rho = 0.6$ g cm^{-3} at a calculated temperature $T =$ 7000 K. The experiments span the predicted pressure range of hydrogen-rich layers in Uranus and Neptune, but not in Jupiter and Saturn. The principal limitation on such shock compression experiments arises from heating of the sample, which becomes so severe as to impose a limit on the increase in density of the sample with single shocking. Use of deuterium, that has a substantially higher initial density than hydrogen, permits achievement of greater compression. Use of a reflected shock onto a preshocked sample also allows greater compression.

The data of Nellis et al. have been used to calibrate an effective pair potential between H_2 molecules, $\phi_{H_2\text{-}H_2}$, valid for the pressure range up to ~0.8 Mbar. The effective potential is intended to include the effects of many-body softening of the true pair potential. As temperatures achieved in the single- and double-shock experiments are comparable to those calculated for deep giant-planet envelopes (see below; Ross et al. 1981), the effective potential can be used to calculate liquid-state thermodynamics of dense molecular hydrogen. Recently, the effective potential derived by Ross has been used by Marley and Hubbard (1986,1988), together with effective potentials $\phi_{He\text{-}He}$ for helium-helium (Nellis et al. 1984) and $\phi_{H_2\text{-}He}$ for hydrogen-helium (Shafer and Gordon 1973) interactions, to compute a model-free energy for mixtures of helium and molecular hydrogen. The helium-helium effective potential has been calibrated by double-shock measurements of helium to 560 kbar and 21,000 K, but the H_2-He potential is so far only available from measurements of dilute gases and may yield significant errors when extrapolated to strongly coupled conditions. In this approximation, the interaction energy Φ of a given configuration of N hydrogen molecules and helium atoms is given by

$$\Phi = \sum_{i<j}^{N} \phi(\mathbf{r}_i - \mathbf{r}_j) \tag{1}$$

where the sum is carried out over all pairs of atoms and/or molecules, located at positions $\mathbf{r}_i$ and $\mathbf{r}_j$. The free energy is then calculated by ensemble-averaging over configuration space with Eq. (1).

In the limit of pure hydrogen (Marley and Hubbard's model), free energy yields calculated Hugoniots which fit the experimental results to within the error bars. Saumon and Van Horn (1986) have also recently calculated an improved H_2 equation of state that is calibrated to the hydrogen shock data.

B. Metallic-Hydrogen Equations of State

1. Methods of Calculation. For interparticle separations on the order of the Bohr radius a_0 and smaller, the fully pressure-ionized phase of hydrogen is expected to be stable. Unfortunately, as we discuss below, the location and nature of the transition from the molecular to the metallic phase of hydrogen remains very uncertain. As discussed by Hubbard et al. (1974), the liquid phase of metallic hydrogen is the only one of interest for planetary interior calculations. A variety of methods for calculating the free energy of metallic hydrogen give very similar results. This is a consequence of the fact that perturbation-theory calculations of the free energy of a pressure-ionized substance with atomic number Z, starting from free electron states, converge rapidly when the pressure exceeds $\sim 300\ Z^5$ Mbar (Kirzhnits 1967). Even at Jovian pressures ($\sim$5 Mbar), differences between theories for hydrogen are not severe.

Zharkov and Trubitsyn (1978) used a quantum-statistical model (QSM) to calculate the equations of state of hydrogen and helium for a zero-temperature crystal lattice. The free energy of liquid metallic hydrogen was then calculated using a Debye-like model for the liquid state. Ross and Seale (1974) and Stevenson (1975) used a perturbation theory for metallic hydrogen and helium wave functions and utilized hard-sphere liquid-state perturbation theory to obtain the liquid-state thermodynamics. Hubbard and DeWitt (1985) used a similar theory for mixtures of hydrogen and helium, but calculated the liquid ensemble averages directly with Monte Carlo techniques. MacFarlane and Hubbard (1983) and Hubbard and MacFarlane (1985) employed an older Thomas-Fermi-Dirac theory (related to the QSM), applicable to mixtures of hydrogen and helium in arbitrary configurations. All of the theories yield similar pressure-density curves for hydrogen, with maximum differences on the order of 10% at Jovian pressures, and convergence to the common limiting model at higher pressures.

2. Need for Accuracy In order to infer the bulk composition of a Jovian planet, it is necessary first to set limits on the mass fraction of hydrogen (either molecular or metallic) that is present in the planet's interior. Such a calculation is feasible in the giant planets because (a) hydrogen is present in substantial proportions in all four giant planets, and (b) hydrogen's compression curve (either isothermal or adiabatic) is well resolved from the compression curves of all other substances. Point (b) is ultimately related to the strong Z-dependence ($\propto Z^5$) of the critical pressure for complete pressure-ionization, as mentioned above.

Suppose that in modeling a giant planet, we deduce an empirical pressure-density profile $P = P(\rho)$ in the planet's interior. For a uniform mixture of metallic hydrogen and another component (say, helium), the $P(\rho)$ relation of the mixture is given to good approximation by the additive-volume law:

$$\frac{1}{\rho(P)} = \frac{1 - Y}{\rho_H} + \frac{Y}{\rho_Y} \tag{2}$$

where ρ_H is the density of pure hydrogen at the same temperature and pressure, and ρ_Y is the density of the other component, which is present with mass fraction Y. Although there were concerns that this law might be a rather crude approximation for hydrogen and helium mixtures in giant planet interiors (Stevenson and Salpeter 1976), Hubbard and MacFarlane (1985) carried out detailed calculations of the pressure-density relation for mixtures of metallic hydrogen and helium using two different thermodynamic models, and showed that the maximum error in Eq. (2) is only 1% or smaller for hydrogen-helium mixtures at pressures greater than 10 Mbar, and that the error is still less than 4% at $P \sim 5$ Mbar. The error is in the sense that the volume of the microscopic mixture is smaller than the volume of a mixture of macroscopic blobs of the two components, by the stated amount. This error is substantially smaller than the probable uncertainty in the hydrogen or helium equations of state. So far, there has not been a careful investigation of the errors in Eq. (2) when it is applied to mixtures of hydrogen and substances other than helium, but this is probably not an urgent item because of the low cosmic abundance of the latter.

We can now address the question of the impact of uncertainties in the hydrogen equation of state on the deduction of the interior composition of a hydrogen-rich planet. Solving Eq. (2) for Y, and taking into account $\rho_H << \rho_Y$ and $\rho \simeq \rho_H$, we find that the uncertainty ΔY in the inferred mass fraction Y is given by

$$\Delta Y \simeq -\delta \tag{3}$$

where $\delta = \Delta\rho_H/\rho_H$ is the relative uncertainty in the hydrogen density.

What is a reasonable value for δ in a planet such as Jupiter? As stated, the maximum difference between metallic-hydrogen equations of state implies $\delta \simeq 0.10$, but one could argue that this range corresponds to the maximum discrepancy between various theories, including some rather crude ones. The largest uncertainty in sophisticated equations of state comes from the contribution of the electron correlation energy (Ross and Shishkevish 1977). On this basis, one might be justified in taking $\delta \simeq 0.03$. On the other hand, there is a substantial thermal contribution to the pressure in the giant planets Jupiter and

Saturn, in the sense that the actual pressure deep in the metallic-hydrogen layers is about 10 to 15% higher than the pressure of layers at the same density but at zero temperature. As we shall discuss further, there is also a substantial uncertainty in the temperature of these layers, with a relative value $\Delta T/T \simeq 0.30$ not unreasonable. It follows, then, that the "thermal" uncertainty in the density would correspond to $\delta(\text{thermal}) \simeq 0.03$. Taking all uncertainties into account, one must conclude that $\delta \simeq 0.05$, which comprises a rather substantial theoretical barrier to detailed modeling of the composition of Jupiter or Saturn.

If, for example, we take the mass fraction of hydrogen in solar composition to be $1 - Y = 0.74$, then the relative uncertainty in deducing the mass fraction of nonhydrogen component in Jupiter, $\Delta Y/Y$, would be about 20%. The uncertainty in the primordial helium abundance is also on this order (Chapter by Gautier and Owen). In matching a theoretical density for a proposed composition with an empirical density (obtained by fitting an interior model to observational constraints), it would therefore be difficult to distinguish between precisely solar composition and a composition where "metals" (elements with $Z > 2$) are enriched over solar by a factor up to ~10.

C. Phase Diagram: Metallic-Molecular Transition in Solid and Liquid Phases

1. Uncertainties in Location and Nature of Transition. The location and nature of the transition from molecular hydrogen H_2 to metallic hydrogen $H+$ is a key issue for understanding the atmospheric abundances of Jupiter and Saturn. The problem affects our understanding in two ways.

(a) The transition, if it is a first-order one, must lead to discontinuities in abundances of chemical constituents across the phase boundary. Thus, in principle, the slow movement of the phase boundary, that might start near the center of the planet and gradually move toward the surface as the planet cools and contracts, would lead to an enrichment or depletion of abundances in the outer layers of the planet with respect to hydrogen. Consider, for example, a molecule such as methane, that we assume is present as a minor species (perhaps in ionized form) in both molecular and metallic hydrogen. Let $x_{1,2}$ be the methane concentration by number in phases 1 and 2, respectively ($1 \Rightarrow H+$; $2 \Rightarrow H_2$). Then in equilibrium, we have

$$\frac{x_1}{x_2} = \exp[(\psi_1 - \psi_2)/kT] \tag{4}$$

where ψ is a function of P and T only, with dimensions of energy, and k is Boltzman's constant (Landau and Lifshitz 1958). So far we do not have calculations of $\psi_1 - \psi_2$ or even estimates of its sign, but it must be nonzero if the phase transition exists. Equation (4) must be supplemented by an equation for conservation of the total amount of methane in the planet. Thus the predicted concentrations of methane in the H_2 envelope and H+ interior become determinate in principle.

In general, if a first-order transition from H_2 to H+ exists in Jupiter and Saturn, one would expect the methane concentrations in their outer layers to be quite different, both because of differing values of P and T at the transition, and because the relative amounts of planetary mass in the molecular and metallic phases are quite different. Obviously the same statement would apply to all other minor species. Yet at the moment, we have evidence that the methane abundances in Jupiter's and Saturn's atmospheres are similar, differing by no more than a factor of about 2, and are enhanced with respect to solar composition by a factor of 2 to 6 (Chapter by Gautier and Owen). We conclude that there is no strong evidence for a first-order phase transition from the methane abundances, but neither can one be ruled out.

(b) According to calculations carried out by Stevenson (1975), a mixture of H+ and He possesses a miscibility gap with a critical temperature of about 10,000 K near pressures of about 5 Mbar. The existence of this gap has profound implications for the relative abundances of hydrogen and helium observed in the atmospheres of Jupiter and Saturn, as we discuss below. The critical temperature is predicted to be a decreasing function of pressure, and thus hydrogen-helium immiscibility should first occur at the lowest pressure for which metallic hydrogen is stable. The planetary mass fraction involved in hydrogen-helium separation is therefore a sensitive function of the transition pressure, particularly in Saturn.

Figure 1 shows a possible phase diagram for hydrogen, together with three planetary adiabats (J,S,U). The latter are, respectively, plausible trajectories for the interior temperature profiles in the hydrogen-rich layers of Jupiter, Saturn and Uranus, and terminate at the approximate maximum pres-

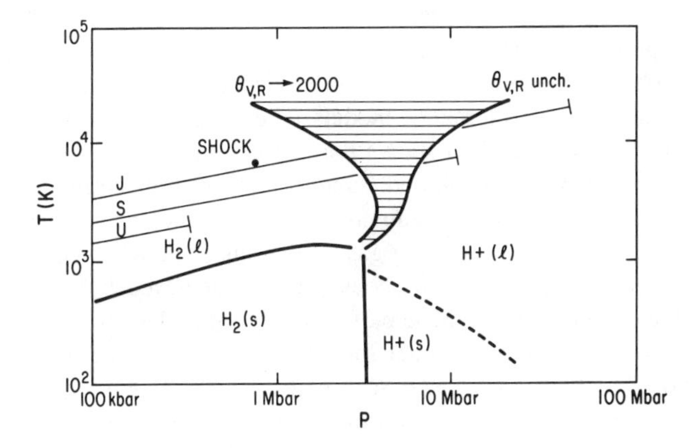

Fig. 1. A possible phase diagram for pure hydrogen, showing liquid (ℓ) and solid (s) phases of H_2 and H+. Also shown are estimated temperature distributions in the giant planets. See text for further explanation of symbols.

sure for these layers in each planet. Also shown ("shock") is the highest shock compression point in the experiments of Nellis et al. The two curves at the top ($\Theta_{V,R} \rightarrow 2000$; $\Theta_{V,R}$ unchanged) are computed boundaries for the $H_2 \rightarrow H+$ phase transition, calculated, respectively, under the assumption that the characteristic vibrational temperature Θ_V and rotational temperature Θ_R for the hydrogen molecule both go to 2000 K in the region of the transition (Hubbard and Stevenson 1984), and under the assumption that these temperatures remain unchanged from their low-density values. The phase boundary is computed using the theory of Marley and Hubbard (1986,1988) to calculate the chemical potential of a hydrogen molecule in the liquid H_2 phase μ_{H_2} and the theory of Hubbard and DeWitt (1985) to compute the chemical potential of a proton in the liquid H+ phase μ_{H+} together with the equation $2\mu_{H+}(T,P) = \mu_{H_2}(T,P)$. The phase boundaries in the liquid region of planetary interest are clearly very sensitive to details about the molecular structure, and there will in general be substantial dependence of the transition pressure on temperature. If the H_2-H+ boundary exists at all in the liquid phase (see Hubbard and Stevenson [1984] for arguments why it may not), it must terminate in a critical point whose value is uncertain.

At lower temperatures, the computed liquid H_2 − liquid H+ phase boundary agrees well with the 3 Mbar transition pressure for the solid phases estimated by Nellis et al., as must necessarily be the case because similar theories are used for both phases. The boundary between solid and liquid phases of H+ (shown dashed) is uncertain, and is not important for planetary interiors because it is located, if it exists at all, at temperatures well below prevailing temperatures in the deep interiors of the giant planets. Ross and Shishkevish (1977) estimate that quantum zero-point oscillations are large enough to cause the metallic-hydrogen lattice to melt at zero temperature for all pressures $\gtrsim$13 Mbar. Since their calculation ignores electron screening, which would reduce melting temperatures further, the sketched boundary probably represents an upper limit for the solid-liquid transition in H+.

To emphasize the uncertainty about the location and nature of the transition, Fig. 2 shows some alternative models. The dashed curve terminating in a critical point on the left (RK) shows the approximate location of the phase boundary calculated by Robnik and Kundt (1983) by considering a model in which hydrogen molecules are ignored, and instead the neutral phase of hydrogen is taken to be an interacting gas of H atoms. The RK phase boundary lies very close to the maximum shock pressure achieved by Nellis et al., although there is no strong evidence of its existence in the latter data.

Another "unorthodox" model for the liquid-state phase transition has been proposed by Ross and Lee (1986). In this model, there is a phase boundary between neutral molecular hydrogen and ionized molecular hydrogen H_2^+), terminating in a critical point (RL). At still higher pressures, there may be an additional phase boundary for the transition from H_2^+ to H+.

Finally, Fig. 2 shows the approximate upper boundary for the instability

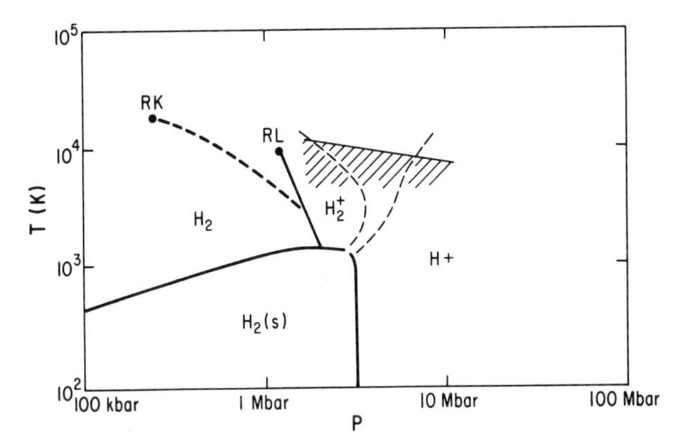

Fig. 2. Phase diagrams for hydrogens showing variants of the boundaries between liquid phases (see text for discussion of the RK and RL phase boundaries). The two dashed curves on the right reproduce the boundaries computed by Marley and Hubbard (1986,1988); see Fig. 1. Also shown is the approximate region for the onset of He-H phase separation in the metallic phase (see text).

region for hydrogen-helium immiscibility in the conducting phases of hydrogen (shaded area). Implications of this region are discussed below.

2. *Uncertainties in Latent Heat.* The temperature distribution in a fully convective giant planet will follow an adiabat except in the vicinity of the H_2-H+ phase transition, as discussed by Stevenson and Salpeter (1976). Because of the discontinuity of entropy across a first-order phase boundary, the temperature distribution will tend to be isothermal in the vicinity of the boundary. Even if the transition is continuous rather than first-order, if it is abrupt enough, the same remark applies.

In the case of the transition curve for $\Theta_{V,R}$ unchanged (Fig. 1), the entropy decreases discontinuously along a Jupiter or Saturn temperature profile as we go into the H+ phase. Thus the actual temperature in the metallic interior will be lower than that for a fully adiabatic planet. On the other hand, for phase diagrams where the phase boundary slopes in the opposite direction (e.g., Fig. 2), the temperature will be higher. Calculations of the Kelvin cooling time for Jovian planets (see, e.g., Stevenson and Salpeter 1976; Hubbard 1977) will therefore be affected by the sign of the latent heat associated with the phase transition.

III. PROPERTIES OF MIXTURES OF HYDROGEN WITH OTHER ELEMENTS

A. Hydrogen-Helium Mixtures: Limited Helium Miscibility

The hydrogen-helium system presents perhaps the most clear-cut situation for interpretation of the composition of giant planet atmospheres. It is generally considered to be true that no process of accumulation of a giant planet could lead to departures of the hydrogen/helium ratio from the primordial value. No plausible process of chemical condensation could segregate the two volatile gases at the temperatures prevailing during the time of formation of the outer planets. Once a massive planetary core is formed, selective distillation would be unimportant. Thus we start from the premise that the bulk proportions of free hydrogen and helium must be identical in all four giant planets, although this assumption may merit further examination.

It follows that any observed deviation from the primordial helium/hydrogen ratio in a giant-planet atmosphere must be attributed to processes of separation within the planet after its formation, processes that thus act to preserve the bulk planetary composition. Stevenson (1975) and previously Salpeter (1973) and Smoluchowski (1967) predicted that helium would not be arbitrarily miscible in metallic hydrogen, and that this phenomenon could lead to altered atmospheric ratios as well as a possible significant energy source via the release of gravitational energy by the sinking of helium toward the center of the planet. According to Stevenson and Salpeter (1976), the process should occur first in the metallic-hydrogen zones of Jupiter and Saturn, while critical temperatures for the onset of immiscibility in the molecular-hydrogen layers are too low to be important in any giant planet.

Calculation of the onset of immiscible behavior in a two-fluid system requires considerable accuracy in the thermodynamic model. Consider the Gibbs free energy $G(P,T,x)$ of a mixture of hydrogen and helium, where x is the helium number fraction. The density-pressure relation $\rho(P,T,x)$ is derived by taking the first partial derivative of this function with respect to P, and for pure hydrogen we estimated the error in the resulting density to be $\sim$5% for the pressure range of interest. The error in the pure-helium equation of state is undoubtedly larger, although no systematic study of its value has been made. Now, in order to find the critical temperature T_c and composition x_c for a given pressure, it is necessary to solve the two equations

$$\left(\frac{\partial^2 G}{\partial x^2}\right)_{T,P} = 0 \tag{5}$$

$$\left(\frac{\partial^2 G}{\partial x^3}\right)_{T,P} = 0 \tag{6}$$

evaluated at a specified P. It is thus not surprising that very good agreement in density-pressure relations for the hydrogen-helium system might not be accompanied by good agreement in predictions of immiscibility behavior, because the latter are calculated from higher derivatives of the Gibbs free energy than are the former.

A model of a fluid of fully ionized nuclei of hydrogen and helium (corresponding to a mixture of metallic hydrogen *and* metallic helium) predicts $T_c \simeq 8000$ K and $x_c \simeq 0.3$ at a pressure of 10 Mbar. The origin of this behavior can be traced to the ion-sphere model of a dense fluid of two species of different charges (DeWitt 1978). Because of the requirement of electrical neutrality, in a strongly coupled two-component plasma, each ion creates a sphere around itself containing only electrons, not ions, and the sphere has a volume which encloses enough of the uniform-density electron fluid to insure overall neutrality within the sphere. In this model, the helium nuclei act as spheres with a larger effective radius than the hydrogen nuclei, and as a result, the mixture minimizes its Gibbs free energy at a sufficiently low temperature by separating into a helium-rich phase and a helium-poor phase.

The situation is more complicated at the pressures prevailing in the giant planets, where helium is not fully pressure ionized. According to calculations by Stevenson (1979), the interactions between a helium nucleus and a surrounding metallic-hydrogen fluid lead to immiscible behavior much as described by the fully ionized model. On the other hand, calculations by Pollack and Alder (1977) led to opposite conclusions. Hubbard and MacFarlane (1985) carried out a three-dimensional Thomas-Fermi-Dirac (TFD3D) calculation which predicted no phase separation of hydrogen-helium mixtures at giant-planet pressures. However, a theory presented by Hubbard and DeWitt (1985), based on the assumption that the electron distribution deviates only weakly from uniform, led to results which confirm those of Stevenson, that helium separates from hydrogen at pressures on the order of 10 Mbar and $T < 10^4$ K. In summary, from a theoretical point of view, the critical temperature for phase separation of helium from hydrogen depends sensitively on the thermodynamic model and is difficult to predict with great accuracy in the pressure range of interest in Jupiter and Saturn.

From the observational point of view, the situation seems relatively clear-cut, and seems to favor the thermodynamic models of Stevenson (1975) and Hubbard and DeWitt (1985). The helium abundance in Saturn's atmosphere is strongly depleted with respect to primordial solar composition (Chapter by Gautier and Owen), and with respect to the Jovian atmosphere. The observational results are in good agreement with the original prediction of Stevenson, and indicate that the temperature distribution in Saturn is well below the critical temperature range for phase separation in metallic hydrogen, while the temperature in Jupiter is only slightly below this range (see Figs. 1 and 2). From the observations, one infers that the temperature in Saturn dropped below the critical value for a solar-composition mixture at a relatively early epoch in the planet's evolution, and that the metallic-hydrogen core began to unmix at this point. Helium enrichment occurred in the deeper layers of this core, while the surrounding molecular-hydrogen envelope was depleted in helium. This process leads to less drastic evolution of the metallic region if the latter is larger in mass, and thus the model is more consistent with a hydrogen phase diagram in which the transition to the metallic phase occurs at a pressure lower than 3 Mbar (Fig. 2; Hubbard and Stevenson 1984; Ross and Lee 1986).

Temperatures in Jupiter are somewhat higher than in Saturn because Jupiter, by virtue of its higher mass, has stored more primordial accumulation energy and retained more of it over the age of the solar system. Thus, separation of helium from hydrogen is less advanced than in Saturn, and a smaller fraction of the planetary mass is involved, so that the atmosphere has only changed its helium fraction slightly, if at all, over the age of the solar system.

In Uranus and Neptune, the maximum pressures in the hydrogen-rich layer do not reach values necessary to produce metallic hydrogen (in any model for the phase diagram), and thus the phase-separation model of Stevenson does not apply to these bodies. One would then predict that the helium/hydrogen ratio in the atmospheres of Uranus and Neptune would be solar (however, see below).

B. Minor Constituents

In principle, the solubility of the minor constituents of a giant-planet atmosphere might also be limited in the deeper layers of the object. However, as pointed out in Hubbard and Stevenson (1984), and in analogy to Eq. (4), if there exists a characteristic energy difference ΔE for the immersion of a minor constituent in metallic hydrogen, then the solubility of the minor constituent in the metallic-hydrogen solute is given by $e^{-\Delta E/kT}$. Since the number fractions of elements other than helium in a solar-composition mixture are smaller than 10^{-3}, the value of ΔE would have to be quite large for this to be a significant effect. Acually, the values of ΔE for abundant species such as C, N and O in H+ are unknown.

IV. STATIC MODELING STUDIES: DETERMINATION OF INTERIOR COMPOSITION

Models of giant planets are calculated by combining a pressure-density relation for a mixture of materials, $P = P(T,\rho,x)$, with equations relating the temperature to the pressure, $T = T(P)$, and the composition to the pressure, $x = x(P)$, ultimately leading to a barotropic relation $P = P(\rho)$ that can be substituted into the equation of hydrostatic equilibrium and solved for the distribution of density within the planet, $\rho(r)$. The latter must satisfy several integral constraints, including the total mass of the planet M, the equatorial radius at a specified pressure a, and the observed even dimensionless multipole moments of the mass distribution $J_2, J_4, \ldots$, as discussed in detail by Hubbard and Stevenson (1984), and Zharkov and Trubitsyn (1978).

As discussed in numerous references (see, e.g., Hubbard and Smoluchowski 1973; Stevenson and Salpeter 1976; Stevenson 1982*b*), comparison of the observed radii and masses of the giant planets with theoretical $a(M)$ curves for various chemical compositions leads to unambiguous conclusions about the different compositions of Uranus and Neptune as compared with Jupiter and Saturn. Because the $T = 0$ compression curve of hydrogen is well separated from that of other elements over the entire pressure range of interest ($P \lesssim 40$ Mbar), the radius of a planet composed primarily of hydrogen will be approximately 2 to 3 times that of a planet of the next nearest composition (helium), and even farther away from that of a planet composed of material with $Z > 2$. Since the radii of Jupiter and Saturn are within a few percent of the theoretical values for solar-composition objects of the same mass, there exists little doubt that both planets are composed largely of hydrogen, with a bulk composition which grossly mirrors the composition of the observable atmosphere, as originally demonstrated by the pioneering work of DeMarcus (1958). Considering the uncertainties in the hydrogen equation of state discussed above, one cannot reliably deduce that Jupiter and Saturn differ from pure solar composition solely from the $a(M)$ curve.

Uranus' and Neptune's radii are, on the other hand, substantially smaller (by a factor of ~ 2) than the radii of solar-composition planets of the same mass. It is thus clear that these bodies are not primarily composed of hydrogen. Although hydrogen is predominant in the atmospheres of these planets, the deep interiors must be composed of denser material. However, there exists great ambiguity as to what this material might be, and further evidence must be adduced to limit the possibilities.

A. The External Gravitational Potential as a Constraint

1. J_2 as a Constraint on Composition. If a planet rotates uniformly at an angular rate ω and is in hydrostatic equilibrium, then the dimensionless quadrupole moment J_2 represents, to lowest order in ω^2, the linear response of the planet to the nonradial gravitational potential perturbation imposed by the rotation. We may define a dimensionless linear response coefficient Λ_2 by

$$\Lambda_2 = \lim_{q \to 0} \frac{J_2}{q} \tag{7}$$

where $q = \omega^2 a^3/GM$, and G is the gravitational constant (Hubbard et al. 1975). In practice, q is not very small in the giant planets (typically about 0.1), and thus the nonlinear response to rotation is also important in J_2. Also, there exists some ambiguity about the correct rotation rate, since giant planet atmospheres rotate with a variety of periods. We adopt the point of view that the quadrupole moment is primarily determined by the response of the planet in its deep, electrically conducting interior layers, whose rotation rate is equal to that of the external magnetic field. Table I lists current results for the Λ_2's of the giant planets, computed in the approximation $\Lambda_2 \simeq J_2/q$, and using the magnetospheric rotation rate (where available), together with the equatorial radius at 1 bar pressure. This table is an updated version of one presented by Hubbard (1984); sources of constants for Jupiter and Saturn are the same as given there. Values for Uranus and Neptune have been updated with the following values.

TABLE I
Quadrupole Response Coefficients of Giant Planets

Planet	Λ_2
Jupiter	0.166
Saturn	0.108
Uranus	0.105
Neptune	0.10–0.18

The rotation period of Uranus' magnetic field (17.24 hr: Warwick et al. 1986) has been used to compute the value of q for this planet. This period is significantly longer than the mean atmospheric rotation period, and leads to a larger value of Λ_2 than the one used before.

Pending arrival of the Voyager 2 spacecraft at Neptune, the best value for Λ_2 for Neptune comes from occultation data (French et al. 1985; Lellouch et al. 1986; Hubbard et al. 1987). The range of uncertainty for Neptune in Table I reflects magnification of errors in using the atmospheric oblateness together with an uncertain value of J_2 (Harris 1984), and is still of limited value for constraining interior models.

The compression curve for hydrogen-rich material is such that the $P(\rho)$ curve roughly resembles the polytropic law $P = K\rho^2$ in the pressure range of interest (Hubbard 1974*a*), and for this law one readily calculates

$$\Lambda_2 = \frac{5}{\pi^2} - \frac{1}{3} = 0.17 \tag{8}$$

in good agreement with the result for Jupiter. On the other hand, Saturn's Λ_2 falls substantially below this theoretical value, implying a much reduced gravitational response to the rotational perturbation. The value given by Eq. (8) represents the approximate response of a homogeneous planet obeying the hydrogen compression curve. If more of the mass of the planet is concentrated in a dense core of higher-Z material, the value of Λ_2 is reduced. In general, Λ_2 is smaller, the greater the degree of central condensation. Thus the primary evidence that Saturn differs substantially from solar composition comes from its value of Λ_2.

The Jovian interior can, according to models, be roughly divided into a hydrogen-rich envelope and a rocky/icy core with a radius of about 10% of the total planetary radius. In Saturn, the same transition is probably farther out, occurring at 15 to 25% of the total planetary radius (Hubbard and Stevenson 1984).

The situation is less clear-cut with Uranus and Neptune. In order to calibrate the extent of disagreement between theory and observation, consider a simple model for these planets, following Hubbard and MacFarlane (1980*a*). The planetary core is composed of iron and magnesium-silicates ("rock") in solar proportions, and is thus in effect a terrestrial planet. Above this core we place an envelope composed of the most abundant ice-forming molecules ("ice"); the latter is composed of CH_4, NH_3 and H_2O in solar proportions. Further, the ratio in mass between the rock core and ice envelope is taken to be solar, or about 1/2.7. If no free hydrogen is included in the model, it has too small a radius for either planet, and in any case atmospheric composition results (see Chapter by Gautier and Owen) show that both planets have atmospheres predominantly composed of hydrogen. Thus it is reasonable to add an outermost hydrogen-rich layer until the radius is in agreement with observation. When this is done, a relatively small amount of hydrogen results in a substantial expansion of the radius, and a model (for either planet) with roughly 4 $M_\oplus$ (where $M_\oplus$ is the Earth's mass) of rock in the core, about 10 $M_\oplus$ of ice in the intermediate envelope, and $\sim$1 $M_\oplus$ of hydrogen-helium in the outer envelope, would agree with the observed radius. The compression curves of these three major components are such that the interfaces are located at roughly equal intervals in radius, i.e., at 1/3 and 2/3 of the total planetary radius.

The above model is reasonable in that it is zoned into the three most abundant chemical categories which differ greatly from each other by volatility. However, it fails to agree with the Λ_2 constraint. The three-layer models of Hubbard and MacFarlane predict $\Lambda_2 = 0.067$ for Uranus and $\Lambda_2 = 0.085$ for Neptune, in substantial disagreement with the observed values (Table I). Apparently, both planets are substantially less differentiated than the three-layer model, and there may even be evidence (French et al. 1985) that Neptune is much less differentiated than Uranus. The latter issue may be definitively resolved if Voyager 2 is able to measure Neptune's magnetospheric rotation period during the planned planetary encounter in 1989.

In principle, the detection of a strong magnetic field at Uranus (Ness et al. 1986) provides further information about the composition and dynamics of possible conducting layers deep within the planet (Torbett and Smoluchowski 1979). The field is in all probability produced in the H-C-N-O-rich intermediate layers or possibly in a metal-rich core, rather than in the hydrogen-rich outer layers.

2. Higher Moments as Constraints on Composition: Uncertainties Due to Differential Rotation/Atmospheric Circulation. In principle, further information about the zoning of material in the deep interior of a giant planet can be obtained by requiring agreement between the computed and observed values of the higher gravitational multipole moments, J_4, J_6, etc. Each moment provides an integral constraint on the density distribution $\rho(r)$ in the planet via

$$Ma^n J_n = -\int_{-1}^{1} d\mu' \int_{0}^{2\pi} d\phi' \int_{0}^{a} dr' r'^{2+n} P_n(\mu')\rho(r',\mu',\phi') \quad (9)$$

where r is the radial distance from the planet's center of mass, θ is the angle from the rotation axis, $\mu = \cos\theta$, ϕ is the azimuthal coordinate, and P_n is a Legendre polynomial. Each J_n measures a moment of the *perturbed* density distribution, and in hydrostatic equilibrium must vanish if the rotational perturbation is absent.

Assuming that the planet rotates as a solid body (unique rotation period), the rotational perturbation directly excites J_2. But higher multipole components are excited indirectly via coupling from lower-order responses, and thus are proportional to higher powers of q. It is obvious from Eq. (9) that they are more sensitive to the density distribution in the outer layers of the planet, and they are strongly correlated with the lower-order multipole components because of the way that they are excited.

If, on the other hand, the planet possesses significant differential rotation, as is evidently the case for the giant planets, the situation becomes more complicated. Unless differential rotation takes place on cylinders, surfaces of constant density, pressure and temperature do not coincide. Furthermore, differential rotation couples directly into high-order multipole responses of the planet rather than indirectly as in the case of uniform rotation, and depending on the magnitude of the differential rotation, could lead to responses that differ substantially from the response to uniform rotation. Thus far, the only detailed studies of this problem have been carried out by Zharkov and Trubitsyn (1978) and by Hubbard (1982). They have found that the lower-degree moments J_2 and J_4 of Jupiter and Saturn are affected only slightly (few percent) by plausible differential rotation on cylinders. However, the higher moments are affected by a significant amount, which reduces their utility for determining interior structure. Instead, the gravitational moments J_6 and higher are likely to be quite sensitive to the dynamics of the outermost planetary layers. This is probably particularly true for Uranus and Neptune.

3. Status of Current Gravitational Potential Constraints: (a) Jupiter. The best current values of equatorial radius a, interior rotation period ω, and the gravitational moments J_2, J_4 and J_6 are presented in Table II (the equatorial radius a at 1 bar is derived from results presented by Lindal et al. [1981]; the magnetospheric rotation period is from Seidelmann and Divine [1977], and the zonal harmonics are from Campbell and Synnott [1985], and are normalized to a radius of 71,398 km rather than a). The value of J_6 is still too uncertain for constraining interior structure, since the strong correlation between this moment and the lower moments J_2 and J_4 allows virtually any interior model which fits the lower moments to fit this moment also. The value of J_4 is consistent with a deep, convective envelope of approximately solar

TABLE II
Constraints on Jupiter Interior Structure

Constraint	Value
Rotation period $2\pi/\omega$	$9^h55^m29.7^s$
a	71492 ± 4 km
J_2	0.014697 ± 0.000001
J_4	−0.000584 ± 0.000005
J_6	0.000031 ± 0.000020

composition (Hubbard 1974*b*). Results of more detailed modeling studies are presented below.

(b) Saturn. There has been no significant improvement of values for Saturn's gravitational potential coefficients since the work of Null et al. (1981). The latter work is based on measurements made by the 1979 Pioneer 11 Saturn flyby, which made a substantially closer approach to Saturn than the subsequent Voyager flybys. There are prospects for tighter constraints on the gravitational harmonics based on analysis of the motions of eccentric ringlets from Voyager imaging data (Nicholson and Porco 1988), but results are not yet available. The rotation period of Saturn's magnetic field (Davies et al. 1983), which is presented in Table III, is measured with some difficulty because of the near-axisymmetry of the field. Also given is the equatorial radius a at 1 bar pressure (Lindal et al. 1985). Note that the gravitational coefficients are normalized to a radius of 60,000 km, not to a.

(c) Uranus. As a result of the successful Voyager II flyby of Uranus in 1987, the status of observational constraints on Uranus' interior structure has been greatly improved. The best source of information about the gravitational harmonics has been observations of the precession of the eccentric Uranian rings via Earth-based stellar occultations (Nicholson et al. 1981; French et al. 1986). These measurements yield J_2 and J_4 and a with a precision which equals or exceeds spacecraft measurements. However, the gravitational harmonics are useless without a determination of q to equal precision, which

TABLE III
Constraints on Saturn Interior Structure

Constraint	Value
Rotation period $2\pi/\omega$	$10^h39^m22.4^s$
a	60268 ± 4 km
J_2	0.016479 ± 0.000018
J_4	0.000937 ± 0.000038
J_6	0.000084 (assumed)

TABLE IV
Constraints on Uranus Interior Structure

Constraint	Value
Rotation period $2\pi/\omega$	$17.24^h \pm 0.01^h$
a	25650 ± 100 km
J_2	0.0033461 ± 0.0000030
J_4	−0.0000321 ± 0.0000037
J_6	0 (assumed)

requires knowledge of the rotation rate of the deep interior. Prior to the Voyager encounter, our only source of information about the planet's rotation period was Earth-based observations of the rotation period of the atmosphere, based on measurements of cloud rotation as determined from photometric variations, and from Doppler-shift measurements (Goody 1982). The periods obtained from these techniques were typically about 16.3 hr. The magnetospheric period of 17.24 hr (Warwick et al. 1986) indicates that substantial differential rotation is present, and that the planet is significantly less centrally condensed than previously believed (Table I). Values of J_2 and J_4 given in Table IV are normalized to a radius of 26,200 km (French et al. 1986). In contrast, the value of a presented in the table is derived from stellar-occultation measurements of the 1-μbar level (Elliot 1982) and then corrected to the "cloudtop" level ($\simeq$1-bar level). The error bar reflects the uncertainty of the latter extrapolation.

(d) Neptune. Neptune's interior structure remains the most poorly constrained of any of the giant planets. There are two reasons for this situation. First, the planet lacks a close-in, continuous ring system like Uranus'. Although there exists evidence from stellar occultations that Neptune possesses ring arcs which vary greatly in azimuth (Hubbard et al. 1986), these partial rings have not been observed enough to yield a model which might be related to Neptune's gravitational field (Goldreich et al. 1986). Further occultation observations may improve the situation.

In the meantime, we are forced to rely for J_2 on observations of the orbital motion of Neptune's massive satellite Triton, which is in an inclined and retrograde orbit (Harris 1984). Because the mass of Triton is not altogether negligible compared with that of Neptune, the pole of Triton's orbit and Neptune's rotational pole precess together about the total angular momentum vector of the system. Uncertainty in the mass of Triton thus affects the value of J_2 which is deduced from the observed regression of Triton's orbit.

Parameters that constrain Neptune's interior structure are given in Table V. The value of J_2 is obtained from the analysis of Harris (1984), normalized to a radius of 25,225 km, with the error bar primarily set by the uncertainty in

TABLE V
Constraints on Neptune Interior Structure

Constraint	Value
Rotation period $2\pi/\omega$	$15.6^h \pm 1.2^h$
a	25269 ± 10 km
J_2	0.0040 ± 0.0005
J_4	0 (assumed)

the mass of Triton. The table lists an improved value for the equatorial radius a obtained by Hubbard et al. (1987), who also observed the distribution of occultation intensity near the center of Neptune's shadow, permitting an independent determination of the planet's oblateness e. Through a knowledge of the value of J_2 from Triton's motion, the value of q and hence the rotation period is inferred from the equation

$$q = 2e - 3J_2 \tag{10}$$

assuming hydrostatic equilibrium and uniform rotation. This calculation involves the difference of two uncertain quantities, which produces amplification of errors even without taking into account possible effects of nonuniform rotation. In fact, the "photometric" rotation period of Neptune, 18.2 hr (Belton et al. 1981), lies substantially outside the range deduced from the occultation data.

B. Deduced Interior Structures, Gross Composition and Related Uncertainties

1. Jupiter. At present, no unique interior model can be found to fit the constraints presented in Table II. There are still too many uncertainties in the phase diagram of hydrogen, too many ways that models can be adjusted to correspond to various conceivable interior temperature distributions and distributions of minor constituents, and too few constraints. However, one may test plausible interior models in the following way. We select a model which has a small number of adjustable parameters, such as the mass of a dense rocky core, and the composition of the hydrogen-rich envelope. These parameters are then varied until agreement is obtained with the constraints listed in Table II. The resulting empirical pressure-density profile for the planet's interior can then be examined for consistency with various assumptions about the distribution of temperature and composition within the planet (Zharkov and Trubitsyn 1978).

(a) What is the Size of the Core? The nucleation model for the formation of giant planets (Perri and Cameron 1974; Mizuno et al. 1978; Harris 1978; Mizuno 1980; Bodenheimer and Pollack 1986; Pollack et al. 1986*a*;

Chapter by Pollack and Bodenheimer) suggests that the critical mass of a rock-ice nucleus is approximately 15 $M_\oplus$, although with considerable uncertainty. That is, the runaway accretion of hydrogen onto an existing nucleus may require a preexisting object of this mass. Many models of Jupiter (see, e.g., Zharkov and Trubitsyn 1978; Hubbard and Horedt 1983) are consistent with such a core. However, the nonuniqueness of the models is such that a substantially smaller core is also admissible. Stevenson and Salpeter (1976) obtained interior models with core masses of ~6 to 13 $M_\oplus$. Recent calculations by Hubbard (1987, unpublished) based on the metallic-hydrogen equation of state of Hubbard and DeWitt (1985) and the improved constraints of Table II yield a model with a rocky core of about 6 $M_\oplus$. Even smaller cores are possible if one uses an extreme metallic-hydrogen equation of state, such as that of Hubbard and MacFarlane (1985), although the latter requires extreme deviations from solar composition in the hydrogen-rich envelope.

If we assume that the composition of Jupiter is precisely solar, that the envelope is composed only of hydrogen and helium, and that all elements with $Z > 2$ are concentrated in a dense core, then the mass of such a core would be about 5 $M_\oplus$. The inferred core in Jupiter is within about a factor of 2 of this number. However, as is evident from atmospheric abundances, the envelope is not pure hydrogen-helium, and if the atmospheric composition is typical of the entire envelope, Jupiter cannot be precisely solar in bulk composition.

(b) How Much Does the Envelope Differ from Solar Composition? The $\rho(P)$ relation for the Jovian envelope lies above that for pure hydrogen; that is, the planet must have a higher density than it would have for pure hydrogen. Utilizing Eq. (2), we may infer the mean value of Y for the Jovian envelope by adjusting Y simultaneously with other parameters (such as the core mass and density) in order to fit the constraints of Table II. The result of this exercise (Hubbard and Horedt 1983) is $Y = 0.30 \pm 0.05$ in the envelope, with a core mass of 11 $M_\oplus$. The more recent calculations of Hubbard (1987, unpublished), which use the more precise gravitational moments of Campbell and Synnott (1985), also find $Y = 0.30$ in the envelope (for the hydrogen equation of state of Hubbard and DeWitt [1985]), but with a core mass of 6 $M_\oplus$, as stated above. The difference in core mass from the earlier calculation to the later one is primarily due to the fact that the Hubbard and Horedt model assumed a central rock core with an envelope of icy material, while the later calculation considers only a rock core, with the remaining icy material assumed to be mixed in the hydrogen-rich envelope. It follows that the detailed structure of Jupiter's core cannot be determined from any constraint currently available.

(c) Comparison of Inferred Envelope Pressure-Density with Theory. The value of Y in the Jovian envelope is of great importance for assessing the

validity of models for the formation of giant-planet envelopes and atmospheres (Pollack et al. 1986*a*). Comparison of the inferred envelope Y with that obtained from observations of the atmospheric composition may provide information about variations of composition with depth.

First, it should be noted that the envelope Y obtained from modeling studies is only a crude average value which pertains to the most massive part of the Jovian envelope (the metallic-hydrogen region). Second, as discussed in Sec. II.B.2, there is an intrinsic uncertainty in the inferred value of Y which is related to the uncertainty in the hydrogen pressure-density relation. It is reasonable to conclude that the inferred envelope Y should lie in the range between 0.25 and 0.35. In contrast, the atmosphere has a helium abundance $Y = 0.18 \pm 0.04$ (Chapter by Gautier and Owen), probably significantly different from the envelope value.

The discrepancy can be partly resolved by noting the probable influence of $Z > 2$ material. There is evidence (Chapter by Gautier and Owen) that abundant elements such as C and N may be enhanced by a factor of 2 to 3 in the atmosphere. Masswise, this is still insignificant and could not account for the discrepancy. Assuming that C, N, and O are all present in the Jovian atmosphere in solar proportions to each other (although present evidence shows that H_2O is *depleted* by a large factor [see Chapter by Gautier and Owen]), but enhanced relative to the H-He component by a factor of up to 4 or 5, this would give a mass fraction of a non-H-He component in the Jovian atmosphere equal to 0.05.

If we assume that the overall helium mass fraction in the envelope is actually 0.24, then a consistent picture emerges. The Jovian envelope could be chemically homogeneous, with a composition nearly identical to the atmosphere, and enhanced in $Z > 2$ material by a modest factor with respect to solar. The outermost envelope layers could be depleted in helium by one of the mechanisms discussed above, although evidence for this is quite marginal. At the center of the planet is a dense core of rocky material which represents the solar complement of the $Z > 2$ material in the envelope.

2. *Saturn: (a) What is the Size of the Core?* It is easier to demonstrate the existence of a dense, massive core in Saturn than in any other giant planet. This is because the large radius of Saturn requires a preponderance of hydrogen, while Saturn's Λ_2 falls substantially below the value for a coreless planet. Hubbard and Stevenson (1984) conclude that, out of a total planetary mass of 95 $M_\oplus$, a mass of 10 to 20 $M_\oplus$ of nonsolar, dense material near the planet's center is required to match the observed Λ_2. Zharkov and Trubitsyn (1978) infer a Saturn core mass of 20 to 30 $M_\oplus$, while Hubbard and Horedt (1983) infer a core mass of 17 to 20 $M_\oplus$. It thus seems very likely that Saturn's core is substantially more massive than Jupiter's, both absolutely and in relation to the total mass of the planet. It follows that Saturn's bulk composition differs from solar by a larger factor than Jupiter's. There is evidence

(Chapter by Gautier and Owen) that $Z > 2$ elements may be present in Saturn's atmosphere by a factor $\sim$2 times the Jovian abundances. This appears to be also true for the planet's bulk composition.

(b) How Much Does the Envelope Differ from Solar Composition? No recent careful modeling study of the Saturnian envelope has been carried out. Because the pressure range is much more limited than in Jupiter, uncertainties in the equation of state in the range 1 to 10 Mbar are much more important. This is precisely the range where the hydrogen equation of state is most uncertain, and the problem is exacerbated by the clear indication of nonuniform distribution of helium in the envelope. In the face of these uncertainties, few reliable conclusions can be drawn about the distribution of $Z > 2$ material in Saturn's envelope. The reader may refer to Hubbard and Stevenson (1984) for a detailed discussion of Saturn's interior models.

3. Uranus. As the improved constraints on Uranus interior structure (Table IV) have only recently become available, detailed studies of possible models have not yet been carried out. In particular, the effect of differential rotation on J_4 has not yet been studied. A model which formally satisfies the new constraints has been presented by Podolak and Reynolds (1987*a*). It resembles the three-layer models of Hubbard and MacFarlane (1980*a*) discussed in Sec. IV.A.1, but with the following important differences. Like the Hubbard and MacFarlane model, there is assumed to be a central core of "rock", consisting of a solar mixture of Fe, Ni, MgO and SiO_2, an intermediate shell of "ice" composed of solar proportions of H_2), CH_4 and NH_3, and an outer envelope of H_2, He and "ice", with the latter enhanced over solar proportions to H_2 and He by a factor A.

The fundamental requirement is to increase the model's moment of inertia so as to obtain a fit to the new, larger Λ_2, and at the same time fit the observed J_4. These two requirements constrain the value of A in the outer envelope and the mass ratio of the ice shell to the central rock core. Interestingly, in order to fit Λ_2, the ice/rock ratio must be 16 or more, substantially greater than the solar ratio. At the same time, A must be small enough to keep the outer envelope density low, as otherwise the absolute value of J_4 becomes too large (see Eq. 9). Podolak and Reynolds find $A < 30$, consistent with the observed composition of Uranus' atmosphere ($A \sim 20$ for CH_4: Pollack et al. 1986*a*). The mass of the rock core is 0.81 $M_\oplus$, the ice shell 11.7 $M_\oplus$ and the hydrogen-rich envelope 2.0 $M_\oplus$.

Clearly the Podolak and Reynolds model is not unique, and it is too early to say whether satisfactory alternative models can be found that have much smaller (and, perhaps, more plausible) ice/rock ratios.

As the mass of the hydrogen-rich envelope is only about 2 $M_\oplus$ (a common feature of all Uranus models), its precise He/H_2 ratio cannot be directly constrained by interior models. However, interior models can be used to provide indirect constraints on the degree to which this ratio might deviate from the solar value. According to Gautier and Owen (see their chapter), Uranus' atmospheric helium mass fraction $Y \simeq 0.26$, quite close to the adopted solar value of 0.24. Unfortunately, the latter number is still somewhat uncertain, and one may reach rather different conclusions depending on the adopted primordial solar value.

The simplest interpretation of the observed He abundance is that it is precisely equal to the primordial solar value. Because the pressure in the H_2-rich layer never becomes great enough to produce He-immiscibility effects (see Fig. 1), there has been no modification of the atmospheric He abundance as there was in the case of Saturn. However, it is possible to conceive of other mechanisms that could either raise or lower the value of Y with respect to the primordial value. If a large number of carbon-bearing planetesimals were accreted into Uranus' atmosphere at a late stage of planetary formation (Pollack et al. 1986*a*), thereby raising the C/H ratio by a factor of about 20, the reduction of these extra carbon atoms to form CH_4 would deplete free hydrogen and thereby increase the He/H_2 ratio by about 6%. On the other hand, Ross and Ree (1980) claim that CH_4 molecules decompose into elemental carbon and molecular hydrogen above pressures of 200 kbar and temperatures of 2000 K. Such conditions almost certainly obtain in much of the hydrogen-rich envelope and throughout the ice shell. If this reaction occurs irreversibly because of the sinking of the dense carbon and the rising of the light hydrogen, the atmospheric He/H_2 ratio would be *decreased* as the extra hydrogen gets to the atmosphere. A limit on this effect can be placed by considering Podolak and Reynolds' extreme model with 11 $M_\oplus$ of ice. Up to ~ 1 $M_\oplus$ of hydrogen could be released from the ice envelope by these reactions, which if mixed into the 2 $M_\oplus$ outer envelope, could greatly reduce its helium mass fraction. It is conceivable that Uranus' atmospheric Y reflects the outcome of competing mechanisms, fortuitously ending up near the primordial value.

4. Neptune. Comparison of Table V with Table IV shows that a much broader range of interior models is possible for Neptune than for Uranus. Nevertheless, it is now apparent that the interiors of these two planets are actually quite dissimilar, which poses interesting challenges to theories for the origin of the outer planets.

Neptune superficially resembles Uranus. It is slightly more massive than Uranus (17.2 $M_\oplus$ vs 14.5 $M_\oplus$), and slightly smaller in radius. As this mass range is still in the ascending part of the radius-mass curve for plausible compositions (see Stevenson and Salpeter 1976, their Fig. 1), the difference in radius is most plausibly interpreted as due to a smaller mass of free hydrogen in Neptune. A hydrogen-helium envelope contributes only a small fraction to the total mass of either planet, but because of the low density of hydrogen, can have a major effect on the radius. Typical results (see, e.g., Hubbard and MacFarlane 1980*a*) obtain a mass for the H_2-He equal to 1.6 $M_\oplus$ in Uranus and 1.1 $M_\oplus$ in Neptune.

As indicated in Table I, there is a distinct possibility that Neptune is substantially less centrally condensed than Uranus. However, in order to establish this result reliably, two key data are required. We need (a) a direct measurement of the mass of Triton in order to reduce the uncertainty in Neptune's J_2 (Harris 1984), and (b) a direct measurement of Neptune's internal rotation period as defined by a magnetic field. Both of these data should be produced by a successful Voyager flyby in 1989.

The other, more certain, difference with Uranus is that Neptune possesses a larger intrinsic heat flow from its interior (Chapter by Conrath and Hanel). After correcting for sunlight which is reradiated from the planet's atmosphere in the thermal infrared, the remaining interior heat flow may be conveniently expressed in terms of a specific luminosity L_i/M, where L_i is the total luminosity of the interior heat flow and M is the planetary mass. For Neptune we have $L_i/M = 2 \times 10^{-7}$ erg g^{-1} s^{-1}, while for Uranus only an upper limit can be determined, $L_i/M < 1.6 \times 10^{-7}$ erg g^{-1} s^{-1}. For comparison, the chondritic $L_i/M = 0.4 \times 10^{-7}$ erg g^{-1} s^{-1}. This result can be interpreted in a variety of ways. Hubbard and MacFarlane (1980*a*) argued that Uranus' greater proximity to the Sun could lead to greater influence of reradiated solar heat on the evolutionary trajectory of Uranus, even if Uranus and Neptune started with the same initial conditions at the time of their formation. With plausible present interior temperatures of ~7000 K and interior compositions similar to that discussed in the preceding section, both planets would have Kelvin cooling times well in excess of the age of the solar system.

Stevenson (1982*b*) points out that even if Uranus or Neptune were formed with a well-segregated interior structure (similar to that proposed by Hubbard and MacFarlane [1980*a*]), the flow of heat from the interior would tend to remix the planet, resulting in less distinct layering and a larger value of Λ_2. The work done in remixing would be deducted from the energy flux to the atmosphere, such that the more homogeneous planet should have a smaller specific luminosity, in contradiction to present indications. However, this model provides a framework for understanding the higher enrichment of $Z > 2$ elements in Neptune's atmosphere as compared with Uranus. With Neptune's less massive H_2-He envelope, convective mixing of material from the ice shell would automatically lead to a greater enhancement compared with Uranus.

If convective stirring results in mixing of ice molecules from the interior into the observable atmosphere, some enhancement in the deuterium/hydrogen ratio in the atmospheres of both Uranus and Neptune is to be expected (Hubbard and MacFarlane 1980*b*).

V. CONCLUSIONS

Despite the great range in the masses of the giant planets, it is likely that each of the four possesses a roughly similar mass of $Z > 2$ material. According to the nucleation model for the origin of giant planets mentioned above, an initial nucleus of nonvolatile material with a mass $\sim$15 $M_{\oplus}$ may be sufficient to capture a much larger mass of gas from the surrounding nebula. Every giant planet appears to contain (at least to within a factor of 2 to 3) about this much non-H_2-He material. However, detailed study of admissible models shows that such material is not sharply confined to a core in any of these planets. The initial state has been obscured by subsequent evolution, which probably involves chemical separation, convective remixing and late accretion. Nevertheless, the initial state can be partially constrained by limits on the bulk composition of the four giant planets. Each of the four has substantially more $Z > 2$ material than solar composition, indicating that the capture of gaseous material onto an initial nucleus was not highly efficient. Moreover, the fraction of $Z > 2$ material increases systematically with heliocentric distance, indicating that the capture efficiency declined with distance from the center of the nebula. The giant planet atmospheres that we see today are relicts of this initial capture process, although they have probably all been modified substantially by subsequent planetary evolution.

THEORIES OF THE ORIGIN AND EVOLUTION OF THE GIANT PLANETS

J. B. POLLACK
NASA Ames Research Center

and

P. BODENHEIMER
University of California at Santa Cruz

The giant planets most likely formed initially by the accretionary growth of solid bodies in the solar nebula. Unlike the situation for their terrestrial planet counterparts, the cores of the giant planets became massive enough to capture gravitationally large quantities of gas from the surrounding solar nebula during the later phases of their growth. In the case of Jupiter and Saturn, but apparently not Uranus and Neptune, the very last phase of gas accretion was characterized by a runaway stage in which most of the H_2 and He of these planets was added in a very short period of time. If almost all of the solid-body growth of Jupiter and Saturn occurred prior to the gas runaway accretion stage, this growth took place in about several times 10^5 to several times 10^6 yr. Following the accretion of solids and gases in the solar nebula, the giant planets contracted to their present sizes over the age of the solar system. At first, this contraction was very rapid, although not hydrodynamic. During this phase, a few planetesimals captured in the outermost part of the envelopes of the giant planets may have remained behind as irregular satellites as the envelope continued to contract. At a later stage, a nebular disk, out of which the regular satellites formed, may have been spun out of the outer envelope of the contracting giant planets, due to a combination of conservation of total angular momentum and the outward transfer of specific angular momentum in the envelope. If the above hypotheses are true, the composition of the irregular satellites directly reflects the composition *of planetesimals from which the giant planets formed, while the composition of the regular satellites is indicative of the composition of the less volatile components of the outermost envelopes of the giant planets. The abundances of elements in the present atmospheres of the giant planets reflect their formation from two distinct phases of the solar nebula: gas and solid.* H_2 *and rare gases were derived entirely from the gas component; C and N from both phases; and rock and water from only the solid component. Since different elements were derived in different proportions from the two components of the solar nebula, since the solid component was preferentially accreted, and since the solid component was partially, but not entirely segregated in a central core region, the ratio of most elements with respect to H can be expected to deviate significantly from the solar value. Furthermore, the partitioning of elements present in accreted planetesimals between the core and envelope was not constant for all elements, since elements present in these planetesimals were introduced into the envelope by vaporization of core material and by the dissolution of planetesimals passing through the envelope on their way to the core. The* C/H *ratios characterizing the atmospheres of all four giant planets are substantially larger than the solar* C/H *ratio. This enhancement and its systematic variation among these planets imply that most of the C present in their atmospheres was derived from the dissolution of planetesimals in their envelopes during the later stages of their growth. Furthermore, it is likely that the* N/C*, the* C/H_2O *and the* $rock/H_2O$ *ratios are subsolar in the atmospheres of the giant planets. Even the* He/H_2 *ratio need not be solar in the atmospheres of the giant planets. It is observed to be distinctly subsolar in the atmospheres of Jupiter and Saturn, most likely due to the partial immiscibility of He in the outer part of the metallic H zone of their envelopes. It may be slightly suprasolar in the atmospheres of Uranus and Neptune, due to the complete reduction of some elements, especially C, present in dissolved planetesimals.*

I. INTRODUCTION

The atmospheres of the giant planets differ in several fundamental ways from those of the terrestrial planets and the larger satellites of the outer solar system. The gaseous envelopes of the giant planets constitute a significant fraction of the planets' total mass, whereas the atmospheres of the other objects represent very minor fractions of their total mass. Thus, the origin and evolution of the atmospheres of the outer planets are intimately tied to the origin and evolution of these planets as a whole.

Hydrogen and helium are the most abundant elements in the atmospheres of the outer planets, whereas other elements are the chief constituents of the atmospheres of the terrestrial planets and the larger satellites of the outer solar system. Thus, the outer planets are singular in that they were able to retain significant quantities of the gases present in the primordial solar nebula that provided the source material for the birth of planets, asteroids and comets.

It has been customary to consider solar elemental abundances as providing a useful first approximation to the composition of the atmospheres of the giant planets in view of the high abundance of hydrogen and helium in these atmospheres. However, there are both theoretical and observational grounds for suspecting that significant deviations from solar elemental abundances may be the rule rather than the exception. From a theoretical viewpoint, the giant planets formed from both the gas and solid phases present in the surrounding solar nebula. Furthermore, as detailed below, they preferentially accumulated the solid component of the solar nebula, with this component being partially segregated to a central core region and partially mixed into the envelope. Since hydrogen and helium were present only in the gas component of the solar nebula, whereas most other elements were present either partly or completely in the solid complement, there is no *a priori* reason to suspect that these other elements should exhibit solar proportions in the current atmospheres of the giant planets. Furthermore, even the helium to hydrogen ratio may not be solar due to helium's limited solubility in hydrogen at low enough temperatures.

The above theoretical arguments are supported by the observed abundances of several elements in the atmospheres of the outer planets, as summarized in Table I. In particular, the C/H ratios of the atmospheres of all four giant planets are larger than the solar C/H ratio, with the magnitude of the enhancement increasing from Jupiter to Saturn to Uranus and Neptune. The He/H ratio in the atmosphere of Saturn is substantially lower than the solar He/H ratio; this ratio in the atmosphere of Jupiter appears to be somewhat less than the solar ratio; and this ratio in Uranus' atmosphere is close to the solar value. A major goal of this chapter is to relate these and other departures from solar elemental abundances to chemical and physical processes that took place during the formation of the giant planets and their subsequent evolution.

The giant planets are encircled by ring systems and regular and irregular satellites. The orbits of the regular satellites generally lie quite close to the

TABLE I.
Composition of the Atmospheres of the Giant Planets

Element	Abundance[a]				References
	Jupiter	Saturn	Uranus	Neptune	
C	2.3 ± 0.2	5.1 ± 2.3	35 ± 15	40 ± 20	Courtin et al. 1984 Lindal et al. 1987 Orton et al. 1987*b*
P[b]	1.4 ± 0.4	2.8 ± 1.6			Courtin et al. 1984
O	0.02[c]				Bjoraker et al. 1986
N	2	3 ± 1			de Pater and Massie 1985
He	0.65 ± 0.15	0.2 ± 0.15	1 ± 0.15		Conrath et al. 1987

[a]Abundances are given as the ratio of the abundance of a given element to that of hydrogen relative to this ratio for the Sun; thus, a value of 1 is equivalent to the solar ratio.
[b]These abundances are lower bounds, based on the abundance of PH_3; see text.
[c]Value applies to the 2 to 6 bar region.

equatorial plane of their parent planet and have low eccentricities. (Triton, the largest satellite of Neptune is an exception to this rule.) Thus, these satellites were probably formed within disks of gas and dust that surrounded their parent planets during their early history. Such disks therefore bear at least a superficial resemblance to the solar nebula. If, as we will argue is likely, such satellite formation disks were created from the outer envelopes of the parent planets, the composition of the satellites can provide useful constraints on the composition of their planets' atmospheres and on conditions in the early solar system.

The irregular satellites of the outer solar system have orbits that lie outside those of the regular satellites. These orbits are highly inclined to the planets' equatorial plane and have a high eccentricity. Such characteristics suggest that they are captured objects, i.e., bodies that formed within the solar nebula and were subsequently placed into orbits about their current planets. The composition of the irregular satellites provide an indication of the nature of the solid material that helped to form the giant planets as well as standards to judge the additional processing undergone by the regular satellites. Comets may also provide useful analogues to the composition of the planetesimals that helped to form the outer planets, although the birthplace of comets is not well constrained at present.

In this chapter, we first review some of the basic properties of the outer planets that constrain their origin and evolution (Sec. II). We next discuss (Sec. III) possible theories of their origin, placing particular emphasis on the so-called core-instability model. According to this model, the giant planets were built initially by the accretion of solid planetesimals, just as their terrestrial counterparts were. But, in their later stages of growth, the giant planets were able to concentrate gravitationally and hence permanently capture large amounts of gas from the surrounding solar nebula. Subsequently, we discuss the evolution of the now fully formed giant planets, with emphasis on the nature and duration of key stages and the formation of the satellite and ring systems (Sec. IV). Using the theoretical framework of this and the previous section, we discuss key processes that introduced, mixed and segregated elements and compounds in the atmospheres of the giant planets. These processes include the division of elements between the gas and solid components of the solar nebula, the preferential accretion of planetesimals, as opposed to gas, by the forming planets, the partitioning of material derived from the planetesimals between the core and envelope, convective mixing of the envelope, and chemical differentiation of the envelope. We then examine the composition of the atmospheres and satellites of the outer planets to elucidate the processes that may have determined them and to constrain chemical conditions in the early solar system. Finally, in Sec. V we conclude this chapter by summarizing our favorite hypotheses for explaining the chemical composition of the atmospheres of the outer planets and citing key problem areas.

There are a number of chapters in this book that are highly relevant to the material discussed here. The chapter by Owen and Gautier provides an in-depth discussion of the chemical composition of the atmospheres of the outer planets, as well as a discussion of some of the implications of these observational results. The chapter by Hubbard is focused on the interior structure of the outer planets. Thus, it gives constraints on their heavy-element (other than H_2 and He) content and its partitioning between the inner and outer regions. It also contains a discussion of the possible segregation of He to the deeper portions of the envelopes of Jupiter and Saturn. The chapter by Conrath et al. discusses the possible importance of He segregation for the internal heat source of the giant planets. A much more detailed description of the formation of satellites than is contained here is given in the chapter by Coradini et al., although they stress an alternative viewpoint concerning the source material for the satellite-forming disks: the solar nebula, rather than the outer envelopes of the giant planets. The chapter by Jessberger et al. summarizes the new compositional data on Comet Halley that may provide a useful analogue to the solid material that helped to form the giant planets. Fianally, chemical processes controlling the composition of the solar nebula and satellite-forming disks are contained in the chapter by Prinn and Fegley.

II. OBSERVATIONAL CONSTRAINTS

Key constraints on the origin and evolution of the giant planets are provided by their composition, axial tilts, internal heat fluxes, orbital and compositional characteristics of their ring and satellite systems, and plausible bounds on the time scale of formation. Below, we summarize these constraints.

Constraints on the bulk composition of the giant planets and its radial distribution can be obtained by fitting models of their interiors to their measured masses, radii and gravitational moments (see, e.g., Hubbard's chapter). For our purposes, it is useful to divide the mass between a H_2/He component (low Z) and a component that encompasses all the other elements (high Z). It is also useful to define the partitioning of the high Z mass between a central core region and the outer envelope. The distinction between low and high Z mass components reflects, to first order, the accreted gas and solid components from which the giant planets formed. Most of the mass in the high Z component presumably is due to rock-forming elements and water, both of which are expected to be almost entirely in their condensed phases in the region of the solar nebula where the giant planets formed (see, e.g., Lewis 1974*a*). Conversely, the temperatures in this region were far above the freezing point of H_2 and He. The division of high Z material between a central core region and an outer envelope region provides a measure of the degree of mixing of the low and high Z material, since the low Z material is located only in the envelope in almost all interior models to date. Note, however, that H_2 may be very soluble in rock and ice at high pressures and, thus, in principle, could be present in nontrivial amounts in the cores of the giant planets (D. Stevenson, personal communication).

Table II summarizes current estimates of the low and high *Z* masses of the giant planets (see, e.g., Hubbard et al. 1980; Hubbard's chapter; Podolak and Reynolds 1984). If the low and high *Z* masses were present in solar elemental proportions, the high *Z* mass would equal about 2% of the low *Z* mass (Cameron 1982). According to Table II, Jupiter, Saturn and Uranus/Neptune contain approximately 5, 25 and 300 times as much high *Z* mass as would be expected from solar elemental abundances. Thus, all four giant planets accreted planetesimals much more efficiently than gas from the surrounding solar nebula. While the low *Z* masses of the giant planets vary by about a factor of 100, their high *Z* masses vary by only a factor of a few. This near constancy of the high *Z* mass may imply that there was a self-regulating process that controlled the magnitude of this component. Finally, high *Z* mass is present in both the core and envelope portions of the giant planets. As we will see in Sec. IV, this distribution of high *Z* material may provide constraints on the relative timing of solid and gas accretion.

Table I contains current estimates of elemental abundance ratios in the atmospheres of the giant planets. In all cases, these values denote the ratio of a given element with respect to H to that expected from solar elemental abundances. As already remarked upon in Sec. I, C/H is suprasolar for all four giant planets, with the degree of enhancement increasing from Jupiter to Saturn to Uranus/Neptune (Pollack et al. 1986*b*; chapter by Owen and Gautier; Lindal et al. 1987; Orton et al. 1987*b*; Courtin et al. 1984). This result implies that some of the C in the outer solar nebula was contained in the solid phase and that some of the solid phase has been dissolved in the envelope (Pollack et al. 1986*a*). As already remarked upon in Sec. I, He/H is subsolar in the atmosphere of Saturn, and, perhaps to a lesser degree, in the atmosphere of Jupiter (Conrath et al. 1984; chapter by Owen and Gautier). Since H and He should have been present entirely in the gas phase of the solar nebula and since there are no obvious mechanisms for effectively fractionating these two elements in the solar nebula, nonsolar He/H ratios in the atmospheres of the giant planets probably reflect a global scale differentiation process (Stevenson and Salpeter 1977*b*; Pollack et al. 1977).

The mildly suprasolar abundance of P in the atmospheres of Jupiter and Saturn is quite interesting in the sense that P is expected to form refractory compounds in the solar nebula (Fegley and Lewis 1980) and hence may have been present almost entirely in the condensed rocky material in the outer solar system. If so, the rock component of the solids that helped to form the giant planets is present in significant quantities in their envelopes.

The abundances of other elements in the atmospheres of the outer planets are less well defined. As shown in Table I, the N/H ratio appears to be somewhat suprasolar in the atmospheres of Jupiter and Saturn, but substantially subsolar in the atmospheres of Uranus and Neptune (Gulkis et al. 1983;

TABLE II
Bulk Properties of the Giant Planets[a]

Property	Jupiter	Saturn	Uranus	Neptune
Total Mass[b]	318.1	95.1	14.6	17.2
Low Z Mass[b,c]	254.1–292.1	72.1–79.1	1.3–3.6	0.7–3.2
High Z Mass[b,c]	26–64	16–23	11–13.3	14–16.5
Axial Inclination[d]	3.1	26.7	98.0	29
Internal Heat Flux[e]	$(1.76 \pm 0.14) \times 10^{-6}$	$(1.52 \pm 0.11) \times 10^{-6}$	$\leq 1.6 \times 10^{-7}$	$(3.4 \pm 1.1) \times 10^{-7}$

[a]Based on Table I of Pollack (1985).
[b]In units of Earth masses.
[c]The Jupiter values were obtained from Hubbard's chapter.
[d]From Blanco and McCuskey (1961); in units of degrees measured from the normal to the planet's orbital plane.
[e]From Hanel et al. (1981*a*,1983) and Pollack et al. (1986*b*); in units of erg g^{-1} s^{-1}—thus, it is the internal luminosity per unit mass of the planet.

dePater and Massie 1985; Courtin et al. 1984; chapter by Owen and Gautier). However, this apparent depletion of N in the atmospheres of Uranus and Neptune may be due to chemical and physical processes occurring at levels deeper than those sensed at radio wavelengths: Dissolution of NH_3 in deep H_2O clouds may strongly deplete the NH_3 abundance at higher levels (Atreya and Romani 1985). Similarly, H_2O has an abundance that is several orders of magnitude smaller than expected from solar abundances in the 2 to 6 bar region of the Jovian atmosphere (Bjoraker et al. 1986*b*), but this depletion may reflect meteorological processes, rather than a true depletion at greater depths (Lunine and Hunten 1987*a*).

The axes of rotation of all the giant planets but Jupiter are tilted significantly with respect to the normal to their orbital planes (cf. Table II). These large obliquities may have been induced by off-center encounters with one or several very large-sized planetesimals during the growth phase of these planets (Safronov 1969). This view is consistent with the overall increase in the obliquity of the giant planets with an increasing ratio of high to low Z mass. Thus, the solids that helped to form the giant planets may have included objects as massive as Mars or the Earth.

Jupiter, Saturn and Neptune radiate to space approximately twice as much energy as the amount of sunlight they absorb, whereas Uranus' thermal output is much closer to its solar input (Hanel et al. 1981*a*,1983; Pollack et al. 1986*b*; Conrath et al. 1987*a*). This difference (cf. Table II) reflects the presence of a strong internal heat source, which feeds on internal heat produced during the planets' formation and early contraction history and during later differentiation processes (Graboske et al. 1975; Pollack et al. 1977; Stevenson 1980; Hubbard 1980 and his chapter). Thus, the present values of the giant planets' internal heat source provide valuable constraints on their origin and evolution.

All four giant planets have ring systems and satellites. The rings or ring arcs lie within or close to the Roche radius, i.e., within several R_p of their primaries, where R_p refers to the planet's radius. The regular satellites begin close to the outer portion of the rings and extend out to several tens of R_p, whereas the irregular ones lie at distances of one to several hundred R_p. Except for the inner satellites of Jupiter, the regular satellites are made primarily of water ice and rock. Interior models of these satellites, constrained to match their observed mean densities, provide estimates of the relative proportion of rock and water ice contained within them. Surprisingly, there appears to be a common mixture characterizing many of the icy satellites of Jupiter, Saturn and Uranus. When the results for individual satellites of a given system are mass weighted, an almost identical rock to ice ratio emerges: A mixture of 55% rock and 45% water ice by mass characterizes all three systems, based on interior models of Ganymede and Callisto for the Jovian system, Titan for the Saturnian system, and Titania and Oberon for the Uranian system (Cassen et al. 1980; Johnson et al. 1987). If the regular satellites formed from high Z material derived from the outer envelopes of their primary, the above relative proportions of rock and water in the satellites may provide useful constraints on the abundances of these materials in the current envelopes of the giant planets. These abundances may also provide constraints on chemical abundances in the early solar system, as discussed in Sec. IV. Note, however, that the regular satellites of Saturn, aside from Titan and perhaps Mimas and Dione, have a significantly higher proportion of water ice than the above system average: about 40% rock and 60% water (Johnson et al. 1987). No useful mean densities have yet been obtained for the satellites of Neptune, a situation that hopefully will be remedied when Voyager flies close to Triton in 1989.

Finally, the visible and near-infrared reflection properties of several of the larger irregular satellites of Jupiter and Saturn's only known irregular satellite, Phoebe, are consistent with their being made of carbonaceous chondritic material, i.e., material containing carbon compounds and hydrated minerals (Degewij et al. 1980). Thus, they appear to be compositionally similar to asteroids prevalent in the outer portion of the asteroid belt and the Trojan asteroids of Jupiter. They, as well as comets, provide compositional analogues to the planetesimals that helped to form the outer planets.

We next consider constraints on the time scale for forming giant planets. An obvious and apparently weakly constraining upper limit on the time scale for assembling each of the four giant planets is the age of the solar system, 4.6 Gyr. However, as will be discussed in Sec. III, some models of the formation of Uranus and Neptune have had difficulties in meeting even this modest constraint. A much more stringent upper limit on the time scale of formation can be obtained from estimates of the lifetime of the solar nebula. Since gas derived from the solar nebula represents a major component of the giant planets, they had to obtain this component prior to the dissipation of the solar nebula. Pre-main sequence stars commonly exhibit infrared excesses that are most readily ascribed to the presence of circumstellar disks (Rydgren and Cohen 1985). Infrared excesses persist for the first several Myr for stars having masses comparable to that of the Sun (Rydgren and Cohen 1985). This time scale is also similar to the lifetime of viscous accretion-disk models of the solar nebula (see, e.g., Lin and Bodenheimer 1982).

A further constraint on the time of formation of the giant planets may be provided by the possible requirement that Jupiter fully formed before accretion in the asteroid belt was completed. It has been suggested that gravitational scattering of planetesimals by Jupiter into the region of the asteroid belt prevented the asteroids from accreting into a single terrestrial-sized body and that this same event may have limited the size to which Mars grew (Safronov 1969).

III. FORMATION

Rival Hypotheses

Since the giant planets accreted significant quantities of both gas and solid from the solar nebula, there are two possible analogues to their initial period of formation. On the one hand, the Sun and other stars represent examples of formation induced by instabilities in large gas clouds. According to this analogue, instabilities, perhaps of a gravitational nature, occurred in the solar nebula, leading to the formation of giant gaseous protoplanets (Cameron 1978). In this event, solids were obtained at later times, perhaps through the capture of stray planetesimals, and/or the precipitation of condensed material inside the planet together with the exchange of material with the solar nebula (Cameron 1978; Pollack et al. 1979). We shall call this model the "gas-instability hypothesis."

Alternatively, the formation of the smaller objects in the solar system may offer a more appropriate analogue for the initial phase of formation of the giant planets. In this event, the accretion of solids through low-velocity encounters dominated the initial growth phase of the giant planets. However, in contrast to asteroids and comets, the giant planets became massive enough to capture large quantities of gas from the surrounding solar nebula (Perri and Cameron 1974; Mizuno et al. 1978; Harris 1978; Stevenson 1982*b;* Bodenheimer and Pollack 1986). We shall call this hypothesis the "core-instability model."

The *gas-instability hypothesis* is consistent with the time scale constraints on the formation of the giant planets. They should form on a dynamical time scale, i.e., roughly the time it takes a sound wave to cross a distance equal to their initial dimensions, say one AU. This time scale is much less than the estimated age of the solar nebula, a few Myr. It also implies that Jupiter would have formed before the asteroids could have accreted into a single body in the asteroid belt, since accretional time scales for the terrestrial planets are typically about 100 Myr (Safronov 1969).

However, this hypothesis has a number of serious problems in accounting for the properties of the giant planets. First, there is no obvious reason why the high Z masses for all four planets should be so similar, even though their low Z masses display a much larger variance. Second, it might be very difficult to form segregated, high Z cores, if the solids were accreted after all the gas had been accreted: the solids would have dissolved and mixed with the gas at the high temperatures and pressures of the deeper portion of the envelopes (Stevenson 1982*a*). Third, this hypothesis seems to be very artificial for Uranus and Neptune, since they are enriched in high Z material by a factor of a few hundred compared to solar abundances. Thus, they would have either had to lose more than 99% of their original low Z mass or had a very efficient means of collecting solids from their environment subsequent to their initial formation. For the above reasons, we and many others do not favor the *gas-instability hypothesis*. Therefore, we will focus the remainder of the chapter on the predictions of the core-instability hypothesis.

The *core-instability hypothesis* is consistent in both a qualitative and a quantitative sense with a number of the key properties of the giant planets. First, the existence of a central core region of high Z material is a natural outcome if giant planet formation were initiated by solid body accretion. Second, the presence of sizeable amounts of high Z material in the envelope can be attributed to the increasing difficulty of late accreting planetesimals in penetrating through an increasingly more massive envelope that develops during the later growth phase (Bodenheimer and Pollack 1986). In this same vein, the suprasolar values of C/H for the four giant planets (cf. Table I) can be approximately reproduced by considering the fate of late accreting planetesimals and allowing for the differences in the low Z masses of the giant planets (Pollack et al. 1986*a*). Third, this hypothesis may account for both the absolute value and near constancy of the high Z masses of the giant planets. When the high Z mass reaches a critical value, gas accretion from the surrounding solar nebula becomes extremely rapid. This critical value is insensitive to the nebular boundary conditions and thus accounts for the similar high Z masses of the four giant planets (Mizuno 1980). Furthermore, a critical value comparable to the current high Z masses results when the accretion occurs on a time scale comparable to the lifetime of the solar nebula (Bodenheimer and Pollack 1986).

The core-instability hypothesis is not without its own problems. First, classical solid-body accretion theory predicts formation times that are greater than the lifetime of the solar nebula for all four giant planets and that are greater than the age of the solar system for Uranus and Neptune (Safronov 1969; Safronov and Ruskol 1982). Second, the ratio of high to low Z masses for Uranus and Neptune imply that they never quite achieved critical high Z masses. Thus, the similarity of their high Z masses to those of Jupiter and Saturn may not be so readily explained by this hypothesis. Below, we discuss the core-instability hypothesis in much greater detail, highlighting the above discussion of its successes and problems.

Critical Core Mass

In their classical paper, Mizuno et al. (1978) constructed an equilibrium series of models of solid cores embedded in the solar nebula. Each such model was characterized by a given value of the core mass and a gaseous envelope in hydrostatic equilibrium that joined smoothly with the temperature and density of the solar nebula at the tidal radius of the protoplanet. Small dust grains provided a major source of the envelope's opacity in its outer, convectively stable region. They found that it was not possible to construct an equilibrium model when the core mass exceeded a critical value of about 10 $M_\oplus$, where $M_\oplus$ is the mass of the Earth. They interpreted the inability to construct an equilibrium model at masses larger than the critical value as indicating that the gaseous envelope underwent a hydrodynamical collapse once the core mass exceeded this bound. Thus, very rapid gas accretion took place at this stage.

Mizuno (1980) found that the critical core mass was very insensitive to the nebula boundary conditions, but was somewhat sensitive to the total opacity due to small grains and gases. Thus, for an opacity similar to that expected from a solar abundance of small grains and gases, this model appeared to be capable of accounting for both the actual value of the high Z masses of the giant planets and its relative constancy among the four giant planets. However, Harris (1978) pointed out that the critical core mass might exhibit a sensitivity to the rate of core accretion. Also, an elegant analytical derivation of the critical core mass by Stevenson (1982*b*) showed that its value decreased significantly when the opacity decreased, the accretion time scale increased, or the mean molecular weight increased.

Thus, it may be possible to alter greatly the critical core mass from the values obtained by Mizuno (1980). In particular, consider the following evolutionary pathway explored by Stevenson (1984*a*). As the core mass increases to values in excess of the mass of the Galilean satellites, the temperature at the core's surface becomes hot enough for water to begin evaporating and to become a significant component of the lower portion of the envelope. With continued growth of the core and even greater amounts of evaporation, water may become a major component of the envelope, thereby significantly elevating its mean molecular weight. Furthermore, if all the grains in the solar nebula had accreted to planetesimals prior to this time and if ablation of planetesimals in their passage through the envelope was negligible, the outer portion of the envelope may have lacked grain opacity. Stevenson (1984*a*) found a critical core mass of 0.2 $M_{\oplus}$ for water-rich envelopes containing no grain opacity. While gas accretion from the solar nebula would have been speeded up once this critical core mass was exceeded, a true runaway gas accretion would not have taken place: gas cannot be added too rapidly or the mean molecular weight would have decreased below the value needed for criticality. Thus a controlled, rapid accretion would have taken place, with the rate of addition of solar nebula gas matching roughly the rate of evaporation of water into the envelope. Stevenson termed the protoplanets as being "superGanymedean puffballs" at this stage in their formation.

SuperGanymedean puffballs have several attractive features. First, they may provide an explanation for the different paths followed by the terrestrial and giant planets. Only in the outer portion of the solar system, where the giant planets formed, were temperatures cold enough for water ice to become a major component of the planetesimals. Thus, it may have been possible to achieve a much smaller value of the critical core mass in the outer solar system than in the inner solar system due to the influence of evaporated water on the mean molecular weight of the envelope. Second, the effective cross sectional area of superganymedean puffballs was much larger than that of the core alone due to the presence of a massive envelope. Consequently, such puffballs may have grown rapidly through encounters with one another. Such a scenario might help to obviate the problem of forming the giant planets quickly enough.

The first evolutionary simulations of the core-instability hypothesis were conducted by Bodenheimer and Pollack (1986), who used an adapted stellar evolution code for this purpose. In their baseline model, they assumed that the envelope had solar elemental abundances and a grain opacity equivalent to high Z elements being present in small grains in solar proportions when their condensed phases were stable at the ambient temperatures and pressures. Thus, this baseline case is antithetical to Stevenson's model. They also assumed constant rates of core accretion.

Figure 1 illustrates the temporal evolution of the masses of the core (solid curves) and envelope (dashed curves) for four models considered by Bodenheimer and Pollack (1986). Cases 1, 6, and 7 correspond to solid accretions rates of 10^{-6}, 10^{-5} and 10^{-7} $M_\oplus$ yr^{-1}, respectively, with a full solar abundance of small grains. Case 5 is the same as case 1, except that the grain abundance has been decreased by a factor of 50. During almost all of the growth phase shown in Fig. 1, the envelope mass is much less than that of the core. However, during the later portion of this growth phase, the envelope's mass grows at a sharply increasing rate and eventually the two masses become the same. The core's mass at this point of equality in the two masses can operationally be equated to a critical core mass, since the envelope mass increases much more rapidly than the core mass past this point.

During the early portions of the growth phase of Fig. 1, the luminosity radiated by the protoplanet to space is supplied almost entirely by the gravitational energy released by core accretion. The luminosity, in turn, controls the rate at which the outer portion of the envelope contracts and hence the rate at which gas is added from the surrounding solar nebula. However, as the core mass approaches its critical value, the gravitational energy released by the contraction of the envelope begins to become the dominant energy source for

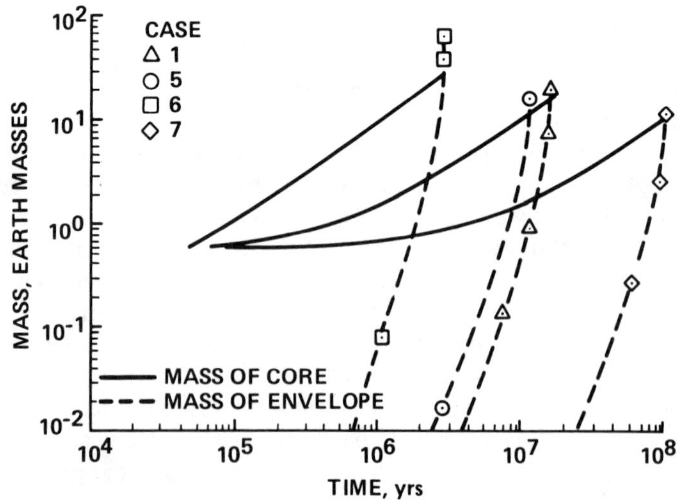

Fig. 1. Envelope mass (dashed) and core mass (solid) as a function of time for several models of Bodenheimer and Pollack (1986). These models simulate the growth of the giant planets for a core-instability scenario of their origin.

the luminosity (cf. Fig. 2). Thus, a strong positive feedback is set up between the rate of gas accretion and the rate of envelope contraction and a true runaway gas accretion results. Although the rate of gas accretion becomes very rapid after the core has achieved its critical mass, the contraction does not become a hydrodynamical collapse.

The value of the critical core mass is very insensitive to the nebular boundary conditions and the amount of grain opacity, in accord with Mizuno's (1980) results, and varies somewhat with the core accretion rate, as proposed by Harris (1978). These dependencies are illustrated in Fig. 1 and shown in numerical detail in Table III. Bodenheimer and Pollack (1986) found that the critical core mass decreased significantly only when both the grain opacity and that due to water vapor were reduced by large factors. Thus, the sensitivity of the critical core mass to total opacity, which was also seen in Mizuno's (1980) calculations, is due chiefly to gaseous and not grain opacity.

If we equate the current high Z mass of Jupiter with its critical core mass, i.e., we neglect planetesimal accretion subsequent to the planet reaching a critical core mass, then an accretion rate of approximately 10^{-4} to 10^{-5} $M_\oplus$ yr^{-1} is implied by the results of Tables II and III. In this case, Jupiter's core achieved a critical mass in several times 10^5 to several times 10^6 yr, in accord with an expected lifetime of a few Myr for the solar nebula, as discussed earlier. Similarly, Saturn may have achieved a critical core mass in several Myr, again in crude accord with the lifetime of the solar nebula. Whether or not such accretion time scales can, in fact, be achieved in a reasonable way will be discussed below.

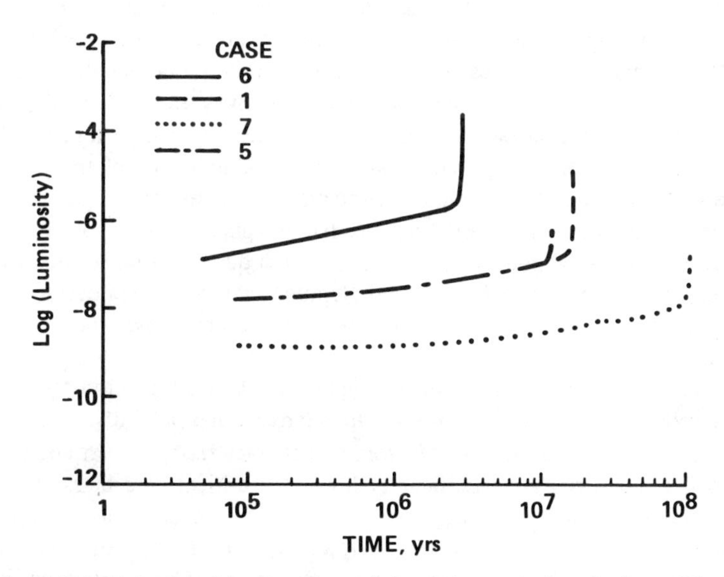

Fig. 2. Intrinsic luminosity of the giant planets during their growth phase for the models shown in Fig. 1.

TABLE III
Critical Core Mass[a]

Case	T (K)	ρ (g cm^{-3})	$\dot{M}$ ($M_\oplus$ yr^{-1})	M_A ($M_\oplus$)	M_{crit} ($M_\oplus$)
1	150	5.8×10^{-11}	10^{-6}	14.5	16.8
2	100	1.2×10^{-10}	10^{-6}	14.5	16.6
3	50	5.8×10^{-11}	10^{-6}	13.9	16.5
4	150	5.8×10^{-10}	10^{-6}	14.3	16.6
5	150	5.8×10^{-11}	10^{-6}	10.0	12.2
6	150	5.8×10^{-11}	10^{-5}	24.5	28.9
7	150	5.8×10^{-11}	10^{-7}	11.0	11.2

[a]T and ρ are the temperature and density of the surrounding solar nebula; $\dot{M}$ is the time independent core accretion rate; M_A is the core mass at the point when the core and envelope accretion rates are the same; and M_{crit} is the critical core mass (by definition it corresponds to the point at which the core and envelope masses are the same). Table from Bodenheimer and Pollack (1986).

Within the context of the calculations illustrated in Fig. 1, Uranus and Neptune differ in a fundamental sense from Jupiter and Saturn. Whereas the low Z mass of Jupiter and Saturn is significantly greater than their high Z mass, the reverse is true for Uranus and Neptune. Hence, Uranus and Neptune never achieved true runaway gas accretion conditions. The reason they did not may lie in the time scale requirements. If core accretion took place more slowly in the zones where Uranus and Neptune formed than in the zones where Jupiter and Saturn formed, as seems likely (Safronov and Ruskol 1982), they may not have reached this point before the solar nebula either had totally dissipated or at least had become much more rarefied. In the latter case, the gas accretion rate may have been limited by a slow transport rate from zones far from the gravitational spheres of influence of the growing planets. Thus, it is to be expected that one giant planet might not quite achieve a critical core mass, but it seems curious that two giant planets evolved almost all the way along the paths of Fig. 1, but did not quite make it. Conceivably, the current gas content of Uranus and Neptune may have been derived from the type of more controlled gas accretion implied by Stevenson's (1984*a*) calculations.

The evolutionary simulations of Fig. 1 and Table III also illustrate some of the requirements needed to make superGanymedian puffballs. Decreasing the grain opacity by a factor of 50 from that characterizing a solar abundance of small grains does not significantly lower the critical core mass. Furthermore, the value of the critical core mass is more sensitive to the opacity supplied by water vapor in the warmer region of the outer envelope (Bodenheimer and Pollack 1986). Thus, making superGanymedian puffballs could be rendered difficult if fresh condensation were occurring in the solar nebula, planetesimals were ablated in the outer envelope, and/or there was a large opacity due to water vapor in the deeper reaches of the envelope.

Limits to Growth

In this subsection, we discuss the factors that controlled the rates at which gas and solids were accreted by the giant planets and the relative timing of the accretion of these two materials. In the previous subsection, we saw that the accretion of solids dominated the early growth phase of the giant planets, but that gas accretion became increasingly rapid as the core mass approached a critical value. Past this point, runaway gas accretion occurred, during which the mass of the envelope rapidly increased and the core mass remained essentially unchanged. Jupiter and Saturn apparently experienced such a runaway accretion, while Uranus and Neptune almost, but not quite achieved such a runaway condition. A key uncertainty in the above scenario is the possible role of a more controlled, but rapid gas accretion phase that may have begun at a much smaller core mass, due to the evaporation of water into the envelope.

Since the ratio of low Z to high Z mass is strongly subsolar for Jupiter and Saturn, there were distinct limits to how much gas they accreted during the runaway gas accretion period. Before considering potential sources of these limits, it is worthwhile to consider where the gas came from. When fully grown, Jupiter's tidal radius had a value of approximately 0.36 AU. Material whose radial distance from the Sun was within several tidal radii of Jupiter's distance would have been strongly gravitationally influenced by it (see, e.g., Lissauer 1987). Using a gas density of 10^{-10} g cc^{-1} and a nebular scale height of 0.1 times the distance to the Sun, values typical of accretion disk models of the solar nebula (Cameron 1978; Lin and Bodenheimer 1982), we obtain a value of about 5 Jovian masses of gas within this strongly perturbed region. Since the tidal radius varies as the cube root of the protoplanet's mass, the mass of readily accessible gas at earlier times during the runaway accretion phase would have been only somewhat smaller than the above value. Thus, the actual gas accretion that took place during runaway gas accretion for Jupiter should not have been limited too severely by the transport of gas from its source regions in the solar nebula to the growing planet if the above nebula models are relevant at this epoch. Similar conclusions can be drawn for Saturn.

The dissipation of the solar nebula by viscous processes and ultimately by a strong solar wind probably limited the amount of gas accreted by Uranus and Neptune and perhaps had some influence on the amount of gas accreted by Saturn. However, it seems unlikely that the dissipation of the nebula limited the amount of gas accreted by Jupiter, since it presumably formed first and the presence of the nebula was necessary for Saturn to obtain its low Z mass. One possible limit to Jupiter's growth and possibly that of Saturn involves its gravitational interaction with the solar nebula. When these planets became massive enough, they may have tidally truncated the solar nebula, i.e., they may have cleared a gap around themselves through the gravitational torque they exerted on the surrounding solar nebula (Lin and Papaloizou 1979). Gap clearing occurred when the torque the planet exerted on the nebula exceeded the torque created by viscous processes in the nebula itself. Tidal truncation could have occurred for plausible values of the turbulent viscosity coefficient when Jupiter achieved its current mass (Lin and Papaloizou 1979*a*).

The rate at which a large object accreted solid planetesimals depended on the space density of planetesimals, the relative velocity between them and the large object, and their orbital period around the Sun (Safronov 1969). When the relative velocity was small compared to the escape velocity from the large object, as it may have been during the initial growth phase, gravitational focusing greatly enhanced the capture cross section over its geometrical value and accretion occurred rapidly. However, once planetesimals were depleted from the initial feeding zone of the protoplanet, larger relative velocities were required for more distant planetesimals to have crossed the orbit of the protoplanet and hence the rate of accretion greatly slowed down (see, e.g., Wetherill 1986; Lissauer 1987). In this sense, the time scale for the giant planets to have reached a critical core mass was determined by the later and not the earlier phases of solid-body accretion.

Using classical accretion theory and a planetesimal space density derived from a "minimum-mass" solar nebula, Safronov and Ruskol (1982) obtained time scales of about 3×10^7 and 2×10^8 yr for Jupiter and Saturn, respectively, to have accreted their high Z mass. Such time scales seem to be incompatible with the expected lifetime of the solar nebula, i.e., no gas would have been available by the time a critical core mass had been achieved. A much lower accretion time scale for the high Z mass would not necessarily have resulted by invoking superganymedean puffballs since such objects might have had low relative velocities with respect to one another and hence might have tended to become prematurely isolated before a protoplanet made from accreting them could have reached a critical core mass for runaway gas accretion to take place.

A potential solution to the above problem has been offered by Lissauer (1987), who suggested that the accretion of the cores of the giant planets took place entirely during the rapid growth phase for planetesimal accretion. Such a sustained rapid growth phase could have occurred if the space density of planetesimals in the region where the outer planets formed was at least 5 to 10 times larger than that implied by a minimum-mass solar nebula. In that event, most of the planetesimals initially present in this region would not have been incorporated into the giant planets, but they may have been gravitationally scattered out of this region by the giant planets, once they had grown to their current masses. In the scenario invoked by Lissauer, the relative velocities of the accreted planetesimals was determined by three-body effects involving the Sun, the protoplanet, and a given planetesimal. Thus, the relative velocities were on the order of the escape velocity at the surface of the protoplanet's Hill sphere and its capture cross section may have been enhanced by a factor of about 1000 over its geometric cross section (Wetherill and Cox 1985). The size of the protoplanet's feeding zone extended for approximately 3 Hill-sphere radii on both sides of its orbit. Lissauer (1987) obtained time scales of about 5×10^5 to 1×10^6 yr for Jupiter to have accreted a high Z mass of 15 to 30 $M_\oplus$.

Very little additional solid body growth occurred during the relatively brief period of runaway gas accretion. If runaway gas accretion was halted by a tidal truncation of the nebula, such a truncation would also have prevented the later accretion of very small planetesimals, which were strongly coupled to the gas phase of the nebula, but would presumably not have prevented bigger planetesimals from reaching the protoplanet. At the end of this gas accretion phase, the protoplanet's envelope would have filled much or all of its Hill sphere. Subsequently, the envelope would have contracted to the present size of the planet, as detailed in the next section. Once the protoplanet had contracted sufficiently, most of its encounters with planetesimals would have resulted in scattering them out of its zone of influence. The amount of high Z accretion that took place between the end of the gas accretion phase and the final contraction epoch would have depended on the relative velocities of the planetesimals and the length of the early contraction phase. Amounts ranging from negligible to significant are possible.

Planetesimal Dissolution

The partial mixing of the high Z material with the low Z material in the envelopes of the giant planets played a major role in determining the current composition of their atmospheres. In an earlier subsection, we saw that the surface of the core became hot enough for significant quantities of water to be evaporated into the envelope once the core mass exceeded about 0.01 $M_\oplus$. As the core mass continued to increase towards the critical value for runaway gas accretion, the envelope became progressively more massive and progressively higher temperatures and pressures characterized its deeper portions. Hence, it became increasingly difficult for planetesimals to reach the core before dissolving in the envelope. In this subsection, we explore this matter.

Figures 3a and b illustrate the temperature and pressure structure of the envelope of one of Bodenheimer and Pollack's (1986) models, case 1, at three epochs during the planet's growth. This model corresponds to a core accretion rate of 10^{-6} $M_\oplus$ yr^{-1} and a solar abundance for the small grain opacity. For this case, the critical core mass equalled 16.8 $M_\oplus$. Both the temperature and pressure near the base of the envelope increase dramatically as the core mass increases towards its critical value.

Planetesimals entering the envelope and falling towards the core experienced several types of interactions with the envelope that affected their passage through it. Gas drag slowed them down, gas dynamical pressure exerted

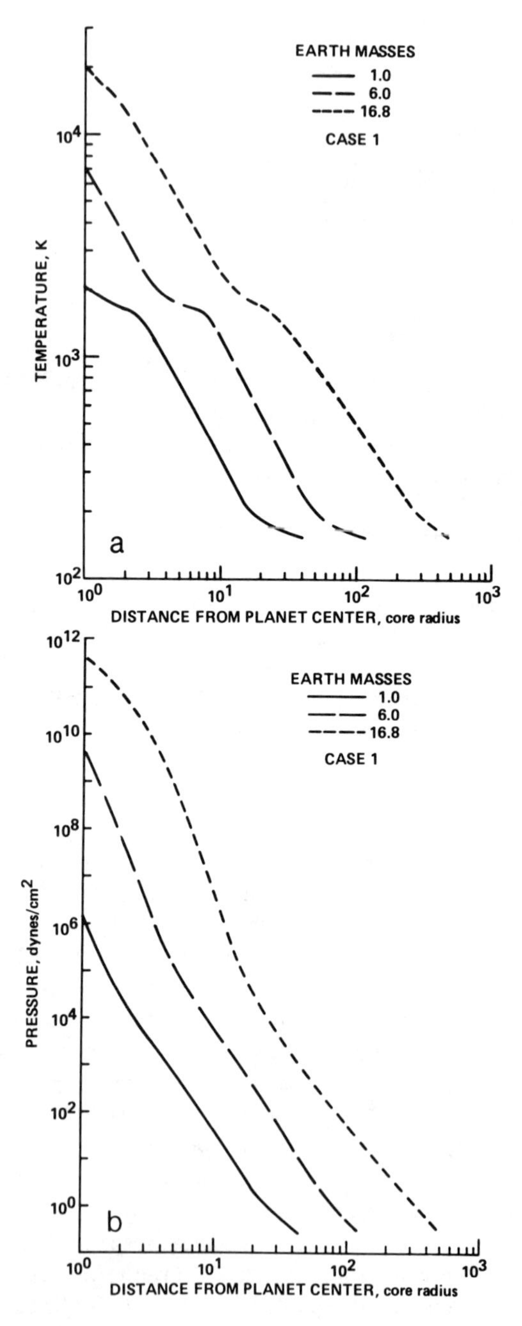

Fig. 3. (a) Temperature within a growing giant planet's envelope as a function of distance from its center for several discrete epochs of case 1 of Fig. 1. These epochs are denoted by the value of the core mass. The distance scale is given in units of the core radius, i.e., a value of 1 refers to the core/envelope interface. The core radius had values of 7.80×10^3, 1.39×10^4 and 1.96×10^4 km when the core mass equalled 1.0, 6.0 and 16.8 $M_\oplus$, respectively. (b) Pressure as a function of distance.

a compressive force acting to fracture them, and radiation from their environment and convective heat exchange with the gas flowing past them tended to warm their surfaces and vaporize their more volatile components (Pollack et al. 1986*a;* Podolak et al. 1988). In the absence of gas drag, the accreted planetesimals traveled at supersonic speeds through the envelope. During the supersonic portion of this traverse, radiation from the shocked region in front of the planetesimals acted as the chief source of surface heating and volatilization.

Figure 4a illustrates the ability of gas drag to slow down planetesimals of varying sizes for case 1 of Bodenheimer and Pollack (1986; Pollack et al. 1986*a*). These planetesimals were made of rocky material and had zero angular momentum. The curves show the distance from the planet's center at which a planetesimal first experienced significant gas drag as a function of the planetesimal's radius. According to this figure, only very tiny planetesimals were significantly slowed down during the early growth phase of the planet, but gas drag should have become very effective for all but the very largest planetesimals (Moon- to Earth-mass objects) as the core mass approaches its critical value. Very similar results were found for planetesimals made of water ice and for other cases of Bodenheimer and Pollack (1986).

The gas dynamical pressure exerted on rocky planetesimals traveling at their free-fall velocity is shown in Fig. 4b (Pollack et al. 1986*a*). Note that this pressure is independent of the size and composition of the planetesimal. A typical value for the compressive strength of planetesimals might have been on the order of 10^8 dyne cm^{-2} (Pollack et al. 1979). Thus, planetesimals probably fragmented near the bottom of the envelope during the later phases of accretion, if they had not been significantly slowed down by gas drag, and if their self-gravity was not strong enough to resist dynamical breakup (self-gravity was important only for very large-sized planetesimals). Such fragmentation would have made it easier to evaporate totally the planetesimal. Note that these conclusions do not depend sensitively on the value chosen for the compressive strength.

Figures 4c and d show the size of the largest planetesimal traveling at its free-fall velocity that was totally evaporated as a function of distance from the planet's center (Pollack et al. 1986*a*). The results of Fig. 4c apply to planetesimals made of rock, while those of Fig. 4d pertain to bodies made solely of water ice. Only the smallest bodies experienced significant volatilization during the early growth phase, but only the largest objects escaped complete evaporation as the core mass approached its critical value. The chief distinction between rocky and icy objects is the ability of small icy objects to evaporate in the outer portion of the envelope.

When all the types of interactions between the envelope and a planetesimal are permitted to operate simultaneously, results very similar to those shown in Figs. 4c and d are found (Pollack et al. 1986*a;* Podolak et al. 1988). If the planetesimals had a broad range of sizes, as seems reasonable, then, over the entire growth phase of the planet, roughly half the

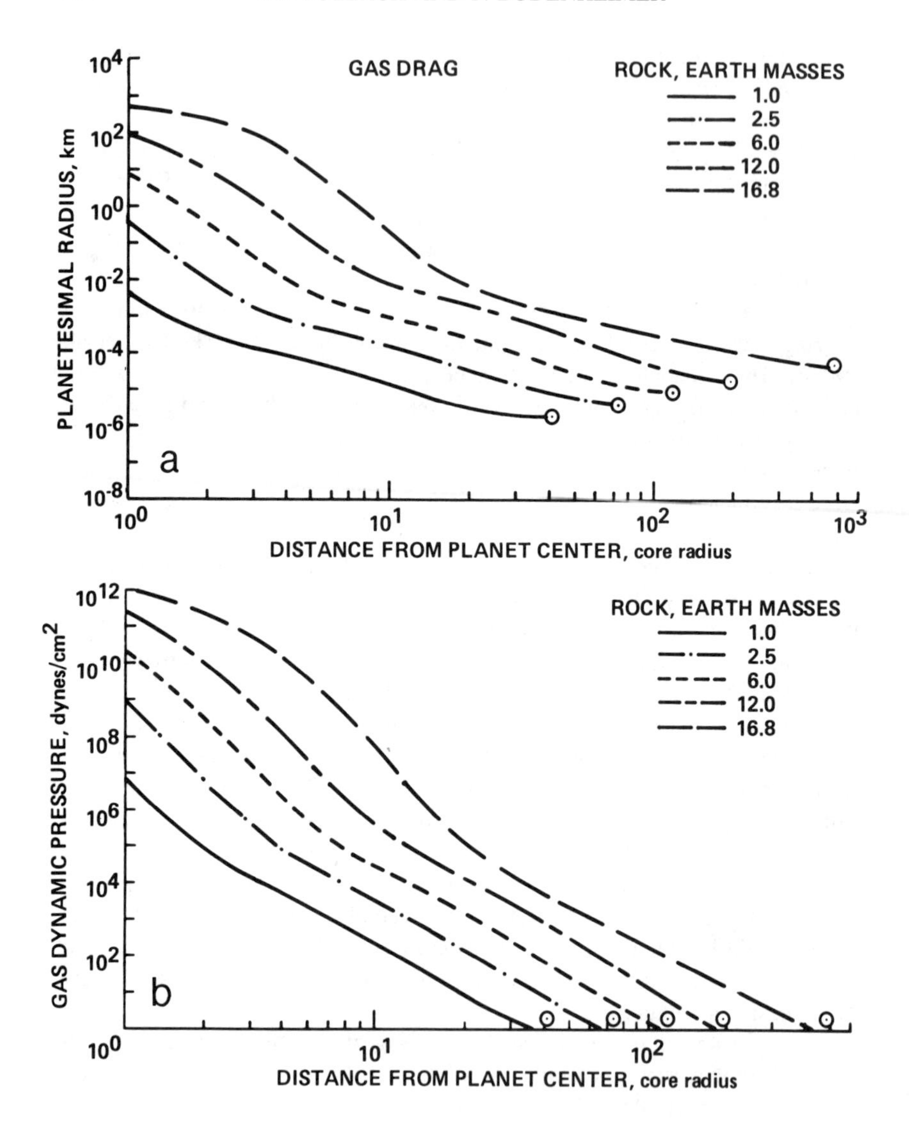

planetesimals dissolved in the envelope and half made it to the core for trajectories having an impact parameter of zero (Pollack et al. 1986*a;* Podolak et al. 1988). A somewhat larger fraction of dissolution would have characterized planetesimals having more oblique trajectories. Much of this dissolution occurred during the later phases of accretion when the envelope was sufficiently massive.

If the planetesimals had a broad range of sizes (meters to hundreds of km), much of their dissolution would have taken place in the deepest portion

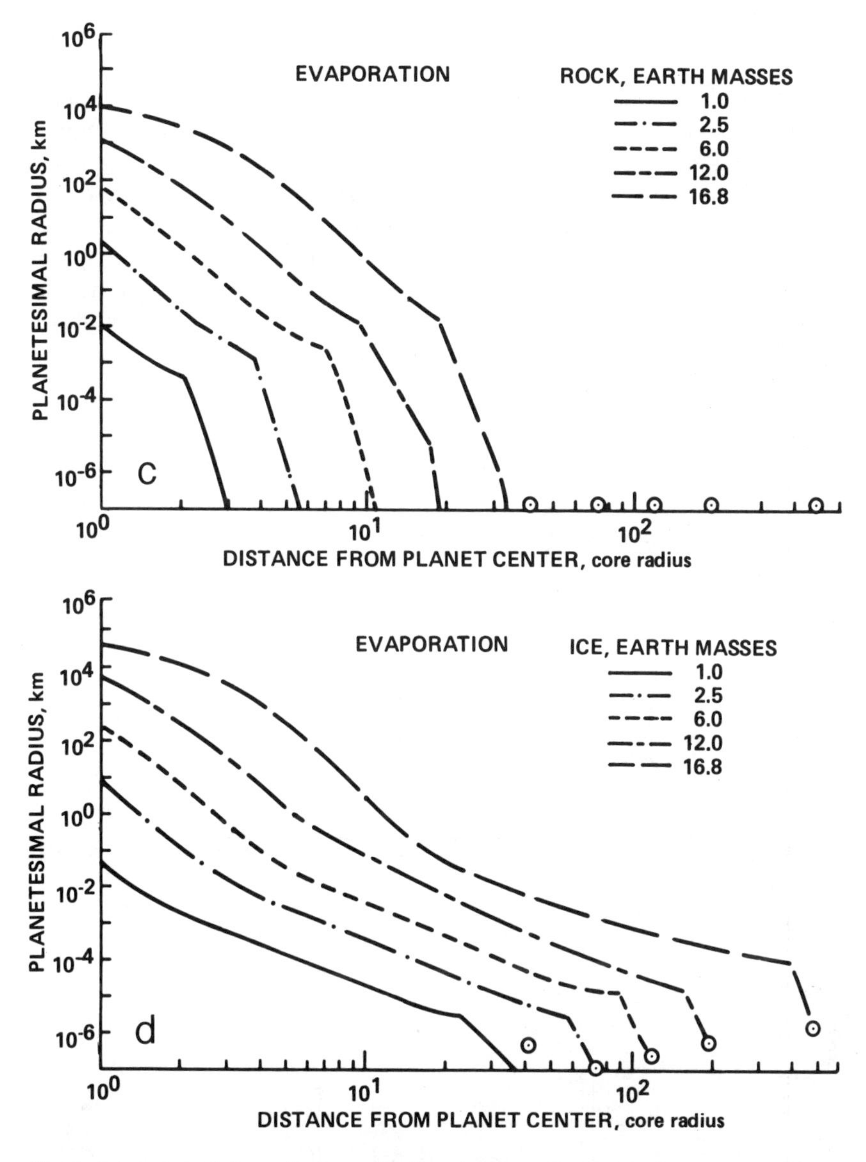

Fig. 4. Interaction of planetesimals passing through the envelopes of the forming giant planets. (a) Size of the largest rocky planetesimal experiencing significant gas drag as a function of distance from the center of the forming giant planet. These results pertain to case 1 of Fig. 1. The various curves correspond to different epochs in the growth of the planet from small core masses to the critical core mass. The circle with the dot inside it on the right-hand side of each curve denotes the location of the interface between the envelope and the solar nebula. (b) Gas dynamic pressure acting on the forward hemisphere of a planetesimal traveling at its free-fall velocity as a function of distance from the center of the growing planet. Note that this pressure is independent of the size and composition of the planetesimal. (c) Size of the largest planetesimal traveling at its free-fall velocity that is completely evaporated as a function of distance in the envelope. These results apply to planetesimals made of rocky material. (d) Same as (c), but for planetesimals made of water.

of the envelope, where most of the gas resided. Thus, the dissolved material would have had a large amount of gas to mix with, but would have initially been largely segregated toward the bottom of the envelope. During the growth phase of the planet, convection zones occurred over only limited regions of the envelope, as illustrated by the vertical bars of Fig. 5a. Thus, dissolved material may not have been well mixed throughout the envelope during this period. As we will see in the next section, complete mixing probably did take place during the early portion of the planet's subsequent contraction phase.

The more volatile species should have been more completely evaporated into the envelopes of the forming planets. Such preferential mixing of the

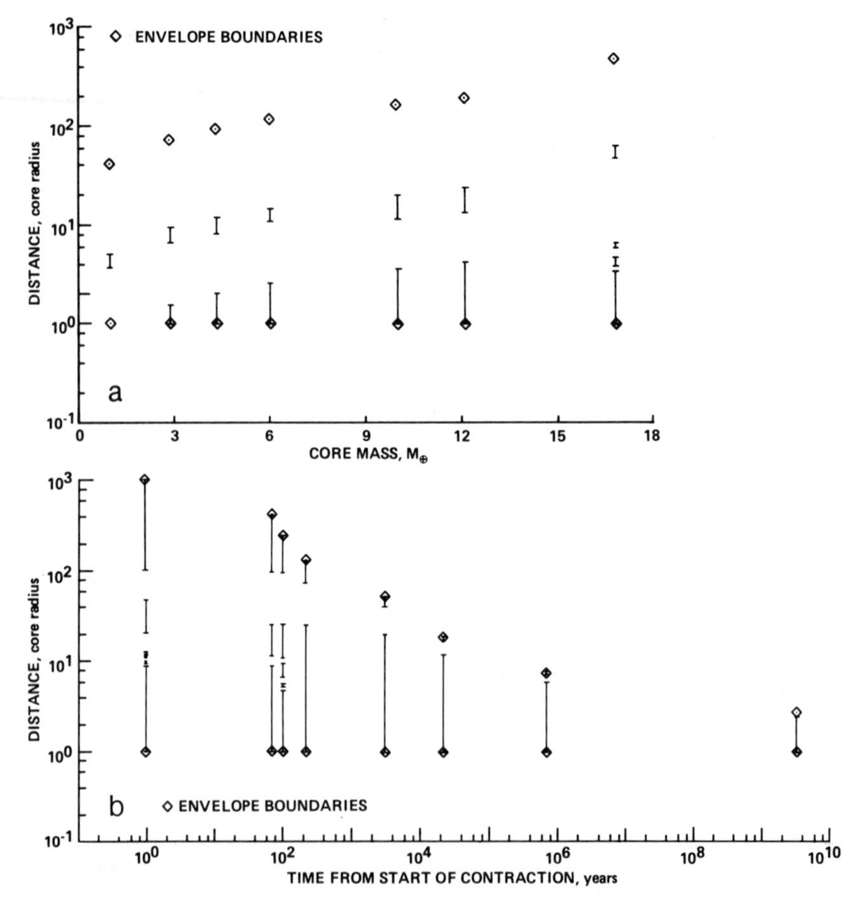

Fig. 5. Location of convection zones within the envelopes of model giant planets is denoted by the vertical lines. Diamonds at the bottom and top of the figure indicate the locations of the core/envelope and envelope/solar nebula interfaces, respectively. These results pertain to case 1 of Fig. 1. (a) Growth phase. Convection zone locations are shown as a function of the core mass. (b) Contraction phase. Convection zone locations are shown as a function of time from the start of the contraction phase.

more volatile species into the envelope would have occurred due to evaporation of material at the core boundary as well as the dissolution of planetesimals traversing the envelope. The results of Figs. 4c and d suggest that the ratio of water to rock dissolved into the envelope may not have greatly exceeded this ratio in the planetesimals. However, greater degrees of enhancement may be possible when allowance is made for planetesimals having oblique trajectories. If carbon and nitrogen were mostly contained in the organic component of planetesimals, the degree of their mixing into the envelope would have been intermediate between the degree of mixing of the water and rock components of the planetesimals. Additional C and N in the envelope would have been derived from the gas phase of the solar nebula.

The potential importance of the mixing of planetesimal material into the envelopes of the giant planets during their growth phase is illustrated by the C/H and P/H ratios characterizing their current atmospheres. As shown in Table I and discussed earlier, C/H is suprasolar in the atmospheres of all four giant planets, with the degree of enhancement increasing from Jupiter to Saturn to Uranus/Neptune. Thus, some of the C in the outer solar nebula was contained in the condensed state and most of the C in the envelopes of the giant planets was derived from this solid component, rather than the gas component.

Figure 6 shows an attempt to reproduce the observed C/H ratio by considering C derived directly from the gas phase of the solar nebula and from the partial dissolution of planetesimals during the planet's growth phase (Pollack et al. 1986*a*). The curves show the predicted C/H ratio in the envelopes of each of the giant planets as a function of the fraction of C contained in the solid phase of the solar nebula α. The fraction β of planetesimals that were assumed to dissolve in the envelope was set equal to 0.5 for all four planets. The C/H ratio is referenced to the solar value, i.e., a value of 1 on the vertical axis of Fig. 6 corresponds to solar elemental abundances. The observed C/H ratios, shown by circles with their associated error bars, can be approximately reproduced by choosing a value of 0.3 for α. More generally, the product of α and β needs to equal about 0.15 for all four giant planets for this model to reproduce the observed C/H ratios. A more refined calculation by Simonelli et al. (1988) indicates that the observed C/H ratios can be reproduced by assuming that about 10% of the C in the outer solar nebula was in the condensed phase. This abundance of C is about 2 times larger than that characterizing the most carbon-rich carbonaceous chondrites (chapter by Lebofsky et al.), but is somewhat less than the abundance of C in Comet Halley (Jessberger et al. 1988). A sizeable fraction of the C in Comet Halley appears to be contained in the so-called CHON particles, that may consist of complex organic polymers (Jessberger et al. 1988).

The calculations illustrated in Fig. 6 provide a simple explanation for the differences in the C/H ratio among the giant planets. Since the high *Z* masses are fairly similar for the giant planets and their low *Z* masses vary consider-

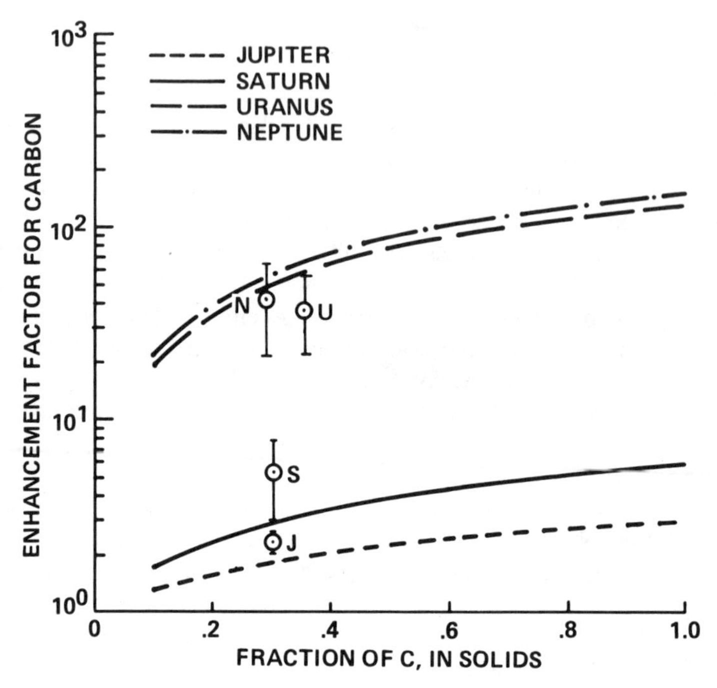

Fig. 6. Enhancement factor for carbon in the atmospheres of the giant planets as a function of the fraction of the C in the outer solar nebula that was in the condensed phase. The enhancement factor is the ratio of C/H in a planetary atmosphere to the solar ratio. Continuous curves show the theoretical results for each giant planet. Half of the C-containing planetesimals were assumed to dissolve in the envelopes of the giant planets during their growth phase and to be well mixed throughout the envelope by the present epoch. Observed values and their uncertainties are shown by the circles and vertical lines for Jupiter (J), Saturn (S), Uranus (U) and Neptune (N).

ably, the variance in the C/H ratio is chiefly due to the variance in the low Z masses for models having constant α and β. The C abundance is determined chiefly by the amount of C-containing planetesimals dissolved in the envelope. For constant α and β, the amount of dissolved C varies only by a factor of a few, with this amount being somewhat larger for Jupiter and Saturn than for Uranus and Neptune. However, the amount of H in the envelope is determined solely by the amount of gas accreted and it thus varies by about a factor of 100 from Jupiter to Uranus/Neptune. An analogy may be drawn here to dissolving a sugar cube of relatively constant size in a beaker of water of greatly varying volume. Much more dilute solutions (low C/H) result for the beakers having the larger volumes (Jupiter). The fact that a model having a constant fraction of C in the condensed phase of the solar nebula and a constant fraction of dissolved planetesimals can approximately fit the observed C/H ratios has some interesting consequences. It suggests that the planetesimals throughout the region of the solar nebula where the giant planets formed had a fairly similar composition. It also suggests that Uranus and Neptune formed in a similar way to the way Jupiter and Saturn did, at least in the epoch prior to their reaching critical core masses.

According to Table I, the P/H ratio in the atmospheres of Jupiter and Saturn are similar to and probably slightly larger than the solar ratio. This ratio was derived from the abundance of PH_3 in these atmospheres. Since additional P may be contained in an unobservable oxidized compound of P at much greater depths in the atmosphere (Barshay and Lewis 1978), the true P/H ratios in the envelopes of Jupiter and Saturn are probably somewhat larger than those given in Table I. We suspect that almost all the P in the outer portion of the solar nebula was contained in the rock component of the planetesimals for the following reasons. First, P readily forms refractory compounds (Fegley and Lewis 1980). Second, such compounds are omni-present in chondritic meteorites (Mason 1962). Third, the P abundance is almost identical in a broad range of meteorites at a value close to its solar abundance (Mason 1962; Cameron 1982). Finally, it seems to us highly unlikely that refractory P compounds formed in the solar nebula or in the interstellar medium (actually in the envelopes of stars losing mass) would be quantitatively converted to volatile P compounds in either location because of kinetic limitations at low temperatures. If P was contained almost entirely in the rock component of planetesimals in the outer solar system, its sizeable abundance in the envelopes of Jupiter and Saturn implies that a significant amount of rock as well as more volatile compounds dissolved into the envelopes of the forming giant planets.

IV. EVOLUTION

After the giant planets had finished accreting, they contracted to their current dimensions. In this section, we first discuss the evolutionary paths followed by these planets subsequent to their formation; consider formation and evolutionary processes that affected the abundance of He in their atmospheres; outline possible scenarios for the origin of their satellite systems within the context of their evolution; and point out the implications of such scenarios for the composition of their atmospheres.

Contraction History

At the end of the runaway gas accretion phase, Jupiter and Saturn had acquired essentially their entire low Z component and had accreted most of their high Z component. The size of the planets at this time depended chiefly on the ability of gas from the surrounding solar nebula to flow into their Roche lobes at a rate commensurate with the rate at which mass was being removed from the outermost part of the Roche lobe by their rapid contraction. Three-dimensional, hydrodynamical calculations indicate that nebular gas existing originally in a ring-like belt about a giant planet can flow into a partially evacuated Hill sphere at a rate of about 10^{-2} $M_\oplus$ yr^{-1} (Hayashi et al. 1985). If this value is adopted as an upper limit to the rate of supply of gas to a protoplanet during the runaway gas accretion phase, then Bodenheimer and Pollack's (1986) simulations imply that Saturn filled all of its Hill sphere during this phase, while the outermost part of Jupiter's Hill sphere was partially evacuated towards the very end of this phase. For both planets, the Hill sphere had a radius of about one thousand times their current radii at the end of this phase. Note that allowance for the angular momentum of the protoplanets would have had very little effect on Bodenheimer and Pollack's simulations of the runaway gas accretion phase; this is because the centrifugal forces near the Hill-sphere boundary would have been much less than the pressure gradient forces for any reasonable choice of the system's total angular momentum, based on the present values for the planet/regular-satellite systems. Throughout the rest of their history, Jupiter and Saturn contracted. Uranus and Neptune followed similar paths to those of Jupiter and Saturn. However, as indicated in the previous section, Uranus and Neptune did not experience a runaway gas accretion phase and, thus, the boundary between their growth phase and their contraction phase may not have been as clean cut as for Jupiter and Saturn.

Figure 7 illustrates the temporal variation of the radius of model planets having the masses of Saturn and Uranus, subsequent to the cessation of accretion (Bodenheimer and Pollack 1986). Initially, both model planets contracted rapidly on a Kelvin-Helmholtz time scale, i.e., one set by the ratio of their

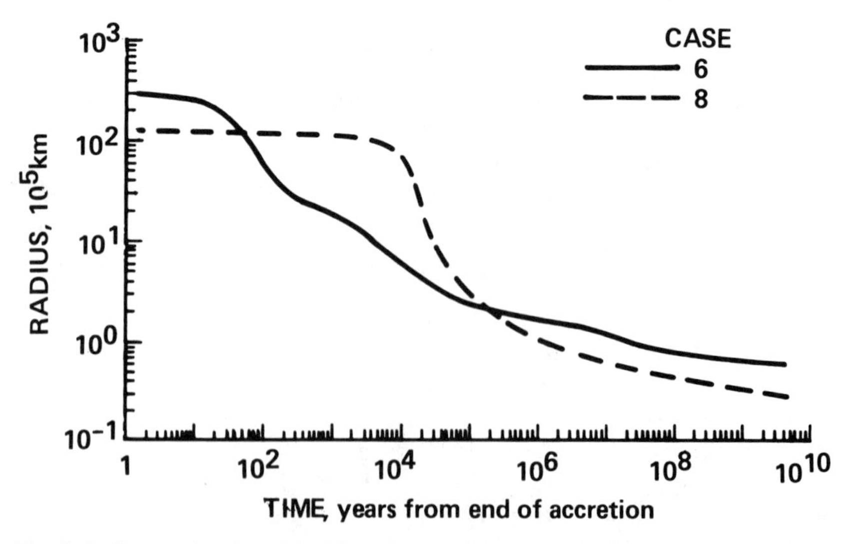

Fig. 7. Radius as a function of time from the end of their growth phase for models having the mass of Saturn (case 6) and Uranus (case 8).

gravitational potential energy to their luminosity. Since the luminosity of Saturn was much higher than that of Uranus during the early contraction epoch (cf. Fig. 8), Saturn contracted more rapidly than Uranus over the first 10^4 yr of this phase. During the later contraction period, an increasing fraction of the fluid envelope attained high densities at which the fluid was much more incompressible. Thus, the rate of contraction progressively slowed down with increasing time from the start of contraction.

This separation into an early and a late contraction phase is also illustrated by the results shown in Fig. 8, in which the luminosities of the Saturn and Uranus models are plotted as a function of the effective temperature of their photospheres. The numbers next to arrows show the time from the onset of contraction (Bodenheimer and Pollack 1986). Except at times close to the present epoch, the internal luminosity, which is the quantity plotted, is much larger than the solar component of the total luminosity. During the early contraction phase, the planet contracted rapidly enough so that temperatures in the envelope increased with increasing time. Thus, the luminosity remained approximately constant, despite the decreasing size of the planet. However, after about 10^4 yr for Saturn and about 2×10^5 yr for Uranus, both the effective temperature and the luminosity decreased with increasing time due to

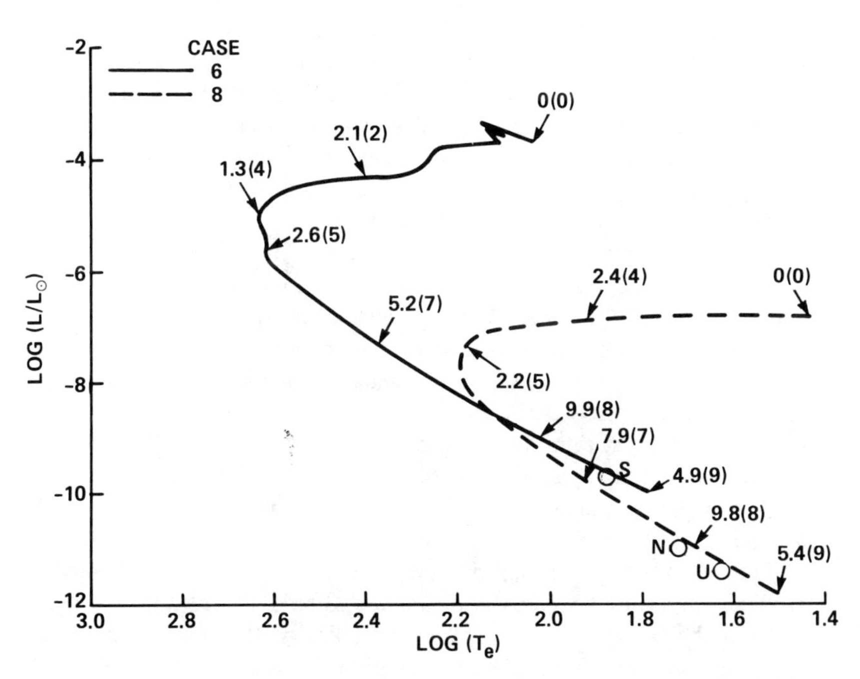

Fig. 8. Intrinsic luminosity as a function of effective temperature for the models of Fig. 7. Times in years are marked on these evolutionary tracks, with the zero point in time referring to the end of accretion. Present observed positions for Saturn, Neptune and an upper limit for Uranus are indicated by the open circles (Pollack et al. 1986*b*).

the increasing incompressibility of the envelope. During this later period, the envelope steadily cooled, with this loss of internal energy becoming a major contributor to the excess thermal energy emitted to space.

The convection zones, which encompassed only portions of the envelope during the growth phase (cf. Fig. 5a), grew to cover an increasing fraction of the planet's envelope, as illustrated in Fig. 5b for the Saturn model of Bodenheimer and Pollack (1986). By the end of the early contraction phase, the convection zone encompassed essentially the entire envelope. Thus, dissolved high Z material may have been well mixed by this point in time. Furthermore, very high convective velocities, on the order of 10^{-2} of the speed of sound, may have ensured that even small condensed grains were well mixed at such early times. However, further calculations are needed to solidify these conclusions, since the simulations of Bodenheimer and Pollack (1986) were based on the envelopes having a solar mixture of elements. Thus, they did not explicitly treat the effects of planetesimal dissolution on the extent of the convection zones.

Helium Abundance

The low Z component of the giant planets contained He and H_2 in solar elemental proportions, as argued earlier. There are two processes that could have led to a nonsolar He/H ratio in their observable atmospheres. First, global scale differentiation is possible due to immiscibility effects. Second, chemical interactions between the dissolved planetesimals and the low Z gas of the envelope could have altered the abundance of H_2 and hence the He/H_2 ratio. These processes are discussed briefly below. Further discussions of these processes, especially the first one, are contained in the Chapters by Conrath et al., Owen and Gautier, and Hubbard.

At high temperatures, He and H_2 are fully miscible in any proportions. However, at low temperatures they are only partially miscible. Such limited miscibility is particularly true of the metallic phase of H, which occurs at pressures in excess of several megabars (Stevenson and Salpeter 1977*b*). Figure 9 shows estimates of the miscibility boundaries for solar proportions of He and H (Stevenson 1982*b*). Also displayed in this figure are estimates of the temperature profiles within Saturn's envelope at several times subsequent to the planet's formation. According to this figure, temperatures inside Saturn's interior are well above the miscibility boundary for molecular hydrogen. Such a conclusion also holds for the other giant planets.

Figure 9 also suggests that the temperature profile of Saturn's interior crossed the miscibility boundary in the outer portion of the metallic hydrogen zone during the last several Gyr of the planet's history. Note that He remained fully miscible in the inner portion of this region. Hence, some He, but not all the He, in the outer portion of the metallic zone would have aggregated into He-rich fluid parcels that would have sunk into the deeper regions of the metallic H zone, where they would have remixed with H (Stevenson and

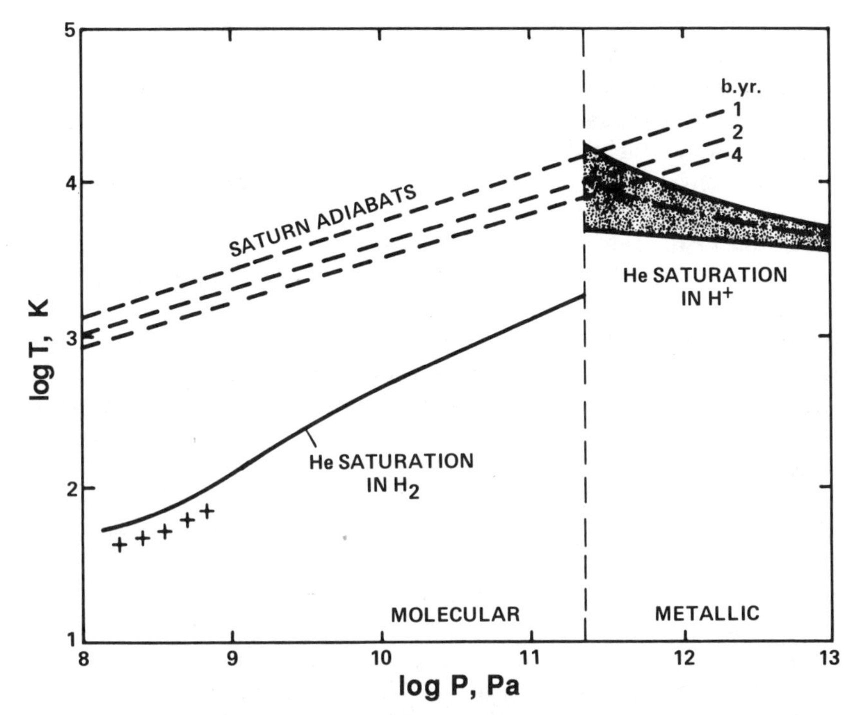

Fig. 9. Saturation temperature of helium in a cosmic mixture of elements as a function of pressure. At temperatures below the saturation temperature helium partially separates from hydrogen. The curves labeled "Saturn adiabats" show the run of temperature through its interior at various times following its formation.

Salpeter 1977*b*). With increasing time from formation, the envelope of Saturn became progressively cooler. Once the miscibility curve was crossed at the metallic H/molecular H interface, a progressively expanding portion of the metallic H region had limited solubility for He and, hence, an increasing amount of He differentiation took place.

Although the molecular H region of the envelope did not directly experience immiscibility, the He abundance here could have been affected by the limited miscibility of He in the outer portion of the metallic H region. In particular, if the envelope was convectively unstable and if He was freely exchanged between the two regions, He would have become progressively less abundant in the molecular envelope and hence in the observable atmosphere with increasing time. The amount of exchange of He across the H phase boundary and whether the He mixing ratio is continuous across the boundary depend on the nature of the phase transition, which is not known (Hubbard's chapter). Nevertheless, the strongly subsolar He/H ratio characterizing Saturn's atmosphere strongly suggests that exchange does occur and that He differentiation has been occurring, in accord with the predictions of Stevenson and Salpeter (1977*b*) and Stevenson (1980).

As helium droplets sunk into Saturn's interior, their viscous interactions with their surroundings acted as a vehicle for converting the change in the planet's gravitational binding energy into heat. Since He is a major constituent of Saturn's envelope, its gravitational segregation could make a major contribution to its internal heat source during its later evolution (Pollack et al. 1977; Stevenson 1980). This conclusion is supported by models of its gravitational contraction history (Pollack et al. 1977). In particular, the predicted magnitude of its internal heat source at the current epoch is a factor of 3 less than the observed value for homogeneous envelope models (Pollack et al. 1977; Grossman et al. 1980; Pollack 1985). The loss of internal heat, built up during the formation and earliest contraction epochs, served as the only heat source in these models. A planet-wide differentiation of He, including exchange between the molecular and metallic regions, could have provided the additional heat source needed to match the observed internal heat flux (Pollack et al. 1977; Stevenson 1980). In effect, this additional heat source slowed down the rate of interior cooling, with this rate acting as a servo mechanism that controlled the rate at which He segregation was occurring.

Since Jupiter is a more massive planet than Saturn, it has cooled more slowly than Saturn. Thus, less, perhaps none, of its metallic H zone has crossed the miscibility boundary for He. This deduction is consistent with the ability of homogeneous cooling models of Jupiter to reproduce approximately its observed internal heat flux at the present epoch (Graboske et al. 1975; Grossman et al. 1980). Nevertheless, the adiabat for Jupiter's interior lies only about 30% higher than the adiabat for Saturn's interior (Conrath et al. 1987*a*). Also, the observed He/H ratio for Jupiter's atmosphere appears to be somewhat subsolar (cf. Table I). Thus, a limited amount of He segregation may have been occurring in Jupiter's envelope.

Pressures as high as several megabars are not realized in the envelopes of Uranus and Neptune. Thus, these envelopes do not contain a metallic H region and so no helium segregation is expected due to immiscibility effects. However, the He/H_2 ratio in their atmospheres may exhibit a small deviation from the ratio derived from solar elemental abundances due to chemical interactions between the low and high Z components of these planets. In particular, a small amount of the molecular hydrogen obtained from the gases of the solar nebula may have been expended in fully reducing chemical compounds derived from dissolved planetesimals. If so, the He/H_2 ratio in the atmospheres of Uranus and Neptune may be slightly suprasolar (Pollack et al. 1986*a*).

Carbon is the key element for altering the He/H_2 ratio due to its high elemental abundance and its proclivity to be almost entirely in its fully reduced form, methane, in the uppermost portion of the envelopes of the giant planets. It seems likely that much of the C in the planetesimals of the outer solar nebula was in the form of organic compounds and CO clathrates, i.e., in compounds that were not fully reduced. Using the observed C/H ratio in Uranus' atmosphere, assuming a neutral reduction/oxidation state for the C in Uranus-forming planetesimals, and allowing for the reduction of N and certain rock-bearing elements, Pollack et al. (1986*a*) estimated that the He/H_2 ratio in its atmosphere may be elevated by about 6% with respect to the solar value. Unfortunately, this deviation is probably too small to be detected, given the uncertainties in the Voyager data and our current knowledge of the solar He/H ratio (see chapter by Owen and Gautier).

If the C/H ratio in Neptune's atmosphere is substantially elevated above that in Uranus' atmosphere, the He/H_2 ratio in its atmosphere would be more suprasolar than is the case for Uranus' atmosphere. Conceivably, measurements taken from the Voyager flyby of Neptune in 1989 will be sensitive enough to detect such a difference between the two planets, as this difference is independent of the solar abundance value. Since the C/H ratios in the atmospheres of Jupiter and Saturn are much less than that for Uranus' atmosphere (cf. Table I), reduction of dissolved planetesimal material is expected to have had only a very minor effect on the He/H_2 ratios in the Jovian and Saturnian atmospheres. Finally, it is worth noting that the He/H_2 ratio may be slightly subsolar in the deeper parts of the envelopes of Uranus and Neptune due to the tendency for elements, such as C, to exist in either their oxidized form or dissociated form at the higher temperatures in these regions.

In summary, He segregation due to its partial solubility in metallic H appears to have greatly lowered the He/H_2 ratio in Saturn's atmosphere, to have somewhat lowered this ratio in Jupiter's atmosphere, and not to have influenced this ratio in the atmospheres of Uranus and Neptune. This ratio may be slightly suprasolar in the latter atmospheres due to the complete reduction of certain elements, especially C, which were derived from the dissolution of planetesimals.

Origin and Properties of Satellites

The composition of the satellites of the outer planets is related to the composition of planetesimals in the solar nebula and directly or indirectly to the composition of the high Z elements in the envelopes of their parent planets. Consequently, it is worthwhile to discuss the origin of these bodies to illuminate such relationships and then to relate their composition to that of their planets. Below, we first explore these matters for the irregular satellites and then for the regular ones.

Gas drag represents one plausible mechanism for capturing the irregular satellites (Pollack et al. 1979). According to this hypothesis, stray planetesimals from the solar nebula passed through the outer portions of the envelopes of the giant planets during their early contraction phase and experienced enough gas drag to be permanently captured. Due to the rapid contraction of the envelope at this epoch, gas was soon eliminated from the orbital location of the captured objects so that they continued to remain in orbit around the planet, rather than to be subjected to continuing gas drag. Had gas drag continued to act on the captured bodies, they would have spiralled into the deeper portions of the planets' envelopes to be incorporated into them. The gas drag model can account approximately for the observed sizes and locations of the irregular satellites (Pollack et al. 1979).

This hypothesis was initially presented at a time when it was thought that the giant planets underwent a hydrodynamical collapse phase just after they finished accreting. As discussed in the first subsection of this section, it now appears that such a collapse did not occur, but, nevertheless, it does seem that the giant planets underwent a rapid contraction shortly after they completed their growth phase (cf. Fig. 7). This very early contraction was more rapid for Saturn than for Uranus. Thus, satellite capture, as opposed to continued planetesimal accretion, may have been easier for Jupiter and Saturn than for Uranus and Neptune. Below, we will assume that this hypothesis is correct and explore some of its implications. Alternative hypotheses are discussed by Pollack (1985) and Stevenson et al. (1986).

Conversion of a captured object into an irregular satellite occurred only during a limited time span (one of rapid contraction), only for certain size ranges, and only for certain impact parameters. In all likelihood, a much larger fraction of captured bodies became incorporated into the planet (Pollack et al. 1979). Thus, this hypothesis for the origin of the irregular satellites implies that some planetesimal accretion occurred after the giant planets had essentially finished accreting gas and began to contract. Unfortunately, it is very difficult to make meaningful quantitative estimates of the amount of planetesimal accretion that occurred during the contraction phase.

The visible albedo and visible and near-infrared spectral characteristics of some of the larger irregular satellites of Jupiter and Phoebe, Saturn's irregular satellite, suggest that they are made of carbonaceous chondritic-like material, as discussed in Sec. II (Degewij et al. 1980). Thus, C, perhaps mostly in the form of organics, is an abundant component of the irregular satellites. This inference is in accord with the conjecture that the C in the atmospheres of the giant planets was derived chiefly from the dissolution of C-containing planetesimals. If the analogy between irregular satellites and carbonaceous chondrites is taken too literally, one would infer that water is present only as water of hydration, and not water ice, in the irregular satellites. But, the very dark appearance of the nucleus of Comet Halley and the very likely large abundance of water ice in it warns against such a facile inference (Keller et al. 1986).

The regular satellites were formed by solid-body accretional processes in nebula disks that surrounded their parent planets in their early histories. There are two possible sources of these disks and hence the source material for the regular satellites: the disks may have been derived from gases, grains and planetesimals present in the nearby solar nebula. In this case, the satellite-forming disks would have been classical viscous accretion disks, in much the same sense that the solar nebula is thought to be (chapter by Coradini et al.). Alternatively, these disks may have been derived from the outer portions of the envelopes of the giant planets themselves (Johnson et al. 1987; Korycansky in preparation). In this latter case, the equatorial regions of the outer envelope may have experienced a steadily increasing angular velocity as the planet contracted due to both conservation of the planet's total angular momentum and the outward transfer of angular momentum within convection zones and their adjacent environs. Eventually, these regions may have rotated fast enough to leave behind disks as the planet continued to contract.

In our opinion, there are several reasons for favoring a planetary source, as opposed to a solar nebula source, for the material in the satellite accretion disk. First, the albedo differences between Phoebe and the small regular satellites of Saturn may indicate different sources for these two sets of objects. Phoebe has a very low albedo, presumably due to a high abundance of organics, whereas the small regular satellites are quite bright and hence have little or no organics on their surfaces (Smith et al. 1982*b*). This difference could be easily understood if Phoebe was a planetesimal captured from the solar nebula and the regular satellites were derived from planetesimals that were evaporated in the envelopes of the giant planets and later recondensed in the outer envelope and nebula disk. Organics need not be produced in the latter situation.

It might be argued that the reflectivity properties of satellites characterize only the top μm of their surfaces and therefore need not be representative of their bulk composition. However, we point out that impact events keep the top km or so of the surface well mixed and therefore prevent a small superficial coating of externally derived material from aliasing greatly the observed surface reflectivity. It might also be argued that near-surface mobilization of volatile ices and ultraviolet/particle irradiation and alteration of these ices could also alias the surface composition. While this is certainly true, such processes are unlikely to lead to the gross apparent differences in surface composition between Phoebe and the small regular satellites of Saturn unless their interiors are also chemically different.

In addition, our current understanding of the early history of the giant planets tends to favor a planet source for the nebular disk. If runaway gas accretion of Jupiter and Saturn were halted by these planets attaining sufficient mass to tidally truncate the solar nebula, as discussed earlier, then they would have filled all or some nontrivial fraction of their Hill spheres prior to truncation and the gas phase of the solar nebula would not have had easy access to these regions thereafter. Hence, it may not have been possible to form a satellite accretion disk from the solar nebula after Jupiter and Saturn had contracted to close to their present dimensions. There may also be a timing problem for this scenario since the solar nebula may have been dissipated before the giant planets contracted sufficiently to leave room for a satellite accretion disk in the region of the regular satellites. This constraint is particularly severe for Uranus, which may not have finished forming before the solar nebula dissipated, let alone contracted to a much smaller size (cf. Fig. 7).

Although we favor a planet source for the satellite forming nebula, we do not consider the matter settled. The alternative viewpoint that the regular satellites formed from material derived from the solar nebula is discussed in the Chapter by Coradini. Below, we consider the implications of each of these hypotheses.

First suppose that the regular satellites formed from material derived from the solar nebula. On the average, the icy Galilean satellites of Jupiter, Titan, Saturn's largest satellite and the outer satellites of Uranus all are made of approximately 55% rock and 45% water by mass, as discussed in Sec. II (Johnson et al. 1987). Therefore, the planetesimals of the outer solar nebula, on the average, had this composition, with the proportion of rock and water being constant throughout the region where Jupiter, Saturn and Uranus formed.

Stevenson (1984*b*) has suggested that the rock-to-ice ratio of Ganymede, Callisto and Titan are enhanced over the ratio for the planetesimals that formed them as a result of strong accretional heating, vaporization of water into their atmospheres, and a strong subsequent loss of water. If so, the rock-to-ice ratios of the smaller, regular satellites of Saturn, 40/60, might be more representative of the plantesimals that formed the regular satellites of Jupiter and Saturn. We tend to favor a rock-to-ice ratio of 55/45 since this ratio also holds for the satellites of Uranus, which are much less massive than the icy satellites of Jupiter and Titan. Therefore, they should have suffered far less water loss during accretion. Also, the mean density of Pluto implies an even somewhat larger rock-to-ice ratio than 55/45 (Simonelli et al. 1988). Pluto presumably formed in the outer solar nebula. As its mass is much less than that of the icy satellites of Jupiter and Titan, water loss during Pluto's accretion should have been much less important than for the more massive satellites. All this leads us to suspect that water loss during accretion was a far less important process than was proposed by Stevenson (1984*b*) and that some other explanation needs to be advanced to explain the differing rock-to-ice ratios for Titan and some of the smaller regular satellites of Saturn.

Dissolution of the water component of these planetesimals in the envelopes of the forming giant planets would have been somewhat more effective than for the rock component. In this case, somewhat more water than rock by mass should characterize the envelopes of Jupiter, Saturn and Uranus than was present in the planetesimals.

The proportions of rock and water incorporated into planetesimals in the outer solar nebula were determined chiefly by the partitioning of O among several chemical species (Johnson et al. 1987). These include oxides, such as SiO_2 and MgO, present in the rock component, CO gas, and, of course, H_2O. If one uses solar elemental abundances and assumes a chondritic assemblage of minerals in the rock, then about 15% of the available O would have been contained in the rock component. The fraction of the remaining 85% of O that was incorporated into water depends strongly on the oxidation state of C in the solar nebular since the solar elemental abundance of C is about 60% of the abundance of O. If none of the C in the outer solar nebula was in the form of gaseous CO, all of the remaining O would have been incorporated into water and the planetesimals would have contained 40% rock and 60% water by mass, if we assume that all the water was condensed in the outer solar nebula, as seems likely. Alternatively, if all the C was in the form of CO, the planetesimals would have been made of 75% rock and 25% water. The satellite-derived value of 55% rock and 45% water implies that some, but not all of the C in the outer solar nebula was in the form of gaseous CO.

Next, suppose that the satellite disks were formed from the outer portions of the envelopes of the giant planets. In that case, the composition of the regular satellites reflects more directly the composition of these envelopes. In particular, the envelopes of the giant planets contain rock and water in proportions of approximately 55 to 45 by mass. Since the planetesimals that dissolved in the envelopes preferentially lost their water component, their rock-to-water mass ratio exceeded 55 to 45. Consequently, CO constituted an even larger fraction of the C compounds in the outer solar nebula than the fraction deduced for the scenario in which satellite disks were derived from the solar nebula. Some evidence for the above conclusions is provided by the rock-to-ice ratio of Pluto, approximately 70/30 (Simonelli et al. 1988).

In summary, gaseous CO and condensed organics were major C-containing species in the outer solar nebula. The large abundance of CO is in accord with Lewis and Prinn's (1980) proposition that kinetic factors prevented CH_4, the thermodynamically favored form of C in the outer solar system, from being the dominant form of C there. Also, the envelopes of the giant planets contain approximately equal masses of dissolved rock and water.

V. SUMMARY

There are strong observational and theoretical reasons for suspecting that deviations from solar elemental abundances are the rule rather than the exception for the atmospheres of the giant planets. As illustrated in Table I, the C/H ratio is distinctly suprasolar in the atmospheres of all four giant planets, while the He/H ratio is distinctly subsolar in the Saturnian atmosphere and probably subsolar in the Jovian atmosphere.

Since the giant planets were assembled from nonsolar mixtures of the gases and solids present in the neighboring solar nebula, it is not surprising that the ratio of elements present in the solid phase, to H, an element present only in the gas phase, is nonsolar in the atmospheres of the giant planets. In general, such ratios can be expected to be suprasolar because the giant planets formed preferentially from the solid component of the solar nebula and because a significant fraction of the planetesimals that formed them dissolved in their envelopes. Furthermore, some elements, such as C and N, were present in both the gas and solid phases of the solar nebula and some elements, such as O, were partitioned among a number of chemical species, not just water. Thus, nonsolar elemental abundance ratios are likely even among the high *Z* elements.

Finally, even though the giant planets accreted He and H in solar proportions from the solar nebula, the atmospheres of the giant planets need not have a solar ratio of He to H_2 for several reasons. In the case of Jupiter and Saturn, the limited miscibility of He in the outer portion of their metallic H zones affected the He content of their atmospheres. Slightly suprasolar He/H_2 ratios are possible in the atmospheres of Uranus and Neptune due to the complete reduction of elements, such as C, derived from dissolved planetesimals. Below, we illustrate these concepts by discussing the atmospheric abundances of several key elements and compounds.

Carbon

Carbon was present in both the gas and solid phases of the solar nebula. Estimates of its partitioning among various compounds in the solar nebula may be derived from the C/H ratio characterizing the atmospheres of the giant planets and the rock-to-water ratios found in their regular satellites. It appears that about 10% of the available C in the outer solar nebula was in the solid phase, with the remainder largely in the form of CO gas. Most likely, the condensed C was contained in organic compounds, as inferred from the spectral properties of the irregular satellites.

The C/H ratio is suprasolar in the atmospheres of all four giant planets, with the degree of enhancement above solar increasing from Jupiter to Saturn to Uranus/Neptune. These data imply that most of the C contained in the atmospheres of the outer planets was derived from dissolved planetesimals, rather than from gas accreted from the solar nebula. The trend in the C/H ratio among the giant planets was determined chiefly by the variation in the ratio of high-to-low *Z* material accreted by them.

Water

Water was present in the outer solar nebula only in the solid phase due to the low temperatures there. However, some of the O in the nebula was present in the form of mineral oxides and CO gas. Thus, water may have had only about half the abundance it could have had if it were the only O reservoir in the solar nebula.

Dissolution of planetesimals in the envelopes of the forming giant planets should have been somewhat more effective for water than for organics or rock, in view of the much greater volatility of water. Consequently, we would expect the water abundance in the atmospheres of the giant planets, at depths below the water condensation level, to be somewhat more suprasolar than the C abundance. This expectation, together with the water-to-rock ratios derived from the regular satellites, strongly indicate that water is not extremely underabundant in the deeper portions of the Jovian atmosphere. Hence, its low abundance in the 2 to 6 bar region of the Jovian atmosphere must have a meteorological rather than a cosmogonical explanation.

Rock Elements

Refractory elements would have been present entirely in the condensed phase of the outer solar nebula. The slightly suprasolar abundance of PH_3 in the atmospheres of Jupiter and Saturn, which provides a lower bound on the P/H ratio in their envelopes, may provide evidence that a significant fraction of the rock component of planetesimals was dissolved in the envelopes of the giant planets during their formation.

The densities of the regular satellites of the outer solar system provide an estimate of the rock-to-water ratio in the envelopes of their planets. If the satellite-forming disks of the outer planets were generated from the outer envelopes of the planets, this ratio equals about 55 to 45 by mass. If these disks formed from material derived directly from the solar nebula, then a ratio of 55 to 45 characterizes the planetesimals in the solar nebula and this ratio is somewhat smaller in the envelopes of the giant planets due to the preferential dissolution of water.

Nitrogen

Like C, N was probably present in both the gas and solid phases of the solar nebula. If CO was the dominant gaseous form of C in the outer solar nebula, then N_2 was probably the dominant N_2-containing gas, again due to kinetic limitations in converting N into its thermodynamically favored form NH_3 at low temperatures. Similarly by analogy to C, the second key reservoir for N in the solar nebula was solid organics.

Since the N/C ratio of organics found in meteorites and comets seems to be subsolar (see, e.g., Kissel and Krueger 1987), the N/C ratio in the planetesimals that helped to form the giant planets and, thus, the N/C ratio of the envelopes of the giant planets may also have been subsolar. Furthermore, since not all the available C in the solar nebula was in the condensed phase, while all the available refractory elements were, the N/S ratio in the envelopes of the giant planets is likely to be substantially subsolar and, in particular, on the order of unity. In such a case, NH_4SH clouds are likely to be more prominent relative to NH_3 clouds in the atmospheres of the giant planets than would be the case if the N/H ratio was solar.

Helium

Significant decreases in the atmospheric He abundance are possible due to the limited miscibility of He in metallic H at sufficiently low temperatures. Such a decrease in the He/H_2 ratio is expected and is observed to be greater for Saturn's atmosphere than for Jupiter's atmosphere due to the more rapid cooling of the interior of less massive Saturn. In order for this process to affect the atmospheric He abundance, there needs to be an exchange of He across the molecular/metallic H interface. Such an exchange implies that there is a much bigger reservoir of He for this differentiation process to encompass than would be the case if only the metallic H region was involved and hence He segregation is a more potent source for the internal heat flux.

Since the envelopes of Uranus and Neptune lack metallic H regions, He differentiation is not expected to occur in their envelopes. In the absence of such differentiation and as a result of the several hundred-fold enrichment of high Z material over solar abundances in the bulk composition of these planets, the complete reduction of dissolved planetesimal material might have an observable effect on the He/H_2 ratio. In particular, slight suprasolar ratios result from this process. The amount of enhancement depends on both the oxidation/reduction state of key elements, such as C, in planetesimals and the fraction of planetesimals that were dissolved in the envelopes of the forming giant planets.

Prospectus

The deviations from solar abundances of elements present in the observable atmospheres of the giant planets is far more than an interesting curiosity. These deviations provide important insights and constraints on the origin and evolution of the giant planets. As these abundances become defined for a wider suite of elements, through *in situ* measurements made from entry probes, a more complete picture will emerge. Properties of satellites, such as their mean densities, provide additional constraints on these fundamental issues.

PART V
Satellites

PRESENT STATE AND CHEMICAL EVOLUTION OF THE ATMOSPHERES OF TITAN, TRITON AND PLUTO

J. I. LUNINE
University of Arizona

S. K. ATREYA
University of Michigan
and

J. B. POLLACK
NASA Ames Research Center

Titan, Triton and Pluto are lunar-sized solid bodies which have substantial atmospheres. The atmosphere of Saturn's satellite Titan has been well characterized by the Voyager 1 flyby; it is mostly molecular nitrogen with an admixture of methane, a surface pressure of 1.5 bars and a surface temperature of 95 K. The atmosphere of Neptune's satellite Triton is inferred from the spectroscopic presence of surface methane and, possibly, nitrogen. Temperatures at Triton's distance from the Sun are substantially lower than at Titan's and hence a thinner atmosphere is expected. Pluto's atmosphere is poorly characterized but spectroscopically identified methane exists partitioned to some unquantified extent between a surface and atmosphere. Preliminary studies of Pluto's satellite Charon show no evidence of methane. The fundamental driving force in the long-term evolution of Titan's atmosphere is the photolysis of methane in the stratosphere to form higher hydrocarbons and aerosols. The current rate of photolysis, together with the inferred undersaturation of methane in the lower troposphere, suggests a reservoir of methane mixed with higher hydrocarbons. This "ethane-methane ocean" serves as both the source and sink of photolysis, and contains a mass of nitrogen comparable to the atmospheric abundance. In the absence of outgassing or external resupply, the ocean composition evolves toward more ethane rich as methane is photolyzed over geologic time. The atmosphere responds to the change in ocean composition with a corresponding change in gaseous composition and spatially averaged cloud composition. The resulting change in the atmospheric thermal structure, and hence climate, is an evolutionary problem which taxes our present understanding of the checks and balances of infrared opacity and incident sunlight. The outstanding problem in the earliest evolution of Titan concerns the origin of atmospheric nitrogen: was it introduced into Titan as molecular nitrogen or ammonia? Measurement of the argon-to-nitrogen ratio in the present atmosphere is a diagnostic test of the competing hypotheses: a large argon abundance suggests molecular nitrogen, while a small abundance favors ammonia. Titan has an atmosphere in contrast to the Galilean satellites because temperatures were low enough in its region of formation to incorporate ammonia into hydrates and methane into clathrates. In the region of Jupiter, methane clathrate was not stable. The absence of ammonia-derived nitrogen, however, is not so easy to explain: perhaps ammonia hydrate was not stable in the Jovian nebula, nitrogen was removed by magnetospheric processes, or the absence of methane prevented a greenhouse environment from being present at the appropriate time to build up a molecular nitrogen atmosphere. Similar speculations on the atmospheres of Triton and Pluto must await the Voyager-Neptune flyby in 1989 and development of advanced Earth-based observational tools for observing the outer solar system.

I. INTRODUCTION

Overview

Three lunar-sized objects in the outer solar system are known or strongly suspected to have atmospheres of mass sufficient to affect surface processes in a measurable way. Saturn's largest satellite Titan is encased in a nitrogen-rich atmosphere with a surface density five times that of the Earth's nitrogen-rich envelope, and a secondary methane component which plays the role of condensable. The outermost planet, Pluto, has an atmosphere which contains at least methane at a partial pressure of perhaps one-fiftieth the total pressure of the Martian atmosphere. Neptune's large satellite Triton has methane and possibly nitrogen present as condensed volatiles on its surface; estimates of the surface atmospheric pressure range over orders of magnitude from microbars to one-tenth of a bar (one bar $\sim$ one atmosphere = 760 torr).

Since the first of these atmospheres was positively identified 44 yr ago (Kuiper 1944), their relationship to the envelopes surrounding the terrestrial planets has been pondered. The work of Urey, his students and scientific descendants in establishing cosmochemistry (the study of the relationship of the chemical composition of planets and their satellites to the origin of the solar system) as a *bona fide* field of study sharpened the questions (Urey 1952; Lewis 1971; Stevenson 1985). The era of exploration of the outer solar system by spacecraft and advanced groundbased techniques provides the basis for making the questions more specific and numerous. Fundamentally, however, we wish to understand *the evolution of the atmospheres of Titan, Triton and Pluto to their present state, the processes during formation of these bodies which initiated their atmospheres, and the characteristics of the atmospheres which are diagnostic of the particular processes by which these objects formed.*

In this chapter we attempt to synthesize the most recent work on the present state and evolution of the atmospheres of Titan, Triton and Pluto, and present in a balanced fashion current ideas on the formation of these objects and their parent systems. Notice that dealing with Titan, Triton and Pluto as a set of objects requires artful rhetoric at times, because the first two objects are satellites of major planets, the last itself a planet with a satellite, Charon. Study of Titan's atmosphere presumably teaches us something about how satellites form around giant planets, because Titan's very regular orbital characteristics strongly suggest an origin associated with the formation of Saturn. Neither Triton nor Pluto have orbits which are sufficiently "regular" (low eccentricity and inclination) that they can be typed so readily. Triton may have formed in place about Neptune, or been captured; Pluto might have been a satellite of one of the outer planets to start with, but no compelling model of its evolution to a solar orbit is available. Pluto's Charon may or may not have been captured. Compounding our difficulty is that we have much more detailed information on Titan than we do on the other objects, because of the Voyager-Saturn encounters in 1980 and 1981. However, the Voyager 2 flyby of Neptune's system in 1989 will provide us with comparable information on Triton.

The difficulties cited above should not be considered a rationale for abandoning comparative studies of the atmospheres of these three objects. The location of Titan, Triton and Pluto in the outer solar system, with consequently low temperatures, affords us study of atmospheres with compositions distinctly different from those of the terrestrial planets. Hypotheses involving steady-state processes and evolution of Titan's atmosphere can be tested on Triton's atmosphere after the 1989 Voyager encounter. Finally, the knowledge acquired of the atmospheres of Titan and Triton will provide us with better notions of what processes and constituents to look for in Pluto's atmosphere, and how to look for them.

In pondering the origin and evolution of atmospheres, we cannot ignore the surfaces and interiors of the bodies in question. Unfortunately, our knowledge of the interior of Titan, based upon bulk density, is rudimentary. For Triton and Pluto, it is essentially nonexistent. The state of our understanding of the surfaces of these bodies is ironically, the reverse: Titan's ubiquitous haze prevents any observation of the surface, whereas Triton's and Pluto's surfaces have to the best of our understanding, been detected by groundbased studies. As we discuss below, our conceptions of the nature of Titan's surface based on inferences from its atmospheric properties are of crucial importance in modeling the evolution and origin of the atmosphere itself.

Organization

We begin this chapter with a subsection containing questions which can be addressed by study of the cryogenic atmospheres of Titan, Triton and Pluto. While not complete, these questions serve as a guide to the types of inquiries which the current state of our knowledge allows. Section II summarizes current understanding of physical processes operating on the atmospheres and surfaces of Titan, Triton and Pluto, to provide a foundation for the discussions of origin and evolution which make up the bulk of the chapter. By necessity, details of the observations and associated controversies are omitted; we emphasize processes. We recommend the excellent review chapters by Hunten et al. (1984), and Cruikshank and Brown (1986), as well as Chapter 7 of Atreya (1986), for more details on the observations. A review chapter by Morrison et al. (1986) contains a brief analysis of issues of the evolution of Titan's atmosphere based upon the available data. Section II also discusses inferred inventories of volatiles based on our understanding of the atmospheres and surfaces. (In this chapter, the term "volatile" is used to denote any material capable of having a substantial vapor pressure under temperature conditions obtained in the outer solar system; anything more volatile than ammonia is included.)

Section III considers atmospheric evolution inferred from physical processes identified as operating in the present-day environment. These include photochemistry and radiation-driven chemistry, atmospheric escape, and surface-atmosphere coupling. An attempt is made to use these processes to "run the clock back" and tie the present-day state to models of the origin and early evolution of Titan, in particular. Section IV considers models for the origin and early evolution of Titan, Triton and Pluto with emphasis on their atmospheres. This section overlaps with the chapters by Coradini et al., and Prinn and Fegley, but is placed in the context of the evolution studies of Sec. III. Section V discusses the relevance of outer solar-system atmospheres to those of the terrestrial planets, in terms of composition and modes of formation. The chapter concludes with a summary of the major points.

Atmospheres of Titan, Triton and Pluto: Questions

Of primary interest is the relationship of the chemical composition of the atmospheres of Titan, Triton and Pluto to that of the gaseous reservoirs (nebulae) out of which they formed. What is the origin of nitrogen in Titan's atmosphere? More specifically, is it primordial, i.e., derived from a gas containing molecular nitrogen, or was it produced by chemical processing of ammonia? How much carbon monoxide was introduced into Titan and its early atmosphere? Was methane brought into Titan as frozen condensate or entrapped in clathrate? The noble gas abundances in Titan's atmosphere, as yet only weakly constrained, contain some of the information needed to answer these questions. Similar questions can be asked for Triton and Pluto, although with less assurance that they are relevant to the poorly determined atmospheric compositions.

A related set of questions arises regarding the physical processes operating during the origin and evolution of these atmospheres. Was Titan formed in an optically thick gaseous nebula or a relatively gas-free environment? What was the primary process responsible for conversion of ammonia to nitrogen: photolysis or shock heating? How hot could a massive primordial atmosphere around Titan have been, and how long did it exist? Was a massive primordial atmosphere lost by extreme ultraviolet (EUV) radiation, a T-Tauri wind or shock heating due to impacts? Why does Titan have an atmosphere while Ganymede and Callisto do not? Could most of Titan's atmosphere have been added by impact of volatile-rich comets after accretion? Were Triton's volatiles extruded onto the surface during a heating episode associated with capture by Neptune? What is the relationship between the atmospheres of Titan, Triton and Pluto and those of the terrestrial planets?

Finally, a set of questions regarding the later evolution of Titan's atmosphere to the present state can be framed. What processes control the overall thermal structure of the present Titan atmosphere? Is it stable against perturbations by volcanic extrusion of volatiles or cometary impacts? Is there a surface liquid hydrocarbon layer (i.e., an ethane-methane ocean)? What is the response of the atmosphere to secular changes in such an ocean due to photolysis of methane? Are there qualitative differences between the evolution of Titan's atmosphere in contact with an ocean as opposed to a solid surface with little or no methane? What is the effect of a large chemical reservoir such as an ocean on the abundance of deuterated species in the atmosphere, and on the atmospheric noble gas ratios? Can the total surface and atmosphere volatile abundance be deduced from an ethane ocean model and our understanding of methane photolysis? For Triton, what is the effect of seasonal variation of insolation on the present-day atmosphere? Is the response of the atmosphere to Triton's seasons, as observed by ground- and space-based observatories, diagnostic of the total volatile abundance on the surface? Is the origin of methane on Pluto related to that of methane on Triton and Titan?

In the remainder of the chapter we address these questions in the context of our current knowledge of Titan, Triton and Pluto.

II. CURRENT UNDERSTANDING OF PHYSICAL PROCESSES IN THE ATMOSPHERES AND SURFACES OF TITAN, TRITON AND PLUTO

Titan

Atmospheric Properties. The most detailed description of Titan's atmosphere based upon the Voyager observations is that of Hunten et al. (1984); significant work has occurred subsequent to the writing of that chapter primarily in the area of thermal structure. The information that we cite in this subsection is derived from the reviews of Hunten et al. (1984), Morrison et al. (1986) and Atreya (1986), unless otherwise noted.

Titan's atmosphere is predominantly molecular nitrogen (N_2), with a surface pressure of 1.5 bars. The air temperature at the surface is 94.5 ± 0.4 K, and drops to 71 K at the tropopause level of ~45 km, 0.1 bar (see subsection on thermal structure, below). This is a conservative lower bound because of opacity effects in the 540 cm^{-1} region of the Voyager Infrared Interferometer Spectrometer (IRIS) data used to determine this value. An upper bound derived from the tropopause temperature and the radio occultation profile is ~102 K (Lellouch et al. 1987). The Voyager radio-occultation experiment, when combined with the temperature measurements of the IRIS experiment, yields a range of atmospheric mean molecular weights which center on 28 AMU. That nitrogen and not carbon monoxide is the dominant gas was determined by Voyager Ultraviolet Spectrometer (UVS) observations of ions of atomic and molecular nitrogen in the upper atmosphere. Atmospheric constituents of different molecular weight are also allowed, so long as they balance in such a way as to yield a mean value near 28 amu. Neon has not been detected; its abundance in the upper atmosphere can be no more than 1%. The upper limit of 6% for argon in the upper atmosphere may still allow for greater amounts in the lower atmosphere; obviously its maximum abundance can be played off against the abundance of methane in the lower atmosphere to yield an appropriate mean molecular weight:

$$28 = 16\, x_{CH_4} + 28\, x_{N_2} + 36\, x_{Ar}; \quad x_{CH_4} + x_{N_2} + x_{Ar} = 1. \tag{1}$$

Here x_y is the mole fraction of component y. Setting the maximum methane abundance near the surface to be its saturated value (but see below) permits up to 25% argon near the surface.

The methane abundance is uncertain, and yet most telling about the atmospheric thermal structure and possible surface states. The maximum physically permissible methane mixing ratio as a function of temperature is given by the saturation vapor pressure (supersaturation may locally permit higher abundances, but, averaged globally and over time, the abundance is expected to be less than its saturated value). Using the vapor pressure relations from Brown and Ziegler (1979) we find the vapor pressure for pure methane over its liquid at 95 K to be 0.19 bars, or 13% mole fraction at the surface. However, the radio occultation temperature profile in the lowermost atmosphere is consistent with a dry adiabat in pure nitrogen gas. Allowing methane condensation at 95 K would drop the adiabatic temperature gradient by a factor of 2 to its moist value. Moreover, variations in mean molecular weight due to methane condensation affect the derivation of temperature from the density determined by the radio occultation results, in such a way as to produce superadiabatic temperature gradients near the surface. Detailed analyses of these constraints have been conducted by Lindal et al. (1983), Eshleman et al. (1983) and Flasar (1983). The argument of Eshleman et al. is that methane condensation cannot occur anywhere that the Voyager-derived temperature gradient exceeds the moist pseudo-adiabat (a moist adiabat in which the condensate is assumed to drop out). They derived an upper limit to methane abundance of 3% at the surface, with condensation occuring above 15 km. Flasar (1983) imposed the constraint that the molecular weight variation due to methane condensation not result in a moist superadiabatic temperature profile, when applied to the T/m data from the radio occultation results. His analysis permits a surface methane abundance of 0.07, or a pressure of 0.11 bars, with condensation beginning at ~3 km above the surface. At the tropopause the methane mixing ratio is limited to 1 to 2% by condensation, consistent with the stratospheric value based on IRIS data. These results set constraints on the atmospheric thermal structure and surface models.

The presence of condensable methane in the troposphere forces consideration of cloud formation, which affects incoming solar and outgoing thermal infrared radiation. The issue of the amount of condensate, morphology of cloud and size of the particles is far from being resolved, because no currently available observational technique is very sensitive to the presence of tropospheric condensate. Recent determinations of limits on cloud optical thickness include Pope and Tomasko (1986) and Toon et al. (1988), which serve as constraints on models of cloud thickness morphology and microphysics. We defer a discussion of cloud models to the subsection on thermal structure of Titan's atmosphere.

Other significant constituents include molecular hydrogen (H_2), at $\sim 2 \times 10^{-3}$ mixing ratio, and hydrocarbons including ethane (C_2H_6), propane (C_3H_8) and acetylene (or ethyne; C_2H_2), in order of abundance. These range from 20 down to 2 parts per million (ppm) in the stratosphere, and are highly oversaturated at the tropopause. A suite of other hydrocarbons and nitriles are present in the stratosphere in varying abundances. Finally, carbon monoxide (CO) has been detected from the Earth at a level of 60 to 150 ppm, and carbon dioxide (CO_2) at 1.5 parts per billion (ppb).

This array of chemical components in Titan's atmosphere carries critical information on current atmospheric processes, surface composition, and weakly constrains models for the origin and evolution of the atmosphere. The essential key to applying the compositional constraints to surface models and evolution lies in the inferred photochemical cycles.

Inferred Photochemical Cycles and Atmospheric Escape. In the present atmosphere of Titan, constituents with different combinations of C and H (such as hydrocarbons CH_4, C_2H_2, C_2H_6, C_2H_4, C_3H_4, C_3H_8 and C_4H_2), C and N (C_2N_2), C, N and H (HCN, HC_3N) and C and O (CO and CO_2) are present. CO and CO_2 could be either indigenous to Titan, or they could have an extraplanetary source. The C-H, C-N-H compounds and hydrogen are all produced as a result of chemical processes operative in the present atmosphere. Although the solar ultraviolet-driven chemistry (photochemistry) plays the dominant role, the precipitation of charged particles may be important at high latitudes, especially when Titan is within the magnetosphere of Saturn. The following discussion on the photochemistry of methane illustrates the manner in which the various C-H, C-N-H and H compounds are formed on Titan. The most extensive photochemical scheme is that of Yung et al. (1984 and references therein).

For the most part, the hydrocarbons on Titan are produced in a manner similar to Jupiter, with one major difference: the Jovian H_2 is replaced by N_2. This results in the production of C-N-H molecules. The CH_4 photochemical scheme, shown in Fig. 1, however, applies to Titan insofar as the production of major hydrocarbon species is concerned. The photodissociation of CH_4 occurs following the absorption of solar photons with wavelengths between 1000 and 1600 Å. For all practical purposes, however, Lyman-alpha photons at 1216 Å account for the bulk (90%) of CH_4 photolysis rate due to the large solar flux at that wavelength. Since the mole fraction of CH_4 is 2% at the tropopause of Titan, the optical depth in CH_4 at Lyman-alpha would be approximately 7×10^6, implying that most of the photolysis would occur several hundred kilometers above the tropopause. The direct photolysis of CH_4 yields a diurnally averaged column-integrated rate of 1.2×10^9 cm^{-2} s^{-1}. The catalytic photolysis (see reaction R18 later), however, could result in a

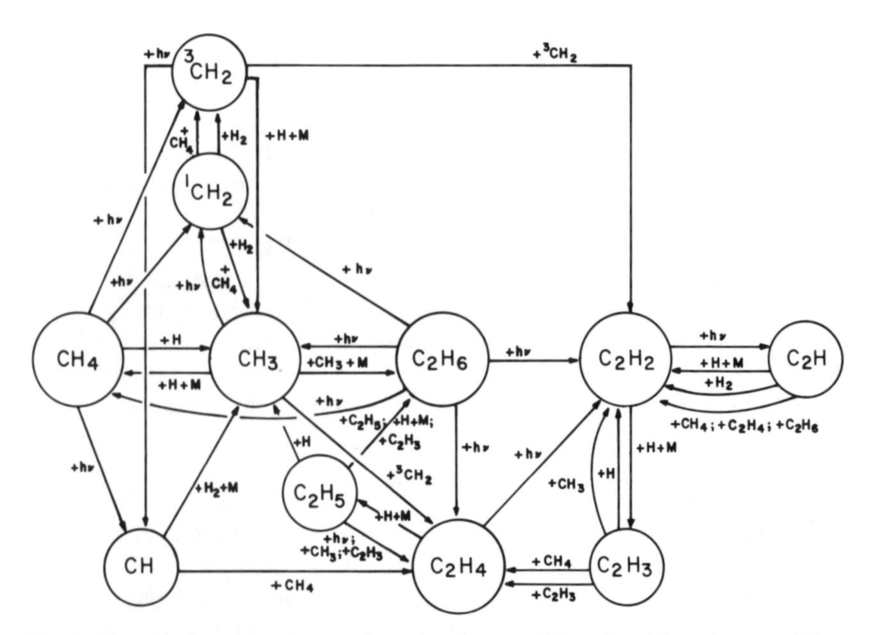

Fig. 1. Photochemistry of methane on the major planets and Titan (figure from Atreya and Romani 1985).

rate 3 times greater (Yung et al. 1984). The photodissociation of CH_4 yields $^1CH_2$, $^3CH_2$, $^3CH_2$ and CH[a] (note the conspicuous absence of CH_3, as it is kinetically disallowed), i.e.

$$CH_4 + h\nu\ (\lambda \lesssim 1450\ \text{Å}) \rightarrow CH + H_2 + H \quad \text{(R1)}$$

$$\rightarrow {}^3CH_2 + 2H \quad \text{(R2)}$$

$$\rightarrow {}^1CH_2 + H_2 \quad \text{(R3)}$$

followed by

$$^1CH_2 + N_2 \rightarrow {}^3CH_2 + N_2. \quad \text{(R4)}$$

Unlike Jupiter, $^1CH_2$ is rapidly quenched by N_2 to $^3CH_2$ (reaction R4).

CH radicals react with CH_4 yielding C_2H_4. The latter is photolyzed below 1700 Å to yield C_2H_2, i.e.

$$CH + CH_4 \rightarrow C_2H_4 + H \quad \text{(R5)}$$

followed by

$$C_2H_4 + h\nu\ (\lambda < 1700\ \text{Å}) \rightarrow C_2H_2 + H_2. \quad \text{(R6)}$$

The most likely fate of the $^3CH_2$ is the conversion to C_2H_2 by the self-reaction

$$^3CH_2 + {}^3CH_2 \rightarrow C_2H_2 + 2H. \quad \text{(R7)}$$

Further reaction of $^3CH_2$ with C_2H_2 can result in the formation of C_3H_4 (methylacetylene or allene):

$$^3CH_2 + C_2H_2 \rightarrow CH_3C_2H\ \text{(methylacetylene)} \quad \text{(R8)}$$

or

$$\rightarrow CH_2 = C = CH_2\ \text{(allene)}. \quad \text{(R9)}$$

Allene has not yet been detected on Titan.

The acetylene produced in the upper atmosphere is transported to the lower atmosphere. It is photolyzed in the stratosphere (at least partially catalytically), and condenses at the tropopause temperature (Yung et al. 1984). The photolysis of C_2H_2 leads to the formation of diacelytene, C_4H_2, and

---

[a]The reader is referred to Yung et al. (1984) for the best currently available data on rate constants, many of which are simply estimates.

possibly higher order polyacetylenes ($C_{2n}H_2$, where n = 3,4,5 . . .) in the following manner (Allen et al. 1980; Romani and Atreya 1986, 1988; Pollack et al. 1987*a;* Glicker and Okabe 1987):

$$C_2H_2 + h\nu \; (2000 \text{ Å} \geq \lambda \geq 1450 \text{ Å}) \rightarrow C_2H + H \quad \text{(R10)}$$

$$C_2H + C_2H_2 \rightarrow C_4H_2 + H \quad \text{(R11)}$$

followed by

$$C_4H_2 + C_2H \rightarrow C_6H_2 + H \quad \text{(R12)}$$

or

$$C_4H_2 + h\nu \rightarrow C_4H + H \quad \text{(R13)}$$

$$\text{or} \rightarrow 2C_2H \quad \text{(R14)}$$

followed by

$$C_4H + C_2H_2 \rightarrow C_6H_2 + H. \quad \text{(R15)}$$

C_8H_2 may be formed on reaction of C_6H_2 with C_2H or on reaction of C_4H with C_4H_2, i.e.

$$C_6H_2 + C_2H \rightarrow C_8H_2 + H \quad \text{(R16)}$$

or

$$C_4H + C_4H_2 \rightarrow C_8H_2 + H. \quad \text{(R17)}$$

At the appropriate proportion of CH_4, the reaction of C_2H with CH_4 is favored, and it leads up to the formation of C_2H_6, i.e.

$$C_2H + CH_4 \rightarrow CH_3 + C_2H_2 \quad \text{(R18)}$$

followed by

$$CH_3 + CH_3 + N_2 \rightarrow C_2H_6 + N_2. \quad \text{(R19)}$$

(It has been suggested [W.R. Thompson, personal communication, 1987] that reactions of CN with the hydrocarbons may be even faster than those of C_2H, resulting in the formation of complex nitriles and other compounds. The abundance of CN and the rate constants and kinetics of the CN reactions at the appropriate atmospheric conditions are poorly known, however.)

As with C_2H_2, the loss of C_2H_6 is by downward transport, followed by photolysis and condensation (Yung et al. 1984). In the subsequent chemistry, propane, butane and other heavier hydrocarbons may be formed, as follows:

$$C_2H_5 + CH_3 + N_2 \rightarrow C_3H_8 + N_2 \qquad (R20)$$

$$C_3H_7 + CH_3 + N_2 \rightarrow C_4H_{10} + N_2 \text{ etc.} \qquad (R21)$$

C_2H_5, C_3H_5, C_3H_6, C_3H_7 etc. are formed in the hydrogen-addition reactions of, respectively, C_2H_4, C_3H_4, C_3H_5, C_3H_6, etc.

Some recycling of CH_4 occurs on reaction of CH_3 with atomic hydrogen, and partially by surface chemistry following the precipitation of the condensible hydrocarbons. The rate of conversion of methane to higher hydrocarbons and the resulting accumulated thickness of material at the surface from Yung et al. (1984) is given in Table I. The values of thickness assume, for simplicity, that the present rate of photolysis was maintained over the age of the

TABLE I
Downward Flux of Products of Methane and Carbon Monoxide Photolysis and Accumulated Depth Over the Age of the Solar System

Species	Flux[a]	Depth[b]
C_2H_6	5.8×10^9	0.6
C_2H_2	1.2×10^9	0.1
C_3H_8	1.4×10^8	0.02
CH_3C_2H	5.7×10^7	0.006
HCN	2.0×10^8	0.02
HC_3N	1.7×10^7	0.002
C_2N_2	6.0×10^6	0.001
CO_2	3×10^5	2×10^{-5}

[a]Flux in $cm^{-2}s^{-1}$ normalized to the surface, from Yung et al. (1984). Discrepancies between their model stratospheric abundance and Voyager IRIS results lead to a factor of 5 uncertainty in flux for propane, a factor of 2 for acetylene and hydrogen cyanide, and less than a factor of 2 for ethane. The model does not include production of aerosols. Loss of methane by photochemistry is 1.5×10^{10} cm^{-2} s^{-1} (Yung et al. 1984).

[b]Depth is given in km for densities of pure substances from Raulin (1985).

solar system. It should also be remembered that they are based upon a model which does not determine the abundance of hydrocarbons converted to the aerosol particles observed in the upper stratosphere of Titan. The results of Pope and Tomasko (1986) indicate that one-fifth of the photolyzed methane is converted to aerosols in the stratosphere (if their mass production rates refer to the surface which is not explicitly stated). A crucial conclusion of our understanding of photochemistry on Titan is that in the absence of resupply of methane, photolysis alone would irreversibly convert all of Titan's methane to heavier hydrocarbons in a time scale ~10 Myr.

The CH_4 photochemistry just discussed is modified due to the presence of large quantities of N_2 on Titan. Being inert, N_2 by itself does not participate directly in the chemistry. However, N_2 dissociates into $N(^2D)$ and $N(^4S)$ by the action of solar ultraviolet radiation below 1000 Å, magnetospheric electrons and galactic cosmic rays. The globally averaged production rate of $N(^2D)$ and (^4S), 1.8×10^9 cm^{-2}s^{-1} (Strobel and Shemansky 1982), by the magnetospheric electrons is by far the largest. It is, respectively, a factor of 3 and 10 greater than that due to the ultraviolet radiation and the cosmic rays. The reactions of $N(^2D)$ and $N(^4S)$ with CH_4, $^3CH_2$ and CH_3 lead to the production of HCN as follows:

$$N(^4S) + CH_3 \rightarrow HCN + H_2 \quad (R22)$$

$$N(^4S) + {}^3CH_2 \rightarrow HCN + H \quad (R23)$$

and

$$N(^2D) + CH_4 \rightarrow HCN + H_2 + H. \quad (R24)$$

The photolysis of HCN gives CN, which produces C_2N_2 in a self reaction, cyanoacetylene (HC_3N) on reaction with C_2H_2 and ethyl cyanide (C_2H_3CN) on reaction with C_2H_4. Formation of polymers is also likely as a result of reactions between CN and other hydrocarbons. The complete chemical scheme involving the above-mentioned reactions in an N_2 - CH_4 atmosphere is depicted in Fig. 2. It is important to recognize that a relatively significant source of HCN lies in the ionosphere of Titan (Atreya 1986). The magnetospheric electrons, as well as the solar EUV photons produce N_2^+ and N^+ ions, following their action on the atmospheric N_2. N_2^+ charge exchanges readily with N, producing N^+ ions. The reaction between N^+ and the upper atmospheric CH_4 gives rise to HCN^+ ions, among other products. HCN^+ reacts with H_2 to give H_2CN^+. The latter is removed by dissociative recombination, yielding HCN in the following manner (Atreya 1986):

$$N^+ + CH_4 \rightarrow HCN^+ + H_2 + H \quad (R25)$$

$$HCN^+ + H_2 \rightarrow H_2CN^+ + H \quad (R26)$$

$$H_2CN^+ + e \rightarrow HCN + H. \quad (R27)$$

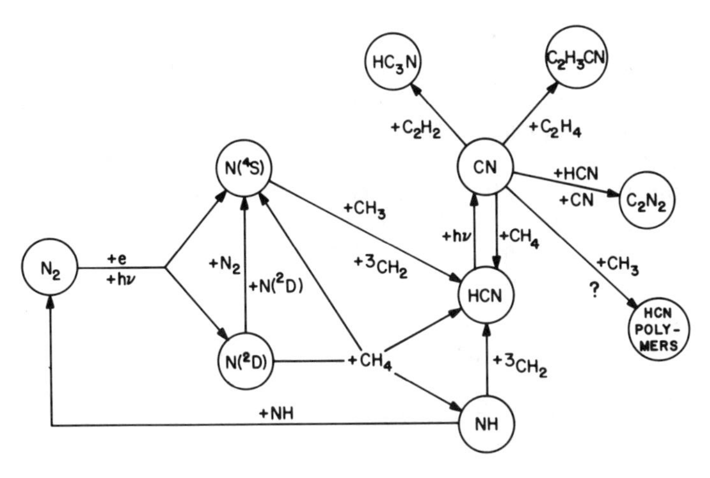

Fig. 2. Photo- and electro-chemistry of methane in the presence of nitrogen (figure from Atreya 1986).

The above-mentioned ionospheric scheme provides a significant source of HCN at ionospheric altitudes. It is also consistent with the results of a Voyager plasma experiment which showed the presence of mass 28 ions in Titan's magnetic tail (Hartle et al. 1982). These ions are presumably $N_2{}^+$ or H_2CN^+ of ionospheric origin picked up by Saturn's rotating magnetosphere.

The photolysis of methane produces large quantities of H and H_2. In fact, the limiting flux of H_2 at the homopause of Titan $(3.8 \pm 2) \times 10^{10}$ cm^{-2} s^{-1} is comparable to the photodecomposition rate of CH_4 (direct as well as catalytic) discussed earlier in this chapter. The limiting flux, φ_l is given by

$$\varphi_l = b_i f_i / H_a \, (1 - m/m_a) \tag{2}$$

where f_i is the constituent mixing ratio, H_a is the atmospheric scale height, b_i is the binary collision parameter, m is the mass of the escaping atom, and m_a is the mass of a nitrogen molecule. b_i can be further expressed in terms of the molecular diffusion coefficient (D_i) and the atmospheric number density n as follows:

$$b_i = nD_i. \tag{3}$$

Also $b_i = AT^a$ where $A = 2.8$ and $a = 0.740$ for an H_2-N_2 system and $A = 4.87$ and $a = 0.698$ for an H-N_2 system (cgs units).

At the critical level of Titan, the thermal velocities of H and H_2 are, respectively, 1.8 and 1.3 km s^{-1}. The critical level signifies the height at which the mean free path of the escaping constituent becomes equal to the scale height—thus collisions become rare. This level for Titan is at a radius of

4175 km where the atmospheric density is 9.3×10^6 cm^{-3}, and the escape velocity is 1.7 km s^{-1}. Thus, escape of both H and H_2 from the exosphere of Titan is assured. The presence of H and H_2 in the present atmosphere of Titan implies continuous supply of their source CH_4.

The primary loss process for nitrogen is formation of atomic nitrogen with a kinetic energy of 0.33 eV or larger, sufficient to escape Titan. Strobel and Shemansky (1982) calculated that 40% of the N atoms produced from molecular nitrogen escape. From the disk-averaged intensity of the molecular nitrogen Rydberg bands, they concluded that 3×10^{26} nitrogen atoms per second are supplied to the magnetosphere, equivalent to loss of 10% of the present abundance of atmospheric nitrogen over the age of the solar system (Hunten et al. 1984).

Finally, the atmospheric CO and CO_2 can be explained if there is an extraplanetary source of an oxygen-bearing constituent. An ideal candidate is water which quickly dissociates to yield OH. The latter in turn reacts with CH_4 photolysis products CH_2, CH_3 and C_2H_2 to eventually produce CO, and CO_2 (Samuelson et al. 1983). The CO_2 condenses out and falls to the surface, with an accumulated yield of a cm thickness over the age of the solar system, assuming the present CO abundance and water influx rate. The rings of Saturn and the carbonaceous chondritic meteorites are well suited for supplying water to the atmosphere of Titan. Another possibility for the presence of CO is that it was captured in a clathrate-hydrate directly from the solar nebula. W. R. Thompson and C. Sagan (personal communication, 1987) surmise that CO may have been produced by shock impact of comets on a surface containing CH_4 and H_2O. Such a mechanism involves a number of uncertainties including the physical state of the surface and size spectrum of impactors which have yet to be quantified. The implications of various evolutionary models for the present abundances of CO and CO_2 are discussed in Sec IV.

Photochemistry and charged-particle chemistry of methane most probably are responsible for the production of the orange haze which uniformly covers the disk of Titan. A suite of laboratory studies have established that a variety of materials, ranging from C_2 hydrocarbons to complex polymers, are produced when methane and nitrogen are subjected to irradiation or an electric discharge (Bar-Nun and Podolak 1979; see also other work referenced in Hunten et al. 1984). More recent work has established that some of these products approximately match the optical constants of the Titan aerosols over a wide range of frequencies (Khare et al. 1984, and references therein). Laboratory experiments do not provide the yield of aerosols relative to simpler hydrocarbons because they cannot simulate condensation in the tropopause region. Photochemical models incorporate condensation but cannot calculate the yield of complex polymeric aerosols (Yung et al. 1984, and references therein).

Observational studies of the aerosols, including the use of Voyager images through 1984, are reviewed in Hunten et al. (1984). Of interest to us here are both the effect of hazes on the thermal structure and the fraction of methane converted to haze vs simpler hydrocarbons and nitriles. The former issue is discussed in the thermal structure subsection. The latter has no firm answer; the most recent estimate of aerosol mass production rate by Pope and Tomasko (1986) is $\sim 8 \times 10^{-14}$ g cm^{-2} s^{-1}, or one-fifth the mass-loss rate of methane. This corresponds to a 0.1 km thick layer of compacted aerosol on the surface if the production rate is integrated over the age of the solar system. Although the photochemical models are constrained to reproduce the column abundances of hydrocarbons in the stratosphere observed by the Voyager IRIS experiment, sufficient uncertainty exists in our understanding of both the aerosols and simpler hydrocarbons that one cannot definitely rule out a model in which most of the products of methane photolysis reaching the surface are complex aerosols. The nominal models would predict, however, that most of the material is in the form of ethane, acetylene and propane. The composition of the material raining down is the primary determinant of surface properties.

Thermal Balance of the Atmosphere. Titan absorbs about 73% of the solar energy incident at the top of its atmosphere (Neff et al. 1985). Thus, it radiates to space at an effective temperature of about 83 K. The temperature profile of Titan's atmosphere is quite different from isothermal because of absorbing haze particles in its stratosphere, and tropospheric gases which absorb in the infrared. Figure 3 illustrates the temperature profile, from Lindal et al. (1983). Hunten (1978) demonstrated the conceptual simplicity of Titan's thermal structure by constructing a temperature profile of the troposphere and stratosphere using sound physical arguments. His temperature profile, truncated at the 1.5 bar level, bears a remarkable resemblance to that derived two years later from the Voyager flyby.

According to Fig. 3, the temperature at Titan's surface is elevated by about 10° above the satellite's effective temperature, due to a modest greenhouse effect (Samuelson 1983). The temperature lapse rate, as discussed above, is close to a dry adiabat in the bottom 3.5 km of the atmosphere. At higher altitudes, a shallower lapse rate occurs, with the temperature achieving a minimum value of about 70 K near an altitude of 40 km. This temperature minimum limits the mixing ratio of methane in the stratosphere to about 2%. Temperatures steadily increase with increasing altitude above the cold trap, attaining a value of 170 K above an altitude of 150 km. This highly elevated asymptotic temperature is believed to be the result of the efficient absorption of sunlight and the inefficient re-emission of thermal energy by haze particles, submicron in size, which populate the lower several hundred kilometers of the stratosphere (Danielson et al. 1973).

Figure 3 also plots the methane mixing ratio along the temperature profile, at which condensation of methane in equilibrium with nitrogen would occur. Clearly, cloud formation is extensive, in the absence of gross supersaturation effects, for methane mixing ratios in excess of 2%. The substantial

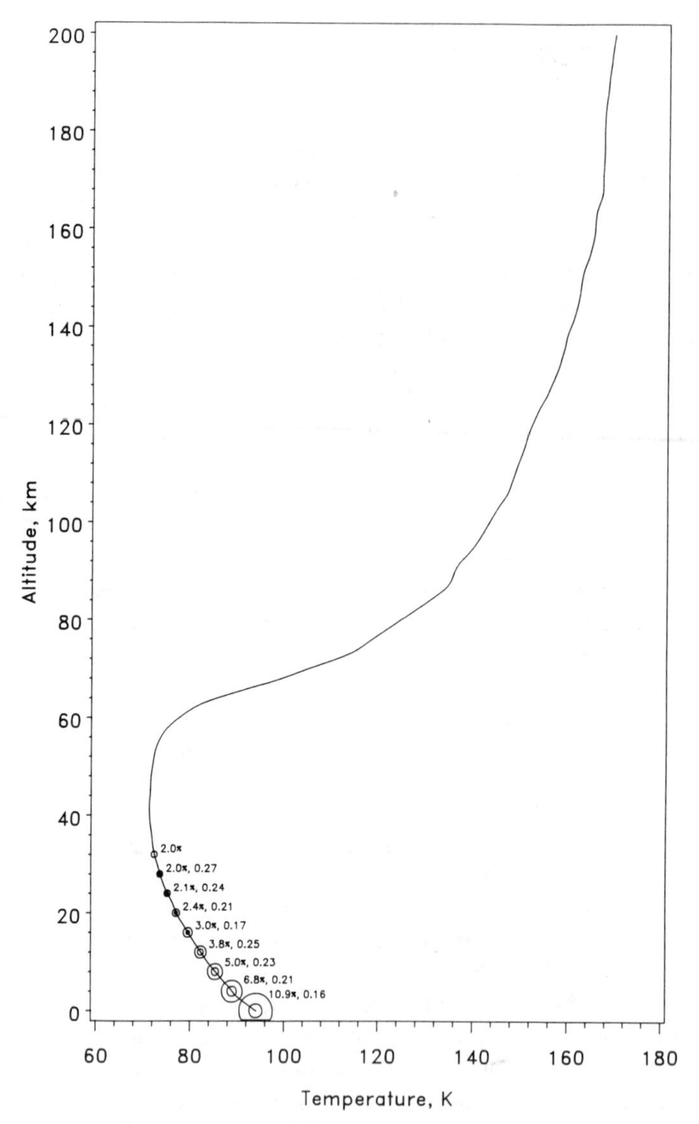

Fig. 3. Saturation equilibrium gas-phase percentages of CH_4 ($X_{M,s}$) and N_2 mole fractions in the condensate (X_N) are given along the altitude-temperature curve for Titan. Outer circle represents uncertainty in $X_{M,s}$ based on thermodynamic data and is assumed proportional to $X_{M,s}$; inner circle represents uncertainty in X_N and is assumed proportional to X_N. Open inner circle represents liquid condensate; filled represents solid. First condensate (base of cloud deck) occurs at the altitude where $X_{M,s}$ equals the surface mole fraction of gaseous methane. Temperature profile is from Lindal et al. (1983); thermodynamic values were calculated by W. R. Thompson who supplied this figure.

solubility of nitrogen in methane implies that cloud particles contain between 15 and 27% nitrogen, and the altitude of condensation is lower than would be obtained assuming pure methane condensate. The results shown here are those of W. R. Thompson (personal communication), who kindly supplied this figure. Kouvaris and Flasar (1985) report similar results. The uncertainties in the figure reflect our incomplete understanding of the thermodynamics of this system at low temperature, as discussed in Thompson (1985).

The surface temperature is determined by a balance between the amount of solar visible and atmospheric thermal radiation absorbed and the amount of thermal radiation emitted, as first described by Samuelson (1983). What follows is a detailed description of the thermal model results of McKay et al. (1988*a*) based upon very recent determinations of collision-induced gas absorption coefficients and constraints of the Voyager IRIS data. Figure 4 illustrates the variation of the net solar flux as a function of altitude in Titan's atmosphere (McKay et al. 1988*a*). It has been normalized to have a value at the top of Titan's atmosphere equal to the fraction of the incident sunlight absorbed by Titan. The calculations were performed assuming a haze model constrained to match the satellite's observed geometric albedo spectrum and to be in a steady state between production and sedimentation. For this calculation, the scattering effects of a methane cloud layer were neglected. The amount of sunlight reaching the ground is on the order of 10% under these

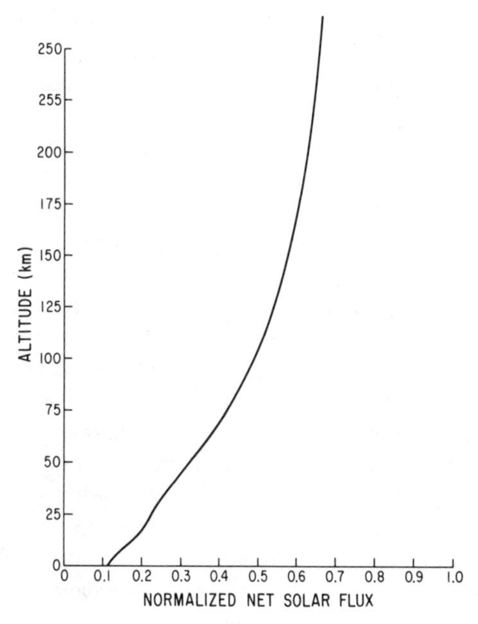

Fig. 4. Net solar flux vs altitude on Titan, normalized as described in text (figure from McKay et al. 1988*a*).

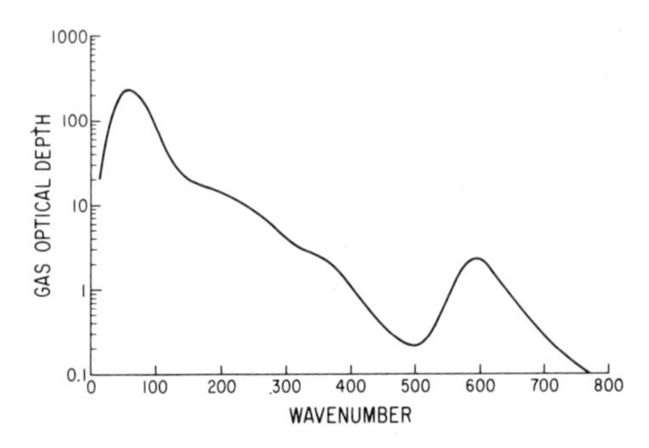

Fig. 5. Infrared optical depth at Titan's surface vs wavelength, due to collisional gas opacity, as calculated in McKay et al. (1988*a*).

assumptions. Absorption of sunlight by the stratospheric haze in the near ultraviolet and blue, and absorption by gaseous methane in the near infrared act to limit severely the amount of sunlight reaching the surface.

Figure 5 plots the pressure-induced gas opacity of Titan's atmosphere as a function of wavenumber in the thermal infrared (McKay et al. 1988*a*). At wavenumbers shortward of about 150 cm^{-1}, the pressure-induced transitions of nitrogen make the atmosphere very opaque; the pressure-induced transitions of methane (in collisions with nitrogen) dominate the region from 150 to 500 cm^{-1}; those of molecular hydrogen are a key between about 500 and 650 cm^{-1}, and the permitted transitions of ethane, acetylene and HCN cause localized elevated opacities between about 650 and 900 cm^{-1}. Thus, the surface is warmed by tropospheric radiation at wavenumbers shortward of 650 cm^{-1} and stratospheric radiation from photochemically produced gases at higher wavenumbers. The surface also receives thermal radiation from the stratospheric haze particles near the 500 cm^{-1} window.

The amount of greenhouse warming of the surface is limited by the window near 500 cm^{-1}. Within this window, the surface can radiate fairly freely to space, while receiving only a small compensation from atmospheric radiation. The pressure-induced transitions of nitrogen, methane and hydrogen play the key role in generating the small but important amount of warming which does occur. Radiation by gases and particles in the hot stratosphere also makes an important contribution to the greenhouse warming.

By way of contrast, a methane cloud can at best have a neutral effect on the greenhouse warming and at worst a cooling effect (McKay et al. 1988*a*). This is because the clouds reduce the amount of sunlight reaching the surface, while their effect in the infrared is limited since they mostly scatter rather than absorb thermal radiation near the 500 cm^{-1} window. This conclusion, reached by McKay et al. (1988*a*), rests upon significant improvements in the absorption coefficients of pressure-induced gas opacity, calculations of both solar and thermal fluxes, and application of available constraints to define the values of free parameters. It differs from earlier modeling results of Thompson and Sagan (1984) and Samuelson (1985). It should be noted that McKay et al. (1988*a*) find a somewhat better fit to the IRIS data if some cloud infrared opacity is added. The constraint that at least a few percent of the incident sunlight reach the surface demands that the clouds be either (1) limited in vertical extent due to a small methane mixing ratio, (2) confined to very large particle sizes or (3) broken in areal coverage (Samuelson 1985). The second possibility is intriguing, since for methane mixing ratios greater than a few percent, the large amount of condensate and lack of condensation nuclei tend to favor formation of raindrops rather than smaller, suspended cloud particles (Toon et al. 1988).

Our understanding of the present state of Titan's thermal structure is crucial to the construction of believable models of the evolution of Titan's atmosphere in response to the photochemical cycle (Sec. III). We can summarize our current understanding of how the observed temperature profile is established as follows:

1. Most sunlight is absorbed before reaching the ground, the blue and ultraviolet by stratospheric haze, the near infrared by methane. Absorption of sunlight by the haze establishes the warm stratosphere.
2. The dominant thermal infrared opacity source is collision-induced absorption, largely by nitrogen-methane collisions. The balance between the 5 to 15% sunlight reaching the ground and the thermal infrared repeatedly absorbed and re-emitted on its way up establishes to first order the tropospheric temperature profile.
3. Clouds play a small, perhaps negligible, role in affecting the heat balance at the surface.

Surface Models. In this section, we show that the model of Titan's surface most consistent with Voyager data is a liquid hydrocarbon ocean with substantial quantities of methane. Difficulties with other surface models are discussed.

The surface of Titan cannot be probed directly except by microwave radio emission and radar. Prior to the Voyager encounter, lack of information on the atmospheric composition hindered models of the surface. Lewis (1971) proposed a surface composed of methane clathrate hydrate and ice, based upon the equilibrium vapor pressure of methane over clathrate at 127 K, the reported spectroscopic abundance of methane on Titan, and reported radio brightness temperatures. The measured surface temperature of 95 K corresponds to a maximum methane pressure over clathrate hydrate of $\sim 2 \times 10^{-6}$ bars (Lunine and Stevenson 1985*a*, and references therein), four orders of magnitude less than the minimum methane surface pressure deduced from the lower-stratospheric mixing ratio of methane of 2%. The predicted maximum vapor pressure of nitrogen over the clathrate is three orders of magnitude lower than Titan's surface pressure. If clathrate is present at the surface of Titan, it is not in thermodynamic equilibrium with the atmosphere; rather, it would be a surface layer of limited thickness determined by the inefficient diffusion of gas molecules through water ice. Good thermodynamic contact requires either that fissures and cracks in the ice penetrate to the submicron level, which is unlikely, or that a regolith be present which is stirred by impacts. Impact causes formation of a regolith as deep as a kilometer (Hartmann 1973), which consists of a wide range of particle sizes (Kawakami et al. 1983); only the upper 100 meters is likely to be stirred sufficiently to be in occasional contact with the gas. A 100-m thick layer of clathrate hydrate can accommodate an equivalent thickness of 15 m of methane and nitrogen, 10 to 100 times less than the total hydrocarbon rainout estimated above, and equivalent to at most 0.1 of the mass of atmospheric nitrogen, or 1 to 10 times the mass of atmospheric methane. (The Voyager determination of Titan's density [Hunten et al. 1984] confirmed that water ice comprises half the mass of the satellite, and so must certainly be a major component of at least the near-surface layers of the satellite.)

A more satisfactory way to infer the nature of the surface is to require consistency with the Voyager determinations of atmospheric composition and thermal structure, and demand that the photolytic destruction of methane be balanced over geologic time by a supply of methane. This approach was considered by Flasar (1983), and in detail by Lunine et al. (1983) and Lunine (1985*a*). The assertions which go into the models are as follows:

1. Methane is present in the atmosphere but at mixing ratios below saturation near the surface. The near-surface temperature gradient is consistent with a nitrogen dry adiabat.

2. Methane is photolyzed in the stratosphere to form higher hydrocarbons, with minimal recycling. The maximum allowed methane in the atmosphere (7% at the surface) would be depleted in 40 Myr in the absence of resupply. The *primary* products of methane photolysis are ethane, acetylene and propane. Ethane and propane are liquid, and acetylene solid, at Titan's surface temperature of 95 K.

The surface state consistent with these constraints is a liquid hydrocarbon layer containing dissolved methane: a so-called ethane-methane ocean. The properties of such a layer which are consistent with the constraints can be summarized as follows:

a. Gaseous methane is in vapor equilibrium with the ocean. Because the ocean is only partially composed of methane, the amount of methane in the atmosphere is below the saturated value for a pure liquid methane surface. Since other hydrocarbons contained in the ocean have vapor pressures orders of magnitude lower than methane at 95 K, the amount of condensable vapor in the lowermost troposphere is too small to alter the temperature gradient from its dry adiabatic value by a measurable amount. If the ocean were pure methane, the near-surface temperature gradient would be significantly smaller than the dry adiabatic value.

b. The mole fraction of methane in the ocean can be estimated if the atmospheric abundance is known. This relationship is illustrated in Fig. 6a. (All figures assume ethane as the liquid heavy hydrocarbon component. The thermodynamic properties of propane are sufficiently similar that the two can be considered identical.) For a methane mole fraction in the atmosphere of 7%, the fraction of methane in the ocean is 60%. For 2% methane in the troposphere, the ocean fraction is 0.15.

c. The depth of the ocean can be calculated from the depth of the ethane layer deposited over the age of the solar system, and augmenting this with enough methane to produce the ocean composition calculated in paragraph b.

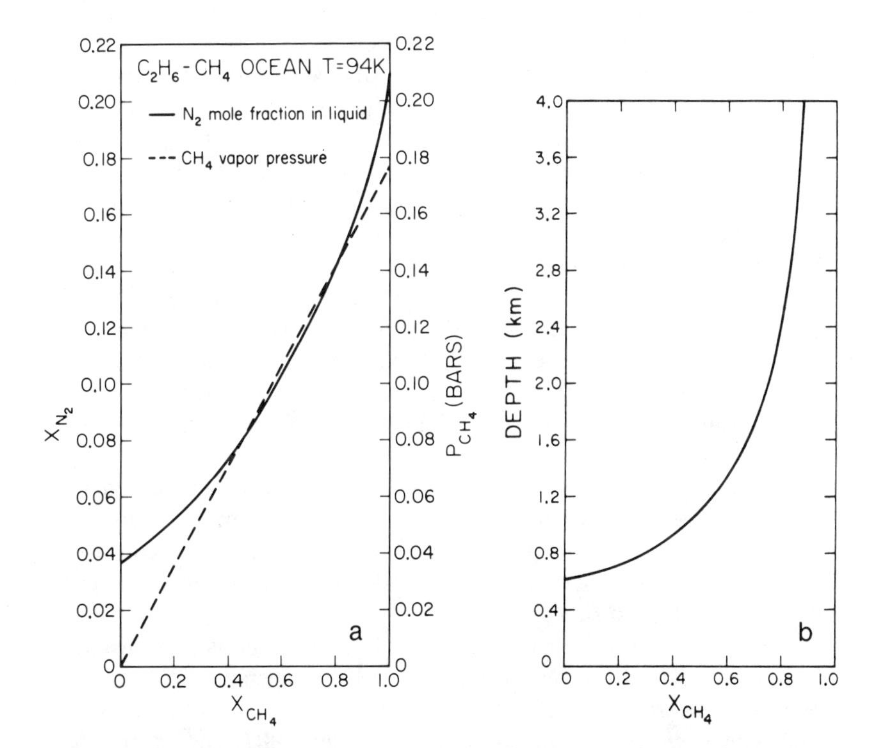

Fig. 6. (a) Partial pressure of methane (bars) at base of atmosphere and mole fraction of molecular nitrogen dissolved in methane-ethane liquid as a function of liquid composition, at 94 K. The abscissa is mole fraction of methane relative to total methane-plus-ethane component. To convert right-hand scale to gaseous methane mole fraction divide by 1.5 bars. The mole fraction of molecular nitrogen dissolved in the liquid is overestimated by as much as 30% for the methane-rich end of the figure, due to the assumption of an ideal solution (see, e.g., Thompson 1985). (b) Ocean depth in km as a function of methane mole fraction in ocean, assuming remainder of ocean is ethane produced photochemically over the age of the solar system at the rate given in Table I (figure based on Lunine 1985*a*).

The ocean depth is given as a function of composition in Fig. 6b. The mass of methane available in the ocean reservoir for photolysis is then determined immediately. For the 60% methane ocean, enough methane is available in the ocean to supply photolysis for 2.5 Gyr beyond the present day; for the 15% case, the ocean methane will last 0.4 Gyr.

Several additional details of the surface state emerge from this model and the constraints given above:

d. Nitrogen is soluble in both methane and ethane. The mole fraction of the ocean which is composed of nitrogen is given in Fig. 6a, as a function of oceanic methane mole fraction for an atmospheric nitrogen pressure of 1.5 bars. The dissolved amount is comparable to the atmospheric mass of nitrogen.

e. Other hydrocarbons produced by methane photolysis either dissolve in the ocean or sink to the bottom, depending upon their solubility in liquid ethane-methane. Acetylene is predicted to sink to the base as a sediment of roughly 100-m thickness, based on Table I. All precipitated hydrocarbons and aerosols are predicted to sink, unless they exhibit substantial porosity. Carbon dioxide is predicted to be soluble; as much as 10^7 times its atmospheric abundance may be dissolved (Raulin 1985). The solubility of noble gases in liquid hydrocarbons will be discussed in the context of origin models, in Sec. IV.

Lunine and Stevenson (1985*b*) and Lunine (1985*b*) consider further implications of the presence of an ocean on the present-day surface and atmosphere, including the possibility of erosion of crater and tectonic water ice topography by an ocean. At least 100 m of water ice vertical topography may be eroded by the hydrocarbon liquid. Thompson (1985) examined the thermodynamics of nitrogen-hydrocarbon systems in more detail, and confirmed the conclusions of the earlier work. The methane-nitrogen cloud structure of the middle and upper troposphere is unchanged by the ethane ocean arguments, because the very small amount of ethane-methane mist produced in the lower troposphere hardly affects the altitude profile of methane mixing ratio. Such a mist, however, if ubiquitous in the lower few km, must be constrained in morphology and particle size by the insolation arguments reviewed in the subsection on thermal structure.

Few tests of the ethane-methane ocean hypothesis are available. Muhleman et al. (1984) measured the brightness temperature of Titan at cm wavelengths to determine a "disk emissivity" of 0.82 ± 0.05. When compared with microwave emissivities of smooth ice (0.87) and liquid ethane (0.91) given in their abstract, the result argues against an ethane ocean at the two standard deviation level. Efforts to detect a radar return from Titan at Goldstone are proceeding. An additional constraint comes from Sagan and Dermott (1982) who argue that the present nonzero free eccentricity of Titan would be destroyed by tidal friction in a surface liquid layer less than a few hundred meters deep. It is not clear how to apply such tidal friction calculations to a km-deep ethane ocean on a satellite for which the amplitude of topography due to cratering is likely to be of order a km (by analogy with Ganymede and Callisto).

Stevenson and Potter (1986) have used the symmetry of northern and southern hemisphere brightness temperatures from Voyager IRIS data at 540 cm^{-1} to argue for dewpoint condensation of methane and molecular nitrogen at polar latitudes. If their model is correct, it requires the methane mole fraction in the lower troposphere to be between 4 and 8%. This would better constrain the composition of a global hydrocarbon ocean, and predict substantial variation in surface composition with latitude. The validity of the model is sensitive to uncertainties in the opacity in the 540 cm^{-1} region, which could be higher than assumed (McKay et al. 1988*a*) and so imply that the symmetry in brightness temperature is atmospheric and not surface in origin.

Alternative models in which the surface is covered largely by solid aerosol or polyacetylenes require that the following assumptions be made:

1. The photochemical models are wrong; most of the products of methane photolysis are acetylene or the complex polymers believed to comprise the aerosols. The alternative, that a significant flux of high-energy particles reaches the surface to convert *most* of the liquid hydrocarbons to solid polymers, is not plausible based on the results of Sagan and Thompson (1984).

2. The absence of liquid methane on the surface implies that methane is resupplied by cometary impact or outgassing over the age of the solar system.

The latter constraint requires some fine tuning of the resupply source, since the atmospheric methane is within a factor of 1.5 to 5 of saturation near the surface. Either we are looking at Titan at a special time, when the surface methane has dried up and the atmospheric methane is within 10 Myr of being depleted, or the resupply mechanism is such that it just avoids saturating the atmosphere with methane.

Figure 7 illustrates the interpretations of Voyager data and our understanding of present-day processes on Titan which lead either to the hydrocarbon ocean model or to the solid polymeric surface case. A clean water ice or water-ammonia hydrate surface requires very recent global resurfacing and is not considered here. The two extreme models have rather different implications for the steady-state evolution of Titan's atmosphere, as we discuss in Sec. III. Figure 8 presents in outline form our current understanding of Titan's atmosphere and surface.

Inferred Volatile Inventory of Titan's Surface. From the hydrocarbon ocean model and an assumed rate of photolysis constant over the age of the solar system, it is possible to estimate the mass of volatiles present on the surface and in the atmosphere of Titan. In Table II, we display the results for a present-day ocean containing 25% methane. From Fig. 6, the same quantities can be estimated for other methane mole fractions in the present-day ocean. Two results are of particular importance: (1) the mass of nitrogen in the ocean is comparable to that in the atmosphere, while the amount of methane in the

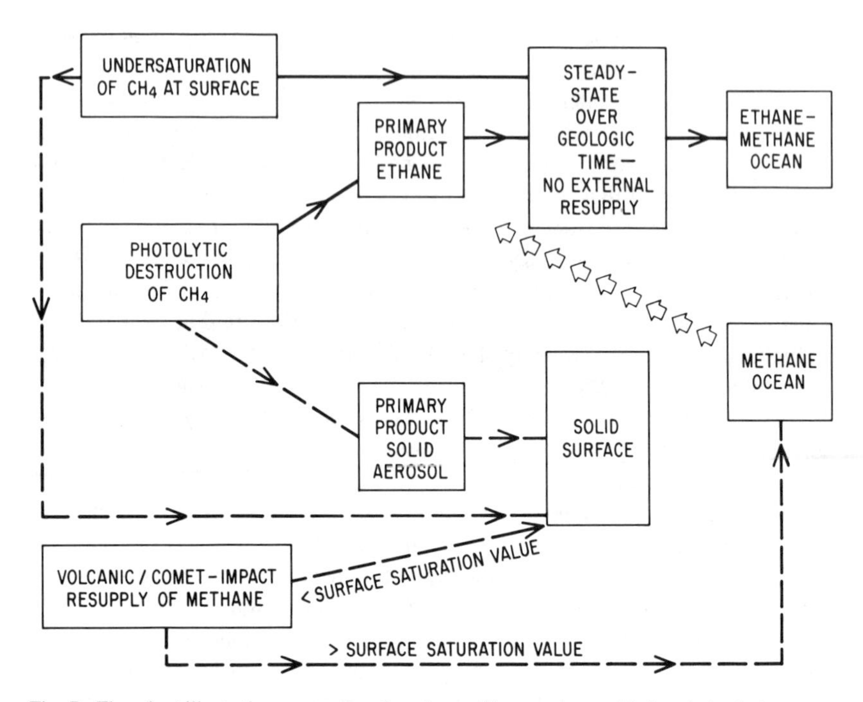

Fig. 7. Flowchart illustrating assumptions based upon Voyager data and inferred physical properties leading to the ethane-methane ocean and solid hydrocarbon surface models.

ocean is orders of magnitude larger than that in the atmosphere, and hence (2) the total methane (including processed hydrocarbons) to nitrogen ratio on Titan's surface is not much greater than unity (and decreases with increasing methane ocean abundance). The volatile complement from the ocean models provides important constraints on the origin and evolution of Titan's atmosphere as described in Secs. III and IV.

Triton and Pluto/Charon

The present review focuses on information acquired since 1986; an excellent review of prior studies is given by Cruikshank and Brown (1986). All of the information available on these objects has been gleaned from groundbased or Earth-orbital observations. We first discuss determinations of the mass, radius and albedo distributions of these objects, and then examine the spectroscopic data on their surface and atmospheric compositions. Finally, models of surface and atmospheric processes based on the data are reviewed.

Physical Properties. The mass of Triton is highly uncertain. It could be as high as 1.3×10^{26} g, based on the barycentric wobble of Neptune (Alden 1943). The most reliable radius determination comes from combining obser-

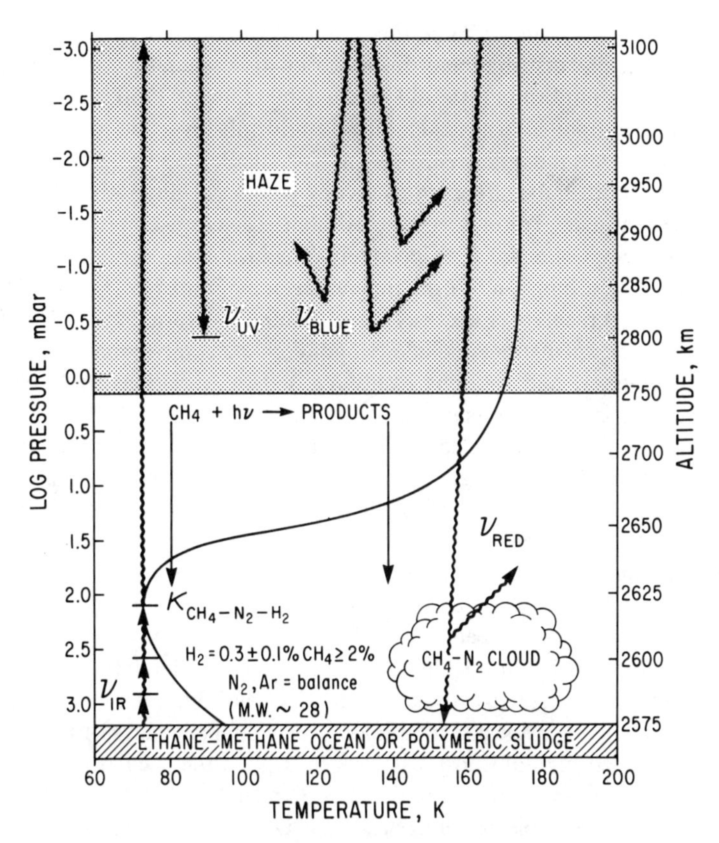

Fig. 8. The atmosphere and surface of Titan based upon Voyager data and inferred physical processes, as discussed in text. Base figure adapted from Thompson and Sagan (1984), with permission. Altitude scale is set for a surface radius of 2575 km.

TABLE II
Approximate Inventory of Volatiles on Titan's Surface[a]

	CH_4 (g)	N_2 (g)
Atmosphere	3×10^{20}	9×10^{21}
Dissolved in ocean	1×10^{22}	3×10^{21}
In ocean and sediment as higher hydrocarbons	6×10^{22}	
Total ocean + atmosphere	7×10^{22}	1×10^{22}

[a]Adapted from Lunine (1985*b*) with modifications from Lunine and Stevenson (1987), for an ocean model with 25% methane.

vations of the visible and infrared flux (Lebofsky et al. 1982; Morrison et al. 1982). For a cosine distribution of infrared radiation around the satellite, i.e., local radiative equilibrium (asteroid model), the radius is 1750 $\pm$ 250 km.; if thermalized sunlight is instead reradiated uniformly over the surface (thick atmosphere model) the radius is 2500 km. The smaller radius yields a density as high as 6 g cm^{-3}; the larger radius gives 2 g cm^{-3}. Unless Alden's mass is a gross overestimate, Triton is as large as Titan. The corresponding albedos in the two cases are 0.4 and 0.2, respectively. The brightness variation of Triton in the *V* filter as a function of orbital phase seen by Franz (1981) in 1977 has been challenged recently: Lark et al. (1987) found no variation at 8900 Å in 1987. Possible explanations include temporal changes on Triton, effects due to choice of filter, and spurious results caused by variations in Neptune's brightness (Franz used the planet as his standard).

The mass of the Pluto-Charon double planet system is well established, being 1.48×10^{25} g (Harrington and Christy 1981). The ratio of masses and the radii are less well constrained. Although observations prior to the mutual eclipses of Pluto and Charon suggested densities <1, the eclipse results establish that the mean density of the system is 1.99 g cm^{-3}, with Pluto's radius being 1123 km, 1.9 times that of Charon (Tholen and Buie 1987). Corresponding blue geometric albedos are 0.51 and 0.37, respectively.

The albedo variation on Pluto may be complex. Marcialis (1988) analyzed photoelectric lightcurves taken over 30 yr and fitted the data with two spots of low albedo, colocated in latitude but on opposite sides of the planet. He also proposed high-albedo polar caps to explain the secular dimming of Pluto with time since 1930. An alternative explanation for the secular dimming is progressive sublimation of bright frosts (Stern et al. 1988).

Additional support for the presence of polar caps comes from an analysis by Sykes et al. (1987) of IRAS pointed observations. Constraining the overall albedos and radii of Pluto and Charon from the eclipse data reported by Tholen et al. (1987), various surface models were tested for consistency with the observed fluxes at 60 and 100 μm. The albedo of Charon was assumed uniform. The best fit involves a Pluto model with polar caps of high albedo (0.3 to 0.8) and low emissivity (0.1 to 0.4), and a low albedo, high emissivity equatorial region. Satisfactory fitting to the IRAS data requires that the temperature decrease with increasing latitude up to the polar caps which are isothermal; completely isothermal models are ruled out. Assuming the polar caps are composed of methane, a maximum methane vapor column abundance of 920 centimeter-amagats (cm-am) is derived.

Analysis of the same IRAS data set by Aumann and Walker (1987) yields a completely isothermal surface for Pluto, in contradiction to the work of Sykes et al. (1987). It has not yet been possible to identify the source of the inconsistency between the two analyses. It should be pointed out that Sykes, et al. used a more sophisticated modeling approach, however. A third IRAS analysis by Tedesco et al. (1987) uses poorer-quality survey-mode data at a single wavelength and should be considered superseded by the pointed observations.

Species Detected Spectroscopically. Information on the volatile components of Triton's and Pluto's surfaces come from near infrared spectrosopy; the visible region has added some information on nonvolatile components. Triton shows methane absorption bands in the near infrared indicative of a condensed phase, generally thought to be fine-grained methane frost (Cruikshank and Apt 1984). An additional spectroscopic feature at about 2.15 μm was tentatively identified in 1984 as being an overtone of pressure-induced absorption in nitrogen (Cruikshank et al. 1984). For plausible surface temperatures (see below), the vapor pressure of gaseous nitrogen is too low to produce the feature; hence a long (meter) path length absorption in condensed nitrogen is required. Because of the required long path length, the feature was ascribed to liquid nitrogen. More recent spectroscopy has either failed to see the liquid nitrogen feature (Rieke et al. 1985) or observed a shape different from laboratory liquid nitrogen (Cruikshank et al. 1987). Moreover, while the spectroscopic data from the early 1980s show substantial variability with orbital phase, suggesting that a nonuniform distribution of volatiles was seen, in the most recent data the effect is apparently masked, and the absorption bands are characteristic of either extremely small grains (microns) or a highly dilute solution of liquid methane (Cruikshank et al. 1987). Both condensed nitrogen and methane would produce significant atmospheres; however, absorption due to gas has not been seen in the infrared data. Spectroscopic data also hint at the presence of fine-grained water frost. Triton's reddish color suggests a component on the surface analogous to the aerosols produced on Titan from photolysis of methane (Cruikshank and Brown 1986).

The near-infrared spectrum of Pluto is dominated by absorption bands most closely resembling those produced by methane ice (Cruikshank and Brown 1986). Shortward of 1 μm, the interpretation of spectral features is more ambiguous. Buie and Fink (1987) find an upper limit column abundance of methane, if all of the absorption is due to gas, of 940 to 1460 cm-am, variable with orbital phase. The variation implies a frost-plus-gas model to be more physically sensible; the best-fit case yields a gaseous abundance of 540 cm-am which is consistent with the IRAS analysis of Sykes et al (1987). There is no evidence for additional spectroscopically active components. Pluto is also reddish, suggesting that methane is being converted to higher hydrocarbons by sunlight or charged particles. A recent experiment separated the infrared signatures of Pluto and Charon, indicating the absence of methane and tentative presence of water ice on Charon's surface (Marcialis et al. 1987).

Although carbon monoxide has not been detected on Triton or Pluto, its signature could be hidden in the 2.3 μm band of methane. It should be possible to ratio the strengths of two methane bands to determine the amount of additional absorption due to carbon monoxide at 2.3 μm, but to date the data are not of sufficiently high quality, nor the laboratory methane analogs sufficiently good matches to the methane features, to be able to derive CO abundance in this fashion.

Surface Models: Physical State of Volatiles and Seasonal Modulations. To understand the nature of the surfaces of Triton and Pluto requires (1) that the spectroscopic data be well characterized, (2) that the thermodynamic properties of the volatiles be understood, and (3) that seasonal modulation of insolation be taken into account. Triton is in a retrograde orbit inclined to the spin axis of Neptune, which is inclined relative to its solar orbit. Harris (1984) determined that the pole of rotation of Triton is probably inclined only about 1° relative to its orbit plane, and that the subsolar latitude of Triton undergoes extreme excursions to ±50° latitude every 600 yr or less, with smaller higher frequency oscillations in between (see figure in Cruikshank and Brown 1986). Trafton (1984) used Harris' results to show that the net deposition of sunlight was greatest at Triton's equator, that volatiles such as methane would tend to migrate toward the poles over geologic time, and that seasonal movement of volatiles from pole to pole by sublimation would occur.

The temperature at the subsolar point of Triton if there is no major surface volatile component is given by

$$T = 72 * [(1 - A) / e]^{0.25} \tag{4}$$

where A = albedo and e = emissivity. For unity emissivity and an albedo of 0.4, Eq. (4) yields a subsolar temperature of 63 K. When a material of sufficient volatility is present on the surface, a significant amount of insolation goes into sublimation; that energy is released upon condensation elsewhere on the body. In effect, the solar energy is redistributed areally by volatile transport; since vapor pressure is an exponential function of inverse temperature, materials are either completely *involatile* or *volatile* in regard to redistribution of solar heat. The ratio of energy going into sublimation to that going into heating is

$$f_{sub} = \frac{L\, P_S}{\sigma T^4} \sqrt{\frac{\mu}{2\pi R_g T}} \tag{5}$$

where L = latent heat of sublimation, P_s = vapor pressure, μ = molecular weight, R_g = universal gas constant and σ = Stephan-Boltzman constant. For methane at 60 K, f_{sub} is 5000; for nitrogen, it is higher by three orders of magnitude. Clearly these materials must alter the temperature distribution on Triton, and by doing so determine the physical states under which they can exist. Trafton (1984) showed that over the age of the solar system the net amount of methane transferred from the equator to the poles could exceed the mass of Triton. Thus, solid (as opposed to liquids that would flow) volatiles are confined to the polar caps. The temperatures of the polar caps, which determine the maximum atmospheric pressure in equilibrium with the surface volatiles (the vapor pressure) are equal to each other due to volatile transport. From Trafton's formulation of the polar-cap temperatures as a function of subsolar point and cap extent, we derive

$$T_{pc} = 67.2\,[(1 - A)/e]^{0.25} \tag{6}$$

for the epoch of the Voyager flyby (mid-1989), when the subsolar latitude will be 47°.6 and assuming for concreteness a polar-cap colatitude of 30°. Figure 9

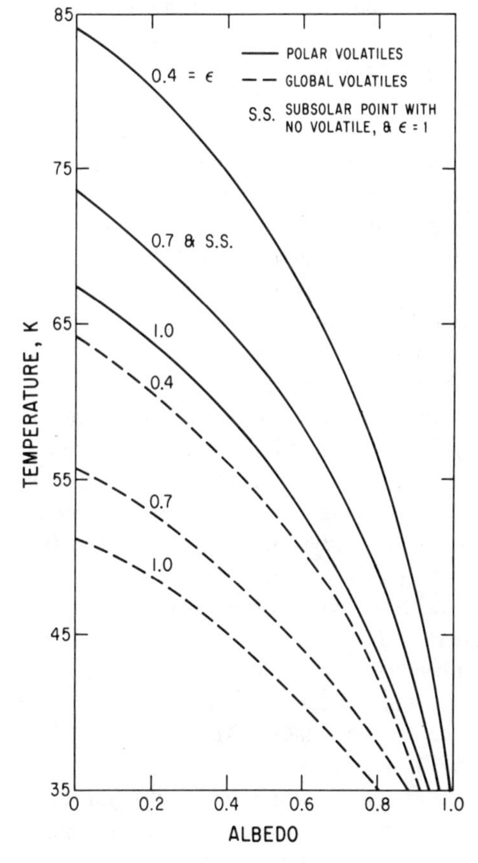

Fig. 9. Surface temperature vs albedo for two different distributions of surface volatiles, and for various emissivities. Polar refers to methane or nitrogen confined to polar caps, and global to uniformly distributed volatiles. SS refers to the temperature at the subsolar point of a non-volatile surface with unity emissivity.

plots temperature as a function of albedo and emissivity for this case. Also shown is the temperature for a liquid volatile, which is assumed to cover uniformly the surface:

$$T_{eq} = 51.[(1 - A)/e]^{0.25}.$$

The proper assignment of emissivity for Triton's surface is difficult. A value for methane ice of 0.4 was favored by Trafton (1984), based on Trafton and Stern's (1983) comparison of the gas abundance of methane on Pluto with a calculated surface temperature. This approach is now questionable, since it was based upon an upper limit for gaseous methane on Pluto by Fink et al. (1980), substantially higher than the most recent determination of Buie and Fink (1987). However, Sykes et al. (1987) find a best fit to IRAS data of Pluto by including methane polar caps with thermal emissivity <0.4. Finally, an emissivity significantly <1 is suggested for liquid nitrogen by the work of Jones (1970). Clearly, determination of the emissivity of condensed nonpolar species is a high priority for laboratory work. Also uncertain is Triton's albedo, since the polar-cap value is likely to be higher than the global average. A global ocean model radiates over 4π steradians (Eq. 7); hence the correct albedo to choose from the radiometric observations is the lower one of 0.2.

Given the temperature range of Fig. 9, nitrogen and methane are most plausibly present in the solid state on Triton's surface, according to the work of Lunine and Stevenson (1985*d*). They assessed a number of surface models in terms of the available thermodynamic and spectroscopic constraints. The primary thermodynamic constraint is the freezing point of nitrogen, 63 K, which is lowered by only one degree through addition of methane. The only cosmochemically abundant material capable of lowering the nitrogen freezing point substantially is neon, which is not expected to be present in significant quantities as discussed in Sec. IV. Figure 9 shows that temperatures on Triton are below 62 K if liquid nitrogen is assumed to be present in a global ocean; hence such a model is thermodynamically inconsistent. If the nitrogen is confined to lakes near the poles, temperatures above 62 K are possible.

The primary spectroscopic constraint is that the 2.3 μm methane absorption band be narrow enough that the 2.16 μm nitrogen feature is visible. Lunine and Stevenson (1985*d*) concluded that for a saturated mole fraction of methane in liquid nitrogen ($\sim$20%), the wing of the methane band could obscure completely the nitrogen feature. This was confirmed experimentally by Piscitelli et al. (1987), who observed the nitrogen feature disappear when the methane mole fraction dissolved in a liquid nitrogen solution reached 1% of the saturated value. Lunine and Stevenson asserted that avoiding methane saturation in liquid nitrogen on Triton would be difficult.

The preferred model for Triton's surface has nitrogen and methane in solid form, perhaps in solid solution, but more probably as a disequilibrium assemblage. The presence of pure nitrogen atop methane or a methane-nitrogen mixture is ensured by seasonal transport and the much higher vapor pressure of nitrogen compared to methane. Lunine and Stevenson estimate that a m-deep layer of nitrogen can be sublimed and deposited elsewhere on a seasonal time scale of 100 yr. The high path length required in the nitrogen solid by the near-infrared observations requires large crystal sizes, on the order of tens of cm, which is plausible only if significant grain metamorphism occurs. A surface temperature below but near the triple point substantially aids this process (Clark et al. 1983). Liquid nitrogen cannot be ruled out absolutely; Cruikshank et al. (1987) assert that the methane absorptions seen in the latest Triton data are well fit by very small amounts of methane dissolved in liquid nitrogen. The same may be true for methane in solid nitrogen, which, however, has not been synthesized for laboratory infrared spectra.

Finally, one can assess the likelihood of a thick, greenhouse atmosphere for Triton, using pressure-induced gas opacities shown to be important on Titan. The low effective temperature of Triton compared to that of Titan limits the vapor pressures which can be raised from the surface, and significant infrared optical depths can be maintained only for a very thick atmosphere (surface temperatures greater than 75 to 80 K). Whether the lack of photometric and spectroscopic variability in recent observations of Triton indicates obscuration by haze in a substantial atmosphere is an issue which will remain open until the Voyager 2 encounter in 1989.

Pluto's rotation axis is highly inclined to the ecliptic (Cruikshank and Brown 1986), and its high orbital eccentricity brings it closer to the Sun than Neptune. Trafton and Stern (1983) examined the dynamics of a methane atmosphere in equilibrium with surface frosts. Because of methane's volatility, Pluto's atmosphere would be expected to be globally uniform in temperature, at least in its near perihelion state. The range of possible temperatures on Pluto near perihelion is the same as that for Triton (for a given assumed model of surface coverage of volatiles). Aphelion temperatures are as much as 20% less than perihelion temperatures (Trafton 1980). Stern et al. (1988) examine the orbitally driven sublimation and freezeout of methane in more detail, and conclude that the process is essential for maintaining bright methane (non-radiation-darkened) on Pluto's surface. The secular dimming of Pluto toward aphelion, in this model, is due to sublimation of methane frosts, in contrast to the fixed-cap model of Marcialis (1988). Their model predicts a lower limit to the atmosphere of Pluto of 16 to 45 cm-am, consistent with data analyses cited above.

Role of Photochemical and Cosmic-Ray Chemistry on Triton and Pluto. In the above-mentioned ranges of temperature, both Triton and Pluto can maintain large column abundances of CH_4 for photolysis. The column abundance of an ideal, isothermal gas in hydrostatic equilibrium is given by $N = P/mg$, where m = mass of gas per molecule, P is pressure and g surface gravity. For a *nominal* surface gravity of 100 cm s^{-2}, methane column abundances range from 3×10^{18} cm^{-2} for $P = 8 \times 10^{-9}$ bar (corresponding to a surface temperature of 40 K if the methane vapor is in equilibrium with surface frost), to 9×10^{23} cm^{-2}, for a vapor pressure of 2.5×10^{-3} bar (corresponding to a surface temperature of 70 K). Since only 6×10^{16} cm^{-2} of CH_4 is required for unit optical depth (where most of the photolysis occurs), it follows that the photolysis regions on Triton and Pluto would lie at least several scale heights above the surface. In the presence of large amounts of N_2 as the background gas, one might then expect the photochemistry on Pluto and Triton to bear some resemblance to that on Titan. The major uncertainty in such a comparison is ignorance of the temperature structure of the atmospheres of Triton and Pluto. Until this issue can be resolved (by the Voyager flyby of Triton in 1989), detailed models of altitude regimes of photolysis and alterations in chemistry due to transport of products vertically can be constructed for neither Triton nor Pluto.

It should be recognized that at the distance of Triton and the perihelion of Pluto's orbit, the solar flux is reduced by a factor of 1000 relative to its value at 1 AU, or by a factor of 10 compared to that at Titan's orbit. This would result in a correspondingly lower production rate of the photolysis products. Since the densities of the products go roughly as the square root of the production rates, they would be lower by about a factor of 3 compared to Titan, all other things being equal. As solar ultraviolet dissociation declines, the charged particle dissociation and ionization of the atmospheric constituents could become more significant. Depending upon the atmospheric composition, pressure and the nature of the magnetic field (intrinsic or induced), galactic cosmic rays with energies greater than 1 GeV, subrelativistic solar-wind protons with energies in the 10 to 100 MeV range, and occasionally solar relativistic protons with energies greater than 10 GeV would interact with Pluto and Triton producing some of the same constituents as in the photochemical processes discussed above. A likelihood also exists for the production of many more complex species, and of polymerization all the way down to the surface (Delitsky and Thompson 1987). Because of the low temperatures prevalent on Triton and Pluto, many of the species produced subsequent to the charged particle induced chemistry are subject to condensation in the form of atmospheric aerosols.

Strazzulla et al. (1984) have measured the cross sections of energetic protons for polymerization on hydrocarbons of relevance to the surfaces of Pluto and Triton. Although the flux of solar cosmic rays at the orbits of Pluto and Triton is not known, a rough estimate may be made by taking the value at 1 AU and assuming a $1/r^2$ dependence in the worst case. Slower variation in flux is indicated by the Pioneer measurements at Jupiter's orbit. McDonald et al. (1974) give for maximum solar activity a flux of 10^{10} cm^{-2}yr^{-1} for protons with energies $\gtrsim$20 MeV, and incident energy spectrum that varies as $E^{-\gamma}$, where $3 \lesssim \gamma \lesssim 6$. Assuming a surface of methane ice, CH_4-vapor pressure of 0.025 mbar (corresponding to a 55 K surface) and McDonald et al.'s energy spectrum with $\gamma = 3$, Strazzulla et al. (1984) calculate that the time constant for polymerizing 50% of the original methane on Pluto would be expected to be 100 Myr at a depth of 0.5 cm, 1 Gyr at a depth of 2.2 cm and 4.6 Gyr at a depth of 5 cm. For Triton, they assumed a monolayer of CH_4 adsorbed on the surface, and calculated polymerization to a depth of 0.1 μm. It is important to recognize that Strazzulla's estimates could easily change by orders of magnitude if a different energy spectrum, e.g., with $\gamma = 6$, and a different flux variation with distance were assumed. Their results should therefore be regarded as tentative. Despite the possibility of polymerization at the surface, it is virtually certain that the surfaces of Triton and Pluto would be continually renewed by seasonal oscillations in the rates of condensation and evaporation, and by gardening. Survival at the surface of some products of Triton's energetic chemistry seem likely, because of the satellite's reddish appearance (Piscittelli et al. 1987).

As for Triton's interaction with Neptune's magnetosphere, it seems at this time to be of little consequence. Unlike Titan, the nitrogen chemistry at Triton is not likely to be controlled by the magnetospheric particles (Atreya 1986). This conclusion is based on the fact that the IUE observations give an "upper limit" of 200 R for Lyman-alpha emission from Neptune (J. T. Clark, personal communication, 1986,1987). Almost all of this intensity can be explained by conventional means: resonance scattering, backscattering of interplanetary/interstellar hydrogen, with perhaps a small contribution from a possible electroglow. Charged-particle input from a magnetosphere is not required. Galactic cosmic rays as well as interplanetary electrons are expected to play an important role in Triton's atmospheric and surface chemistry.

Stability of Methane on Triton and Pluto. Trafton (1980) considered the hydrodynamic escape (blowoff) of methane from Pluto. He argued that if the average thermal energy per molecule is comparable to the gravitational potential energy at the exobase, then free escape of the molecules can occur. The essential condition is that

$$\lambda = GMm/RkT = 1.5 \text{ at the exobase} \qquad (8)$$

where G = gravitational constant, M = mass of Pluto, m = mass of methane molecule, k = Boltzmann's constant, T = temperature, R is radial distance from Pluto's center. At Pluto's surface, for the parameters given earlier, λ = 25 at T = 50 K. An atmosphere supplied by methane sublimation at the surface and supported hydrostatically would extend up to altitudes at which λ = 1. Trafton showed that for temperatures in excess of 47 K, the blowoff flux would be $> 2 \times 10^{13}$ cm^{-2}s^{-1} (referred to the surface), enough to vaporize a mass of methane equal to the mass of Pluto. The result suggested that a heavier gas, e.g. nitrogen, is present in the atmosphere of Pluto to limit the escape of methane which must diffuse to the exobase.

Hunten and Watson (1982) applied the analytic scheme of Watson et al. (1981) to calculate a more realistic escape flux accounting for energy transport in the atmosphere. Escape at the exobase is driven by solar EUV heating, which is absorbed many scale heights above the surface. The efficiency of escape of gas is primarily limited by the conduction of heat from the level of absorption of ultraviolet to the region where the bulk of the atmosphere is located. The energy limited escape flux is then determined by allowing the temperature minimum to equal 0. Using this model, Hunten and Watson (1982) calculated an escape flux for methane on Pluto of 4×10^{10} cm^{-2}s^{-1} referred to the surface, implying that the amount of methane lost over the age of the solar system is less than 0.5% of the mass of Pluto, i.e., equivalent to a layer 8 km deep. Although this is a large quantity of methane, it is much smaller than the maximum amount of methane which could have been incorporated in Pluto based on cosmochemical models (see Sec. V), which in terms of mass fraction is in the range of 5 to 15%.

The absence of methane on Charon (Marcialis et al. 1987) combined with its smaller mass, suggests that hydrodynamic escape may have depleted that object of its methane. Application of the Watson et al. (1981) formalism supports that conclusion (Trafton et al. 1988).

Because Triton's mass is probably substantially greater than that of Pluto, the same conclusion holds more strongly for that satellite. Other loss processes for methane such as tidal stripping by Neptune have not been demonstrated to be important. It appears that, as for Titan, the dominant loss of methane on Triton is photochemical. On Pluto, energy-limited escape may indeed dominate over photolysis as the primary sink of methane, but it is not rapid enough at present to require invoking a heavier secondary gas to maintain methane on Pluto. The role of enhanced EUV flux from the early Sun in atmospheric escape is discussed in Sec. V.

III. EVOLUTION OF TITAN'S ATMOSPHERE DRIVEN BY PHOTOCHEMISTRY

We have laid the foundation for discussing the primary processes operating at present which drive the evolution of the atmosphere of Titan. The coupling of the irreversible photochemical conversion of methane with the thermal structure of the troposphere is achieved through the postulated existence of an ethane-methane-nitrogen ocean. In this section we explore the interactions between these systems, and describe how the results of such models test our understanding of the present Titan atmosphere, as well as constrain models for its formation and early evolution. Couplings in the absence of a massive ocean are also discussed. We further consider the effect of a massive ocean and the photochemical cycle on the atmospheric D to H ratio. Finally, we describe the additional information on Triton and Pluto required for similar modeling of those bodies.

Coupled Evolution of Titan's Ocean and Atmosphere

Figure 10 illustrates the conceptual coupling between the photochemical cycle, atmospheric thermal structure and hydrocarbon ocean described in Lunine and Stevenson (1985*b*). The model is a closed box for nitrogen, which is partitioned between ocean and atmosphere but does not escape. The loss of methane is entirely photochemical. To first order, two methane molecules derived by evaporation from the ocean are photolyzed to form one molecule of ethane, which condenses and falls back to the ocean. The primary correction to this scheme is the production of acetylene and aerosols, decreasing the ethane production rate by a factor of perhaps 10 to 30%. Hydrogen escapes from the system, with the consequence that the process is irreversible. As a function of time, then, assuming no resupply of methane by outgassing or cometary impact, the ocean (1) becomes more ethane rich, and (2) decreases in number of molecules (volume).

The effects of these changes on the atmosphere are determined by the thermodynamic properties of the ocean. Since the ethane-methane liquid system is nearly ideal in the sense of Raoult's law (Lunine et al. 1983), the partial pressure of methane in the atmosphere is proportional to the mole fraction of

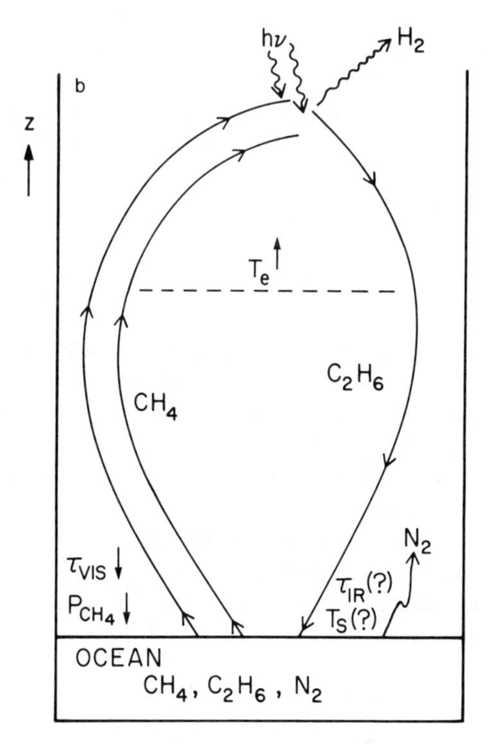

Fig. 10. Illustration of the physical processes which couple the atmospheric thermal structure to the evolution of the ocean, as described in the text.

methane in the ocean; hence at a fixed temperature, the amount of atmospheric methane decreases as the ocean becomes more ethane rich. The nitrogen dissolved in the ocean tends to outgas to the atmosphere with time, both because its solubility decreases for increasing ethane mole fraction (Fig. 6) and because the ocean is decreasing in volume.

The atmospheric thermal structure is determined largely by the amount of sunlight reaching the surface and the infrared opacity contributed by pressure-induced absorption in gaseous nitrogen, hydrogen and methane. The decreasing methane pressure determined by ocean evolution decreases the infrared opacity and tends to lower the surface temperature. The increasing nitrogen pressure has the opposite effect. Which effect dominates depends upon the relative contributions of the two gases to the infrared opacity, and the relative rates at which the gas pressures change with changing ocean composition. In addition, the decrease in methane mole fraction in the ocean increases the altitude at which methane cloud formation occurs, which probably acts to increase the amount of solar flux reaching the ground. Finally, the change in the thermal structure of the troposphere changes the mole fraction of methane admitted to the stratosphere, which under some conditions could alter the photolysis rates and composition of products.

Analytic Model of Climate Evolution. The rather complex couplings cited above were explored in a preliminary way by Lunine and Stevenson (1985*b*). An analytic representation for the change in Titan's tropospheric climate with ocean evolution illustrates the interplay between the various physical processes. The infrared optical depth at Titan's surface due to pressure-induced gas opacity, suitably averaged over frequency, is

$$\tau = \left(P_N^2 \frac{\alpha_{NN}}{2} + P_N P_H \alpha_{NH} + P_N P_C \alpha_{NC} \right) \frac{1}{\eta^2 kT\, mg} \quad (9)$$

where α_{xy} = absorption coefficient for collision of molecule x with molecule y, in cm^{-1} amagat^{-2}, P_x is partial pressure of molecule x, η = Loschmidt's number, m is mass of primary atmospheric molecule, g is the surface gravitational acceleration, k is Boltzmann's constant and T is temperature. We exclude other collisional combinations such as methane-methane because their contribution is negligible.

We wish to find the change in surface temperature with time, or equivalently, with decreasing ocean methane mole fraction x_C, since the methane conversion rate is assumed constant with time. We use the grey atmosphere solution to the equation of radiative transfer in the Eddington limit, i.e.,

$$T = T_e\,(0.5 + 0.75\,\tau)^{0.25} \quad (10)$$

where T_e is an "effective" temperature which is tied to the amount of solar flux reaching Titan's lower troposphere. To make further progress, we must

(a) find expressions for the change in nitrogen, methane and hydrogen pressures with ocean composition and temperature, and (b) decide on the relative importance, and temperature dependences of, absorption due to collisions between various pairs of molecules.

The methane pressure follows Raoult's law:

$$P_C = x_C\, e_C \tag{11}$$

where e_c is the saturation vapor pressure of methane at the surface temperature T (deviations from Eq. [10] for methane can be as large as 40% in pressure; W.R. Thompson, personal communication). The solubility of nitrogen in methane and ethane is given by Henry's law (Lunine and Stevenson 1985*b*):

$$P_{\rm N} = K\, x_{\rm N} \tag{12}$$

where K is the Henry's law constant and $x_{\rm N}$ is the mole fraction of nitrogen in the ocean. We evaluate this constant for a mixed ethane-methane liquid, and conserve the total amount of nitrogen in ocean and atmosphere:

$$M_{\rm N} = 4\,\pi\, R_T^2 P_{\rm N}/g + 28\, P_{\rm N} M_{\rm o}/K\mu_{\rm o} \tag{13}$$

where $M_{\rm N}$ is the total amount of nitrogen in the atmosphere and ocean, $M_{\rm o}$ is the mass of ethane and methane in the ocean, R and g are the radius and surface gravity of Titan and $\mu_{\rm o}$ is the ocean mean molecular weight. This expression is strictly valid for dilute nitrogen solutions only ($K >> P_{\rm N}$).

The effect of the ocean on the hydrogen partial pressure has not been explored. Two reasonable limiting cases are (1) constant atmospheric mixing ratio of hydrogen and nitrogen (implying similar thermodynamic interactions with the ocean) and (2) constant partial pressure of hydrogen (no interaction of hydrogen with the ocean).

We first assume that collisions between nitrogen molecules dominate the infrared opacity. In this case, the dependence on methane and hydrogen pressure disappears, and optical depth goes as the square of the nitrogen pressure in Eq. (9). Evaluating Eqs. (9) through (13) yields the result that the optical depth increases as methane mole fraction decreases, i.e., surface temperatures were colder in the past. The physical argument is that as the ocean becomes more ethane rich and decreases in volume, nitrogen comes out of solution to increase the atmospheric pressure and hence infrared optical depth. Figure 11 summarizes the quantitative results of such a model. Two curves are shown, one for the solubility of N_2 constant with temperature as considered above, and the other for a solubility which decreases with increasing temperature. Work by Thompson (1985) suggests that the latter is more reasonable. The dashed lines are temperature boundaries for the freezing out of the ocean.

Recent work by Dagg et al. (1984,1986) and Courtin (1988) confirms

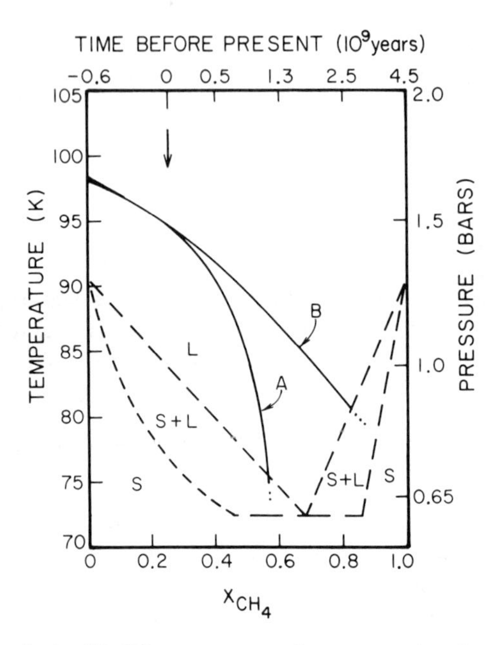

Fig. 11. Evolution of a simplified Titan ocean-atmosphere system, plotted as surface temperature vs mole fraction of methane in ocean, for a present-day ocean containing 25% methane. The sole atmospheric opacity source in this model is nitrogen-nitrogen molecular collisions; as discussed in the text more detailed atmosphere models yield different results. Curve *A* assumes a temperature-dependent solubility of nitrogen in the ocean; curve *B* assumes temperature independence. Two auxiliary scales are plotted: atmospheric pressure in bars corresponding to temperature, and time before present corresponding to ocean methane mole fraction assuming constant photolysis rate. Dotted lines are ethane-methane solid-liquid phase diagram from Moran (1959), with L = all liquid, $S + L$ = coexisting liquid and solid and S = all solid (figure from Lunine and Stevenson 1985*b*).

what was suspected for several years (Courtin 1982), that the dominant gas opacity in Titan's atmosphere is contributed by collisions between nitrogen and methane, and nitrogen and hydrogen molecules. Although the absorption coefficients for the various collision pairs differ only by a factor of several below 100 cm^{-1} (and hence the much larger mole fraction of nitrogen would result in its opacity dominating), the nitrogen-nitrogen collisional opacity drops rapidly to zero for higher frequencies. The higher multipole processes leading to methane and hydrogen absorption provide substantial opacity out to 400 wavenumbers. Since the peak of Titan's Planck function is at ~200 wavenumbers for the present surface temperature, and opacity averaging schemes tend to weight this or higher frequencies (Mihalas 1978), clearly for the present conceptual model the appropriate choice of opacity is that due to nitrogen-methane and nitrogen-hydrogen collisions. Thompson (1985) has discussed possible effects of including the nitrogen-methane collisions.

Lunine (1988) considered the evolution of Titan's atmosphere for these two sources of opacity, using the grey atmosphere formalism described above, and absorption coefficients calculated as in Courtin (1988). The coefficients were frequency averaged using a Rosseland-mean scheme. Hence, the nitrogen-hydrogen and nitrogen-methane cases had to be considered separately. The two alternative assumptions for the behavior of the hydrogen vapor pressure lead to an optical depth in Eq. (9) either (1) proportional to the square of the nitrogen pressure, or (2) linearly proportional to the nitrogen pressure. Case (1) is formally identical to the results in Fig. 11; for case (2), the evolution is in the same direction but more gradual. If the nitrogen-methane opacity dominates, Eq. (9) becomes more complex because the nitrogen and methane vapor pressures change in opposite directions with ocean evolution. The methane pressure dominates, and the sense of the evolution is toward warmer temperatures in the past.

The frequency-averaged absorption coefficient for nitrogen-hydrogen was significantly larger than for nitrogen-methane, and on this basis one would argue that Titan's surface was colder in the past. However, as Lunine (1988) points out, the averaging technique probably underestimates the methane contribution, and a non-grey scheme may be required to determine the evolution of the surface for both opacity sources.

The analysis presented shows that the present state of Titan's troposphere, and the evolution of its thermal structure, must be understood in the context of the surface physical state and composition. Additional effects, not included above, are the change in stratospheric thermal balance due to changing tropospheric temperatures (and hence changing methane mole fraction admitted to the stratosphere), the increase in solar luminosity over geologic time, and changes in cloud deck thickness and opacity. To first order, the effects of cloud infrared and visible opacity tend to cancel each other out (McKay et al. 1988*a*). However, the play-off between methane vapor pressure and cloud infrared optical depth is complex, and includes changes in altitude of the lifting condensation level, higher cloud density for larger temperatures, possible change in cloud particle size distribution for different cloud densities, and possible changes in cloud morphology and areal homogeneity for large changes in surface temperature. These may alter the rough balance between the visible and infrared optical depths due to clouds, and hence contribute to the evolution of the thermal structure.

The surface-atmosphere coupling presented here does not work if an ocean containing large quantities of exchangeable methane and nitrogen is not present. The solid polymer model decouples the atmosphere from any steady-state interaction with the surface. Sporadic resupply of atmospheric methane by volcanic or impact sources would perturb the atmospheric thermal balance, but analysis of such effects is rendered difficult by their stochastic nature.

Clearly, then, observational evidence for or against the presence of a hydrocarbon ocean carries implications both for the long-term photochemical cycle on Titan *and* long-term climate evolution. Our understanding of this satellite's evolution will increase enormously when such evidence can be obtained.

Effect of Photochemical Cycle on D/H Ratio. Recent observations of deuterated methane in the reflection spectrum of Titan near the 1.6-μm region lead to a D/H ratio in the atmosphere of $1.65^{+1.65}_{-0.8} \times 10^{-4}$, 8 times higher than the value derived for Jupiter and Saturn (and generally assumed to be primordial) and comparable to the value found on Uranus (de Bergh et al. 1987*a*) and on the Earth. It is compatible with an analysis of Voyager IRIS data of Titan (Kim and Caldwell 1982). Pinto et al. (1986) explored the effects of a hydrocarbon ocean, and the photochemical cycle, on the deuterated methane abundance relative to total atmospheric methane. Fractionation into the ocean by vapor-pressure differences turns out to be negligible, since the deuterated methane molecule turns out to have a nearly identical vapor pressure to that of normal methane (the result of a fortuitous cancellation between inter- and intra- molecular effects on the vapor pressure). Other solubility effects lead to at most an atmospheric deuterium enhancement of 1.3.

Also considered was the possible photochemical partitioning of deuterium in methane. The slightly higher binding energy of deuterium in methane compared to normal hydrogen results in preferential dissociation of CH_4. The deuterated methane enhancement factor is then dependent upon the ratio of the amount of methane photolyzed to the total available methane reservoir, i.e., ocean composition. For a 25% methane ocean, Pinto et al. (1986) find $CH_3D/CH_4 = 1.7$. Thus, at most a factor of 2 enrichment in deuterated atmospheric methane may be caused by surface and atmospheric photochemical process. The remainder may be a signature of the original materials out of which Titan was assembled; a discussion of the source of this deuterium-rich material is given in Sec. IV.

Evolution of Surfaces of Triton and Pluto

At present we have no information on the abundance of methane and nitrogen on the surfaces of Triton and Pluto. Establishment by Voyager 2 of the areal coverage of volatile material on Triton, and the atmospheric pressures of methane and nitrogen, would aid in understanding how much volatile material is present. The seasonal transport of volatiles on Triton amounts to meters thickness; Earth-based observations as solstice is approached in the next 20 yr could establish whether volatiles become depleted from the summer hemisphere. The photochemical processing of methane on Triton and Pluto also puts constraints on the amount of surface methane; since the photolysis rate is down by a factor of 10 from Titan to Triton, the depth of processed organics is 10s to 100s of meters on the latter body. Clearly the sort of modeling presented in this section must await the results of the Voyager flyby of

Triton in 1989; for Pluto improved groundbased techniques will remain the primary tool for some time to come.

IV. ORIGIN AND EARLY EVOLUTION OF THE ATMOSPHERES OF TITAN, TRITON AND PLUTO

As with any atmosphere, the atmospheres of Titan, Triton and Pluto contain information on the environments within which these bodies formed and underwent early evolution driven by atmospheric escape, enhanced EUV radiation from the Sun, bombardment by debris and internal processes. In this section we examine these processes and identify their chemical signatures in the present atmospheres. Of greatest interest are the origin of nitrogen and methane on Titan and Triton, for which noble gas abundances may be diagnostic, and the implications of the deuterium abundance in the atmosphere of Titan for its environment of formation. The role of nebular chemistry in the formation of atmospheres is contained in the Chapter by Prinn and Fegley; some of the discussion below overlaps with material in that chapter.

Regular satellites such as Titan undoubtedly originated in a nebula of gas and dust that surrounded their parent outer planets during the latter's early history. The origins of Triton and Pluto are much less clear. Given the large inclination of Triton's orbital plane to Neptune's equatorial plane, it is possible alternatively to consider Triton to be a regular satellite formed in a nebula around Neptune, with this system subsequently severely disturbed by a passing large planetesimal; or to suppose that it was formed in the solar nebula and later captured by Neptune. The latter hypothesis is advocated by McKinnon (1984); earlier capture hypotheses are reviewed by Cruikshank and Brown (1986).

Pluto/Charon represents a fully tidally evolved system (Burns 1986*a*) and hence no information on the mechanism of producing this double planet system is available from the orbit of Charon. Although a number of hypotheses for ejection of Pluto from orbit around Neptune have been advanced, some involving Triton, they all suffer from significant problems when compared to the present state (Burns 1986*a*). Burns advocates the view that Triton, Pluto and Charon were leftover planetesimals; Triton was captured by gas drag into Neptune orbit while Pluto-Charon remained in solar orbit because their 3:2 resonance with Neptune prevented close encounters with the outermost giant planet.

The rest of this section is organized to provide an overview of the formation and early evolution of satellite atmospheres first (focusing necessarily on Titan) from models of the nebula, through condensation and accretion of planetesimals, to processes affecting post-accretional evolution of the primordial Titan atmosphere. Following this, the two competing scenarios for the origin of nitrogen in Titan's atmosphere are discussed in terms of diagnostic observations. The issue of the total surface abundance of methane is then considered. We close with an assessment of the significance of the deuterium abundance in Titan's atmosphere, and the relationship of the origin of Titan's volatiles to those on Triton.

An Overview of the Formation and Early Evolution of Titan's Atmosphere: Nebula Models. The giant planets formed from the gaseous and solid components of the primordial solar nebula, a flattened disk of material derived from a collapsing molecular cloud fragment in association with the formation of the Sun. Solid-body accretion processes operating in the solar nebula led to the creation of progressively larger-sized planetesimals that served as the source material for the cores of the giant planets (see, e.g., Pollack 1985). Extensive treatments of nebula models are available in the chapters by Coradini et al., Morfill et al. and Pollack and Bodenheimer. Briefly, the solar nebula underwent perhaps four major stages of evolution. First was a growth phase dominated by the accretion of material from the collapsing molecular cloud fragment. By the end of this stage the solar nebula may have been quite massive, perhaps several tenths of a solar mass. The next stage was marked by turbulence and consequent outward transfer of angular momentum and inward transfer of nebular mass (Lynden-Bell and Pringle 1974). Following this, grain formation would have increased the nebular opacity to the extent that thermal convection would have been dominant in some regions. Planetesimal formation would have reduced the grain opacity and, in the final stage, gas and small grains were eliminated from the solar system by a strong wind emanating from the Sun. Based on the properties of T-Tauri stars, the formation and dissipation of the nebula took place in 10 Myr (see discussion in Lissauer 1987).

The radial temperature profile in such a nebula is complicated by the variation in opacity due to grain formation (Ruden and Lin 1986), and does not follow a simple power law. However, the general trend defined by planet composition (Lewis 1974*a*) is maintained in sophisticated numerical models. The same type of physics as pertained to the solar nebula controlled the qualitative behavior and evolution of satellite-forming accretion disks around giant planets. However, there are a number of key differences between the two, detailed in Stevenson et al. (1986). Primary among them is that pressures in the satellite-forming nebulae were 5 to 8 orders of magnitude larger than the pressures in the outer solar nebula. As a consequence the satellite-nebulae of Jupiter and Saturn, at least, were optically thick throughout and may have had temperature profiles inversely proportional to radial distance from the planet at some point during their evolution (Lunine and Stevenson 1982*a*). Temperatures were higher in the inner regions of these disks than in the surrounding solar nebula. Satellite accretion disks also evolved much faster than the solar nebula, with dynamic time scales of perhaps 10^2 to 10^4 yr.

Planetestimal Composition. The composition of the building blocks of satellites is determined by a combination of the temperature and oxidation state of the nebula from which they formed, and also possibly by an inheritance of small amounts of solid material from earlier times. At temperatures expected in the outer solar nebula during most of its history, and in a nebula around Saturn, both water ice and rock-forming materials would have been entirely in the condensed state. The relative proportions of these two key ingredients depends strongly on the partitioning of elemental oxygen (O) among various chemical species (Lewis and Prinn 1980; Prinn and Fegley 1981*a*: Johnson et al. 1987; Pollack and Bodenheimer, personal communication). In particular, about 15% of the O available in the solar nebula would have gone into oxides of the rock component. The remainder would have been partitioned between CO and water. CO can be a potent sink for O since the solar abundance of C is 60% that of O. Thus, the relative proportion of rock and water in planetesimals depends primarily on the oxidation state and molecular speciation of C in the solar and giant planet nebulae.

As quantified in detail by Lewis and Prinn (1980) and Prinn and Fegley (1981*a*), methane is the thermodynamically favored form of C at the temperatures characterizing the outer solar nebula. However, kinetic factors may prevent this equilibrium from being realized. If there is effective exchange of material between the inner and outer parts of a nebula, CO derived from the warmer regions, where reactions proceeded rapidly, may not have been converted to methane in the outer regions before remixing to warmer regions occurs (or even within the lifetime of the nebula). A similar conclusion holds for gas derived from the molecular cloud, which is dominated by CO (see chapter by Irvine et al.). A giant planet (protosatellite) nebula, with hydrogen pressures orders of magnitude higher than in the solar nebula, *would* be dominated by methane (Prinn and Fegley 1981*a*). The oxidation state of nitrogen tends to follow the same direction as carbon (see chapter by Prinn and Fegley).

Because the oxidation state of carbon so strongly controls the abundance of water, and because CO is too volatile to form a pure condensate, the densities of the satellites are diagnostic of the oxidation state of the nebulae from which they formed. First, suppose that methane was the dominant form of C in the outer solar nebula. In that case, the rock to water-ice ratio in planetesimals would have been approximately 40/60 by mass. Conversely, if CO were the dominant C species, this ratio would have equalled about 75/25. The mass ratios derived from large satellites of the outer solar system provide an *upper bound* on this ratio for nebula planetesimals, since it is possible to lose water by vaporization during accretion (see below). (On the other hand, the rock-to-ice ratio of planetesimals formed in a satellite-nebula around Saturn might have been altered by preferential dissolution of water in the Saturnian atmosphere, and loss of rock to the core.) The observed density of Titan implies a mass ratio of 55/45, suggesting either (a) some CO in the nebula, or

(b) a CH_4-rich nebula with loss of water during Titan's accretion. Interestingly, the Uranian satellites' densities fall within the same rock-to-ice range (Johnson et al. 1987); since they are not large enough to lose significant quantities of water during accretion (see below), they probably formed from a gas containing significant quantities of CO. This gas could have been the solar nebula (Johnson et al. 1987) or a disk of material spun out from Uranus after a giant impact, in which CO was produced from shock heating (Stevenson 1984*b*). Until we learn Triton's density from the Voyager flyby, we cannot say anything about the oxidation state of its formation region.

Condensation of planetesimals is a powerful fractionation process which largely determines the bulk composition of the assembled body. Incorporation of a gaseous molecular species into solids is accomplished by (a) direct condensation, (b) clathration and (c) physical adsorption. Considering elements more volatile than water ice, temperature profiles for nebulae around giant planets (Prinn and Fegley 1981*a*; Lunine and Stevenson 1982*a*; Pollack and Consolmagno 1984) allow for direct condensation of ammonia hydrates and methane ice in the Saturnian nebula. In the solar nebula, the *T-P* profile also may allow for condensation of carbon dioxide if it comprises as much as 1% of the total carbon. Direct condensation of molecular nitrogen, argon and carbon monoxide are unlikely in the solar nebula, and certainly very implausible at Saturn's distance from the Sun.

Clathrate formation involves a structural alteration in water ice to accommodate more volatile species in voids, or cages. Application of the thermodynamics of clathrate formation to astrophysical environments and a review of earlier work is presented in Lunine and Stevenson (1985*a*). As with condensation, clathrate formation requires a threshold partial pressure of volatile gases as a function of temperature. Once water ice is induced to form clathrate, all gases present in the surrounding nebula will be incorporated to varying degrees, largely determined by their volatility. The total volatile-to-ice ratio in clathrate is 1 to 6. Essentially all of the condensed ice can be converted to clathrate, excluding that which formed ammonia hydrates (Lewis 1972*a*; Lunine and Stevenson 1985*a*), so long as good physical contact exists between the gas and ice. This requires frequent collisions between planetesimals to expose fresh ice to the gas, plausible in a nebula around Saturn but perhaps less so in the solar nebula, where orbital times are longer.

The chemical partitioning of gases into clathrate has important consequences for the proportions of nitrogen, carbon and noble gas species incorporated into satellites. Figure 12 illustrates the fractionation. The left panel shows that methane much more readily incorporates in clathrate than does nitrogen, as shown by experimental data. As confirmed by recent data on CO clathrate (Davidson et al. 1987), CO is predicted by Lunine and Stevenson to incorporate more readily than nitrogen but less so than methane. This figure makes the generous assumption that all of the nitrogen is molecular; it thus yields the maximum amount of nitrogen incorporated in the clathrate. As an

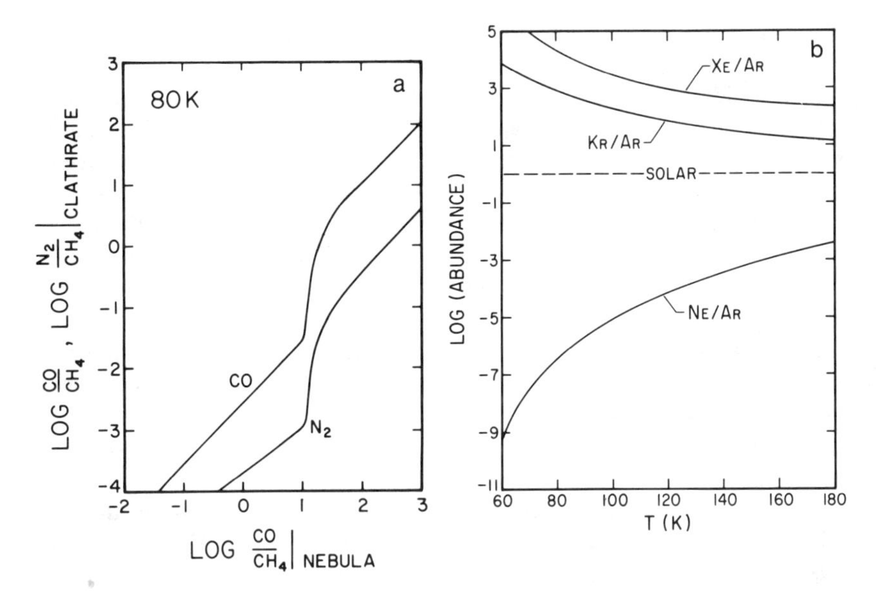

Fig. 12. (a) Ratio of CO and N_2 to CH_4 incorporated in clathrate as a function of CO-to-CH_4 ratio in a solar-composition nebula, at 80 K. All nitrogen is assumed to be in the form of N_2; total carbon-to-oxygen ratio in H_2O + CO + CH_4 is solar. (b) Noble gas abundance ratios in CH_4-dominated clathrate relative to abundance ratios in solar composition gas, as a function of temperature (figure from Lunine and Stevenson 1985*a*).

illustrative example, assume the nitrogen-to-methane ratio in the present Titan surface-atmosphere system is primordial and equals the ratio of nitrogen to methane incorporated in clathrate (see discussion of the origin of nitrogen below). This ratio is ~0.1 to 1 (Table II). Selecting a value of unity for concreteness, and reading across Fig. 12a from the ordinate value log N_2/CH_4 = 0 shows that such a value in the clathrate requires a CO-to-CH_4 ratio in the nebular gas of ~500. Physically, the methane ratio in the nebula gas must be small, so that methane does not overwhelm nitrogen and occupy almost all of the cage sites. We also see from the figure that the CO-to-N_2 ratio in the clathrate is 10; for the example cited, CO is the dominant molecule in the clathrate (because the methane nebular abundance is so small), and by the starting assumption in this example, should dominate the present Titan atmosphere (see discussion of CO photolysis later in this section). Note that the CO-dominated nebula required in this example produces a Titan which is too rock rich.

Alternatively, if we assume the CO-to-CH_4 ratio in the nebula was less than unity, as seems to be required by Titan's density, we find that the nitrogen-to-methane ratio in the clathrate is $<10^{-3}$, and an additional source of nitrogen to supply the present-day atmosphere is required. As discussed below, the likely source is ammonia, brought into Titan in the frozen chemical compounds ammonia monohydrate, ammonia dihydrate, or pure condensed ammonia.

One additional piece of information not contained in Fig. 12 is that nitrogen and argon partition into clathrate with roughly equal ease. Thus argon becomes a major indicator of the origin of nitrogen in Titan, as discussed below.

Physical adsorption, in which volatile molecules stick directly to ice grain surfaces, produces a similar chemical partitioning effect. However, the amount of area available for adsorption is limited, and hence the amount of volatile material which can be incorporated in the ice is much less than the maximum which can be incorporated in clathrate. Amorphous ice grains have a much higher specific adsorption area (Mayer and Pletzer 1986), and if volumetric contact between ice and gas is inhibited, adsorption could be more important than clathration.

Accretion of Large Satellites. Accretion of a satellite provides the initial source of energy for creation of a primordial atmosphere. Although a wide range of accretion models have been explored, which differ in the manner in which energy is deposited in the subsurface, surface or gaseous envelope of a protosatellite, accretional energy input is an important process only for satellites as large or larger than Rhea and Titania. Above this mass threshold, enough heating is generated to melt and/or vaporize volatiles (see the extended discussion in Stevenson et al. 1986). For bodies the size of the Galilean satellites and Titan, accretional heating quickly raises a substantial atmosphere, that overwhelms the ambient hydrogen envelope from the surrounding nebula (Lunine and Stevenson 1982). This atmosphere is sufficiently thick that planetesimals deposit their energy largely in the envelope; hence even if accretion starts off gas free, it conforms to the gaseous scenarios during the later stages. If Triton is as massive as Titan, then the same considerations apply; Pluto's reported mass makes accretional heating a marginal process for raising a massive primordial atmosphere (although methane would be volatilized to some extent).

No models for the accretion of Titan have been published, except for the thesis of Lunine (1985*a*) which assumes the planetesimals to be largely rock, methane clathrate hydrate and ammonia hydrate. To some extent the results apply to ammonia-poor planetesimals as well. The heating during accretion is sufficient to devolatilize the clathrate and release most of the methane into the primordial atmosphere. The only methane intact after accretion is buried in the innermost 5 to 20% of the mass of Titan (Lunine and Stevenson 1987). Ammonia and water form a massive envelope together with the methane and any other volatiles previously clathrated. The ammonia and water are in vapor pressure equilibrium with a surface ammonia-water "magma" ocean. Some of the atmosphere escapes back into the nebula by a hydrodynamic flow which taps part of the energy of planetesimal infall (Stevenson et al. 1986); the net effect is to increase the rock-to-ice ratio of Titan relative to that of the original planetesimals.

The final state is shown in Fig. 13. The primordial atmosphere consists of up to several hundred bars of methane, and up to 200 bars of ammonia and water. The mass of the primordial atmosphere due to gaseous accretion is 10^3 times larger than the present atmosphere.

Escape of Primordial Atmosphere. Energy sources available for escape include solar EUV radiation, T-Tauri wind, and post-accretional impact. (The primordial atmosphere does not adiabatically expand away, or "uncork", as the surrounding nebula is dissipated, because many of the processes occurring during accretion were thermodynamically irreversible.) Enhanced EUV from the Sun can drive atmospheric escape at the energy-limited rate described by Watson et al. (1981) and Hunten et al. (1987). Analysis of pre-main-sequence stars by International Ultraviolet Explorer (IUE) satellite indicates that before the Sun reached the stellar main sequence at an age of 50 Myr, its flux shortward of 2000 Å may have been up to 10,000 times greater than the present value (Zahnle and Walker 1982; Canuto et al. 1982). Lunine (1985*a*) applied the atmospheric escape model of Watson et al. (1981) (discussed in Sec. II in the context of the present atmosphere of Pluto) to the primordial Titan atmosphere. For an EUV flux 100 times the present-day, the escape flux is 10^{15} g yr^{-1}, insufficient to permit the primordial atmosphere to escape within the age of the solar system. EUV fluxes greater than this were not considered as the model formalism breaks down; whether higher fluxes persisted sufficiently long to remove the primordial atmosphere remains unclear.

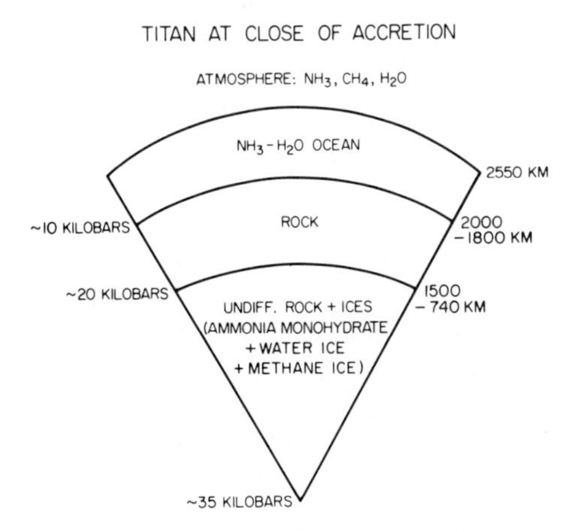

Fig. 13. Schematic model of the interior structure and atmospheric composition of Titan immediately after accretion. Radii are indicated on the right side of the figure; approximate pressure on the left (figure from Lunine and Stevenson 1987).

Loss of a massive Titan atmosphere by T-Tauri wind from the Sun (or an analogous process at Saturn) has never been quantified. Uncertainties include the coupling of the wind to the atmosphere (is it direct or through the Saturnian nebula?), and the timing and duration of the wind.

Impact of solar orbiting debris represents a potentially large source of energy for escape. This loss process is qualitatively different from hydrodynamical escape at the exobase level powered by EUV. It involves accelerating air in the lower atmosphere to escape velocity by shock waves accompanying the impactor during descent and after it hits the surface. The impact process and net loss of solid target debris have been considered by O'Keefe and Ahrens (1982*c*); atmospheric loss for the case of Mars has been considered by Watkins (1983; see also Lewis and Prinn 1984). Most recently, Walker (1986*b*) analyzed theoretically the atmospheric loss due to impacts as a function of the impactor velocity and the escape velocity of the planet. He concluded that (1) escape occurs only if the impactor velocity is greater than the planetary escape velocity, (2) impact erosion does not occur from planets with escape velocities >10 km s^{-1}, and (3) the amount of gas which escapes is roughly the amount intercepted by the impactor.

Titan's escape velocity is 3 km s^{-1}. Impact of Saturn-orbiting planetesimals with relative velocities lower than this would have led to net accretion of material. Solar-orbiting debris, on the other hand, would have eroded the atmosphere and surface upon impact. The epoch of the accretion of Titan and its primordial atmosphere from Saturn-orbiting material could have been followed by a period when projectiles external to the Saturn system removed most of the gas. The Saturnian satellite cratering record has not yet been definitively interpreted in terms of solar- or planet-orbiting projectiles (Chapman and McKinnon 1986). However, if the Shoemaker and Wolfe (1984) analysis of the Jovian and Saturnian cratering record is correct, several hundred Earth masses worth of debris existed in the outer solar system at the close of planet formation. These objects could have been of singular importance in the escape and chemical modification of a primordial Titan atmosphere.

Structure and Cooling of Primordial Titan Atmosphere. Figure 14 illustrates a model for an ammonia-rich primordial Titan atmosphere. The atmospheric ammonia and water, in gaseous and cloud form, maintain the high surface temperatures by virtue of their high infrared opacities. Giant impacts cause escape of the atmosphere and hence reduction in the infrared opacity, resulting in cooling and consequent condensation of ammonia and water at the surface. Impact heating also converts ammonia to nitrogen, and methane to complex hydrocarbons. Photochemical conversion of ammonia to nitrogen also takes place in the upper atmosphere. These processes are detailed in the following subsection. As temperatures drop below 200 to 250 K, clathrate

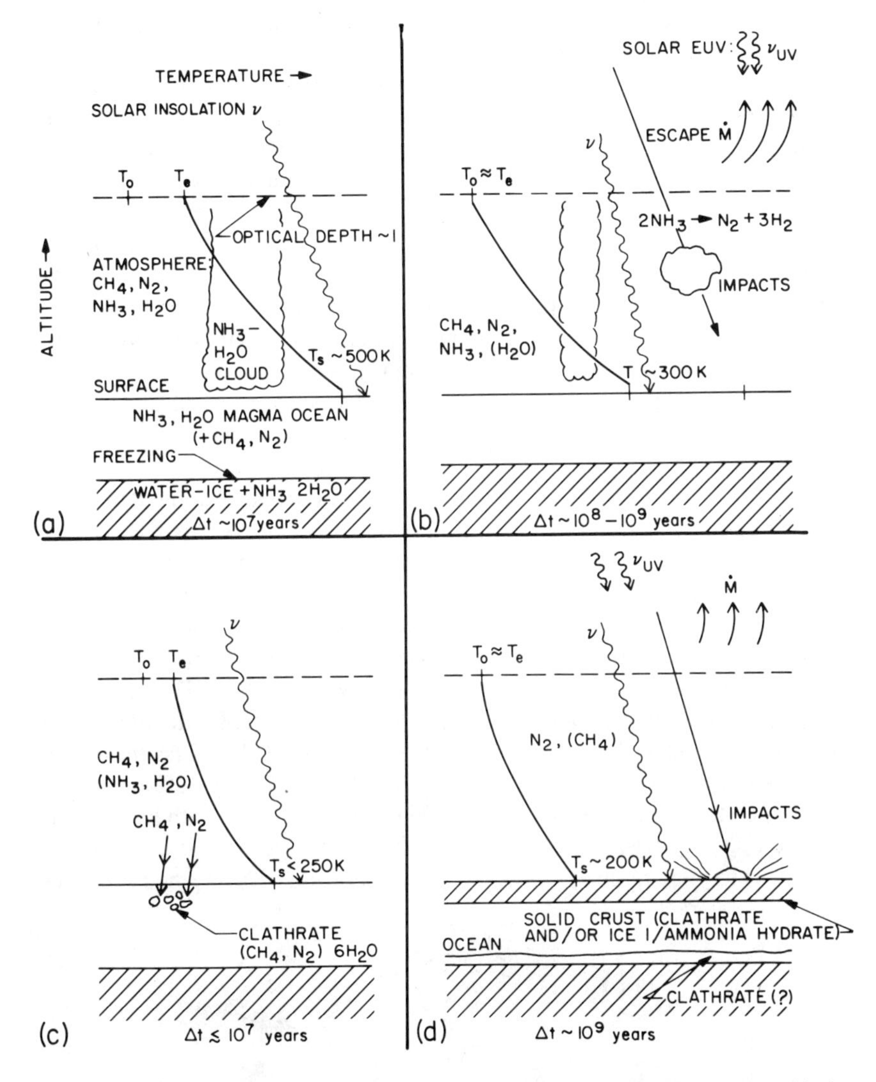

Fig. 14. Illustration of phases of atmospheric evolution after accretion. Temperature-altitude scale is schematic; time scales are typical for a model with atmospheric mass-loss rate of 10^{14} g yr^{-1} and starting mass 10^{23} to 10^{24} g. Minor constituents are indicated in parentheses. Although some processes may occur in all panels, only those important to the particular stage are shown in a given panel. (a) Hot, post-accretional atmosphere cools down. (b) Atmosphere achieves thermal steady state in which effective temperature T_e is given by the insolation value T_o. Cooling rate is now determined by mass-loss rate. Photochemical and impact-induced $NH_3 \rightarrow N_2$ conversion may occur here (and in panel d). (c) Base of atmosphere enters clathrate stability field, dropping surface pressure rapidly and forcing $T_e > T_o$, increasing the cooling rate. (d) Ocean freezeover terminates clathrate formation; atmosphere returns to steady state (figure from Lunine 1985*a*).

formation at the base of the atmosphere may occur, causing a further drop in surface pressure and temperature. Eventually the surface freezes over.

The end state of such an evolution is hard to specify. If bombardment is efficient enough to remove 10^{25} g of atmosphere over 100 Myr, it is too much to ask that it leave behind 10^{22} g of nitrogen and methane. As discussed in Lunine and Stevenson (1987) the present surface methane complement was likely to have been outgassed from the interior during the overturning of Titan's core. If this occurred as late as 500 Myr after formation, the epoch of primordial atmospheric erosion would likely have been concluded. Was nitrogen outgassed with the methane, was it produced during the terminal bombardment and retained, or was it produced after the bombardment and methane outgassing? To address these questions we must consider the two processes by which ammonia can be converted to nitrogen in the early atmosphere of Titan: photolysis and shock heating due to impacts.

N_2 from NH_3 Photolysis. Ammonia can be efficiently converted to nitrogen photochemically (Atreya et al. 1977). The photolysis of NH_3 occurs following the absorption of solar ultraviolet photons with wavelengths between 1600 and 2300 Å. Neither CH_4 nor any other gas in significant proportion on Titan absorbs in this region. As a consequence, the photolysis of NH_3 would proceed unhindered by the presence of other constituents. Figure 15 shows a schematic of NH_3 photochemistry relevant to Titan. The photolysis of NH_3 yields amidogen radicals (NH_2) in the ground state. Production of NH ($a^1\Delta$) is also likely, but with a low quantum yield of 4%. The maximum photolysis rate at Titan is on the order of 10^{-6} s^{-1}. Nearly 30% of NH_3 is recycled by the reaction of NH_2 with H. The reaction of NH_2 with itself yields hydrazine, N_2H_4. Hydrazine reacts with atomic hydrogen and also undergoes photolysis to yield hydrazyl, N_2H_3. The self-reaction of N_2H_3 then produces

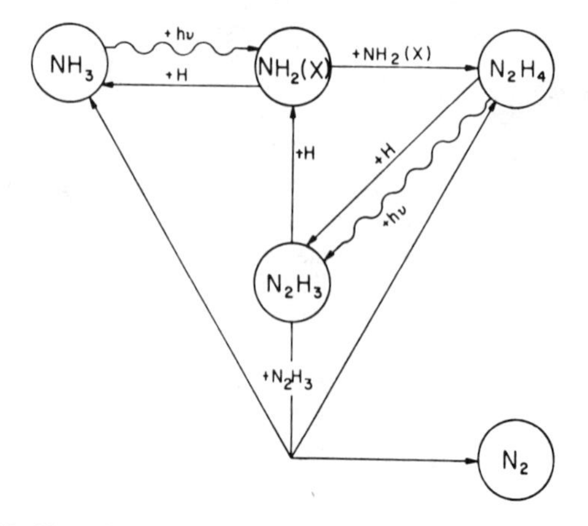

Fig. 15. Photochemistry of ammonia (figure from Atreya et al. 1978).

N_2. Once N_2 is produced, it accumulates in the atmosphere, as it is inert and highly stable. A fraction (approximately 10%) of N_2 would, however, be lost due to the thermal and nonthermal escape mechanisms.

The scenario described works provided that the gas phase abundances of both NH_3 and N_2H_4 are appreciable. At the present-day surface temperature of Titan, 94 K, the vapor pressure and the corresponding mixing ratio (by volume) of NH_3 are, respectively, 2.7×10^{-8} mbar and 0.02 ppb. The low column abundance of NH_3 has not allowed detection of this species on Titan. The present abundance of NH_3 is too low to produce substantial amounts of N_2 by photolysis. If NH_3 photolysis were the source of Titan's N_2, a warm temperature in the primordial atmosphere would be needed. Temperatures cannot be too high, however. Above approximately 300 K, comparable amounts of water and ammonia would be present in the vapor phase. The photolysis of H_2O would produce OH, whereas the photolysis of NH_3 would yield NH_2. The reaction between NH_3 and H_2O and their radical products would entirely choke off the production of N_2H_4 and subsequently N_2. Below 200 K, however, the saturation vapor pressure of NH_3 dominates that of H_2O. At the melting point of NH_3 (195.3 K), the ratio is nearly 100,000. Below 150 K, the saturated vapor density of the key intermediate constituent, N_2H_4, drops dramatically with a corresponding drop in the yield of N_2 (Atreya et al. 1977). The photolysis of NH_3 leading up to the production of N_2 can thus proceed quite readily at temperatures between 200 and 150 K.

According to a pre-Voyager model of Atreya et al. (1978), the photolysis of NH_3 on Titan would proceed at a diurnally averaged rate of approximately 6×10^{10} cm^{-2} s^{-1} assuming a temperature of 150 K. This implies that it could take up to 100 Myr, or 2% of the geologic time, to account for the present-day surface pressure of 1500 mbar on Titan. An epoch of enhanced solar ultraviolet flux would result in an increase in the rate of photolysis, and a corresponding decrease in the time constant for producing the present-day surface pressure of N_2 to as low as 10,000 yr (Atreya 1986). This estimate is based upon IUE observations of pre-main-sequence stars which suggest that the solar flux shortward of 2000 Å could have been as much as 10,000 times its present value (Zahnle and Walker 1982). Allowing for nonlinearities in the N_2 production rate due to the escape of N_2, the decreasing vapor pressures of NH_3 and N_2H_4 as Titan cooled, and the effect on photolysis rate of the change with time of the solar ultraviolet flux, the above time constant could approach 50,000 to 100,000 yr—still short compared to geologic time scales (Atreya 1986).

The necessity for invoking enhanced solar flux to drive the conversion goes away if a long time period of ~100 Myr existed when Titan was warm. Even after the terminal bombardment phase erodes a primordial atmosphere, this is possible. Outgassing of all the methane required to produce the present ethane-methane ocean corresponds to a surface pressure of 20 bar, assuming no condensation. Pollack (1973) has calculated that a 0.5 bar surface pressure of CH_4 alone would have led to enough greenhouse effect to raise Titan's surface temperature to 150 K. Rapid conversion of NH_3 to N_2 may still be necessary because of possible evolution of hazes and clouds that could prevent penetration of the photolyzing flux to the region where NH_3 was present.

Impact Production of Titan's N_2. Projectiles passing through an atmosphere at supersonic velocities generate a shock in front of themselves. Atmospheric gas encountering the shock front is strongly compressed and heated to very high temperatures, on the order of thousands of degrees Kelvin. At these high temperatures thermodynamic equilibrium is very quickly realized, and a gas composition very different from the ambient atmosphere is produced. Reaction rates decrease very rapidly as the gas parcel subsequently cools by radiation, and the high-temperature chemistry is "quenched" into the parcel. In this manner, meteoroids passing through an early Titan atmosphere containing significant amounts of ammonia vapor may have converted a fraction of it to N_2. Independent estimates of the amount of N_2 formed in this way during Titan's early history indicate that the shock-induced conversion of NH_3 to N_2 could easily have generated the satellite's current inventory of N_2 in ocean and atmosphere (Jones and Lewis 1987; Anderson and Stevenson 1987; McKay et al. 1988*b*).

The amount of nitrogen produced during passage of meteoroids through Titan's early atmosphere depends on the velocity and mass of the projectile, the temperature history of gas parcels, chemical reaction rates and the abundance of ammonia in the atmosphere. A convenient means for estimating the total amount of nitrogen generated by meteoroid passage through Titan's atmosphere is provided by the following equation:

$$[N_2] = E_K * F_K * Y \tag{14}$$

where E_K is the total kinetic energy of the projectiles participating in the conversion, F_K is the fraction of the kinetic energy going into shock heating, and Y is the number of moles of N_2 produced per unit of energy of shock heating. Projectiles with sizes between m and km are most effective at producing shock conversion since larger bodies pass through the atmosphere without losing much of their kinetic energy and smaller bodies dissipate the bulk of their kinetic energy above the cold trap, where little ammonia would be present (McKay et al. 1988*b*).

The yield Y depends on the quench temperature, the pressure of the shocked gas, and the mole fraction of ammonia in the ambient atmosphere. Jones and Lewis (1987) show that the value of Y is fairly insensitive to shock temperature and pressure over a wide range of reasonable values. Experimental simulations of the shock conversion of NH_3 to N_2 have been carried out by McKay et al. (1988*b*) who used a focused laser to generate a shock. They find that Y varies approximately linearly with ammonia mole fraction for fractions above about 0.1, but that it sharply decreases with decreasing mole fraction below this value. Thus, significant shock conversion of NH_3 to N_2 occurred when ammonia was a major constituent of the lower atmosphere of Titan. This condition was most likely met during the satellite's accretion when heat deposited by impact raised the surface temperature well above its current value (McKay et al. 1988*a,b*).

Jones and Lewis (1987) estimated the shock production of N_2 during Titan's early history, subsequent to formation. Impactors derived from an early breakup of the neighboring satellite Hyperion and planetesimals scattered from the zone where Uranus and Neptune formed were studied as possible generators of Titan's N_2 atmosphere as well as a number of organic compounds. Using an estimate of the latter flux provided by Shoemaker and Wolfe (personal communication), they concluded that it was the dominant producer of shocked N_2 and that Titan's current N_2 volatile inventory could be generated by this source. In making this calculation, they assumed that the NH_3 mole fraction in the lower atmosphere was 0.5; this requires the hot gaseous accretion model described above.

Anderson and Stevenson (1987) and McKay et al. (1988*b*) advocated generating N_2 from NH_3 during accretion. In this case, the planetesimals that were adding mass to the satellite also chemically reprocessed some of the earliest atmosphere. Anderson and Stevenson (1987) obtained estimates of Y by using a laboratory-based reaction rate for the interconversion of N_2 and NH_3 and by calculating the temperature history of a gas parcel passing through a shock front. McKay et al. (1988*b*) derived Y from their laboratory data. Both groups concluded that Titan's N_2 volatile inventory could be generated by this shock processing.

Table III (McKay et al. 1988*b*) shows that a significant yield of organic compounds can be obtained in a mixture of NH_3 and CH_4 subjected to shock heating. For a laboratory simulation in which the starting materials were equal molar quantities of NH_3 and CH_4, the yields of HCN and C_2H_2 are comparable to that of N_2. The observed yields can be approximately reproduced by setting the quench temperature equal to 2000 K, as shown by the third column of the table. However, the much larger yield of C_2H_6 than expected on the basis of this quench temperature may reflect the role played by ultraviolet photolysis in the simulation (ultraviolet light is produced in the initial plasma discharge). Thus, copious quantities of organics, especially HCN and C_2H_2, may have been produced by meteoroids passing through Titan's early atmosphere, ultimately to sediment out of the atmosphere onto the surface.

Photochemical Destruction of Primordial Carbon Monoxide. Incorporation of molecular nitrogen into Titan by accretion of clathrate brings in equal or greater amounts of carbon monoxide, as Fig. 12a shows. To reproduce the present atmospheric abundance of CO relative to N_2, destruction of CO by photochemical or shock chemical processes must be invoked. Sam-

TABLE III
Gases Produced in a Laboratory Simulation of Shock Heating[a]

Gas	Experimental Yield	Theoretical Yield[b]	Detection Method[c]
N_2	1.2×10^{17}	5.7×10^{17}	1
H_2	2.1×10^{18}	4.0×10^{18}	1
HCN	2.9×10^{17}	2.6×10^{17}	2
C_2H_2	1.1×10^{17}	4.8×10^{17}	2
C_2H_2	1.5×10^{17}	4.8×10^{17}	3
C_2H_6	8.9×10^{15}	9.2×10^{13}	3
$CH_3C{\equiv}CH$	7.3×10^{15}	1.4×10^{15}	3
C_2H_4	1.2×10^{15}	2.2×10^{16}	3
C_3H_8	1.0×10^{14}	4.7×10^{13}	3

[a]Results are from McKay et al. 1988*b*. The initial gases were CH_4 and NH_3 in equal molar proportions.
[b]Based on thermodynamic equilibrium at a quench temperature of 2000 K.
[c]1 = Thermocouple Gas Chromatography; 2 = Infrared Spectroscopy; 3 = Flame Ionization Detector Gas Chromatography.

uelson et al. (1983) formulated a photochemical scheme in which water from particulate material, raining down on Titan's atmosphere, could convert CO to CO_2, and reproduce the abundances of both in the atmosphere. Given a sufficient water influx rate, Samuelson et al. estimate that about 0.1 of a Titan atmosphere of CO could be destroyed over the age of the solar system. Whether the scheme is capable of destroying 100 times that amount (a conservative estimate of the amount of primordial CO brought in with nitrogen) has not been assessed.

Origin of Nitrogen on Titan: Observational Tests

We have enumerated two models for the origin of nitrogen on Titan: introduction of primordial nitrogen in clathrate hydrate formed in the nebula, and incorporation of ammonia into Titan followed by chemical conversion to molecular nitrogen. Both the bulk density of Titan and the atmospheric argon-to-nitrogen ratio provide tests of these models.

Test 1: Bulk Density of Titan. As noted above, Titan's density falls between that for a rock-water ice body formed from a methane-rich and carbon monoxide-rich gaseous nebula. (Again, the difference lies in the amount of oxygen available to form water after sequestration into rock and carbon species.) Since it is easier to envision the loss of water rather than rock during accretion, the measured density implies one of the following simple scenarios: (a) Titan formed in a relatively methane-rich environment and lost some water during accretion. Because CO is not the dominant carbon species in this model, most of the clathrate cages would contain methane, and nitrogen would be largely excluded from the clathrate (Fig. 12). Hence the only source of nitrogen readily incorporated in Titan would be ammonia. (b) Titan formed in an environment containing comparable amounts of methane and carbon monoxide with a corresponding rock-to-ice ratio directly reflected in the present bulk density (corrected for compression). Again, because of the strong propensity for methane to occupy most of the cage sites, the amount of molecular nitrogen brought into Titan in clathrate is probably insufficient to account for the present-day nitrogen-to-(methane + processed hydrocarbons) ratio. Of course, more complex models are possible, in which the rock-to-ice ratio of planetesimals which formed Titan is decoupled from the amount of CO in the nebular gas (for example, due to processes involved in the spinout of a Saturnian nebula). The silicate mass fractions for various satellite systems, and corresponding amounts of CO and CH_4 in the primordial nebulae are quantified in Johnson et al. 1988).

Test 2: Argon to Nitrogen Ratio. Owen (1982*b*) first pointed out that if Titan's nitrogen were primordial, the argon-to-nitrogen ratio in the atmosphere should be approximately solar. He also showed that the low upper limit to neon in the atmosphere (1%) is inconsistent with an atmosphere captured directly from a solar composition gas. Thus the atmosphere was brought into Titan in ice phases condensed from the nebular gas out of which Titan formed. Lunine and Stevenson (1985*a*) quantified a compositional test of the origin of nitrogen in Titan's atmosphere using clathrate thermodynamics. If the present-day nitrogen is derived entirely from clathrate, then the Ar-to-N_2 ratio in Titan's atmosphere would be 10^{-2} to 10^{-1} (Lunine and Stevenson 1985*a*). The molecular properties of Ar and N_2 are such that their outgassing histories should be similar. If the N_2 were derived from photochemistry or shock heating, and any primitive N_2 transported in clathrate, then Ar/N_2 would depend on the amount of N_2 produced photochemically and on the argon outgassing and escape history. Since the clathrate is dominated by CH_4 for all but the most oxidized-nebula scenarios (Fig. 12*a*), the amount of argon brought in depends upon its propensity for incorporation in clathrate, which is similar to that of N_2. Then the Ar/CH_4 ratio in the clathrate is about 10^{-5}. If argon outgassed to a similar extent as the CH_4 brought into Titan in clathrate, then the argon-to-N_2 ratio in the present atmosphere would be $\sim 10^{-5}$, assuming (for example) a methane-to-nitrogen ratio in the current atmosphere plus ocean of order unity. The maximum possible argon abundance brought into Titan in methane-dominated clathrate could be as high as 1 to 10% of the atmospheric nitrogen abundance, but would require that all of the argon be outgassed and retained by the atmosphere, while only a small fraction of the methane did likewise. We regard this as unlikely.

Thus, a low value of argon in Titan's atmosphere ($<<1\%$) would suggest nitrogen was derived from ammonia; a large argon abundance ($>1\%$) would suggest that much of the nitrogen was brought into Titan as N_2.

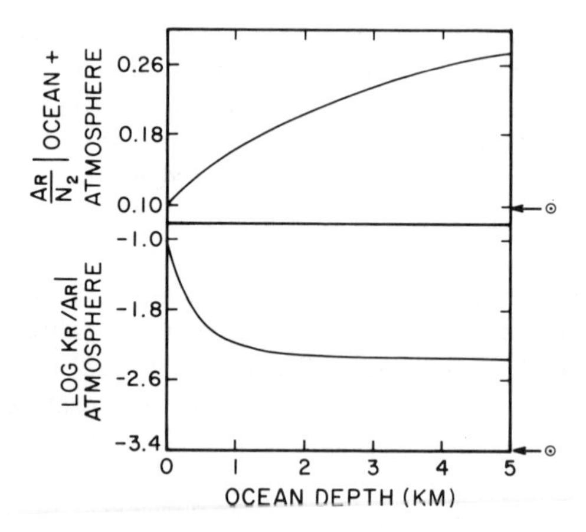

Fig. 16. Abundance patterns in Titan's atmosphere and ocean, as a function of ocean depth, predicted from solubility considerations. See text for details (figure from Lunine and Stevenson 1985*a*).

The above test will be affected slightly by the presence of the ethane ocean. Figure 16 plots the argon-to-nitrogen ratio in ocean and atmosphere, assuming the atmospheric value is measured at 10%. This figure also plots the logarithmic Kr-to-Ar ratio in the atmosphere assuming an ocean-plus-atmosphere ratio of 10%. Both fractionation effects are due to differential solubility of the gases in liquid hydrocarbons. Clearly, interpretation of atmospheric mixing ratios in Titan's atmosphere could require surface sampling if an ocean is present.

Origin of Methane on Titan

Lunine and Stevenson (1987) pursued a model in which methane is incorporated into Titan as clathrate hydrate, then outgassed during core overturn of the satellite induced by radiogenic heating. The model is described in the chapter by Schubert et al., and we therefore do not elaborate. However, the implications of the model are twofold: (1) Most of the methane brought into Titan is lost during accretion—only about 10% is retained in the deep interior beyond accretion; (2) Outgassing of methane from the interior must occur late, after $\sim 5 \times 10^8$ yr, to avoid trapping of methane in an aqueous ammonia upper mantle and to avoid loss during hydrodynamic escape. The subsequent transition to the present epoch has not been studied, but could involve complex climate changes as a methane-nitrogen ocean is formed and evolves to the present state.

Figure 17 presents two schematic timelines for the origin and evolution of Titan's atmosphere, based upon the primitive nitrogen and ammonia-converted nitrogen hypotheses.

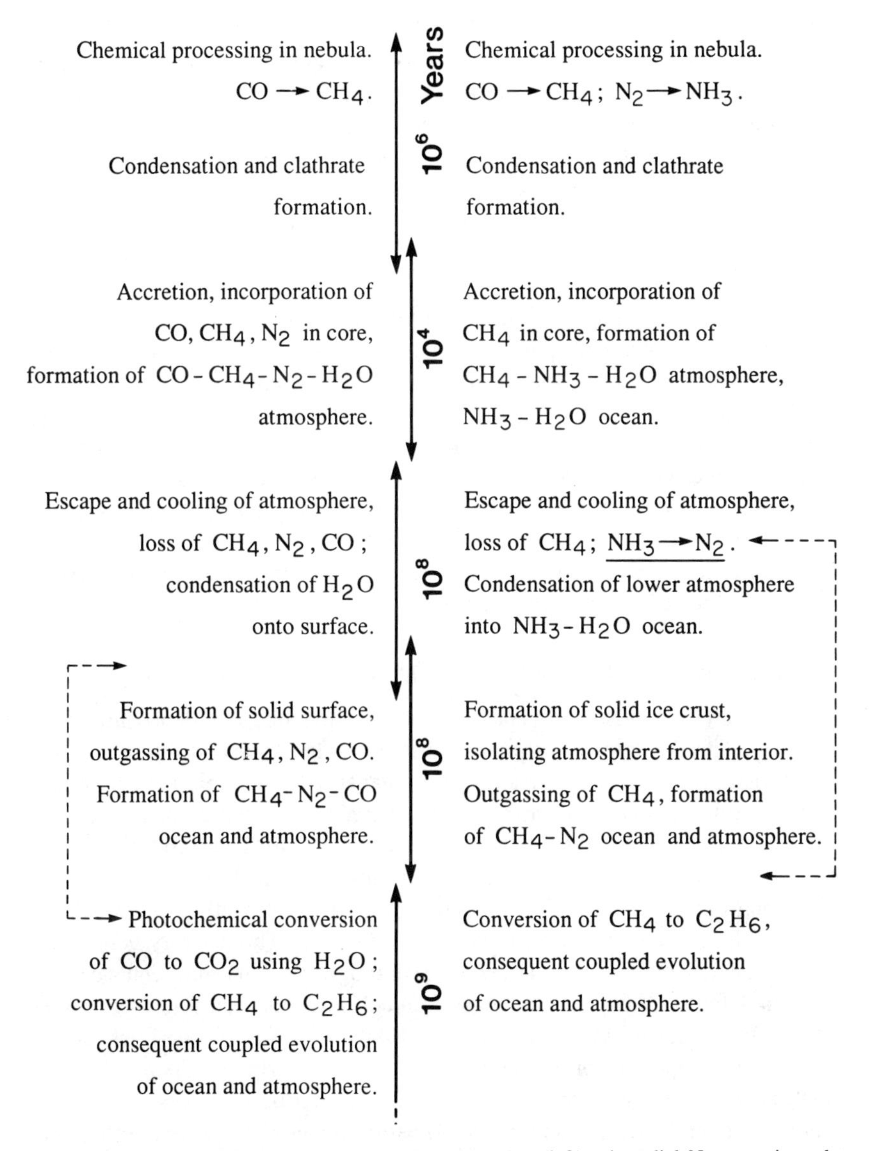

Fig. 17. Schematic timelines of Titan evolution, assuming (left) primordial N_2 scenario and (right) N_2 derived from ammonia.

Why do not Ganymede and Callisto have Nitrogen Atmospheres?

The absence of detectable atmospheres on Ganymede and Callisto, physically near-twins of Titan, requires explanation. The nebula around Jupiter was too hot to allow formation of methane clathrate (Lunine and Stevenson 1982), but could have allowed condensation of ammonia hydrates. We can only speculate on the essential differences between the evolution of the giant Jovian satellites and Titan, but three key differences suggest themselves: (1) Ganymede and Callisto never accreted significant quantities of ammonia, hence no source for a nitrogen atmosphere existed. (2) Ganymede and Callisto incorporated ammonia; a nitrogen atmosphere was raised but was eroded away by intense magnetospheric bombardment and/or a solid body bombardment flux more intense or longer in duration than at Saturn. (3) Ganymede and Callisto incorporated ammonia but not methane; as a consequence no methane outgassing occurred, a post-bombardment greenhouse phase never occurred and ammonia was converted to N_2 only in very small amounts.

Whether any of these scenarios is right requires spectroscopic observations of the poles of Ganymede and Callisto by Galileo, since molecular nitrogen and other volatiles could be cold-trapped at extremely high latitudes on those satellites (D. J. Stevenson, personal communication).

Implications of Deuterium Abundance for the Origin of Titan

Owen (1987) assessed the significance of the enhanced deuterium abundance in Titan's atmosphere for its formation. He concluded that it was unlikely that processes in the solar or giant-planet nebulae could have produced such an enhancement, and that some of Titan's atmosphere must have been contributed by interstellar material which never fully equilibrated with the hydrogen gas forming the solar nebula. Grinspoon and Lewis (1987*a*) have also found that it is difficult to obtain significant fractionation in the nebula, even in the presence of surface catalysis on grains. The delivery vehicle for the interstellar deuterium is uncertain. Although the deuterium enhancement seen in Titan's atmospheric methane is much less than that found in organic phases in meteorites (Yang and Epstein 1983), it is close to the D/H ratio in the water component of Halley's comet (1 to 5 $\times$ 10^{-4}; Eberhardt et al. 1987*a*). The hypothesis that comets supplied deuterium-rich ices to Titan has yet to be quantified; the net gain of cometary material to the atmosphere depends upon a number of factors such as impactor velocity, size distribution and volatile content, as well as on the atmospheric mass of Titan at the time. In any event, Owen's assertion that the deuterium enhancement is interstellar cannot be tested until the origin of cometary material itself is determined. This will require *in situ* measurement of the D-to-H ratio in the ice component of a comet nucleus, together with analyses of molecular abundances of species trapped in the ice. Such a test awaits the development of the Comet Rendezvous Asteroid Flyby (CRAF) Mission. Observations of comets from the next generation of infrared observatories, e.g., Stratospheric Observatory for Infrared Astronomy (SOFIA), Infrared Space Observatory (ISO) and Space Infrared Telescope Facility (SIRTF), will also be required.

Origin of Volatiles on Triton and Pluto

Questions which we can address about Titan's volatiles cannot be addressed for Triton until certain measurements are made. These include determination of bulk density and measurement of (or upper limits on) abundances of atmospheric argon and neon. These tests will be performed during Voyager 2's flyby in 1989.

The density derived for the Pluto-Charon system (Tholen et al. 1987; Tholen and Buie 1987) strongly suggests an origin in a water-depleted region; i.e., a CO-rich gas such as the solar nebula. However, the presence of methane shows that the carbon in the nebula was not purely in CO. Trapping of nebular gases in water ice, either by adsorption or clathration, necessarily results in chemical fractionation effects. Relative to carbon monoxide, methane is preferentially incorporated in water ice (Lunine and Stevenson 1985*a*). Hence, an origin for Pluto in a CO-rich nebula with traces of methane could lead to (a) a high bulk density, and (b) incorporation of some methane along with carbon monoxide during accretion. Lunine (1987) concluded that measurement by *Giotto* of comparable amounts of methane and carbon monoxide coming from the nucleus of Halley's comet is consistent with that comet's origin in just such a nebula, assuming the gases were trapped in clathrate hydrate.

Finally, scenarios for origin and evolution of Triton and Pluto's atmospheres must also consider the thermal effects of catastrophes such as orbital ejection or capture. This issue has been examined by McKinnon (1984).

V. RELATIONSHIP TO TERRESTRIAL-PLANET ATMOSPHERES

We consider the relationship of the atmospheric composition of Titan, Triton and Pluto to those of the terrestrial planets. Nitrogen- and carbon-containing gas species represent the major constituents of both types of atmospheres. For all of these atmospheres, except Venus', the abundance of water vapor is limited by the surface temperature, although this limitation is more severe in the case of the outer solar system objects.

Perhaps the most significant difference between the inner and outer solar system atmospheres is their state of oxidation. The Earth represents an extreme case in this regard due to the high abundance of molecular oxygen. This apparently reflects the Earth's unique position in the solar system as an abode of life. However, C is present in its fully oxidized state CO_2 in all the terrestrial atmospheres and in its fully reduced state in the outer solar-system objects. This difference could reflect the ability of water to buffer the oxidation state of the carbon species in the atmospheres of the terrestrial planets, while the oxidation state of the C in the outer solar system atmospheres may be primordial, preserved because water buffering is not possible at the low temperatures obtained. Much of the terrestrial buffering is due to dissociated water in magma chambers.

Size may also have played an important role in the differing evolutionary history of these two classes of atmospheres. On the one hand, the terrestrial planets may be large enough so that their atmospheres evolved significantly over their lifetime due to volcanic and tectonic changes with time. Thus, for example, the Martian atmosphere may have been much denser during its early history and a clement climate may have prevailed due to more vigorous outgassing (Pollack et al. in preparation). By way of contrast, the outer solar system objects may have undergone early dramatic changes followed by tectonic quiescence; subsequent evolution would be externally driven by photochemical processes. (Since an ammonia-water magma beneath the surface of Titan could be maintained to the present day, that satellite may be tectonically active [Lunine and Stevenson 1987].)

Another parallel between the inner and outer solar system atmospheres has to do with partitioning of carbon and nitrogen species between the atmosphere and lithosphere. In the case of the Earth and probably Mars, much more C resides in the lithosphere than in the atmosphere at present, due to weathering processes that readily convert CO_2 gas into carbonate rocks, and in the case of Mars, to the large buffering capacity of the regolith. In the outer solar system objects, this buffering is due largely to the low surface temperatures which put most of the methane into surface frosts or oceans. Nitrogen is more volatile so that a much larger fraction of it is present in the atmospheres of Earth, Mars and Titan. Venus is, of course, a special case since its high surface temperature prevents the accumulation of large near-surface reservoirs of C and N.

VI. CONCLUSIONS

We have reviewed the current state of knowledge of the atmospheres of Titan, Triton and Pluto and theoretical models for their origin and evolution. All three atmospheres contain methane, while Titan and probably Triton also have nitrogen. The fundamental driving process for the evolution of Titan's atmosphere is the irreversible photolysis of methane. If a surface reservoir of liquid methane exists to resupply the atmosphere, then it is subject to enrichment in ethane due to the long-term photolysis of methane. The ocean in turn drives slow climate changes as the atmospheric composition changes in response to the ocean evolution.

The key issue in the origin and early evolution of Titan's atmosphere is the source of molecular nitrogen: was it brought into Titan as molecular nitrogen or ammonia? A diagnostic experiment to test the two models is measurement of atmospheric argon: a low ($<<1\%$) value of the atmospheric argon-to-nitrogen ratio implies the present molecular nitrogen was derived from ammonia, while a high (>1%) value would suggest an origin as N_2. Two schemes for conversion of ammonia to nitrogen are photolysis and shock chemistry during projectile impacts. Coupling of the early evolution of Titan to the slow evolution of atmosphere and (if existent) hydrocarbon ocean remains a major challenge.

Little can be said about the evolution of Triton's and Pluto's atmospheres until definitive compositional and physical measurements can be made. The Voyager 2 encounter with Triton in 1989 is expected to raise our understanding of that object's atmosphere to the point that useful comparative studies with Titan's can be made. The seasonal modulations of Triton's atmosphere may set limits on the abundance and distribution of volatiles on its surface; long-term Earth-based observations are required to observe such effects.

The parallels and differences between these outer solar system atmospheres and their terrestrial cousins challenge our understanding both of their origin, and of the operation of current physical processes in the atmospheres. Most challenging, and perhaps most rewarding, is to establish the interaction between the atmospheres and surface reservoirs. In every case except Venus, the atmospheres we observe on the "small" bodies of the solar system are in intimate contact with surface and subsurface volatiles, which in the long term control the composition and evolution of these thin gaseous envelopes.

Acknowledgments. We thank B. Rizk, T. Owen, T. Jones, C. McKay, L. Lebofsky, and W.R. Thompson for useful discussions and permission to use certain material. We are grateful to D.P. Cruikshank, W. R. Thompson, D. M. Hunten, and R. Marcialis for their reviews of the manuscript. JIL was supported by a NASA grant from the Planetary Atmospheres, and Geology and Geophysics Programs. SKA acknowledges support from the NASA Planetary Atmospheres Program and the Voyager Project.

IO'S TENUOUS ATMOSPHERE

T. V. JOHNSON and D. L. MATSON
Jet Propulsion Laboratory

Among the satellites which might possess tenuous atmospheres, Io is the best known and studied. The nature of Io's atmosphere is, however, still the subject of considerable debate despite many different types of observations which relate, directly or indirectly, to atmospheric characteristics or processes. This chapter reviews the observational evidence for Io's atmosphere and the various theoretical approaches to modeling it. Significant elements of our current understanding include the following: (1) SO_2 and its various photochemical products play a major role in the atmosphere although constraints on other species are not strong in some cases. Atomic species such as Na and K are present in smaller amounts and may play an important role in the ion chemistry of the atmosphere. (2) Vapor pressure equilibrium between condensed SO_2 on the surface and atmospheric gas is an important process in controlling the distribution of SO_2 gas globally, but cannot account for the neutral densities at the terminator required for most ionospheric models, given realistic surface temperature distributions. (3) The density of the atmosphere probably varies greatly with local time and geographic position and may be strongly influenced by temporal changes in volcanic activity and/or magnetospheric interactions.

I. INTRODUCTION

There is no "official" definition of a tenuous atmosphere. Previous discussions of planetary atmospheres have used the term to describe various types of gaseous envelopes ranging from the relatively robust atmosphere of Mars to cases, such as cometary comae, where the gases have no significant gravitational tie to the body or are composed mainly of gases which are not indigenous to the planet, such as the solar-wind-implanted atmosphere of the Moon. A working definition used by atmospheric modelers is to call an atmosphere tenuous if most of the incident solar energy reaches the surface and most of the reflected and emitted energy from the surface escapes back to space. For the purposes of this chapter, considering satellite atmospheres, tenuous atmospheres are assumed to be anything from essentially vacuum to a fraction of a millibar.

An early discussion of the probabilities of various solar system bodies being able to support some form of atmosphere was given by Kuiper (1952*a*), discoverer of the first satellite atmosphere (Titan's). In his book, *The Atmospheres of the Earth and Planets,* Kuiper listed the planets and satellites in decreasing order of their ability to retain an atmosphere, using as a figure of merit a quantity proportional to the ratio of escape speed to thermal molecular velocity at the top of the atmosphere. Kuiper (1952*a*) realized that this was only a first approximation and that many potentially important factors were ignored; also, we now have significantly improved values for the physical constants (radii, masses, albedos, temperatures) of many of the smaller bodies in his list. Nevertheless, Kuiper's list, shown in Table I, remains a useful

TABLE I
Ability of Planetary Objects to Retain Atmospheres[a]

Object[b]	Atmosphere detected?	Comments
Jupiter	Yes	H, He
Saturn	Yes	H, He
Neptune	Yes	H, He, CH_4
Uranus	Yes	H, He, CH_4
Earth	Yes	N_2, O_2
Venus	Yes	CO_2
Pluto	Yes	CH_4
Triton	Yes (?)	CH_4
Mars	Yes	CO_2
Titan	Yes	N_2, CH_4
Ganymede	No	H_2O frost surface
Io	Yes	SO_2, ?
Callisto	No	H_2O frost
Europa	No (?)	H_2O frost
Mercury	No (?)	Na seen in emission
Moon	No (?)	Na seen in emission

[a]Table after Kuiper (1952*a*).
[b]Listed in order of decreasing ratio of escape speed to thermal molecular velocity at the top of the atmosphere, using physical properties known to Kuiper (1952*a*).

ranking for small-body atmospheres. All the objects listed in rank down to Io have detectable atmospheres except for Ganymede (assuming that Triton and Pluto possess at least atmospheres of methane at vapor pressures corresponding to their surface temperatures; see the chapter by Lunine et al.). Collisionally thick atmospheres have not been detected for any of the objects below Io in the list, although a very tenuous atomic atmosphere of sodium and potassium has been detected around Mercury (see Hunten et al. 1988*b*). There is little hard data for the smaller icy satellites, and theoretical arguments similar to Kuiper's are still the only available guide to these bodies' atmospheres, suggesting that most of them possess only cold cometary-type envelopes of water molecules and related dissociation products escaping their surfaces through the processes of evaporation by sputtering of the surface and charged particles, and ultraviolet light. Triton and the Pluto-Charon system are covered in the chapter by Lunine, Atreya and Pollack and so Io remains as the focus for our following discussion.

II. IO'S ATMOSPHERE: A BRIEF HISTORY OF OBSERVATIONS

Io has long been known to be an unusual satellite. Its very high albedo compared with that of the Moon and planets led to speculation about frost deposits on its surface early in the century. The first modern measurements suggesting an Io atmosphere were reported by Binder and Cruikshank (1964). Their measurements showed that Io was brighter by about 10% during an interval of ~15 min immediately following the satellite's emersion from eclipse. One possible explanation advanced for this effect was the condensation of an atmospheric gas during eclipse and its subsequent evaporation. It is indicative of the frustrating state of our knowledge of this satellite that the relation of post-eclipse brightening to an Io atmosphere and even the reality of the phenomena are still the subject of debate more than twenty years after the first report.

In 1971, observations of the occultation of β Scorpii C by Io provided an upper limit of $\sim 10^{-6}$ bar to the atmosphere, placing Io's atmosphere in the tenuous category (Smith and Smith 1972). This was followed by what is usually regarded as the first detection of Io's atmosphere when Pioneer 10 performed a radio wavelength occultation experiment. The radio data were interpreted as showing an ionosphere with relatively high peak electron densities on both the entry and exit from occultation; preliminary estimates of the neutral atmosphere required to explain the ionospheric data (by analogy to the Earth and Mars) yielded surface pressures of $\sim 10^{-7}$ bar (Kliore et al. 1974,1975).

Subsequent to the Pioneer 10 flyby, a large number of observations from groundbased telescopes provided a wealth of new data (not always understandable) related to Io and its atmosphere. Atomic spectral line emissions from neutral sodium were discovered by Brown (1974) coming from Io. Observations and theoretical analyses by a number of groups quickly established that the observed sodium was concentrated in a neutral *cloud* of atoms about Io, which had escaped Io but were still orbiting Jupiter and that resonant scattering of sunlight was the emission mechanism (Trafton et al. 1974; Matson et al. 1974; Bergstralh et al. 1975,1977). Neutral potassium was also detected (Trafton 1975). A number of lines of evidence suggested that these neutral atoms are escaping from Io by nonthermal, energetic means, either from the surface or from Io's atmosphere (see reviews by Pilcher and Strobel [1982] and Cheng et al. [1986]).

The situation was further complicated by the discovery of emission from ionized sulfur in the magnetosphere, also associated with Io (Kupo et al. 1976). The important role of sulfur compounds on Io and in the magnetosphere became evident when the Voyager flybys showed that sulfur and oxygen were major components of the magnetospheric plasma and energetic particle populations (Broadfoot et al. 1979; Bridge et al. 1979; Krimigis et al. 1979; Vogt et al. 1979) with a huge plasma torus of sulfur and oxygen circling Jupiter at Io's orbital distance. Analyses of telescopic reflection spectra identified sulfur dioxide frost on Io's surface (Fanale et al. 1979; Smythe et al. 1979), and the Voyager IRIS (infrared interferometer spectrometer) experiment identified gaseous SO_2 over one of the major volcanic hotspots discovered by Voyager (Hanel et al. 1979*a*; Pearl et al. 1979). The IRIS also provided temperature measurements at various locations on Io. The Voyager imaging system returned high-quality data on the location of volcanic regions, and the distribution of albedo markings on the surface, particularly the lack of prominent bright polar caps (Smith et al. 1979); it also detected auroral emissions associated with eruptive plume activity on the dark side of Io (Cook et al. 1981).

Most post-Voyager observations related to Io's atmosphere have been primarily concerned with characterization of the torus properties. Two recent observations that give information about atmospheric species in the near vicinity of Io are those of sodium near Io carried out by observing mutual eclipses (Schneider et al. 1987) and the detection of neutral sulfur and oxygen emissions near Io with the IUE (International Ultraviolet Explorer) (Ballester et al. 1987).

The brief history above summarizes most of the observational evidence bearing on Io's atmosphere. The details of the observations and numerous theoretical analyses are described in a number of excellent reviews in recent years; these reviews also contain extensive references to the original research papers (Kumar and Hunten 1982; Fanale et al. 1982*b*; Pilcher and Strobel 1982; Cheng et al. 1986; Nash et al. 1986; Schneider et al. 1988). In this chapter, we will attempt to review critically the major pieces of observational evidence, discuss theoretical approaches to modeling Io's atmosphere and highlight key unresolved issues.

III. OBSERVATIONS

A fundamental problem in dealing with Io's atmosphere is that each observation is associated with a specific time and place on Io. One of the few things most observers agree upon is that phenomena on this satellite are likely to be highly variable, both spatially and temporally. In addition, most of the observations are not straightforward, but require considerable, often difficult, interpretation in order to produce a result useful for studies of atmospheric pressure or composition. Unfortunately this situation creates many possibilities for confusion and error. Given this situation, it is perhaps not surprising that there is no single model or theory that adequately explains all the observations to the satisfaction of all workers and that the total set of data often seems overconstrained or even self-contradictory. Without going into exhaustive detail concerning each observation, this section notes some of the key factors associated with the observations most frequently used to constrain Io atmospheric models.

Sulfur Compounds

Sulfur in some form is believed to be an important constituent of Io's surface and atmosphere. The evidence for large amounts of sulfur being supplied to the magnetosphere in one form or another is very strong. The rates implied by the magnetospheric observations, however, do not come close to depleting Io of a cosmic inventory of sulfur over geologic time, and a number of lines of argument suggest that the sulfur escaping is only a small fraction of the sulfur supplied to the surface (Johnson et al. 1979; Johnson and Soderblom 1982). This suggests in turn that there is a large source of sulfur compounds being supplied to the surface of Io. The optical properties of the surface also have been used to infer the presence of sulfur in the form of SO_2 and S (see review by Nash et al. 1986). Finally, thermodynamic models of plume eruptions on Io rely on SO_2 and S liquid to explain the characteristics of the plumes (Kieffer 1982). In summary, the amount of sulfur in the crust of Io is not known accurately, but all evidence points to a surface with at least surficial deposits of sulfur compounds and the strong likelihood that extensive, thick deposits may occur in many places.

Pioneer 10 Occultation

The results from the radio occultation experiment in 1973 by the Pioneer 10 spacecraft are used as the prime input for most aeronomy models of the atmosphere. The radio data provide a measurement of the integrated electron density along the line of sight to the spacecraft as it went behind Io as seen from the Earth. After considerable processing, including assuming spherical geometry for the electrons around Io, these data were used to infer the electron number density as a function of altitude over two regions on Io, approximately over the evening and morning terminators. The high ($\sim 10^4$ cm^{-3})

peak electron densities from these interpretations form the basic rationale for virtually all the models with relatively high neutral densities (see Kumar and Hunten 1982).

There are several properties of the Pioneer data which must be kept in mind, however. First, the radius of Io determined from the occultation does not agree with stellar occultation or Voyager image determinations. The sense of the discrepancy is that the entry occultation occurred "on time" for the best radius data, while re-emergence was ~75 km "late." It is possible that this problem results from the geometry of the occultation (which was distant and with a considerable projection angle into the sky plane) and small errors in either the spacecraft or satellite emphemeris. The data have not been thoroughly re-analyzed with more recent ephemeris information (A. J. Kliore, personal communication, 1987), so better results are not available. Second, the exit occultation profile (electron density vs altitude) is quite dissimilar from the entry profile; not only is the electron density at the peak smaller, but the shape is distinctly more flattened, and does not follow the type of pattern expected for a simple scale-height model. As a result, most atmospheric models have dealt with the entry data alone and have either ignored or provided *ad hoc* explanations for the exit data. Finally, as seen in Fig. 1, the exit occulta-

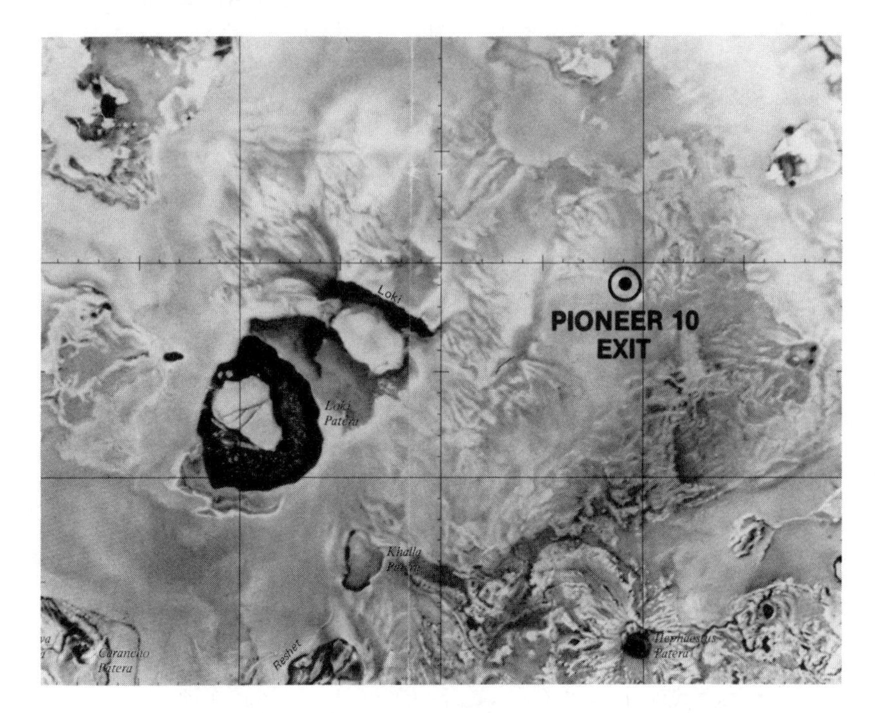

Fig. 1. Portion of the shaded relief map for the Loki Patera region of Io showing the location of the Pioneer 10 radio occultation exit point. Map prepared by U.S.G.S. for NASA.

tion occurred very close to the Loki region on Io (Kliore 1980). Loki was the most active volcanic area observed by Voyager and had eruptive plumes several hundred km high associated with it during each encounter (as opposed to Pele, which was observed during the Voyager 1 encounter but not the Voyager 2). Groundbased infrared observations strongly suggest that the Loki region has continued to be one of Io's most important centers of volcanic activity (Johnson et al. 1984*a,b*; Johnson et al. 1988). It seems likely that at least the Pioneer exit occultation data, occurring 6 yr before the Voyager encounters, were affected by plume material to some degree, although no detailed modeling of such a situation has been attempted.

Voyager IRIS Data

The detection of SO_2 gas by IRIS (Pearl et al. 1979*a*) is the only direct measurement of the composition of Io's atmosphere. All current models assume that SO_2 is the major atmospheric constituent, with minor amounts of photochemically derived S and O products, Na and K. However, this measurement relates to a unique place on Io, Loki Patera, once again. The gas was detectable only because the volcanically heated surface of the Loki region provided a strong continuum source of infrared flux allowing the observation of the absorption spectrum of the tenuous, cold gas above it. The column density of the observed gas due to the ambient atmosphere vs the gas directly associated with the volcanic activity cannot be determined from the data. Within the uncertainties in the amount of material involved in the eruptive plumes, the relative fraction of atmospheric compared with volcanic gas could range from 0 to 1 (Johnson et al. 1979; Collins 1981; Pearl et al. 1979). Additionally, the IRIS data provide very little information on SO_2 elsewhere, since there were no other infrared sources large enough and warm enough to allow absorption spectra to be obtained. Because of the circumstances of observation, the lack of detections of gas near other hotspots does not constrain either the atmospheric or the plume abundances for SO_2 gas (Pearl et al. 1979; Pearl and Sinton 1982). Thus, direct evidence for SO_2 gas as the major constituent of a global ambient atmosphere comes from a single observation of a very atypical, active region on Io; inferences from less direct observations and theoretical arguments are required to extend the IRIS data to a model of the whole atmosphere.

IRIS data are also important for measurements of surface temperatures. The major volcanic hotspots were easily detected and in some cases good measurements of the spectral flux distributions allowed temperatures to be derived (Hanel et al. 1979*a*; Pearl and Sinton 1982). Surface temperatures in other places were also measured, although signal-to-noise and pointing uncertainties limit the usefulness of the data, particularly for the night-side observations. Of particular importance for this discussion of the atmosphere is the estimate of 110 K which has been set as an upper bound to the near-terminator temperatures (Pearl et al. 1979). This temperature has been used frequently as if it were an actual measurement and not just an upper bound. Terminator temperatures lower by 10 deg or more are more realistic (J. C. Pearl, personal communication, 1987).

Dark-Side Aurora

An observation of apparent auroral activity associated mostly with plume areas on the night side of Io was reported by Cook et al. (1981). However, these observations do not constrain closely the composition or pressure of the atmosphere. The localized nature of the emissions indicate that on the night side at least the gases responsible for the glow are concentrated around plumes and not distributed widely as in an ambient atmosphere.

IUE Observations

An upper bound on ultraviolet absorptions associated with SO_2 was derived from IUE observations by Butterworth et al. (1980). They interpreted these data as an upper limit on the integrated column density of gas of 0.008 cm-amagat. However, the behavior of these absorptions under Io-like conditions has not been studied in the laboratory. Calculations of the theoretical effects of the curve of growth for a particular band model have been carried out by Belton (1982*a*). These suggest that the limiting atmospheric pressure may be very poorly constrained for a cold atmospheric model. How well the actual Io atmosphere matches any of the assumptions used in the various interpretations is not known, and the constraints from these data must be regarded as ambiguous pending the availability of further information needed to improve the modeling.

A subsequent search for SO_2 absorptions in IUE data with higher spectral resolution was also unsuccessful (Ballester et al. 1988). Since interpretation of these data may be less affected by the uncertainties discussed by Belton, a more stringent upper limit on global SO_2 on the day side may be possible when these observations are fully analyzed.

IV. ATMOSPHERIC MODELS

Aeronomy Models

Some of the most detailed models of Io's atmosphere have been those designed to match the ionospheric profiles from Pioneer 10. Early models used a variety of compositions and both solar ultraviolet and charged-particle ionization sources. Since the Voyager encounters, virtually all models have assumed SO_2 as the major constituent of the atmosphere, and Voyager data have been used to calculate the ionization from electron impact. These models are thoroughly reviewed in Kumar and Hunten (1982) and in more recent work by Kumar (1985). Photochemical-type models of this sort generally explain the electron-density profile (usually only the entry profile as noted above) for a given neutral surface pressure and scale height. Among the problems encountered by this class of model are difficulties in explaining the relatively high sodium flux into the neutral cloud surrounding Io and matching the details of escape rates for S and O required by various interpretations of the magnetospheric observations (see Kumar 1985). Table II gives typical parameters for SO_2 models from Kumar and Hunten (1982) and Kumar (1985); both models use ultraviolet energy and electron precipitation as ionization sources, with the electron contribution being required to match the observed peak electron densities. A representative model electron-density profile, compared with the Pioneer 10 occultation data is shown in Fig. 2.

Buffering Models: the "Leighton-Murray Io"

Immediately following the Voyager encounters in 1979, the results from a number of observations of Io, its atmosphere, its surface and the magnetosphere were synthesized by several groups into what amounts to the current Io paradigm. The key elements in this synthesis were the identification of SO_2 frost absorptions in telescopic reflection spectra of Io (Fanale et al. 1979; Smythe et al. 1979) and the measured abundance of SO_2 gas near the subsolar point by the IRIS experiment. Pearl et al. (1979) noted that the inferred surface pressure in the vicinity of Loki ($\sim 10^{-7}$ bar) was very close to the vapor pressure of solid SO_2 at the temperature of the surrounding surface ($\sim$130 K). They proposed an SO_2 atmosphere with pressures near the local equilibrium vapor pressure. This picture requires large variations in local pressure across the surface due to the strong variation of vapor pressure with temperature for SO_2, leading to essentially no atmosphere on the night side (Pearl et al. 1979; Ip and Axford 1979); qualitatively this type of behavior also accounted for the lower pressures required by the photochemical models ($\sim 10^{-9}$ bar) to support the ionosphere over the cooler terminator regions.

The buffered SO_2 model for Io's atmosphere is similar in concept to an idea first proposed for Mars by Leighton and Murray (1966). They noted that the then recently determined low pressure of the Martian atmosphere was understandable in terms of an atmosphere that was tied to the vapor pressure of CO_2 at the polar caps. Although analysis of a large amount of Mars data since then has resulted in many modifications and extensions of the original idea, the

TABLE II
Atmospheric Models[a]

Source	n_o, cm^{-3}	E, eV	Flux, cm^{-2} s^{-1}
Kumar and Hunten (1982)	1.2×10^{11}	20	1.2×10^{10}
Kumar (1985)	4.0×10^{10}	600	2.0×10^{8}

[a] n_o is surface number density of atmosphere, E is the energy of the electrons assumed in the ionization model, and Flux is the required flux of the electrons with energy E.

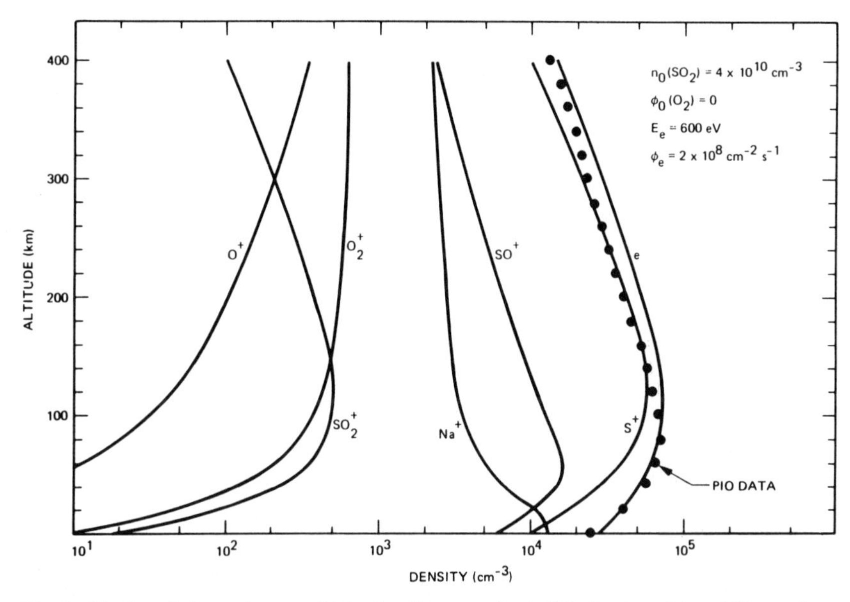

Fig. 2. The best-fit ionosphere model for Io with atmospheric SO_2 density of 4×10^{10} cm^{-3} at the surface and mono-energetic incident electron flux of 2×10^8 cm^{-2} s^{-1} at 600 eV as the source of ionization. The Pioneer 10 radio occultation data on the downstream side (ingress) are shown for comparison. Profiles for various ions are indicated (Figure after Kumar 1985).

Leighton-Murray buffering concept remains a key element in our understanding of the atmosphere/surface interactions on Mars. Application of the same ideas to Io is complicated by a number of factors. First, the apparent vapor pressure vs temperature coincidence relating the IRIS measurement of daytime gas pressure and the pressures required by photochemical models at the terminator is not that strong when realistic temperature distributions are carefully examined, particularly considering the special conditions of the IRIS observation noted above. Second, Io has no bright polar caps; the buffering model described above assumes pressure equilibrium with *local* SO_2 frost (which is stable at the equator and believed to cover a reasonable fraction of the surface) and not with a limited region at the coldest planetary temperatures, such as Mars' CO_2 polar deposits. Third, the large pressure differences implied by local equilibrium are almost certain to have meteorological consequences, including winds, that will modify the pressure distribution from the "pure" equilibrium model. Finally, there are nonthermal processes, such as sputtering by energetic particles, which may play a significant role in transporting volatile materials and in the escape of material from Io's surface and atmosphere.

Aeronomical models such as those listed in Table II have difficulty accounting for the electron densities in the Pioneer 10 entry profile for surface atmospheric pressures of $SO_2 \lesssim 6 \times 10^{-10}$ bar. Figure 3 is a plot of SO_2 vapor pressure vs temperature from Wagman (1979) and Dean (1973) after

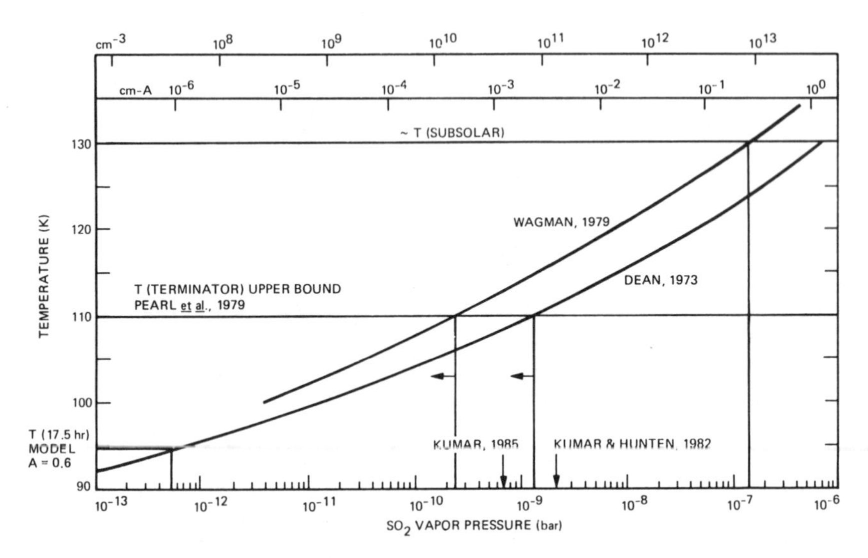

Fig. 3. SO_2 vapor pressure vs temperature. Corresponding scales for surface density (cm^{-3}) and atmospheric column (cm-A) are also shown. The temperature range is characteristic of Io's surface. The pressure corresponding to three temperatures are shown: subsolar, terminator upper bound and a model temperature for 17.5 hr. Terminator pressures required by two atmospheric models are also indicated.

Kumar and Hunten (1982). The *upper bound* terminator temperature of 110 K given by Pearl et al. results in a vapor pressure of only $\sim 2 \times 10^{-10}$ bar for the Wagman pressure curve (which is more recent and cited as more reliable by Kumar and Hunten), a factor of three below the lowest model pressure. Even if the older estimate of Dean is used, the vapor pressure is just barely above 1×10^{-9} bar, sufficient for the Kumar model but still below the Kumar and Hunten model values in Table II. More realistic terminator temperatures range from 95 K (derived from a standard thermal inertia model) to 100 K (J.C. Pearl, personal communication, 1987). Because of the strong temperature dependence of vapor pressure on temperature, these lower values result in predicted pressures 2 to 3 orders of magnitude below the ionospheric model values. There are effects that may result in the actual pressures being either higher or lower than given by the vapor pressure curve, which are discussed in more detail below, but the simple, local equilibrium vapor pressure model does not seem to produce the pressures required by the photochemical model of the ionospheric electron densities.

One factor which could alter the pressure distribution is wind, driven by the large pressure gradients that result from equilibrium models. The most complete treatment to date of this effect has been given by Ingersoll et al. (1985). They found that supersonic flow of sublimation-driven winds could result under some conditions. They investigated a range of different models and assumptions about surface temperature and sticking coefficients, and found that wind flow did tend to distribute gas away from the subsolar point as expected and raise the terminator pressures over the equilibrium values. This effect only amounted to about a factor of 2 enhancement in pressure in most cases, however, and the final pressures were still controlled strongly by the assumed terminator temperature; they concluded that only in the more extreme models (again with terminator temperatures at the upper-limit value of 110 K) could they get terminator pressures in the range required by the photochemical models. Another variant of wind-related models has been proposed where the source of the pressure difference is a new volcanic eruption of large scale on a relatively atmosphere-less Io; these calculations show that for reasonable choices of volcanic gas outflow, substantial temporary atmospheres can be built up through the flow of the volcanic gases across the surface (Baumgardner et al. 1987; Moreno et al. 1988*a*). Whether such effects are important only for short periods of time or whether enough gas outflow activity is present to make this a major contributor to the "average" atmosphere is uncertain. However, the scale of possible disturbance by large volcanic events should be kept in mind when trying to reconcile atmospheric data taken at different epochs.

Most elaborations of the buffering model have been concerned with the issue of the most appropriate temperature and location of the regions that control the equilibrium pressure. Fanale et al. (1982*b*) pointed out that SO_2 gas will preferentially condense in colder regions. On a surface consisting of a mixture of bright, cold areas, and darker, warmer areas, they proposed that the local atmospheric pressure should be controlled by the coldest regions in any given area. For local cold traps with temperatures 8 to 9 K lower than the average, Fanale et al. calculated that the pressure contours on the surface would be lowered by a factor of 15, with pressure falling below 10^{-9} bar only 40 to 60 deg from the subsolar point, far from the terminator.

There are several possibilities for producing local temperature variations and subsequent local cold trapping. Topographic effects, particularly shaded regions at low solar illumination angles, may be important in some cases. For instance, "permanently" shaded regions at high latitudes have been proposed as long-term cold traps for lunar volatiles (Watson et al. 1961*a,b*). While Io's level of geologic activity mitigates against geologically long-lived topographic cold traps, such high-latitude areas could be a significant reservoir of condensed SO_2. Another local cold-trap source, investigated by Matson and Nash (1983) is the subsurface layers of the surface. They note that if the upper surface is as porous as many models imply, the temperatures only a few cm below the surface will be very low, $<\sim 90$ K. If gas molecules have relatively free access to these regions through the porous surface, then the effective buffering temperatures may be tens of degrees below the calculated surface temperatures and the local equilibrium temperatures, correspondingly less (from $\sim 10^{-12}$ subsolar to $\sim 10^{-15}$ bar at the terminator). This particular regolith cold-trap model is a low-temperature extreme for cold trapping schemes which have been investigated quantitatively, and there are still many unknowns concerning the thermophysics of Io's surface that might cause the actual subsurface temperatures to be somewhat higher (e.g., the solid-state greenhousing models for bright icy surfaces discussed by Brown and Matson [1987] and Matson and Brown [1988]).

A quantitative assessment of albedo cold trapping has been carried out by McEwen et al. (1988) based on photometric analysis of Voyager imaging data. They found that actual albedo contrast on Io's surface should produce local temperature variations far greater than the modest 8 to 9 K assumed by Fanale et al. Local temperature-contrasts for surfaces under the same solar illumination as large as 40 K are predicted. The brighter, colder regions should have noontime temperatures <120 K (or >10 K colder than "average" Io) with a corresponding reduction of local pressures in equilibrium with these regions at all illuminations of more than an order of magnitude (see Fig. 3). They also noted the heterogeneous pattern of bright material, with the leading hemisphere having bright, cold material concentrated in the equatorial regions. This pattern results in the mid-latitudes having higher mean temperatures than the equator in this hemisphere. Flow of SO_2 driven by sublimation pressure differences will definitely not be symmetric about the subsolar point in this case. McEwen et al. suggest that flow to equatorial, *cooler* regions might be part of the explanation of the lack of polar cap deposits. They conclude that the actual albedo range on Io should result in albedo cold trapping as proposed by Fanale et al. and that the pressure reduction from this effect is probably even larger than originally suggested.

Nonthermal Models

Both the aeronomical models and the various buffering models assume that the atmospheric gas (SO_2 in most cases) pressure and distribution are controlled by surface and atmospheric temperature and gravity. Io's position deep within the radiation belts of Jupiter, surrounded by a dense corotating plasma of heavy ions, leads to a wide variety of nonthermal processes involving charged particles which may affect the surface and atmosphere. Sputtering of material from Io's surface by charged-particle impact was suggested by Matson et al. (1974) well before the Voyager encounters as an important mechanism for escape of atoms and ions from the satellite. Since that time, the role of charged-particle interactions with Io has been extensively studied by many authors; Voyager particle measurements and many laboratory studies now support this work, which has been reviewed by Cheng et al. (1986) (see also chapter by Chang and Johnson). A key parameter in discussing such effects is the column abundance of atmospheric gas. Below about 10^{15} cm^{-2} the atmosphere is collisionally thin and material sputtered with velocities exceeding the escape velocity of $\sim$2.5 km s^{-1} will escape directly to the magnetospheric environment; above this value sputtered atoms are thermalized in the atmosphere and the important escape processes relate to charged-particle interactions with the atmosphere. Figure 4 shows the various estimates of gas pressure on Io in terms of column abundance, number density at the surface and surface pressure in bars.

Most of the models discussed result in situations where the thermal equilibrium atmospheric pressure is too high in some regions to allow direct sputtering from the surface, but allow it in other places, particularly in the night hemisphere. The expected column density of sputtered material will itself form a type of atmosphere even in the absence of an ambient or local equilibrium vapor-pressure component. Such "sputter-corona" atmospheres have

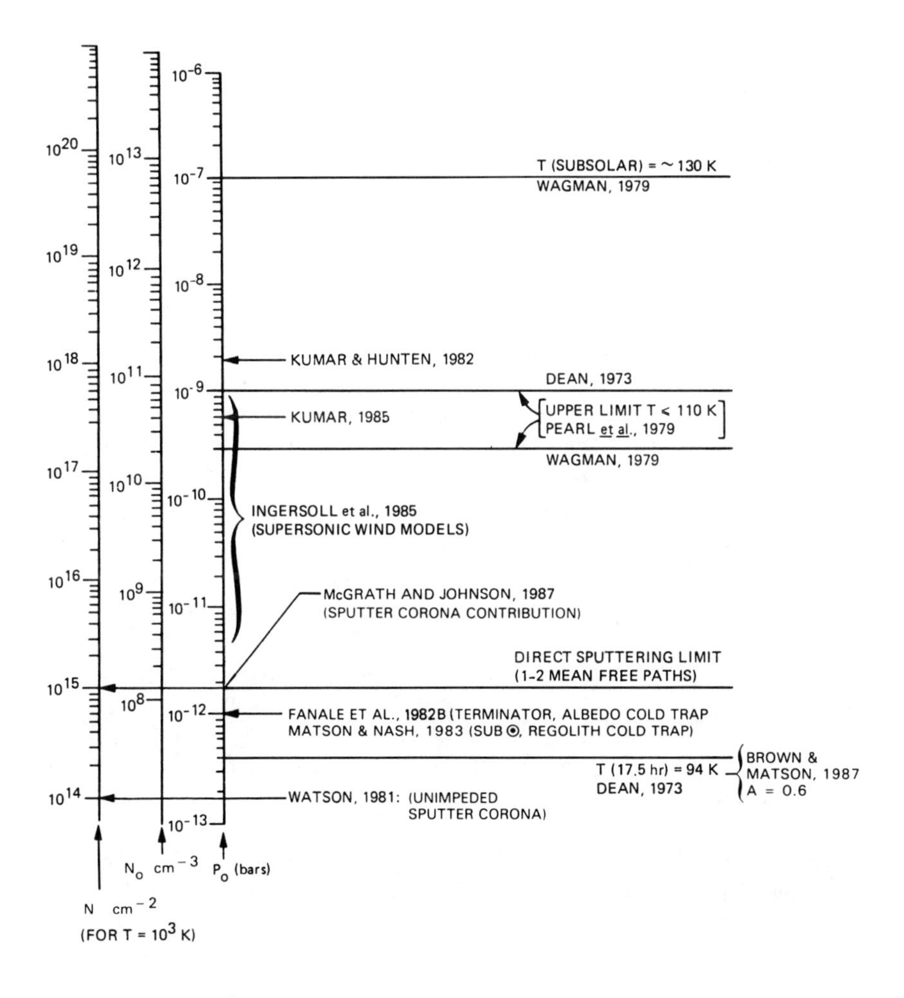

Fig. 4. Comparison of various values for SO_2 vapor-pressure and atmospheric models.

been investigated by Watson (1981), Sieveka and Johnson (1985,1986) and McGrath and Johnson (1987); column densities of $\sim 10^{15}$ cm^{-2} (corresponding to about $\sim 10^{-12}$ bar from a SO_2 thermal atmosphere) result from these models. Fully integrated models of atmospheric density as function of longitude and local time, including vapor-pressure equilibrium, cold traps and sputtering effects, have not yet been produced; however, the sputter corona results suggest that when the local buffering temperature is below ~95 K (on the night side, and in high latitude and terminator regions), sputter-produced coronas may be the dominant atmospheric component.

V. CONCLUSIONS

Despite more than a decade of modern observations and two Voyager flybys, the nature of Io's atmosphere remains an enigma. There are some areas of agreement but no general model that satisfies all of the currently accepted constraints. Given the nature of the incompatabilities between various models and data sets, it seems likely that we are still missing some crucial datum or insight into the problem. The following summarizes what we believe to be the status of some of the key issues related to the atmosphere.

Composition. SO_2 is almost certainly a constituent of both the surface and atmosphere, since its detection is relatively unambiguous in both cases. Current models assume that this compound, its photochemical products, and trace amounts of Na and K are the only species present. It is possible that this is a misconception. Limits on some other gases are surprisingly weak, in many cases based on the β Scorpii C stellar occultation data rather than direct spectroscopic data (Smith and Smith 1972; Pearl et al. 1979). Consideration of some other species might be useful in resolving some of the apparent contradictions in the current models. Among the possible candidates of interest is H_2S, a common volcanic gas along with SO_2 which could have a vapor pressure below the stellar occultation limit if buffered at polar temperatures and still be an important constituent relative to SO_2. Interestingly, H_2S has been observed to darken and change color under ultraviolet irradiation (Lebofsky and Fegley 1976), making any H_2S polar deposits more compatible with the optical data than caps of SO_2.

Pressure. The biggest problem with the atmospheric data set 15 yr after the Pioneer 10 occultation remains the interpretation of the derived electron-density profiles. The data and analyses reviewed above suggest several things: (a) that whatever is supporting the electron densities in the ionosphere, it is probably not merely SO_2 in local vapor-pressure equilibrium with the surrounding surface temperature and (b) that the actual distribution of gas across the surface of Io is most likely highly variable with geographic location and local time of day, and may change on both long and short time scales under the influence of changing volcanic activity and changes in the interaction with Io's magnetospheric environment.

Prospects for improving our understanding of Io's atmosphere are good. Advances in instrumentation and techniques for ground and space-based astronomy offer many opportunities for measurements related to atmospheric problems. The next period of mutual satellite-satellite occultations and eclipses should yield significant new results. More extensive use of the innovative techniques pioneered by Schneider and his colleagues (Schneider et al. 1987) will be capable of probing the sodium atmosphere very close to Io's surface and can perhaps be extended to other species. Observations of the volcanic hotspot emission at multiple wavelengths also has proven to be a powerful technique for studying the distribution and intensity of eruptive activity during the last mutual event sequence (Goguen et al. 1988). New space-based instrumentation may also allow spatially resolved spectroscopic and imaging studies of eruptive plumes and atmospheric absorptions. In 1995, the Galileo spacecraft is scheduled to make a 1000 km altitude flyby of this interesting satellite and to continue studying it and the entire Jovian system for a period of $\sim$2 yr, including the opportunities to make one or more radio occultation measurements of Io. Combined with a better understanding of the temporal behavior of Io phenomena from ground and space-based programs, the Galileo observations may finally solve some of the current difficulties and, undoubtedly will produce both new insights and new problems to solve.

Acknowledgments. This work was carried out at the Jet Propulsion Laboratory, California Institute of Technology, under a contract with the National Aeronautics and Space Administration.

EFFECTS OF MAGNETOSPHERE INTERACTIONS ON ORIGIN AND EVOLUTION OF ATMOSPHERES

A. F. CHENG
The Johns Hopkins University

and

R. E. JOHNSON
University of Virginia

Interactions with planetary magnetospheres can affect the origin and evolution of atmospheres, especially tenuous ones in which the exosphere comprises a significant portion of the whole. Magnetospheric plasma incident on an atmosphere deposits energy, causes ejection of atmospheric species into space, and produces chemical modifications of atmospheric species. For sufficiently tenuous atmospheres, energetic particles and photons can reach the surface of the planet or satellite, causing sputtering as well as physical and chemical modifications of the surface. This can result in the ejection of new species into the atmosphere, affecting its composition. Magnetospheric interactions can be the dominant source and loss processes for tenuous atmospheres and cause formation of spatially extended "sputter coronae" and/or neutral clouds in planetary magnetospheres. Magnetospheric interactions are discussed for the atmospheres of Mercury, Io and the icy satellites of Jupiter, Saturn and Uranus.

I. INTRODUCTION

The origin and the evolution of atmospheres are governed both by interactions with condensed bodies (planets, their satellites and rings, comets, or meteoroids) and by interactions with energetic photons, particles and electromagnetic fields of the space environment. We present below several examples of tenuous atmospheres whose interactions with the space environment are essential for understanding atmospheric sources and losses, composition, and thermal structure. Specifically we shall discuss the atmosphere of Mercury, the neutral and plasma tori of Io at Jupiter, and the atmospheres of the icy satellites of Jupiter, Saturn and Uranus, including the rings of Saturn. We shall emphasize recent advances and attempt to minimize overlap with earlier reviews of magnetosphere-satellite interactions (Cheng et al. 1986; Johnson et al. 1985), the Io torus (Pilcher and Strobel 1982; Sullivan and Siscoe 1982; Belcher 1983; Brown et al. 1983) and Titan's atmosphere (Hunten et al. 1984; Neubauer et al. 1984). Complementary discussion on magnetospheric interaction is contained in two other chapters of this book: one by Johnson and Matson (mainly Io), and another by Lunine, Atreya and Pollack (Titan, Triton and Pluto). In the remainder of this section, we present a general overview of atmosphere interactions with the space environment, beginning with a discussion of energetic photon bombardment, then discussing effects of charged particle bombardment, and concluding with consequences for planetary magnetospheres.

Irradiation by energetic photons from the Sun leads to important sources and losses for the atmospheres of planets and their satellites. Ionization and dissociation of atmospheric species, excitation of electromagnetic radiation, and initiation of photochemistry are familiar consequences of bombardment by solar ultraviolet radiation. Less familiar are the effects of solar ionizing photons incident directly on planetary or satellite surfaces, as is possible for objects with very tenuous atmospheres. Then absorption of solar ultraviolet photons can result in ejection of atoms or molecules into the atmosphere. Such photon-induced desorption processes may be important sources for the very tenuous atmospheres of Mercury and the icy satellites of the outer solar system. Of course, heating of the surface by absorption of longer wavelength solar radiation also causes sublimation of the more volatile surface species. However, sublimation decreases in importance for the cold, icy satellites of the outer solar system.

Charged particle bombardment, which also causes ionization and dissociation, makes important contributions to sources and losses of atmospheric species. Ionization in the presence of electromagnetic fields can be an important atmospheric loss mechanism when it occurs near or above the exobase (which can be the surface of the parent body for very tenuous atmospheres). In this case, newly ionized species are generally swept away by ambient electric and magnetic fields of the solar wind or the planetary magnetosphere. Then ionization, whether by ultraviolet photons or charged particles, leads immediately to loss of the atmospheric species. This loss process is particularly important for the tenuous atmospheres of Mercury and the icy satellites of the outer solar system. By way of contrast, in the terrestrial ionosphere most of the ionization occurs far below the exobase, so newly created ions are not swept away immediately.

Energetic ion bombardment also contributes to atmospheric loss by collisional ejection of atmospheric species through a variety of processes: two-body collisions between an incident ion and an atmospheric neutral, a cascade of two-body collisions ("atmospheric sputtering") initiated by the incident ion, and charge exchange reactions. These processes occurring near or above the exobase can be the dominant atmospheric loss mechanisms, e.g., at Io.

Likewise fast electron bombardment contributes to atmospheric escape. Exospheric molecules can be excited to a repulsive, dissociative state by electron impact so as to create fast atoms above the escape velocity. Electron impact dissociation of N_2 is the dominant escape mechanism for N atoms from Titan. Production of fast H atoms from H_2 electron impact dissociation may also be important for the exospheres of Saturn and Uranus.

Atmospheric neutrals that are ejected from the exospheres of Io and Titan generally remain below the escape speeds from Jupiter and Saturn, respectively, if ejection occurs following sputtering or glancing two-body collisions. The same is true for escape following electron impact dissociation and thermal escape. These escaping neutrals remain gravitationally bound to their respective planets, forming neutral clouds near the orbits of Io and Titan, respectively. These neutral clouds can be regarded as part of an extended exosphere at both Jupiter and Saturn. The neutrals in these clouds occupy orbits analogous to the so-called "satellite orbits" in Chamberlain models of Earth's exosphere. These neutral clouds can have profound implications for magnetospheres, since ionization and dissociation of these clouds can yield important plasma sources (dominant in the case of Jupiter) as well as important mass-loading effects.

In the case of sufficiently tenuous atmospheres, charged particle bombardment is not only an atmospheric loss mechanism as discussed above, but also an important source mechanism. Fast charged particles incident on the surface of the parent body can eject neutrals into the atmosphere by a variety of processes collectively known as sputtering. Sputtering can result from a cascade of collisions with and among the atomic cores of the target ("nuclear" or "collision cascade" sputtering) or it can result from electronic excitation ("electronic sputtering"). Sputtering by charged particle impacts can be the dominant atmospheric source for the icy satellites of the outer solar system.

Atmospheres can be subjected to charged particle bombardment by being immersed either in the solar wind or in a planetary magnetosphere. The solar wind is a magnetized plasma expanding hypersonically away from the Sun. A sufficiently magnetized planet presents an obstacle to the solar wind flow, creating a cavity around which the solar wind flow is deflected. Within this cavity, called the magnetosphere, the planetary magnetic field traps charged particles, allowing a substantial magnetospheric particle population to accumulate. Four planets (Earth, Jupiter, Saturn and Uranus) are currently known to have sufficiently strong magnetic fields that the solar wind is always deflected far above the atmospheric exobase. In the case of Mercury, the exobase is the surface, and its magnetic field is strong enough to deflect the solar wind well above the surface most of the time, but not all of the time. The solar wind can directly impact the surface of Mercury during rare periods of unusually high dynamic pressure. All five planets, including Mercury, are said to have "intrinsic" magnetospheres. When the solar wind is deflected far above the exobase, charged particles in the solar wind have no direct access to the atmospheres of these planets except in very small regions near the magnetic poles (threaded by open field lines). Only a small fraction of incident solar wind particles can gain entry to an intrinsic planetary magnetosphere via processes that are not yet well understood. On the other hand, the magnetic fields of Venus and Mars are sufficiently weak that the solar wind impinges directly on their atmospheres, creating a comet-like interaction that has been reviewed elsewhere (Russell et al. 1982; Ip and Axford 1982; Luhmann 1986) and will not be pursued further here.

Planetary satellites whose orbits lie well within the magnetosphere are exposed to magnetospheric charged-particle fluxes rather than solar wind fluxes. The Galilean satellites of Jupiter and the icy satellites of Saturn and Uranus fall in this category. The orbit of Titan, however, is sufficiently far from Saturn that the dayside portion of the orbit occasionally lies in the solar wind as the size of Saturn's magnetosphere varies.

The incident solar wind consists of a hydrogen and helium plasma in approximately solar proportions ($\sim$ 4% He by number), with incident proton energies $\lesssim$ 1 keV and with the flux varying approximately as the inverse square of distance from the Sun. Factor-of-two temporal variations in the flux are present. Energetic particle events, such as those associated with solar flares, also occur. On the other hand, magnetospheric charged-particle populations exhibit an important or dominant heavy ion component, with the notable exception of Uranus, and particle fluxes at energies up to tens of MeV are far greater than those encountered in the solar wind. Temporal variations exceeding two orders of magnitude are found in the magnetospheric charged particle fluxes at Mercury, Earth and Uranus.

Perhaps the most surprising aspect of the particle populations found in planetary magnetospheres is their unusual composition. The particle populations do not exhibit solar abundance ratios in any of the five known intrinsic planetary magnetospheres. At Jupiter, the plasma mass density, charge density, and pressure are all dominated by sulfur and oxygen ions. As both the solar wind and the Jovian atmosphere are dominated by hydrogen and helium, it is evident that the bulk of the magnetospheric plasma found its origin elsewhere. It is now clear that the sulfur and oxygen must have escaped from Jupiter's satellite Io, with its SO_2 volcanoes and SO_2 frosts (Fanale et al. 1982*c*; Kumar and Hunten 1982; Pilcher and Strobel 1982; Sullivan and Siscoe 1982; Nash et al. 1986; Cheng et al. 1986; Belcher 1983; Brown et al. 1983). Likewise, the magnetospheric plasma at Saturn is dominated by oxygen and nitrogen ions. The ultimate source of the oxygen ions is the water ice surfaces of the icy satellites and rings, while the source of the nitrogen is the atmosphere of Titan (Scarf et al. 1984; Cheng et al. 1986; Hunten et al. 1984; Neubauer et al. 1984). The magnetosphere of Mercury must be populated by sodium ions which may dominate the mass and energy densities (Cheng et al. 1987). Proton and alpha populations are also injected following complex interactions between the solar wind, magnetosphere, surface and atmosphere of Mercury. The Earth's magnetosphere is generally dominated by protons, whose origin may be either the solar wind or the atmosphere, but O^+ ions can occasionally become the dominant species during disturbed periods, for energies below roughly 100 keV (see, e.g., Krimigis et al. 1985; Shelley et al. 1985, and references therein). The O^+, and many other heavy ions found in the magnetosphere, clearly come from the Earth's atmosphere. Finally, Uranus is exceptional in that the magnetospheric ion population is totally dominated by protons; no heavy ions of mass > 2 have been detected in the inner magnetosphere (Bridge et al. 1986; Krimigis et al. 1986). The absence even of the alpha population argues against the solar wind as the source of this plasma (McNutt et al. 1987), and the most likely sources are the ionosphere and the atomic hydrogen corona (Shemansky and Smith 1983, 1986; Broadfoot et al. 1986; Cheng 1987).

The unique plasma compositions found in planetary magnetospheres are the clear signatures of magnetospheric interactions with the surfaces and atmospheres of planets and their satellites. Below we first describe the general physical processes and then discuss particular cases.

II. INTERACTION OF IONS WITH GASES AND SOLIDS

When energetic photons are absorbed in a molecular gas, a large fraction of the absorbed energy goes eventually into dissociation. As seen in Fig. 1, this leads to heating of the gas if the photon is absorbed by a molecule well below the exobase (process *c*). Given typical photon ionization cross sections of 10^{-18} cm^2 to 10^{-17} cm^2, only a small fraction of ionizing photons can be absorbed above the exobase. The same is true for photodissociation. When energy is deposited near the exobase (by photon absorption or other processes), and when this energy is thermalized, the high energy tail of the Maxwell-Botzmann distribution is populated and Jeans escape can occur (process *b*). Charged particle bombardment can be more important than photon absorption as an atmospheric heat source, particularly for the outer planets. If a molecule is dissociated in the exosphere, by photon absorption or, more likely, by electron impact, the dissociation can lead to atomic or molecular loss (process *a*). The energizing processes described for incident photons will also

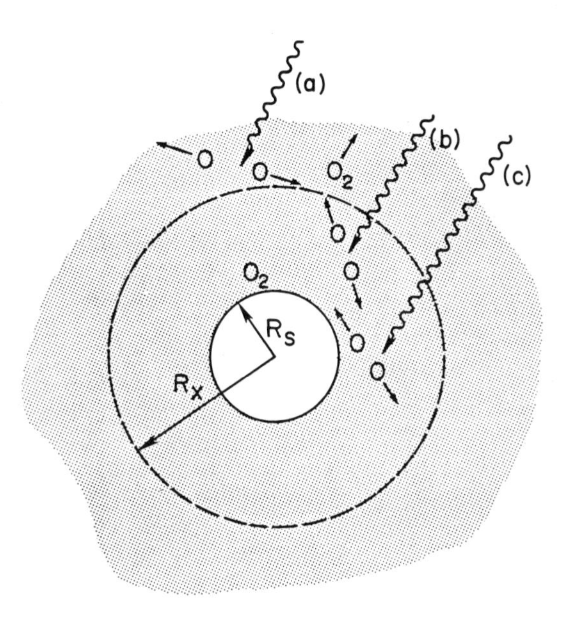

Fig. 1. Schematic of photodissociation in an atmosphere of exobase radius R_x around a satellite of radius R_s. Dissociation products can escape directly (*a*) or heat the atmosphere (*c*), contributing to Jeans escape (*b*).

occur when the molecular species is condensed on the surface. In this case, the species ejected from the surface must overcome the surface binding energy (sublimation energy). This process is referred to as photon-stimulated desorption, or, sometimes, photosputtering. In insulating materials, the typical cross section for photo-absorption is $\sim 10^{-18}$ to $\sim 10^{-17}$ cm^2 for photons of energy above the threshold for dissociation of molecules. It is not certain in the solid state how efficiently this energy is converted into energetic dissociation (see e.g., Tolk et al. 1982; Brenig and Menzel 1985). However, desorption can be an important process when dissociation occurs in the outermost monolayer. The species desorbed from an object's surface can be incorporated within the atmosphere, affecting its composition. Alternatively, if the atmosphere is collisionally thin, and if the molecule (atom) has sufficient energy, it can escape from the gravitational field of the object.

Processes similar to the above occur when plasma ions bombard an object. However, energetic ion bombardment differs in several ways. A single ion carries considerable momentum and can ionize or excite a large number of target species in a given volume. The resulting chemical processes depend on excitation density. Ions can also charge exchange with neutrals and, thereby, become disconnected from the magnetic field. The high momentum and the possibility of charge exchange are important for determining the plasma ion's ability to penetrate an atmosphere. The latter determines whether interactions occur above or below the exobase and whether interactions occur with the surface.

In Fig. 2 are shown processes that can occur when a plasma is incident on an object (satellite or planet) with a gaseous envelope. An incident ion can strike an atom or molecule in the exosphere, directly ejecting it or sending it into a larger bound orbit. It can also undergo electron capture (charge exchange) becoming a fast neutral which may or may not be deflected. When charge exchange occurs a "slow" ion is left behind. This slow ion can be swept away by the moving interplanetary or magnetospheric magnetic field if this field penetrates the atmosphere. Ions penetrating the exobase can energize a number of molecules, setting up a cascade of collisions that can result in molecules escaping or following ballistic trajectories in the exosphere. The energy deposited also heats the atmosphere. As is customary, we use the word sputtering for the nonthermal motion of the initial cascade of collisions. When

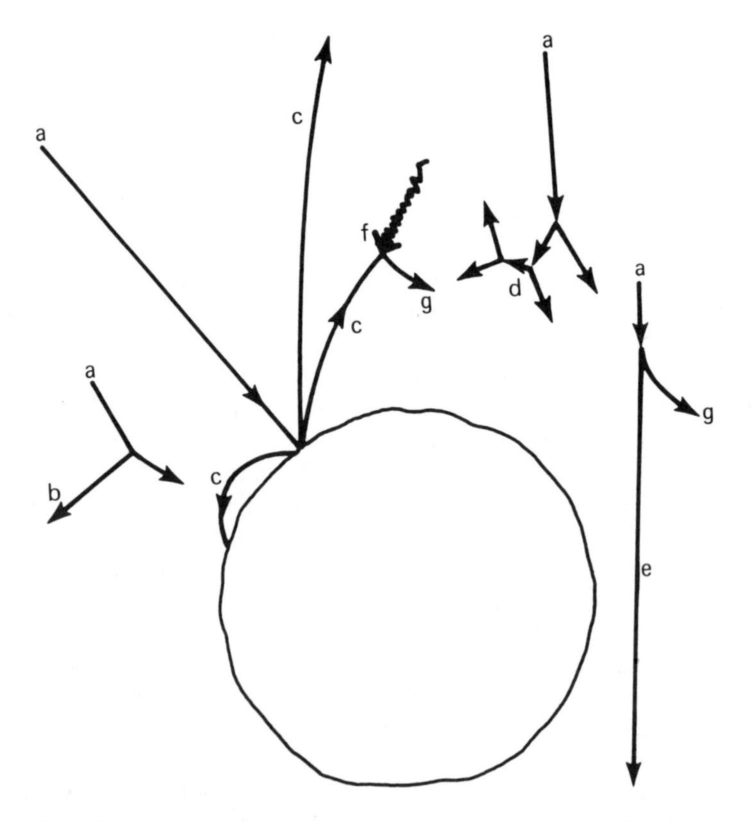

Fig. 2. Plasma bombardment of surface and atmosphere, showing incident ions (*a*), scattered neutrals (*b*), sputtered neutrals (*c*), atmospheric sputtering (*d*), charge exchange (*e*), ionization (*f*) and pickup of new ions (*g*).

the energy is thermalized, we refer to plasma heating of the gas. This can lead to loss via Jean's escape. As with the photon case, if the atmosphere is thin enough to be penetrated by the energetic particle, then sputtering can occur directly from the surface.

Collision cascade sputtering in gaseous atmospheres (Haff and Watson 1979) is very similar to that in solids, which has been extensively studied and is well understood in principle (Sigmund 1969; Andersen and Bay 1975). It is required only that the atmosphere be thick enough to allow full development of the collision cascades. Sputtering by collision cascades in solids is the dominant sputtering mechanism for incident ion velocities less than that corresponding to a few keV per nucleon. Collision cascade sputtering yields are roughly $\lesssim 1$ molecule per incident ion for protons incident on water ice. For heavy ions like O^+ the collisional energy transfer to the target species is more efficient, and collision cascade sputtering yields are typically an order of magnitude greater. The sputtering yield roughly scales as U^{-1}, where U is the sublimation energy of the target species ($U = 0.53$ eV for H_2O ice). In the case of atmospheric sputtering, the same U^{-1} scaling applies, except that the gravitational binding energy at the exobase replaces the sublimation energy (Haff and Watson 1979; McGrath and Johnson 1987).

When a fast ion penetrates a solid, it loses energy by causing electronic excitations (mainly ionizations) as well as by collisional energy transfer to target atoms that created the collision cascades described above. The electronic energy loss leads to a second sputtering mechanism in solids, "electronic sputtering," which tends to be most important for velocities between a few keV per nucleon and a few MeV per nucleon. In this velocity range, and particularly in weakly bonded insulators like ices, electronic sputtering yields for incident heavy ions can exceed several thousand molecules per incident ion, orders of magnitude greater than collision cascade sputtering yields. This may happen because a high density of ionizations can be created in a small volume, from which the deposited energy can escape only slowly via thermal conduction, allowing greatly enhanced sublimation from the strongly heated "thermal spike" region. It is important to note that this thermal spike effect does not occur in gaseous atmospheres, where individual ionization events are widely separated so that electron-ion recombination energizes the gas by individual molecular dissociations, as was the case for photon bombardment.

Via the processes described above, plasma bombardment can lead to loss of material from an atmosphere, formation of an expanded neutral corona (exosphere), heating of the atmosphere, and additions of surface material to the atmosphere. In addition, the incident ions can be neutralized and become part of the atmosphere. This has all been known for some time. What has changed recently is that laboratory data on ion interaction with outer solar system materials have been used to explain quantitatively some of the important spacecraft and groundbased observations.

A most intriguing consequence of ion bombardment is the production of gas-phase (atmospheric) species from the solid. Because water ice is found on the surfaces of many of the outer solar system objects, the water ice sputtering data, reported in Cheng et al. (1986) and Johnson et al. (1984*a*), are important for describing the gaseous envelopes of many objects. In addition, experimental data have been obtained for the sputtering of condensed SO_2 and sulfur (of interest at Io), condensed NH_3, CO and CO_2 (see, for example, Johnson et al. 1984*b*; Lanzerotti et al. 1985; Johnson et al. 1985; Brown and Johnson 1986; Chrisey et al. 1986). Finally, there has been considerable interest in condensed methane. Methane ice and/or methane clathrate in water ice may be relevant to the satellites of Uranus, Triton, Pluto and Charon, and is thought to behave under irradiation like many more complex organics (Strazzulla et al. 1984, 1985*a*; Strazzulla 1986; Lanzerotti et al. 1987).

In addition to sputtering, the principal effect of ion irradiation of condensed molecular gases is their chemical alteration. O_2 and H_2 are produced from H_2O (Brown et al. 1984; Reimann et al. 1984). These more volatile species can form a gaseous envelope above a very low temperature water ice surface. Similarly, irradiation of condensed NH_3 results in the production of H_2 and N_2, while SO_2 and CO_2 produce O_2 efficiently (Haring et al. 1983). Of considerable importance is the fact that all sulfur-containing species and carbon-containing species lose volatiles rapidly under irradiation and leave behind nonvolatile sputter-resistant residues (Strazzulla et al. 1985*a*; Chrisey et al. 1985; Lanzerotti et al. 1987). As these residues are dark, this process is of interest for a number of outer solar system objects (Cheng and Lanzerotti 1978; Strazzulla et al. 1984). Although the data base for these important species has increased considerably over the past few years, there are many important gaps in our knowledge. One major gap is the sputtering (desorption) of trace elements of semivolatiles from refractory solids. This is a process of considerable interest for Mercury, the Moon and other "airless" rocky bodies. It is also of interest for producing free sodium from the surface of Io. The sputtering of NaS and Na from condensed sodium sulfide has recently been measured (Chrisey et al. 1988).

When describing the interaction of the plasma with the gaseous envelope, the important processes are, in principle, well understood. The processes in Fig. 2 that lead to loss of molecules from the atmosphere give the escaping species characteristic energy and angular distributions (Sieveka and Johnson 1984). In charge exchange collisions between identical or nearly identical species, the neutralized plasma ion is forward directed along its initial motion, as shown in Fig. 2. On the other hand, in noncharge exchanging collisions an initially slow O atom receives preferentially a small fraction of the collision energy and leaves the collision predominantly at right angles to the fast incident ion's direction of motion. These two processes have very different effective radii of interactions (cross sections) as charge exchange between species with identical or nearly identical ionization energies can occur at large separations (many atomic radii) whereas collisional energy transfer requires the nuclei to approach each other closely. On the other hand, at very low velocities ($\lesssim$ 1 km s^{-1}) a slow ion can orbit a polarizable neutral forming, briefly, a molecular complex. The ion and neutral from such an encounter will share the center of mass motion and emerge with random orientations with respect to the initial relative motion. The differences in the nature of collisions at high and low velocities are important for determining the fate of the neutrals.

In contrast to the single collision cases above, sputtered species, that are produced after a collision cascade and that are not equilibrated with the ambient gas, have distinct energy spectra. At high energies, the energy spectrum of the sputter ejecta decays roughly as $1/E^2$, where E is the energy of the sputtered atom or molecule (see, e.g., Anderson and Bay 1975; Haring et al. 1983; Reimann et al. 1984). Consequently, the energy spectrum of sputter ejecta is much harder than that due to plasma heating (Jeans escape), which is roughly Maxwellian and decays exponentially at high energy (Sieveka and Johnson 1984). The angular distributions for sputtering and Jeans escape are, however, similar and directed preferentially normal to the surface (exobase). These processes will be considered when describing specific satellites or planets below.

The major uncertainty in describing the plasma interaction with an atmosphere is our limited ability to determine the induced currents, electromagnetic fields and actual plasma flow profile. In the following, we concentrate on the atomic and molecular physics aspects of the plasma modification of atmospheres rather than poorly understood details of the plasma flow.

III. MERCURY

The planet Mercury has a tenuous atmosphere whose nature is largely determined by interactions with the planetary surface and magnetosphere. The atmosphere is an exosphere, meaning that the exobase is at Mercury's surface. Thus an atmospheric neutral will typically fall back to the surface before colliding with another atmospheric particle. The neutral can then scatter off the surface, perhaps undergoing multiple collisions if the surface is microscopically rough, or it can be adsorbed on the surface and shortly thereafter be re-emitted into the atmosphere. These interactions are clearly critical for determining the atmospheric velocity distribution, thermal structure and distribution over Mercury's surface. For a recent review, see Hunten et al. (1988*b*).

Interactions with the magnetosphere are also important for determining the atmospheric sources and losses. Two of the primary constituents of the atmosphere are hydrogen and helium, both of which may find their ultimate origins in the solar wind (Goldstein et al. 1981; Hunten et al. 1988*b*). Protons and alpha particles in the solar wind enter the magnetosphere and can be implanted within Mercury's surface, whereupon they are neutralized and ultimately released back into the atmosphere. The magnetospheric interactions with the solar wind therefore determine in large part the sources of atmospheric hydrogen and helium, although outgassing of radiogenic helium may also be an important source (Goldstein et al. 1981).

Magnetospheric interactions are likewise important for atmospheric losses. For three of the known atmospheric species—helium, sodium and potassium—ionization is believed to be an important or dominant loss mechanism (Hunten et al. 1988*b*; Cheng et al. 1987). While the ionization is caused primarily by solar ultraviolet radiation, the magnetosphere nevertheless plays a key role. This is because a newly created ion is immediately swept up by the magnetospheric electric and magnetic fields, and can either be driven back into the surface of Mercury or out into deep space. If re-implanted within the surface, the ion is neutralized again and eventually recycled into the atmosphere. The efficiency of this recycling process therefore determines the net atmospheric loss rate due to ionization. Since the atmospheric scale heights are small compared to Mercury's radius, most of the ionization occurs close to the surface and a significant fraction of the ions will be recycled.

Observations

Hydrogen, helium and oxygen were discovered in the atmosphere of Mercury by the Mariner 10 airglow spectrometer (Broadfoot et al. 1976). A well-measured scale height was determined only for helium, and the oxygen detection was reported as tentative. The measured helium scale height corresponds to a temperature of 575 K, roughly consistent with the subsolar surface temperature. The hydrogen altitude distribution is not well-fitted by a single scale height, and Shemansky and Broadfoot (1977) propose that both a hot component and a cold component are present, with kinetic temperatures characteristic of the day side and the night side, respectively.

Sodium and potassium were recently discovered by groundbased optical spectrometers (Potter and Morgan 1985, 1986). The originally reported sodium column density of 8×10^{11} cm^{-2} has been revised downward to (1 to 2) $\times 10^{11}$ cm^{-2} (Hunten et al. 1988*b*). The potassium column density is roughly 1% of the sodium column. Both the sodium and potassium column densities are observed to be variable.

Table I summarizes the available data on the abundances of the five known species in Mercury's atmosphere. Additional species of comparable or greater abundance may also be present. Only upper limits from the Mariner 10 ultraviolet occultation experiment (Broadfoot et al. 1976) are available for the column densities of Ar, O_2, H_2, N_2, CO_2 and H_2O. These typically lie more than 3 orders of magnitude above the observed values for the known atmospheric species.

Solar radiation pressure can have a substantial effect on the trajectories of atmospheric neutrals; for sodium, the resulting acceleration can exceed half the surface gravity of Mercury, but varies strongly with radial velocity (owing

TABLE I
Column Density and Surface Number Density of Atmospheric Species at Mercury

Species	N (cm^{-2})	n (cm^{-3})	Notes
H		8, 82	1, 2
He	1.5×10^{11}	4.5×10^{3}	1
O	5.7×10^{10}	7.1×10^{3}	1, 3
Na	$(1–2) \times 10^{11}$	$(2–4) \times 10^{4}$	3, 4
K	$(0.5–1.5) \times 10^{9}$	$(1.5–5) \times 10^{2}$	3, 4

(1) Near subsolar point (Broadfoot et al. 1976).
(2) Thermal and nonthermal components, respectively, of surface number density.
(3) Assumed temperature 575 K as measured for He (Potter and Morgan 1985; Hunten et al. 1988*b*); considerable variability observed.

to the deep solar Fraunhofer line) and distance to the Sun (Smyth 1986). However, recent observations (Hunten et al. 1988*b*) have shown that the sodium emission is closely confined to the visible disk of Mercury, and there is no evidence for an extended sodium tail on the night side as predicted by Ip (1986*a*) and Smyth (1986). The absence of an antisolar sodium tail is consistent with a sodium velocity distribution corresponding to typical dayside surface temperatures of ~ 500 K and a sodium atmospheric scale height ~ 50 km for the observed population.

Hydrogen and Helium

The implications of magnetospheric processes at Mercury for hydrogen and helium source and loss rates have been investigated by many authors (see, e.g., Hunten et al. 1988*b*; Goldstein et al. 1981, and references therein). The most important source mechanisms are outgassing of radiogenic helium from the interior of the planet and impact of solar wind protons and alpha particles onto the surface, followed by neutralization and release into the atmosphere. The most important loss mechanisms are Jeans escape and photoionization.

For helium, Goldstein et al. (1981) conclude that the combined radiogenic source and solar wind source could approximately balance the losses due to photoionization and Jeans escape, within the order-of-magnitude uncertainties. The radiogenic source of helium is estimated as (0.6 to 7) $\times 10^{22}$ s^{-1}, while the solar wind source is estimated as (0.6 to 2) $\times 10^{23}$ s^{-1}. Major uncertainties in the solar wind source arise because the mechanisms responsible for entry of solar wind plasma into the magnetosphere are not well understood. Solar wind plasma can impact the surface directly during the infrequent periods when the solar wind dynamic pressure is great enough to push the magnetospheric boundary down to the surface (this never occurs for the more strongly magnetized planets Earth, Jupiter, Saturn and Uranus). Solar wind plasma can also enter the magnetosphere: (1) via dayside magnetic field reconnection, quasi-steady or sporadic; (2) in magnetospheric cusps over the magnetic poles; or (3) in the nightside magnetotail so as to enter the plasma sheet and be transported or precipitated to the surface. Goldstein et al. (1981) conclude that direct impact of solar wind on the surface occurs only ~ 6% of the time and is relatively unimportant. None of the remaining mechanisms is well understood, even in Earth's magnetosphere. The relative importance of these various entry mechanisms is an important issue, because it affects not only the absolute magnitude of the solar wind source, but also where on Mercury's surface the source is greatest—day side or night side, high latitude or equatorial regions.

The helium loss rate due to photoionization is estimated as $(1-\gamma)\ 7 \times 10^{22}\ s^{-1}$ (Hunten et al. 1988*b*), where γ is the fraction of newly created ions recycled into the surface. This fraction is determined by magnetospheric electromagnetic fields and is not well known at present. Numerical simulations suggest a value of $\gamma \lesssim 0.5$ (Ip 1987*a*). The atmospheric helium loss rate due to photoionization also depends marginally on the estimate that the nightside helium density is ~ 50 times the observed value on the day side (see below). The Jeans escape rate may be less important, because the atmospheric velocity distribution function may be strongly depleted at high velocities (Shemansky and Broadfoot 1977). The magnitude of this depletion depends on the nature of interactions with the surface and is still controversial (Hodges 1980; Shemansky 1980). A key question is the degree of energy accommodation of helium and hydrogen in collisions with the surface, which determines the rate at which the velocity distributions equilibrate to the surface temperature. It is clear that the energy accommodation is not efficient, but a quantitative understanding is not available (Shemansky and Broadfoot 1977). Inefficient energy accommodation is inferred from the observed distribution of helium across the day side to the terminator (Broadfoot et al. 1976). The observed terminator density suggested a nightside density ~ 50 times the subsolar density, whereas with efficient accommodation there would be a greater concentration of helium toward the night side (Hunten et al. 1988*b*).

Similar uncertainties apply to the hydrogen atmosphere. The solar wind source of hydrogen would exceed that for helium by a factor of ~ 22, the average abundance ratio in the solar wind, if hydrogen and helium have equal probabilities for entry into the magnetosphere, implantation in the surface, and release into the atmosphere (Goldstein et al. 1981, Hunten et al. 1988*b*). The assumption of equal probabilities may be a reasonable first approximation but has not been justified in detail. The total loss rate of hydrogen is dominated by Jeans escape and is estimated as $3 \times 10^{23}\ s^{-1}$ (Hunten et al. 1988*b*). This estimated loss rate is highly uncertain. In the first place, the high velocity tail of the velocity distribution may be depleted, and the extent of this depletion depends on poorly understood surface interactions. In the second place, the loss rate estimate assumes equal column densities on day side and night side. This assumption was justified by noting the presence of a cold hydrogen component in the subsolar distribution, which suggested very rapid transport between day side and night side.

In summary, the solar wind sources of hydrogen and helium appear to balance the estimated losses of these species within the order-of-magnitude uncertainties, but a quantitative understanding is not available.

Sodium and Oxygen

The situation for sodium and oxygen differs from that for hydrogen and helium. This is because energy accommodation with the surface will be relatively efficient for these heavy species (Hunten et al. 1988*b*), and because simple estimates suffice to show that the sodium source to the atmosphere must come from within Mercury itself, with external sources like the solar wind and meteoroid infall being relatively unimportant (Ip 1986*b*; McGrath et al. 1986; Cheng et al. 1987). Photon-induced desorption and thermal desorption are probably the dominant mechanisms for sodium ejection to the atmosphere, and Mercury's surface is not expected to be anomalously enriched in sodium (see Fig. 3). Efficient energy accommodation with the surface leads to a velocity distribution well characterized by the surface temperature. This is consistent with a recent observation of the hyperfine structure of the sodium lines, which implied a kinetic temperature of $\sim$ 500 K in good agreement with the local surface temperature (Potter and Morgan 1985; Hunten et al. 1988*b*).

As will be discussed below, there is rapid exchange of sodium between the surface and the atmosphere. If we consider the combined surface - atmosphere system, then sodium exchange between surface and atmosphere does not contribute to the net loss from the system. This net loss rate of sodium is dominated by photoionization (Smyth 1986; Ip 1986*b*; McGrath et al. 1986) and is estimated by Cheng et al. (1987) as $7\,(1-\gamma) \times 10^{24}\ s^{-1}$. Here $\gamma \lesssim 0.5$ is, as before, the poorly known fraction of ions recycled into the surface. This net loss rate of sodium is balanced by upward transport of sodium within Mercury's surface regolith, followed by ejection into the atmosphere. Meteoritic and other exogenic sources are relatively unimportant (Ip 1986*b*; McGrath et al. 1986; Cheng et al. 1987).

Sodium atoms released into the atmosphere fall back to the surface under gravity and solar radiation pressure after a short free flight time, typically $< 10^3$ s. Upon impact the sodium atoms generally stick to the surface, although it is not yet clear whether they are more likely to be adsorbed or chemically bound there. If sodium is mainly adsorbed on the surface, its sticking lifetime may be as short as minutes on the night side and much less on the day side (Hunten et al. 1988*b*). If, on the other hand, most of the surface Na is chemically bound with binding energy $\sim$ 2 eV, then the dominant mechanism for ejection from the surface will be photon-stimulated desorption. The lifetime against such desorption for chemically bound sodium on Mercury's surface is estimated as 10^4 to 10^5s (McGrath et al. 1986; Cheng et al. 1987), by

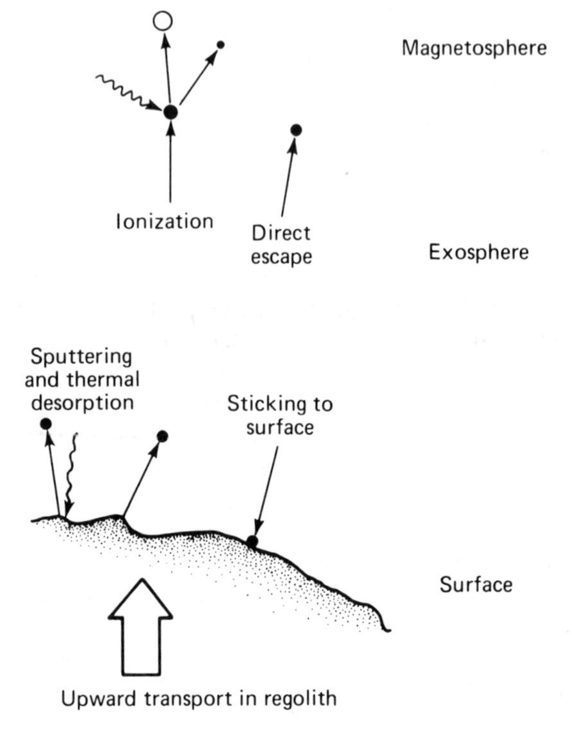

Fig. 3. Interactions among magnetosphere, exosphere and surface of Mercury. Atmospheric species (e.g. Na) fall to the surface and can stick there, becoming either chemically bound or adsorbed. They can be released from the surface by photon or particle bombardment or by thermal desorption. Atmospheric species can escape directly or be ionized, thereafter being swept away or re-implanted into the surface.

comparison with the vapor phase photodissociation lifetimes of NaO_2 and NaOH (see, e.g., Kirchhoff 1983) after scaling to Mercury. These estimates are roughly consistent with Na ejection rates from Na C1, the only system for which data are available. Additional laboratory measurements are clearly needed. Even with this lifetime for fresh removal of chemically bound sodium from the surface, maintenance of the sodium atmosphere does not require Mercury's surface to be anomalously enriched in sodium; a sodium surface abundance of $\sim 10^{-3}$ would suffice. If most of the sodium is adsorbed rather than chemically bound, the required surface abundance would be even less, because thermal desorption would yield a much shorter sticking lifetime on the surface.

Charged particle sputtering is expected to be less important (McGrath et al. 1986; Cheng et al. 1987). Magnetospheric ion fluxes were not well measured *in situ* at Mercury, but an upper limit to the ion fluxes can be placed by assuming that the plasma β (the ratio of particle pressure to magnetic pressure) satisfies $\beta \lesssim 1$. This constraint is obeyed at least approximately by all the other known planetary magnetospheres. If this constraint is imposed, charged particle sputtering becomes inadequate to maintain the sodium atmosphere unless the surface of Mercury were largely or entirely sodium, which is implausible.

For oxygen, the dominant net loss mechanism from the combined surface-atmosphere system is again photoionization. Oxygen chemically binds to the surface upon impact. The oxygen net loss rate is much less than that of sodium, by a factor (4 to 9) $\times 10^{-3}$ (Cheng et al. 1987), primarily because of the longer photoionization time for oxygen, roughly 60 times that for sodium, and smaller atmospheric column density. The recycling fractions γ should be similar for oxygen and sodium ions. Likewise, the net potassium loss rate is much less than that of sodium, by roughly a factor of 60 (Cheng et al. 1987).

The rapid ionization loss of sodium from the atmosphere, and to a much smaller extent ionization of other heavy species, is expected to cause formation of a heavy ion magnetosphere (Cheng et al. 1987). Sodium ions from the atmosphere may typically be accelerated to keV energies, and may dominate the mass and time-averaged energy budgets of the magnetosphere, which are estimated as 300 g s^{-1} and 3×10^9 W, respectively.

IV. IO ATMOSPHERE AND PLASMA TORUS

It is by now well known that sulfur and oxygen ions are the most abundant species by number in Jupiter's magnetosphere (Belcher 1983; Brown et al. 1983). At thermal energies (about 50 eV in the Io torus, and higher energies at larger radii), they dominate both the mass and charge densities in the magnetosphere. Sulfur and oxygen ions continue to be important or dominant in the composition at energies up to several MeV (Krimigis and Roelof 1983). These results are confirmed not only by Voyager ultraviolet spectroscopy but also by *in situ* plasma and energetic particle measurements on Voyager. Observations by groundbased optical spectrometers and the International Ultraviolet Explorer (IUE, an Earth-orbiting satellite) have also confirmed the sulfur and oxygen composition of the magnetosphere (see, e.g., Moos et al. 1985, and references therein). That composition is unique in the known universe.

Jupiter's satellite Io is the ultimate source of the sulfur and oxygen in the Jovian magnetosphere. There are high abundances of sulfur and oxygen on Io's surface and in its atmosphere. Gaseous SO_2 with a surface pressure of 10^{-7} bar (column density$= 5 \times 10^{18}$ cm^{-2}) was detected by the Voyager 1 infrared spectrometer near one of the volcanic plumes; this pressure was consistent with the SO_2 vapor pressure at the local surface temperature 130 K (Pearl et al. 1979). SO_2 frost has also been detected (see review by Nash et al. 1986).

However, the escape of sulfur and oxygen from Io remains a nontrivial issue (see reviews by Fanale et al. 1982*c*; Kumar and Hunten 1982; Nash et al. 1986; Cheng et al. 1986). The volcanic eruptions themselves are not energetic enough to drive sulfur and oxygen into space against Io's gravity (Kieffer 1982), and the atmosphere is too cold for thermal escape to be adequate to maintain the plasma torus. Voyager ultraviolet observations also imply that the total supply rate of oxygen, atomic and/or molecular, required to maintain the Io plasma torus is comparable to or up to 4 times that of sulfur (Shemansky 1987); a contrary view has been expressed by Moreno et al. (1985). Finally, short-lived mass 64 ions, either SO_2^+ or S_2^+, have been observed in the torus far from Io (Bagenal and Sullivan 1981), implying the presence of the parent neutral molecule (SO_2 or S_2) far from Io (Cheng 1984*a*).

The escape of molecular SO_2 from Io is further suggested by recent models of the cold, inner portion of the Io torus. This is the portion found well within Io's orbit (radii $\lesssim 5.5$ R_J). Barbosa and Moreno (1988) have modeled the temperature and the composition of the cold torus by considering inward radial diffusion of hot plasma from outside Io's orbit through neutral clouds containing molecular SO_2 and its dissociation product SO. The presence of neutral SO appears to resolve difficulties found in earlier work (Richardson and Siscoe 1983; Moreno and Barbosa 1986).

Finally, the sources of free sodium and potassium at Io are still uncertain (see reviews by Fanale et al. 1982*c*; Nash et al. 1986; Cheng et al. 1986; Brown et al. 1983). Charged particle sputtering from Io's surface and/or atmosphere is probably important (Matson et al. 1974), but the nature of materials containing Na and K on Io's surface is unknown, as is the nature of Na and K species incorporated into the atmosphere. Chrisey et al. (1988) have recently measured the sputtering of Na and NaS from condensed sodium sulfide.

It is now widely agreed that the magnetosphere itself drives the escape of sulfur and oxygen from Io, by plasma impacting onto the surface and/or exobase. The magnetospheric plasma also ionizes the neutrals ejected from Io to create the very plasma that dominates its own composition. Figure 4 shows the chain of processes that links the Jovian magnetosphere (specifically the Io plasma torus), the surface and atmosphere of Io, and the sulfur and oxygen clouds in the plasma torus. The chain is shown as being closed on itself, meaning that the Io plasma torus is self-regenerating. Each ion impact on Io's surface or atmosphere results in loss of the ion from the plasma torus. But many neutrals (tens to thousands) are sputtered into the neutral clouds, and a large fraction of them are eventually ionized and re-injected into the torus, resulting in a net increase in the ion population of the torus. Indeed, the plasma torus density would grow indefinitely except that other processes intervene.

One widely held view is that the plasma torus density is stabilized by flux-tube centrifugal interchange instability (Huang and Siscoe 1986, 1987). The centrifugal interchange instability is an MHD instability analogous to the

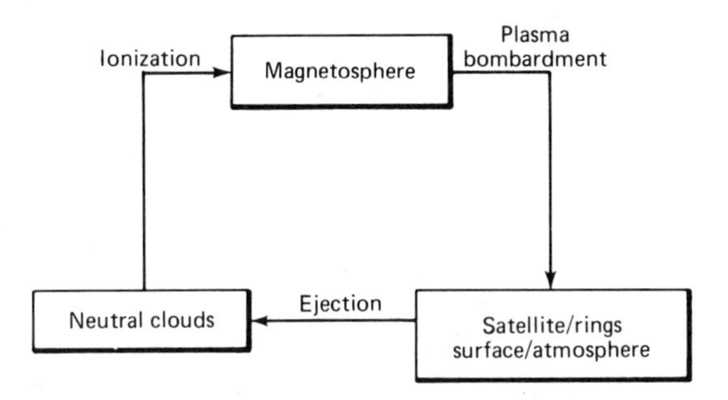

Fig. 4. Closed chain of interactions among the magnetosphere, satellites and neutral clouds.

Rayleigh-Taylor instability in ordinary fluids that occurs when a dense fluid overlies a light one. At Jupiter, the rapid plasma rotation causes centrifugal force to dominate gravity, and interchanges are centrifugally driven. The interchanges cause a rapid transport of plasma out from the torus, allowing sulfur and oxygen ions to fill the Jovian magnetosphere and limiting the growth of the Io torus. This is because centrifugal interchange instability causes the plasma transport lifetime in the torus to decrease as the plasma density increases. Hence, the plasma losses increase more rapidly than plasma sources as the torus density increases, stabilizing the torus.

However, the torus density can also be stabilized in the absence of centrifugal interchange instability. Presumably, in this case the plasma transport lifetime would be determined externally, e.g., by Jovian atmospheric wind-driven interchanges (Coroniti 1974), and would not depend on torus density. When the torus density is sufficiently high, an increase in density now causes a decrease in electron temperature (Barbosa et al. 1983; Smith and Strobel 1985), thereby limiting the plasma source and stabilizing the torus. In these models, the electron temperature is maintained by Coulomb collisional energy transfer from hot ions, the ions having been originally energized by pickup in the rotating Jovian magnetic field (see, e.g., Sullivan and Siscoe 1982; Cheng et al. 1986). An additional unknown energy source, of strength comparable to that of ion pickup, may be required in the Io torus (Shemansky 1988). Provided that unknown energy sources do not dominate ion pickup, and given a transport lifetime roughly independent of torus density, the electron cooling provides an upper limit to the torus density that is only a few times greater than typical present values (Cheng 1988).

Hence the average torus plasma density can find a stable equilibrium with or without centrifugal interchange instability (Cheng 1988). However, the temporal variability of the torus, and specifically its response to changes in input parameters, can depend significantly on the nature of plasma transport. For example, changes in Io's volcanic activity may lead to changes in its atmosphere density, thereby changing the sputtering yield of neutrals from plasma bombardment. If the transport lifetime is independent of torus density, then even arbitrarily large sputtering yields cannot cause the torus density to rise above several times present-day values, owing to electron cooling (again, barring dominant unknown energy sources). However, if the transport lifetime is inversely proportional to density, as may be roughly true for centrifugal interchange instability (Huang and Siscoe 1986), then the electron cooling effect does not occur and the torus density may become much greater than present-day values (Cheng 1988). On the other hand, a significant drop in average torus density is permitted by either transport model.

Voyager plasma observations (Richardson and McNutt 1987) place a significant constraint on centrifugal interchange transport models in the Io torus. Namely, there are no plasma density fluctuations exceeding 10% within and near the Io torus over distance scales as short as 20 km. The presence of nearly empty flux tubes in the torus as suggested by Pontius et al. (1986) can be ruled out. However, it is not clear that centrifugal interchange requires density fluctuations exceeding 10%. In the Siscoe and Summers (1981) model of centrifugal interchange diffusion, the diffusion coefficient at $L \gtrsim 6$ is roughly $D_{LL} \approx 2.2 \times 10^{-5} (\Delta L)^2 \text{ s}^{-1}$, where ΔL is the characteristic eddy size. Given the measured gradient in flux tube content (Bagenal et al. 1985; Bagenal and Sullivan 1981), the requirement of density fluctuations $< 10\%$ leads to $\Delta L < 0.22$ or $D_{LL} \lesssim 1.1 \times 10^{-6} \text{ s}^{-1}$ near $L = 6$ which is quite acceptable (see, e.g., Sullivan and Siscoe 1982; Pilcher and Strobel 1982; Cheng 1986*b*).

The observational situation regarding temporal variability of the torus is not entirely settled. There are monthly variations by a factor of ~ 2 in electron temperature and ion densities, according to groundbased optical observations (Morgan 1985), Voyager extreme ultraviolet observations (Sandel et al. 1979) and Earth-orbital ultraviolet observations from IUE (Moos et al. 1985). Thirteen IUE observations over a 5 yr period following the Voyager encounters showed the torus to be stable over the long term (Moos et al. 1985). However, there are indications of order-of-magnitude density variations from optical observations of sulfur line ratios, although the interpretation of these observations is unclear (Morgan 1985). Finally, Pioneer 10 apparently detected torus densities an order of magnitude less than during the Voyager encounters (see e.g., Mekler and Eviatar 1980); this discrepancy has never been explained in terms of instrumental effects and may indicate a true temporal variation. However, the near-Io sodium cloud at the time of Voyager 1 encounter was generally similar to that observed around the time of Pioneer encounter (Goldberg et al. 1980), arguing against drastic variations of sodium escape rate or electron density.

In summary, factor-of-2 temporal variations in the torus are well established, but it is not clear whether more extreme variations have been observed. The long-term stability of Io's atmosphere, therefore, remains an open question.

Io's Surface

Nash et al. (1986) and Clark et al. (1986) recently reviewed the physical properties and chemical composition of Io's surface. With its active volcanoes and its high abundances of elemental sulfur and condensed SO_2, Io's surface is unique among known bodies in the solar system. Io's surface is spectrally bland and characterized by very high reflectivity in the visible and near-infrared, with a strong 4 μm absorption band and a steep absorption edge shortward of 0.6 μm. There is no evidence for H_2O ice bands in the near-infrared (Pilcher 1979), contrasting with the spectra of the other three Galilean satellites of Jupiter as well as the icy satellites of Saturn and Uranus. The 4 μm absorption band has been identified as due to condensed SO_2 (Fanale et al. 1979; Smythe et al. 1979). This identification is strengthened by the detection of gaseous SO_2 near one of the volcanic plumes (Pearl et al. 1979) and ubiquitous sulfur and oxygen ions in the plasma torus and Jovian magnetosphere (Belcher 1983; Brown et al. 1983; Krimigis and Roelof 1983). However, it remains controversial whether the condensed SO_2 is present mainly as frost or adsorbate (see, e.g., Matson and Nash 1983; Howell et al. 1984). On the other hand, the high spectral reflectance of Io's surface in the visible and near infrared and the steep absorption below 0.6 μm are consistent with elemental sulfur (Wamsteker et al. 1974). The spectral reflectance of a sulfur surface depends on its granularity, temperature and thermal history (Gradie and Veverka 1984), and eclipse observations show that no more than $\sim$ 50% of Io's surface can be covered with elemental S_8. The remarkable colors of Io's surface, pastel yellow with orange, green and grey tints, are consistent with a variety of sulfur allotropes on the surface. The volcanic plumes also appear to fall into two classes, those driven by SO_2 and those driven by sulfur (McEwen and Soderblom 1983).

Other species must be present that have not been identified spectrally. Io's high bulk density and the presence of 10 km scale surface relief imply the presence of silicates near the surface. The silicate rocks may be largely covered by a thin layer of sulfur-rich materials. In addition, sodium and potassium must be present on the surface in some form, possibly as $Na_2 S_x$ and $K_2 S_x$ ($x \approx 1$ to 5) (Lunine and Stevenson 1985). These semi-volatiles must be present to maintain the atomic Na and K clouds that coorbit with Io in the plasma torus.

The considerable volcanic activity of Io yields a relatively young surface with no evidence for impact craters due to rapid resurfacing. Voyager photographs indicate widespread vents, plumes and surface flows. Estimates of the resurfacing rate vary from $\sim 10^{-1}$ cm yr^{-1} (based on cratering rates) to ~ 10 cm yr^{-1} (based on heat loss) (Nash et al. 1986). The visible reflectance spectra of vent source regions, plume deposits, and the caldera floors of Pele and Surt are generally consistent with various heated sulfur and sulfur-basalt mixtures (McEwen and Soderblom 1983). Sulfur and SO_2 are probably intimately mixed (Howell et al. 1984), and SO_2 frost is not simply associated with the obvious bright areas. Synoptic observations of Io in the mid-ultraviolet and near-infrared further indicate that condensed SO_2 is globally distributed, but the SO_2 is more abundant on the bright leading hemisphere of Io and less abundant on the darker and redder trailing hemisphere.

In addition to the darkness of the trailing hemisphere compared to the leading hemisphere, Io is characterized by dark polar regions and anomalous post-eclipse brightening. The anomalous eclipse observations are apparently explained as due to emission from hot spots unaffected by eclipse and nonuniformly distributed over Io's surface (see review by Veverka et al. 1986). In addition, the surface is inferred to have very low thermal inertia, probably indicating a very high porosity. A porous surface may arise from sublimation of S_8 from sulfur flows (Nash 1987) which would leave behind a fluffy, polymeric sulfur structure. Matson and Nash (1983) proposc that cold-trapping may occur in a porous, subsurface regolith, yielding an SO_2 vapor pressure above the surface (away from hot spots, plumes, etc.) much lower than that above an SO_2 frost in thermal equilibrium at the surface temperature. The dark polar regions also suggest that the SO_2 atmosphere is, at least locally, collisionally thin (Matson and Nash 1983).

Io's Atmosphere

Our understanding of Io's atmosphere has advanced significantly, but many uncertainties remain. While earlier work considered models that were everywhere collisionally thin (see, e.g., Matson and Nash 1983) or everywhere thick (Kumar 1984), more recent work has included atmospheric fluid dynamics as well as condensation and sublimation (Ingersoll et al. 1985; Thomas 1987). These models are collisionally thick over part, but not all, of Io's surface. ("Thick" implies radial column densities exceeding roughly 10^{15} cm^{-2}, more than in an exosphere.)

Constraints on the nature of the atmosphere can be inferred from direct observations of the atmosphere or from observations and modeling of the plasma torus. Ballester et al. (1987) recently observed oxygen and sulfur emissions from Io's atmosphere. Owing to the low local electron temperature ($\sim$ 2 eV), a significant fraction of these emissions must have arisen from a dense, collisionally thick region. Schneider et al. (1987) have mapped sodium emissions very near Io by using a Europa occultation (see below) and find results consistent with a very low exobase.

In addition, the Io plasma torus imposes constraints on the escape of oxygen and sulfur from Io (Cheng 1984*a*; McGrath and Johnson 1987, and references therein). The total oxygen escape rate from Io amounts to roughly (0.3 to 3) $\times$ 10^{28} s^{-1} in the form of atoms or molecules. The oxygen escape rate is comparable to that of sulfur, or up to 4 times greater (Shemansky 1987). Finally, there is significant escape of SO_2 and/or S_2, as inferred from observations of SO_2^+ and/or S_2^+ and suggested by cold torus models. These requirements significantly constrain atmospheric models. Models that are everywhere collisionally thin have difficulty yielding an adequately high oxygen escape rate (Cheng 1984*a*); even after formation of a sputter corona (Sieveka and Johnson 1985) the escape rate can only approach the lower end of the estimated requirement ($\sim 3 \times 10^{27}$ s^{-1}). However, once the atmospheric exobase lies above the surface, so atmospheric sputtering takes over, then adequately high escape rates are easily obtained even with the exobase much less than an Io radius above the surface (McGrath and Johnson 1987).

However, thick atmosphere models are also strongly constrained. In one class of these models (Kumar 1984), photochemical processes and diffusive separation cause O to be the dominant species in the exosphere even though SO_2 is the primary constituent near the surface. The escape rate of O then tends to be far greater than that for all other species, conflicting with observations. Processes like strong dynamical mixing or mutual drag effects (Cheng 1984*a*; Hunten 1985) appear to be needed to obtain more nearly equal sulfur and oxygen escape rates.

More realistically, Io's atmosphere will be collisionally thick over a significant fraction of Io's surface, but probably not the entire surface. The thick portions will include regions above large dayside frost deposits and any active volcanic plumes (see, e.g., Ingersoll et al. 1985; Thomas 1987). This view is directly supported by the Voyager infrared observations (Pearl et al. 1979) and by IUE observations of sulfur and oxygen emissions from Io's atmosphere (Ballester et al. 1987) that originate in a collision-dominated region. A thick atmosphere over a major portion of the surface is also supported by the Pioneer observations of an ionosphere at Io (Kliore et al. 1975; Kumar 1985). The nightside atmosphere may be thin or may contain abundant O_2 (Kumar and Hunten 1982). The presence of dark polar caps, however, suggests that the atmosphere is collisionally thin over the poles (Matson and Nash 1983). The poles may have been darkened by plasma bombardment (Chrisey et al. 1987).

Sputter Coronae

If the atmosphere is collisionally thin, then incident corotating ions (O^{n+} at 260 eV, S^{n+} at 520 eV, total incident flux (0.7-1) $\times$ 10^{10} cm^{-2} s^{-1}) impact the surface and sputter ejecta follow ballistic trajectories. If, on the other hand, the atmosphere is thick and a conducting ionosphere forms, the corotating plasma flow is deflected above the surface by shielding of the motional electric field (Schulz and Eviatar 1977; Ip 1982). However, the incident plasma flow still reaches the exobase (Wolf-Gladrow et al. 1987) and atmospheric sputtering will occur. In all cases, a sputter corona (Sieveka and Johnson 1984; McGrath and Johnson 1987*a*) is expected to form.

As discussed above, plasma bombardment produces energetic sputtered atoms and molecules and causes heating of the region penetrated. For bombardment of a surface with significant conductivity, the heating can generally be ignored, but this is not the case in an atmosphere.

Sieveka and Johnson (1984) have calculated the sputtered neutral corona produced by ions impinging on a surface covered with condensed SO_2. They show that a giant extended corona is produced. As seen in Fig. 5, the sputter corona has of the order of an exosphere of gas. Since the charge exchange cross sections of the incident ions with the sputter-ejected SO_2 and O_2 (Johnson et al. 1984*b*) are of the order of 10^{-15} cm^2 (Johnson and Strobel 1982; McGrath and Johnson 1988), fresh ions are produced in the disk subtended by Io. The ions produced by charge exchange can be accelerated by the penetrating field into the surface producing additional sputtering, but this process is self limiting. First, any excessive production of pickup ions will lead to deflection of the incident plasma flow by shielding of the electric field. Second, excessive gas production will limit the penetration of the ions. As the predominant corotating oxygen and sulfur have penetration depths of 10^{15} to 10^{16} SO_2 cm^{-2}, with an equivalent spread in penetration depths, Sieveka and Johnson (1985) have pointed out that the sputter limit is an atmospheric column density $\sim 10^{16}$ SO_2 cm^{-2}, which can exceed an exospheric component.

McGrath and Johnson (1987) have considered the sputter-produced corona above a collisionally thick atmosphere (exobase above the surface). The corotating ions deposit their energy in a column of gas $\sim 10^{16}$ cm^{-2}, and the energy deposited below the exobase (or conducted there) can cause significant

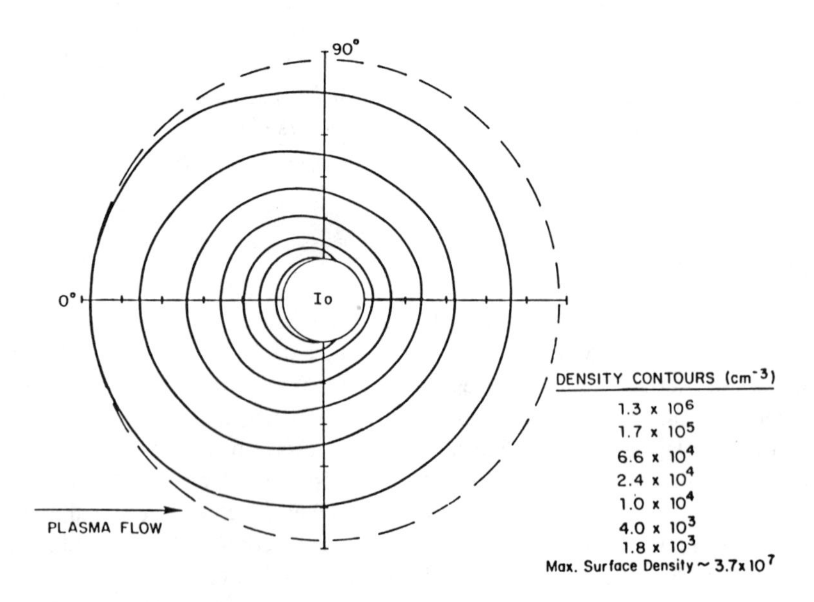

Fig. 5. Model of SO_2 sputter corona formed by sputtering of Io's surface (figure from Sieveka and Johnson 1985).

heating of the atmosphere. The sputter corona contains energetic molecules in the tail of the sputter energy distribution, so it extends more than an Io radius above the surface even if the exobase is close to the surface. Molecular and atomic species are lost, therefore, to sputtering, collisions of the incident plasma with neutrals in the corona, and charge exchange or electron impact ionization of corona species followed by sweeping. Rates for each of these processes were calculated. Figure 6 shows the resulting net supply of new species to the torus region. In this calculation, the exobase temperature is 1000 K, the plasma is assumed to be undeflected above the exobase, and re-impact by freshly produced ions is ignored. The escape rates from atmospheric sputtering and heating are so large that an exobase altitude well above the surface may be consistent only with significant plasma deflection to limit the interaction with the corona and limit the plasma heating. Ip (1982) has previously estimated that plasma deflection reduces the effective cross section of Io for absorption of incident plasma by a factor ~ 0.17, and similar results are obtained from numerical MHD simulations (Wolf-Gladrow et al. 1987; Linker et al. 1987). With this reduction factor to account for plasma deflection and ignoring re-impact of fresh ions, we find from Fig. 6 a maximum exobase altitude $< 0.5\ R_{Io}$ for a supply rate of 10^{30} amu s^{-1}.

It is seen that the oxygen and sulfur supply rates at the time of Voyager are easily obtained with an exobase not far above the surface and an SO_2

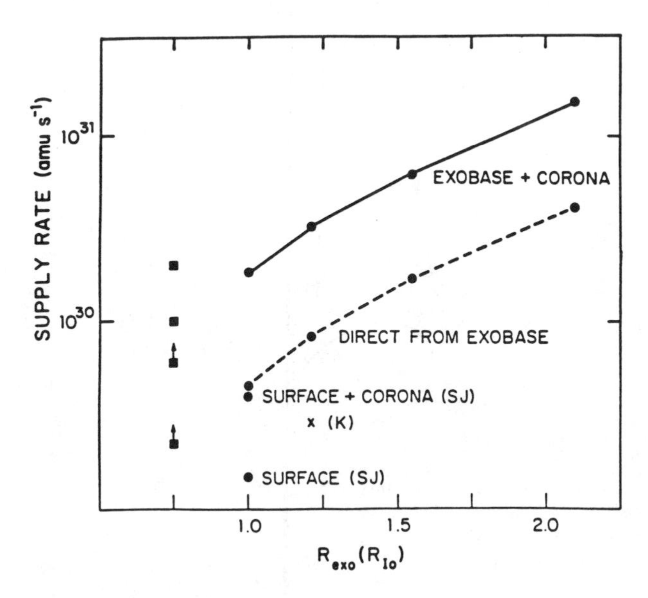

Fig. 6. Combined escape rates of sulfur and oxygen from Io, assuming atmospheric sputtering and an incident ion flux of 10^{10} cm^{-2} s^{-1} consisting of sulfur and oxygen ions in equal proportion.

atmosphere. While up to two oxygen ions are supplied to the torus per sulfur ion in this picture, the resulting oxygen ion population in the torus is nearly equal to the sulfur ion population because charge exchange preferentially replaces oxygen ions by sulfur ions (Johnson and Strobel 1982; Brown et al. 1983). (See McGrath and Johnson [1988] for new cross sections.)

The above has primarily emphasized SO_2 sputtering and ejection. Recently Chrisey et al. (1987) measured the sputtering of sulfur in the ion-energy region of interest for Io, and Torrisi et al. (1986) gave yields for high-energy ions. The sputter yields at low temperature for corotating ions are significant ($\sim$ 1/3 those of SO_2), and sulfur sputters predominantly as S_2, which has a mass identical to SO_2. The escape fraction for sputtered sulfur is less than that for SO_2 sputtering, because of softer energy spectra. With a 50/50 coverage of SO_2 and sulfur, Chrisey et al. found that ejected sulfur contributed about 10% of the total neutral mass loss from Io. They also showed that sulfur changed color after the bombardment, with the color change reversible by annealing. As the color of sulfur surfaces reflects the thermal history (Nash

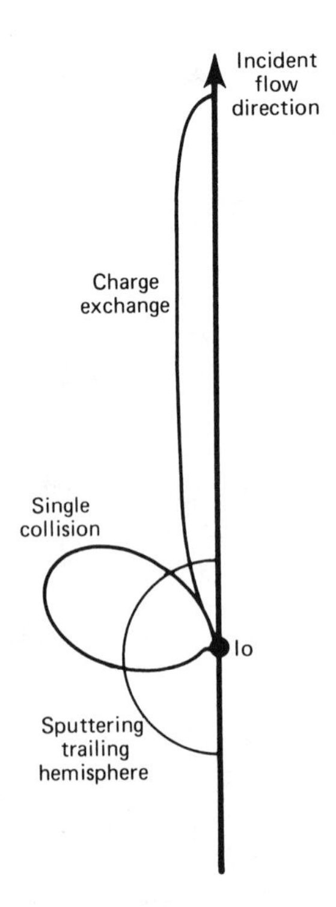

Fig. 7. Schematic of energy and angular distributions of ejecta from charge exchange, single collision and sputtering.

1987), the polar caps on Io are likely to be indicative of an ion-bombarded surface which has not been annealed.

Sodium Cloud

Aspects of the above picture of the plasma interaction are nicely confirmed by the observations of neutral Na near Io. The ionization rates of escaping S and O are low, so that the neutral oxygen and sulfur clouds extend far from Io (Brown 1981; Smyth and Shemansky 1983). In contrast, the Na ionization rate is high, so that the Na cloud extends only partially around Io's orbit.

Three components are conventionally identified in the sodium cloud

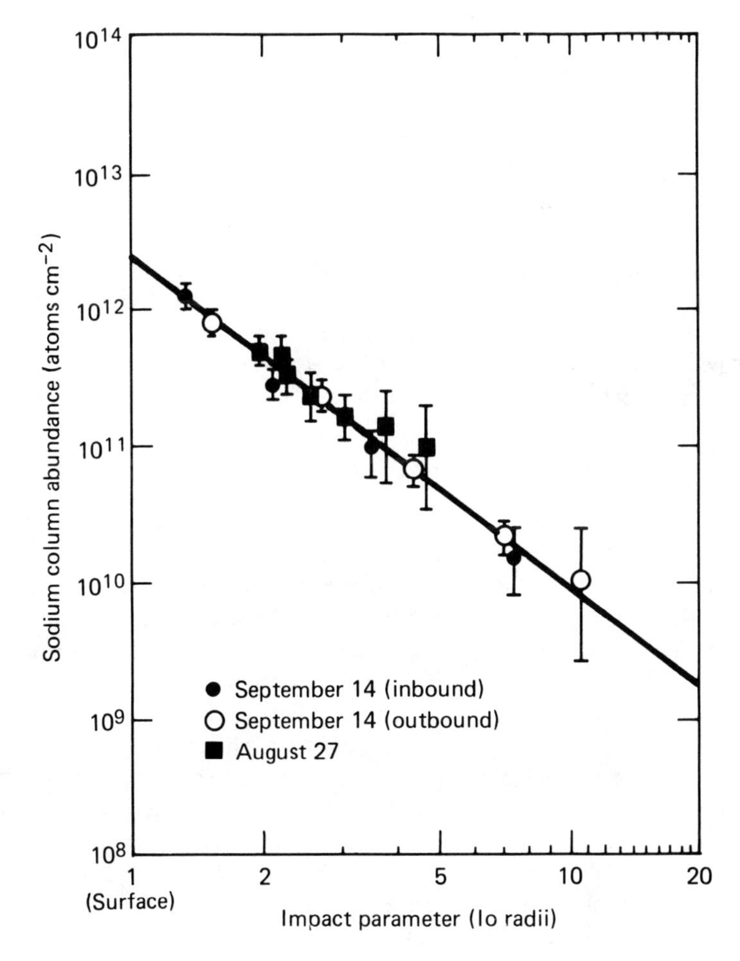

Fig. 8. Observed radial structure of Io's sodium atmosphere, from Europa occulation (Schneider et al 1987).

(Brown and Yung 1976): (1) the so-called *A* cloud very close to Io's visible disk and containing sodium gravitationally bound to Io; (2) the *B* cloud extending 5 to 15 arcsec from Io, containing sodium escaping from Io; and (3) the extremely tenuous *C* cloud, consisting of much faster sodium permeating the inner Jovian magnetosphere.

Pilcher et al. (1984) first correlated *B*- and *C*-cloud features with the sputter energy distributions and single collision energy distributions calculated by Sieveka and Johnson (1984; see Fig. 7). Trauger (1985) Doppler analyzed the very energetic *C*-cloud species and identified very fast charge exchange Na atoms. This is a very weak signal as the Na^+ ions in the torus only charge exchange effectively with atomic Na (McGrath and Johnson 1987). As neutral Na may be present in the corona at the few % level, this process is much less probable than the single collision ejection by the more numerous sulfur and oxygen ions.

Recently, Schneider et al. (1987) observed the neutral Na within ~ 0.5 R_{Io} of the surface using Io to occult Europa. The deduced atmospheric density profile (see Fig. 8) is consistent with the sputter corona of McGrath and Johnson (1987) if the electron impact ionization lifetime within ~ 10 R_{Io} is $\gtrsim 2 \times 10^4$ s (McGrath and Johnson 1987*a*). Assuming a sodium abundance of the order of a few % places the exobase below ~ 0.5 R_{Io} above the surface. Comparison of such measurements with detailed models (Smyth 1983, 1986) is very exciting, as are the recent extreme ultraviolet observations of Ballester et al. (1987) in which emissions from excited O and S were observed close to Io. These S and O emissions can be due in large part to locally electron-excited atoms from electron impact dissociation of SO_2.

V. ICY SATELLITES OF JUPITER

The icy satellites of the outer planets are exposed to radiations which can produce atmospheres. In the case of the satellites of Saturn and Uranus, the gas phase species produced by sputtering readily escape the satellite's gravitational field. This is not the case for Jupiter's icy satellites.

The only surface species clearly identified on these satellites is water ice (see, e.g., Clark et al. 1986; McKinnon and Parmentier 1986; Malin and Pieri 1986). The purity of the water ice, at least at the optical surface, is very high. If intimately mixed with the water ice, any dark contaminant cannot have a globally averaged abundance of more than a few wt. % at Europa or more than 10 wt. % at Ganymede. The water ice on Callisto may be less pure. There is also strong evidence, from water ice infrared band depths, for a fine-grained ice surface, with the grain size affected by charged particle sputtering. Magnetospheric plasma bombardment can preferentially destroy the smaller grains and thereby reduce the visual reflectance and increase the infrared band depths. This process appears to be important for the visible albedos and infrared band depths of the leading and trailing hemispheres of Europa, Ganymede and Callisto (Clark et al. 1983). The trailing hemispheres of Europa and Ganymede are darker and have deeper bands than the leading hemisphere, with the hemispheric asymmetry less pronounced for Callisto. In fact, Callisto's leading hemisphere is actually darker in the blue and near ultraviolet. The average visible brightness also decreases from Europa to Callisto. These trends are consistent with the expectation that sputtering effects should be more important for Europa, less so for Ganymede, and still less for Callisto. The bright polar regions of Ganymede actually appear to be covered by frost (Squyres 1980).

In addition to alteration of grain sizes, plasma bombardment leads to implantation of magnetospheric ions as well as dissociation and chemical modifications of the surface. Implantation of Jovian magnetospheric sulfur ions has been observed on the trailing hemisphere of Europa (Lane et al. 1981). Other exogenic processes like meteoroid bombardment or endogenic processes can also produce dark contaminants in the water ice surfaces of the icy Galilean satellites or create dark, ice-free regions (McKinnon and Parmentier 1986; Malin and Pieri 1986; Spencer 1987). Meteoroid bombardment may play a role in darkening the leading hemisphere of Callisto. Callisto's surface is heavily cratered with little or no evidence for resurfacing, while there is considerable evidence for resurfacing and/or internal activity on Ganymede, and still more for Europa (Crawford and Stevenson 1987). Little is known concerning the nature of the dark material on Ganymede and Callisto, except that it is spectrally red.

Initially, there was some question whether or not the satellites had magnetic fields that would deflect the incident plasma. However, Lane et al. (1981) discovered ultraviolet spectral features attributed to S-O bonds from sulfur ion implantation in Europa's surface. Johnson et al. (1988*a*) recently showed a direct correlation of the longitudinal dependence of the absorption seen through the 0.35 μm filter on Voyager with longitudinal dependences of the flux of the magnetospheric ions bombarding Europa. Recent laboratory measurements (O'Shaughnessy et al. 1988) have confirmed that plasma bombardment alters reflectance spectra, although the Europa spectra have yet to be reproduced in the laboratory. In addition, Clark et al. (1985) have interpreted the differences in the band depths of the 1.04 μm feature between the leading and trailing hemispheres as being due to ion bombardment. Therefore, it is clear that ions do strike the surface of this object, implying both the absence of a significant magnetic field and average atmospheric column densities less than the mean penetration depth of the ions. For the principal plasma ions at Europa, this implies column densities less than $\sim 10^{16}$ to 10^{17} H_2O cm^{-2}. Ions above 10 keV can penetrate even greater column densities, and sputtering by energetic ions yields a lower limit to the atmosphere (Brown et al. 1982; Haff et al. 1981; Johnson et al. 1983). Absorption features suggestive of ion bombardment are seen at Ganymede, suggesting that it too has a tenuous atmosphere. Voyager ultraviolet occultation measurements (Broadfoot et al.

1981), set an upper limit of 1.5×10^9 cm^{-3} at the surface of Ganymede for the densities of H_2O or O_2. At Callisto the plasma flux is low and the surface exhibits considerably more dark material which is not associated with plasma bombardment.

Because ions reach the surface, sputtering occurs. Therefore, gas-phase species are produced, creating a water-product neutral corona. The plasma flux onto Europa is such that corotating ions bombard the trailing hemisphere almost exclusively, but highly energetic ions can reach the leading hemisphere. On all of the satellites the sputter yield depends on temperature for temperatures above ~ 100 K. Therefore, the coronal densities are enhanced when the trailing hemisphere faces the Sun. In Fig. 9 are shown the expected sputtering rates compared to sublimation rates, assuming the plasma to be all H^+ or all O^+ (Johnson et al. 1981, 1983; Cheng et al. 1986). In fact, at Europa both S^+ and O^+ are principal components. Since the surface is bright

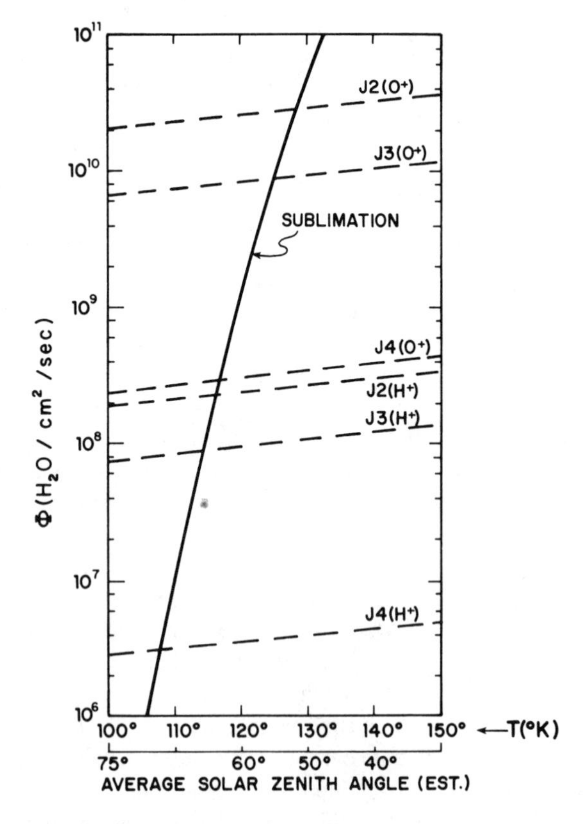

Fig. 9. Sublimation vs sputtering at icy Galilean satellites of Jupiter, assuming incident ions to be protons (curves labeled H^+) or oxygen ions (labeled 0^+) (figure from Johnson et al. 1982 and Cheng et al. 1986).

and the effective temperature is low, the sputter corona is dominant over most of the bombarded disk. At Callisto, the opposite is the case.

The sputtering of water ice also results in the production of H_2 and O_2. This H_2 readily escapes as it would in a sublimation water atmosphere (Yung and McElroy 1977). O_2 cannot escape and, in fact, does not condense over most of the surface as efficiently as H_2O. Therefore, the atmosphere produced by sputtering of water becomes dominated by O_2. Voyager ultraviolet observations set an upper limit of $< 2 \times 10^{15}$ cm^{-2} for the column density of gaseous O_2 at Ganymede (Broadfoot et al. 1981). This O_2 will eventually condense out at the poles. It is then modified by incident ions and recycled into the surface. Therefore, a much larger column density, possibly of the order of 10^{16} O_2 cm^{-2}, may be condensed on the surfaces of Ganymede and Europa (Johnson et al. 1982).

The presence of sputter coronae due to plasma bombardment results in the transport of water products over the surfaces of the satellites (Sieveka and Johnson 1982) and in the production of a water product plasma torus by the same means as described for Io. Energetic corona particles directly escape or are removed from the corona by collisions and charge exchange. A plasma source of the order of 10^{27} amu s^{-1} is expected from both Europa and Ganymede. Cook et al. (1987) have suggested the possibility of an active surface on Europa. If present, an active surface would enhance the associated plasma source. The ion bombardment, which leads to a light scattering surface (Johnson 1985*b*), and the transport and redeposition of material can alter the reflecting properties of the surface. This is possibly responsible for the brightness of Europa and the poles at Ganymede.

VI. SATURN

Several of Saturn's satellites have important interactions with the magnetosphere. Mimas, Enceladus, Tethys, Dione and Rhea are icy bodies with radii up to about half that of Earth's Moon, and all have substantial fractions of their surfaces covered by water ice. These icy satellites are all located in the inner magnetosphere within 3 to 9 Saturn radii (R_S). In addition, Saturn's moon Titan interacts strongly with the outer magnetosphere. Titan is larger than Mercury and has a dense atmosphere consisting mainly of nitrogen and methane plus other hydrocarbons. It is far enough from Saturn (20.2 R_S) that it occasionally finds itself outside Saturn's magnetosphere as the magnetospheric boundaries move in and out. Titan was within the magnetosphere during the Voyager 1 encounter and perhaps part of the Voyager 2 encounter (Neubauer et al. 1984).

The interaction of the icy satellites with the magnetosphere can be summarized by Fig. 4. Charged particle sputtering and meteoroid bombardment eject water products into the magnetosphere, forming vast neutral water-product clouds. The clouds are ionized and dissociated to maintain an oxygen-ion heavy plasma torus that was observed both by Pioneer and by the Voyager spacecraft (Frank et al. 1980; Sittler et al. 1983; Lazarus and McNutt 1983). While the Voyager plasma instrument was initially unable to identify the heavy ion species, subsequent modeling work using the full response function of the instrument yielded strong evidence in favor of O^+ rather than N^+ as the dominant heavy species in the inner magnetosphere (Richardson 1986). The O^+ composition favors the water-product clouds, rather than Titan, as the source of Saturn's inner plasma torus. The ion densities show a broad peak between the orbits of Tethys (4.9 R_S) and Dione (6.3 R_S), and the maximum density is about 50 ions cm^{-3}. Electron temperatures there are variable and range from < 10 to > 100 eV.

The presence of a neutral water-product cloud in the inner magnetosphere has been inferred from Voyager low energy charged particles (LECP) observations of charge-exchange neutrals from Saturn (Kirsch et al. 1981; Cheng 1986*a*). These observations are consistent with charge exchanges in a water-product cloud extending from 3 to 11 R_S, with a total thickness of 1.5 R_S and an average density of 20 cm^{-3}, plus charge exchanges in the atomic hydrogen cloud of the outer magnetosphere (the Titan torus). Alternatively, the LECP energetic neutral observations can be explained without such a water-product density if molecular hydrogen at a density of $\sim$ 10 to 20 cm^{-3} is present together with atomic hydrogen in the Titan torus (Ip 1984). However, H_2^+ at thermal energies has not been detected in the outer magnetosphere. The H_2 density is probably less than that of atomic H in the Titan torus (Eviatar and Podolak 1983; Cheng and Krimigis 1988). No electromagnetic emissions have been observed from either a water-product neutral cloud or a molecular hydrogen cloud at wavelengths from the infrared to the ultraviolet. Likewise, no optical or ultraviolet emissions from ions in the plasma torus have been observed.

The lack of independent information from optical and ultraviolet spectroscopy of Saturn's water-product cloud and inner plasma torus has greatly hindered our understanding of those objects. Detailed models (see, e.g., Richardson et al. 1986) are poorly constrained by observations. In contrast with the situation at Jupiter, the importance of interactions between Saturn's magnetosphere and its icy satellites for the magnetospheric mass and energy budgets is not yet clear.

The interaction between Titan and Saturn's magnetosphere resembles that of Venus with the solar wind. Titan, like Venus, does not have a strong intrinsic magnetic field, so the corotating plasma of Saturn's magnetosphere impacts directly on Titan's dense, collisional atmosphere. A long tail of protons and heavy ions (mainly nitrogen ions) is extracted and carried downstream, analogous to the ion tail of comets; the resulting loss rate of heavy ions from Titan is about 10^{24} s^{-1}. The ion tail may wrap several times around Saturn (Eviatar et al. 1982); for a contrary view, see Goertz (1983). In addition, about 3×10^{27} hydrogen atoms and molecules and about 3×10^{26}

nitrogen atoms escape from Titan each second. For recent reviews of Titan's atmosphere and interactions with Saturn's magnetosphere, see Hunten et al. (1984) and Neubauer et al. (1984).

Hydrogen atoms and molecules and nitrogen atoms escape into the magnetosphere and form the Titan torus, a vast neutral gas cloud orbiting Saturn. The existence of an atomic hydrogen torus around Saturn supplied by Titan was actually predicted well before the Pioneer and Voyager encounters (McDonough and Brice 1973). The Voyager ultraviolet spectrometer observed that the atomic hydrogen cloud of the Titan torus extended radially from 8 to 25 R_S with an average density of 10 to 20 cm^{-3} (Broadfoot et al. 1981; Sandel et al. 1982). Escape of atomic N from Titan is predicted to maintain a somewhat smaller average atomic N density in the Titan torus of ~ 0.4 cm^{-3} (Eviatar and Podolak 1983). Shemansky and Smith (1983) have suggested that Saturn's atmosphere is the dominant source of H for the Titan torus, but Richardson and Eviatar (1988) have criticized this idea on the grounds that a high H density in the Dione-Tethys-Rhea region would be incompatible with the observed plasma there. Ionization of neutrals in the Titan torus is clearly an important plasma source for the magnetosphere, but may not be dominant. The Titan torus also differs from the Io torus in that the escape of neutrals from Titan is not driven primarily by the impact of magnetospheric ions.

Saturn's magnetosphere interacts strongly with the rings as well as the satellites. Of particular interest here is the E ring that extends from 3 to 8 R_S with peak density near the orbit of Enceladus at 4 R_S (Pang et al. 1984; Mendis et al. 1984). The E ring therefore lies within the inner heavy ion plasma torus. It is detectable by groundbased telescopes but only when the rings are edge-on to the Earth. It consists of 2 μm diameter grains, probably water ice, but its total surface area is small compared to that of the satellites. Sputtering of the E ring is therefore not an important source for the water cloud in the inner magnetosphere.

Atmosphere of Saturn's Satellites

The inner satellites of Saturn have predominantly water ice surfaces identified both by the reflectance spectra (see, e.g., Clark et al. 1984, 1986) and the density (Morrison et al. 1986). These objects exhibit leading/trailing hemispherical differences like the icy Galilean satellites again indicative of some exogenically induced darkening agent. In this region of the solar system, other volatiles (such as NH_3) might be expected to condense as well as water ice and it has been suggested that volcanism could have occurred even on these small objects (Stevenson 1982*c*). In fact, the surface of Enceladus appears to be covered uniformly by a fresh frost that may be due to impacts or volcanism. In addition to cratering, leading/trailing differences in ice grain size attributed to plasma bombardment have been observed (see, e.g., Clark et al. 1986). Such bombardment would also liberate any more volatile species from the ice grains in relatively short times (Lanzerotti et al. 1984), making their presence difficult to detect. Therefore, in the following we presume water ice to be the only important surface species.

The icy inner satellites of Saturn are small so that plasma bombardment results in direct gravitational escape of most of the sputtered species. Hence these satellites act to supply water products to the Saturnian magnetosphere. Rates of sputter ejection are given in Table II (Cheng et al. 1986; Johnson et al. 1988*b*). At the temperature of these satellites the predominant ejected species is H_2O (Brown et al. 1984). As these species orbit Saturn they may be dissociated by photons or plasma electrons. This process results in the production, primarily, of a fast H atom, while the orbit of the heavier OH is changed very little by this ejection. Dissociation of OH leaves an orbiting O. Eventually these species are ionized by photons, plasma electrons and charge transfer. Once ionized they are swept up by the rotating magnetic field and contribute to the local plasma.

Prior to being ionized, the neutral H_2O, OH and O orbit Saturn, forming a neutral water-product exosphere for Saturn. Because the ionization lifetimes are longer than orbital periods, this exosphere is in the form of a torus centered on the satellite orbital plane, which also closely coincides with the equatorial plane of the magnetic field. The water-product torus so produced is shown in Fig. 10. This is calculated using the local ionization rates (Johnson et al. 1988*b*). The neutral cloud is seen to be peaked in density about the satellite sources and is an atmosphere of enormous proportions. Its column density is nevertheless too low to be seen from the Earth. However, it exhibits

TABLE II
Satellite Sputter Fluxes of H_2O by Incident O^+ and Net Satellite Rates[a]

	Pioneer[a]	Voyager I[a] cold	Voyager I[a] hot	Voyager II[a] cold	Voyager II[a] hot	Source Rates J_S (molecules s^{-1})
	Fluxes[b] (10^8 cm^{-2} s^{-1})					
Mimas	0	0		0		0
Enceladus	0.03	0		0	1.0[b]	8×10^{23}
Tethys	0.7	0.3	4[b]	0.6	0.6	1.6×10^{25}
Dione	2[b]	1.3	5[b]	1.3	0.6	3.4×10^{25}
Rhea	0.07	0.5	2[b]	1.3[b]	0.2	2.5×10^{25}
Total						7.6×10^{25}

[a]Pioneer data (Frank et al. 1980); Voyager I and II hot plasma (Lanzerotti et al. 1982); this plasma data was not extrapolated to the plane of satellite orbits, but for each satellite, erosion rate measurements closest to this plane were used; hot plasma was assumed predominantly 0^+; cold plasma obtained using ion densities from Sittler et al. model of Voyager I outbound and Voyager II inbound/outbound average.
[b]10^8 H_2O cm^{-2} s^{-1} = μm/10^3 yr.

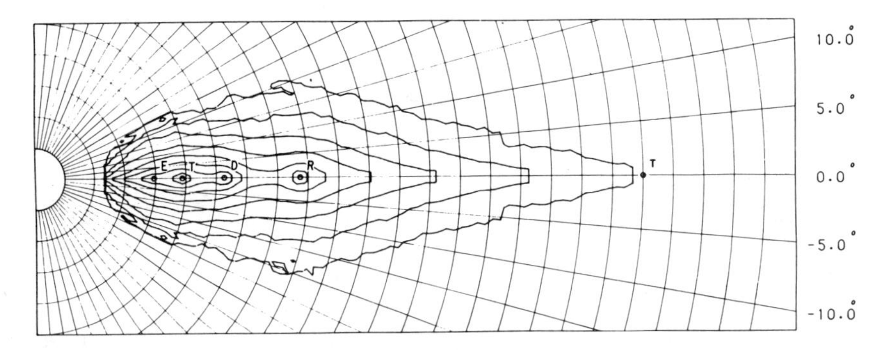

Fig. 10. Model of water-product neutral cloud at Saturn. Outer large contour is at a density of 10^{-2} cm^{-3}, and two contours are given per decade, spaced by $10^{0.5}$ (Johnson et al. 1988*b*). Isotropic ejection of neutrals from the satellites is assumed.

other observable effects. First, it is the source of the heavy ion plasma in the satellite orbital plane and, therefore, it can be thought of as having an ionosphere. Second, fast radiation belt ions in the region can undergo electron capture (charge-exchange) reactions and become fast neutrals that escape from the magnetosphere. Such charge-exchange neutrals were observed by Voyager (Kirsch et al. 1981; Ip 1984; Cheng 1986*a*). Finally, it coexists with the tenuous E ring which may be a precipitate of the water-product exosphere (see below).

The plasma source from ionization of the water-product torus is very nonuniform. It is seen from Fig. 11 that the Pioneer and Voyager measurements of the plasma density generally exhibit the structure observed due to the plasma source except at Rhea. Therefore, the plasma lifetime is relatively independent of radius. Similarly, the scale height of the atmosphere reasonably describes the scale height of the plasma. The main plasma loss processes are radial transport, charge exchange, and, in the inner region, electron-ion dissociative recombination (Eviatar 1984; Richardson et al. 1986).

When the relative collision velocities between ions and neutrals are low, ion-neutral orbiting collisions can occur. After such collisions the neutrals have lower average velocities than they would for charge exchange, in which the fresh neutral has the ion's initial speed. In the same region of the magnetosphere, the electrons are cool so that ion-neutral recombination becomes feasible. These processes can, in principle, result in accumulation of large densities of neutrals (Johnson et al. 1988*b*). This is not described in the calculation of neutral densities in Fig. 10.

One of the interesting features of the region occupied by Saturn's icy satellites is the presence of the tenuous E ring. This is a ring of ~ 2 μm size particles with peak density in the vicinity of Enceladus extending out to ~ 8 R_S. The ring also increases in its thickness above and below the satellite

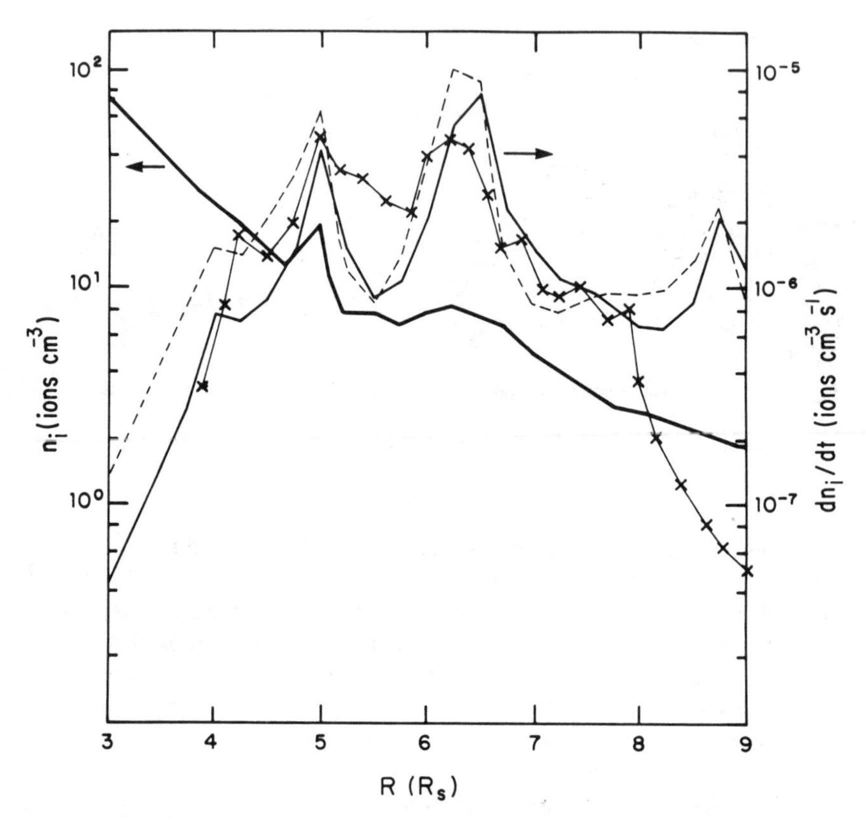

Fig. 11. Observed heavy ion density in Saturn's inner magnetosphere compared with ionization source from water-product neutral cloud (Johnson et al. 1988*b*). Heavy solid line (referred to left-hand axis) gives equatorial plane heavy-ion density from Voyager 2 data as extrapolated by Sittler. Solid line with crosses (left axis) gives the same from Pioneer 11 data. Light solid line (referred to right-hand axis) gives the calculated ion source for isotropic ejection of neutrals. Dashed line (right axis) gives the ion source for trailing hemisphere ejection.

orbital plane with increasing distance from Enceladus, roughly resembling the neutral cloud distribution in this region. It has been proposed that this ring is the product of particle ejection from Enceladus (Haff et al. 1983; Pang et al. 1984) or that it is a steady precipitate of the neutral cloud (Cheng et al. 1986). The difficulty in understanding the E ring is that lifetimes against destruction and removal of the E ring grains by plasma interactions are less than 10^4 yr (Cheng et al. 1982; Morfill et al. 1983; Haff et al. 1983; Johnson et al. 1984*a*).

The E ring particles are eroded by plasma ions but also accrete neutrals from the neutral cloud. In the vicinity of Mimas, the ion velocities are low enough that they do not sputter but also accrete onto ring particles. The relative rates of the ion erosion and material accretion are difficult to determine, since the charge of the small grains and their fluffiness are not well known. A porous structure can greatly reduce sputtering yields (Hapke 1986). Johnson et al. (1988*b*) recently examined these processes and found that circumstances could be imagined in which grain growth could occur in the Mimas/Enceladus region with grain destruction occurring at larger radii. Accretion of water clusters from the main rings of Saturn may also contribute to maintenance of the E ring (Ip 1986*b*). Alternatively, the E ring can be maintained by ejection of particulate material from Enceladus followed by gradual destruction and outward radial transport.

VII. URANUS

The first *in situ* observations of Uranus leading to discovery of the Uranus magnetosphere were made by Voyager 2 in January, 1986. There were many surprising results and their analysis is far from complete. Chief among the surprises are the very large tilt angle of 60° between the magnetic and rotation axes and the nearly pure proton composition of the magnetosphere (Ness et al. 1986; Bridge et al. 1986; Krimigis et al. 1986). Prior to encounter, it was predicted that darkening and polymerization of methane and other organic ices under particle irradiation may explain the low albedos of the Uranus rings and the dark matter on the satellites (Cheng and Lanzerotti 1978). A heavy ion plasma torus from ionization and dissociation of water products sputtered from the icy Uranian satellites was also predicted (Cheng 1984*b*; Eviatar and Richardson 1986).

However, no heavy ions were detected in the Uranian magnetosphere, except for a small abundance ($\sim 10^{-3}$) of H_2^+ at MeV energies (Krimigis et al. 1986). The charged particle fluxes were nevertheless large enough to blacken and polymerize organic ice surfaces in geologically short times (Krimigis et al. 1986; Lanzerotti et al. 1987). The blackening and polymerization are still expected even if the organic species are intimately mixed into a water ice matrix, as would be the case, for example, in a methane-water clathrate (Lanzerotti et al. 1987).

Two distinct neutral populations are maintained in the magnetosphere of Uranus. The first is an atomic hydrogen exosphere (also termed the hydrogen "corona" in the literature, by analogy with Earth's geocorona). The Uranian hydrogen corona has two components, a thermal component with an exospheric temperature of 750 K, plus a nonthermal component which extends beyond 5 R_U (Broadfoot et al. 1986). The second neutral population is a theoretically predicted heavy neutral torus consisting of water products sputtered from the icy Uranian satellites (Cheng 1984*b*; Eviatar and Richardson 1986). The Uranian hydrogen corona was detected in Lyman-α emissions by Voyager, although the detection of a nonthermal component is still preliminary. The heavy neutral torus has not been observed in any electromagnetic emissions, and no heavy ions were detected. The failure to observe a heavy ion torus at Uranus was initially surprising, since such tori are observed at Jupiter and Saturn. We discuss below why an observable heavy ion torus did not form at Uranus.

Satellite Surfaces

As with the Galilean and inner Saturnian satellites, water ice is the principal component of the surfaces of the Uranian satellites. Whereas the other icy satellites exhibit a darkening agent which absorbs more strongly at shorter wavelength (i.e. it is 'red'), the dark component of the Uranian satellites is spectrally flat (Johnson et al. 1987; Veverka et al. 1987). In addition, there are no clear trends on these satellites (such as leading/trailing brightness differences) that would indicate whether the principal darkening agent is exogenic in origin. This could be due, in part, to the large dipole tilt giving a more uniform bombardment profile for these objects. Therefore, there has been considerable discussion on the role of the local radiation in forming this observed dark material (see, e.g., Lanzerotti et al. 1987; Thompson et al. 1987). This discussion centers on whether or not other volatile species such as CH_4, CO, etc might be present (e.g., as an ice clathrate). Such species would be rapidly darkened and polymerized by the local plasma bombardment (Lanzerotti et al. 1987). Recent analysis of satellite densities, however, suggests that an organic material is mixed with the ice throughout the volume of the satellite (see, e.g., Johnson et al. 1987). If this is the case, plasma bombardment also preferentially removes the volatiles, enhancing the concentration of the dark matter.

Heavy Neutral Torus

Water products from the icy satellites of Uranus are ejected into the magnetosphere by several mechanisms, forming a heavy neutral torus in the shape of a thin disk confined near the satellite orbital plane. This plane is inclined 60° relative to the magnetic equatorial plane (Cheng 1987). (By "water products" is meant the total of H_2O, OH, O and O_2.) These heavy neutrals can be ejected from icy surfaces by micrometeoroid impact, photosputtering and charged particle sputtering. Of these ejection mechanisms, meteoroid impact and photosputtering appear to be the most important. Charged particle sputtering is much less important at Uranus than at Jupiter or Saturn, principally for three reasons: (1) the ion fluxes there are relatively low to begin with; (2) the ions are mainly protons which have much smaller sputtering yields than heavy ions; and (3) the large magnetic dipole tilt angle (see below). This tilt angle causes a severe dilution of the heavy ion source, because newly created ions are not confined within the thin disk of heavy neutrals but rapidly fill the magnetic flux tube on which they are injected.

Photosputtering at Uranus by the mechanism of Harrison and Schoen (1967) may yield an OH source estimated at 10^6 cm^{-2} s^{-1} (Eviatar and Richardson 1986). This estimate is highly uncertain owing to the lack of laboratory data. For comparison, the charged-particle sputtering rates are at most a few times 10^5 cm^{-2} s^{-1} at Miranda and less than 10^5 cm^{-2} s^{-1} at the remaining satellites (Lanzerotti et al. 1987). These sputtering rates were appropriately averaged over the satellite orbits to include effects of the large magnetic dipole tilt relative to the spin axis, which causes the satellites to spend much of the time at high magnetic latitudes where charged-particle fluxes are very small. Finally the micrometeoroid flux and the resulting rate of vapor production in the outer solar system are uncertain by orders of magnitude (see, e.g., Haff and Eviatar 1986), but the hypervelocity impact source of vapor may reach 10^7 cm^{-2} s^{-1} for water ice surfaces on the Uranian satellites.

While the combined water product source from micrometeoroid impact, photosputtering and charged-particle sputtering may reach 10^7 cm^{-2} s^{-1} for pure water ice surfaces, the relatively low albedo of the Uranian satellites indicates that most of their surfaces contain a dark, probably carbonaceous substance. A nominal water-product source of 10^6 cm^{-2} s^{-1} is therefore adopted as an average over the satellites' surfaces.

Table III shows expected properties of the Uranian heavy neutral torus (Cheng 1987), assuming a water-product source of 10^6 cm^{-2} s^{-1}. Each satellite is assumed to maintain a steady state, homogeneous heavy neutral torus. The loss mechanisms for heavy neutrals are electron impact ionization, charge exchange and photoionization. Electron impact and charge exchange are both important within the inner magnetosphere (radius <6 R$_U$) but photoionization becomes dominant in the outer magnetosphere. In this model, the total heavy neutral density within the inner magnetosphere becomes 0.3 cm^{-3} after summing over the contributions from Miranda, Ariel and Umbriel. The neutral densities are actually expected to be strongly concentrated toward the satellite orbits, as was the case for the Saturn heavy neutral tori.

TABLE III
Neutral Tori of Uranian Satellites

	R_s (km)	r_o (R$_U$)	r_+ (R$_U$)	r_- (R$_U$)	ΔZ (R$_U$)	n_o (cm^{-3})
Miranda	240	5.1	9.9	2.9	0.38	0.06
Ariel	585	7.5	17	3.8	0.68	0.11
Umbriel	595	10.4	29	4.7	1.1	0.12
Titania	795	17.0	73	6.1	2.3	0.02
Oberon	775	22.8	150	6.9	3.6	0.002

R_s = satellite radius; r_o = satellite orbit radius; r_+ = apoapsis and r_- periapsis for neutrals ejected at 1 km s^{-1} in the forward and backward directions, respectively; ΔZ = torus half thickness from orbital plane; n_o = heavy neutral density.

Hydrogen Corona

The source of heavy ions in the inner magnetosphere is assumed to be ionization of the heavy neutral torus. Voyager observations indicate a proton density near 5 R_U of roughly 1 cm^{-3}, while the heavy ion density is less than 0.01 cm^{-3} (McNutt et al. 1987). The heavy neutral torus model of Table III is consistent with these observations (Cheng 1987).

Two proton sources are considered for the inner magnetosphere, ionization of the atomic hydrogen corona and ionospheric protons driven by photoelectron escape. Only the nonthermal component of the hydrogen corona is relevant here, and its density is highly uncertain. Shemansky (1986) suggested a hydrogen density of $\sim$ 70 cm^{-3} near 5 R_U. The H density near 5 R_U cannot be much greater than 60 cm^{-3} without conflicting with Voyager LECP observations (Krimigis et al. 1988), but it could be much less. The Voyager LECP upper limit on the H density was obtained from the failure to observe energetic charge-exchange neutrals outside the Uranian magnetosphere. The photoelectron-driven proton source was scaled from Jupiter (Swartz et al. 1975) to yield a ionospheric proton source of 7×10^5 cm^{-2} s^{-1} at Uranus.

Since ions move within a given magnetic flux tube, the various ion sources must be compared for a given magnetic flux tube in the inner magnetosphere. The ionospheric photoelectron-driven source injects protons at the feet of the flux tube, while ionization of the corona injects them all along the flux tube. Heavy ions are injected only within the volume occupied by the heavy neutral torus, which is only a small fraction of the total flux tube volume, because of the thinness of the neutral disk and its 60° tilt from the magnetic equator. This geometric effect causes the heavy ion source to be greatly diluted relative to the proton source. The two proton sources turn out to be roughly comparable ($\sim 2 \times 10^{-7}$ $cm^{-3}s^{-1}$ near 5 R_U) for a corona density of 70 cm^{-3} and can maintain the observed proton density of $\sim$ 1 cm^{-3} for a plasma transport loss time of $\sim$ 30 days. Even if the nonthermal coronal density near 5 R_U turns out to be $<<$ 70 cm^{-3}, the total proton source would be reduced only by a factor $\sim$ 2, and the model would not be greatly affected.

The question of how to maintain a nonthermal H corona is one of great interest. One possibility is that it may be maintained as a consequence of the same processes that excite the electroglow (Broadfoot et al. 1986). The electroglow consists of nonauroral hydrogen emissions detected from the dayside upper atmospheres of Uranus, Jupiter and Saturn. The electroglow at wavelengths less than 1110 Å is directly excited by solar radiation, as its brightness scales accurately as the inverse square of distance from the Sun, and it may be explained by fluorescence of solar ultraviolet (Yelle et al. 1987). The situation at longer wavelengths is controversial; it may also be solar excited directly, or it may involve soft electron excitation and magnetosphere-ionosphere interactions. Ip (1987*b*) has suggested that bombardment of the upper atmosphere by charge-exchange neutrals above 30 keV may contribute to the excitation of electroglow. However, this idea suffers from the difficulty that the ultraviolet emissions excited by energetic neutral bombardment will be very similar to those excited by energetic proton precipitation, and the resulting emitted spectrum would be incompatible with electroglow.

VIII. CONCLUSIONS

Magnetosphere interactions are important for the origin and evolution of tenuous atmospheres. We have seen how magnetosphere interactions affect the sources, losses, thermal structure and composition of atmospheres at Mercury, Io and the icy satellites of Jupiter, Saturn and Uranus. We have restricted the scope of the present discussion to these particular cases of important magnetosphere interactions, for which significant advances in our understanding have recently occurred. Of course, magnetosphere interactions also directly affect the ionosphere and exosphere of substantial (as opposed to tenuous) atmospheres. Chamberlain and Hunten (1987), for example, have reviewed effects of magnetosphere-induced losses over cosmogonic time scales on the atmospheric evolution of terrestrial planets. The upper atmospheres and ionospheres of the outer planets have been reviewed by Atreya (1986).

The recent discovery of a sodium atmosphere at Mercury has transformed our understanding of its magnetosphere, whose mass and energy budgets may be dominated by sodium ions. Groundbased observations of the sodium atmosphere may yield important new insights into the interactions among the surface, sodium atmosphere and magnetosphere of Mercury. Many uncertainties remain, however, concerning the hydrogen and helium exospheres of Mercury and their interactions with Mercury's surface.

Our understanding of Io's atmosphere has advanced rapidly, and direct optical and ultraviolet observations of the atmosphere have been made. More detailed models of Io's atmosphere should include formation of a sputter corona and deflection of the incident plasma flow above the surface by the conducting ionosphere. Io's atmosphere is collisionally thick over part, but probably not all, of Io's surface.

The hydrogen and water-product neutral clouds of Saturn and their relationships to Saturn's icy satellites and rings also remain controversial. Is the E ring a condensate from the water-product cloud? What is the role of micrometeoroid impacts on the main rings and icy satellites? How does the water product cloud compare with the Titan torus in terms of dynamical consequences for the magnetosphere?

The first *in situ* observations of the Uranian magnetosphere revealed it to be totally dominated by protons, with no trace of the predicted heavy ions. An unobservably small heavy-ion density can be understood in the context of a simple model for the water-product neutral cloud maintained by the Uranian satellites, taking into account the large tilt angle between the magnetic dipole and rotation axes. Many questions, nevertheless, remain open; for example, what is the nature of the electroglow? What is the structure of the nonthermal hydrogen corona, and how is it maintained?

Acknowledgments. A. F. Cheng acknowledges support under a NASA contract between the Johns Hopkins University and the Department of the Navy. R. E. Johnson acknowledges support by NASA's Planetary Atmospheres Division.

FORMATION OF THE SATELLITES OF THE OUTER SOLAR SYSTEM: SOURCES OF THEIR ATMOSPHERES

A. Coradini, P. Cerroni, G. Magni
Istituto di Astrofisica Spaziale

and

C. Federico
Università degli Studi, Perugia

We address here the origin of regular satellite systems, highlighting the key processes leading to their present configuration. When atmospheres are present on these bodies, they are probably connected with the final phases of satellite formation, although thermodynamic conditions prevailing in the early stages of formation may have been crucial in determining the amount of volatiles present in the forming body. There appear to have been four major stages of formation. (1) The disk phase *linked the formation of the primary body to that of the satellites. During this phase the conditions leading to the observed differences among different satellite systems were probably established. (2) The formation* phase of intermediate-sized bodies *probably occurred next. In this phase, the mass infall onto the satellite disk stopped, the disk relaxed and turbulence decayed. Also, in this phase the opacity probably decreased because of the increased mean grains' size. During this period, grains separated from the gas phase, resulting in a two-layer configuration, which may have become unstable. As a result of such instability, local mass concentrations may have formed planetesimal bodies. (3) After phase (2), the* collisional evolution *of planetesimals probably resulted in larger bodies (satellite embryos) capable of gravitationally capturing smaller-sized bodies. (4) Finally, a series of subsequent evolutionary phases link the* presently observed state *of satellite systems to the primordial phases described above. The key to understanding the formation of satellite atmospheres are the internal composition and the thermal histo-*

ry of the satellites. The thermal history of the satellites is not discussed in detail here, but rather the potential sources of an atmosphere are identified. Such sources are discussed in relation to the composition, the primordial thermal evolution and the internal structure of the satellites.

I. INTRODUCTION

Regular satellite systems around giant planets closely resemble the planetary system, as underlined by many authors (see, e.g., Black 1971; Alfvén and Arrhenius 1976; Coradini et al. 1981*b*; Safronov and Ruskol 1982). On the other hand, satellite systems also present some interesting distinctive features, most of which can be justified by their not having originated as isolated systems. Thus, for example, both the gravitational and the radiation field in which satellite disks formed depended on the combined effects of the presence of the Sun and the protoplanet. As a consequence, the gravitational field was anisotropic and weakly dependent on time, while the radiation field depended on the luminosity of the Sun and the protoplanet, both of which are known to have varied during the evolution of these bodies. In particular, extreme ultraviolet (EUV) and ultraviolet radiation from the Sun could have been much stronger in the past, while the contribution to the radiation field from the protoplanet was strongly dependent on its evolutionary phase (Bodenheimer and Pollack 1986), and might have been extremely important when satellite disks were not dominated by viscous dissipation processes (Coradini and Magni 1984).

Furthermore, protosatellite disks were embedded in the solar nebula, the density and composition of which strongly affected the absorbtion of ultraviolet radiation. In particular, gradients in the concentrations of H_2 and water vapor molecules may have caused significant differences among satellite disks.

Differences between protoplanetary and satellite disks could depend also on the amount of solid material present, and on its mass distribution. In fact, grains in protoplanetary disks were probably similar in composition and mass to those of the interstellar medium; in satellite disks, the population of solid bodies had probably evolved since it was composed both of small-sized grains and of fragments of the planetesimals which had survived the formation of giant planet cores.

A full understanding of the formation process of regular satellite systems around giant planets cannot be reached without a detailed study of the formation and evolution of the central body. It is therefore interesting to give here a brief resumè of the present state of knowledge in this field.

The formation and evolution of giant planets has been modeled in the framework of two different scenarios. In one model, giant protoplanets are formed by gravitational instabilities in the solar nebula (Cameron 1978;

Bodenheimer et al. 1980*a*), and are characterized by solar chemical composition and masses probably larger than the present masses of the giant planets. The solid core could be formed by sedimentation of solid material to the center of the structure, or by capture of solid planetesimals (Pollack et al. 1979).

In another model, the cores are formed first, through an accumulation mechanism similar to that generally accepted for the formation of the terrestrial planets (Safronov 1969; Safronov and Ruskol 1982). As the core grows larger, more and more nebular gas is concentrated in its sphere of influence, until an envelope large and massive enough is formed to start a rapid contraction phase, or a collapse onto the core. The critical core mass to trigger the instability is a crucial parameter in these models; its values have been estimated starting from static equilibrium models of core/envelope configurations and successively increasing the core masses up to values beyond which no purely static solution is possible (Perri and Cameron 1974). According to these authors, the major drawback of this model is to predict critical core masses of 70 $M_{\oplus}$, which are not in agreement with the actual values of the cores of the giant planets. This result is mainly due to the assumed purely adiabatic structure throughout the envelope (Harris 1978).

In a more recent model by Mizuno et al. (1980), the evaluation of the temperature structure during planet evolution leads to critical core masses not too far from the deduced giant-planet cores (15 to 25 $M_{\oplus}$). [It should be mentioned that Stevenson (1984*a*) has found, according to his own model, a substantially lower critical core mass (0.1 $M_{\oplus}$).] In such a scenario, the giant planets are produced by the subsequent accretion of "superganymedean puffballs"; this idea is intriguing, and deserves further study.

One of the major achievements of Mizuno's model is to explain in a self-consistent way the cores of the giant planets being comparable in mass, despite formation at different distances in the primordial solar nebula. Mizuno (1980) claims that the gas collapsing onto the core could have been depleted in grains by the formation process of giant-planet cores, which should have contained most of the solid material originally present in the zone. He also assumes that solid material in the form of planetesimals should not have contributed largely to the gas opacity (see Fig. 1).

More recently, the accretionary growth of the core has been modeled in a self-consistent way by Bodenheimer and Pollack (1986), who attempted to solve the problems left open by Mizuno (1980) and Safronov and Ruskol (1982) in their models of the early stages of formation of the Jovian planets. Their effort was addressed in particular to an explanation of the long accretion time and to the requirement of high opacity in order to obtain the proper values of core masses. According to their results, the accretion of both core and envelope terminates, and the planet contracts and cools to its present state, on a time scale of 5 Gyr. The problem of the opacity is addressed particularly in a companion paper by Pollack et al. (1986*a*), where it is con-

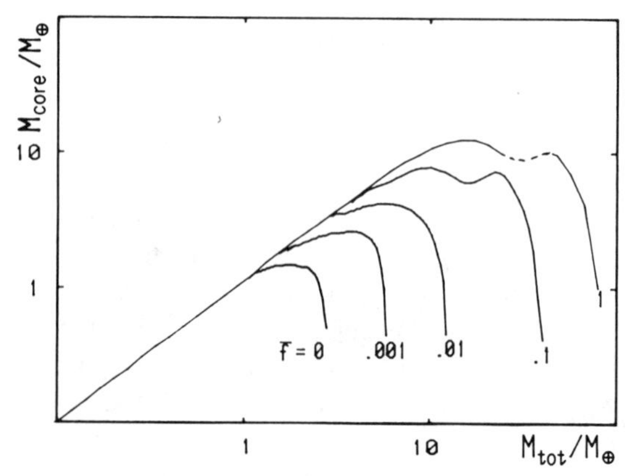

Fig. 1. Core mass as a function of total mass (core plus envelope) for conditions of hydrostatic equilibrium. The parameter $\bar{f}$ denotes the ratio of assumed grain opacity to that expected in a cold region of the solar nebula. Note that no equilibrium solution exists for core masses exceeding certain critical values (figure after Mizuno 1980).

cluded that planetesimals in the feeding zone of the giant planets could penetrate into the satellite disk in different stages of the formation process of the planet, possibly contributing with their dissolution to the gas opacity. This process could also justify opacity values high enough to allow the formation of a core of the correct dimensions to trigger the gas instability leading to the formation of the giant planets and affect the thermodynamical processes of the accretion disk.

As a consequence, some of the properties of satellite disks, such as, for example, the grain opacity and the gas density of the protosolar nebula at the time of satellite formation, depend on the formation process of the central body, although these disks have only been formed at the end of this process.

Once the accretion on the disk ends, and the central body is formed, the satellite disk cools and relaxes. During this process, the formation of the regular satellites and possibly the capture of the irregular ones takes place. It is clear therefore that the initial phases are crucial in determining the early characteristics of the regular satellites as well as their further evolution. For example, the pressures and temperatures of the disk will permit the condensation of ices with high-volatile content. These volatile-rich ices can be absorbed by the forming satellites and their percentage will affect greatly the rheology of the substances constituting the satellites and their thermal history.

The formation of an atmosphere around a satellite, which may originate by degassing of the satellite, by nebular gas capture or by bombardment of the satellite's surface by external impactors, is also related to these primordial phases of satellite evolution. For all these reasons, self-consistent models of the primitive phases of disk formation are important to the study and understanding of the formation of atmospheres in satellites.

II. THE DISK PHASE

The origin of the solar system is generally believed to involve a disk of gas and dust swirling around the proto-Sun and formed almost contemporarily with it. Thus, in recent years, the hydrodynamics of the growth of protostellar/protosolar disks from the collapse of a rotating gas cloud has been extensively studied (Cameron 1978; Lin and Papaloizou 1980; Lin 1981; Lin and Bodenheimer 1982; Cassen and Pettibone 1976; Cassen and Moosman 1981; Cassen and Summers 1983). A complete solution of this problem though, requiring a three-dimensional hydrodynamical scheme, presents overwhelming difficulties, and it has not been obtained hitherto. Two different approaches are usually followed: the first is to study analytically the process, introducing simplifying assumptions in outlining its basic physics; the second is to develop appropriate numerical models.

A. Analytical Models

In what follows, disk models by Cassen and Moosman (1981) and Cassen and Summers (1983), modified to account for the peculiarities of satellite disks, will be used to describe an accretion disk forming around a giant planet in the final stages of the primary formation process. In this framework, a symmetrical disk with respect to the central plane is formed around a central body of mass M_C, in the origin of a spherical coordinate system (see Fig. 2). The mass infall rate from the primordial solar nebula, where both the disk and the central body are embedded, is assumed to be small, 10^{-2} $M_{\oplus}$ yr^{-1} (Hayashi et al. 1985), and the specific angular momentum of the infalling matter is also assumed to be small.

A physical description of an accretion disk requires the mass infall rate $\dot{M}_C$, the specific angular momentum Γ of matter infalling through the Hill lobe

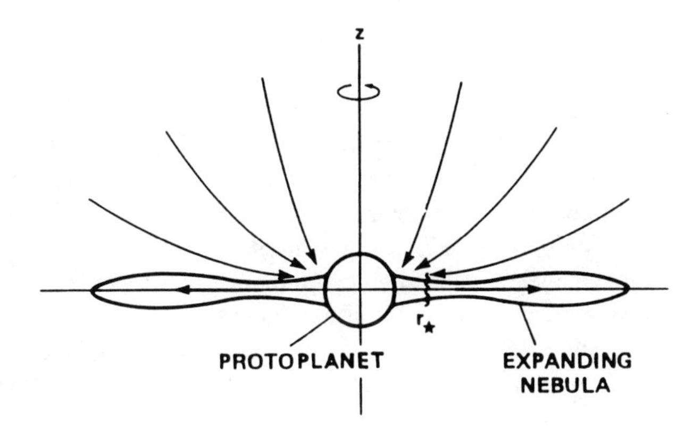

Fig. 2. Schematic illustration of an accretion disk. The region $r < r_*$ is that of direct deposition of angular momentum due to the infalling matter (figure after Cassen and Summers 1983).

at distance R_H, the accretion time τ_a, and the viscous and centrifugal radii R_v and R_{cf} to be specified (where R_v and R_{cf} are defined, respectively, as the size reached by the disk due to viscous angular momentum transfer and the distance of Keplerian rotation of infalling matter, if angular momentum is conserved). These parameters can be defined as follows:

$$\begin{aligned} \dot{M}_C &= 4\,\pi\,\beta\,R_H{}^2\,\rho_{SN} V_{ff}\,(R_H) \\ \Gamma &= \Omega_K R_H{}^2 \\ \tau_a &= M_C/\dot{M}_C \\ R_\mathrm{v} &= (\nu\,\tau_a)^{1/2} \\ R_{cf} &= \Gamma^2/GM_C \end{aligned} \tag{1}$$

where ρ_{SN} is the density of the solar nebula, V_{ff} the free-fall velocity at the Hill radius, which can be reasonably evaluated by assuming that the matter at the boundary of the Hill lobe has Keplerian rotation about the Sun with angular velocity Ω_K, ν is the viscosity, here assumed to have a typical value of 10^{15} cm^2 s^{-1} in agreement with numerical calculations by Cabot et al. (1987*a,b*) and Coradini and Magni (1986), and $\beta < 1$ is a parameter, which accounts for the perturbation induced on the accretion process by the presence of the Sun.

The structure of the disk further depends on two adimensional parameters, ϵ and p, where ϵ, defined as

$$\epsilon = R_{cf}/R_H \tag{2}$$

is related to the amount of specific angular momentum of infalling matter, while p, defined as

$$p=(R_\mathrm{v}/R_{cf})^2 \tag{3}$$

is a measure of the importance of viscosity effects. $p >> 1$ implies strong viscosity effects, outward flux of angular momentum, disk size $R_D = R_\mathrm{v}$, and low mass, while $p << 1$ implies negligible angular momentum transport, disk size essentially determined by the specific angular momentum and equal to R_{cf}, and large and time-dependent disk mass.

Using the parameters defined above, the physical properties of the disk, such as the disk radius R_D, mass M_D, surface density σ_D, overall density ρ_D, effective temperature T_D, pressure P_D and scale height z_D can be derived in a simple dimensional way following Cassen and Summers (1983) and Coradini and Magni (1986):

$$R_D=\max(R_\mathrm{v},R_{cf})$$

$$M_D = M_C \qquad for\ p << 1$$

$$M_D = M_C (R_{cf}/R_v)^{1/2} \qquad \text{for } p >> 1$$

$$\sigma_D = M_D / \pi R_D^2$$

$$T_D = (9G\dot{M}\sigma_D\ \nu\tau_a / 8R_D^3\sigma_S)^{1/4}$$

$$z_D = (2k_B T_D R_D^3 N_A / GM_C)^{1/2}$$

$$\rho_D = \sigma_D / 2z_D$$

$$P_D = k_B T_D N_A \sigma_D / \mu \tag{4}$$

where k_B is Boltzmann's constant, N_A Avogadro's number, μ is the molecular weight of the gas and σ_S is Stefen's constant.

In this approximation, an accretion disk will be assumed to survive when its mean pressure P_D exceeds both the dynamical pressure P_{TT} due to a T Tauri wind from the young Sun, and the pressure in the solar nebula P_{SN}, expressed as in Coradini and Magni (1986) by

$$P_{SN} = 30\ r^{-3.7}\text{AU g cm}^{-1}\ \text{s}^{-2}. \tag{5}$$

P_{TT} has been computed assuming for the mass loss and asymptotic escape velocity of the T Tauri wind, respectively: $M_{TT} = 7 \times 10^{-6} M_\odot\ \text{yr}^{-1}$, and $V = 30\ \text{km s}^{-1}$ (Torbett 1984) and imposing mass and energy conservation (Coradini and Magni 1986).

Table I shows the main physical properties obtained for Jupiter and Saturn disks by Coradini and Magni (1986), while in Fig. 3 the values obtained for P_D, P_{SN} and P_{TT} are plotted for the first eight planets. It can be seen that pressure constraints allow the formation of stable accretion disks only around the giant planets, with Uranus and Neptune being located at the boundary of the stability region. The angular momentum of these disks is low, thus indicating a viscous evolution; the mass of the disk is in agreement with the "effective mass" of the regular satellites, obtained by adding to the present mass of the satellites the amount of H and He needed to reconstruct solar abundances. Finally, the shock region, where gas first infalls, is inside the present position

TABLE I

Main Characteristics of Jupiter and Saturn Disks[a]

	R_{cf}/R_H	R_v/R_H	p	ϵ	M_D	T_D	z_D/R_H	$2z_D/R_D$
Jupiter	3×10^{-4}	0.2	6×10^5	10^{-3}	7×10^{28}	30 K	0.06	0.5
Saturn	10^{-4}	0.3	10^7	3×10^{-4}	10^{28}	10 K	0.08	0.8

[a]Units are cgs throughout.

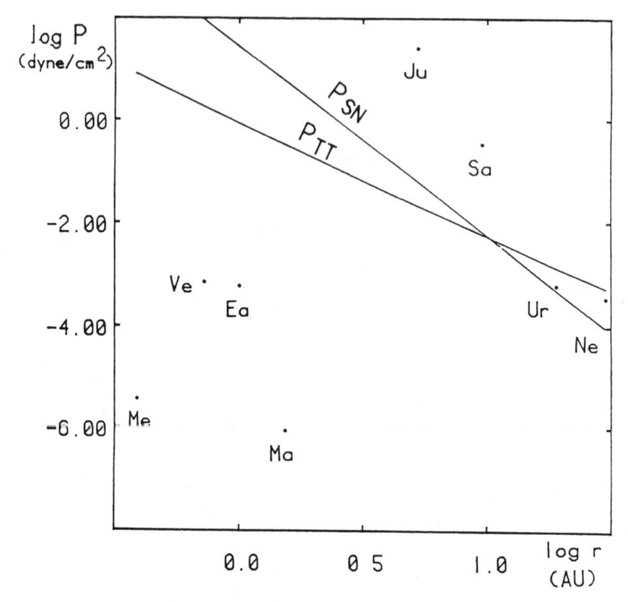

Fig. 3. Characteristic pressures from analytical models of accretion disks. P_{SN} refers to the pressure at the midplane of the solar nebula, P_{TT} to the T Tauri pressure (figure after Coradini and Magni 1986).

of the regular satellites. A large fraction of infalling solid particles thus experienced evaporation, and subsequent recondensation when transported outwards by the viscous evolution. It should also be noted that temperatures in the outer regions of the disk are low enough to allow the condensation of ices.

B. Numerical Models

We shall now describe in greater detail the disk structure, taking into account the main physical processes affecting it, such as, for example, energy transport, time evolution of the accretion, viscous shear and turbulence. These problems cannot be approached through the simple analytical model discussed in Sec. II.A, and numerical models must be developed. In the presence of mass infall, generating supersonic flows and shock waves, the accretion problem requires hydrodynamic and time-dependent techniques; furthermore, the gravitational force of the Sun, in the reference frame adopted, is time dependent, which also requires the use of time-dependent techniques.

A fully physically consistent dynamical model, requiring three-dimensional hydrodynamics, has been hitherto ruled out by the extreme complexity of the problem, and the amount of computer time required. Two alternative approaches can be used, for which a satisfactory approximate treatment can be developed: (1) a *large-scale* description—here, distances are comparable to the Hill radius, and a polytropic gas law and two-dimensional hydrodynamics can be used; (2) a *small-scale* approximation, where the region containing the regular satellites is described—here, equilibrium between mass infall to the planet and mass motions in the disk holds, and mass infall is assumed to be governed by viscosity. Models can thus be computed using the general scheme for the treatment of accretion disks developed by Shakura and Sunyaev (1973), and Lynden-Bell and Pringle (1974).

These equilibrium models have been improved by introducing a detailed treatment of the vertical turbulent convection and radiative transfer, taking into account both gas and grain opacities in determining self-consistent values of turbulent viscosity. The strength of turbulent viscosity is computed taking into account the local values of physical parameters such as density, pressure and temperature, and is directly related to the growth factor of the convective instability (Cabot et al. 1987*a,b*) (Coradini and Magni 1986). The most important results will be described in what follows, omitting the theoretical details.

Time-dependent large-scale model. Time-dependent models are crucial to determine the evolutionary stage in the history of giant planets when a disk structure was formed. Sekiya et al. (1984) studied the hydrodynamics of gas inflow by means of a three-dimensional computation with a "smoothed

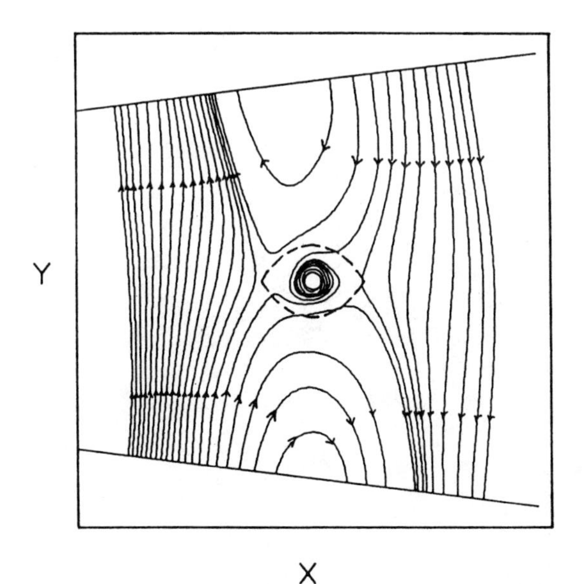

Fig. 4. Equatorial streamlines of nebula gas elements around proto-Jupiter with 1/10 of the present mass. The dashed curves denote the Hill lobe of proto-Jupiter, and x and y are the heliocentric coordinates in a rotating frame where both the Sun and proto-Jupiter are at rest (figure after Sekiya et al. 1984).

particle method." In Fig. 4 are shown the streamlines of a fluid element at a stage when the mass of the protoplanet is 1/10 of its present mass. Assuming the orbit of proto-Jupiter to be circular, and the gas not able to escape after having reached the planetary surface, Sekiya et al. (1984) found an inflow rate for the nebular gas of the order of $10^{-2} M_{\oplus}$ yr^{-1}.

More recently, Coradini and Magni (1986) studied the time evolution of the accretion process, using Eulerian two-dimensional hydrodynamics. In order to avoid complications due to three-dimensional hydrodynamics, two different, but internally consistent, two-dimensional hydrodynamic schemes have been developed. In the first scheme (flat model), isotropy in the vertical direction is assumed; the force field contains gravitational interactions with Sun and planet, and centrifugal and Coriolis forces. By means of this model, a satisfactory description of the situation in the midplane of the disk can be given, but no information can be supplied on its vertical structure. In the second scheme (axisymmetric model), the vertical structure is described in detail, but the force field and other physical parameters in the midplane are averaged over the azimuthal angle.

Since the protoplanet growth was not followed in detail, the mass distribution around the central body at the starting point of the model evolution is unknown. The accretional evolution of the central planet in a solar nebula depleted by the solid material already accumulated in the solar core, can be

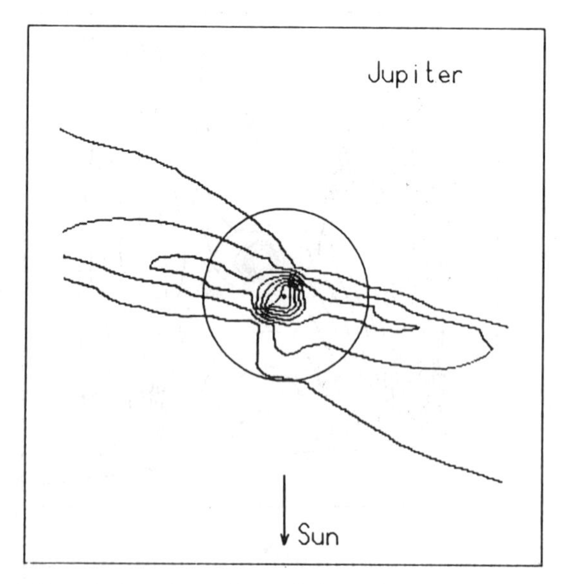

Fig. 5. "Flat model" for Jupiter, after 1.1 free-fall times. The central body contains 95% of its present mass. The spacing of density contours corresponds to $\Delta\rho = 10^{0.4}$, i.e., about 2.5. The outer contour corresponds to the solar nebula density ρ_{SN} at the present distance of the planet, the inner contour to a value 100 ρ_{SN}.

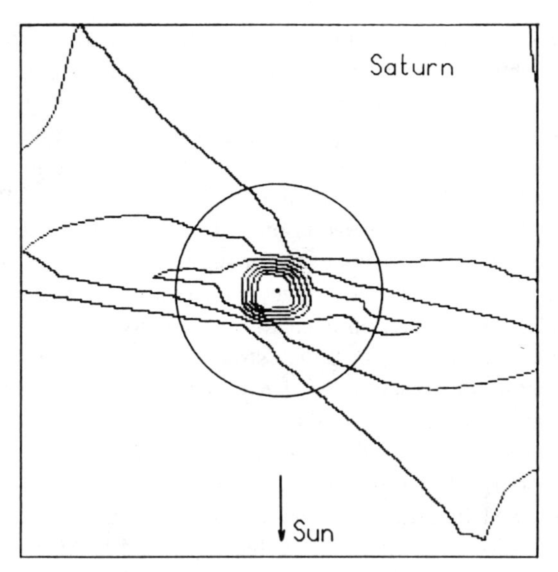

Fig. 6. Same as in Fig. 5, for Saturn.

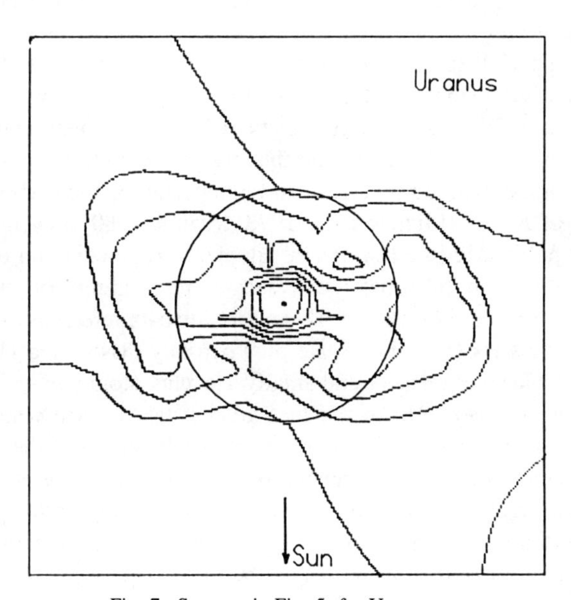

Fig. 7. Same as in Fig. 5, for Uranus.

followed for several free-fall times, in order to allow the matter in the Hill lobe to relax, until a slowly varying, quasi-equilibrium state is reached.

In Figs. 5, 6 and 7, the flat model is shown, for Jupiter, Saturn and Uranus, respectively. The three models can be compared directly, the evolution time being 1.1 free-fall times. Here, the free-fall time is defined as the ratio between the Hill lobe radius and the free-fall velocity. Whatever the model of planet formation, the evolution is started with a central body of mass equal to 95% of its present mass; e.g., for Jupiter, this yields a mass of 1.8×10^{30} g. At this early evolutionary stage, some common features are exhibited by these models. The accretion is clearly asymmetrical, in contrast with the results by Sekiya et al. (1984); a disk structure can be identified, extending from the planet to a distance of $\sim 0.1\ R_H$, denser for Jupiter than for Saturn and Uranus, as a consequence of the larger Jovian gravity, and of the higher density of the protosolar nebula at the distance of Jupiter.

These models allow a prediction to be made about the mass infall rate and the short- and long-range variations of the accretion, the angular momentum distribution, and the size and shape of the disk. Moreover, a knowledge of the average values of the mass infall rate for several values of the central-body mass, together with the assumption that this mass rate coincide with the growth rate $\dot{M}_C$ of the central body allows the function $M_C(t)$ to be integrated, by means of the function $\dot{M}_C(M_C)$. A qualitative behavior of the time growth of the planet mass can thus be obtained. Finally, at the end of the mass infall, the mass loss from the disk, due to the tidal influence of the Sun, can also be estimated.

The case of Jupiter is treated here as an example, in the framework of the "nucleated accretion" model (Stevenson 1982*a*). The overall accretion process is supposed to run through three phases: (1) core formation by planetesimals collisions (time span = 10^7 to 10^8 yr); (2) hydrodynamic collapse and gas infall onto the core (10^2 to 10^3 yr); (3) formation of a low-mass accretion disk around the nearly completely grown protoplanet (10^3 yr).

Although radiative processes can be important in the early phases of the formation of Jupiter (Mizuno et al. 1978; Mizuno 1980; Stevenson 1982*a*), a simplified polytropic-like law can be adopted for the equation of state of the gas. The polytropic index has four values corresponding, for growing values of density, to an isotherm ($\gamma = 1$), and to adiabats referring to a mixture of neutral He plus H_2 ($\gamma = 1.43$), He plus partially dissociated H_2 ($\gamma = 1.15$; Magni and Mazzitelli 1979) and finally He plus atomic H ($\gamma = 5/3$). The change from the isotherm to the first adiabat takes place when the optical depth is comparable with the mean density scale length of the infalling gas, while the other changes of γ depend on the dissociation of H_2. The overall shape of the adiabats depend on the opacity, and mainly on the grain opacity.

When the protoplanet mass is $M_C = 1.8 \times 10^{30}$ g, the parameters of the solar nebula are assumed to be $T_{SN} = 600\ r^{-1}$; $\rho_{SN} = f^* 1.28 \times 10^{-9} r^{-2.7}$, (standard solar nebula), where f^* is a depletion factor derived by Safronov and

Ruskol (1982), which accounts for the decrease in gas density in the protoplanet feeding zone as the protoplanet grows. This happens if there is a low velocity of refilling in the rarefied regions, caused by viscosity not high enough and absence of mixing (Safronov and Ruskol 1982).

The evolution of the gas in the protoplanet's Hill lobe is largely independent of the previous thermal evolution in the solar nebula; thus, viscous stresses and thermal time scales are different from those of the surrounding nebula, and a different adiabat has to be adopted.

The gas adiabat was constrained to fit the solar nebula thermodynamical quantities, but two different scenarios were tested: (1) the gas is adiabatic when compressed with respect to the solar nebula background; (2) the gas is isothermal up to a density 100 ρ_{SN}, and adiabatic for larger densities. These two cases correspond, respectively, to high (HO) and low (LO) opacity cases, and are essentially related to a different amount of small solid particles in the accreting gas and to the thermal structure of the protoplanet.

In Fig. 8, the adiabats from this model are compared with those derived by Mizuno (1980) and Lunine and Stevenson (1982). It can be seen that both Mizuno's, and Lunine and Stevenson's adiabats are cooler; it should be remembered, however, that the curve by Mizuno applies to a region very close to the protoplanet, while that by Lunine and Stevenson applies to an evolved disk, just before Jupiter's satellites formed. It can thus be expected that by increasing the evolution time, the $\rho - T$ relation obtained in the equilibrium model will approach that computed by Lunine and Stevenson, overlapping that by Mizuno in the inner part of the disk.

Some interesting results of the time-dependent model for the Jupiter disk are shown in Figs. 9 through 15. The time integration has been carried out for

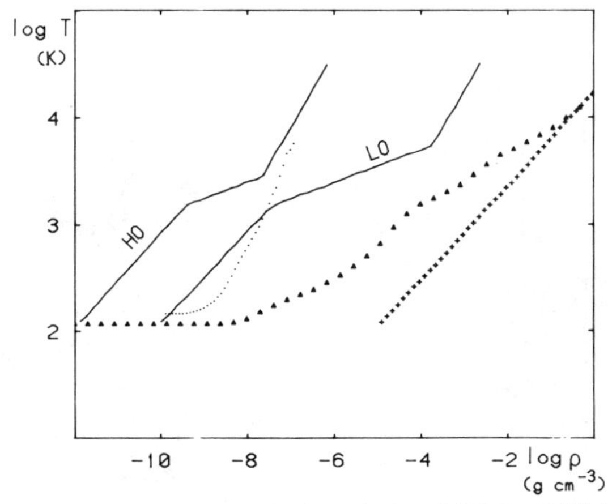

Fig. 8. Solid lines showing adiabats from the dynamical model (LO and HO indicate, respectively, low- and high-opacity models), and from the equilibrium model (dotted line). Crosses refer to the model by Lunine and Stevenson (1982*a*); triangles to that by Mizuno (1980).

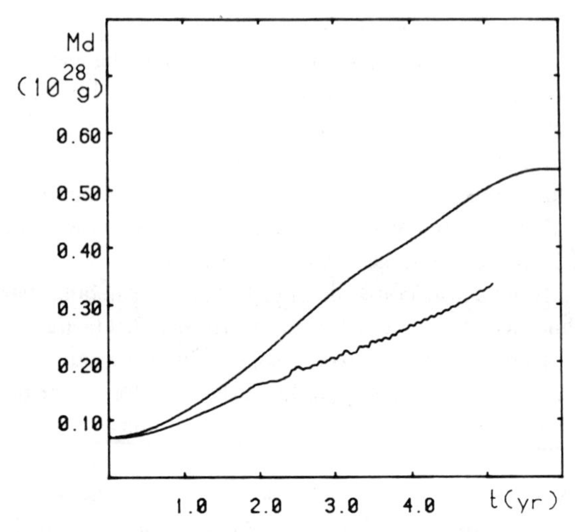

Fig. 9. Mass inside the Hill lobe M_H vs time for axisymmetric (upper curve) and flat (lower curve) models (figure after Coradini and Magni 1986).

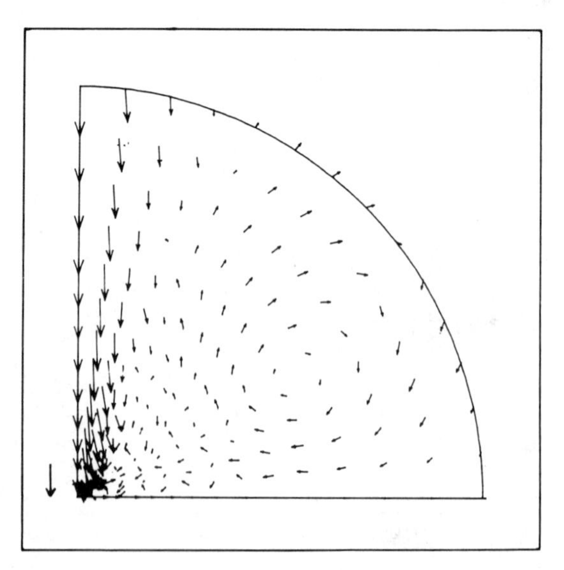

Fig. 10. Stream velocity from the polar region in the axisymmetric Jupiter model ($t \cong 8$ yr). The velocity is normalized with respect to the free-fall velocity. The scale is given by the single arrow, corresponding to unit normalized velocity.

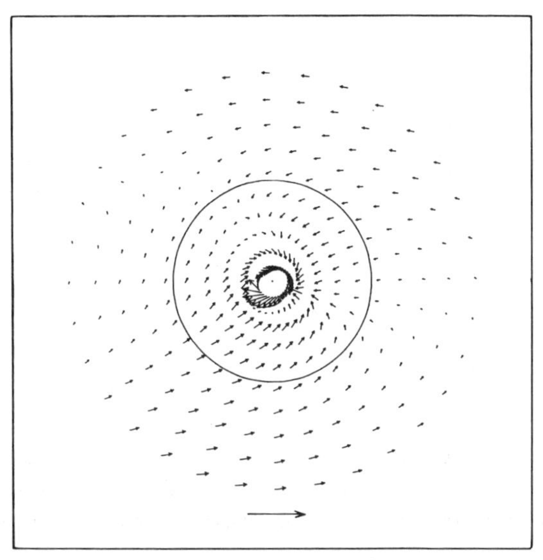

Fig. 11. Accretion on the central plane, after ~8 yr. The arrows correspond to the velocity vector for the fluid. A change in direction is evident, after the gas has been steadily captured. The scale is given by the single arrow (10 km s^{-1}).

~8 yr, corresponding to 10 free-fall times. Figure 9 shows a comparison of the mass inside the Hill lobe as a function of time for flat and axisymmetric models, with a central-body mass of 1.8×10^{30} g. The slope of the axisymmetric model is steeper, as accretion is also possible here from the polar regions of the Hill lobe (Fig. 10); on the disk midplane, accretion is hindered by the largest values of the specific angular momentum for the infalling matter (Fig. 11). In any case, the two mean accretion values do not differ significantly.

Figure 12 allows a comparison to be made of the accreted mass inside the Hill lobe for low- and high-opacity axisymmetric models; the different slopes are due to different values of the gas compressibility in the inner regions of the Hill lobe. A rebound in the accretion is evident in these figures, strongly dependent on the opacity; this is due to the excess in the pressure gradient in the inner region of the Hill lobe, which in turn is due to the viscous shear being neglected in this dynamical model. The presence of viscous shear ensures a sufficient rate of mass inflow in the inner region of the disk, and a consequent depletion of matter, which allows new matter to be accreted. The maximum value for the mass in the Hill lobe is comparable to the effective satellite mass (Lunine and Stevenson 1982; Coradini and Magni 1984).

The structure of the disk at the end of the integration time (~8 yr) in the axisymmetric and flat geometries, is shown in Figs. 13 and 14, respectively. Figure 14 also shows the prograde region of the disk (crosses), i.e., the region

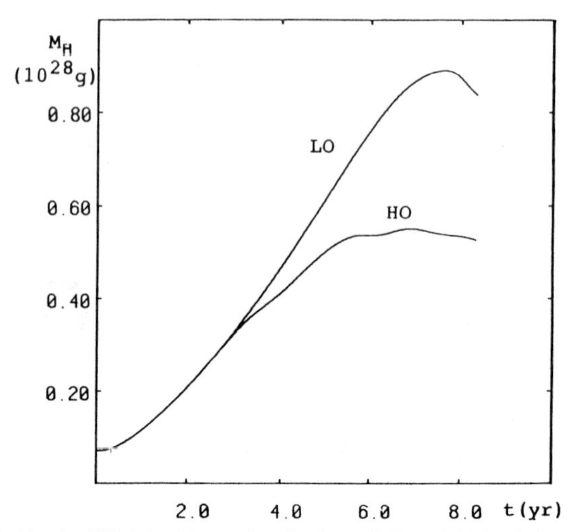

Fig. 12. Mass inside the Hill lobe M_H vs time for low- (LO) and high- (HO) opacity models.

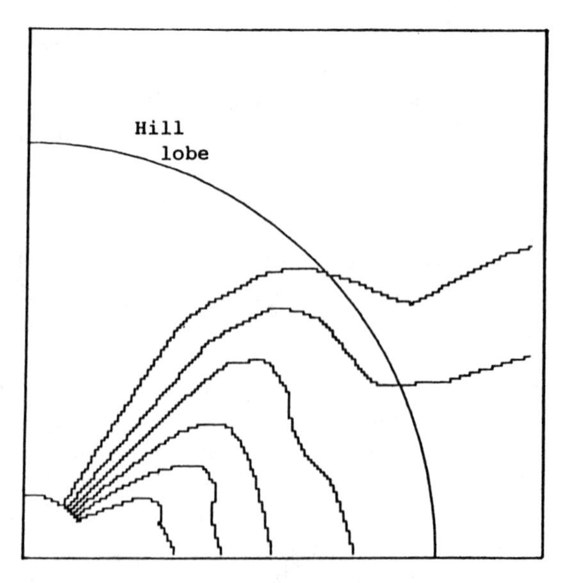

Fig. 13. Jupiter disk at t=8.3 yr for the axisymmetric model (low opacity). The spacing of density contours corresponds to $\Delta\rho = 10^{0.4}$, i.e., ~2.5. The outer contour corresponds to the solar nebula density ρ_{SN} at the present distance of Jupiter; the inner contour corresponds to a value of 100 ρ_{SN}.

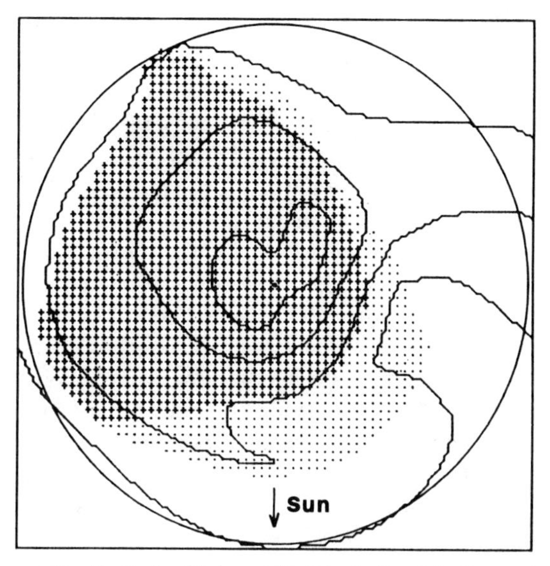

Fig. 14. Same as in Fig. 13, for the "flat model" of the Jupiter disk. Crosses indicate the region rotating as the regular satellites; dots and crosses, the gravitationally bound region.

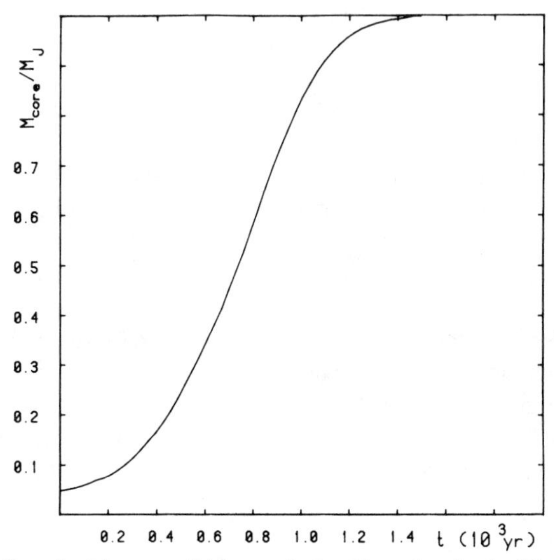

Fig. 15. Growth of the mass of the proto-Jupiter M_C vs time for the "flat model."

rotating as the regular satellites, and the gravitationally bound region (dots and crosses), defined as the region where the energy is lower than the escape energy from the Hill lobe. Only a qualitative estimate of the extent of this last region can be given, as energy is not strictly conserved in the reference frame adopted. Furthermore, it is clear from the figures that regions of the disk with scale length comparable to that of the Hill lobe are thick in the vertical direction, and exhibit strong anisotropies.

The time evolution of the protoplanet mass, obtained by integrating the function $\dot{M}_C$ (M_C), is shown in Fig. 15; the mass captured in the Hill lobe is supposed to be totally accreted onto the central core. The mass of the central body at the end of the collapse is uncertain (Mizuno 1980; Stevenson 1982*a*). However, it can be seen in Fig. 15 that the slow accretion phase, assumed to start after the very rapid hydrodynamic collapse, occurs on a time scale of 1.4×10^3 yr, a time shorter by one order of magnitude than that obtained by Safronov and Ruskol (1982). This is a lower limit for the accretion time, since the accretion is driven also by radiative and convective losses, and the Kelvin-Helmholtz time scale may become important. The accretion time of the planet has been found to be quite insensitive to the initial value of the central mass.

On the basis of these results, a region of the disk can be isolated, where the angular momentum has the same orientation as the regular satellite systems, and where the gas is gravitationally bound to the protoplanet. This region will be called the satellite accretion disk. The possibility of Kelvin-Helmholtz instabilities setting in the planet accretion region, as a consequence of the tangential velocity gradients (responsible also, due to the Coriolis forces, for the transition from the retrograde to the prograde direction; Figs. 11 and 14) has also been studied. Following Canuto et al. (1984), this instability mechanism can be used to determine the value of turbulent kinematic viscosity; a typical value obtained for this parameter is $\sim 10^{15}$ cm^2 s^{-1}.

Equilibrium Models. A more detailed description of the inner regions of satellite accretion disks can be obtained, once the mass infall rate into the inner regions has been determined, and stationary accretion has been reached. A great deal of work has been devoted in recent years to the physics of protosolar disks (see, e.g., Cabot et al. 1987*a,b*). We report here the results obtained, by applying a similar approach to satellite disks (Coradini and Magni 1986).

If the disk is reasonably thin, it can be divided into contiguous vertical columns, the structure of each being independent of that of adjacent ones, so that vertical gradients can be determined in each column independently.

In a thin disk, the viscous stress acting between different radial sections of the disk can be determined globally (Lynden-Bell and Pringle 1974). Although in satellite accretion disks the ratio of thermal to gravitational energy is at times substantial, hydrostatic equilibrium can be assumed to hold, and the pressure gradient to be balanced by the vertical component of the gravitational attraction, with most of the disk mass being confined in regions where the disk geometry is not too distorted. The self-gravitation of the disk can be neglected, being nowhere larger than 1/100 of the vertical component of the central-body gravity.

A more detailed description of satellite accretion disks can be obtained assuming stationary accretion and axisymmetry. Of particular importance are the physical and thermodynamical properties of these disks, both to investigate the primordial stages of satellite formation, and to determine the reservoirs of volatiles, which are the principal components of an atmosphere.

For stationary disks, in the thin-disk approximation, the vertical structure can be determined once the mass infall rate, the turbulent viscosity and the energy transfer mechanisms are known. Generally, in the astrophysical approximation, the turbulent viscosity is expressed in terms of the sound speed and of the characteristic scale height of the model, scaled by an *ad hoc* parameter α (Shakura and Sunyaev 1973). Following Canuto et al. (1984), α can be expressed in terms of the growth rate of the unstable modes of the physical mechanisms generating turbulence. For thick satellite disks, convection has been found to be the dominant instability mechanism in the inner regions (Coradini and Magni 1984). In the outer radiative regions of the disk, the Kelvin-Helmoltz instability has been introduced as that feeding turbulence (Coradini and Magni 1986).

In Fig. 16, values of the disk central temperature (a) and the disk surface density (b) are shown for the accretion disks of Jupiter and Saturn. For both

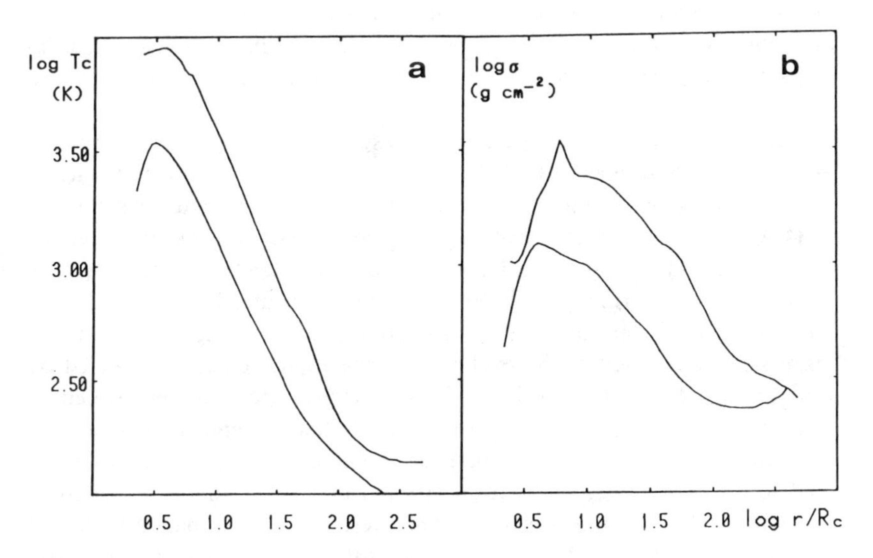

Fig. 16. Radial distribution of temperature at the midplane (a) and surface density (b), for the equilibrium model of Jupiter (upper curve) and Saturn (lower curve) disks.

models, an accretion time $\tau_a \cong 250$ yr and a mass infall rate per unit surface $\dot{M}_C \propto r^{-1.2}$ have been used, obtained as an extrapolation of the results of the two dimensional axisymmetric model. It should be pointed out that this is a parametric time scale adopted to obtain the best possible fit between the total mass of the disk and the effective satellite mass, compatible with the convergence of the model, while the actual accretion time for this model is 1.4×10^3 yr.

The values obtained for the total mass of the disk are 4.4×10^{28} g for Jupiter and 9.2×10^{28} g for Saturn. These masses are larger than the effective satellite masses, respectively equal to 2.8×10^{28} and 1.1×10^{28} g. A large fraction of this mass is contained in the outer part of the disk, which requires an inward drift of the solid component in order to form the satellites where they are today, while the gas mass excess can be removed by tidal effects. In both cases, the convective zone overlaps the area where the regular satellites are present: for Jupiter, this zone is located at a distance of 5.7 R_J, and at a height of 4.1 R_J on the central plane, while for Saturn it is located at a distance of 0.5 R_S and at a height of 0.6 R_S above the central plane. Moving outwards, the Fe grains start to condense, again first in the upper part of the disk where the temperature is lower, and later on the central plane. The icy particles appear only at very large distances from the protoplanet, and simultaneously within the whole column. Disks obtained by this refined model are more turbulent, and have a larger mass, than those obtained by previous calculations, on the basis of an empirical mass law and an approximate turbulence model (Coradini and Magni 1984).

In Fig. 17 a and b, the vertical sections of the two disks are plotted; in this figure the convective zones, and the condensation and sublimation zones of refractory and icy particles are also represented. In the calculations described above, the presence of ices other than water ice was not taken into account.

In order to assess the occurrence of these ices in satellite disks, following Pollack and Consolmagno (1984), we have plotted the equation of state of various disk models (Lunine and Stevenson 1982*a*; Pollack and Consolmagno 1984; Coradini and Magni 1986) against the regions of stability of various ices derived by Lewis (1972*a*). Lewis examined in detail the chemical species existing in local thermodynamic equilibrium in a gas of solar elemental composition, at low temperature and a variety of pressures. In Fig. 18a are shown the primordial phases of the evolution of the Jupiter disk, as modeled by Coradini and Magni (1986); it can be seen that temperatures in the central plane of the disk, are everywhere too high to allow methane clathrate and ammonia hydrate to condense. In later evolutionary stages, such as those modeled by Lunine and Stevenson (1982*a*) after mass infall has stopped, water ice and ammonia hydrates can be present in the formation region of Callisto only although it has to be verified that enough mass is left. The case of Saturn is different. The outer region of the disk overlaps the condensation

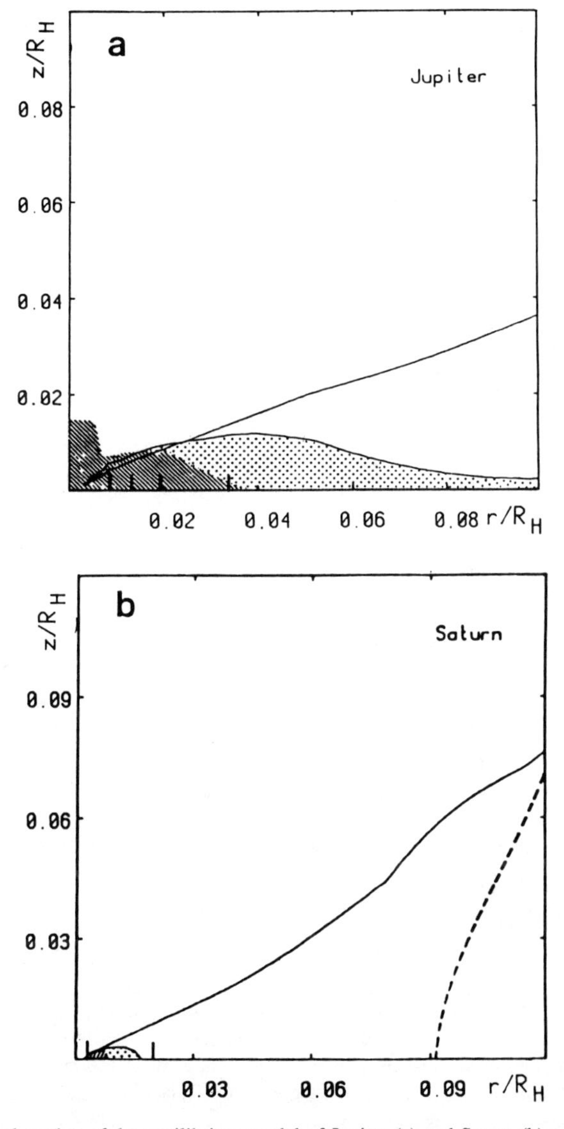

Fig. 17. Vertical section of the equilibrium model of Jupiter (a) and Saturn (b) accretion disks. The upper curve indicates where 99% of the surface density is reached. The dotted region indicates the convective zone, the hatched region, the zone depleted of refractory elements. Bars indicate the position of regular satellites. The dashed line indicates in (b) the inner boundary of the condensation zone of water ice.

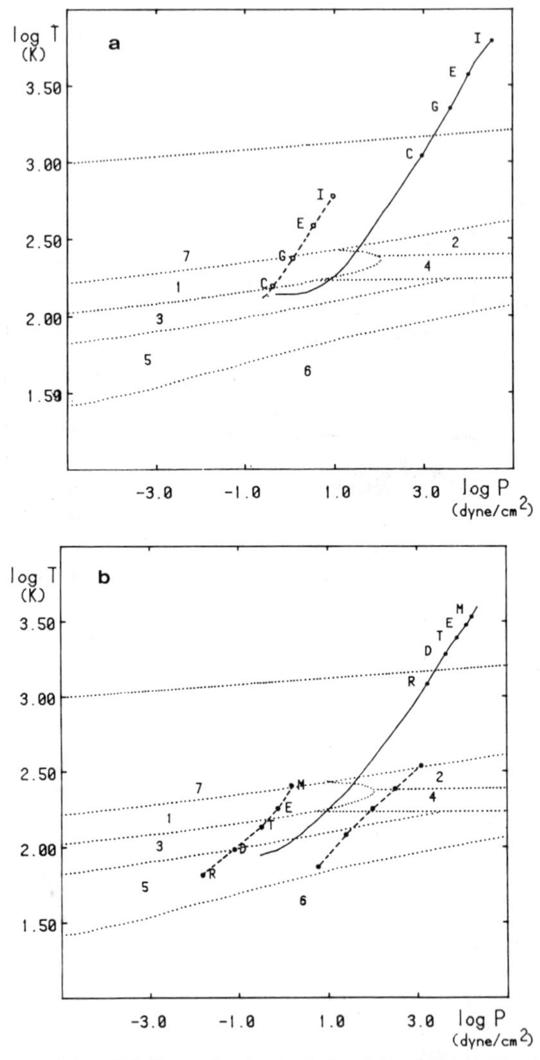

Fig. 18. Comparison of the adiabatic profiles for the Jupiter (a) and Saturn (b) nebulae against the region of stability of various ices in the log P − log T plane (after Lewis 1972; Pollack and Consolmagno 1984). The stability regions correspond to: (1) water ice; (2) liquid water; (3) ammonia hydrate; (4) liquid − ice; (5) methane − clathrate; (6) methane − ice; (7) hydrous silicates. Pressure is in dyne cm^{-2} and temperature in Kelvin. The solid line in (a) and (b) indicates the model by Coradini and Magni (1986), which refers to a very primordial phase. The dashed lines indicate the model by Lunine and Stevenson (1982*a*), in (a) and in (b) a minimum- and a maximum-mass nebula (respectively left and right), after Pollack and Consolmagno (1984). Dots indicate the position of regular satellites.

region of water ice, ammonia hydrate and clathrates even in the primordial phases (Fig. 18b); in this case, the adiabatic profiles of both the small- and large-mass nebulae (Pollack and Consolmagno 1984) lie in the stability region of ammonia hydrates and methane clathrate. A detailed study of the disk relaxation from the primordial phases discussed here should be carried out, in order to increase our understanding of the complex chemistry of satellite formation.

The equilibrium model by Coradini and Magni (1986) predicts values for the disk mass inside the Hill lobe which for Jupiter are in good agreement with the effective satellite mass as computed by Lunine and Stevenson (1982*a*), while for Saturn they are larger. Furthermore, the regular satellites region, as described by these models for both Jupiter and Saturn, contains an amount of mass too small to account for the effective satellite mass, and is characterized by temperatures too high to allow the condensation of ice. Thus, to obtain realistic conditions for the satellite formation phase, it should be assumed that mass infall decay in the late accretion stage is accompanied by turbulence decay and cooling of the disk in a few hundred years; so that ice could con-

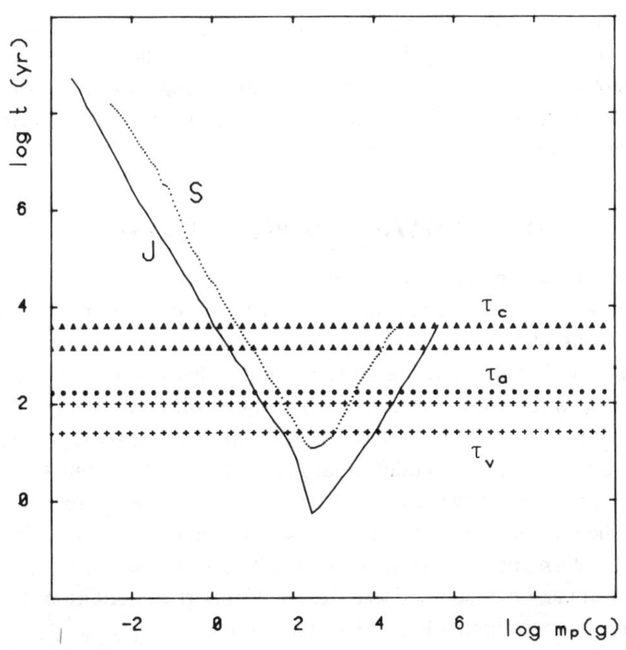

Fig. 19. Critical infall time of solid particles of mass m_p for Jupiter (J) and Saturn (S). The region bounded by triangles represent the mean values of the collisional growth time of solid particles in a disk τ_c; crosses indicate the dissipative viscous time τ_v of the disk, and τ_a = 250 yr is the accretion time of the disk, i.e., the time required to add the last 1% of the mass to the protoplanet. Note that the region corresponding to times longer than the viscous dissipation time has no physical meaning.

dense in the satellite formation region, and inward migration through aereodynamic drag could bring a significant fraction of solid particles from outside to inside the satellite region.

Figure 19 shows the migration times of solid particles of mass m_p from an upper boundary of the disk where its mass is equal to the effective satellite mass, to a distance twice that of the farthest regular satellite in order to estimate the upper boundary of the outer feeding zone. The disk model referred to in these calculations is the equilibrium model (Coradini and Magni 1986) shown in Fig. 16. τ_v is the mean value of the dissipative viscous time scale of the disk, defined as $4\Gamma/(3\pi\nu\Omega_K)$; τ_a is the accretion time of the disk; and τ_c is the *e*-folding time of collisional growth of an ensemble of particles of equal mass at temperature and density corresponding to those of the standard solar nebula (Coradini et al. 1981*b*). The range of values assumed by τ_v and τ_c in the region of the disk where the migration times have been calculated are shown in the hatched regions. It can be seen that only particles ranging in mass from 10^2 to 10^4g for Jupiter, and from 10^2 to 10^3g for Saturn, will be able to enter the formation region of the satellites. Whether the total mass present in these mass ranges will be large enough to allow the formation of the satellites is a key question, as yet unsolved.

Finally, it should be mentioned that our approach cannot be easily extended to model the Uranus disk. This disk, in fact, cannot be considered to be flat, and the approximations described above fail to apply; thus, two-dimensional calculations will be needed to model the equilibrium accretion for the Uranian system.

III. FORMATION OF PLANETESIMALS

Detailed disk models allow pressure and temperature conditions in the phases characterized by a constant mass infall rate to be determined, as shown in the previous section.

In the models discussed above, the mass distribution per unit surface, obtained with a refined treatment of viscosity, is different from that observed in the present satellite systems. For example, most of the mass is confined in a small region, while in the models a large percentage of the mass is located in the external regions of the disk. Furthermore, the temperature in the internal region of the disk is too high to allow condensation of ices, while the satellites of Jupiter, Saturn and Uranus are known to be composed of a large fraction of ices. Thus, it can be concluded that the most important conditions for satellite evolution are determined when mass infall stops.

As Stevenson et al. (1986) has pointed out, when mass infall stops, the temperature in the disk subsequently decreases. In the cooling phase, nucleation of condensates occurs from the gas phase. This nucleation can be either heterogeneous (on the surfaces of pre-existing refractory seed nuclei) or homogeneous. Growth of these condensation centers to μm-sized grains is by diffusion and rapid coagulation ($<10^3$ yr; Stevenson et al. 1986).

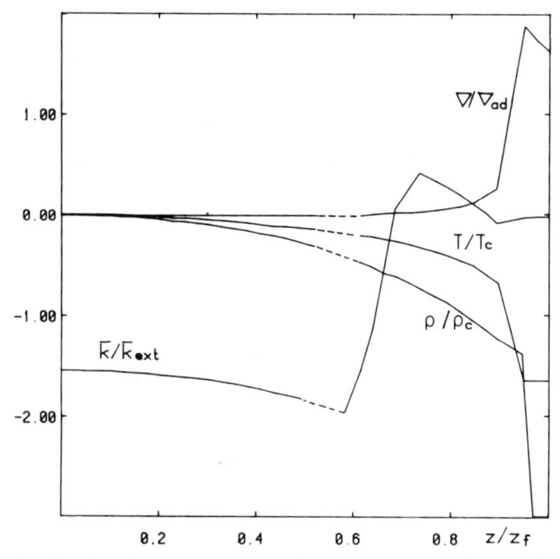

Fig. 20. Vertical distribution of temperature, density, temperature gradient and opacity at $r = 3.7$ R_C for a low-viscosity model. Temperature and density are normalized with respect to the central values, the temperature gradient to the adiabatic one, and the opacity to its value at the top of the column. The height z is normalized to the final scale height of the disk z_f. The dashed line indicates a radiative region.

Subsequent growth can be slower, but it is aided by the relative motion between gas, possibly turbulent and grains, or by the possible sticking of grains during collisions (Coradini et al. 1980; Nakagawa et al. 1981; Völk et al. 1980; Morfill 1983; Weidenschilling 1984; Nakagawa et al. 1986). Condensation is likely to begin near the disk photosphere, where the temperature is lowest, and cause the formation of a cloud of μm-sized grains. Unlike the gas, which is supported by pressure gradients, the dust grains settle toward the midplane of the disk. In Fig. 20, the vertical distribution of temperature, density, temperature gradients and opacity at $r = 3.7\ R_S$ (i.e., between Mimas and Enceladus) is shown as an example, for a low-viscosity disk model, i.e., characterized by a high critical Reynolds number of the order of 150. Temperature and density are normalized with respect to the central values: the temperature gradient with respect to the adiabatic one, and the opacity to its value k_{ext} at the top of the column. Dashes indicate a radiative region. Ice particles are present in the upper part of the column. The two convective zones are separated by a small radiative region; therefore evaporation of already formed grains is inhibited (Coradini and Magni 1984).

A situation like that described in Fig. 20 is possible only in the presence of mass infall. When mass accretion onto the disk stops, convection in the disk is not energetically sustained and a higher degree of quiescence is reached. The solid component, driven by the viscous coupling with the gas, settles toward the midplane of the disk. The density of the central plane increases until the configuration becomes gravitationally unstable. In this phase only radial perturbations can grow, and tangential perturbations when arising are destroyed by differential rotation and tidal forces.

The contraction of the ring under its self-gravitation continues until the density reaches such a value that the time scale of development of a tangential perturbation is shorter than the time scales of disrupting effects (differential rotation and tidal forces). The ring can then break up in fragments distributed in circular orbits.

The fragments can collapse until they reach an equilibrium configuration, i.e., attain solid-body density or become virialized structures. Ward (1976), following a previous work (Goldreich and Ward 1973), developed a model of satellite formation from "local" planetesimals resulting from the contraction and fragmentation of an infinitely thin dust disk. The influence of the gas component on the contraction and fragmentation was not taken into account in this model. More recently, Coradini et al. (1981*a,b*) and Sekiya (1983) studied the onset of gravitational instabilities in a two-fluid system (gas and solid particles coupled by viscosity). In the model by Coradini et al., a time-dependent system is simulated by following the sedimentation of the solid component which is in disequilibrium with the gas until the critical value of the grain density on the disk midplane is reached and gravitational instabilities can set in. Due to gas particle coupling, fragmentation and formation of the planetesimals proceed in quite a different way from the Ward model. Another possibility, suggested by Weidenshilling (1984), is that grains may not totally settle to the disk midplane due to turbulence. In this case, grains will grow mainly by binary collisions and the planetesimals obtained are significantly smaller in size (of the order of meters).

The evolution of a satellite disk with particles of equal mass has been computed by Coradini et al. (1981*b*) for Jupiter and Saturn, for three different values of the particle mass: 1, 10^4 and 10^8 g. This interval does not strictly coincide with the mass range of particle survival previously identified, but contains it. This enlargement can be justified, on the grounds of the large uncertainties still affecting this type of model.

In Fig. 21, the masses M_P of planetesimals produced by fragmentation of the unstable ring are shown. These values are computed according to Coradini et al. (1981). Masses range from 4×10^{15} g to 2×10^{17} g. An increase in the particle mass by 8 orders of magnitude results in M_P increasing at most 50 times and $M_{\rm ring}$ increasing at most 2 times. More massive particles imply a less efficient viscous drag and, consequently, shorter evolutionary times of the unstable ring and the fragments (Coradini et al. 1981*a,b*). The values obtained for planetesimal masses do not differ greatly from those obtained by Ward (1976).

The scenario previously described implies high values of the particle density on the central plane of the disk and, consequently, a high degree of

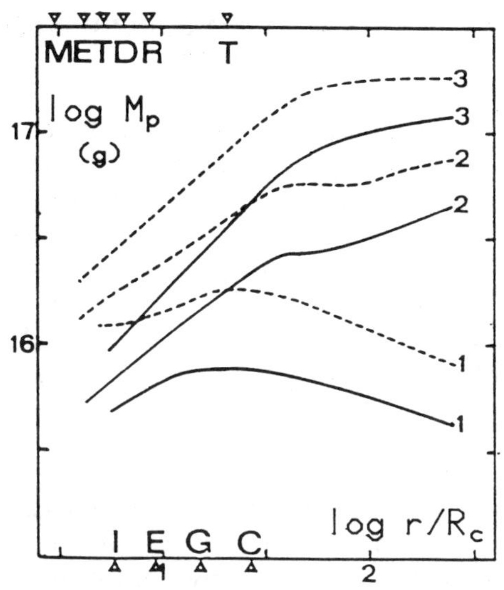

Fig. 21. Planetesimal masses for particle masses equal to 1 g (1), 10^4 g (2), and 10^8 g (3), for the Jupiter system (solid lines) and the Saturn system (dashed lines). Triangles indicate the position of regular satellites.

quiescence in the disk. Harris and Kaula (1975) and Harris (1978) suggested that satellites have been formed from the same circumsolar planetesimals responsible for the growth of the solid core of the planet and captured in the satellite accretion disk during the last phases of the formation of the primary body. In the above mentioned models, the accumulation of debris in orbit about an accreting planet is a consequence of mutual collisions of planetesimals inside the sphere of influence of the planet. The debris give rise to a disk, that, when the incident flux of planetesimals decreases, flattens and becomes gravitationally unstable. The main difference of these models from the previous ones by Ruskol (1960,1972,1973) consists in the inclusion in the model of tidal effects. Safronov et al. (1986) present another scenario for the formation of satellite bodies. In a preliminary phase, prior to the beginning of gas accretion, a swarm of solid material begins to form, due to inelastic collisions and subsequent capture of particles by the planet's gravitational field. The presence of gas decreases the particle velocities thereby increasing the probability of constructive collisions, but it increases also the time scale of planetesimals formation. Smaller heliocentric planetesimals can be captured by gas drag interactions with the disk, while larger bodies can be captured by collisions. In this stage, gas drag was also present leading to contraction of the orbits and losses of the inner part of the swarm to the central body. This complicated scenario is probably closer to reality than the simplified one previously presented; however, detailed disk models are needed to investigate the possibilities suggested by this approach.

IV. EVOLUTION OF PLANETESIMALS AND FORMATION OF SATELLITES

The further evolution of planetesimals is characterized by collisional accretion and orbital decay due to dissipative effects such as gas drag on planetesimals and mutual collisions (Safronov 1969; Safronov and Vityazev 1985*b*). In a first phase, planetesimals can be considered as colliding bodies with relative velocities given mainly by gravitational close encounters (Safronov 1969). The presence of gas drag has been taken into account through the values of the Safronov number, as in Safronov (1969). The time to double the mass of a planetesimal is given by

$$\tau_c = (4\vartheta\bar{\rho}/3\pi G)^{1/2}\ \ln 2/NM_p \tag{6}$$

where $N = \sigma_p/2h_pM_p$ is the number density of planetesimals, σ_p the surface density of particles in the disk and $h_p = \pi\bar{v}/4\Omega_K$ the half height of vertical dispersion and $\bar{v}^2 = GM_p/R_p$ the mean relative velocity of planetesimals with bulk density ρ and radius R_p. ϑ is the Safronov number (Safronov 1969), depending on impact dissipation, collisional cross section, mass distribution of planetesimals and gas drag. Values of ϑ ranging from 5 to 50 seem to take into account all these effects. The collisional accretion of planetesimals goes on until an embryo is formed, which prevails and captures mass from its own feeding zone. The accretion time of the embryo, taking into account gravitational focusing of the swarm of planetesimals, can be evaluated (Safronov 1969) as follows:

$$\tau_f = 2\pi(M_f\bar{\rho}^2/36\pi)^{1/3}/(1+2\vartheta)\sigma_p\Omega_K. \tag{7}$$

M_f, the final mass of the embryo, can be obtained from the width of the feeding zone (Vityazev et al. 1978) that enlarges as ϑ decreases.

The orbital decay of planetesimals is due to the drag exerted by the gas on individual bodies. When planetesimals reach a sufficiently large mass, radial decay can be modified also by their tidal effects on the surrounding gas. By this mechanism, angular momentum and consequently gas are pushed away, with a gas-depleted gap forming inside the disk along planetesimals orbits (Lin and Papaloizou 1980). The minimum mass of the body creating the tunnel was computed at first to be 10^{17} g (Weidenschilling 1982), and later, by a different method, to be 10^{25} g (Lunine and Stevenson 1982). Weidenschilling (1982) assumed that the tidal torque, at the distance of 1 scale height away from the satellite, must balance the viscous torque, that tends to eliminate a gap of the order of the Hill lobe. If the disk is quiescent and the viscosity is molecular, the minimum mass obtained to open a tunnel is of the order of 10^{17} g. Lunine and Stevenson (1982) noted that a gaseous disk behaves differently from a particulate one because the pressure effects are important. In this situation, the impulse approximation cannot be used for the gas element in the Hill lobe. Moreover, they state that the molecular viscosity seems inappropriate to obtain reasonable evolutive times (10^3 yr) for the redistribution of angular momentum in the disk. This more realistic treatment of the problem leads to a value of the minimum opening-gap body of the order of 10^{25} g. Safronov et al. (1986) note that the presence of numerous large bodies with crossing orbits makes the probability of formation of stable tunnels very low. Ward (1986) compares the results by Weidenschilling (1982) and by Lunine and Stevenson (1982), and points out that in both cases the calculations are too approximate and that probably the nonlinearity and three dimensionality of the problem need to be taken into account.

Mutual encounters among planetesimals also supply a dissipative mechanism, which does not play an important role, as shown by Fig. 22 (Coradini et al. 1981*a*). This figure shows the evolutive times of the swarm of planetesimals obtained by gravitational instabilities in a highly sedimentated disk. The mass of planetesimals varies as a function of the distance from the primary; at the same planetocentric distance, planetesimals are considered to have equal mass. The surface density of the planetesimal swarm is assumed to be equal to that of the original disk owing to the rapidity of the planetesimal formation compared to the evolutive times of the disk. This can be verified using the characteristic time scale by Hayashi et al. (1977). Figure 22 shows as well that the collisional growth of planetesimals is fast enough to allow the formation of embryos in spite of orbital decay, at least in the region of the regular satellites. Only planetesimals lying beyond a radius $r > 100\ R_c$ can spiral without any appreciable growth.

Table II gives the masses of the embryos for Jupiter and Saturn at the end of the accretion. For massive satellites ($M_s > 10^{24}$ g), the agreement with the present mass is reasonable for $\vartheta < 5$; this condition implies high relative velocities of planetesimals and hardly agrees with the assumption of a dense circumplanetary gas disk.

One possibility to increase the relative velocity is to assume that, at a certain stage of the satellite history, resonant configurations among the accreting bodies are reached. In a resonant system, such as the Galilean satellites, the smaller bodies experience induced eccentricities, so that the widths of the feeding zones can be enlarged.

For the planetary system, Patterson (1987) suggested that planet embryos could have formed at the two-body exterior resonance due to capture and accretion of planetesimals. Similar mechanisms can probably be applied to the satellite systems.

The inner feeding zones may be replenished by the orbital decay of outer slowly growing planetesimals, the amount of mass in the outer regions being comparable to that of the regular satellites. This feeding mechanism, if efficient, has a lower-limit time scale of about 10^3 yr, so that the growth time scale of embryos may be significantly increased. Shorter time scales of forma-

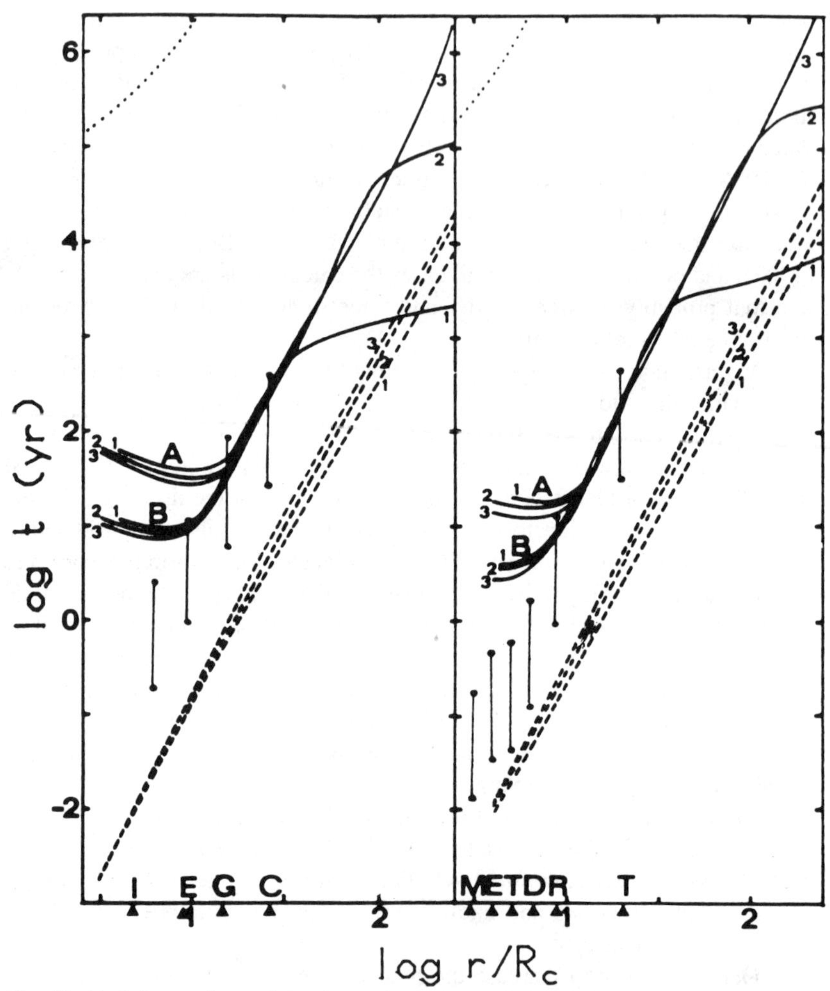

Fig. 22. Evolutionary times of the planetesimals around Jupiter (left) and Saturn (right), with particle masses equal to 1 g (1), 10^4 g (2), and 10^8 g (3). Solid lines refer to radial infall time of planetesimals. The infall time has been evaluated for $\vartheta = 50$ (A) and $\vartheta = 5$ (B). Dashed lines refer to the collisional accretion time τ_a. Dotted lines represent the dissipative infall time due to mutual encounters among planetesimals. Bars represent the growth time τ_f of embryos when $\vartheta = 5$ and $\vartheta = 50$.

TABLE II
Mass of the Embryo for Jupiter and Saturn Satellites, for Different Values of ϑ and No Overlapping of the Feeding Zones

Satellite	M_f/M_s		
	$\vartheta = 1$	$\vartheta = 5$	$\vartheta = 50$
Io	0.3	0.1	0.04
Europa	0.5	0.2	0.07
Ganymede	0.3	0.2	0.04
Callisto	0.6	0.2	0.07
Mimas	28.3	12.7	4.0
Enceladus	18.3	8.2	2.6
Tethys	53.9	24.2	7.5
Dione	50.0	22.6	7.1
Rhea	2.9	1.3	0.4
Titan	0.6	0.2	0.04

tion can be obtained if the combined effects of radial drift and orbital resonances are taken into account. In fact, Weidenschilling and Davis (1985) have shown that resonant perturbations oppose the gas drag and can cause trapping of bodies in stable orbits. Induced eccentricities are large, causing overlap of orbits in different resonances. This effect might speed again the accretion.

An important effect that needs investigating is the role of density waves in the accretion of satellites. Density waves can produce angular momentum transfer between an accreting embryo and the surrounding material (gas and/or small particles). Ward (1986) stresses the fact that density waves can be important as a coupling mechanism between gas and solid components, and that the combination of gas drag and density-wave torques may furnish an important degree of radial mobility for objects of virtually any size range. Ward (1986) attempts to quantify the effect of these processes on the accretion of Titan and concludes that for reasonable values of α, Titan can open a gap, but the damping length and hence width of this gap is comparable to the orbital radius. Narrow gaps are possible for $\alpha \cong 0.1$ to 1, but for such vigorous turbulence, a mass larger than Titan's is required for gap formation.

When comparing the results by Ward (1986) and Coradini and Magni (1984), it can be seen that:

1. "Traditional" accretion in the presence of gas does not allow the formation of large-sized satellites;
2. More satisfactory results could be reached assuming that the satellites grew at the expenses of material formed in the external region of the disk and subsequently migrated inward (Fig. 22);

3. The combined effect of gas drag and resonant trapping can speed the accretion and prevent the isolation of embryos, making the gas-rich scenario more plausible;
4. Radial migration of the satellites could have modified the position of satellites in the past; it would then be extremely important to establish when this migration process stopped.

V. SUBSEQUENT EVOLUTION: SOURCES OF THE ATMOSPHERE.

In the framework of satellite evolution by accretion processes, further collisional growth can take place if the relative velocities of embryo-impactors do not exceed a few km s^{-1}. In fact, the embryo will be catastrophically fragmented when the specific energy of the impact exceeds a threshold value, depending on the material strength, on the self-gravitation of the embryo and on the size of the impactor. In the strength-dominated regime, experiments performed by Lange and Ahrens (1982*c*) on icy targets indicate that this threshold impact specific energy is of the order of 10^2 J kg^{-1}. This seems to exclude cometary impactors, as the high relative velocities would most probably lead to the shattering of the target although the size ratio between impactor and embryo must be taken into account. Moreover, in order to have mass gain in the impact of a planetesimal on the embryo, Horedt (1980) argued that the impact velocity must be <3.2 times the escape velocity.

During an impact, a fraction of the impactor's kinetic energy will be irreversibly trapped as heat in the target, with the depth of burial of the heat depending on the size of the impactor. Following Wetherill (1976), a critical impactor size can be determined, above which heat fails to reach the surface by thermal conduction due to the growth of the satellite. This critical size is given by

$$R_{\mathrm{crit}} > k/(8\,\dot{R}) \tag{8}$$

where k is the thermal diffusivity and $\dot{R}$ the radius accretion rate. For Ganymede, this critical radius is approximately 2 cm, when normal lattice diffusivity of ice is used and $\dot{R}$ is about 1 km yr^{-1} (see Fig. 23); under the conditions of accretion, though, the dominant mode of heat transfer will be by mixing the surface material with subsequent impacts rather than by ordinary lattice conduction. Safronov (1969) computed values >100 times larger for k when this effect is taken into account; when this correction is applied, the critical radius becomes 2 m for Ganymede and 25 m for Callisto, the difference being due to an accretion time ~ 10 times smaller for Callisto, computed according to Sec. IV (see also Coradini et al. 1982).

Thus, it seems reasonable to assume that most impactors had sizes larger than this threshold value, and were able to deposit a substantial fraction of their kinetic energy beneath the accreting surface. This, in our opinion, rules out the possibility of planetary bodies starting their evolution as cold bodies, in the framework of the gas-free accretion scenario (Coradini et al. 1982; Federico and Lanciano 1983). The thermal and internal evolution of the satellites will depend on satellite size, composition, heat sources and initial conditions; it has been recently extensively reviewed by Stevenson et al. (1986) and Schubert et al. (1986). We will not follow here the same pattern, but rather try to identify the potential sources of an atmosphere in relation to composition and initial thermal structure of the satellites. We will not deal with the subsequent thermal evolution; the interested reader is referred to Schubert et al. (1986) and to Kirk and Stevenson (1987).

To assess the ability of planetary bodies to retain a stable atmosphere, Owen (1982*a*) proposed a very simple thermal stability parameter, defined on the basis of the escape velocity from the body, on its albedo and its distance from the Sun, according to which Titan and Triton rank just higher than Ganymede and Callisto, for a given gas species. Other nonthermal escape mechanisms discussed by Hunten et al. (see their chapter), act in similar directions for Ganymede, Callisto and Titan. The observed differences will therefore depend largely on the molecular weight of the released volatiles: released hydrogen will always escape, but N_2 and CH_4 released, for example, on Titan, will be stable under the conditions prevailing on this satellite. In what follows, we discuss briefly evolutionary routes for large satellites, assessing similarities and dissimilarities possibly leading to the release of different volatiles, and to the formation of an atmosphere. The evolution of small satellites will not be specifically dealt with here since, with the possible exception of satellites where tidal heating provides an important additional heat source, it is not relevant to the discussion of the possible formation and retention of an atmosphere.

The accretional temperature profiles derived by Lanciano et al. (1981) and Coradini et al. (1982) for Ganymede and Callisto are described as follows. The main energy sources to be accounted for are the conversion of accretional gravitational energy into heat and the energy produced by the decay of long-lived radioactive isotopes (U,K,Th), the amount of which is based on chondritic abundances in the silicate component of the ice-silicates mixture. This model has been developed in the framework of an accretion theory by means of large colliding objects depositing, as shown above, heat deep inside the target body. Furthermore, it is assumed that the interval between two subsequent impacts by planetesimals larger than 1 km in radius on the same area, is long enough to allow radiative cooling of the impact melt. The surface temperature in this model is therefore constant, and has been taken to be 100 K.

Solid-state convection has been simulated through a parameterized scheme in this model, a temperature-dependent viscosity has been assumed,

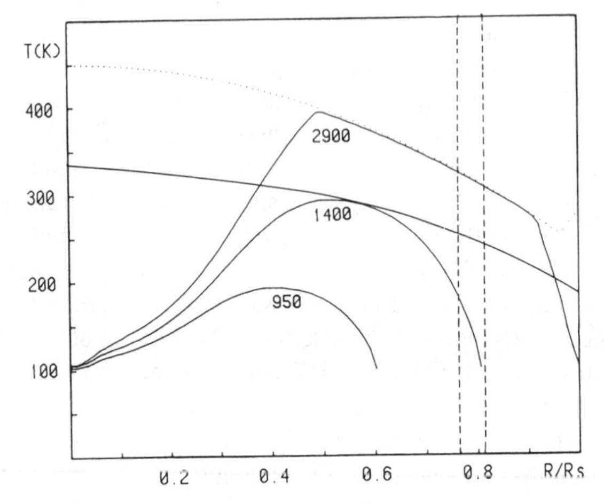

Fig. 23. Early thermal profiles for Ganymede. On each curve is labeled the time in years. The dotted line is the melting curve for water ice, the solid line the melting curve for $NH_3 - H_2O$. solution (after Johnson et al. 1985), and the dashed lines indicate the calculated boundary between clathrate and water plus methane (after Lunine and Stevenson 1985*a*). The region bounded by vertical dashed lines indicates the uncertainty in the calculated high-pressure boundary.

and the rheological properties of the water ice-silicate mixture have been assumed to be dominated by the viscosity of ice.

The accretional thermal profiles computed in this model for Ganymede and Callisto are shown in Figs. 23 and 24. The accumulation time interval for both satellites, based on the expression given in Sec. IV, is short enough ($<10^4$ yr) for the radioactive energy release to be neglible. For each impact, the parameter giving the initial partitioning of impact kinetic energy into

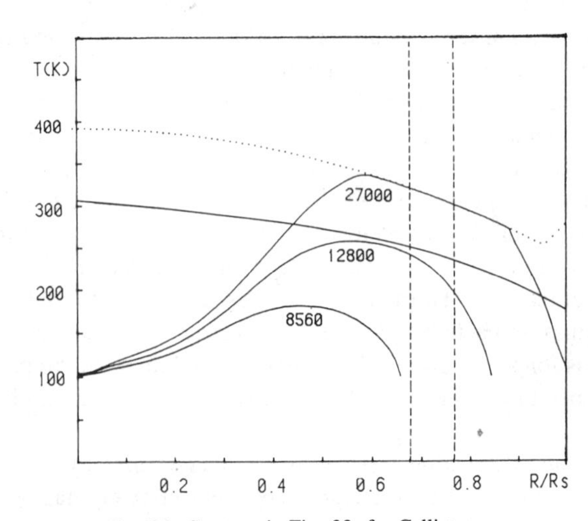

Fig. 24. Same as in Fig. 23, for Callisto.

waste heat was computed following the theory developed by Kieffer and Simmonds (1980) suitably modified in Coradini et al. (1982); for projectile and target materials and impact velocities typical of the accumulation process of Ganymede and Callisto, h has been found to range between 0.1 and 0.2. The melting curve of water ice, drawn for comparison in Figs. 23 and 24, indicates the presence of a large melted convective zone at the end of the accumulation in both models. Radiative cooling at the surface of the satellites, together with the strongly temperature-dependent rheology of water ice, accounts for the presence of an external rigid conductive shell in the models of both satellites. The thickness of this lithosphere is different in the two models, being larger for Callisto ($\sim$360 km) than for Ganymede ($\sim$210 km); this is due to convection being $\sim$10 times more vigorous in Ganymede than in Callisto. It should be noted that the presence of a crust of different thickness could allow a different subsequent thermal evolution for the two satellites, which is borne out by the widely different geologic surface features.

The choice of the parameter h, the fraction of the accretional gravitational energy irreversibly trapped as heat in the growing body, deserves some further discussion. In our model, a value of h as small as 0.2 is sufficient for the satellites to reach the end of the accumulation phases as internally hot and partially molten bodies. It could be argued that *a fortiori* this will be true, when values as large as those proposed by Ahrens and O'Keefe (1985) and Lange and Ahrens (1986*a*) ($0.85 < h < 0.9$) on the basis of impact experiments on water ice-silicate mixtures are used.

A second highly uncertain parameter, deserving some discussion, is the silicate mass fraction f. This parameter is crucial to the thermal evolution of icy satellites. In fact, the amount of long-lived radioactive isotopes, accounting for the subsequent thermal evolution after the accumulation is completed, is proportional to the silicate mass fraction. Moreover, the possibility of efficiently transferring heat through convection is also directly connected to the silicate mass fraction f. A larger f will increase the viscosity of the mixture; it has been computed (Friedson and Stevenson 1983) that an increase of viscosity of an order of magnitude over that of clean ice should be appropriate for the silicate mass fraction of the icy satellites. A larger viscosity will cause convection to set in at higher temperatures, with the final state being thus hotter and the melted region larger than what is derived by assuming convection is driven by the rheology of pure ice. According to the analysis performed by Friedson and Stevenson (1983), this effect will be relevant for a value of the silicate volume fraction appropriate to Ganymede.

Unfortunately, the observational constraints on satellite composition permit only the computation of an overall density, which does not allow a definite estimate to be made for the silicate mass fraction f, affected by uncertainties both in the percentage of anhydrous vs hydrate silicates and in the occurrence of phase transitions in water ice. In Fig. 25 the silicate mass fraction f is plotted vs the distance from the primary body for the satellites of Jupiter, Saturn and Uranus. The values of f have been computed from:

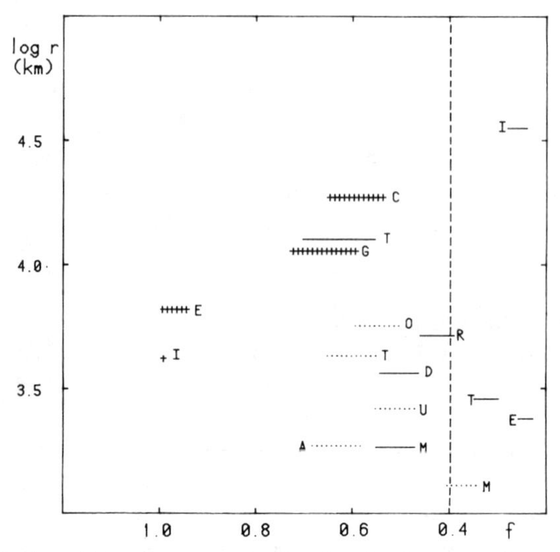

Fig. 25. Distribution of the silicate mass fraction f vs the distance to the central body for the satellites of Jupiter (crosses), Saturn (solid lines), and Uranus (dots). The dashed line indicates cosmic abundances.

$$f=(1-\rho_{ice}/\rho_{sat})/(1-\rho_{ice}/\rho_{sil}) \tag{9}$$

where the subscripts "sat," "sil" and "ice" refer to the entire satellite and to its silicate and ice components, respectively. The effect on f of the uncertainties in silicate density have been quantified, computing f for the two extreme values 2.5 g cm^{-3} (hydrated silicates) and 3.5 g cm^{-3} (anhydrous silicates); these values are indicated as error bars on the graph. It can be seen that in the Jovian system the silicate mass fraction seems to be correlated with the distance from the primary body, decreasing with increasing distances, while no regularities can be found for the Saturn and Uranus systems. A possible explanation for this lack of regularity has been put forward (Pollack and Consolmagno 1984, and reference therein) based on the role of collisional evolution. Present satellites could be the result of random reaccretion of collisionally disrupted planetesimals, which would explain the random distribution of silicate mass fraction with radial distance, possibly superimposed on a radial variation such as that exhibited by the Jupiter system.

Although thermal evolutionary models of icy satellites are still affected by the large uncertainties discussed above, it is interesting to extrapolate the results obtained for Ganymede and Callisto to Titan, for which no complete thermal model has been proposed to date.

It is interesting to note that Titan is intermediate in mass, radius and computed density between Ganymede and Callisto. On the basis of these parameters alone, one would be justified in inferring for Titan the same bulk composition as for Ganymede and Callisto; however, thermodynamic conditions prevailing in the Saturn nebula were possibly such to justify the occurrence of water ice in solution with ammonia as ammonia monohydrate, and incorporation of methane, nitrogen and argon as a clathrate in water ice (Figs. 18a,b). The dense atmosphere surrounding Titan, composed mainly of N_2 and CH_4, further supports different bulk compositions for these satellites. Notwithstanding these differences, the concentration of NH_3, if present, will not exceed the bulk ratio $NH_3/(NH_3 + H_2O) = 0.15$, based on cosmic abundances of oxygen and nitrogen (Consolmagno 1985). It can thus be confidently assumed that the thermal evolution will still be controlled by the rheology of water ice, until the melting temperature of ammonia monohydrate is reached. In Figs. 23 and 24 on the thermal profiles of Ganymede and Callisto, respectively, are superimposed as well the solid-liquid transition curve for a solution with the eutectic point at $T = 173$ K (after Johnson et al. 1985; Johnson and Nicol 1987), and a clathrate decomposition curve at high pressures, computed by Lunine and Stevenson (1985*a*). The presence of ammonia monohydrate in solution with water ice allows a larger melted convective zone to be formed inside the satellite at the end of the accumulation, which for Titan will be completed in a time intermediate between Ganymede and Callisto. On a very qualitative basis, the thickness of the lithosphere of Titan could be estimated from the intersection of the ammonia monohydrate melting curve with the conductive profile in the lithosphere. From Fig. 24 this thickness is about 100 km, and from Fig. 25 it is about 140 km, therefore thinner with respect to the lithosphere of Ganymede and Callisto. The presence of large convective motions beneath a thin lithosphere will favor the cracking of the lithosphere and the degassing of NH_3 from beneath it. Photochemical calculations indicate that, if NH_3 is outgassed from Titan, it may have been converted to a dense N_2 atmosphere during the lifetime of the satellite (Atreya et al. 1978). In a different scenario, a recent model by Lunine and Stevenson (1987) proposes a similar internal source for Titan's atmospheric N_2.

If most of the nitrogen is not in the form of ammonia, it could be incorporated as N_2 in clathrates. The supposed occurrence of clathrate decomposition at high pressures (Lunine and Stevenson 1985*a*) would provide a reservoir of clathrate guest molecules (N_2, CH_4, Ar) which could rise buoyantly through cracks to the surface and then into the atmosphere.

It emerges clearly from this qualitative discussion, that more detailed models are needed to account for the presence and the evolution of an atmosphere on Titan, in the framework of a gas-free accretion scenario. In particular, the thermal evolution of the satellite, the coupled evolution of its atmosphere and its interaction with a solid surface need to be further investigated. This is particularly true when recent experimental work by Johnson and Nicol (1987) is taken into account, which indicates the unexpected importance of dihydrate phases of water ammonia mixtures at pressures of interest for Titan's interior.

An alternative mechanism to explain the present atmosphere of Titan in a gas-rich accretion scenario has been proposed by Lunine and Stevenson (1982*b*), and discussed further in Lunine and Stevenson, 1985*a,b*). According to this model, incoming planetesimals formed mainly of water ammonia ice and methane clathrate hydrate will be fragmented in the primordial atmosphere and will gradually heat up the atmosphere and the satellite surface; at the same time, the increasing amount of volatiles released into the atmosphere through vaporization will change the composition of the atmosphere from the initial cosmic mixture of H_2 and He to a mixture dominated by CH_4 and NH_3. The subsequent evolution, assuming NH_3 is photolytically dissociated to N_2 in the primordial warm period as suggested by Atreya et al. (1978), has been outlined by Lunine and Stevenson (1985a). Briefly, the study of the dissociation pressure curve for a N_2-CH_4 mixture predicts that clathrate starts to be formed at 200 K; because of the greater propensity of CH_4 to be incorporated in clathrates, the composition of the atmosphere will change during the cooling phase, resulting in its being enriched in N_2 with decreasing temperature. Although the results depend on initial gas abundances, Lunine and Stevenson (1985*a*) for comparable initial values of N_2 and CH_4, find final abundances in agreement with the present observed atmospheric ratio N_2/CH_4 for Titan.

The subsequent evolution of the atmosphere thus formed has been modeled by Lunine and Stevenson (1985*b*) through the interaction with a surface methane-ethane-nitrogen ocean; the details of this work are beyond the scope of this review, which is aimed at the assessment of similarities and dissimilarities in the primordial thermal evolution of these satellites, possibly leading to the formation of an atmosphere.

The presence of a substantial atmosphere has been inferred for Triton as well, from groundbased infrared spectral measurements; this is not surprising since Triton, on the basis of the simple criterion for the stability of an atmosphere (as derived by Owen [1982*a*]) is more favored than Titan for the retention of a stable atmosphere. The hypothesized composition of this atmosphere is mainly nitrogen, with methane contributing only 10^{-3} of the total pressure. The surface composition of this satellite has been the object of some debate. The spectral features have been interpreted by Cruikshank and Brown (1986) as indicating the presence of liquid nitrogen, methane ice and water ice while Lunine and Stevenson (1985*a*), on the grounds of thermodynamical theoretical considerations, argue for the presence of solid nitrogen and methane. The large uncertainties in the values of virtually all the crucial parameters for this satellite prevent further speculations on the origin of its atmosphere, which must be postponed until a closer look of the satellite is obtained by Voyager 2 in 1989.

No other satellite in the outer solar system is known to date to possess a bound, stable, thick atmosphere. A tenuous atmosphere seems to be present on Io, while Enceladus seems to be the source of the tenuous observed E-ring of Saturn. In both cases, gases are vented during volcanic eruptions, the energy for which is supplied by a complex tidal evolution.

An alternative mechanism proposed for the E-ring is the expulsion of impact-produced fragments, although the observed spherical shape of these particles seems to favor a volcanic origin. It should be noted that a similar interaction mechanism has been proposed to explain the homogeneous dark material coating the surface of Umbriel (Stone and Miner 1986). This model, when applied to Umbriel, would require the existence of dark material orbiting with Umbriel and being accreted by it. Umbriel should also be the source of this material, either by the expulsion of impact ejecta, or through volcanic eruptions (although there is no evidence of volcanic activity on this satellite; nor does it present the complex tidal evolution typical of Enceladus).

VI. CONCLUSIONS

The sequence of events leading to the formation of regular satellite systems has been reviewed here, with particular interest to the primordial phases of formation of satellite accretion disks. Possible implications for the presence of atmospheres have been discussed as well, on the basis of the initial composition and thermal evolution of satellites.

Our conclusions about primordial satellite accretion disks can be summarized as follows:

1. The formation of satellite accretion disks is a natural consequence of the physical conditions in the solar nebula, and a by-product of the formation mechanism of the giant planets;
2. Accretion times do not exceed some 10^3 yr, and are only slightly dependent on the critical value of the core mass, but strongly dependent on gas and grain opacity;
3. Disks are turbulent, at least in the accretion phase, due to convective and/or Kelvin-Helmholtz instabilities;
4. Disk masses are large enough to feed satellite formation regions;
5. Matter is spread outside the present regular satellite region, but radial drift due to gas-particle friction may produce inward solid-particle migration (Weidenschilling 1977*b*; Coradini et al. 1980);
6. Disks have low angular momentum, with a significant mass infall in the hot region, so that solid particles experience evaporation and subsequent recondensation.
7. Ice is not present initially in the regular satellite region; it can condense afterwards, at the end of the accretion phase if turbulence decays and the disk cools down.

It should be noted that no quantitative model of this cooling phase has as yet been carried out. An alternative approach to determine the thermodynamic conditions prevailing when satellites were formed is to extrapolate backwards from the present state. For example, Prinn and Fegley (see their chapter) discuss in detail possible scenarios for the formation of satellites and the reservoirs of volatiles. This approach is based on chemical and thermodynamical considerations, rather than on physical and dynamical ones. It would be of the utmost importance to try to integrate these two different approaches into a single model for the formation of satellites. This has been attempted for the protosolar cloud by Morfill and Völk (1984), who discuss the transport of dust, trace gases and vapors in a turbulent protostellar accretion disk, giving a rough idea of the gross chemical structure and solid matter distribution in the planetary region. A similar approach has never been attempted to model satellite disks, which are not isolated systems and are characterized by large gradients in the physical quantities. These properties make it even more difficult to perform a coupled numerical simulation, which would treat simultaneously the dynamical and chemical evolution of satellite disks. Detailed studies of the disk relaxation from the primordial stages discussed here should be performed as well, thus allowing a more complete understanding of the complex chemistry of satellite formation, and of its implications for the formation of atmospheres.

Color Section

Plates 1, 2, 3, 4a, 4b and 5 give example spectra from PUMA 1 mode 0. In the lower part the original time based spectrum is plotted (yellow). In the same box the spectrum is inverted (red) and plotted with the correct mass scale (bars indicate mass decades). In the upper part a relevant portion of the spectrum is enlarged (yellow). The (blue) overlain spectrum is superimposed assuming singly charged atomic ions and normal isotopic composition of the elements to aid interpretation of the measured spectra. Vertical bars indicate integer mass positions, horizontal bars give with factors of 10 the intensities (log scale). The number in the upper left is the sequence number of the spectrum. The spectra are discussed in the text.

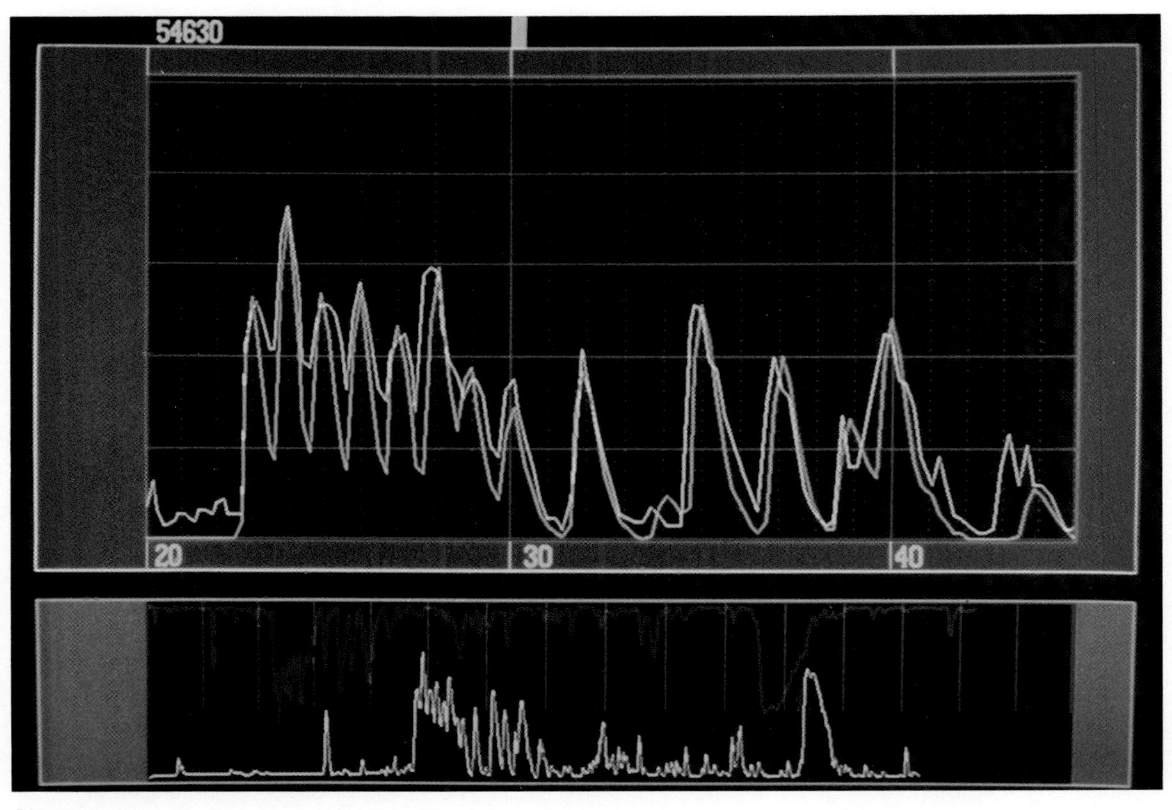

Plate 1.

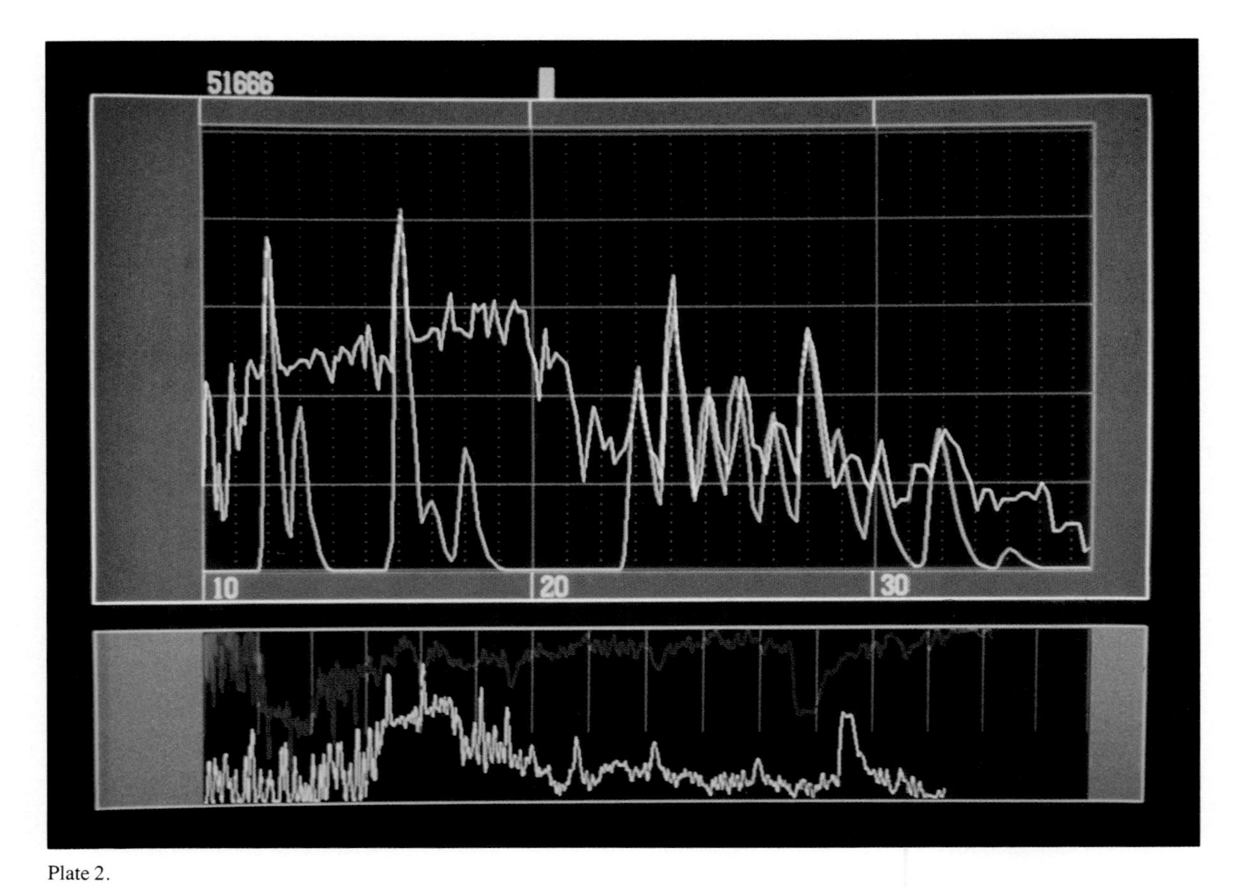

Plate 2.

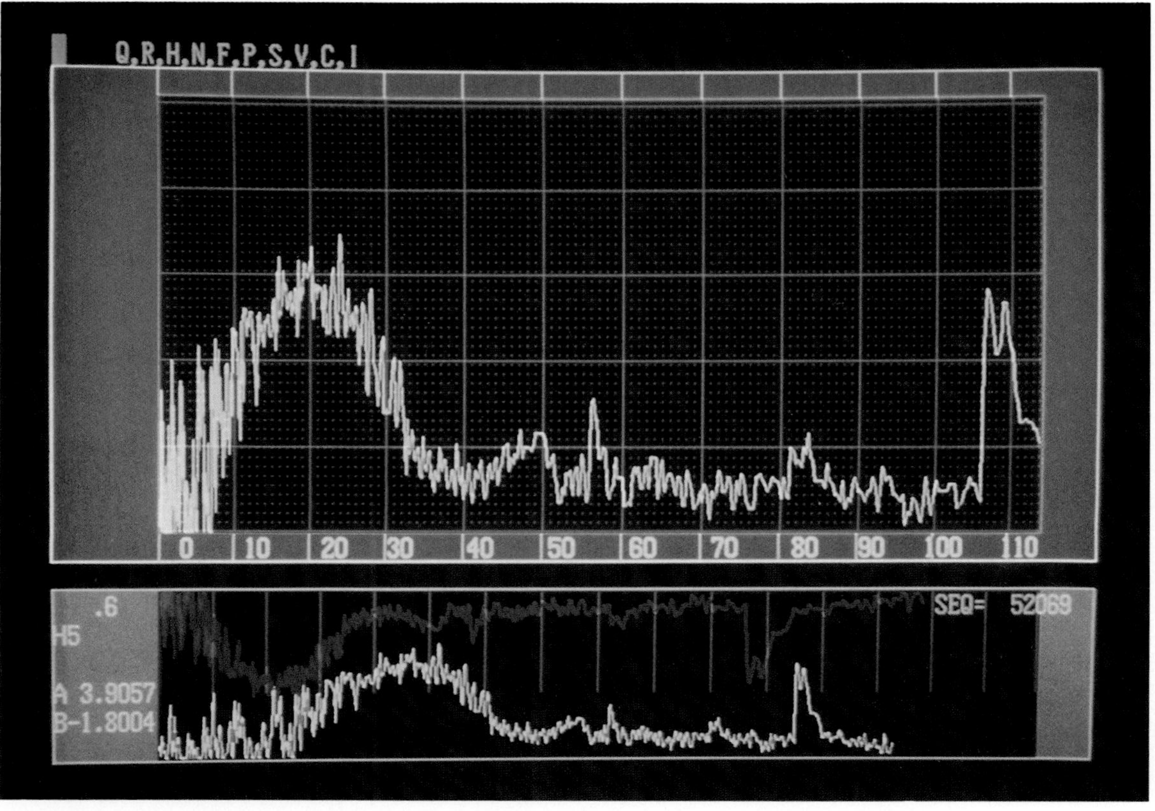
Q,R,H,N,F,P,S,V,C,I
0
10
20
30
40
50
60
70
80
90
100
110
.6
H5
A 3.9057
B-1.8004
SEQ= 52069

Plate 3.

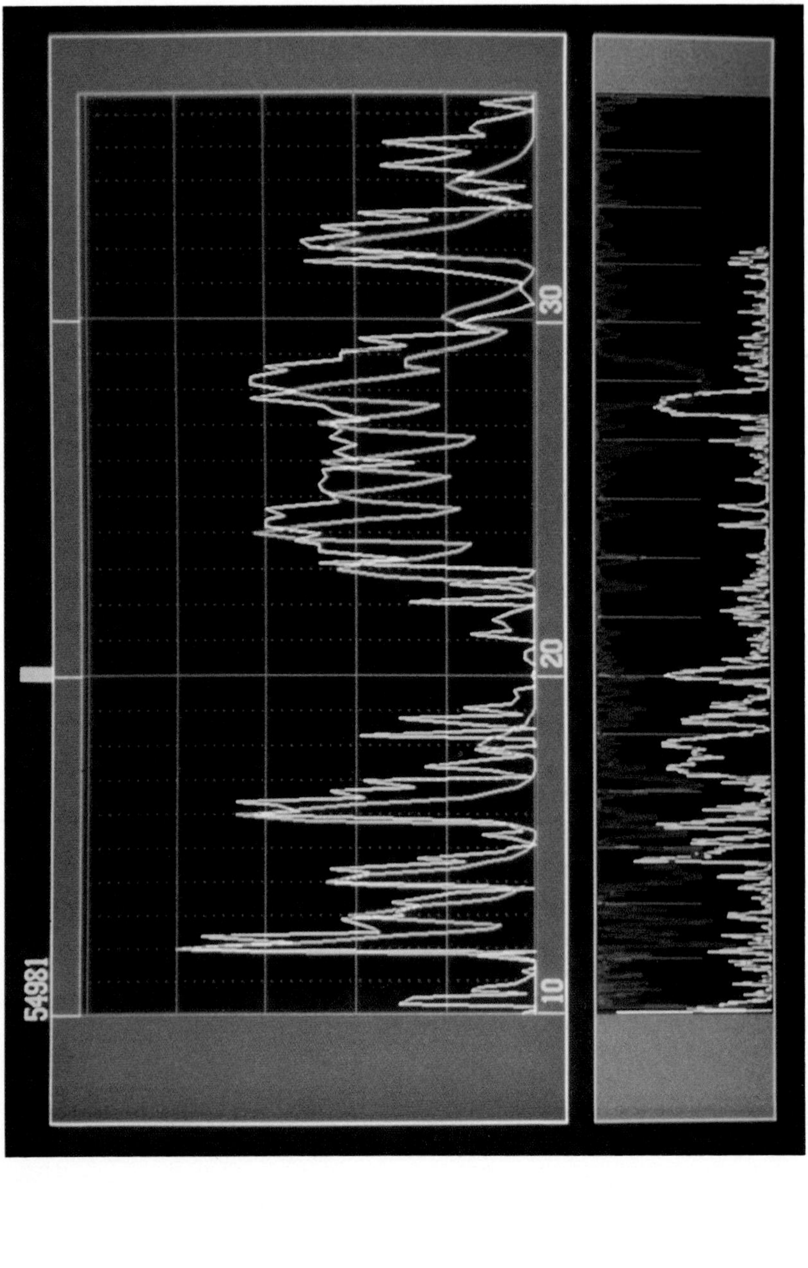

Plate 4a.

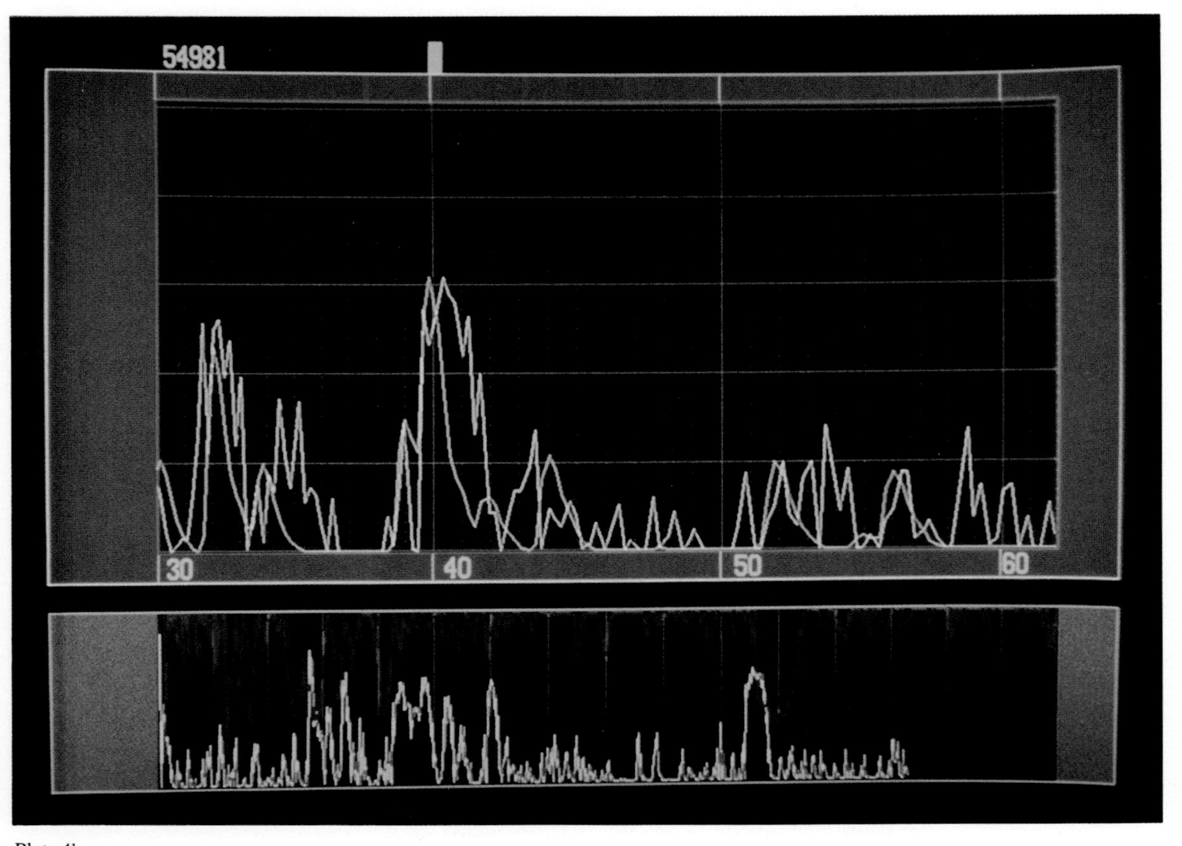

Plate 4b.

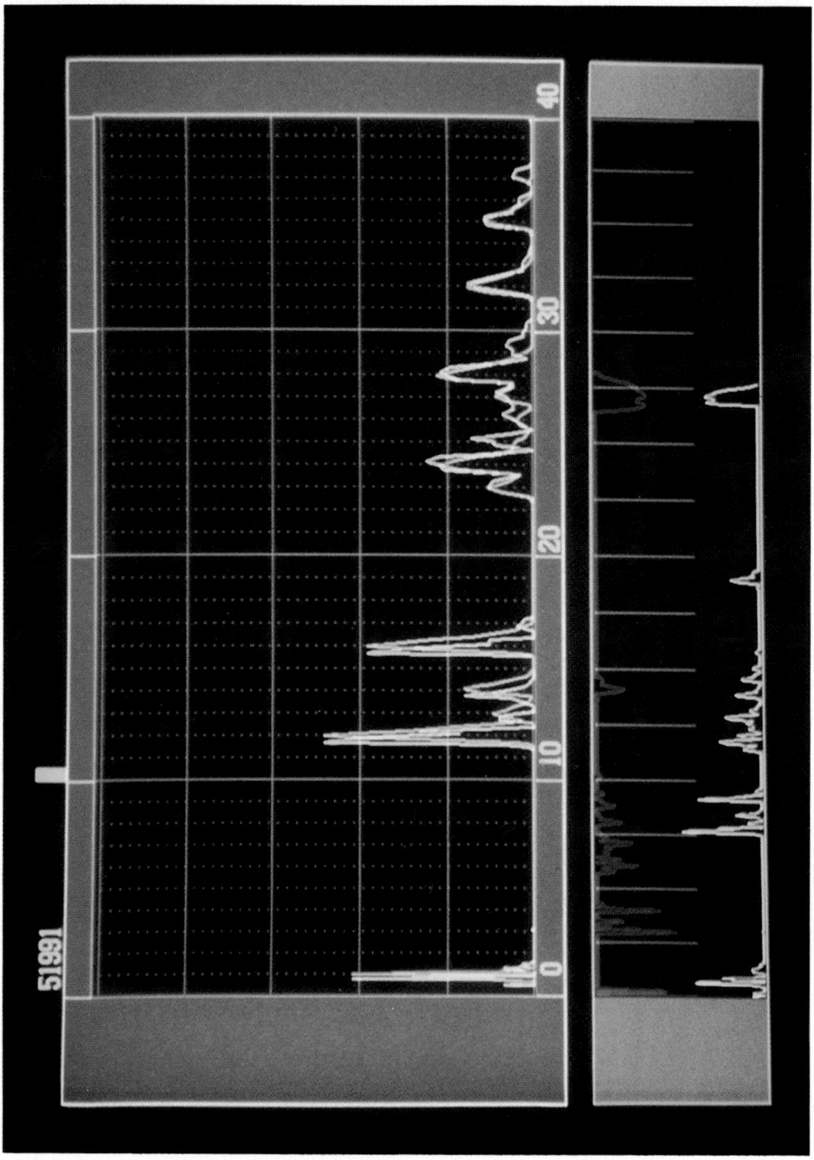

Plate 5.

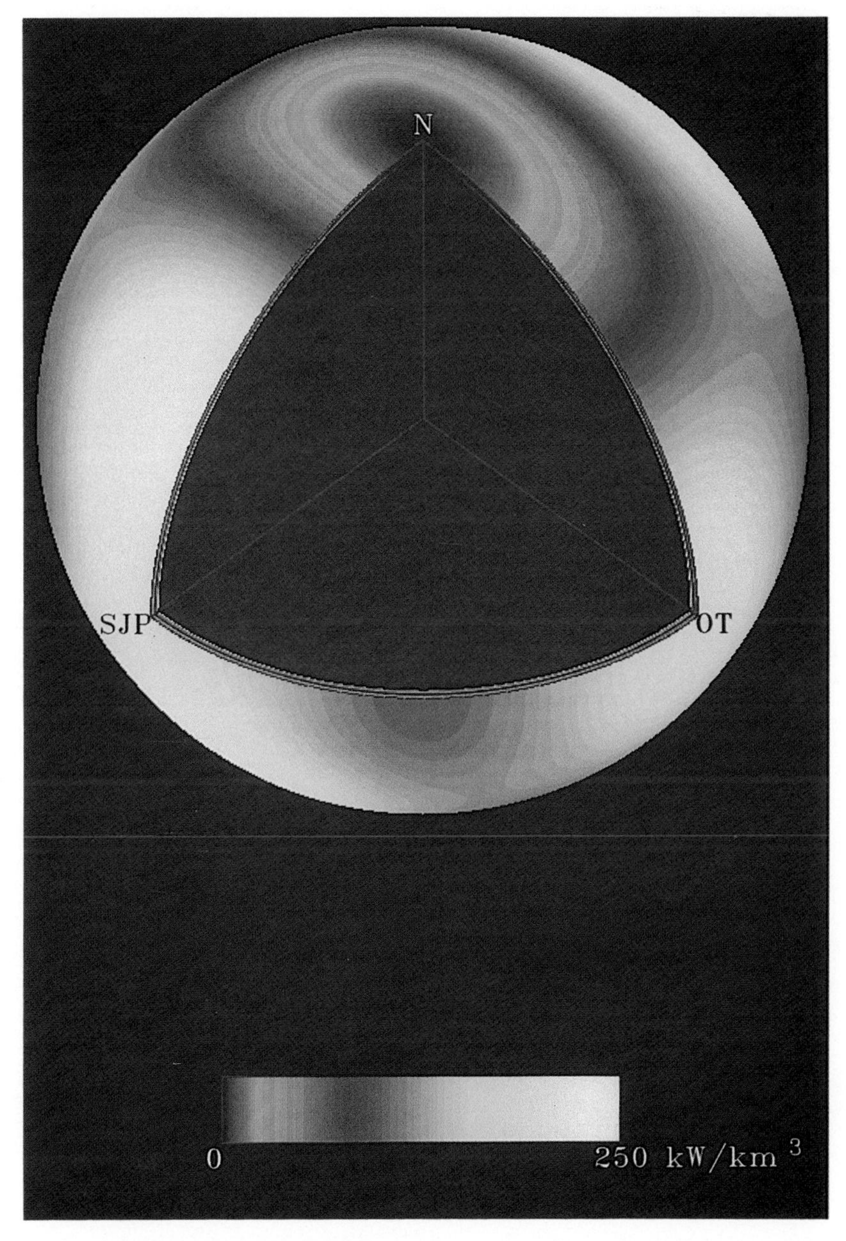

Plate 6. Spatial distribution of the rate of internal heat production (kWkm^{-3}) by tidal dissipation in a model of Io in which tidal heating occurs in a thin, near-surface asthenosphere (after Segatz et al. 1988). Heat production is highest near the equator. OT is the orbit tangent point and SJP is the sub-Jupiter point. N is the north pole. The triangular cut through the satellite shows heating only in a thin shell (asthenosphere) near the surface.

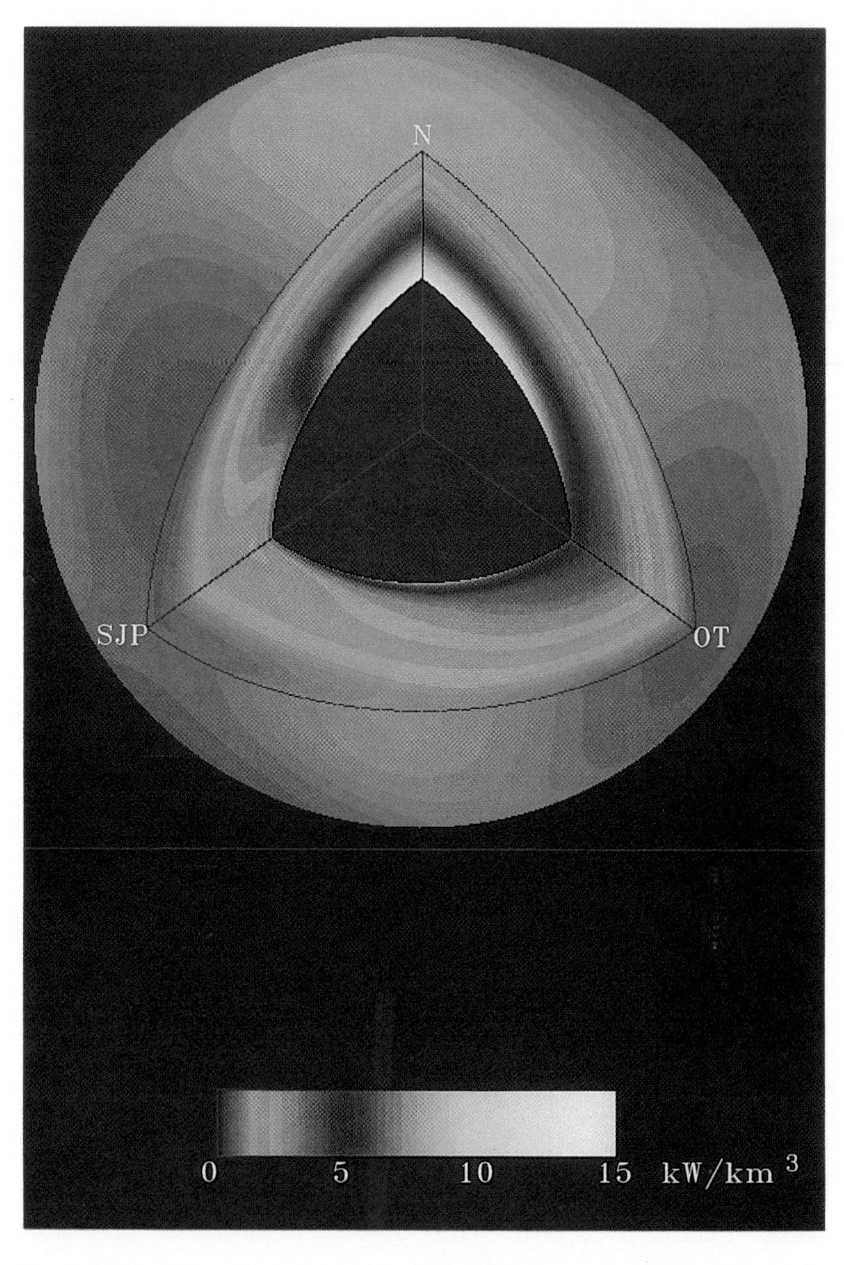

Plate 7. Similar to Plate 6 for an Io model in which tidal heating occurs throughout Io's mantle (after Segatz et al. 1988). Mantle dissipation concentrates heat production at the poles and near the core-mantle interface. The triangular cut through the satellite shows heating throughout the mantle. The inner spherical boundary is the outer surface of the core.

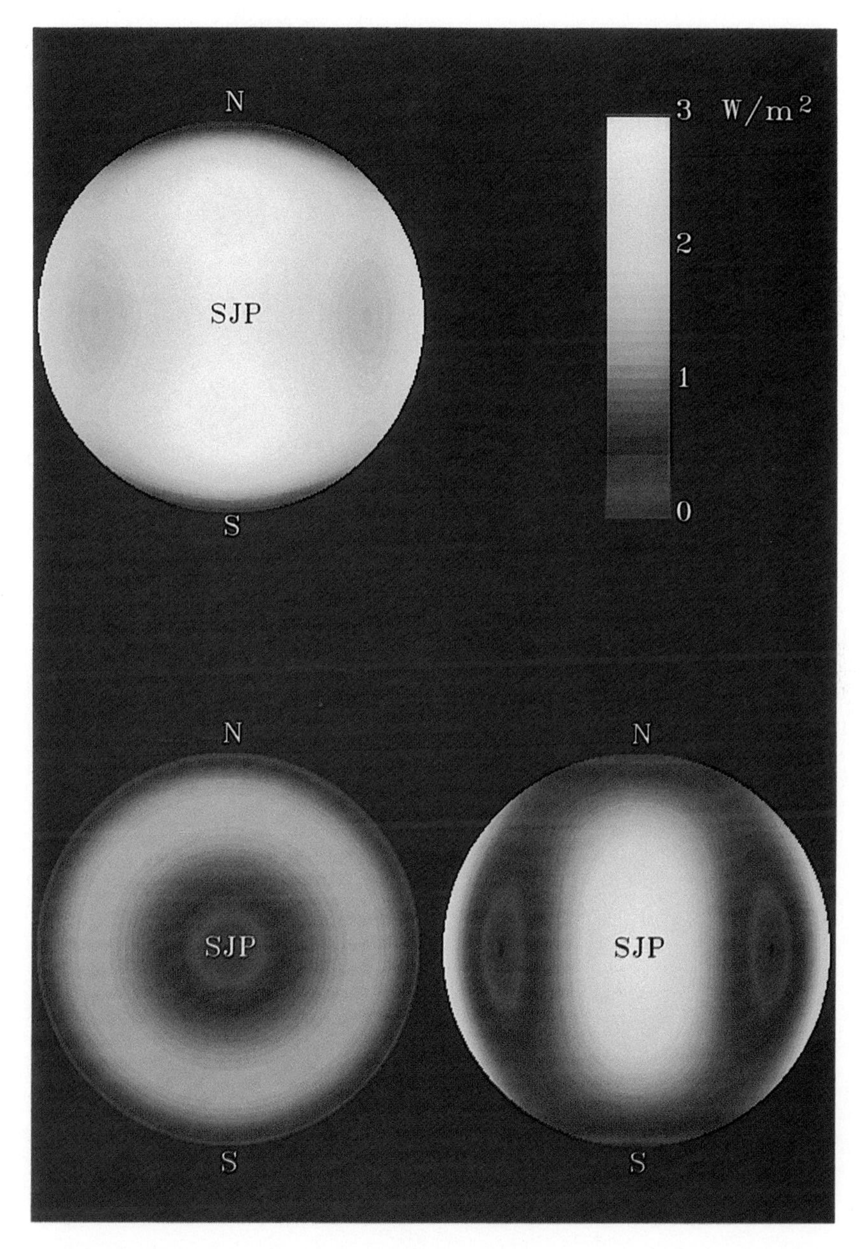

Plate 8. Surface heat flow (W m^{-2}) corresponding to the heat production distribution of Plate 6. Total surface heat flux is shown in upper left. The bottom panels show separate contributions to the surface heat flow from the tidal forcing due to Io's varying orbital distance from Jupiter (left) and due to Io's libration (right) (after Segatz et al. 1988). SJP is the sub-Jupiter point and N and S are the north and south poles. Surface heat flow is a maximum at the equator when tidal heating occurs in a thin asthenosphere.

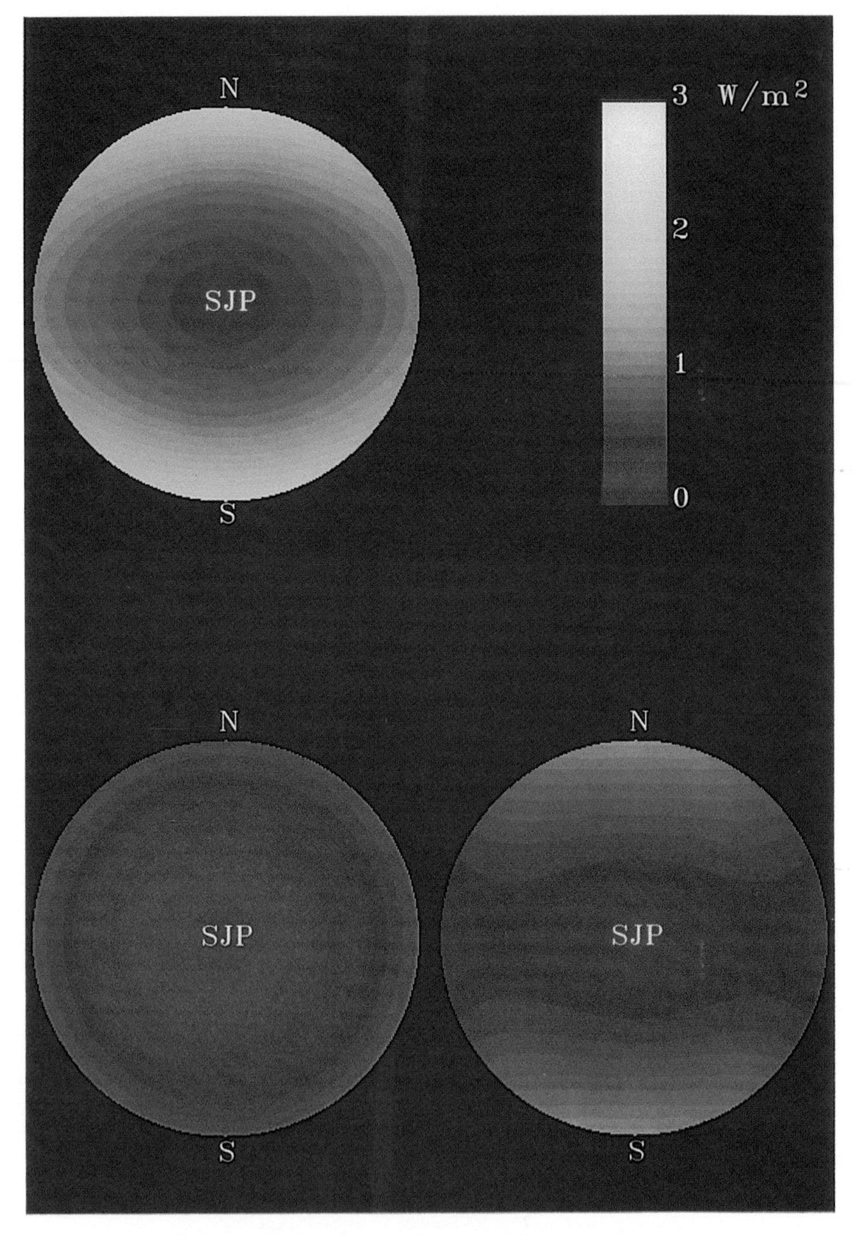

Plate 9. Similar to Plate 8 for the heat production distribution shown in Plate 7 (after Segatz et al. 1988). Surface heat flow maximizes at the poles when tidal heating occurs throughout Io's mantle.

Glossary

GLOSSARY*

Compiled by Melanie Magisos

accretion	the formation of planets or other bodies by aggregation of smaller bodies.
accretion, heterogeneous or inhomogeneous	the growth of planets from material whose composition varies during accretion.
accretion, homogeneous	the growth of planets from material whose composition stays constant throughout the accretion period.
adiabat	the trajectory in pressure-temperature space followed by a parcel of matter which undergoes changes in volume without exchanging heat with its surroundings.
adiabatic temperature gradient	the vertical temperature gradient, often approximated by convective regions of atmospheres, nebulae or solid planets, where adiabatic cooling and heating dominate over other thermal exchange mechanisms.
albedo, bond	ratio of the total reflected power to the incident solar power intercepted by a planet.

*We have used some definitions from *Glossary of Astronomy and Astrophysics* by J. Hopkins (by permission of the University of Chicago Press, copyright 1980 by the University of Chicago), from *Astrophysical Quantities* by C. W. Allen (London: Athlone Press, 1973), and from *Glossary of Geology,* edited by M. Gary, R. McAfee, and C. L. Wolff (Washington, D.C.: American Geological Institute, 1972). We also acknowledge definitions and helpful comments from various chapter authors and especially from D.H. Grinspoon.

albedo, geometric	ratio of the mean disk intensity at zero planetary phase angle to that for a perfect Lambert disk.
Alfvén waves	a transverse magnetohydrodynamic wave generated by tensions in lines of magnetic force.
Amor asteroids	asteroids having perihelion distance 1.017 AU $< q <$ 1.3 AU.
amu	atomic mass unit; AMU is also used.
Apollo asteroids	asteroids having semimajor axis $a >$ 1.0 AU and perihelion distance $q <$ 1.017 AU.
Aten asteroids	asteroids having semimajor axis $a <$ 1.0 AU.
atmospheric cratering	the ejection of atmospheric gases from a planet by impacts of planetesimals, asteroids or comets.
AU	astronomical unit = 1.496×10^{13} cm.
bar	a unit of atmospheric pressure; 1 bar = 10^6 dyne cm^{-2}= 0.987 atm.
Bohr radius	radius of the first orbital in the hydrogen atom, 0.529×10^{-8} cm.
blanketing effect	same as the *greenhouse effect,* except that the energy source is assumed to be located at the planet's surface.
Bonnor-Ebert sphere	an isothermal equilibrium state of a self-gravitating sphere of gas. Such a sphere has a finite radius and nonzero surface pressure.
Brownian motion	random motion of microscopic particles suspended in a gas due to collisions with gas molecules.
carbonaceous chondrite	a type of meteorite, rich in carbon, which appears to have undergone very little processing since its formation and thus is believed to be the most primitive type of meteorite.
carbonate metamorphism	*see* silicate reconstitution.

carbonate-silicate cycle	the process by which atmospheric carbon dioxide exchanges with CO_2 trapped in carbonate rocks.
C asteroid class	low albedo (mean p_v: 0.039) objects with featureless visual and near-infrared spectra. They comprise the majority of belt asteroids and dominate the central portion around 2.5 to 3.0 AU.
clathrates	a structure formed by the systematic inclusion of certain molecules in cavities within a crystal lattice. Structure I clathrate hydrates have 46 water molecules in the unit cell which form two pentagonal 12-sided polyhedra and six 14-sided polyhedra. Structure II clathrate hydrates have 136 water molecules in the unit cell which consists of sixteen pentagonal 12-sided polyhedra and eight 16-sided polyhedra.
clay mineral	a hydrous aluminum silicate, with Mg or Fe substituting for Al in the structure.
collisional evolution	change in size distribution of planetesimals due to collisions.
continuously habitable zone (CHZ)	the range of distances from a star in which a planet could maintain liquid water on its surface for a period of several billion years.
Copernicus	a satellite observatory launched by NASA to observe the ultraviolet spectral region. Also known as OAO-3 (Orbiting Astronomical Observatory-3).
cryosphere	The polar ice caps and permafrost-laden regions of Earth and Mars. On Venus, a term used to describe the cold upper atmosphere on the night side.
Eddington limit	the stellar core mass above which radiation pressure exceeds gravity, making the star's structure unstable; roughly 1.44 $M_\odot$.
emissivity	ratio of the radiation emitted by a body to that emitted by a blackbody at the same temperature.
energy balance	ratio of the emitted planetary energy to the absorbed solar energy.

Fischer-Tropsch synthesis production of organic molecules by the hydrogenation of carbon monoxide in the presence of a suitable catalyst.

fractionation the physical separation of one phase, element or isotope from another.

Galilean satellites the four largest satellites of Jupiter—Io (J I), Europa (J II), Ganymede (J III) and Callisto (J IV)—discovered by Galileo in 1610. All are locked in synchronous rotation with Jupiter.

giant molecular cloud (GMC) the most massive objects in our Galaxy (*see* molecular cloud), the formation sites for massive stars.

Gibbs free energy the function $G = E - TS + PV$, where E is the internal energy, T is the temperature, S is the entropy, P is the pressure and V is the volume. The Gibbs free energy reaches a minimum at equilibrium.

greenhouse effect the warming of a planet's surface caused by the trapping of outgoing infrared energy by the planet's atmosphere.

HAC hydrogenated amorphous carbon, a postulated constituent of interstellar grains.

Henry's law at sufficiently high dilution in a liquid solution, the partial pressure of a solute is proportional to its concentration.

Herbig-Haro objects semi-stellar, emission-line nebulae which are produced by shock waves in the supersonic outflow of material from young stars; also referred to as Herbig-Haro nebulae.

Hill sphere an approximately spherical region within which a planet, rather than the Sun, dominates the motion of particles.

Hubble time a time equal to one over the Hubble constant H, representing approximately the age of the universe.

Hugoniot the locus of points describing the pressure-volume-energy relations or states that may be achieved within a material by shocking it from a given initial state.

hydrocarbon	organic molecule composed solely of hydrogen and carbon atoms.
ion-molecule chemistry	chemistry based upon binary, gas phase reactions between positive molecular ions and neutral molecules, ultimately powered by cosmic ray and ultraviolet ionization; predictions of ion molecule chemistry provide a reasonable match to observed abundances for small interstellar molecules.
IRAS	Infrared Astronomical Satellite, launched by the United States.
IRIS	Infrared Interferometer Spectrometer, a Voyager instrument.
Jeans (or thermal) escape	the process by which planets and satellites lose the low mass molecular species in their atmosphere, usually hydrogen and helium due to species in the upper atmosphere achieving thermal velocities in excess of the escape velocity.
Kelvin-Helmholtz instability	the tendency of waves to grow on a shear boundary between two regions in relative motion parallel to the boundary.
Keplerian relative velocity	the velocity difference between two objects on Keplerian orbits.
Keplerian shear	shearing motion of an ensemble of particles, each on a nearly circular, Keplerian orbit. Orbital velocity decreases with orbital radius, yielding the shear. Viscous drag on such shear, due to collisions, plays a key role in ring processes.
Kirkwood gaps	regions in the asteroid zone which have been swept clear of asteroids by the perturbing effects of Jupiter. They were named for the American astronomer Daniel Kirkwood, who explained them in 1866.
Lagrangian points	the five equilibrium points in the restricted three-body problem. Two of the Lagrange points (L_4 and L_5) are located at the vertices of equilateral triangles formed by

	the two primaries (e.g., Sun and Saturn, or Saturn and satellite) and are stable; the other three are unstable and lie on the line connecting the two primaries.
Large Deployable Reflector	one in the NASA series of "great observatories," a space telescope for the far infrared and submillimeter spectral region.
libration	oscillation of an angular quantity about a fixed value. It often describes a stable orbital or rotational configuration.
$M_\odot$	solar mass = 1.989×10^{33}g.
mass extinction coefficient	optical thickness per unit mass.
Maxwell distribution	an expression for the statistical distribution of velocities among the molecules of a gas at a given temperature in equilibrium.
Miller-Urey synthesis	formation of organic molecules by the passage of an electric discharge, or energetic radiation, through a mixture of methane and ammonia over refluxing water.
moist greenhouse	a warm, humid atmosphere that is in equilibrium with liquid water at the surface.
molecular cloud	a condensation of interstellar gas with a density that is sufficiently high so that most of the hydrogen is in molecular form. These objects vary in mass from the nearby cold, dark clouds of a few hundred solar masses to the giant molecular clouds which may contain 10^5 to 10^6 solar masses.
nitriles	any organic compound containing the group $—C \equiv N$, which on hydrolysis yields an acid with elimination of ammonia.
noble gases	the gases He, Ar, Kr, Ne, Xe, Rn which rarely undergo chemical reactions, also known as inert gases and rare gases.
opacity	a loosely defined term referring to the ability of a medi-

	um to extinguish radiation of any given wavelength. In various applications, opacity has been used to mean: (a) optical thickness divided by physical thickness; (b) optical or radio thickness; or (c) mass extinction coefficient.
optical thickness	a measure of the ability of an absorbing or scattering medium to extinguish radiation defined so that $I/I_o = e^{-\tau}$ when I and I_o are the final and incident intensities and τ is the optical thickness.
PAH	polycyclic aromatic hydrocarbon, organic molecules containing tens of carbon atoms in linked ring structures, a postulated constituent of the interstellar medium.
paleosols	fossilized soils. They may provide information on the O_2 and CO_2 content of the Earth's early atmosphere.
parahydrogen	a molecule of hydrogen (H_2) in which the proton spins are opposite.
parsec	1 parsec = the distance at which 1 AU subtends 1 arcsec = 206,265 AU = 3.26 lightyear = 3.086×10^{18} cm.
planetary phase angle	the angle between the sunward direction from an object and the direction from which an observation is being made.
planetary phase function	the function that describes the variation in intensity of reflected light with phase angle for a given object.
planetesimal	small rocky or icy body formed in the primordial solar nebula.
polymerization	joining molecules together to form a larger, macromolecular complex. Originally the joined units were identical to each other (monomers); now the term commonly includes heteromolecular complexes.
ppb	parts per billion.
ppm	parts per million.

prograde motion	orbital or rotational motion in the same direction as the prevailing sense of rotation.
QCC	quenched carbonaceous composite, the name given by the investigators to the products formed by an electrical discharge in a tube containing methane; a poorly characterized, complex organic material, sometimes referred to as "tholin."
Raoult's law	in an ideal solution, the vapor pressure of a dissolved substance equals the product of its mole fraction in solution and the vapor pressure of the pure substance.
Rayleigh criterion	governing criterion for stability of an axisymmetric, rotating, inviscid fluid. Stability occurs only when the angular momentum per unit mass of the fluid increases monotonically outward.
Rayleigh scattering	the scattering law appropriate when the scattering molecules or particles are small with respect to the wavelength of light being scattered. The scattering is inversely proportional to the fourth power of the wavelength.
Rayleigh-Taylor instability	a type of hydrodynamic instability for static fluids (e.g., cold dense gas above hot rarefied gas).
regolith	surface layer of loose, fragmental debris produced by impacts on the surface of a body.
retrograde motion	orbital or rotational motion opposite to the prevailing sense of rotation.
Reynolds number	a dimensionless number ($R = L\mathrm{v}/\nu$, where L is a typical dimension of the system, v is a measure of the velocities that prevail, and ν is the kinematic viscosity) that governs the conditions for the occurrence of turbulence in fluids.
Roche limit	the minimum distance at which a fluid satellite influenced by its own gravitation and that of a central mass can be in mechanical equilibrium. For a satellite of zero tensile strength, and the same mean density as its pri-

mary, in a circular orbit around its primary, this critical distance is 2.46 times the radius of the primary.

Roche (Hill's) lobe — the largest closed zero-velocity potential surface surrounding a secondary body (satellite) in orbit about a primary. This surface passes through the Lagrange L_2 point which lies between the primary and secondary bodies.

runaway greenhouse — an atmosphere in which the infrared absorbing species are evaporated from surface reservoirs. The greenhouse blanketing of the infrared species heats the planet, leading to more evaporation and thus more heating.

runoff channels — one of several types of channels observed on the Martian surface. These channels appear to have been formed over a long period of time by liquid water flowing on the surface.

Safronov number — a parameter θ relating the random velocity in a swarm of particles to the escape velocity of a characteristic size particle (usually the largest in the swarm). θ = (escape velocity/random velocity)^2/2.

Schwarzschild criterion — stability against convective motions of a heated fluid requires that the temperature gradient be smaller in absolute magnitude than the adiabatic temperature gradient.

silicate reconstitution — the process in which silicate minerals are reformed from carbonate minerals plus silica, thereby releasing gaseous CO_2. This process occurs when carbonate minerals on the seafloor are subducted downwards at plate margins. Also referred to as "carbonate metamorphism."

silicate weathering — the process in which acids dissolved in groundwater react chemically with silicate minerals, releasing cations such as Ca^{++}, Mg^{++}, and Fe^{++}, along with silica and bicarbonate.

SNC meteorites — Shergottites, Nakhlites and Chassigny, a group of eight differentiated meteorites for which Mars is suggested as parent body.

solar EUV — extreme ultraviolet, short wavelength ultraviolet radiation from the Sun.

steradian	the solid angle which, having its vertex in the center of a sphere, cuts out an area of the surface of the sphere equal to that of a square with sides of length equal to the radius of the sphere. A complete sphere contains 4π steradians.
stratosphere	an upper portion of a planetary atmosphere, above the troposphere and below the ionosphere, characterized by a vertical temperature gradient which is stable against convection. On the Earth, its lower limit varies from about 8 to about 20 km; its upper limit lies at about 25 km.
thermal blanketing	the thermal regime of an early planetary atmosphere in which the energy of impact is trapped by a co-accreting radiatively opaque atmosphere.
thermal wind shear	vertical gradient in the horizontal wind speed caused by a horizontal temperature gradient assuming vertical hydrostatic balance and horizontal geostrophic balance.
tholins	organic solids produced by the irradiation of mixtures of cosmically abundant reducing gases. Term derived from the Greek for "muddy," and proposed by C. Sagan and B. Khare but not yet universally accepted in the literature.
Titius-Bode law	a mnemonic device discovered by Titius in 1766 and advanced by Bode in 1772, used for remembering the distances of the planets from the Sun. Take the series 0, 3, 6, 12, . . .; add 4 to each member of the series, and divide by 10. The resulting sequence 0.4, 0.7, 1.0, 1.6, . . . gives the approximate distance from the Sun (in AU) of Mercury, Venus, Earth, Mars, . . ., out to Uranus. The law fails for Neptune and beyond.
TMC-1	Taurus Molecular Cloud 1, a nearby, cold dark cloud (*see* molecular cloud).
torr	a unit of pressure equal to 1/760 of an atmosphere, or about 1 mm Hg.
tropopause	upper boundary of the troposphere (about 15 km), where the temperature gradient goes to zero.

troposphere	lowest level of the Earth's atmosphere, from zero altitude to about 15 km above the surface. This is the region where most weather occurs. Its temperature decreases from about 290 K to 240 K. More generally, the convective region of a planet's atmosphere.
T Tauri stars	young, late-type stars that are precursors to solar-mass stars characterized by emission-line spectra, infrared excesses, and irregular variability. The prototype for this class of stars is T Tau.
T Tauri wind	outflow from a T Tauri star.
UIR	unidentified infrared bands, a series of emission features seen in the infrared spectra between three and twelve microns of many planetary nebulae and H II regions.
valley networks	networks of branching runoff channels, or gullies, on the Martian surface that resemble river systems on the Earth.

Bibliography

BIBLIOGRAPHY

Compiled by Mary Guerrieri and Melanie Magisos

Abe, Y., and Matsui, T. 1985. The formation of an impact-generated H_2O atmosphere and its implications for the early thermal history of the Earth. *Proc. Lunar Planet. Sci. Conf.* 15, *J. Geophys. Res.* 90:C545–C559.

Abe, Y., and Matsui, T. 1986. Early evolution of the Earth: Accretion, atmosphere formation and thermal history. *Proc. Lunar Planet. Sci. Conf.* 17, *J. Geophys. Res.* 91:E291–E302.

Abe, Y., and Matsui, T. 1988. Evolution of an impact-generated H_2O-CO_2 atmosphere and formation of a hot proto-ocean on Earth. *J. Atmos. Sci.*, in press.

Abelson, P. H. 1966. Chemical events on the primitive Earth. *Proc. Natl. Acad. Sci. (USA)* 55:1365–1372.

Abt, H. A. 1983. Normal and abnormal binary frequencies. *Ann. Rev. Astron. Astrophys.* 21:343–372.

Adachi, I., Hayashi, C., and Nakazawa, K. 1977. The gas drag effect on the elliptic motion of a solid body in the primordial solar nebula. *Prog. Theor. Phys.* 56:1756–1771.

Adadurov, G. A. D., Balashov, D. B., and Dremin, A. N. 1961. A study of the volumetric compressibility of marble at high pressures. *Bull. Acad. Sci., USSR Geophys. Ser.* 5:463–466.

Adams, F. C., Lada, C. J., and Shu, F. H. 1987. Spectral evolution of young stellar objects. *Astrophys. J.* 312:788–806.

Adel, A., and Slipher. V. M. 1934. The constitution of the atmospheres of the giant planets. *Phys. Rev.* 46:902–906.

A'Hearn, M. F. Are cometary nuclei like asteroids? 1986. In *Asteroids, Comets, Meteors II,* eds. C.-I. Lagerkvist, B. A. Lindblad, H. Lundstedt and H. Rickman (Uppsala: Uppsala University), pp. 187–190.

A'Hearn, M. F., and Feldman, P. F. 1985. S_2: A clue to the origin of cometary ice? In *Ices in the Solar System,* eds. J. Klinger, D. Benest, A. Dollfus and R. Smoluchowski (Dordrecht: D. Reidel), pp. 463–471.

A'Hearn, M. F., and Millis, R. L. 1980. Abundance correlations among comets. *Astron. J.* 85:1528–1537.

A'Hearn, M. F., Feldman, P. D., Schleicher, D. G. 1983. The discovery of S_2 in Comet IRAS-Araki-Alcock 1983d. *Astrophys. J.* 274:L99–L103.

A'Hearn, M. F., Birch, P. V., and Klinglesmith, D. A., III. 1986. Gaseous jets in comet P/Halley. In *20th ESLAB Symposium on the Exploration of Halley's Comet,* vol. 1, ESA-SP 250 (Noordwijk: ESA), pp. 483–486.

Ahrens, T. J. 1987. Shock wave techniques for geophysics and planetary physics. In *Methods of Experimental Physics,* eds. C. Sammis and T. L. Henyey (New York: Academic Press), pp. 185–255.

Ahrens, T. J., and Gregson, V. G., Jr. 1964. Shock compression of crustal rocks: Data for quartz, calcite, and plagioclase rocks. *J. Geophys. Res.* 69:4839–4874.

Ahrens, T. J., and O'Keefe, J. D. 1972. Shock melting and vaporization of lunar rocks and minerals. *Moon* 4:214–249.

Ahrens, T. J., and O'Keefe, J. D. 1977. Equation of state and impact-induced shock-wave attenuation on the Moon. In *Impact and Explosion Cratering,* eds. D. J. Roddy, R. O. Pepin and R. B. Merrill (Elmsford, N.Y.: Pergamon Press), pp. 639–656.

Ahrens, T. J., and O'Keefe, J. D. 1985. Shock vaporization and the accretion of the icy satellites of Jupiter and Saturn. In *Ices in the Solar System,* eds. J. Klinger, D. Benest, A. Dollfus and R. Smoluchowski (Dordrecht: D. Reidel), pp. 631–654.

Ahrens, T. J., and O'Keefe, J. D. 1987. Impact on the Earth, ocean, and atmosphere. *J. Impact Eng.* 5:13–32.

Alaerts, L., Lewis, R. S., and Anders, E. 1979. Isotopic anomalies of noble gases in meteorites and their origins. IV. C3 (Ornans) carbonaceous chondrites. *Geochim. Cosmochim. Acta* 43:1421–1432.

Alden, H. L. 1943. Observations of the satellite of Neptune. *Astron. J.* 50:110–111.

Alexander, E. C., Jr. 1975. ^{40}Ar/^{39}Ar studies of Precambrian chert: An unsuccessful attempt to measure the time evolution of the atmospheric ^{40}Ar/^{36}Ar ratio. *Precambrian Res.* 2:239–344.

Alfvén, H., and Arrhenius, G. 1976. *Evolution of the Solar System,* NASA SP-345.

Allamandola, L. J., Tielens, A. G. G. M., and Barker, J. R. 1987. The IR emission features: Emissions from PAH molecules and amorphous carbon particles. In *Polycyclic Aromatic Hydrocarbons and Astrophysics,* eds. A. Leger, L. d'Hendecourt and N. Boccara (Dordrecht: D. Reidel), pp. 255–272.

Allègre, C. J., Staudacher, T., Sarda, P., and Kurz, M. 1983. Constraints on evolution of Earth's mantle from rare gas systematics. *Nature* 303:762–766.

Allègre, C. J., Staudacher, T., and Sarda, P. 1986/87. Rare gas systematics: Formation of the atmosphere, evolution and structure of the Earth's mantle. *Earth Planet. Sci. Lett.* 81:127–150.

Allen, C. W. 1976. *Astrophysical Quantities,* 3rd ed. (London: Athlone Press).

Allen, M., Pinto, J. P., and Yung, Y. L. 1980. Aerosol photochemistry and variations related to the sunspot cycle. *Astrophys. J.* 242:L125–L128.

Allen, M., Delitsky, M., Huntress, W., Yung, Y., Ip, W.-H., Schwenn, R., Rosenbauer, H., Shelley, E., Balsiger, H., and Geiss, J. 1987. Evidence for methane and ammonia in the coma of comet Halley. *Astron. Astrophys.* 187:502–512.

Allison, M. 1983. Planetary waves in Jupiter's equatorial atmosphere. *Bull. Amer. Astron. Soc.* 15:836 (abstract).

Alt, J. C., Honnorez, J., Laverne, C., and Emmermann, R. 1986. Hydrothermal alteration of a 1-km section through the upper oceanic crust, Deep Sea Drilling Project Hole 504B: Mineralogy, chemistry, and evolution of seawater-basalt interactions. *J. Geophys. Res.* 91:10309–10336.

Alvarez, L. W., and Muller, R. A. 1984. Evidence from crater ages for periodic impacts on the Earth. *Nature* 308:718–720.

Alvarez, W., Alvarez, L. W., Asaro, F., and Michel, H. V. 1980. Extra-terrestrial cause for the Cretaceous-Tertiary extinction: Experimental results and theoretical interpretation. *Science* 208:1095–1108.

Amari, S., and Ozima, M. The extra-terrestrial noble gases in deep sea sediments. 1988. *Geochim. Cosmochim. Acta.* 52:1087–1095.

Anders, E. 1964. Origin, age, and composition of meteorites. *Space Sci. Rev.* 3:583–714.

Anders, E. 1968. Chemical processes in the early solar system, as inferred from meteorites. *Acc. Chem. Res.* 1:289–298.

Anders, E. 1986. What can meteorites tell us? In *The Comet Nucleus Sample Return Mission,* ed. O. Meliter, ESA SP-249, pp. 31–39.

Anders, E. 1987. Local and exotic components of primitive meteorites, and their origin. *Phil. Trans. Roy. Soc. A* 323:287–304.

Anders, E. 1988. Circumstellar material in meteorites: Noble gases, carbon, and nitrogen. In *Meteorites and the Early Solar System,* eds. J. F. Kerridge and M. S. Matthews (Tucson: Univ. of Arizona Press), pp. 927–955.

Anders, E., and Ebihara, M. 1982. Solar-system abundances of the elements. *Geochim. Cosmochim. Acta* 46:2363–2380.

Anders, E., and Owen, T. 1977. Mars and Earth: Origin and abundance of volatiles. *Science* 198:453–465.

Andersen, H., and Bay, H. 1975. Sputtering yield measurements. In *Sputtering by Particle Bombardment. I. Physical Sputtering of Single Element Solids* (Heidelberg: Springer-Verlag), pp. 145–218.

Anderson, A. T. 1974. Chlorine, sulfur, and water in magmas and oceans. *Geol. Soc. Amer. Bull.* 85:1485–1492.

Anderson, A. T. 1975. Some basaltic and andesitic gases. *Rev. Geophys. Space Phys.* 13:37–55.

Anderson, D. L. 1972. The internal composition of Mars. *J. Geophys. Res.* 77:789–795.

Anderson, D. L. 1983. Chemistry of the primitive mantle. *Lunar Planet. Sci.* 14:5 (abstract).

Anderson, W. W., and Stevenson, D. J. 1987. Origin of nitrogen of Titan and Triton by accretion shock processing of primordial ammonia-rich atmospheres. Origin and Evolution of Planetary and Satellite Atmospheres, Abstract Booklet, 10-14 March, Tucson, AZ.

Appleby, J. F. 1986. Radiative convective equilibrium models of Uranus and Neptune. *Icarus* 65:383–405.

Appleby, J. F., and Hogan, J. S. 1984. Radiative-convective equilibrium models of Jupiter and Saturn. *Icarus* 59:336–366.

Armstrong, K. R., Harper, D. A., Jr., and Low, F. J. 1972. Far-infrared temperature of the planets. *Astrophys. J.* 178:L89–L92.

Arquilla, R., and Goldsmith, P. F. 1985. Density distributions in dark clouds. *Astrophys. J.* 29:436–454.

Arrhenius, G., De, B. R., and Alfvén, H. 1974. Origin of the ocean. In *The Sea,* ed. E. Goldberg (New York: Wiley), pp. 839–861.

Atreya, S. K. 1986. *Atmospheres and Ionospheres of the Outer Planets and Their Satellites* (New York: Springer-Verlag), pp. 145–197.

Atreya, S. K., and Romani, P. N. 1985. Photochemistry and clouds of Jupiter, Saturn, and Uranus. In *Recent Advances in Planetary Meteorology,* ed. G. E. Hunt (Cambridge: Cambridge Univ. Press), pp. 17–68.

Atreya, S. K., Donahue, T. M., and Kuhn, W. R. 1977. The distribution of ammonia and its photochemical products on Jupiter. *Icarus* 31:348–355.

Atreya, S. K., Donahue, T. M., and Kuhn, W. R. 1978. Evolution of a nitrogen atmosphere on Titan. *Science* 201:611–613.

Atreya, S. K., Donahue, T. M., Sandel, B. R., Broadfoot, A. L., and Smith, G. R. 1979. Jovian upper atmospheric temperature measurement by the Voyager 1 UV spectrometer. *J. Geophys. Res. Lett.* 6:795–798.

Audouze, J. 1977. The importance of CNO isotopes in astrophysics. In *CNO Isotopes in Astrophysics,* ed. J. Audouze (Dordrecht: D. Reidel), pp. 3–11.

Aumann, H. H., and Walker, R. G. 1987. IRAS observations of the Pluto-Charon system. *Astron. J.* 94:1088–1091.

Aumann, H. H., Gillespie, C. M., Jr., and Low, F. J. 1969. The internal powers and effective temperatures of Jupiter and Saturn. *Astrophys. J.* 157:L69–L72.

Aumann, H. H., Gillett, F. C., Beichman, C. A., de Jong, T., Houck, J. R., Low, F., Neugebauer, G., Walker, R. G., and Wasselius, P. R. 1984. Discovery of a shell around Alpha Lyrae. *Astrophys. J.* 278:L23–L27.

Bach, G. G., Kuhl, A. L., and Oppenheim, A. K. 1975. On blast waves in exponential atmospheres. *J. Fluid Mech.* 71:105–122.

Bachet, G., Cohen, E. R., Dore, P., and Birnbaum. 1983. The translational rotational absorption spectrum of hydrogen. *Can. J. Phys.* 61:591–603.

Bagenal, F., McNutt, R., Belcher, J., Bridge, H., and Sullivan, J. 1985. Revised ion temperatures for Voyager plasma measurements in the Io plasma torus. *J. Geophys. Res.* 90:1755–1757.

Bahcall, J. N., and Bahcall, S. 1985. The Sun's motion perpendicular to the galactic plane. *Nature* 316:706–708.

Baker, V. R. 1978. The Spokane flood controversy and the Martian outflow channels. *Science* 202:1240–1256.

Ballester, G. E., Moos, H. W., Feldman, P. D., Strobel, D. F., Summers, M. E., Bertaux, J.-L., Skinner, T. E., Festou, M. C., and Lieske, J. H. 1987. Detection of neutral oxygen and sulfur emissions near Io using IUE. *Astrophys. J.* 319:L33–L38.

Ballester, G. E., Feldman, P. D., and Moos, H. W. 1988. SO_2 on the atmosphere of Io: A near UV, high resolution IUE spectrum. *Eos: Trans. AGU* 69:394–395.

Bally, J. 1986. Interstellar molecular clouds. *Science* 232:185–193.

Balsiger, H., Altwegg, K., Bühler, F., Geill, J., Ghielmetti, A. G., Goldstein, B. E., Goldstein, R., Huntress, W. T., Ip, W.-H., Lazarus, H. J., Meier, A., Neugebauer, M., Rettenmund, U., Rosenbauer, H., Schwenn, R., Sharp, R. D., Shelley, E. G., Ungstrup, E., and Young, D. T. 1986. Ion composition and dynamics at comet Halley. *Nature* 321:330–334.

Banks, P. M., and Kockarts, G. 1973. *Aeronomy: Part A* (New York: Academic Press).

Barber, D. J. 1985. Phyllosilicates and other layer-structured minerals in stony meteorites. *Clay Min.* 20:415–454.

Barbosa, D. D., and Moreno, M. A. 1988. A comprehensive model of ion diffusion and charge exchange in the cold Io torus. *J. Geophys. Res.* 93:823–836.

Barbosa, D. D., Coroniti, F. V., and Eviatar, A. 1963. Coulomb thermal properties and stability of the Io torus. *Astrophys. J.* 274:429–442.

Barker, J. R., Allamandola, L. J., and Tielens, A. G. G. M. 1987. Anharmonicity and the interstellar polycyclic aromatic hydrocarbon infrared emission spectrum. *Astrophys. J.* 315:L61–L65.

Bar-Nun, A. 1979. Acetylene formation on Jupiter: Photolysis or thunderstorms? *Icarus* 38:180–191.

Bar-Nun, A., and Podolak, M. 1979. The photochemistry of hydrocarbons in Titan's atmosphere. *Icarus* 38:115–122.

Bar-Nun, A., Herman, G., Laufer, D., and Rappaport, M. L. 1985. Trapping and release of gases by water ice and implications for icy bodies. *Icarus* 63:317–332.

Bar-Nun, A., Dror, J., Kochavi, E., and Laufer, D. 1987. Amorphous water ice and its ability to trap gases. *Phys. Rev. B* 35:2427–2435.

Barshay, S. S. 1981. Combined Condensation-Accretion Models of the Terrestrial Planets. Ph.D. Thesis, Massachusetts Inst. of Technology.

Barshay, S. S., and Lewis, J. S. 1975. Chemistry of solar material. In *The Dusty Universe,* eds. G. B. Field and A. G. W. Cameron (New York: Neale Watson Academic Publ.), pp. 33–40.

Barshay, S. S., and Lewis, J. S. 1976. Chemistry of primitive solar material. *Ann. Rev. Astron. Astrophys.* 14:81–90.

Barshay, S. S., and Lewis, J. S. 1978. Chemical structure of the deep atmosphere of Jupiter. *Icarus* 33:593–611.

Barsukov, V. L., Basilevsky, A. T., Burba, G. A., Bobinna, N. N., Kryuchkov, V. P., Kuzmin, R. O., Nikolaeva, O. V., Pronin, A. A., Ronca, L. B., Chernaya, I. M., Shashkina, V. P., Garanin, A. V., Kushky, E. R., Markov, M. S., Sukhanov, A. L., Kotelnikov, V. A., Rzhiga, O. N., Petrov, G. M., Alexandrov, Yu. N., Sidorenko, A. I., Bogomolv, A. F., Skrypkik, G. I., Bergman, M. Yu., Kudrin, L. V., Bokshtein, I. M., Kronrod, M. A., Chochia, P. A., Tyuflin, Yu. S., Kadnichansky, S. A., and Akim, E. L. 1986. The geology and geomorphology of the Venus surface as revealed by the radar images obtained by Veneras 15 and 16. *Proc. Lunar Planet. Sci. Conf.* 16, *J. Geophys. Res.* 91:D378–D398.

Barth, C. A., Stewart, A. I., Hord, C. W., and Lane, A. L. 1972. Mariner 9 ultraviolet spectrometer experiment: Mars airglow spectroscopy and variations in Lyman alpha. *Icarus* 17:457–468.

Basaltic Volcanism Study Project. 1981*a*. Thermal histories of the terrestrial planets. In *Basaltic Volcanism on the Terrestrial Planets* (New York: Pergamon Press), pp. 1129–1234.

Basaltic Volcanism Study Project. 1981*b*. Table 4.3.2.d. In *Basaltic Volcanism on the Terrestrial Planets* (New York: Pergamon Press), p. 642.

Basford, J. R., Dragon, J. C. Pepin, R. O., Coscio, M. R., Jr., and Murthy, V. R. 1973. Krypton and xenon in lunar fines. *Proc. Lunar Sci. Conf.* 4:1915–1955.

Bates, D. R., and McDowell, M. R. C. 1957. Atmospheric helium. *J. Atmos. Terr. Phys.* 11:200–208.

Bates, D. R., and McDowell, M. R. C. 1959. Escape of helium. *J. Atmos. Terr. Phys.* 16:393–395.

Bates, D. R., and Patterson, T. N. L. 1981. Hydrogen atoms and ions in the thermosphere and exosphere. *Planet. Space Sci.* 5:257–273.

Baumgardner, J., Moreno, M. A., Schubert, G., and Kivelson, M. G. 1987. Two classes of volcanic eruptions and their corresponding atmospheres on Io. *Bull. Amer. Astron. Soc.* 19:856 (abstract).

Becker, R. H., and Pepin, R. O. 1984. The case for a martian origin of the shergottites: Nitrogen and noble gases in EETA 79001. *Earth Planet. Sci. Lett.* 69:225–242.

Becker, R. H., Pepin, R. O., Rajan, R. S., and Rambaldi, E. R. 1986*a*. Light noble gases in Weston metal grain surfaces. *Meteoritics* 21:331–332 (abstract).

Becker, R. H., Rajan, R. S., and Rambaldi, E. R. 1986*b*. Solar wind helium, neon and argon released by oxidation of metal grains from the Weston chondrite. In *Workshop on Past and Present Solar Radiation: The Record in Meteoritic and Lunar Regolith Material,* eds. R. O.

Pepin and D. S. McKay, LPI Tech. Rept. 86-02 (Houston: Lunar and Planetary Inst.), pp. 12–13.

Beckwith, S., Sargent, A. I., Scoville, N. Z., Masson, C. R., Zuckerman, B., and Phillips, T. G. 1986. Small-scale structure of the circumstellar gas of HL Tauri and R Monocerotis. *Astrophys. J.* 309:755–761.

Beer, R., and Taylor, F. W. 1973. The abundance of CH_3D and the D/H ratio in Jupiter. *Astrophys. J.* 179:309–327.

Beer, R., and Taylor, F. W. 1978. The D/H and C/H ratios in Jupiter from the CH_3D phase. *Astrophys. J.* 219:763–767.

Begelman, M. C., and Rees, M. J. 1976. Can cosmic clouds cause climatic catastrophes? *Nature* 261:298–299.

Beichman, C. A., Myers, P. C., Emerson, J. P., Harris, S., Mathieu, R., Benson, P. J., and Jennings, R. E. 1986. Candidate solar-type protostars in nearby molecular cloud cores. *Astrophys. J.* 307:337–349.

Belcher, J. 1983. The low energy plasma in the Jovian magnetosphere. In *Physics of the Jovian Magnetosphere,* ed. A. J. Dessler (New York: Cambridge Univ. Press), pp. 68–105.

Bell, J. F. 1986. Mineralogical evolution of the asteroid belt. *Lunar Planet. Sci.* XVII:985–986 (abstract).

Bell, J. F., Cruikshank, D. P., and Gaffey, M. J. 1985. The composition and origin of the Iapetus dark material. *Icarus* 61:192–209.

Bell, M. B., Avery, L. A., Matthews, H. E., Feldman, P. A., Watson, J. K. G., Madden, S. C., and Irvine, W. M. 1988. A study of C_3HD in cold interstellar clouds. *Astrophys. J.* 326:924–930.

Bell, P. M., Mao, H. K., and Hemley, R. J. 1986. Observations of solid H_2, D_2, and N_2 at pressures around 1.5 Mbar at 25 °C. *Physica* 139 & 140B:16–20.

Bellamy, C. J. 1975. *The Infrared Spectra of Complex Molecules* (London: Chapman & Hall).

Belton, M. J. S. 1982*a*. An interpretation of the near-ultraviolet absorption spectrum of SO_2: Implications for Venus, Io, and laboratory measurements. *Icarus* 52:149–165.

Belton, M. J. S. 1982*b*. An introductory review of our present understanding of the structure and composition of Uranus' atmosphere. In *Uranus and the Outer Planets,* ed. G. Hunt (Cambridge: Cambridge Univ. Press), pp. 155–172.

Belton, M. J. S., Wallace, L., and Howard, S. 1981. The periods of Neptune: Evidence for atmospheric motions. *Icarus* 46:263–274.

Benlow, A., and Meadows, A. J. 1977. The formation of the atmospheres of the terrestrial planets by impact. *Astrophys. Space Sci.* 46:293–300.

Bergstralh, J. T., Matson, D. L., and Johnson, T. G. 1975. Sodium D-line emission from Io: Synoptic observations from Table Mountain Observatory. *Astrophys. J.* 195:L131–L135.

Bergstralh, J. T., Young, J. W., Matson, D. L., and Johnson, T. V. 1977. Sodium D-line emission from Io: A second year of synoptic observation from Table Mountain Observatory. *Astrophys. J.* 211:L51–L55.

Bernatowicz, T. J., and Podosek, F. A. 1978. Nuclear components in the atmosphere. In *Terrestrial Rare Gases,* eds. E. C. Alexander, Jr., and M. Ozima (Tokyo: Japan Scientific Society Press), pp. 99–135.

Bernatowicz, T., Podosek, F., Honda, M., and Kramer, F. 1984. The atmospheric inventory of xenon and noble gases in shales: The plastic bag experiment. *J. Geophys. Res.* 89:4597–4611.

Bernatowicz, T. J., Kennedy, B. M., and Podosek, F. A. 1985. Xe in glacial ice and the atmospheric inventory of noble gases. *Geochim. Cosmochim. Acta* 49:2561–2564.

Bernatowicz, T., Fraundorf, G., Ming, T., Anders, E., Wopenka, B., Zinner, E., and Fraundorf, P. 1987. Evidence for interstellar SiC in the Murray carbonaceous meteorite. *Nature* 330:728–730.

Berner, R. A., Lasaga, A. C., and Garrels, R. M. 1983. The carbonate-silicate geochemical cycle and its effect on atmospheric carbon dioxide over the past 100 million years. *Amer. J. Sci.* 283:641–683.

Bertaux, J. L., and Belton, M. J. S. 1979. Evidence of SO_2 on Io from UV observations. *Nature* 282:813–815.

Bézanger, C., Bézard, B., and Gautier, D. 1986. Spatial variation of the thermal structure of Jupiter's atmosphere. In *The Jovian Atmospheres,* eds. M. Allison and L. D. Travis, NASA CP-2441, pp. 79–82.

Bézard, B. 1985. Thermal structure of Saturn's atmosphere. In *The Atmospheres of Saturn and Titan,* ESA SP-241, pp. 21–31.

Bézard, B., and Gautier, D. 1985*a*. Proceedings of the Conference on the Jovian Atmospheres, 6-8 May, New York, NY.

Bézard, B., and Gautier, D. 1985*b*. A seasonal climate model of the atmospheres of the giant planets at the Voyager encounter time. I. Saturn's stratospheres. *Icarus* 61:296–310.

Bézard, B., and Gautier, D. 1986. A model of the spatial and temporal variation of the Uranus thermal structures. In *The Jovian Atmospheres,* eds. M. Allison and L. D. Travis, NASA CP-2441, pp. 254–260.

Bézard, B., Marten, A., Baluteau, J. P., Gautier, D., Flaud, J.-M., and Camy-Peyret, C. 1983. On the detectability of H_2S in Jupiter. *Icarus* 55:259–271.

Bézard, B., Gautier, D., and Conrath, B. 1986*a*. A seasonal model of the Saturnian upper troposphere: Comparison with Voyager infrared measurements. *Icarus* 60:274–288.

Bézard, B., Gautier, D., and Marten, A. 1986*b*. Detectability of HD and non-equilibrium species in the upper atmospheres of the giant planets from their submillimeter spectrum. *Astron. Astrophys.* 161:387–402.

Bézard, B., Drossart, P., Maillard, J. P., Tanago, G., Lacome, N., Pounigue, G., Levy, A., and Gudachvilli, G. 1987. High-resolution spectroscopy of Saturn at 5 μm. II. Cloud structure and gaseous composition. *Bull. Amer. Astron. Soc.* 19:849 (abstract).

Bibring, J. P., and Rocard, F. 1984. Organic chemistry by irradiation in space. *Adv. Space Res.* 4:103–106.

Biermann, L., Giguere, P. T., and Huebner, W. F. 1982. A model of comet coma with interstellar molecules in the nucleus. *Astron. Astrophys.* 108:221–226.

Biloen, P., and Sachtler, W. M. H. 1981. Mechanism of hydrocarbon synthesis over Fischer-Tropsch catalysts. In *Advances in Catalysis,* eds. D. D. Eley, H. Pines and P. B. Weisz (New York: Academic Press), pp. 165–216.

Binder, A. B., and Cruikshank, D. P. 1964. Evidence for an atmosphere on Io. *Icarus* 3:299–305.

Bischoff, J. L., Rosenbauer, R. J., Aruscavage, P. J., Baedecker, P. A., and Crock, J. G. 1983. Seafloor massive sulfide deposits from 21°N, East Pacific Rise: Juan de Fuca Ridge; and Galapagos Rift: Bulk chemical composition and economic implications. *Econ. Geol.* 78:1711–1720.

Bjoraker, G. L., Larson, H. P., and Kunde, V. G. 1986*a*. The abundance and distribution of water vapor in Jupiter's atmosphere. *Astrophys. J.* 311:1058–1072.

Bjoraker, G. L., Larson, H. P., and Kunde, V. G. 1986*b*. The gas composition of Jupiter derived from 5-μm airborne spectroscopic observations. *Icarus* 66:579–609.

Black, D. C. 1971. On the equivalence of the planet-satellite formation processes. *Icarus* 15:115–119.

Black, D. C. 1972. On the origins of trapped helium, neon, and argon isotopic variations in meteorites—I. Gas-rich meteorites, lunar soil and breccia. *Geochim. Cosmochim. Acta* 36:347–375.

Black, D. C., and Bodenheimer, P. 1976. Evolution of rotating interstellar clouds. II. The collapse of protostars of 1, 2, and 5 $M_\odot$. *Astrophys. J.* 206:138–149.

Black, J. H., and Willner, S. P. 1984. Interstellar absorption lines in the infrared spectrum of NGC 2024 IRS2. *Astrophys. J.* 279:673–678.

Blake, G. A., Sutton, E. C., Masson, C. R., and Phillips, T. G. 1987. Molecular abundances in OMC-1: The chemical composition of interstellar molecular clouds and the influence of massive star formation. *Astrophys. J.* 315:621–645.

Blanco, V. M., and McCuskey, S. W. 1961. *Basic Physics of the Solar System* (Reading, Mass.: Addison-Wesley Publ. Co.).

Boato, G. 1954. The isotopic composition of hydrogen and carbon in the carbonaceous chondrites. *Geochim. Cosmochim. Acta* 6:209–220.

Bochsler, P., and Geiss, J. 1977. Elemental abundances in the solar wind. *Trans. IAU* 16B:120–123.

Bochsler, P., Geiss, J., and Kunz, S. 1986. Abundances of carbon, oxygen, and neon in the solar wind during the period from August 1978 to June 1982. *Solar Phys.* 103:177–201.

Bockelée-Morvan, D., Crovisier, J., Despois, D., Forveille, T., Gérard, E., Schraml, J., and Thum, C. 1986. A search for HCN and other parent molecules in Comets P/Giacobini-Zinner

1984e and P/Halley 1982i. In *Proceedings of the NRAO Workshop on Cometary Radio Astronomy,* eds. W. M. Irvine, F. P. Scloerb and L. E. Tacconi-Garman (Green Bank: NRAO), pp. 59–63.

Bodenheimer, P. 1968. The evolution of protostars of 1 and 12 solar masses. *Astrophys. J.* 153:483–494.

Bodenheimer, P., and Pollack, J. B. 1986. Calculations of the accretion and evolution of giant planets: The effects of solid cores. *Icarus* 67:391–408.

Bodenheimer, P., Grossman, A. S., DeCampli, W. M., Marcy, G., and Pollack, J. B. 1980*a*. Calculations of the evolution of the giant planets. *Icarus* 41:293–308.

Bodenheimer, P., Tohline, J. E., and Black, D. C. 1980*b*. Criteria for fragmentation in a collapsing rotating cloud. *Astrophys. J.* 242:209–218.

Boesgaard, A. M., and Steigman, G. 1985. Big bang nucleosynthesis: Theories and observations. *Ann. Rev. Astron. Astrophys.* 23:319–379.

Bogard, D. D., and Johnson, P. 1983. Martian gases in an Antarctic meteorite? *Science* 221:651–654.

Bond, G. C. 1962. *Catalysis by Metals* (London: Academic Press).

Borghese, A., Bussoletti, E., and Colangeli, L. 1987. Amorphous carbon and the unidentified infrared bands. *Astrophys. J.* 314:422–428.

Borucki, W., and Chameides, W. 1984. Lightning: Estimates of the rates of energy dissipation and nitrogen fixation. *Rev. Geophys. Space Phys.* 22:353–372.

Borucki, W., McKay, C., and Whitten, R. 1984. Possible production by lightning of aerosols and trace gases in Titan's atmosphere. *Icarus* 60:260–273.

Boss, A. P. 1982. Hydrodynamical models of presolar nebula formation. *Icarus* 51:623–632.

Boss, A. P. 1983. Fragmentation of a nonisothermal protostellar cloud. *Icarus* 55:181–184.

Boss, A. P. 1984*a*. Protostellar formation in rotating interstellar clouds. IV. Nonisothermal collapse. *Astrophys. J.* 277:768–782.

Boss, A. P. 1984*b*. Angular momentum transfer by gravitational torques and the evolution of binary protostars. *Mon. Not. Roy. Astron. Soc.* 209:543–567.

Boss, A. P. 1985. Three dimensional calculations of the formation of the presolar nebula from a slowly rotating cloud. *Icarus* 61:3–9.

Boss, A. P. 1986. Protostellar formation in rotating interstellar clouds. V. Nonisothermal collapse and fragmentation. *Astrophys. J. Suppl* 62:519–552.

Boss, A. P. 1987*a*. Theory of collapse and protostar formation. In *Interstellar Processes,* eds. D. J. Hollenbach and H. A. Thronson (Dordrecht: D. Reidel), pp. 321–348.

Boss, A. P. 1987*b*. Bipolar flows, molecular gas disks, and the collapse and accretion of rotating interstellar clouds. *Astrophys. J.* 316:721–732.

Boss, A. P. 1987*c*. Protostellar formation in rotating interstellar clouds. VI. Nonuniform initial conditions. *Astrophys. J.* 319:149–161.

Boss, A. P., and Bodenheimer, P. 1979. Fragmentation in a rotating protostar: A comparison of two 3-D computer codes. *Astrophys. J.* 234:289–295.

Boss, A. P., and Haber, J. G. 1982. Axisymmetric collapse of rotating, interstellar clouds. *Astrophys. J.* 255:240–244.

Boulanger, F., Baud, B., and van Albada, G. D. 1985. Warm dust in the neutral interstellar medium. *Astron. Astrophys.* 144:L9–L12.

Boulos, M. S., and Manuel, O. K. 1971. The xenon record of extinct radioactivities in the Earth. Science 174:1334–1336.

Bowell, E., Chapman, C. R., Gradie, J. C., Morrison, D., and Zellner, B. 1978. Taxonomy of asteroids. *Icarus* 35:313–335.

Bowers, T. S., and Taylor, H. P., Jr. 1985. An integrated chemical and stable-isotopic model of the origin of midocean ridge hot spring system. *J. Geophys. Res.* 90:12583–12606.

Brace, W. F., and Kohlstedt, D. L. 1980. Limits on lithospheric stress imposed by laboratory experiments. *J. Geophys. Res.* 85:6248–6252.

Bradley, J. P., and Brownlee, D. E. 1986. Cometary particles: Thin sectioning and electron beam analysis. *Science* 231:1542–1544.

Bradley, J. P., Brownlee, D. E., and Fraundorf, P. 1984. Carbon compounds in interplanetary dust: Evidence for formation by heterogeneous catalysis. *Science* 223:56–58.

Braterman, P. S., Cairns-Smith, A. G., and Sloper, R. W. 1983. Photo-oxidation of hydrated Fe^{+2}: Significance for banded iron formations. *Nature* 303:163–164.

Bratton, R. J., and Brindley, G. W. 1965. Kinetics of vapor phase hydration of magnesium oxide. Part 2: Dependence on temperature and water vapor pressure. *Trans. Faraday Soc.* 61:1017–1025.

Bregman, J. D., Campins, H., Witteborn, F. C., Wooden, D. H., Rank, D. M., Allamandola, L. J., Cohen, M., and Tielens, A. G. G. M. 1987. Airborne and groundbased spectrophotometry of Comet P/Halley from 5 to 13 micrometers. *Astron. Astrophys.* 187:616–620.

Breig, E. L., Hanson, W. B., Hoffman, J. H., and Kayser, D. C. 1976. In situ measurements of hydrogen concentration and flux between 160 and 300 km in the thermosphere. *J. Geophys. Res.* 81:2677–2686.

Brenig, W., and Menzel, D. 1985. *Desorption Induced by Electronic Transitions DIET II* (Berlin: Springer-Verlag).

Bridge, H. S., Belcher, J. W., Lazarus, A. J., Sullivan, J. D., McNutt, R. L., Bagenal, F., Scudder, J. D., Sittler, E. C., Siscoe, G. L., Vasyliunas, V. M., Goertz, C. K., and Yeates, C. M. 1979. Plasma observations near Jupiter: Initial results from Voyager 1. *Science* 204:987–990.

Bridge, H. S., Belcher, J. W., Coppi, B., Lazarus, A. J., McNutt, R. L., Jr., Richardson, J. D., Sands, M. R., Selesnick, R. S., Sullivan, J. D., Hartle, R. E., Ogilivie, K. W., Sittler, E. C., Jr., Bagenal, F., Wolff, R. S., Vasyliunas, V. M., Siscoe, G. L., Goertz, C. K., and Eviatar, A. 1986. Plasma observations near Uranus: Initial results from Voyager 2. *Science* 233:89–93.

Brinkmann, R. T. 1971. More comments on the validity of Jeans escape rate. *Planet. Space Sci.* 19:791–794.

Broadfoot, A. L., Shemansky, D., and Kumar, S. 1976. Mariner 10: Mercury atmosphere. *Geophys. Res. Lett.* 3:577–580.

Broadfoot, A. L., Belton, M. J. S., Takacs, P. Z., Sandel, B. R., Shemansky, D. E., Holberg, J. B., Ajello, J. M., Atreya, S. K., Donahue, T. M., Moos, H. W., Bertaux, J. L., Blamont, J. E., Strobel, D. F., McConnell, J. C., Calgarno, A., Goody, R., and McElroy, M. B. 1979. Extreme ultraviolet observations from Voyager 1 encounter with Jupiter. *Science* 204:979–982.

Broadfoot, A. L., Sandel, B. R., Shemansky, D. E., McConnell, J. C., Smith, G. R., Holberg, J. B., Atreya, S. K., Donahue, T. M., Strobel, D. F., and Bertaux, J. L. 1981. Overview of the Voyager ultraviolet spectrometry results through Jupiter encounter. *J. Geophys. Res.* 86:8259–8284.

Broadfoot, A. L., Herbert, F., Holberg, J. B., Hunten, D. M., Kumar, S., Sandel, B. R., Shemansky, D. E., Smith, G. R., Yelle, R. V., Strobel, D. F., Moos, H. W., Donahue, T. M., Atreya, S. K., Bertaux, J. L., Blamont, J. E., McConnell, J. C., Dessler, A. J., Linick, S., and Springer, R. 1986. Ultraviolet spectrometer observations of Uranus. *Science* 233:74–79.

Bronshtehn, V. A. 1983. *Physics of Meteoric Phenomena* (Dordrecht: D. Reidel).

Brooke, T. Y., and Knacke, R. F. 1986. The nucleus of Comet P/Arend-Rigaux. *Icarus* 67:80–87.

Brown, G. N., Jr., and Ziegler, W. T. 1979. Vapor pressure and heats of vaporization and sublimation of liquids and solids of interest in cryogenics below 1-atm pressure. *Adv. Cryogenic Eng.* 25:662–670.

Brown, H. 1949. Rare gases and the formation of the Earth's atmosphere. In *The Atmospheres of the Earth and Planets,* ed. G. P. Kuiper (Chicago: Univ. of Chicago Press), pp. 260–268.

Brown, H. 1952. Rare gases and formation of the Earth's atmosphere. In *The Atmospheres of the Earth and Planets,* 2nd ed., ed. G. P. Kuiper (Chicago: University of Chicago Press), pp. 258–266.

Brown, R. A. 1974. Optical line emission from Io. In *Exploration of the Planetary System,* eds. A. Woszczyk and C. Iwaniszewska (Dordrecht: D. Reidel), pp. 527–531.

Brown, R. A. 1981. The Jupiter hot plasma torus: Observed electron temperature and energy flows. *Astrophys. J.* 244:1072–1080.

Brown, R. A., and Yung, Y. 1976. Io: Its atmosphere and optical emission. In *Jupiter,* ed. T. Gehrels (Tucson: Univ. of Arizona Press), pp. 1102–1145.

Brown, R. A., Pilcher, C., and Strobel, D. 1983. Spectrophotometric studies of the Io torus. In *Physics of the Jovian Magnetosphere,* ed. A. J. Dessler (New York: Cambridge Univ. Press), pp. 197–225.

Brown, R. D., and Rice, E. 1981. Interstellar deuterium chemistry. *Phil. Trans. Roy. Soc. London* A303:523–533.
Brown, R. H., and Matson, D. L. 1987. Thermal effects of insolation propagation into the regoliths of airless bodies. *Icarus* 72:84–94.
Brown, W. L., and Johnson, R. E. 1986. Sputtering of ices: A review. *Nucl. Instrum. Meth.* B13:295–303.
Brown, W. L., Lanzerotti, L. J., and Johnson, R. E. 1982. Fast ion bombardment of ice and its astrophysical implications. *Science* 198:103–105.
Brown, W. L., Augustyniak, W. M., Marcantonio, K. J., Simmons, E. H., Boring, J. W., Johnson, R. E., and Reimann, C. T. 1984. Electronic sputtering of low temperature molecular solids. *Nucl. Instrum. Meth.* B1:307–314.
Brownlee, D. E. 1987. Interstellar grains in the solar system. In *Interstellar Processes,* eds. D. J. Hollenbach and H. Thronson (Dordrecht: D. Reidel), pp. 513–530.
Brownlee, D. E., Wheelock, M. M., Temple, S., Bradley, J. P., and Kissel, J. 1987. A quantitative comparison of comet Halley and carbonaceous chondrites at the submicron level. *Lunar Planet. Sci.* XVIII:133–134 (abstract).
Bryan, W. B., and Moore, J. G. 1977. Compositional variations of young basalts in the Mid-Atlantic Ridge rift valley near lat. 36°49′N. *Geol. Soc. Amer. Bull.* 88:556–570.
Buie, M. W., and Fink, U. 1987. Methane absorption variations in the spectrum of Pluto. *Icarus* 70:483–498.
Bunch, T. E., and Chang, S. 1980. Carbonaceous chondrites—II. Carbonaceous chondrite phyllosilicates and light element geochemistry as indicators of parent body processes and surface conditions. *Geochim. Cosmochim. Acta* 44:1543–1577.
Burgess, R., Wright, I. P., and Pillinger, C. T. 1987. Evidence for hydrothermal alteration in meteorites of high petrologic type. *Lunar Planet. Sci.* XVIII:137–138 (abstract).
Burghele, A., Dreibus, G., Palme, H., Rammensee, W., Spettel, B., Weckwerth, G., and Wänke, H. 1983. Chemistry of shergottites and the Shergotty parent body (SPB): Further evidence for the two-component model of planet formation. *Lunar Planet. Sci.* XIV:80–81 (abstract).
Buriez, J. C., and de Bergh, C. 1981. A study of the atmosphere of Saturn based on methane line profiles near 1.1 μm. *Astron. Astrophys.* 94:382–390.
Burns, J. A. 1986*a*. The evolution of satellite orbits. In *Satellites,* eds. J. A. Burns and M. S. Matthews (Tucson: Univ. of Arizona Press), pp. 117–158.
Burns, J. A. 1986*b*. Some background about satellites. In *Satellites,* eds. J. A. Burns and M. S. Matthews (Tucson: Univ. of Arizona Press), pp. 1–38.
Burns, L. E. 1985. The Border Ranges ultramafic and mafic complex, south-central Alaska: Cumulate fractionates of island-arc volcanics. *Can. J. Earth Sci.* 22:1020–1038.
Bussoletti, E. 1985. Interstellar grains. *Rev. Nuovo Cimento* 8:1–57.
Butchart, I., McFadzean, A. D., Whittet, D. C. B., Geballe, T. R., and Greenberg, J. M. 1986. Three micron spectroscopy of the galactic centre source IRS 7. *Astron. Astrophys.* 154:L5–L7.
Butler, W. A., Jeffery, P. M., Reynolds, P. M., and Wasserburg, G. J. 1963. Isotopic variations in terrestrial xenon. *J. Geophys. Res.* 68:3283–3291.
Butterworth, P. S., Caldwell, J., Moore, V., Owen, T., Rivolo, A. R., and Lane, A. L. 1980. An upper limit to the global SO_2 abundance on Io. *Nature* 285:308–309.
Byerlee, J. 1978. Friction of rocks. *Pure Appl. Geophys.* 16:615–626.
Cabot, W., and Savedoff, M. P. 1982. Meridional circulation in optically thick accretion disks. *Astron. Astrophys.* 112:L1–L2.
Cabot, W., Canuto, V. M., Hubickyj, O., and Pollack, J. B. 1987*a*. The role of turbulent convection in the primitive solar nebula. I. Theory. *Icarus* 69:387–422.
Cabot, W., Canuto, V. M., Hubickyj, O., and Pollack, J. B. 1987*b*. The role of turbulent convection in the primitive solar nebula. II. Results. *Icarus* 69:423–457.
Cahen, S. 1986. About the standard solar model. In *Proceedings of the 2nd Institut d'Astrophysique de Paris Workshop: Advances in Nuclear Astrophysics,* eds. E. Vangioni-Flam, J. Audouze, M. Casse and J. P. Chieze (Gif-sur-Yvette: Editions Frontières), pp. 97–104.
Cadogan, P. H. 1977. Paleoatmospheric argon in Rhynie chert. *Nature* 268:38–41.
Cameron, A. G. W. 1962. The formation of the sun and planets. *Icarus* 1:13–69.
Cameron, A. G. W. 1973*a*. Accumulation processes in the primitive solar nebula. *Icarus* 18:407–450.

Cameron, A. G. W. 1973*b*. Abundances of the elements in the solar system. *Space Sci. Rev.* 15:121–146.

Cameron, A. G. W. 1975. Cosmogonical considerations regarding Uranus. *Icarus* 24:280–284.

Cameron, A. G. W. 1978. Physics of the primitive solar accretion disk. *Moon and Planets* 18:5–40.

Cameron, A. G. W. 1979. The interaction between giant gaseous protoplanets and the primitive solar nebula. *Moon and Planets* 21:173–183.

Cameron, A. G. W. 1982. Elemental and nuclidic abundances in the solar system. In *Essays in Nuclear Astrophysics,* eds. C. A. Barnes, D. D. Clayton and D. N. Schramm (Cambridge: Cambridge Univ. Press), pp. 23–43.

Cameron, A. G. W. 1983. Origin of the atmospheres of the terrestrial planets. *Icarus* 56:195–201.

Cameron, A. G. W. 1984. Star formation and extinct radioactivities. *Icarus* 60:416–427.

Cameron, A. G. W. 1985*a*. Formation and evolution of the primitive solar nebula. In *Protostars & Planets II,* eds. D. C. Black and M. S. Matthews (Tucson: Univ. of Arizona Press), pp. 1073–1099.

Cameron, A. G. W. 1985*b*. Formation of the prelunar accretion disk. *Icarus* 62:319–327.

Cameron, A. G. W. 1986. The impact theory for the origin of the Moon. In *Origin of the Moon,* eds. W. K. Hartmann, R. J. Phillips and G. J. Taylor (Houston: Lunar and Planetary Inst.), pp. 609–616.

Cameron, A. G. W., and Benz, W. 1987. Planetary collision calculations: Origin of Mercury. *Lunar Planet. Sci.* XVIII:151–152 (abstract).

Cameron, A. G. W., and Pine, M. R. 1973. Numerical models of the primitive solar nebula. *Icarus* 18:377–406.

Cameron, A. G. W., and Pollack, J. B. 1976. On the origin of the solar system and of Jupiter and its satellites. In *Jupiter,* ed. T. Gehrels (Tucson: Univ. of Arizona Press), pp. 61–84.

Cameron, A. G. W., and Truran, J. W. 1977. The supernova trigger for formation of the solar system. *Icarus* 30:447–461.

Cameron, A. G. W., and Ward, W. R. 1976. The origin of the Moon. *Lunar Planet. Sci.* VII:120–122 (abstract).

Cameron, A. G. W., DeCampli, W. M., and Bodenheimer, P. 1982. Evolution of giant gaseous protoplanets embedded in the primitive solar nebula. *Icarus* 49:298–312.

Campbell, D. B., and Burns, B. A. 1980. Earth-based radar imagery of Venus. *J. Geophys. Res.* 85:8271–8281.

Campbell, D. B., Head, J. W., Harmon, J. K., and Hine, A. A. 1983. Venus: Identification of banded terrain in the mountains of Ishtar Terra. *Science* 221:644–647.

Campbell, D. B., Head, J. W., Harmon, J. K., and Hine, A. A. 1984. Venus: Volcanism and rift formation in Beta Regio. *Science* 226:167–170.

Campbell, J. K., and Synnott, S. P. 1985. Gravity field of the Jovian system from Pioneer and Voyager tracking data. *Astron. J.* 90:364–372.

Campins, H., and Ryan, E. V. 1987. The structure of the silicate emission in comet Halley. *Bull. Amer. Astron. Soc.* 19:839 (abstract).

Campins, H., and Ryan, E. V. 1988. Identification of crystalline silicates in cometary dust. *Bull. Amer. Astron. Soc.* 20:732 (abstract).

Canalas, R., Alexander, E., and Manuel, O. 1968. Terrestrial abundance of noble gases. *J. Geophys. Res.* 73:3331–3334.

Canuto, V. M., and Goldman, I. 1985. Analytical model for large-scale turbulence. *Phys. Rev. Lett.* 54:430–433.

Canuto, V. M., Levine, J., Augustsson, T., and Imhoff, C. 1982. UV radiation from the young Sun and oxygen levels in the pre-biological palaeoatmosphere. *Nature* 296:816–820.

Canuto, V. M., Levine, J. S., Augustsson, T. R., Imhoff, C. L., and Giampapa, M. S. 1983. The young Sun and the atmosphere and photochemistry of the early Earth. *Nature* 305:281–286.

Canuto, V. M., Goldman, I., and Hubickyj, O. 1984. A formula for the Shakura-Sunyaev turbulent viscosity parameter. *Astrophys. J.* 280:L55–L58.

Capps, R. W., Gillett, F. C., and Knacke, R. F. 1978. Infrared observations of the OH source W33A. *Astrophys. J.* 226:863–868.

Carey, W. C., Walker, R. M., and Bradley, J. P. 1987. Interplanetary dust particles and comet Halley: A comparative study. *Meteoritics* 22:348–349 (abstract).

Carlson, R. W., and Judge, D. L. 1976. Pioneer 10 ultraviolet photometer observations of Jupiter: The helium to hydrogen ratio. In *Jupiter,* ed. T. Gehrels (Tucson: Univ. of Arizona Press), pp. 418–440.
Carr, M. H. 1986. Mars: A water-rich planet. *Icarus* 68:187–216.
Carr, M. H. 1987. Water on Mars. *Nature* 326:30–35.
Carr, M. H., Saunders, R. W., Strom, R. G., and Wilhelms, D. E. 1984. *The Geology of the Terrestrial Planets,* NASA SP-469.
Carter, N. L. 1976. Steady state flow of rocks. *Rev. Geophys. Space Phys.* 14:301–360.
Cassen, P. M., and Moosman, A. 1981. On the formation of protostellar disks. *Icarus* 48:353–376.
Cassen, P., and Pettibone, D. 1976. Steady accretion of a rotating fluid. *Astrophys. J.* 208:500–511.
Cassen, P. M., and Summers, A. 1983. Models of the formation of the solar nebula. *Icarus* 53:26–40.
Cassen, P., Peale, S. J., and Reynolds, R. T. 1980. On the comparative evolution of Ganymede and Callisto. *Icarus* 41:232–239.
Cassen, P. M., Smith, B. F., Miller, R. H., and Reynolds, R. T. 1981. Numerical experiments on the stability of preplanetary disks. *Icarus* 48:377–392.
Cassen, P., Peale, S. J., and Reynolds, R. T. 1982. Structure and thermal evolution of the Galilean satellites. In *Satellites of Jupiter,* ed. D. Morrison (Tucson: Univ. of Arizona Press), pp. 93–128.
Cassen, P., Shu, F., and Terebey, S. 1985. Protostellar discs and star formation: An overview. In *Protostars & Planets II,* eds. D. C. Black and M. S. Matthews (Tucson: Univ. of Arizona Press), pp. 448–483.
Cernicharo, J., and Guélin, M. 1987. Metals in IRC+10216: Detection of NaC1, A1C1, and KC1, and tentative detetection of A1F. *Astron. Astrophys.* 183:L10–L12.
Cernicharo, J., Guélin, M., Menten, K. M., and Walmsley, C. M. 1987. C_6H: Astronomical study of its fine and hyperfine structure. *Astron. Astrophys.* 181:L1–L4.
Cernicharo, J., Kahane, C., Guélin, M., and Gomez-Gonzalez, J. 1988. Tentative detection of CH_3NC towards Sgr B2. *Astron. Astrophys.* 189:L1–L2.
Cess, R. D., and Caldwell, J. 1979. A model of Saturn's seasonal stratosphere at the time of the Voyager encounters. *J. Atmos. Sci.* 37:1883–1885.
Cess, R. D., Ramanathan, V., and Owen, T. 1980. The martian paleoclimate and enhanced carbon dioxide. *Icarus* 41:159–165.
Cess, R. D., Carlson, B. E., Caldwell, J., Nolt, I. G., Gillett, F. C., and Tokunaga, A. T. 1981. Latitude variation in Jovian stratospheric temperature. *Icarus* 46:249–255.
Chamberlain, J. W. 1963. Planetary coronae and atmospheric evaporation. *Planet. Space Sci.* 11:901–960.
Chamberlain, J. W., and Hunten, D. M. 1987. *Theory of Planetary Atmospheres* (Orlando: Academic Press).
Chapman, C. R. 1976. Asteroids as meteorite parent bodies: The astronomical perspective. *Geochim. Cosmochim. Acta* 40:701–719.
Chapman, C. R. 1979. The asteroids: Nature, interrelations, origins, and evolution. In *Asteroids,* ed. T. Gehrels (Tucson: Univ. of Arizona Press), pp. 25–60.
Chapman, C. R. 1987. The asteroid belt: Compositional structure and size distributions. *Bull. Amer. Astron. Soc.* 19:839 (abstract).
Chapman, C. R., and McKinnon, W. B. 1986. Cratering of planetary satellites. In *Satellites,* eds. J. A. Burns and M. S. Matthews (Tucson: Univ. of Arizona Press), pp. 492–580.
Chapman, C. R., Morrison, D., and Zellner, B. 1975. Surface properties of asteroids: A synthesis of polarimetry, radiometry, and spectrophotometry. *Icarus* 25:104–130.
Chapple, W. M. 1978. Mechanics of thin-skinned fold and thrust-belts, *Geol. Soc. Amer. Bull.* 89:1189–1198.
Cheng, A. F. 1984*a*. Escape of sulfur and oxygen from Io. *J. Geophys. Res.* 89:3939–3944.
Cheng, A. F. 1984*b*. Magnetosphere, rings, and moons of Uranus. In *Uranus and Neptune,* ed. J. T. Bergstralh, NASA CP-2330, pp. 541–546.
Cheng, A. F. 1986*a*. Energetic neutral particles from Jupiter and Saturn. *J. Geophys. Res.* 91:4524–4530.

Cheng, A. F. 1986*b*. Radial diffusion and ion partitioning in the Io torus. *Geophys. Res. Lett.* 13:517–520.
Cheng, A. F. 1987. Proton and oxygen plasmas at Uranus. *J. Geophys. Res.* 92:15309–15314.
Cheng, A. F. 1988. Two classes of models for temporal variability of the Io torus. *J. Geophys. Res.* 93, in press.
Cheng, A. F., and Krimigis, S. M. 1988. Energetic neutral particle imaging of Saturn's magnetosphere. In *Yosemite Conference Proceedings.*
Cheng, A. F., and Lanzerotti, L. J. 1978. Ice sputtering by radiation belt protons and the rings of Saturn and Uranus. *J. Geophys. Res.* 83:2597–2602.
Cheng, A. F., Lanzerotti, L. J., and Pironello, V. 1982. Charged particle sputtering of ice surfaces in Saturn's magnetosphere. *J. Geophys. Res.* 87:4567–4570.
Cheng, A. F., Haff, P., Johnson, R. E., and Lanzerotti, L. J. 1986. Interactions of planetary magnetospheres with icy satellite surfaces. In *Satellites,* eds. J. A. Burns and M. S. Matthews (Tucson: Univ. of Arizona Press), pp. 403–436.
Cheng, A. F., Johnson, R. E., Krimigis, S. M., and Lanzerotti, L. J. 1987. Magnetosphere, exosphere, and surface of Mercury. *Icarus* 71:430–440.
Chopra, P. N., and Paterson, M. S. 1984. The role of water in the deformation of dunite. *J. Geophys. Res.* 89:7861–7876.
Chrisey, D. B., BorLng, J. W., Phipps, J. A., Johnson, R. E., and Brown, W. L. 1986. Sputtering of molecular gas solids by keV ions. *Nucl. Instrum. Meth.* B13:360–364.
Chrisey, D. B., Johnson, R. E., Phipps, J., McGrath, M., and Boring, J. 1987. Sputtering of sulfur by keV ions: Application to the magnetospheric plasma interaction with Io. *Icarus* 70:111–123.
Chrisey, D. B., Johnson, R. E., Boring, J. W., and Phipps, J. A. 1988. The ejection of sodium from sodium sulfide by the sputtering of the surface of Io. *Icarus* 75:233–244.
Chyba, C. 1987. The cometary contribution to the oceans of primitive Earth. *Nature* 330:632–635.
Clark, B. C. 1986. Comets, volcanism, the salt rich regolith and cycling of volatiles on Mars. In *Symposium on Mars: Evolution of Its Climate and Atmosphere,* LPI Contrib. 599 (Houston: Lunar and Planetary Inst.), pp. 15–17.
Clark, B. C., Mason, L. W., and Kissel, J. 1987. Systematics of the "CHON" and other light element particle populations in Comet Halley. In *20th ESLAB Symp. on the Exploration of Halley's Comet,* vol. 3, ESA SP-250, eds. B. Battrick, E. J. Rolfe and R. Reinhard (Noordwijk: ESA), pp. 353–358.
Clark, R. N., Fanale, F. P., and Zent, A. P. 1983. Frost grain metamorphism: Implications for remote sensing of planetary surfaces. *Icarus* 56:233–245.
Clark, R. N., Brown, R. H., Owensby, P. D., and Steele, A. 1984. Saturn's satellites: Near-infrared spectrophotometry (0.65–2.5 μm) of the leading and trailing sides and compositional information. *Icarus* 58:265–281.
Clark, R. N., Fanale, F. P., and Gaffey, M. J. 1986. Surface composition of the natural satellites. In *Satellites,* eds. J. A. Burns and M. S. Matthews (Tucson: Univ. of Arizona Press), pp. 437–491.
Clayton, D. D. 1978. The cloudy state of interstellar matter. In *Protostars and Planets,* ed. T. Gehrels (Tucson: Univ. of Arizona Press), pp. 13–42.
Clayton, D. D. 1984. ^{26}Al in the interstellar medium. *Astrophys. J.* 280:144–149.
Clayton, D. D. 1985. Aluminum clues to the formation of the solar system. *Nature* 315:633–634.
Clayton, R. N., and Mayeda, T. K. 1983. Oxygen isotopes in eucrites, shergottites, nakhlites, and chassignites. *Earth Planet. Sci. Lett.* 62:1–6.
Clayton, R. N., and Mayeda, T. K. 1984. The oxygen isotope record in Murchison and other carbonaceous chondrites. *Earth Planet. Sci. Lett.* 67:151–161.
Clayton, R. N., and Thiemens, M. H. 1980. Lunar nitrogen: Evidence for secular change in the solar wind. In *The Ancient Sun: Fossil Record in the Earth, Moon and Meteorites,* eds. R. O. Pepin, J. A. Eddy and R. B. Merrill (New York: Pergamon Press), pp. 463–473.
Clayton, R. N., Onuma, N., and Mayeda, T. K. 1976. A classification of meteorites based on oxygen isotopes. *Earth Planet. Sci. Lett.* 30:10–18.
Coats, R. R. 1962. Magma type and crustal structure in the Aleutian Arc. *The Crust of the Pacific Basin,* Amer. Geophys. Union Monograph 6, pp. 92–109.
Cogley, J. G., and Henderson-Sellers, A. 1984. The origin and earliest state of the Earth's hydrosphere. *Rev. Geophys. Space Phys.* 22:131–175.

Cohen, M., and Kuhi, L. V. 1979. Observational studies of pre-main-sequence evolution. *Astrophys. J. Suppl.* 41:743–843.

Cohen, M., Tielens, A. G. G. M., and Allamandola, L. J. 1985. A new emission feature in IRAS spectra and the polycyclic aromatic hydrocarbon spectrum. *Astrophys. J.* 299:L93–L97.

Cohen, R. E., Kornakci, A. S., and Wood, J. A. 1983. Mineralogy and petrology of chondrules and inclusions in the Mokoia CV3 chondrite. *Geochim. Cosmochim. Acta* 47:1739–1757.

Cohen, R. E., Kornakci, A. S., and Wood, J. A. 1983. Mineralogy and petrology of chondrules and inclusions in the Mokoia CV3 chondrite. *Geochim. Cosmochim. Acta* 47:1739–1757.

Colin, L. 1983. Basic facts about Venus. In *Venus,* eds. D. M. Hunten, L. Colin, T. M. Donahue, and V. I. Moroz (Tucson: Univ. of Arizona Press), pp. 10–26.

Combes, F., Gérin, M., Wootten, A., Wlodarczak, G., Clausset, F., and Encrenaz, P. J. 1987. Acetone in interstellar space. *Astron. Astrophys.* 180:L13–L16.

Combes, M., Moroz, V. I., Crifo, J. F., Lamarre, J. M., Charra, J., Sanko, N. F., Soufflot, A., Bibring, J. P., Cazes, S., Coron, N., Crovisier, J., Emerich, C., Encrenaz, T., Gispert, R., Grigoryev, A. V., Guyot, G., Krasnopolsky, V. A., Nikolsky, Yu. V., and Rocard, F. 1986. Infrared sounding of Comet Halley from Vega 1. *Nature* 321:266–268.

Conrath, B. J., and Gautier, D. 1980. Thermal structure of Jupiter's atmosphere obtained by inversion of Voyager 1 infrared measurements. In *Remote Sensing of Atmospheres on Oceans,* ed. A. Deepak (New York: Academic Press), pp. 611–630.

Conrath, B. J., and Gierasch, P. J. 1984. Global variations of the *para* hydrogen fraction in Jupiter's atmosphere and implications for dynamics on the outer planets. *Icarus* 57:184–204.

Conrath, B. J., and Pirraglia, J. A. 1983. Thermal structure of Saturn from Voyager infrared measurements: Implications for atmospheric dynamics. *Icarus* 53:286–292.

Conrath, B. J., Flasar, F. M., Pirraglia, J. A., Gierasch, P. J., and Hunt, G. E. 1981. Thermal structure and dynamics of the Jovian atmosphere. 2. Visible cloud features. *J. Geophys. Res.* 86:8769–8775.

Conrath, B. J., Gautier, D., Hanel, R. A., and Hornstein, J. S. 1984. The helium abundance of Saturn from Voyager measurements. *Astrophys. J.* 282:807–815.

Conrath, B., Gautier, D., Hanel, R., Lindal, G., and Marten, A. 1987. The helium abundance of Uranus from Voyager measurements. *J. Geophys. Res.* 92:15003–15010.

Consolmagno, G. J. 1985. Resurfacing Saturn's satellites: Models of partial differentiation and expansion. *Icarus* 64:401–413.

Consolmagno, G. J., and Drake, M. J. 1977. Composition and evolution of the eucrite parent body: Evidence from rare earth elements. *Geochim. Cosmochim. Acta* 41:1271–1282.

Consolmagno, G. J., and Jokipii, J. 1978. Al and the partial ionization of the solar nebula. *Moon and Planets* 19:253–259.

Consolmagno, G. J., and Lewis, J. S. 1978. The evolution of icy satellite interiors and surfaces. *Icarus* 34:280–293.

Cook, A. F., Shoemaker, E. M., Smith, B. A., Danielson, G. E., Johnson, T. V., and Synnott, S. P. 1981. Volcanic origin of the eruptive plumes on Io. *Science* 211:1419–1422.

Cook, A. F., Burratti, B., Mullins, K. H., Shoemaker, E. M., and Fielder, R. 1987. Venting through the ice on Europa. *Icarus,* in press.

Cook, T. L., and Harlow, F. H. 1978. Three-dimensional dynamics of protostellar evolution. *Astrophys. J.* 225:1005–1020.

Coradini, A., and Magni, G. 1984. Structure of the satellitary accretion disk of Saturn. *Icarus* 59:376–391.

Coradini, A., and Magni, G. 1986. Protosatellitary accretion disks. In *The Solid Bodies of the Outer Solar System,* ESA SP-242, pp. 5–15.

Coradini, A., Federico, C., and Magni, G. 1980. Time evolution of grains in the protosolar nebula. *Moon and Planets* 22:47–61.

Coradini, A., Federico, C., and Magni, G. 1981*a*. Formation of planetesimals in an evolving protoplanetary disk. *Astron. Astrophys.* 98:173–185.

Coradini, A., Federico, C., and Magni, G. 1981*b*. Gravitational instabilities in satellite disks and formation of regular satellites. *Astron. Astrophys.* 99:255–261.

Coradini, A., Federico, C., and Lanciano, P. 1982. Ganymede and Callisto: Accumulation heat content. In *The Comparative Studies of the Planets,* eds. A. Coradini and M. Fulchignoni (Dordrecht: D. Reidel), pp. 61–70.

Corliss, J. B., Dymond, J. B., Gordon, L. I., Edmond, J. M., von Herzen, R. P., Ballard, R. D.,

Green, K., Williams, D., Bainbridge, A., Crane, K., and van Andel, T. H. 1979*a*. Submarine thermal springs on the Galapagos rift. *Science* 203:1073–1083.

Corliss, J. B., Gordon, L. I., and Edmond, J. M. 1979*b*. Some implications of heat/mass ratios in Galapagos rift hydrothermal fluid for models of seawater-rock interaction and the formation of oceanic crust. In *Deep Drilling Results in the Atlantic Ocean: Ocean Crust,* eds. C. E. Harrison and D. E. Hayes (Washington, D.C.: Amer. Geophys. Union), pp. 391–402.

Coroniti, F. V. 1974. Energetic electrons in Jupiter's magnetosphere. *Astrophys. J. Suppl.* 27:261–281.

Courtin, R. 1982. The spectrum of Titan in the far-infrared and microwave regions. *Icarus* 51:466–475.

Courtin, R. 1988. Pressure-induced absorption coefficients for radiative transfer calculations in Titan's atmosphere. *Icarus,* 75:245–254.

Courtin, R., Gautier, D., and Lacombe, A. 1979*a*. Indications of supersaturated stratospheric methane on Neptune from its atmospheric thermal profile. *Icarus* 37:236–248.

Courtin, R., Lena, P., De Muizon, M., Rouzn, D., Nicollier, C., and Wijnbergen, J. 1979*b*. Far infrared photometry of planets: Saturn and Uranus. *Icarus* 38:411–419.

Courtin, R., Gautier, D., Marten, A., Bézard, B., and Hanel, R. 1984. The composition of Saturn's atmosphere at northern temperate latitudes from Voyager IRIS spectra: NH_3, PH_3, C_2H_2, C_2H_6, CH_3D, CH_4, and the Saturnian D/H isotopic ratio. *Astrophys. J.* 287:899–916.

Cowling, T. G. 1951. The condition for turbulence in rotating stars. *Astrophys. J.* 114:272–286.

Cox, P., and Leene, A. 1987. IRAS observations of the Pleiades. In *Star Formation in Galaxies,* ed. C. J. Lonsdale Persson, NASA CP-2466, pp. 117–121.

Crabb, J., and Anders, E. 1981. Noble gases in E-chondrites. *Geochim. Cosmochim. Acta* 45:2443–2464.

Craig, H., and Lupton, J. E. 1976. Primordial neon, helium, and hydrogen in oceanic basalts. *Earth Planet. Sci. Lett.* 31:369–385.

Craig, H., Clarke, W. B., and Beg, M. A. 1975. Excess ^{3}He in deep water on the East Pacific Rise. *Earth Planet. Sci. Lett.* 26:125–132.

Crater Analysis Techniques Working Group. 1979. Standard techniques for presentation and analysis of crater size-frequency data. *Icarus* 37:467–474.

Crawford, G. D., and Stevenson, D. J. 1987. Gas-driven water volcanism and the resurfacing of Europa. *Icarus* 73:66–79.

Creager, K. C., and Jordan, T. H. 1984. Slab penetration into the lower mantle. *J. Geophys. Res.* 89:3031–3049.

Croswell, K., Hartmann, L., and Avrett, E. H. 1987. Mass loss from FU Orionis objects. *Astrophys. J.* 312:227–242.

Crowell, J. C. 1983. Ice ages recorded on Gondwanan continents. *Trans. Geol. Soc. S. Africa* 86:237–262.

Crowley, T. J. 1983. The geologic record of climatic change. *Rev. Geophys. Space Phys.* 21:828–877.

Cruikshank, D. P. 1986. Dark matter in the solar system. Paper presented at COSPAR, Toulouse, July 1986.

Cruikshank, D. P., and Apt, J. 1984. Methane on Triton: Physical state and distribution. *Icarus* 58:306–311.

Cruikshank, D. P., and Brown, R. H. 1986. Satellites of Uranus and Neptune, and the Pluto-Charon system. In *Satellites,* eds. J. A. Burns and M. S. Matthews (Tucson: Univ. of Arizona Press), pp. 836–873.

Cruikshank, D. P., Brown, R. H., and Clark, R. N. 1984. Nitrogen on Triton. *Icarus* 58:293–305.

Cruikshank, D. P., Brown, R. H., Tokunaga, A. T., Smith, R. G., and Piscitelli, J. R. 1988. Volatiles on Triton: The infrared spectroscopic evidence, 2.0 to 2.5 micrometers. *Icarus* 74:413–423.

Dagg, I. R., Anderson, A., Yan, S., Smith, W., and Read, L. A. 1984. Collision-induced absorption in nitrogen at low temperatures. *Can. J. Phys.* 63:625–631.

Dagg, I. R., Anderson, A., Yan, S., Smith, W., Joslin, C. G., and Read, L. A. A. 1986. Collision-induced absorption in gaseous mixtures of nitrogen and methane. *Can. J. Phys.* 64:7–15.

Dalgarno, A., and Lepp, S. 1984. Deuterium fractionation mechanisms in interstellar clouds. *Astrophys. J.* 287:L47–L50.

Danielson, R. E., Caldwell, J. J., and Larach, D. R. 1973. An inversion in the atmosphere of Titan. *Icarus* 20:437–443.
Davidson, D., Desando, M., Gough, S., Handa, Y., Ratcliffe, C., Ripmeester, J., and Tse, J. 1987. A clathrate hydrate of carbon monoxide. *Nature* 328:418–419.
Davies, G. F., and Arvidson, R. E. 1981. Martian thermal history, core segregation, and tectonics. *Icarus* 45:339–346.
Davies, M. E., Abalakin, V. K., Lieske, J. H., Seidelmann, P. K., Sinclair, A. T., Sinzi, A. M., Smith, B. A., and Tjuflin, Y. S. 1983. Report of the IAU Working Group on cartographic coordinates and rotational elements of the planets and satellites: 1982. *Celestial Mech.* 29:309–321.
Davis, D. R., Chapman, C. R., Greenberg, R., Weidenschilling, S. J., and Harris, A. W. 1979. Collisional evolution of asteroids. In *Asteroids,* ed. T. Gehrels (Tucson: Univ. of Arizona Press), pp. 528–557.
Davis, D. R., Suppe, J., and Dahlen, F. A. 1983. Mechanics of fold-and-thrust belts and accretionary wedges. *J. Geophys. Res.* 89:1153–1172.
Davis, D. R., Chapman, C. R., Weidenschilling, S. J., and Greenberg, R. 1985. Collisional history of asteroids: Evidence from Vesta and the Hiriyama families. *Icarus* 62:30–53.
Dayhoff, M. O., Eck, R., Lippincott, E. R., and Sagan, C. 1967. Venus: Atmospheric evolution. *Science* 155:556–557.
Dean, J. A., ed. 1973. *Lange's Handbook of Chemistry* (New York: McGraw-Hill).
de Bergh, C., Lutz, B. L., Owen, T., Brault, J., and Chauville, J. 1984. Monodeuterated methane in the outer solar system. II. Its detection on Uranus at 1.6 microns. *Astrophys. J.* 311:501–510.
de Bergh, C., Lutz, B. L., Owen, T., and Maillard, J. P. 1987. Neptune: Short term variations in the spectrum at 1.6 microns and the search for deuterated methane. *Bull. Amer. Astron. Soc.* 19:864 (abstract).
de Bergh, C., Lutz, B. L., Owen, T., and Chauville, J. 1988. Monodeuterated methane in the outer solar system. III. Its abundance on Titan. *Astrophys. J.* 329:951–955.
de Bergh, C., Lutz, B. L., Maillard, J. P., and Owen, T. 1989. Monodeuterated methane in the outer solar system. IV. Its detection and abundance on Neptune. *Astrophys. J.*, submitted.
DeCampli, W. M. 1981. T-Tauri winds. *Astrophys. J.* 244:124–146.
Degewij, J., Cruikshank, D. P., and Hartmann, W. K. 1980. Near-infrared colorimetry of J6 Himalia and S9 Phoebe: A summary of 0.3 to 2.2 μm reflectances. *Icarus* 44:541–547.
Deines, P., and Wickman, F. E. 1985. The stable carbon isotopes in enstatite chondrites and Cumberland Falls. *Geochim. Cosmochim. Acta* 49:89–95.
Delaney, J. R., Muenow, D. W., and Graham, D. G. 1978. Abundance and distribution of water, carbon and sulfur in the glassy rims of submarine pillow basalts. *Geochim. Cosmochim. Acta* 42:581–594.
Delbourgo-Salvador, P., Gry, C., Malinie, G., and Audouze, J. 1985. Effects of nuclear uncertainties and chemical evolution on the standard big bang nucleosynthesis. *Astron. Astrophys.* 150:53–61.
Delitsky, M. L., and Thompson, W. R. 1987. Chemical processes in Triton's atmosphere and surface. *Icarus* 70:354–365.
Delsemme, A. H., ed. 1977. *Comets, Asteroids, and Meteorites: Interrelations, Evolution and Origin* (Toledo: Univ. of Toledo Press).
Delsemme, A. H. 1982. Chemical composition of cometary nuclei. In *Comets,* ed. L. L. Wilkening (Tucson: Univ. of Arizona Press), pp. 85–130.
Delsemme, A. H. 1985. The nature of the cometary nucleus. *Publ. Astron. Soc. Pacific* 97:861–870.
DeMarcus, W. C. 1958. The constitution of Jupiter and Saturn. *Astron. J.* 63:2–28.
de Muizon, M., Geballe, T. R., d'Hendecourt, L. B., and Baas, F. 1986. New emission features in the infrared spectra of two IRAS sources. *Astrophys. J.* 306:L105–L108.
DePaolo, D. J. 1980. Source of continental crust: Neodymium isotope evidence from the Sierra Nevada and the Peninsular Ranges. *Science* 209:684–687.
de Pater, I. 1986. Jupiter's zone-belt structure at radio wavelengths. II. Comparison of observations with model atmosphere calculations. *Icarus* 68:344–365.
de Pater, I., and Massie, S. T. 1985. Models of the millimeter-centimeter spectra of the giant planets. *Icarus* 62:143–171.
DesMarais, D. J. 1985. Carbon exchange between the mantle and the crust, and its effect upon

the atmosphere: Today compared to Archean time. In *The Carbon Cycle and Atmospheric CO_2: Natural Variations Archean to Present,* eds. E. T. Sundquist and W. S. Broecker, Geophysical Monograph 32 (Washington, D.C.: Amer. Geophys. Union), pp. 602–611.

DesMarais, D. J., and Moore, J. G. 1984. Carbon and its isotopes in mid-oceanic basaltic glasses. *Earth Planet. Sci. Lett.* 69:43–57.

Detrick, R. S., Buhl, P., Vera, E., Mutter, J., Orcutt, J., Madsen, J., and Brocher, T. 1987. Multi-channel seismic imaging of a crustal magma chamber along the East Pacific Rise. *Nature* 326:35–41.

DeWitt, H. E. 1978. Equilibrium statistical mechanics of strongly coupled plasmas by numerical simulation. In *Strongly Coupled Plasmas,* eds. G. Kalman and P. Carini (New York: Plenum), pp. 81–115.

d'Hendecourt, L. B., and Allamandola, L. J. 1986. Time dependent chemistry in dense molecular clouds. III. Infrared band cross sections of molecules in the solid state at 10 K. *Astron. Astrophys. Suppl.* 64:453–467.

Dictor, R. A., and Bell, A. T. 1986. Fischer-Tropsch synthesis over reduced and unreduced iron oxide catalysts. *J. Catalysis* 97:121–136.

Dietrich, W. F., and Simpson, J. A. 1979. The isotopic and elemental abundances of neon nuclei accelerated in solar flares. *Astrophys. J.* 231:L91–L95.

Dodd, R. T. 1981. *Meteorites: A Petrologic-Chemical Synthesis* (Cambridge: Cambridge Univ. Press).

Dohnanyi, J. S. 1969. Collisional model of asteroids and their debris. *J. Geophys. Res.* 74:2531–2554.

Dole, S. H. 1962. The gravitational concentration of particles in space near the Earth. *Planet. Space Sci.* 9:541–553.

Donahue, T. M. 1986. Fractionation of noble gases by thermal escape from accreting planetesimals. *Icarus* 66:195–210.

Donahue, T. M., and Pollack, J. B. 1983. Origin and evolution of the atmosphere of Venus. In *Venus,* eds. D. M. Hunten, L. Colin, T. M. Donahue and V. I. Moroz (Tucson: Univ. of Arizona Press), pp. 1003–1036.

Donahue, T. M., Hoffman, J. H., and Hodges, R. R., Jr. 1981. Krypton and xenon in the atmosphere of Venus. *Geophys. Res. Lett.* 8:513–516.

Donahue, T. M., Hoffman, J. H., Hodges, R. R., Jr., and Watson, A. J. 1982. Venus was wet: A measurement of the ratio of deuterium to hydrogen. *Science* 216:630–633.

Donn, B. 1963. The origin and structure of icy cometary nuclei. *Icarus* 2:396–402.

Donn, B. 1968. Polycyclic hydrocarbons, Platt particles, and interstellar extinction. *Astrophys. J.* 152:L129–L133.

Donn, B. 1976. A comparison of the composition of new and evolved comets. In *Comets, Asteroids, Meteorites: Interrelations, Evolution and Origin,* ed. A. H. Delsemme (Toledo: Univ. of Toledo), pp. 15–24.

Donn, B. 1988. Formation and structure of a cometary nucleus. *Astron. Astrophys.,* in press.

Donn, B., and Rahe, J. 1982. Structure and origin of cometary nuclei. In *Comets,* ed. L. L. Wilkening (Tucson: Univ. of Arizona Press), pp. 203–226.

Donn, B., Wickramasinghe, N. C., Hudson, J. P., and Stecher, T. P. 1966. On the formation of graphite grains in cool stars. *Astrophys. J.* 153:451–463.

Donn, B., Khanna, R., Salsbury, D., Allen, J., and Moore, J. 1987. Problems with polycyclic aromatic hydrocarbons in interstellar grains. *Bull. Amer. Astron. Soc.* 18:1031 (abstract).

Draine, B. T. 1985. Grain evolution in dark clouds. In *Protostars & Planets II,* eds. D. C. Black and M. S. Matthews (Tucson: Univ. of Arizona Press), pp. 621–640.

Draine, B. T., and Anderson, N. 1985. Temperature fluctuations and infrared emission from interstellar grains. *Astrophys. J.* 292:494–499.

Drake, M. J. 1979. Geochemical evolution of the eucrite parent body: Possible nature and evolution of asteroid 4 Vesta? In *Asteroids,* ed. T. Gehrels (Tucson: Univ. of Arizona Press), pp. 765–782.

Drapatz, S., Larson, H. P., and Davis, D. S. 1986. Search for methane in Comet Halley. *Proc. 20th ESLAB Symposium on the Exploration of Halley's Comet,* vol. 1, eds. B. Battrick, E. J. Rolfe and R. Reinhard, ESA SP-250 (Noordwijk: ESA), pp. 347–352.

Dreibus, G., and Wänke, H. 1979. On the chemical composition of the Moon and the eucrite parent body and a comparison with the composition of the Earth. *Lunar Planet. Sci.* X:315–317 (abstract).

Dreibus, G., and Wänke, H. 1980. The bulk composition of the eucrite parent asteroid and its bearing on planetary evolution. *Z. Naturforsch.* 35a:204–216.

Dreibus, G., and Wänke, H. 1984. Accretion of the Earth and the inner planets. In *Proc. 27th Intl. Geol. Cong.*, vol. 11, (Utrecht: VNU Science Press), pp. 1–20.

Dreibus, G., and Wänke, H. 1985. Mars, a volatile-rich planet. *Meteoritics* 20:367–381.

Dreibus, G., and Wänke, H. 1987. Volatiles on Earth and Mars: A comparison. *Icarus* 71:225–240.

Dreibus, G., Spettel, B., and Wänke, H. 1979. Halogens in meteorites and their primordial abundances. In *Origin and Distribution of the Elements,* ed. L. H. Ahrens (New York: Pergamon Press), pp. 33–38.

Dreibus, G., Palme, H., Rammensee, W., Spettel, B., and Wänke, H. 1981. Chemistry of the Shergotty parent body. *Meteoritics* 16:310–311 (abstract).

Dreibus, G., Palme, H., Rammensee, W., Spettel, B., Weckwerth, G., and Wänke, H. 1982. Composition of Shergotty parent body: Further evidence of a two component model of planet formation. *Lunar Planet. Sci.* XIII:186–187 (abstract).

Dreibus, G., Rieder, R., and Wänke, H. 1988. Can we relate the volatiles on Earth and Mars to those contained in C1-chondrites? *Lunar and Planet. Sci.* XIX:283–284 (abstract).

Drell, S. D., Foley, H. M., and Ruderman, M. A. 1965. Drag and propulsion of large satellites in the ionosphere: An Alfvén propulsion engine in space. *J. Geophys. Res.* 70:3131–3145.

Dry, M. E. 1981. The Fischer-Tropsch synthesis. In *Catalysis Science and Technology,* eds. J. R. Anderson and M. Budart (Berlin: Springer-Verlag), pp. 159–255.

Duba, A. G., and Boland, J. N. 1984. High temperature electrical conductivity of the carbonaceous chondrites Allende and Murchison. *Lunar Planet. Sci.* XV:232–233 (abstract).

DuFresne, E. R., and Anders, E. 1962. On the chemical evolution of the carbonaceous chondrites. *Geochim. Cosmochim. Acta* 26:1085–1114.

Dufton, P. L., Keenan, F. P., and Hibbert. A. 1986. The abundance of phosphorus in the interstellar medium. *Astron. Astrophys.* 164:179–183.

Duley, W. W. 1984. Evidence against biological grains in the interstellar medium. *Q. J. Roy. Astron. Soc.* 25:109–113.

Duley, W. W. 1987. Formation, destruction and extinction of carbon grains and PAH molecules. In *Polycyclic Aromatic Hydrocarbons and Astrophysics,* eds. A. Leger, L. d'Hendecourt and N. Boccara (Dordrecht: D. Reidel), pp. 373–386.

Duley, W. W., and Williams, D. A. 1981. The infrared spectrum of interstellar dust: Surface functional groups on carbon. *Mon. Not. Roy. Astron. Soc.* 196:269–274.

Duley, W. W., and Williams, D. A. 1986. PAH molecules and carbon dust in interstellar clouds. *Mon. Not. Roy. Astron. Soc.* 219:859–864.

Durand, B., ed. 1980. *Kerogens* (Paris: Editions Technip).

Eberhardt, P., Geiss, J., Graf, H., Grogler, N., Mendia, M. D., Morgeli, M., Schwaller, H., and Stettler, A. 1972. Trapped solar wind gases in Apollo 12 lunar fines 12001 and Apollo 11 breccia 10046. *Proc. Lunar Sci. Conf.* 2:1821–1856.

Eberhardt, P., Eugster, O., and Marti, K. 1965. A redetermination of the isotopic composition of atmospheric neon. *Z. Naturforsch.* 21a:623–624.

Eberhardt, P., Krankowsky, D., Shulte, W., Dolder, U., Lämmerzahl, P., Berthelier, J. J., Woweries, J., Stubbeman, U., Hodges, R. R., Hoffman, J. H., and Illiano, J. M. 1986. On the CO and N_2 abundance in comet Halley. In *20th ESLAB Symp. on the Exploration of Halley's Comet,* vol. 1, eds. B. Battrick, E. J. Rolfe and R. Reinhard, ESA SP-250, (Noordwijk: ESA), pp. 383–386.

Eberhardt, P., Dolder, U., Schulte, W., Krankowsky, D., Lämmerzahl, P., Hoffman, J. H., Hodges, R. R., Berthelier, J. J., and Illiano, J. M. 1987*a*. The D/H ratio in water from Halley. *Astron. Astrophys.* 187:435–437.

Eberhardt, P., Hodges, R. R., Krankowsky, D., Berthelier, J. J., Schulte, W., Dolder, U., Lämmerzahl, P., Hoffman, J. H., and Illiano, J. M. 1987*b*. The D/H and $^{18}O/^{16}O$ isotopic ratios in comet Halley. *Lunar Planet. Sci.* XVIII:252–253 (abstract).

Eberhardt, P., Krankowsky, D., Schulte, W., Dolder, U., Lämmerzahl, P., Berthelier, J. J., Woweries, J., Stubbemann, U., Hodges, R. R., Hoffman, J. H., and Illiano, J. M. 1987*c*. The CO and N_2 abundance in comet P/Halley. *Astron. Astrophys.* 187:481–484.

Edmond, J. M., Corliss, J. B., and Gordon, L. I. 1979*a*. Ridge crest-hydrothermal metamorphism at the Galapagos spreading center and reverse weathering. In *Deep Drilling Results* *in the Atlantic Ocean: Oceanic Crust,* eds. M. Talwani, C. G. Harrison and D. E. Hayes (Washington, D.C.: Amer. Geophys. Union), pp. 383–390.

Edmond, J. M., Measures, C., McDuff, R. E., Chan, L. H., Collir, R., Grand, D., Gordon, L. I., and Corliss, J. B. 1979*b*. Ridge crest hydrothermal activity and balances of major and minor elements in the ocean: The Galapagos data. *Earth Planet. Sci. Lett.* 45:1–18.

Ellder, J., Friberg, P., Hjalmarson, Å., Höglund, B., Irvine, W. M., Johansson, L. E. B., Olofsson, H., Rydbeck, G., Rydbeck, O. E. H., and Guélin, M. 1980. On methyl formate, methane, and deuterated ammonia in Orion A. *Astrophys. J.* 242:L93–L97.

Elliot, J. L. 1982. Rings of Uranus: A review of occultation results. In *Uranus and the Outer Planets,* ed. G. E. Hunt (Cambridge: Cambridge Univ. Press), pp. 237–256.

Ellsworth, K., and Schubert, G. 1983. Saturn's icy satellites: Thermal and structural models. *Icarus* 54:490–510.

Elmegreen, B. G. 1978. On the interaction between a strong stellar wind and a surrounding disk nebula. *Moon and Planets* 19:261–277.

Elmegreen, B. G. 1979. On the disruption of a protoplanetary disk nebula by a T Tauri-like solar wind. *Astron. Astrophys.* 80:77–78.

Encrenaz, T. 1984. Primordial matter in the solar system: A study of its chemical composition from remote spectroscopic analysis. *Space Sci. Rev.* 38:35–87.

Encrenaz, T., Hardorp, J., Owen, T., and Woodman, J. H. 1974. Observational constraints on model atmospheres for Uranus and Neptune. In *Exploration of the Planetary System,* eds. A. Woscyk and I. Iwaniszewska (Dordrecht: D. Reidel), pp. 487–496.

Erickson, E. F., Goorvitch, D., Simpson, J. P., and Stecker, D. W. 1978. Far infrared brightness temperature of Jupiter and Saturn. *Icarus* 35:61–73.

Eshleman, V. R., Lindal, G. F., and Tyler, G. L. 1983. Is Titan wet or dry? *Science* 221:53–55.

Esposito, L. W. 1984. Sulfur dioxide: Episodic injection shows evidence for active Venus volcanism. *Science* 223:1072–1074.

Etique, P., Signer, P., and Wieler, R. 1981. An in-depth study of neon and argon in lunar soil plagioclase, revisited: implanted solar flare noble gases. *Lunar Planet. Sci.* XII:265–267 (abstract).

Eugster, O., Eberhardt, P., and Geiss, J. 1967. The isotopic composition of krypton in unequilibrated and gas rich chondrites. *Earth Planet. Sci. Lett.* 2:385–393.

Everhart, E. 1967. Intrinsic distributions of cometary perihelia and magnitudes. *Astron. J.* 72:1002–1011.

Eviatar, A. 1984. Plasma in Saturn's magnetosphere. *J. Geophys. Res.* 89:3821–3828.

Eviatar, A., and Podolak, M. 1983. Titan's gas and plasma torus. *J. Geophys. Res.* 88:833–840.

Eviatar, A., and Richardson, J. 1986. Predicted satellite tori in the magnetosphere of Uranus. *Astrophys. J.* 300:L99–L102.

Eviatar, R., Siscoe, G. L., Scudder, J., Sittler, E., and Sullivan, J. 1982. The plumes of Titan. *J. Geophys. Res.* 87:8091–8103.

Ezer, D., and Cameron, A. G. W. 1965. A study of solar evolution. *Can. J. Phys.* 43:1497–1517.

Fallick, A. E., Hinton, R. W., Mattey, D. P., Norris, S. J., Pillinger, C. T., Swart, P. K., and Wright, I. P. 1983. No unusual compositions of the stable isotopes, of nitrogen, carbon and hydrogen in SNC meteorites. *Lunar Planet. Sci.* XIV:183–184 (abstract).

Fan, C. Y., Gloeckler, G. and Horestadt, D. 1984. The composition of heavy ions in solar energetic particle events. *Space Sci. Rev.* 38:143–178.

Fanale, F. P. 1971. A case for catastrophic early degassing of the Earth. *Chem. Geol.* 8:79–105.

Fanale, F. P. 1976. Martian volatiles: Their degassing history and geochemical fate. *Icarus* 28:179–202.

Fanale, F. P., and Cannon, W. A. 1971. Physical adsorption of rare gas on terrigenous sediments. *Earth Planet. Sci. Lett.* 11:362–368.

Fanale, F. P., and Cannon, W. A. 1972. Origin of planetary primordial rare gas: The possible role of adsorption. *Geochim. Cosmochim. Acta* 36:319–328.

Fanale, F. P., and Jakosky, B. M. 1982. Regolith-atmosphere exchange of water and carbon dioxide on Mars: Effects on atmospheric history and climate change. *Planet. Space Sci.* 30:819–831.

Fanale, F. P., Brown, R. H., Cruikshank, D. P., and Clark, R. N. 1979. Significance of absorption features in Io's IR reflectance spectrum. *Nature* 280:760–763.

Fanale, F. P., Salvail, J. R., Banerdt, W. B., and Saunders, R. S. 1982*a*. Mars: The regolith-atmosphere-cap system and climate change. *Icarus* 50:381–407.

Fanale, F. P., Banerdt, W., Elson, L., Johnson, T. V., and Zurek, R. 1982*b*. Io's surface: Its phase composition and influence on Io's atmosphere and Jupiter's magnetosphere. In *Satellites of Jupiter,* ed. D. Morrison (Tucson: Univ. of Arizona Press), pp. 756–781.

Fazio, G. G., Traub, W. A., and Wright, E. L. 1976. The effective temperature of Uranus. *Astrophys. J.* 209:633–637.

Federico, C., and Lanciano, P. 1983. Thermal and structural evolution of four satellites of Saturn. *Ann. Geophys.* 1:469–476.

Fegley, M. B., Jr. 1983. Primordial retention of nitrogen by terrestrial planets and meteorites. *Proc. Lunar Planet. Sci. Conf.* 13, *J. Geophys. Res.* 88:A853–A868.

Fegley, M. B., Jr., and Lewis, J. S. 1979. Thermodynamics of selected trace elements in the Jovian atmosphere. *Icarus* 38:166–179.

Fegley, M. B., Jr., and Lewis, J. S. 1980. Volatile element chemistry in the solar nebula: Na, K, F, C1, Br, and P. *Icarus* 41:439–455.

Fegley, M. B., Jr., and Prinn, R. G. 1985. Equilibrium and non-equilibrium chemistry of Saturn's atmospheres: Implications for the observability of PH_3, N_2, CO, and GeH_4. *Astrophys. J.* 299:1067–1078.

Fegley, M. B., Jr., and Prinn, R. 1986. Chemical models of the deep atmosphere of Uranus. *Astrophys. J.* 307:852–865.

Fegley, M. B., Jr., and Prinn, R. G. 1988*a*. Chemical constraints on the water and total oxygen abundances in the deep atmosphere of Jupiter. *Astrophys. J.* 324:621–625.

Fegley, M. B., Jr., and Prinn, R. G. 1988*b*. The predicted abundances of deuterium-bearing gases in the atmospheres of Jupiter and Saturn. *Astrophys. J.* 326:490–508.

Fegley, M. B., Prinn, R. G., Hartman, H., and Watkins, G. H. 1986. Chemical effects of large impacts on the Earth's primitive atmosphere. *Nature* 319:305–308.

Fehn, U., Green, K. E., von Herzen, R. P., and Cathles, L. M. 1983. Numerical models for the hydrothermal field at the Galapagos spreading center. *J. Geophys. Res.* 88:1033–1048.

Feierberg, M. A., and Drake, M. J. 1980. The meteorite-asteroid connection: The infrared spectra of eucrites, shergottites, and Vesta. *Science* 209:805–807.

Feierberg, M. A., Lebofsky, L. A., and Larson, H. P. 1981. Spectroscopic evidence for aqueous alteration products on the surface of low-albedo asteroids. *Geochim. Cosmochim. Acta* 95:971–981.

Feierberg, M. A., Lebofsky, L. A., and Tholen, D. J. 1985*a*. The nature of C-class asteroids from 3-μm spectrophotometry. *Icarus* 63:183–191.

Feierberg, M. A., Lebofsky, L. A., and Tholen, D. J. 1985*b*. Are T, P, and D asteroids really ultraprimitive? *Bull. Amer. Astron. Soc.* 17:730 (abstract).

Feldman, P. D., Festou, M. C., A'Hearn, M. F., Arpigny, C., Butterworth, P. S., Cosmovici, C. B., Danks, A. C., Gilmozzi, R., Jackson, W. M., McFadden, L. A., Patriarchi, P., Schleicher, D. G., Tozzi, G. P., Wallis, M., Weaver, H. A., and Woods, T. N. 1986. IUE observations of comet Halley: Evolution of the UV spectrum between September 1985 and July 1986. In *20th ESLAB Symposium on the Exploration of Halley's Comet,* eds. B. Battrick, E. J. Rolfe and R. Reinhard, vol. 1, ESA SP-250 (Noordwijk: ESA), pp. 325–328.

Ferlet, R. 1981. Abundance of interstellar nitrogen. *Astron. Astrophys.* 98:L1–L3.

Fernández, J. A. 1980. Evolution of comet orbits under the perturbing influence of the giant planets and nearby stars. *Icarus* 42:406–421.

Fernández, J. A. 1985. The formation and dynamical survival of the comet cloud. In *Dynamics of Comets: Their Origin and Evolution,* eds. A. Carusi and G. B. Valsecchi (Dordrecht: D. Reidel), pp. 45–70.

Fernández, J. A., and Ip, W.-H. 1981. Dynamical evolution of a cometary swarm in the outer planetary region. *Icarus* 47:470–479.

Fernández, J. A., and Jockers, K. 1983. Nature and origin of comets. *Rep. Prog. Phys.* 46:665–772.

Fertl, W. H. 1976. *Abnormal Formation Pressures* (New York: Elsevier).

Fink, U., and Larson, H. 1978. Deuterated methane observed on Saturn. *Science* 201:343–345.

Fink, U., Smith, B. A., Benner, D. C., Johnson, J. R., Reitsema, H. J., and Westphal, J. A. 1980. Detection of a CH_4 atmosphere on Pluto. *Icarus* 44:62–71.

Fisher, D. E. 1975. Trapped helium and argon and the formation of the atmosphere by degassing. *Nature* 256:113–114.

Fisher, D. E. 1985. Noble gases from oceanic island basalts do not require an undepleted mantle source. *Nature* 316:716–718.

Flasar, F. M. 1973. Gravitational energy sources in Jupiter. *Astrophys. J.* 186:1097–1106.

Flasar, F. M. 1983. Oceans on Titan? *Science* 221:55–57.

Flasar, F. M. 1986. Para-hydrogen equilibrium and superadiabatic lapse rates on Uranus: Evidence of methane condensation? *Bull. Amer. Astron. Soc.* 18:757 (abstract).

Flasar, F. M., and Birch, F. 1973. Energetics of core formation: A correction. *J. Geophys. Res.* 78:6101–6103.

Flasar, F. M., Conrath, B. J., Gierasch, P. J., and Pirraglia, J. A. 1987. Voyager infrared observations of Uranus' atmosphere: Thermal structure and dynamics. *J. Geophys. Res.* 92:15011–15018.

Forrest, W. J., McCarthy, J. F., and Houck, J. R. 1979. 16-39 micron spectroscopy of oxygen-rich stars. *Astrophys. J.* 233:611–620.

Frakes, L. A. 1979. *Climates Throughout Geologic Time* (New York: Elsevier).

Francis, P. W., and Wadge, G. 1983. The Olympus Mons aureole: Formation by gravitational spreading. *J. Geophys. Res.* 88:8333–8334.

Francis, P. W., and Wood, C. A. 1982. Absence of silicic volcanism on Mars: Implications for crustal composition and volatile abundance. *J. Geophys. Res.* 87:9881–9889.

François, L. M., and Gérard, J.-C. 1986. A numerical model of the evolution of ocean sulfate and sedimentary sulfur during the last 800 million years. *Geochim. Cosmochim. Acta* 50:2289–2302.

Frank, L. A., Burek, L., Ackerson, K., Wolfe, J., and Mihalov, J. 1980. Plasmas in Saturn's magnetosphere. *J. Geophys. Res.* 85:5695–5708.

Franz, O. G. 1981. UBV photometry of Triton. *Icarus* 45:602–606.

Freer, R. 1981. Diffusion in silicate minerals and glasses: A data digest and guide to the literature. *Contrib. Mineral Petrol.* 76:440–454.

French, R. G., Melroy, P. A., Baron, R. L., Dunham, E. W., Meech, K. J., Mink, D. J., Elliot, J. L., Allen, D. A., Ashley, M. C. B., Freeman, K. C., Erickson, E. F., Goguen, J., and Hammel, H. B. 1985. The 1983 June 15 occultation by Neptune. II. The oblateness of Neptune. *Astron. J.* 90:2624–2638.

French, R. G., Elliot, J. L., and Levine, S. E. 1986. Structure of the Uranian rings. II. Ring orbits and widths. *Icarus* 67:134–163.

Frerking, M. A., Wilson, R. W., Linke, R. A., and Wannier, P. G. 1980. Isotopic abundance ratios in interstellar carbon monsulfide. *Astrophys. J.* 240:65–73.

Frerking, M. A., Langer, W. D., and Wilson, R. W. 1982. The relationship between CO abundance and visual extinction in interstellar clouds. *Astrophys. J.* 262:590–605.

Frick, U., and Pepin, R. O. 1981. Study of solar wind gases in a young lunar soil. In *Lunar Planet. Sci.* XII:303–305 (abstract).

Frick, U., Mack, R., and Chang, S. 1979. Noble gas trapping and fractionation during synthesis of carbonaceous matter. *Proc. Lunar Planet. Sci. Conf.* 10:1961–1973.

Frick, U., Becker, R. H., and Pepin, R. O. 1988. Solar wind record in the lunar regolith: Nitrogen and noble gases. *Proc. Lunar Planet. Sci. Conf.* 18:87–120.

Fricker, P. E., and Reynolds, R. T. 1968. Development of the atmosphere of Venus. *Icarus* 9:221–230.

Friedson, A. J., and Ingersoll, A. P. 1987. Seasonal meridional energy balance and thermal structure of the atmosphere of Uranus: A radiative-convective dynamical model. *Icarus* 69:135–156.

Friedson, A. J., and Stevenson, D. J. 1983. Viscosity of rock-ice mixtures and applications to the evolution of icy satellites. *Icarus* 56:1–14.

Frisch, P. C., and York, D. G. 1985. Interstellar clouds near the Sun. In *The Galaxy and the Solar System,* eds. R. Smoluchowski, J. N. Bacall and M. S. Matthews (Tucson: Univ. of Arizona Press), pp. 83–100.

Gaffey, M. J. 1987. Instrumental requirements and observational strategies for spectrophotometric data acquisition during a CRAF-type asteroid flyby. *Lunar Planet. Sci.* XVIII:308–309 (abstract).

Gaffey, M. J., and Lazarewicz, A. R. 1988. The thermal and mineralogical evolution of small, planetary objects. In preparation.

Gaffey, M. J., and McCord, T. B. 1979. Mineralogical and petrological characterization of asteroid surface materials. In *Asteroids,* ed. T. Gehrels (Tucson: Univ. of Arizona Press), pp. 688–723.
Garcia, M. O., Liu, N. W. K., and Muenow, D. W. 1979. Volatiles in submarine volcanic rocks from the Mariana Island arc and trough. *Geochim. Cosmochim. Acta* 43:305–312.
Garrels, R. M., and Lerman, A. 1984. Coupling of the sedimentary sulfur and carbon cycles—An improved model. *Amer. J. Sci.* 284:989–1007.
Gaskell, R., Synnott, S. P., McEwen, A. S., and Schaber, G. G. 1988. Large-scale topography of Io: Implications for internal structure and heat transfer. *Geophys. Res. Lett.* 15:581–584.
Gatley, I., and Kaifu, N. 1987. Infrared observations of interstellar molecular hydrogen. In *Astrochemistry,* eds. M. S. Vardya and S. P. Tarafdar (Dordrecht: D. Reidel), pp. 153–166.
Gautier, D. 1985. Isotopic ratios in giant planets. In *Isotopic Ratios in the Solar System,* ed. D. Gautier (Toulouse: Cepadues-Editions), pp. 181–190.
Gautier, D., and Courtin, R. 1979. Thermal properties of the giant planets. *Icarus* 39:28–45.
Gautier, D., and Grossman, K. 1972. A new method for the determination of the mixing ratio hydrogen to helium in the giant planets. *J. Atmos. Sci.* 29:788–792.
Gautier, D., and Owen, T. 1983*a*. Cosmological implications of helium and deuterium abundances on Jupiter and Saturn. *Nature* 302:215–218.
Gautier, D., and Owen, T. 1983*b*. Cosmogonical implications of elemental and isotopic abundances in atmospheres of the giant planets. *Nature* 304:691–694.
Gautier, D., and Owen, T. 1985. Observational constraints on models for giant planet formation. In *Protostars & Planets II,* eds. D. C. Black and M. S. Matthews (Tucson: Univ. of Arizona Press), pp. 832–846.
Gautier, D., Conrath, B., Flasar, M., Hanel, R., Kunde, V., Chedin, A., and Scott, N. 1981. The helium abundance of Jupiter from Voyager. *J. Geophys. Res.* 86:8713–8720.
Gautier, D., Bézard, B., Marten, A., Baluteau, J. P., Scott, N., Chedin, A., Kunde, V., and Hanel, R. 1982. The C/H ratio in Jupiter from the Voyager infrared investigation. *Astrophys. J.* 257:901–912.
Gautier, T. N. 1986. Observations of infrared cirrus. In *Light on Dark Matter,* ed. F. P. Israel (Dordrecht: D. Reidel), pp. 49–54.
Geballe, T. R. 1986. Absorption by solid and gaseous CO towards obscured infrared objects. *Astron. Astrophys.* 162:248–252.
Geballe, T. R., Baas, F., Greenberg, J. M., and Schutte, W. 1985. New infrared absorption features due to solid phase molecules containing sulfur in W33A. *Astron. Astrophys.* 146:L6–L8.
Gehrels, T. 1957. Indiana expedition to South Africa, April-June, 1957. *Astron. J.* 62:244.
Geiss, J. 1982. Processes affecting abundances in the solar wind. *Space Sci. Rev.* 33:201–217.
Geiss, J. 1987. Composition measurements and the history of cometary matter. *Astron. Astrophys.* 187:859–866.
Geiss, J., and Boschler, P. 1981. On the abundances of rare ions in the solar wind. In *Solar Wind IV,* ed. H. Rosenbauer (Lindau: Max-Planck-Institut für Aeronomie), pp. 403–413.
Geiss, J., and Bochsler, P. 1982. Nitrogen isotopes in the solar system. *Geochim. Cosmochim. Acta* 46:529–548.
Geiss, J., and Bochsler, P. 1985. Ion composition in the solar wind in relation to solar abundances. In *Proc. of the Intl. Conf. on Isotopic Ratios in the Solar System* (Paris: CNES), in press.
Geiss, J., and Reeves, H. 1981. Deuterium in the solar system. *Astron. Astrophys.* 93:189–199.
Gerin, M., Wootten, H. A., Combes, F., Boulanger, F., Peters, W. L., Kuiper, T. B. H., Encrenaz, P. J., and Bogey, M. 1987. Deuterated C_3H_2 as a clue to deuterium chemistry. *Astron. Astrophys.* 173:L1–L4.
Gierasch, P. J., and Conrath, B. J. 1987. Layered convection on the outer planets. *J. Geophys. Res.* 92:15019–15029.
Gierasch, P. J., Conrath, B. J., and Magalhaes, J. A. 1986. Zonal mean properties of Jupiter's upper troposphere from Voyager infrared observations. *Icarus* 67:456–483.
Gillett, F. C. 1986. IRAS observations of cool excess around main sequence stars. In *Light on Dark Matter,* ed. F. P. Israel (Dordrecht: D. Reidel), pp. 61–69.
Gillett, F. C., and Rieke, G. H. 1977. 5-20 micron observations of Uranus and Neptune. *Astrophys. J.* 218:L141–L144.

Gillett, F. C., Jones, T. W., Merrill, K. M., and Stein, W. A. 1975. Anisotropy of constituents of interstellar grains. *Astron. Astrophys.* 45:77–81.

Gilra, D. P. 1972. Collective excitations and dust particles in space. In *The Scientific Results from the Orbiting Astronomical Observatory (OAO-2),* ed. A.D. Code, NASA SP-130, pp. 295–319.

Glicker, S., and Okabe, H. 1987. Photochemistry of diacetylene. *J. Phys. Chem.,* in press.

Goguen, J. D., Sinton, W. M., Matson, D. L., Howell, R. R., Dyck, H. M., Johnson, T. V., Brown, R. H., Veeder, G. J., Lane, A. L., Nelson, R. M., and McClaren, R. A. 1988. Multicolor IR photometry of occultations of Io by Ganymede and Callisto during 1985: Temperatures, areas and locations of major active hot spots and relative satellite astrometry. *Icarus,* submitted.

Gold, T. 1964. Outgassing processes on the moon and Venus. In *The Origin and Evolution of Atmospheres and Oceans,* eds. P. J. Brancazio and A. J. W. Cameron (New York: Wiley), pp. 249–256.

Goldberg, B. A., Mekler, Y., Carlson, R., Johnson, T. V., and Matson, D. L. 1980. Io's sodium emission cloud and the Voyager 1 encounter. *Icarus* 44:305–317.

Goldreich, P., and Ward, W. R. 1973. The formation of planetesimals. *Astrophys. J.* 183:1051–1061.

Goldreich, P., Tremaine, S., and Borderies, N. 1986. Toward a theory for Neptune's arc rings. *Astron. J.* 92:490–494.

Goldsmith, P. F., and Arquilla, R. 1985. Rotation in dark clouds. In *Protostars & Planets II,* eds. D. C. Black and M. S. Matthews (Tucson: Univ. of Arizona Press), pp. 137–149.

Goldsmith, P. F., Snell, R. L., Erickson, N. R., Dickman, R. L., Schloerb, F. P., and Irvine, W. M. 1985. Search for molecular oxygen in dense interstellar clouds. *Astrophys. J.* 289:613–617.

Goldsmith, P. F., Irvine, W. M., Hjalmarson, Å., and Ellder, J. 1986. Variations in the HCN/HNC abundance ratio in the Orion molecular cloud. *Astrophys. J.* 310:383–391.

Goldstein, B. E., Suess, S., and Walker, R. J. 1981. Mercury: Magnetospheric processes and the atmospheric supply and loss rates. *J. Geophys. Res.* 86:5485–5499.

Goody, R. M. 1982. The rotation of Uranus. In *Uranus and the Outer Planets,* ed. G. E. Hunt (Cambridge: Cambridge Univ. Press), pp. 143–153.

Goody, R. M., and Walker, J. C. G. 1972. *Atmospheres* (Englewood Cliffs: Prentice-Hall).

Gottlieb, C. A., Ball, J. A., Gottlieb, E. W., and Dickinson, D. F. 1979. Interstellar methyl alcohol. *Astrophys. J.* 227:422–432.

Gough, D. O. 1981. Solar interior structure and luminosity variations. *Solar Phys.* 74:21–34.

Graboske, H. C., Jr., Pollack, J. B., Grossman, A. S., and Olness, R. J. 1975*a*. The structure and evolution of Jupiter: The fluid contraction stage. *Astrophys. J.* 199:265–281.

Graboske, H. C., Jr., Olness, R. J., and Grossman, A. S. 1975*b*. Thermodynamics of dense hydrogen-helium fluids. *Astrophys. J.* 199:255–264.

Gradie, J., and Tedesco, E. F. 1982. The compositional structure of the asteroid belt. *Science* 216:1405–1407.

Gradie, J., and Veverka, J. 1980. The composition of the Trojan asteroids. *Nature* 283:840–842.

Gradie, J., and Veverka, J. 1984. Photometric properties of powdered sulfur. *Icarus* 58:227–245.

Gradie, J. C., Chapman, C. R., and Williams, J. G. 1979. Families of minor planets. In *Asteroids,* ed. T. Gehrels (Tucson: Univ. of Arizona Press), pp. 359–390.

Gradie, J., Hayashi, J., Zuckerman, B., Epps, H., and Howell, R. 1987. Physical properties of the Beta Pictoris circumstellar disk. *Lunar Planet. Sci.* XVIII:351–352 (abstract).

Gradie, J., Mouginis-Mark, P., Hayashi, J., and Flynn, L. 1988. Surface temperature and variability of an active lava lake: Lessons to be applied to Io. In *Lunar Planet. Sci.* XIX:407–408 (abstract).

Grady, D. E., and Moody, R. L. 1985. Shock and release equations of state of calcite. *Sandia Report SAND85-0947* (Albuquerque: Sandia Natl. Labs.), p. 25.

Grady, M. M., Wright, I. P., Carr, L. P., and Pillinger, C. T. 1986. Compositional differences in enstatite chondrites based on carbon and nitrogen stable isotope measurements. *Geochim. Cosmochim. Acta* 50:2799–2813.

Graham, J. A., and Frogel, J. A. 1985. An FU Orionis star associated with Herbig-Haro Object 57, *Astrophys. J.* 289:331–341.

Grasdalen, G. L., Strom, S. E., Strom, K. M., Capps, R. W., Thompson, D., and Castelaz, M.

1984. High spatial resolution IR observations of young stellar objects: A possible disk surrounding HL Tauri. *Astrophys. J.* 283:L57–L61.

Gray, C. M., and Compston, W. 1987. Excess ^{26}Mg in the Allende meteorite. *Nature* 251:495–497.

Greeley, R. 1987. Release of juvenile water on Mars: Estimated amounts and timing associated with volcanism. *Science* 236:1653–1654.

Greenberg, J. M. 1974. The interstellar depletion mystery, or where have all those atoms gone? *Astrophys. J.* 189:L81–L85.

Greenberg, J. M. 1976. Radical formation, chemical processing, and explosion of interstellar grains. *Astrophys. Space Sci.* 39:9–18.

Greenberg, J. M. 1982. What are comets made of? A model based on interstellar dust. In *Comets,* ed. L. L. Wilkening (Tucson: Univ. of Arizona Press), pp. 131–163.

Greenberg, J. M., and Chlewicki, G. 1983. A far-ultraviolet extinction law: What does it mean? *Astrophys. J.* 272:563–578.

Greenberg, J. M., Grim, R., and van IJzendoorn, L. 1986. Interstellar S_z in comets. In *Asteroids, Comets, Meteors II,* eds. C.-I. Lagerqvist, B. A. Lindblad, H. Lundstedt and H. Rickman (Uppsala: Uppsala Univ.), pp. 225–227.

Greenberg, R. 1978. Orbital resonance in a dissipative medium. *Icarus* 33:62–73.

Greenberg, R. 1979. Growth of large, late-stage planetesimals. *Icarus* 39:141–150.

Greenberg, R. 1980. Numerical simulation of planet growth. *Lunar Planet. Sci.* XI:365–367 (abstract).

Greenberg, R. 1988. Particle properties and the large-scale structure of planetary rings. *Icarus* 75:527–539.

Greenberg, R., and Rizk, B. 1987. Incipient runaway growth of planetesimals. *Lunar Planet. Sci.* XVIII:362–363 (abstract).

Greenberg, R., and Scholl, H. 1979. Resonances in the asteroid belt. In *Asteroids,* ed. T. Gehrels (Tucson: Univ. of Arizona Press), pp. 310–333.

Greenberg, R., Wacker, J. F., Hartmann, W. K., and Chapman, C. R. 1978*a*. Planetesimals to planets: Numerical simulation of collisional evolution. *Icarus* 35:1–26.

Greenberg, R., Hartmann, W. K., Chapman, C. R., and Wacker, J. F. 1978*b*. The accretion of planets from planetesimals. In *Protostars and Planets,* ed. T. Gehrels (Tucson: Univ. of Arizona Press), pp. 599–622.

Greenberg, R., Weidenschilling, S. J., Chapman, C. R., and Davis, D. R. 1984. From icy planetesimals to outer planets and comets. *Icarus* 59:87–113.

Greenberg, R., Carusi, A., and Valsecchi, G. B. 1986. Two-body models for planetary encounters: Failure criteria. *Lunar Planet. Sci.* XVII:287–288 (abstract).

Greenberg, R., Carusi, A., and Valsecchi, G. B. 1988. Outcome of planetary close encounters: A systematic comparison of methodologies. *Icarus* 75:1–29.

Grevesse, N. 1984*a*. Abundances of elements in the Sun. In *Frontiers of Astronomy and Astrophysics,* ed. R. Pallavicini (Florence: Italian Astronomical Society), p. 71.

Grevesse, N. 1984*b*. Accurate atomic data and solar photospheric spectroscopy. *Phys. Scripta* T8:49–58.

Grevesse, N., Sauval, A. J., and van Dishoeck, E. F. 1984. An analysis of vibration—Rotation lines of OH in the solar infrared spectrum. *Astron. Astrophys.* 141:10–16.

Grieve, R. A. F. 1982. The record of impact on Earth: Implications for a major Cretaceous-Tertiary impact event. In *Conference on Large Body Impacts and Terrestrial Evolution: Geological, Climatological, and Biological Implications,* eds. L. T. Silver and P. H. Schultz, Geol. Soc. Amer. SP 190 (Boulder: Geol. Soc. of Amer.), pp 25–37.

Grieve, R. A. F., and Dence, M. R. 1979. The terrestrial cratering record. II. The crater production rate. *Icarus* 38:230–242.

Grigorýev, F. V., Kormer, S. B., Mikhailova, O. L., Tolochko, A. P., and Urlin, V. D. 1972. Experimental determination of the compressibility of hydrogen at densities 0.5-2 g/cm^3. Metallization of hydrogen. *Sov. Phys. JETP (Letters)* 16:201–204.

Grinspoon, D. H. 1987. Was Venus wet? Deuterium reconsidered. *Science* 238:1702–1704.

Grinspoon, D. H., and Lewis, J. S. 1987*a*. Deuterium fractionation in the pre-solar nebula: Kinetic limitations on surface catalysis. *Icarus* 72:430–436.

Grinspoon, D. H., and Sagan, C. 1987*b*. Was the early Earth shrouded in impact generated dust? *Bull. Amer. Astron. Soc.* 19:872 (abstract).

Grinspoon, D. H., and Lewis, J. S. 1988. Cometary water on Venus: Implications of stochastic comet impacts. *Icarus* 74:21–35.

Gross, S. 1972. On the exospheric temperature of hydrogen-dominated planetary atmospheres. *J. Atmos. Sci.* 29:214–218.

Grossman, A. S., Pollack, J. B., Reynolds, R. T., Summers, A. L, and Graboske, H. C., Jr. 1980. The effect of dense cores on the structure and evolution of Jupiter and Saturn. *Icarus* 42:358–379.

Grossman, L. 1972. Condensation in the primitive solar nebula. *Geochim. Cosmochim. Acta* 36:597–619.

Grossman, L., and Larimer, J. W. 1974. Early chemical history of the solar system. *Rev. Geophys. Space Phys.* 12:71–101.

Grover, R., and Hardy, J. W. 1966. The propagation of shocks in exponentially decreasing atmospheres. *Astrophy. J.* 148:48–60.

Guélin, M. 1985. Chemical composition and molecular abundances of molecular clouds. In *Molecular Astrophysics: State of the Art and Future Directions,* eds. G. H. F. Dierckson, W. F. Huebner and P. W. Langhoff (Dordrecht: D. Reidel), pp. 23–44.

Guélin, M. 1987. Radio and millimeter observations of less complex molecules. In *Astrochemistry,* eds. M. S. Vardya and S. P. Tarafdar (Dordrecht: D. Reidel), pp. 171–181.

Guélin, M., Langer, W. D., and Wilson, R. W. 1982. The state of ionization in dense molecular clouds. *Astron. Astrophys.* 107:107–127.

Gulkis, S., and de Pater, I. 1984. A review of the millimeter and centimeter observations of Uranus. In *Uranus and Neptune,* ed. J. T. Bergstralh, NASA Conf. Pub. 2330, pp. 225–262.

Gulkis, S., Olsen, E. T., Klein, M. J., and Thompson, T. J. 1983. Uranus: Variability of the microwave spectrum. *Science* 221:453–455.

Gusten, R., and Ungerechts, H. 1985. Constraints on the sites of nitrogen nucleosynthesis from $^{15}NH_3$-observations. *Astron. Astrophys.* 145:241–250.

Haas, M. R., Erickson, E. F., McKibbin, D. D., Goorvitch, D., and Caroff, L. J. 1982. Far-infrared spectrophotometry of Saturn and its rings. *Icarus* 51:476–490.

Haff, P. K., and Eviatar, A. 1986. Micrometeoroid impact on planetary satellites as a magnetospheric mass source. *Icarus* 66:258–269.

Haff, P. K., and Watson, C. C. 1979. The erosion of planetary and satellite atmospheres by energetic atomic particles. *J. Geophys. Res.* 84:8436–8442.

Haff, P. K., Watson, C., and Yung, Y. L. 1981. Sputter ejection of matter from Io. *J. Geophys. Res.* 86:426–438.

Haff, P. K., Eviatar, A., and Siscoe, G. 1983. Ring and plasma: The enigmae of Enceladus. *Icarus* 56:426–438.

Hagen, W., Allamandola, L. J., and Greenberg, J. M. 1980. Infrared absorption lines by molecules in grain mantles. *Astron. Astrophys.* 85:L3–L6.

Hale, L. D., Morton, C. J., and Sleep, N. H. 1982. Reinterpretation of seismic reflection data over East Pacific Rise. *J. Geophys. Res.* 87:7707–7718.

Hamano, Y., and Ozima, M. 1978. Earth-atmosphere evolution model based on Ar isotopic data. In *Adv. Earth Planet. Sci.,* eds. E. C. Alexander, Jr., and M. Ozima (Tokyo, Japan: Acad. Publ.) 3:155–171.

Hammel, H. B., and Buie, M. W. 1987. An atmospheric rotation period of Neptune determined from methane-band imaging. *Icarus* 72:62–68.

Hanel, R., Conrath, B., Flasar, M., Kunde, V., Lowman, P., Maguire, W., Pearl, J., Pirraglia, J., Samuelson, R., Gautier, D., Gierasch, P., Kumar, S., and Ponnamperuma, C. 1979*a*. Infrared observations from the Jovian system from Voyager 1. *Science* 204:972–976.

Hanel, R., Conrath, B., Flasar, M., Herath, L., Kunde, V., Lowman, P., Maguire, W., Pearl, J., Pirraglia, J., Samuelson, R., Gautier, D., Gierasch, P., Horn, L., Kumar, S., and Ponnamperuma, C. 1979*b*. Infrared observations of the Jovian system from Voyager 2. *Science* 206:952–956.

Hanel, R. A., Crosby, D., Herath, L., Vanous, D., Collins, D., Creswick, H., Harris, C., and Rhodes, M. 1980. Infrared spectrometer for Voyager. *Appl. Opt.* 19:1391–1400.

Hanel, R. A., Conrath, B. J., Herath, L., Kunde, V. G., and Pirraglia, J. A. 1981*a*. Albedo, internal heat, and energy balance of Jupiter: Preliminary results of the Voyager infrared investigation. *J. Geophys. Res.* 86:8705–8712.

Hanel, R., Conrath, B., Flasar, F. M., Kunde, V., Maguire, W., Pearl, J., Pirraglia, J., Sam-

uelson, R., Herath, L., Allison, M., Cruikshank, D., Gautier, D., Gierasch, P., Horn, L., Koppany, R., and Ponnamperuma, C. 1981*b*. Infrared observations of the Saturnian system from Voyager 1. *Science* 212:192–200.
Hanel, R., Conrath, B., Flasar, F. M., Kunde, V., Maguire, W., Pearl, J., Pirraglia, J., Samuelson, R., Cruikshank, D., Gautier, D., Gierasch P., Horn, L., and Ponnamperuma, C. 1982. Infrared observations of the Saturnian system from Voyager 2. *Science* 215:544–548.
Hanel, R. A., Conrath, B. J., Kunde, V. G., Pearl, J. C., and Pirraglia, J. A. 1983. Albedo, internal heat flux, and energy balance of Saturn. *Icarus* 53:262–285.
Hanel, R., Conrath, B., Flasar, F. M., Kunde, V., Maguire, W., Pearl, J., Pirraglia, J., Samuelson, R., Cruikshank, D., Gautier, D., Gierasch, P., Horn, L., and Schulte, P. 1986. Infrared observations of the Uranian system. *Science* 233:70–74.
Hanks, T. C., and Anderson, D. L. 1969. The early thermal history of the Earth. *Phys. Earth Planet. Int.* 2:19–29.
Hanner, M. S. 1981. On the detectability of icy grains in the comae of comets. *Icarus* 47:342–350.
Hapke, B. 1986. On the sputter alteration of regoliths of outer solar system bodies. *Icarus* 66:270–279.
Haring, R. A., Haring, A., Klein, F. W., Kummel, A. C., and deVries, A. E. 1983. Reaction sputtering of simple condensed gases by keV heavy ion bombardment. *Nucl. Instrum. Meth.* 24:529–533.
Harmon, J. K., Campbell, D. B., Hine, A. A., Shapiro, I. I., and Marsden, B. G. 1988. Radar observations of comet IRAS-Araki-Alcock 1983. *Astrophys. J.*, in press.
Harper, D. A., Loewenstein, R. F., and Davidson, J. A. 1984. On the nature of the material surrounding Vega. *Astrophys. J.* 285:808–812.
Harrington, R. S., and Christy, J. W. 1981. The satellite of Pluto. III. *Astron. J.* 86:442–443.
Harris, A. W. 1978. The formation of the outer planets. *Lunar Planet. Sci.* IX:459–461 (abstract).
Harris, A. W. 1984. Physical properties of Neptune and Triton inferred from the orbit of Triton. In *Uranus and Neptune,* ed. J. T. Bergstralh, NASA CP-2330, pp. 357–373.
Harris, A. W., and Kaula, W. M. 1975. A co-accretional model of satellite formation. *Icarus* 24:516–524.
Harris, S. A. 1977. The aureole of Olympus Mons. *J. Geophys. Res.* 82:3099–3107.
Harrison, H., and Schoen, R. 1967. Evaporation of ice in space: Saturn's rings. *Science* 157:1175–1176.
Hart, M. H. 1978. The evolution of the atmosphere of the Earth. *Icarus* 33:23–39.
Hart, M. H. 1979. Habitable zones around main sequence stars. *Icarus* 37:351–357.
Hart, R. 1973. A model for chemical exchange in the basalt-seawater system of oceanic layer II. *Can. J. Earth Sci.* 10:799–816.
Hart, S. R., Dymond, J., and Hogan, L. 1979. Preferential formation of the atmosphere-sialic crust system from the earth. *Nature* 278:156–159.
Harteck, P., and Jensen, J. H. D. 1948. Uber den Sauerstoffgehalt der Atmosphare. *Z. Naturforsch.* 3a:591–595.
Hartle, R. E., and Taylor, H. A., Jr. 1983. Identification of deuterium ions in the ionosphere of Venus. *Geophys. Res. Lett.* 10:965–968.
Hartle, R. E., Sittler, E. C., Ogilvie, K. W., Scudder, J. D., Lazarus, A. J., and Atreya, S. K. 1982. Titan's ion exosphere observed from Voyager 1. *J. Geophys. Res.* 87:1383–1394.
Hartmann, L., and Kenyon, S. J. 1985. On the nature of FU Orionis objects. *Astrophys. J.* 299:462–478.
Hartmann, W. K. 1973. Ancient lunar mega-regolith and subsurface structure. *Icarus* 18:634–646.
Hartmann, W. K. 1978. Planet formation: Mechanism of early growth. *Icarus* 33:50–61.
Hartmann, W. K. 1980. Surface evolution of two-component stone/ice bodies in the Jupiter region. *Icarus* 44:441–453.
Hartmann, W. K. 1984. Does crater "saturation equilibrium" occur in the solar system? *Icarus* 60:56–74.
Hartmann, W. K. 1986. Moon origin: The impact-trigger hypothesis. In *Origin of the Moon,* eds. W. K. Hartmann, R. J. Phillips and G. J. Taylor (Houston: Lunar and Planetary Inst.), pp. 579–608.

Hartmann, W. K., and Davis, D. R. 1975. Satellite-sized planetesimals and lunar origin. *Icarus* 24:504–515.

Hartmann, W. K., Strom, R. G., Weidenschilling, S. J., Blasius, K. R., Woronow, A., Dence, M. R., Grieve, R. A. F., Diaz, J., Chapman, C. R., Shoemaker, E. M., and Jones, K. L. 1981. Chronology of planetary volcanism by comparative studies of planetary cratering. In *Basaltic Volcanism on the Terrestrial Planets* (New York: Pergamon Press), pp. 1050–1127.

Hartmann, W. K., Cruikshank, D. P., and Tholen, D. J. 1985. Outer solar system materials: Ices and color systematics. In *Ices in the Solar System,* eds. J. Klinger, D. Benest, A. Dollfus and R. Smoluchowski (Dordrecht: D. Reidel), pp. 169–181.

Hartmann, W. K., Tholen, D. J., and Cruikshank, D. P. 1987. The relationship of active comets, "extinct" comets, and dark asteroids. *Icarus* 69:33–50.

Hashimoto, A., and Grossman, L. 1987. Alteration of Al-rich inclusions inside amoeboid olivine aggregates in the Allende meteorite. *Geochim. Cosmochim. Acta* 51:1685–1704.

Hawkins, I., and Jura, M. 1987. The $^{12}C/^{13}C$ ratio of the interstellar medium in the neighborhood of the Sun. *Astrophys. J.* 317:926–950.

Hayashi, C. 1981. Structure of the solar nebula, growth and decay of magnetic fields, and effects of magnetic and turbulent viscosities on the nebula. *Prog. Theor. Phys. Suppl.* 70:35–53.

Hayashi, C., and Nakano, T. 1965. Thermal and dynamical properties of a protostar and its contraction to the stage of quasi-static equilibrium. *Prog. Theor. Phys.* 34:754–775.

Hayashi, C., Nakazawa, K., and Adachi, I. 1977. Long-term behavior of planetesimals and the formation of the planets. *Publ. Astron. Soc. Japan* 29:163–197.

Hayashi, C., Nakazawa, K., and Nakagawa, Y. 1985. Formation of the solar system. In *Protostars & Planets II,* eds. D. C. Black and M. S. Matthews (Tucson: Univ. of Arizona Press), pp. 1100–1153.

Hayatsu, R., and Anders, E. 1981. Organic compounds in meteorites and their origins. In *Topics in Current Chemistry, Cosmo-and Geochemistry* (Berlin: Springer-Verlag) 99:1–39.

Hayes, W. D. 1968. The propagation upward of the shock wave from a strong explosion in the atmosphere. *J. Fluid Mech.* 32:317–331.

Head, J. W. 1974. Lunar dark-mantle deposits: Possible clues to the distribution of early mare deposits. *Proc. Lunar Sci. Conf.* 5:207–222.

Head, J. W., and Wilson, L. 1986. Volcanic processes and landforms on Venus: Theory, predictions, and observations. *J. Geophys. Res.* 91:9407–9446.

Heisler, J., Tremaine, S., and Alcock, C. 1987. The frequency and intensity of comet showers from the Oort cloud. *Icarus* 70:269–288.

Henkel, C., Wilson, T. L., and Bieging, J. 1982. Further ($^{12}C/^{13}C$) ratios from formaldehyde: A variation with distance from the galactic center. *Astron. Astrophys.* 109:344–351.

Hennecke, E. W., and Manuel, O. K. 1975. Noble gases in an Hawaiian xenolith. *Nature* 257:778–780.

Herbert, F., and Sonett, C. P. 1978. Primordial metamorphism of asteroids via electric induction on a T Tauri-like solar wind. *Astrophys. Space Sci.* 55:227–239.

Herbert, F., and Sonett, C. P. 1979. Electromagnetic heating of minor planets in the early solar system. *Icarus* 40:484–496.

Herbert, F., and Sonett, C. P. 1980. Electromagnetic inductive heating of the asteroids and moon as evidence bearing on the primordial solar wind. In *The Ancient Sun: Fossil Record in the Earth, Moon and Meteorites,* eds. R. O. Pepin, J. A. Eddy and R. B. Merrill (New York: Pergamon Press), pp. 563–576.

Herbert, F., Sonett, C. P., and Wiskerchen, M. J. 1977. Model "zero-age" lunar thermal profiles resulting from electrical induction. *J. Geophys. Res.* 82:2054–2060.

Herbig, G. H. 1958. T Tauri stars, flare stars, and related objects as members of stellar associations. In *Stellar Populations,* ed. D. J. K. O'Connell (Amsterdam: North Holland), pp. 129–142.

Herbig, G. H. 1977. Eruptive phenomena in early stellar evolution. *Astrophys. J.* 217:693–715.

Herbig, G. H. 1982. Stars of low to intermediate mass in the Orion nebula. In *Symposium on the Orion Nebula to Honor Henry Draper,* eds. A. E. Glassgold, P. J. Huggins and E. L. Schucking (New York: New York Acad. Sci.), pp. 64–78.

Herbst, E., Adams, N. G., and Smith, D. 1984. Theoretical reinvestigation of hydrocarbon and cyanoacetylene abundances in TMC-1. *Astrophys. J.* 285:618–621.

Herbst, E., Adams, N. G., Smith, D., and DeFrees, D. J. 1987. Ion-molecule calculation of the abundance ratio of CCD to CCH in dense interstellar clouds. *Astrophys. J.* 312:351–357.

Herring, J., and Kyle, L. 1961. Density in a planetary exosphere. *J. Geophys. Res.* 66:L1980–L1982.

Herron, T. L., Ludwig, W. J., Stoffa, P. L., Kan, T. K., and Buhl, P. 1978. Structure of the East Pacific Rise crest from multichannel seismic data. *J. Geophys. Res.* 83:798–804.

Hertogen, J., Vizgirda, J., and Anders, E. 1977. Composition of the parent body of eucrite meteorites. *Bull. Amer. Astron. Soc.* 9:458–459 (abstract).

Herzberg, G. 1952. Laboratory absorption spectra obtained with long paths. In *The Atmospheres of the Earth and Planets,* ed. G. P. Kuiper (Chicago: Univ. of Chicago Press), pp. 406–416.

Heyer, M. H. 1988. The magnetic evolution of the Taurus molecular clouds. II. A reduced role of the magnetic field at high densities. *Astrophys. J.* 324:311–320.

Heyer, M. H., Vrba, F. J., Snell, R. L., Schloerb, F. P., Strom, S. E., Goldsmith, P. F., and Strom, K. M. 1987. The magnetic evolution of the Taurus molecular clouds. I. Large scale properties. *Astrophys. J.* 321:855–876.

Heymann, D., and Mazor, E. 1968. Noble gases in unequilibrated ordinary chondrites. *Geochim. Cosmochim. Acta* 32:1–19.

Hibbert, A., Dufton, P. L., and Keenan, F. P. 1985. Oscillator strengths for transitions in NI and the interstellar abundance of nitrogen. *Mon. Not. Roy. Astron. Soc.* 213:721–734.

Hildebrand, A. R., Jones, T. D., and Lebofsky, L. A. 1987. Is Ceres differentiated? *Meteoritics* 22:410–411 (abstract).

Hildebrand, R. H., Loewenstein, R. F., Harper, D. A., Orton, G. S., Keen, J., and Withcomb, S. E. 1985. Far-infrared and submillimeter brightness temperatures of the giant planets. *Icarus* 64:64–87.

Hiyagon, H., and Ozima, M. 1986. Partition of noble gases between olivine and basalt melt. *Geochim. Cosmochim. Acta* 50:2045–2057.

Hiyagon, H., Hunemohr, H., Kennedy, B. M., Reynolds, J. H., and Smith, S. P. 1986. Noble gases from Alberta, Canada, natural gas fields. *Terra Cognita* 6:104–105 (abstract).

Hjalmarson, Å. 1985. Astrochemistry—Observational aspects. In *(Sub)-Millimeter Astronomy,* eds. P. A. Shaver and K. Kjär, ESO Conf. Workshop Proc. No. 22, pp. 285–326.

Hobbs, L. M. 1986. Observations of gaseous circumstellar disks. III. *Astrophys. J.* 308:854–858.

Hobbs, L. M., Vidal-Jadjar, A., Ferlet, R., Albert, C. E., and Gry, C. 1985. The gaseous component of the disk around Beta Pictoris. *Astrophys. J.* 293:L29–L33.

Hodges, R. R. 1980. Methods for Monte Carlo simulation of the exospheres of the moon and Mercury. *J. Geophys. Res.* 85:164–170.

Hodges, R. R., Jr., and Donahue, T. M. 1987. Water vapor in the atmosphere of Venus. *Icarus,* to be submitted.

Hoffman, J. H., Hodges, R. R., Jr., Donahue, T. M., and McElroy, M. B. 1980. Composition of the Venus lower atmosphere from the Pioneer Venus mass spectrometer. *J. Geophys. Res.* 85:7882–7890.

Holland, H. D. 1963. On the chemical evolution of the terrestrial and cytherion atmospheres. In *The Origin and Evolution of Atmospheres and Oceans,* eds. P. J. Brancasio and A. G. W. Cameron (New York: Wiley), pp. 86–101.

Holland, H. D. 1978. *The Chemistry of the Atmosphere and Oceans* (New York: Wiley).

Holland, H. D. 1984. *The Chemical Evolution of the Atmosphere and Oceans* (Princeton: Princeton Univ. Press).

Holland, H. D., and Zbinden, E. A. 1986. Paleosols and the evolution of the atmosphere, Part I. In *Physical and Chemical Weathering in Geochemical Cycles,* eds. A. Lerman and M. Meybeck (NATO ASI Institute).

Hollis, J. M., Snyder, L. E., Blake, D. H., Lovas, F. J., Suenram, R. D., and Ulich, B. L. 1982. New interstellar molecular transitions in the 2 millimeter range. *Astrophys. J.* 251:541–548.

Holsapple, K. A., and Schmidt, R. M. 1982. On the scaling of crater dimensions. 2. Impact processes. *J. Geophys. Res.* 87:1849–1870.

Holton, J. R. 1972. *An Introduction to Dynamic Meteorology* (New York: Academic Press).

Honda, M., Reynolds, J. H., Roedder, E., and Epstein, S. 1987. Noble gases in gem-class diamonds from known localities: occurrences of solar-like helium and neon. *J. Geophys. Res.* 92:12531–12538.

Horedt, G. P. 1978. Blow-off of the protoplanetary cloud by a T-Tauri like solar wind. *Astron. Astrophys.* 64:173–178.

Horedt, G. P. 1980. Accretional heating as the major cause of compositional differences among the meteorite parent bodies, the Moon and the Earth. *Icarus* 43:215–222.

Horedt, G. P. 1982*a*. Blow-off of planetary atmospheres and of the protoplanetary nebula. *Phys. Earth Planet. Int.* 29:252–260.

Horedt, G. P. 1982*b*. Mass loss from the protoplanetary nebula. *Astron. Astrophys.* 110:209–214.

Hourigan, K., and Ward, W. R. 1987. Radial migration of preplanetary material: Implications for the accretion time scale problem. *Icarus* 60:29–39.

Housley, R. M. 1978. Modelling lunar eruptions. *Proc. Lunar Planet. Sci. Conf.* 9:1473–1484.

Howell, R. R., Cruikshank, D. P., and Fanale, F. P. 1984. Sulfur dioxide on Io: Spatial distribution and physical state. *Icarus* 57:83–92.

Hoyle, F. 1955. *Frontiers in Astronomy* (London: William Heinemann), pp. 68–72.

Hoyle, F., and Wickramasinghe, N. C. 1962. On graphite particles as interstellar grains. *Mon. Not. Roy. Astron. Soc.* 124:417–433.

Hoyle, F., Wickramasinghe, N. C., Al-Mufti, S., Olavosen, A. H., and Wickramasinghe, D. T. 1982. Infrared spectroscopy over the 2.9–3.9 micron waveband in biochemistry and astronomy. *Astrophys. Space Sci.* 83:405–409.

Huang, T., and Siscoe, G. 1986. Stability of the Io torus. *J. Geophys. Res.* 91:10164–10166.

Huang, T., and Siscoe, G. 1987. Types of planetary tori. *Icarus* 70:366–378.

Hubbard, W. B. 1974*a*. Gravitational field of a rotating planet with a polytropic index of one. *Astron. Zh.* 51:1052–1058.

Hubbard, W. B. 1974*b*. Inversion of gravity data for giant planets. *Icarus* 21:157–161.

Hubbard, W. B. 1977. The Jovian surface condition and cooling rate. *Icarus* 30:305–310.

Hubbard, W. B. 1978. Comparative thermal evolution of Uranus and Neptune. *Icarus* 35:177–181.

Hubbard, W. B. 1980. Intrinsic luminosities of the giant planets. *Rev. Geophys. Space Phys.* 18:1–9.

Hubbard, W. B. 1982. Effects of differential rotation on the gravitational figures of Jupiter and Saturn. *Icarus* 52:509–515.

Hubbard, W. B. 1984*a*. Interior structure of Uranus. In *Uranus and Neptune,* ed. J. T. Bergstralh, NASA CP-2330, pp. 291–325.

Hubbard, W. B. 1984*b*. Uranus and Neptune. In *Planetary Interiors* (New York: Van Nostrand Rheinhold), pp. 282–290.

Hubbard, W. B., and DeWitt, H. E. 1985. Statistical mechanics of light elements at high pressure. VII. A perturbative free energy for arbitrary mixtures of H and He. *Astrophys. J.* 290:388–393.

Hubbard, W. B., and Horedt, G. P. 1983. Computation of Jupiter interior models from gravitational inversion theory. *Icarus* 54:456–465.

Hubbard, W. B., and Marley, M. S. 1988. Optimized Jupiter, Saturn and Uranus interior models. *Icarus,* in press.

Hubbard, W. B., and MacFarlane, J. J. 1980*a*. Structure and evolution of Uranus and Neptune. *J. Geophys. Res.* 85:225–234.

Hubbard, W. B., and MacFarlane, J. J. 1980*b*. Theoretical predictions of deuterium abundances in the Jovian planets. *Icarus* 44:676–682.

Hubbard, W. B., and MacFarlane, J. J. 1985. Statistical mechanics of light elements at high pressure. VIII. Thomas-Fermi-Dirac theory for binary mixtures of H with He, C, and O. *Astrophys. J.* 297:133–144.

Hubbard, W. B., and Smoluchowski, R. 1973. Structure of Jupiter and Saturn. *Space Sci. Rev.* 14:599–662.

Hubbard, W. B., and Stevenson, D. J. 1984. Interior structure of Saturn. In *Saturn,* eds. T. Gehrels and M. S. Matthews (Tucson: Univ. of Arizona Press), pp. 47–87.

Hubbard, W. B., Zharkov, V. N., and Trubitsyn, V. P. 1974. Significance of gravitational moments for interior structure of Jupiter and Saturn. *Icarus* 21:147–151.

Hubbard, W. B., Slattery, W. L., and DeVito, C. L. 1975. High zonal harmonics of rapidly rotating planets. *Astrophys. J.* 199:504–516.

Hubbard, W. B., MacFarlane, J. J., Anderson, J. D., Null, G. W., and Biller, E. D. 1980. The structure of Saturn inferred from Pioneer 11 gravity data. *J. Geophys. Res.* 85:5909–5916.

Hubbard, W. B., Avey, H. P., Carter, B., Frecker, J., Fu, H. H., Gehrels, J.-A., Gehrels, T.,

Hunten, D. M., Kennedy, H. D., Lebofsky, L. A., Mottram, K., Murphy, T., Nielsen, A., Page, A. A., Reitsema, H. J., Smith, B. A., Tholen, D. J., Varnes, B., Vilas, F., Waterworth, M. D., Wu, H. H., and Zellner, B. 1985. Results from observations of the 15 June 1983 occultation by the Neptune system. *Astron. J.* 90:655–667.

Hubbard, W. B., Brahic, A., Sicardy, B., Elicer, L.-R., Roques, F., and Vilas, F. 1986. Occultation detection of a neptunian ring-like arc. *Nature* 319:636–640.

Hubbard, W. B., Nicholson, P. D., Lellouch, E., Sicardy, B., Brahic, A., Vilas, F., Bouchet, P., McLaren, R. A., Millis, R. L., Wasserman, L. H., Elias, J. H., Matthews, K., McGill, J. D., and Perrier, C. 1987. Oblateness, radius, and mean stratospheric temperature of Neptune from the 1985 August 20 occultation. *Icarus* 72:635–646.

Hubbert, M. K., and Rubey, W. W. 1959. Role of fluid pressure in mechanics of overthrust faulting, I. Mechanics of fluid-filled solids and its application to overthrust faulting. *Geol. Soc. Amer. Bull.* 70:115–166.

Hudson, G. B., Kennedy, B. M., and Podosek, F. A. 1988. The early solar system abundance of ^{244}Pu as inferred from the St. Severin chondrite. *Proc. Lunar Planet. Sci. Conf.* 19, in press.

Huebner, W. F. 1987. First polymer in space identified in comet Halley. *Science* 237:628–630.

Huffman, D. R. 1977. Interstellar grains: The interaction of light with a small-particle system. *Adv. Phys.* 26:129–230.

Hunt, J. M. 1972. Distribution of carbon in crust of Earth. *Bull. Amer. Assoc. Petrol. Geol.* 56:2273–2277.

Hunten, D. M. 1973. The escape of light gases from planetary atmospheres. *J. Atmos. Sci.* 30:1481–1494.

Hunten, D. M. 1978. A Titan atmosphere with a surface temperature of 200K. In *The Saturn System,* eds. D. M. Hunten and D. Morrison, NASA CP-2068, pp. 127–140.

Hunten, D. M. 1979. Capture of Phobos and Deimos by protoatmospheric drag. *Icarus* 37:113–123.

Hunten, D. M. 1982. Thermal and nonthermal escape mechanisms for terrestrial bodies. *Planet. Space Sci.* 30:773–783.

Hunten, D. M. 1984. Atmospheres of Uranus and Neptune. In *Uranus and Neptune,* ed. J. T. Bergstralh, NASA CP-2330, pp. 27–54.

Hunten, D. M. 1985. Blow-off of an atmosphere and possible application to Io. *Geophys. Res. Lett.* 12:271–273.

Hunten, D. M., and Donahue, T. M. 1976. Hydrogen loss from the terrestrial planets. *Ann. Rev. Earth Planet. Sci.* 4:265–292.

Hunten, D. M., and Watson, A. J. 1982. Stability of Pluto's atmosphere. *Icarus* 51:665–667.

Hunten, D. M., Tomasko, M. G., Flasar, F. M., Samuelson, R. E., Strobel, D. F., and Stevenson, D. J. 1984. Titan. In *Saturn,* eds. T. Gehrels and M. S. Matthews (Tucson: Univ. of Arizona Press), pp. 671–759.

Hunten, D. M., Pepin, R. O., and Walker, J. C. G. 1987. Mass fractionation in hydrodynamic escape. *Icarus* 69:532–549.

Hunten, D. M., Pepin, R. O., and Owen, T. 1988*a*. Planetary atmospheres. In *Meteorites and the Early Solar System,* eds. J. F. Kerridge and M. S. Matthews (Tucson: Univ. of Arizona Press), pp. 565–591.

Hunten, D., Morgan, T., and Shemansky, D. 1988*b*. The Mercury atmosphere. In *Mercury,* eds. F. Vilas, C. R. Chapman and M. S. Matthews (Tucson: Univ. of Arizona Press), pp. 561–611.

Hunter, J. H., Jr. 1969. The collapse of interstellar gas clouds and the formation of stars. *Mon. Not. Roy. Astron. Soc.* 142:473–498.

Hut, P., and Weissman, P. R. 1985. Dynamical evolution of cometary showers. *Bull. Amer. Astron. Soc.* 17:690 (abstract).

Hut, P., Alvarez, W., Elder, W. P., Hanson, T., Kauffman, E. G., Keller, G., Shoemaker, E. M., and Weissman, P. R. 1987. Comet showers as a cause of stepwise extinctions. *Nature* 329:118–126.

Igarashi, G., and Ozima, M. 1988. Origin of isotopic fractionation of terrestrial xenon. *Proc. of the NIPR Symp. on Antarctic Meteorites, No. 1,* pp. 315–320.

Imbrie, J., and Imbrie, K. P. 1981. *Ice Ages: Solving the Mystery* (Short Hills, N.J.: Enslow).

Ingersoll, A. P. 1969. The runaway greenhouse: A history of water on Venus. *J. Atmos. Sci.* 26:1191–1198.

Ingersoll, A. P., and Porco, C. C. 1978. Solar heating and internal heat flow on Jupiter. *Icarus* 35:27–43.

Ingersoll, A. P., Münch, G., Neugebauer, G., Diener, D. J., Orton, G. S., Schupler, B., Schroeder, M., Chase, S. C., Ruiz, R. D., and Trafton, L. M. 1975. Pioneer 11 infrared radiometer experiment: The global heat balance of Jupiter. *Science* 188:472–473.

Ingersoll, A. P., Summers, M. E., and Schlipf, S. G. 1985. Supersonic meterology of Io: Sublimation driven flow of SO_2. *Icarus* 64:375–390.

Innanen, K. A., Patrick, A. T., and Duley, W. W. 1978. The interaction of the spiral density wave and the Sun's galactic orbit. *Astrophys. Space Sci.* 57:511–515.

Ip, W.-H. 1980. Condensation and agglomeration of cometary ice: The HDO/H_2O ratio as tracer. In *Ices in the Solar System,* eds. J. Klinger, D. Benest, A. Dollfus and R. Smoluchowski (Dordrecht: D. Reidel), pp. 389–396.

Ip, W.-H. 1982. On charge exchange and knock-on processes in the exosphere of Io. *Astrophys. J.* 262:780–785.

Ip, W.-H. 1984. An estimate of the H_2 density in the atomic hydrogen cloud of Titan. *J. Geophys. Res.* 89:2377–2379.

Ip, W.-H. 1986*a*. The sodium exosphere and magnetosphere of Mercury. *Geophys. Res. Lett.* 13:423–426.

Ip, W.-H. 1986*b*. Plasmatization and recondensation of the saturnian rings. *Nature* 320:143–145.

Ip, W.-H. 1987*a*. Dynamics of electrons and heavy ions in Mercury's magnetosphere. *Icarus* 70:440–445.

Ip, W.-H. 1987*b*. Magnetopheric charge exchange effect on the electroglow of Uranus. *Nature* 326:775–777.

Ip, W.-H., and Axford, W. I. 1980. A weak interaction model for Io and the Jovian magnetosphere. *Nature* 283:180–183.

Ip, W.-H., and Axford, W. I. 1982. Theories of physical processes in the cometary comae and ion tails. In *Comets,* ed. L. L. Wilkening (Tucson: Univ. of Arizona Press), pp. 588–634.

Ip, W.-H., and Fernández, J. A. 1988. Exchange of condensed matter among the outer and terrestrial proto-planets and the effect on surface impact and atmospheric accretion. *Icarus* 74:47–61.

Ip, W.-H., and Huntress, W. T. 1987. Gas phase chemistry in the coma of comet Halley. Origin and Evolution of Planetary and Satellite Atmospheres, Abstract Booklet, 10–14 March, Tucson, AZ.

Irvine, W. M., and Hjalmarson, Å. 1984. The chemical composition of interstellar molecular clouds. *Origins of Life* 14:15–23.

Irvine, W. M., and Hjalmarson, Å. 1986. Observational astrochemistry. *Adv. Space Res.* 6:227–236.

Irvine, W. M., and Schloerb, F. P. 1984. Cyanide and isocyanide abundances in the cold dark cloud TMC-1. *Astrophys. J.* 282:516–521.

Irvine, W. M., Good, J. C., and Schloerb, F. P. 1983. Observations of SO_2 and HCS^+ in cold molecular clouds. *Astron. Astrophys.* 127:L10–L13.

Irvine, W. M., Schloerb, F. P., Hjalmarson, Å., and Herbst, E. 1985. The chemical state of dense interstellar clouds: An overview. In *Protostars & Planets II,* eds. D. C. Black and M. S. Matthews (Tucson: Univ. of Arizona Press), pp. 579–620.

Irvine, W. M., Goldsmith, P. F., and Hjalmarson, Å. 1987*a*. Chemical abundances in molecular clouds. In *Interstellar Processes,* eds. D. J. Hollenbach and H. A. Thronson (Dordrecht: D. Reidel), pp. 561–609.

Irvine, W. M., Avery, L. W., Friberg, P., Matthews, H. E., and Ziurys, L. M. 1987*b*. Newly detected molecules in dense interstellar clouds. In *Interstellar Matter,* eds. J. Moran and P. T. P. Ho (New York: Gordon and Breach), pp. 15–28.

Isotomin, V. G., Grechnev, K. V., and Kochnev, V. A. 1980. Mass spectrometry of the lower atmosphere of Venus: Krypton isotopes and other recent results of the Venera -11 and -12 data processing. Preprint D-298, Space Research Inst., Academy of Sciences, USSR.

Ito, E., Harris, D. M., and Anderson, A. T. 1983. Alteration of oceanic crust and the geologic cycling of chlorine and water. *Geochim. Cosmochim. Acta* 47:1613–1624.

Jackson, M. J., and Pollack, H. N. 1987. Mantle devolatilization and convection: Implications for the thermal history of the Earth. *Geophys. Res. Lett.* 14:737–740.

Jagoutz, E. 1987. New light on shergottites: ALHA 77005, the shock age. *Meteoritics* 22:417–418 (abstract).

Jagoutz, E., and Wänke, H. 1986. Sr and Nd isotopic systematics of shergotty meteorite. *Geochim. Cosmochim. Acta* 50:939–953.

Jagoutz, E., Baddenhausen, H., Palme, H., Blum, K., Cendales, J., Dreibus, G., Spettel, B., Lorenz, V., and Wänke, H. 1979. The abundances of major, minor and trace elements in the Earth's mantle as derived from primitive ultramafic nodules. *Proc. Lunar Planet. Sci.* 11:2031–2050.

Jagoutz, E., Carlson, R. W., and Lugmair, G. W. 1980. Equilibrated Nd-unequilibrated Sr isotopes in mantle xenoliths. *Nature* 286:708–710.

Jakosky, B. M., and Ahrens, T. J. 1979. The history of an atmosphere of impact origin. *Proc. Lunar Planet. Sci. Conf.* 10:2727–2739.

Jansa, L. F., and Pe-Piper, G. 1987. Identification of an underwater extraterrestrial impact crater. *Nature* 327:612–614.

Javoy, M., Pineau, F., and Allègre, C. J. 1982. Carbon geodynamic cycle. *Nature* 300:171–173.

Javoy, M., Pineau, E., and Delorme, H. 1986. Carbon and nitrogen isotopes in the mantle. *Chem. Geol.* 57:41–62.

Jeans, J. H. 1916. *The Dynamical Theory of Gases* (Cambridge: Cambridge Univ. Press).

Jenkins, E. B. 1987. Element abundances in the interstellar atomic material. In *Interstellar Processes,* eds. D. J. Hollenbach and H. A. Thronson (Dordrecht: D. Reidel), pp. 533–559.

Jenkins, E. B., Jura, M., and Loewenstein, M. 1983. *Copernicus* observations of CI: Pressures and carbon abundances in diffuse interstellar clouds. *Astrophys. J.* 270:88–104.

Jenkins, W. J., Edmond, J. M., and Corliss, J. B. 1978. Excess ^3He and ^4He in Galapagos hydrothermal waters. *Nature* 272:156–158.

Jessberger, E. K., and Kissel, J. 1987. Bits and pieces from Halley's Comet. In *Lunar Planet. Sci.* XVIII:466–467 (abstract).

Jessberger, E. K., Kissel, J., Fechtig, H., and Krueger, F. R. 1986. On the average chemical composition of cometary dust. In *The Comet Nucleus Sample Return Mission,* ESA-SP-249, pp. 27–30.

Jessberger, E. K., Kissel, J., Fechtig, H., and Kissel, J. 1987. On the average chemical composition of cometary dust. In *Physical Processes in Comets, Stars, and Active Galaxies,* eds. W. Hildebrandt, E. Meyer-Hofmeister and H.-C. Thomas (Berlin: Springer-Verlag), pp. 26–33.

Jessberger, E. K., Christoforidis, A., and Kissel, J. 1988. Aspects of the major element composition of Halley's dust. *Nature* 332:691–695.

Johansson, L. E. B., Andersson, C., Ellder, J., Friberg, P., Hjalmarson, Å., Höglund, B., Irvine, W. M., Olofsson, H., and Rydbeck, G. 1984. Spectral scan of Orion A and IRC+10216 from 72 to 91 GHz. *Astron. Astrophys.* 130:227–256.

Johnson, M. L., and Nicol, M. 1987. The ammonia-water phase diagram and its implications for icy satellites. *J. Geophys. Res.* 92:6339–6349.

Johnson, M. L., Schawke, A., and Nicol, M. 1985. Partial phase diagram for the system NH_3-H_2: The water rich region. In *Ices in the Solar System,* eds. J. Klinger, D. Benest, A. Dollfus and R. Smoluchowski (Dordrecht: D. Reidel), pp. 39–47.

Johnson, R. E. 1985*a*. Comment on the evolution of interplanetary particles. In *Ices in the Solar System,* eds. J. Klinger, D. Benest, A. Dollfus and R. Smoluchowski (Dordrecht: D. Reidel), pp. 337–339.

Johnson, R. E. 1985*b*. Polar frost in Ganymede. *Icarus* 62:344–347.

Johnson, R. E., and Strobel, D. 1982. Charge exchange in the Io torus and exosphere. *J. Geophys. Res.* 87:10385–10393.

Johnson, R. E., Lanzerotti, L. J., Brown, W. L., and Armstrong, T. P. 1981. Erosion of Galilean satellite surfaces by Jovian magnetospheric particles. *Science* 212:1027–1030.

Johnson, R. E., Lanzerotti, L. J., and Brown, W. L. 1982. Planetary applications of ion-induced erosion of condensed gas frosts. *Nucl. Instrum. Meth.* 198:147–157.

Johnson, R. E., Boring, J. W., Reimann, C. T., Barton, L. A., Sieveka, E. M., Garrett, J. W., Farmer, K. R., Brown, W. L., and Lanzerotti, L. J. 1983. Plasma ion induced molecular ejection of the Galilean satellites: Ejected molecule energies. *Geophys. Res. Lett.* 10:892–895.

Johnson, R. E., Lanzerotti, L., and Brown, W. L. 1984*a*. Sputtering processes: Erosion and chemical change. *Adv. Space Res.* 4:41–51.

Johnson, R. E., Garrett, J., Boring, J. W., Barton, L., and Brown, W. L. 1984*b*. Erosion and modification of SO_2 ice by ion bombardment of the surface of Io. *Proc. Lunar Planet. Sci. Conf.* 14, *J. Geophys. Res. Suppl.* 89:B711–B715.

Johnson, R. E., Barton, L. A., Boring, J. W., Jesser, W. A., Brown, W. L., and Lanzerotti, L. J. 1985. Charged particle modification of ices in the Jovian and Saturnian systems. In *Ices in the Solar System,* eds. J. Klinger, A. Dollfus and R. Smoluchowski (Dordrecht: D. Reidel), pp. 301–316.

Johnson, R. E., Nelson, M. L., McCord, T. B., and Gradie, J. C. 1988*a*. Analysis of Voyager images of Europa: Plasma bombardment. *Icarus,* 75:423–436.

Johnson, R. E., Pospieszalska, M., Sieveka, E. M., Cheng, A. F., Lanzerotti, L. J., and Sittler, E. C. 1988*b*. The neutral cloud and heavy ion inner torus at Saturn. *Icarus,* submitted.

Johnson, T. V., and Soderblom, L. A. 1982. Volcanic eruptions on Io: Implications for surface evolution and mass loss. In *Satellites of Jupiter,* ed. D. Morrison (Tucson: Univ. of Arizona Press), pp. 634–646.

Johnson, T. V., Cook, A. F., II, Sagan, C., and Soderblom, L. A. 1979. Volcanic resurfacing rates and implications for volatiles on Io. *Nature* 280:746–750.

Johnson, T. V., Morrison, D., Matson, D. L., Veeder, G. J., Brown, R. H., and Nelson, R. M. 1984*a*. Io volcanic hot spots: Stability and longitudinal distribution. *Science* 226:134–137.

Johnson, T. V., Matson, D. L., and Veeder, F. J. 1984*b*. Volcanic distribution on Io: Constraints from IRTF data and Voyager imaging. *Bull. Amer. Astron. Soc.* 16:655 (abstract).

Johnson, T. V., Brown, R. H., and Pollack, J. B. 1987. Uranus satellites: Densities and composition. *J. Geophys. Res.* 92:14884–14894.

Johnson, T. V., Veeder, G. F., Matson, D. L., Brown, R. H., Nelson, R. M., and Morrison, D. 1988. Io: Leading side volcanism in 1986. *Lunar Planet. Sci.* XIX:559–560 (abstract).

Jones, C. M. 1982. Paleoatmospheric Argon in Cherts and the Degassing of the Earth. Ph.D. Thesis, Sheffield Univ.

Jones, M. C. 1970. *Far Infrared Absorption in Liquified Gases,* NBS Tech. Note 390 (Washington, D.C.).

Jones, T. D., and Lewis, J. S. 1987. Estimated impact shock production of N_2 and organic compounds on early Titan. *Icarus* 72:381–393.

Jones, T. D., Lewis, J. S., and Lebofsky, L. A. 1986. Mid-IR reflectance of low albedo surface analogs. *Bull. Amer. Astron. Soc.* 18:763 (abstract).

Jones, T. D., Lebofsky, L. A., Lewis, J. S., and Marley, M. S. 1988. The composition and origin of the C, P, and D asteroids. *Icarus,* in press.

Joseph, C. L., Snow, T. P., Seab, C. G., and Crutcher, R. M. 1986. Interstellar abundances in dense, moderately reddened lines of sight. I. Observational evidence for density-dependent depletion. *Astrophys. J.* 309:771–782.

Kahn, R. 1985. The evolution of CO_2 on Mars. *Icarus* 62:175–190.

Kaneoka, I., Takaoka, N., and Aoki, K. 1977. Rare gases in a phlogopite nodule and phlogopite-bearing peridotite in South African Kimberlites. *Earth Planet. Sci. Lett.* 36:181–186.

Kaneoka, I., and Takaoka, N. 1978. Excess ^{129}Xe and high $^3He/^4He$ ratios in olivine phenocrysts of Kapuho lava and xenolithic dunites from Hawaii. *Earth Planet. Sci. Lett.* 39:382–386.

Kaneoka, I., and Takaoka, N. 1980. Rare gas isotopes in Hawaiian ultramafic nodules and volcanic rocks: Constraint on genetic relationships. *Science* 208:1366–1368.

Karato, S. 1986. Does partial melting reduce the creep strength of the upper mantle? *Nature* 319:309–310.

Karhu, J., and Epstein, S. 1986. The implication of the oxygen isotope records in coexisting cherts and phosphates. *Geochim. Cosmochim. Acta* 50:1745–1756.

Kasting, J. F. 1982. Stability of ammonia in the primitive terrestrial atmosphere. *J. Geophys. Res.* 87:3091–3098.

Kasting, J. F. 1987. Theoretical constraints on oxygen and carbon dioxide concentrations in the Precambrian atmosphere. *Precambrian Res.* 34:205–228.

Kasting, J. F. 1988. Runaway and moist greenhouse atmosphere and the evolution of Earth and Venus. *Icarus* 74:472–494.

Kasting, J. F., and Ackerman, T. P. 1986. Climatic consequences of very high CO_2 levels in the Earth's early atmosphere. *Science* 234:1383–1385.

Kasting, J. F., and Pollack, J. B. 1983. Loss of water from Venus. I. Hydrodynamic escape of hydrogen. *Icarus* 53:479–508.

Kasting, J. F., and Walker, J. C. G. 1981. Limits on oxygen concentration in the prebiological atmosphere and the rate of abiotic fixation of nitrogen. *J. Geophys. Res.* 86:1147–1158.

Kasting, J. F., Zahnle, K. J., and Walker, J. C. G. 1983. Photochemistry of methane in the Earth's early atmosphere. *Precambrian Res.* 20:121–148.

Kasting, J. F., Pollack, J. B., and Crisp, D. 1984*a*. Effects of high CO_2 levels on surface temperature and atmospheric oxidation state on the early Earth. *J. Atmos. Chem.* 1:403–428.
Kasting, J. F., Pollack, J. B., and Ackerman, T. P. 1984*b*. Response of Earth's surface temperature to increases in solar flux and implications for loss of water from Venus. *Icarus* 57:335–355.
Kaula, W. M. 1963. Tidal dissipation in the Moon. *J. Geophys. Res.* 68:4959–4965.
Kaula, W. M. 1979. Thermal evolution of Earth and Moon growing by planetesimal impacts. *J. Geophys. Res.* 84:999–1008.
Kaula, W. M., and Beachey, A. E. 1986. Mechanical models of close approaches and collisions of large protoplanets. In *Origin of the Moon,* eds. W. K. Hartmann, R. J. Phillips and G. J. Taylor (Houston: Lunar and Planetary Inst.), pp. 567–576.
Kawakami, S.-I., Mizutani, H., Takagi, Y., Kato, M., and Kumazawa, M. 1983. Impact experiments on ice. *J. Geophys. Res.* 88:5806–5814.
Keene, J., Blake, G. A., Phillips, T. G., Huggins, P. J., and Beichman, C. A. 1985. The abundance of atomic carbon near the ionization fronts in M17 and S140. *Astrophys. J.* 299:967–980.
Keil, K. 1968. Mineralogical and chemical relationships among enstatite chondrites. *J. Geophys. Res.* 73:6945–6976.
Keller, H. U., Arpigny, C., Barbieri, C., Bonnet, R. M., Cazes, S., Coradini, M., Cosmovici, C. B., Delamere, W. A., Huebner, W. F., Hughes, D. W., Jamar, C., Malaise, D., Reitsema, H. J., Schmidt, H. U., Schmidt, W. K. H., Seige, P., Whipple, F. L., and Wilhelm, K. 1986. First Halley multicolour camera imaging results from Giotto. *Nature* 321:320–325.
Kelley, S., Turner, G., Butterfield, A. W., and Shepherd T. J. 1986. The source and significance of argon isotopes in fluid inclusions from areas of mineralization. *Earth Planet. Sci. Lett.* 79:303–318.
Kerridge, J. F. 1980*a*. Secular variations in composition of the solar wind: Evidence and causes. In *The Ancient Sun: Fossil Record in the Earth, Moon and Meteorites,* eds. R. O. Pepin, J. A. Eddy and R. B. Merrill (New York: Pergamon Press), pp. 475–489.
Kerridge, J. F. 1980*b*. Isotopic clues to organic synthesis in the early solar system. *Lunar Planet. Sci. Conf.* XI:538–540 (abstract).
Kerridge, J. F. 1982. Whence so much ^{15}N? *Nature* 295:643–644.
Kerridge, J. F. 1983. Isotopic composition of carbonaceous-chondrite kerogen: Evidence for an interstellar origin of organic matter in meteorites. *Earth Planet. Sci. Lett.* 64:186–200.
Kerridge, J. F. 1985. Carbon, hydrogen and nitrogen in carbonaceous chondrites: Abundances and isotopic compositions in bulk samples. *Geochim. Cosmochim. Acta* 49:1707–1714.
Kerridge, J. F., and Bunch, T. E. 1979. Aqueous activity on asteroids: Evidence from carbonaceous meteorites. In *Asteroids,* ed. T. Gehrels (Tucson: Univ. of Arizona Press), pp. 745–764.
Kerridge, J. F., and Chang, S. 1985. Survival of interstellar matter in meteorites: Evidence from carbonaceous material. In *Protostars & Planets II,* eds. D. C. Black and M. S. Matthews (Tucson: Univ. of Arizona Press), pp. 738–754.
Kerridge, J. F., Haymon, R. M., and Kaster, M. 1983. Sulfur isotope systematics at the 21°N site, East Pacific Rise. *Earth Planet. Sci. Lett.* 66:91–100.
Khare, B. N., Sagan, C., Arakawa, E. T., Suits, F., Callcott, T. A., and Williams, M. W. 1984. Optical constants of organic tholins produced in a simulated Titanian atmosphere: From soft x-ray to microwave frequencies. *Icarus* 60:127–137.
Kieffer, S. W. 1982. Dynamics and thermodynamics of volcanic eruption: Implications for the plumes on Io. In *Satellites of Jupiter,* ed. D. Morrison (Tucson: Univ. of Arizona Press), pp. 647–723.
Kieffer, S. W., and Simonds, C. H. 1980. The role of volatiles and lithology in the impact cratering process. *Rev. Geophys. Space Phys.* 18:143–181.
Kiehl, J. T., and Dickinson, R. E. 1987. A study of the radiative effects of enhanced atmospheric CO_2 and CH_4 on early Earth surface temperatures. *J. Geophys. Res.* 92:2991–2998.
Kiess, C. C., Corliss, C. H., and Kiess, H. K. 1960. High dispersion spectra of Jupiter. *Astrophys. J.* 132:221–231.
Kim, S.-J., and Caldwell, J. 1982. The abundance of CH_3D in the atmosphere of Titan, derived from 8- to 14- micron thermal emission. *Icarus* 52:475–482.
Kipp, M. W., and Melosh, H. 1986. Origin of the Moon: A preliminary numerical study of colliding planets. *Lunar Planet. Sci.* XVII:420–421 (abstract).

Kir'choff, V. 1983. Atmospheric sodium chemistry and diurnal variations: An update. *Geophys. Res. Lett.* 10:721–724.

Kirk, R. L., and Stevenson, D. J. 1987. Thermal evolution of a differentiated Ganymede and implications for surface features. *Icarus* 69:91–134.

Kirsch, E., Krimigis, S. M., Ip, W.-H., and Gloeckler, G. 1981. X-ray and energetic neutral particle emission from Saturn's magnetosphere. *Nature* 292:718–721.

Kirzhnits, D. A. 1967. *Field Theoretical Methods in Many-Body Systems* (Oxford: Pergamon Press).

Kissel, J. 1986. The Giotto particulate analyser. In *The Giotto Mission—Its Scientific Investigations,* eds. R. Reinhard and B. Battrick, ESA SP-1077, pp. 67–68.

Kissel, J., and Krueger, F. R. 1987. The organic component in dust from comet Halley as measured by the PUMA mass spectrometer on board Vega 1. *Nature* 326:755–760.

Kissel, J., Sagdeev, R. Z., Bertaux, J. L., Angarov, V. N., Audouze, J., Blamont, J. E., Büchler, K., Evlanov, E. N., Fechtig, H., Formenkova, M. N., von Hoerner, H., Inogamov, N. A., Khromov, V. N., Knabe, W., Krueger, F. R., Langevin, Y., Leonas, V. B., Levasseur-Regourd, A. C., Manadze, G. G., Podkolzin, S. N., Shapiro, V. D., Tabaldyev, S. R., and Zubkov, B. V. 1986*a*. Composition of comet Halley dust particles from Vega observations. *Nature* 321:280–282.

Kissel, J., Büchler, K., Clark, B. C., Fechtig, H., Grün, E., Hornung, K., Ingenbergs, E. B., Jessberger, E. K., Krueger, F. R., Kuczera, H., McDonnell, J. A. M., Morfill, G. M., Rahe, J., Schwehm, G. H., Sekanina, Z., Utterback, N. G., Völk, H. J., and Zook, H. A. 1986*b*. Composition of comet Halley dust particles from Giotto observations. *Nature* 321:336–337.

Kley, D. 1984. Lyman-α absorption cross-section of H_2O and O_2. *J. Atmos. Chem.* 2:203–210.

Kliore, A. J. 1980. The Pioneer 10 Io occultation in the light of Voyager. Paper presented at The Satellites of Jupiter, IAU Colloquium No. 57, Kailua-Kona, Hawaii, May.

Kliore, A. J., and Woiceshyn, P. M. 1976. Structure of the atmosphere of Jupiter from Pioneer 10 and 11 radio occultations. In *Jupiter,* ed. T. Gehrels (Tucson: Univ. of Arizona Press), pp. 216–237.

Kliore, A. J., Cain, D. L., Fjelbo, G., Seidel, B. L., Sykes, M., and Rasool, S. I. 1974. Preliminary results on the atmospheres of Io and Jupiter from Pioneer 10 S-band occultation experiment. *Science* 183:323–324.

Kliore, A. J., Fjeldbo, G., Siedel, B. L., Sweetnam, D. N., Sesplaukis, T. T., and Woiceshyn, P. M. 1975. Atmosphere of Io from Pioneer 10 radio occultation measurements. *Icarus* 24:407–419.

Kliore, A. J., Patel, I. R., Lindal, G. F., Sweetnam, D. N., Hotz, H. B., Waite, J. H., and McDonough, T. R. 1980. Structure of the ionosphere and atmosphere of Saturn from Pioneer 11 Saturn radio occultation. *J. Geophys. Res.* 85:5857–5870.

Knacke, R. F. 1977. Carbonaceous compounds in interstellar dust. *Nature* 269:132–134.

Knacke, R. F. 1987. Sampling the stuff of a comet. *Sky and Telescope* 73:246–250.

Knacke, R. F. 1978. Mineralogical similarities between interstellar dust and primitive solar system material. In *Protostars and Planets,* ed. T. Gehrels (Tucson: Univ. of Arizona Press), pp. 112–131.

Knacke, R. F., and McCorkle, S. 1987. Spectroscopy of the Kleinmann-Low nebula: Scattering in an absorption band. *Astron. J.* 94:972–976.

Knacke, R. F., McCorkle, S., Puetter, R. C., Erickson, E. F., and Kratschmer, W. F. 1982. Observations of interstellar ammonia ice. *Astrophys. J.* 260:141–146.

Knacke, R. F., Geballe, T. R., Noll, K. S., and Tokunaga, A. T. 1985*a*. Search for interstellar methane. *Astrophys. J.* 298:L67–L69.

Knacke, R. F., Puetter, R. C., Erickson, E., and McCorkle S. 1985*b*. Interstellar dust spectra between 2.5 and 3.5 microns: A search for hydrated silicates. *Astron. J.* 90:1828–1831.

Knacke, R. F., Brooke, T. Y., and Joyce, R. R. 1987. The 3.2-3.6 micron emission features in comet Halley: Spectral identifications and similarities. *Astron. Astrophys.* 187:625–628.

Knauth, L. P., and Epstein, S. 1976. Hydrogen and oxygen isotope ratios in nodular and bedded cherts. *Geochim. Cosmochim. Acta* 40:1095–1108.

Kockarts, G., and Nicolet, M. 1962. Le problème aéronomique de l'hélium et de l'hydrogène neutres. *Ann. Geophys.* 18:269–290.

Kolodny, Y., Kerridge, J. F., and Kaplan, I. R. 1980. Deuterium in carbonaceous chondrites. *Earth Planet. Sci. Lett.* 46:149–158.

Kompaneets, A. S. 1960. A point explosion in an inhomogeneous atmosphere. *Soviet Phys.-Doklady* 5:46–48.

Kondo, K. E., and Ahrens, T. J. 1983. Heterogeneous shock-induced thermal radiation in minerals. *Phys. Chem. Minerals* 9:173–181.

Kouvaris, L. C., and Flasar, F. M. 1985. Determination of the saturated methane vapor profile for a nitrogen-methane Titan atmosphere. *Bull. Amer. Astron. Soc.* 17:740 (abstract).

Kowal, C. 1971. Discussion. In *Physical Studies of Minor Planets,* ed. T. Gehrels, NASA SP-267, pp. 185–186.

Kowal, C. T. 1979. Chiron. In *Asteroids,* ed. T. Gehrels (Tucson: Univ. of Arizona Press), pp. 436–439.

Kozak, R. C., and Schaber, G. G. 1986. Gravity-spreading origin of the Venusian tesserae. *Lunar Planet. Sci.* XVII:444–445 (abstract).

Kramers, J. D. 1979. Lead, uranium, strontium, potassium, and rubidium in inclusion-bearing diamonds and mantle-derived xenoliths from southern Africa. *Earth Planet. Sci. Lett.* 48:58–70.

Krankowsky, D., and Eberhardt, P. 1988. Evidence for the composition of ices in the nucleus of comet Halley. In *Comet Halley 1986: World-Wide Investigations, Results and Interpretations* (Chichester: Ellis Horwood), in press.

Krankowsky, D., Lämmerzahl, P., Herrwerth, I., Woweries, J., Eberhardt, P., Dolder, U., Herrmann, U., Schulte, W., Berthelier, J. J., Illiano, J. M., Hodges, R. R., and Hoffman, J. H. 1986. In situ gas and ion measurements at comet Halley. *Nature* 321:326–329.

Krasnopolsky, V. A. 1985. Total injection of water vapor into the Venus atmosphere. *Icarus* 62:221–229.

Kratschmer, W., and Huffman, D. R. 1979. Infrared extinction of heavy ion irradiated and amorphous olivine, with applications to interstellar dust. *Astrophys. Space Sci.* 61:195–203.

Krebs, H. J., Bonzel, H. P., and Gafner, G. 1979. A model study of the hydrogenation of CO over polycrystalline iron. *Surface Sci.* 88:269–283.

Kresák, L. 1981. Evolutionary aspects of the splits of cometary nuclei. *Bull. Astron. Inst. Czech.* 24:264–283.

Krimigis, S. M., and Roelof, E., 1983. Low-energy particle population. *Physics of the Jovian Magnetosphere,* ed. A. J. Dessler (New York: Cambridge Univ. Press), pp. 106–156.

Krimigis, S. M., Armstrong, T. P., Axford, W. E., Bostrom, L. O., Fan, C. Y., Gloeckler, G., Lanzerotti, L. J., Keath, E. P., Zwickl, R. D., Carbary, J. F., and Hamilton, D. C. 1979. Low-energy charged particle environment at Jupiter: A first look. *Science* 204:998–1003.

Krimigis, S. M., Gloeckler, G., McEntire, R., Potemra, T., Scarf, F., and Shelley, E. 1985. Magnetic storm of September 4, 1984: A synthesis of ring current spectra and energy densities measured with AMPTE/OCE. *Geophys. Res. Lett.* 12:329–332.

Krimigis, S. M., Armstrong, T. P., Axford, W. I., Cheng, A. F., Gloeckler, G., Hamilton, D. C., Keath, E. P., Lanzerotti, L. J., and Mauk, B. H. 1986. The magnetosphere of Uranus: Hot plasma and radiation environment. *Science* 233:97–102.

Krimigis, S. M., Keath, E., Mauk, B., Cheng, A. F., Lanzerotti, L. J., Lepping, R. P., and Ness, N. F. 1988. Observations of energetic ion enhancements and fast neutrals upstream and downstream of Uranus' bow shock by the Voyager 2 spacecraft. *Planet. Space Sci.* 36:311–328.

Krüger, F. R. 1984. Ion emission from solid surfaces: Comparison of dust impact with other excitations. In *The Giotto Spacecraft Impact-Induced Plasma Environment, Proc. Giotto PEWG Meeting,* eds. E. Rolfe and B. Battrick, ESA-SP 224, pp. 49–54.

Krüger, F. R., and Kissel, J. 1987. The chemical composition of the dust of comet P/Halley as measured by "PUMA" onboard Vega-1. *Naturwissenschaften* 74:312–316.

Kuhi, L. V. 1964. Mass loss from T-Tauri stars. *Astrophys. J.* 140:1409–1433.

Kuhn, W. R., and Atreya, S. K. 1979. Ammonia photolysis and the greenhouse effect in the primordial atmosphere of the Earth. *Icarus* 37:207–213.

Kuhn, W. R., and Kasting, J. F. 1983. The effects of increased CO_2 concentrations on surface temperature of the early Earth. *Nature* 301:53–55.

Kuiper, G. P. 1944. Titan: A satellite with an atmosphere. *Astrophys. J.* 100:378–383.

Kuiper, G. P. 1951. On the origin of the solar system. In *Astrophysics,* ed. J. A. Hynek (New York: McGraw-Hill), pp. 357–424.

Kuiper, G. P. 1952*a*. *The Atmospheres of the Earth and Planets,* ed. G. P. Kuiper (Chicago: Univ. of Chicago Press).

Kuiper, G. P. 1952*b*. Planetary atmospheres and their origin. In *The Atmospheres of the Earth and Planets,* ed. G. P. Kuiper (Chicago: Univ. of Chicago Press), pp. 306–405.

Kumar, S. 1984. Sulfur and oxygen escape from Io and a lower limit to atmospheric SO_2 at Voyager 1 encounter *J. Geophys. Res.* 89:7399–7406.

Kumar, S. 1985. The SO_2 atmosphere and ionosphere of Io: Ion chemistry, atmospheric escape, and models corresponding to the Pioneer 10 radio occultation measurements. *Icarus* 61:101–123.

Kumar, S., and Hunten, D. M. 1982. The atmospheres of Io and other satellites. In *Satellites of Jupiter,* ed. D. Morrison (Tucson: Univ. of Arizona Press), pp. 782–806.

Kumar, S., Hunten, D. M., and Taylor, H. A. 1981. H_2 abundance in the atmosphere of Venus. *Geophys. Res. Lett.* 8:237–240.

Kumar, S., Hunten, D. M., and Pollack, J. B. 1983. Nonthermal escape of hydrogen and deuterium from Venus and implications for loss of water. *Icarus* 55:369–389.

Kunde, V. G., Hanel, R. A., Maguire, W. C., Gautier, D., Baluteau, J.-P., Marten, A., Chedin, A., Husson, N., and Scott, N. 1982. The tropospheric gas composition of Jupiter's north equatorial belt (NH_3, PH_3, CH_3D, GeH_4, H_2O) and the Jovian D/H isotopic ratio. *Astrophys. J.* 263:443–467.

Kung, C.-C., and Clayton, R. N. 1978. Nitrogen abundances and isotopic compositions in stony meteorites. *Earth Planet. Sci. Lett.* 38:421–435.

Kupo, I., Mekler, Yu., and Eviatar, A. 1976. Detection of ionized sulfur in the Jovian magnetosphere. *Astrophys. J.* 205:L51–L53.

Kurz, M. 1986. Cosmogenic helium in a terrestrial igneous rock. *Nature* 320:435–439.

Kurz, M., Jenkins, W. J., Schilling, J. G., and Hart, S. R. 1982. Helium isotopic variations in the mantle beneath the central North Atlantic Ocean. *Earth Planet. Sci. Lett.* 58:1–14.

Kusaka, T., Nakano, T., and Hayashi, C. 1970. Growth of solid particles in the primordial solar nebula. *Prog. Theor. Phys.* 44:1580–1596.

Kushiro, I., Syono, Y., and Akimoto, S. 1968. Melting of a peridotite nodule at high pressures and high water pressures. *J. Geophys. Res.* 73:6023–6029.

Kyser, T. K., and Rison, W. 1982. Systematics of rare gas isotopes in basic lavas and ultramafic xenoliths. *J. Geophys. Res.* 87:5611–5630.

Kyser, T. K., and O'Neil, J. R. 1984. Hydrogen isotope systematics of submarine basalts. *Geochim. Cosmochim. Acta* 48:2123–2133.

Kyte, F. T., and Wasson, J. T. 1986. Accretion rate of extraterrestrial matter: Iridium deposited 33 to 67 million years ago. *Science* 232:1225–1229.

Lacy, J. H., Baas, F., Allamandola, L. J., Persson, S. E., McGregor, P. J., Lonsdale, C. J., Geballe, T. R., and van de Bult, C. E. P. 1984. 4.6 micron absorption features due to solid phase CO and cyano-group molecules toward compact infrared sources. *Astrophys. J.* 276:533–543.

Lada, C. J. 1985. Cold outflows, energetic winds, and enigmatic jets around young stellar objects. *Ann. Rev. Astron. Astrophys.* 23:267–317.

Lal, M., and Ramanathan, V. 1984. The effects of moist convection and water vapor radiative processes on climate sensitivity. *J. Atmos. Sci.* 41:2238–2249.

Lambert, D. L. 1978. The abundances of the elements in the solar photosphere-VIII. Revised abundances of carbon, nitrogen and oxygen. *Mon. Not. Roy. Astron. Soc.* 182:249–272.

Lancet, M. S., and Anders, E. 1970. Carbon isotope fractionation in the Fischer-Tropsch synthesis and in meteorites. *Science* 170:980–982.

Lanciano, P., Federico, C., and Coradini, A. 1981. Primordial thermal history of growing planetary objects. *Lunar Planet. Sci.* XII:586–588 (abstract).

Landau, L. D., and Lifshitz, E. M. 1958. *Statistical Physics* (London: Pergamon Press).

Lane, A. L., Nelson, R. M., and Matson, D. L. 1981. Evidence for sulfur implantation in Europa's UV absorption data. *Nature* 292:38–39.

Lange, M. A., and Ahrens, T. J. 1980. The evolution of an impact-generated atmosphere. *Lunar Planet. Sci.* XI:596–598 (abstract).

Lange, M. A., and Ahrens, T. J. 1982*a*. The evolution of an impact generated atmosphere. *Icarus* 51:96–120.

Lange, M. A., and Ahrens, T. J. 1982*b*. Impact-induced dehydration of serpentine and the evolution of planetary atmospheres. *Proc. Lunar Planet. Sci. Conf.* 13, *J. Geophys. Res.* 87:A451–A456.

Lange, M. A., and Ahrens, T. J. 1982*c*. Impact fragmentation of ice-silicate bodies. *Lunar Planet. Sci.* XII:417–418 (abstract).
Lange, M. A., and Ahrens, T. J. 1983. Shock-induced CO_2-production from carbonates and a proto-CO_2-atmosphere on the Earth. *Lunar Planet. Sci.* XIV:419–420 (abstract).
Lange, M. A., and Ahrens, T. J. 1984. FeO and H_2O and the homogeneous accretion of the Earth. *Earth Planet. Sci. Lett.* 71:111–119.
Lange, M. A., and Ahrens, T. J. 1986*a*. Impact vaporization and the accretion of icy satellites. *Lunar Planet. Sci.* XVII:458–459 (abstract).
Lange, M. A., and Ahrens, T. J. 1986*b*. Shock-induced CO_2 loss from $CaCO_3$: Implications for early planetary atmosphere. *Earth Planet. Sci. Lett.* 77:409–418.
Lange, M. A., and Ahrens, T. J. 1987*a*. Atmospheric blow-off during accretion of the terrestrial planetary atmospheres. *Icarus,* submitted.
Lange, M. A., and Ahrens, T. J. 1987*b*. Accretion of the icy Jovian and Saturnian satellites. *Icarus,* to be submitted.
Lange, M. A., Lambert, P., and Ahrens, T. J. 1985. Shock effects on hydrous minerals and implications for carbonaceous meteorites. *Geochim. Cosmochim. Acta* 49:1715–1726.
Langer, W. D., Frerking, M. A., Linke, R. A., and Wilson, R. W. 1979. Detection of deuterated formaldehyde in interstellar clouds. *Astrophys. J.* 232:L169–L173.
Langer, W. D., Goldsmith, P. F., Carlson, E. R., and Wilson, R. W. 1980*a*. Evidence for isotopic fractionation of carbon monoxide in dark clouds. *Astrophys. J.* 235:L39–L44.
Langer, W. D., Schloerb, F. P., Snell, R. L., and Young, J. S. 1980*b*. Detection of deuterated cyanoacetylene in the interstellar cloud TMC-1. *Astrophys. J.* 239:L125–L128.
Langer, W. D., Graedel, T. E., Frerking, M. A., and Armentrout, P. B. 1984. Carbon and oxygen isotope fractionation in dense interstellar clouds. *Astrophys. J.* 277:581–604.
Langevin, Y., Kissel, J., Bertaux, J. L., and Chassefière, E. 1987. Impact ionization mass spectrometry of cometary grains on board Giotto, Vega 1 and Vega 2 spacecrafts: Preliminary statistical analysis of spectra in compressed modes. *Lunar Planet. Sci.* XVIII:533 (abstract).
Lanzerotti, L., Brown, W. L., Augustyniak, W., Johnson, R. E., and Armstrong, T. 1982. Laboratory studies of charged particle erosion of SO_2 ice and applications to the frosts of Io. *Astrophys. J.* 259:920–929.
Lanzerotti, L. J., Brown, W. L., Marcantonio, K. J., and Johnson, R. E. 1984. Production of ammonia-depleted surface layers in Saturnian satellites by ion sputtering. *Nature* 312:139–140.
Lanzerotti, L. J., Brown, W. L., and Johnson, R. D. 1985. Laboratory studies of ion irradiations of water, sulfur dioxide and methane ices. In *Ices in the Solar System,* ed. J. Klinger, D. Benest, A. Dollfus and R. Smoluchowski (Dordrecht: D. Reidel), pp. 317–336.
Lanzerotti, L. J., Brown, W. L., and Marcantonio, K. J. 1987. Experimental study of erosion of methane ice by energetic ions and some considerations for astrophysics. *Astrophys. J.* 313:910–919.
Larimer, J. W. 1967. Chemical fractionations in meteorites—I. Condensation of the elements. *Geochim. Cosmochim. Acta* 31:1215–1238.
Larimer, J. W. 1971. Composition of the Earth: Chondritic or achondritic? *Geochim. Cosmochim. Acta* 35:769–786.
Larimer, J. W., and Anders, E. 1967. Chemical fractionations in meteorites. II. Abundance patterns and their interpretation. *Geochim. Cosmochim. Acta* 31:1239–1270.
Larimer, J. W., and Bartholomay, M. 1979. The role of carbon and oxygen in cosmic gases: Some applications to the chemistry and mineralogy of enstatite chondrites. *Geochim. Cosmochim. Acta* 43:1455–1466.
Lark, N. L., Hammel, H. B., Rigler, M. A., Cruikshank, D. P., and Tholen, D. J. 1987. Triton: Orbital brightness variations, June 1987. *Bull. Amer. Astron. Soc.* 19:858 (abstract).
Larson, H. P. 1980. Infrared spectroscopic observations of the outer planets, their satellites, and the asteroids. *Ann. Rev. Astron. Astrophys.* 18:43–75.
Larson, H. P., Fink, U., Treffers, R., and Gautier, T. N. 1975. Detection of water vapor on Jupiter. *Astrophys. J.* 197:L137–L140.
Larson, H. P., Feierberg, M. A., Fink, U., and Smith, H. A. 1979. Remote spectroscopic identification of carbonaceous chondrite mineralogies: Applications to Ceres and Pallas. *Icarus* 39:257–271.
Larson, H. P., Feierberg, M. A., and Lebofsky, L. A. 1983. The composition of asteroid 2 Pallas and its relation to primitive meteorites. *Icarus* 56:398–408.

Larson, H. P., Davis, D. S., Hofmann, R., and Bjoraker, G. L. 1984. The Jovian atmospheric window at 2.7 μm: A search for H_2S. *Icarus* 60:621–639.

Larson, H. P., Davis, D. S., Black, J. H., and Fink, U. 1985. Interstellar absorption features toward the compact infrared source W33A. *Astrophys. J.* 299:873–880.

Larson, R. B. 1969. Numerical calculations of the dynamics of a collapsing proto-star. *Mon. Not. Roy. Astron. Soc.* 145:271–295.

Larson, R. B. 1972. The collapse of a rotating cloud. *Mon. Not. Roy. Astron. Soc.* 156:437–458.

Larson, R. B. 1980. The FU Orionis mechanism. *Mon. Not. Roy. Astron. Soc.* 190:321–335.

Larson, R. B. 1984. Gravitational torques and star formation. *Mon. Not. Roy. Astron. Soc.* 206:197–207.

Lasaga, A. C., Berner, R. A., and Garrels, R. M. 1985. An improved geochemical model of atmospheric CO_2 fluctuations over the past 100 million years. In *The Carbon Cycle and Atmospheric CO_2: Natural Variations Archean to Present,* Geophysical Monograph 32, pp. 397–411.

Latimer, W. M. 1950. Astrochemical problems in the formation of the Earth. *Science* 112:101–104.

Laul, J. C. 1986. The Shergotty consortium and SNC meteorites: An overview. *Geochim. Cosmochim. Acta* 50:875–887.

Laul, J. C., Smith, M. R., Wänke, H., Jagoutz, E., Dreibus, G., Palme, H., Spettel, B., Burghele, A., Lipschutz, M. E., and Verkouteren, R. M. 1986. Chemical systematics of the Shergotty meteorite and the composition of its parent body (Mars). *Geochim. Cosmochim. Acta* 50:909–926.

Laumbach, D. D., and Probstein, R. F. 1968. A point explosion in a cold exponential atmosphere. *J. Fluid Mech.* 32:53–75.

Layden, G. K., and Brindley, G. W. 1963. Kinetics of vapor-phase hydration of magnesium oxide. *J. Amer. Ceramic Soc.* 46:518–522.

Lazarus, A., and McNutt, R. 1983. Low energy plasma ion observations in Saturn's magnetosphere. *J. Geophys. Res.* 88:8831–8846.

Lebofsky, L. A. 1978. Asteroid 1 Ceres: Evidence for water of hydration. *Mon. Not. Roy. Astron. Soc.* 182:17P–21P.

Lebofsky, L. A. 1980. Infrared reflectance spectra of asteroids: A search for water of hydration. *Astron. J.* 85:573–585.

Lebofsky, L. A., and Fegley, M. B., Jr. 1976. Laboratory reflection spectra for the determination of chemical composition of icy bodies. *Icarus* 28:379–387.

Lebofsky, L. A., Feierberg, M. A., Tokunaga, A. T., Larson, H. P., and Johnson, J. R. 1981. The 1.7- to 4.2-μm spectrum of asteroid 1 Ceres: Evidence for structural water in clay minerals. *Icarus* 48:453–459.

Lebofsky, L. A., Rieke, G. H., and Lebofsky, M. 1982. The radii and albedos of Triton and Pluto. *Bull. Amer. Astron. Soc.* 14:766 (abstract).

Lebofsky, L. A., Sykes, M. V., Nolt, I. G., Radostitz, J. V., Veeder, G. J., Matson, D. L., Ade, P. A. R., Griffin, M. J., Gear, W. K., and Robson, E. I. 1985. Submillimeter observations of the asteroid 10 Hygiea. *Icarus* 63:192–200.

Lebofsky, L. A., Sykes, M. V., Tedesco, E. F., Veeder, G. J., Matson, D. L., Brown, R. H., Gradie, J. C., Feierberg, M. A., and Rudy, R. J. 1986. A refined "standard" thermal model for asteroids based on observations of 1 Ceres and 2 Pallas. *Icarus* 68:239–251.

Lebreton, Y., and Maeder, A. 1986. The evolution and helium content of the sun. *Astron. Astrophys.* 161:119–124.

Lecar, M. 1987. The surface density of the solar nebula. *Bull. Amer. Astron. Soc.*, in press.

Lecar, M., and Franklin, F. A. 1973. On the original distribution of the asteroids. I. *Icarus* 20:422–436.

Lee, T., Papanastassiou, D. A., and Wasserburg, G. J. 1976. Demonstration of ^{26}Mg excess in Allende and evidence for ^{26}Al. *Geophys. Res. Lett.* 3:41–44.

Léger, A., and d'Hendecourt, L. 1987. Identifications of PAHs in astronomical IR spectra—Implications. In *Polycyclic Aromatic Hydrocarbons and Astrophysics,* eds. A. Léger, L. d'Hendecourt and N. Boccara (Dordrecht: D. Reidel), pp. 223–254.

Léger, A., and Puget, J. L. 1984. Identification of the "unidentified IR emission features" of interstellar dust? *Astron. Astrophys.* 137:L5–L8.

Léger, A., d'Hendecourt, L., and Boccara, N., eds. 1987. *Polycyclic Aromatic Hydrocarbons and Astrophysics* (Dordrecht: D. Reidel).

Leighton, R. B., and Murray, B. C. 1966. Behavior of carbon dioxide and other volatiles on Mars. *Science* 153:136–144.

Leitch, C. A., and Smith, J. V. 1981. Mechanical aggregation of enstatite chondrites from an inhomogeneous debris cloud. *Nature* 290:228–230.

Lellouch, E., Hubbard, W. B., Sicardy, B., Vilas, F., and Bouchet, P. 1986. Occultation determination of Neptune's oblateness and methane stratospheric mixing ratio. *Nature* 324:227–231.

Lellouch, E., Coustenis, A., Gautier, D., Raulin, F., Dubouloz, N., and Frere, C. 1988. Titan's atmospheric temperature profile: A reanalysis of the Voyager 1 radio occultation and IRIS 7.7 micron data. *Icarus,* in press.

Leung, C. M., Herbst, E., and Huebner, W. F. 1984. Synthesis of complex molecules in dense interstellar clouds via gas-phase chemistry: A pseudo time-dependent calculation. *Astrophys. J. Suppl.* 56:231–256.

Levin, B. J. 1972*a*. Origin of the Earth. In *The Upper Mantle,* ed. A. R. Ritsema (New York: Elsevier), pp. 7–30.

Levin, B. J. 1972*b*. Revision of initial size, mass, and angular momentum of the solar nebula and the problems of its origin. In *Origin of the Solar System,* ed. H. Reeves (Paris: CNRS), pp. 341–360.

Levin, B. J. 1978. Relative velocities of planetesimals and the early accumulation of planets. *Moon and Planets* 19:289–296.

Lewis, J. S. 1969. Observability of spectroscopically active compounds in the atmosphere of Jupiter. *Icarus* 10:393–409.

Lewis, J. S. 1970. Venus: Atmospheric and lithospheric composition. *Earth Planet. Sci. Lett.* 10:73–80.

Lewis, J. S. 1971. Satellites of the outer planets: Their physical and chemical nature. *Icarus* 15:174–185.

Lewis, J. S. 1972*a*. Low temperature condensation from the solar nebula. *Icarus* 16:241–252.

Lewis, J. S. 1972*b*. Metal/silicate fractionation in the solar system. *Earth Planet. Sci. Lett.* 15:286–290.

Lewis, J. S. 1973*a*. Chemistry of the outer solar system. *Space Sci. Rev.* 14:401–410.

Lewis, J. S. 1973*b*. Origin and composition of the terrestrial planets and satellites of the outer planets. In *The Origin of the Solar System,* ed. H. Reeves (Paris: CNRS), pp. 202–205.

Lewis, J. S. 1974*a*. The temperature gradient in the solar nebula. *Science* 186:440–443.

Lewis, J. S. 1974*b*. The chemistry of the solar system. *Sci. Amer.* 230(3):50–65.

Lewis, J. S. 1980*a*. Lightning synthesis of organic compounds on Jupiter. *Icarus* 43:85–95.

Lewis, J. S. 1980*b*. Lightning on Jupiter: Rate, energetics, and effects. *Science* 210:1351–1352.

Lewis, J. S., and Fegley, M. B., Jr. 1984. Vertical distribution of disequilibrium species in Jupiter's troposphere. *Space Sci. Rev.* 39:163–192.

Lewis, J. S., and Prinn, R. G. 1980. Kinetic inhibition of CO and N_2 reduction in the solar nebula. *Astrophys. J.* 238:357–364.

Lewis, J. S., and Prinn, R. G. 1984. *Planets and Their Atmospheres: Origin and Evolution* (New York: Academic Press).

Lewis, J. S., Barshay, S. S., and Noyes, B. 1979. Primordial retention of carbon by the terrestrial planets. *Icarus* 37:190–206.

Lewis, J. S., Watkins, G. H., Hartmann, H., and Prinn, R. G. 1982. Chemical consequences of major impact events on Earth. In *Geological Implications of Impacts of Large Asteroids and Comets on the Earth,* eds. L. T. Silver and P. H. Schultz (Boulder: Geological Soc. Amer.), Spec. Paper 190, pp. 215–221.

Lewis, R. S., Srinivasan, B., and Anders, E. 1975. Host phase of a strange Xe component in Allende. *Science* 190:1251–1262.

Lewis, R. S., Ming, T., Wacker, J. F., Anders, E., and Steel, E. 1987. Interstellar diamonds in meteorites. *Nature* 326:160–162.

Lin, D. N. C. 1981. Convective-accretion disk model for the primordial solar nebula. *Astrophys. J.* 246:972–984.

Lin, D. N. C., and Bodenheimer, P. 1982. On the evolution of convective accretion disk models of the primordial solar nebula. *Astrophys. J.* 262:768–779.

Lin, D. N. C., and Papaloizou, J. 1979*a*. Tidal torques on accretion disks in binary systems with extreme mass ratios. *Mon. Not. Roy. Astron. Soc.* 186:799–812.

Lin, D. N. C., and Papaloizou, J. 1979*b*. On the structure of circumbinary accretion disks and the tidal evolution of commensurable satellites. *Mon. Not. Roy. Astron. Soc.* 188:191–201.

Lin, D. N. C., and Papaloizou, J. 1980. On the structure and evolution of the primordial solar nebula. *Mon. Not. Roy. Astron. Soc.* 191:37–48.

Lin, D. N. C., and Papaloizou, J. 1985. On the dynamical origin of the solar system. In *Protostars & Planets II,* eds. D. C. Black and M. S. Matthews (Tucson: Univ. of Arizona Press), pp. 981–1072.

Lin, D. N. C., and Pringle, J. E. 1987. A viscosity prescription for a self-gravitating accretion disk. *Mon. Not. Roy. Astron. Soc.* 225:607–613.

Lindal, G. F., Wood, G. E., Levy, G. S., Anderson, J. D., Sweetnam, D. N., Hotz, H. B., Buckles, B. J., Holmes, D. P., Doms, P. E., Eshleman, V. R., Tyler, G. L., and Croft, T. A. 1981. The atmosphere of Jupiter: An analysis of the Voyager radio occultation measurements. *J. Geophys. Res.* 86:8721–8727.

Lindal, G. F., Wood, G. E., Hotz, H. B., Sweetnam, D. N., Eshleman, V. R., and Tyler, G. L. 1983. The atmosphere of Titan: An analysis of the Voyager 1 radio occultation measurements. *Icarus* 53:348–363.

Lindal, G. F., Sweetnam, D. N., and Eshleman, V. R. 1985. The atmosphere of Saturn: An analysis of the Voyager radio occultation measurements. *Astron. J.* 90:1136–1146.

Lindal, G. F., Lyons, J. R., Sweetnam, D. N., Eshleman, V. R., Hinson, D. P., and Tyler, G. L. 1986. The atmosphere of Uranus: Results of the Voyager 2 radio occultation measurements. *Bull. Amer. Astron. Soc.* 18:756 (abstract).

Lindal, G. F., Lyons, J. R., Sweetnam, D. N., Eshleman, V. R., Hinson, D. P., and Tyler, G. L. 1987. The atmosphere of Uranus: Results of radio occultation measurements with Voyager 2. *J. Geophys. Res.* 92:14987–15001.

Lindzen, R. S., Hou, A. Y., and Farrell, B. F. 1982. The role of convective model choice in calculating the climatic impact of doubling CO_2. *J. Atmos. Sci.* 39:1189–1205.

Linke, R. A., Guélin, M., and Langer, W. D. 1983. Detection of $H^{15}NN^+$ and $HN^{15}N^+$ in interstellar clouds. *Astrophys. J.* 271:L85–L88.

Linker, J., Kivelson, M. G., and Walker, R. J. 1987. Three-dimensional MHD simulations of plasma flow past Io: Mass loading effects. *EOS: Trans. AGU* 68:1428 (abstract).

Lissauer, J. J. 1987. Timescales for planetary accretion and the structure of the protoplanetary disk. *Icarus* 69:249–265.

Liszt, H. S., and Vanden Bout, P. A. 1985. Upper limits on the O_2/CO ratio in two dense interstellar clouds. *Astrophys. J.* 291:178–182.

Liu, L.-G. 1979. On the 650-km discontinuity. *Earth Planet. Sci. Lett.* 42:202–208.

Lockwood, G. W., Lutz, B. L., Thompson, D. T., and Warnock, A., III. 1983. The albedo of Uranus. *Astrophys. J.* 266:402–414.

Loewenstein, R. F., Harper, D. A., and Moseley, H. 1977. The effective temperature of Neptune. *Astrophys. J.* 218:L145–L146.

Low, F. J. 1965. Planetary radiation at infrared and millimeter wavelength. *Lowell Obs. Bull.* 129(9):184–187.

Low, F. J., Beintema, D. A., Gautier, T. N., Gillette, F. C., Beichman, C. A., Neugebauer, G., Young, E., Aumann, H. H., Boggess, N., Emerson, J. P., Habing, H. J., Hauser, M. G., Houck, J. R., Rowan-Robinson, M., Soifer, B. T., Walker, R. G., and Wesselius, P. R. 1984. Infrared cirrus: New components of the extended infrared emission. *Astrophys. J.* 278:L19–L22.

Lubimova, H. 1958. Thermal history of the Earth with consideration of the variable thermal conductivity of its mantle. *Geophys. J. Roy. Astron. Soc.* 1:115–134.

Luhmann, J. 1986. The solar wind interaction with Venus. *Space Sci. Rev.* 44:241–306.

Lunine, J. I. 1985*a*. Volatiles in the Outer Solar System. Ph.D. Thesis, California Inst. of Technology.

Lunine, J. I. 1985*b*. Titan's surface: Implications for Cassini. In *The Atmospheres of Saturn and Titan,* Proc. Intl. Workshop, Alpbach, Austria, ESA SP-241, pp. 83–88.

Lunine, J. I. 1987. Trapping of gases in water ice and the volatile inventory of Halley's comet. *Bull. Amer. Astron. Soc.* 19:887 (abstract).

Lunine, J. I. 1988. Thermal evolution of Titan's atmosphere. *Lunar Planet. Sci.* XIX:705–706 (abstract).

Lunine, J. I., and Hunten, D. M. 1987. Moist convection and the abundance of water in the troposphere of Jupiter. *Icarus* 69:566–570.

Lunine, J. I., and Stevenson, D. J. 1982*a*. Formation of the Galilean satellites in a gaseous nebula. *Icarus* 52:14–39.
Lunine, J. I., and Stevenson, D. J. 1982*b*. Post-accretional evolution of Titan's surface and atmosphere. *Bull. Amer. Astron. Soc.* 14:713 (abstract).
Lunine, J. I., and Stevenson, D. J. 1985*a*. Thermodynamics of clathrate hydrate at low and high pressures with application to the outer solar system. *Astrophys. J. Suppl.* 58:493–531.
Lunine, J. I., and Stevenson, D. J. 1985*b*. Evolution of Titan's coupled ocean-atmosphere system and interaction of ocean with bedrock. In *Ices in the Solar System,* eds. J. Klinger, D. Benest, A. Dollfus and R. Smoluchowski (Dordrecht: D. Reidel), pp. 741–757.
Lunine, J. I., and Stevenson, D. J. 1985*c*. Physics and chemistry of sulfur lakes on Io. *Icarus* 64:345–367.
Lunine, J. I., and Stevenson, D. J. 1985*d*. Physical state of volatiles on the surface of Triton. *Nature* 317:238–240.
Lunine, J. I., and Stevenson, D. J. 1987. Clathrate and ammonia hydrates at high pressure: Application to the origin of methane on Titan. *Icarus* 70:61–77.
Lunine, J. I., Stevenson, D. J., and Yung, Y. L. 1983. Ethane ocean on Titan. *Science* 222:1229–1230.
Lupton, J. E. 1983*a*. Terrestrial rare gases: Isotope tracer studies and clues to primordial components in the mantle. *Ann. Rev. Earth Planet. Sci.* 11:371–414.
Lupton, J. E. 1983*b*. Fluxes of helium-3 and heat from submarine hydrothermal systems: Guaymas Basin versus 21°N EPR. *EOS: Trans. AGU* 64:723 (abstract).
Lüst, R. 1952. Die Entwicklung einer um einen zentralkörper rotierenden Gasmasse. I. Lösungen der hydrodynamischen Gleichungen mit turbulenter Reibung. *Z. Naturforsch.* 7a:87–98.
Lutz, B. L., Owen, T., and Cess, R. D. 1976. Laboratory band strengths of methane and their application to the atmospheres of Jupiter, Saturn, Uranus, Neptune, and Titan. *Astrophys. J.* 203:541–551.
Lutzky, M., and Lehto, D. L. 1968. Shock propagation in spherically symmetric exponential atmospheres. *Phys. Fluid* 11:1466–1472.
Lynden-Bell, D., and Pringle, J. E. 1974. The evolution of viscous disks and the origin of the nebular variables. *Mon. Not. Roy. Astron. Soc.* 168:603–637.
MacDonald, G. J. F. 1959. Calculations on the thermal history of the Earth. *J. Geophys. Res.* 64:1967–2000.
MacDonald, G. J. F. 1963. The deep structure of continents. *Rev. Geophys.* 1:587–665.
MacFarlane, J. J., and Hubbard, W. B. 1982. Internal structure of Uranus. In *Uranus and the Outer Planets,* ed. G. Hunt (New York: Cambridge Univ. Press), pp. 111–124.
MacFarlane, J. J., and Hubbard, W. B. 1983. Statistical mechanics of light elements at high pressure. V. Three-dimensional Thomas-Fermi-Dirac theory. *Astrophys. J.* 272:301–310.
MacKinnon, I. D. R., and Rietmeijer, F. J. M. 1987. Mineralogy of chondritic interplanetary dust particles. *Rev. Geophys.* 25:1527–1553.
MacLeod, J. M., Avery, J. W., and Broten, N. W. 1981. Detection of deuterated cyanodiacetylene (DC_5N) in Taurus Molecular Cloud 1. *Astrophys. J.* 251:L33–L36.
Macy, W. W., Jr., and Sinton, W. M. 1977. Detection of methane and ethane emission on Neptune but not on Uranus. *Astrophys. J.* 218:L79–L81.
Macy, W. W., Jr., and Smith, W. H. 1978. Detection of HD on Saturn and Uranus and the D/H ratio. *Astrophys. J.* 222:L137–L140.
Magni, G., and Mazzitelli, I. 1979. Thermodynamical properties and equation of state for hydrogen and helium in stellar conditions. *Astron. Astrophys.* 72:134–147.
Malin, M. C., and Pieri, D. C. 1986. Europa. In *Satellites,* eds. J. A. Burns and M. S. Matthews (Tucson: Univ. of Arizona Press), pp. 689–718.
Mamyrin, B. A., and Tolstikhin I. N. 1984. *Helium Isotopes in Nature* (Amsterdam: Elsevier Science Publishers).
Manabe, S., and Wetherald, R. T. 1967. Thermal equilibrium of the atmosphere with a given distribution of relative humidity. *J. Atmos. Sci.* 24:241–259.
Mange, P. 1961. Diffusion of the thermosphere. *Ann. Geophys.* 17:277–291.
Manuel, O. K., and Sabu, D. D. 1981. The noble gas record of the terrestrial planet. *Geochem. J.* 15:245–267.
Marcialis, R. L. 1988. A two-spot model for the surface of Pluto. *Astron. J.* 95:941–977.
Marcialis, R. L., Rieke, G. H., and Lebofsky, L. A. 1987. The surface composition of Charon: Tentative identification of water ice. *Science* 237:1349–1351.

Marley, M. S., and Hubbard, W. B. 1986. Calculation of the molecular-metallic hydrogen transition. *Bull. Amer. Astron. Soc.* 18:780 (abstract).

Marley, M. S., and Hubbard, W. B. 1988. Thermodynamics of dense molecular hydrogen-helium mixtures at high pressure. *Icarus* 73:536–544.

Marsden, B. G. 1970. On the relationship between comets and minor planets. *Astron. J.* 75:206–217.

Marsh, S. P. 1980. *LASL Shock Hugoniot Data* (Berkeley: Univ. of California Press).

Marten, A., Courtin, R., Gautier, D., and Lacombe, A. 1980. Ammonia vertical density profiles on Jupiter and Saturn from their radioelectric and infrared emissivities. *Icarus* 41:410–422.

Marti, K., and Craig, H. 1987. Cosmic-ray-produced neon and helium in the summit of Maui. *Nature* 325:335–337.

Marti, K., Lugmair, G. W., and Sheinin, N. B. 1977. Sm-Nd-Pu systematics in the early solar system. *Lunar Sci.* VIII:619–621 (abstract).

Martin, P. G. 1978. *Cosmic Dust* (Cambridge: Oxford Univ. Press).

Martin, P. G. 1987. Carbon in the interstellar medium. In *Polycyclic Armomatic Hydrocarbons and Astrophysics,* eds. A. Leger, L. d'Hendecourt and N. Boccara (Dordrecht: D. Reidel), pp. 215–222.

Marty, B., and Jambon, A. 1987. C/^{3}He in volatile fluxes from the solid Earth: Implications for carbon geodynamics. *Earth Planet. Sci. Lett.* 83:16–26.

Mason, B. 1962. *Meteorites* (New York: John Wiley).

Mason, B. 1971. *Handbook of Elemental Abundances in Meteorites,* ed. B. Mason (New York: Gordon and Breach), pp. 209–213.

Mason, B. 1979. Cosmochemistry Part 1. Meteorites. In *Data of Geochemistry,* ed. M. Fleischer. U.S. Geol. Survey Prof. Paper 440-B-1.

Masursky, H. 1973. An overview of geological results from Mariner 9. *J. Geophys. Res.* 78:4009–4030.

Mathis, J. S. 1986*a*. IRAS cirrus observations and the nature of dust. In *Light on Dark Matter,* ed. F. P. Israel (Dordrecht: D. Reidel), pp. 171–176.

Mathis, J. S. 1986*b*. Observations and theories of interstellar dust. In *Interrelationships among Circumstellar, Interstellar, and Interplanetary Dust,* eds. J. A. Nuth III and R. E. Stencel, NASA CP-2403, pp. 29–36.

Mathis, J. S., Rumpl, W., and Nordsieck, K. H. 1977. The size distribution of interstellar grains. *Astrophys. J.* 217:425–433.

Matson, D. L. 1986. Infrared Astronomical Satellite asteroid and comet survey. Preprint Version No. 1. JPL Internal Document D-3698.

Matson, D. L., and Brown, R. H. 1988. Solid state greenhouses and their implications for icy satellites. *Icarus,* in press.

Matson, D. L., and Nash, D. 1983. Io's atmosphere: Pressure control by regolith cold trapping and surface venting. *J. Geophys. Res.* 88:4771–4783.

Matson, D. L., Johnson, T. V., and Fanale, F. P. 1974. Sodium D-line emission from Io: Sputtering and resonant scattering hypothesis. *Astrophys. J.* 192:L43–L46.

Matsuda, J., Lewis, R. S., Takahashi, H., and Anders, E. 1980. Isotopic anomalies of noble gases in meteorites and their origins—VII. C3V carbonaceous chondrites. *Geochim. Cosmochim. Acta* 44:1861–1874.

Matsui, T., and Abe, Y. 1986*a*. Evolution of an impact-induced atmosphere and magma ocean on the accreting Earth. *Nature* 319:303–305.

Matsui, T., and Abe, Y. 1986*b*. Impact-induced atmospheres and oceans on Earth and Venus. *Nature* 322:526–528.

Matsui, T., and Abe, Y. 1986*c*. Formation of a "magma ocean" on the terrestrial planets due to the blanketing effect of an impact-induced atmosphere. *Earth, Moon, Planets* 34:223–230.

Matsui, T., and Abe, Y. 1987. Evolutionary tracks of the terrestrial planets. *Earth, Moon, Planets* 39:207–214.

Matthews, H. E., and Irvine, W. M. 1985. The hydrocarbon ring C_3H_2 is ubiquitous in the Galaxy. *Astrophys. J.* 217:L61–L65.

Matthews, H. E., Feldman, P. A., and Bernath, P. F. 1987. Upper limits to interstellar PO. *Astrophys. J.* 312:358–362.

Mayer, E., and Pletzer, R. 1986. Astrophysical implications of amorphous ice: A microporous solid. *Nature* 319:298–301.

Mayor, M., and Mazeh, T. 1987. The frequency of triple and multiple stellar systems. *Astron. Astrophys.* 171:157–177.
Mazor, E., Heymann, D., and Anders, E. 1970. Noble gases in carbonaceous chondrites. *Geochim. Cosmochim. Acta* 34:781–820.
McAlister, H. A., Hartkopf, W. I., Hutter, D. J., Shara, M. M., and Franz, O. G. 1987. ICCD speckle observations of binary stars. I. A survey for duplicity among the bright stars. *Astron. J.* 92:183–194.
McCauley, J. F., Carr, M. H., Cutts, J. A., Hartmann, W. K., Masursky, H., Milton, D. J., Sharp, R. R., and Wilhelms, D. E. 1972. Preliminary Mariner 9 report on the geology of Mars. *Icarus* 17:289–327.
McCord, T. B., Adams, J. B., and Johnson, T. V. 1970. Asteroid Vesta: Spectral reflectivity and compositional implications. *Science* 168:1445–1447.
McCord, T. B., Charette, M. P., Johnson, T. V., Lebofsky, L. A., Pieters, C., and Adams, J. B. 1972. Lunar spectral types. *J. Geophys. Res.* 77:1349–1359.
McDonald, F. F., Fichtel, D. C., and Fisk, L. A. 1974. Solar particles (observations, relationship to the sun, acceleration, interplanetary medium). In *High Energy Particles and Quanta in Astrophysics,* eds. F. F. McDonald and C. E. Fichtel (Cambridge, Mass.: MIT Press), pp. 212–272.
McDonnell, J. A. M., Kissel, J., Grün, E., Grard, R. J. L., Langevin, Y., Olearcyk, R. E., Perry, C. H., and Zarnecki, J. C. 1986. Giotto's dust impact detection system didsy and particulate impact analyser PIA: Interim assessment of the dust distribution and properties within the coma. In *20th ESLAB Symposium on the Exploration of Halley's Comet,* vol. 2, eds. B. Battrick, E. J. Rolfe and R. Reinhard, ESA SP-250, pp. 25–38.
McDonough, T., and Brice, N. 1973. A Saturnian gas ring and the recycling of Titan's atmosphere. *Icarus* 20:136–145.
McDuff, R. E., and Edmond, J. M. 1982. On the fate of sulfate during hydrothermal circulation at mid-ocean ridges. *Earth Planet. Sci. Lett.* 57:117–132.
McElroy, M. B. 1972. Mars: An evolving atmosphere. *Science* 175:443–445.
McElroy, M. B., and Donahue, T. M. 1972. Stability of the Martian atmosphere. *Science* 177:986–988.
McElroy, M. B., and Prather, M. J. 1981. Noble gases in the terrestrial planets: Clues to evolution. *Nature* 293:535–539.
McElroy, M. B., and Yung, Y. L. 1976. Oxygen isotopes in the Martian atmosphere: Implications for the evolution of volatiles. *Planet. Space Sci.* 24:1107–1113.
McElroy, M. B., Yung, Y. L., and Nier, A. O. 1976. Isotopic composition of nitrogen: Implications for the past history of Mars' atmosphere. *Science* 194:70–72.
McElroy, M. B., Kong, T.-Y., and Yung, Y. L. 1977. Photochemistry and evolution of Mars' atmosphere: A Viking perspective. *J. Geophys. Res.* 82:4379–4388.
McElroy, M. B., Prather, M. J., and Rodriguez, J. 1982*a*. Escape of hydrogen from Venus. *Science* 215:1614–1615.
McElroy, M. B., Prather, M. J., and Rodriguez, J. M. 1982*b*. Loss of oxygen from Venus. *Geophys. Res. Lett.* 9:649–651.
McEwen, A. S., and Soderblom, L. A. 1983. Two classes of volcanic plumes on Io. *Icarus* 55:191–218.
McEwen, A. S., Johnson, T. V., Matson, D. L., and Soderblom, L. A. 1988. The global distribution, abundance, and stability of SO_2 on Io. *Icarus,* 75:450–478.
McFadden, L. A. 1983. Spectral Reflectance of Near-Earth Asteroids: Implications for Composition, Origin and Evolution. Ph.D. Thesis, Univ. of Hawaii.
McFadden, L. A., and Vilas, F. 1987. The 3 : 1 Kirkwood gap as sources of ordinary chondrites: Perspectives from spectral reflectance. *Lunar Planet. Sci.* XVIII:614–615 (abstract).
McFadden, L. A., Gaffey, M. J., and McCord, T. B. 1985. Near-Earth asteroids: Possible sources from reflectance spectroscopy. *Science* 229:193–195.
McFarlane, J. J., and Hubbard, W. B. 1983. Statistical mechanics of light elements at high pressure. V. Three-dimensional Thomas-Fermi-Dirac theory. *Astrophys. J.* 272:301–310.
McGovern, P., and Schubert, G. 1988. Thermal evolution of the Earth: Effects of volatile exchange between atmosphere and interior. In preparation.
McGrath, M., and Johnson, R. E. 1987. Magnetospheric plasma sputtering of Io's atmosphere. *Icarus* 69:519–531.

McGrath, M., and Johnson, R. E. 1988. Cross sections for the Io plasma torus. *J. Geophys. Res.*, submitted.

McGrath, M., Johnson, R. E., and Lanzerotti, L. 1986. Sputtering of sodium on the planet Mercury. *Nature* 323:694–696.

McKay, C. P., and Thomas, G. E. 1978. Consequences of a past encounter of the Earth with an interstellar cloud. *Geophys. Res. Lett.* 5:215–218.

McKay, C. P., Pollack, J. B., and Courtin, R. 1988*a*. The thermal structure of Titan's atmosphere. *Icarus*, in press.

McKay, C. P., Scattergood, T. W., Pollack, J. B., Borucki, W. J., and Van Ghysegahm, H. T. 1988*b*. High temperature shock formation of N_2 and organics on primordial Titan. *Nature* 332:520–522.

McKinnon, W. B. 1984. On the origin of Triton and Pluto. *Nature* 311:355–358.

McKinnon, W., and Parmentier, E. M. 1986. Ganymede and Callisto. In *Satellites,* eds. J. A. Burns and M. S. Matthews (Tucson: Univ. of Arizona Press), pp. 717–763.

McNally, D. 1964. The collapse of interstellar gas clouds. I. The effect of cooling on clouds having initially polytropic density distributions. *Astrophys. J.* 140:1088–1099.

McNaughton, N. J., Borthwick, J., Fallick, A. E., and Pillinger, C. T. 1981. Deuterium/hydrogen ratios in unequilibrated ordinary chondrites. *Nature* 294:639–641.

McNutt, R., Selesnick, R., and Richardson, J. 1987. Low energy plasma observations in the magnetosphere of Uranus. *J. Geophys. Res.* 92:4399–4410.

McSween, H. Y. 1979*a*. Are carbonaceous chondrites primitive or processed? A review. *Rev. Geophys. Space Phys.* 17:1059–1078.

McSween, H. Y. 1979*b*. Alteration in Cl carbonaceous chondrites inferred from modal and chemical variations in matrix. *Geochim. Cosmochim. Acta* 43:1761–1770.

McSween, H. Y. 1987. Matrix compositions in antarctic and non-antarctic CM carbonaceous chondrites. *Lunar Planet. Sci.* XVIII:631–632 (abstract).

McSween, H. Y., and Richardson, S. M. 1977. The composition of carbonaceous chondrite matrix. *Geochim. Cosmochim. Acta* 41:1145–1161.

McSween, H. Y., Jr., and Stolper, E. M. 1980. Basaltic meteorites. *Sci. Amer.* 242(6):44–53.

Melnick, G., Russell, R. W., Gosnell, T. R., and Harwit, M. 1983. Spectrophotometry of Saturn and its rings form 60 to 180 microns. *Icarus* 53:310–318.

Melosh, H. J. 1984. Impact ejection, spallation, and the origin of meteorites. *Icarus* 59:234–260.

Melosh, H. J., and Sonett, C. P. 1986. When worlds collide: Jetted vapor plumes and the Moon's origin. In *Origin of the Moon,* eds. W. K. Hartmann, R. J. Phillips and G. J. Taylor (Houston: Lunar and Planetary Inst.), pp. 621–642.

Mendis, D. A., Hill, J., Ip, W.-H., Goertz, C., and Grün, E. 1984. Electrodynamic processes in the ring system of Saturn. In *Saturn,* eds. T. Gehrels and M. S. Matthews (Tucson: Univ. of Arizona Press), pp. 546–592.

Mendis, D. A., Houpis, H. L. F., and Marconi, M. L. 1985. *Fund. Cosmic Phys.* 10:1–30.

Menzel, D. H., Coblentz, W. W., and Lampland, C. O. 1926. Planetary temperatures derived from water-cell transmissions. *Astrophys. J.* 63:177–187.

Merrill, K. M. 1974. 8-13 micron spectropolarimetry of Comet Kohoutek. *Icarus* 23:566–567.

Merrill, K. M., and Stein, W. A. 1976. 2-14 micron stellar spectrophotometry I. Stars of the conventional spectral sequence. *Publ. Astron. Soc. Pacific* 88:285–293.

Mewaldt, R. A., Spalding, J. D., and Stone, E. C. 1984. A high-resolution study of the isotopes of solar flare nuclei. *Astrophys. J.* 280:892–901.

Meyer, J. P. 1985*a*. The baseline composition of solar energetic particles. *Astrophys. J. Suppl.* 57:151–171.

Meyer, J. P. 1985*b*. Solar-stellar outer atmospheres and energetic particles and galactic cosmic rays. *Astrophys. J. Suppl.* 57:173–204.

Michard, G., Abarede, F., Michard, A., Minster, J.-F., Charlou, J.-L., and Tan, N. 1984. Chemistry of solutions from the 13°N East Pacific Rise hydrothermal site. *Earth Planet. Sci. Lett.* 67:297–307.

Mihalas, D. 1978. *Stellar Atmospheres* (San Francisco: W. H. Freeman), p. 57.

Millar, T. J., and Freeman, A. 1984. Chemical modelling of molecular sources. II. L183. *Mon. Not. Roy. Astron. Soc.* 207:425–432.

Millar, T. J., Leung, C. M., and Herbst, E. 1987. How abundant are complex interstellar molecules? *Astron. Astrophys.* 183:109–117.

Miller, S. L. 1955. Production of some organic compounds under possible primitive Earth conditions. *J. Amer. Chem. Soc.* 77:2351–2361.
Miller, S. L. 1961. The occurrence of gas hydrates in the solar system. *Proc. Nat. Acad. Sci. USA* 47:1798–1808.
Miller, S. L. 1985. Clathrate hydrates in the solar system. In *Ices in the Solar System,* eds. J. Klinger, D. Benest, A. Dollfus and R. Smoluchowski (Dordrecht: D. Reidel), pp. 59–79.
Milton, D. J. 1973. Water and processes of degradation in the martian landscape. *J. Geophys. Res.* 78:4037–4047.
Minh, Y. C., Irvine, W. M., and Ziurys, L. M. 1987. Hocot observations of interstellar clouds. In *Molecular Clouds in the Milky Way and External Galaxies,* eds. R. L. Dickman and J. S. Young (Berlin: Springer-Verlag), in press.
Misener, D. J. 1974. Cationic diffusion in olivine to 1400 C and 35 kbar. In *Geochemical Transport and Kinetics,* eds. A. W. Hofmann, B. J. Giletti, H. S. Yoder, Jr., and R. A. Yund (Washington, D.C.: Carnegie Institution), pp. 117–129.
Mittlefehldt, D. W. 1979. The nature of asteroidal differentiation processes: Implications for primordial heat sources. *Proc. Lunar Planet. Sci. Conf.* 10:1975–1993.
Mizuno, H. 1980. Formation of the giant planets. *Prog. Theor. Phys.* 64:544–557.
Mizuno, H., and Boss, A. P. 1985. Tidal disruption of dissipative planetesimals. *Icarus* 63:109–133.
Mizuno, H., Nakazawa, K., and Hayashi, C. 1978. Instability of a gaseous envelope surrounding a planetary core and formation of giant planets. *Prog. Theor. Phys.* 60:699–710.
Mizuno, H., Hayashi, C., and Nakazawa, K. 1980. Dissolution of primordial rare gases into the molten Earth's material. *Earth Planet. Sci. Lett.* 50:202–210.
Mizuno, H., Markiewicz, W. J., and Völk, H. J. 1988. Grain growth in turbulent protoplanetary accretion disks. *Astron. Astrophys.* 195:183–192.
Moore, J. G., and Fabbi, B. P. 1971. An estimate of the juvenile sulfur content of basalt. *Contrib. Mineral. Petrol.* 33:118–127.
Moos, H. W., Skinner, T. E., Durrance, S. T., Feldman, P., Festou, M. C., and Bertaux, J. L. 1985. Long term stability of the Io high temperature plasma torus. *Astrophys. J.* 294:369–383.
Moran, D. W. 1959. Low Temperature Equilibria in Binary Systems. Ph.D. Thesis, Univ. of London.
Morbey, C. L., and Griffin, R. F. 1987. On the reality of certain spectroscopic orbits. *Astrophys. J.* 317:343–352.
Moreno, M. A., and Barbosa, D. 1986. Mass and energy balance of the cold Io torus. *J. Geophys. Res.* 91:8993–8997.
Moreno, M. A., Kivelson, M. G., and Schubert, G. 1988*a*. Io: An atmosphere formed by volcanic and hotspot sublimation. *27th COSPAR,* Espsoor, Finland, July 1988 (abstract).
Moreno, M. A., Schubert, G., Baumgardner, J., and Kivelson, M. 1988*b*. Volcanic eruptions and the atmosphere of Io. *Icarus,* submitted.
Morfill, G. E. 1983. Some cosmochemical consequences of a turbulent protoplanetary cloud. *Icarus* 53:41–54.
Morfill, G. E. 1985. Physics and chemistry in the primitive solar nebula. In *Birth and Infancy of Stars,* eds. R. Lucas and A. Omont (Amsterdam: North Holland), pp. 693–794.
Morfill, G. E. 1988. Protoplanetary accretion discs with coagulation and evaporation. *Icarus* 75:371–380.
Morfill, G. E., and Clayton, R. N. 1986. Oxygen isotope fractionation in the primitive solar nebula: Theory. *Lunar Planet. Sci.* XVII:569–570 (abstract).
Morfill, G. E., and Stenholm, L. G. 1980. Structure of molecular clouds III. Effects of MHD waves in collapsing fragments. *Astron. Astrophys.* 90:134–139.
Morfill, G. E., and Völk, H. J. 1984. Transport of dust and vapour, and chemical fractionation in the early protosolar cloud. *Astrophys. J.* 287:371–395.
Morfill, G. E., Grün, E., and Johnson, T. V. 1983. Saturn's E, F, and G rings: Modulated by the plasma sheet? *J. Geophys. Res.* 88:5573–5579.
Morfill, G. E., Tscharnuter, W., and Völk, H. J. 1985. Dynamical and chemical evolution of the protoplanetary nebula. In *Protostars & Planets II,* eds. D. C. Black and M. S. Matthews (Tucson: Univ. of Arizona Press), pp. 493–533.

Morfill, G. E., Goertz, C. K., and Havnes, O. 1987. Thermal cycling and fluctuations in the protoplanetary nebula. *Icarus* 76:391–403.

Morgan, J. S. 1985. Temporal and spatial variation in the Io torus. *Icarus* 62:389–414.

Morgan, J. W., and Anders, E. 1979. Chemical composition of Mars. *Geochim. Cosmochim. Acta* 43:1601–1610.

Morgan, J. W., and Anders, E. 1980. Chemical composition of Earth, Venus and Mercury. *Proc. Natl. Acad. Sci. USA* 77:6973–6977.

Morgan, J. W., Higuchi, H., Takahashi, H., and Hertogen, J. 1978. A "chondritic" eucrite parent body: Interference from trace elements. *Geochim. Cosmochim. Acta* 42:27–38.

Moroz, V. I., Golovin, Yu. M., Ekonomov, A. P., Moshkin, B. E., Parfent'ev, N. A., and San'ko, N. F. 1980. Spectrum of the Venus day sky. *Nature* 284:243–244.

Morrison, D. M., ed. 1983. *Planetary Exploration through Year 2000* (Washington, D.C.: Superintendent of Documents).

Morrison, D. M., and Lebofsky, L. A. 1979. Radiometry of asteroids. In *Asteroids,* ed. T. Gehrels (Tucson: Univ. of Arizona Press), pp. 184–205.

Morrison, D. M., Cruikshank, D. P., and Brown, R. H. 1982. Diameters of Triton and Pluto. *Nature* 300:425–427.

Morrison, D. M., Owen, T., and Soderblom, L. A. 1986. The satellites of Saturn. In *Satellites,* eds. J. A. Burns and M. S. Matthews (Tucson: Univ. of Arizona Press), pp. 764–801.

Morton, J. L., and Sleep, N. H. 1985*a*. Seismic reflections from a Lau Basin magma chamber. In *Geology and Offshore Resources of Pacific Island Arcs—Tonga Regions,* eds. D. W. Scholl and T. L. Vallier (Houston: Circum-Pacific Council for Energy and Mineral Resources), pp. 441–453.

Morton, J. L., and Sleep, N. H. 1985*b*. A mid-ocean ridge thermal model: Constraints on the volume of axial hydrothermal heat flux. *J. Geophys. Res.* 90:11345–11353.

Morton, J. L., Sleep, N. H., Normark, W. R., and Tompkins, D. H. 1987. Structure of the southern Juan de Fuca ridge from seismic reflection records. *J. Geophys. Res.* 92:11315–11326.

Moseley, S. H., Conrath, B. J., and Silverberg, R. F. 1985. Atmospheric temperature profiles of Uranus and Neptune. *Astrophys. J.* 292:L83–L86.

Mouginis-Mark, P. J., Wilson, L., and Head, J. W. 1982. Explosive volcanism on Hecates Tholus, Mars: Investigation of eruption conditions. *J. Geophys. Res.* 87:9890–9904.

Mouginis-Mark, P. J., Wilson, L., and Zimbelman, J. W. 1988. Polygenic eruptions of Alba Patera, Mars: Evidence of channel erosion on pyroclastic flows. *Bull. Volcanology,* 50(6).

Mouschovias, T.Ch. 1977. A connection between the rate of rotation of interstellar clouds, magnetic fields, ambipolar diffusion, and the periods of binary stars. *Astrophys. J.* 211:147–151.

Mouschovias, T.Ch., and Paleologou, E. V. 1986. The effect of ambipolar diffusion on magnetic braking of molecular cloud cores: An exact, time-dependent solution. *Astrophys. J.* 308:781–790.

Muehlenbachs, K., and Clayton, R. 1976. Oxygen isotope composition of the oceanic crust and its bearing on seawater. *J. Geophys. Res.* 81:4365–4369.

Muhleman, D., Berge, G., and Rudy, D. 1984. Microwave emission from Titan and the Galilean satellites. *Bull. Amer. Astron. Soc.* 16:686 (abstract).

Mukhin, L. M., Evlanov, E. N., Fomenkova, M. N., Khromov, V. N., Kissel, J., Prilutsky, O. F., Zubkov, B. V., and Sagdeev, R. Z. 1987. Different types of dust particles in Halley's Comet. In *Lunar Planet. Sci.* XVIII:674–675 (abstract).

Mumma, M. J., Weaver, H. A., and Larson, H. P. 1986. The ortho/para ratio of water vapor in comet Halley. In *20th ESLAB Symposium on the Exploration of Halley's Comet,* vol. 1, eds. B. Battrick, E. J. Rolfe and R. Reinhard, ESA SP-250 pp. 341–346.

Murphy, R. E., and Fesen, R. A. 1974. Spatial variations in the Jovian 20 micrometer flux. *Icarus* 21:42–46.

Murray, B. C., and Wildey, R. L. 1963. Stellar and planetary observations at 10 microns. *Astrophys. J.* 137:692–693.

Myers, P. C., and Benson, P. J. 1983. Dense cores in dark clouds. II. NH_3 observations and star formation. *Astrophys. J.* 266:309–320.

Myers, P. C., and Goodman, A. A. 1988. Magnetic molecular clouds: Indirect evidence for magnetic support and ambipolar diffusion. *Astrophys. J.* 329:392–405.

Mysen, B. O., and Boettcher, A. L. 1975. Melting of a hydrous mantle: I. Phase relations of natural peridotite at high pressures and temperatures with controlled activities of water, carbon dioxide, and hydrogen. *J. Petrol.* 16:520–548.

Nagahara, H., and El Goresy, A. 1984. Yamato-74370: A new enstatite chondrite (EH4). *Lunar Planet. Sci.* XV:583–584 (abstract).

Nakagawa, Y. 1978. Statistical behavior of planetesimals in the primitive solar system. *Prog. Theor. Phys.* 59:1834–1851.

Nakagawa, Y., Nakazawa, K. and Hayashi, C. 1981. Growth and sedimentation of dust grains in the primordial solar nebula. *Icarus* 45:517–528.

Nakagawa, Y., Hayashi, C., and Nakazawa, K. 1983. Accumulation of planetesimals in the solar nebula. *Icarus* 54:361–376.

Nakagawa, Y., Sekiya, M., and Hayashi, C. 1986. Settling and growth of dust particles in a laminar phase of a low-mass solar nebula. *Icarus* 67:375–390.

Nakano, T. 1987. Formation of planets around stars of various masses. I. Formulation and a star of one solar-mass. *Mon. Not. Roy. Astron. Soc.* 224:107–130.

Nautiyal, C. M., Padia, J. T., Rao, M. N., and Venkatesan, T. R. 1981. Solar flare neon: Clues from implanted noble gases in lunar soils and rocks. *Proc. Lunar Planet. Sci. Conf.* 12:627–637.

Nautiyal, C. M., Rao, M. N., and Venkatesan, T. R. 1983. Ancient solar flare neon and solar cosmic ray proton fluxes from gas-rich meteorites. *Lunar Planet. Sci.* XIV:544–545 (abstract).

Nash, D. B. 1987. Sulfur in a vacuum: Sublimation effects on frozen melts and applications to Io's surface and torus. *Icarus* 72:1–34.

Nash, D. B., Carr, M., Gradie, J., Hunten, D., and Yoder, C. 1986. Io. In *Satellites,* eds. J. A. Burns and M. S. Matthews (Tucson: Univ. of Arizona Press), pp. 629–688.

Neff, J. S., Humm, D. C., Bergstralh, J. T., Cochran, A. L., Cochran, W. D., Barker, E. S., and Tull, R. G. 1984. Absolute spectrophotometry of Titan, Uranus, and Neptune: 3500–10,500λ. *Icarus* 60:221–235.

Neff, J. S., Ellis, T. A., Apt, J., and Bergstralh, J. T. 1985. Bolometric albedos of Titan, Uranus, and Neptune. *Icarus* 62:425–432.

Nellis, W. J., Ross, M., Mitchell, A. C., van Thiel, M., Young, D. A., Ree, F. H., and Trainor, R. J. 1983. Equation of state of molecular hydrogen and deuterium from shock-wave experiments to 760 kbar. *Phys. Rev.* A27:608–611.

Nellis, W. J., Holmes, N. C., Mitchell, A. C., Trainor, R. J., Governo, G. H., Ross, M., and Young, D. A. 1984. Shock compression of liquid helium to 56 Gpa (560 kbar). *Phys. Rev. Lett.* 53:1248–1251.

Ness, N., Acuna, M., Behannon, K., Burlaga, L., Connerney, J., Lepping, R., and Neubauer, F. 1986. Magnetic fields at Uranus. *Science* 233:85–89.

Neubauer, F. M., Gurnett, D., Scudder, J., and Hartle, R. 1984. Titan's magnetospheric interaction. In *Saturn,* eds. T. Gehrels and M. S. Matthews (Tucson: Univ. of Arizona Press), pp. 760–787.

Neukum, G., König, B., Fechtig, H., and Storzer, D. 1975. Cratering in the Earth-moon system: Consequences for age determination by crater counting. *Proc. Lunar Planet. Sci. Conf.* 6:2597–2620.

Newman, M. J., and Rood, R. T. 1977. Implications of solar evolution for the Earth's early atmosphere. *Science* 198:1035–1037.

Newton, R. C. 1987. Late Archean/early Proterozoic CO_2 streaming through the lower crust and geochemical segregation. *Geophys. Res. Lett.* 14:287–290.

Ney, E. P. 1977. Star dust. *Science* 195:541–546.

Nicholson, P. D., and Porco, C. C. 1988. A new constraint on Saturn's zonal gravity harmonics from Voyager observations of an eccentric ringlet. *J. Geophys. Res.*, in press.

Nicholson, P. D., Matthews, K., and Goldreich, P. 1981. The Uranus occultation of 10 June 1979. I. The rings. *Astron. J.* 86:596–606.

Niederer, F. R., and Papanastassiou, D. A. 1984. Ca isotopes in refractory inclusion. *Geochim. Cosmochim. Acta* 48:1279–1793.

Niemeyer, S., and Marti, K. 1981. Noble gas trapping by laboratory carbon condensates. *Proc. Lunar Planet. Sci. Conf.* 12:1177–1188.

Nier, A. O. 1950. A redetermination of the relative abundances of the isotopes of carbon, nitrogen, oxygen, argon and potassium. *Phys. Rev.* 77:789–793.

Nier, A. O., and McElroy, M. B. 1977. Composition and structure of Mars' upper atmosphere: Results from the neutral mass spectrometers on Viking 1 and 2. *J. Geophys. Res.* 82:4341–4349.

Nier, A. O., McElroy, M. B., and Yung, Y. L. 1976. Isotopic composition of the Martian atmosphere. *Science* 194:68–70.

Nishida, S. 1983. Collisional processes of planetesimals with a protoplanet under the gravity of the protosun. *Prog. Theor. Phys.* 70:93–105.

Noll, K. S., Knacke, R. F., Geballe, T. R., and Tokunaga, A. T. 1986. Detection of carbon monoxide in Saturn. *Astrophys. J.* 309:L91–L94.

Noll, K. S., Knacke, R. F., Geballe, T. R., and Tokunaga, A. T. 1988. The origin and vertical distribution of carbon monoxide on Jupiter. *Astrophys. J.* 324:1210–1218.

Nolt, I. G., Radostitz, J. V., Donnelly, R. J., Murphy, R. E., and Ford, H. C. 1974. Thermal emission of Saturn's rings and disc at 35 μm. *Nature* 248:659–660.

Nordyke, M. D. 1977. Nuclear cratering experiments: United States and Soviet Union. In *Impact and Explosion Cratering: Planetary and Terrestrial Implications,* eds. D. J. Roddy, R. O. Pepin and R. B. Merrill (New York: Pergamon Press), pp. 103–124.

Norman, C. A., and Silk, J. 1980. Clumpy molecular clouds: A dynamic model self-consistently regulated by T Tauri star formation. *Astrophys. J.* 238:158–174.

Norman, M. L., and Wilson, J. R. 1978. The fragmentation of isothermal rings and star formation. *Astrophys. J.* 224:497–511.

Norman, M. L., Wilson, J. R., and Barton, R. T. 1980. A new calculation on rotating protostar collapse. *Astrophys. J.* 239:968–981.

Null, G. W., Lau, E. L., Biller, E. D., and Anderson, J. D. 1981. Saturn gravity results obtained from Pioneer 11 tracking data and Earth-based Saturn satellite data. *Astron. J.* 86:456–468.

Nuth, J. A. 1985. Meteoritic evidence that graphite is rare in the interstellar medium. *Nature* 318:166–168.

Nuth, J. A., Donn, B., and Nelson, R. 1986. Refractory grain processing in circumstellar shells: Diagnostic infrared signatures. *Astrophys. J.* 310:L83–L86.

Nyquist, L. E., Bogard, D. D., Wooden, J. L., Wiesmann, H., Shih, C.-Y., Bansal, B. M., and McKay, G. 1979. Early differentiation, late magmatism, and recent bombardment of the Shergottite parent planet. *Meteoritics* 14:502 (abstract).

O'Keefe, J. D., and Ahrens, T. J. 1977*a*. Impact induced energy partitioning, melting, and vaporization on terrestrial planets. *Proc. Lunar Sci. Conf.* 8:3357–3374.

O'Keefe, J. D., and Ahrens, T. J. 1977*b*. Meteorite impact ejecta: Dependence of mass and energy lost on planetary escape velocity. *Science* 198:1249–1251.

O'Keefe, J. D., and Ahrens, T. J. 1982*a*. The interaction of the Cretaceous-Tertiary extinction bolide with the atmosphere, ocean and solid Earth. In *Proc. Conf. on Large Body Impacts and Terrestrial Evolution,* Geol. Soc. Amer. SP 190, pp. 103–120.

O'Keefe, J. D., and Ahrens, T. J. 1982*b*. Impact mechanics of the Cretaceous-Tertiary extinction bolide. *Nature* 298:123–127.

O'Keefe, J. D., and Ahrens, T. J. 1982*c*. Cometary and meteorite swarm impact on planetary surfaces. *J. Geophys. Res.* 87:6668–6689.

O'Keefe, J. D., and Ahrens, T. J. 1988. Large scale impact on the Earth with an atmosphere. *Lunar Planet. Sci.* XIX:887–888 (abstract).

Olberg, M., Bester, M., Rau, G., Pauls, T., Winnewisser, G., Johansson, L. E. B., and Hjalmarson, Å. 1985. A new search for and discovery of deuterated ammonia in three molecular clouds. *Astron. Astrophys.* 142:L1–L4.

Olofsson, H. 1984. Deuterated water in Orion KL and W51M. *Astron. Astrophys.* 134:36–44.

Omont, A. 1986. Physics and chemistry of interstellar polycyclic aromatic molecules. *Astron. Astrophys.* 164:159–178.

O'Nions, R. K., and Oxburgh, E. R. 1983. Heat and helium in the Earth. *Nature* 306:429–431.

Oort, J. 1950. The structure of the cloud of comets surrounding the solar system and a hypothesis concerning its origin. *Bull. Astron. Inst. Netherlands* 11:91–110.

Öpik, E. J. 1963. Selective escape of gases. *Geophys. J. Roy. Astron. Soc.* 7:490–509.

Öpik, E. J. 1973. Comets and the formation of planets. *Astrophys. Space Sci.* 21:307–398.

Oró, J., and Kimball, A. P. 1961. Synthesis of purines under possible primitive Earth conditions. I. Adenine from hydrogen cyanide. *Arch. Biochem. Biophys.* 94:217–227.

Orton, G. S. 1975. The thermal structure of Jupiter I. Implications of Pioneer 10 infrared radiometer data. *Icarus* 26:125–141.

Orton, G. S., and Ingersoll, A. P. 1977. Pioneer 10 and 11 and ground based infrared data on Jupiter: The thermal structure and He-H_2 ratio. In *Jupiter,* ed. T. Gehrels (Tucson: Univ. of Arizona Press), pp. 207–215.

Orton, G. S., and Ingersoll, A. P. 1980. Saturn's atmospheric temperature structure and heat budget. *J. Geophys. Res.* 85:5871–5881.

Orton, G. S., Griffin, M. J., Ade, P. A. R., Nolt, I. G., Radostitz, J. V., Robson, E. I., and Gear, W. K. 1986. Submillimeter and millimeter observations of Uranus and Neptune. *Icarus* 67:289–304.

Orton, G. S., Aitken, D. K., Smith, C., Roche, P. F., Caldwell, J., and Snyder, R. 1987*a*. The spectra of Uranus and Neptune at 8-14 and 17-23 μm. *Icarus* 70:1–12.

Orton, G. S., Baines, K. H., Bergstralh, J. T., Brown, R. H., Caldwell, J., and Tokunaga, A. T. 1987*b*. Infrared radiometry of Uranus and Neptune at 21 and 32 μm. *Icarus* 69:230–238.

O'Shaughnessy, D. J., Boring, J. W., and Johnson, R. E. 1988. Measurements of reflectance spectra of ion-bombarded ice and application to surfaces in the outer solar system. *Nature* 333:240–241.

Ostro, S. J., Campbell, D. B., and Shapiro, I. I. 1985. Mainbelt asteroids: Dual-polarization radar observations. *Science* 229:442–446.

Owen, T. 1982*a*. A brief survey of the solar system. In *Formation of Planetary Systems,* ed. A. Brahic (Toulouse: Cepadues), pp. 585–650.

Owen, T. 1982*b*. The composition and origin of Titan's atmosphere. *Planet. Space Sci.* 30:833–838.

Owen, T. 1986. Update of the Anders-Owen model for martian volatiles. In *Workshop on the Evolution of the Martian Atmosphere,* eds. M. Carr, P. James, C. Leovy, R. Pepin and J. Pollack, LPI Tech. Rept. 86-07 (Houston: Lunar and Planetary Inst.), pp. 31–32.

Owen, T. 1987. How primitive are the gases in Titan's atmosphere? *Adv. Space Res.* 7:51–54.

Owen, T., Biemann, K., Rushneck, D. R., Biller, J. E., Howarth, D. W., and Lafleur, A. L. 1977. The composition of the atmosphere at the surface of Mars. *J. Geophys. Res.* 82:4635–4639.

Owen, T., Cess, R. D., and Ramanathan, V. 1979. Early Earth: An enhanced carbon dioxide greenhouse to compensate for reduced solar luminosity. *Nature* 277:640–642.

Owen, T., Lutz, B. L., and de Bergh, C. 1986. Deuterium in the outer solar system: Evidence for two distinct reservoirs. *Nature* 320:244–246.

Owen, T., Maillard, J. P., de Bergh, C., and Lutz, B. 1988. Deuterium on Mars: The abundance of HDO and the value of D/H. *Science* 240:1767–1770.

Oxburgh, E. R., and O'Nions, R. K. 1987. Helium loss, tectonics, and the terrestrial heat budget. *Science* 237:1583–1588.

Oyama, V. I., Carle, G. C., Woeller, F., Pollack, J. B., Reynolds, R. A., and Craig, R. A. 1980. Pioneer Venus gas chromatography of the lower atmosphere of Venus. *J. Geophys. Res.* 85:7891–7901.

Ozima, M. 1975. Ar isotopes and Earth-atmosphere evolution models. *Geochim. Cosmochim. Acta* 39:1127–1134.

Ozima, M., and Kudo, K. 1972. Excess argon in submarine basalts and an Earth-atmosphere evolution model. *Nature Phys. Sci.* 239:23–24.

Ozima, M., and Nakazawa, K. 1980. Origin of rare gases in the Earth. *Nature* 284:313–316.

Ozima, M., and Podosek, F. A. 1983. *Noble Gas Geochemistry* (Cambridge: Cambridge Univ. Press).

Ozima, M., and Zashu, S. 1983*a*. Primitive helium in diamonds, *Science* 219:1067–1068.

Ozima, M., and Zashu, S. 1983*b*. Noble gases in submarine pillow volcanic glasses. *Earth Planet. Sci. Lett.* 62:24–40.

Ozima, M., and Zashu, S. 1988. Solar-type Ne in Zaire cubic diamonds. *Geochim. Cosmochim. Acta* 52:19–25.

Ozima, M., Zashu, S., Mattey, D. P., and Pillinger, C. T. 1985*a*. Helium, argon and carbon isotopic compositions in diamonds and their implications in mantle evolution. *Geochem. J.* 19:127–134.

Ozima, M., Podosek, F. A., and Igarashi, G. 1985*b*. Terrestrial xenon constraints on the early history of the Earth. *Nature* 315:471–474.

Pagel, B. E. J. 1982. Abundances of elements of cosmological interest. *Phil. Trans. Roy. Soc. London* A307:19–37.

Palme, H., Suess, H. E., and Zeh, H. D. 1981. Abundances of the elements in the solar system.

In *Landolt-Bornstein,* vol. 2, *Astronomy and Astrophysics,* eds. K. Schaifers and H. H. Voight (Berlin: Springer-Verlag), pp. 257–272.

Pang, K. D., Voge, C. C., Rhoads, J., and Ajello, J. 1984. The E-ring of Saturn and satellite Enceladus. *J. Geophys. Res.* 89:9459–9470.

Paresce, F., and Bowyer, S. 1986. The Sun and the interstellar medium. *Sci. Amer.* 225(3):92–99.

Parker, E. N. 1963. Hydrostatic properties of a coronal atmosphere. In *Interplanetary Dynamical Processes,* ed. R. E. Marshak (New York: Wiley), pp. 41–50.

Parkhomenko, E. I. 1967. *Electrical Properties of Rocks,* ed. G. V. Keller (New York: Plenum Press). Trans. from Russian.

Patterson, C. W. 1987. Resonance, capture and evolution of the planets. *Icarus* 70:319–333.

Peale, S. J., and Cassen, P. 1978. Contribution of tidal dissipation to lunar thermal history. *Icarus* 36:245–269.

Peale, S. J., Cassen, P., and Reynolds, R. T. 1979. Melting of Io by tidal dissipation. *Science* 203:892–894.

Pearl, J. C., and Sinton, W. M. 1982. Hot spots of Io. In *Satellites of Jupiter,* ed. D. Morrison (Tucson: Univ. of Arizona Press), pp. 724–755.

Pearl, J. C., Hanel, R., Kunde, V., Maguire, W., Fox, K., Gupta, S., Ponnamperuma, C., and Raulin, F. 1979. Identification of gaseous SO_2 and new upper limits for other gases on Io. *Nature* 288:757–758.

Pearl, J. C., Conrath, B. J., Hanel, R. A., Pirraglia, J. A., Bézard, B., and Coustenis, A. 1988. Albedo, internal heat flux, and energy balance of Uranus. *Icarus,* submitted.

Peck, J. A. 1983. An SEM petrographic study of C3(V) meteorite matrix. *Lunar Planet. Sci.* XIV:598–599 (abstract).

Peck, J. A. 1984. Origin of the variation in properties of CV3 meteorite matrix and matrix clasts. *Lunar Planet. Sci.* XV:635–636 (abstract).

Penzias, A. A. 1983. Isotopic fractionation and mass motion in giant molecular clouds. *Astrophys. J.* 273:195–201.

Pepin, R. O. 1985. Evidence of martian origins. *Nature* 317:473–475.

Pepin, R. O. 1987. Volatile inventories of the terrestrial planets. *Rev. Geophys.* 25:293–296.

Pepin, R. O., and Phinney, D. 1978. Components of xenon in the solar system. *Earth, Moon, Planets,* submitted.

Pepin, R. O., and Signer, P. 1965. Primordial rare gases in meteorites. *Science* 149:253–265.

Perri, F., and Cameron, A. G. W. 1974. Hydrodynamic instability of the solar nebula in the presence of a planetary core. *Icarus* 22:416–425.

Phillips, R. J., Kaula, W. M., McGill, G. E., and Malin, M. C. 1981. Tectonics and evolution of Venus. *Science* 212:879–887.

Phillips, T. G., Scoville, N. Z., Kwan, J., Huggins, P. J., and Wannier, P. G. 1978. Detection of $H_2{}^{18}O$ and an abundance estimate for interstellar water. *Astrophys. J.* 222:L59–L62.

Phillips, T. G., Kwan, J., and Huggins, P. J. 1980. Detection of submillimeter lines of CO (0.65 mm) and H_2O (0.79 mm). In *Interstellar Molecules,* ed. B. H. Andrew (Dordrecht: D. Reidel), pp. 21–24.

Phinney, D., Tennyson, J., and Frick, U. 1978. Xenon in CO_2 well gas revisited. *J. Geophys. Res.* 83:2313–2319.

Pieters, C. M., Head, J. W., Patterson, W., Pratt, S., Garvin, J., Barsukov, V. L., Basilevsky, A. T., Khodakovsky, I. L., Selivanov, A. S., Panfilov, A. S., Getkin, Y. M., and Narayeva, Y. M. 1986. The color of the surface of Venus. *Science* 234:1379–1383.

Pilcher, C. B. 1979. The stability of water on Io. *Icarus* 37:559–574.

Pilcher, C., and Strobel, D. 1982. Emissions from neutrals and ions in the Jovian magnetosphere. In *Satellites of Jupiter,* ed. D. Morrison (Tucson: Univ. of Arizona Press), pp. 807–845.

Pilcher, C., Smyth, W., Combi, M., and Fertel, J. 1984. Io's directional sodium features: Evidence for a magnetospheric wind-driven gas escape mechanism. *Astrophys. J.* 287:427–444.

Pillinger, C. T. 1984. Light element stable isotopes in meteorites—from grams to picograms. *Geochim. Cosmochim. Acta* 48:2739–2766.

Pinto, J. P., and Holland, H. D. 1987. Paleosols and the evolution of the atmosphere, Part II. In preparation.

Pinto, J. P., Lunine, J. I., Kim, S-J., and Yung, Y. L. 1986. D to H ratio and the origin and evolution of Titan's atmosphere. *Nature* 319:388–390.

Pirronello, V., Brown, W. L., Lanzerotti, L. J., Marcantonio, K. J., and Simmons, E. H. 1982.

Formaldehyde formation in a H_2O/CO_2 ice mixture under irradiation by fast ions. *Astrophys. J.* 262:636–640.

Piscitelli, J. R., Cruikshank, D. P., and Bell, J. F. 1987. Laboratory studies of irradiated nitrogen-methane mixtures: Application to Triton. *Icarus,* in press.

Platt, J. R. 1956. On the properties of interstellar dust. *Astrophys. J.* 123:486–490.

Podolak, M., and Cameron, A. G. W. 1974. Models of the giant planets. *Icarus* 22:123–148.

Podolak, M., and Reynolds, R. T. 1984. Consistency tests of cosmogonic theories for models of Uranus and Neptune. *Icarus* 57:102–111.

Podolak, M., and Reynolds, R. T. 1987. The rotation rate of Uranus, its internal structure, and the process of planetary accretion. *Icarus* 70:31–36.

Podolak, M., Pollack, J. B., and Reynolds, R. T. 1988. Interactions of planetesimals with protoplanetary atmospheres. *Icarus* 73:163–179.

Podosek, F. A. 1972. Gas retention chronology of Petersburg and other meteorites. *Geochim. Cosmochim. Acta* 36:755–772.

Pollack, E. J., and Alder, B. J. 1977. Phase separation for a dense fluid mixture of nuclei. *Phys. Rev.* A15:1263–1268.

Pollack, J. B. 1971. A nongrey calculation of the runaway greenhouse: Implications for Venus' past and present. *Icarus* 14:295–306.

Pollack, J. B. 1973. Greenhouse models of the atmosphere of Titan. *Icarus* 19:43–58.

Pollack, J. B. 1979. Climatic change on the terrestrial planets. *Icarus* 37:479–553.

Pollack, J. B. 1984. Origin and history of the outer planets: Theoretical models and observational constraints. *Ann. Rev. Astron. Astrophys.* 22:389–424.

Pollack, J. B. 1985. Formation of giant planets and their satellite-ring systems: An overview. In *Protostars & Planets II,* eds. D. C. Black and M. S. Matthews (Tucson: Univ. of Arizona Press), pp. 791–831.

Pollack, J. B., and Black, D. C. 1982. Noble gases in planetary atmospheres: Implications for the origin and evolution of atmospheres. *Icarus* 51:169–198.

Pollack, J. B., and Consolmagno, G. 1984. Origin and evolution of the Saturn system. In *Saturn,* eds. T. Gehrels and M. S. Matthews (Tucson: Univ. of Arizona Press), pp. 811–866.

Pollack, J. B., and Yung, Y. L. 1980. Origin and evolution of planetary atmospheres. *Ann. Rev. Earth Planet. Sci.* 8:424–487.

Pollack, J. B., Grossman, A., Moore, R., and Graboske, H. 1976. The formation of Saturn's satellites and rings as influenced by Saturn's contraction history. *Icarus* 29:35–48.

Pollack, J. B., Grossman, A. S., Moore, R., and Graboske, H. C., Jr. 1977. A calculation of Saturn's gravitational contraction history. *Icarus* 30:111–128.

Pollack, J. B., Burns, J. A., and Tauber, M. E. 1979. Gas drag in primordial circumplanetary envelopes: A mechanism for satellite capture. *Icarus* 37:587–611.

Pollack, J. B., Toon, O. B., and Boese, R. 1980. Greenhouse models of Venus' high surface temperature, as constrained by Pioneer Venus measurements. *J. Geophys. Res.* 85:8223–8231.

Pollack, J. B., McKay, C. P., and Christofferson, B. M. 1985. A calculation of the Rosseland mean opacity of dust grains in primordial solar system nebulae. *Icarus* 64:471–492.

Pollack, J. B., Podolak, M., Bodenheimer, P., and Christofferson, B. 1986*a*. Planetesimal dissolution in the envelopes of the forming, giant planets. *Icarus* 67:409–443.

Pollack, J. B., Rages, K., Baines, K. H., Bergstralh, J. T., Wenkert, D., and Danielson, G. E. 1986*b*. Estimates of the bolometric albedos and radiation balance of Uranus and Neptune. *Icarus* 65:442–466.

Pollack, J. B., Rages, K., Pope, S. K., Tomasko, M. G., Romani, P. N., and Atreya, S. K. 1987. Nature of the stratospheric haze on Uranus: Evidence for condensed hydrocarbons. *J. Geophys. Res.* 13:15037–15065.

Pollack, J. B., Kasting, J. F., Richardson, S. M., and Poliakoff, K. 1987*b*. The case for a wet, warm climate on early Mars. *Icarus* 71:203–224.

Pontius, D., Hill, T. W., and Rassbach, M. 1986. Steady state plasma transport in a corotation dominated magnetosphere. *Geophys. Res. Lett.* 11:1097–1100.

Pope, S. K., and Tomasko, M. G. 1986. Titan's aerosols and the presence of a methane cloud. *Bull. Amer. Astron. Soc.* 18:816 (abstract).

Poreda, R., and di Brozolo, F. R. 1984. Neon isotope variations in Mid-Atlantic Ridge basalts. *Earth Planet. Sci. Lett.* 69:277–289.

Postawko, S. E., and Kuhn, W. R. 1986. Effect of the greenhouse gases (CO_2, H_2O, SO_2) on Martian paleoclimate. *Proc. Lunar Planet. Sci. Conf.* 16, *J. Geophys. Res.* 91:D431–D438.

Potter, A., and Morgan, T. 1985. Discovery of sodium in the atmosphere of Mercury. *Science* 229:651–653.

Potter, A., and Morgan, T. 1986. Potassium in the atmosphere of Mercury. *Icarus* 67:336–340.

Prasad, S. S., and Huntress, W. T. 1982. Sulfur chemistry in dense interstellar clouds. *Astrophys. J.* 260:590–598.

Prentice, A. J. R. 1978. Towards a modern La Placian theory for the formation of the solar system. In *The Origin of the Solar System,* ed. S. F. Dermott (Chichester, N.Y.: Wiley), pp. 111–161.

Prinn, R. G. 1979. On the possible roles of gaseous sulfur and sulfanes in the atmosphere of Venus. *Geophys. Res. Lett.* 6:807–810.

Prinn, R. G. 1982. Origin and evolution of planetary atmospheres: An introduction to the problem. *Planet. Space Sci.* 30:741–753.

Prinn, R. G. 1988. On neglect of non-linear momentum terms in solar nebula accretion disk models. *Astrophys. J.,* submitted.

Prinn, R. G., and Barshay, S. 1977. Carbon monoxide on Jupiter and implications for atmospheric convection. *Science* 198:1031–1034.

Prinn, R. G., and Fegley, B. 1981. Kinetic inhibition of CO and N_2 reduction in circumplanetary nebulae: Implications for satellite composition. *Astrophys. J.* 249:308–317.

Prinn, R. G., and Fegley, B. 1987*a*. The atmospheres of Venus, Earth, and Mars: A critical comparison. *Ann. Rev. Earth Planet. Sci.* 15:171–212.

Prinn, R. G., and Fegley, B. 1987*b*. Bolide impacts, acid rain, and biospheric traumas at the Cretaceous-Tertiary boundary. *Earth Planet. Sci. Lett.* 83:1–15.

Prinn, R. G., and Olaguer, E. P. 1981. Nitrogen on Jupiter: A deep atmospheric source. *J. Geophys. Res.* 86:9895–9899.

Prinn, R., and Owen, T. 1976. Chemistry and spectroscopy of the Jovian atmosphere. In *Jupiter,* ed. T. Gehrels (Tucson: Univ. of Arizona Press), pp. 319–371.

Prinn, R. G., Larson, H. P., Caldwell, J. J., and Gautier, D. 1984. Composition and chemistry of Saturn's atmosphere. In *Saturn,* eds. T. Gehrels and M. S. Matthews (Tucson: Univ. of Arizona Press), pp. 88–149.

Puget, J. L., Léger, A., and Boulanger, F. 1985. Contribution of large polycyclic aromatic molecules to the infrared emission of the interstellar medium. *Astron. Astrophys.* 142:L19–L22.

Quijano-Rico, M., and Wänke, H. 1969. Determination of boron, lithium and chlorine in meteorites. In *Meteorite Research,* ed. P. M. Millman (Dordrecht: D. Reidel), pp. 132–145.

Rahe, J. 1980. Ultraviolet spectroscopy of comets. In *Proc. 2nd European IUE Conf.,* ESA SP-157, p. 125.

Rahe, J. 1983. Cometary spectroscopy and implications to the chemical composition, physical properties and origin of comets. In *Highlights of Astronomy,* vol. 6, ed. R. M. West (Dordrecht: D. Reidel), pp. 333–345.

Rahe, J., and Donn, B. 1968. Formation and development of cometary Type I tails. *Astron. J.* 73:S32 (abstract).

Rahe, J., and Donn, B. 1969. Ionization and ray formation in comets. *Astron. J.* 74:256–258.

Rao, M. N., Nautiyal, C. M., Padia, J. T., and Venkatesan, T. R. 1983. Confirmation of solar flare neon composition using lunar pyroxenes and SCR proton fluxes in last hundred million years. *Lunar Planet. Sci.* XIV:632–633 (abstract).

Rasool, S. I., and de Bergh, C. 1970. The runaway greenhouse and accumulation of CO_2 in the Venus atmosphere. *Nature* 226:1037–1039.

Raulin, F. 1985. Chimie organique dans Titan: Expériences de simulation en laboratoire et spéculations. In *The Atmospheres of Saturn and Titan: Proc. Intl. Workshop,* Alpbach, Austria, ESA SP-241, pp. 161–173.

Raup, D. M., and Sepkoski, J. J., Jr. 1984. Periodicity of extinctions in the geologic past. *Proc. Natl. Acad. Sci. USA* 81:801–805.

Ray, J., and Heymann, D. 1980. A model for nitrogen isotopic variations in the lunar regolith: Possible solar system contributions from a nearby planetary nebula. In *The Ancient Sun: Fossil Record in the Earth, Moon and Meteorites,* eds. R. O. Pepin, J. A. Eddy and R. B. Merrill (New York: Pergamon Press), pp. 491–512.

Read, P. L. 1986. Super-rotation and diffusion of axial angular momentum: II. A review of quasi-axisymmetric models of planetary atmospheres. *J. Roy. Met. Soc.* 112:243–272.

Regev, O., and Shaviv, G. 1981. Formation of protostars in collapsing, rotating, turbulent clouds. *Astrophys. J.* 245:934–959.
Reimann, C. T., Boring, J. W., Johnson, R. E., Garrett, J. W., Farmer, K. R., Brown, W. L., Marcantonio, K. J., and Augustyniak, W. M. 1984. Ion-induced molecular ejection from D_2O ice. *Surf. Sci.* 147:227–240.
Reimers, C. E., and Komar, P. D. 1979. Evidence for explosive volcanic density currents on certain Martian volcanoes. *Icarus* 39:88–110.
Reipurth, B., Bally, J., Graham, J. A., Lane, A. P., and Zealey, W. J. 1986. The jet and energy source of HH34. *Astron. Astrophys.* 164:51–66.
Revercomb, H. E., Sromovsky, L. A., Suomi, V. E., and Boese, R. W. 1985. Net thermal radiation in the atmosphere of Venus. *Icarus* 61:521–538.
Richardson, J. 1986. Thermal ions at Saturn: Plasma parameters and implications. *J. Geophys. Res.* 91:1381–1389.
Richardson, J., and Eviatar, A. 1988. Limits on the extent of Saturn's hydrogen cloud. *J. Geophys. Res.*, in press.
Richardson, J., and McNutt, R. 1987. Observational constraints on interchange models at Jupiter. *Geophys. Res. Lett.* 14:64–67.
Richardson, J., and Siscoe, G. 1983. The problem of cooling the cold Io torus. *J. Geophys. Res.* 88:2001–2009.
Richardson, J., Eviatar, A., and Siscoe, G. L. 1986. Satellite tori at Saturn. *J. Geophys. Res.* 91:8749–8755.
Richardson, S. H., Gurney, J. J., Erlank, A. J., and Harris, J. W. 1984. Origin of diamonds in old enriched mantle. *Nature* 310:198–202.
Richet, P., Bottinga, Y., and Javoy, M. 1977. A review of hydrogen, carbon, nitrogen, oxygen, sulphur, and chlorine stable isotope fractionation among gaseous molecules. *Ann. Rev. Earth Planet. Sci.* 5:65–110.
Ridgway, S. T. 1974. Jupiter: Identification of ethane and acetylene. *Astrophys. J.* 187:L41–L43.
Ridgway, S. T., Larson, H. P., and Fink, U. 1976. The infrared spectrum of Jupiter. In *Jupiter*, ed. T. Gehrels (Tucson: Univ. of Arizona Press), pp. 384–417.
Rieke, G. H., Lebofsky, L. A., and Lebofsky, M. J. 1985. A search for nitrogen on Triton. *Icarus* 64:153–155.
Ringwood, A. E. 1958. The constitution of the mantle III: Consequences of the olivine-spinel transition. *Geochim. Cosmochim. Acta* 15:195–212.
Ringwood, A. E. 1959. On the chemical evolution and density of the planets. *Geochim. Cosmochim. Acta* 15:257–283.
Ringwood, A. E. 1966*a*. The chemical composition and origin of the Earth. In *Adv. Earth Sci.*, ed. P. M. Hurley (Boston: MIT Press), pp. 287–356.
Ringwood, A. E. 1966*b*. Mineralogy of the mantle. In *Adv. Earth Sci.*, ed. P. Hurley (Boston: MIT Press), pp. 357–398.
Ringwood, A. E. 1977*a*. Basaltic magmatism and the bulk composition of the Moon. I. Major and heat-producing elements. *Moon* 16:389–423.
Ringwood, A. E. 1977*b*. Composition of the core and implications for origin of the Earth. *Geochem. J.* 11:111–135.
Ringwood, A. E. 1979*a*. *On the Origin of the Moon* (New York: Springer-Verlag).
Ringwood, A. E. 1979*b*. *Origin of the Earth and the Moon* (New York: Springer-Verlag).
Ringwood, A. E. 1986. Composition and origin of the Moon. In *Origin of the Moon*, eds. W. K. Hartmann, R. J. Phillips and G. J. Taylor (Houston: Lunar and Planetary Inst.), pp. 673–698.
Ringwood, A. E., and Kesson, S. E. 1977. Basaltic magmatism and the bulk composition of the Moon II. Siderophile and volatile elements in Moon, Earth and chondrites: Implications for lunar origin. *Moon* 16:425–464.
Robert, F., and Epstein, S. 1982. The concentration and isotopic composition of hydrogen, carbon, and nitrogen in carbonaceous meteorites. *Geochim. Cosmochim. Acta* 46:81–95.
Robert, F., Merlivat, L., and Javoy, M. 1979. Deuterium concentration in the early solar system: Hydrogen and oxygen isotope study. *Nature* 282:785–789.
Robert, F., Javoy, M., Halbout, J., Dimon, B., and Merlivat, L. 1987*a*. Hydrogen isotope abundances in the solar system. Part I: Unequilibrated chondrites. *Geochim. Cosmochim. Acta* 51:1787–1805.
Robert, F., Javoy, M., Halbout, J., Dimon, B., and Merlivat, L. 1987*b*. Hydrogen isotope abun-

dances in the solar system. Part II: Meteorites with terrestrial-like D/H ratio. *Geochim. Cosmochim. Acta* 51:1807–1822.
Robnik, M., and Kundt, W. 1983. Hydrogen at high pressures and temperatures. *Astron. Astrophys.* 120:227–233.
Romani, P. N., and Atreya, S. K. 1986. Polyacetylene photochemistry and condensation of Uranus. *Bull. Amer. Astron. Soc.* 18:758 (abstract).
Romani, P. N., and Atreya, S. K. 1988. Methane photochemistry and haze production on Neptune. *Icarus* 74:424–445.
Ronov, A. B., and Yaroshevsky, 1967. Chemical structure of the Earth's crust. *Geochemistry* 11:1041–1066.
Ronov, A. B., and Yaroshevskiy, A. A. 1976. A new model for the chemical structure of the Earth's crust. *Geochim. Intl.* 13(6):89–91.
Ross, J. E., and Aller, L. H. 1976. The chemical composition of the Sun. *Science* 191:1223–1229.
Ross, M. N., and Lee, Y. 1986. The insulator-conductor transition in dense molecular hydrogen at high temperature. Preprint.
Ross, M., Graboske, H. C., Jr., and Nellis, W. J. 1981. Equation of state experiments and theory relevant to planetary modeling. *Phil. Trans. Roy. Soc. London* A303:303–313.
Ross, M. N., and Ree, F. H. 1980. Repulsive forces of simple molecules and mixtures at high density and temperature. *J. Chem. Phys.* 73:6146–6152.
Ross, M. N., and Schubert, G. 1985. Tidally forced viscous heating in a partially molten Io. *Icarus* 64:391–400.
Ross, M. N., and Schubert, G. 1987. Tidal heating in an internal ocean model of Europa. *Nature* 325:133–134.
Ross, M. N., and Seale, D. 1974. Perturbation approximation to the screened coulomb gas. *Phys. Rev.* A9:396–399.
Ross, M., and Shishkevish, C. 1977. Molecular and metallic hydrogen. Defense Advanced Research Projects Agency, Rept. R-2046-ARPA.
Rossow, W. B., Henderson-Sellers, A., and Weinreich, S. K. 1982. Cloud feedback: A stabilizing effect for the early Earth? *Science* 217:1245–1247.
Rubey, W. W. 1951. Geologic history of sea water: An attempt to state the problem. *Geol. Soc. Amer. Bull.* 62:1111–1147.
Ruden, S., and Lin, D. N. C. 1986. The global evolution of the primordial solar nebula. *Astrophys. J.* 308:883–901.
Ruskol, E. L. 1960. Origin of the Moon I. *Soviet Astron. AJ* 4:657–668.
Ruskol, E. L. 1963. Origin of the Moon II. *Soviet Astron. AJ* 7:221–227.
Ruskol, E. 1972. Origin of the Moon III. *Soviet Astron. AJ* 15:646–654.
Russell, C. T., Luhmann, J., Elphic, R., and Neugebauer, M. 1982. Solar wind interaction with comets: Lessons from Venus. In *Comets,* ed. L. L. Wilkening (Tucson: Univ. of Arizona Press), pp. 561–587.
Rydbeck, O. E. H., and Hjalmarson, Å. 1985. Radio observations of interstellar molecules, of their behaviour, and of their physics. In *Molecular Astrophysics: State of the Art and Future Directions,* eds. G. H. F. Dierckson, W. F. Huebner and P. W. Langhoff (Dordrecht: D. Reidel), pp. 45–175.
Rydberg, A. E., and Cohen, M. 1985. Young stellar objects and their circumstellar dust: An overview. In *Protostars & Planets II,* eds. D. C. Black and M. S. Matthews (Tucson: Univ. of Arizona Press), pp. 371–385.
Safronov, V. S. 1969. *Evolution of the Protoplanetary Cloud and Formation of the Earth and the Planets* (Moscow: Nauka Press). In Russian. Trans. NASA TTF-677, 1972.
Safronov, V. S. 1972. Accumulation of the planets. In *Origin of the Solar System,* ed. H. Reeves (Paris: CNRS), pp. 89–113.
Safronov, V. S. 1979. On the origin of asteroids. In *Asteroids,* ed. T. Gehrels (Tucson: Univ. of Arizona Press), pp. 975–991.
Safronov, V. S., and Ruskol, E. L. 1982. On the origin and initial temperature of Jupiter and Saturn. *Icarus* 49:284–296.
Safronov, V. S., and Ruzmaikina, T. V. 1978. On the angular momentum transfer and the accumulation of solid bodies in the solar nebula. In *Protostars and Planets,* ed. T. Gehrels (Tucson: Univ. of Arizona Press), pp. 545–564.

Safronov, V. S., and Ruzmaikina, T. V. 1985. Formation of the solar nebula and the planets. In *Protostars & Planets II*, eds. D. C. Black and M. S. Matthews (Tucson: Univ. of Arizona Press), pp. 959–980.

Safronov, V. S., and Vityazev, A. V. 1985. Origin of the solar system. *Sov. Sci. Rev. Astrophys. Space Phys.* 4:1–98.

Safronov, V. S., Pechernikova, G. V., Ruskol, E. L., and Vitjazev, A. V. 1986. Protosatellite swarms. In *Satellites*, eds. J. A. Burns and M. S. Matthews (Tucson: Univ. of Arizona Press), pp. 89–116.

Sagan, C. 1960. The radiation balance of Venus. JPL Tech. Rept. No. 32-34.

Sagan, C., and Dermott, S. F. 1982. The tide in the seas of Titan. *Nature* 300:731–733.

Sagan, C., and Khare, B. N. 1979. Tholins: Organic chemistry of interstellar grains and gas. *Nature* 277:102–107.

Sagan, C., and Mullen, G. 1972. Earth and Mars: Evolution of atmospheres and surface temperatures. *Science* 177:52–56.

Sagan, C., and Thompson, W. R. 1984. Production and condensation of organic gases in the atmosphere of Titan. *Icarus* 59:133–161.

Sagdeev, R. Z., Kissel, J., Evlanov, E. N., Mukhin, L. M., Zubkov, B. V., Prilutskii, O. F., and Fomenkova, M. N. 1986. Elemental composition of the dust component of Halley's Comet: Preliminary analysis. In *20th ESLAB Symposium on the Exploration of Halley's Comet*, vol. 3, eds. B. Battrick, E. J. Rolfe and R. Reinhard, ESA SP-250, pp. 349–352.

Saito, K., Basu, A. R., and Alexander, E. C. Jr., 1978. Planetary-type rare gases in an upper mantle-derived amphibole. *Earth Planet. Sci. Lett.* 39:274–280.

Saito, S., Kawaguchi, K., Yamamoto, S., Ohishi, M., Suzuki, H., and Kaifu, N. 1987. Laboratory detection and astronomical identification of a new free radical, CCS ($^3\Sigma^-$). *Astrophys. J.* 317:L115–L118.

Sakai, H., Casadevall, T. J., and Moore, J. G. 1982. Chemistry and isotope ratios of sulfur in basalts and volcanic gases at Kilauea Volcano, Hawaii. *Geochim. Cosmochim. Acta* 46:729–738.

Sakata, A., Setsuko, W., Okutsu, Y., Shintani, H., and Nakada, Y. 1983. Does a 2,200 Å hump observed in artificial carbonaceous composite account for UV interstellar extinction? *Nature* 301:493–494.

Sakata, A., Wada, S., Tanabe, T., and Onaka, T. 1984. Infrared spectrum of the laboratory-synthesized quenched carbonaceous composite (QCC): Comparison with the infrared unidentified emission bands. *Astrophys. J.* 287:L51–L54.

Sakata, A., Wada, S., Onaka, T., and Tokunaga, A. T. 1987. Infrared spectrum of quenched carbonaceous composite (QCC). II. A new identification of the 7.7 and 8.6 micron unidentified infrared emission bands. *Astrophys. J.* 320:L63–L67.

Salpeter, E. E. 1973. On convection and gravitational layering in Jupiter and in stars of low mass. *Astrophys. J.* 181:L183–L186.

Samuelson, R. E. 1983. Radiative equilibrium model of Titan's atmosphere. *Icarus* 53:364–387.

Samuelson, R. E. 1985. Titan's methane clouds. *Bull. Amer. Astron. Soc.* 17:740 (abstract).

Samuelson, R. E., Maguire, W. C., Hanel, R. A., Kunde, V. G., Jennings, D. E., Yung, Y. L., and Aiken, A. C. 1983. CO_2 on Titan. *J. Geophys. Res.* 88:8709–8715.

Sandel, B. R., Belton, M. J. S., Takacs, P. Z., Shemansky, D. E., Holberg, J. B., Ajello, J. M., Atreya, S. K., Donahue, T. M., Moos, H. W., Bertaux, J. L., Blamont, J. E., Strobel, D. F., McConnell, J. C., Dalgarno, A., Goody, R., and McElroy, M. B. 1979. Extreme ultraviolet observations from the Voyager 2 encounter with Jupiter. *Science* 286:962–966.

Sandel, B. R., Shemansky, D., Broadfoot, A., Bertaux, J., Blamont, J., Belton, M. J. S., Ajello, J. M., Holberg, J. B., Atreya, S. K., Donahue, T. M., Moos, H. W., Strobel, D. F., McConnell, J. C., Dalgarno, A., Goody, R., McElroy, M. B., and Takacs, P. Z. 1982. Extreme ultraviolet observations from the Voyager 2 encounter with Saturn. *Science* 215:548–553.

Sandford, S. A., and Walker, R. M. 1985. Laboratory infrared transmission spectra of individual interplanetary dust particles from 2.5 to 25 microns. *Astrophys. J.* 291:838–851.

Sarda, P., Staudacher, T., and Allègre, C. J. 1985. ^{40}AR/^{36}Ar in MORB glasses: Constraints on atmosphere and mantle evolution. *Earth Planet. Sci. Lett.* 72:357–375.

Sargent, A. I., and Beckwith, S. 1987*a*. Kinematics of the circumstellar gas of HL Tauri and R Monocerotis. *Astrophys. J.* 323:294–305.

Sargent, A. I., and Beckwith, S. 1987*b*. Velocity structure in the circumstellar gas of HL Tau. *Bull. Amer. Astron. Soc.* 18:959 (abstract).

Sasaki, S., and Nakazawa, K. 1986. Terrestrial Xe fractionation due to escape of primordial H_2-He atmosphere. In *Abstracts for Japan-U.S. Seminar on Terrestrial Rare Gases,* Yellowstone Natl. Park, September, pp. 68–71 (abstract).

Sasaki, S., and Nakazawa, K. 1988. Origin of the isotopic fractionation of terrestrial xenon. *Earth Planet. Sci. Lett.* 89:323–334.

Sato, M. 1976. Oxygen fugacity and other thermochemical parameters of Apollo 17 high Ti basalts and their implications of the reduction mechanism. *Proc. Lunar Sci. Conf.* 7:1323–1344.

Saumon, D., and Van Horn, H. M. 1986. Toward an improved pure hydrogen EOS for astrophysical applications. In *Strongly Coupled Plasmas,* eds. F. J. Rogers and H. E. DeWitt (New York: Plenum), pp. 173–177.

Saunders, R. S., Bills, T. G., and Johansen, L. 1981. The ridged plains of Mars. *Lunar Planet. Sci.* XII:924–925 (abstract).

Savage, B. D., and Matthis, J. S. 1979. Observed properties of interstellar dust. *Ann. Rev. Astron. Astrophys.* 17:73–111.

Scarf, F. L., Gurnett, D. A., Kurth, W. S., Anderson, R. R., and Shaw, R. R. 1981. An upper bound to the lightning flash rate in Jupiter's atmosphere. *Science* 213:684–685.

Scarf, F. L., Frank, L. A., Gurnett, D., Lanzerotti, L. J., Lazarus, A., and Sittler, E. 1984. Measurements of plasma, plasma waves, and superthermal charged particles in Saturn's inner magnetosphere. In *Saturn,* eds. T. Gehrels and M. S. Matthews (Tucson: Univ. of Arizona Press), pp. 318–353.

Schidlowski, M., Hayes, J. M., and Kaplan, I. R. 1983. Isotopic inferences of ancient biochemistries: Carbon, sulfur, hydrogen, and nitrogen. In *The Earth's Earliest Biosphere: Its Origin and Evolution,* ed. J. W. Schopf (Princeton: Princeton Univ. Press), pp. 149–186.

Schleicher, D. G., and A'Hearn, M. F. 1986. Limits on OD in comet P/Halley. *Bull. Amer. Astron. Soc.* 18:795 (abstract).

Schloerb, F. P. 1987. Millimeter-wave experiments for cometary missions. In *Workshop on the Multi-Comet Mission,* NASA, in press.

Schloerb, F. P., Snell, R. L., Langer, W. D., and Young, J. S. 1981. Detection of deuteriocyanobutadiyne (DC_5N) in the interstellar cloud TMC-1. *Astrophys. J.* 251:L37–L42.

Schloerb, F. P., Friberg, P., Hjalmarson, Å., Höglund, B., and Irvine, W. M. 1983. Observations of sulfur dioxide in the Kleinmann-Low nebula. *Astrophys. J.* 264:161–171.

Schmidt, O. J. 1957. *Four Lectures on the Theory of the Earth's Origin,* 3rd ed. (Moscow: Izdatel'stvo AN).

Schmidt, R. M., and Housen, K. R. 1987. Some recent advances in the scaling of impact and explosion cratering. *Intl. J. Impact Eng.* 5:543–560.

Schneider, N. M., Hunten, D. M., Wells, W. K., and Trafton, L. J. 1987. Eclipse measurements of Io's sodium atmosphere. *Science* 238:55–58.

Schneider, N. M., Smyth, W. H., and McGrath, M. A. 1988. Paper to appear in *Time Variable Phenomena in the Jovian System.* Proceedings of the conference held in Flagstaff, AZ, August, 1987.

Schneider, S. H., and Londer, R. 1984. *The Coevolution of Climate and Life* (San Francisco: Sierra Club Books).

Schofield, J. T., Taylor, F. W., and McCleese, D. J. 1982. The global distribution of water vapor in the middle atmosphere of Venus. *Icarus* 52:263–278.

Schubart, J., and Matson, D. L. 1979. Masses and densities of asteroids. In *Asteroids,* ed. T. Gehrels (Tucson: Univ. of Arizona Press), pp. 84–97.

Schubert, G. 1979. Subsolidus convection in the mantles of terrestrial planets. *Ann. Rev. Earth Planet. Sci.* 7:289–342.

Schubert, G., and Reymer, A. P. S. 1985. Continental volume and freeboard through geologic time. *Nature* 316:336–339.

Schubert, G., Cassen, P., and Young, R. E. 1979. Subsolidus convective cooling histories of terrestrial planets. *Icarus* 38:192–211.

Schubert, G., Stevenson, D. J., and Ellsworth, K. 1981. Internal structures of the Galilean satellites. *Icarus* 47:46–59.

Schubert, G., Spohn, T., and Reynolds, R. T. 1986. Thermal histories, compositions, and inter-

nal structures of the moons of the solar system. In *Satellites,* eds. J. A. Burns and M. S. Matthews (Tucson: Univ. of Arizona Press), pp. 224–292.
Schubert, G., Ross, M. N., Stevenson, D. J., and Spohn, T. 1988. Mercury's thermal history and the generation of its magnetic field. In *Mercury,* eds. F. Vilas, C. R. Chapman and M. S. Matthews (Tucson: Univ. of Arizona Press), pp. 429–460.
Schultz, P. H. 1987. Possible non-random impact fluxes on the Moon in recent time (< 100 my). *Eos: Trans. AGU* 68:344 (abstract).
Schulz, M., and Eviatar, A. 1988. Charged particle absorption by Io. *Astrophys. J.* 211:L149–L154.
Schwarcz, H. P., Hoefs, J., and Welte, D. 1969. Carbon. In *Handbook of Geochemistry,* vol. II/1 (Berlin: Springer-Verlag), pp. 6-B-1–6-O-3.
Schwartz, P. R., Simon, T., and Campbell, R. 1986. The T Tauri radio source. II. The winds of T Tauri. *Astrophys. J.* 303:233–238.
Schwartzman, D. W. 1973. Argon degassing model of the Earth. *Nature Phys. Sci.* 245:20–21.
Schwerer, F. C., Nagata, T., and Fisher, R. M. 1971. Electrical conductivity of lunar surface rocks and chondrite meteorites. *Moon* 2:408.
Scott, D. H., and Tanaka, K. L. 1982. Ignimbrites of Amazonis Planitia region of Mars. *J. Geophys. Res.* 87:1179–1190.
Scoville, N. Z., Kleinmann, S. G., Hall, D. N. B., and Ridgway, S. T. 1983. The circumstellar and nebular environment of the Becklin-Neugebauer object: 2.5 micron spectroscopy. *Astrophys. J.* 275:201–224.
Segatz, M., Spohn, T., Ross, M., and Schubert, G. 1988. Tidal dissipation, surface heat flow, and figure of viscoelastic models of Io. *Icarus* 75:187–206.
Seidelmann, P. K., and Divine, N. 1977. Evaluation of Jupiter longitudes in System III (1965). *Geophys. Res. Lett.* 4:65–68.
Sekanina, Z. 1976. A probability of encounter with interstellar comets and the likelihood of their existence. *Icarus* 27:123–133.
Sekanina, Z. 1977. Relative motions of fragments of the split comets. I. A. new approach. *Icarus* 30:574–594.
Sekanina, Z. 1980. Physical similarities between dissipating comets and short-lived fragments of the split comets. *Bull. Amer. Astron. Soc.* 12:511 (abstract).
Sekiya, M. 1983. Gravitational instabilities in a dust-gas layer and formation of planetesimals in the solar nebula. *Prog. Theor. Phys.* 69:1116–1130.
Sekiya, M. K., Nakazawa, K., and Hayashi, C. 1980*a*. Dissipation of the rare gases contained in the primordial Earth's atmosphere. *Earth Planet. Sci. Lett.* 50:197–201.
Sekiya, M., Nakazawa, K., and Hayashi, C. 1980*b*. Dissipation of the primordial terrestrial atmosphere due to irradiation of the solar EUV. *Prog. Theor. Phys.* 64:1968–1985.
Sekiya, M., Hayashi, C., and Nakazawa, K. 1981. Dissipation of the primordial terrestrial atmosphere due to irradiation of the solar far-UV during T-Tauri stage. *Prog. Theor. Phys.* 66:1301–1316.
Sekiya, M., Miyama, S. M., and Hayashi, C. 1987. Gas flow in the solar nebula leading to the formation of Jupiter. *Earth, Moon, Planets* 39:1–15.
Sellgren, K. 1984. The near-infrared continuum emission of visual reflection nebulae. *Astrophys. J.* 277:623–633.
Seyfried, W. E., Janecky, D. R., and Mottl, M. J. 1984. Alteration of oceanic crust: Implications for geochemical cycles of lithium and boron. *Geochim. Cosmochim. Acta* 48:557–569.
Schloerb, F. P. 1987. Millimeter-wave experiments for cometary missions. In *Workshop on the Multi-Comet Mission,* NASA, in press.
Schloerb, F. P., Snell, R. L., Langer, W. D., and Young, J. S. 1981. Detection of deuteriocyanobutadiyne (DC_5N) in the interstellar cloud TMC-1. *Astrophys. J.* 251:L37–L42.
Shafer, R., and Gordon, R. G. 1973. Quantum scattering theory of rotational relaxation in H_2-He mixtures. *J. Chem. Phys.* 58:5422–5443.
Shakura, N. I., and Sunyaev, R. A. 1973. Black holes in binary systems. Observational appearance. *Astron. Astrophys.* 24:337–355.
Shapley, H. 1921. Note on a possible factor in changes of geologic climate. *J. Geol.* 29:502–504.
Sharpton, V. L., Grieve, R. A. F., and Goodacre, A. K. 1987. Periodicity in the Earth's cratering record? *Eos: Trans. AGU* 68:344 (abstract).
Shelley, E. G., Klumpar, D. M., Peterson, W., Ghielmetti, A., Balsiger, H., Geiss, J., and

Rosenbauer, H. 1985. AMPTE/OCE observations of the plasma composition below 17 keV during the September 4 1984 magnetic storm. *Geophys. Res. Lett.* 12:321–324.

Shemansky, D. 1980. Comment on "Methods of Monte Carlo simulation of the exospheres of the Moon and Mercury" by R. Hodges. *J. Geophys. Res.* 85:221.

Shemansky, D. 1987. Ratio of oxygen to sulfur in the Io plasma torus. *J. Geophys. Res.* 92:6141–6146.

Shemansky, D. 1988. Energy branching in the Io torus: The failure of neutral cloud theory. *J. Geophys. Res.* 93:1773–1784.

Shemansky, D., and Broadfoot, A. L. 1977. Interaction of the surfaces of the Moon and Mercury with their exospheric atmospheres. *Revs. Geophys. Space Phys.* 15:491–499.

Shemansky, D., and Smith, G. R. 1983. Whence comes the "Titan" torus. *Eos: Trans. AGU* 63:1019 (abstract).

Shemansky, D., and Smith, G. R. 1986. The implications for the pressure of a magnetosphere on Uranus in the relationship of EUV and radio emission. *Geophys. Res. Lett.* 13:2–5.

Shih, C. V., Nyquist, L. E., Bogard, D. D., McKay, G. A., Wooden, J. L., Bansal, B. M., and Wiesman, H. 1982. Chronology and petrogenesis of young achondrites, Shergotty, Zagami, and ALHA 7705: Late magmatism on a geologically active planet. *Geochim. Cosmochim. Acta* 46:2323–2344.

Shizgal, B., and Blackmore, R. 1986. A collisional kinetic theory of a plane parallel evaporating planetary atmosphere. *Planet. Space Sci.* 34:279–291.

Shklovskii, I. S. 1951. On the possibility of explaining the difference in chemical composition of the Earth and Sun by thermal dissipation of light gases. *Astron. Zh.* 28:234–243.

Shoemaker, E. M. 1977*a*. Astronomically observable crater forming projectiles. In *Impact and Explosion Cratering: Planetary and Terrestrial Implications,* eds. D. J. Roddy, R. O. Pepin and R. B. Merrill (New York: Pergamon), pp. 617–628.

Shoemaker, E. M. 1977*b*. Why study impact craters? In *Impact and Explosion Cratering: Planetary and Terrestrial Implications,* eds. D. J. Roddy, R. O. Pepin and R. B. Merrill (New York: Pergamon Press), pp. 1–9.

Shoemaker, E. M. 1982. The collision of solid bodies. In *The New Solar System,* eds. J. K. Beatty, B. O'Leary and A. Chaikin (Cambridge: Cambridge Univ. Press), pp. 33–44.

Shoemaker, E. M., and Wolfe, R. F. 1982. Cratering time scales for the Galilean satellites. In *Satellites of Jupiter,* ed. D. Morrison (Tucson: Univ. of Arizona Press), pp. 277–339.

Shoemaker, E. M., and Wolfe, R. F. 1984. Evolution of the Uranus-Neptune planetesimal swarm. *Lunar Planet. Sci.* XV:780–781 (abstract).

Shoemaker, E. M., and Wolfe, R. F. 1986. Mass extinctions, crater ages, and comet showers. In *The Galaxy and the Solar System,* eds. R. Smoluchowski, J. N. Bahcall and M. S. Matthews (Tucson: Univ. of Arizona Press), pp. 338–386.

Shoemaker, E. M., Williams, J. G., Helin, E. F., and Wolfe, R. F. 1979. Earth-crossing asteroids: Orbital classes, collision rates with Earth, and origin. In *Asteroids,* ed. T. Gehrels (Tucson: Univ. of Arizona Press), pp. 253–282.

Shu, F. H. 1977. Self-similar collapse of isothermal spheres and star formation. *Astrophys. J.* 214:488–497.

Shu, F. H., and Tereby, S. 1984. The formation of cool stars from cloud cores. In *Cool Stars, Stellar Systems, and the Sun,* eds. S. Baliunas and L. Hartmann (Berlin: Springer-Verlag), pp. 78–89.

Shu, F. H., Adams, F. C., and Lizano, S. 1987. Star formation in molecular clouds: Observation and theory. *Ann. Rev. Astron. Astrophys.* 25:23–72.

Shull, J. M., and Beckwith, S. 1982. Interstellar molecular hydrogen. *Ann. Rev. Astron. Astrophys.* 20:163–190.

Sieveka, E. M., and Johnson, R. E. 1982. Thermal and plasma-induced molecular re-distribution on icy satellites. *Icarus* 51:528–540.

Sieveka, E., and Johnson, R. E. 1984. Ejection of atoms and molecules from Io by plasma ion impact. *Astrophys. J.* 287:418–426.

Sieveka, E., and Johnson, R. E. 1985. Non-isotopic coronal atmosphere on Io. *J. Geophys. Res.* 90:5327–5331 and *J. Geophys. Res.* 91:4608 (erratum).

Sigmund, P. 1969. Theory of sputtering. I. Sputtering yields of amorphous and polycrystalline targets. *Phys. Rev.* 184:383–416.

Simakov, G. V. M. N., Pavlovskiy, N. M., Kalashnikov, N. G., and Turnin, R. F. 1974. Shock compressibility of twelve minerals. *Izv. Phys. Solid Earth* 8:488–492.

Simonelli, D. P., Pollack, J. B., and McKay, C. P. 1987. Carbon in the outer solar system. *Bull. Amer. Astron. Soc.* 19:818 (abstract).

Siscoe, G. L., and Summer, D. 1981. Centrifugally driven diffusion of Iogenic plasma. *J. Geophys. Res.* 96:8471–8479.

Sittler, E. C., Ogilvie, K., and Scudder, J. 1983. Survey of low energy plasma electrons in Saturn's magnetosphere: Voyagers 1 and 2. *J. Geophys. Res.* 88:8847–8870.

Sleep, N. H. 1979. Thermal history and degassing of the Earth: Some simple calculations. *J. Geol.* 87:671–686.

Sleep, N. H., and Wolery, T. J. 1978. Thermal and chemical constraints on venting of hydrothermal fluids at mid-ocean ridge crests. *J. Geophys. Res.* 83:5913–5922.

Smith, B. A. 1984. Near-infrared imaging of Uranus and Neptune. In *Uranus and Neptune,* ed. J. T. Bergstralh, NASA CP-2330, pp. 213–223.

Smith, B. A., and Smith, S. A. 1972. Upper limits for an atmosphere on Io. *Icarus* 17:218–222.

Smith, B. A., and Terrile, R. J. 1984. A circumstellar disk around β Pictoris. *Science* 226:1421–1424.

Smith, B. A., Soderblom, L. A., Johnson, T. V., Ingersoll, A. P., Collins, S. A., Shoemaker, E. M., Hunt, G. E., Masursky, H., Carr, M. H., Davies, M. E., Cook, A. R. II, Boyce, J., Danielson, G. E., Owen, T., Sagan, C., Beebe, R. F., Veverka, J., Strom, R. G., McCauley, J. F., Morrison, D., Briggs, G. A., and Suomi, V. E. 1979. The Jupiter system through the eyes of Voyager 1. *Science* 204:951–972.

Smith, B. A., Soderblom, L., Batson, R., Bridges, P., Inge, J., Masursky, H., Shoemaker, E., Beebe, R., Boyce, J., Briggs, G., Bunker, A., Collins, S. A., Hansen, C. J., Johnson, T. V., Mitchell, J. L., Terrile, R. J., Cook, A. F., Cuzzi, J., Pollack, J. B., Danielson, G. E., Ingersoll, A. P., Davies, M. E., Hunt, G. E., Morrison, D., Owen, T., Sagan, C., Veverka, J., Strom, R., and Suomi, V. E. 1982. A new look at the Saturn system. *Science* 215:504–537.

Smith, B. A., Soderblom, L. A., Beebe, R., Bliss, D., Boyce, J. M., Brahic, A., Briggs, G. A., Brown, R. H., Collins, S. A., Cook, A. F. II, Croft, S. K., Cuzzi, J. N., Danielson, G. E., Davies, M. E., Dowling, T. E., Godfrey, D., Hansen, C. J., Harris, C., Hunt, G. E., Ingersoll, A. P., Johnson, T. V., Krauss, R. J., Masursky, H., Morrison, D., Owen, T., Plescia, J., Stoker, C., Strom, R. G., Suomi, V. E., Synnott, S. P., Terrile, R. J., Thomas, P., Thompson, W. R., and Veverka, J. 1986. Voyager 2 in the Uranian system: Imaging science results. *Science* 233:43–64.

Smith, D., Adams, N. G., and Alge, E. 1982. Some H/D exchange reactions involved in the deuteration of interstellar molecules. *Astrophys. J.* 263:123–129.

Smith, J. V. 1982. Hetergeneous growth of meteorites and planets, especially the Earth and Moon. *J. Geol.* 90:1–48.

Smith, L. L., and Gross, S. H. 1972. The evolution of water vapor in the atmosphere of Venus. *J. Atmos. Sci.* 29:173–178.

Smith, R. A., and Strobel, D. F. 1985. Energy partitioning in the Io plasma torus. *J. Geophys. Res.* 90:9469–9493.

Smith, R. G., Shemansky, D. E., Holberg, J. B., Broadfoot, A. L., Sandel, B. R., and McConnell, J. C. 1983. Saturn's upper atmosphere from Voyager 2 EUV solar and stellar occultations. *J. Geophys. Res.* 88:8667–8678.

Smith, R. G., Sellgren, K., and Tokunaga, A. T. 1986. Moderate spectral resolution observations of 3 micron absorption features in highly obscured objects. In *Summer School on Interstellar Processes,* eds. D. J. Hollenbach and H. A. Thronson, NASA TM-88342, pp. 127–128.

Smith, S. P., and Reynolds, J. H. 1981. Excess ^{129}Xe in a terrestrial sample as measured in a pristine system. *Earth Planet. Sci. Lett.* 54:236–238.

Smoluchowski, R. 1967. Internal structure and energy emission of Jupiter. *Nature* 215:691–695.

Smoluchowski, R. 1973. Dynamics of the Jovian interior. *Astrophys. J.* 185:L95–L99.

Smyth, W. H. 1983. Io's sodium cloud: Explanation of the east-west asymmetries. II. *Astrophys. J.* 234:1148–1153.

Smyth, W. H. 1986. Nature and variability of Mercury's sodium atmosphere. *Nature* 323:696–699.

Smyth, W. H., and Shemansky, D. 1983. Escape and ionization of atomic oxygen from Io. *Astrophys. J.* 271:865–875.

Smythe, W. D., Nelson, R. M., and Nash, D. B. 1979. Spectral evidence for SO_2 frost or adsorbate on Io's surface. *Nature* 280:766–767.

Snell, R. L. 1981. A study of nine interstellar dark clouds. *Astrophys. J. Suppl.* 45:121–175.
Snow, T. P., and Joseph, C. L. 1985. A study of depletions within the Rho Ophiuchi cloud based on IUE observations of HD 147889. *Astrophys. J.* 288:277–283.
Soifer, B. T., Puetter, R. C., Russell, R. W., Willner, S. P., Harvey, P. M., and Gillett, F. C. 1979. The 4-8 micron spectrum of the infrared source W33A. *Astrophys. J.* 232:L53–L57.
Soifer, B. T., Willner, S. P., Capps, R. W., and Rudy, R. J. 1981. 4-8 micron spectrophotometry of OH 0739-14. *Astrophys. J.* 250:631–635.
Šolc, M., Vanýsek, V., and Kissel, J. 1986. Carbon stable isotopes in comets after encounters with P/Halley. In *20th ESLAB Symposium on the Exploration of Halley's Comet,* vol. 1, eds. B. Battrick, E. J. Rolfe and R. Reinhard, ESA SP-250, pp. 373–376.
Šolc, M., Jessberger, E. K., Hsiung, P., and Kissel, J. 1987. Halley dust composition. In *Proc. 10th Regional Meeting of the IAU,* eds. Z. Ceplecha and P. Pecina (Prague: Astron. Inst. Czech. Acad. Sci.), pp. 47–50.
Solomon, S. C., and Head, J. W. 1984. Venus banded terrain: Tectonic models for band formation and their relationship to lithospheric thermal structures. *J. Geophys. Res.* 89:6885–6897.
Sonett, C. P., Colburn, D. S., and Schwartz, K. 1968. Electrical heating of meteorite parent bodies and planets by dynamo induction from a pre-main sequence T Tauri "solar wind." *Nature* 219:924–926.
Sonett, C. P., Colburn, D. S., Schwartz, K., and Kiel, K. 1970. The melting of asteroidal-sized bodies by unipolar dynamo induction from a primordial T Tauri sun. *Astrophys. Space Sci.* 7:446–488.
Sonett, C. P., Colburn, D. S., and Schwartz, K. 1975. Formation of the lunar crust: An electrical source of heating. *Icarus* 24:231–255.
Sparks, R. S. J. 1978. The dynamics of bubble formation and growth in magmas: A review and analysis. *J. Vol. Geotherm. Res.* 3:1–37.
Spencer, J. R. 1987. Icy Galilean satellites reflectance spectra: Less ice on Ganymede and Callisto? *Icarus* 70:99–110.
Spinrad, H. 1987. Comets and their composition. *Ann. Rev. Astron. Astrophys.* 25:231.
Spitzer, L., Jr. 1952. The terrestrial atmosphere above 300 km. In *The Atmospheres of the Earth and Planets,* ed. G. P. Kuiper (Chicago: Univ. of Chicago Press), pp. 211–247.
Spitzer, L. 1985. Average density along interstellar lines of sight. *Astrophys. J.* 290:L21–L24.
Squyres, S. W. 1980. Surface temperature and retention of H_2O frost on Ganymede and Callisto. *Icarus* 4:502–510.
Squyres, S. W. 1984. The history of water on Mars. *Ann. Rev. Earth Planet. Sci.* 12:83–106.
Squyres, S. W., and Carr, M. H. 1986. Geomorphic evidence for the distribution of ground ice on Mars. *Science* 231:249–252.
Squyres, S. W., Reynolds, R. T., Cassen, P. M., and Peale, S. J. 1983*a*. Liquid water and active resurfacing on Europa. *Nature* 301:225–226.
Squyres, S. W., Reynolds, R. T., Cassen, P. M., and Peale, S. J. 1983*b*. The evolution of Enceladus. *Icarus* 53:319–331.
Squyres, S. W., Reynolds, R. T., Summers, A. L., and Shung, F. 1988. Accretional heating of the satellites of Saturn and Uranus. *J. Geophys. Res.* 93:8779–8794.
Srinivasan, B., Gros, J., and Anders, E. 1977. Noble gases in separated meteoritic minerals: Murchison (C2), Ornans (C3), Karoonda (C5), and Abee (E4). *J. Geophys. Res.* 82:762–778.
Stacey, F. D. 1977. *Physics of the Earth,* 2nd ed. (New York: Wiley).
Stahler, S. W. 1983*a*. The birthline for low-mass stars. *Astrophys. J.* 274:822–829.
Stahler, S. W. 1983*b*. The equilibria of rotating, isothermal clouds. II. Structure and dynamical stability. *Astrophys. J.* 268:165–184.
Stahler, S. W. 1984. The cyanopolyynes as a chemical clock for molecular clouds. *Astrophys. J.* 281:209–218.
Stahler, S. W., Shu, F. H., and Taam, R. E. 1981. The evolution of protostars III. The accretion envelope. *Astrophys. J.* 298:727–737.
Staudacher, T. 1987. Upper mantle origin for Harding County well gases. *Nature* 352:605–607.
Staudacher, T., and Allègre, C. J. 1982. Terrestrial xenology. *Earth Planet. Sci. Lett.* 60:389–406.
Steel, T. M., and Duley, W. W. 1987. A 217.5 nm absorption feature in the spectrum of small silicate particles. *Astrophys. J.* 315:337–339.
Stern, S. A. 1986. The effects of mechanical interaction between the interstellar medium and comets. *Icarus* 68:276–283.

Stern, S. A. 1987. Extra-solar Oort cloud encounters and planetary impact rates. *Icarus* 69:185–188.
Stern, S. A., and Shull, J. M. 1988. The thermal evolution of comets in the Oort cloud by stars and supernovae. *Nature* 332:407–411.
Stern, S. A., Trafton, L. M., and Gladstone, G. R. 1988. Why is Pluto bright? Implications of the albedo and lightcurve behavior of Pluto. *Icarus* 75:485–498.
Stevenson, D. J. 1975. Thermodynamics and phase separation of dense, fully ionized hydrogen-helium fluid mixtures. *Phys. Rev.* B12:3999–4007.
Stevenson, D. J. 1979. Solubility of helium in metallic hydrogen. *J. Phys. F. (Metal Phys.)* 9:791–801.
Stevenson, D. J. 1980. Saturn's luminosity and magnetism. *Science* 208:746–748.
Stevenson, D. J. 1982*a*. Formation of the giant planets. *Planet. Space Sci.* 30:755–764.
Stevenson, D. J. 1982*b*. Interiors of the giant planets. *Ann. Rev. Earth Planet. Sci.* 10:257–295.
Stevenson, D. J. 1982*c*. Volcanism and igneous processes in small icy satellites. *Nature* 298:142–144.
Stevenson, D. J. 1983. Structure of the giant planets: Evidence for nucleated instabilities and post-formational accretion. *Lunar Planet. Sci.* XIV:770 (abstract).
Stevenson, D. J. 1984*a*. On forming giant planets quickly (superganymedean puffballs!). *Lunar Planet. Sci.* XV:822–823 (abstract).
Stevenson, D. J. 1984*b*. Composition, structure, and evolution of Uranian and Neptunian satellites. In *Uranus and Neptune,* NASA CP-2330, pp. 405–423.
Stevenson, D. J. 1985*a*. Cosmochemistry and structure of the giant planets and their satellites. *Icarus* 62:4–15.
Stevenson, D. J. 1985*b*. Partition of noble gases at extreme pressures within planets. *Lunar Planet. Sci.* XVI:821–822 (abstract).
Stevenson, D. J. 1987*a*. Origin of the Moon: The collisional hypothesis. *Ann. Rev. Earth Planet. Sci.* 15:271–315.
Stevenson, D. J. 1987*b*. Chemical processing of the solar nebula by leaky protoplanets (Where did all the CO go?). Origin and Evolution of Planetary and Satellite Atmospheres, Abstract Booklet, 10-14 March, Tucson, AZ.
Stevenson, D. J. 1987*c*. Uranus. *Bull. Amer. Astron. Soc.* 19:851 (abstract).
Stevenson, D. J., and Lunine, J. I. 1986. Mobilization of cryogenic ice in outer solar system satellites. *Nature* 323:46–48.
Stevenson, D. J., and Lunine, J. I. 1988. Rapid formation of Jupiter by diffusive redistribution of water vapor in the solar nebula. *Icarus* 75:146–155.
Stevenson, D. J., and Potter, B. E. 1986. Titan's latitudinal temperature distribution and seasonal cycle. *Geophys. Res. Lett.* 13:93–96.
Stevenson, D. J., and Salpeter, E. E. 1976. Interior models of Jupiter. In *Jupiter,* ed. T. Gehrels (Tucson: Univ. of Arizona Press), pp. 85–112.
Stevenson, D. J., and Salpeter, E. E. 1977*a*. The dynamics and helium distribution properties for hydrogen-helium fluid planets. *Astrophys. J. Suppl.* 35:239–261.
Stevenson, D. J., and Salpeter, E. E. 1977*b*. The phase diagram and transport properties of hydrogen-helium fluid planets. *Astrophys. J. Suppl.* 35:221–237.
Stevenson, D. J., Harris, A. W., and Lunine, J. I. 1986. Origins of satellites. In *Satellites,* eds. J. A. Burns and M. S. Matthews (Tucson: Univ. of Arizona Press), pp. 39–88.
Stier, M. T., Traub, W. A., Fazio, G. G., Wright, E. L., and Low, F. J. 1978. Far infrared observations of Uranus, Neptune and Ceres. *Astrophys. J.* 226:347–349.
Stone, E. C., and Miner, E. D. 1986. The Voyager 2 encounter with the Uranian system. *Science* 233:1–64.
Strazzula, G. L. 1985. Modification of grains by particle bombardment in the early solar system. *Icarus* 61:48–56.
Strazzulla, G. L. 1986*a*. Organic material from Phoebe to Iapetus. *Icarus* 66:397–400.
Strazzulla, G. L. 1986*b*. "Primitive" galactic dust in the solar system? *Icarus* 67:63–70.
Strazzulla, G. L., Calcagno, L., and Foti. G. 1984. Build-up of carbonaceous material by fast protons on Pluto and Triton. *Astron. Astrophys.* 140:441–444.
Strazzulla, G. L., Calcagno, L, and Foti, G. 1985*a*. Dark materials by fast protons of frozen methane: II. Astrophysical implications. *Il Nuovo Cimento* 8:63–75.
Strazzulla, G. L., Calcagno, L., Foti, G., and Sheng, K. L. 1985*b*. Interaction between solar

energetic particles and interplanetary grains. In *Ices in the Solar System,* eds. J. Klinger, D. Benest, A. Dollfus and R. Smoluchowski (Dordrecht: D. Reidel), pp. 273–285.

Strens, M. R., and Cann, J. R. 1982. A model of hydrothermal circulation in fault zones at mid-oceanic ridge crests. *Geophys. J. Roy. Astron. Soc.* 71:225–240.

Strobel, D. F. 1982. Chemistry and evolution of Titan's atmosphere. *Planet. Space Sci.* 30:839–848.

Strobel, D. F., and Shemansky, D. E. 1982. EUV emission from Titan's upper atmosphere: Voyager 1 encounter. *J. Geophys. Res.* 87:1361–1368.

Strom, R. G. 1987. The solar system cratering record: Voyager 2 results at Uranus and implications for the origin of impacting objects. *Icarus* 70:517–535.

Strom, R. G., Trask, N. J., and Guest, J. E. 1975. Tectonism and volcanism on Mercury. *J. Geophys. Res.* 80:2478–2507.

Studier, M. H., Hayatsu, R., and Anders, E. 1968. Origin of organic matter in early solar system I. Hydrocarbons. *Geochim. Cosmochim. Acta* 32:151–173.

Suess, H. E. 1949. Die Haufigkeit der Edelgase auf der Erde und in Kosmos. *J. Geol.* 57:600–607.

Sullivan, J., and Siscoe, G. L. 1982. In-situ observations of Io torus plasma. In *Satellites of Jupiter,* ed. D. Morrison (Tucson: Univ. of Arizona Press), pp. 846–871.

Sun, S.-S. 1982. Chemical composition and origin of the Earth's primitive mantle. *Geochim. Cosmochim. Acta* 46:179–192.

Sutton, E. C., Blake, G., Masson, C., and Phillips, T. G. 1985. Molecular line survey of Orion A from 215 to 247 GHz. *Astrophys. J. Suppl.* 58:341.

Suzuki, H., Ohishi, M., Kaifu, N., Ishikawa, S., and Kasuga, T. 1986. Detection of the interstellar C_6H radical. *Publ. Astron. Soc. Japan* 38:911–917.

Swartz, W. E., Reed, R. W., and McDonough, T. 1975. Photoelectron escape from the ionosphere of Jupiter. *J. Geophys. Res.* 88:495–500.

Swindle, T. D., Caffee, M. W., and Hohenberg, C. M. 1986. Xenon and other noble gases in shergottites. *Geochim. Cosmochim. Acta* 50:1001–1015.

Sykes, M. V., Cutri, R. M., Lebofsky, L. A., and Binzel, R. P. 1987. Observations of Pluto and Charon by IRAS. *Science* 237:1336–1340.

Takaoka, N. 1972. An interpretation of general anomalies of xenon and the isotopic composition of primitive xenon. *Mass Spectrometry* 20:287–302.

Talbot, R. J., and Newman, M. J. 1977. Encounters between stars and dense interstellar clouds. *Astrophys. J. Suppl.* 34:295–308.

Taylor, H. P., Jr. 1983. Oxygen and hydrogen isotope studies of hydrothermal interactions at submarine and subaerial spreading centers. In *Hydrothermal Processes at Seafloor Spreading Centers,* eds. P. A. Rona, K. Bostrom, L. Laubier and K. L. Smith (New York: Plenum Press), pp. 83–139.

Taylor, R. C. 1979. Pole orientations of asteroids. In *Asteroids,* ed. T. Gehrels (Tucson: Univ. of Arizona Press), pp. 480–493.

Tedesco, E. F. 1984. Asteroid compositional "rings": Clues to the compositions of primordial planetesimals in the middle solar system. In *Planetary Rings,* ed. A. Brahic (Toulouse: Cepadues-Editions), pp. 703–712.

Tedesco, E. F., and Gradie, J. 1987. Discovery of M class objects among the near-Earth asteroid populations. *Astron. J.* 93:738–746.

Tedesco, E. F., Veeder, G. J., Dunbar, R. S., and Lebofsky, L. A. 1987. IRAS constraints on the sizes of Pluto and Charon. *Nature* 327:127–129.

Terebey, S., Shu, F. H., and Cassen, P. 1984. The collapse of the cores of slowly rotating isothermal clouds. *Astrophys. J.* 286:529–551.

Terrile, R. J., and Westphal, J. A. 1977. The vertical cloud structure of Jupiter from 5 μm measurements. *Icarus* 30:274–281.

Tholen, D. J. 1984. Asteroid Taxonomy from Cluster Analysis of Photometry. Ph.D. Thesis, Univ. of Arizona.

Tholen, D. J., and Buie, M. W. 1987. Pluto and Charon: Radii, density and orbital elements from mutual event photometry through 1987. *Bull. Amer. Astron. Soc.* 19:859–860 (abstract).

Tholen, D. J., Buie, M. W., Binzel, R. P., and Frueh, M. L. 1987. Improved physical parameters for the Pluto-Charon system. *Science* 237:512–514.

Thomas, G. E. 1978. The interstellar wind and its influence on the interplanetary environment. *Ann. Rev. Earth Planet. Sci.* 6:173–204.

Thomas, N. 1987. Condensation and sublimation of Io—I. *Mon. Not. Roy. Astron. Soc.* 226:195–207.
Thomas, P., Veverka, J., and Dermott, S. 1986. Small satellites. In *Satellites,* eds. J. A. Burns and M. S. Matthews (Tucson: Univ. of Arizona Press), pp. 802–835.
Thompson, W. R. 1985. Phase equilibria in N_2-hydrocarbon systems: Applications to Titan. *The Atmospheres of Saturn and Titan: Proc. Intl. Workshop,* Alpbach, Austria, ESA SP-241, pp. 109–119.
Thompson, W. R., and Sagan, C. 1984. Titan: Far infrared and microwave remote sensing of methane clouds and organic haze. *Icarus* 60:236–259.
Thompson, W. R., Reid, B., Murray, B., Khare, B., and Sagan, C. 1987. Coloration and darkening of methane clathrate and other ices by charged particle irradiation: Applications to the outer solar system. *J. Geophys. Res.* 92:14933–14948.
Thomsen, L. 1981. ^{129}Xe on outgassing of the atmosphere. *J. Geophys. Res.* 85:4374–4378.
Thorne, L. R., Anicich, V. G., Prasad, S. S., and Huntress, W. T. 1984. The chemistry of phosphorus in dense interstellar clouds. *Astrophys. J.* 280:139–143.
Tielens, A. G. G. M. 1983. Surface chemistry of deuterated molecules. *Astron. Astrophys.* 119:177–184.
Tielens, A. G. G. M., and Allamandola, L. J. 1987*a*. Composition, structure, and chemistry of interstellar dust. In *Interstellar Processes,* eds. D. J. Hollenbach and H. A. Thronson (Dordrecht: D. Reidel), pp. 397–470.
Tielens, A. G. G. M., and Allamandola, L. J. 1987*b*. Evolution of interstellar dust. In *Physical Processes in Interstellar Clouds,* eds. G. E. Morfill and M. Scholer (Dordrecht: D. Reidel), pp. 333–376.
Tinsley, B. A. 1974. Hydrogen in the upper atmosphere. *Fund. Cosmic Phys.* 1:201–300.
Töksoz, M. N., Hsui, A. T., and Johnston, D. H. 1978. Thermal evolutions of the terrestrial planets. *Moon and Planets* 18:281–321.
Tokunaga, A. T., Orton, G. S., and Caldwell, J. 1983. New observational constraints on the temperature inversions of Uranus and Neptune. *Icarus* 53:141–146.
Tolk, N. H., Traum, M. M., Tully, J. C., and Madey, T. E., eds. 1983. *Desorption Induced by Electronic Transitions, DIET I,* vol. 24, *Springer Series in Chemical Physics* (Berlin: Springer-Verlag).
Tolstikhin, I. N. 1978. A review: Some recent advances in isotope geochemistry of light rare gases. In *Terrestrial Rare Gases* eds. E. C. Alexander, Jr., and M. Ozima (Tokyo: Japan Scientific Society Press), pp. 33–62.
Tomasi, C. 1979*a*. Non-selective absorption by atmospheric water vapor at visible and near infrared wavelengths. *Quart. J. Roy. Meteor. Soc.* 105:1027–1040.
Tomasi, C. 1979*b*. Weak absorption by atmospheric water vapor in the visible and near-infrared spectral region. *Il Nuovo Cimento* 2:511–526.
Tomasko, M. G., West, R. A., and Castillo, N. D. 1978. Photometry and polarimetry of Jupiter at large phase angles. *Icarus* 33:558–592.
Tomeoka, K., and Buseck, P. R. 1985. Indicators of aqueous alteration in CM carbonaceous chondrites: Microtextures of a layered mineral containing Fe, S, O, and Ni. *Geochim. Cosmochim. Acta* 49:2149–2163.
Toon, O. B., Pollack, J. B., Ward, W., Burns, J. A., and Bilski, K. 1980. The astronomical theory of climatic change on Mars. *Icarus* 44:552–607.
Toon, O. B., Pollack, J. B., Ackerman, T. P., Turco, R. P., McKay, C. P., and Kiu, M. S. 1982. Evolution of an impact generated dust cloud and the effects on the atmosphere. In *Large Body Impacts and Terrestrial Evolution: Geological, Climatological, and Biological Implications,* eds. L. T. Silver and P. H. Schultz, Geol. Soc. Amer. SP 190 (Boulder: Geol. Soc. of Amer.), pp. 187–200.
Toon, O. B., McKay, C. B., Courtin, R., and Ackerman, T. P. 1988. Methane rain on Titan. *Icarus* 75:255–284.
Torbett, M. V. 1984. Hydrodynamic ejection of bipolar flows from objects undergoing disk accretion: T Tauri stars, massive pre-main-sequence objects, and cataclysmic variables. *Astrophys. J.* 278:318–325.
Torbett, M. V., and Smoluchowski, R. 1979. The structure and magnetic field of Uranus. *Geophys. Res. Lett.* 6:675–676.
Torrisi, L., Coffa, S., Foti, G., and Strazzulla, G. 1986. Sulfur erosion of 1.0 MeV helium ions. *Rad. Eff.* 100:61–69.

Tozer, D. C. 1967. Towards a theory of thermal convection in the mantle. In *The Earth's Mantle*, ed. T. F. Gaskell (London: Academic Press), pp. 305–353.
Trafton, L. M. 1967. Model atmospheres of the major planets. *Astrophys. J.* 147:765–781.
Trafton, L. M. 1975. Detection of a potassium cloud near Io. *Nature* 258:690–692.
Trafton, L. M. 1980. Does Pluto have a substantial atmosphere? *Icarus* 44:53–61.
Trafton, L. M. 1981. The atmospheres of the outer planets and satellites. *Rev. Geophys. Space Phys.* 19:43–89.
Trafton, L. M. 1984. Large seasonal variations in Triton's atmosphere. *Icarus* 58:312–324.
Trafton, L. M., and Stern, S. A. 1983. On the global distribution of Pluto's atmosphere. *Astrophys. J.* 267:872–881.
Trafton, L. M., and Wildey, C. 1974. Jupiter: A comment on the 8 to 14 micron limb darkening. *Astrophys. J.* 194:499–502.
Trafton, L. M., Parkinson, T., and Macy, W., Jr. 1974. The spacial extent of sodium emission around Io. *Astrophys. J.* 190:L85–L89.
Trafton, L. M., Stern, S. A., and Gladstone, G. R. 1988. The Pluto-Charon system: The escape of Charon's primordial atmosphere. *Icarus* 74:108–120.
Trauger, J. 1985. Doppler-resolved spectral imagery of Io sodium phenomena. *Bull. Amer. Astron. Soc.* 17:694 (abstract).
Trauger, J., Roesler, F., and Michelson, M. 1977. The D/H ratio on Jupiter, Saturn, and Uranus based on new HD and H_2 data. *Bull. Amer. Astron. Soc.* 9:516 (abstract).
Treffers, R., and Cohen, M. 1974. High-resolution spectra of cool stars in the 10- and 20-micron regions. *Astrophys. J.* 188:545–552.
Troutman, W. W., and Davis, C. W. 1965. The two-dimensional behavior of the shocks in the atmosphere. *Air Force Weapons Lab. Report,* AFWL-TR-65-151.
Tscharnuter, W. M. 1978. Collapse of the presolar nebula. *Moon and Planets* 19:229–236.
Tscharnuter, W. M. 1987. Models of star formation. In *Physical Processes in Comets, Stars, and Active Galaxies (Lecture Notes in Physics),* eds. E. Meyer-Hofmeister, H.-C. Thomas and W. Hillebrandt (Berlin: Springer-Verlag).
Tullis, T. 1979. High temperature deformation of rocks and minerals. *Rev. Geophys. Space Phys.* 17:1137–1154.
Turcotte, D. L., and Schubert, G. 1988. Tectonic implications of radiogenic noble gases in planetary atmospheres. *Icarus* 74:36–46.
Turekian, K. K., and Clark, S. P., Jr. 1969. Inhomogenous accumulation model of the Earth from the primitive solar nebula. *Earth Planet. Sci. Lett.* 6:346–348.
Turekian, K. K., and Clark, S. P., Jr. 1975. The non-homogeneous accumulation model for terrestrial planet formation and the consequences for the atmosphere of Venus. *J. Atmos. Sci.* 32:1257–1261.
Turner, B. E., and Bally, J. 1987. Detection of interstellar PN: The first identified phosphorus compound in the interstellar medium. *Astrophys. J.* 321:L75–L80.
Turner, B. E., and Rickard, L. J. 1987. C_3H_2 in infrared cirrus. *NRAO Quart. Rept.*, Jan.-March 1987, pp. 20–21.
Tyburczy, J. A., and Ahrens, T. J. 1986. Dynamic compression and volatile release of carbonates. *J. Geophys. Res.* 91:4730–4744.
Tyburczy, J. A., Frisch, B., and Ahrens, T. J. 1986. Shock-induced volatile loss from a carbonaceous chondrite: Implications for planetary accretion. *Earth Planet. Sci. Lett.* 80:201–207.
Tyburczy, J. A., Krishnamurthy, R. V., Epstein, S., and Ahrens, T. J. 1987. Impact induced devolatilization and hydrogen isotopic fractionation of serpentine; possible implication for planetary accretion. *Earth and Planet. Sci. Lett.*, in press.
Tyler, G. L., Eshleman, V. R., Anderson, J. D., Levy, G. S., Lindal, G. F., Wood, G. E., and Croft, T. A. 1982. Radio science with Voyager 2 at Saturn: Atmosphere and ionosphere and the masses of Mimas, Tethys, and Iapetus. *Science* 215:553–558.
Unni, C. K., and Schilling, J. G. 1978. Cl and Br degassing by volcanism along the Reykjanes Ridge and Iceland. *Nature* 272:5648–5652.
Urey, H. C. 1947. The thermodynamic properties of isotopic substances. *J. Chem. Soc.* 562–581.
Urey, H. C. 1952. *The Planets, Their Origin and Development* (New Haven, Conn.: Yale Univ. Press).

Urey, H. C. 1953. Chemical evidence regarding the Earth's origin. In *Intl. Conf. Pure and Applied Chem. and Plenary Lectures* 13:188–217.
Urey, H. C. 1959*a*. The atmospheres of the planets. *Handbuch der Physik*, ed. S. Flugge (Berlin: Springer-Verlag), pp. 363–418.
Urey, H. C. 1959*b*. Primary and secondary objects. *J. Geophys. Res.* 64:1721–1737.
Vanajakshi, T. C., and Jenkins, A. W., Jr. 1985. Effect of turbulent viscosity on the isothermal collapse of a rotating protostellar cloud. *Astrophys. J.* 294:502–512.
van Dishoeck, E. F., and Black, J. H. 1987. The abundance of interstellar CO. In *Physical Processes in Interstellar Clouds*, ed. M. Scholer (Dordrecht: D. Reidel), in press.
Vannice, M. A. 1982. Catalytic activation of carbon monoxide on metal surfaces. In *Catalysis Science and Technology*, eds. J. R. Anderson and M. Boudart (Berlin: Springer-Verlag), pp. 139–198.
Vanýsek, V. 1987. Isotopic abundances in comets. In *Astrochemistry*, eds. M. S. Vardya and S. P. Tarafdar (Dordrecht: D. Reidel), pp. 461–467.
Vanýsek, V., and Rahe, J. 1978. The $^{12}C/^{13}C$ isotope ratio in comets, stars and interstellar matter. *Moon and Planets* 18:441–446.
Vardavas, I. M., and Carver, J. H. 1985. Atmospheric temperature response to variations in CO_2 concentration and the solar-constant. *Planet. Space Sci.* 33:1187–1207.
Veeder, G. J. 1986. The IRAS asteroid data. In *IRAS Asteroid and Comet Survey*, Preprint Version No. 1, ed. D. L. Matson, 2:1–2:54, JPL Internal Document D-3698.
Veizer, J. 1983. Geologic evolution of the Archean-early Proterozoic Earth. In *The Earth's Earliest Biosphere: Its Origin and Evolution*, ed. J. W. Schopf (Princeton: Princeton Univ. Press), pp. 240–259.
Veizer, J., Holser, W. T., and Wilgus, C. K. 1980. Correlation of $^{13}C^{12}C$ and $^{34}S/^{32}S$ secular variations. *Geochim. Cosmochim. Acta* 44:579–587.
Veverka, J., and Thomas, P. 1979. Phobos and Deimos: A preview of what asteroids are like? In *Asteroids*, ed. T. Gehrels (Tucson: Univ. of Arizona Press), pp. 628–651.
Veverka, J., Thomas, P., Johnson, T. V., Matson, D., and Housen, K. 1986. The physical characteristics of satellite surfaces. In *Satellites*, eds. J. A. Burns and M. S. Matthews (Tucson: Univ. of Arizona Press), pp. 342–402.
Veverka, J., Thomas, P., Helfenstein, D., Brown, R. H., and Johnson, T. V. 1987. Satellites of Uranus: Disk integrated photometry from Voyager imaging. *J. Geophys. Res.* 92:14895–14904.
Vidal-Madjar, A. 1987. Deuterium formation and abundance. In *Space Astronomy and Solar System Exploration*, ESA SP-268, pp. 73–84.
Vidal-Madjar, A., Laurent, C., Gry, C., Bruston, P., Ferlet, R., and York, D. G. 1983. The ratio of deuterium to hydrogen in interstellar space. *Astron. Astrophys.* 120:58–62.
Vilas, F., and Smith, B. A. 1985. Reflectance spectrophotometry (0.5-1.0 μm) of outer belt asteroids: Implications for primitive, organic solar system material. *Icarus* 64:503–516.
Vinogradov, A. P., Surkov, Y. A., and Kirnozov, F. F. 1973. The content of uranium, thorium and potassium in the rocks of Venus as measured by Venera. *Icarus* 20:253–259.
Vityazev, A. V., Pechernikova, A. V., and Safronov, V. S. 1978. Limiting masses, distances and times for the accumulation of the planets of the terrestrial group. *Sov. Astron. AJ* 22:60–63.
Vizgirda, J., and Ahrens, T. J. 1982. Shock compression of aragonite and implications for the equation of state of carbonates. *J. Geophys. Res.* 87:4747–4758.
Vogt, R. E., Cook, W. R., Cummings, A. C., Garrard, T. L., Gehrels, N., Stone, E. C., Trainor, J. H., Schardt, A. W., Conion, T., Lal, N., and McDonald, F. B. 1979. Voyager 1: Energetic ions and electrons in the jovian magnetosphere. *Science* 204:1003–1006.
Völk, H. J., Jones, F. C., Morfill, G. E., and Rosen, S. 1980. Collisions between grains in a turbulent gas. *Astron. Astrophys.* 85:316–325.
von Damm, K. L., Grant, B., and Edmond, J. M. 1983. Preliminary report on the chemistry of hydrothermal solution at 21°N, East Pacific Rise. In *Hydrothermal Processes at Seafloor Spreading Centers*, eds. P. A. Rona, K. Bostrom, L. Laubier and K. L. Smith (New York: Plenum Press), pp. 369–390.
Von Damm, K. L., Edmond, J. M., Grant, B., Measures, C. I., Walden, B., and Weiss, R. F. 1985. Chemistry of submarine hydrothermal solution at 21°N, East Pacific Rise. *Geochim. Cosmochim. Acta* 49:2197–2220.

von Weizssäcker, C. F. 1948. Die Rotation Kosmischyer Gasmassen. *Z. Naturforsch.* 3a:524–539.

von Zahn, U., and Moroz, V. I. 1984. Composition of the Venus atmosphere below 100 km altitude. Paper presented at the 25th Plenary Meeting of COSPAR, June 28–30, Graz, Austria.

von Zahn, U., Kumar, S., Niemann, H., and Prinn, R. 1983. Composition of the Venus atmosphere. In *Venus,* eds. D. M. Hunten, L. Colin, T. M. Donahue and V. I. Moroz (Tucson: Univ. of Arizona Press), pp. 299–430.

Wacker, J. F., and Anders, E. 1984. Trapping of xenon in ice and implications for the origin of the Earth's noble gases. *Geochim. Cosmochim. Acta* 48:2372–2380.

Wacker, J. R., and Anders, E. 1986. Trapping of noble gases by carbon: Applications to meteorites and planets. In *Abstracts for Japan-U.S. Seminar on Terrestrial Rare Gases,* Yellowstone Natl. Park, September, pp. 89–92 (abstract).

Wagman, D. D. 1979. *Sublimation Pressure and Enthalpy of* SO_2 (Washington, D.C.: Thermodynamics Data Center, Natl. Bur. Standards).

Walker, J. C. G. 1975. Evolution of the atmosphere of Venus. *J. Atmos. Sci.* 32:1248–1256.

Walker, J. C. G. 1977. *Evolution of the Atmosphere* (New York: Macmillan).

Walker, J. C. G. 1982*a*. Climatic factors on the Archean Earth. *Paleogeogr. Paleoclimat. Paleoecol.* 40:1–11.

Walker, J. C. G. 1982*b*. The earliest atmosphere of the Earth. *Precambrian Res.* 17:147–171.

Walker, J. C. G. 1986*a*. Carbon dioxide on the early Earth. *Origins of Life* 16:117–127.

Walker, J. C. G. 1986*b*. Impact erosion of planetary atmospheres. *Icarus* 68:87–98.

Walker, J. C. G., Turekian, K. K., and Hunten, D. M. 1970. An estimate of the present-day deep-mantle degassing rate from data on the atmosphere of Venus. *J. Geophys. Res.* 75:3558–3561.

Walker, J. C. G., Hays, P. B., and Kasting, J. F. 1981. A negative feedback mechanism for the long-term stabilization of Earth's surface temperature. *J. Geophys. Res.* 86:9776–9782.

Walker, J. C. G., Klein, C., Schidlowski, M., Stevenson, D. J., and Walter, M. R. 1983. Environmental evolution of the Archean-Early Proterozoic Earth. In *Earth's Earliest Biosphere: Its Origin and Evolution,* ed. J. W. Schopf (Princeton: Princeton Univ. Press), pp. 260–290.

Wallace, D., and Sagan, C. 1979. Evaporation of ice in planetary atmospheres: Ice-covered rivers on Mars. *Icarus* 39:385–400.

Wallace, L. 1980. The structure of the Uranus atmosphere. *Icarus* 43:231–259.

Wallace, L. 1984. The seasonal variation of the thermal structure of the atmosphere of Neptune. *Icarus* 59:367–375.

Walmsley, C. M., Hermsen, W., Henkel, C., Mauersberger, R., and Wilson, T. L. 1987. Deuterated ammonia in the Orion hot core. *Astron. Astrophys.* 172:311–315.

Walter, F. M. 1986. X-ray sources in regions of star formation. I. The naked T Tauri stars. *Astrophys. J.* 306:573–586.

Wamsteker, W., Kroes, R. L., and Fountain, J. A. 1974. On the surface composition of Io. *Icarus* 10:1–7.

Wänke, H. 1981. Constitution of terrestrial planets. *Phil. Trans. Royal Soc. London* A303:287–302.

Wänke, H., and Dreibus, G. 1986. Geochemical evidence for the formation of the Moon by impact-induced fission of the proto-Earth. In *Origin of the Moon,* eds. W. K. Hartmann, R. J. Phillips and G. J. Taylor (Houston: Lunar and Planetary Inst.), pp. 649–672.

Wänke, H., Dreibus, G., and Jagoutz, E. 1984. Mantle chemistry and accretion history of the Earth. In *Archaean Geochemistry,* eds. A. Kroner, G. N. Hanson and A. M. Goodwin. (Berlin: Springer-Verlag), pp. 1–24.

Wannier, P. G., Linke, R. A., and Penzias, A. A. 1981. Observations of $^{14}N/^{15}N$ in the galactic disk. *Astrophys. J.* 247:522–529.

Ward, D. B. 1977. Far infrared spectral observations of Saturn and its rings. *Icarus* 32:437–442.

Ward, W. R. 1974. Climatic variations on Mars. 1. Astronomical theory of insolation. *J. Geophys. Res.* 79:3375–3386.

Ward, W. R. 1976. Some remarks on the accretion problem. In *Fisica e Geologia Planetaria* (Roma: Accademia nazionale dei Lincei), pp. 225–239.

Ward, W. R. 1984. The solar nebula and the planetary disk. In *Planetary Rings,* eds. R. Greenberg and A. Brahic (Tucson: Univ. of Arizona Press), pp. 660–686.

Ward, W. R. 1986. Density waves in satellite precursor disks. In *The Solid Bodies of the Outer Solar System,* ESA SP 242, pp. 1–4.

Warnatz, J. 1984. Rate coefficients in the C/H/O system. In *Combustion Chemistry,* ed. W. C. Gardiner, Jr. (New York: Springer-Verlag), pp. 197–360.

Warwick, J. W., Evans, D. R., Romig, J. H., Sawyer, C. B., Desch, M. D., Kaiser, M. L., Alexander, J. K., Carr, T. D., Staelin, D. H., Gulkis, S., Poynter, R. L., Aubier, M., Boischot, A., Leblanc, Y., Lecacheux, A., Pedersen, B. M., and Zarka, P. 1986. Voyager 2 radio observations of Uranus. *Science* 233:102–106.

Wäsch, R. 1986. An approach to the nature of the silicatic component of the comet Halley. In *20th ESLAB Symposium on the Exploration of Halley's Comet,* vol. 2., eds. B. Battrick, E. J. Rolfe and R. Reinhard, ESA SP-250 pp. 265–267.

Wasserman, I., and Weinberg, M. D. 1987. Theoretical implications of wide binary observations. *Astrophys. J.* 312:390–401.

Wasson, J. T., and Wetherill, G. W. 1979. Dynamical, chemical and isotopic evidence regarding the formation locations of asteroids and meteorites. In *Asteroids,* ed. T. Gehrels (Tucson: Univ. of Arizona Press), pp. 926–974.

Waters, J. W., Gustinic, J. J., Kahar, R. K., Kuiper, T. B. H., Roscoe, H. K., Swanson, P. N., Rodriquez, E. N. R., Kuiper, E. N., Kerr, A. R., and Thaddeus, P. 1980. Observations of interstellar H_2O emission at 183 gigahertz. *Astrophys. J.* 235:57–62.

Watkins, G. H. 1983. The Consequences of Cometary and Asteroidal Impacts on the Volatile Inventory of the Terrestrial Planets. Ph.D. Thesis, Massachusetts Inst. of Technology.

Watkins, H., and Lewis, J. S. 1984. Loss of volatiles from Mars as the result of energetic impacts. In *Workshop on Water on Mars* (Houston: Lunar and Planetary Inst.), pp. 91–93 (abstract).

Watson, A. J., Donahue, T. M., and Walker, J. C. G. 1981. The dynamics of a rapidly escaping atmosphere: Applications to the evolution of Earth and Venus. *Icarus* 48:150–166.

Watson, A. J., Donahue, T. M., and Kuhn, W. R. 1984. Temperatures in a runaway greenhouse on the evolving Venus: Implications for water loss. *Earth. Planet. Sci. Lett.* 68:1–6.

Watson, C. C. 1981. The sputter-generation of planetary coronae: Galilean satellites of Jupiter. *Proc. Lunar Planet. Sci. Conf.* 12:1569–1583.

Watson, D. M. 1984. Far infrared spectroscopy of molecular clouds. In *Galactic and Extragalactic Infrared Spectroscopy,* eds. M. F. Kessler and J. P. Phillips (Dordrecht: D. Reidel), pp. 195–219.

Watson, K., Murray, B. C., and Brown, H. 1961*a*. On the possible presence of ice on the Moon. *J. Geophys. Res.* 66:1598–1600.

Watson, K., Murray, B. C., and Brown, H. 1961*b*. The behavior of volatiles on the lunar surface. *J. Geophys. Res.* 66:3033–3045.

Watson, W. D., and Walmsley, C. M. 1982. Chemistry relevant to molecular clouds near H II regions. In *Regions of Recent Star Formation,* eds. R. S. Roger and P. E. Dewdney (Dordrecht: D. Reidel), pp. 357–377.

Watt, G. D. 1985. Time-dependent chemistry. II. Dependence of the chemistry on the initial [C]/[O] abundance ratio. *Mon. Not. Roy. Astron. Soc.* 212:93–103.

Wedepohl, K. H. 1981. Der primare Erdmantel (Mp) und die durch die Krustenbildung verarmte Mantelzusammensetzune (Md). *Fortschr. Miner.* 59:203–205.

Weidenschilling, S. J. 1975. Mass loss from the region of Mars and the asteroid belt. *Icarus* 26:361–366.

Weidenschilling, S. J. 1977*a*. The distribution of mass in the planetary system and solar nebula. *Astrophys. Space Sci.* 51:153–158.

Weidenschilling, S. J. 1977*b*. Aerodynamics of solid bodies in the solar nebula. *Mon. Not. Roy. Astron. Soc.* 180:57–70.

Weidenschilling, S. J. 1978. A constraint on pre-main-sequence mass loss. *Moon and Planets* 19:279–287.

Weidenschilling, S. J. 1980. Dust to planetesimals: Settling and coagulation in the solar nebula. *Icarus* 44:172–189.

Weidenschilling, S. J. 1982. Origin of regular satellites. In *The Comparative Studies of the Planets,* eds. A. Coradini and M. Fulchignoni (Dordrecht: D. Reidel), pp. 49–59.

Weidenschilling, S. J. 1984. Evolution of grains in a turbulent solar nebula. *Icarus* 60:553–567.

Weidenschilling, S. J., and Davis, D. R. 1985. Orbital resonances in the solar nebula: Implications for planetary accretion. *Icarus* 62:16–29.

Weidenschilling, S. J., and Davis, D. R. 1987. Orbital resonances in the solar nebula: Timescales and resonance widths. *Lunar Planet. Sci.* XVIII:1068–1069 (abstract).

Weidenschilling, S. J., and Lewis, J. S. 1973. Atmospheric and cloud structures of the Jovian planets. *Icarus* 20:465–476.

Weiland, J. L., Blitz, L., Dwek, E., Hauser, M. G., Magnani, L., and Rickard, L. J. 1986. Infrared cirrus and high latitude molecular clouds. *Astrophys. J.* 306:L101–L104.

Weissman, P. R. 1982. Dynamical history of the Oort cloud. In *Comets,* ed. L. L. Wilkening (Tucson: Univ. of Arizona Press), pp. 637–658.

Weissman, P. R. 1983. Cometary impacts on the terrestrial planets. In *Conference on Planetary Volatiles,* eds. R. O. Pepin and R. O'Connell, LPI Technical Report 83-01 (Houston: Lunar and Planetary Inst.), pp. 192–193.

Weissman, P. R. 1985*a*. Dynamical evolution of the Oort cloud. In *Dynamics of Comets: Their Origin and Evolution,* eds. A. Carusi and G. B. Valsecchi (Dordrecht: D. Reidel), pp. 87–96.

Weissman, P. R. 1985*b*. Cometary dynamics. *Space Sci. Rev.* 41:299–349.

Weissman, P. R. 1986. The Oort cloud and the galaxy: Dynamical interactions. In *The Galaxy and the Solar System,* eds. R. Smoluchowski, J. N. Bahcall and M. S. Matthews (Tucson: Univ. of Arizona Press), pp. 204–237.

Weissman, P. R. 1987. The cratering history in the inner solar system: No evidence for an enhanced cometary flux. *Bull. Amer. Astron. Soc.* 19:861 (abstract).

Weissman, P. R. 1988. Physical processing of cometary nuclei since their formation. In *Comet Halley: Worldwide Investigations, Results, and Interpretations,* in press.

Weissman, P. R., and Wetherill, G. W. 1974. Periodic Trojan-type librations in the Earth-Sun system. *Astron. J.* 79:404–412.

Welhan, J. A., and Craig, H. 1983. Methane, hydrogen, and helium in hydrothermal fluids at 21°N on the East Pacific Rise. In *Hydrothermal Processes at Seafloor Spreading Centers,* eds. P. A. Rona, K. Bostrom, L. Laubier and K. L. Smith (New York: Plenum Press), pp. 391–410.

Werner, M. W., Crawford, M. K., Genzel, R., Hollenbach, D. J., Townes, C. H., and Watson, D. M. 1984. Detection of shocked atomic gas in the Kleinmann-Low nebula. *Astrophys. J.* 282:L81–L84.

Wernicke, B., and Burchfiel, B. C. 1982. Modes of extensional tectonics. *J. Struc. Geol.* 4:105–115.

Westphal, J. A. 1969. Observations of localized 5-micron radiation from Jupiter. *Astrophys. J.* 157:L63–L64.

Wetherill, G. W. 1954. Variations in the isotopic abundances of neon and argon extracted from radioactive minerals. *Phys. Rev.* 96:679–683.

Wetherill, G. W. 1967. Collisions in the asteroid belt. *J. Geophys. Res.* 72:2429–2444.

Wetherill, G. W. 1971. Cometary versus asteroidal origin of chondritic meteorites. In *Physical Studies of Minor Planets,* ed. T. Gehrels, NASA-SP-267, pp. 447–459.

Wetherill, G. W. 1975. Late heavy bombardment of the moon and terrestrial planets. *Lunar Sci. Conf.* 6:1539–1561.

Wetherill, G. W. 1976. The role of large bodies in the formation of the Earth and Moon. *Proc. Lunar Sci. Conf.* 7:3245–3258.

Wetherill, G. W. 1977. Evolution of the Earth's planetesimal swarm subsequent to the formation of the Earth and Moon. *Proc. Lunar Sci. Conf.* 8:1–16.

Wetherill, G. W. 1978. Accumulation of the terrestrial planets. In *Protostars and Planets,* ed. T. Gehrels (Tucson: Univ. of Arizona Press), pp. 565–598.

Wetherill, G. W. 1980. Formation of the terrestrial planets. *Ann. Rev. Astron. Astrophys.* 18:77–113.

Wetherill, G. W. 1981*a*. Multi-ring basins. *Proc. Lunar Planet. Sci. Conf.* 12:1–18.

Wetherill, G. W. 1981*b*. Solar wind origin of ^{36}Ar on Venus. *Icarus* 46:70–80.

Wetherill, G. W. 1985. Occurrence of giant impacts during the growth of the terrestrial planets. *Science* 228:877–879.

Wetherill, G. W. 1986. Accumulation of the terrestrial planets and implications concerning lunar origin. In *Origin of the Moon,* eds. W. K. Hartmann, R. J. Phillips and G. J. Taylor (Houston: Lunar and Planetary Inst.), pp. 519–550.

Wetherill, G. W. 1987. Dynamical relationships between asteroids, meteorites, and Apollo-Amor objects. *Phil. Trans. Roy. Soc. London* A323:323–337.

Wetherill, G. W. 1988. Where do the Apollo objects come from? *Icarus* 76:1–18.

Wetherill, G. W., and Cox, L. P. 1985. The range of validity of the two-body approximation in models of terrestrial planet accumulation: II. Gravitational cross-sections and runaway accretion. *Icarus* 63:290–303.

Wetherill, G. W., and Stewart, G. R. 1986. The early stages of planetary accretion. *Lunar Planet. Sci.* XVII:939 (abstract).

Wetherill, G. W., and Stewart, G. R. 1987. Factors controlling early runaway growth of planetesimals. *Lunar Planet. Sci.* XVIII:1077 (abstract).

Wetherill, G. W., and Williams, J. G. 1979. Origin of differentiated meteorites. In *Origin and Distribution of the Elements,* ed. L. H. Ahrens (Oxford: Pergamon Press), pp. 19–31.

Wheelock, M. N., and Brownlee, D. E. 1986. *Elemental Composition of Submicron Volumes of Meteoritic Material: Analogues to Cometary Dust Grains.* (Seattle: Univ. of Washington).

Whipple, F. L. 1951. A comet model. I. The acceleration of Comet Encke. *Astrophys. J.* 11:375–394.

Whipple, F. L. 1963. On the structure of the cometary nucleus. In *The Moon, Meteorites, and Comets,* vol. 4, eds. B. M. Middlehurst and G. P. Kuiper (Chicago: Univ. of Chicago Press), pp. 639–664.

Whipple, F. L. 1964. The history of the solar system. *Proc. Natl. Acad. Sci.* 51:711–718.

Whipple, F. L. 1975. Do comets play a role in galactic chemistry and x-ray bursts? *Astron. J.* 80:525–531.

Wieler, R., Baur, H., and Signer, P. 1986. Noble gases from solar energetic particles revealed by closed system stepwise etching of lunar soil minerals. *Geochim. Cosmochim. Acta.* 50:1997–2017.

Wiens, R. C., Becker, R. H., and Pepin, R. O. 1986. The case for a Martian origin of the shergottites, II. Trapped and indigenous gas components in EETA 79001 glass. *Earth Planet. Sci. Lett.* 77:149–158.

Wildt, R. 1932. Absorptions Spektren und Atmosphären der Grossen Planeten. *Veröff Univ. Sternwarte Göttingen* 2 22:171–180.

Wilkening, L. L. 1977. Meteorites in meteorites: Evidence for mixing among the asteroids. In *Comets, Asteroids, Meteorites: Interrelations, Evolution, and Origins,* ed. A. H. Delsemme (Toledo: Univ. of Toledo), pp. 389–396.

Wilkening, L. L. 1978. Carbonaceous chondritic material in the solar system. *J. Naturwissenschaften* 65:73–79.

Williams, J. G. 1969. Secular Perturbations in the Solar System. Ph.D. Thesis, Univ. of California, Los Angeles.

Williams, J. G., and Faulkner, J. 1981. The positions of secular resonance surfaces. *Icarus* 46:390–399.

Willner, S. P. 1984. Observed spectral features of dust. In *Galactic and Extragalactic Infrared Spectroscopy,* eds. M. F. Kessler and J. P. Phillips (Dordrecht: D. Reidel), pp. 37–57.

Willner, S. P., Russell, R. W., Puetter, R. C., Soifer, B. T., and Harvey, P. M. 1979. The 4 to 8 micron spectrum of the galactic center. *Astrophys. J.* 229:L65–L68.

Willner, S. P., Gillett, F. C., Herter, T. L., Jones, B., Krassner, J., Merrill, K. M., Pipher, J. L., Puetter, R. C., Rudy, R. J., Russell, R. W., and Soifer, B. T. 1982. Infrared spectra of protostars: Composition of the dust shells. *Astrophys. J.* 253:174–187.

Willson, L. A., Bowen, G. H., and Struck-Marcell, C. 1987. Mass loss on the main sequence. *Comm. Astrophys.,* in press.

Wilson, L., and Head, J. W. 1981. Ascent and eruption of basaltic magma on the Earth and Moon. *J. Geophys. Res.* 86:2971–3001.

Wilson, L., and Head, J. W. 1983. A comparison of volcanic eruption processes on Earth, Moon, Mars, Io and Venus. *Nature* 302:663–669.

Wilson, L., and Mouginis-Mark, P. J. 1987. Volcanic input to the atmosphere from Alba Patera on Mars. *Nature* 330:354–357.

Wilson, L., Sparks, R. S. J., and Walker, G. P. L. 1980. Explosive volcanic eruptions. IV. The control of magma properties and conduit geometry on eruption column behavior. *Geophys. J. Roy. Astron. Soc.* 63:117–148.

Wilson, R. W., Langer, W. D., and Goldsmith, P. F. 1981. A determination of the carbon and oxygen isotope ratios in the local interstellar medium. *Astrophys. J.* 243:L47–L52.

Winkler, K.-H., and Newman, M. J. 1980. Formation of solar-type stars in spherical symmetry. II. Effects of detailed constitutive relations. *Astrophys. J.* 238:311–325.

Wisdom, J. 1983. Chaotic behavior and the origin of the 3/1 Kirkwood gap. *Icarus* 56:51–74.

Wisdom, J. 1985*a*. Meteorites may follow a chaotic route to Earth. *Nature* 315:731–733.

Wisdom, J. 1985*b*. A perturbative treatment of motion near the 3/1 commensurability. *Icarus* 63:272–289.

Wise, D. U. 1974. Continental margins, freeboard, and the volumes of continents and oceans through time. In *Geology of Continental Margins,* eds. C. A. Burk and C. L. Drake (New York: Springer-Verlag), pp. 45–58.

Witt, A. N., and Schild, R. E. 1986. Hydrogenated carbon grains in reflection nebulae. *Bull. Amer. Astron. Soc.* 18:1030 (abstract).

Witt, A. N., Bohlen, R. C., and Stecher, T. P. 1986. Colors of reflection nebulae. III. Ultraviolet scattering by small particles in reflection nebulae associated with 17, 20, and 23 Tauri. *Astrophys. J.* 302:421–431.

Witteborn, F. C., Bregman, J. D., Lester, D. F., and Rank, D. M. 1982. A search for fragmentation debris near Ursa Major stream stars. *Icarus* 50:63–71.

Wolery, T. J. 1978. Some Chemical Aspects of Hydrothermal Processes at Mid-Oceanic Ridges—A Theoretical Study. I. Basalt-Sea Water Reactions and Chemical Cycling Between the Oceanic Crust and Oceans. II. Calculation of Chemical Equilibrium Between Aqueous Solution and Minerals. Ph.D. Thesis, Northwestern Univ., Evanston, Illinois.

Wolery, T. J., and Sleep, N. H. 1976. Hydrothermal circulation and geochemical flux at mid-ocean ridges. *J. Geol.* 84:249–275.

Wolf-Gladrow, D., Neubauer, F. M., and Lussem, M. 1987. Io's interaction with the plasma torus: A self-consistent model. *J. Geophys. Res.* 92:9949–9961.

Wood, C. A., and Amsbury, D. L. 1986. Salt tectonics on Venus. *Lunar Planet. Sci.* XVII:954–955 (abstract).

Wood, J. A. 1979. Review of the metallographic cooling rates of meteorites and a new model for the planetesimals in which they formed. In *Asteroids,* ed. T. Gehrels (Tucson: Univ. of Arizona Press), pp. 849–891.

Wood, J. A., and Ashwal, L. D. 1981. SNC meteorites: Igneous rocks from Mars? *Proc. Lunar Planet. Sci.* 12:1359–1375.

Wood, J. A., and Chang, S. 1985. *The Cosmic History of the Biogenic Elements and Compounds,* NASA SP-476.

Wood, J. A., and Morfill, G. E. 1988. A review of solar nebula models and the relationship of meteorites to the nebula. In *Meteorites and the Early Solar System,* eds. J. F. Kerridge and M. S. Matthews (Tucson: Univ. of Arizona Press), pp. 329–347.

Woods, T. N., Feldman, P. D., and Dymond, K. F. 1986. The atomic carbon distribution in the coma of comet Halley. In *20th ESLAB Symposium on the Exploration of Halley's Comet,* vol. 1, eds. B. Battrick, E. J. Rolfe and R. Reinhard, ESA SP-250, pp. 431–435.

Wootten, A. 1987. Deuterated molecules in interstellar clouds. In *Astrochemistry,* eds. M. S. Vardya and S. P. Tarafdar (Dordrecht: D. Reidel), pp. 311–320.

Wootten, A., Loren, R. B., and Snell, R. L. 1982. A study of DCO+ emission regions in interstellar clouds. *Astrophys. J.* 255:160–175.

Woronow, A. 1978. A general cratering history model and its implications for the lunar highlands. *Icarus* 34:76–88.

Wright, E. L. 1976. Recalibration of the far infrared brightness temperatures of the planets. *Astrophys. J.* 210:250–253.

Wright, I. P., Carr, R. H., and Pillinger, C. T. 1986. Carbon abundance and isotopic studies of Shergotty and other shergottite meteorites. *Geochim. Cosmochim. Acta* 50:983–991.

Wyckoff, S., Lindholm, E., Wehinger, P. A., Peterson, B. A., Zucconi, J. M., and Festou, M. 1988. The carbon isotopes in comet P/Halley. *Astrophys. J.*, in press.

Wyllie, P. J. 1980. The origin of kimberlite. *J. Geophys. Res.* 85:6902–6910.

Yabushita, S., and Allen, A. J. 1985. On the effect of interstellar matter on terrestrial climate. *Observatory* 103:249–252.

Yamamoto, S., Saito, S., Kawaguchi, K., Kaifu, N., Suzuki, H., and Ohishi, M. 1987*a*. Laboratory detection of a new carbon-chain molecule C_3S and its astronomical identification. *Astrophys. J.* 317:L119–L121.

Yamamoto, S., Saito, S., Ohishi, M., Suzuki, H., Ishikawa, S.-I., Kaifu, N., and Murakami, A.

1987*b*. Laboratory and astronomical detection of the cyclic H_3H radical. *Astrophys. J.* 322:L55–L58.

Yamamoto, T. 1987. Chemical composition of cometary ice and grains, and the origin of comets. In *Astrochemistry,* eds. M. S. Vardya and S. P. Tarafdar (Dordrecht: D. Reidel), pp. 565–575.

Yang, J., and Anders, E. 1982*a*. Sorption of noble gases by solids, with reference to meteorites. II. Chromite and carbon. *Geochim. Cosmochim. Acta* 46:861–875.

Yang, J., and Anders, E. 1982*b*. Sorption of noble gases by solids, with reference to meteorites. III. Sulfides, spinels, and other substances; on the origin of planetary gases. *Geochim. Cosmochim. Acta* 46:877–892.

Yang, J., and Epstein, S. 1983. Interstellar organic matter in meteorites. *Geochim. Cosmochim. Acta* 47:2199–2216.

Yang, J., and Epstein, S. 1985. A study of stable isotopes in Shergotty meteorite. *Lunar Planet. Sci.* XVI:25–26 (abstract).

Yang, J., Lewis, R. S., and Anders, E. 1982. Sorption of noble gases by solids, with reference to meteorites. I. Magnetite and carbon. *Geochim. Cosmochim. Acta* 46:841–860.

Yelle, R., McConnell, J. C., Sandel, B., and Broadfoot, A. 1987. The dependence of electroglow on the solar flux. *J. Geophys. Res.* 92:15110–15124.

York, D. G., Spitzer, L., Bohlin, R. C., Hill, J., Jenkins, E. B., Savage, B. D., and Snow, T. P. 1983. Interstellar abundances of oxygen and nitrogen. *Astrophys. J.* 266:L55–L59.

Yuan, C., and Cassen, P. 1985. Protostellar angular momentum transport by spiral density waves. *Icarus* 64:435–447.

Yung, Y. L., and McElroy, M. B. 1977. Stability of an oxygen atmosphere on Ganymede. *Icarus* 30:97–103.

Yung, Y. L., and McElroy, M. B. 1979. Fixation of nitrogen in the prebiotic atmosphere. *Nature* 203:1002–1004.

Yung, Y. L., and Pinto, J. P. 1978. Primitive atmosphere and implications for the formation of channels on Mars. *Nature* 273:730–732.

Yung, Y. L., Allen, M., and Pinto, J. P. 1984. Photochemistry of the atmosphere of Titan: Comparison between model and observations. *Astrophys. J. Suppl.* 55:465–506.

Yung, Y. L., Pinto, J., and Friedl, R. 1987*a*. Kinetic isotopic fractionation in the solar nebula. Origin and Evolution of Planetary and Satellite Atmospheres, Abstract Booklet, 10-14. March, Tucson, AZ.

Yung, Y. L., Wen, J. S., Allen, M., Pierce, K. K., Paulson, S., and Pinto, J. P. 1987*b*. HDO in the martian atmosphere: Implications for the abundance of crustal water. *Bull. Amer. Astron. Soc.* 19:818 (abstract).

Zadnik, M. G., and Jeffery, R. M. 1985. Radiogenic neon in an archean anorthsite. *Chem. Geol.* 52:119–125.

Zahnle, K. J., and Kasting, J. F. 1986. Mass fractionation during transonic hydrodynamic escape and implications for loss of water from Venus and Mars. *Icarus* 68:462–480.

Zahnle, K. J., and Walker, J. C. G. 1982. The evolution of solar ultraviolet luminosity. *Rev. Geophys. Space Phys.* 20:280–292.

Zahnle, K., Kasting, J. F., and Pollack, J. B. 1988. Evolution of a steam atmosphere during Earth's accretion. *Icarus* 74:62–97.

Zähringer, J. 1962. Isotopie-Effekt und Haufigkeiten der Edelgase in Steinmeteoriten und auf der Erde. *Z. Naturforsch.* 17a:460–471.

Zähringer, J. 1968. Rare gases in stony meteorites. *Geochim. Cosmochim. Acta* 32:209–237.

Zaikowski, A., and Knacke, R. F. 1975. Infrared spectra of carbonaceous chondrites and the composition of interstellar grains. *Astrophys. Space Sci.* 37:3–9.

Zel'dovich, Y. B., and Raizer, Yu.P. 1967. *Physics of Shock Waves and High-Temperature Hydrodynamic Phenomena* (New York: Academic Press).

Zellner, B., Tholen, D. J., and Tedesco, E. F. 1985*a*. The eight-color asteroid survey: Results for 589 minor planets. *Icarus* 61:355–416.

Zellner, B., Thirunagari, A., and Bender, A. 1985*b*. The large-scale structure of the asteroid belt. *Icarus* 62:505–511.

Zharkov, V. N., and Trubitsyn, V. P. 1978. *Physics of Planetary Interiors* (Tucson: Pachart Press).

Zinner, E., and Epstein, S. 1988. Heavy carbon in individual oxide grains from the Murchison meteorite. *Earth Planet. Sci. Lett.* 84:359–368.

Zinner, E., McKeegan, K. D., and Walker, R. M. 1983. Laboratory measurements of D/H ratios in interplanetary dust. *Nature* 305:119–121.

Ziurys, L. M. 1987. Detection of PN, the first interstellar phosphorus-containing molecule. *Astrophys. J.* 321:L81–L85.

Zook, H. A. 1981. On a new model for the generation of chondrites. *Lunar Planet. Sci.* XII:1242–1243 (abstract).

Zvyagina, E. V., and Safronov, V. S. 1972. Mass distribution of protoplanetary bodies. *Soviet Astron. J.* 15:810–817.

Zvyagina, E. V., Pechernikova, G. V., and Safronov, V. S. 1974. Qualitative solution of the coagulation equation with allowance for fragmentation. *Soviet Astron. J.* 17:793–800.

ACKNOWLEDGMENTS

The editors acknowledge the support of the National Aeronautics and Space Administration, through Grant NAGW-716; the University of Arizona; and the University of Arizona Press for the preparation of this book. They also acknowledge the extensive help of referees and of J. E. Frecker, one of the proofreaders for this book. The following authors wish to acknowledge specific funds involved in supporting the preparation of their chapters.

Ahrens, T. J.: NASA Grants NSG-7129 *and* NGL-05-002-105
Atreya, S. K.: NASA Grant NSG-7404
Bodenheimer, P.: NSF Grant AST 8521636
Boss, A. P.: NASA Grant NAGW-398 *and* NSF Grant AST 8515644
Cheng, A. F.: NASA Grant under Task 1 of Contract between Johns Hopkins University and the Department of the Navy N00039-87-C-5301
Donahue, T. M.: NASA Grant NAS2-12314 *and* NSF Grant ATM 8520653
Fegley, B., Jr.: NASA Grants NAG 9-108, NAGW-997 *and* NAGW-821
Gautier, D.: CNES Grant *and* ATP Planétologie contracts 57.47, 67.43 *and* 67.44
Greenberg, R.: NASA Grant NAGW-1029
Hubbard, W. B.: NASA Grant NSG-7045
Hunten, D. M.: NASA Grant NSG-7558 *and* NSF Grant AST 85-14520
Irvine, W. M.: NASA Grant NAGW-436 *and* NSF Grant AST 8512903
Johnson, R. E.: NASA Grant NAGW-461
Jones, T. D.: NASA Grants NAGW-340, NAGW-1146 *and* NGT-50186
Knacke, R. F.: NASA Grant NSG-7286 *and* NSF Grants AST 8414825
Lebofsky, L. A.: NASA Grant NSG-7114
Lunine, J. I.: NASA Grant NAGW-1039
Owen, T.: NASA Grant NGR 33-015165
Pepin, R. O.: NASA Grant NAG 9-60
Prinn, R. G.: NSF Grant ATM 8710102 *and* NASA Grant NAG9-108
Schubert, G.: NASA Grants NSG 7315 *and* NSG 7164
Sleep, N. H.: NSF Grant EAR 87-19278
Solomon, S. C.: NASA Grants NSG 7297 *and* NAGW-1077
Turcotte, D. L.: NASA Grant NAGW-797
Walker, J. C. G.: NASA Grant NAGW-176

Index

Index